ENCYCLOPEDIA OF

BIOPROCESS TECHNOLOGY: FERMENTATION, BIOCATALYSIS, AND BIOSEPARATION

VOLUME 5

WILEY BIOTECHNOLOGY ENCYCLOPEDIAS

Encyclopedia of Bioprocess Technology: Fermentation, Biocatalysis, and Bioseparation
Edited by Michael C. Flickinger and Stephen W. Drew

Encyclopedia of Molecular Biology
Edited by Thomas E. Creighton

Encyclopedia of Cell Technology
Edited by Raymond E. Spier

Encyclopedia of Ethical, Legal, and Policy Issues in Biotechnology
Edited by Thomas J. Murray and Maxwell J. Mehlman

ENCYCLOPEDIA OF BIOPROCESS TECHNOLOGY: FERMENTATION, BIOCATALYSIS, AND BIOSEPARATION
EDITORIAL BOARD

ENCYCLOPEDIA OF BIOPROCESS TECHNOLOGY: FERMENTATION, BIOCATALYSIS, AND BIOSEPARATION

VOLUME 5

Michael C. Flickinger
University of Minnesota
St. Paul, Minnesota

Stephen W. Drew
Merck and Co., Inc.
Rahway, New Jersey

A Wiley-Interscience Publication
John Wiley & Sons, Inc.
New York / Chichester / Weinheim / Brisbane / Singapore / Toronto

This book is printed on acid-free paper. ♾

Published simultaneously in Canada.

For ordering and customer service, call 1-800-CALL-WILEY.

Library of Congress Cataloging-in-Publication Data:

Flickinger, Michael C.
The encyclopedia of bioprocess technology : fermentation, biocatalysis, and bioseparation / Michael C. Flickinger, Stephen W. Drew.
p. cm.
Includes index.
ISBN 0-471-13822-3 (alk. paper)
1. Biochemical engineering--Encyclopedias. I. Drew, Stephen W., 1945- . II. Title.
TP248.3.F57 1999
660.6′03--dc21 99-11576
CIP

Printed in the United States of America.

10 9 8 7 6 5 4 3 2 1

ENCYCLOPEDIA OF

BIOPROCESS TECHNOLOGY:
FERMENTATION, BIOCATALYSIS,
AND
BIOSEPARATION

VOLUME 5

PYRUVATE, PRODUCTION USING DEFECTIVE ATPASE ACTIVITY

ATSUSHI YOKOTA
SEIGO AMACHI
FUSAO TOMITA
Hokkaido University
Sapporo, Japan

KEY WORDS

Energy metabolism
Escherichia coli
F_1-ATPase
Futile cycle
Glycolytic pathway
Hemoglobin
Phosphotransferase system
Phosphoglycerate kinase
Pyruvate kinase
Pyruvic acid

OUTLINE

INTRODUCTION

Much work has been done on the development of producers of useful metabolites. Metabolic engineering has enabled more scientific and strategic breeding than the conventional method based on mutagenesis followed by selection. In general, the targets for breeding are biosynthetic pathways of the metabolites, and the best example is the metabolic regulatory fermentation of various kinds of amino acids. In contrast, few studies consider energy metabolism as the target of breeding. However, it seems that manipulation of energy metabolism opens up a new dimension in breeding, because the energy status of the cell deeply concerns both regulation of energy production and biosynthesis and assembly of cell materials, although it does not necessarily directly affect the biosynthetic pathways of the desired metabolites. In this article, the development of producers of useful metabolites by the manipulation of energy metabolism is described with special emphasis on our work, which is improvement of the pyruvic acid productivity of a pyruvic acid–producing mutant of *Escherichia coli* K-12 by a defect in F_1-ATPase activity.

ENHANCEMENT OF CENTRAL METABOLISM BY A DEFECT IN THE ACTIVITY OF F_1F_0-ATP SYNTHASE

This section mainly describes our work, which has demonstrated both the enhancement of glucose metabolism and the improvement of pyruvic acid productivity by the introduction of a defect in an energy-producing reaction in an *E. coli* strain.

Improvement of Pyruvic Acid Production by Strain TBLA-1, an F_1-ATPase-Defective Mutant of *E. coli* W1485*lip2*

For the development of producers of useful metabolites, it seems important to consider points both upstream and downstream of the metabolic flow as well as the biosynthetic pathway itself, such as increases in the supply of precursor metabolites that flow into the biosynthetic pathways and the efficient transport of substrates and metabolites across the membranes. Under these concepts, we have focused on the improvement of fermentative production by enhancement of the glycolytic pathway. That is, it was expected that the enhancement of the glycolytic pathway would lead to the improvement of producers through smooth assimilation of substrate carbohydrates.

As a model system to assess this idea, we have established pyruvic acid production by *E. coli* W1485*lip2* (ATCC 25645) (1), a lipoic acid auxotroph of *E. coli* K-12 (2). This system was employed for several reasons. First of all, it was expected that the effect of enhancement of the glycolytic pathway would be obvious because pyruvate is the end product of glycolysis. Second, *E. coli* strain K-12 is suitable for characterization because much information about its genetics and physiology is available. Strain W1485*lip2* requires lipoic acid for its growth and produces a large amount of pyruvic acid in cultures in which lipoic acid is limiting, where the activity of the pyruvate dehydrogenase complex becomes almost negligible (1).

Then, how can we enhance the glycolytic pathway? Much work has revealed that flow in the glycolytic pathway is enhanced by a decrease in the energy level of the cell. That is, under those conditions, phosphofructokinase (PFK) I, one of the isozymes of *E. coli* PFK, and pyruvate kinase (PYK) II, one of the isozymes of *E. coli* PYK, have been shown to be activated by ADP and AMP, respectively (3,4). Therefore, decrease in the energy level of the cell seemed to be effective for enhancement of the glycolytic pathway. To decrease the energy level, it seemed appropriate to impair the generation of ATP through oxidative phosphorylation. For these reasons, we have introduced a defective gene of the α-subunit of F_1-ATPase, *atpA401* (5), into the strain W1485*lip2* by transduction, and obtained an F_1-ATPase-defective mutant, TBLA-1 (6). Both strains were cultured aerobically in the pyruvic acid production medium, Medium 7 (6), containing 50 g/L of glucose, using jar fermentors with pH controlled at 6.0. As shown in Figure 1, strain TBLA-1 was found to produce pyruvic acid much more effectively than the parental strain in spite of some decreases both in growth rate and in growth level (6). That is, the parental strain, W1485*lip2*, consumed 50 g/L of glucose and produced 25.5 g/L of pyruvic acid after culture for 32 h; under the same conditions, strain TBLA-1 consumed glucose and produced more than 30 g/L of pyruvic acid after culture for 24 h.

Table 1 shows F_1-ATPase activities, fermentation data, and fermentation parameters of various strains. The fermentation data were obtained after culture for 16 h, at which time maximum growth was attained in the strains listed. As seen in Table 1, the level of growth in strain TBLA-1 decreased to 67% of the parental strain because of impairment of oxidative phosphorylation. However, in strain TBLA-1, both the amount of glucose consumed per cell ([B]/[A] in Table 1) and the amount of pyruvic acid produced per cell ([C]/[A] in Table 1) increased by factors of 1.9 and 2.8, respectively, compared with strain W1485*lip2* (6). In strain TBLA-1, the conversion yield of pyruvic acid from consumed glucose ([C]/[B] in Table 1) also increased by a factor of about 1.5 (6). Furthermore, as all of these changes have not been observed in a revertant from strain TBLA-1 (no. 63-1) having a normal level of F_1-ATPase activity (Table 1), it was concluded that the enhancement of pyruvic acid productivity was brought about by the defect in F_1-ATPase activity (6).

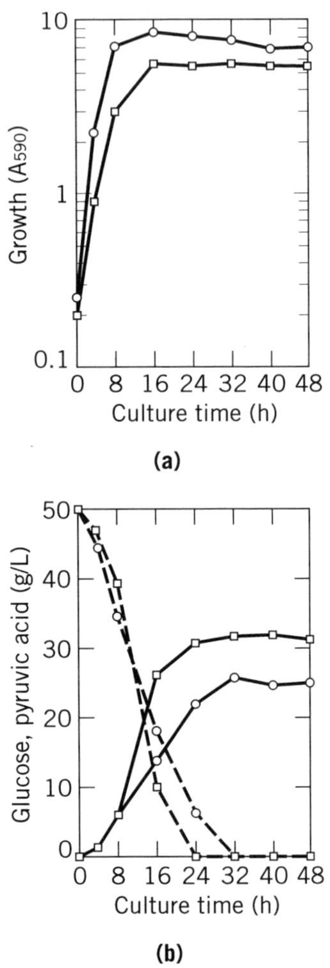

Figure 1. Time courses of pyruvic acid production by strains W1485*lip2* and TBLA-1 (6) showing data for growth (**a**) and pyruvic acid production (**b**). In **b**, *dashed lines* indicate glucose consumption and *solid lines* indicate pyruvic acid production. ○, strain W1485*lip2*; □, strain TBLA-1.

Table 1. F_1-ATPase Activities and Fermentation Profiles of Various Strains

Strain	W1485*lip2*	TBLA-1	No. 63-1
F_1-ATPase activity[a]	6.3	0.09	5.7
Growth (A_{590}) (A)	8.5	5.7	7.2
Glucose consumed (g/L) (B)	32.1	40.4	35.6
Pyruvic acid produced (g/L) (C)	13.5	25.9	14.8
(B)/(A)	3.8	7.1	4.9
(C)/(A)	1.6	4.5	2.1
(C)/(B)	0.42	0.64	0.42

[a] μmol min^{-1} mg $protein^{-1}$.

These results seem to be the first example of the application of a mutant defective in energy metabolism in the field of biotechnology. Although we have not optimized culture conditions for pyruvic acid production, the productivity of pyruvic acid in strain TBLA-1 in terms of grams of pyruvic acid produced per liter per hour is quite efficient among the pyruvic acid producers so far reported (7,8). The highest accumulation of pyruvic acid has been reported for the wild-type *Torulopsis glabrata* IFO 0005 (8), which requires nicotinic acid, thiamine, pyridoxine, and biotin for its growth. This yeast strain is capable of producing 57 g/L of pyruvic acid from 100 g/L of glucose in 59-h culture. The commercial importance of pyruvic acid includes its use as a starting material for the synthesis of many drugs and agrochemicals. Especially, pyruvic acid has been used as one of the substrates for the enzymatic production of L-dihydroxyphenylalanine (L-DOPA).

Enhancement of Glucose Metabolism in Strain TBLA-1

In the course of these studies, Jensen and Michelsen (9) reported the physiological characteristics of *E. coli* mutants having deletions in the *atp* operon coding for H^+-ATPase. In the mutants, they demonstrated increases in the oxygen consumption rate and in the level of *b*-type cytochromes by factors of 1.4 and 1.8, respectively, compared with the wild-type strain. It was also found in the *atp* mutant that the flow through the glycolytic pathway increased, and twice as much acetate was produced as a main by-product from consumed glucose. The increased oxygen consumption rate in the *atp* mutants is considered to contribute to the enhanced operation of the glycolytic pathway and to enhanced ATP production by substrate-level phosphorylation through the smooth reoxidation by the electron transport chain of NADH generated from the glycolytic pathway. Also, increase in the content of *b*-type cytochromes seems to contribute to enhanced respiration.

In the cells of strain TBLA-1 growing in pyruvic acid–producing medium, oxygen consumption rates (Fig. 2a), and the content of *b*-type cytochromes were also found to increase by factors of 1.7 and 1.4, respectively, compared with strain W1485*lip2* (10). These changes suggested increases in the amount of some components of the respiratory chain.

In the *E. coli* respiratory chain, two types of NADH dehydrogenases, designated NDH-1 and NDH-2, have been found (11). The expression of the *nuo* operon coding for NDH-1 has been shown to be repressed anaerobically by ArcA, a gene product from *arcA* (*arc* for aerobic respiration control) (12). NDH-2 is specified by the *ndh* gene. The expression of this gene is subject to *fnr* (fumarate nitrate reduction) gene product (Fnr) -mediated anaerobic repression (13). Thus, these two dehydrogenase genes are expressed aerobically and repressed anaerobically. Two types of terminal oxidases, Cyt *o* and Cyt *d*, are known in *E. coli* (14); the former is coded by the *cyo* operon and the latter by the *cyd* operon. The *cyo* operon is expressed aerobically and is repressed anaerobically by both ArcA and Fnr (15,16). On the other hand, the expression of the *cyd* operon has been shown to be regulated by both ArcA and Fnr in such a way that the expression is maximum under microaerobic conditions (15,17,18). To clarify the mechanism(s) for the enhanced respiratory activities observed in strain TBLA-1, it seemed necessary to measure and compare the amounts of the NADH dehydrogenases and terminal oxidases of strains TBLA-1 and W1485*lip2*. This kind of approach might allow us to consider how the energy status of the cells can affect the activities of the oxygen-responding regulators ArcA and Fnr.

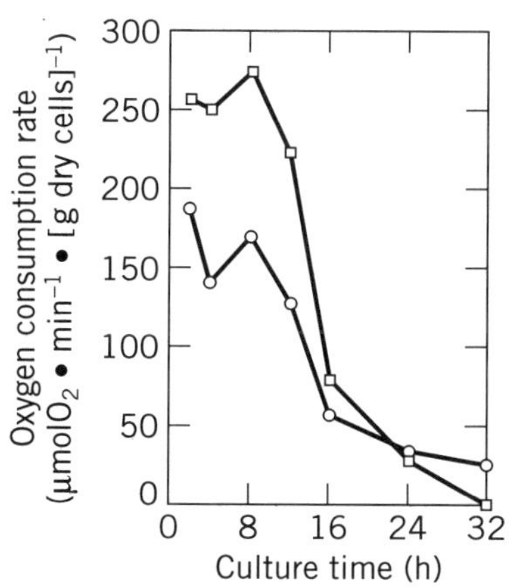

(a)

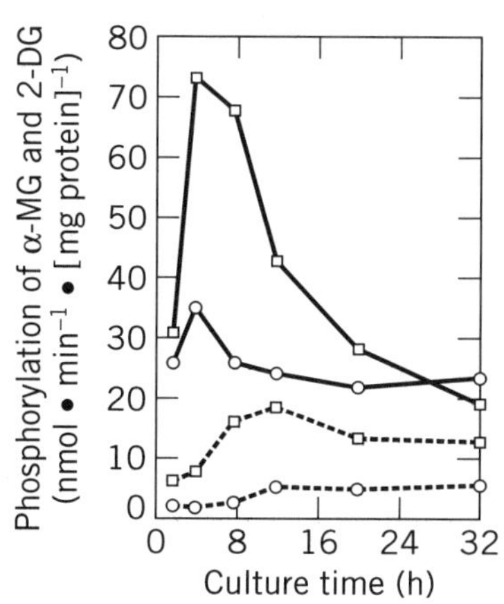

(b)

Figure 2. Oxygen consumption rates (**a**) and phosphotransferase (PTS) activities (**b**) of strains W1485*lip2* and TBLA-1 during pyruvic acid production (10) were measured using the cells harvested from the cultures for pyruvic acid production presented in Figure 1. In **b**, *solid lines* and *dashed lines* indicate PEP-dependent phosphorylation of methyl α-D-glucopyranoside (α-MG) and 2-deoxy-D-glucose (2-DG), respectively. ○, strain W1485*lip2*; □, strain TBLA-1.

To elucidate the enhancement of glucose consumption in strain TBLA-1, activities of the phosphotransferase system (PTS) for glucose uptake were measured by phosphoenolpyruvate (PEP) dependent phosphorylation of methyl α-D-glucopyranoside and 2-deoxy-D-glucose using the toluenized cells. We found that these activities were higher in strain TBLA-1 than in strain W1485*lip2* (Fig. 2b) (10). These changes were considered to be important explanations for the higher activities of glucose consumption per cell in strain TBLA-1. PTS not only catalyzes the formation of glucose-6-phosphate on uptake of glucose but also catalyzes the conversion of PEP to pyruvate, seemingly the same reaction as that of pyruvate kinase. Therefore, increase in the activity of PTS found in strain TBLA-1 is considered to be equivalent to the enhancement of glycolysis. Enhancement of glucose consumption has been reported in some *atp* mutants (9,19). However, PTS activities of these mutants under glycolytic conditions have not yet been reported. Therefore, our results are the first implication of the importance of PTS activities in the regulation of glycolysis in *atp* mutants. Furthermore, the increase in PTS activity is considered to lead to increased yield of pyruvate from consumed glucose in strain TBLA-1 because PEP is converted to pyruvate much more effectively with the increased operation of PTS. Rohwer et al. (20) studied PTS activities of a strain of *E. coli* K-12 growing on succinate by measuring initial uptake rates of methyl α-D-glucopyranoside. They found that the uptake rate of the strain decreased when the expression level of the *atp* operon decreased and suggested that the PTS activity decreases under energy-deficient conditions. Thus, it seems quite interesting to examine PTS activities of *atp* mutants in more detail under various culture conditions.

Measurements of the activities of all the glycolytic enzymes revealed that those of phosphoglycerate kinase and PYK-I were higher in strain TBLA-1 than in strain W1485*lip2* during the logarithmic growth phase by factors of about 1.5 at their maxima (10). These differences were considered to be not so distinct as those found in PTS activities. Moreover, so far as the activities of PYK-I were concerned, differences were not observed after the culture had continued for 12 h. However, as these two enzymes are involved in substrate-level phosphorylation, increased activities of these enzymes seem to contribute to the supply of ATP required for the synthesis of cell materials during the logarithmic growth phase. Also, the increases seemed to contribute to increase in the rate of glucose consumption per cell during the logarithmic growth phase in strain TBLA-1. Because activities of all the glycolytic enzymes were measured with crude enzymes that had been gel-filtered to eliminate low molecular weight materials, the increases in these enzyme activities were not related to activation by low molecular weight ligands such as ADP and AMP.

These interesting changes found so far in strain TBLA-1 are summarized in Figure 3 (21). It became apparent that various kinds of changes that seem to compensate for energy shortage take place in strain TBLA-1, which is defective in F_1-ATPase activity. At present, it is not clear if these changes are indigenous to *atp* mutants. These points need to be investigated using continuous cultures in which the physiological state of each cell is homogeneous and well defined. Investigation of the mechanisms of these changes seems to be quite important and will shed light on the energy metabolism of *E. coli*.

Applicability of a Defect in F_1F_o-ATP Synthase Activity to Other Industrially Important Organisms

It seems interesting to consider whether the enhancement of glucose metabolism and the concomitant improvement

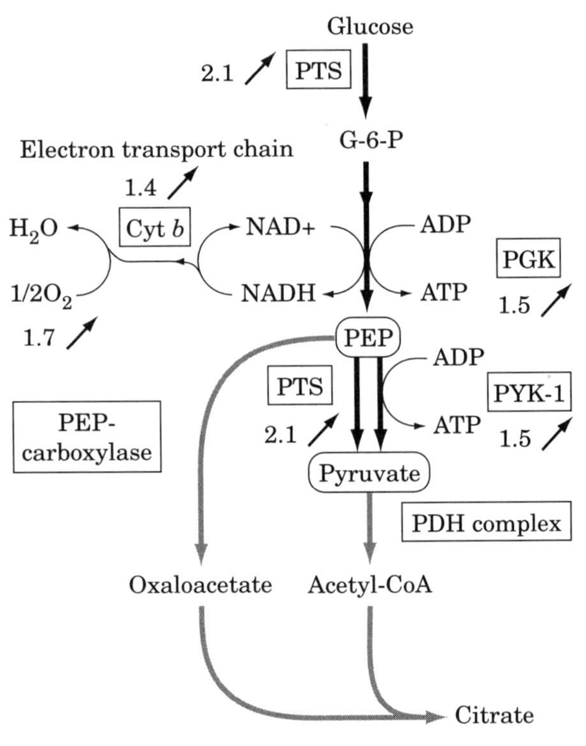

Figure 3. Changes in activities of glucose metabolism in strain TBLA-1 (from 21, with permission of Japan Society for Bioscience, Biotechnology, and Agrochemistry). Each number with an *arrow* shows an increase of the indicated reaction compared with strain W1485*lip2*. The increase for PTS is the average value for PEP-dependent phosphorylation of α-methylglucoside after culture for 4–12 h (Fig. 2b). The increase for oxygen consumption rate is also the average value for 4 to 12 h in Figure 2a. G-6-P, glucose-6-phosphate; Cyt b, b-type cytochromes; PDH, pyruvate dehydrogenase; PGK, phosphoglycerate kinase. Hatched lines indicate weak reactions.

of the productivity of a metabolite by the introduction of a gene defective for F_1-ATPase observed in *E. coli* can be applicable to other industrially important microorganisms. To date, other than *E. coli*, physiological changes similar to that observed in *E. coli* mutants (6,9), the enhancement of glucose metabolism and increase in respiratory activity, have also been reported in *atp* mutants of *Bacillus subtilis* (19). It was also demonstrated that the expression of genes coding for some of the respiratory enzymes has also been increased by disruption of the *atp* operon (19). From these observations, it seems that increases in the activities of both rate-limiting enzymes of the glycolytic pathway and respiration for the reoxidation of NADH generated via glycolysis are prerequisite for the enhancement of glucose metabolism. Here, for simplicity, let us consider the possibility of the enhancement of glucose metabolism only from the former aspect, increased activity of rate-limiting enzymes of the glycolytic pathway. Generally in bacteria, PFK and PYK, both of which catalyze the irreversible reactions in glycolysis, have been shown as the control points for the glycolytic flux. In many cases these enzymes are reported to be activated or inhibited allosterically by metabolites of the glycolytic pathway and by adenine nucleotides, and activation is normally observed in response to decrease in the energy level of the cell. Thus, it is expected that a defect in the activity of F_1F_o-ATP synthase lowers the energy level of the cell, thereby leading to the enhancement of glycolysis from the activation of PFK or PYK under those conditions.

Table 2 summarizes effectors for PFKs and PYKs from some of the industrially important bacteria. As reported (6,9,19), defects in F_1F_o-ATP synthase activity resulted in the enhancement of glucose metabolism in *E. coli* and in *B. subtilis*. Consistently, PFK-1 and PYK-2 of *E. coli* are activated by ADP and AMP, respectively (3,4). In strain TBLA-1 defective in F_1F_o-ATP synthase activity, an increase in the activity of PYK-1 that is different from allosteric activation by fructose-1,6-diphosphate (FDP) (22) was observed (10), suggesting synergistic effect for the increase of PYK activity under glycolysing conditions. On the other hand, the control point of glycolysis has not been established in *B. subtilis* (23). A PFK from *B. stearothermophilus* (24) and PYKs from *B. licheniformis* (25) and *B. stearothermophilus* (26) have been shown to be activated by AMP or ADP, suggesting the possibility of enhancement of glycolytic flux by a defect in F_1F_o-ATP synthase activity. In the case of glutamic acid–producing bacteria, *Brevibacterium flavum* no. 2247 (28,29) and *Corynebacterium glutamicum* (30), the control point for the regulation of glycolysis appeared to be the PYK reaction. A mutant of *B. flavum* no. 2247 defective in ATPase activity has already been isolated in our laboratory. In this mutant enhancement of glucose metabolism has also been observed as in the case of *E. coli* mutants (Yokota et al., unpublished data). Little information is available concerning the regulation of glycolysis in industrially important actinomycetes that produce antibiotics. Purification and characterization of PFK from *Streptomyces coelicolor* A3 (2) have been reported to indicate the enzyme as a regulatory step in the glycolytic pathway (31).

Table 2. Effectors for Phosphofructokinase (PFKs) and Pyruvate Kinase (PYKs) from Industrially Important Bacteria

Organisms	PFK	PYK	References
Escherichia coli K-12	PFK-I Cooperativity (Yes) ADP (+) PEP (−) PFK-II Cooperativity (No)	PYK-I Cooperativity (Yes) FDP (+) GTP (−) ATP + succinyl-CoA (−) PYK-II Cooperativity (Yes) AMP (+)	3,4,22
Bacillus subtilis 60015	ND	Cooperativity (Yes)	23
Bacillus licheniformis A-5	Cooperativity (No) PEP (−)	Cooperativity (Yes) ADP (+), AMP (+) ATP (−), Pi (−)	25,27
Bacillus stearothermophilus NCA1503 (ATCC7954)	Cooperativity (No) ADP (+) (+), GDP (+) PEP (−)	Cooperativity (Yes) AMP (+), GMP (+) R5P (+) Pi (−)	24,26
Brevibacterium flavum No. 2247 (ATCC14067)	Cooperativity (No)	Cooperativity (Yes) AMP (+) ATP (−), GTP (−)	28,29
Corynebacterium glutamicum M2	ND	Cooperativity (Yes) AMP (+) ATP (−)	30
Streptomyces coelicolor A3(2)	Cooperativity (No) PEP (−)	ND	31
Streptococcus lactis ML3	ND	Cooperativity (Yes) FDP (+) Pi (−)	32

Presence (Yes) or absence (No) of homotropic cooperativity (cooperativity) in the substrate saturation curve for fructose-6-phosphate in the case of PFK and for PEP in the case of PYK. (+), positive effector; (−), negative effector; ND, not determined; FDP, fructose-1,6-diphosphate; Pi, inorganic phosphate; R5P, ribose-5-phosphate; PEP, phosphoenolpyruvate.

The control point for the regulation of glycolytic flux in lactic acid bacteria has extensively been investigated using *Lactococcus* (formerly *Streptococcus*) *lactis* (32). In *L. lactis*, glycolytic flux appeared to be controlled in PYK that has FDP as an activator and inorganic phosphate as an inhibitor. In a glycolysing cell of *L. lactis*, a high intracellular concentration of FDP and a low intracellular concentration of inorganic phosphate have been reported (33), suggesting the presence of an already fully activated form of PYK in the cell. Moreover, in the case of lactic acid bacteria, F_1F_o-ATP synthase has been shown to work in the direction of pumping out protons on hydrolysis of ATP for keeping intracellular pH at neutral under acidic culture conditions. In fact, our work (34) showed that a mutant of *L. lactis* ssp. *lactis* C2 with reduced membrane-bound ATPase activity had almost the same productivity of lactic acid from glucose in a pH-controlled fermentor as that of the parental strain. Thus, glucose consumption per cell did not increase as a result of the mutation. On the other hand, the mutant showed higher acid sensitivity than that of the parental strain, indicating the importance of F_1F_o-ATP synthase in the pH homeostasis of the cell. In acidic culture conditions, higher transcriptional levels of the gene encoding the β-subunit of F_1F_o-ATP synthase and higher amounts of the β-subunit protein than those observed under neutral culture conditions were detected in both strains (35). In contrast, no increase in the transcriptional level of the *atp* operon encoding F_1F_o-ATP synthase was found in *E. coli* in response to acidic culture conditions (36), suggesting the difference of the function of F_1F_o-ATP synthase in different microorganisms. The industrial importance of the mutant defective in the activity of F_1F_o-ATP synthase in the prevention of postacidification of yogurt and fresh cheese has been proposed (34).

From all these examples, mutations in energy metabolism caused by defects in the activity of F_1F_o-ATP synthase are quite interesting for the breeding of industrially important microorganisms. Because of the pleiotropic nature of the mutation, the development of a novel type strain can be expected.

OVEREXPRESSION OF GENES OF ENZYMES INVOLVED IN THE FORMATION OF FUTILE CYCLES

Futile cycles are cyclic metabolic pathways that involve biosynthesis and degradation of certain metabolic intermediates with energy spilling. For example, the coexistence of PFK and fructose bisphosphatase and of PYK and PEP synthase forms futile cycles. The activities of these enzymes are under kinetic control by adenine nucleotides. That is, when the energy level of the cell decreases, glycolytic flow is enhanced, while under energy-sufficient conditions, gluconeogenetic flow is enhanced and thus both enzymes are not activated simultaneously.

Patnaik et al. (37) cloned the PEP synthase gene on a multicopy plasmid, and overexpressed it in *E. coli* K-12 to form a futile cycle as shown in Figure 4. When the PEP synthase activity was overexpressed 15-fold, an increase in the oxygen consumption rate per cell was observed. Furthermore, an increase in the glucose consumption rate per cell and the excretion of pyruvate and acetate were observed when the activity was overexpressed 30-fold. Patnaik et al. inferred that the increase in PEP concentration due to the overexpression of PEP synthase gene caused the increase in PTS activity, resulting in the enhancement of glucose consumption rate. Similarly, Chao and Liao (38) overexpressed genes of PEP carboxylase and PEP carboxykinase simultaneously in *E. coli* K-12 to form a futile cycle (Fig. 4). Also in this case, with the incremental induction of both enzymes, an increase in the oxygen consumption rate per cell was first observed, and an increase in the glucose consumption rate per cell and excretion of pyruvate and acetate then followed. Chao and Liao concluded that the *arc* system had no relationship to the increase in the oxygen consumption rate.

These physiological changes caused by the overexpression of the genes of the enzymes involved in futile cycles resemble those observed in *atp* mutants (9,19). The operation of futile cycles causes waste of ATP. Therefore, these cells might be suffering from shortage of energy as in *atp* mutants. Although no information is available as to the changes in the activities of PTS and glycolytic enzymes in these cells, it seems interesting to use these futile cycles to improve the fermentative production of useful compounds through enhancement of glucose metabolism.

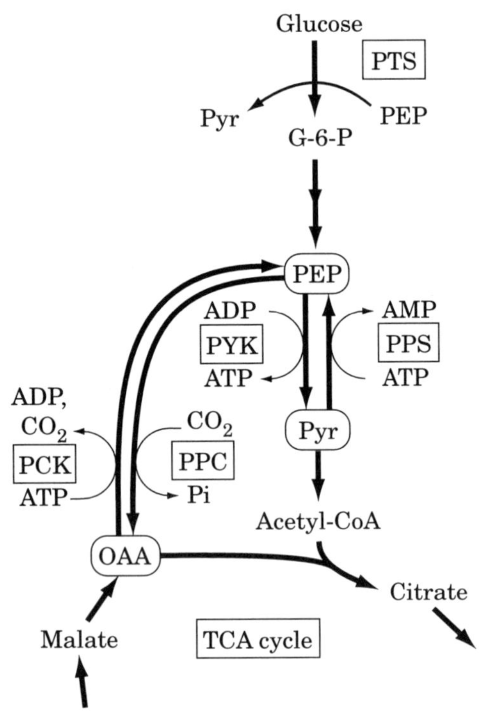

Figure 4. PEP pyruvate and PEP oxaloacetate futile cycles formed in *E. coli* K-12 (from 21, with permission of Japan Society for Bioscience, Biotechnology, and Agrochemistry). Pyr, pyruvate; OAA, oxaloacetate; PPS, PEP synthase; PPC, PEP carboxylase; PCK, PEP carboxykinase, G-6-P, glucose-6-phosphate.

IMPROVEMENT OF PRODUCTION OF USEFUL COMPOUNDS BY THE EXPRESSION OF THE HEMOGLOBIN GENE FROM *VITREOSCILLA*

Finally, in contrast to our concept, we discuss examples in which production of various kinds of useful compounds was improved by increased efficiency of energy production. Bacteria of the genus *Vitreoscilla* are Gram-negative chemoorganotrophs that move by gliding. They are aerobes and use molecular oxygen as the terminal electron acceptor when they live in hypoxic conditions. Under these conditions, a strain of *Vitreoscilla* was found to synthesize a kind of hemoprotein (*Vitreoscilla* hemoglobin, VHb) similar to that of the eukaryotic hemoglobin (39,40). VHb is considered to contribute to the growth of the organism under hypoxic conditions through enhanced diffusion of oxygen into the cytoplasm (41).

When the VHb gene containing its promoter region was introduced into *E. coli* K-12, active VHb protein was produced in response to the decrease in oxygen concentration in the culture medium, as in *Vitreoscilla* cells (42). The regulation of the gene expression was found to be at the transcription level (42). By expression of the VHb gene, various interesting physiological changes were observed in *E. coli* K-12 cells, including increases in growth rate, final growth level, and oxygen consumption rate per cell (43). Moreover, expression of the VHb gene improves various kinds of fermentative production under hypoxic conditions. For example, production of enzymes by *E. coli* (44,45) and by *B. subtilis* (46), production of actinorhodin by *Streptomyces coelicolor* (47), and production of cephalosporin C by *Acremonium chrysogenum* (48) were found to be improved. It should be noted that in contrast to *atp* mutants, enhancement of glucose consumption has not been observed in the *E. coli* K-12 strain expressing the VHb gene (44). Efficient production of these metabolites under hypoxic conditions has much merit in industrial-scale production using large fermentors in which the supply of oxygen frequently becomes a limiting factor.

Figure 5a shows an *E. coli* K-12 cell expressing the VHb gene. It has been assumed that VHb enhances diffusion of oxygen from periplasm to cytoplasm across the cell membrane by reversible association with free oxygen and increases the concentration of effective oxygen ([Oc] + [HbOc]) in the cytoplasm (41,49). In cells of *E. coli* K-12 expressing the VHb gene, a twofold increase in ATP content was found compared with cells without the VHb gene (49). In general, the main terminal oxidase of *E. coli* cells is believed to be Cyt *o* under aerobic conditions; in contrast, under oxygen-limited conditions Cyt *d* becomes the main oxidase. The cells exhibiting increased ATP production with expression of VHb gene were found to possess both types of cytochromes. It has been found that Cyt *d* dose not have proton-pumping activity, although it has a higher affinity to oxygen than Cyt *o*. On the other hand, Cyt *o* has been reported to have proton-pumping activity (50). Based on this information, it has been assumed that in cells without the VHb gene, Cyt *d* works as a main terminal oxidase owing to the low oxygen concentration in the cells (Fig. 5b), whereas in the cells expressing the VHb gene, Cyt *o* as well as Cyt *d* work under the conditions of increased effective

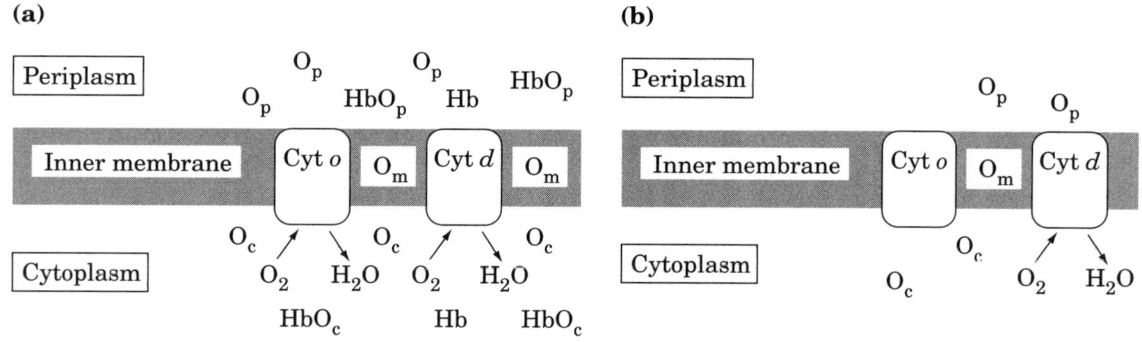

Figure 5. Effect of VHb gene expression on energy metabolism of *E. coli* K-12 (from 21, with permission of Japan Society for Bioscience, Biotechnology, and Agrochemistry). (**a**) Cells expressing VHb gene. (**b**) Cells without VHb gene. O, free oxygen (subscripts p, m, and c denote periplasm, inner membrane, and cytoplasm, respectively); Hb, deoxyhemoglobin; HbO, oxyhemoglobin.

oxygen in the cells (Fig. 5a). Therefore, in the latter case, ATP production by F_1F_o-ATP synthase would be enhanced under the hypoxic conditions caused by the increase in proton motive force. Under these conditions, various kinds of fermentative production would be improved through the enhancement of biosynthetic reactions (49).

CONCLUSIONS

In this article, we have described recent work on the energy metabolism and strain improvement with special emphasis on our work on pyruvic acid production by an *atp* mutant. Improvement of producers of useful metabolites by manipulation in energy metabolism provides not only a new direction in strain improvement but also provides useful information in the basic research of energy metabolism. From the work described, it can be concluded that defects of energy production from the *atp* mutation cause enhancement of glycolysis and thereby are suitable for the improvement of the production of primary metabolites such as pyruvate or amino acids, while the enhancement of energy production attained by the expression of the VHb gene seems to be favorable for the improvement of production of such products as protein and secondary metabolites, which require more energy for biosynthesis.

BIBLIOGRAPHY

1. A. Yokota, H. Shimizu, Y. Terasawa, N. Takaoka, and F. Tomita, *Appl. Microbiol. Biotechnol.* **41**, 638–643 (1994).
2. A.A. Herbert and J.R. Guest, *J. Gen. Microbiol.* **53**, 363–381 (1968).
3. J. Babul, *J. Biol. Chem.* **253**, 4350–4355 (1978).
4. D. Kotlarz, H. Garreau, and H. Buc. *Biochim. Biophys. Acta* **381**, 257–268 (1975).
5. S.D. Dunn, *Biochem. Biophys. Res. Commun.* **82**, 596–602 (1978).
6. A. Yokota, Y. Terasawa, N. Takaoka, H. Shimizu, and F. Tomita, *Biosci. Biotechnol. Biochem.* **58**, 2164–2167 (1994).
7. A. Yokota and S. Takao, *Agric. Biol. Chem.* **48**, 2663–2668 (1984).
8. T. Yonehara and R. Miyata, *J. Ferment. Bioeng.* **78**, 155–159 (1994).
9. P.R. Jensen and O. Michelsen, *J. Bacteriol.* **174**, 7635–7641 (1992).
10. A. Yokota, M. Henmi, N. Takaoka, C. Hayashi, Y. Takezawa, Y. Fukumori, and F. Tomita, *J. Ferment. Bioeng.* **83**, 132–138 (1997).
11. K. Matsushita, T. Ohnishi, and H.R. Kaback, *Biochemistry* **26**, 7732–7737 (1987).
12. J. Bongaerts, S. Zoske, U. Weidner, and G. Unden, *Mol. Microbiol.* **16**, 521–534 (1995).
13. J. Green and J.R. Guest, *Mol. Microbiol.* **12**, 433–444 (1994).
14. L. Thöny-Meyer, *Microbiol. Mol. Biol. Rev.* **61**, 337–376 (1997).
15. S. Iuchi, V. Chepuri, H.-A. Fu, R.B. Gennis, and E.C.C. Lin, *J. Bacteriol.* **172**, 6020–6025 (1990).
16. P.A. Cotter and R.P. Gunsalus, *FEMS Microbiol. Lett.* **91**, 31–36 (1992).
17. H.-A. Fu, S. Iuchi, and E.C.C. Lin, *Mol. Gen. Genet.* **226**, 209–213 (1991).
18. P.A. Cotter, S.B. Melville, J.A. Albrecht, and R.P. Gunsalus, *Mol. Microbiol.* **25**, 605–615 (1997).
19. M. Santana, M.S. Ionescu, A. Vertes, R. Longin, F. Kunst, A. Danchin, and P. Glaser, *J. Bacteriol.* **176**, 6802–6811 (1994).
20. J.M. Rohwer, P.R. Jensen, Y. Shinohara, P.W. Postma, and H.V. Westerhoff, *Eur. J. Biochem.* **235**, 225–230 (1996).
21. A. Yokota, *Nippon Nógeikagaku Kaishi* **71**, 11–16 (1997) (in Japanese).
22. E.B. Waygood and B.D. Sanwal, *J. Biol. Chem.* **249**, 265–274 (1974).
23. M. Diesterhaft and E. Freese, *Biochim. Biophys. Acta* **268**, 373–380 (1972).
24. B.C. Valdez, S.H. Chang, and E.S. Younathan, *Biochem. Biophys. Res. Commun.* **156**, 537–542 (1988).
25. F.W. Tuominen and R.W. Bernlohr, *J. Biol. Chem.* **246**, 1746–1755 (1971).
26. H. Sakai, K. Suzuki, and K. Imahori, *J. Biochem. (Tokyo)* **99**, 1157–1167 (1986).
27. C.K. Marschke and R.W. Bernlohr, *Arch. Biochem. Biophys.* **156**, 1–16 (1973).
28. S. Sugimoto and I. Shiio, *Agric. Biol. Chem.* **53**, 1261–1268 (1989).

29. H. Ozaki and I. Shiio, *J. Biochem.* (*Tokyo*) **66**, 297–311 (1969).
30. M.S.M. Jetten, M.E. Gubler, S.H. Lee, and A.J. Sinskey, *Appl. Environ. Microbiol.* **60**, 2501–2507 (1994).
31. A.M.C.R. Alves, G.J.W. Euverink, M.J. Bibb, and L. Dijkhuizen, *Appl. Environ. Microbiol.* **63**, 956–961 (1997).
32. L.B. Collins and T.D. Thomas, *J. Bacteriol.* **120**, 52–58 (1974).
33. J. Thompson and D.A. Torchia, *J. Bacteriol.* **158**, 791–800 (1984).
34. A. Yokota, S. Amachi, S. Ishii, and F. Tomita, *Biosci. Biotechnol. Biochem.* **59**, 2004–2007 (1995).
35. S. Amachi, K. Ishikawa, S. Toyoda, Y. Kagawa, A. Yokota, and F. Tomita, *Biosci. Biotechnol. Biochem.* **62**, 1574–1580 (1998).
36. E. Kasimoglu, S.-J. Park, J. Malek, C.P. Tseng, and R.P. Gunsalus, *J. Bacteriol.* **178**, 5563–5567 (1996).
37. R. Patnaik, W.D. Roof, R.F. Young, and J.C. Liao, *J. Bacteriol.* **174**, 7527–7532 (1992).
38. Y.P. Chao and J.C. Liao, *J. Biol. Chem.* **269**, 5122–5126 (1994).
39. S.J. Boerman and D.A. Webster, *J. Gen. Appl. Microbiol.* **28**, 35–43 (1982).
40. S. Wakabayashi, H. Matsubara, and D.A. Webster, *Nature* **322**, 481–483 (1986).
41. C. Khosla and J.E. Bailey, *J. Mol. Biol.* **210**, 79–89 (1989).
42. C. Khosla and J.E. Bailey, *J. Bacteriol.* **171**, 5995–6004 (1989).
43. C. Khosla and J.E. Bailey, *Nature* **331**, 633–635 (1988).
44. C. Khosla, J.E. Curtis, J. DeModena, U. Rinas, and J.E. Bailey, *Biotechnology* **8**, 849–853 (1990).
45. M. Khosravi, D.A. Webster, and B.C. Stark, *Plasmid* **24**, 190–194 (1990).
46. P.T. Kallio and J.E. Bailey, *Biotechnol. Prog.* **12**, 31–39 (1996).
47. S.K. Magnolo, D.L. Leenutaphong, J.A. DeModena, J.E. Curtis, J.E. Bailey, J.L. Galazzo, and D.E. Hughes, *Biotechnology.* **9**, 473–476 (1991).
48. J.A. DeModena, S. Gutiérrez, J. Velasco, F.J. Fernández, R.A. Fachini, J.L. Galazzo, D.E. Hughes, and J.F. Martin, *Biotechnology* **11**, 926–929 (1993).
49. P.T. Kallio, D.J. Kim, P.S. Tsai, and J.E. Bailey, *Eur. J. Biochem.* **219**, 201–208 (1994).
50. A. Puustinen, M. Finel, T. Haltia, R.B. Gennis, and M. Wikström, *Biochemistry* **30**, 3936–3942 (1991).

R

REVERSE MICELLES, ENZYMES

FRAGISKOS N. KOLISIS
National Technical University of Athens
Athens, Greece

KEY WORDS

Biocatalysis
Enzyme extraction
Enzymes
Lipases
Reverse micelles

OUTLINE

REVERSE MICELLES

Reverse micelles are fine dispersions of water in organic solvent stabilized by a surfactant molecule. The main property of this system is the coexistence of two completely different microphases in an apparently macrohomogeneous solution. The nature of the surfactant molecules, known as amphiphiles, induces molecular organizates, among which reverse micelles are thermodynamically stable and are formed spontaneously. The solutions are optically isotropic because one solvent is finely dispersed into the other, forming microdroplets that have a diameter on the order of 100 Å.

Formation of Reverse Micelles

The formation of the reverse micelles depends on the relative concentrations of water, surfactant, and organic solvent and can be explained by the energy change caused by dipole–dipole interactions between the polar head groups of the surfactant molecule. This mechanism can be understood by taking into consideration the values of the interfacial tension between water and oil (the term *oil* refers to any hydrophobic solvent). This interfacial tension (with values varying from 30 to 60 dyn/cm) causes a reduction of the surface between the two solvents, which results in their separation. The addition of surfactants induces the formation of micelles because of the reduction of the interfacial tension down to very low values (10^{-5} dyn/cm). The amphiphilic nature of the surfactant molecule causes its self-association in solution. In a reverse micellar solution, the polar head of the surfactant is directed toward the aqueous core of the micelle, and the hydrophobic tail extends into the continuous organic phase (Fig. 1). Depending on the hydrophilic–lipophilic balance (HLB) of the surfactant (i.e, the rate of the weight percentage of hydrophilic groups to the weight percentage of hydrophobic groups in the surfactant molecule), the interface will curve spontaneously toward water or the oil phase, favoring the formation of water-in-oil or oil-in-water structures, respectively. Various types of surfactants have been used as amphiphiles to formulate reverse micellar systems, including natural membrane lipids and artificial surfactants. The most commonly used surfactants are the anionic AOT [bis(2-ethylhexyl) sodium sulfosuccinate]; the cationic CTAB (cetyltrimethyl ammonium bromide); various nonionic surfactants of the Brij, Triton and Tween series; and the zwitterionic biological membrane-forming phospholipids. They can be grouped in three main categories: anionic, cationic, and nonionic, depending on the nature of their polar moiety (Fig. 2).

Cationic micelles are very small ($[H_2O]/[surfactant] \leq 3$). To increase the micelle size, and thus the amount of water solubilized, a cosurfactant is included in the system, usually an aliphatic alcohol, the cosurfactant being partitionated between the interface and the apolar phase. Anionic surfactants form micellar aggregates of large size ($[H_2O]/[surfactant] \leq 60$). There is a surfactant concentration under which the reversed micelle formation does not occur the critical micellar concentration (CMC). The CMC depends on the chemical structure of the surfactant, solvent, temperature, and pressure.

The solubilization properties of surfactants are often expressed by a three- or four-component phase diagram, af-

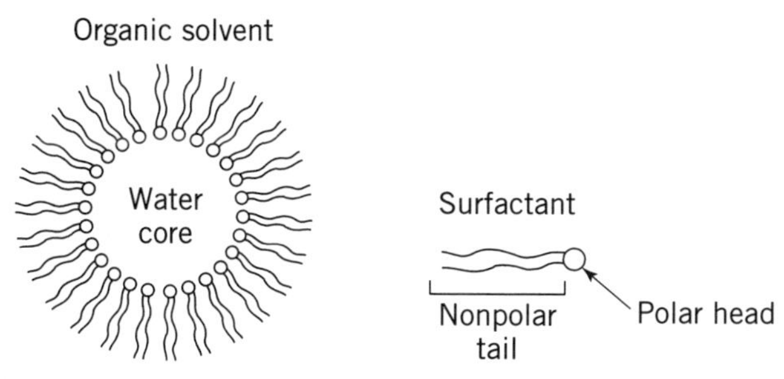

Figure 1. Reverse micelle.

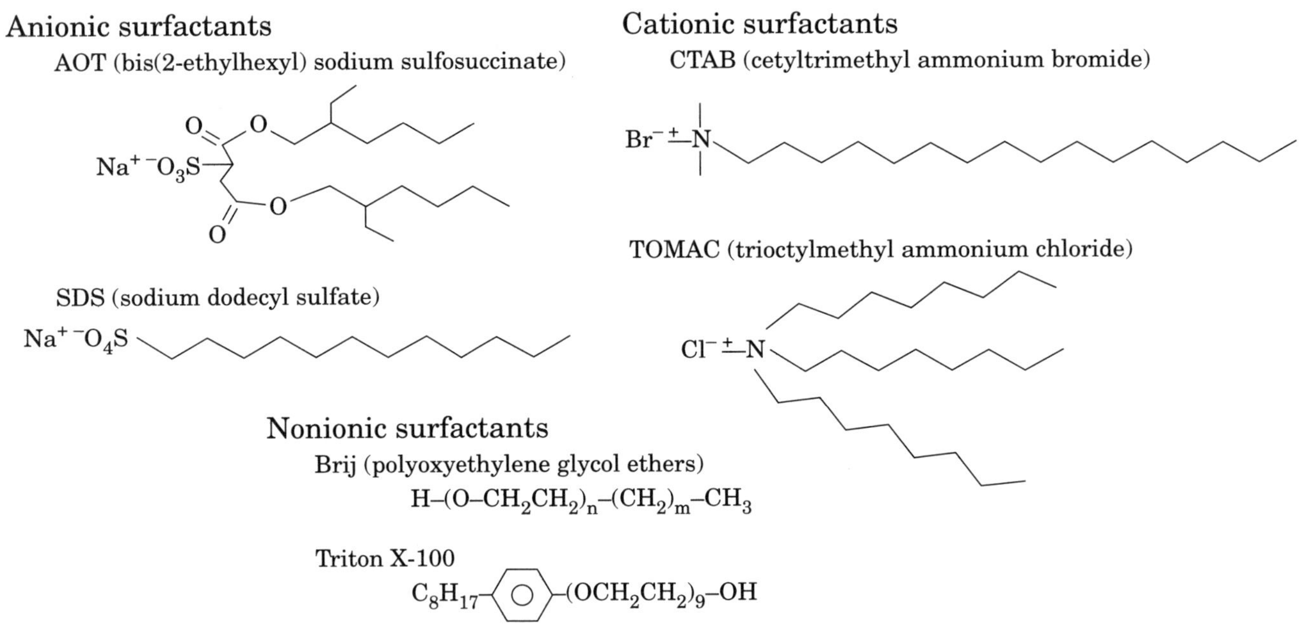

Figure 2. Some surfactants used for reverse micelles.

ter determination of regions of optical transparency. From these diagrams, it is easy to calculate the volume of each component and mix them by microinjection of the aqueous buffer solution into a stirred solution of surfactant dissolved in organic solvent in order to obtain a transparent solution. Figure 3 depicts the phase diagram of AOT/isooctane/water, showing the reverse micelles region.

Micelle Size and Structure

The size, shape, and solubilizing ability of the reverse micelles depend strongly on the nature of the amphiphile molecule. The structure of microemulsions has been studied using various techniques such as NMR, electron paramagnetic resonance (EPR), IR, fluorescence spectroscopy, chemical relaxation, and various scattering techniques such as dynamic light scattering, small-angle X-ray scattering, and small-angle neutron scattering.

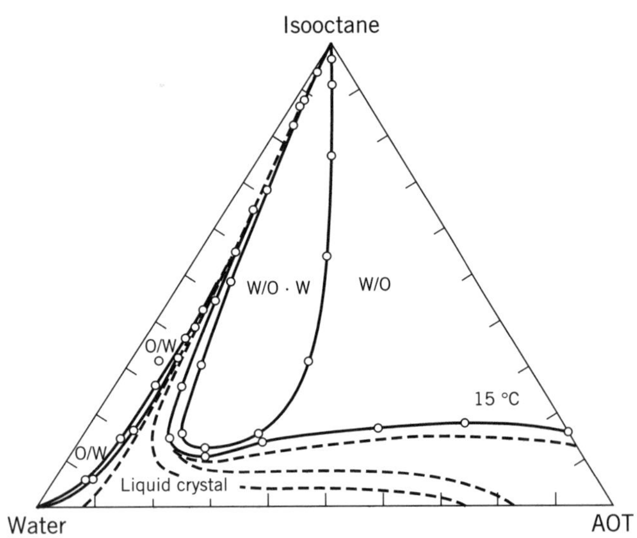

Figure 3. Phase diagram of AOT/isooctane/water, showing the reverse micelles region. O = oil; W = water.

In the majority of the works performed with reverse micelles in biological systems, sodium(bis-2-ethylhexyl)-sulfosuccinate (Aerosol-OT or AOT) has been used as surfactant. This anionic surfactant has the ability to form reverse micelles in a great number of hydrophobic solvents (oils), incorporating considerable amounts of water. The water pools that are formed can have radii from 15 to >100 æ, allowing the solubilization of hydrophilic biomolecules of various sizes.

The structure of the AOT reverse micelles strongly depends on the amount of incorporated water. The water content can be expressed by the molar ratio w_0, with $w_0 = [H_2O]/[AOT]$, reflecting the sizes of the reverse micelles. Many experimental studies using scattering techniques have shown the existence of empirical relations between w_0 and the radius of the reverse micelles (whole microdroplet), for example, $r(nm) = 1.5 + 0.175\, w_0$ (1). Also, for a given w_0 value, the increase of the respective quantities of water and AOT lead to an increase of the number of micelles.

It has been well established from various structural studies that the AOT reverse micelles are spheres formed by a monolayer of AOT molecules surrounding a water core. The water molecules that are located in the close vicinity of the AOT membrane are quite immobilized and possess different properties as compared to bulk water. This is because these water molecules serve to hydrate the polar heads of the surfactant. On the contrary, the water molecules that are situated in the center of the sphere have almost the same properties as bulk water. By increasing

w_0, that is, increasing the amount of water, the overall behavior of the water phase tends to that of normal water.

In general, the physical characteristics of the water in the reverse micelles strongly depends on the value of w_0 and on the nature of the surfactant, and the presence of these differently organized water molecules can be categorized in three different domains in every type of reverse micellar system: *free water*, whose properties and structure become closer to those of bulk water as w_0 increases; *bound water*, whose properties are qualitatively different than those of bulk water because of hydrophilic interactions with the polar head groups of the surfactant; and *surfactant tails*, which penetrate the apolar solvent. When the amount of water is not sufficient to hydrate the surfactant molecules, there is not free water and reverse micelles have only two domains: bound water and surfactant.

Reverse Micelles Dynamics

Reverse micelles are not rigid, impenetrable spheres; they are dynamic entities. In addition to the normal Brownian motions, a number of relaxation processes characterize the monomer, that is, the individual surfactant molecules participating in the micellar aggregate and the water molecules. In contrast to micelles formed in the aqueous solution, the reverse micelles are electrically neutral in their exterior shell, so that no Coulombic repulsions occur, and collisions are frequent. The collision of reverse micelles can be predicted from simple diffusion theory to be on the order of nanoseconds. Not all collisions result in mixing of the aqueous pool. When micelles collide with each other, they may build a transient dimer with a communication channel, which permits the exchange of materials.

Exchange of the contents of reverse micelles is believed to occur by means of a collision–fusion–fission process. The characteristic time for exchange of solubilizates between water pools in AOT reverse micelles has been found to be 10^{-4} to 10^{-6} seconds. The exchange is a function of droplet size and temperature, decreasing with increasing droplet size at a given temperature, and increasing with increasing temperature at a given w_0. The rate of exchange for reverse micelles stabilized by the surfactant alkylarylsulfonate is much slower, of the order of hours, whereas that of CTAB reverse micelles is on the order of microseconds.

pH and Effect of the Ionic Strength in the Reverse Micelles

The pH of the solubilized water in reverse micelles can differ form that of the bulk water, as a consequence of the high concentration of its electric charges caused by the ionization of the polar head groups of the surfactant molecule. For example, in an anionic surfactant reverse micellar system, a decrease the apparent pH of the water within the reverse micelles is observed, and this difference is larger for low w_0.

In studies concerning the role of salts, when salts are present in the aqueous pool of the micelles, it has been found that the ionic strength, the type of salt, and the type of buffer used can affect the size as well as the shape of the reversed micelles.

These two factors, pH and ionic strength, are of particular importance for the encapsulation of enzymes inside reverse micelles in their application in biocatalysis and liquid–liquid extraction of proteins.

ENZYMES, IN REVERSE MICELLES

Enzymes represent novel routes to a number of important products, ranging from β-blockers to synthetic cocoa butter (2). Although they have outstanding catalytic properties, including regioselectivity, stereoselectivity, and enantioselectivity, which can be applied for the preparation of compounds not available by chemical means, their use in organic synthesis is quite limited. The main reason is that most organic compounds of commercial interest are very sparingly soluble in aqueous media used in conventional enzymology. Water is a poor solvent for nearly all applications in industrial chemistry.

Biocatalysts—enzymes, microbial cells, and others—are traditionally applied to aqueous reaction systems, including water-soluble reactants. Since the recent development of biotechnology, much wider applications of biocatalysts have been demanded, for example, bioconversions of biological and xenobiotic compounds having lipophilic or hydrophobic characters. In such bioconversion reaction systems, it is indispensable to introduce organic solvents into the system to improve the poor solubility of the reactants in water. Furthermore, in an attempt to apply biocatalysts for the conversion of petrochemicals, the substrates and products are themselves organic solvents in such biochemical reactions.

The recent realization that enzymes can function also in organic solvents has given the synthetic potential of enzymes a powerful boost for many processes. Other uncommon biocatalysis media are supercritical gases such as CO_2(CO_2SF), liquid membranes, and reverse micelles. The activity and specificity of the enzymes in these uncommon environments are changed. The enantiomeric excesses are normally increased. Several substrates (e.g, lipophilic ones) are better soluble in these media and downstream processing is often easier. Various enzymes (such as alcohol dehydrogenase and esterases, to mention some of them), catalyzed asymmetric redox reactions, stereospecific transesterification, regioselective esterification of sugars, and others have been studied in nonaqueous systems.

Enzymes Inside Reverse Micelles

The use of reverse micelles to solubilize enzymes in organic solvents has attracted considerable interest in the past decade (3–5). A large variety of enzymes, including cofactor-dependent enzyme, have been successfully encapsulated in reverse micellar systems, retaining their catalytic activity. The enzyme molecules can be entrapped in the reverse micelles, avoiding direct contact with the organic medium, potentially limiting their denaturation. The interior of the reverse micelles acts like a microreactor that provides a favorable aqueous microenvironment for enzyme activity as well as an enormous interfacial area through which the conversion of hydrophobic substrates can be catalyzed. Increasing the interfacial area is of great technological interest since this results in the increase of the number of substrate molecules available to react.

When biomolecules such as enzymes are solubilized in reverse micelles, they can be localized in the different microcompartments, depending on their chemical structure or relative solubility in these microphases. The degree of hydrophobicity of the biomolecule plays an important role to the preferential localization among the various microenvironments (water core, bound water, surfactant membrane, and organic solvent). Hydrophilic enzymes are entrapped in the water core, whereas others with an amphiphilic character can be anchored to the surfactant layer, penetrating even in the continuous organic phase (Fig. 4). Recent studies relate the enzyme localization in the different domains of the reverse micelles with their selectivity for various substrates with a different degree of polarity. This hypothesis at least stands for lipase-catalyzed esterification reactions. In addition, this interaction between the enzyme molecule and the interface can explain the different enzyme behavior found in several cases with respect to bulk water.

Solubilization of Enzymes in Reverse Micelles. The most common technique used for the encapsulation of enzymes in the polar core of reverse micelles is the injection method. The protein already solubilized in a concentrated stock aqueous (or buffered) solution is added to the surfactant-containing organic solution upon stirring. With this method, the solubilization is almost instantaneous, and it is possible to control the amount of water present in the system. Another method is to contact the protein solution with the micellar solution upon stirring. Phase transfer will occur, depending on the charge of distribution of the enzyme, surfactant charge, and pH and ionic strength of the aqueous phase. With this method, the enzyme solubilization is slow, but a large amount can be solubilized with minimal values of w_0. A third way of introducing the enzyme is to add the dry enzyme into the empty reverse micelles while aggitating. This technique causes denaturation of the enzyme to a degree. The injection and dry addition methods are commonly used in biocatalytic applications, the latter being suited to hydrophobic proteins. The phase-transfer technique is the basis for extraction of proteins from aqueous solutions. In addition to the method of protein inclusion to the reverse-micelles, other variables that can influence protein solubilization are the size and isoelectric point of the protein, the size of the reverse micelles, the nature of the surfactant used, and the pH and ionic strength of the aqueous solution.

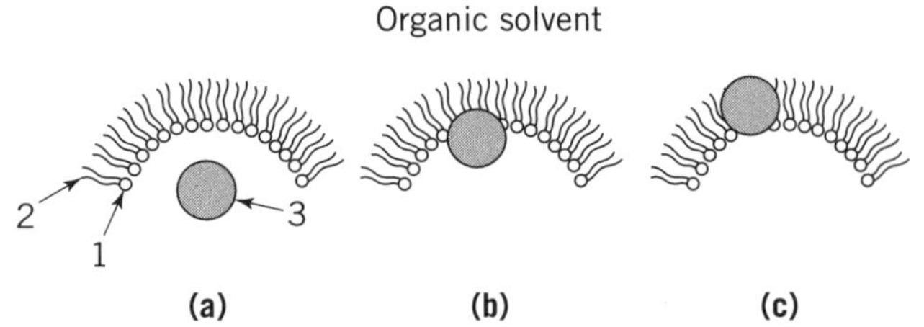

Figure 4. Enzyme localization in reverse micelles depending on the degree of hydrophobicity of the molecule. (**a**) Hydrophilic; (**b**) Amphiphilic, and (**c**) Hydrophobic enzyme. *1*, hydrophilic head of the surfactant; *2*, hydrophobic tail of the surfactant; and *3*, enzyme molecule.

Factors Affecting Enzyme Solubilization. For both the injection and phase contact methods, proteins with a net charge opposite to the surfactant charge and a low ionic strength will favor solubilization. As far as the size of protein molecule is concerned, their solubilization may occur even when the radius of the molecule is larger than the micellar polar core radius. The protein itself probably stabilizes the enlarged micelle, constructing a protective surfactant shell with a certain amount of water. Data support the possibility that there is not a definite maximal molecular size upon which encapsulation in surfactant aggregates does not occur.

The hydrophilic character of large proteins may have a significant role on the formation of reversed micelles. It seems that the hydrophilic proteins will favor the formation of surrounding surfactant shell. This fact could account for the solubilization of large entities, even when the total charge of the macromolecules is identical to the charge of the surfactant heads, thus excluding the formation of micelles via direct electrostatic interactions between the protein and the surfactant layer. This type of interaction within a micellar system can be particularly important for the phase transfer of proteins, applied to the liquid–liquid extraction of proteins and other molecules via reverse micelles. The various parameters affecting this process are discussed in more detail in the section on liquid–liquid extraction.

Enzyme Conformation in Reverse Micelles. Enzyme-containing reverse micellar solutions are optically clear. Thus, spectroscopic techniques, such as ultraviolet (UV), fluorescence, circular dichroism (CD), EPR, and NMR, are able to be used for monitoring the conformation changes that may occur in enzyme molecules upon their solubilization in reverse micelles. Moreover, in this system it possible to precisely control the range of water microenvironments (interfacial/bulk), the size of the micelles by varying the water/surfactant molar ratios, and the role of the chemical structure of the surfactant.

Generally, a more ordered structure has been identified by CD for myelin basic proteins (6) and poly(α-amino acids) (7) in AOT reverse micelles and trypsin and α-chymotrypsin in lecithin reverse micelles (8). UV spectra of enzymes in reverse micelles show little change from aqueous solutions. Fluorescence spectra in reverse micelles are red shifted for some enzymes and blue shifted for others (9,10). An increase in quantum yield has been observed with α-chymothrypsin (10), trypsin (11), pepsin (11), lysozyme (10), and a decrease in quantum yield has been observed for lactate dehydrogenase (12). ^{1}H-NMR studies of some peptides and small proteins show that met-encephalin became more folded, and pancreatic secretory trypsin inhibitor became highly flexible (13). Furthermore, EPR studies have been performed for investigation of structural changes to α-chymotrypsin (14), serum albumin (15), and cutinase (16) upon solubilization in reverse micelles.

The information derived by several studies shows that proteins can be categorized according to the perturbation of their structure induced by reverse micelles: conformations that are not changed or only slightly changed after incorporation into reverse micelles; conformations that are altered in reverse micelles but unaffected by the amount of water present in the system; and conformations that are altered and that also change as a function of the water/surfactant molar ratio.

Biocatalysis in Reverse Micelles

The first report on enzymes retaining their catalytic activity in reverse micellar systems was in 1978 (2,17,18). Today the number of active enzymes described in reverse micelles has increased to more than 50, including hydrolases, dehydrogenases, and oxidases used in single or multienzyme catalyzed reactions. Enzymatic catalysis in microemulsions has been used for a variety of reactions, such as synthesis of esters, peptides and sugar acetals, transesterifications, various hydrolysis reactions, and steroid transformations. Several applications with potential interest for biotechnological conversions have been developed; some of them are summarized in Table 1.

Enzymatic reactions in reverse micelles are found to obey normal Michaelis-Menten kinetics, but some attention should be given to the way of expressing the kinetic constant K_m. K_m is the substrate concentration where the reaction rate is half the maximum rate. The substrate concentration in a reversed micellar system can be based on the total volume of the system or the water-pool volume of the micelles, but the latter is very difficult to measure. Other factors that have to be taken under consideration is the compartmentalization of the enzyme and substrates between the various microdomains of the reverse micelles and the degree of hydrophobicity of substrate, product, and enzyme molecule, which is related to its localization in the micelles. Several theoretical models have been developed to explain enzyme kinetics in reverse micelles. They are based on different assumptions, such as all droplets in the medium are equal, and there are not any changes in the enzymes kinetic parameters (K_m and V_{max}) during their entrapment in the reverse micelles (the diffusional models). The nondiffusional models are based on the fact that the differences in the kinetic constants with respect to bulk water are caused by conformational changes produced in the reverse micellar microenvironment. Another important parameter obtained from Michaelis-Menten kinetics is the K_{cat}, the turnover number, which is related to maximum reaction velocity. Because K_{cat} is concentration independent, its determination is more straightforward. Turnover numbers for enzymes in reverse micelles are usually dependent on the amount of water present in the system, which is expressed by the molar water-to-surfactant ratio, w_0. This dependence is often a bell-shaped curve, explained by the existence of an optimal inner cavity for catalytic activity, which corresponds to the size of the enzyme. Maximum K_{cat} is sometimes higher than that in aqueous solutions, giving rise to the concept of superactivity (31). This effect can be explained by a greater reactivity of the structured water in the micelle, increased rigidity of the enzyme (32), and enhanced substrate concentration at the enzyme surface (33). Thus, the first parameter that should be considered when studying the enzyme activity in reverse micelles is the amount of the water in the system (w_0). The electric properties of the substrate and surfactant can often have a marked effect on the K_m and K_{cat} constants. When the charges of substrate and surfactant are the same, the differences of K_{cat} and K_m values with those found in water are not significant, but when the charges are different, partition effects are usually observed because of the increase of the substrate concentration near the vicinity of the surfactant layer, hampering their access to the enzyme, which is surrounded by water. α-Chymotrypsin and trypsin encapsulated in AOT and CTAB show a decrease in the second-order rate constant (K_{cat}/K_m). The specificity of the enzymes can also be altered because of the distribution of the substrate between the system phases or interactions with the surfactant layer. An example is the oxidation of aliphatic primary alcohols, catalyzed by alcohol dehydrogenase encapsulated in an AOT/octane system (24). In the reverse micellar system, octanol ($n = 8$) is the optimum substrate, whereas in water the preferred substrate is butanol ($n = 4$) (17). It is profound that these parameters are more accurately determined using the active enzyme concentration, which can be obtained typically by active site titration, for example, based on fluorescence techniques for the active site titration of α-chymotrypsin and lactate dehydrogenase (1). Another parameter to take into account is the pH. When the enzyme is entrapped in reverse micelles, significant shifts of the pH profile usually occur, because of the heterogeneous dis-

Table 1. Examples of Bioconversions in Reversed Micellar Systems

Type of reaction	Enzyme	Reverse micelles
Reduction of decanal	Alcohol dehydrogenase	AOT/isooctane (19)
Polymer synthesis	Horseradish peroxidase	AOT/isooctane (20)
Reduction of ketosteroids	Alcohol dehydrogenase	AOT/isooctane (21)
Oxidation of polyunsaturated fatty acids, linoleic acid	Lipoxidase	AOT/isooctane (22)
Cholesterol oxidation	Cholesterol oxidase	CTAB/octanol/octa ne (23)
Oxidation of aliphatic alcohols	Alcohol dehydrogenase	AOT/octane (24)
Peptide synthesis	α-Chymotrypsin	AOT/isooctane (25)
Hydrolysis of olive oil	Lipase	AOT/isooctane (26)
Inter- and *trans*-esterification	Lipase	AOT/isooctane (27)
Bioluminescent assay	Luciferase	Brij-96/octane (28)
Fatty acid esterification with alcohols	Lipase	AOT/isooctane (29)
Polyol–fatty acid esters	Lipase	AOT/isooctane (30)

tribution of protons inside the polar core of the micelle. Larger shifts of optimal pH occur for small values of w_0. The shifts reported in literature for systems with anionic surfactants are of one to two unities (3).

Lipases in Reverse Micelles

A particular case of enzymatic studies in microemulsions is that of lipases (34). Among the enzymes so far investigated, lipase is one of the most advantageous because it is stable, inexpensive, and widely used in the development of various applications in the detergents, oils and fats, dairy, and pharmaceutical industries. In addition to hydrolysis of triglycerides to glycerol and free fatty acids, lipases can be used in esterification and transesterification reactions in low-water content media. This catalytic process is heterogeneous and can be favored by the use of microemulsion. Lipases are active almost exclusively near interfaces in a classical heterogeneous procedure. There are various reports in the literature concerning studies on the hydrolytic and synthetic action of lipases in different types of microemulsions. Table 2 lists examples of the studies performed so far with lipases from various sources in water-in-oil microemulsions.

Hydrolytic Reactions. Lipase catalyzes the hydrolysis of long-chain aliphatic esters to glycerol and free fatty acids. Several hydrolytic reactions performed by lipases in microemulsion systems have been reported, such as the hydrolysis of triolein and tributyrin by pancreatic lipase in an AOT/octane microemulsion system (35). The same system has been used for the hydrolysis of palm kernel olein by *Rhizopus arrhizus* lipase (36). Hydrolysis of triolein and tributyrin has been reported with the use of *Rhizopus delemar* lipase in AOT/isooctane, CTAB/isooctane, and $C_{12}E_4$/decane systems (37–41). Olive oil hydrolysis by *Candida rugosa* lipase and glycerolysis of triolein by *Chromobacterium viscosum* lipase in AOT microemulsion systems have been performed (42,43). A bioreactor system has been designed that is suitable for continuous glycerolysis of fats and oils, using liposomes as a matrix for adsorbing enzyme and AOT reverse micellar systems, and as a supplier of interfacial area for glycerolysis (45). Palm oil and trimyristine hydrolysis by *Rhizopus* sp. lipases for the production of monoglycerides and diglycerides with a good yield has been also reported in AOT and $C_{12}E_3$ based systems (40,41).

Table 2. Lipase-Catalyzed Reactions in Microemulsions

Lipase source	Microemulsion system
	Hydrolytic reactions
Pancreas	AOT/octane (35)
R. arrhizus	AOT/octane (36)
R. delemar	AOT/isooctane (37)
	CTAB/isoctane (38)
	$C_{12}E_4$/decane (38)
	Lecithin/isooctane (39)
Rhizopus sp.	AOT/isooctane (40)
	$C_{12}E_3$/isooctane (41)
Candida rugosa	AOT/isooctane (42)
Chromobacterium viscosum	AOT/isooctane (45)
Candida cylinracea	Lecithin/cyclooctane (44)
	Synthetic reactions
Rhizopus delemar	Phosphatidylcholine/hexane (27)
	$C_{12}E_3$/nonane (47)
	$C_{12}E_4$/decane (49)
	AOT/isooctane (30)
	AOT/hexane (47)
Rhizopus arrhizus	AOT/isooctane (50)
Chromobacterium viscosum	AOT/heptane (46)
Candida cylindracea	Brij-35/olive oil (48)
	AOT/isooctane (30)
Penicillium simplicissimum	AOT/isooctane (50)

Several parameters play an important role in lipase-catalyzed hydrolytic reactions in microemulsion systems. The reverse micellar structural characteristics, such as the size, shape, and solubilizing ability, which depend strongly on the nature of the polar moiety of the amphiphile used, strongly affect the enzyme activity. Anionic, cationic, and nonionic types of surfactants have been used in various hydrolytic reactions. Among the different systems studied, the anionic surfactants seem to be more suitable for the hydrolytic activity of lipase, as compared to the cationic ones, whereas nonionic systems do not seem adequate for this type of reaction.

Other important parameters affecting lipase activity that have been studied in detail by several groups are pH, temperature, water content (w_0), fatty acid chain length of the triglycerides used as substrates, and enzyme stability. In studies concerning the hydrolysis of triolein by *Rhizopus delemar* lipase, the pH optima were reported to be 6.5 in the anionic systems and 5.8 in the cationic systems, considering as pH values the values of the buffered enzyme solutions used for the preparation of the various microemulsion systems. The pH optimum for the same reaction performed in classic heterogeneous biphasic systems was found to be 5.6. This difference is obviously related to the fact that the pH values within the water-dispersed phase do not necessarily correspond to that in the bulk water. The difference observed between AOT and CTAB systems is possibly the result of the different amounts of bound water and the nature of the ions present in the water pools of each system. In the case of AOT systems, it has been reported that 8 to 10 water molecules are indispensable for hydrating each Na^+ and sulphonate group, whereas for CTAB, only 1 to 3 water molecules per Br^- and ammonium head are enough. In other studies on AOT-formulated microemulsion systems, a pH optimum of 7.1 was reported for lipase from *Candida rugosa* 7.0 for lipase from *Rhizopus* sp.; and a maximum in the range 6.0–7.5 for lipase from *Chromobacterium viscosum*. Temperature also plays an important role in lipase activity in microemulsions; in the case of anionic systems, the optimum temperature is reported to be 30 °C; in cationic systems, 22.5 °C. However, at temperatures lower than 15 °C and higher than 40 °C, the structure of these systems may be considerably altered.

Water content in microemulsions appears to be the most important factor that determines the reaction rates. The molar ratio w_0 (w_0 = $[H_2O]/[\text{surfactant}]$) values are used to express the water content of the system. Several reports concerning the activity of lipase from various sources as a function of w_0 shows a maximum in the range of 9 to 11. The same pattern is also observed for other enzymes studied in the same microemulsion systems, such as chymotrypsin and pyrophosphatase. However, when reverse micelles based on lecithin were used, the maximum reaction rate was observed at an extremely low water content ($w_0 = 2.2$).

In studies concerning the effect of the fatty acid chain length as well as the type of microemulsion system on the substrate conversion rate, it was found that for microbial origin lipases, triolein was a slightly better substrate as compared to tributyrin, but pancreatic lipase cleaves triolein 15 times more efficiently than tributyrin. Also, the enzyme in anionic microemulsion systems shows a higher affinity as compared to the cationic systems, whereas nonionic systems do not seem adequate for hydrolytic reactions.

One serious problem for the application of lipase to triglyceride hydrolysis is the stability of the enzyme in the microemulsion systems. The size of the microemulsion microdroplet seems to play an important role on the enzyme stability. It has been reported that lipase stability is improved as w_0 decreases. Also, the presence of various additives, such as $CaCl_2$ and casein, considerably increases the stability of lipase in AOT microemulsion systems.

Synthetic Reactions. The hydrolytic reaction catalyzed by lipase is reversible because the enzyme can also catalyze the synthesis and transesterification of triglycerides, under certain conditions. This catalytic process is of great technological interest and may be favored by the use of microemulsions. The lipophilic substrates and products can be solubilized in the continuous organic phase, whereas lipase remains active in the water core of the microemulsion droplets. In this case, specific synthetic reactions can be catalyzed, under mild conditions. The synthesis of triacylglycerol by *Rhizopus delemar* lipase has been reported in a phosphatidylcholine reverse micellar system (27). The esterification of glycerol with oleic acid, catalyzed by *Chromobacterium viscosum* lipase in glycerol-in-oil–type microemulsions of AOT in heptane, and the synthesis of various fatty acid esters in water-in-oil microemulsions, stabilized by CTAB, have also been reported (46). Similar systems have been used for the transesterification of palm oil with stearic acid (47). Synthesis of heptyloleate and the esterification of various triglycerides by *Candida cylindracea* lipase in a microemulsion-like system has been reported using Brij-35 as surfactant (48). The esterification of various alcohols with fatty acids by *Rhizopus delemar* lipase has been performed using nonionic microemulsion systems stabilized by $C_{12}E_4$, with significantly fast reaction rates also reported (49). In this work, it was shown that pH ranging from 3.45 to 8.55 does not affect the reaction rates in this type of microemulsion system. Also, studying the effect of the system's water content showed that when $w_0 < 5$ the conversion is considerably slow, but for $w_0 > 7$ a 90% conversion is attained in 4 h. Esterification reactions of alcohols with fatty acids and synthesis of polyols have been reported (30) using *Rhizopus delemar* and *Candida cylindracea* lipase in AOT microemulsion systems. Recent studies have been reported on the activity of lipases from *Rhizopus delemar, Rhizopus arrhizus*, and *Penicillium simplicissimum* in esterification reactions of various aliphatic alcohols with fatty acids in microemulsions formulated by AOT in isooctane (50). The results showed a remarkable selectivity of the lipases regarding the chain length and the structure of the substrates. *P. simplicissimum* lipase presented higher reaction rates in the esterification of long-chain alcohols as well as secondary alcohols. Primary alcohols had a low reaction rate, and tertiary alcohols had a very slow rate of esterification. Long-chain fatty acids were better catalyzed as compared to the shorter ones. *R. delemar* and *R. arrhizus* lipases showed a preference for the esterification of short-chain primary alcohols, but the secondary alcohols had a low rate of esterification and the tertiary could not be catalyzed. Medium-chain fatty acids were also better catalyzed than the long ones. Fluorescence quenching measurements suggest that there are not any significant changes of the shape and size of the microemulsions caused by the substrates, which could have been related to the enzyme selectivity. In extensive studies using a large number of substrates with different degrees of polarity as well as spectroscopic techniques, this selectivity appears to be related to the localization of the enzyme molecule within the micellar microstructure, because of the hydrophobic/hydrophilic character of the protein and not of a specificity of the enzyme itself. This hypothesis can further be supported if one takes into consideration the structural differences of these lipases.

Kinetic Studies. Lipase-catalyzed hydrolysis of triglycerides in reverse micelles follows Michaelis-Menten kinetics. The release of fatty acids as a function of reaction time at different enzyme concentrations and fixed substrate concentrations (20–40%, v/v, where $[S] \gg K_m$) is proportional to the enzyme concentration. The K_m and V_{max} parameters of the Michaelis-Menten kinetic equation can be determined from the Lineweaver-Burk double-reciprocal plot of the lipase activity only as apparent ones. To find a true K_m, the substrate concentration at the enzyme surface must be used. This requires knowledge of the substrate partitioning between various microphases, as the water pool, the interface, and the bulk organic solvent. The partition coefficient is difficult to be determined experimentally. Another constant obtained from Michaelis-Menten kinetics is the turnover number, K_{cat}, which is related to the maximum reaction velocity. Determination of K_{cat} depends on accurate measurement of active-enzyme concentration in the microemulsion system. Usually, the total enzyme concentration and not the active-enzyme concentration is used for the calculation of K_{cat}. The latter can be achieved by active site titration studies.

Kinetic studies of the esterification reaction on the reaction of ethylene glycol and fatty acids (30) and of lauric acid and (-)menthol (29), catalyzed by lipase in AOT/isooctane/water microemulsion systems, have been described. In the second case, the authors have shown that the kinetics of this synthesis follow a ping-pong Bi-Bi mechanism. According to this mechanism, lipase reacts with lauric acid to form the lipase–lauric acid complex. The complex is then transformed to a carboxylic–lipase intermediate and water is released. This is followed by an attack of (-)menthol to this intermediate, forming the (-)menthyl laureate. The values of all apparent kinetic parameters were determined, and a model concerning this type of reaction involving two hydrophobic substrates was proposed. According to this, the partitioning of the two hydrophobic substrates between the continuous organic phase and the reverse micelles must be taken into account to determine the real kinetic constants.

Lipase Applications in Reverse Micelles

The potential applications of lipases in biotechnological processes have been the subject of increased scientific, economic, and industrial interest and efforts in recent years. Regioselectivity, stereospecificity, substrate specificity, and low energy consumption are some of the features that make lipase-catalyzed processes an alternative to conventional nonenzymatic reactions. Special attention is given to the transformation of fats and oils as well as the preparation of synthetically useful enantiomers. Among the systems under investigation, the encapsulation of lipases in reverse micelles seems to present several advantages such as greater interfacial area, lower mass transfer limitations, and enhancement of catalytic activity.

The development of a bioreactor that enables reactions in a semibatch or continuous mode allowing the separation of substrates, products, and enzymes is a critical demand in reversed micellar technology today. Recently, two types of reactors have been developed for the glycerolysis of olive oil by lipases. In the first case, the bioreactor, abbreviated as LRM system, uses liposomes for adsorbing the enzyme and reversed micelles as a good supplier of interfacial area for the glycerolysis reaction (45). According to this system, the lipase was adsorbed on the liposomes prepared in a simple manner, and reverse micelles containing the immobilized enzyme were formed by injecting appropriate amounts of glycerol and predetermined amounts of water in AOT/isooctane system. Olive oil in isooctane is fed at controlled flow rates into the bioreactor by pressurizing the reservoir with nitrogen gas. The reaction takes place at 37 °C. The products are obtained from the outlet in a single phase, but the microemulsions are not broken. A microporous membrane is used to separate products formed (oleic acid, monoolein, and diolein) and unreacted substrate (triolein) from the microdroplets containing immobilized lipase. The membrane, made of polysulfone, is about 12.5 μm thick, and the micropores have a maximum diameter of 25 nm. This system presents a long operational stability. The second bioreactor system is a membrane bioreactor for simultaneous lipolysis and bioproducts separation in a reversed micellar system (51). The main purpose of this type of bioreactor is to retain lipase behind the membrane where reaction occurs and to recover the products on the permeate side. The membrane is a ceramic one, resistant to organic solvents, made of a zirconium oxide layer over a porous carbon support with a 10,000 molecular weight cutoff. The inside diameter is 6 mm; the wall thickness, 2 mm; and the effective length, 20 cm. The internal permeation area is 38 cm^2. The reaction mixture is recirculated from the thermostated vessel into the reactor by means of a gear pump flowing tangentially to the membrane. Part of the solution is permeated through the wall of the semipermeable membrane (by connective flow) as a result of the hydrostatic pressure difference, and it is recycled to the reactor. All connecting piping is solvent resistant.

Enzymatic catalysis in reverse micelles has also been performed mostly in batch-type reactors in various other processes. Some examples are the synthesis of peptides by α-chymotrypsin in reverse micelles in a hollow fiber reactor (25), or the production of tryptophan by tryptophanase in a continuous reversed micellar reactor (52).

Liquid–Liquid Extraction of Enzymes Using Reverse Micelles

The application of reverse micelles in the extraction and isolation of enzymes is a promising technique at the downstream biotechnological processes, having interesting advantages over existing processes for large-scale recovery of extracellular enzymes (53). The traditional process of transfer proteins into organic solvents often causes irreversible denaturation of proteins and loss of biological function. A reverse micellar extraction combines the potential for concentration and purification of the enzymes through transfer between aqueous and organic micellar phases. Enzymes can be extracted without loss of activity. Examples of enzyme extraction using reverse micellar systems are summarized in Table 3.

The recovery of enzymes from the aqueous phase involves two steps: forward and back extraction. In the first step, the desired enzyme is transferred selectively from the aqueous phase to the reversed micellar phase, and in the second step, the enzyme is subsequently recovered from the reversed micellar phase by extraction with a second aqueous phase, the micellar phase being recovered for further extractions. The rates of forward transfer from the aqueous to the micellar phase are extremely fast. Upon steering, for example, only a few seconds are necessary for the transfer of cytochrome *c* into a system of AOT/isooctane. The process of back extraction into the new aqueous phase is much slower and requires constant agitation. In ideal conditions, the back extraction of cytochrome *c* can be achieved in several minutes. The distribution of an enzyme between the reversed micellar and the aqueous phase is influenced by several factors such as pH, ionic strength, and surfactant type (anionic, cationic, or nonionic). The pH determines the ionization state of the surface-charged groups of the protein molecule. The solubilization is favored when attractive electrostatic interactions occur between the protein molecule and the surfactant head groups; in this case, the overall charge of the surfactant is opposite to the overall charge of the protein. This implies that, for cationic surfactants, solubilization of the protein in reverse micelles is favored at pH values above the isoelectric point (pl) of the protein, whereas the opposite is true for anionic surfactants. Together with the pH, the ionic strength plays a very important role in protein recovery. The presence of salts in the aqueous phase can alter the equilibrium size of the surfactant aggregates.

Table 3. Examples of Enzyme Extraction Using Reversed Micelles

Enzyme	Reversed micellar system
Trypsin (54)	AOT/isooctane, octane, heptane
Extracellular alkaline protease (55)	AOT/isooctane
α-Amylase (56)	TOMAC/isooctane
Separation of a mixture of cytochrome *c*, ribonuclease, lysozyme (57)	AOT/isooctane
Lysozyme, α-chymotrypsin, pepsin (58)	AOT/isooctane
Isocitrate dehydrogenase, β-hydroxybutyrate dehydrogenase, glucose-6-phosphate dehydrogenase (59)	CTAB/octane/exanol
α-Chymotrypsin (60,61)	NR_4^+ /cycloexane
	AOT/octane/water/glycerol
Ribonuclease A (62)	AOT/isooctane

A higher ionic strength will decrease the size of the micelle, and this effect can be stronger for larger ions, which causes shorter electrostatic interactions. Thus, it is possible to achieve a selective transfer based on size exclusion effects. The attraction between the surfactant and the protein is weaker, and this contributes to exclude the protein from the micellar core. When the ionic strength is very low, phase transfer does not occur. The transfer of proteins from an aqueous to a micellar phase requires a minimum of ionic strength in the aqueous phase. In addition to the surfactant charge, other surfactant-dependent parameters have to be taken under consideration, such as the size of the reversed micelles that are formed, the energy required to enlarge the reverse micelles, and the charge density on the inner surface of the reverse micelles.

The liquid–liquid extraction technique in general is well known, and apparatus and scale-up rules have been established for numerous applications, including the use in recovery processes for antibiotics and organic acids from fermentation broths. The forward and back extractions can be performed in a continuous mode, with the reverse micellar phase circulating between the two extraction units. No specialized equipment is required. The process developed for conventional liquid–liquid extraction is, in principle, also suitable for this application. The most important types of equipment are mixer/settler, agitated column, centrifugal, and membrane extractors. At the present time, process development has centered on the use of mixer/settlers and membrane extractors.

CONCLUDING REMARKS

So far, the results derived from studies with reversed micelles show that the use of these systems can reach a considerable biotechnological relevance and be extended to a wide range of applications. In the field of bioconversion, the enzyme solubilization in reversed micelles clearly offers advantages over other techniques, including the very fast reaction rates or the small quantity of enzyme to be used in comparison to other approaches of enzymology in extreme conditions. However, there are a few problems that must be solved in order to take full advantage of this system, such as the need to remove the excess surfactant from the products and to develop appropriate reactors, which will allow the continuous production and in situ recovery of the bioproducts applying new separation techniques, for example, membrane polymers. Another promising field for reverse micellar systems application is in the downstream processing for the separation of other bioproducts besides proteins and peptides, such as amino acids, nucleotides, nucleic acids, solvents, organic acids, antibiotics and steroids.

BIBLIOGRAPHY

1. J.D. Nicholson and J.H.R. Clarke, in K.L. Mittal and B. Lidman eds., *Surfactants in Solution*, vol. 3, Plenum, New York, 1984, p. 1663.
2. H.W. Blanch, in F.N. Kolisis, B. Macris, and D. Kekos, eds., *Biocatalysis in Organic Solvents*, EEC Advanced Workshops on Biotechnology, 1992.
3. M.J.M. Castro and J.M.S. Cabral, *Biotech. Adv.* **6**, 151–167 (1988).
4. M. Waks, *Proteins: Struct. Funct. Genet.* **1**, 4–15 (1986).
5. P.L. Luisi and R. Wolf, in Mittal and Fedler eds., *Solution Behavior of Surfactants*, vol. 2, Plenum, New York, 1982, pp. 887–905.
6. C. Nicot, M. Vachet, M. Vincent, J. Gallay, and M. Waks, *Biochemistry* **24**, 7024–7032 (1985).
7. M. Seno, H. Noritomi, Y. Kuroyanagi, K. Iwamamoto, and G. Ebert, *Colloid Polym. Sci.* **262**, 727–733 (1984).
8. O. Peng and P.L. Luisi, *Eur. J. Biochem.* **188**, 471–480 (1990).
9. S. Barbaric and P.L. Luisi, *J. Am. Chem. Soc.* **103**, 4239–4244 (1981).
10. C. Grandi, R.E. Smith, and P.L. Luisi, *J. Biol. Chem.* **256**, 837–843 (1981).
11. P.L. Luisi, F. Henninger, M. Joppich, A. Dossena, and G. Casnati, *Biochem. Biophys. Res. Comm.* **74**, 1384–1389 (1977).
12. G.B. Strambini and M. Gonnelli, *J. Phys. Chem.* **92**, 2850–2853 (1988).
13. A. de Marco, L. Zetta, E. Menegatti, and P.L. Luisi, *J. Biochem. Biophys. Methods* **12**, 335–347 (1986).
14. P. Marzola, C. Forte, C. Pincino, and C.A. Veracini, *FEBS Lett.* **289** (1991).
15. L.V. Belonova, R.M. Davydov, and V.P. Timofeev, *Russ. J. Phys. Chem.* **57**, 1670–1673 (1983).
16. V. Papadimitriou, A. Xenakis, C.T. Cazianis, H. Stamatis, M. Egmont, and F.N. Kolisis, *Ann. N.Y. Acad. Sci.* **799**, 275–280 (1996).
17. A. Sanchez-Ferrer and F. Garcia-Ramona, *Enzyme Microb. Technol.* **16**, 409–415 (1994).

18. K. Martinek, I.V. Berezin, Y.L. Khmelnitski, N.L. Klyachko, and A.V. Levashov, *Biocatalysis* **1**, 9–15 (1987).
19. K. Martinek and A.N. Semenov, *Biochim. Biophys. Acta* **659**, 90–101 (1981).
20. A.M. Rao, V.T. John, R.D. Gonzalez, J.A. Akkara, D.L. Kaplan, *Biotechnol. Bioeng.* **41**, 531–540 (1993).
21. R. Hillhorst, C. Laane, and C. Veeger, *FEBS Lett.* **159**, 225–228 (1983).
22. J.F.G. Vliegenhart, G.A. Veldink, and J. Boldingh, *Agric. Food Chem.* **27**, 623–626 (1979).
23. C. Laane and J. Spruijt, in J.M.C. Duarte ed., *Recent Developments in Biotechnology*, Plenum, pp. 142–150.
24. K. Martinek, A.V. Levashov, Y.L. Khmelnissky, N.L. Klyachko, and I.V. Berezin, *Science* **218**, 889–891 (1982).
25. P. Luthi and P.L. Luisi, *J. Chem. Soc.* **106**, 7285–7286 (1984).
26. D. Han and J.S. Rhee, *Biotechnol. Lett.* **7**, 615–621 (1985).
27. S. Morita, H. Narita, T. Matoba, and M. Kito, *J. Am. Oil Chem. Soc.* **61**, 1571–1574 (1984).
28. E.I. Belyaeva, L. Brovko, N.N. Ugarova, N.L. Klyachko, A.V. Levashov, K. Martinek, and I.V. Berezin, *Doklady Akad. Nauk. SSSR* **273**, 494–497 (1983).
29. H. Stamatis, A. Xenakis, U. Menge, and F.N. Kolisis, *Biotechnol. Bioeng.* **42**, 103–110 (1993).
30. D.G. Hayes and E. Gulari, *Biotechnol. Bioeng.* **40**, 110–118 (1992).
31. N.L. Klyachko, A.V. Levashov, and K. Martinek, *Molekulyarnaya Biologiya (Russ)* **18**, 1019–1032 (1984).
32. K. Martinek, I.V. Berezin, Y.L. Khmelnissky, N.L. Klyachko, and A.V. Levashov, *Coll. Czech. Chem. Commun.* **52**, 2589–2602 (1987).
33. E. Ruckenstein and P. Karpe, *J. Colloid Inter. Sci.* **139**, 408–436 (1990).
34. A. Ballesteros, U. Bornscheuer, A. Capewell, D. Combes, J.S. Condoret, K. Koening, F.N. Kolisis, A. Marty, U. Menge, T. Scheper, H. Stamatis, A. Xenakis, *Biocatalysis* **13**, 1–42 (1995).
35. E.A. Malakhova, B.I. Kurganov, A.V. Levashov, I.V. Berezin, and K. Martinek, *Doklady Akad. Nauk. SSSR* **270**, 474–477 (1983).
36. T. Kim and K. Chung, *Enzyme Microb. Technol.* **11**, 528–532 (1989).
37. A. Xenakis, T.P. Valis, and F.N. Kolisis, *Progr. Colloid Polym. Sci.* **79**, 88–93 (1989).
38. T.P. Valis, A. Xenakis, and F.N. Kolisis, *Biocatalysis* **6**, 267–279 (1992).
39. R.K. Schmidli and P.L. Luisi, *Biocatalysis* **3**, 367–376 (1990).
40. K. Holmberg and E. Osterberg, *Progr. Colloid Polym. Sci.* **82**, 181–189 (1990).
41. K. Holmberg and E. Osterberg, *J. Am. Oil Chem. Soc.* **65**, 1544–1548 (1988).
42. S.W. Tsai, C.L. Chiang, *Biotechnol. Bioeng.* **38**, 206–211 (1991).
43. K. Holmberg, B. Lassen, and M.B. Stark, *J. Am. Oil Chem. Soc.* **66**, 1796–1800 (1989).
44. R. Scartazzini and P.L. Luisi, *Biocatalysis* **3**, 377–380 (1990).
45. P.S. Chang, J.S. Rhee, and J.J. Kim, *Biotechnol. Bioeng.* **38**, 1159–1165 (1991).
46. P.D.I. Fletcher, R.B. Freedman, B.H. Robinson, G.D. Rees, and R. Schomacker, *Biochim. Biophys. Acta* **912**, 278–282 (1987).
47. K. Holmberg and E. Osterberg, *Progr. Colloid Polym. Sci.* **74**, 98–102 (1987).
48. M. Bello, D. Thomas, and M.D. Legoy, *Biochem. Biophys. Res. Commun.* **146**, 361–367 (1987).
49. F.N. Kolisis, T.P. Valis, and A. Xenakis, *Ann. N. Y. Acad. Sci.* **613**, 674–680 (1990).
50. H. Stamatis, A. Xenakis, M. Provelegiou, and F.N. Kolisis, *Biotechnol. Bioeng.* **42**, 931–937 (1993).
51. D.M.F. Prazeres, F.A.P. Garcia, and J.M.S. Cabral, *Biotechnol. Bioeng.* **41**, 761–770 (1993).
52. D.K. Eggers and H.W. Bloch, *Bioproc. Eng.* **3**, 83–91 (1988).
53. M. Dekker, R. Hilhorst, and C. Laane, *Anal. Biochem.* **178**, 217–226 (1989).
54. Q. Chang and J. Chen, *Biotechnol. Bioeng.* **46**, 172–174 (1995).
55. R.S. Rahaman, J.Y. Chee, J.M.S. Cabral, and T.A. Hatton, *Biotechnol. Prog.* **4**, 218–225 (1988).

RHEOLOGY OF FILAMENTOUS MICROORGANISMS, SUBMERGED CULTURE

S.M. Heydarian
A.P. Ison
P. Ayazi Shamlou
University College London
London, United Kingdom

KEY WORDS

Culture viscosity
Filamentous microorganisms
Mixer-viscometers
Morphology-viscosity
Non-Newtonian flow
Rheology
Rheometers

OUTLINE

INTRODUCTION

Many biotechnology products are manufactured using microorganisms and mammalian cells that behave in a Newtonian manner. In other cases, the product of the culture or the cellular material itself may cause the culture to be highly non-Newtonian. It is the processing of non-Newtonian materials that can cause major design and operational problems. For example, in most filamentous fermentations in submerged cultures, at the start of fermentation the broth rheology is Newtonian, not too different from water. However, often the microbiological requirements of the fermentation are such that the prevailing operating conditions lead to the development of diverse and complex non-Newtonian broth rheology, and extreme changes in the apparent viscosity of the broth can occur during fermentation (1–3). These changes can cause practical difficulties in maintaining an adequate power input per unit volume in the bioreactor. Subsequently, significant heterogeneity in dissolved oxygen tension and shear rate distribution can occur in the vessel contents (4–6). Consequently, transport processes such as heat and mass transfer and metabolic activity of the cells can suffer, cell damage can occur, and a fall in productivity can result (7–9). The extent of non-Newtonian behavior of the broth is closely related to the growth of the biomass and the development of a three-dimensional network of highly branched hyphae in the case of mycelia microorganisms (10–12). Considerable work has been done in recent years to understand these complex structures by measuring the mycelia morphology in dispersed (filamentous) and pelleted (aggregated) forms (13–17). However, there are as yet no general predictive constitutive equations that can be used confidently for design and scale-up, and even for the same strain there is no consensus of opinion among researchers on the precise effects of mycelia morphology on broth rheology or how the rheology of such systems should be described. The uncertainties associated with the use of predictive rheological equations are to a large extent understandable considering the wide range of parameters that are thought to influence the rheology of the cultures. These include the mode of fermentation (e.g., batch, fed-batch, or continuous), growth rate, level of the dissolved oxygen concentration and pH levels of the culture, level of mixing and shear intensity in the fermentor, and presence of solid debris, impurities, surfactants, and additives. These parameters critically affect the properties of the individual hyphae, for example, their length and length distribution, diameter, and mechanical integrity (strength, rigidity, and flexibility), as well as the physicochemical interactions between the hyphae (compactness, roughness) and between the hyphae and the continuous liquid phase (3,15,18–22). The analysis of rheological data is further complicated by the degree of hyphae aggregation (i.e., formation of clumps) and by the flocculation of clumps (16). The superimposed effects of all these parameters on broth rheology can be difficult to simulate.

Few publications are available on the rheology of non-filamentous cultures. In cases where data are published (e.g., for yeast suspensions), there is considerable disagreement between the results of different researchers (11,23–25). Fermentation cultures and biological solutions that do not exhibit mycelia growth but can show non-Newtonian behavior include high concentrations of single-cell suspensions (e.g., yeasts), high molecular weight fermentation products (e.g., polysaccharides), solutions containing globular proteins, supercoiled plasmid DNA and macromolecules (e.g., membrane protein complexes), chromosomal DNA, and animal cells in solutions in high concentrations (11,26).

This review article focuses largely on the rheology of the microbial suspensions in order to examine some of the key factors that are perceived to affect their flow properties. The intention is to highlight the results of some of the recent research on the interactions between mycelia morphology and culture rheology without emphasizing the effect of rheology on, for example, transport processes such as heat and mass transfer. For these aspects, the reader is referred to other relevant reviews on the subject (4–6, 27–29).

VISCOMETRY

In addition to biomass concentration, the extent of non-Newtonian behavior of fermentation broths is determined largely by the primary properties of the hyphae, which include hyphal morphology and dimensions, hyphal flexibility, and ability to form clumps. The primary properties of the hyphae in turn are critically dependent on the detailed fermentation conditions, making it difficult to develop a satisfactory theoretical simulation of the non-Newtonian behavior of broths.

An alternative engineering approach to the study of flow properties of a fermentation culture is to experimentally develop the flow curve, or shear diagram, which is a plot of shear stress, τ, against shear rate, γ, in the laminar region (Fig. 1). This information is usually obtained in a conventional laboratory viscometer such as the cup-and-

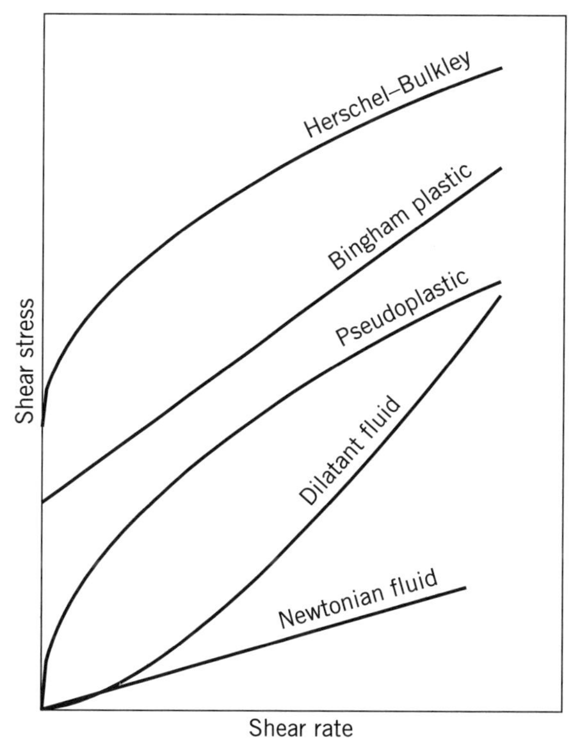

Figure 1. Rheological models.

bob (coaxial cylinder), capillary tube, or a cone-and-plate viscometer. However, most cultures are highly heterogeneous and contain large amounts of insoluble debris and particulate material, which can be fragile and can separate and sediment during rheological measurements. This renders many of the conventional rheometers unsuitable for characterization of biological suspensions. In such cases useful rheological data can be generated indirectly in laminar flow by using a calibrated, scaled-down, "engineering" viscometer, for example, an instrumented pipe flow viscometer or impeller viscometer (16–34). The former is widely used in the biochemical industry for flow characterization and is essentially a cup-and-bob viscometer except that the rotating inner cylinder is replaced by a mixer, (e.g., a turbine, paddle, disc, or helical ribbon) (Fig. 2). From the rheological viewpoint, the basic difficulty in using engineering viscometers is that these instruments do not provide a well-defined and uniform shear field, and therefore, individual elements of fluid are likely to undergo periods of exposure to different levels of shear during measurements. In the case of the impeller viscometer, the evaluation of an average shear rate in the cup is based on an empirical technique originally developed for mechanically agitated vessels by Metzner and Otto (35) and later modified for relatively simple rotating units (e.g., a disc spindle) by Mitschka (36). These techniques have been fully described elsewhere (31,37) and are therefore only briefly reviewed here. The estimation of the average shear rate, γ_{av}, in the cup of an impeller viscometer is based on the assumption that the average shear rate, γ_{av}, is directly proportional to the rotational speed of the impeller, N; the constant of proportionality, k_s, which is generally known as the impeller shear rate constant, is uniquely dependent on the geometry of the impeller viscometer but is independent of the fluid rheology. Thus, $\gamma_{av} = k_s N$. Figure 3 summarizes the experimental procedure used in evaluating the shear rate constant, k_s, for an impeller (viscometer) based on the Metzner and Otto technique. The key steps in the determination of k_s are as follows.

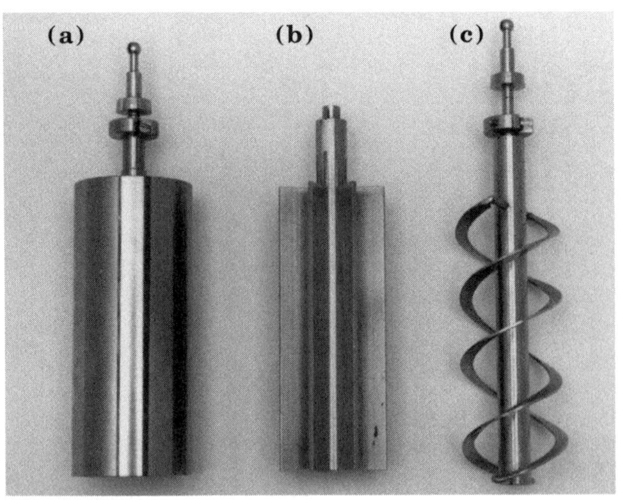

Figure 2. Mixer for rheometers. (**a**) solid cylinder; (**b**) turbine; (**c**) helical ribbon.

1. The laminar Newtonian power curve is obtained for the impeller-cup-and-bob viscometer from a series of measurements of torque and speed using one or more Newtonian liquids.
2. The flow curve is developed for a (model) non-Newtonian liquid, for example, carboxymethycellulose (CMC) solution using a conventional cup-and-bob viscometer.
3. The cup of the impeller viscometer is filled with the (model) non-Newtonian liquid.
4. The impeller speed is set at a predetermined value, N_1, and the corresponding torque, T_1, is measured. These values are used to calculate the corresponding dimensionless power number, $(P_o)_1$, $[P_o = P/(\rho N^3 D^5)]$.
5. From the Newtonian power curve, the Reynolds number, $(Re)_1$ corresponding to $(P_o)_1$ is obtained.
6. Using the definition of the general Reynolds number, $(Re = \rho N D^2/\mu_a)$ and the value of the Re_1 from step 5, the apparent viscosity, $(\mu_a)_1$, is calculated.
7. From the non-Newtonian flow curve the value of the shear rate, $(\gamma_a)_1$, is obtained corresponding to $(\mu_a)_1$.
8. Steps 4 through 7 are repeated for a range of speeds (N_2, N_3, etc.), and these values are plotted against the corresponding values of shear rates ($(\gamma_a)_2$, $(\gamma_a)_3$, etc.).

The plot obtained in step 8 is often a straight line through the origin and has a slope equal to the shear rate constant, k_s. It should be noted that the above procedure is valid only if the measurements of speed and torque (power) fall in the laminar flow region of the power curve, this being the region where the power number is a function of the Reynolds number. For most turbine impellers, the Metzner and Otto technique gives a value of the shear rate constant, k_s, in the range between 10 and 13 (35), whereas for helical ribbon and large paddle impellers its value falls in the range of 26 to 30 (38).

An important practical consideration is the selection of an appropriate rheometer (e.g., a rotating viscometer or a capillary flow rheometer). Different methods of measuring the rheological parameters of a non-Newtonian liquid do not necessarily yield the same result. This is not to say that one method is better than another but merely that it reflects the fact that the liquid flow field is different in the different devices, and the response of a non-Newtonian liquid depends on the flow field to which it is exposed. Therefore, in specifying a value for the apparent viscosity, it is important to quote the means by which the parameter has been measured. Moreover, the method of measurement should be closely related to the application of the final product. For example, for the design of a mechanically agitated bioreactor, an appropriate viscometer is the cup-and-bob or the turbine impeller viscometer, whereas the capillary tube or the instrumented pipeflow viscometer can be used effectively to gather rheological data for pipeline transportation of non-Newtonian biological suspensions. The following discussion is based on rheological data obtained in the former group of viscometers.

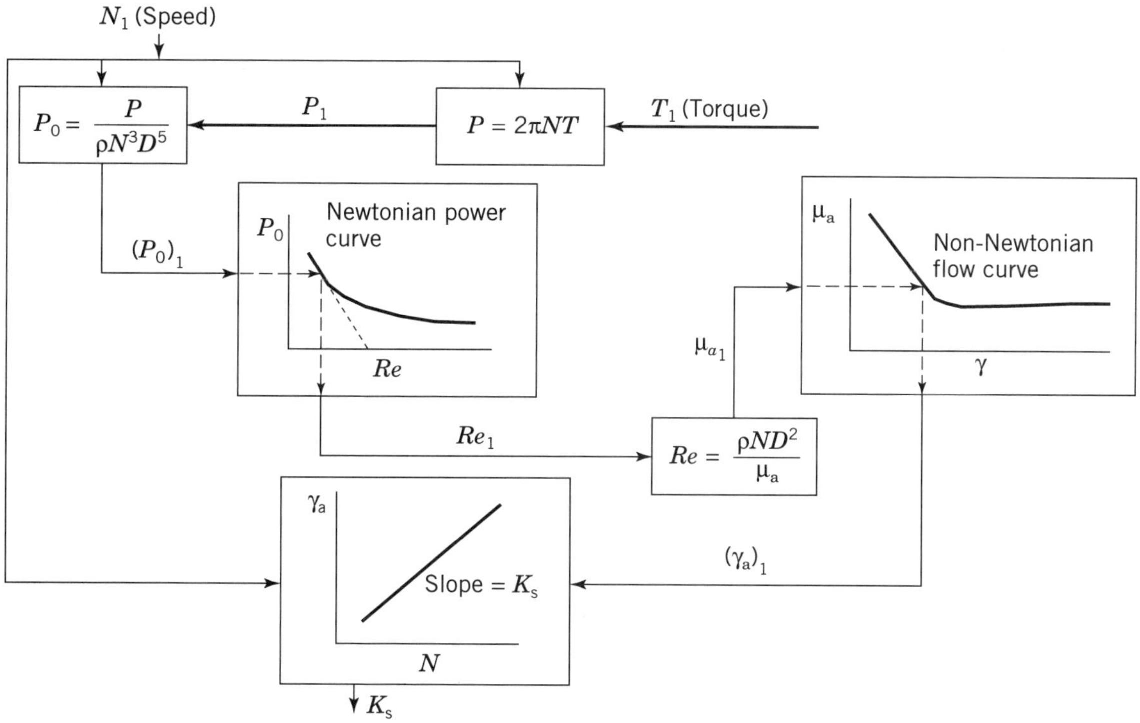

Figure 3. Experimental procedures for calculation of K_s from Metzner and Otto equation.

Commercial Viscometers

Bohlin, Haake, Brookfield, Rheomat, Carrimed, and Rheomatrics are just a few of the more commonly used devices used for the rheological characterization of liquids (39). The first four viscometers are available in both laboratory and portable (battery-operated) form, and some, such as the Brookfield, can also be used on-line. The state of technology has vastly improved over the past two decades, and current instruments such as the Bohlin can be fitted with different heads (e.g., coaxial cylinders, parallel plates, or cone-and-plate geometry) and can perform a variety of measurements including controlled stress and controlled strain measurements. Additionally, the Bohlin rheometer can accurately determine the response of the liquid to more complex types of motion imposed on the sample, including oscillatory motion from which rheological information can be obtained on the viscoelastic properties of the liquid.

These rheometers can be used to develop the non-Newtonian flow curve directly over a wide range of shear rates (10^{-6}–10^{7} s^{-1}) of interest to most process engineering operations (39). The low shear rates represent processes involving, for example, sedimentation of fine particles (10^{-6}–10^{-4} s^{-1}), leveling due to surface tension (10^{-2}–10^{-1}), and gravity drainage (10^{-1}–10^{1}). The medium range of shear rates covers processes such as extrusion, chewing and swallowing (10^{0}–10^{2} s^{-1}), mixing in mechanically agitated vessels, and pipe flow (10^{1}–10^{3} s^{-1}). Shear rates of 10^{3} s^{-1} to 10^{7} s^{-1} occur in operations such as spraying, centrifugation, and high-pressure homogenization.

RHEOLOGICAL MODELS

For design purposes it is often desirable to describe mathematically the information contained in a flow curve by fitting empirical equations to experimental measurements. Numerous flow equations have been proposed to describe the shape of the flow curves, including the Bingham plastic, Herschel–Bulkley, Casson, and power law equations. There is, however, no theoretical justification in recommending any of these equations a priori, and the selection of an appropriate flow equation should be based on practical considerations; the model equation should give an accurate fit of the experimental flow curve, it should have a minimum number of independent constants that are readily evaluated, and preferably the constants should have some physical basis.

A few researchers have reported time-dependent rheology (6,40,41) and viscoelastic behavior (42) for filamentous cultures. However, in the majority of cases the rheology of the cultures appears to exhibit time-independent shear thinning behavior, which can be described adequately by a two-parameter power law equation (31,43,44). This is characterized by a flow curve that, over a reasonable range of shear rate, is a straight line on a logarithmic plot of shear stress against shear rate (Fig. 4). Thus the relation between shear stress, τ, and shear rate, γ, can be written as $\tau = K\gamma n$, in which K is the consistency index and n is known as the flow behavior index and has a value less than unity for most cultures. The value of the flow behavior index gives a measure of the extent of the non-Newtonian behavior of the broth. For $n = 1.0$, the culture has Newtonian flow properties, with $K = \mu$. As n deviates from 1.0, the broth becomes increasingly non-Newtonian.

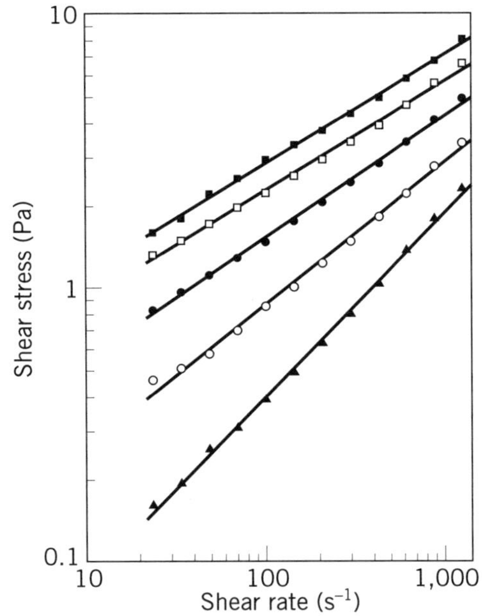

Figure 4. Shear stress against shear rate in log-log scale. Samples from *S. erythraea* (CA 340), fermentation in 7-L stirred bioreactor. Biomass concentration of the samples was between 3 and 12 g L^{-1}. Details of the fermentation areas in Ref. 45.

Flow equations have also been recommended for fermentation cultures based on the Bingham plastic, Casson, and Herschel–Bulkley models, all of which contain a yield stress parameter that needs to be evaluated from experimental measurements (16,42,46). The flow curves of many biological suspensions rise rapidly and then flatten out at relatively low shear rate. Under these conditions it is difficult to differentiate between the power law model and a model containing a yield stress, and very accurate data at quite low shear rates would be required to prove or disprove the existence of "true" yield stress (42,46). In the absence of such information, it seems prudent to use the power law model to describe the rheological data of biological materials over a specified range of shear rate.

RHEOLOGY OF MICROBIAL CULTURES

Filamentous cultures exhibit two extreme types of morphology in submerged culture, highly dense pelleted and free mycelia, with loosely packed mycelium forming a transition between the two types (Fig. 5). The freely dispersed mycelia and its entanglement to form clumps can cause the fermentation to become highly viscous and non-Newtonian. Past studies on morphology have attempted to distinguish between the different types by using various morphological parameters including compactness, roughness, and fractal dimension. These studies show that the compactness of *Penicillium chrysogenum* varies between about 0.3 for highly filamentous structure to close to unity for pellets (16). The fractal of *P. chrysogenum, Streptomyces griseus*, and *Streptomyces tendae* have been measured by Patankar et al. (47) and in our laboratory (45) and seem to vary between 1.9 and 2.7 in filamentous form and between 1.45 and 2.00 in pelleted form (47).

The pelleted type of growth consists of compact discrete masses of hyphae that generally lead to Newtonian behavior of the broth (48). However, in many cases as the size of the pellets increases, the diffusion of nutrients, particularly O_2, into the center of the pellets becomes limited, resulting in low productivity (6,49). Moreover, there are reported cases of cultures consisting of pelleted mycelium that exhibit clear non-Newtonian behavior (50). Great caution must therefore be exercised in generalizing the flow behavior of microbial cultures.

The flow curves presented later are intended to serve as typical rheological examples of real biological microorganisms extensively studied in the past. The discussion examines the effects of some of the important parameters that are perceived to influence broth rheology. For these reasons, the details of fermentation conditions are not included, but where necessary, the reader is referred to other publications and references where further information on the relevant aspects of the work can be found. Additionally, with the exception of the information given in Table 1, a conscious decision has been made to avoid recommending specific constitutive equations because the absolute values of the constitutive parameters K and n are highly system specific and ideally should always be based on experimental flow curves obtained under the actual fermentation conditions for which they are required.

Filamentous Mold Cultures

Numerous publications are available on the rheology of mold fermentation cultures; mostly the mycelium is in the filamentous form because these cultures show strong non-Newtonian behavior. The organisms most widely studied are *P. chrysogenum* and *Aspergillus niger*. Table 1 gives details on the range of parameters used in the rheological characterization of these materials. The non-Newtonian behavior of most of the cultures investigated has been modeled either in terms of a flow behavior index, n, and consistency coefficient, K, of a power law shear thinning equation, or in terms of a yield stress, τ_0, and a plastic viscosity, μ_p, of the Bingham plastic model. In a few cases the Casson model is used, in which case the yield stress, τ_0, and the shear rate coefficient, K_c, are given as the two model parameters. In these studies, plastic viscosity, μ_p, and yield stress, τ_0, in the Bingham plastic model and the consistency coefficient, K, in the power law equation were found to increase and the flow behaviour index, n, decrease with increasing biomass concentration. The increase was found to be nonlinear, and its extent was highly strain dependent (Table 1).

An examination of Table 1 shows that equations are also available in which the rheological parameters have been related to the morphological properties of the individual hyphae. However, recent measurements have shown that in many cases as much as 90% of the mycelium exists in flocculated (clump) form and not in a freely dispersed form, suggesting that rheological parameters are more likely to be related to clump properties and not to the de-

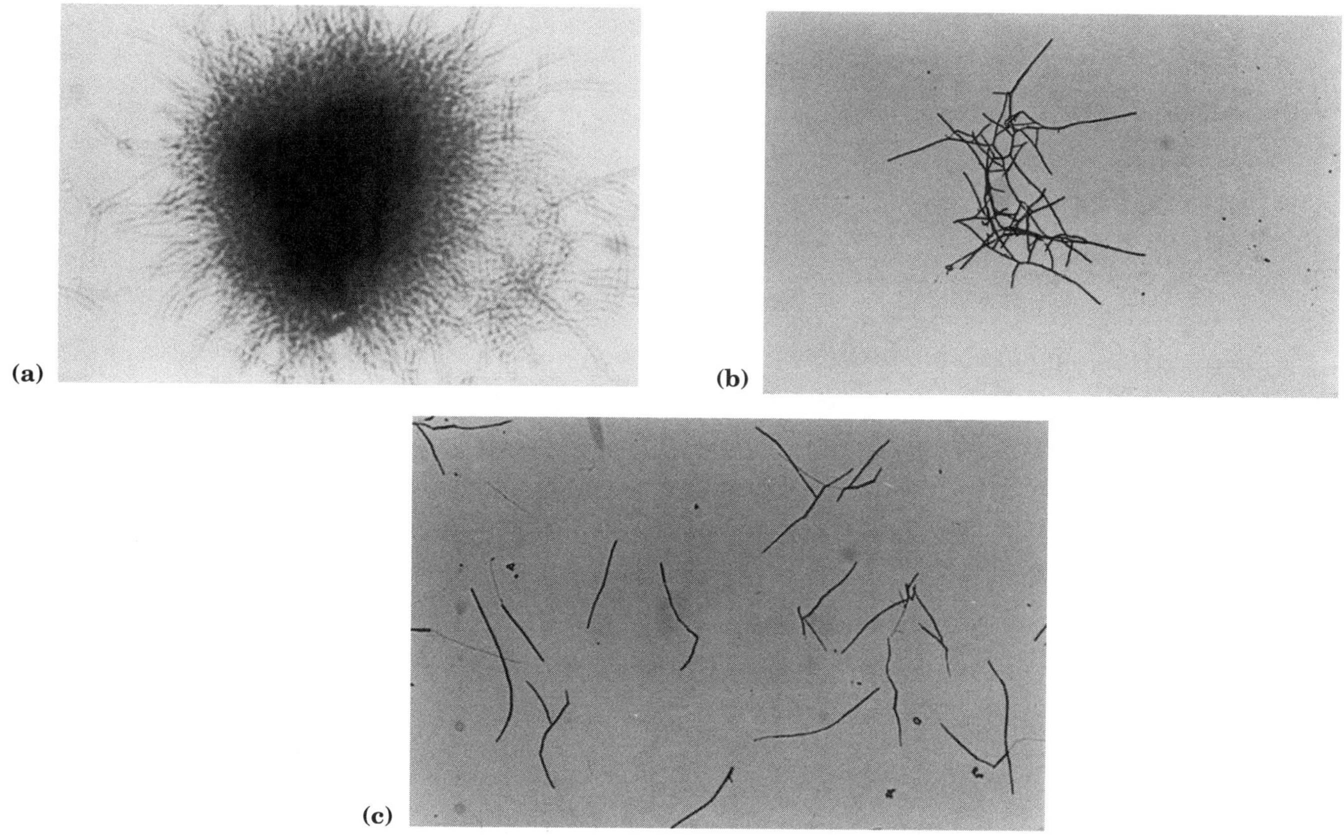

Figure 5. Example of *S. erythraea* (CA 340) morphology. (**a**) Pellet; (**b**) clump; (**c**) dispersed mycelium.

tailed characteristics of the individual hyphae. Recent measurements carried out by Tucker and Thomas (56) and Olsvik et al. (16) appear to support this conclusion and suggest that, for example, the compactness and the roughness of hyphae flocs can be successfully related to the rheological parameter (Table 1).

Filamentous Actinomycetes Cultures

In filamentous form, the sizes of the individual hyphae in the mycelium are generally less than 100 μm, and the broths exhibit considerable non-Newtonian flow behavior (31,46). However, recent experimental evidence obtained in our laboratory using three industrial strains of Actinomyces ranging in size from 15 μm to 25 μm indicated that the properties of the individual hyphae have little effect on broth rheology (3,44). The results of these experiments, discussed in more detail later, also revealed that there was little change in the size of the mean main length of the hyphae, and furthermore very few clumps were observed during hyphal characterization.

Figure 6 shows the variations of n and K of *Saccharopolyspora erythraea* (strain CA 340) mycelia cultures as a function of time of fermentation (45). The data were obtained in a standard coaxial cylinder viscometer using fresh samples of the culture that were removed during fermentation in a 7-L stirred bioreactor (45). The plots shown in Figure 6 are effectively snap shots of a developing rheology, showing the dramatic changes in the non-Newtonian behavior of the fermentation broth during the growth period.

Similar rheological data are shown in Figures 7, 8, and 9, respectively, for the fermentation of industrial strains of *S. erythraea, Aspergillus roseorufa*, and *Streptomyces rimosus* (3,44). The data for *S. rimosus* were obtained from an industrial-scale mechanically agitated bioreactor; in the case of *S. erythraea* and *A. roseorufa*, samples were taken during fermentation in a small, laboratory-scale mechanically stirred unit. In all three cases, the broths contained a significant amount of particulate matter, which made the use of a standard cup-and-bob viscometer inappropriate. The rheological parameters plotted in Figures 7–9 were therefore obtained by using a Brookfield 2HA-TDVII Synchro-Lectric rheometer fitted with a disc-spindle impeller, for which the average shear rate at each rotational speed was obtained by using Mitschka's technique, which is described fully elsewhere (36).

Figure 10 shows experimental data on the mean main hyphal length as a function of fermentation time. The width of the plotted lines indicates the 95% confidence limits of the data. Variations of the power law coefficient, K, with time are also shown in Figure 10 for comparison. The mean main hyphal length fluctuates throughout the fermentation time, but no observable trend in the data can be detected for any of the microorganisms. Additionally these plots indicate no correlation between the mean main hyphal length and broth rheology. The mean main hyphal

Table 1. Rheological Correlations for Biological Cultures

Microorganism	Rheological model	Variation of theological parameters[a]	Shear range (s^{-1})	Viscometer	Comments	References
A. niger (dispersed mycelia)	Power law	$K = 0.308X^{3.15}d_p^{0.229}$ $n = 0.559X^{-0.058}d_p^{-0.082}$	30–200	Pipe flow	$2.3 < X < 11.2$	51
A. niger (dispersed mycelia)	Power law Bingham Casson	$K = 4.3 \times 10^{-4}X^{3.3}$ $\tau_0 = 0.02X^{2.3}$ $K_c = 0.048\,X^{0.26}$ $\tau_0 = 2.5 \times 10^{-3}X^3$	50–650 for all three models	Coaxial cylinder	Power law was preferred; morphology changes were not considered.	31
A. niger (dispersed mycelia)	Power law	$K = -0.56 + 1.8 \times 10^{-3}RX^{1.7}$ for continuous culture and $K = 0.38 + 4.8 \times 10^{-5}RX^{2.9}$ for fed-batch culture	10–50	Turbine impeller	–	16
A. niger (dispersed mycelia)	Power law for early stage ($x < 4$ g/L), casson for later stage	–	50–150	Turbine impeller and pipe flow	Results for two different viscometers were not consistent.	32
A. niger (dispersed mycelia)	Herschel–Bulkley	$\tau_0 = 4.2 \times 10^{-6} \times X^{2.6}L_{hgu}^{0.65}$ $\tau_0 = 4.8 \times 10^{-7}X^{2.5}R^{3.2}$	–	Haake Rotovisco RV12	–	46
A. awamori	Power law	$K = 6.9 \times 10^{-5}X^{3.96}$	–	Coaxial cylinder	Morphological changes were not considered.	31
P chrysogenum (dispersed mycelia)	Casson	$\tau_0 \alpha X^{2.5}$ $K_c = 1.4 \times 10^{-3}X^1L_{hgu}^{0.6}$	–	Turbine impeller	Flexibility can affect rheology.	20
P. chrysogenum (dispersed mycelia)	Power law Bingham Casson	$K = 3.6 \times 10^{-3}X^{2.5}$ $\tau_0 = 0.043X^{2.1}$ $\mu_p = 2.2 \times 10^{-3}X^{0.78}$ $\tau_0 = 8.3 \times 10^{-3}X^{2.5}$ $K_c = 0.047X^{0.19}$	50–650 for all three models	Coaxial cylinder	Morphology changes not considered; power law was preferred.	31
P. chrysogenum (dispersed mycelia)	Herschel–Bulkley Casson	$K = aX^{2.8}R^{0.7}Com^{1.2}$ $1 - n = bX^{0.7}R^{0.6}Com^{0.9}$ $\tau_0 = cX^{3.2}R^{0.6}Com^{1.5}$ $\tau_0 = eX^{2.9}R^{0.6}Com^{1.0}$ $K_c = fX^{1.2}R^{0.1}Com^{1.0}$	3.5–75.8 for both models	Turbine impeller	Morphological measurement was for clump, which is more than 90% of total.	16

P. chrysogenum (dispersed mycelia)	Casson	$\tau_0 = aX^2$	–	Turbine impeller	–	41
P. Chrysogenum (dispersed mycelia)	Bingham	$\tau_0 = aX$	–	MacMichael viscometer	–	52
P. chrysogenum (pellet)	Casson	$K_c = aE_x + b$ $\tau_0 = c(E_x/d)^3$	<20	–	E_x = pellet volume fraction	50
S. levoris	Power law Bingham Casson	$K = 0.27X^{0.7}$ $\tau_0 = 1.1X^{0.75}$ $\mu_p = 5.9 \times 10^{-3} X^{0.32}$ $\tau_0 = 0.64X^{0.79}$ $K_c = 0.042X^{0.2}$	50–650 for all three models	Coaxial cylinder	Morphological changes not considered; power law was preferred.	31
S. levoris	Herschel–Bulkley	$\tau_0 = 4.2 \times 10^{-6} \times X^{2.6} L_{hgu}^{0.65}$	–	Haake Rotovisco RV12	–	46
S. fradiae	Herschel–Bulkley and Casson	Apparent viscosity, consistency index, and yield stress change with morphology.	<1300	Coaxial cylinder	Mathematical relation was not shown; Casson was preferred.	53
S. grisens	Bingham	τ_0 and plastic viscosity varied sinousidally with fermentation age.	–	–	–	54
S. erythraea *S. rimoscus* *A. roseourufa*	Power law	Rheology is related to biomass, not morphology.	<250	Disk spindle impeller	Mean hyphal length was between 15 and 25 μm.	44
S. erythraea	Power law	$K \alpha X^{2.14}$	24–880	Coaxial cylinder	–	45
Baker's yeast	Power law	$K \alpha e^{3.43E_c}$; no relation for n	10^4–10^5	Coaxial cylinder	Acidic pH lowered the viscosity.	23
Baker's yeast, *S. cerevisiae* and *C. utilis*	Newtonian	$\mu/\mu_0 = 1/1 - (h_s E_c)^a$; $a = 0.5$ for *S. cerevisiae*, $X < 150$ g/L	–	Falling sphere viscometer	*C. utilis* is considerably smaler than *S. cerevisiae* and had a higher value of viscosity.	25
Baker's yeast	Newtonian	$\mu = \mu_0(1 + 2.5E_c + 7.25 \times E_c^2)$	–	–	–	55
X. campestris	Power law and Herschel–Bulkley	The non-Newtonian behavior is controlled by molecular properties of dissolved polysaccharide.	<200	–	Thixotrophy was shown.	5

[a] a, b, c, d, e and f are constant parameters.

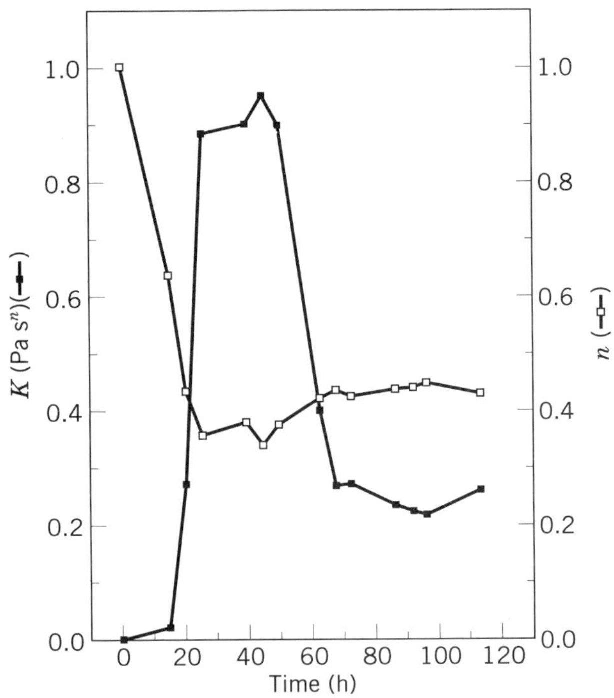

Figure 6. Variation of consistency index, K, and flow behavior index, n, of *S. erythraea* (strain CA 340) culture as a function of time of the fermentation.

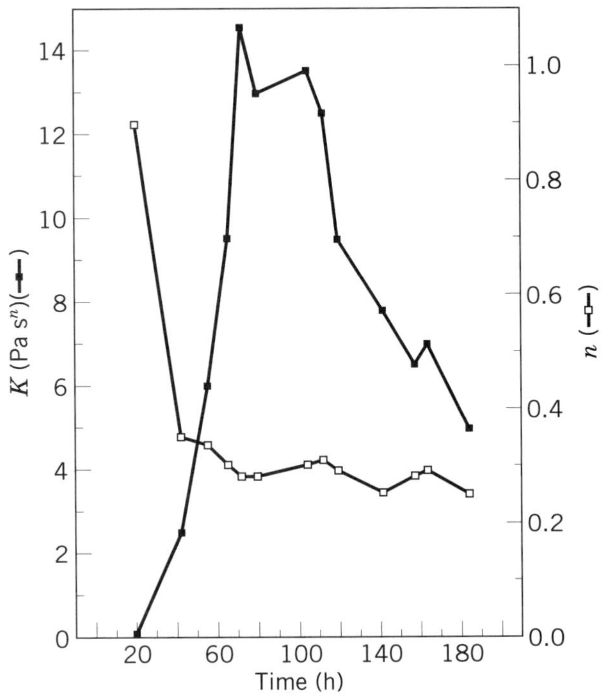

Figure 8. Variation of consistency index, K, and flow behavior index, n, of *S. roseorufa* industrial strain culture as a function of time of the fermentation.

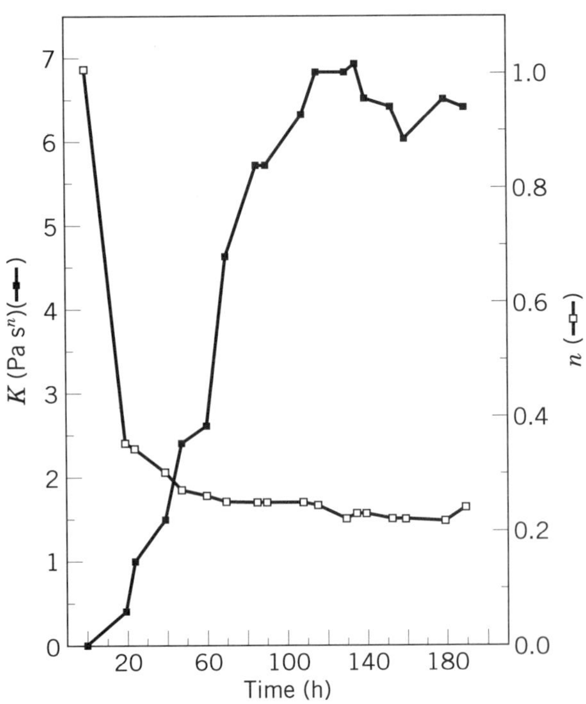

Figure 7. Variation of consistency index, K, and flow behavior index, n, of *S. erythraea* industrial strain culture as a function of time of the fermentation.

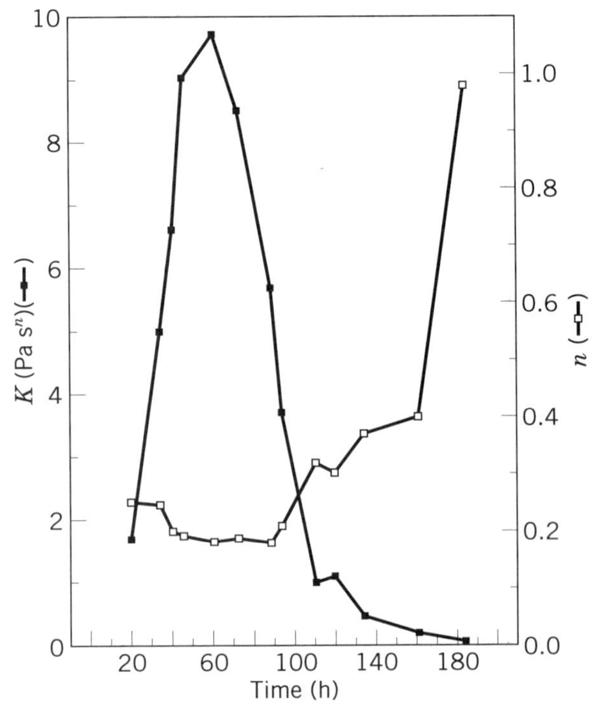

Figure 9. Variation of consistency index, K, and flow behavior index, n, of *S. rimosus* industrial strain culture as a function of time of the fermentation.

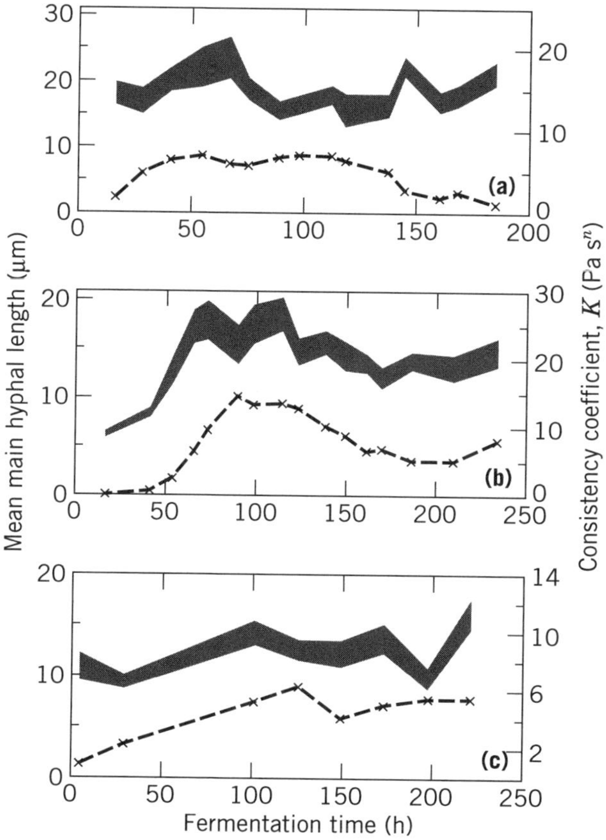

Figure 10. Variation of mean main length and consistency index, K, of fermentation cultures of (**a**) *S. erythraea*, (**b**) *A. roseorufa*, and (**c**) *S. rimoscus*.

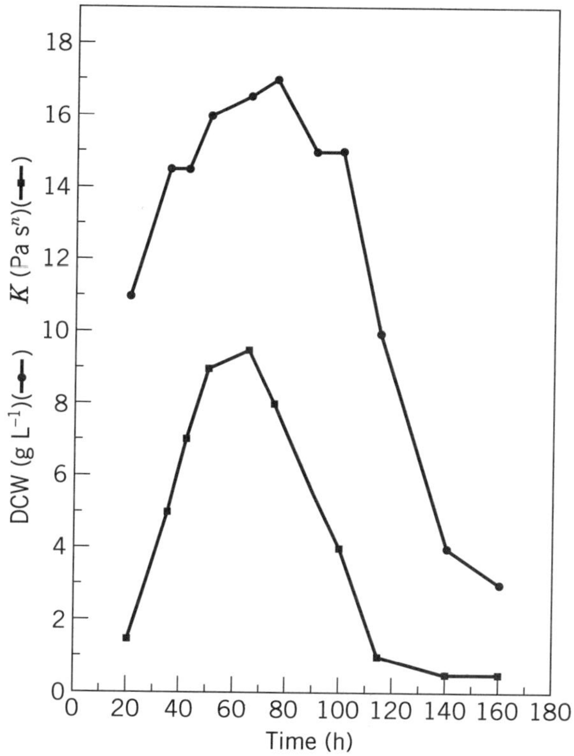

Figure 11. Variation of dry cell weight and consistency index for *S. erythraea* culture.

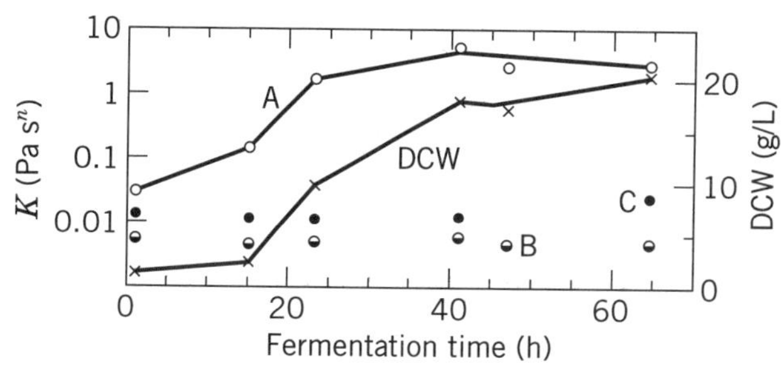

Figure 12. Effect of biomass, insoluble particulate matter, and soluble compounds on rheology of *S. erythraea* culture.

length of the three microorganisms shown in Figure 10 are significantly smaller than those reported previously for *Streptomyces clavuligerus* (45–120 mm), and *P. chrysogenum* (e.g., 130–420 μm and 60–120 μm).

An examination of the information given in Table 1 shows that, compared with the fungal cultures, the rheological parameters, K, τ_o, μ_p, and K_c, for example, are less sensitive to changes in the biomass concentration. The effect of dry cell weight (DCW) on the consistency coefficient, K, of *S. erythraea* culture is shown in Figure 11. The similarity of the two profiles in Figure 11 is evident from these plots, and a series of experiments were conducted to examine the factors responsible for the observed dependency of the broth consistency coefficient, K, on DCW. Two fermentation batches were prepared and run in the same way except that only one batch was inoculated with *S. erythraea*. Several samples were taken periodically from each batch for rheological measurements and DCW evaluation. Some of the samples were also centrifuged prior to analysis to remove *S. erythraea* cells and/or medium particulate and insoluble debris. Figure 12 shows the results of these experiments and indicates that for the inoculated batch (A) the profiles of the consistency coefficient, K, and the DCW are similar and increase with fermentation time. However, once the *S. erythraea* biomass was removed by centrifugation (curve B) the broth rheology changed dramatically, the variation in K disappeared, and the flow behavior index increased from its value of about 0.2 to a value close to unity, indicating a change from a highly shear thinning to an almost Newtonian behavior. Samples from the uninoculated batch showed very little change in their rheology, which remained Newtonian practically throughout (curve C). Similar results shown in Figure 13 were observed for the fermentation of *A. roseorufa*, confirming that the high apparent viscosity of the filamentous cultures depends on the presence and concentration of the organisms, and that other constitutive components (e.g., insoluble particulate matter) appear to have little effect on broth rheology, at least for the cases presented here.

The effect of pH of the culture on broth rheology was also investigated, and it was found to have a strong

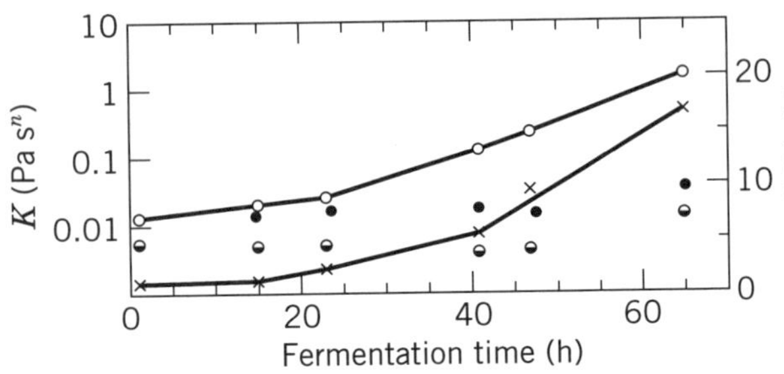

Figure 13. Effect of biomass, insoluble particulate matter, and soluble compounds on rheology of *S. roseorufa* culture.

effect on the extent of non-Newtonian behavior of the culture. This effect was somewhat unexpected and was further investigated for one of the microorganisms, *S. rimosus*. A large sample of the broth was removed from the bioreactor and divided into several specimens. Using concentrated hydrochloric acid and concentrated sodium hydroxide (60% by weight) the pH of each sample was altered to cover the range of pH between 3 and 10, in increments of 0.5 pH. The rheological properties of each sample were then measured using the Brookfield viscometer. Figure 14 shows the effect of pH on both the consistency coefficient, K, and the flow behavior index, n, of the broth. Measurements of the hyphal morphology (not shown) revealed no effect of pH on the mean main hyphal length of the individual hyphae. An important implication of the data in Figure 14, however, is that alteration in charges on and/or close to the surface of the microorganisms, arising from vari ations in pH, could play a role in determining the observed changes in broth rheology. The zeta potential characterizes the nature and the extent of surface charges, and the zeta potential–pH profiles of the microorganisms should reflect these effects. No electrophoretic measurements are available in the literature concerning the zeta potential–pH profiles of microorganisms. These profiles should be developed experimentally and studied to examine how they are affected by broth conditions. The results could provide valuable information on the interrelationship between broth rheology, hyphal morphology, and biomass concentration.

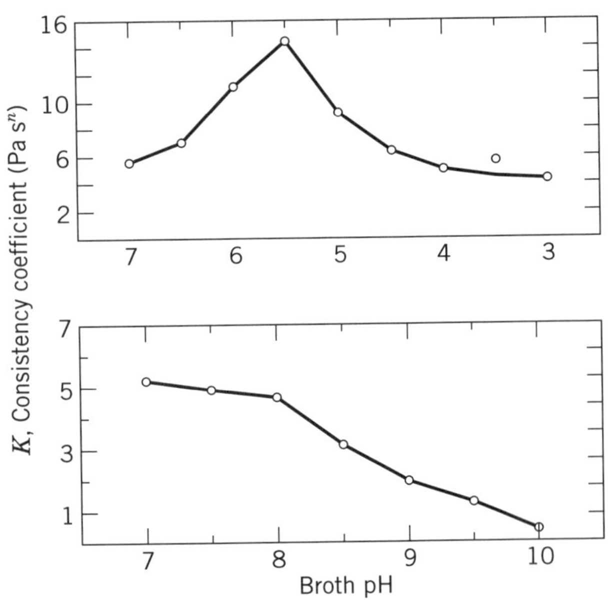

Figure 14. Effect of pH on consistency index of *S. rimosus* culture.

CONCLUSION

Considerable experimental work has been published on the rheology of biological suspensions since the 1960s. Advances made in image analysis techniques in the early 1990s resulted in further work on relating the rheological properties to the morphological characteristics of filamentous microorganisms. No consistent opinion has yet emerged on the interaction between the two. Recent developments on the measurement of fractal properties of mycelium may provide a useful basis for further elucidation of these complex interactions.

NOMENCLATURE

Com	Compactness of mycelial aggregate
D	Tank diameter (m)
DCW	Dry cell weight (g L^{-1})
d_p	Diameter of mycelial aggregate (m)
E_c	Volume fraction of wet cells
h_s	Packing factor
K	Consistency index (N sn m^{-2})
K_c	Casson viscosity (N s $m^{-2})^{1/2}$
K_s	Impeller shear rate constant
L_e	Length of the hyphae (μm)
L_{hgu}	Length of hyphal growth unit (μm)
n	Flow behavior index
N	Rotational speed of impeller (s^{-1})
P	Power (W)
P_0	Impeller power number
R	Roughness of mycelial aggregates
T	Torque (N m)
X	Concentration of biomass (g L^{-1})
γ	Shear rate (s^{-1})
γ_{ave}	Average shear rate (s^{-1})
μ_0	Viscosity of cell-free medium (N m^{-2} s)
μ	Viscosity of culture (N m^{-2} s)
μ_a	Apparent viscosity (N m^{-2} s)
μ_p	Plastic viscosity (Pa s)
ρ	Density (kg m^{-3})
τ	Shear stress (N m^{-2})
τ_0	Yield stress (N m^{-2})

BIBLIOGRAPHY

1. M. Borovic, A. Cimerman, W. Steiner, and T. Koloini, *Appl. Microbiol. Biotechnol.* **34**, 579–581 (1991).
2. P. Ayazi Shamlou, D.J. Pollard, A.P. Ison, and M.D. Lilly, *Chem. Eng. Sci.* **49**, 303–312 (1994).
3. S.J. Warren, E. Keshavarz-More, P.A. Shamlou, M.D. Lilly, C.R. Thomas, and K. Dixon, *Bioprocess Eng.* **13**, 45–48 (1995).
4. C.M. Tuffile and F. Pinho, *Biotechnol. Bioeng.* **12**, 849–871 (1970).
5. M. Charles, *Adv. Biochem. Eng.* **8**, 1–62 (1978).
6. H.W. Blanch and S.M. Bhavaraju, *Biotechnol. Bioeng.* **18**, 745–790 (1976).
7. F. Vardar and M.D. Lilly, *Appl. Microbiol. Biotechnol.* **14**, 203–211 (1982).
8. G. Larsson and S.O. Enfors, *Bioprocess Eng.* **3**, 123–127 (1988).
9. H. Funahashi, H. Harada, H. Taguchi, and H. Yoshida, *J. Ferment. Technol.* **20**, 277–282 (1987).
10. W. Ruohang and C. Webb, *Biotechnol. Tech.* **9**, 55–58 (1995).
11. M. Reuss, D. Debus, and G. Zoll, *German Biochem. Eng.* 233–236 (June 1982).
12. J.C. Van Suijdam and J.C. Metz, *Biotechnol. Bioeng.* **23**, 111–148 (1981).
13. H.L. Packer and C.R. Thomas, *Biotechnol. Bioeng.* **35**, 870–881 (1990).
14. K.G. Tucker, T. Kelly, P. Delgrazia, and C.R. Thomas, *Biotechnol. Prog.* **4**, 353–359 (1992).
15. E. Olsvik, and B. Kristiansen, *Biotechnol. Adv.* **12**, 1–39 (1994).
16. E. Olsvik, K.G. Tucker, C.R. Thomas, and B. Kristiansen, *Biotechnol. Bioeng.* **42**, 1046–1052 (1993).
17. S.K. Treskatis, V. Oregeldinger, H. Wolf, and E.D. Gilles, *Biotechnol. Bioeng.* **53**, 191–201 (1997).
18. N. Ohta, Y.S. Park, K. Yahiro, and M. Okabe, *J. Ferment. Bioeng.* **79**, 443–448 (1995).
19. L. Ju, C.S. Ho, and J.F. Shanaham, *Biotechnol. Bioeng.* **38**, 1223–1232 (1991).
20. B. Metz, N.W.F. Kossen, and J.C. Van Suijdam, *Adv. Biochem. Eng.* **11**, 103–156 (1979).
21. J.C. van Suijdam and B. Metz, *Biotechnol. Bioeng.* **23**, 111–148 (1981).
22. H. Eiki, H. Ishida, H. Saito, Y. Oka, and T. Osono, *J. Ferment. Bioeng.* **67**, 345–349 (1989).
23. S. El-Temtamy, L. farahat, A. Nour el-din, and A. Gaber, *Eur. J. Appl. Microbiol. Biotechnol.* **5**, 156–160 (1982).
24. B.W. Shimmons, W.Y. Svrcek and J.E. Zajic, *Biotechnol. Bioeng.* **18**, 1793–1805 (1976).
25. M. Reuss, M. Popovic and W.K. Bronn, *Appl. Microbiol. Biotechnol.* **8**, 167–175 (1979).
26. R.P. Tomkiewicz, A. Biviji, and M. King, *Biorheology* **31**, 511–520 (1994).
27. B. McNeil and L.M. Harvey, *Crit. Rev. Biotechnol.* **13**, 275–304 (1993).
28. K. Gbewonyo, G. Hunt, and B. Buckland, *Bioprocess Eng.* **8**, 1–7 (1992).
29. G. Li, H. Qiu, Z. Zheng, Z. Cai, and S. Yang, *J. Chem. Tech. Biotechnol.* **42**, 385–391 (1995).
30. U. Bjorkman, *Biotechnol. Bioeng.* **29**, 114–129 (1987).
31. D.G. Allen and C.W. Robinson, *Chem. Eng. Sci.* **45**, 37–48 (1990).
32. N. Blakebrough, W.J. McManamey, and K.R. Tart, *J. Appl. Chem. Biotechnol.* **78**, 453–461 (1978).
33. E.S. Olsvik and B. Kristiansen, *Biotechnol. Bioeng.* **40**, 375–387 (1992).
34. N. Blakebrough, W.J. McManamay, and W.J. Tart, *J. Appl. Chem. Biotechnol.* **28**, 453–461 (1978).
35. A.B. Metzner and R.E. Otto, *AIChEJ.* **3**, 3–10 (1957).
36. P. Mitschka, *Rheological Acta* **21**, 207–209 (1982).
37. Z. Kemblowski and B. Kristiansen, *Biotechnol. Bioeng.* **28**, 1474–1483 (1986).
38. P. Ayazi Shamlou and M.F. Edwards, *Chem. Eng.* **40**, 1773–1781 (1985).
39. H. Barnes, *Chem. Eng.* **24**, 17–24 (1993).
40. J.J.T.M. Bongenaar, N.W.F. Metz, and F.W. Meijboom, *Biotechnol. Bioeng.* **15**, 201–206 (1973).
41. J.A. Roels, J. Van Den Berg, and R.M. Voncken, *Biotechnol. Bioeng.* **16**, 181–208 (1974).
42. M. Mohseni, H. Kautola, and D.G. Allen, *J. Ferment. Bioeng.* **83**, 281–286 (1977).
43. T. Oalman and H.W. Blanch, *Crit. Rev. Biotechnol.* **4**, 133–184 (1986).
44. S.J. Warren, E. Keshavarz-Moore, P. Ayazi Shamlou, M.D. Lilly, C.R. Thomas, and K. Dixon, *Biotechnol. Bioeng.* **45**, 80–85 (1995).
45. S.M. Heydarian, Ph.D. Thesis, University of London, London, U.K. (1998).
46. M. Mohseni and D.G. Allen, *Biotechnol. Bioeng.* **48**, 257–265 (1995).
47. D.B. Patankar, T. Liu, and T. Oolman, *Biotechnol. Bioeng.* **42**, 571–578 (1993).
48. J.H. Kim, J.M. Lebeault, and M. Reuss, *Eur. J. Appl. Microbiol. Biotechnol.* **18**, 11–16 (1983).
49. E. Ujokova, Z. Fencl, M. Muzilkova, and L. Seichert, *Biotechnol. Bioeng.* **22**, 227–241 (1980).
50. R. Witter, R. Matthes, and K. Schugerl, *Appl. Microbiol. Biotechnol.* **18**, 17–23 (1983).
51. I.A. Fatile, *Appl. Microbiol. Biotechnol.* **21**, 60–64 (1985).
52. F.H. Deindoerfer and E.L. Gaden, *Appl. Microbiol.* **3**, 253–257 (1970).
53. N.P. Ghildyal, M.S. Thakur, S. Srikanta, S.A. Jaleel, S.G. Prapulla, M.S. Prasad, P.N. Devi, and B.K. Lonsane, *J. Chem. Tech. Biotechnol.* **38**, 221–234 (1987).
54. J.W. Richards, *Prog. Ind. Microbiol.* **3**, 253–260 (1961).
55. B.W. Shimmons, W.Y. Svrcek, and J.E. Zajic, *Biotechnol. Bioeng.* **18**, 1793–1805 (1976).
56. K.G. Tucker and C.R. Thomas, *Food Bioprod. Proc.* **71**, 111–117 (1993).

See also *ASPERGILLUS*; SECONDARY METABOLITE PRODUCTION, ACTINOMYCETES, OTHER THAN STREPTOMYCES; XANTHAN GUM.

ROLLER BOTTLE CULTURE, MIXING

D.R. Unger
M. Liu
P.D. Swanson
F.J. Muzzio
Rutgers University
Piscataway, New Jersey
Joye Bramble
James Searles
John Aunins
Merck and Company
West Point, Pennsylvania

KEY WORDS

Cell propagation
Mixing
Particle settling
Roller bottle
Virus infection

OUTLINE

INTRODUCTION

The settling behavior of biological cells with densities different than the fluid is a key factor affecting the performance of roller bottle bioreactors. The fully three-dimensional velocity field in a roller bottle bioreactor was simulated using a finite-element method, and the settling behavior of cells was simulated for both steady and time-dependent bottle rotation rates using particle dynamics algorithms. These simulations were validated experimentally using fluorescent particle tracers. Under steady flow conditions the flow is divided into two regions, one in which the particles settle to the wall and one in which the particles remain suspended indefinitely, but unsteady flow conditions (characterized by repeated reversal of the flow direction) lead to complete cell settling. These observations showed qualitative agreement with culture infection kinetic data in which changes in the roller bottle rotation rate resulted in enhanced infection. The main mixing limitation observed experimentally in roller bottles is very slow mixing in the axial direction, caused by flow segregation at the central plane of symmetry of the bottle. This limitation is readily overcome by imposing a small-amplitude vertical rocking motion to the bottle, forcing fluid to flow through the segregation plane and leading to a completely homogeneous mixture in just a few revolutions of the bottle.

BACKGROUND

Manual and automated roller bottle systems have been used for more than 40 years in the pharmaceutical, biochemical, and medical fields for processes such as cell growth and infection, heterologous glycoprotein production, vaccine preparation, and high-density plant cell cultivation (1–10). Despite efforts by numerous investigators to develop unit operations such as microcarrier culture for the production of anchorage-dependent cells or cell products (11–19), roller bottle systems still persist in research and industry. Additionally, for industrial-scale cell culture production (e.g., vaccines), cells are frequently expanded in roller bottles before transfer to microcarrier cultures for the final growth phase even when unit operation-based systems are utilized (17).

Widespread use of the roller bottle has several reasons. Most notably, the process relies on very simple technology: a horizontal cylindrical vessel is filled approximately one-fourth full and rotated axially; increased production is accomplished simply by increasing the number of bottles. Thus, scale-up development is straightforward, resulting in reduced developmental time lines for industry and faster introduction to market for new products. The system allows constant fluid–gas contact and easy addition of nutrients without interruption of the process. In addition, the process is capable of maintaining sterile conditions for prolonged times, contamination of one or more roller bottles does not necessarily result in the failure of an entire lot, precise control of nutrient and waste product levels is possible, and the direct monitoring of cell growth and metabolism is relatively simple (6).

On the other hand, roller bottles are limited in surface area available for growth and in the volume of harvest fluid obtained. Manpower and facility space requirements are higher than for unit operation systems such as microcarriers because hundreds or thousands of roller bottles can be operated for a single production run, although new automation systems are addressing the manpower issue (10,20). In addition, cell growth and infection are believed to be suboptimal in certain vaccine products because of flow and mixing dynamics, perhaps by preventing infected cells from attaching to host cells attached to the bottle walls (4). Although these issues point toward an obvious need for flow analysis and process design criteria, there have been no published results to date on either of these topics.

The following sections detail computational and experimental characterizations of the fluid flow profiles within rotating roller bottles, including particle trajectories, fluid mixing patterns, and unsteady-state flow strategies. The results point to ways in which cell proliferation and infection can be optimized by simple modifications to roller bottle rotation. This article is organized as follows. First, the numerical method for obtaining the three-dimensional flow solution is described. The flow solution is then validated by comparison with experimental measurements obtained via particle imaging velocimetry. Computationally predicted mixing patterns are compared with mixing patterns obtained by visualization of fluorescent dye streaklines. The model is then used to examine cell motion in the flow field. Computationally predicted cell motion patterns are compared with experimentally observed particle motion patterns obtained via long-time exposure photography of fluorescent particles. Then, particle settling results for the roller bottle under steady flow conditions are discussed and related to cell growth results obtained for various bottle rotation speeds. Finally, attention turns to improving the performance of roller bottles. The use of unsteady motions on the cylinder to enhance cell settling is explored, and the use of rocking flow perturbations to enhance axial mixing is demonstrated. The final portion of this chapter is devoted to conclusions and recommendations for future research.

DEVELOPMENT AND EXPERIMENT VALIDATION OF A COMPUTATIONAL MODEL OF FLUID AND CELL MOTION IN A ROLLER BOTTLE

Flow Field Solution

The three-dimensional flow field for a standard 850-cm^2 roller bottle geometry was obtained using a commercial CFD software package, FLUENT. This and similar software packages have been used by several researchers in the past to solve fluid dynamics problems (21–26). The equations of motion for the flow are the well-known Navier–Stokes equations, which, for a two-dimensional flow, include the momentum equation in two components and the continuity equation:

$$\boldsymbol{u}/t + \boldsymbol{u}\nabla\boldsymbol{u} = -\nabla p + (1/\mathrm{Re})\nabla^2\boldsymbol{u} + (1/\mathrm{Fr}) \qquad (1)$$

$$\nabla\boldsymbol{u} = 0 \qquad (2)$$

where $\boldsymbol{u}$ is the dimensionless velocity and p is the dimensionless pressure. Re in Equation 1 is the Reynolds number, $\mathrm{Re} = \rho v L/\mu$, where ρ and $\boldsymbol{\mu}$ are the fluid density and viscosity, and v and L are the bottle's linear rotational velocity and the liquid height, respectively; Fr is the Froude number, $\mathrm{Fr} = gL/v^2$, where g is gravity. The rationale for selecting liquid height as the natural length scale for the flow is that the liquid height is the diffusional length scale in the system. Viscous forces can be viewed as diffusive transport of momentum, with the kinematic viscosity (μ/ρ) playing the role of a momentum diffusivity. Hence, the liquid depth is the natural length scale for the action of viscous forces.

The experimental geometry is shown in Figure 1a. To obtain the velocity field, it was assumed that (*1*) the flow is steady, (*2*) the flow is in the Stokes regime (creeping flow conditions), (*3*) the free boundary is both planar and horizontal (effects of surface tension and viscous drag on the boundary are neglected). For the conditions used in the biotechnology industry, Re = 10, confirming the validity of the creeping flow assumption.

The physical domain was discretized using a curvilinear mesh; although the bottle has a total length to diameter ratio of 2:1, because of its symmetry only one-half of the system needed to be simulated, giving a length to diameter ratio of 1:1. The liquid height in the bottle was one-fourth the total diameter, which is typical of industrial practice. A 80 × 20 × 94 structured computation grid was used; Figure 1b represents a schematic of the z-plane grid (for the computation a higher node density was used). An acceptable node density for the computational mesh was found by creating several meshes with different node densities. The number of nodes in the mesh was successively increased by 25% until an average velocity difference of less than 3% was achieved for two successive meshes. The lower node density was then used to generate the computational mesh.

Velocity and pressure fields were obtained via iteration using a Sun SPARC 20 workstation. Approximately 140 Mb of RAM and 60 h of CPU time were required for solution convergence. To determine convergence, a criterion of 10^{-6} was used for each normalized residual velocity and pressure component. Residuals for each iteration were normalized versus the residual values obtained after the second iteration of the solver. To test the sufficiency of the convergence criterion, the simulation was run using convergence criteria ranging from 10^{-5} to 10^{-6} with no significant change in the velocity field when the convergence criteria were varied over this range (<1.0% change in velocity magnitude, on average). Sensitivity analysis (i.e., trajectory closing) showed that such a mesh was sufficiently accurate. These flow data were then used for all subsequent particle tracking simulations.

The condition adopted in the base case simulation was $\Omega = 0.25$ rpm, $\mu = 0.125$ Pa s at room temperature, giving a Reynolds number of Re 1.32. At this Re, the flow can be considered to be in the creeping flow regime. The flow was assumed to be sufficiently slow so that it could be assumed that the free surface was always horizontal and no surface tension effects were present at the bottle walls. The free surface was therefore modeled as a frictionless wall (i.e., a surface that exerts no stresses on the fluid). Several assumptions were also made regarding the particles simulated in the postprocessing program. The particles in the flow were sufficiently dilute so that they did not interact with each other or affect the fluid flow in any way. The particles were assumed large enough so that Brownian forces could be ignored. All the assumptions used here are valid for the majority of industrial uses of roller bottles.

The Fluid Velocity Field in a Roller Bottle

Two flow visualization experiments were performed to validate the simulations. The experiments were conducted in

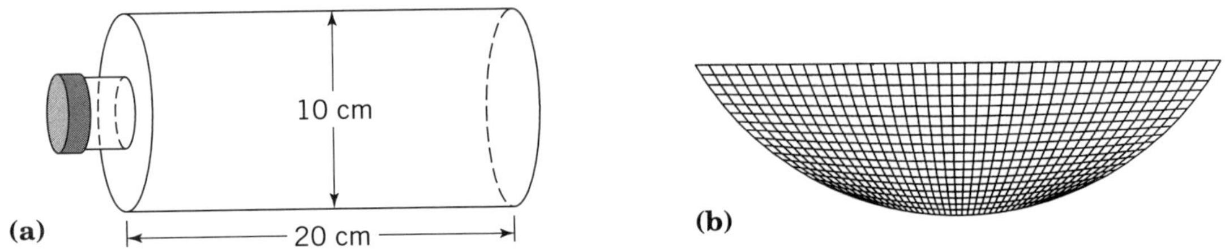

Figure 1. (**a**) Schematic diagram of the roller bottle. (**b**) Schematic diagram of the computational grid used for the simulations.

a glass roller bottle 10 cm in diameter and 20 cm in length. The working fluid was glycerin, which has a density of 1.26 g/cm^3 and a viscosity of 1.25 Pa s. The bottle was rotated using an apparatus that consisted of a set of rollers whose rotation speed and direction were accurately controlled by a computer. The rotation rate for the two experiments performed was 0.25 rpm, giving a Re of 0.13. This flow condition can be assumed to be representative of all conditions in the creeping flow regime, because in this regime the velocity field becomes independent of Re. The velocity field was first validated via comparison with particle imaging velocimetry (PIV) results. The use of particle imaging techniques to study fluid flow problems is well documented in the literature and has been reviewed by Adrian (27). Information that is pertinent to our specific system is provided here. The flow was seeded with 12-μm-diameter silver-coated hollow glass spheres (Potter's Industries) that have a specific gravity of 1.17. These particles are sufficiently small as to provide accurate measurements of the instantaneous velocity field. A 10-mJ New Wave Research pulsed mini-YAG laser was used as the illumination source. The laser beam passed through a series of optical components to produce a sheet of light 1 mm thick (at a focal distance of 1 m). A 32 × 32 grid was analyzed using cross-correlation data acquisition, with 10 ms between laser pulses. Resultant data from PIV was in the form of velocity direction and magnitude at the center of each interrogation area.

The experimental velocity field was measured in the vertical plane normal to the bottle axis in the middle of the bottle. Figure 2 shows the experimental (Fig. 2a) and computational (Fig. 2b) velocity vector fields. The length and color of the arrows represent the velocity magnitude in the plane, ranging from low (dark blue) to high (red). As seen in this figure, the velocity fields are qualitatively identical, with the exception of some small experimental scattering near the free surface of the liquid.

The computational velocity field was further validated by comparing computational and experimental mixing patterns obtained for passive tracers at the center plane of the bottle. This first step for simulating mixing patterns is to determine the trajectories followed by point particles as a result of the equation of motion:

$$dx/dt = v \quad (3)$$

where x is the particle position and v is the fluid velocity. Figure 3 shows the flow pathlines predicted using 4th-order Runge–Kutta integration of equation 3, which are also the flow streamlines because the flow is steady. A stagnation point is located on the symmetry line at 0.6 unit lengths below the free surface (the unit length is the height of the liquid). A similar approach was used to simulate the evolution of a vertical filament initially placed at the center of the bottle (Fig. 4a). The "line" consisted of 10,000 particles moving at the flow velocity. Figures 4b–4e show the configuration adopted by the line, as revealed by the particle positions, at 4, 8, 16, and 32 bottle revolutions, respectively. As the particles are moved by the flow, the line is twisted around the stagnation point. The line stretch in this steady two-dimensional flow is linear with time. The evolution of a fluid line was also observed ex-

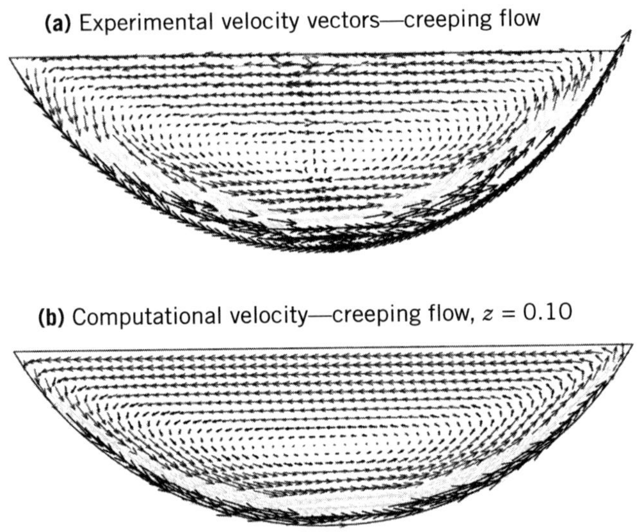

Figure 2. (**a**) Particle imaging velocimetry determination of the velocity field at the center plane of the bottle. (**b**) Finite volume results of velocity field at the center plane of the bottle.

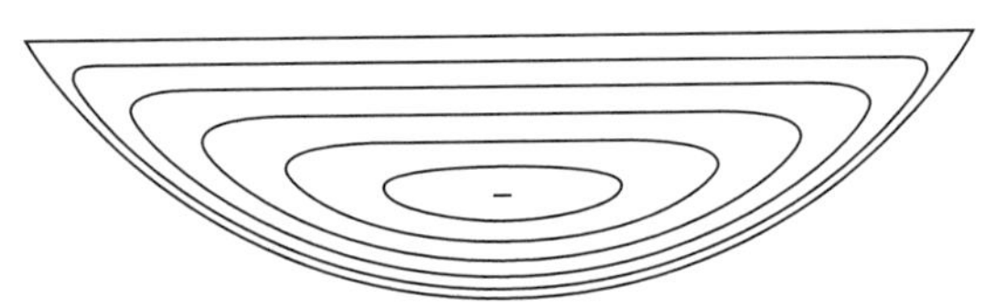

Figure 3. Computational stream lines of the flow at the center plane of the bottle under steady flow conditions.

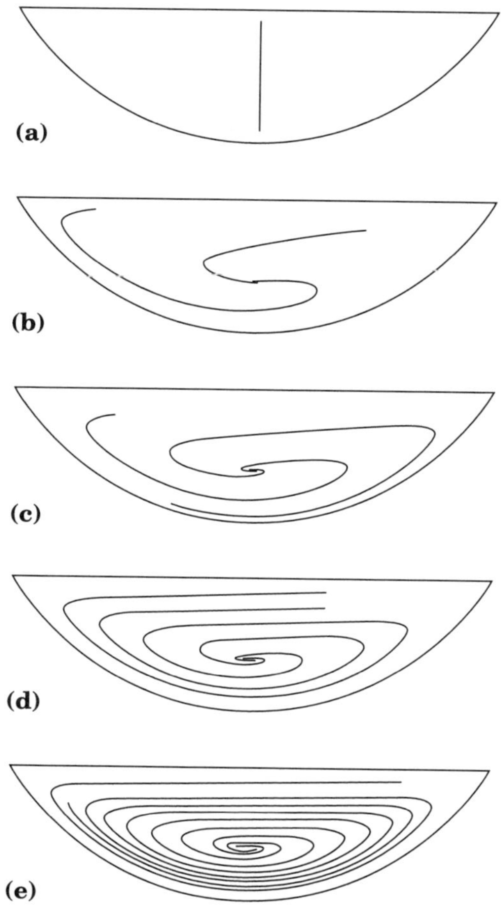

Figure 4. Evolution of a vertical line placed in the flow field. (a) Initial condition. (b–e) The line-after 4, 8, 16, and 32 bottle revolutions, respectively.

Figure 5. Initial condition for the flow visualization experiment. The fluorescent dye was placed in the center of the bottle axially so that end effects would cancel.

Figure 6. Experimental validation of the evolution of a vertical line placed in the flow field.

perimentally to verify the simulations. Figure 5 represents the initial condition, which consisted of a vertical streak of dyed fluid placed in the center plane of the roller bottle, and Figure 6 shows the evolution of the streak of dye after 20 revolutions. Once again, excellent qualitative agreement between experiments and simulations was obtained, and it was concluded that the simulations indeed provide an accurate representation of the velocity field in the roller bottle.

Once the computational velocity field was validated, it was used to obtain a detailed characterization of fluid motion in the bottle. To provide a clear illustration of a three-dimensional velocity vector field is not a trivial task. This is accomplished by using two-dimensional cuts of the velocity field, where the in-plane components are shown using a vector and the total magnitude of the velocity field is shown using color contours, with red signifying high velocity and blue indicating low velocity.

Figure 7 shows vertical cross sections at 2-cm intervals through the bottle (i.e., $z = 0.0$ is 0.0 m from the end wall, and $z = 0.10$ cm is at the symmetry plane in the center of the roller bottle). Figure 7f corresponds to the center of the bottle and is therefore identical to Figure 2b. Moreover, Figures 7d and 7e are also identical to Figure 7f, indicating that end effects play little or no role in the fluid flow past 5 cm from the end of the bottle. The velocity field shows both reflection symmetry relative to the center plane (expected from the symmetry of the boundary) and antisymmetry between the left and right sides of the bottle (expected from the creeping flow condition and the rotational symmetry of the bottle). As discussed next, this observation can have important implications because the flow induced by the end walls of the bottle is the only means of axial mixing.

Figure 7a represents the velocity field at the end of the roller bottle. Due to the no-slip condition at the bottle wall, all fluid follows the bottle wall in a counterclockwise direction. When the fluid reaches the free surface (at the right side), it detaches from the wall and flows outward in the z-direction. At the left side, the fluid immediately adjacent to the wall is drawn downward; this motion is "fed" by an inward flow from the z-direction. Figures 7b and 7c show the presence of a high-velocity component near the surface of the liquid, caused by the end-wall effects. However, detailed examination of the velocity vectors displayed

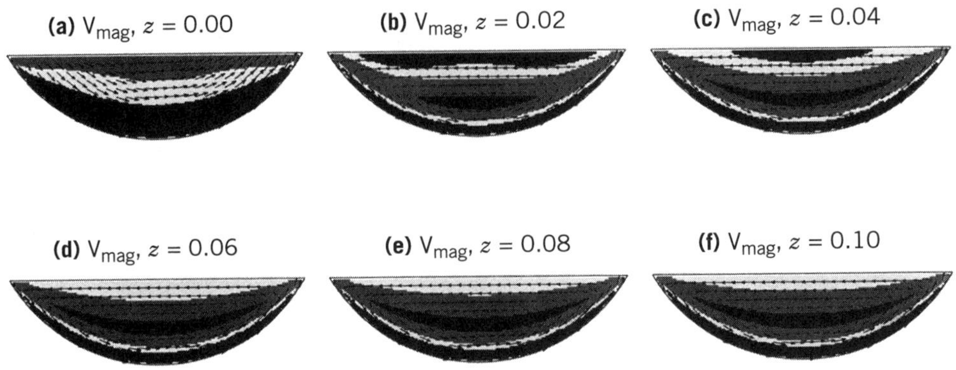

Figure 7. Finite-volume results. Velocity field at several vertical planes spaced along the axis of the bottle.

in Figures 7b and 7c indicates that the x- and y-components of the velocity field are largely unaffected by end-wall effects; such effects appear primarily in the axial (z) component of the velocity field.

The end-wall effects are more clearly shown by a top view of the bottle. Figure 8 shows a velocity contour plot just below the liquid free surface. The end of the bottle is at the bottom of this figure, and the symmetry plane is at the top. Flow in the axial (z) direction is clearly visible in this figure. The fluid flows upward from the end wall in the lower right-hand corner, travels in a counterclockwise loop, then flows downward at the lower right-hand corner as a result of the end-wall flow. There is a stagnation point 0.46 cm from the end of the bottle, as shown by the dark blue color. Moving toward the center of the bottle, end effects have less influence on the flow; there is less than 2% difference in the velocity field between the two-dimensional symmetry plane and the plane at 5 cm from the end wall. Similar flow patterns were observed for other values of Re of less than 10.

An alternative illustration of the three-dimensional flow pattern in the roller bottle is obtained by computing the pathlines of the flow. Figure 9 shows six pathlines for fluid particles initially placed 0.2 cm from the bottom of the bottle. The center of the bottle is at the left-hand side of the figure, and the end wall is at the right. Examination of the pathlines from left to right reveals the effect of the end wall on the flow field. The pathline at the center plane is a two-dimensional vertical loop identical to the pathlines observed in Figure 3. Toward the end of the bottle, the pathlines begin to bend near the free surface toward the center of the bottle. All the pathlines shown are closed loops. The pathlines at the center plane of the bottle are two-dimensional loops, and the pathlines away from the symmetry plane are three-dimensional loops. In either situation, however, the same result ensues: fluid particles in the flow remain trapped in the closed periodic orbits and do not contact other portions of the flow. Such regular motion patterns can pose significant barriers to the mixing process. Experimental evidence of this obser-

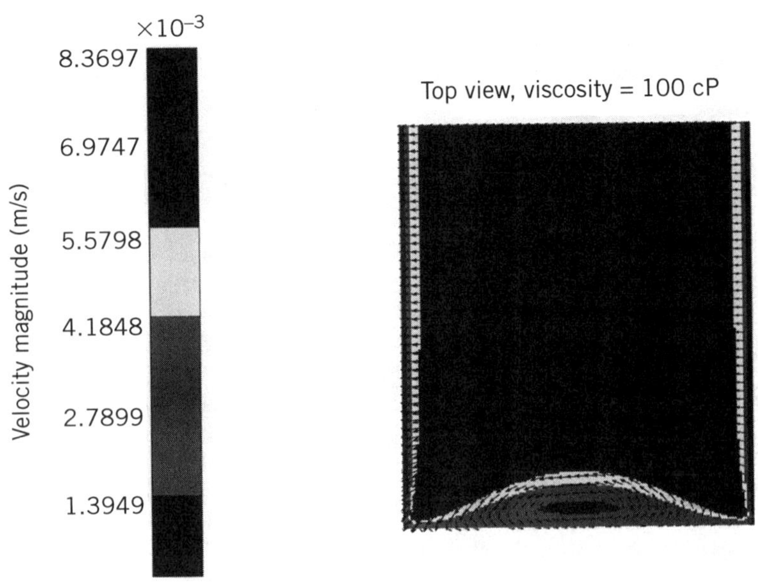

Figure 8. Finite-volume results. Velocity field (*left*) at a horizontal plane near the upper surface of the fluid. *Right*, top view.

vation and approaches for overcoming mixing limitations follow.

The Motion of Cells in a Roller Bottle

The initial point for the modeling of small particles in the flow field is given by the equations of motion proposed by Maxey and Riley (28):

$$m_p \frac{d\mathbf{v}}{dt} = (m_p - m_F)\boldsymbol{g} + m_F \frac{D\mathbf{u}}{Dt} - 6\pi a^2 \mu \int [\pi\nu(t - \tau)]^{-1/2} \left(\frac{d\mathbf{X}}{dt}\right) d\tau - \left(\frac{1}{2}\right) m_F \frac{d[\mathbf{u} - v - (1/10)a^2\nabla^2\mathbf{u}]}{dt} - 6\pi a\mu \boldsymbol{X} \quad (4)$$

where $\boldsymbol{X} = \boldsymbol{u} - \boldsymbol{v} - (1/6)a^2\nabla^2\boldsymbol{u}$.

In these equations, $\boldsymbol{u}$ is the (dimensional) fluid velocity vector, $\boldsymbol{v}$ is the (dimensional) cell velocity vector, a is the cell radius, ν is the fluid kinematic viscosity; and m_p and m_F correspond to the weight of the cell and that of the fluid displaced by the cell, respectively. The five terms on the right-hand side of the equation correspond to buoyancy forces, pressure forces, flow history effects, added mass effect, and Stokes drag forces. This equation can be significantly simplified for the roller bottle problem. Because $\boldsymbol{u}$ is known from the numerical solution of the flow, the relative magnitude of each term can be evaluated. Analysis has shown that the leading order terms are inertia, buoyancy, and drag forces, thus the equation becomes:

$$\frac{d\boldsymbol{v}}{dt} = G\boldsymbol{g} + \frac{1}{S_{tk}}(\boldsymbol{u} - \boldsymbol{v}) \quad (5)$$

where $G = 2(\gamma - 1)gL/[(2\gamma + 1)U^2]$, and $S_{tk} = (2\gamma+1)a^2U/(9\nu L)$; γ is the particle density relative to the fluid density, L is the depth of the fluid, and U is the bottle rotation velocity. If we assume that the particles are cells and the rotation speed and liquid are typical of those used in the biotechnology industry, this equation can be further simplified. Under such conditions, $G \sim O(10^{+3})$ and $S_{tk} \sim O(10^{-6})$, so that the inertial term can be neglected, greatly simplifying the problem. The particle velocity is obtained immediately as

$$\boldsymbol{v} = \boldsymbol{u} + \mathbf{V_s}, \quad \mathbf{V_s} = \frac{2a^2(\gamma - 1)}{9\nu U}\boldsymbol{g} \quad (6)$$

where $\boldsymbol{v}$, the total particle velocity, is simply the sum of $\boldsymbol{u}$, the fluid velocity, and $\boldsymbol{V}_S$, the particle terminal settling velocity. Once $\boldsymbol{v}$ is known, particle positions are found by numerically integrating the equations

$$\frac{d\boldsymbol{x}}{dt} = \boldsymbol{v} \quad (7)$$

using the standard 4th-order Runge–Kutta method. To understand small particle behavior in a steady rotating roller bottle, equations 6 and 7 are used to simulate the motion of particles on the center plane of the bottle. Figure 10 shows the particle trajectories for a settling velocity $V_s = 0.05\ U$, where U is the nondimensional tangential velocity of the bottle wall. Six particles were initially placed on the vertical center line. The particles showed different behaviors: while the two particles nearest to the free surface reached the rotating wall in a fraction of a bottle rotation, the remaining four particles moved in closed orbits. Such particles will never reach the bottom wall. This simulation indicates that whether a particle settles or not depends not only on its settling velocity, but also on its initial location within the flow field. Moreover, as the settling velocity used in this simulation is much larger than those corresponding to real systems, Figure 10 actually indicates that a substantial fraction of the cells in a roller bottle could actually remain trapped indefinitely in recirculating trajectories within the bottle flow, never reaching the bottle wall.

This simulation result was investigated experimentally by using long-exposure-time photography to reveal the pathline followed by a fluorescent particle with a settling velocity similar to the one corresponding to Figure 9. A 200-μm fluorescent polystyrene particle was placed in the center plane of the roller bottle. This particle had a terminal settling velocity of approximately 0.5 cm/min, or 0.06 U, where U is the linear rotation velocity of the roller bottle. Using long-exposure photography, the particle pathline was photographed for one bottle revolution under fluorescent light (Fig. 11). The original position of the particle was approximately 0.6 length units from the free surface (where the unit length is the height of the fluid). The pathline revealed in Figure 11 closely resembles the pattern shown in Figure 10, which is a simulation for particles of roughly similar settling velocity. When the particle was originally placed in a different location, closer to the bottle center, the particle followed a different closed pathline. When placed close enough to the upper surface, the particle eventually settled and remain attached to the bottle wall.

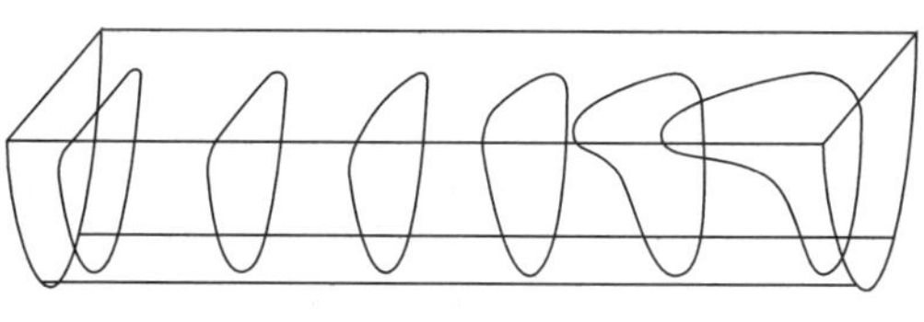

Figure 9. Pathlines of passive tracer particles starting from different positions inside the bottle.

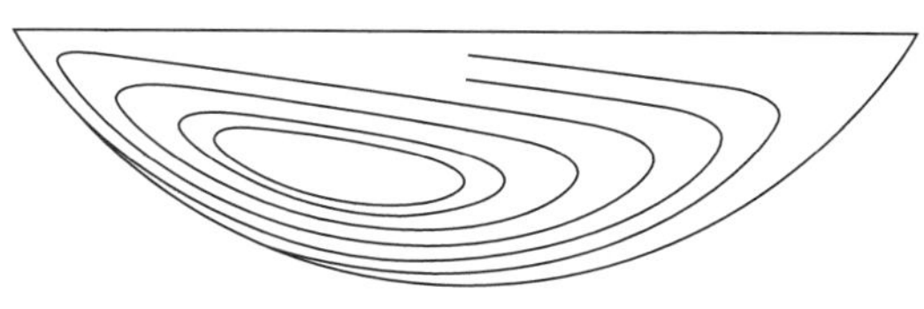

Figure 10. Pathlines of six particles with settling velocity of V_s = 0.05 U placed in the flow.

Figure 11. Experimental validation of a particle pathline in the flow. The particle had a settling velocity of approximately 0.06 U.

DETAILED ANALYSIS OF CELL SETTLING

Particle Settling Results

A more complete picture of cell behavior can be obtained by following a large number of particles over a long period of time. Such simulations were conducted by placing 20,000 particles uniformly distributed across the entire central cross section of the bottle and following them for 20 rotations of the bottle. Whenever a particle touched the bottle wall, it was (arbitrarily) assumed that the particle had "settled" and was removed from the simulation. Figure 12 shows Poincaré diagrams obtained from such a simulations for settling velocities V_s = 0, 0.02 U, 0.05 U, 0.1 U, and 0.2 U. For these simulations, the bottle was rotating counterclockwise at a nondimensional linear velocity of U = 1.0. For V_s = 0 (Fig. 12a), the structure of the particle positions is identical to those of the flow streamlines, because such particles are in fact fluid particles. As V_s increases, the particle paths are perturbed away from streamlines, but as it is shown in Figure 12b, for V_s = 0.02 U most particles remain trapped in closed periodic orbits and never reach the bottle walls; only those particles initially very close to the free surface and the bottle wall are able to reach the bottle boundary. As settling velocity increases, more particles settle, but only for large values of V_s (i.e., V_s = 0.2 U; Fig. 12e) are most particles able to reach the bottle walls.

Although such simulations give information about the dynamic behavior of particle motion, the settling rate cannot be inferred from them. To obtain the spatial distribution of particle settling rate, the settling time of each particle was recorded and plotted as a function of initial particle location. The settling time distribution in the bottle is shown in Figure 13. The bottle is rotated clockwise at a dimensionless linear velocity of U = 1.0 at the wall. Results correspond to V_s = 0.02 (Fig. 13a), 0.05 (Fig. 13b), 0.1 (Fig. 13c), and 0.2 (Fig. 13d). The colors from dark gray to white represent the particle residence time from 0 to the maximum computing time. The smooth core region in dark gray represents the particles that do not settle at all. For small V_s = 0.02 (Fig. 13a), only a narrow strip of particles

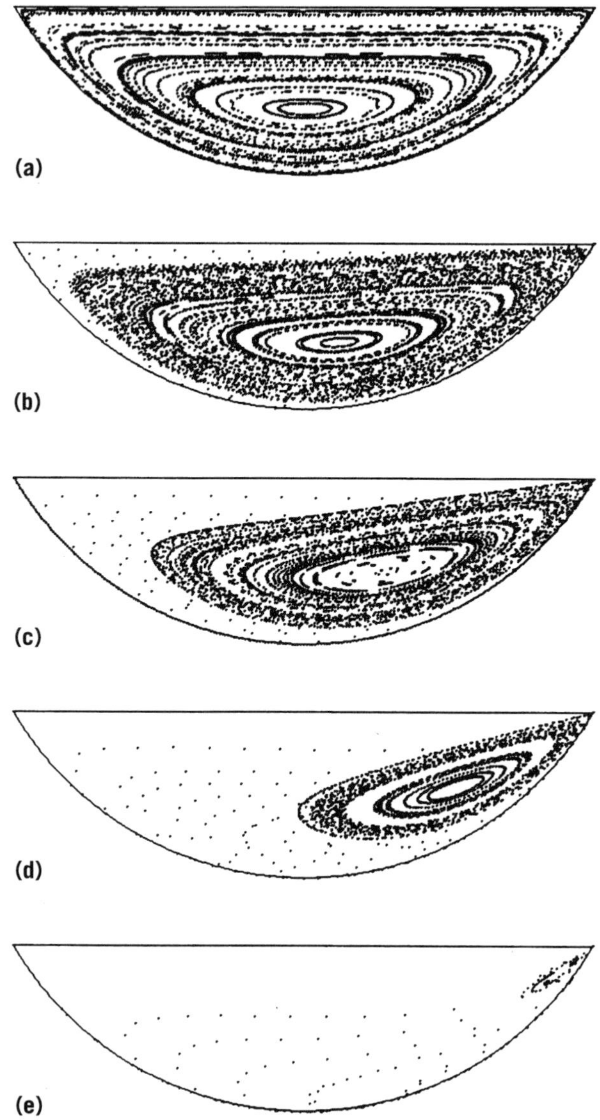

Figure 12. Poincaré diagram of particles, initially distributed uniformly in the central vertical cross section of the bottle, for settling velocities of V_s = 0, 0.02 U, 0.05 U, 0.1 U, and 0.2 U, respectively, from top to bottom of figure (**a–e**).

under the top free surface and near the bottom on the right side of the bottle settle. The width of this strip increases as V_s increases and the size of the red core decreases. The tip of this core region is always at the upper-left corner of the flow and leans to the left side of the bottom of the bottle. At V_s = 0.1 (Fig. 13b), most of the particles in the bottle will settle. When V_s is increased to 0.2 (Fig. 13c), almost all the particles settle to the wall. The spatial distribution of particle settling time in the flow is shown clearly by the shades of gray. Short-lived particles are always on the right side of the bottle for this flow direction. If the flow direction were reversed, Figure 13 would appear as a mirror image because of the symmetry of the system. These results clearly reveal that the particle residence time

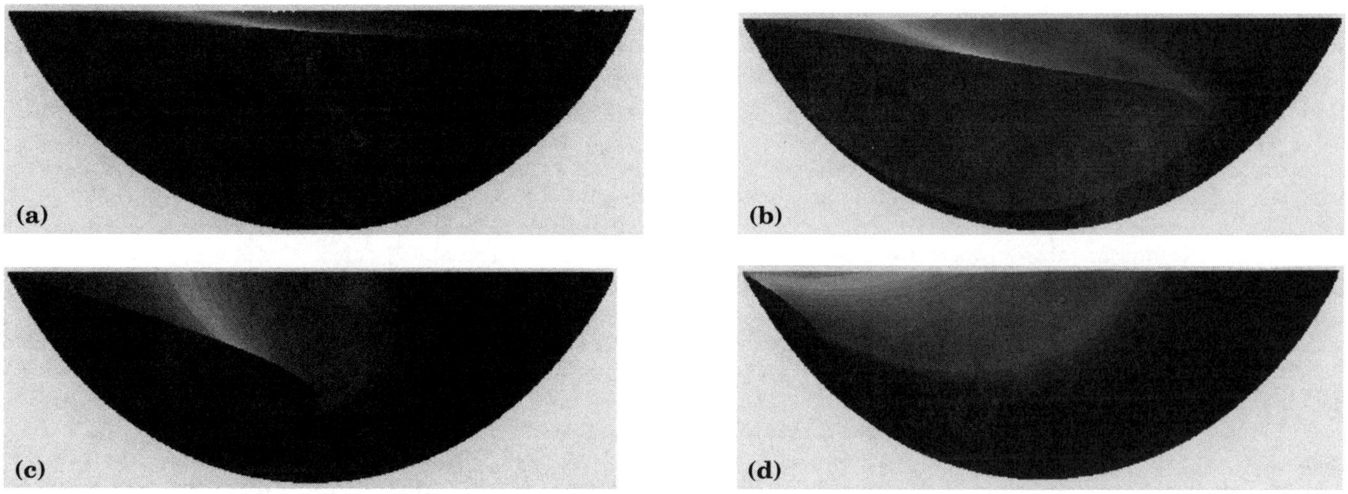

Figure 13. Contour plots of residence time of particles against particle initial position for V_s = 0.02 U, (**a**) 0.05 U, (**b**) 0.1 U, (**c**), and 0.2 U (**d**) after 20 bottle revolutions. The shades in this figure range from dark gray to white, corresponding to zero residence time to infinite residence time.

distribution in suspension is strongly affected by the recirculation of the flow.

The rate of particle settling is shown in Figure 14; the fraction of particles suspended in the flow at a given time is plotted against the dimensionless time, where one bottle revolution corresponds to 12.56 time units. All the settling processes are nonlinear (except for $V_s = 0.2$). Particles settle to the wall relatively fast at the initial stage, and then the rate of settling (the slope of the curves) levels off in a relatively short time, about 8 to 12 time units. With large settling velocity ($V_s = 0.2$, the heaviest particles examined), the particles settle at almost a constant rate.

Effects of Bottle Rotation Speed on Cell Culture Experiments

Previous experiments demonstrated that the growth of anchorage-dependent cells and cell culture products (e.g., varicella virus) in roller bottles was much less efficient than that observed in stationary cultures such as T-flasks. It was hypothesized that these differences were caused in part by decreased cell (uninfected and infected cells) sedimentation rates present in the roller bottle cultures as the result of circulating liquid flows that trapped the particles in circular orbits. The decreased sedimentation times resulted in lower cell plating efficiencies for uninfected cells and increased degradation times for virus-infected cells, that were unable to reach the cell culture surface for attachment or infection. To further investigate this hypothesis, the kinetics of virus-infected cell sedimentation and attachment were studied at different rotation rates. The attachment to and growth of cells on the roller bottle surface were also studied to determine the primary mechanism controlling the differences observed in cell growth rates as a function of rotation rate.

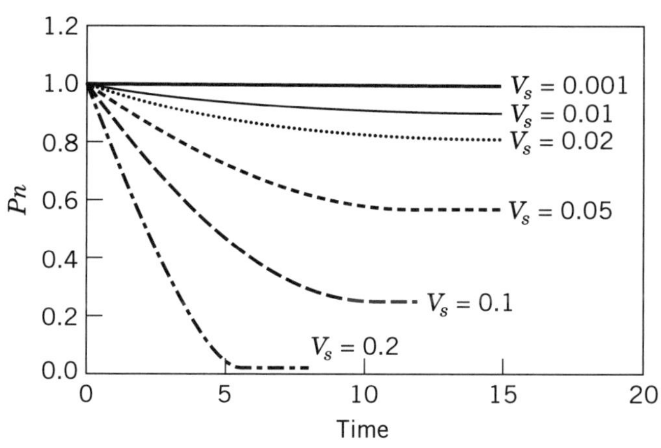

Figure 14. Fraction of particles (Pn) suspended in the flow as a function of time for varied settling velocities (V_s).

Materials and Procedures. Oka strain of varicella was obtained from the Biken Institute, Japan. MRC-5 cells were obtained from ATCC or NIBSC and used to generate a manufacturer's working cell bank (MWCB) at Merck. MEM was manufactured by Merck and supplemented at the time of use with fetal bovine serum (FBS), neomycin, and glutamine. Eagle's basic medium (BME) was purchased from Gibco. Horseradish peroxidase–conjugated donkey antigoat antibody (affinity-purified F[ab']2 fragment) was obtained from Jackson Immunoresearch Labs (West Grove, Pa. [Cat# 705-036-147]). Goat antivaricella antiserum was obtained from Lampire (Lot 2200). 3,3-Diaminobenzidine (DAB) was obtained from Sigma, prepared as a 25 mg/mL solution in water, and stored at −20 °C. Immediately before use, DAB stock solution was thawed and diluted with a peroxide/nickel enhancing solution (Amersham) in phosphate buffer.

A freshly trypsinized, concentrated cell suspension was placed in roller bottles with MEM growth medium supplemented with 10% FBS, 50 μg/mL neomycin, and 2 mM glutamine. The bottles were incubated at 37 °C and rotated at the desired rate. After 4–6 days (depending on the process used), the roller bottles were refed with fresh nutrient medium. Roller bottle cultures with an established cell monolayer on the roller bottle surface were infected by the addition of freshly trypsinized, varicella virus-infected MRC-5 cells. The infected cells were added to the liquid nutrient medium in the roller bottle 6–8 days after the initial roller bottle cell plant. Particle counting and size analysis utilized an Elzone particle size analyzer model with a 48-μm orifice.

An in situ immunostaining procedure was used. The medium was aspirated from the culture vessel (T-flask or roller bottle) using vacuum. The culture vessel was then rinsed three times with PBS with 1% BSA. After the final rinse with PBS, a fixing solution of 5% acetic acid, 5% water, and 90% methanol was added to the culture vessel and exposed to all surfaces for 15–30 min. The methanol is then aspirated from the culture vessel with vacuum. The culture vessels are then rinsed three more times with PBS, and stored with a PBS overlay at 4–8 °C until stained. Immunostaining of the fixed bottles was carried out by a dual antibody method. First, 2–3 mL of a 1:500 dilution of goat antivaricella antiserum in PBS with 1% BSA was added to the culture vessels, covering the entire growth surface of the T-flask and a 3- to 4-cm-wide strip along one side of the roller bottle. The vessels were then placed on a rocker platform and rocked gently at room temperature for 1–2 h to allow binding of the primary antibody. The culture vessels were then rinsed three times with PBS to remove unbound antibody, and 3 ml of a 1:500 dilution of the anti-goat antibody was added. The vessels were again incubated at room temperature with gentle rocking for 30–45 min. Unbound secondary antibody was then removed by three PBS rinses. Then, 10 mL of peroxidase substrate solution was added to each culture vessel and the vessels were incubated at 37 °C until foci of infected cells appeared. The peroxidase solution was then aspirated from the vessels with vacuum and the vessels again rinsed three times with PBS. The stained areas were left under a PBS overlay, and the vessels were stored at 4–8 °C. Infectious foci were counted using a microscope at 10× magnification.

Varicella plaque assay was conducted in the following manner: MRC-5 cells were grown in 60-mm plates with 5 mL/plate of BME with 10% FBS, 0.2% glutamine, and 0.05% neomycin; 10 plates were prepared per sample with 5 plates per dilution. The plates were then incubated at 35.5 °C with a 3% CO_2 gas environment for 2 days. Before addition of the samples, the cell growth medium was removed and 5 mL/plate of MEM with 2% FBS, 0.2% glutamine, and 0.05% neomycin was added to each plate. Diluted virus samples of 0.1 mL were added to each plate. Samples were diluted to ensure a plaque counting range of 15–80 plaques per plate. The plates were mixed well and incubated for 7 days at 35.5 °C with a 3% CO_2 gas environment. Following incubation, the medium was removed by aspiration and 2 mL of 0.2% Coomassie blue stain in 100% ethanol containing 1% v/v glacial acetic acid was added to each plate. After 15 min at room temperature, the stain was removed and the plate was rinsed with cold water. The plates were inverted to dry. Plaques were then counted in each plate using a light box to assist in visualizing the plaques.

Results. In the first study, the kinetics of virus-infected cell sedimentation and attachment were studied at three rotation rates: 0 (T-flask stationary culture), 1/4 rpm (roller bottle culture), and 1/2 rpm (roller bottle). A monolayer of MRC-5 cells was established on the surface of the T-flasks and roller bottles. Following growth of these cells to confluence, varicella-infected MRC-5 cells in nutrient medium were added to the T-flasks or roller bottles and rotated at 0, 1/4, and 1/2 rpm. Samples of the supernatant were collected at various times from 0 to 36 h and were measured for infected cell number remaining. At the same time intervals, flasks and roller bottles were sacrificed and stained with a varicella-specific monoclonal antibody marker to determine the number of infectious foci on the cell monolayer. Infectious foci appear when an infected cell has attached to, fused with, and infected a cell attached to the roller bottle surface followed by infection of neighboring cells. Thus, the disappearance of infected cells from the supernatant could be correlated to the appearance of infectious foci on the cell monolayer (where it is assumed that one infected cell equals one infectious focus).

The infectivity of varicella-infected cells degrades exponentially at culture temperatures. To optimize the use of this material and to provide for as synchronous an infection as possible, the time that the infected cells remain in solution is to be minimized. Figure 15 demonstrates that as the rotation rate is reduced the fraction of infected cells remaining in the supernatant decreases at a given time. This finding is further supported by the increased appearance of infectious foci on the cell monolayer (Fig. 16) as the rotation rate is decreased, thus establishing the correlation between number of infected cells disappearing from the supernatant with the increase in infectious foci. Additionally, a mass balance was performed to demonstrate that as a function of time the appearance of infectious foci equaled the initial number of cells in the supernatant minus the intrinsic degradation rate in solution (degradation rates were determined from independent experiments).

In the second study, lower rotation rates were used to further optimize the roller bottle rotation rate for the infection of the cell monolayer. Rotation rates lower than 1/8 rpm were not selected because the cell monolayer remains without nutrient medium coverage for too long, resulting in poor cell growth and viability. Thus, the following three rotation rates were selected in the second study:

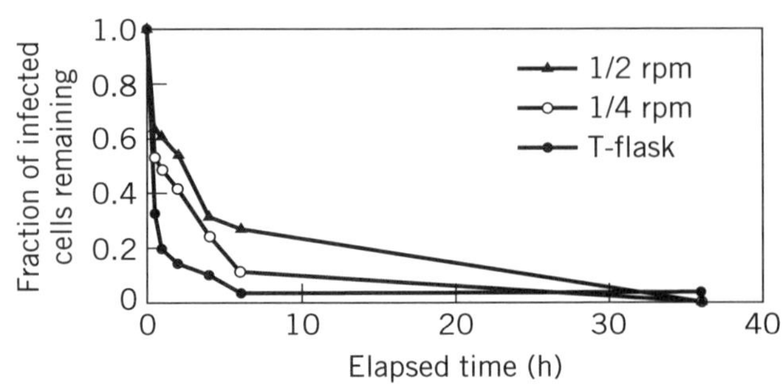

Figure 15. Fraction of cells remaining in the supernatant as a function of roller bottle rotation rate (rpm).

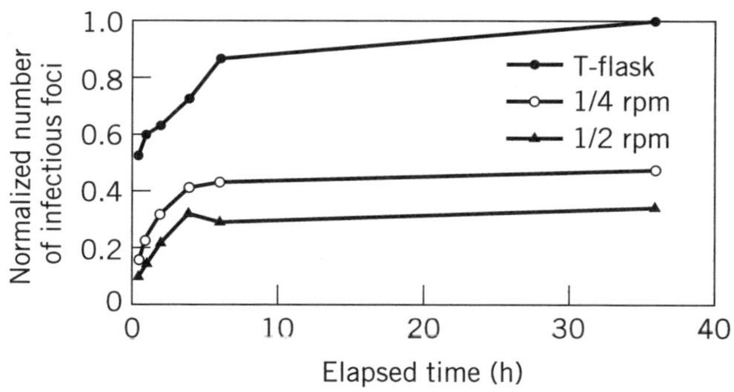

Figure 16. Quantity of infectious foci on the cell monolayer as a function of rotation rate (rpm).

0 (T-flask stationary culture), 1/8 rpm (roller bottle culture), and 1/4 rpm (roller bottle). In Figure 17, the disappearance of infected cells correlates with a decrease in rotation rate, further supporting the hypothesis.

In the final study performed, cell growth rates and final cell densities were determined as a function of roller bottle rotation rate to evaluate if rotation rate was a critical parameter in achieving high cell densities once the cells attached to the culture vessel surface. Two rotation rates were evaluated in this study: 1/4 and 1/2 rpm. Higher cell numbers were achieved when higher rotation rates were utilized (Fig. 18). The higher cell densities as rotation rate is increased may be attributed to improved transport of nutrients, gases, and waste products to and from the bulk

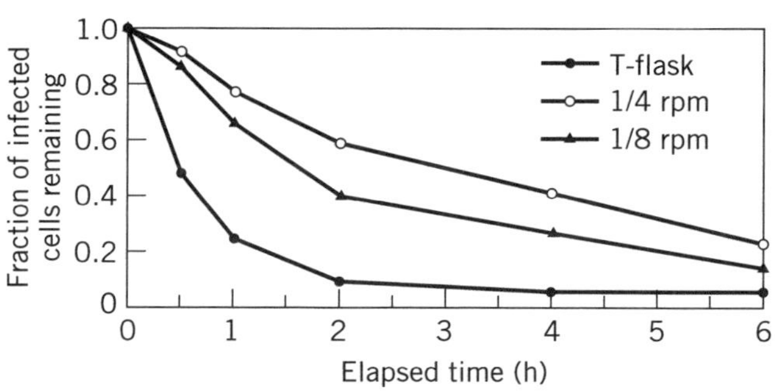

Figure 17. Fraction of infectious foci on the cell monolayer as a function of rotation rate, for lower rotation rates.

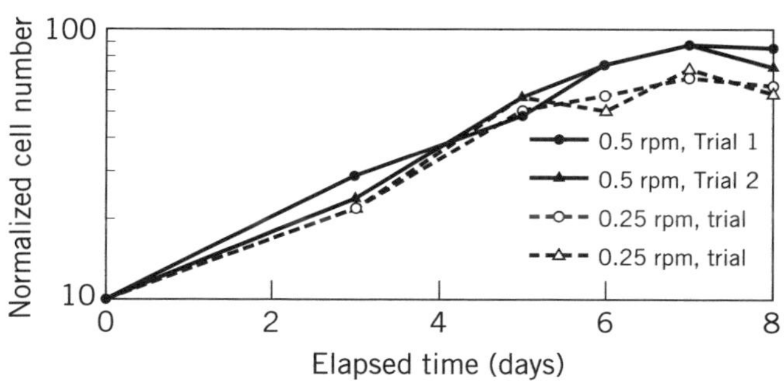

Figure 18. Cell growth rates and final cell densities as a function of rotation rate.

solution to and from the cell monolayer. The results of this experiment pointed to the need to further optimize the cell proliferation steps of the process. These studies were carried out, and the resulting process has shown much greater productivity (data not shown).

The liquid flow modeling shows good qualitative agreement with the experimental data. In both cases, the settling dynamics are nonlinear, in which the rate at which particles are removed from the suspension starts out high and decreases thereafter. In addition, the model correctly predicted that decreases in rotation rate resulted in improved sedimentation of cells onto the cell monolayer.

IMPROVING THE PERFORMANCE OF ROLLER BOTTLES

Using Unsteady Rotation to Enhance Settling

Comparison of Figure 3 and Figure 12 indicates that incomplete settling is caused by the strong recirculation present in the flow, which in two-dimensional cuts of the flow field causes the flow to behave as sets of nested closed loops. When a particle is in the lower part of the flow (below the stagnation point), it falls through loops toward an outer loop. However, if the particle does not reach the wall, it then moves to the upper part of the flow and then falls toward an inner loop. The combination of these two effects causes the particle to move up and down in an closed orbit. In fact, without Brownian motion a particle can exhibit only one of two phenomena: settle to the bottom of the bottle or move in a closed orbit.

Intuitively, such closed orbits can be disrupted by perturbing the flow in a time-dependent manner, perhaps resulting in enhanced settling. This hypothesis was tested by examining settling in time-periodic flows, which are the simplest and most easily controllable time-dependent flows. There are different ways of introducing time periodicity into the flow. A simple approach is to reverse the flow direction periodically. For the creeping flow used in this study, transient effects that occur when the flow reverses direction are negligible because inertial terms in the Navier–Stokes equations are very small; therefore, the periodic flow can be simulated using two identical steady flows that are mirror images of each other, each one acting for a predetermined amount of time in an alternating fashion.

Although mixing of passive tracers will not improve at all in such a periodic flow because the tracers would simply move back and forth along the same streamlines, settling of heavy particles is greatly enhanced by periodic reversal of the direction of rotation. The settling behavior of particles depended strongly on the value of the flow period T, defined as the total displacement of the bottle wall (measured in dimensionless units, where one unit is equal to half the depth of the liquid) between a pair of reversals (i.e., a flow period). For example, if the bottle completes a rotation between reversals, the flow period is two total rotations and one obtains a value of $T = 25.13$. For some values of T, complete settling can be achieved in short times even for a much smaller settling velocity than those studied for steady flow.

The particle settling rate in the periodic flows was studied by placing 6,540 particles uniformly in the flow domain.

As before, particles that hit the wall were considered to be settled at the wall. Two small values of settling velocity were studied, $V_s = 0.002$ and 0.005. The particles were followed for 30 periods for each case, and the fraction of particles suspended in the flow was plotted against time. The results for $V_s = 0.005$ are shown in Figure 19, which demonstrates that the settling processes for all values of T are linear. At small values of T, $T < 25$, almost all particles settle very fast, and the smaller the value of T, the faster the settling. However, at $T = 25$, almost no settling occurs, and the same behavior is observed for $T = 50$. The particle settling rate increases from small T to $T = 25$ and then decreases until $T = 35$. The settling rate for $T = 35$ and 40 are about the same during most of the flow process, except at the later stage when settling in the $T = 40$ case continues at a linear rate while settling in the $T = 35$ case slows down. Therefore, the fastest settling is achieved at small T or between $T = 35$ and $T = 40$. Of those two conditions, the latter is preferable because for $T < 25$, the bottle does not complete a rotation between reversals, and therefore some regions of the bottle wall never touch the liquid. Such a condition would lead to drying and lysis of the cell monolayer in such regions of the bottle wall and therefore to reduced process efficiency.

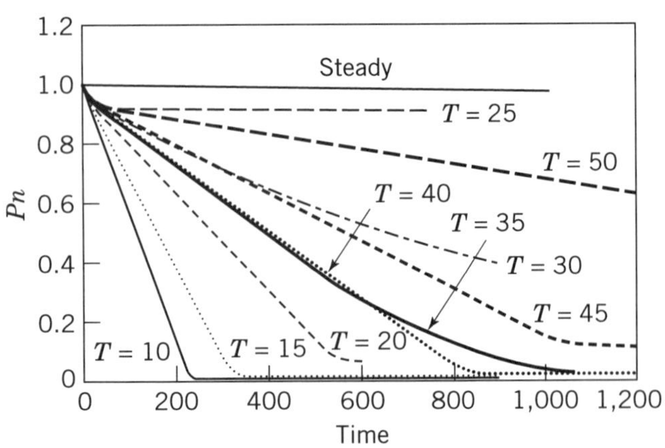

Figure 19. Fraction of particles (Pn) suspended in the flow as a function of time for varied period lengths. The settling velocity for all figures was the same, $V_s = 0.003$ U.

Using Rocking Motion to Enhance Axial Mixing

As mentioned, the flow patterns revealed by simulations suggest slow fluid mechanical mixing, especially in the axial direction. This expectation was confirmed experimentally by using a pH indicator (bromothymol blue) and an acid–base neutralization to study mixing behavior (29). The experiments were carried out on an apparatus that consisted of a set of rollers whose rotation speed and direction were accurately controlled by a computer (Fig. 20). To achieve flows within the creeping regime, glycerin was used as the working fluid. The glycerin initially was mixed with the indicator and 0.5 mL 1 M NaOH, causing the glycerin to turn a deep blue color. With the roller bottle on the apparatus, approximately 20 mL of glycerin was then removed from the bottle with a syringe and mixed with 1.0 mL of 1 M HCl, turning the fluid yellow. This fluid was then injected back into the roller bottle near the back wall (the bottom of the bottle when upright) with a syringe and carefully mixed with the surrounding fluid at the base of the bottle using a specially modified spatula.

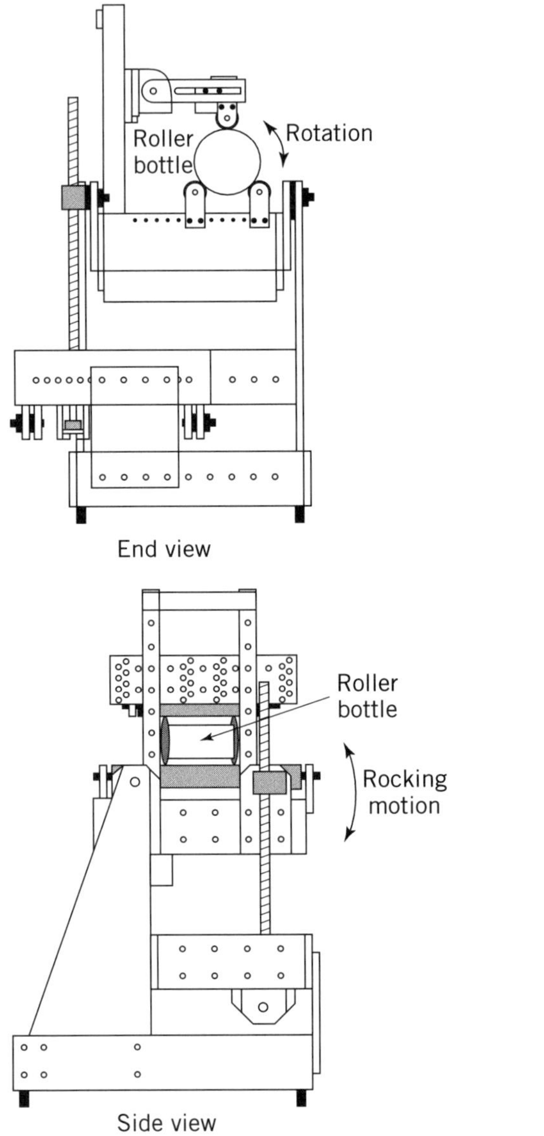

Figure 20. Experimental apparatus used to investigate mixing enhancement by introduction of rocking motion.

After the acidic glycerin was injected back into the roller bottle, the mixer was turned on and the bottle was rotated at constant speed. Figure 21a–c shows the results of this experiment. The initial condition is shown in Figure 21a, mixing after 32 revolutions in Figure 21b, and 64 revolutions in Figure 21c. These figures clearly reveal the effect of the symmetry plane located in the center of the bottle. After 32 revolutions, mixing within each half of the bottle is complete. This mixing results from the axial motion of fluid caused by the end-wall effects. However, even after 64 bottle revolutions, mixing is minimal between the two halves of the bottle; neutralization of the base on the right-hand side will occur only through the action of diffusion.

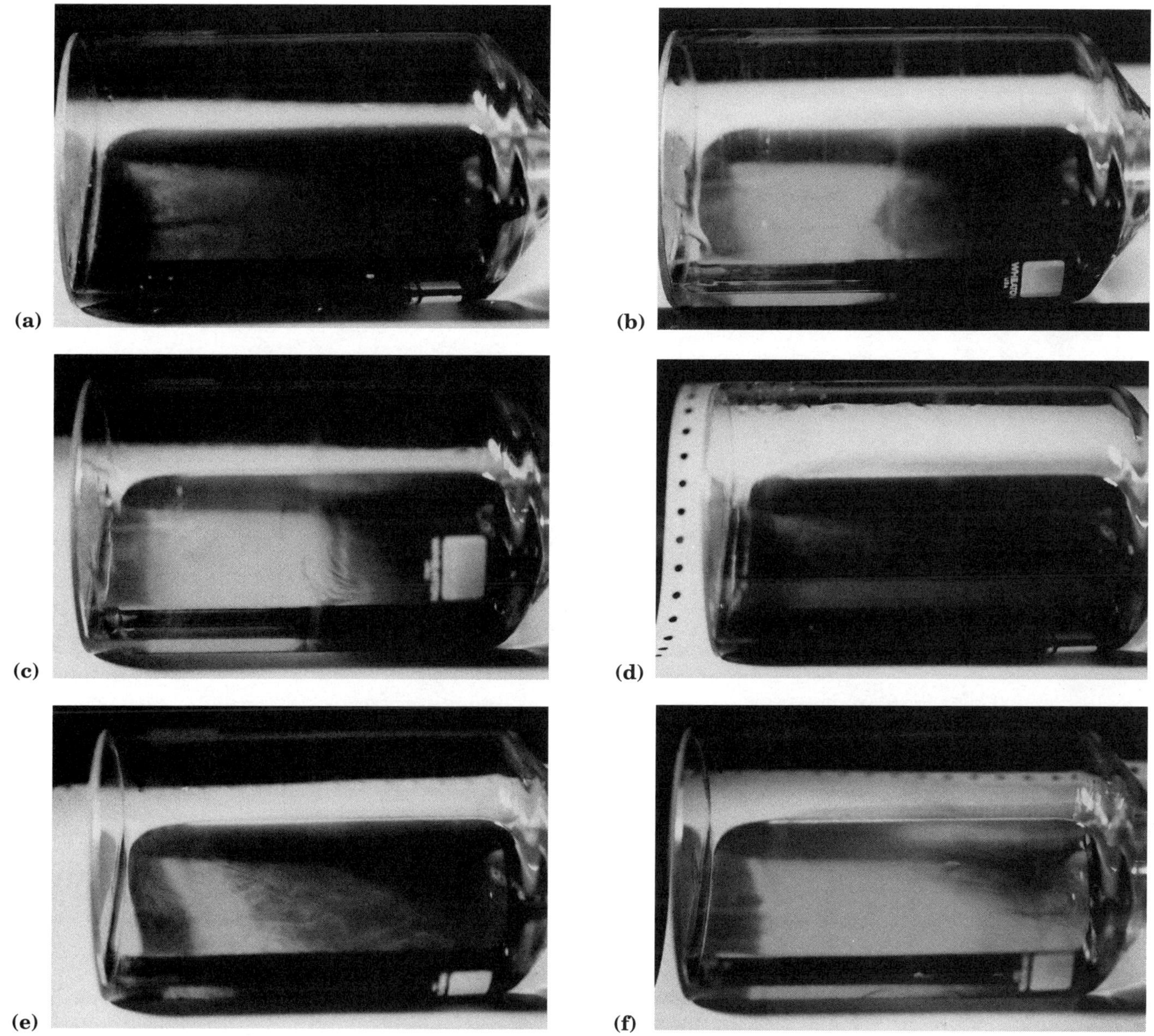

Figure 21. Results from mixing experiments performed (**a–c**) without rocking motion, corresponding to (**a**) the initial condition, (**b**) 32 revolutions, and (**c**) 64 revolutions; (**d–f**) with rocking motion at amplitude of 12 degrees, frequency equal to 1.6 revolutions per rock, corresponding to (**d**) the initial condition, (**e**) 32 revolutions, and (**f**) 64 revolutions; and (**g–i**) with rocking motion at amplitude of 12 degrees, frequency equal to 3.2 revolutions per rock, corresponding to (**g**) the initial condition, (**h**) 32 revolutions, and (**i**) 64 revolutions.

Recent experiments dealing with mixing of dry powders in a rotating and rocking cylinder have shown that a small-amplitude rocking motion in the vertical direction can greatly enhance mixing in the axial direction (30). This approach was implemented here to attempt to enhance mixing in the roller bottle. Two conditions were examined: (*1*) mixing at a rocking frequency of 1.6 bottle revolutions per rocking cycle and (*2*) mixing at a rocking frequency of 3.2 bottle revolutions per rocking cycle.

Figure 21d–f represents the effect of rocking on mixing in the system. The roller bottle was rocked approximately 12 degrees at a rocking rate of 1.6 bottle revolutions per rock, and mixing improved substantially in the system as a result of the rocking. Similarly, Figure 21g–i represents mixing for a rocking rate of 3.2 bottle revolutions per rock; rocking indeed enhanced mixing, which is essentially complete after 64 revolutions. Moreover, because rocking disrupts the recirculating flow patterns described earlier, it is also likely to disrupt cyclical cell motions and enhance settling. It is important, however, to realize that if rocking were used to enhance mixing in a cell growth process, the choice of rocking frequency and amplitude should take into

Figure 21. Continued.

account the need to keep the cells submerged in the nutrient medium as uniformly as possible. If an integer number of revolutions per rocking cycle were chosen, then the cells at the top of the bottle when the rocking reached its maximum angle of inclination would always be the same and would be exposed to nutrients for a substantially shorter fraction of time than those at the bottom. The rocking frequencies of 1.6 and 3.2 revolutions per rocking cycle were chosen because these frequencies should keep the cells fairly uniformly exposed to the nutrients, although the cells closer to the neck of the bottle will still be exposed less to the nutrients.

CONCLUSIONS

The three-dimensional flow in a roller bottle was simulated using a finite-volume method and used to study the particle settling behavior in the bioreactor. From the steady flow simulations of cell motion in two-dimensional cuts of the velocity field, it was shown that a large number of particles in the flow remain trapped in circulating loops and never reach the walls of the roller bottle. Only for large settling velocities (i.e., $V_s = 0.2\ U$) do most particles reach the bottle walls. Thus, in many situations, poor mixing and settling of cells could occur.

It was shown that settling could be enhanced by destroying the closed-orbit motion of the cells. A simple time-dependent flow that periodically reversed the rotation direction was an effective approach for this purpose. Particles were found to settle faster for small values of the flow period T, but such short periods are undesirable because such conditions fail to wet the entire bottle wall, which is essential to keep the cell monolayer intact and viable. Longer periods were also shown to be effective for enhancing settling, and nearly optimal values were identified.

Major limitations in mixing performance in the axial direction were caused by the presence of a vertical plane of symmetry at the center of the bottle. Such limitations, however, can be readily overcome by introducing a small-amplitude vertical rocking motion that disrupts the plane of symmetry, leading to much faster and complete axial mixing. The frequency of such motion was shown to have a significant effect on mixing rate. Additional work is needed to identify optimum values of frequency and amplitude.

BIBLIOGRAPHY

1. H. Tanaka, F. Nishijima, M. Suwa, and T. Iwamoto, *Biotechnol. Bioeng.* **25**, 2359–2365 (1983).
2. H. Tanaka, *Process Biochem.* **22**, 106–115 (1987).
3. C.Y. Hong, T.P. Labuza, and S.K. Harlander, *Biotechnol. Prog.* **5**, 137–143 (1989).
4. Y.A. Elliot, *Bioprocess Technol.* **10**, 207–216 (1990).
5. V.G. Kalthod, *Novel Carrier and Reactor for Culture of Attachment Dependent Mammalian Cells*, D.Sc. Thesis, Washington University, St. Louis, Mo., 1991.

6. E.I. Tsao, M.A. Bohn, D.R. Omstead, and M.J. Munster, *Ann. N.Y. Acad. Sci.* **665**, 127–136 (1992).
7. R. Pennell and C. Milstein, *J. Immunol. Methods* **146**, 43–48 (1992).
8. E. Olivas, B.B.D.-M. Chen, and W.S. Walker, *J. Immunol. Methods* **182**, 73–81 (1995).
9. R. Singhvi, J.F. Markusen, B. Ky, B.J. Horvath, and J.G. Aunins, *Cytotechnology* **22**, 79–87 (1996).
10. R. Kunitake, A. Suzuki, H. Ichihashi, S. Matsuda, O. Hirai, and K. Morimoto, *J. Biotechnol.* **52**, 289–294 (1997).
11. E. Van Hemert, D.G. Kilburn, and A.L. Van Wezel, *Biotechnol. Bioeng.* **11**, 875–885 (1969).
12. C. Horng and W. McLimans, *Biotechnol. Bioeng.* **17**, 713–732 (1975).
13. R.E. Spier and J.P. Whiteside, *Biotechnol. Bioeng.* **18**, 649–659 (1976).
14. D.W. Levine, D. Wang, and W.G. Thilly, *Biotechnol. Bioeng.* **2**, 821–845 (1979).
15. J.J. Clark and M.D. Hirtenstein, *J. Interferon Res.* **1**, 391–400 (1981).
16. B.J. Montagnon, B. Fanget, and A.J. Nicolas, *Dev. Biol. Stand.* **47**, 55–56 (1981).
17. V.G. Edy, *Adv. Exp. Med. Biol.* **172**, 169–176 (1984).
18. E. Rivera, C.G. Sjosten, R. Bergman, and K.A. Karlsson, *Res. Vet. Sci.* **41**, 391–400 (1986).
19. R.M. Gallegos Gallegos, E.L. Espinosa Larios, L.R. Ramirez, R.K. Schmid, and A.G. Setien, *Arch. Med. Res.* **26**, 59–63 (1995).
20. R. Archer and L. Wood, *Proceedings of the 11th Annual Meeting of the European Society for Animal Cell Technology*, Brighton, U.K., September 2–6, 1991.
21. A.K. Majumdar, R.H. Whitesides, S.L. Jenkins, and D.L. Bacchus, *J. Propulsion and Power* **6**, 5–10 (1990).
22. V. Dupont, M. Pourkashanian, A. Williams, and R. Wooley, *Fuel* **72**, 497–504 (1993).
23. J. Seth and W.R. Wilcox, *J. Cryst. Growth* **114**, 357–363 (1991).
24. D.Y. Tang, J.J. Ou, R.H. Heist, S.H. Chen, and A.J. Dukat, *Ind. Eng. Chem. Res.* **32**, 1727–1733 (1993).
25. P.D. Swanson, F.J. Muzzio, A. Annapragada, and A. Adjei, *Int. J. Pharmaceutics* **142**, 33–52 (1996).
26. D.M. Hobbs and F.J. Muzzio, *Chem. Eng. Sci.* **53**, 3199–3214 (1998).
27. R.J. Adrian, *Annu. Rev. Fluid Mech.* **23**, 261–304 (1991).
28. M.R. Maxey and J.J. Riley, *Fluids* **26**, 883–889 (1983).
29. D.J. Lamberto, F.J. Muzzio, and A.L. Tondovich, *Chem. Eng. Sci.* **51**, 733 (1996).
30. C. Wightman, R.M. Paul, F.J. Muzzio, R.E. Riman, and E.K.Gleason, *Powder Technol.* **84**, 231–240 (1995).

See also AMMONIA TOXICITY, ANIMAL CELLS; CULTURE MEDIA, ANIMAL CELL, LARGE SCALE PRODUCTION.

S

SAMPLING METHODS (REACTORS, CONTAMINATION)

M.P. NANDAKUMAR
H. HÅKANSON
B. MATTIASSON
Lund University
Lund, Sweden

KEY WORDS

Biosensors
Coaxial catheter
Expanded bed
Flow injection analysis
On-line analysis
On-line cell disintegration
Sample handling
Sampling
Sampling probes

OUTLINE

INTRODUCTION

To optimize fermentation processes, that is, to ensure maximum growth and/or productivity, the fermentation conditions should be controlled and maintained at their optimum values. However, the media for fermentations and other bioprocesses are very complex in nature. These are often multiphase systems. The liquid phase contains soluble nutrients (e.g., sugar, proteins, vitamins, salts, and products of metabolism such as amino acids, alcohols, and organic acids). Furthermore, there may be a solid phase of suspended material (insoluble substrate/product, carrier materials, etc.). The gas phase is in equilibrium with dissolved gases in the medium. Analysis of any of these compounds by traditional techniques is tedious and time-consuming and cannot provide on-line information of the process. Recently, the increasing demand of high-value products has focused interest on well-controlled, optimized processes. This has initiated the development of new on-line measurement techniques (1–3). The ideal situation would be to have specific probes that can operate in situ during the fermentation. However, at present there is only a limited number of such probes available for measuring pH, dissolved oxygen, and carbon dioxide. There is still no suitable technology for operating biosensors containing a biological macromolecule. In the last few years research on biosensors for bioprocess control has been growing. These sensors offer the ability to measure concentrations of various biomolecules, as well as biological parameters, very specifically using immobilized enzymes or cells or other biochemical/biological entities. Unfortunately, these sensors are inactivated by steam sterilization (4). Because sensor technology is not yet sufficiently advanced to permit this type of monitoring, the majority of analysis is still carried out outside of the bioreactor, a process requiring sampling and sample-handling strategies. This review highlights the importance of sampling and sample handling and the techniques used.

STRATEGIES FOR SAMPLING

For analysis of any bioprocess, off-line or on-line sampling is a prerequisite and is the basis for the final analytical result. The following strategies/criteria are important when designing a sampling device (5).

1. The sterility of the process must be maintained during sampling.
2. The delay between sampling and analysis may result in changes in the composition of the sample. Precautions must be taken to prevent or compensate for that.
3. The analysis must be accurate and reproducible.
4. When constructing the sampling and sample treatment unit, the selection of material must be governed by compatibility criteria such that the process being monitored is not negatively influenced.
5. Protein adsorption and cell colonization on the sampling and sample-handling device may lead to an imparted mass transport and formation of unwanted immobilized catalytic activities that may severely influence the composition of the sample before analysis can be carried out.

Considering these parameters, a sampling system for on-line monitoring of fermentation processes must meet the following criteria.

1. The system can take representative samples from fermenter but should not disturb the fermentation process.
2. The system should be sterilizable and provide a sterile barrier to avoid contamination.
3. Separation of cells, solids, or macromolecules should be carried out with as short a delay as possible within the sampling devices (6).

There are many different types of sampling units available to achieve these goals (7). This review highlights a detailed description of some of the sampling devices used for on-line monitoring of bioprocesses, especially when biosensors are used in the analytical step. It should, however, be stressed that the sampling and sample handling as described here will be applicable when setting up any kind of analysis to monitor a bioprocess. Figure 1 presents a schematic description of different sampling modes.

NATURE OF THE SAMPLE

The character of the reaction mixture to be sampled determines the choice of sampling method. Different kinds of analytes require different sampling protocols.

Volatile Compounds and Compounds Present in Gaseous Phase

Volatile compounds may be present in the head space of a fermentation or be dissolved in the liquid medium. At equilibrium, the concentration in the head space reflects that in the liquid. The composition of the gas phase is usually less complex than that of the aqueous phase; therefore, when the metabolites are volatile, it is probably simpler to monitor the concentration in the gas phase. The risks of infection of the fermentation are lower when sampling from the gas phase. This has made it attractive to use analysis of the gas phase when monitoring industrial fermentations.

There are two main modes described in the literature for the sampling of gases and volatile compounds:

1. Analysis of the outlet gases or of the headspace gases.
2. Analysis of the volatile gaseous compounds dissolved in the culture medium.

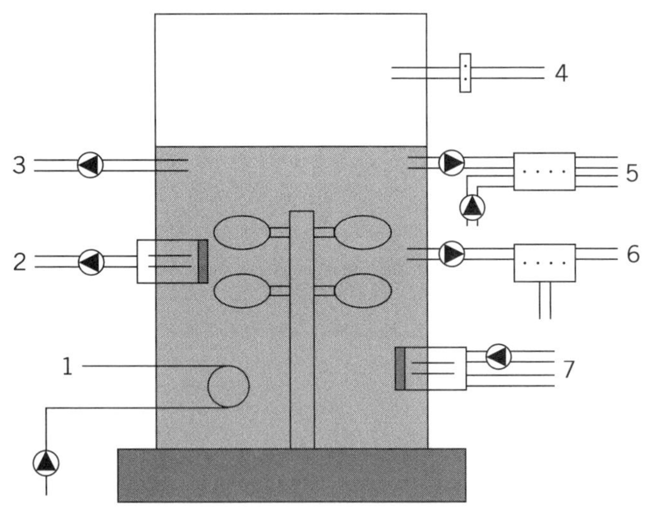

Figure 1. Schematic presentation of different modes of sampling from a bioreactor. The different concepts are (1) dialysis, (2) a membrane unit for removing a particulate-free sample from the turbulent liquid around the impeller, (3) direct sampling, (4) sampling of headspace, (5) sampling and filtering the sample outside the fermenter using a receiving buffer to catch the permeate, (6) sampling with filtering where the permeate is collected, (7) a membrane dialysis probe with the receiving buffer being pumped through the unit.

A sampling method is applied that is based on a technology for supplying oxygen to bioprocesses that are disturbed by bubbling oxygen. Oxygen is supplied in a tubing made of hydrophobic material. The oxygen penetrates the walls of the tubing and is released to the cultivation broth. Similar phenomena can take place for gaseous compounds with an opposite concentration gradient (8). Tubings made from materials that are permeable to organic solvent molecules are immersed in the fermentation broth, and an inert gas that flows through the lumen of the tubings transports the solvent molecules to an analyzer (e.g., a gas-sensitive semiconductor sensor placed downstream of the tubing) (9,10).

Recently a new approach for monitoring carbon dioxide during fermentation was reported based on a piezoelectric crystal (11). A change in the oscillation frequency of the crystal occurs due to selective sorption of a gaseous compound to the surface, and the change in frequency is directly proportional to the mass deposited. Fast response, simple use, and low cost may lead to a more extensive use of this technique. More details on sampling of gases during fermentations are described in the literature (12,13).

Analysis of Nonvolatile Compounds

Analytes Dissolved in the Liquid Medium. Automatic sampling, either continuously or by a frequently repeated procedure, is a prerequisite for on-line or "at-line" monitoring. The sampling system normally serves also as a sterile barrier between the bioreactor and the analytical system. Different analytical systems can be used. Flow injection analysis (FIA) is a common and convenient way to carry out the analysis (14). The samples are heterogeneous, with compounds dissolved in the medium, some forming complexes with other constituents, some not fully soluble, some particulate, and so forth. Manual sampling and off-line analysis are still in practice; however, they result in delays in analysis, risks of infection, and less reproducible sample handling (15). Sampling can be done using sampling units mounted either inside or outside the bioreactor. Several systems of both kinds have been reported recently (16,17). The advantages and disadvantages of internal and external sampling modules are presented in Table 1.

Table 1. Advantages and Disadvantages of Internal and External Sampling Modes

Mode of sampling	Advantages	Disadvantages
In situ modules	Short answering time, simple design, less contamination, suited for coupling on on-line analyzers	Problem of long-term stability, not exchangeable, adhesion of microorganisms and fouling, positioning is crucial, size limitation
External modes	Exchangeable, suited for coupling of on-line analysers	Complex design, longer answering time, fouling, shear stress

Direct Sampling. Direct sampling from a bioreactor is possible through a catheter with a small inner diameter. It is possible to connect this sample stream directly to a FIA system or any other analytical device. However, for precise measurements, the metabolic activity in the sample must be stopped immediately after sampling, otherwise unwanted reactions may take place during the period after sampling but before analysis. If a deactivation is carried out with addition of an inhibitory agent (18), one has to examine the possibility of interference from the reagent on the analysis (19).

One device that addresses these challenges is the coaxial catheter unit (20,21), which is connected to the bioreactor for direct sample withdrawal. It consists of two concentrically placed tubings, an inner and an outer (Fig. 2); the outer protrudes beyond the inner a few millimeters. The sampling is done via the inner tubing. An inhibitor is fed into the space between the tubings, and the broth plus the inhibitor are mixed in the small mixing zone at the tip of the catheter. Inhibition thus takes place at the moment of sampling. A major advantage of the direct sampling system is that a complete sample is obtained, and it is therefore possible to measure the biomass concentration on-line (22) and even perform analysis on intracellular compounds (23,24). The coaxial catheter has not yet been used for sample withdrawal from cultures of filamentous fungi and highly viscous fluids. This method is so far limited to studies of unicellular organisms.

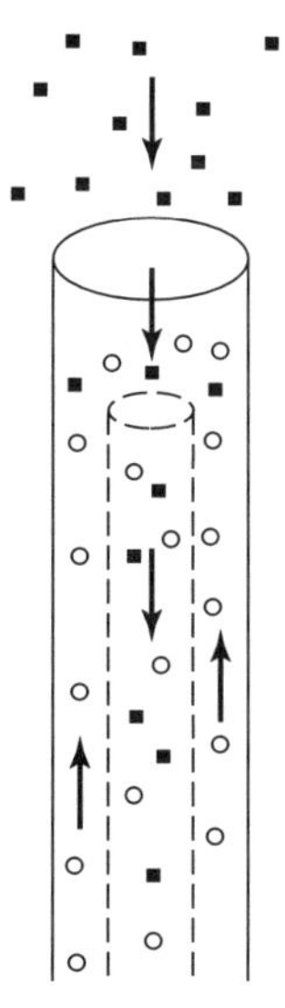

Figure 2. The coaxial catheter. The unit consists of two concentrically placed catheters, the inner being a few millimeters shorter that the outer. Sample is sucked into the inner and is transported away for analysis. The space between the two catheters is used to transport an inhibitor solution toward the tip of the coaxial catheter, where it is mixed with the sample and then withdrawn. If the inner catheter sucks at, for example, double the rate of the inhibitor feed, one volume inhibitor is mixed with one volume of sample and then transported to analysis.

Membrane-Based Sampling Techniques. The sampling system acts as an aseptic barrier between the bioreactor and the analytical equipment. The basic idea behind this type of sampling device is to utilize a membrane for separation during the sampling process. In most cases separations based on differences in size are carried out. The membrane-based sampling techniques can also be applied to cultures of filamentous fungi and in other viscous media. The major disadvantage of the membrane module is clogging of the membrane, mainly due to deposition of proteins on the membrane surface. Reviews on membrane modules are present in the literature (3,25,26).

When using a membrane sampling module, there are three types of separation forces that can be utilized: pressure-driven difference, concentration difference, and electrical potential difference. Applying any of these forces, transport over the membrane will occur and may serve several purposes:

1. Cells are removed from the sample, thus stopping the metabolic activity.
2. Compounds that may interfere with the analysis are retained.
3. Membrane sampling acts as a barrier between the analytical system and the bioreactor.

The details of four major types of membrane separation processes, used in combination with FIA, are presented in Table 2.

Membrane modules can be categorized into two groups: those placed in a recycle loop connected to the bioreactor (filtration outside the reactor), and those placed in situ (as with dialysis filtration). When the membrane unit is placed outside the reactor, fermentation broth is usually pumped through the membrane unit and then recycled to the fermentor. Some characteristics both pro and con of such a device are listed in Table 3.

What is special with the external loop membrane unit is that upon passage of the fermentation broth through the loop, special conditions may prevail. Thus, oxygen limitations may be severe, and, damaging shear forces may appear. This type of device is thus less suited to dealing with high-cell-density fermentations of aerobic cells or with

Table 2. Membrane Separation Procedures Used for Sampling

Mode of separation	Particles separated	Driving force
Filtration	Water and dissolved species	Gravity
Microfiltration	Water and dissolved species (possible to separate unicellular organisms)	Pressure difference
Ultrafiltration	Water and salts (separates biologicals, colloids, and macromolecules)	Pressure difference
Electrodialysis	Charged molecules and ions	Electrical potential
Dialysis	Low molecular weight species	Concentration difference

Table 3. Advantages and Disadvantages in Using a Membrane Unit Placed Outside the Bioreactor

Pro	Con
Replacement of membrane possible	Oxygen depletion may appear in the external loop
Quick response	Shear forces
	Risk of infection

shear-stress-sensitive cells. The great advantage to using an external membrane is the greater flexibility. Thus, a membrane may be easily replaced when needed.

There are two basic modes of operation: the permeate may be collected and used for analysis, or alternatively, a receiving buffer is pumped through the unit on the collection side of the membrane to collect substances passing the membrane. It is important that analyte collection is carried out under sterile conditions. This is usually secured by intermittent sterilization (27–29).

The membranes used for external filtration are made of a variety of materials (e.g., nylon, polypropylene, acrylnitril copolymer, cellulose acetate, cellulose nitrate, polyvinylpyrroliden fluoride) and are available in different pore sizes. Normally, membranes with pore sizes of 0.2 μm are preferred because they guarantee the separation of medium from the biomass and other suspended material (30). Sampling units with external filters are available commercially. The Bio-Pem (B. Braun, Melsungen, Germany) consists of a chamber (vol. 175 mL) containing a magnetically stirred rod. The stirrer ensures a tangential flow across the membrane, which is placed in the bottom of the chamber (31). This unit was successfully used for sample withdrawal during a cultivation process involving unicellular microorganisms but was found unsuitable for sampling cultures of fungi. A major draw back of the Bio-Pem is the large dead volume in the membrane module, which may cause substrate depletion in the loop as well as high response times. An alternative method involves a cross-flow filtration module in which the dead volume is less, resulting in a much shorter response time (<10 s). This module is applied efficiently in fed-batch penicillin fermentations on complex media (32) and for wastewater treatment (33).

There is still another unit based on a planar membrane (Fig. 3). The membrane is clamped between two metal blocks, which are then screwed together. During filtration, the incoming broth (feed) is pumped into grooves in the upper block, and the filtrate is collected in the lower chamber. The broth is separated into two streams: the particular-free filtrate (permeate), which passes the membrane, and the particulate-enriched retentate, which is returned to the fermentor. The filtrate is then led to an injection loop for later injection into the analytical system. After passing the loop, two possibilities arise: the filtrate is either returned to the bioreactor or sent to the waste collector. The latter would quickly empty a small fermentor. To avoid this, the filtrate can be recycled to the fermentor, but this increases the risk of contamination. Placing a sterile barrier (e.g., a microfilter) between the injection loop and the bioreactor reduces the risk (34). This may be acceptable in microbial fermentations, but in certain types of processes, such as human cell line production, such a strategy cannot be applied because such cells are very slowly growing and must be cultivated during long time periods and at the same time have a high value. An alternative approach is called discontinuous filtration. The filtrate line is closed between injections and opened shortly before the injection to flush the tubings and the loop (35).

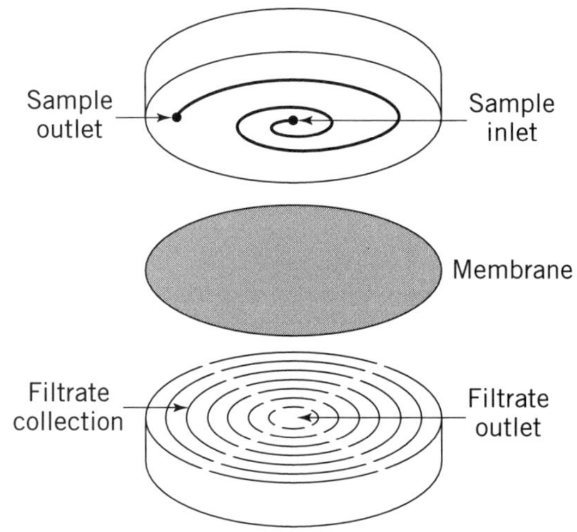

Figure 3. Schematic presentation of the flux membrane dialysis module. The membrane is mounted between the two holders, and sample is introduced on one side. The filtrate is collected and transported for future analysis.

There are several units with planar membranes available on the market. These units have been used for different applications. The A-SEP TM module (Applicon, Schiedam, The Netherlands) has been used for monitoring acetate and phosphate during *Escherichia coli* fermentations (36), utilizing a standard 47-mm-wide 0.45-μm MF membrane. The Minitan TM system (Millipore, Bedford, Mass.), using two membranes, has been employed for the determination of glucose in *Candida rugosa* cultivations

(37), followed by several others for the determination of glucose, diols, and ethanol (38,39).

Membrane filters may also be available in the form of hollow fibers of different diameter and porosity (Fig. 4). Depending on the intended application, the cartridge contains different numbers of hollow fibers. In accordance with the cross-flow filtration principle, the medium is pumped at high velocity (1 m s^{-1}) through the fiber. A pressure is applied, and the filtrate is collected on the outer side of the fibers. It is also possible to pump the broth on the outer side of the fiber(s) and collect the filtrate inside (40). However, depending on the membrane geometry, this latter may be less attractive. When dealing with anisotropic membranes it is advisable to apply the complex medium to the smooth side of the membrane.

A range of different applications are described in the literature. A stainless steel module with only one fiber inside has been reported (41). The fiber had an inner diameter of 5.5 mm and a length of about 200 mm. The dead volume of the module was 2.5 cm^3; and it was successfully used for sampling in conjunction with determination of ammonium, glucose, and phosphate. Reports on the use of several in-house built hollow-fiber modules can be found in the literature, covering vastly different applications in mammalian cell culture (42), lignocellulose hydrolysate fermentations (43), and biodegradation of naphthalene sulphonic acid (44). Systems are also available commercially from Bio-FIO (Glasgow, U.K.) and used for monitoring different types of fermentation components (45,46).

Despite the potential complications associated with operating an external loop filtration system, it remains a popular approach. The great advantage is that when membrane clogging occurs, it is possible to intervene without ruining the fermentation.

In dialysis, a membrane is used to separate a complex solution from a more well defined solution. Pores in the

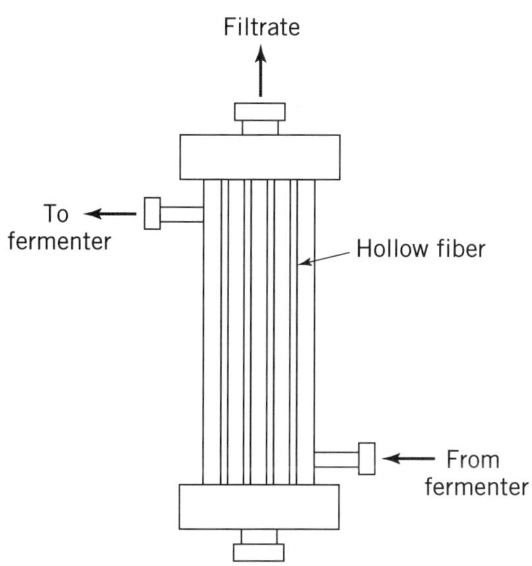

Figure 4. Schematic presentation of the hollow-fiber modules for sampling. A perfusion buffer is passed through the lumen of the fibers, and the fermentation broth is introduced on the shell side of the unit.

membrane allow exchange of small molecules between the two liquid volumes. The driving force is concentration difference. The different modes of dialysis originate from whether the donor and acceptor phases are stagnant or moving. In equilibrium dialysis both are stagnant and, at equilibrium, the concentration in both phases will be the same of those molecules able to pass the membrane. In continuous dialysis, compounds passing the membrane are continuously removed. This results in a larger concentration gradient, and a higher flux is maintained.

The membranes used for dialysis are classified according to their molecular weight cut-off. Most dialysis membranes are made from cellulose materials, but because many organisms are able to hydrolyze cellulosic materials, this limits their applicability. There are several sampling units that incorporate dialysis membranes. They are usually placed in the fermentor, and a receiving dialysis carrier buffer is circulated on the permeate side of the membrane. The carrier buffer is thus pumped into the bioreactor and, after having received the dialysate molecules, removed again. The major problem associated with this approach is the deposition of cells and fouling of the membrane on the fermentor side. The accumulated layer of cells may reduce the mass transfer from the fermentor to the accepting solution and also cause decreases in concentration of metabolites passing through the membrane because they run the risk of being metabolized by the cell layer before entering the dialysis membrane. To help reduce fouling it is possible to place the dialysis probe into the zone of the fermentor where the impeller creates the greatest turbulence (47).

Another means of obtaining similar results, but with fewer design restrictions, is to construct a dialysis probe that besides containing the membrane unit also has a stirring function, so that a high tangential liquid flow is generated over the membrane. This approach was used by Mandenius et al. (48) in constructing a system for use in dense particulate suspensions. The membrane is placed in the front of the probe, and membrane fouling is avoided by means of the rotating magnet driven by an electrical motor inside the probe (Fig. 5).

Zabriskie and Humprey (49) developed a dialysis probe situated within one baffle of a stirred fermentor. They were able to take samples for glucose analysis during cultivation of yeast cells on a synthetic medium. The disadvantage of this dialyzer is that it can be used only in this type of bioreactor. Another sampling device was presented by Glaxo Laboratories, Ltd. (50). This unit consist of an elongated support with a milled helical groove. A tubular membrane made of cellulosic materials or polytetrafluoroethylene was fixed to the support using ring seals.

An in situ dialysis membrane module was developed at the University of Hannover and is marketed by Advanced Biotechnology Corporation (ABC). The module consists of a unit supporting the membrane, and one or more extension units. The membrane module is usually inserted through a standard 19- or 25-mm port in the top of the bioreactor or, alternatively, through a 25-mm port below the liquid surface. When inserted from the top of the bioreactor, one or more extension tubes are needed to place the membrane below the liquid surface. Liquid passing the

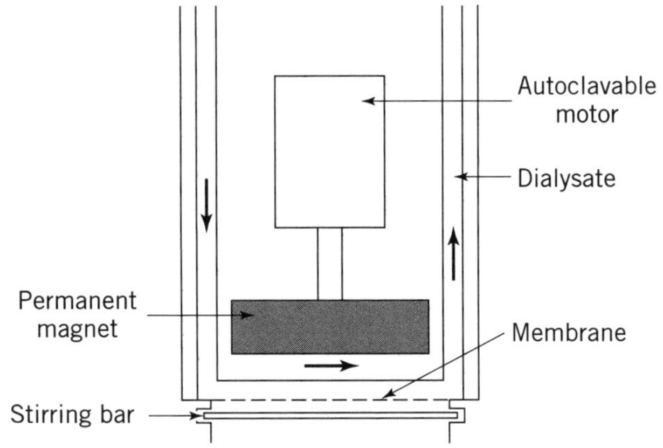

Figure 5. Cross section of the dialysis probe. The stirring bar is rotated by the rotating magnetic field when the motor is on. Receiving solution is pumped over the inner side of the membrane where dialysate is received and removed for analysis *Source*: From Ref. 48.

membrane is collected in 16 grooves, and these 16 permeate streams are joined in four openings placed evenly along the membrane holding unit. Through these four openings the permeate enters the central channel (10 mm i.d.), and a continuous sample stream to the analyzer leaves the unit from the top of the probe (51).

Gibson and Woodward (52) have also described an internal dialysis module, based on the same principle but without a stirrer. Ethanol and glucose could be sampled with this unit, and the results from the subsequent analyses agreed well with the on-line data.

Another modified technique is the microdialysis used for continuous in vivo sampling, which has been described in reviews (3,53,54). The microdialysis sampling unit is a probe, and it is a miniaturized form of a dialysis unit. It is manufactured in several different configurations, one is seen in Figure 6.

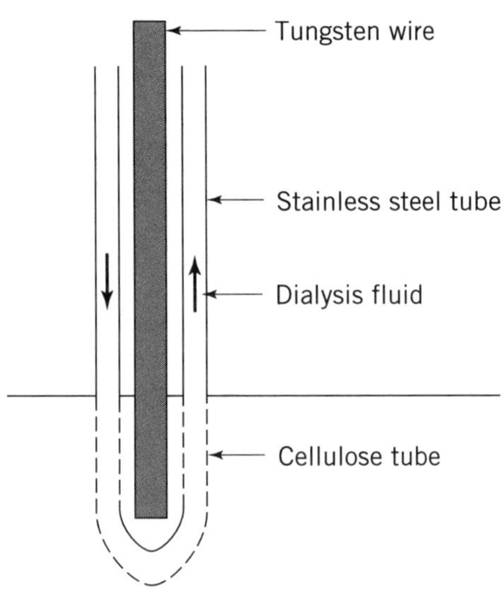

Figure 6. The microdialysis unit.

Typical flow rates of the perfusion liquid are 0.5–25 μL/min, and to avoid pulses in the flow, a peristaltic pump or preferably a syringe pump should be used. Due to its ease of operation and the general approach, the number of applications increases rapidly. It can easily be coupled on-line to a number of separation and detection techniques. This technique was used in fermentation processes for the production of ethanol, penicillin, and so on (55). However, the slow flux of liquid and thus the long response time from an analysis based on this sampling technique, compared with the conventional sensing devices, should be considered.

The major problem in dialysis units is the fouling and clogging by macromolecules on the membrane surface. This can be overcome by adding molecules that compete for the binding site of the protein, adding a displacer (56), changing the pH (57), or pumping the solution tangentially over the membrane surface. An excellent review on dialysis for flow injection analysis has been published (26).

MODES THAT AVOID OR AT LEAST REDUCE THE NEED FOR SAMPLE TREATMENT

As already noted a major concern in sample handling is clogging. Quite often packed-bed systems are used, and then particulate matter constitutes a major obstacle. In these cases, particulate matter has to be removed prior to analysis. However, with the introduction of expanded-bed chromatography, it is possible to handle large-scale chromatographic processes without separating out the cells (58). This made it tempting to investigate the possibility of using the same approach in an analytical system.

Enzyme-based analysis was studied using glucose oxidase immobilized to Streamline® particles. These particles are characterized by being of heterogeneous size and density distribution, such that when used in an expanded bed, a stable expanded bed can be obtained. Glucose was monitored using this technique (59). A more challenging issue is whether the same system would allow binding reactions to take place. If so, one should be able to run immunoassays in whole broth and maybe even in cell homogenates. Assays of human serum albumin in the presence of increasing amounts of heat-killed yeast cells was used as a model system. Antibodies against albumin were present on the Streamline® (60). The assays seemed successful, and therefore the model was also studied to see if the intracellular enzyme β-galactosidase could be quantified in a cell suspension or a cell homogenate. The results were so encouraging that it is safe to say that this technique allows binding assays without the preceding separation of particulate matter from the sample.

PRETREATMENT OF SAMPLES

Another vital part of the analysis is the pretreatment of the samples. This includes dilution, centrifugation, extraction, and filtration, depending on the nature of the samples. Certain samples can be analyzed without pretreatment; however, for complex samples extensive clean-up is required before introduction into the flow system. This topic has been discussed in several reviews (61,62). The reproducibility and rapidity of flow injection analysis promotes its applicability for on-line pretreatment of samples with less manual intervention (63–65) (Fig. 8). Additionally pretreatment steps by FIA are rapid. The pretreatment and analysis, including chemical conversion, usually require less than a minute. Furthermore, when samples are toxic or pathogenic, FIA offers the advantage of removing, or at least minimizing, human exposure. An excellent review on sample pretreatment by FIA was published by Clark et al. (66) and includes a description of the theoretical background of sample pretreatment.

Disintegration of Cells

Monitoring levels of intracellular compounds has so far been possible only via off-line analyses. However, after the recent development of an on-line disintegration system that can be used together with expanded-bed adsorption in a flow system, a new possibility seems to have emerged. The important issues are that the whole broth is taken and treated and after that the whole homogenate can be processed in, for example, a binding assay.

The whole broth is introduced into an FIA system where it first passes an expanded bed containing Streamline® with immobilized lysozyme. On passage of *Micrococcus lysodeiticus* cells it has been proven that a high degree of disintegration is achieved, as judged from the amount of protein released (23). When adding a subsequent step involving sonication using a flow cell, the disintegration is even more efficient. The experimental set-up is shown in Figure 7. A culture of *E. coli* was studied and its level of β-galactosidase production was monitored. The detection step consisted of an expanded bed containing Streamline® with immobilized antibodies to β-galactosidase.

Upon introduction of the whole broth, homogenization takes place by treatment in the two units; and afterward liberated enzyme is trapped by the immunosorbent. The enzyme is quantified by reading the amount of *o*-nitrophenol liberated after passing a pulse of *o*-nitrophenyl-β-galactopyranoside over the sorbent. After the assay the immunocomplex is broken and the system is ready for another assay (24).

Centrifugation

Centrifugation is commonly used for separating cells and particulate materials from the fermentation broth. It is possible to include a small-scale centrifuge in an FIA system, with recycling of the sample to the reaction vessel (67).

Liquid–Liquid Extraction

Liquid–liquid extraction is based on the partitioning of the analytes between an aqueous and an organic phase and is widely used for gas chromatographic analysis. Integrating this technique into FIA is difficult because it is difficult to automate, laborious, and expensive. Solid-phase extraction (SPE) is based on sorption of analytes in solution to a sorbent packed in a cartridge or reactor, typically the size of a precolumn (10–20 × 2–4.6 mm in length and i.d., respectively) (68). While the analyte is trapped into the packing material, compounds with little or no affinity for the sorbent pass through the column. After washing the column to further remove the interfering substances, the analytes are then desorbed and eluted by a small volume of an appropriate solvent. Recently, Vreuls et al. (69) demonstrated the coupling of on-line SPE to gas chromatography. A precolumn packed with mixed anion/cation exchange materials was used on-line in a flow system for removal of ionic substances in aliquots of a filtrate from fermentation.

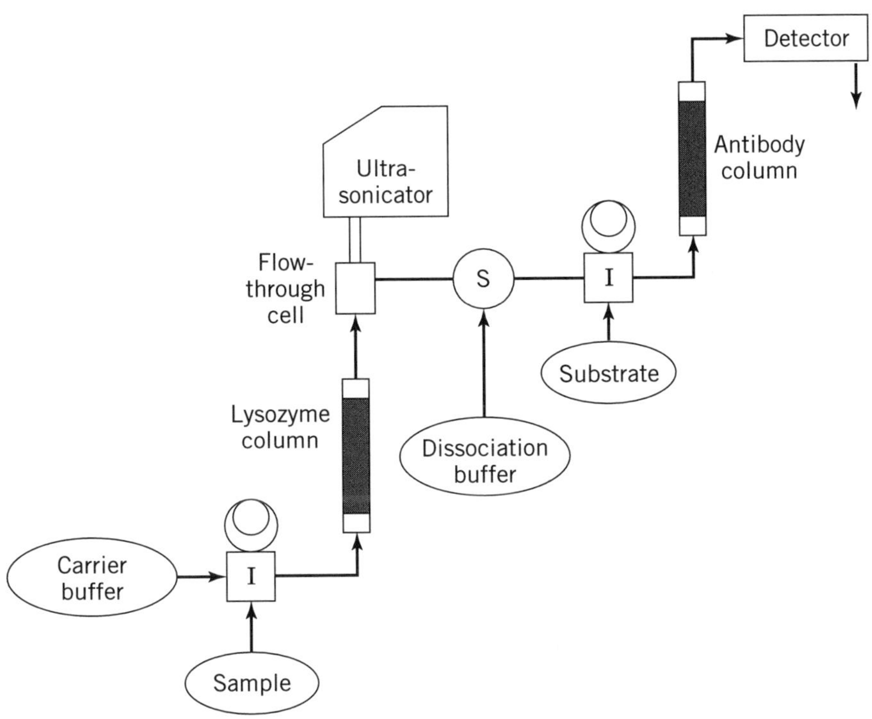

Figure 7. Schematic presentation of a complete flow system, including an expanded flow column with immobilized lysozyme and ultrasonicator flow cell. An expanded column with immobilized antibodies is also included for the subsequent ELISA.

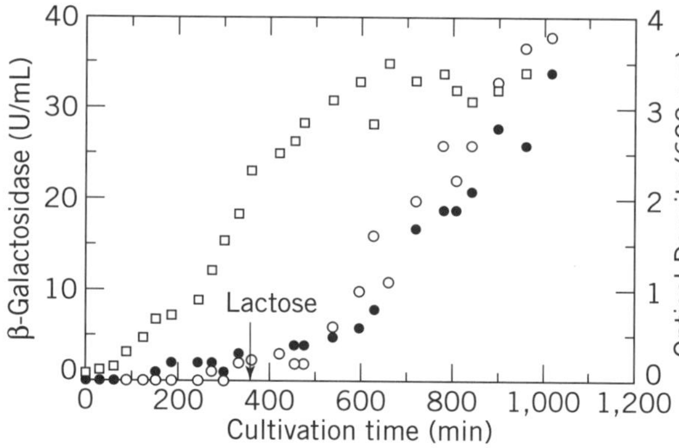

Figure 8. Monitoring of a shake-flask experiment. On-line (●) and off-line (○) β-galactosidase activity and o.d. (□) measurements. At the arrow the inducer lactose was added.

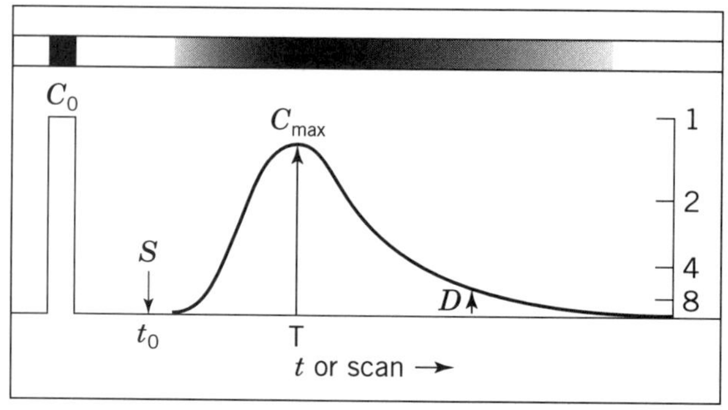

Figure 9. In addition to peak height a dispersed sample zone yields an infinite number of vertical readouts. The initial concentration, C_0, is injected at time t_0, and the pulse appears as a continuum of concentrations as exemplified by the peak. At time T, C_{max} is the highest value read, but a reading could be done anywhere along the peak (e.g., a diluted value at point D).

Preconcentration Techniques

If the analyte concentration is less than the concentration range of the detection method, the sample must be concentrated. Most of the manual pretreatment methods are time-consuming and expensive. Microcolumns with ion exchange materials can be used to significantly reduce the volume of the analyte by adsorbing the sample into a solid support in the column and by a further elution (70).

Dilution of Samples

If the analyte concentration is more than that of the detectable range of the analytical system, the sample must be diluted. The same goes for a sample in which inhibiting or otherwise disturbing substances are present in high concentrations. Manual dilution is used for off-line analysis. On-line dilution is now widely applied for fermentation processes (66) and includes the use of gradient dilution, zone sampling, and cascade dilution.

In the gradient dilution, an originally homogeneous sample zone disperses during its movement through the FIA conduit. The zone undergoes a change from an initial plug shape with a concentration C_0, to an exponentially modified gaussian shape with a maximum concentration in the zone of C_{max}. If the peak maximum does not lie in the calibration range of the measurement, the zone can be made at a time on the profile that yields a decreased response. By selection of measurement times longer than the time of peak maximum (T in Fig. 9), an element with a decreased concentration is measured, yielding a proportionally decreased absorbance. This measurement is reproducible from sample to sample and can be calibrated.

In zone sampling a small portion of the sample zone injected into a second flowing stream results in dilution (71,72). In cascade dilution the sample volume is reduced through stream splitting using differential pumping coupled with a true dilution by merging with a subsequent carrier stream.

Sample Injection

Sample handling and pretreatment can be achieved by using many techniques in FIA. The method of injection is equally important. The purpose of sample injection in FIA is to deliver a well-defined sample bolus reproducibly into a continuously flowing carrier stream. Ruzicka and Hansen (73) described that changing the injected sample volume is a powerful way to change the dispersion. An increase in peak height and sensitivity of measurement is best achieved by increasing the volume of the injected sample solution. Conversely, dilution of overly concentrated sample material is achieved by reducing the sample volume.

There are two methods of injection: time-based injection (a sample loaded for a precise time at a given flow rate) and volume-based injection (the sample fills a geometrically defined volumetric cavity). This well-defined sample bolus is then inserted into the carrier stream. Volume-based injections are performed using rotary valves, commutators, and hydrodynamic injections. A detailed methodology used for injection has been published (74).

Calibration of Sampling System

For any kind of analysis, whether it is on-line or off-line, the system needs to be calibrated to ensure that the compound of interest is being measured reliably. It is easy to calibrate an off-line system. However, the continuous systems described so far commonly have no possibility for calibration and recalibration on-line without disrupting the monitoring procedure. Thus, any changes occurring in the sample during the recalibration, are not registered (5). Another important limitation is the fact that the calibration solution and the sample solution are usually not treated in the same manner.

In-line calibrating systems are used in medical monitoring. Attempts were made for in-line calibration during fermentation with *E. coli* using a double-lumen catheter. The calibration was done by closing the sample inlet from the fermentor and substituting it with a sample of known analyte concentration (20) (Fig. 10). A further development of this technology was the use of the coaxial catheter for sampling. In on-line monitoring of glucose and lactate in blood, the calibration solution was introduced as a pulse

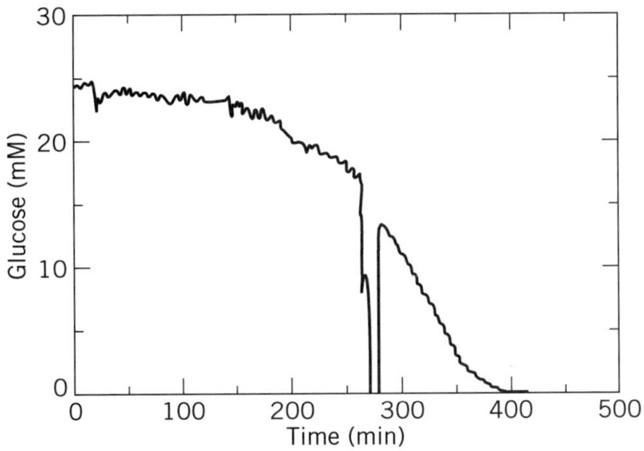

Figure 10. Monitoring of glucose in batch cultivation of *Escherichia coli*. After approx. 270 min a calibration solution is introduced, and the reading is back approx. 10 min later. *Source*: From Ref. 20.

in the heparin stream fed via the outer space in the coaxial catheter. This led to an on-line monitoring of the process without having to interrupt the reading (Fig. 11). Furthermore, by introducing the calibration solution at the tip of the sampling catheter, the whole system is calibrated. Thus a more true calibration value is obtained (75,76).

CONCLUSION

Sampling and sample treatment are extremely important when applying good analytical technique to the monitoring and control of biotechnological processes. Too often much effort is spent on developing a sensitive assay, but very little or no concern has been given to sampling and sample handling. The examples given in this article clearly demonstrate that many good techniques can be applied. Which method to use will be governed by the system studied, the equipment available, and also the earlier experience of the researcher.

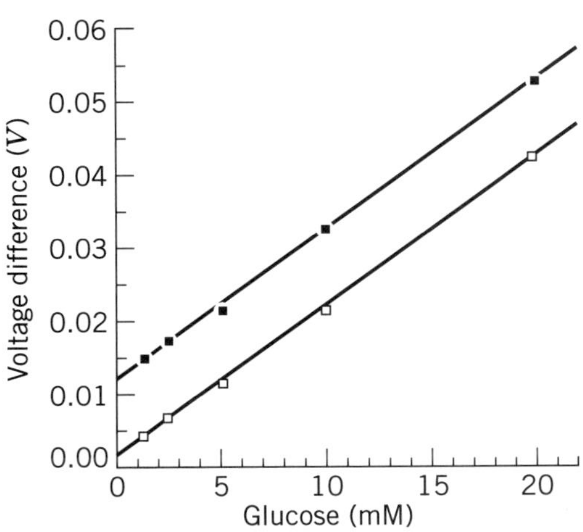

Figure 11. On-line outlet signal (V) as a function of glucose concentrations (mM); (■) glucose standard solution with added internal standard (5 mM), $y = 1.14 \times 10^{-2} + 2.05 \times 10^{-3}x$, $r^2 = 1.000$; (□) glucose standard solution, $y = 1.09 \times 10^{-3} + 2.04 \times 10^{-3}x$, $r^2 = 1.000$.

BIBLIOGRAPHY

1. D.J. Clarke, B.C. Blake-Coleman, M.R. Calder, R.J.G. Carr, and S.C. Moody, *J. Biotechnol.* **1**, 135–158 (1984).
2. R.T. Hatch, *Annu. Rep. Ferment. Proc.* **5**, 231–235 (1982).
3. N.C. van de Merbel, H. Lingeman, and U.A. Th. Brinkman, *J. Chromatogr. A* **725**, 13–27 (1996).
4. I. Karube, T. Matsunaga, S. Suzuki, T. Asano, and S. Hob, *J. Biotechnol.* **1**, 286–291 (1984).
5. B. Mattiasson and H. Håkanson, *Trends Biotechnol.* **11**, 136–142 (1993).
6. Hun.-C. Shu, Ph.D. Thesis Lund University, Lund, Sweden, 1994, ISRN LUTKDH/TKBT-94/1020-SE.
7. T. Buttler, Ph.D. Thesis Lund University, Lund, Sweden, 1996, ISRN LUNDKL/(NDKAK-1034).
8. K.D. Vorolop, J.W. Becke, and J. Klein, *Biotechnol. Lett.* **5**(8), 509–512 (1983).
9. J.H. Lee, J.C. Woodard, R.J. Pagan, and P.L. Rojers, *Biotechnol. Lett.* **3**, 351–356 (1981).
10. C.F. Mandenius and B. Mattiasson, *Eur. J. Appl. Microbiol. Biotechnol.* **18**, 197–200 (1983).
11. O.F. Fiho, J.F. de Andrade, A.A. Suleiman, and G.G. Guilbault, *Anal. Chem.* **61**, 746–748 (1989).
12. T. Buttler, C. Nilsson, L. Gorton, G. Marko-Varga, and T. Laurell, *J. Chromatogr. A* **725**, 41–56 (1996).
13. T. Lorenz, W. Schimidt, and K. Schugerl, *Chem. Eng. J.* **35**, B15–B22 (1987).
14. B. Mattiasson and B. Danielsson, (1996) in R.F. Taylor and J.S. Schultz eds., *Handbook of Chemical and Biological Sensors*, IOP Publishing, Philadelphia, pp. 533–552.
15. J. Nielson, *Process Qual. Control.* **2**, 371–384 (1982).
16. N. Vriezen and J.P. Van Dijken, *Biotechnol. Tech.* **11**(6), 427–430 (1997).
17. T. Kullick, U. Bock, J. Schubert, T. Scheper, and K. Schügerl, *Anal. Chim. Acta* **300**, 25–31 (1995).
18. K. Nikolajsen, J. Nielsen and J. Villadsen, *Anal. Chim. Acta* **214**, 137–145 (1988).
19. J. Nielsen, K. Nikolajsen, and J. Villadsen, *Biotechnol. Bioeng.* **33**, 1127–1134 (1989).
20. O. Holst, H. Håkanson, A. Miyabayashi, and B. Mattiasson, *Appl. Microbiol. Biotechnol.* **28**, 32–36 (1988).
21. H. Håkanson, M. Nilson, and B. Mattiasson, *Anal. Chim. Acta* **249**, 61–65 (1991).
22. S. Benthin, J. Nielsen, and J. Villadsen, *Anal. Chim. Acta* **247**, 45–50 (1991).
23. M.P. Nandakumar, A. Tokaj, and B. Mattiason, *Bioseparation*, in press.
24. A. Tokaj, M.P. Nandakumar, O. Holst, and B. Mattiasson, *Bioseparation*, in press.
25. K. Schügerl, in H.J. Rehm, G. Reed, A. Puhler, and P. Stadler eds., *Biotechnology*, 2nd ed., vol. 4, VCH, Weirheim, 1991, pp. 152–156.
26. J.F. van Staden, *Fresenius J. Anal. Chem.* **352**, 271–302 (1995).
27. K.H. Kroner and N. Papamichel, *Process Biochem.* **23**(10), 3–4 (1991).

28. A. Reckenwald, K.H. Kroner, and M.R. Kula, *Enzyme Microb. Technol.* **7**, 607–612 (1985).
29. M. Garn, M. Gasin, C. Thommen, and P. Cevey, *Biotechnol. Bioeng.* **34**, 423–428 (1989).
30. T. Lorenz, W. Schimidt, and K. Schügerl, *Chem. Eng. J.* **35**, B15–B22 (1987).
31. K.H. Kroner and M.R. Kula, *Anal. Chim. Acta* **163**, 3–15 (1984).
32. L.H. Christensen, J. Nielsen, and J. Villadsen, *Chem. Eng. Sci.* **46**(12), 3304–3307 (1991).
33. K.M. Pedersen, M. Kummel, and H. Soeberg, *Anal. Chim. Acta* **238**, 191–199 (1990).
34. N.C. van de Merbel, I.M. Kool, H. Lingeman, U.A.Th. Brinkman, A. Kolhorn, and L.C. de Rijke, *Chromatographia* **33**, 525–532 (1992).
35. T.A. Butler, A.J. Kristina Johanson, L. Gorton, and G.A. Markovarga, *Anal. Chem.* **65**(19), 2628–2636 (1993).
36. L.F. Forman, B.D. Thomas, and F.S. Jacobson, *Anal. Chim. Acta* **249**, 101–111 (1991).
37. F. Valero, J. Lafuente, M. Poch, C. Sola, A.N. Aranjo, and J.L.F.C. Lima, *Biotechnol. Bioeng.* **36**, 647–651 (1990).
38. M.J. Hayward, T. Kotiano, A.K. Lister, R.G. Cooks, G.D. Austin, R. Narayan, and G.T. Tsao, *Anal. Chem.* **62**, 1978–1804 (1990).
39. U. Brand, B. Reinhardt, F. Ruther, T. Scheper, and K. Schügerl, *Sensors Actuators* **4**, 315–318 (1991).
40. H. Hoffman, Ph.D. Thesis, University of Hannover, Hannover, Germany, 1985.
41. K. Schügerl, *Anal. Chim. Acta* **213**, 1–9 (1988).
42. S.S. Ozturk, J.C. Thrift, J.D. Blackie, and D. Nareh, *Biotechnol. Bioeng.* **48**, 201–208 (1995).
43. C.F. Mandenius, *Biotechnol. Bioeng.* **32**, 123–129 (1988).
44. J. Meschke, H. Bennemann, M. Herbst, S. Dormeier, and D.C. Hempel, *Bioprocess Eng.* **3**, 151–157 (1988).
45. N.C. van de Merbel, H. Lingman, U.A.Th. Brinkman, A. Kolhorn, and L.C. de Rijke, *Anal. Chim. Acta* **279**, 39–50 (1993).
46. N.C. van de Merbel, P. Zuur, Frijlink, J.J.M. Holthuis, H. Lingeman, and U.A.Th. Brinkman, *Anal. Chim. Acta* **303**, 175–186 (1995).
47. T. Lorenz, W. Schimidt, and K. Schügerl, *Chem. Eng. J.* **35**, B15–B22 (1987).
48. C.F. Mandenius, B. Danielsson, and B. Mattiasson, *Anal. Chim. Acta* **163**, 135–141 (1984).
49. D.W. Zabriskie and A.E. Humphery, *Biotechnol. Bioeng.* **20**, 1295–1301 (1978).
50. U.S. Pat. 3,830,106 (1974), Glaxo Laboratories, Ltd.
51. L.H. Christensen, J. Nielsen, and J. Villadsen, *Anal. Chim. Acta* **249**, 123–136 (1991).
52. T.D. Gibson and J.R. Woodward, *Anal. Chim. Acta* **213**, 61–68 (1988).
53. H. Benreniste and P.C. Huttemeier, *Prog. Neurobiol.* **35**, 365–372 (1991).
54. T.E. Robinson and J.B. Justice, Jr., Eds., *Microdialysis in the Neurosciences*, Elsvier Science, Amsterdam, 1991.
55. G. Marko-Varga, T. Buttler, L. Gorton, and C. Gronstervall, *Chromatographia* **35**, 285–289 (1993).
56. N.C. van de Marbel, J.M. Teule, H. Lingeman, and U.A.T. Britman, *J. Pharm. Biomed. Anal.* **10**, 225–231 (1992).
57. J.D.H. Cooper, D.C. Turnell, B. Green, and F. Verillon, *J. Chromatogr.* **456**, 53–69 (1988).
58. H. Chase, *Trends Biotechnol.* **12**, 296–303 (1994).
59. M.P. Nandakumar, A. Lali, and B. Mattiason, (1998) Submitted.
60. B. Mattiason and M.P. Nandakumar, (1998) Submitted.
61. M.C. Hennion, *Trends. Anal. Chem.* **10**, 317–322 (1991).
62. U.A.Th. Brinkman, *J. Chromatogr. A* **665**, 217–231 (1994).
63. S. Olsen, J. Ruzicka, and E.H. Hansson, *Anal. Chim. Acta* **136**, 101–112 (1982).
64. E. Martins, M. Bengtsson, and G. Johansson, *Anal. Chim. Acta* **169**, 31–42 (1985).
65. B. Karlsberg and S. Thelander, *Anal. Chim. Acta* **98**, 1–7 (1978).
66. G.D. Clark, D.A. Whitman, G.D. Christian, and J. Ruzicka, *Crit. Rev. Anal. Chem.* **21**(5), 357–375 (1990).
67. N.F. Thornhill and N.B. Fish, *Biotechnol. Tech.* **7**, 19–24 (1993).
68. G.R. van der Hoff, R.A. Banmann, U.A.Th. Brinkman, and P. van Zoonen, *J. Chromatogr.* **644**, 367–373 (1993).
69. J.J. Vreulls, G.J. De Jong, R.T. Ghijsen, and U.A.Th. Brinkman, *J. AOAC Int.* **77**, 306–327 (1994).
70. J. Ruzicka and E.H. Hansen *Anal. Chim. Acta* **214**, 1–27 (1983).
71. B.F. Reis, A.O. Jacintho, J. Moratti, F.J. Krug, E.A.G. Zagatto, F.H. Bergamin, and L.C.R. Pessenda, *Anal. Chim. Acta* **123**, 221–228 (1981).
72. J. Ruzicka and E.H. Hansen, *Anal. Chim. Acta* **145**, 1–15 (1983).
73. J. Ruzicka and E.H. Hansen, *Flow Injection Analysis*, 2nd ed., Wiley, New York, 1988.
74. F.J. Krag, F.F. Bergamin, and E.A.G. Zagatto, *Anal. Chim. Acta* **179**, 103–107 (1986).
75. M. Kyrölainen, H. Håkanson, and B. Mattiasson, *Biotechnol. Bioeng.* **45**, 122–128 (1995).
76. R. Kok and P. Hoasi, *Biosensors* **3**, 89–100 (1987/88).

SCALE-UP. See DIMENSIONAL ANALYSIS, SCALE-UP; MAMMALIAN CELL CULTURE REACTORS, SCALE-UP; SCALE-UP, STIRRED TANK REACTORS.

SCALE-UP, STIRRED-TANK REACTORS

D.C. HEMPEL
H. DZIALLAS
Technische Universität Braunschweig
Braunschweig, Germany

KEY WORDS

Dispersing capacity
Mass transport
Mixing time
Newtonian fluids
Non-Newtonian fluids
Power input
Pumping capacity
Scale-up

Shearing stress
Stirred-tank reactor

OUTLINE

INTRODUCTION

The scale-up of a biotechnological process developed in the laboratory often presents problems that, owing to the many parameters involved and their complexity, do not permit any generalized solution. As a result, criteria for scale-up have always been developed on the basis of specific processes. The main problem in all considerations of scale-up is that no strictly defined criteria exist for conducting biotechnological processes and, hence, their scaled translation. Of the biological and physical factors important in scale-up, generally valid relationships can be established only for the latter. Thus in the following, the experimentally proven physical fundamentals of scale-up for conventional stirred-tank reactors are brought together in a way that reflects their importance in practical operation. Such considerations have been published in the past but were based on the scientific knowledge available at that time (1–5). This article is an updated short version. For further information about scale-ups of fermentation processes and extended literature see Ref. 5.

SCALE-UP METHODS

Balance Equation

A scale-up using a solution of the complete equations for momentum, energy, and mass balances is exceedingly difficult to determine for a complex microbiological process. Thus for a scale-up on the basis of mathematical models, there are still no practicable equations available at present that take into account the kinetics and transport processes in the stirred-tank reactor. Nonetheless, the application of simplified balance equations is often helpful in order to understand the processes involved (6,7).

Dimensional Analysis

Dimensional analysis, introduced successfully into chemical technology and already a classical method of scale-up, is also helpful when dealing with biotechnological processes. Using it, a number of dimensionless quantities (π-quantities) are derived from the relevant process-influencing quantities, which using the mathematical interrelationships of the similarity principle, provide criteria for scale-up (see also Refs. 8 and 9). The invariance of all dimensionless π-quantities relevant to the problem provides the simplest rule for scale-up, but the rule is dependent upon complete similarity. Whereas physical processes often do behave similarly upon change in scale, similarity in the presence of a chemical or microbial process is often unattainable. Even for purely physical processes, a complete similarity can often not be attained. As shown in the section "Performance Characteristics of the Stirred-Tank Reactor," the translation rule Re = idem (idem means that the numerical value of Re is held constant in scale translation) is valid for the energy input in scaling-up stirred-tank reactors, without formation of vortices. But if, nonetheless, a vortex should form, as in unbaffled vessels, then both the Reynolds number and the Froude number must retain their numerical value. From the definition of Re and Fr, it immediately becomes obvious that the requirements Re = idem and Fr = idem cannot be fulfilled because $n \cdot d^2$ = idem cannot equal $n^2 \cdot d$ = idem.

Because it is rarely possible to maintain all dimensionless characteristic quantities at the same numerical value, it becomes necessary to restrict oneself to partial similarities in which only some of the π-quantities remain fixed. To achieve this, the rate-determining step of the reaction for the complex microbial system must be known. That means that from the many process parameters, such as nutrient requirements, physicochemical conditions of growth, or transport processes, the limiting (or inhibiting) quantities have to be determined by careful analysis. Problems then arise when there is a shift in the effect of the individual limiting factors when the scale is altered. Such an approach to scale-up, in which each rate-determining step, both of the microbial reaction and of the transport processes, is mathematically correlated, has been put forward from time to time (for example, see Ref. 10) and has also been introduced highly successfully (11–14).

Maintenance of Identical Operating Conditions

A further simplification of the scale-up criteria, which follows from accepting the notion of partial similarity, is the maintenance of identical operating conditions (usually assuming that there is geometrical similarity). The most common examples of scale-up rules that have been published in the specialized biotechnological literature are based on this concept; extrapolation is often performed ac-

Table 1. Effects of Different Criteria in Linear Scaling-Up by a Factor of 5

	Laboratory scale	Scale-up criterion			
		P/V (kW/m^3)	n (s^{-1})	$n \cdot d$ (m/s)	Re
Diameter D (m)	1	5	5	5	5
Specific power input P/V (kW/m^3)	1	1	25	0.2	1.6×10^{-3}
Power input P (kW)	1	125	3125	25	0.21
Rotational speed of stirrer n (s^{-1})	1	0.34	1	0.2	0.04
Tip speed of stirrer $n \cdot d$ (m/s)	1	1.71	5	1	0.2
Reynolds number	1	8.55	25	5	1

Source: From Ref. 20.

cording to empirical rules of thumb that are still used today in the fermentation industry. Typical examples of this are the following:

- Identical volumetric power input P/V
- Identical mixing time or adaption of the mixing time to different stirred-tank reactor sizes according to purely empirical translation rules
- Identical tip speed of the agitator
- Identical volumetric mass transfer coefficient k_La
- Identical Reynolds number, Re

In fact, according to Ref. 15, about one third of plants employed the translation rules P / V = idem and k_La = idem, and about 20% used the tip speed of the stirrer (i.e., assuming shear stress). Another 20% of industrial plants employ a scale-up on the basis of mixing time and Reynolds number. The remainder scaled-up on the basis of a limiting or inhibiting substrate or product component—most commonly on the basis of the concentration of dissolved oxygen.

A scale-up founded on a single operating condition has diverse effects on the other operating quantities, as shown in Table 1 (see also Refs. 16–19). Using a linear scaling factor of 5 (corresponding to a volume scaling factor of 125) each column shows the change in the operating quantity that results from the respective scale-up criterion. Table 1 shows that it is not possible to simultaneously fulfill several of the scale-up criteria mentioned (20).

PHYSICAL FUNDAMENTALS OF SCALE-UP CONDITIONS

Performance Characteristics of the Stirred-Tank Reactor

The stirring power, P, which produces the three-dimensional flow field can be calculated from the dimensionless power number or Newton number:

$$Ne \equiv \frac{P}{\rho \cdot n^3 \cdot d^5} \tag{1}$$

From considerations of the similarity principle, the formal correlation of the performance characteristic for a given geometry is described by

$$Ne = f(Re, Fr, Q) \tag{2}$$

where the Froude number is $Fr \equiv n^2 \cdot d/g$, the Reynolds number is $Re \equiv n \cdot d^2/\nu$, and the dimensionless gasflow number is $Q \equiv q_G/n \cdot d^3$.

Power Input in Nongassed Systems of Newtonian Fluids. In systems that have no gas input ($Q = 0$) and no vortex formation (as is effectively the case in a vessel with baffles), the effect of the Froude number is negligible. The performance characteristics of the nongassed system of Newtonian fluids, which is then solely a function of the Reynolds number (21–23) is depicted in Figure 1 for a selection of stirrers. The geometrical arrangement and the working range of the stirrers have been standardized in some countries (e.g., DIN 28131 in Germany). Except for the helical and anchor stirrers which extend right to the wall, stirred vessels are equipped with baffles to avoid vortex formation. The reduction in the number of baffles from 4 to 3 or 2 generally lowers the energy input by 10 or 20% (23). In a completely turbulent system (a condition fulfilled with most stirrers at $Re > 10^4$), the performance characteristic is independent of the Reynolds number, that is, Ne = constant. If the scale-up considerations are based on a constant volumetric power input, $P/V \sim P/d^3$ = idem, then the following is valid for the scaled translation:

$$n^3 \cdot d^2 = \text{idem} \tag{3}$$

Power Input in Nongassed Systems of Non-Newtonian Fluids. In many fermentation broths, for example, those containing polysaccharides or mycelium-forming microor-

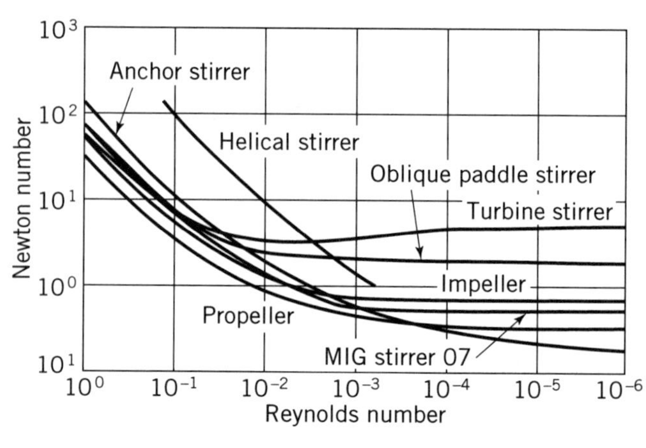

Figure 1. Performance characteristics of some types of stirrer with Newtonian fluids.

ganisms, non-Newtonian flow behavior is encountered: there is a complicated nonlinear relationship between the shearing stress, τ, and shear rate, $\dot{\gamma}$. For non-Newtonian fluids, the viscosity property can be defined only in the form of an apparent viscosity

$$\eta_a \equiv \frac{\tau}{\dot{\gamma}} \tag{4}$$

A number of models have been put forward to describe the non-Newtonian flow behavior. The most efficacious originate from the Ostwald-De Waele law for pseudoplastic fluids

$$\tau = K' \cdot \dot{\gamma}^{m'} \tag{5}$$

and from Herschel–Bulkley law for plastic fluids

$$\tau = \tau_0 + K'' \cdot \dot{\gamma}^{m''} \tag{6}$$

In order to determine the energy input in such fluids, the approach adopted by Metzner and Otto (24) and Metzner et al. (25) has proved itself. According to this concept, there is a direct proportionality between the rate of shear and the rotational speed of the stirrer

$$\dot{\gamma} = K \cdot n \tag{7}$$

whereby the apparent viscosity, η_a, is obtained from the rheological flow curve of the fluid involved. The coefficient K depends only slightly on the rheological behavior of the fluid but mainly on the stirring system, for example, $K \approx$ 11 to 13 for turbine agitators and $K \approx$ 15 to 25 for anchor agitators (15,26). In Figure 2 the performance characteristic of the turbine stirrer for pseudoplastic fluids is plotted versus the Reynolds number which has been modified by η_a (24,27).

A similar performance characteristic was measured by Sanchez et al. (28) and Velasco et al. (29) for a single and a dual Rushton turbine when mixing a mycelial broth under nonaerated conditions. As with Newtonian fluids, a constant Newton number is attained here at sufficiently high Reynolds numbers (i.e., the performance characteristic is independent of the viscosity). Whereas this independence of the viscosity is unequivocally valid in a nongassed system, considerable dependence on the viscosity occurs in a gassed system of non-Newtonian fluids, as is illustrated in a following section.

The apparently simple approach advocated by Metzner and Otto (24) is made more complicated by the fact that the rheological behavior of fermentation broths often changes drastically during the fermentation, as is shown in Figure 3 for a decrease of the apparent viscosity during the cultivation of *Cellulomonas uda* grown on printed newspaper (26).

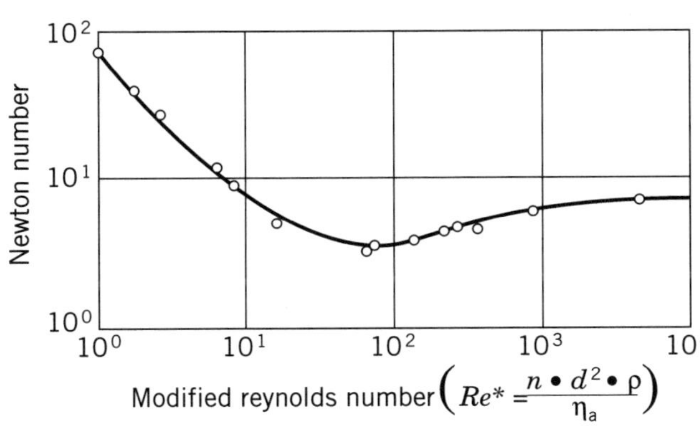

Figure 2. Performance characteristic of turbine stirrer with pseudoplastic fluid.

Power Input in Gassed Systems at Low Viscosities. In gassed systems of aqueous-like fluids, a decrease in the stirring power input can be observed with increasing gas flow. This decrease is caused by gas cushions that form behind the stirrer blades and diminish the rotational resistance of the stirrer. The effect was first studied by Oyama and Endoh (30); the results are shown in Figure 4. Since then a series of measured values have been published; particularly extensive are the measurements made by Zlokarnik (31) and Judat (32). Judat (32) gives a relationship valid for a turbine stirrer in a turbulent region of flow ($Re > 10^4$) in water. This relationship is also applicable, in a somewhat modified form, to large vessels of up to $V = 900$ m^3:

$$Ne_G = Z\frac{Ne_0 + 187QFr^{-0.32}(d/D)^{1.53} - 4.6Q^{1.25}}{1 + 136Q(d/D)^{1.14}} \tag{8}$$

Here $Ne_0 \approx 4.9$ is the Newton number of the turbine stir-

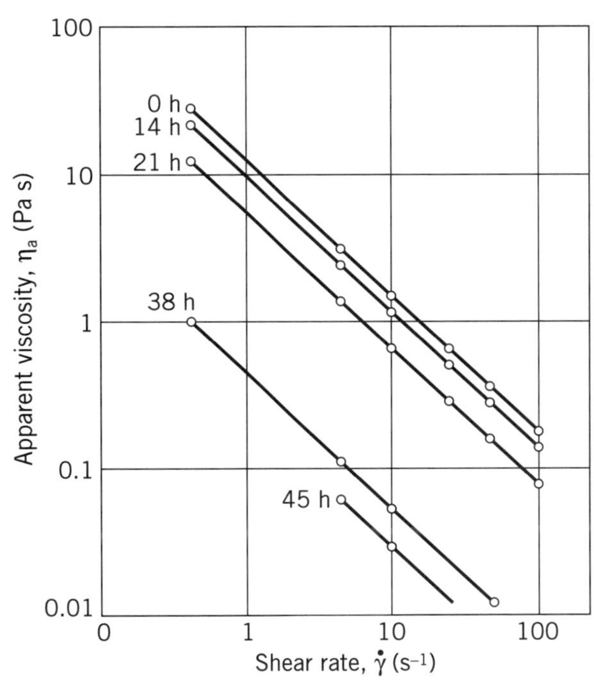

Figure 3. Time course of changes in apparent viscosity in the fermentation of cellulose.

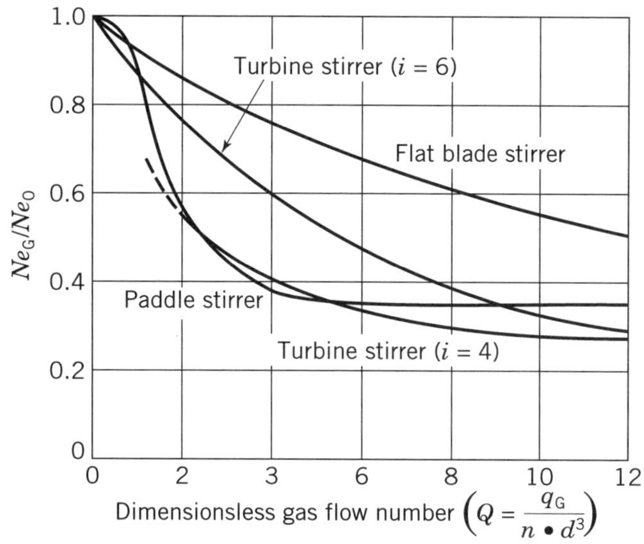

Figure 4. Power input of several types of stirrer in Newtonian fluids, dependent on the gas throughput.

ring in a turbulent region of flow without gassing, and Z is the number of stirrer blades. Equation 8 is valid for $Re > 10^4$, $Fr \leq 0.07(d/D)^3$, $0.2 \leq d/D \leq 0.42$, and $h/d > 0.75$.

Detailed investigations with other types of stirrers and multiple impellers were carried out by Schlueter and Deckwer (33). Their experimental results were well described by equation 8 with a maximum deviation of only 30%. Further investigations are necessary to describe properly the power input in stirred vessels if the impeller position is varied and multiple stirrer arrangements are used.

The influence of the Reynolds number on the power input in a gassed system has been studied by Zlokarnik (31) and is given, for the turbine stirrer, by

$$\frac{Ne_G}{Ne_0} = 1 + [(3.9Re^{0.12} + 6 \cdot 10^{-12}Re^{3.45}) \cdot (0.22Q^{0.1} + 6.25Q^3)]^{-1} \qquad (9)$$

which is valid for $Q < 0.05$ and $800 \leq Re \leq 10^4$. Ne_G is the Newton number of the turbine stirrer gassed with Q at $Re > 10^4$.

Power Input in Gassed Systems of Newtonian and Non-Newtonian Fluids at Higher Viscosities. Whereas the influence of the gas on the energy input in aqueous-like fluids is about the same for the different types of stirrer, each stirrer type displays a very different effect when the viscosity increases. Detailed measurements on this have been published by Hoecker and Langer (34) and Hoecker (35). To illustrate the results, the form completely equivalent to equation 2 was chosen,

$$Ne = f(Q, Fr, Ga) \qquad (10)$$

with the Galilei number

$$Ga \equiv \frac{Re^2}{Fr} = \frac{d^3 \cdot g}{\nu^2}$$

With turbine stirrers, a decrease in the power input is observed with increasing viscosity (Fig. 5). As in the correlation of the power input for nongassed systems of non-Newtonian fluids, the results with sparging can also be expressed by introducing an apparent viscosity, η_a, as results from the approach taken by Metzner and Otto (24) (i.e., equations 4 and 7). As Figure 5 shows, even the results for Newtonian glycerine–water solutions and for pseudoplastic carboxymethylcellulose–water solutions can be expressed by the same form of curve (34). With viscoelastic polyacrylamide–water solutions (Figure 5b), the decrease in the power input with viscosity is even more pronounced than with Newtonian and pseudoplastic fluids; this can probably be traced back to the development of orthogonal tension in viscoelastic fluids, which tends to damp the turbulence. The illustration of the performance characteristic for turbine stirrers in aqueous glycerine solutions in terms of the gas flow in Figure 6 clearly shows that at high viscosities a decrease in power input with the gas throughput hardly makes itself felt, but rather the Froude number exerts a more marked effect (34,35). The slight influence of the dimensionless gasflow number is due to the formation of gas cushions behind the stirrer blades, which is independent of the gas throughput.

The relationship given in equation 9 to calculate the influence of the Reynolds number on the power input with gassed turbine stirrers is roughly applicable to non-Newtonian fluids, too. In Figure 7 the results obtained by

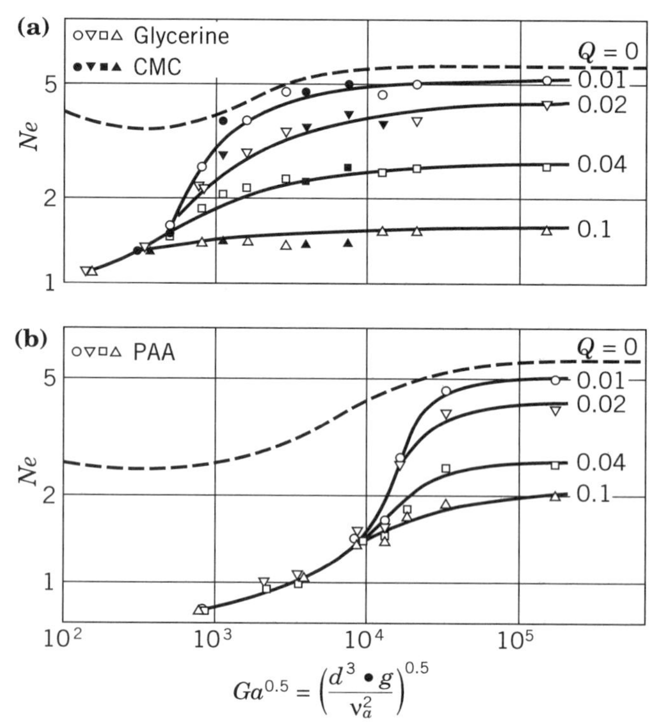

Figure 5. Power input with turbine stirrer (d/D = 0.33) in (**a**) Newtonian (glycerine) and pseudoplastic (CMC) and (**b**) viscoelastic (PAA) fluids.

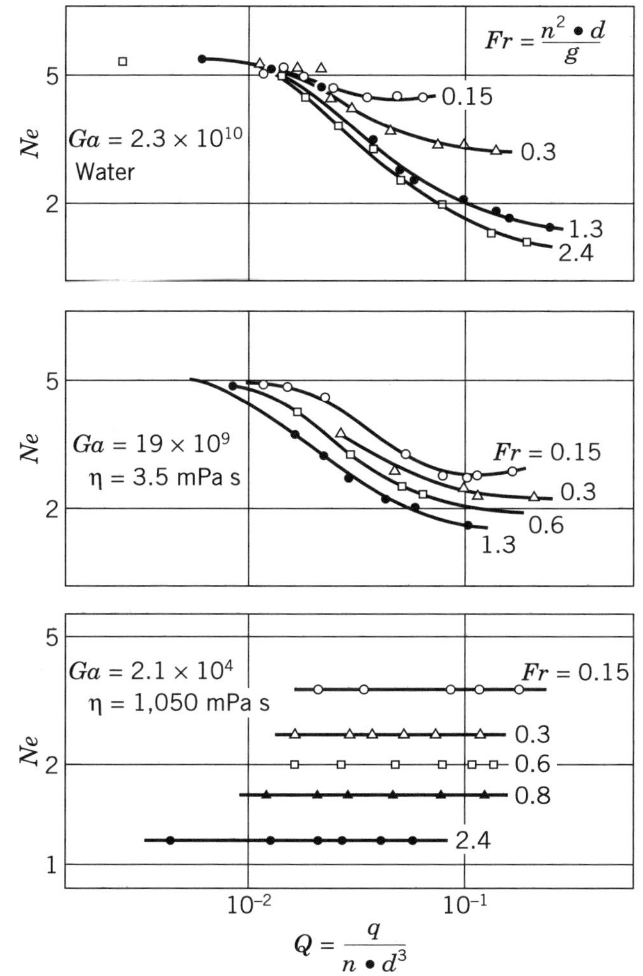

Figure 6. Power input with turbine stirrer ($d/D = 0.33$) in glycerine–water solutions.

Hoecker and Langer (34) are depicted for water, glycerine, carboxymethylcellulose (CMC), and polyacrylamide (PAA) solutions, plotted as $Ne \cdot f(Re, Q)$ corresponding to equation 9 as a function of $Q(1 + 38(d/D)^5)$. The dotted curve shows the results obtained by Zlokarnik (31). From the values taken from the line of the curve, the performance characteristics can be calculated exactly enough from Figure 7 with $Ne = Ne_G/f(Re, Q)$ with equation 9; this is the case for Newtonian and pseudoplastic fluids in the region $Ga > 10^6$ ($Re > 10^3$) and for viscoelastic fermentation broths with $Ga > 10^9$ (34,36).

Completely different behavior from that seen with a turbine stirrer is observed with a multistage impulse countercurrent (MIG) agitator: In Figure 8 the results obtained by Hoecker and Langer (34) and Hoecker (35) are plotted and are analogous to those in Figure 5. Here a decrease in the power input comparable in magnitude to that with the turbine stirrer cannot be established; this is due to the less marked formation of gas cushions with a MIG agitator. Because the interference multistage impulse countercurrent (INTERMIG) and the oblique arm paddle stirrer both exhibit hydrodynamic behaviour similar to that with a MIG stirrer, this result is to be expected in these cases, too.

Dispersing Capacity of the Stirrer

The dispersing capacity of a stirrer is limited by the so-called flooding point, which gives the maximum gas throughput that can be distributed by the stirrer, $q_{G,max}$. If this maximum gas throughput is exceeded, the stirrer is completely "flooded" by the gas, and the circulatory flow provided by the agitator breaks down. This flooding point exists in the transitional zone leading to turbulent flow and is particularly marked in the turbulent flow region. It manifests itself in a sudden decrease in the power input and mass transfer capacity. In the turbulent area of flow ($Re > 10^4$) the measured data can be correlated using the dimensionless form

$$Q = f(Fr, d/D) \qquad (11)$$

Judat (32) reports that the flooding point for turbine stirrers for $Re > 10^4$ and $0.2 \leq d/D \leq 0.42$ is

$$Q_{max} = \frac{0.21Fr^{2.1d/D}}{[(d/D)^{-1} - 2.04]^{1.3}} + \frac{0.14Fr^{7.54d/D}}{[(d/D)^{-1} - 2.25]^{1.5}} \qquad (12)$$

The flooding characteristic for different stirrers is depicted in Figure 9 (32). Because these results have all been acquired in laboratory experiments, comparison measurements must be carried out in order to achieve a dependable scale-up. A model to describe the flooding characteristic of stirrers by means of dimensional analysis was suggested by Zehner (37).

For fermentation processes that require low rotational speeds of the stirrer at high sparging rates or at very high gas throughput, Breucker et al. (38) made an interesting observation. For different types of turbine, propeller, or INTERMIG stirring systems they showed that sparge rings near the wall prevent the introduced gas flow from flooding the impeller. In the rotational speed range in which the stirrer is normally flooded, flooding did not occur. This extension of the operation range results in greater safety in vessel design. From the observations available for media of higher viscosity and for non-Newtonian fluids, conservative design considers the dispersing capacity of the agitator as lower than with Newtonian fluids.

Mixing Time Characteristics of Stirred-Tank Reactors

The mixing time is commonly defined as the time necessary to reduce concentration differences in the volume in question to a minimum. The mixing time should thus always be related to the method of measurement used and the degree of homogenity desired. From theoretical considerations of similarity, the mixing time characteristic in the case of Newtonian fluids (with negligible density and viscosity differences) is given by

$$n \cdot \theta = f(Re) \qquad (13)$$

with the dimensionless characteristic mixing number $n \cdot \theta$.

In Figure 10 the mixing time characteristics of some types of agitator are illustrated. Those conditions of geometry (which deviate from the German norm DIN 28131) are given in the diagram. In the range of turbulent flow

Symbol				Ga	Fr
Water	Glycerine	CMC	PAA		
●	▲	△	○	$1.3 \times 10^9 - 2.3 \times 10^{10}$	
	▼			$1.37 \times 10^8 - 4.6 \times 10^8$	
	■	□		$10^7 - 6.8 \times 10^7$	> 0.6
	×	+		$1.3 \times 10^6 - 6.4 \times 10^6$	
	◄			$7.1 \times 10^5 - 9.06 \times 10^5$	
------------------				$2.2 \times 10^6 - 7.15 \times 10^7$	> 0.9

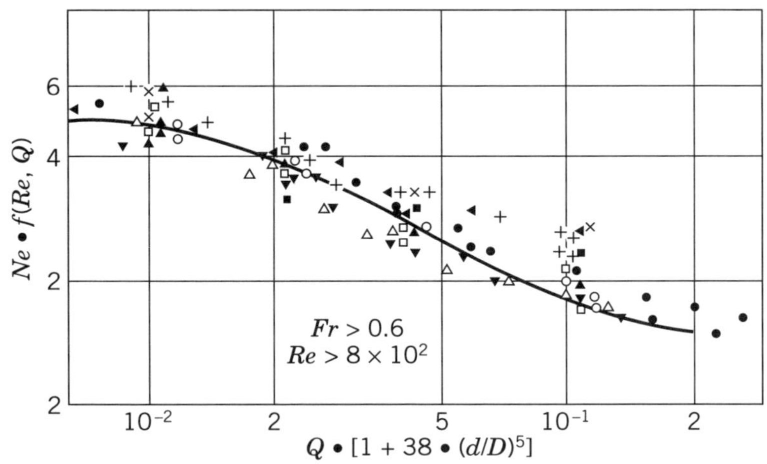

Figure 7. Power input with turbine stirrer (d/D = 0.33): Influence of Reynolds number and gas throughput [$f(Re, Q)$] according to equation 9.

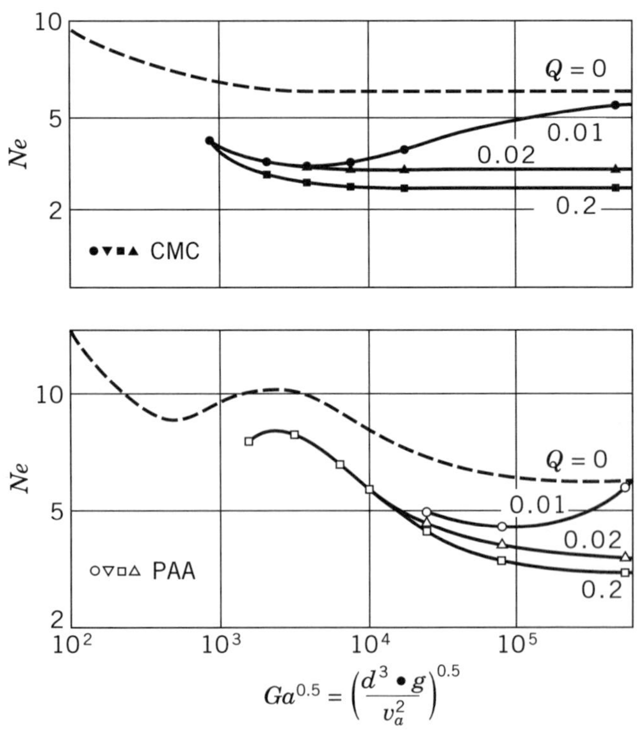

Figure 8. Power input with MIG stirrer (d/D = 0.6, 3 stirrers, H/D = 2), substance systems as in Figure 5.

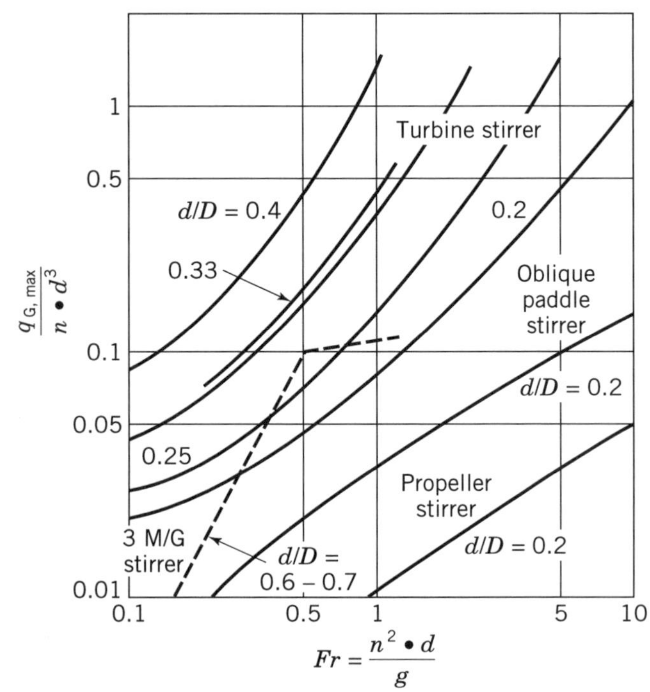

Figure 9. Dispersing capacity of several types of stirrer, $Re > 10^4$.

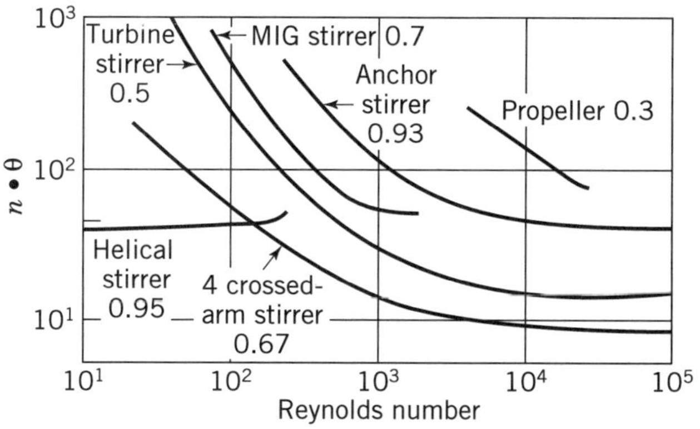

Figure 10. Mixing time characteristics of several types of stirrer in Newtonian fluids.

($Re > 10^4$) the characteristic mixing number is practically independent of the Reynolds number (39–41). Figure 11 shows a plot of the power input that is required to achieve those mixing times in the turbulent region given in Figure 10. The characteristic dimensionless numbers used in this diagram, which were formulated by Zlokarnik (42) on a theoretical consideration of similarity, permit the stirring power input to be determined for a given mixing time, or vice versa (43). For the most favorable conditions of geometry and operation, the stirring power input required in the region of turbulent flow (Ne = constant, $n \cdot \theta$ = constant) is obtained from the relationship

$$\frac{P \cdot \theta^3}{\rho \cdot D^5} = 300 \tag{14}$$

For a scale-up with P/V = idem there is, from equation 14, a corresponding increase in the mixing time in the region of turbulent flow by

$$\frac{\theta}{D^{2/3}} = \text{idem} \tag{15}$$

for example,

$$\frac{\theta}{V^{2/9}} = \text{idem} \tag{16}$$

A scale-up on the basis of constant mixing times (θ = idem) requires, according to equation 14, an increase in the power input by

$$\frac{P/V}{D^2} = \text{idem} \tag{17}$$

which leads to required energies that cannot be defended either on a technical or an economic basis. Bearing in mind the need for an economical stirring power input, it is indeed meaningful to allow an even longer mixing time when scaling-up than that given by equation 16.

The influence of non-Newtonian flow behavior on the mixing time characteristics can be derived from Figure 12

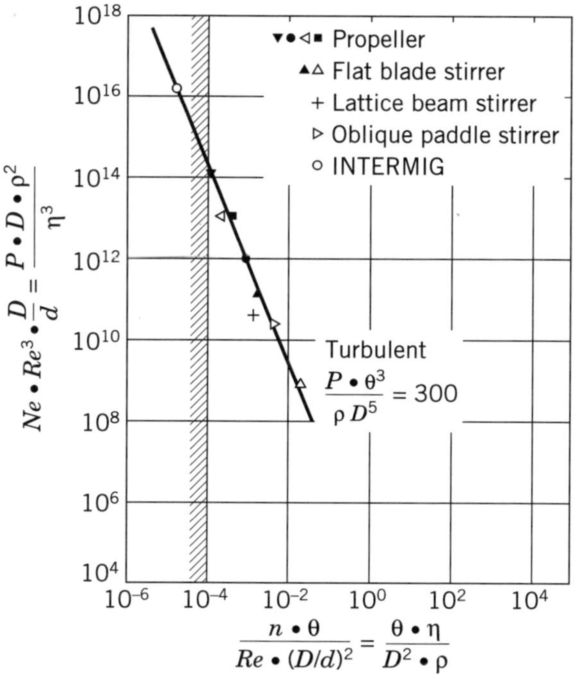

Figure 11. Power input to obtain given mixing times with favorable geometrical and operating conditions.

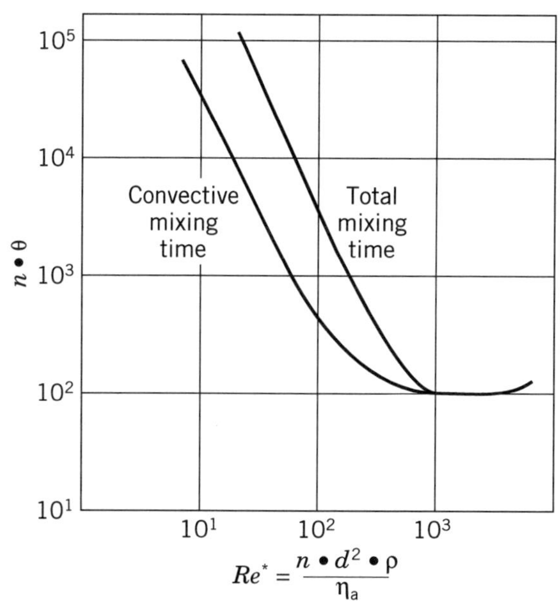

Figure 12. Mixing time characteristics in non-Newtonian fluids (crossed-arm stirrer in CMC solution).

(44). The Reynolds number here has again been modified by the apparent viscosity, as determined according to the method of approach used by Metzner and Otto (24) (equations 4 and 7).

For pseudoplastic fluids Opara (44) observed that in the laminar flow region, about 95% of the liquid is distributed very rapidly by convection, whereas the remainder is mixed very slowly by diffusion. This convective mixing time corresponds roughly to the mixing time in Newtonian fluids (Fig. 10). The total mixing time in non-Newtonian fluids can be as much as a factor of 10 longer than the convective mixing time.

Mass Transport between Gas and Liquid Phase

Apart from the stirring power, mixing time, and dispersing capacity of the stirrer, mass transfer between gas and liquid is often the prime influence during an aerobic fermentation. The oxygen demand of the aerobic culture has to be the same as the oxygen transfer rate (OTR $= \dot{n}_{O_2}$) in the steady-state condition

$$\dot{n}_{O_2} = \frac{\dot{N}_{O_2}}{V} = k_L a (c^*_{O_2} - c_{O_2})_{av} \tag{18}$$

In this, it has been assumed that because oxygen is such a poorly soluble gas, the overall rate of mass transport is determined essentially by the mass transfer resistance in the liquid phase. In fact the transport coefficient in the gas phase is larger than that in the liquid phase by a factor of 100. The liquid-phase mass transfer coefficient, k_L, multiplied by the interfacial exchange area per unit volume, a, is termed the *volumetric mass transfer coefficient*.

The averaged concentration difference at the phase boundary $(c^*_{O_2} - c_{O_2})_{av}$ in the whole reaction vessel is, in the case of an ideally mixed liquid phase and complete segregation of the gas phase, equal to the logarithmic mean concentration

$$(c^*_{O_2} - c_{O_2})_{av} = \frac{c^*_{O_2,in} - c^*_{O_2,out}}{\ln\left(\dfrac{c^*_{O_2,in} - c_{O_2}}{c^*_{O_2,out} - c_{O_2}}\right)} \tag{19}$$

In stirred tank reactors the average concentration difference can be calculated from equation 19 with sufficient accuracy.

Mass Transfer in Aqueous Fluids. Many equations have been published for the calculation of the volumetric mass transfer coefficient, $k_L a$, in low-viscosity aqueous fluids (see Ref. 5). Regardless of the measuring system used, these equations give differences in the $k_L a$ value of up to a factor of 10. In particular, the influence of the coalescence behavior of the fluid system on $k_L a$ has not been elucidated. Consequently the $k_L a$ values found with the model system of sulfite oxidation in water must be viewed critically when applying such values to fermentation broths.

Theoretical considerations of the similarity of mass transfer between gas and liquid phase made by Zlokarnik (45) led to the formal relationship

$$k_L a = f\left(\frac{P}{V}, \frac{q_G}{V}, \ldots\right) \tag{20}$$

with a gas throughput per unit volume q_G/V. However, Judat (46) showed, by evaluating a series of published $k_L a$ values, that in order to fit the measured data better, the gas throughput related to the cross-sectional area of the reactor (i.e., the superficial gas velocity)

$$u_G = \frac{q_G}{A} \tag{21}$$

should be taken as the characteristic gassing quantity. This finding is in agreement with the empirical equation of the type

$$k_L a = K_1 \left(\frac{P}{V}\right)^{\alpha} u_G^{\beta} \tag{22}$$

which was compiled by Moo-Young and Blanch (3) for coalescent and noncoalescent aqueous solutions; their findings are listed in Table 2 (SI units, i.e., $k_L a$ in s^{-1}, P/V in W/m^3, and u_G in m/s). According to Van't Riet (47), differences of up to 35% arise in the $k_L a$ values in coalescent and noncoalescent systems. Van't Riet (47) gives the following correlations for low-viscosity fermentation broths from a series of published data (SI units as in equation 22):

$$\text{Coalescing medium:} \quad k_L a = 0.026 \left(\frac{P}{V}\right)^{0.4} u_G^{0.5} \tag{23}$$

$$\text{Noncoalescing medium:} \quad k_L a = 0.002 \left(\frac{P}{V}\right)^{0.7} u_G^{0.3} \tag{24}$$

In modifying the dimensionless quantities proposed by Zlokarnik there follows, for the calculation of the volumetric mass transfer coefficient, $k_L a$, the functional relationship

$$k_L a \left(\frac{\nu}{g^2}\right)^{1/3} = \left(\frac{P}{V \cdot \rho \cdot (g^4 \cdot \nu)^{1/3}}, \frac{u_G}{(g \cdot \nu)^{1/3}}, Sc, \sigma^*, S_i\right) \tag{25}$$

In this, Sc $\equiv \nu/\mathcal{D}$ is the Schmidt number, $\sigma^* \equiv \sigma/\rho(g \cdot \nu^4)^{1/3}$ is the dimensionless surface tension, and S_i represents characteristic quantities that are intended to describe the coalescence behavior of the fluid system but whose composition, as already mentioned, still remains largely unclear, though Breucker et al. (38) found that qualitatively the mass transfer coefficient for the noncoalescing system is up to a factor of 4 higher than for coalescing systems at comparable conditions. In the absorption of oxygen in fermentation broths, σ^* and Sc are practically determined only by the viscosity. The surface tension itself changes only very slightly and is not significant. The diffusion coefficient of aqueous solutions exhibiting Newtonian flow behavior can be correlated with the viscosity. On the other hand, the diffusion coefficient in non-Newtonian aqueous solutions is independent of ν and has the same value as water (48,49).

In the mass transfer of gases other than O_2, the Schmidt number and surface tension must also be considered in some cases. However the transport of CO_2, which is also of interest in fermentations, can be calculated without much error by using the same $k_L a$, owing to the very slight differences in diffusion coefficients between O_2 and CO_2. Indeed the overall transport rate is greater for CO_2 because it is 40 times more soluble in water than O_2.

Table 2. Sorption Characteristics of Aqueous Systems for Turbine Stirrers

				Range of validity	
Liquid	K_1	α	β	P/V (kW/m^3)	u_G (cm/s)
Water	0.024	0.4	0.5	0.26–5.3	0.3–1.8
Electrolyte solution	0.018	0.74	0.26	0.26–5.3	0.3–1.8
Water	–	0.4	0.35	0.03–18	0.1–0.5
KCl solution					
0.22 N	–	0.71	0.36	0.03–18	0.1–0.5
0.1N	–	0.63	0.62	0.03–18	0.1–0.5
Water	0.0275	0.42	0.43	0.44–10	0.37–1.11
Electrolyte solution (Na_2SO_4 + KOH)	0.017	0.52	0.43	0.44–10	0.37–1.11

Note: $Z = 6$, $d/D = 0.33$; according to Equation 22. For more information see Ref. 5.

By evaluating published data, Judat (46) proposed a correlation that for aqueous systems covers the entire region of technical interest:

$$k_La\left(\frac{\nu}{g^2}\right)^{1/3} = \frac{9.8 \cdot 10^{-5}\left(\dfrac{P}{V \cdot \rho(g^4 \cdot \nu)^{1/3}}\right)^{\alpha}}{(B^{-\beta} + 0.81 \cdot 10^{-0.65/B})^{1.05}} \qquad (26)$$

The dimensionless gassing number is here given by

$$B \equiv \frac{q_G/D^2}{(\nu \cdot g)^{1/3}} \qquad (27)$$

The exponents are given as $\alpha = 0.42$ and $\beta = 0.6$. Although the exponents are dependent on the fluid system, the addition of the exponents, $\alpha + \beta \approx 1$, is approximately valid for all substances. Henzler (4) used this circumstance to evaluate the experimental results available with a relationship of the form

$$k_La\left(\frac{\nu}{g^2}\right)^{1/3} = K_2\left(\frac{P}{V \cdot \rho(g^4 \cdot \nu)^{1/3}}\right)^{\alpha}\left(\frac{u_G}{(g \cdot \nu)^{1/3}}\right)^{\beta} \qquad (28)$$

Taking into account $\beta \approx 1 - \alpha$, the simplified form was derived:

$$\frac{k_La}{u_G}\left(\frac{\nu^2}{g}\right)^{1/3} = K_2\left(\frac{P}{V \cdot u_G \cdot \rho \cdot g}\right)^{\alpha} \qquad (29)$$

which reflects the measured values with almost the same accuracy as equation 26.

In Figure 13 the sorption characteristics according to equation 29 are given for water and aqueous salt solutions. The results were obtained in reactor volumes of between 2.51 and 900 m^3. They lie close together, independent of the vessel volume. Equation 29 can thus be taken as a translation rule in scale-up on the basis of the oxygen transfer. In Table 3 the constant K_2 and exponent α are given for coalescent and noncoalescent fluid systems for turbine stirrers having $d/D = 0.14 - 0.5$.

Mass Transfer in Fluids of Higher Viscosity. For fluids of higher viscosity, in particular those with non-Newtonian flow behavior, there have been few systematic investigations published that deal with the determination of the volumetric mass transfer coefficient, k_La. Figure 14 shows a plot of several sorption characteristics for millet pulp, aqueous glucose solutions, and aqueous CMC solutions using the same correlation (equation 29) as for low-viscosity fluids (4,35,49). Thereby the translation rule given by equation 29 is confirmed for all fluid systems, even though the viscosity and the diffusion coefficient change by a factor of 1,000 and 30, respectively. To illustrate the experimental results for non-Newtonian fluids it is necessary, however, as with the determination of the power input, to calculate the apparent viscosity with equations 4 and 7 according to the concept of Metzner and Otto (24).

In Table 3 all sorption characteristics compiled by Henzler (4) are shown. Because these sorption characteristics are valid only for the particular rheological behavior and coalescence properties of each system, such properties have to be measured in the fermentation medium before an exact calculation of the mass transport is possible. Perez and Sandall (50) proposed the following correlation to calculate the volumetric mass transfer coefficient for pseudoplastic non-Newtonian fluids according to an approach of Sideman et al. (51):

$$k_La\left(\frac{d^2}{\mathcal{D}}\right) = 21.24 Re^{111} Sc^{0.5}\left(\frac{\eta_a \cdot u_G}{\sigma}\right)^{0.447}\left(\frac{\eta_G}{\eta_a}\right)^{0.694} \qquad (30)$$

The range of the flow behavior indices according to equation 5 is $m' = 0.916 - 1.00$ and $K' = (0.009 - 0.04)$ Pa $\cdot$ s$^{m'}$ with a vessel diameter of $D = 0.1524$ m. Kawase and Moo-Young (52) showed that the proposed model correlates reasonably well with the relatively wide variety of data measured by different authors for a range of vessel diameters of $D = 0.15 - 0.6$ m and for the flow behavior indices $m' = 0.59 - 0.95$ and $K' = (0.00355 - 10.8)$ Pa $\cdot$ s$^{m'}$.

A completely different correlation for the calculation of the volumetric mass transfer coefficient, k_La, was suggested by Mooyman (53):

$$k_La = C(270\varphi_G + 135\varphi_G \log(P_a/V)) \qquad (31)$$

Equation 31 expresses the overall transfer coefficient as an enhancement formula dependent on two variables: the gas holdup fraction, φ_G, that is given by

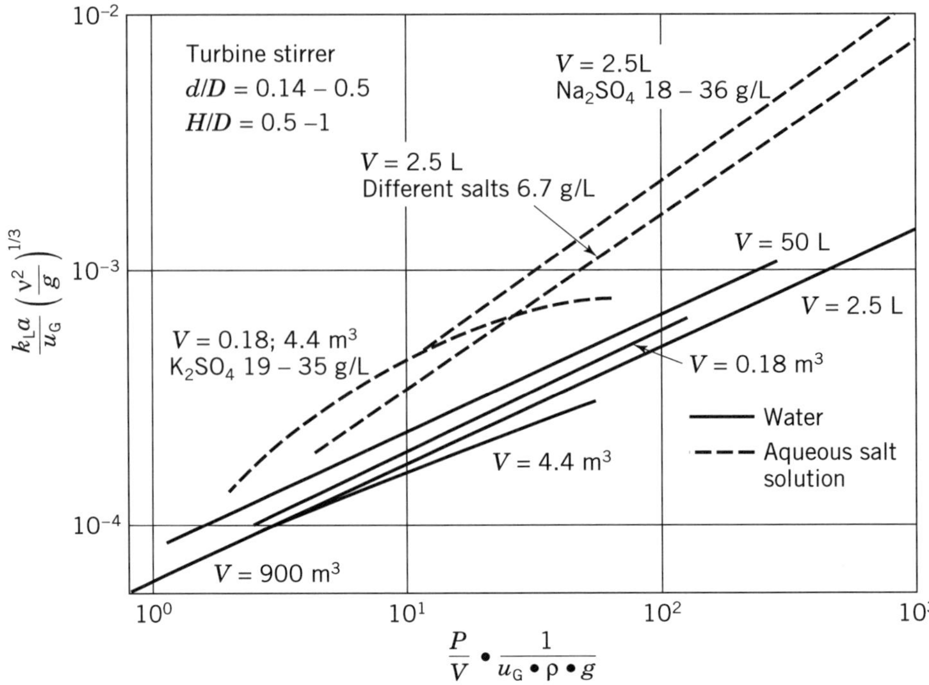

Figure 13. Sorption characteristics of aqueous systems for turbine stirrers.

Table 3. Sorption Characteristics of Different Fluid Systems for Turbine Stirrers

			Range of validity		
Liquid	Liquid	α	P/V (kW/m^3)	u_G (cm/s)	η (mPa s)
Water	7.5×10^{-2}	0.43	0.01–16.7	0.07–1.8	1
Aqueous salt solution					
6.7 g/L	7.2×10^{-5}	0.68	0.15–16	0.12–0.47	1
18–36 g/L	8.3×10^{-5}	0.71	0.23–18	0.12–1.8	1
Aqueous glucose solution	2.65×10^{-5}	0.7	–	–	12–267
Millet pulp	9.0×10^{-5}	0.7	0.23–2.2	0.19–0.78	1.3–70.2
Aqueous CMC solution, m = 0.4–0.82, equation 5	3.6×10^{-4}	0.55	0.06–6.2	0.22–1.9	16–1500

Note: Z = 6, d/D = 0.14–0.5; according to equation 29. For more information see Ref. 5.

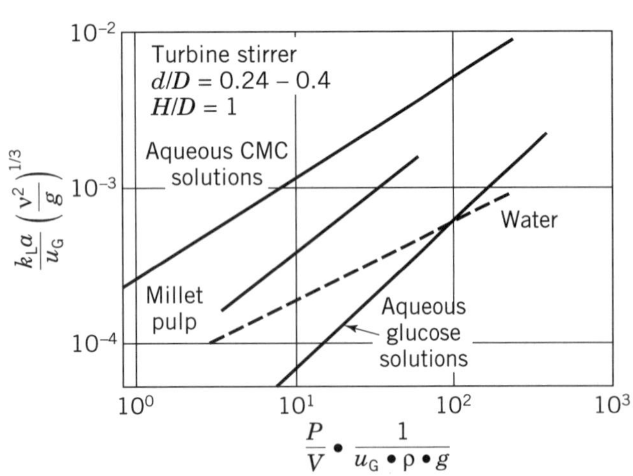

Figure 14. Sorption characteristics for millet pulp, glucose solutions, and CMC solutions.

$$\varphi_G = 0.1\left(\log\left(10.915 u_G \frac{p}{p_{av}}\right) - 0.9\right) \qquad (32)$$

and the mechanical plus the adiabatic gas expansion power input per unit of volume, P_a/V; C is a multiplication factor required to correct published or calculated $k_L a$ data to actual values observed in industrial fermentations with operating volumes above 250 L (53).

The effects of the suspension of solids on the mass transfer in sparged tank reactors have been investigated by various authors (e.g., Mills et al. [54] and Oguz et al. [55]). Of particular note is the dimensionless correlation developed by Oguz et al. (55) for the $k_L a$ in air-sparged agitated tanks containing slurries of different materials:

$$k_L a \frac{V}{q_G} = 2.173 \cdot 10^{-3}\left[\frac{P/q_G}{\rho^{1/3}(\eta_a \cdot g)^{2/3}}\right]^{0.607} \qquad (33)$$

The influence of the solid concentration becomes evident

in the calculation of the apparent viscosity of the slurries, which all showed non-Newtonian behavior. Consequently the viscosity was calculated with equations 4 and 7 according to the concept of Metzner and Otto (24). Oguz et al. (55) showed that an increase in the solid concentration leads to an increase of the slurry viscosity that corresponds to a decrease in $k_L a$. The deviation of the measured data from the values calculated with equation 33 was in the range of 12.5%.

For a scale-up on the basis of constant volumetric mass transfer coefficients, there follows from the generally valid equation 29

$$\left(\frac{P}{V}\right) u_G^{(1-\alpha)/\alpha} = \text{idem} \quad (34)$$

for the power input required. The exponent α is approximately 0.67 for noncoalescing aqueous systems as well as for more highly viscous Newtonian fermentation fluids. From equation 34 it follows that

$$\left(\frac{P}{V}\right) u_G^{0.5} = \text{idem} \quad (35)$$

For example, with $u_G \sim q_G/D^2$

$$\left(\frac{P}{V}\right) \frac{q_G^{0.5}}{D} = \text{idem} \quad (36)$$

If surface-active substances, such as chemical antifoam agents, are added to the fermentation medium, the volumetric mass transfer coefficient at the interfacial surface usually drops, and the mass transfer behavior tends toward a coalescing fluid system. In a coalescing fluid system and for non-Newtonian fluids, the exponent α is approximately 0.5. Consequently it follows from equation 29 that

$$\left(\frac{P}{V}\right) u_G = \text{idem} \quad (37)$$

and

$$\left(\frac{P}{V}\right) \frac{q_G}{D^2} = \text{idem} \quad (38)$$

In applying this translation rule (equations 35–38), the dispersing capacity of the stirrer, as given by equation 12, also has to be taken into consideration.

Shearing Stress and Pumping Capacity of the Stirrer

The mixing of the fermentation broth, the distribution and dispersion of the gases, and the suspending of solid constituents of the culture medium and microorganism pellets are effected in the stirred tank by a complex three-dimensional flow field. In only a few isolated cases (those that have a defined flow field) can the shearing stress and the pump capacity of the stirrer be calculated. With turbulent mixing it is possible, however, only to estimate both the longitudinal mixing by the pumping around of the vessel contents, and the shear stress on the fluid caused by the fluctuations in turbulence.

In a theoretical consideration of turbulent flow, Reynolds split up the measured local liquid velocity, u_L, into the sum of a mean (deterministic) velocity, $\bar{u}$, and a superposed stochastic fluctuation velocity, u'. According to this hypothesis, the turbulent Reynolds shear stress produced by the stirrer is proportional to the square of the local fluctuation velocity of the turbulence and is approximately given by the tip speed of the stirrer

$$\tau_R \sim (n \cdot d)^2 \quad (39)$$

With the Reynolds shear stress, τ_R, the effects of the stochastic liquid fluctuation velocity, u', are mathematically seized. τ_R can be interpreted as a virtual shear stress, that is, a consequence of the impulse transport caused by the turbulent motions.

Hoffmann et al. (56) determined the shear stress in stirred-tank reactors by means of a model system consisting of flocculated clay. Their investigations showed that the proportionality between shear stress of the clay floc system and the tip speed of the stirrer is of little influence; this is caused by the distinctness of the length scale of the vortices resulting from the Reynolds shear stress and the size of the relatively small flocs. The clay floc system is suitable for describing the influence of shear stress on tissue cell cultures (57).

Therefore, the model of Cherry and Kwon (58) is proposed for the calculation of shear stress. It is based on Kolmogoroff's theory of local homogenous isotropic turbulence: the length scale of the microeddies in the dissipative area (with laminar flow of the eddies) yields

$$\lambda = \left(\frac{\nu^3}{\varepsilon}\right)^{0.25} \quad (40)$$

where ε is the energy dissipation rate (i.e., the energy input to the system that is completely dissipated by the formation of vortices)

$$\varepsilon = \frac{P}{\rho \cdot V} \quad (41)$$

To employ this model by Cherry and Kwon (58), the particles exposed to shear have to be smaller than 12 times the length scale, λ, a condition that is usually fulfilled in fermentation processes. Under these circumstances the microeddies induced by the energy input or the energy dissipation rate, respectively, cause the shear stress. The particles are put into rotation by the velocity gradients in the microeddies, which are significantly larger than the gradients caused by the liquid fluctuation velocities in Reynolds hypothesis. The pulsating shear stress that acts upon the flocs during the contact with the microeddies leads to

$$\tau = 5.33\rho(\varepsilon \cdot \nu)^{0.5} \quad (42)$$

that is,

$$\tau = 5.33\left(\frac{P}{V}\nu \cdot \rho\right)^{0.5} = K_3\left(\frac{P}{V}\right)^{0.5} \qquad (43)$$

The influence of the liquid fluctuation velocities (i.e., the influence of the shear stress caused by the microeddies) has its maximum value if the dimension of the particles is close to the length scale of the microeddies. In Figure 15 the relationship between the experimentally determined shear stress of the clay floc system and the volumetric energy input for different types of stirrers is shown, using a relationship in analogy to equation 43

$$\tau = K_3\left(\frac{P}{V}\right)^{\alpha} \qquad (44)$$

The experimentally determined maximum shear stress deviates from the data calculated with the model of Cherry and Kwon (58) by two to six times. This can be explained by the fact that, according to the type of impeller system, there exist areas with increased energy dissipation rates in the reactor in the area close to the stirrer. This effect was not taken into consideration in the model, which uses a mean energy dissipation rate over the entire reactor volume that can be up to 30 times smaller than the maximum value of the energy dissipation rate close to the stirrer (59). Because the main part of the energy input is dissipated near the impeller, the model has to be extended by a supplementary corrective term containing information about the type of stirrer and its geometry (i.e., about the area with increased dissipation rates). By analogy with the work of Liepe (60), who studied the ratio between the maximum and the mean dissipation rate, $\varepsilon_{max}/\varepsilon$, of different impeller systems for dispersion tasks, Hoffmann et al. (56) suggested a corrective function

$$f = f\left(\frac{V}{V_S \cdot Ne}\right)^{\beta} \qquad (45)$$

with the effective energy dissipation volume, V_S, that is proportional to the volume of the stirrer ($V_S \sim d^2 \cdot h_B$). The values of V/V_S and Ne are given in Figure 15. By fitting the experimental data, Hoffmann et al. (56) finally determined the following correlation for the calculation of shear stress in stirred-tank reactors:

$$\tau = 0.367\left(\frac{P}{V}\left(\frac{V}{V_S \cdot Ne}\right)^{0.42}\right)^{0.55} \qquad (46)$$

Figure 16 shows the calculated shear stress according to equation 46 compared with the experimentally determined data (56). The deviation of the model from the measured data is in a range of up to 30%, which means the model is suitable for a scale-up of the reactor performance on the basis of constant shear stress. In a sparged stirred-tank reactor the shear stress is also dependent on the gas input. Experimental investigations of Henzler and Biedermann (61) showed that there is a qualitative dependency of the shear stress on the pneumatic power input and the type of sparging system. They observed that with constant volumetric power input the shear stress increases with the increasing portion of pneumatic power input and with increasing superficial gas velocity. They described some of

	Stirrer	d/D	V/V_S	Ne
●	Propeller	0.33	200.7	0.31
■	INTERMIG	0.65	40.7	0.65
▲	Turbine	0.33	310.1	5.0
△	Turbine	0.65	42.5	5.5
□	4-blade	0.33	135.7	9.6
—— Model according to Ref. 58				

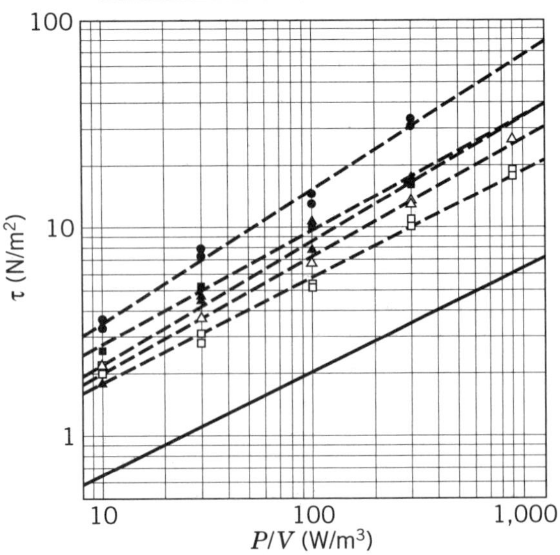

Figure 15. Experimentally determined maximum shear stress versus the volumetric power input, according to equation 44.

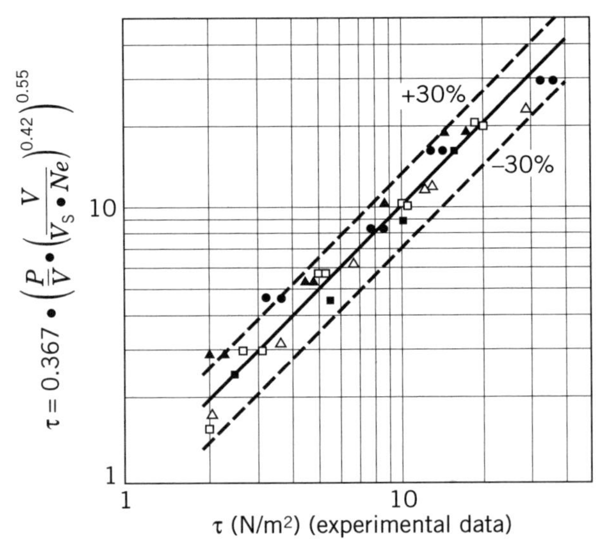

Figure 16. Experimentally determined shear stress compared with calculated data, according to equation 46; for symbols, see Figure 15.

the observed effects causing the increased shear stress, such as the turbulence in the lag zone of rising bubbles, coalescence or dispersion of bubbles close to the sparger, and the bursting of bubbles at the liquid surface (as is observed in bubble columns) (62). Because a model or correlations to describe these effects quantitatively are not yet known, further investigations are necessary to describe the effects of sparging and bubbles on the mechanical shear stress in stirred vessels.

A relationship for the circulating volumetric flow, which is produced by the pumping capacity of the stirrer, can be derived in a manner analogous to the principle that applies to rotary pumps. In this way Van de Vusse (39) calculated the circulating volumetric flow of propellers with the assumption that counterpressure is "zero" and obtained

$$\dot{V} = 0.5\,\frac{\pi}{4}\,n \cdot d^2 \cdot h_S \tag{47}$$

where h_S is the pressure head produced by the propeller. With $h_S \sim d$, the generalized relationship

$$\dot{V} = K_4 \cdot n \cdot d^3 \tag{48}$$

follows from equation 47. K_4 is a constant that is essentially dependent on the type of stirrer (63) (see also Refs. 16, 19, and 64).

Varying details have been published regarding the suspension of solids in stirring apparatuses. Most authors say that in achieving the same suspension result in a larger vessel, the power input can be diminished (43). The circulating volumetric flow given by equation 48 is sufficiently large under the conditions of stirring that generally prevail in stirred-tank reactors so that a uniform distribution of cell pellets and solid constituents in the culture medium is guaranteed. This uniform distribution is supported by the very slight density differences between solid and liquid in the fermentation medium, and the relatively small solid particles and the small fraction of solid material ($d_P < 2$ mm, $\varphi_S < 20\%$) in the fermentation medium. Consequently the suspension process in general takes on no particular significance in scaling up stirred-tank reactors.

Mass Transport Processes in Mycelium Agglomerates

Mycelium-forming microorganisms can build up many morphological forms while submerged fermentation, depending on the conditions of fermentation. Forms include loose hyphae that are distributed homogenously throughout the medium or collect in agglomerates that appear as compact pellets. Whereas the filamentous forms result in highly viscous, non-Newtonian fermentation fluids, the mycelial pellets are usually present in Newtonian fluid systems at a considerably lower viscosity (65). An adequate shearing of the mycelium-forming microorganisms thereby leads to the preferred formation of pellets rather than a filamentous mycelial structure.

Regardless of whether the formation of pellets and compact cell agglomerates leads to fluids of lower viscosity (and thus to more turbulent conditions), problems regarding the oxygen supply to the cells can arise, in particular with large, compact pellets. When the inner and outer mass transport resistances are too large, the danger of autolysis exists, and only the external pellet layer is involved in the biotechnological process.

The effective diffusion coefficient of oxygen in mycelial pellets of *Aspergillus niger* was found to be about $\mathcal{D}_{Pe} = 10^{-4}$ cm^2/s (2). This value is larger by a factor of 10 than the molecular diffusion coefficient, $\mathcal{D}$, in the culture medium. Because the diffusion coefficient determined by Miura (2) also varied with the volumetric power input, this result must be interpreted as indicating that the mycelium forms a loose pellet structure that is penetrated by the turbulent microflow. Thus in such cultures an adequate but not excessive turbulence (i.e., shear stress) is required in order for pellets with a loose structure to form.

Whereas the mass transport inside mycelial aggregates has still not been widely researched, reliable equations exist for determining the mass transfer from fluid to pellet. The mass transfer coefficient k_{Pe} on the surface of the pellet of diameter d_{Pe} can be calculated from the formal relationship

$$Sh = f(Re_{Pe}, Sc) \tag{49}$$

where $Sh \equiv k_{Pe} \cdot d_{Pe}/\mathcal{D}$ is the Sherwood number, $Re_{Pe} \equiv u_{Pe} \cdot d_{Pe}/\nu$ is the Reynolds number formed with the relative velocity between pellet and fluid, and $Sc \equiv \nu/\mathcal{D}$ is the Schmidt number. The relative velocity at the pellet is determined solely by local turbulent fluctuation velocities, u', which are effected by the turbulent eddies of the scale of the pellets. According to the theory of isotropic turbulence put forward by Kolmogoroff, it follows for this velocity that

$$\sqrt{\overline{u'^2}} \sim (\varepsilon \cdot d_{Pe})^{1/3} \tag{50}$$

where ε is the mass-related energy of dissipation calculated with equation 41. Then the Reynolds number, formed with the relative velocity

$$u_{Pe} \sim \sqrt{\overline{u'^2}}$$

is

$$Re_{Pe} = \left(\frac{P}{V}\,\frac{d_{Pe}^4}{\rho \cdot \nu^3}\right)^{1/3} \tag{51}$$

In Figure 17 the experimental results according to equation 49, from different authors, are shown; the experimental data, following the theory, have been plotted as

$$\frac{Sh - 2}{Sc^{1/3}} = f(Re_{Pe}) \tag{52}$$

The only published equations that are mentioned here are those relationships reported by Liepe and Moeckel (66); with the relative velocity defined here they are given by

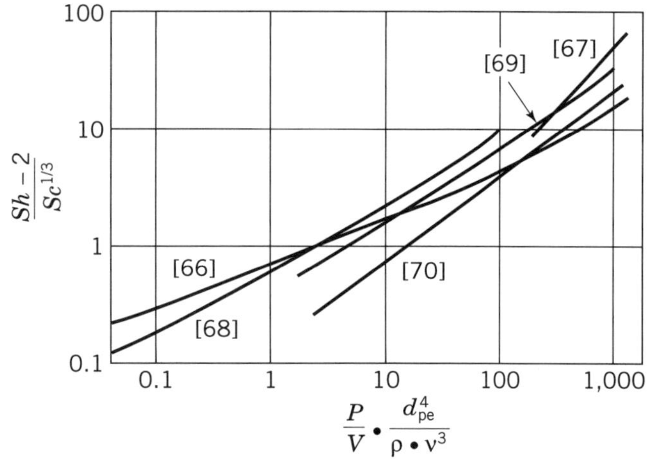

Figure 17. Mass transfer at the pellet surface as dependent on the particle Reynolds number. Numbers in brackets indicate reference.

$$\frac{Sh - 2}{Sc^{1/3}} = 0.35\left(\frac{\Delta\rho}{\rho}\right)^{1/3}\left(\frac{P}{V}\frac{d_{\mathrm{Pe}}^4}{\rho \cdot \nu^3}\right) \qquad (53)$$

The curve given by equations 52 and 53 is illustrated in Figure 17, marked according to Liepe and Moeckel (66). Apart from the measurements made by Sicardi et al. (67), all measurements were performed in a nongassed system. Nevertheless the comparison of results shows that the presence of gas has virtually no influence on the mass transfer from fluid to pellet.

SCALE-UP EXAMPLES OF BIOTECHNOLOGICAL PROCESSES BY MAINTENANCE OF IDENTICAL OPERATING CONDITIONS

Scale-Up on the Basis of the Oxygen Transfer Rate

A scale-up of aerobic biotechnological processes is often performed successfully using the criterion of identical oxygen transfer rate (OTR = idem), or the same volumetric mass transfer coefficient (k_La = idem). The reason is that in the aerobic culture of yeasts and bacteria, which have high rates of respiration, the rate-limiting reaction step is often located in the oxygen transfer resistance at the gas–liquid interface. In the fermentation of mycelium-forming microorganisms there arises, under certain conditions, a further limiting mass transport resistance at the surface of or inside the mycelial pellet. That is the reason why a scale-up on the basis of both mass transport resistances is more successful.

Karow et al. (71) took the oxygen transfer rate as a scale-up criterion for the production of penicillin and streptomycin. In Figure 18 the relative penicillin concentrations are shown as a function of the oxygen transfer rate. The results for the production of streptomycin are very similar. The scale-up covers 4 orders of magnitude. At oxygen transfer rates of more than 0.5 mol/(L h) the product yield is maximal and nearly constant. The successful translation from the laboratory to a production scale, performed by

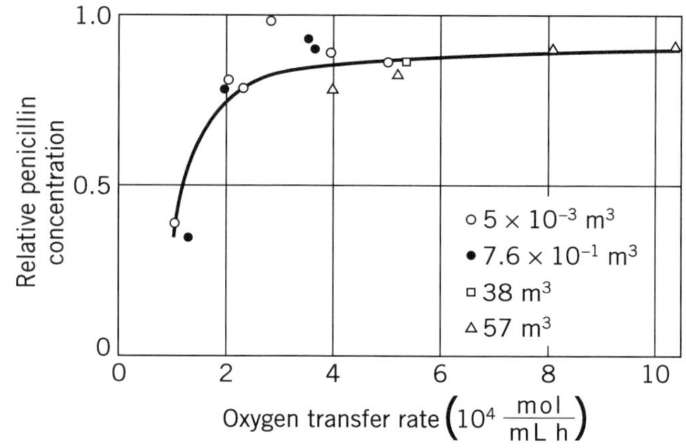

Figure 18. Production of penicillin as a function of the oxygen transfer rate.

Karow et al. (71), was also confirmed by Bylinkina et al. (72) for the fermentation of penicillin and streptomycin. They carried out a scale-up from the 15-L fermentor via a 100-L and 3-m^3 fermentor up to a 63-m^3 production fermentor on the basis of the mass transfer coefficient at the gas–liquid interface. Wegrich and Shurter (73) also described a successful scale-up of penicillin production. The same yields of penicillin were attained in an 8-m^3 fermentor as in a 100-m^3 fermentor. They did not only keep the superficial gas velocity constant but also the volumetric power input of the stirrer. According to equation 37, holding these parameters constant in non-Newtonian fluids corresponds to the scale-up rule k_La = idem.

A further, now "historical," example of scale-up on the basis of the oxygen transfer is shown in Figure 19, in which the production of baker's yeast is depicted (74). Although the oxygen transfer rate in the model system (sulfite oxidation), used as the scale-up criterion, was measured, no significant difference in the production of yeast could be

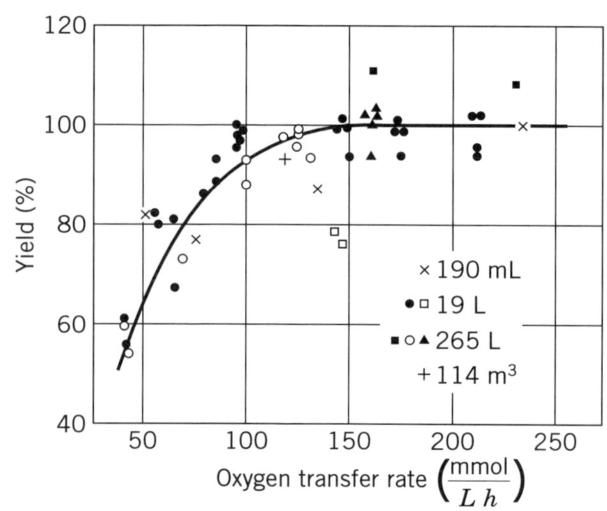

Figure 19. Scale-up of fermentation of baker's yeast based on oxygen transfer rate (sulfite oxidation).

established, which was dependent only on the size of the reactor. Those points plotted in Figure 19 were obtained using shaker flasks with a 190-mL liquid content and in 19-L, 265-L, and 114-m^3 fermentors. Jari (75) was able to show for the production of the primary metabolites alkaline and acid proteases and the secondary metabolites nystatin and fumagillin that a scale-up can be carried out using the same oxygen transfer rate (k_La = idem).

In the bacterial production of vitamin B_{12}, Bartholomew (76) established that the yield at high oxygen transfer rates is negatively influenced by a not very clearly described effect (Fig. 20). Since, moreover, this effect is dependent on the fermentor size, the oxygen transfer rate in this region cannot be used as a scale-up criterion. The dotted curve in Figure 20 applies to the yield only when small-scale equipment is used, and the continuous line represents the yield achieved in a production fermentor. In other words, the production fermentor is oversized. Takei et al. (77) studied protease production by *Streptomyces* sp. in a 0.03-m^3- and a 0.2-m^3-sized fermentor and found the same effect. The productivity of the cells in synthesizing protease is indeed determined by the oxygen transfer rate but is nonetheless diminished at higher oxygen transfer rates by a fermentor size–dependent effect.

In the production of glucoamylase by *Endomyces* sp., the enzyme yield is also not solely determined by the oxygen transport. A scale-up on the basis of identical k_La values also leads to oversizing of the production fermentor. The yields depicted in Figure 21, expressed in relative enzyme activities as a function of the volumetric mass transfer coefficients, were measured in fermentors of 0.06 m^3, 3 m^3, and 30 m^3 (78). In the latter three examples of fermentation described, no adequate scale-up among the fermentor sizes was achieved by using solely the oxygen transfer rate (i.e., the k_La value). The reasons for such deviations can be diverse. For example, in applying the same k_La values as a scale-up criterion, the influence of the surface aeration has to be taken into consideration. Fuchs et al. (79) showed that this effect makes itself felt at fermentor volumes smaller than 200 L.

The translation of the oxygen transfer rate from a laboratory fermentor to a production scale can still lead to

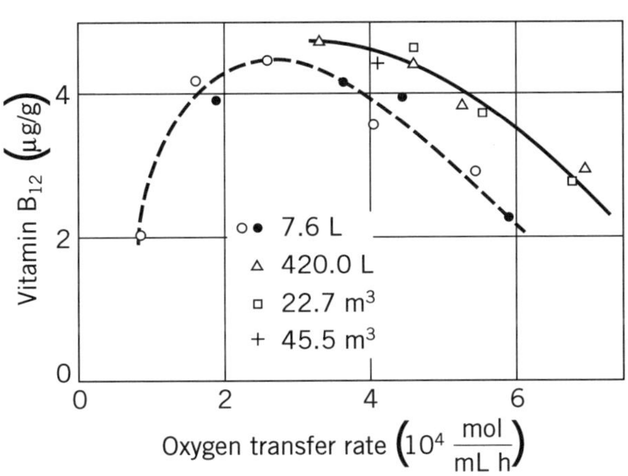

Figure 20. Production of vitamin B_{12} as function of the oxygen transfer rate.

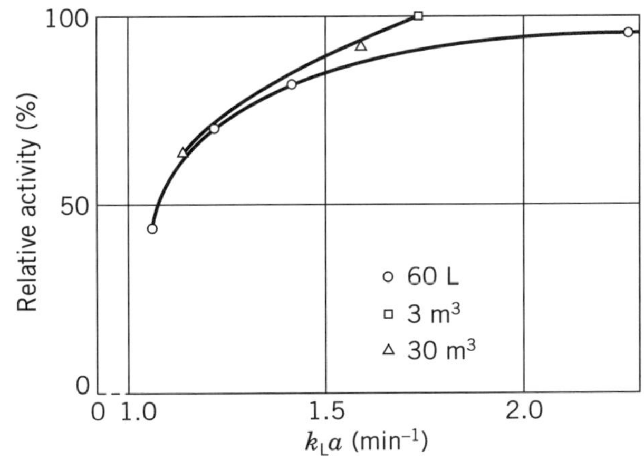

Figure 21. Relative glucoamylase activity as function of the k_La value.

considerable errors for other reasons. Thus, with the same k_La value, the oxygen transfer rate in the production fermentor can be larger than in the laboratory fermentor because, owing to the higher hydrostatic pressure, the driving force (concentration difference) will be larger. But with the same bubble size (identical rising velocity of the bubbles) the average residence time of the air bubble in the larger apparatus is longer; it is thus deficient in oxygen, and consequently, the positive effect of hydrostatic pressure, just cited, is diminished.

Scale-Up on the Basis of the Volumetric Power Input

One-third of the scale-up translations performed in the fermentation industry use the scale-up criterion "identical volumetric power input," as Margarites and Zajic (15) reported. Apart from the simplicity of the measurement in applying this scale-up rule, the reason for its industrial success probably lies in the fact that almost all hydrodynamic and mass transport phenomena can be correlated using the dissipated energy (and thus using P/V). Nevertheless, or maybe simply because of it, this criterion provides an extremely rough rule of thumb, which permits a scale-up only on an empirical basis.

Maxon (80) studied the influence of the volumetric power input on the production of novobiocin using *Streptomyces niveus*. In Figure 22 the novobiocin concentration produced is plotted against the volume-related power input for three different sizes of stirrer. Above $P/V = 1$ kW/m^3 the yields of the antibiotic are maximal and largely independent of P/V. Scale-up based on this criterion is not dependable because it has been established that there is a dependance on the diameter of the stirrer. According to Wang and Fewkes (81) a scale-up of this process is possible only owing to the partial similarity of the turbulent shearing stress and the pumping capacity of the stirrer.

In the fermentation of penicillin, the volumetric power input, P/V, was often used in the past as a criterion for scale-up. In Figure 23 the penicillin concentration is plotted as a function of the volume-related stirring power, as described by Gaden (82), who gave P/V values between 1.5

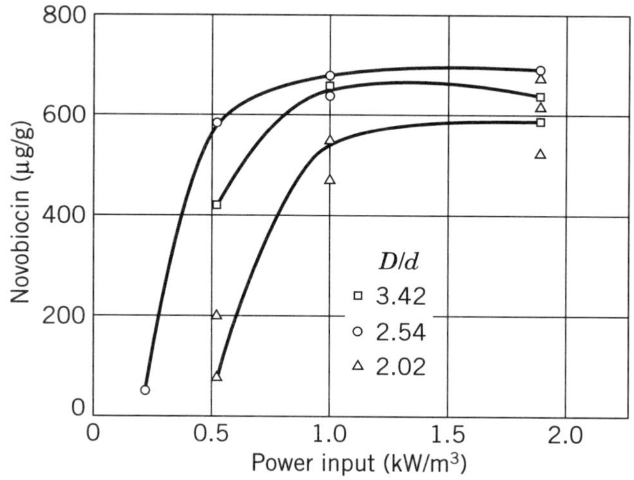

Figure 22. Novobiocin concentration after 115 h of fermentation duration, dependent upon the volumetric power input.

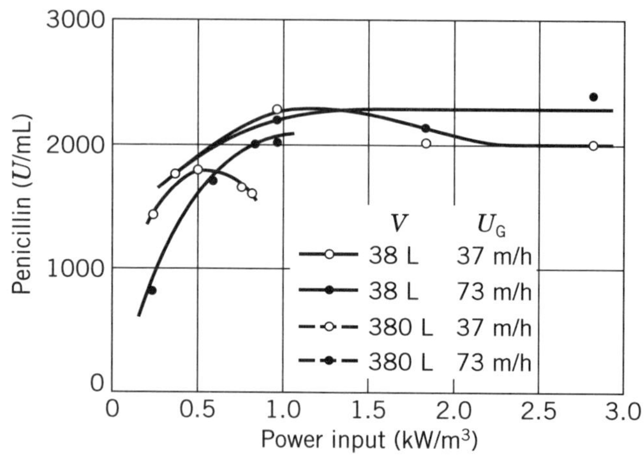

Figure 23. Influence of the volumetric power input on the synthesis of penicillin.

and 3 kW/m^3 as a scale-up rule. Similar results were achieved by Humphrey (83,84), who established a correlation between penicillin production and the volumetric power input and carried out a scale-up at a volume-related stirring power between 1 and 2.5 kW/m^3. Humphrey (84) reports a volumetric power input of 1.5 kW/m^3 for a scale-up of streptomycin production.

NOMENCLATURE

A	Reactor cross-sectional area (m^2)
a	Interfacial exchange area per unit volume (m^2/m^3)
$B \equiv (q_G/D^2)/(\nu \cdot g)^{1/3}$	Dimensionless gassing number
C	Factor
$c^*_{O_2}$	Saturation concentration of O_2 in fermentation medium (mg/L)
c_{O_2}	Concentration of O_2 in fermentation medium (mg/L)
$c^*_{O_2,in}$, $c^*_{O_2,out}$	Saturation concentration of O_2 at entrance and exit of reactor (mg/L)
D	Diameter of stirred-tank reactor (m)
$\mathcal{D}$	Diffusion coefficient of the liquid phase (m^2/s)
$\mathcal{D}_{Pe}$	Effective diffusion coefficient in mycelial pellets (m^2/s)
d	Diameter of the stirrer (m)
d_P	Diameter of solid particles (m)
d_{Pe}	Diameter of mycelial pellets (m)
$Fr \equiv n^2 \cdot d/g$	Froude number
$Ga \equiv d^3 \cdot g/\nu$	Galilei number
g	Acceleration due to gravity (m/s^2)
H	Liquid filling height in the reactor (m)
h	Distance between stirrers (m)
h_B	Height of stirrer blade (m)
h_S	Pressure head of the stirrer (m)
i	Number of stirrer blades
K	Proportionality constant
K'	Consistency index [(Pa s)m']
K''	Consistency index [(Pa s)m'']
K_1, K_2, K_3, K_4	Constants
k_L	Liquid-side mass transfer coefficient at gas–liquid interface (m/s)
k_{Pe}	Mass transfer coefficient at mycelium pellet surface (m/s)
k_La	Volumetric mass transfer coefficient at gas–liquid interface (s^{-1})
m', m''	Flow behavior parameters
$Ne \equiv P/\rho \cdot n^3 \cdot d^5$	Newton number
Ne_G	Newton number in gassed reactor at $Re > 10^4$
Ne_0	Newton number without gassing
$\dot{N}_{O_2}$	Oxygen mass flow (kg/s)
n	Rotational speed (s^{-1})
$\dot{n}_{O_2}$	Oxygen transfer rate (mass flow density) [kg/(s · m^3)]
P	Stirring power (kW)
P_a	Mechanical plus adiabatic gas expansion power input (kW)
p	Atmospheric pressure (N/m^2)
p_{av}	Average pressure at half the liquid height (N/m^2)
$Q \equiv q_G/n \cdot d^3$	Dimensionless gassing number
Q_{max}	Dimensionless gas flow number at flooding point of stirrer
q_G	Gas throughput (m^3/s)
$q_{G,max}$	Maximum gas throughput (m^3/s)

$Re \equiv n \cdot d^2/\nu$	Reynolds number
$Re^* \equiv n \cdot d^2 \cdot \rho/\eta_a$	Modified Reynolds number
$Re_{Pe} \equiv u_{Pe} \cdot d_{Pe}/\nu$	Pellet Reynolds number
S_i	Characteristic quantities for coalescence of fluids
$Sc \equiv \nu/\mathcal{D}$	Schmidt number
$Sh \equiv k_{Pe} \cdot d_{Pe}/\mathcal{D}$	Pellet Sherwood number
u'	Turbulent fluctuation velocity of the fluid (m/s)
$\bar{u}$	Mean velocity of the fluid (m/s)
u_G	Gas velocity based on empty cross section area of the reactor (m/s)
u_L	Local liquid velocity (m/s)
u_{Pe}	Relative velocity between pellet and fluid (m/s)
V	Reactor volume (m^3)
$\dot{V}$	Circulating volumetric flow by stirrer pumping (m^3/s)
V_S	Volume stirred up by the stirrer blades (m^3)
Z	Number of stirrer blades
α	Exponent
β	Exponent
$\dot{\gamma}$	Shear rate (s^{-1})
ε	Mass-related energy dissipation rate (kW/kg)
ε_{max}	Maximum energy dissipation rate (kW/kg)
η	Dynamic viscosity (Pa s)
η_a	Apparent dynamic viscosity (Pa s)
η_G	Dynamic viscosity of the gas phase (Pa s)
θ	Mixing time (s)
λ	Length scale of microeddies (m)
ν	Kinematic viscosity (m^2/s)
ρ	Density of fermentation fluid (kg/m^3)
σ	Surface tension (N/m)
$\sigma^* \equiv \sigma/[\rho(g \cdot \nu^4)^{1/3}]$	Dimensionless surface tension
τ	Shear stress (Pa)
τ_R	Reynolds shear stress (Pa)
τ_0	Yield stress (Pa)
φ_G	Gas holdup
φ_S	Solid holdup

BIBLIOGRAPHY

1. D.I.C. Wang and A.E. Humphrey, in D.J.D. Huckenhull ed., *Progress in Industrial Microbiology*, vol. 8, Temple Press Books, London, 1968, pp. 1–34.
2. Y. Miura, in T.K. Ghose, A. Fiechter, and N. Blakebrough eds., *Advances in Biochemical Engineering*, vol. 4, Springer-Verlag, Berlin, 1976, pp. 3–40.
3. M. Moo-Young and H.W. Blanch, in A. Fiechter ed., *Advances in Biochemical Engineering*, vol. 19, Springer-Verlag, Berlin, 1981, pp. 1–69.
4. H.-J. Henzler, *Chem.-Ing.-Tech.* **54**, 461–476 (1982).
5. D.C. Hempel, in R.K. Finn, P. Praeve, M. Schlingmann, W. Crueger, K. Esser, R. Thauer, and F. Wagner eds., *Biotechnology Focus 1*, Hanser Publishers, Munich, 1988, pp. 51–94.
6. J. Bryant, in T.K. Ghose, A. Fiechter, and N. Blakebrough eds., *Advances in Biochemical Engineering*, vol. 5, Springer-Verlag, Berlin, 1977, pp. 101–124.
7. R. Mann, P.P. Mavors, and J.C. Middleton, *Trans. Inst. Chem. Eng.* **59**, 271–278 (1981).
8. J. Pawlowski, *Die Ähnlichkeitstheorie in der physikalisch-technischen Forschung: Grundlagen und Anwendung*, Springer-Verlag, Berlin, 1971.
9. J. Stichlmaier, *Chem.-Ing.-Tech.* **63**, 38–51 (1991).
10. F. Kudrewizki, *Acta Biotechnol.* **4**, 89–104, 247–254 (1984).
11. M. Reuss, R.K. Bajpal, D. Debus, R. Lenz, H. Niebelschuetz, and A. Papalexiou, *6th Int. Ferment. Symp.*, London, Ontario, Canada, 1980.
12. R.K. Bajpal and M. Reuss, *Can. J. Chem. Eng.* **60**, 384–392 (1982).
13. N.M.G. Oosterhuis, N.M. Groesbeek, A.P.C. Olivier, and N.W.F. Kossen, *Biotechnol. Lett.* **5**, 141–146 (1983).
14. N.M.G. Oosterhuis, *Scale-Up of Bioreactors*, Proefschrift, Techn. Hogeschool, Delft, Netherlands, 1984.
15. A. Margarites and J.E. Zajic, *Biotechnol. Bioeng.* **20**, 939–1001 (1978).
16. J.Y. Oldshue, *Fluid Mixing Technology*, McGraw-Hill, New York, 1983.
17. J.Y. Oldshue, *Chem. Eng. Prog.* **5**, 33–42 (1989).
18. J.Y. Oldshue, *AIChE Symp. Ser.* **89**, 158–163 (1993).
19. J.Y. Oldshue, *Chem. Eng. Prog.* **93**, 70–73 (1997).
20. J.Y. Oldshue, *Biotechnol. Bioeng.* **8**, 3–24 (1966).
21. J.H. Rushton, E.W. Costich, and H.J. Everett, *Chem. Eng. Prog.* **46**, 395–404, 467–476 (1950).
22. M. Zlokarnik, in E. Bartholome, E. Biekert, H. Hellmann, H. Ley eds., *Ullmanns Enzyklopaedie der Technischen Chemie*, vol. 2, Verlag Chemie GmbH, Weinheim/Bergstrasse, 1972, pp. 259–281.
23. K. Kipke, *GVC-Vortragstagung: Verfahrenstechnische Fortschritte beim Mischen, Dispergieren und bei der Wärmeuebertragung in Fluessigkeiten*, VDI-Verlag, Duesseldorf, 1978, 21–36.
24. A.B. Metzner and R.E. Otto, *AIChE J.* **3**, 3–11 (1957).
25. A.B. Metzner, R.H. Feehs, H.L. Ramos, R.E. Otto, and J.D. Toothill, *AICHE J.* **7**, 3–9 (1961).
26. R. Rapp, H. Reng, D.C. Hempel, and F. Wagner, *Biotechnol. Bioeng.* **26**, 1167–1175 (1984).
27. P.H. Calderbank and M.B. Moo-Young, *Trans. Inst. Chem. Eng.* **37**, 26–33 (1959).
28. A. Sanchez, A. Martinez, L.G. Torres, and E. Galindo, *Process Biochem.* **27**, 351–365 (1992).
29. D. Velasco, A. Martinez, L.G. Torres, and E. Galindo, in A.W. Nienow ed., *3rd International Conference on Bioreactor and Bioprocess Fluid Dynamics, BHR Group Conference Series Publication No. 5*, Mechanical Engineering Publications Limited, London, 1993, pp. 101–116.
30. Y. Oyama and K. Endoh, *J. Chem. Eng. Jpn.* **19**, 2–11 (1955).

31. M. Zlokarnik, *Chem.-Ing.-Tech.* **45**, 689–692 (1973).
32. H. Judat, Dissertation, Universitaet Dortmund, Germany, 1976.
33. V. Schlueter and W.D. Deckwer, *Chem.-Ing.-Tech.* **64**, 644–645 (1992).
34. H. Hoecker and G. Langer, *Rheol. Acta* **16**, 400–412 (1977).
35. H. Hoecker, Dissertation, Universitaet Dortmund, Germany, 1979.
36. H.-J. Henzler, *GVC-Vortragstagung: Verfahrenstechnische Fortschritte beim Mischen, Dispergieren und bei der Wärmeuebertragung in Fluessigkeiten*, VDI-Verlag, Duesseldorf, 1978, pp. 91–110.
37. P. Zehner, *Chem.-Ing.-Tech.* **60**, 531–539 (1988).
38. C. Breucker, A. Steiff, P.-M. Weinspach, *Proceedings of the European Conference on Mixing and Centrifugal Separation / Fluid Engineering*, vol. 6, British Hydromechanics Research Association, Cranfield, Bedford, 1988, pp. 399–406.
39. J.G. Van de Vusse, *Chem.-Ing.-Tech.* **31**, 583–587 (1959).
40. S. Nagata, *Mixing: Principles and Applications*, Kodansha, Tokyo, and Wiley, New York, 1975.
41. H.-J. Henzler, *Untersuchungen zum Homogenisieren von Fluessigkeiten oder Gasen, VDI-Forschungheft*, vol. 587, VDI-Verlag, Duesseldorf, 1978.
42. M. Zlokarnik, *Chem.-Ing.-Tech.* **39**, 539–548 (1967).
43. A. Mersmann, W.D. Einenkel, and M. Kaeppel, *Chem.-Ing.-Tech.* **47**, 953–996 (1975).
44. M. Opara, *Verfahrenstechnik* **9**, 446–449 (1975).
45. M. Zlokarnik, in T.K. Ghose, A. Fiechter, and N. Blakebrough eds., *Advances in Biochemical Engineering*, vol. 8, Springer-Verlag, Berlin, 1978, pp. 133–151.
46. H. Judat, *Chem.-Ing.-Tech.* **54**, 520–521 (1982).
47. K. Van't Riet, *Ind. Eng. Chem. Process Des. Dev.* **18**, 357–375 (1979).
48. K. Akita and F. Yoshida, *Ind. Eng. Chem. Process Des. Dev.* **13**, 84–91 (1974).
49. H. Yagi and F. Yoshida, *Ind. Eng. Chem. Process Des. Dev.* **14**, 488–493, (1975).
50. J.F. Perez and O.C. Sandall, *AIChE J.* **20**, 770–775 (1974).
51. S. Sideman, O. Hortacsu, and J.W. Fulton, *Ind. Eng. Chem. (Int. Ed.)* **58**, 32–47 (1966).
52. Y. Kawase and M. Moo-Young, *Chem. Eng. Res. Des.* **66**, 285–288 (1988).
53. J.G. Mooyman, *Biotechnol. Bioeng.* **29**, 180–186 (1987).
54. D.B. Mills, R. Bar, and D.J. Kirwan, *AIChE J.* **33**, 1542–1549 (1987).
55. H. Oguz, A. Brehm, and W.-D. Deckwer, *Chem. Eng. Sci.* **42**, 1815–1822 (1987).
56. J. Hoffmann, K. Buescher, and D.C. Hempel, *Chem.-Ing.-Tech.* **67**, 210–214 (1995).
57. J. Hoffmann, J. Tralles, and D.C. Hempel, *Chem.-Ing.-Tech.* **64**, 953–956 (1992).
58. R.S. Cherry and K.-Y. Kwon, *Biotechnol. Bioeng.* **36**, 563–571 (1990).
59. R.K. Geisler, Dissertation, Technische Universitaet Munich, Germany, 1991.
60. F. Liepe, *Verfahrenstechnische Berechnungsmethoden, vol. 4: Stoffvereinigen in fluiden Phasen*, VCH, Weinheim, 1988.
61. H.-J. Henzler and A. Biedermann, *Chem.-Ing.-Tech.* **68**, 1546–1561 (1996).
62. K. Buescher and D.C. Hempel, *Chem.-Ing.-Tech.* **68**, 1452–1455 (1996).
63. B.K. Revill, *4th Eur. Conf. on Mixing*, Paper B1, Noordwijkerhout, Netherlands, 1982.
64. J.Y. Oldshue, in C.S. Ho and J.Y. Oldshue eds., *Biotechnology Processes: Scale-Up and Mixing*, AIChE, New York, 1987, pp. 3–5.
65. B. Metz, N.W.F. Kossen, J.C. Van Suijdam, in T.K. Ghose, A. Fiechter, and N. Blakebrough eds., *Advances in Biochemical Engineering*, vol. 11, Springer-Verlag, Berlin, 1979, pp. 103–156.
66. F. Liepe, H.O. Moeckel, *Chem. Tech. (Leipzig)* **28**, 205–209 (1976).
67. S. Sicardi, R. Conti, G. Baldi, L. Franzino, *Chem.-Ing.-Tech.* **53**, MS 870/81 (1981).
68. P.L.T. Brian, H.B. Hales, T.K. Sherwood, *AIChE J.* **15**, 727–733 (1969).
69. P. Harriot, *AIChE J.* **8**, 93–102 (1962).
70. P.H. Calderbank, M.B. Moo-Young, *Chem. Eng. Sci.* **16**, 39–54 (1961).
71. E.O. Karow, W.H. Bartholomew, M.R. Sfat, *J. Agric. Food. Chem.* **1**, 302–312 (1953).
72. E.S. Bylinkina, V.V. Birukov, S.A. Voroshilova, in H. Dellweg ed., *Abstracts 5th International Fermentation Symposium*, Berlin, 1976.
73. O.G. Wegrich, R.A. Shurter, *Ind. Eng. Chem.* **45**, 1153–1160 (1953).
74. J. Strohm, R.F. Dale, H.J. Peppler, *Appl. Microbiol.* **7**, 235–238 (1959).
75. M. Jarai, in G. Terni ed., *4th International Fermentation Symposium Proceedings*, Fermentation Technology Today, Osaka, 1972, pp. 97–105.
76. W.H. Bartholomew, in Umbreit ed., *Advances in Applied Microbiology 2*, Academic Press, New York, 1960, pp. 289–300.
77. H. Takei, K. Mizusawa, and F. Yoshida, *J. Ferment. Technol.* **53**, 151–158 (1975).
78. H. Taguchi, T. Imomanaka, S. Teramoto, M. Takatsu, and M. Sato, *J. Ferment. Technol.* **46**, 823–828 (1968).
79. R. Fuchs, D.D.Y. Ryu, A.E. Humphrey, *Ind. Eng. Chem. Process Des. Dev.* **10**, 190–196 (1971).
80. W.D. Maxon, *J. Biochem. Microbiol. Tech. Eng.* **1**, 311–324 (1959).
81. D.I. Wang, R.C.J. Fewkes, in L.A. Underkofler ed., *Developments in Industrial Microbiology*, American Institute of Biological Science, Washington, D.C., 1977, pp. 39–56.
82. E.J. Gaden, Jr., *Sci. Rep. Inst. Superore di Sanita* **1**, 161–176 (1961).
83. A.E. Humphrey, *J. Ferment. Technol.* **42**, 265–278 (1964).
84. A.E. Humphrey, *J. Ferment. Technol.* **42**, 334–345 (1964).

See also BIOREACTORS, AIR-LIFT REACTORS; BIOREACTORS, FLUIDIZED-BED; BIOREACTORS, CONTINUOUS STIRRED-TANK REACTORS; DIMENSIONAL ANALYSIS, SCALE-UP; FERMENTATION MONITORING, DESIGN AND OPTIMIZATION; MAMMALIAN CELL CULTURE REACTORS, SCALE-UP; MASS TRANSFER.

SECONDARY METABOLITE PRODUCTION, ACTINOMYCETES, OTHER THAN STREPTOMYCES

Ann C. Horan
Schering Plough Research Institute
Kenilworth, New Jersey

KEY WORDS

Actinomycetes
Cell wall composition
Chemotaxonomy
Fatty acids
Fermentation
Menaquinones
Phospholipids
Secondary metabolites

OUTLINE

INTRODUCTION

Actinomycetes are an industrially important group of microorganisms, producing, in fermentation, numerous therapeutically relevant natural products. In the pharmaceutical and biotechnology industries, they continue to be the focus of many natural product screening programs, especially in the search for novel antibiotics active against

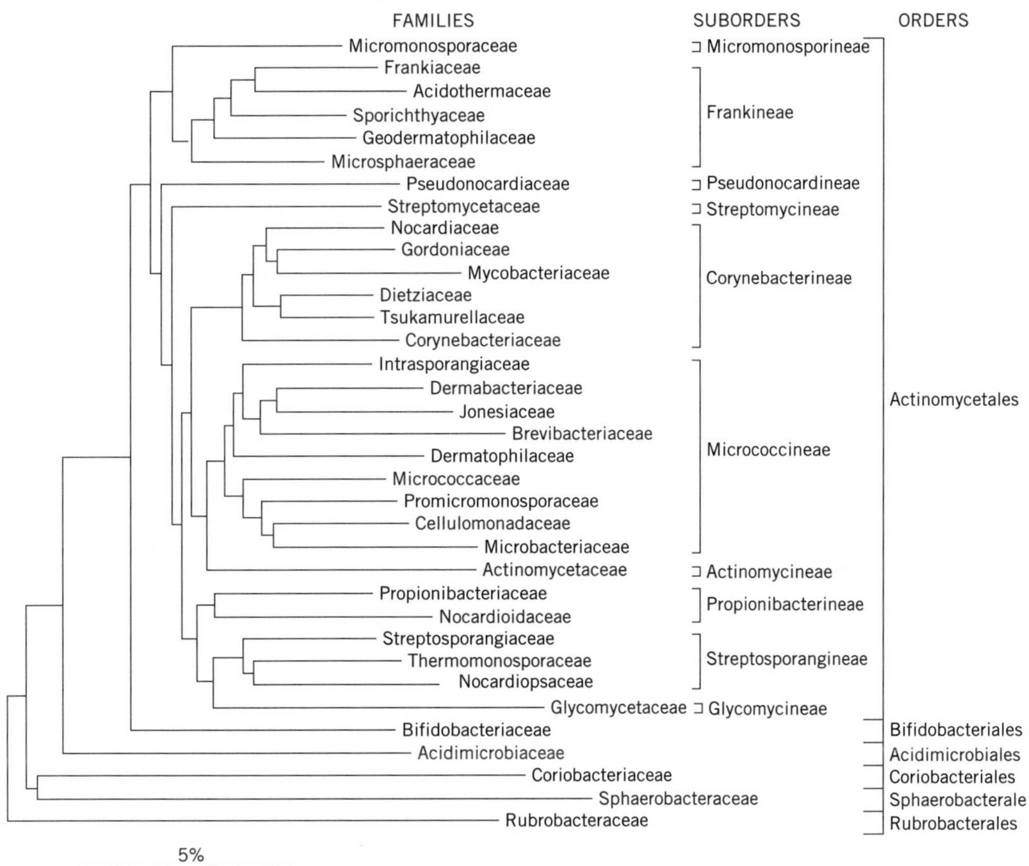

Figure 1. Relatedness of actinobacteria showing the presence of 6 orders and 10 suborders of the Actinomycetales based on 16S rDNA/rRNA sequence comparison. The scale bar represents 5 nucleotide substitutions per 100 nucleotides. *Source*: From Stackebrandt et al. (2).

Table 1. Morphological Characteristics of Members of the Families Streptosporangiaceae, Nocardiopsaceae, Thermomonosporaceae, Pseudonocardiaceae, Micromonosporaceae, and Glycomycetaceae

Family/genus	Substrate mycella	Aerial mycelia	Spore chains	Spores
		Streptosporangiaceae		
Streptosporangium	Branching, nonfragmenting; yellow, brown, orange, red	Present; white, pink to greenish-gray; spore vesicles borne on long or short sporangiophores, singly or clustered	Spore vesicles contain a coiled chain of arthrospores formed by septation of an unbranched, spiral hyphae within an expanded sporangiophore sheath.	Nonmotile; smooth, spherical, oval or rod-shaped
Herbidospora	Branching, nonfragmenting; colorless, tan to yellow-brown	Absent; spore chains, in mass, white	Straight, short chains, 10 to 30 spores, borne on tips of branching sporophores, in clusters, directly from the vegetative mycelium	Nonmotile; smooth, oval
Microbispora[a]	Branching, nonfragmenting; yellow, brown, orange, violet	Present; pink to white; bearing longitudinal pairs of spores often closely arranged along the hyphae, either sessile or on short sporophores	Spores in pairs	Nonmotile; smooth, spherical to oval
Microtetraspora	Branching, nonfragmenting; yellow, brown, orange, violet	Present; white, pink, blue-gray; short, sparsely branched	Spores in chains of 4	Nonmotile; smooth, warty, spherical to oval
Planobispora	Branching, nonfragmenting; colorless or rose	Present; white to light rose; bearing cylindrical to clavate sporangia, singly or in bundles, on short sporangiophores	Longitudinal pairs of spores in sporangia	Motile; peritrichous flagella; smooth, straight to slightly curved
Planomonospora	Branching, nonfragmenting; grayish-yellow, pink, reddish, orange	Present; white to rose; bearing cylindrical to clavate sporangia, sessile or on sporangiophores	Single spore in each sporangium	Motile; peritrichous flagella; straight to curved.
		Nocardiopsaceae		
Nocardiopsis	Branching, fragmenting into rods and coccal elements; yellow, yellowish-brown, olive	Present; white, yellowish-gray; long, moderately branched; straight to flexous, zigzag fragmentation	Fragmenting into long (greater than 50) chains of spores	Nonmotile; smooth, irregular
		Thermomonosporaceae		
Thermomonospora[b]	Branching, nonfragmenting; colorless to pale yellow	Present; white; differentiates into a single, heat sensitive spore	Single, sessile, or on the end of short branched or unbranched sporophores	Nonmotile; smooth to ridged
Actinomadura	Branching, nonfragmenting; colorless to yellow, orange, brown, red-brown, red, violet, gray-green	Present; white, yellow, pink, blue, green, gray; branching, carrying short to long chains of arthrospores	Straight, flexous, hooked, open loops, irregular spirals of 1 to 4 turns	Nonmotile; folded, irregular, smooth, spiny, or warty
Spirillospora[c]	Branching, nonfragmenting; white, pale yellow, pale pink, red to reddish brown	Present; white; bearing spherical sporangia on sporangiophores	Each spore vesicle contains coiled, branched aerial hyphae that differentiate into spores.	Motile, 1 to 7 subpolar flagella; smooth, rod-shaped, curved

		Micromonosporaceae		
Micromonospora	Branching, well developed; tan to orange, brown, blue-green, purple	Absent or a gray to white bloom	Borne singly, sessile or on short sporophores, in branched clusters	Nonmotile; smooth to warty, oval
Actinoplanes	Branching, nonfragmenting; tan to red, brown, violet	Absent or scanty, sporangia, spherical, subspherical, clavate, irregular, arising from the substrate mycelia, sessile or on short sporangiophores	Within the sporangia	Motile, polar flagella; spherical or short rods
Catellatospora	Branching, nonfragmenting; tan, yellow, mustard gold, orange, reddish brown.	Absent	Short, straight to flexous, branched chains of spores directly from the substrate mycelia	Nonmotile; cylindrical, ovoid, smooth to slightly rough
Couchioplanes	Branching, nonfragmenting; dark blue	Present; branching	Short spore chains in irregular spirals, directly from the substrate mycelia or on short aerial mycelia	Motile, polar flagellation; oval, short rods, smooth
Catenuloplanes	Branching, nonfragmenting; orange, red, brown	Present; sparse, branching	Short spore chains in spirals of 1 to 2 turns, from the substrate or aerial mycelia	Motile, peritrichous flagella; rod-shaped, straight to curved, smooth
Dactylosporangium	Branching, nonfragmenting; orange, rose, or wine.	Absent; sporangia present, borne on short sporangiospores, finger-shaped to claviform	Within the sporangia a single row of 3 to 4 spores	Motile, polar flagella; oblong, ovoid
Pilimelia	Branching, nonfragmenting; pale lemon-yellow, golden yellow, orange	Absent; sporangia present, spherical ovoid, pyriform, cylindrate, borne on sporangiospores	Within the sporangia, in parallel or swirl rows, and as conidia, in chains	Motile, sporangiospores, laterally inserted tuft of flagella; nonmotile, rod-shaped conidiospores
Spirilliplanes	Present, nonfragmenting, branching, yellow to orange	Absent	Short chains of spores arranged in spirals and clusters directly from the substrate mycelia	Motile, oval to short rods, smooth, flagella type not given
		Pseudonocardiaceae		
Pseudonocardia	Zigzag fragmentation and long chains of spore-like structures; mycelial wall with a characteristic electron dense outer layer	Present; white; fragmenting	Spore chains long, 50 or more per chain	Nonmotile; smooth, spiny, irregular
Actinopolyspora	Branching, occasionally fragments; buff	Present; white, with an outer sheath	Long chains of spores	Nonmotile; smooth
Actinosynnema	Fragmenting	Present; form synnemata (dome-like structures)	Chains of arthrospores at the tip of the synnemata	Motile arthrospores (flagellation not given)
Amycolatopsis	Fragments into long chains of squarish to ellipsoid structures	Rarely formed; when present, white; may fragment into chains of spores	Chains of squarish to oval fragments	Nonmotile; oval, smooth, warty
Kibdelosporangium	Zigzag fragmentation	Present; white; sporangia-like structures that do not contain spores but germinate when placed on agar	Long chains	Nonmotile; smooth, irregular

Table 1. Morphological Characteristics of Members of the Families Streptosporangiaceae, Nocardiopsaceae, Thermomonosporaceae, Pseudonocardiaceae, Micromonosporaceae, and Glycomycetaceae *(continued)*

Family/genus	Substrate mycella	Aerial mycelia	Spore chains	Spores
Kutzneria	Branching, nonfragmenting	Present; white, olive gray; branching with globose sporangia	Long chains formed by separation of coiled, unbranched hyphae	Nonmotile; spherical, rod-shaped or oval
Lentzea	Branching, nonfragmenting; yellow to yellow brown	Present; white to whitish-yellow; fragmenting into rod-shaped elements	Chains of rod-like structures	Nonmotile; smooth, rod-like
Saccharomonospora	Branching, rarely fragments; beige, lilac, green	Present; white becoming green; with an outer sheath	Single densely packed, on the aerial mycelia	Nonmotile; oval, smooth, warty
Saccharopolyspora	Fragmenting; spores often present	Present; white, pink to brownish gray, often sparse; with an outer sheath	Long and short chains	Nonmotile; spore sheath ornamentation hairy, spiny, smooth
Saccharothrix	Zigzag fragmentation; yellow to yellow brown	Present; white to yellowish-gray, zigzag fragmentation	Chains of ovoid to irregular elements	Nonmotile; smooth, ovoid
Streptoalloteichus	Branching, bearing spore-like vesicles containing one to several spores	Present; spore chains in clusters and sclerotia	Spore chains on the aerial mycelia, sporangiospores on the substrate mycelia	Motile sporangiospores by a single long polar flagellum
Thermocrispum	Branching, nonfragmenting; yellow to light brown	Present; white; aggregating into clusters that fragment into rod-like structures	Chains of rod-like structures	Nonmotile; smooth, rod-like
		Glycomycetaceae		
Glycomyces	Branching, nonfragmenting; pale yellow to tan	Present; white; forming spores	Short chains	Nonmotile; square-ended
		Genera of Uncertain Affiliation		
Actinobispora	Branching, nonfragmenting; bearing paired spores; light orange	Present; white; sparse; bearing paired spores	Longitudinal pairs	Nonmotile; smooth, spherical
Actinocorallia	Branching, nonfragmenting; ivory to pale yellow	Absent	Coralloid sporophores arising from the substrate mycelia bearing long chains (30+) of spores	Nonmotile; smooth, cylindrical
Actinokineospora	Branching, nonfragmenting; (on agar); colorless to yellow brown	Present; white; bearing chains of spores in a sheath	Irregularly curved; less than 50 per chain	Motile, peritrichous flagella; smooth, ovoid to squarish
Planotetraspora	Branching, nonfragmenting; ivory	Present; whitish gray; cylindrical sporangia on short sporangiophores	Single row of 4 spores per sporangium	Motile, a single polar flagellum
Thermobispora	Branching, nonfragmenting; yellow to yellow brown	Present; white; monopodially branched; bearing pairs of spores	In longitudinal pairs on the aerial hyphae	Nonmotile; smooth, spherical to oval

[a] *Microbispora bispora* has been removed from the genus to form a new, thermophilic single species genus, *Thermobispora* (7).
[b] The genus *Thermomonospora*, as defined by Kroppenstedt and Goodfellow (8), contains two species, *T. curvata* and *T. formosensis*.
[c] The genus *Spirillospora* was last placed in the family Streptosporangiaceae (9) based on 16S rDNA but has now been moved to the family Thermonomosporaceae (2).

Table 2. Cell Wall Types and Corresponding Whole-Cell Sugar Patterns of Aerobic Actinomycetes Containing *meso*-Diaminopimelic Acid

Cell Wall		Whole-cell sugar pattern		
Type	Distinguishing major constituents	Type	Diagnostic sugars	Representative genera
II	Glycine	D	Xylose, arabinose	*Micromonospora*
III	None	B	Madurose[a]	*Actinomadura*
		C	None	*Nocardiopsis*
IV	Arabinose, galactose	A	Arabinose, galactose	*Amycolatopsis*

Source: According to Lechevalier and Lechavalier (13).
[a]Madurose is 3-*O*-methyl-D-galactose.

Muramic acid
↓
1 L-Alanine (Gly, L-Ser)
↓
2 D-Glutamic Acid (D-AlaNH$_2$, Gly, GlyNH$_2$)
↓
3 *meso*-Diaminopimelic acid
(LL-Dap, L-Ala, L-Dab, L-Glu, L-Lys, L-Orn, L-Hsr)
↓
4 D-Alanine

Figure 2. The peptide stem of the actinomycete (and other Gram-positive bacteria) peptidoglycan and the possible variations in amino acid content (in parentheses). Ala, alanine; Dab, diaminobutyric acid; Glu, glutamic acid; Gly, glycine; Hsr, homoserine; Lys, lysine; Orn, ornithine; Ser, serine. *Source*: From Suzuki et al. (15).

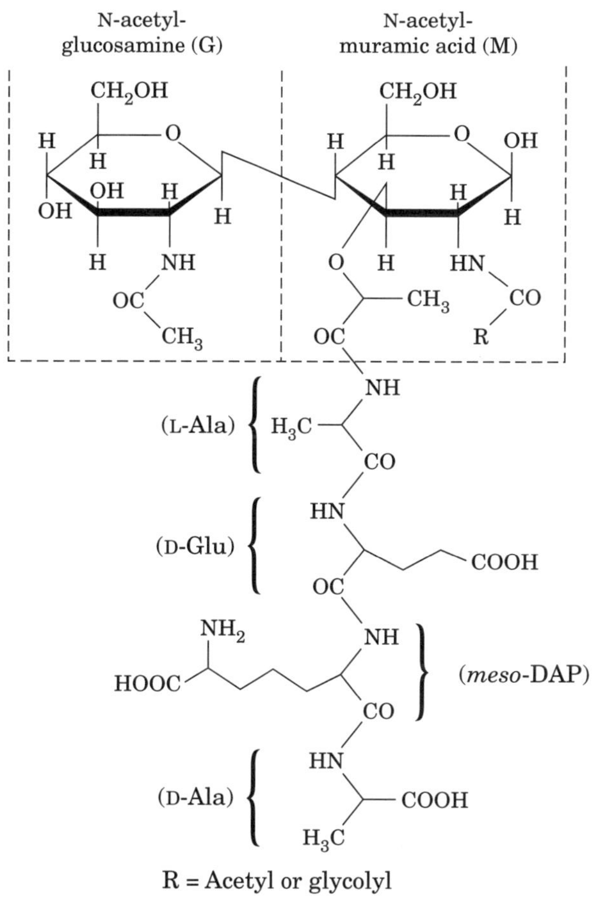

Figure 3. Complete structure of a single subunit of the peptidoglycan showing the linkage between the two amino sugars that make up the glycan strand and between muramic acid and the four amino acids in the peptide. Also indicated is the position of acetyl/glycolyl on muramic acid. *Source*: Modified from Suzuki et al. (15).

highly resistant, infectious Gram-positive pathogens. In an actinomycete-based natural product screening program, the organisms must be isolated from the environment, grown in pure state, preserved, and fermented, and the fermentation products must be tested for biological activity. Essential to this process is a knowledge of actinomycete ecology, taxonomy, and physiology, all of which contribute to the isolation of new strains with the potential of producing novel compounds.

Actinomycetes are Gram-positive, filamentous bacteria with a mol % G + C content of DNA greater than 50%. They can be divided into two broad groups: the fermentative organisms that are found in the natural cavities of man and animals and the larger group of oxidative organisms found in soil (1). The majority of oxidative soil actinomycetes can be further subdivided into two large groups based on the isomer of diaminopimelic acid in the peptidoglycan: the streptomycetes, containing LL-diaminopimelic acid (DAP) and the remainder containing *meso*-DAP. Further subdivisions to family, genus, and species relies on analysis of actinomycete morphology, determination of chemical constituents of cell walls and whole cells, biochemical characteristics, and resistance patterns. Sequence data generated from the analysis of small subunit RNA (16S) have been applied to the understanding of actinomycete phylogeny as well as molecular identification. A new hierarchical classification based on 16S rDNA/rRNA sequences (Fig. 1), proposed by Stackebrandt et al. (2), will be followed as phenotypic data are presented. Aerobic, saprophytic, filamentous actinomycetes containing *meso*-DAP in their cell

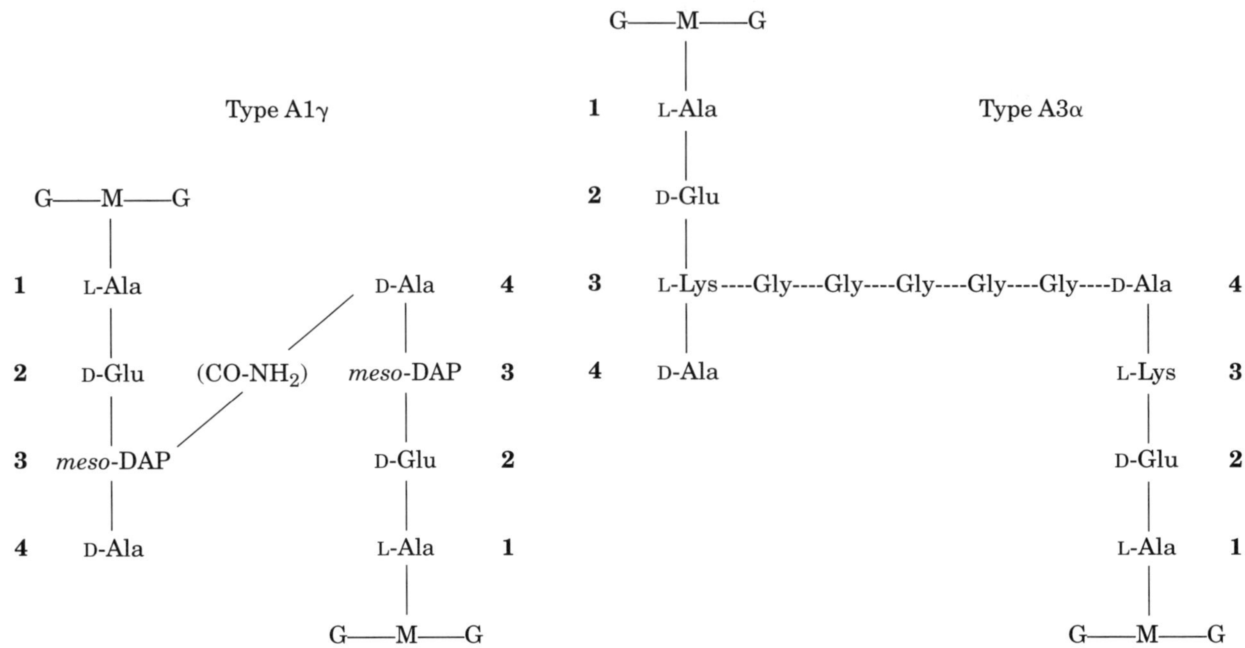

Figure 4. Example of a directly cross-linked *meso*-DAP–containing peptidoglycan of type A (cross-linkage between positions 3 and 4 of two peptide subunits), 1 (no peptide bridge), and γ (*meso*-DAP in position 3), and an example of cross-linking through a peptide bridge, A (as above), 3 (the bridge is an oligopeptide), and α (L-lysine in position 3. Ala, alanine; Dap, diaminopimelic acid; G, *N*-acetylglucosamine; Glu, glutamic acid; Gly, glycine; Lys, lysine; M, *N*-acetyl- or *N*-glycolylmuramic acid. *Source*: Modified from Suzuki et al. (15).

walls, which have not been placed in any of the actinomycete families defined by 16S rDNA sequence data, including Actinobispora (3), Actinocorallia (4), Actinokineospora (5), Planotetraspora (6), and Thermobispora (7) will also be described. The suborders Actinomycineae, Micrococcineae, Corynebacterineae, Propionibacterineae and Frankineae (2) will not be discussed in this article.

IDENTIFICATION OF NONSTREPTOMYCETE ACTINOMYCETES

Morphological Characterization

Meso-DAP–containing soil actinomycetes have a complex morphology. When grown on agar, a substrate mycelia forms that may penetrate the agar and that can bear spores, sporelike structures, or sporangia. Aerial mycelia may arise from the substrate mycelia and can be branching. The aerial mycelia may be in a sheath or in a sporangium and may fragment into motile or nonmotile spores. Morphological characteristics can be observed by growing the organisms on minimal media, for example, water agar (crude agar, 15 g; tap water, 1,000 mL), dilute Bennett's agar (agar, 15 g; glucose, 1 g; pancreatic digest of casein, 0.2 g; yeast extract, 0.1 g; beef extract, 0.1 g; distilled water, 1,000 mL), or dilute yeast–dextrose agar (agar, 15 g; yeast extract, 1 g; dextrose, 1 g; tap water, 1000 mL). After 7–28 days of incubation at 28 °C, plates can be examined directly under a light microscope using a long working distance, 40× objective. Detailed morphological analysis, such as spore and sporangiophore ornamentation, location and attachment of flagella on motile spores, and sporangial structures and spore formation within the sporangia will require either scanning or electron microscopic observations. In some cases, actinomycetes can be identified to genus based on their characteristic morphology (Table 1).

Chemical Characterization

Although the majority of actinomycetes readily form morphological structures, many do not, making preliminary identification based on this characteristic impossible. The laboratories of Huber and Mary Lechevalier at the Waksman Institute, Rutgers University, New Jersey, pioneered the extensive use of chemical markers to delineate genera of actinomycetes. The first large-scale, systematic application of chemical criteria to taxonomy focused on the analysis of cell-wall and whole-cell sugars and amino acids. Representatives of all the described actinomycete genera were analyzed, and distinct patterns were discerned (11,12) (Table 2). In conjunction with the data generated by Schleifer and Kandler (14) delineating the mode of cross-linking between subunits of the Gram-positive peptidoglycan (Figs. 2 to 4), a clearer understanding of the relationship of wall structure to wall chemotype was obtained (15) (Tables 3 and 4). The presence of either glycolyated or acetylated muramic acid in the cell wall (17,18) and lipids such as mycolic acids (19), phospholipids (20), menaquinones (21), and fatty acids (22,23) resulted in additional taxonomic discrimination. Phospholipid patterns are presented in Table 5, and the structure of typical ac-

Table 3. Peptidoglycan Classification

Position of cross-link	Peptide bridge	Amino acid at position 3
	Peptidoglycan A	
Cross-linkage between positions 3 and 4 of two peptide subunits	1. None	α-L-Lysine β-L-Ornithine γ-*meso*-DAP
	2. Polymerized subunits	α-L-Lysine
	3. Monocarboxylic L-amino acid or glycine or oligopeptides	α-L-Lysine β-L-Ornithine γ-LL-DAP
	4. Contains a dicarboxylic amino acid	α-L-Lysine β-L-Ornithine γ-*meso*-DAP δ-L-DAB
	Peptidoglycan B	
Cross-linkage between positions 2 and 4 of two peptide subunits	1. Contains an L-amino acid	α-L-Lysine β-L-Homoserine γ-L-Glutamic acid δ-L-Alanine
	2. Contains a D-amino acid	α-L-Ornithine β-L-Homoserine γ-L-DAB

Source: From Suzuki et al. (15) based on Schleifer and Kandler (14).
Note: DAB, diaminobutyric acid.

Table 4. Relationship Between Wall Chemotype and Peptidoglycan Type in Actinomycetes

Wall type[a]	Major wall amino acids[a]	Other distinguishing characteristics[b]	Peptidoglycan type[c]	Representative genera
I	Glycine, LL-DAP	None	A3γ	*Streptomyces*
II	Glycine, *meso*- or OH-DAP	None	A1γ	*Actinoplanes* *Micromonosporaa*
III	*meso*-DAP	Madurose in whole cells	A1γ	*Actinomadura* *Streptosporangium*
IV	*meso*-DAP	Arabinose and galactose in cell wall	A1γ	*Amycolatopsis* *Nocardia*
V	Lysine, ornithine	None	A5β	*Actinomyces*
VI	Lysine (aspartic acid, var.)	Galactose (var.)	A3α, A4α	*Couchioplanes, Oerskovia*
VII	Diaminobutyric acid, glycine (lysine, var.)	None	B2α	*Agromyces*
VIII	Ornithine	None	B2β	*Bifidobacterium*

Source: From Minnikin and Goodfellow (16).
Note: var. = variable
[a]Major constituents of actinomycete cell walls according to Lechevalier and Lechevalier (13).
[b]All cell wall preparations contain major amounts of alanine, glutamic acid, glucosamine, and muramic acid.
[c]From Schleifer and Kandler (14).

tinomycete fatty acids and the patterns obtained from their analysis are presented in Tables 6 and 7. Chemotaxonomic markers aid in the identification of actinomycetes to the genus level. Further discrimination to the species level can be achieved using physiological tests, including utilization of various carbon and nitrogen sources; hydrolysis of complex carbohydrates and macromolecules; and resistance to chemicals, including antibiotics (24,25). Sequence analysis, particularly of the 16S rDNA, may play a significant role in future definitions of actinomycete species. However, because of the highly conserved nature of small subunit RNA, very high similarity levels (greater than 97%) among related strains will be observed. In order to delineate species within these groups, DNA–DNA reassociation may be required (26).

The chemotaxonomic characteristics for each genus of the families described in the next section are presented in Table 8.

CHARACTERISTICS OF ACTINOMYCETE FAMILIES

Streptosporangiaceae

Streptosporangiaceae (9,27) are aerobic, Gram-positive, non-acid-fast, chemoorganotrophic actinomycetes that form stable substrate and aerial mycelia. Aerial mycelia

Table 5. Phospholipid Patterns of Aerobic Actinomycetes

```
             CH2 – OO C – R′
              |
R – COO – CH          O
              |          ||
             CH2 – O – P – OY
                         |
                        OH
```

R, R′ = Long chain alkyl

Y = Polar head groups, PG, DPG,[a] PI,[a] PIM, PE, PC, PME

Phospholipid Pattern	Phosphatidyl-ethanolamine (PE)	Phosphatidylmethyl-ethanolamine (PME)	Phosphatidyl-choline (PC)	Phosphatidyl-glycerol (PG)	GluNU[a]
PI	Absent	Absent	Absent	Variably present	Absent
PII	Present	Absent	Absent	Absent	Absent
PIII	Variably present	Variably present	Present	Variably present	Absent
PIV	Variably present	Variably present	Absent	Absent	Present
PV	Absent	Absent	Absent	Present	Present

Source: According to Lechevalier et al. (20).
[a]DPG, diphosphatidy glycerol; PI, phosphatidylinositol; GluNU, phospholipids of unknown structure containing glucosamine.

Table 6. Structure of Representative Actinomycete Fatty Acids

Fatty acid type	Common example and structure
Straight-chain	*Hexadecanoic* $CH_3\text{-}(CH_2)_{14}\text{-}COOH$
Cis-unsaturated	*Oleic* $CH_3\text{-}(CH_2)_7\text{-}CH{=}CH\text{-}(CH_2)_7\text{-}COOH$ *cis*
Tuberculostearic	*10-Methyloctadecanoic* CH_3 | $CH_3\text{-}(CH_2)_7\text{-}CH\text{-}(CH_2)_8\text{-}COOH$
Cis-cyclopropane	*11,12-Methyleneoctadecanoic* CH_2 /\\ $CH_3\text{-}(CH_2)_5\text{-}CH\text{-}CH\text{-}(CH_2)_9\text{-}COOH$ *cis*
Iso	*13-Methyltetradecanoic* CH_3 | $CH_3\text{-}CH\text{-}(CH_2)_{11}\text{-}COOH$
Anteiso	*12-Methyltetradecanoic* CH_3 | $CH_3\text{-}CH_2\text{-}CH\text{-}(CH_2)_{10}\text{-}COOH$

Source: From Minnikin and O'Donnell (16).

may carry chains of two or more spores or be differentiated into sporangia (spore vesicles) containing one, two, or more spores. Spores are either motile or nonmotile. Cell walls contain *meso*-DAP and N-acetylated muramic acid, but lack characteristic sugars; whole cells contain madurose. The peptidoglycan is of the A1γ type, having a cross-linkage between positions 3 and 4 of adjacent peptide subunits, with *meso*-DAP at position 3 of the peptide stem (14). Lipid profiles contain saturated, *iso*- and *anteiso*-fatty acids; major proportions of tetrahydrogenated menaquinones, with nine isoprene units saturated at sites III and VIII; and glucosamine-containing polar lipids. Mycolic acids are absent. The mol % G + C content of the DNA ranges from 66 to 74.

Nocardiopsaceae

Nocardiopsaceae (28,29) are aerobic, Gram-positive, non-acid-fast, chemoorganotropic actinomycetes that form well-developed, branching substrate mycelia that may fragment into coccoid to bacillary forms. Aerial mycelia are long and moderately branching, with zigzag fragmentation into irregular spores. The family is defined on the basis of chemotaxonomic markers because the morphological characteristics are the same for the genus *Saccharothrix* (family Pseudonocardiaceae). Cell walls contain *meso*-DAP, N-acetylated muramic acid, and no characteristic sugars. The peptidoglycan is A1γ. Lipid profiles contain phosphatidylcholine and phosphatidylmethylethanolamine; major portions of variably saturated menaquinones with 10 isoprene units; *iso*-branched, *anteiso*-branched, and 10-methyl-branched fatty acids; and high levels of octadecanoic acid. The combination of 15 to 20% *anteiso*-C17:0 (14-methylhexadecanoic acid) with 20 to 25% 10-methyl-C18:0 (tuberculostearic acid) or its precursor, oleic acid, is diagnostic and unique among bacteria. Mycolic acids are absent. The G + C content of the DNA is 64 to 69 mol %. The family contains only one genus, *Nocardiopsis*.

Thermomonosporacea

Thermomonosporacea (8) are aerobic, Gram-positive, non-acid-fast, chemoorganotrophic actinomycetes producing branched substrate mycelia bearing aerial mycelia that differentiate into single or short chains of arthrospores. Cell walls contain *meso*-DAP, N-acetylated muramic acid, and no characteristic sugars. The peptidoglycan type is

Table 7. Fatty Acid Types

Type	*Iso-* 15:0	*Anteiso* 15:0	*Iso-* 16:0	16:0	10-Methyl 16:0	*Iso-* 17:0	*Anteiso-* 17:0	17:0	17:1	10 Methyl 17:0	*Iso-* 18:0	18:0	18.1	10-Methyl 18:0	OH_2	Cyclo-19
1a	–	–	–	+ + +	–	–	–	–	–	–	–	+	+ + +	–	–	–
1b	–	–	–	+ + +	+ +	–	–	–	–	+	–	+	+ +	+ + +	–	–
1c	–	–	–	+ + +	–	–	–	–	–	–	–	+	+ +	–	–	+ +
2a	+	–	+ + +	+ +	–	+	+	+	+	–	+	–	–	–	+ +	–
2b	+ + + +	+ +	+ +	+	–	+ + +	+	–	–	–	–	–	–	–	–	–
2c	+ +	+ + +	+ + +	+	–	+	+ +	–	–	–	–	–	–	–	(V)	–
2d	+ + +	+ +	+ + + +	+ +	–	+ +	+ +	+	+	–	+	+	+ +	–	–	–
3a	–	–	+ +	+ + +	+	–	–	+	+	+	+	+	+ +	+ +	–	–
3b	+ + + +	+	+ +	–	–	+ +	+ +	+	+ +	+	–	–	+	–	–	–
3c	+ +	+	+ + +	+	+	+	+	+ +	+	+ +	+	+	–	–	+ +	–
3d	+	+	+ + +	+	–	+	+ + +	+	+	+ +	+	+	+ +	+ +	–	–
3e	+ +	–	+ + + +	+	–	+	+	–	+	+	–	–	–	–	–	–
3f	+ +	+	+ + + +	+	–	+	+	+ +	+ +	+	–	–	–	–	+	–
3g	+ +	+	+ + + +	+	–	+	+ +	+	+	+	–	–	–	–	+	–

Source: From Kroppenstedt (23).
Note: + = 1.5%; + + = 5–15%, + + + = 15–25%, + + + + = 25%; (V) = variable, usually less than 2% for one component.

A1γ. Whole cells contain madurose. Lipid profiles contain mixtures of straight- and branched-chain fatty acids; hydrogenated menaquinones with nine isoprene units; and major amounts of phosphatidylglycerol, phosphatidylinositol, and phosphatidylmannosides, some strains contain minor amounts of phosphatidylglycerol. The G + C content of the DNA is mol % 66 to 72.

Micromonosporaceae

Micromonosporaceae (10,30,31) are aerobic, Gram-positive, and non-acid-fast, forming nonfragmenting, branching substrate mycelia that, in the genus *Micromonospora*, form nonmotile spores, singly either sessile or on short sporophores. Aerial mycelia are rarely formed, but when present are scanty. Sporangia, which may be present and can contain two to five or numerous spores, are formed on sporangiophores arising directly from the substrate mycelia. Globose, subglobose, or rod-shaped spores are formed in the sporangia or in chains arising from the substrate mycelia. In water the spores become motile by means of polar or lateral tufts of flagella. Cell walls contain *meso-* or 3-hydroxy-DAP, or both. The peptidoglycan is of the A1γ type; glycine is the first amino acid (acetate in *Pilimelia*) in the stem peptide attached to a glycolated muramic acid. Whole-cell hydrolysates contain arabinose and xylose. Lipid profiles have saturated, *iso-*, and *anteiso*-fatty acids, and diphosphatidylglycerol, phosphatidylethanolamine, phosphatidylglycerol, and phosphatidylinositol as characteristic phospholipids; menaquinones are of varying lengths. Mycolic acids are absent. The mol % G + C content of the DNA ranges from 71 to 73.

Pseudonocardiaceae

Pseudonocardiaceae (32,33) are a heterogeneous grouping of organisms based on 16S rDNA/rRNA sequence data. Aerobic, Gram-positive, mesophilic or thermophilic, and chemoorganotrophic, one genus is halophytic. Heterogeneous morphologies include stable or fragmenting substrate and aerial mycelia; single or short chains of nonmotile spores on both the aerial and substrate mycelia; sporangial-like bodies not containing spores; and long chains of interwoven spores with motile spores at the tip. The peptidoglycan contains *meso*-DAP and is acetylated; many genera contain arabinose and galactose in their cell walls and whole cells, others do not. Rhamnose in combination with galactose in whole-cell hydrolysates appears to be diagnostic for some genera. Lipid profiles include mono-methyl-branched, straight-chain saturated and unsaturated, and hydroxy fatty acids and are genera specific; tetrahydrogenated menaquinones with 8, 9, or 10 isoprene units; phospholipid patterns II, III and IV are represented; mycolic acids are not present. The mol % G + C content of the DNA ranges from 64 to 79.

Glycomycetaceae

Glycomycetaceae (34) are aerobic, Gram-positive, non-acid-fast, and chemoorganotrophic, forming stable substrate mycelia and aerial mycelia that fragment into short chains of square-ended spores. The cell wall contains *meso*-DAP and glycine and N-glycolated muramic acid; the whole-cell sugar pattern is xylose and arabinose. The peptidoglycan is A1γ. Lipid profiles contain diphosphatidylglycerol and phosphatidylinositol; nitrogenous phospholipids are not found, nor are mycolic acids; tetra- and hexahydrogenated menaquinones with 10 isoprene units predominate. The mol % G + C content of the DNA is 71. The family contains only one genus, *Glycomyces*.

SCREENING ACTINOMYCETES FOR NATURAL PRODUCTS

Isolation from Natural Habitats

Studies have been performed (35–38) to discover appropriate conditions for isolating various groups of actinomycetes from natural habitats, thereby enhancing diversity and increasing the chances of discovering novel natural products. A testing paradigm was described by Horan (39) that evaluated the effect of soil type, media

Table 8. Chemotaxomic Characteristics of Members of the Families Streptosporangiaceae, Nocardiopsaceae, Thermomonosporaceae, Micromonosporaceae, Pseudonocardiaceae, and Glycomycetaceae

Family/genus	Diagnostic whole-cell sugars	Diagnostic cell-wall components[a]	Acyl type of muramic acid	Menaquinones	Phospholipid type	Fatty acid type[b]	mol % G+C
			Streptosporangiaceae				
Streptosporangium	Madurose	None	Acetyl	MK-9($H_{0,2,4}$)	IV	3c	69–71
Herbidospora	Madurose	None	Acetyl	MK-10($H_{2,4,6}$)	IV	3c	69–71
Microbispora	Madurose	None	Acetyl	MK-9($H_{0,2,4}$)	IV	3c	67–74
Microtetraspora	Madurose	None	Acetyl	MK-9($H_{0,2,4}$)	IV	3c	64–69
Planobispora	Madurose	None	Acetyl	MK-9($H_{0,2,4}$)	IV	3c	70–71
Planomonospora	Madurose	None	Acetyl	MK-9($H_{0,2,4}$)	IV	3c/a	72
			Nocardiopsaceae				
Nocardiopsis	None	None	Acetyl	MK-10($H_{4,6}$)	III	3d	64–69
			Thermomonosporaceae				
Thermomonospora	+/− Madurose	None	Acetyl	MK-9($H_{2,4,6,8}$)	I	3a/c	72
Actinomadura	Madurose	None	Acetyl	MK-9($H_{4,6,8}$)	I	3a	66–70
Spirillospora	Madurose	None	Acetyl	MK-9($H_{4,6}$)	I/II	3a	71–73
			Micromonosporaceae				
Micromonospora	Arabinose Xylose	OH-DAP Glycine	Glycolyl	MK-9(H_4) MK-10($H_{4,6}$) MK-12	II	3b	71–73
Actinoplanes	Arabinose Xylose +/− Galactose	OH-DAP Glycine	Glycolyl	MK-9(H_4) MK-10(H_4)	II	2d/2c	72–73
Catellatospora	Arabinose Xylose	OH-DAP Glycine 3-*O*-Methyl-rhamnose	Glycolyl	MK-10($H_{2,6}$) MK-9($H_{4,6}$)	III	2c	70.6 71.5
Couchioplanes	Arabinose Xylose Galactose	L-Lysine glycine Serine	Glycolyl	MK-9($H_{4,6}$)	II	2c	69.9 72.1
Catenuloplanes	Xylose +/− Galactose	L-Lysine Serine Glycine	Glycolyl	MK-9(H_8) MK-10(H_8)	III	2c	71–72
Dactylosporangium	Arabinose Xylose	Glycine	Glycolyl	MK-9($H_{2,6,8}$)	II	3b/2d	72–73
Pilimelia	Arabinose Xylose	Glycine	Glycolyl	MK-9($H_{2,4}$)	II	2d/2b	NR
Spirilliplanes	3-*O*-Methylmannose Xylose Galactose	Glycine	Glycolyl	MK-10(H_4)	II	2d	69
			Pseudonocardiaceae				
Pseudonocardia	Arabinose Galactose	Arabinose	Acetyl	MK-8(H_4)	III	3f/3e	68–79
Actinopolyspora	Arabinose Galactose Ribose	Arabinose Galactose	Acetyl	MK-9($H_{4,6}$)	III	2e/2c	64
Actinosynnema	None	None	Acetyl	MK-9($H_{4,6}$)	II	3d	71–73
Amycolatopsis	Arabinose Galactose	Arabinose Galactose	Acetyl	MK-9($H_{2,4}$)	II	3f	66–69
Kibdelosporangium	Arabinose Galactose tr-Madurose Rhamnose	Trace galactose	Acetyl	MK-9(H_4)	II	3f	66
Kutzneria	Galactose Rhamnose	None	Acetyl	MK-9(H_4)	II	3c/3e	70.3 70.7
Lentzea	None	None	Acetyl	MK-9($H_{0,2}$)	II	3d	68.6
Saccharomonospora	Arabinose Galactose	Arabinose Galactose	Acetyl	MK-9(H_4) MK-8(H_4)	II	2a	66–70

Table 8. Chemotaxomic Characteristics of Members of the Families Streptosporangiaceae, Nocardiopsaceae, Thermomonosporaceae, Micromonosporaceae, Pseudonocardiaceae, and Glycomycetaceae *(continued)*

Family/genus	Diagnostic whole-cell sugars	Diagnostic cell-wall components[a]	Acyl type of muramic acid	Menaquinones	Phospholipid type	Fatty acid type[b]	mol % G+C
Saccharopolyspora	Arabinose Galactose	Arabinose Galactose	Acetyl	MK-9($H_{2,4}$) MK-10(H_4)	III	3c	70–71.5
Saccharothrix	Galactose Rhamnose	None	Acetyl	MK-9(H_4)	II	3f	70–73
Streptoalloteichus	Galactose Rhamnose	Galactose	Acetyl	MK-9(H_6) MK-10(H_6)	II	NR	NR
Thermocrispum	Arabinose Trace galactose	None	Acetyl	MK-9(H_4)	II	3f	69–73
			Glycomycetaceae				
Glycomyces	Xylose Arabinose	Glycine	Glycolyl	MK-10($H_{2,4}$)	I	2c	71–73
			Genera of Uncertain Affiliation				
Actinobispora	Arabinose Galactose Xylose	None	Acetyl	MK-7(H_2) MK-9(H_2)	IV	NR	71
Actinocorallia	None	None	Acetyl	MK-9($H_{4,6}$)	II	1a	73
Actinokineosporia	Arabinose Galactose Rhamnose	Arabinose Galactose	Acetyl	MK-9(H_4)	II	3c	69–72
Planotetraspora	Arabinose Galactose Xylose Ribose	Glutamic acid	NR	NR	NR	NR	NR
Thermobispora	Madurose Galactose	None	Acetyl	MK-9($H_{0,2,4}$)	IV	3c	70

Note: NR, not reported.
[a]All organisms contain *meso*-DAP.
[b]Growth conditions may affect the profile (15).

constituents, and selective pressure on the isolation of actinomycetes. Resistance to antibiotics and physiological data, derived from taxonomic analysis of the various species of actinomycetes, formed the basis for the choice of selective pressures (antibiotics) and soil isolation media constituents. Soil samples were suspended in distilled water and serially diluted. The suspensions plated onto the surface of agar media containing various carbon and nitrogen sources and antibiotics, including the antifungal agent nystatin at 50 μg/mL. Resulting colonies were isolated to purity and identified microscopically to either genus or broad morphological group. Antibiotics in the agar medium had the most profound effect on the types of actinomycetes isolated, followed by soil type and media constituents. Novobiocin and spectinomycin at 20 μg/mL selectively enhanced the isolation of micromonosporae: up to 90% of isolates belonged to this group. Rifamycin (15 μg/mL), rosaramicin (10 μg/mL), everninomicin (10 μg/mL), and gentamicin (5 μg/mL) significantly reduced the streptomycete population, allowing diverse, *meso*-DAP containing actinomycetes to proliferate. Pretreatment of the soil sample, particularly air drying for 48 hours before plating, and rifamycin in the agar positively affected the isolation of nocardioform (fragmenting aerial and substrate mycelia) actinomycetes. Table 9 summarizes selective isolation techniques for non-streptomycete actinomycetes.

Growth and Fermentation of Isolates

Organisms are isolated from soil plates using sterile toothpicks, streaked onto the surface of a rich medium, usually American Type Culture Collection (ATCC) medium 72 (53); and grown at 28 °C for 5 to 7 days. Isolated colonies are transferred to broth (ATCC medium 172) in 25-mm tubes containing 10 mL of media. After 3 to 5 days' incubation at 28 °C on a rotary shaker at 200 to 250 rpm, the tubes are removed. Sterile glycerol is added to a final concentration of 10%, and 3 mL aliquots are distributed into sterile vials. The contents of a frozen vial are used as the inoculum that initiates the fermentation process. One milliliter of a thawed suspension is added to 10 mL of germination medium in a 25-mm tube stoppered with a Morton closure, and the tube is incubated for 3 to 5 days at 28 °C and shaken (200 to 250 rpm) on rotary shaker. After incubation, 1 mL of the germination tube growth (G1) is transferred into 10 mL of fermentation medium in 25-mm tubes and incubated as above. For primary screening, large numbers of diverse actinomycetes should be fermented in me-

Table 9. Methods for the Isolation of Actinomycetes from Natural Habitats

Genus	Habitat	Method
Streptosporangium *Kurtzneria*	Soil, possibly slightly acidic	Air dry soil, plate on AV agar containing penicillin G (0.8 μg/mL) and/or polymyxin (4 μg/mL) or gentamicin (2–5 μg/mL) (38,40)
Microbispora and *Microtetraspora*	Soil	Air dry soil, dry heat 120 °C/h, treat with 1.5% phenol, plate on AV, MGA-SE, and HV agars alone or on minimal media with rubromycin or streptomycin without phenol treatment (9,41)
Dactylosporangium, *Planobispora*, *Planomonospora*, *Spirillospora*	Soil	Baiting using organic substrates including grass, centrifugation for motile spored organisms, and plating on egg white agar containing nystatin (50 μg/mL) (42–44)
Herbidospora	Soil and plant material	Plant samples desiccated at 28 °C for 1 week, ground in blender with water, suspension incorporated into yeast extract agar containing nystatin and cyclohexamide; pour plates incubated 28 °C >2 weeks (45)
Thermonomonospora	Soil, mesophiles; overheated substances, bagasse, compost, fodders and manure, thermophiles	Sample heated to 100 °C, dilution plate onto media containing rifamycin (12 μg/mL) and/or kanamycin (25 μg/mL) (8,35)
Actinomadura	Soil	Air dry soil, dilution plate on media containing rifamycin (15 μg/mL), mefoxitin (10 μg/mL), or gentamicin (2–5 μg/mL) (38,39)
Micromonospora	Soil, lake muds, sediments	Air dry soil, dilution plate on soluble starch-yeast extract agar plus novobiocin (20 μg/mL), spectinomycin (20 μg/mL), or gentamicin (5–50 μg/mL) (39)
Actinoplanes *Couchioplanes* *Dactylosporangium*	Soil, sediments	Air dry soil, dilution plate on water agar, colloidal chitin agar or HV agar containing 0.1% potassium tellurite (30), or novobiocin (25 μg/mL) (27), or baiting and centrifugation (42)
Pilimelia	Soil, rarely encountered	Baiting using hair (30)
Catenuloplanes	Soil	Dilution plating on starch-casein agar plus nalidixic acid (25 μg/mL), kanamycin (12.5 μg/mL), cefsulodin (5 μg/mL), and kabicidin (6.25 μg/mL) (46)
Spirilliplanes	Soil	Dry heat 120 °C/h, dilution plating onto HV agar (10)
Amycolatopsis	Soil	Dilution plate on AV agar plus vancomycin (1–10 μg/mL) and polymyxin B (5 U/mL) (38); some species isolated from human specimens (47)
Saccharomonospora	Soil, thermophilic strains isolated from leaf litter, manure, compost	Isolate using Anderson Sampler, sedimentation chamber with penicillin G and polymyxin (5 μg/mL) added to half-strength nutrient agar (35)
Saccharopolyspora	Soil and decaying plant material	Dilution plate on yeast extract (1 g/L), glucose (1 g/L), agar (15 g/L) plus novobiocin (5 μg/mL) (A. Horan, unpublished data)
Kibdelosporangium	Soil	Ampicillin (4 μg/mL) and nalidixic acid (10 μg/mL) in complex medium incubated at 43 °C (48)
Saccharothrix	Soil	AV or starch-casein agars plus penicillin G (5–10 μg/mL) and nalidixic acid (15 μg/mL) or starch-casein-nitrate agar plus rifamycin (10 μg/mL) (38,39)
Streptoalloteichus	Soil, arid	Dilution plate on ISP-2 or Bennett's agars plus gentamicin or kanamycin (10 μg/mL), incubate 43 °C (49)
Thermocrispum	Municipal waste and mushroom composts	Dilution plate and Anderson Sampler, TSA plus rifamycin (10 μg/mL), erythromycin and oleandomycin (100 μ/mL), novobiocin (15 μg/mL), 50 °C (50)
Glycomyces	Soil	Dilution plate on Czapek-sucrose agar plus novobiocin (25 μg/mL) and streptomycin (15 μg/mL) (30)
Actinocorallia	Soil	Dilution plate on colloidal chitin-vitamin agar plus kabicidin (10 μg/mL) (4)
Actinokineospora	Soil, fallen leaves	Dilution plates plus kanamycin and/or nalidixic acid (50 μg/mL) or baiting and centrifugation (5,51)

Note: Detailed selective procedures are not reported for the following general: *Nocardiopsis, Actinoployspora* (not isolated from a natural habitat, appeared as a contaminant), *Catellatospora, Pseudonocardia, Actinobispora*, and *Planotetraspora*. *Lentzea* has been isolated from human specimens (52).
[a]All media contain nystatin at 50–100 μg/mL.

dia that support the production of a wide variety of secondary metabolites. More than one medium is often used, three to four are optimal. The fermentation is harvested after 4 to 6 days and can either be extracted with solvent (butanol, methyl isobutyl ketone, or ethyl acetate) or filtered (paper or 10,000 mw cutoff filters). The extract or filtrate is tested for secondary metabolite production using biological activity or chromatography. Primary fermentations of wild-type strains, under the conditions outlined, can yield from 0.01 μg/mL to as much as 400 μg/mL of

Table 10. Methods for the Production of Actinomycete Secondary Metabolites

Parameter	Example 1 (54): saccharocarcins by *Saccharothrix aerocolonigenes*	Example 2 (55): megalomicin by *Micromonospora megalomicea*
Germination medium (per liter)	Glucose, 10 g; trehalose, 10 g; hydrolyzed casein, 5 g; soy flour, 5 g; yeast extract, 5 g; calcium carbonate, 2 g; tap water	Dextrose, 1 g; potato starch, 24 g; beef extract, 3 g; tryptone, 5 g; yeast extract, 5 g; calcium carbonate, 2 g; tap water
Germination conditions	G1 48 hr at 28 °C, 250 rpm; G2 48 h at 28 °C, 250 rpm	G1 72 h at 35 °C, 300 rpm
% inoculum	10	5
Fermentation medium (per liter)	Dextrin, 30 g; molasses, 10 g; soluble starch, 5 g; tap water	Dextrose, 10 g; starch, 20 g; yeast extract, 5 g; casein hydrolysate, 5 g; calcium carbonate, 4 g; tap water
Fermentation conditions	5 days at 30 °C, 250 rpm	60–70 hr at 31 °C, 300 rpm
Extraction procedure	Whole broth with ethyl acetate	Adjust whole broth to pH 9.5 and extract with ethyl acetate

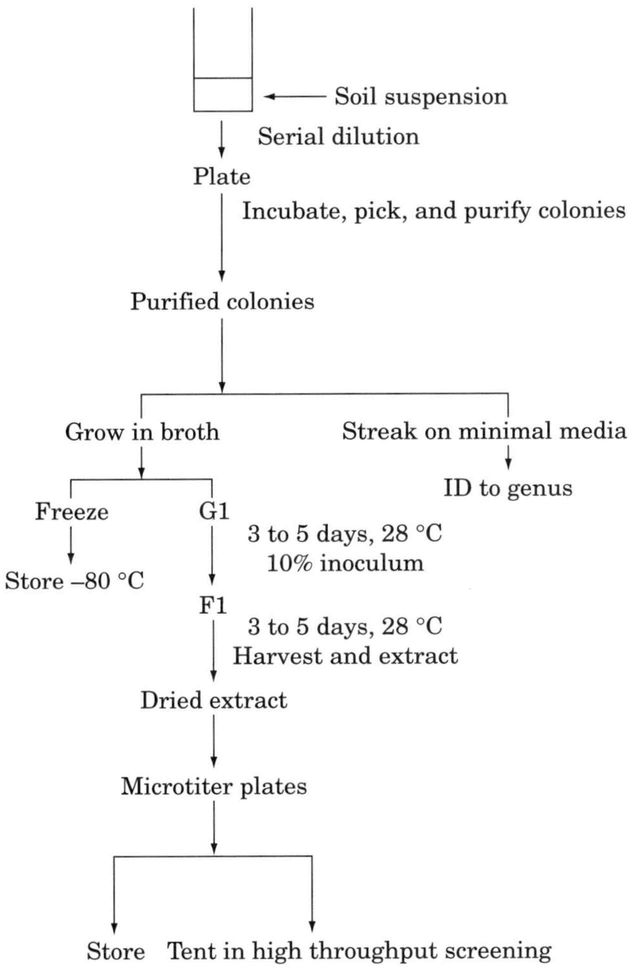

Figure 5. Outline of the primary screen process.

product. Examples of the production of secondary metabolites by two different genera of actinomycetes are presented in Table 10.

The advent of high throughput screening has resulted in a dramatic increase in the number of samples assayed, including natural products, with a concurrent decrease in the time required to test samples. In addition, an assay's lifetime has also decreased because all available samples can be tested in a relatively short time span (one month). For natural products to effectively compete with synthetic compounds and combinatorial libraries for available assay slots, a readily available source of samples is required. High throughput screening requires small amounts of test sample; a 10-mL fermentation results in enough extract for as many as 100 assays. Many natural product screening programs have addressed the issue of sample availability by developing libraries of extracts stored in 96-well microtiter plates. For each extract that is stored, the corresponding producing culture must also be preserved and its location catalogued. The nonstreptomycete actinomycetes described in this section can be stored at −80 °C in 10% glycerol for many years. However, data on the stability of natural product extracts stored over a period of years as well as the ability of preserved nonstreptomycete actinomycetes to maintain productivity have yet to appear in the literature. Figure 5 outlines the steps in a typical primary screen. Once activity has been detected in an extract or filtrate, the producing culture is fermented in a large enough volume to isolate enough material for characterization and structure elucidation. Many of the secondary metabolites produced by the *meso*-DAP containing actinomycetes are listed in Table 11.

SUMMARY

Nonstreptomycete actinomycetes continue to be a source of novel, therapeutically relevant natural products. Procedures described to effectively isolate a wide variety of genera rely on (*1*) collecting ecologically and geographically distinct soil samples, (*2*) using selective pressures (incorporating antibiotics and other chemicals into isolation media) that suppress streptomycetes while allowing diverse genera to survive, and (*3*) incorporating media constituents that favor the growth of rare actinomycetes. Evaluating the effectiveness of soil isolation procedures in obtaining diverse nonstreptomycete actinomycetes can quickly be achieved by identifying isolates macro- and microscopically. More definitive identification currently re-

Table 11. Secondary Metabolites Produced by *meso*-DAP-Containing Actinomycetes

Genus	Secondary metabolites[a]
Streptosporangium	Anthracyclines (figaroic acids, fragilomycins, carminomycin, carborubicin, sibiromycin; ansamycins (ansamitocin); bleomycins (platomycins, victomycin); glycopeptides (A-84575); macrolides (aculeximycin); peptides; phenazines; sugars (sporacuracins, 21906A); selenomycin; WS79089
Microbispora	Peptides (cochinmicins); phenazines; EV22 (SCH31828)
Microtetraspora	Amino acid containing (azaserine); talisomycin B; kistamicins; pyralomicins; macrolactam (fluvirucins); Bu-2313
Planobispora	Peptides; GE 2270
Planomonospora	Peptides (sporangiomycin); polyene
Thermomonospora	Naphthaquinone (SCH38519)
Actinomadura	Anthracyclines (akrobomycins, barminomycins, benanomicins, carminomycins, pradimicin, rubeomycins, rubstomycins); ansamycin (rifamycins); amino acid containing (madumycin); anthraquinones (oxanthromicin, maduramycin); indolocarabozole (AT2433); nucleosides (chloropentastatin, adenosine, coformycin, pentostatin); macrolactam (SCH38516, SCH38518, fluvirucin); macrocyclic lactone (kijanimicin); prodigiosins; polyethers (cationomycin, hidamicin, kijimycin); peptide related (formestin, luzopeptin, parvodicin), sugar containing (veractamycins, esperamicins); tetracyclin related (ES-119, SCH33256, SCH36969)
Spirillospora	Peptide macromolecular (Spirillomycin); polyene.
Micromonospora	Aminoglycosides (gentamicin, fortimicin, kanamycin, neomycin, sagamicin, sisomicin, verdamicin, G-52, G-418, J120); anthracyclines (daunorubicin, sibanomicin); anthraquinone (dynemicin); ansamycins (halomycins, rifamycins); benzodiazepine (sibanomicin); macrocyclic lactones (antlermicin, tetrocarcins); macrolides (clostomicin, erythromycin, juvenimicin, megalomicin, mycinamicin, izenamicin, rustmicin, primycin, rosaramicin); nucleosides (astromycin, dapiramicin); orthosomicins (everninomicins); peptides and amino acid containing (actinomycin, bottromycin, chalcidin, micromonosporin, 68-1147, glycine, negamicin); pigments (genistein, rubradirin, nocardiorubin); sugar containing (calicheamicin); xanthone (citreamicins); LL-E-19085
Actinoplanes	Glycopeptides (actaplanin, teicoplanin, ramoplanin); macrolides (lipiarmycin); peptides and amino acid containing (gardimycin, mycoplanecin, planothiocin, taitomycin, azetidine, azaserine, plauracins, virginiamycin, thristreptone); orthosomicin (SE737/74D); polyenes (octamycin, SE293); quinone (purpuromycin); chuangxinmycin; A/15104Y
Couchioplanes	Polyene (67-121)
Catenuloplanes	CP-54715
Dactylosporangium	Aminoglycosides (dactimicin, G-367); macrolides (clostomicin, tiacumin); orthosomicins (SF-2107); peptide containing (capreomycin, SF-2329); polyether (AC-7230, CP44161); quinone (DK-7814); tetracycline (SCH34164)
Pseudonocardia	Glycopeptides (azureomycins, boxazomycins, helvedardins); peptide (41043); polyene; quinones
Actinosynnema	Ansamycins (macbecins, C-33196); β-lactams (nocardicins)
Amycolatopsis	Ansamycins (rifamycins, kanglemycin); glycopeptides (avoparcin, vancomycin, orienticins, decaplanin, ristocetin, balhimycin); macrocyclic lactones (quartromycins); macrolides (A-59770); quinones (benzathrins, epoxyquinomicins, azicemicins); sugar (efrotomycin type); tetracycline (cetocycline); octacosamicins
Kibdelosporanglum	Glycopeptides (aridicins, kibdelins, decaplanin, A-80407); cycloviracins
Kutzneria	Chloramphenicol; glycoside macrolides (sporaviridins)
Saccharomonospora	Glycopeptide (AB-65)
Saccharopolyspora	Aminoglycosides (nebramycin, destomycin, saccharocin, sporaricin); alkaloids (hatomamicin); cinodines (coumamidines); lactones (belactins, nodusmicin); macrolides (erythromycin, sporeamicin); peptides (vanoxonin); CL307-24
Saccharothrix	Alkaloids (tetrazomine); glycopeptides (galacardins); indolocarbazoles (staurosporin, K-252a, rebeccamycins, SCH43228); macrolides (kayamycin, sekothrixide, N-814-103 factors); macrocyclic lactones (saccharocarcins); peptides (dopsisamine, madumycin, polynitroxin, nocamycin, sandramycin); polyether (portmicin); sugar (LL-C-19004); phosphonothrixin; karnamicins; tetrazomine
Streptoalloteichus	Aminoglycosides (kanamycins); bleomycins (talisomycins); siderochelin
Thermocrispum	Undescribed antifungal antibiotic (50)
Glycomyces	Amino acid containing (azaserine); LL-D05139
Actinocorallia	Peptide (azaserine related)

Note: Secondary metabolites not reported from *Herbidospora, Catellatospora, Pilimelia, Spirilliplanes, Actinopolyspora, Lentzea, Actinobispora*, and *Planotetraspora*.

[a]Data from Actinomycetes Antibiotic Data Base, Technomics Inc., C. P. O. Box 882, Tokyo, 00-91, Japan.

quires extensive chemical analysis of cell-wall and whole-cell components. Future identification of actinomycetes from natural habitats will rely on molecular approaches, particularly sequencing the gene encoding the small subunit of RNA (16S). This gene is a likely target because it is relatively small (1.5 kb), many actinomycete-type strains have been sequenced, and the data are readily available on-line for comparison to soil isolated, unknown strains. Numerous isolates can be fermented in small volumes. The fermentations are extracted or filtered, and the resulting samples are stored for testing against novel targets using high throughput screening methodologies. And so the quest for new metabolites with therapeutic potential from new organisms continues.

BIBLIOGRAPHY

1. M.P. Lechevalier and H.A. Lechevalier, in A.L. Demain and N.A. Solomon eds. *Biology of Industrial Microorganisms*, Benjamin/Cummins, Menlo Park, California, 1985, pp. 315–358.
2. E. Stackebrandt, F.A. Rainey, and N.L. Ward-Rainey, *Int. J. Syst. Bacteriol.* **47**, 479–491 (1997).
3. C. Jiang, L. Xu, Y. Yang, G. Guo, J. Ma, and Y. Liu, *Int. J. Syst. Bacteriol.* **41**, 526–528 (1991).
4. S. Iinuma, A. Yokota, T. Hasegawa, and T. Kanamaru, *Int. J. Syst. Bacteriol.* **44**, 230–234 (1994).
5. T. Hasegawa, *Actinomycetologica* **2**, 31–45 (1988).
6. H. Rumano, W. Guizhen, and L. Junying, *Int. J. Syst. Bacteriol.* **43**, 468–470 (1993).
7. Y. Wang, Z. Zhang, and J. Ruan, *Int. J. Syst. Bacteriol.* **46**, 933–938 (1996).
8. R.M. Kroppenstedt and M. Goodfellow, in A. Ballows, H.G. Trüper, M. Dworkin, W. Harder, K.H. Schleifer eds. *The Prokaryotes, Volume 2: A Handbook on the Biology of Bacteria: Ecophysiology, Isolation, Identification, Application*, 2nd ed., Springer-Verlag, New York, 1992, pp. 1085–1114.
9. M. Goodfellow, in A. Ballows, H.G. Trüper, M. Dworkin, W. Harder, K.H. Schleifer eds. *The Prokaryotes, Volume 2: A Handbook on the Biology of Bacteria: Ecophysiology, Isolation, Identification, Application*, 2nd ed., Springer-Verlag, New York, 1992, pp. 1115–1138.
10. T. Tamura, M. Hayakawa, and K. Hatano, *Int. J. Syst. Bacteriol.* **47**, 97–102 (1997).
11. M.P. Lechevalier, *J. Lab. Clin. Med.* **71**, 934–944 (1968).
12. M.P. Lechevalier and H.A. Lechevalier, *Int. J. Syst. Bacteriol.* **20**, 435–443 (1970).
13. H.A. Lechevalier and M.P. Lechevalier, in M.P. Starr, H. Stolp, H.G. Trüper, A. Balows, and H.G. Schlegel eds. *The Procaryotes: A Handbook on Habitats, Isolation and Identification of Bacteria*, Springer-Verlag, New York, 1981, pp. 1915–1922.
14. K.H. Schleifer and O. Kandler, *Bacteriol. Rev.* **36**, 407–477 (1972).
15. K. Suzuki, M. Goodfellow, and A.G. O'Donnell, in M. Goodfellow and A.G. O'Donnell eds. *Handbook of New Bacterial Systematics*, Academic Press, New York, 1993, pp. 195–238.
16. D.E. Minnikin and A.G. O'Donnell, in M. Goodfellow, M. Mordarski, and S.T. Williams eds. *The Biology of the Actinomycetes*, Academic Press, New York, 1984, pp. 337–388.
17. K. Uchida and K. Aida, *J. Gen. Appl. Microbiol.* **23**, 249–260 (1977).
18. K. Uchida and A. Seino, *Int. J. Syst. Bacteriol.* **47**, 182–190 (1997).
19. M.P. Lechevalier, A.C. Horan, and H.A. Lechevalier, *J. Bacteriol.* **105**, 313–318 (1971).
20. M.P. Lechevalier, C. DeBievre, and H.A. Lechevalier, *Biochem. Syst. Ecol.* **5**, 249–260 (1977).
21. M.D. Collins, in M. Goodfellow and D.F. Minnikin eds. *Chemical Methods in Bacterial Systematics*, Soc. for Applied Biotech., Tech. Series No. 20, Academic Press, London, 1985, pp. 267–287.
22. R.M. Kroppenstedt, in M. Goodfellow and D.F. Minnikin eds. *Chemical Methods in Bacterial Systematics*, Soc. for Applied Biotech., Tech. Series No. 20, Academic Press, London, 1985, pp. 173–199.
23. R.M. Kroppenstedt, in A. Ballows, H.G. Trüper, M. Dworkin, W. Harder, K.H. Schleifer eds. *The Prokaryotes, Volume 2: A Handbook on the Biology of Bacteria: Ecophysiology, Isolation, Identification, Application*, 2nd ed., Springer-Verlag, New York, 1992, pp. 1139–1156.
24. E.B. Shirling and D. Gottlieb, *Int. J. Syst. Bacteriol.* **16**, 313–340 (1966).
25. A.C. Horan and B.C. Brodsky, *Int. J. Syst. Bacteriol.* **32**, 195–200 (1982).
26. E. Stackebrandt and B.M. Goebel, *Int. J. Syst. Bacteriol.* **44**, 846–849 (1994).
27. M. Goodfellow, L.J. Stanton, K.E. Simpson, and D.E. Minnikin, *J. Gen. Microbiol.* **136**, 19–34 (1990).
28. J. Meyer, in S.T. Williams, M.E. Sharpe, and J.G. Holt eds. *Bergey's Manual of Systematic Bacteriology, Volume 4*, Williams & Wilkins, Baltimore, 1989, pp. 2562–2568.
29. F.A. Rainey, N. Ward-Rainey, R.M. Kroppenstedt, and E. Stackebrandt, *Int. J. Syst. Bacteriol.* **46**, 1088–1092 (1996).
30. G. Vobis, in A. Ballows, H.G. Trüper, M. Dworkin, W. Harder, K.H. Schleifer eds. *The Prokaryotes, Volume 2: A Handbook on the Biology of Bacteria: Ecophysiology, Isolation, Identification, Application*, Springer-Verlag, 2nd ed., New York, 1992, pp. 1029–1060.
31. C. Koch, R.M. Kroppenstedt, F.A. Rainey, and E. Stackebrandt, *Int. J. Syst. Bacteriol.* **46**, 765–768 (1996).
32. M.T. Embley, J. Smida, and E. Stackebrandt, *Syst. Appl. Microbiol.* **11**, 44–52 (1988).
33. M.T. Embley, in A. Ballows, H.G. Trüper, M. Dworkin, W. Harder, K.H. Schleifer eds. *The Prokaryotes, Volume 2: A Handbook on the Biology of Bacteria: Ecophysiology, Isolation, Identification, Application*, 2nd ed., Springer-Verlag, New York, 1992, pp. 996–1027.
34. D.P. Labeda, R.T. Testa, M.P. Lechevalier, and H.A. Lechevalier, *Int. J. Syst. Bacteriol.* **35**, 417–421 (1985).
35. T. Cross, in M.P. Starr, H. Stolp, H.G. Trüper, A. Balows, and H.G. Schlegel eds. *The Procaryotes: A Handbook on Habitats, Isolation and Identification of Bacteria*, Springer-Verlag, New York, 1981, pp. 2091–2102.
36. M. Goodfellow and S.T. Williams, *Ann. Rev. Microbiol.* **37**, 189–216 (1983).
37. S.T. Williams and E.M.H. Wellington, in J.D. Bu'lock, L.J. Nisbet, and D.J. Winstanley eds. *Bioactive Microbial Products: Search and Discovery*, Academic Press, London, 1982, pp. 9–26.
38. M.C. Shearer, *Dev. Ind. Microbiol.* **28**, 91–97 (1987).
39. A.C. Horan, in V.P. Gullo ed. *The Discovery of Natural Products with Therapeutic Potential*. Butterworth-Heinemann, Boston, 1994, pp. 3–30.
40. T.S. Whitham, M. Athalye, D.E. Minnikin, and M. Goodfellow, *Antonie Van Leeuwenhoek* **64**, 387–429 (1993).
41. M. Hayakawa and H. Nonomura, *J. Ferment. Technol.* **65**, 501–509 (1987).
42. C.E. Bland and J.N. Couch, in M.P. Starr, H. Stolp, H.G. Trüper, A. Balows, and H.G. Schlegel eds. *The Procaryotes: A Handbook on Habitats, Isolation and Identification of Bacteria*, Springer-Verlag, New York, 1981, pp. 2004–2010.
43. N.S. Makkar and T. Cross, *J. Appl. Bacteriol.* **52**, 209–218 (1982).
44. F.P. Mertz, *Int. J. Syst. Bacteriol.* **44**, 274–281 (1994).
45. T. Kudo, T. Ioth, S. Miyadoh, T. Shomura, and A. Seino, *Int. J. Syst. Bacteriol.* **43**, 319–328 (1993).
46. A. Yokota, T. Tamura, T. Hasegawa, and L.H. Huang, *Int. J. Syst. Bacteriol.* **43**, 805–812 (1993).
47. A.F. Yassin, B. Haggene, H. Budzikiewicz, and K.P. Schaal, *Int. J. Syst. Bacteriol.* **43**, 414–420 (1993).
48. K. Tomita, Y. Hoshino, and T. Miyaki, *Int. J. Syst. Bacteriol.* **43**, 297–301 (1993).
49. K. Tomita, in S.T. Williams, M.E. Sharpe, and J.G. Holt eds. *Bergey's Manual of Systematic Bacteriology, Volume 4*, Williams & Wilkins, Baltimore, 1989, pp. 2569–2572.

50. F. Korn-Wendisch, F. Rainey, R.M. Kroppenstedt, A. Kempf, A. Majazza, H.J. Kutzner, and E. Stackebrandt, *Int. J. Syst. Bacteriol.* **45**, 67–77 (1995).
51. T. Tamura, M. Hayakawa, H. Nonomura, A. Yokota, and K. Hatano, *Int. J. Syst. Bacteriol.* **45**, 371–378 (1995).
52. A.F. Yassin, F.A. Rainey, H. Brzezinka, K.D.-Jahnke, H. Weissbrodt, H. Budzikiewica, E. Stackebrandt, and K.P. Schaal, *Int. J. Syst. Bacteriol.* **45**, 357–363 (1995).
53. R. Gherna, P. Pienta, and R. Cote eds. *ATCC Catalogue of Bacteria and Bacteriophages*, 18th ed., American Type Culture Collection, Rockville, Md., 1992.
54. A.C. Horan, M.C. Shearer, V. Hegde, M.L. Beyazova, B.C. Brodsky, A. King, R. Berrie, K. Cardaci, and M. Nimeck, *J. Antibiot.* **50**, 119–125 (1997).
55. M.J. Weinstein, G.H. Wagman, J.A. Marguez, R.T. Testa, E.M. Oden, and J.A. Waitz, *J. Antiobiot.* **22**, 253–258 (1969).

See also CEPHALOSPORINS; FERMENTATION MONITORING, DESIGN AND OPTIMIZATION; MEDIUM FORMULATION AND DESIGN, *E. COLI* AND *BACILLUS* SPP.; METABOLITES, PRIMARY AND SECONDARY; MUTAGENESIS.

SECONDARY METABOLITES, ANTIBIOTICS

WILLIAM R. STROHL
Merck Research Laboratories
Rahway, New Jersey

KEY WORDS

Antibacterial drugs
Antibiotic drug discovery
Antibiotic markets
Antibiotic resistance
Anticancer drugs
Antifungal drugs
Beta-lactams
Macrolides
Natural products
Polyketides

OUTLINE

INTRODUCTION

Definition of Natural Product Antibiotics

The field of natural products owes its origins to the discovery of penicillin and its ability to kill bacteria. Since Fleming's original discovery in 1928 of the antibacterial effects of *Penicillium chrysogenum* and Flory and Chaim's efforts more than a decade later to bring that discovery to production, the field of natural products has undergone enormous changes.

Vuillemin first introduced the concept of antibiotic activity in 1889, using the term *influences antibiotiques* (antibiotic influences) to describe negative interactions between plants and animals (1). Waksman later coined the term *antibiotic* to describe "a chemical substance derived from microorganisms which has the capacity of inhibiting growth, and even destroying, other microorganisms in dilute solutions" (2). Natural product antibiotics are almost universally secondary metabolites, compounds generally produced at low growth rates or after growth has ceased, nonessential for growth of the producing organisms in pure culture, and typically possessing unusual structures not found in primary metabolites of central metabolism. Although natural product antibiotic biosynthesis is not essential to the producing organism in pure culture, these compounds are critical to the producing organisms in their natural environment, both for survival and competitive advantage (3).

In more recent years, the term *antibiotics* sometimes has been used more broadly to define anti-infective agents, essentially in line with Waksman's definition, as well as *bioactive natural products*. This latter group includes immunomodulatory agents, statins, antitumor drugs, antiparasitic compounds, insecticides, and other bioactive natural products, many of which also possess weak antibacterial or antifungal activities. Indeed, the immunosuppressive drug cyclosporin was originally discovered in an antifungal screening program. Additionally, the biosynthesis of bioactive antibiotics is often very similar to that of other bioactive natural products. As a result, cyclosporin, a weak antifungal antibiotic but an excellent immunosuppressive agent, is usually included in discussions of other peptide antibiotics. Similarly, when reviews are written about macrolide antibiotics, other bioactive natural products such as tacrolimus (formerly called FK-506) and rapamycin, both of which are immunosuppressive agents (4), are typically included in discussions with erythromycin, a bacteriostatic antibiotic; avermectin, a potent antiparasitic compound; and tylosin, an agricultural feed additive.

This certainly makes sense because all these bioactive natural products possess similar modes of biosynthesis.

An extensive table including all marketed antibiotics (both natural product-derived and chemosynthetic) as well as all other marketed bioactive natural products, as of December 1996, was recently compiled for the book entitled *Biotechnology of Antibiotics* (5). Additionally, a recent review by Shu (6) highlighted most of the recent important discoveries in the field of natural products, ranging from new antifungal activities to natural products active as neuromodulators. Thus, this review will focus on natural product anti-infective and antitumor agents.

Overview of the Natural Product Antibiotics Field

Cragg et al. (7) published in 1997 a well-cited review of the natural products field in which they analyzed the percentage of natural products in various fields. Of 520 new approved drugs between 1983 and 1994, 55.6% were chemically synthesized drugs, 24.4% were semisynthetic compounds, 5.8% were unmodified natural products, 8.8% were chemically synthesized drugs patterned after natural products, and 5.4% were "biologics" (e.g., recombinant proteins) (7). Thus, approximately 40% of all newly approved drugs over that period were derived from natural product leads. In breaking down the categories, however, natural products dominated the antibacterial markets (63% were derived from natural product leads) and the anticancer market, in which 61% of the new drugs were derived from natural product leads. Natural product leads also strongly contributed to the antihyperintensive market (48% were derived from natural product leads) and the anti-inflammatory market, in which one-third of the new drugs approved between 1983 and 1994 were derived from natural product leads. According to Cragg et al., however, natural products did not contribute in the period of 1983–1994 to newly approved drugs in the antifungal, analgesic, anxiolytic, antihistamine, antidepressant, cardiotonic, or hypnotic fields (7).

Strohl (5) conducted a similar analysis of bioactive natural products with antibiotic activities, identifying a list of 252 natural and synthetic compounds used clinically and agriculturally in the United States with selected compounds included from other countries. Thus, although this analysis was not all-inclusive, it represented the majority of commercially important antibiotics and antimetabolic drugs. Of those compounds identified, 75 (30%) were natural products, another 75 (30%) were semisynthetic compounds derived from natural products, and the remaining 40% were chemically synthesized compounds (5).

The discovery of new, novel natural products continues to rise dramatically (5). Based on an analysis of papers published in *Journal of Antibiotics*, Bérdy (8) indicated that about 11,900 antibiotics and bioactive natural products had been discovered through 1994, with increasing rates of discovery every year through 1990. He projected that the total of new, novel natural product structures discovered by the year 2000 would top 16,000 (8). The annual rate of new natural product descriptions published is now more than 500 per year, up from 200 to 300 per year 20 years ago (5,8). Nevertheless, there is the widespread perception in both the scientific literature and the popular press that the discovery of new natural products and natural product antibiotics is on the wane (5). This anomaly, however, can easily be explained. Because most new natural product drugs still are concentrated in the fields of anti-infectives and anticancer drugs, and natural product leads have been around long enough in both the anti-infectives and anticancer areas, the law of diminishing returns is most prevalent in these fields. For example, from 1962, when nalidixic acid was developed as the first quinolone antibiotic, to 1996, no new classes of clinical natural-product-derived antibacterials have been introduced to the market. In 1993, only one new natural product antibiotic was approved by the Food and Drug Administration for use in the United States, none in 1994, and only a few in 1995 and 1996. Overall, from 1980 to 1990, the worldwide discovery of more than 4,600 new bioactive natural products was published, yet only 20 new natural-product-derived drugs (3 natural products and 17 semisynthetic products) were commercialized in that period. Finally, in the period spanning from World War II to 1975, approximately 1.6% of all natural product antibiotics reported (either by publication or patent) were eventually commercialized. Since 1975, this number has dropped to about 0.11%. Thus, fewer antibiotics are being commercialized from an increasingly larger pool of isolated candidate compounds. These data suggest that the Golden Era of antibiotics, generally recognized as having begun shortly after World War II ended, lasted nearly 30 years before the law of diminishing returns caught up to it. Another major contributing factor for the dearth of commercialized natural product antibiotics from 1975 to the present is that the U.S. government, academic researchers, and pharmaceutical companies alike turned their efforts toward the defeat of cancer and heart disease, believing that infectious diseases were under control (5,9).

On the other hand, in pharmacological areas in which natural products are making new inroads, many discoveries of bioactive natural products have resulted in compounds that are still in preclinical or clinical trials. Since the late 1960s, Umezawa and colleagues in Japan showed interest in discovering non-anti-infective bioactive natural products, particularly in the area of protease inhibitors (10). Although researchers in U.S. pharmaceutical companies showed interest in developing natural products as bioactive molecules in areas other than the anti-infective and anticancer markets (i.e., whole cell assays), many viewed the Japanese mechanism-based assay approach more as a curiosity than a serious discovery effort. It was not until the publication of Merck's patent and paper in *Proceedings of the National Academy of Science* (11) in 1980 on lovastatin as a remarkable inhibitor of hydroxymethylglutaryl-CoA reductase, and ultimately an outstanding cholesterol-lowering drug, that the inclusion of natural products in mechanism-based screens was considered seriously.

Since that time much has changed, and although some pharmaceutical companies have minimized natural product screening efforts because of the perceived cost-to-benefit ratio, others have sought to expand the scope of natural products in their screening programs. As evidenced by publications in journals such as *Journal of Antibiotics*, patent applications, and clinical trial candidates (6), it has now become apparent that natural products can significantly augment synthetic chemical approaches for the discovery of new drug entities in a wide range of pharmacologically important areas. The two most significant recognized drawbacks of natural products in a broad-based screening program are the inclusion of crude extracts (which usually include salts, proteins, fatty acids and other potentially interfering materials) in in vitro assays (e.g., receptor-based, enzyme inhibition, or protein–protein interaction assays) and the time it takes to turn a crude-extract hit into a chemical structure. Despite these drawbacks, natural products now have been found to contribute to significant leads in most therapeutic areas (6).

The sources of natural products are also likely to change. Bérdy (8) showed that of the 11,900 antibiotics that had been discovered through 1994, approximately 6,600 (55%) were produced by *Streptomyces* spp. The remainder of producing organisms included the filamentous fungi, which produced 2,600 (22%); nonactinomycete bacteria, which produced 1,400 (12%); and non-*Streptomyces* strains of actinomycetes, which produced 1,300 (11%) (8). In the broader pharmacological field, it is widely recognized that a great percentage of pharmacologically derived products in use today are derived from or patterned after plant products (6). It can be expected that in the future a wider range of microorganisms, including well-known but underused groups such as the filamentous fungi, myxobacteria, bacilli, and pseudomonads, will contribute to the pool of natural products in larger ways (5).

Market for Natural Product Antibiotics

The total world sales for prescription pharmaceuticals was estimated in a September 1998 news release to be $308 billion. In a recent report of the 200 most-prescribed medicines in the United States (12), natural products accounted for 10 of the top 35 prescribed drugs (Table 1). All the top-prescribed natural products or semisynthetic drugs derived from natural products are anti-infective agents except for three statins: simvastatin, lovastatin and pravastatin. Although it is difficult to identify the entire contribution of natural products to the world pharmaceutical market economy, it should easily total more than $30 billion, including the natural-product-derived anti-infective agents (>$20 billion), the natural-product-derived anticancer agents (in which the anthracyclines and taxanes alone represent more than $3 billion in market size), the statins (with a market size estimated at more than $6 billion and rising at a rate of more than 20% annually), and the large pool of plant-derived natural products or compounds patterned after plant products.

The total world market for anti-infective agents in 1996 was reportedly $23 billion (13). The 1995 U.S. anti-infectives market alone was greater than $8 billion, with cephalosporins (45%), penicillins (15%), fluoroquinolones (11%), tetracyclines (6%), and macrolides (5%) comprising the majority of sales. Of these top-selling anti-infective agents, only the fluoroquinolones lack the distinction of being natural products, and even they are patterned after the original quinolone, nalidixic acid. Thus, in deference to their origins, antibacterial drugs are still natural products, semisynthetic compounds derived from fermentation products, or patterned after natural products.

The world antifungal market in 1995 was approximately $3 billion. Although this represents a much smaller market than the antibacterial market, the antifungal market is projected to grow at 20% annually. This rapid growth in antifungals unfortunately is driven by the dramatic increase in life-threatening fungal diseases, mostly as a result of opportunistic fungal pathogens infecting immunocompromised patients.

The antiviral market was $1.8 billion in 1995, and the projected 1998 market is $3 billion. Although the top product in 1995 was acyclovir (50% of the market [14]), more than 17 million people were infected with HIV, a number that is expected to reach 30 to 40 million by year 2000 (14). Because of the development of successful new anti-HIV drugs such as the HIV-proteinase inhibitors, it is expected that the antiviral market should increase substantially within the next few years.

Biogenesis of Natural Product Antibiotics

Since the discovery and elucidation of the structure of penicillin, there has been an enormous effort placed into finding new natural product antibiotics, particularly those of the antibacterial, antifungal, and anticancer therapeutic classes. Natural product antibiotics, for the most part, fall into several major classes of structures, as shown in Table 2, based on the precursors from which they are assembled and the mechanisms of assembly. Precursors common for the synthesis of natural products include short chain acyl-CoAs such as acetyl-CoA, malonyl-CoA, and propionyl-CoA, which are used to synthesize polyketides; activated amino acids, which are used to build peptides and β-lactams; various pyrimidines, which are the building blocks for nucleoside natural products; isoprene units, which form the basis of terpene, alkaloid, and substituted steroid natural products; and sugars and amino sugars, which are the building blocks for aminoglycosides as well as for the sugar moieties that decorate a wide variety of natural products. Many natural products result from the fusion of moieties arising from two or more different biogenic classes. For example, 6-deoxyerythronolide B (DEB), the macrolide backbone of erythromycin, is synthesized by a type I modular polyketide synthase, but the molecule is not biologically active until the sugars desosamine, attached at C-5, and cladinose, attached at C-3, are added. Several other macrolides, including spiromycin and tylosin, have this same characteristic. Similarly, anthracyclines are comprised of the aglycone portions, derived from the activity of a type II iterative polyketide synthase system, and daunosamine, a 2,3,6-trideoxy-3-aminohexose synthesized from TDP-glucose (15). Glycopeptides such as vancomycin also contain core units derived from the activity of a nonribosomal peptide synthetase that are decorated by sugar units to make bioactive molecules. An exaggerated example of this is the antifungal compound pradimicin, which is synthesized by a type II iterative polyketide synthase and then decorated both with sugar moi-

Table 1. Ranking of Natural Product Pharmaceuticals as Prescription Medicines in the United States (1997)

Natural product[a]	Brand name	Company	Rank[b]
Amoxicillin	Trimox	Apothecon	1
Azithromycin	Zithromax	Pfizer	9
Amoxicillin-clavulanate	Augmentin	SmithKline Beecham	14
Simvastatin	Zocor	Merck	15
Clarithromycin	Biaxin	Abbott	24
Amoxicillin	Amoxil	SmithKline Beecham	25
Cephalexin	Cephalexin	Teva	26
Penicillin V (K^+)	Veetids	Apothecon	32
Amoxicillin	Amoxicillin	Teva	33
Pravastatin	Pravachol	Bristol-Meyers Squibb	34
Cephalexin	Cephalexin	Apothecon	54
Lovastatin	Mevacor	Merck	57
Cefprozil	Cefzil	Bristol-Meyers Squibb	58
Cefuroxime	Ceftin	Glaxo Wellcome	72
Erythromycin	Ery-Tab	Abbott	82
Neomycin/polymyxin B/HC	Neomycin/polymyxin B/HC	Schein	122
Mupirocin	Bactroban	SmithKline Beecham	130
Cefaclor	Cefaclor	Mylan	139
Doxycycline	Doxycycline	Zenith	168
Amoxicillin	Amoxicillin	Warner-Chilcott	184
Clarithromycin	Biaxin suspension	Abbott	192
Cefixime	Suprax	Wyeth-Ayerst	196

[a]Includes both natural products and semisynthetic compounds derived from natural products.
[b]Ranking based on number of prescriptions sold.
Source: Data from Ref. 12.

eties and D-amino acid moieties (16). Other molecules, such as the rifamycins, contain a unique building block such as 3-amino-5-hydroxybenzoic acid (AHBA), which is used as the primer with which to initiate the synthesis of the macrocycle (17). Thus, many natural products are derived from mixed biogenic origins. For more detailed information on specific antibiotics, recent books edited by Vining and Stuttard (18) and by Strohl (19) cover the biogenesis and biosynthesis of virtually all the major classes of antibiotics and bioactive natural products.

ANTIBACTERIAL AGENTS

Introduction

The most widely prescribed, best known, best characterized, and perhaps also the most widely disparaged natural products are the antibacterials. After the discovery of penicillin by Fleming in 1928 and its subsequent development as a drug during World War II, the pharmaceutical industry and well-known academics such as Waksman of Rutgers University worked feverishly to discover new antibacterials that would eliminate bacterial infections of all kinds. The natural product antibiotics gramicidin (1939), streptomycin (1944), chlortetracycline (1948), chloramphenicol (1948), cephalosporin C (1948), neomycin (1949), oxytetracycline (1950), erythromycin (1952), tetracycline (1953), vancomycin (1955), rifamycins (1959), lincomycin (1962), pristinamycin (1962), and nalidixic acid (1962), the predecessor to fluoroquinolones, were all discovered by 1962 (22). No new, truly novel classes of natural product antibacterial drugs have been marketed since then. Even in the cases of modern synthetic antibacterials, all successful synthetic antibacterials have been biomimetically derived from natural product structures (the best example being the fluoroquinolones, which are based on nalidixic acid). There are new classes of candidate antibacterials, such as the oxazolidinones (23), which are not derived from natural products, but these have yet to break into the market, in part because of associated toxicities exhibited by this class of compounds. The targets for these drugs, including cell wall biosynthesis, protein biosynthesis, and nucleic acid biosynthesis, all have been targeted many times over by analogues of them, as well as by other compounds, both synthetic and naturally derived. As we enter the 21st century, there is a critical need for new, potent antibacterials to which resistance is not easily gained.

Needs and Markets

With the emergence of new infectious diseases as well as the emergence of bacterial strains resistant to existing antibiotics, there is an enormous challenge to pharmaceutical companies, researchers, and governments to develop new methods for treating existing infectious diseases and those emerging as new health threats. As recently reviewed by Strohl (5), there are several reasons for the need to develop new antibiotics, preferably with novel structures and activities.

First and foremost, pathogenic bacteria are acquiring or developing resistance to existing antibiotics and classes of antibiotics in direct correlation with the use of those antibiotics to treat them (24). As Julian Davies stated, "it is frightening to realize that one single base change in a gene encoding a bacterial β-lactamase can render useless \$100 million worth of pharmaceutical research effort" (25). Although this is an oversimplification of the problem, it

Table 2. Biogenic Classes of Natural Product Antibiotics

Biogenic class and subclass[a]	Example compounds or groups	Example activities
	Nonribosomal peptide synthetases	
Diketopiperazines	Gliotoxin	Antibiotic
Cyclic peptides	Cyclosporin	Immunosuppressive
	Gramicidin	Antibacterial
Depsipeptide	Enniatin, valinomycin	Ionophores
Lipopeptides	Echinocandins	Antifungal
Glycopeptides	Vancomycin	Antibacterial
	Bleomycins	Antitumor
β-Lactams	Penicillins, cephalosporins	Antibacterial
Lantibiotic	Nisin	Antibacterial
Peptidolactones	Actinomycin D	Antitumor
Phosphinopeptide	Bialaphos	Phytotoxic
	Polyketides	
Type I modular		
Macrolides	Erythromycin, tylosin	Antibacterial
Pentacyclic lactones	Avermectin, milbemycin	Insecticidal acaricidal
Ascomycins	Tacrolimus	Immunosuppressive
Ansamycins	Rifamycin, ansatrienins	Tuberculostatic, antibacterial
Polyenes	Nystatin, amphotericin B	Antifungal
Polyethers	Monensin	Growth promotants
Type I iterative		
Statins	Lovastatin, pravastatin	Cholesterol-lowering
Type II iterative		
Tetracyclines	Oxytetracycline	Antibacterial
Anthracyclines	Doxorubicin, aclarubicin	Cytotoxic antitumor
	Aminoglycosides	
Aminoglycosides	Streptomycin, kanamycin	Antibacterial
Aminocyclitols	Spectinomycin	Antibacterial
Cyclitol	Kasugamycin	Antibacterial
	Isoprenoids	
Steroidal	Squalamine	Antitumor, antibacterial
Sesquiterpene	Artemisinin (Qinghao), epolone	Antimalarial
Diterpene	Paclitaxel (Taxol)	Antitumor
Triterpene	Fusidic acid	Antibacterial
Indole alkaloid	Vinblastine	Antitumor
	Others	
Lincosamides	Lincomycin, clindamycin	Antibacterial
Chorismic acid-derived	Chloramphenicol	Antibacterial
Amino acids	Cycloserine	Antibacterial
Quinolones	Nalidixic acid, fluoroquinolones	Antibacterial
Nucleosides	Nikkomycins, polyoxins	Antifungal
Mitosanes	Mitomycin C	Antitumor, cytotoxic
Oligosaccharides	Everninomycin, avilamycin	Antibacterials

Source: Data from Refs. 5, 18–21.

succinctly sums up the problem with the development of antibiotic resistance by pathogenic bacteria. The most critical reason for the development of new antibiotics is the race to remain a step ahead of antibiotic-resistance development, which has proceeded at an alarming rate (26). Even since the earliest use of antibacterial agents, the emergence of resistance to the drugs used has been observed. For example, within 4 years after the introduction of penicillin during World War II, several strains of penicillin-resistant bacteria were isolated from infected patients (27). After the war, outbreaks of dysentery in Japan, caused by sulfonamide-resistant *Shigella* spp., presented a serious health hazard to the recovering population. Within just a few years after introduction in 1950 of the newly developed antibiotics, streptomycin, tetracycline, and chloramphenicol to help quell the problem, resistance to each of these antibiotics was observed (28). By 1969, 69% of shigellae were resistant to sulfonamides, streptomycin, tetracycline, and chloramphenicol (28). Thus, in fewer than 20 years, nearly 70% of all *Shigella* spp. in Japan became resistant to four antibiotics to which there was minimal resistance before their use (28). Widespread antibiotic use has resulted in the rapid spread of multidrug resistant pathogens, and it has become both a local and global health hazard of epidemic proportions. The incidence of methicillin-resistant *Staphylococcus aureus* (MRSA), vancomycin-resistant enterococci (VRE; particularly by *Enterococcus faecium* and *Enterococcus faecalis*), and β-lactam-resistant *Streptococcus pneumoniae* are among the most serious problems, but are by no means the only ones (29). The incidence of nosocomial infections (i.e., hospital-acquired) is on the rise, and of those, an increasingly greater percentage of the infectious agents (typically MRSA and VRE) are drug-resistant (30). The U.S. Centers for Disease Control and Prevention (CDC) estimates that of the 40 million hospitalizations in the United States each year, 2 million patients (5%) contract nosocomial infections, and that as many as 58,000 of them (0.15%) die of them (31). Unfortunately, VRE are nearly untreatable with current antibiotics, with the possible exceptions of the experimental drugs Synercid or teichoplanin, which have been used successfully to treat some cases of VRE (32). The mortality rate of patients infected with multiple-drug resistant enterococci is 70%, making this a serious health care problem (32).

Second, new pathogens are evolving at an alarming rate. For example, today there are more than 30 new infectious diseases (e.g., AIDS, Ebola, Legionnaires disease, hantavirus, Lyme disease, and food-borne *E. coli* O157:H7 infections) that were unknown 20 years ago (5,33). Third, even with the development of remarkable anti-infective drugs with outstanding biological activities, there are still certain pathogens that naturally defy even the best efforts to defeat them. Perhaps the best example of this is our current inability to treat successfully *Pseudomonas aeruginosa* infections resulting from cystic fibrosis (5).

Finally, many potentially important antibiotics have associated toxicities that limit their use. The best example of this is gentamycin and other aminoglycosides, which are limited in effectiveness because of their associated nephrotoxicity and ototoxicity (5). Thus, critical challenges lay ahead for the entire anti-infectives field, many of which may be met by the discovery and development of new natural product leads.

There are three critical needs in the antibacterial market. First, new parenteral drugs that can successfully treat MRSAs and VREs are absolutely required. This need will grow dramatically in the coming years as the once-rare incidence of vancomycin-resistant *S. aureus* strains, the single greatest worry of infectious disease experts (34), expands. Second, even with a seemingly crowded pediatric market, there is still a critical need (and market place) for an orally absorbed, pediatric-safe (which eliminates fluoroquinolones, which are not considered safe for children) antibacterial agent that can successfully treat upper respiratory tract infections (primarily otitis media caused by β-lactam resistant streptococci, moraxellae, and *Haemophilus influenzae*) and that is not compromised by β-lactam, aminoglycoside, or MLS (macrolide-lincosamine-streptogramin) resistance or degraded by extended spectrum β-lactamases (35).

The pediatric antibacterial market, which has grown sharply since the early 1980s, has become one of the most important anti-infective markets today. Pediatric antibacterials typically are prescribed for middle-ear infections (otitis media) caused by *H. influenzae, Moraxella* spp., and *Streptococcus pneumoniae*. By the year 2000, it is expected that the U.S. pediatric market alone will be $2.2 billion, with annual increases of approximately 15%. The current leading pediatric antibacterials are amoxicillin, Augmentin (amoxicillin plus clavulanic acid), azithromycin, and clarithromycin. It is expected that the ketolides may make a significant impact on this market. The final critical need is for a new tuberculocidal drug. Antibiotic resistance in *Mycobacterium tuberculosis* is considered to be a major growing problem. Currently, a combination treatment with isoniazid and rifamycin (both tuberculostats) is the method of choice, but many strains of *M. tuberculosis* have recently been found that are resistant to one or both of these antibiotics, and approximately 14% of all tuberculosis cases reported today involve *M. tuberculosis* strains that are resistant to one or more antibiotics (36). The problem facing doctors today is that there is no proved, effective alternative yet on the market to treat isoniazid- and rifamycin-resistant *M. tuberculosis* (36).

Antibiotic-Resistant Bacteria

Antibiotic resistance is a major clinical problem that will only get worse with time. In the case of bacteria, resistance can be mediated by several different types of mechanisms, including modification of the target to make it unsusceptable to the antibiotic, enzymatic modification of the antibiotic to render it inactive, exclusion of the antibiotic by modification of membrane permeability, active export of the antibiotic, and bypass of the critical physiological mechanisms or structures inhibited by the antibiotic. Resistance to some these mechanisms, encoded by chromosomal mechanisms, is attained at a rate of 1×10^{-8}, and others are obtained at much higher frequencies through horizontal plasmid transfer.

Plasmid-encoded TEM- and SHV-type β-lactamases are the most common mechanisms responsible for β-lactam resistance in Gram-negative bacteria (37). These β-lactamases have generally broad substrate specificities for penicillins as well as some first- and second-generation cephalosporins and are responsible for much of the resistance to β-lactams observed in the clinics. It is well documented that as new β-lactams have been developed to resist plasmid-borne β-lactamases, the β-lactamases themselves have evolved (25,37). In an extraordinary example of natural directed evolution, TEM- and SHV-type β-lactamases recently have been found containing specific point mutations, all falling in close proximity to the active site cavity, which confer resistance to third-generation cephalosporins and monobactams (37).

Vancomycin has long been the last line of defense against antibiotic-resistant bacteria, particularly MRSAs and β-lactam- and aminoglycoside-resistant enterococci. Just a little over a decade ago, it was thought that microorganisms would never become vancomycin resistant because of its unusual mechanism of action (38). This illusion was shattered, however, when VREs were reported in England and France in 1987 (39). VREs now are being discovered at an alarming rate, with no new last line of defense substitute firmly entrenched on the market (40).

Not all clinically significant antibiotic resistance is plasmid- or transposon-mediated. *S. pneumoniae*, the primary causative agent in pediatric otitis media and pneumonia as well as bacterial meningitis and bacteremia, remained sensitive to penicillin for many years. Recently, however, an alarming number of penicillin-resistant *S. pneumoniae* strains have recently been isolated from infected patients, correlating with the use of penicillins to treat them (41). These penicillin-resistant *S. pneumoniae* were not found to produce β-lactamases, the common plasmid-borne mechanism for resistance. Instead, they had incurred chromosomal mutations giving rise to altered penicillin-binding proteins (PBPs), the target for penicillin activity (41). This discovery has a touch of irony, because *S. pneumoniae* is the bacterium with which plasmid-mediated transformation was first observed. Similarly, the Oxford strain of *S. aureus*, isolated 50 years ago, is sensitive to nearly every antibiotic tested against it (42). The staphylococci, however, acquired resistance to aminoglycosides very quickly, and by 1992, 32% of *S. aureus* strains tested were found to be resistant to methicillin, the drug of choice to treat staphylococcal infections, up from 2% in 1975 (42). Like pneumococci, MRSAs have altered PBPs, giving them a resistance mechanism that cannot be countered easily. Thus, vancomycin has become the drug of last resort to treat MRSAs. Ever since *Enterococcus faecalis* was shown to transfer vancomycin resistance to *S. aureus* in the laboratory in 1992 (43), one of the greatest fears of health professionals universally has been the emergence of vancomycin-resistant, methicillin-resistant staphylococci (44). Recently, such doubly resistant, nearly untreatable strains have appeared in sporadic cases, confirming the greatest fears of infectious disease professionals (45).

A new mechanism of antibiotic resistance recently was discovered in meropenem-resistant *Pseudomonas aeruginosa* strains (46). *P. aeruginosa* is an opportunistic pathogen that contributes to nosocomial infections, eye infections, and opportunistic infections of burn patients and those with cystic fibrosis. The mechanism was found to be based on the alteration of outer membrane proteins, causing a reduction in permeability of several antibiotics, including meropenem, through the outer membrane (46). This is a significant finding, because carbapenems (e.g., imipenem) have been considered the antibiotics of choice for the treatment of opportunistic pseudomonads.

β-Lactams

As is well known, the discovery of penicillin was the genesis of modern pharmaceutical efforts to produce antibiotics, the magic bullets that at one time were thought to cure the world of disease. β-Lactams function by binding to PBPs, especially transpeptidases, that normally function in peptidoglycan biosynthesis. Because peptidoglycan is novel to bacteria, these drugs specifically act on a bacterial target. Since bacteria cannot grow in normal environments without their cell wall, inhibition of cell wall biosynthesis results in killing of the bacteria; thus, all β-lactams by definition are bacteriocidal drugs.

The total antibacterial market tops $20 billion annually, with nearly all the top-selling antibacterial drugs being natural products, semisynthetics derived from natural products, or compounds patterned after natural products. Even more than a half century after the first widespread use of penicillin during World War II to treat bacterial infections, β-lactams as a whole still make up 60%, or more than $12 billion worth, of the total antibacterial market. The commercial β-lactam antibiotics are comprised of many types of several different classes, ranging from penicillin G to the monobactam biomimetic drug, aztreonam (Table 3).

The natural product β-lactams can be separated into five separate biogenic subclasses (47): penams (e.g., penicillins), cephems (cephamycin C, cephalosporins), clavams (clavulanic acid, clavamycins), carbapenems (thienamycin), and monobactams (aztreonam) (Table 4). The ring structures of the various β-lactams differ from the classical penicillin core as follows: cephems have an expanded 6-membered ring containing a double-bond but retaining the sulfur atom; clavams have an oxygen atom replacing the sulfur of penicillins, and they tend to differ at C-2 and C-6; in carbapenems, the sulfur is replaced by a carbon and a double bond is present; and the monobactams are monocyclic (48).

β-Lactams of the penicillin, cephamycin, and cephalosporin subclasses share a common biogenic origin, i.e., the condensation of three activated amino acids, L-α-aminoadipate, L-cysteine, and L-valine by aminoadipyl-cysteinyl-valine (ACV) synthase (*pcbAB* gene products) to form the tripeptide δ-L-(α-aminoadipyl)-L-cysteinyl-D-valine (ACV), which is then cyclized by the enzyme isopencillin N (IPN) synthase (*pcbC* gene product) to form isopenicillin N (49). ACV synthases comprise a subset of a larger class of nonribosomal peptide synthases, which have been characterized in detail, both biochemically and molecularly (49,50). Both ACV synthases (49,50) and IPN synthases (49) of all β-lactam producers are highly conserved, indicating common ancestral origins. IPN is either directly converted to penicillin G via IPN acyltransferase (*penDE* gene product), or via IPN epimerase (*cefD* gene product) to penicillin N, which is converted via deacetoxycephalosporin C (DAOC) synthase (*cefE* gene product; also called expandase because it expands the 5-membered ring of penicillins to the 6-membered ring of cephalosporins) to deacetoxycephalosporin C (DAOC) (49).

The clavam β-lactamase inhibitor, clavulanic acid, is produced by *Streptomyces clavuligerus*, which also produces cephamycins (49). Interestingly, the biogenesis of the bicyclic nucleus of clavulanic acid occurs via a completely different mechanism than does the biosynthesis of other β-lactams (49,51). In *S. clavuligerus*, the clavulanic acid biosynthesis genes are just downstream of the cephamycin C biosynthesis genes (51).

Not all β-lactam-producing taxonomic groups of organisms synthesize all subclasses of β-lactams, as depicted in Table 4. Filamentous fungi, for example, produce β-lactams only of the penam and cephem classes, whereas various actinomycetes produce β-lactams of all subclasses

Table 3. Important Clinical and Experimental β-Lactams

Subclass	Drugs on market[a]
Penicillins	Ampicillin, *amoxicillin*, bacampicillin, cloxacillin, floxacillin, mezlocillin, nafcillin, oxacillin, penicillin G, penicillin V
Penicillinase-resistant penicillins	*Methicillin*, dicloxacillin
Antipseudomonal penicillins	Carbenicillin indanyl, piperacillin, *ticarcillin*
First-generation cephalosporins	Cefadroxil, cefazolin, *cephalexin*, cephalothin, cephapirin, cephradine
Second-generation cephalosporins	*Cefaclor*, cefmetazole, cefonicid, cefotetan, *cefuroxime*
Third-generation cephalosporins	Cefamandole, *cefixime*, cefoperazone, cefotaxime, cefpodoxime proxetil, *cefprozil*, ceftazidime, *ceftibuten*, ceftizoxime, ceftriaxone
Oxycephams (third generation)	Flomoxef, latamoxef
Fourth-generation cephalosporins	*Cefepime*
Cefam	*Cefoxitin*
Carbacefem	*Loracarbef*
Carbapenems	*Imipemen / cilistatin*, meropenem, MK-826, panipenem (Japan only)
Monobactams	Aztreonam
Clavams (β-lactamase inhibitors)	*Clavulanate*, sulbactam, tazobactam
Penicillins/β-lactamase inhibitors	*amoxicillin / clavulanate (Augmentin)*, ampicillin/sulbactam (Unasyn), pipericillin/tazobactam (Zosyn), ticarcillin/clavulante (Timentin)

[a]Compounds in italics are most important commercial antibiotics.

Table 4. Biogenic Classes of β-Lactams

Subclass	Example compounds	β-Lactams produced by[a]		
		Fungi	G^+ bacteria	G^- bacteria
Penams	Penicillin G	*Penicillium, Aspergillus*	—	—
Cephems	Cephalosporin C	*Cephalosporium*	—	—
	DAOC[b]	—	—	Various strains
	Cephabacins	—	—	*Lysobacter*
	Cephamycin C	—	*Streptomyces, Nocardia*	—
Carbapenems	Thienamycin	—	*Streptomyces cattleya*	*Serratia, Erwinia*
Monobactams	Aztreonam[c]	—	*Nocardia*	*Pseudomonas*
Clavams	Clavulanic acid	—	*Streptomyces*	—
	Clavamycins	—	*Streptomyces*	—

Note: See Cohen and Aharonowitz (47) for more information. Dash means not produced by this group.
[a]Example taxonomic groups are given by genus names rather than an exhaustive listing of producing organisms.
[b]DAOC = deacetoxycephalosporin C.
[c]Example given is the biomimetic synthetic compound of the class.

except the penams and cephalosporins of the cephem subclass. The cephems in general are produced by the taxonomically widest range of microorganisms, including filamentous fungi, Gram-negative bacteria (e.g., lysobacters), and actinomycetes. The cephem products accumulated by each group differs, however, likely indicating evolutionary branches in terms of enzymes present. The Gram-negative bacteria, *Flavobacterium* sp. SC 12.154 and *Lysobacter lactamgenes* accumulate a 7-formylamino analogue of cephalosporin called cephabacin (49,52), an unusual natural β-lactam structure that is highly resistant to β-lactamases (49). The actinomycetes *Streptomyces clavuligerus* and *Nocardia lactamdurans* produce cephamycin C, which is formed from DAOC by the enzymes (gene names in parentheses) DAOC hydroxylase (*cefF*), deacetylcephalosporin C (DAC) *O*-carbamoyltransferase (*cmcH*), cephalosporin 7-α-hydroxylase (*cmcI*), and 7-α-hydroxy-*O*-carbamoyldeacetoxycephalosporin C methyltransferase (*cmcJ*) (49). The filamentous fungus, *Cephalosporium acremonium*, on the other hand, converts DAOC to cephalosporin C in two steps: DAOC hydroxylase (*cefF*), and DAC acetyltransferase (*cefG*) (49).

As is well known, penicillin was the first widely used clinical antibiotic. The first oral penicillin, used primarily to treat Gram-positive bacteria, was the fermentation product phenoxymethylpenicillin (penicillin V), still sold today under the name Veetids (5). Second-generation penicillins include the aminopenicillin group of ampicillin, amoxicillin, and bacampicillin, which exhibited much better gastrointestinal tract absorption and possess longer half lives. These aminopenicillins also exhibit greater activity against Gram-negative bacteria at the expense of activity against Gram-positives. Unfortunately, these penicillins are highly susceptible to β-lactamases. Development of β-lactamase inhibitors such as clavulanic acid, sulbactam, and tazobactam, however, have greatly extended the use of these broad-spectrum penicillins (47,49). In fact, Augmentin (oral amoxicillin/clavulanic acid), Unasyn (injectable ampicillin/sulbactam), Timentin (injectable ticarcillin/clavulanic acid), and Zosyn (injectable pipericillin/tazobactam) are widely used antibiotics. Carbenicillin indanyl, the first oral antipseudomonal β-lactam, and ticarcillin, a potent parenterally administered drug, are third-generation penicillins (53). The development of antipseudomonal fluoroquinolones, however, has rendered carbenicillin indanyl obsolete. Pipericillin, azlocillin, and mezlocillin are parenteral fourth-generation penicillins that have not found wide usage clinically (53).

First-generation cephalosporins typically inhibited Gram-positive bacteria other than the enterococci. Second- and third-generation cephalosporins typically inhibit Gram-negative bacteria at the expense of activity against Gram-positives. A new fourth-generation parenteral cephalosporin, cefepime (Maxipime), possesses activity against both Gram-positives and Gram-negatives (54). Additionally, cefipime penetrates bacteria well, is not susceptible to β-lactamases, and has a lower propensity to induce β-lactamases (55). Unfortunately, this broad-spectrum cephalosporin does not effectively treat either VREs or MRSAs (54). Ceftriaxone was very recently approved as a one-time-only injectable for treatment of pediatric otitis media. The concept behind the development of this drug is that the greatest problem, and reason for failure, with treatment of otitis media is patient compliance; the option of a one-time-only injectable removes this as an obstacle.

Carbapenems are very broad spectrum, β-lactamase-resistant β-lactams that contain a carbon atom instead of a sulfur atom in the 5-membered ring (47). Imipenem/cilistatin (Primaxin) was the first clinical carbapenem antibiotic, developed in the mid-1970s and made available commercially since 1986, as a synthetic biomimetic analogue of the natural product thienemycin (56). Imipenem is administered with cilistatin, an inhibitor of human renal dehydropeptidase, to protect the antibiotic from enzymatic degradation. Imipenem has broad-spectrum activity against Gram-negatives and anaerobic bacteria and moderate activity against Gram-positive bacteria (53). Unfortunately, imipenem also is limited by potential renal toxicity (56). Meropenem (Merrem) (57) is a new carbapenem antibiotic approved in 1996 for use as a broad-spectrum injectable antibiotic for treatment of intra-abdominal infections and for pediatric bacterial meningitis. The advantage of meropenem over imipenem is that meropenem is stable in the presence of renal dehydropeptidase, thereby bypassing the need for cotherapy with cilistatin (53). A third carbapenem called panipenem is available in Japan but not in the United States (58). Merck currently has a new, very potent, broad-spectrum, long-acting carbapenem, MK-826, in phase-II clinical trials (59). This new carbapenem drug candidate has been demonstrated to be highly resistant to β-lactamases, including plasmid-encoded extended spectrum β-lactamases (ESBLs) (59). Nevertheless, certain rarely pathogenic bacteria (particularly *Acinetobacter baumannii*) contain specific carbapenem-degrading enzymes known as carbapenemases (60), which may, through horizontal gene transfer, enter the broader pathogen field in the future.

An interesting new parenteral β-lactam drug candidate has recently been described. The new Glaxo-Wellcome experimental compound, GV129606, called a trinem, is a highly potent antibacterial with efficacy against Gram-positives, Gram-negatives, aerobes and anaerobes alike, plus marked resistance to extended spectrum, clinically relevant β-lactamases (61). This compound may mark a new path to be followed in the β-lactam field.

Macrolides

The macrolide antibiotics are a group of compounds produced by actinomycetes via type I modular polyketide synthases (62). The core aglycone of the prototype macrolide, erythromycin, is synthesized by the huge multifunctional enzyme, 6-deoxyerythronolide B synthase (DEBS), which consists of three, multidomain subunits, each subunit with a molecular weight around 300,000 Da. The macrolide chain is synthesized by a set of enzymatic functional domains within the proteins in a linear, stepwise fashion. Interestingly, the enzymatic active sites are arranged in the polypeptides in the same order as their apparent function (62).

Although macrolides theoretically can have ring sizes ranging from as few as 6-membered to over 30-membered, most of the clinically relevant macrolides possess 14-membered or 16-membered rings (Table 5). Current clinical antibacterial macrolides are mostly derived from erythromycin, although a wide variety of biologically active macrolides are used for several purposes across several therapeutic areas (Tables 5 and 6).

The macrolide antibiotic, erythromycin, was first discovered in 1952, the year the double-helical structure of DNA was first announced. Macrolides, which are bacteriostatic agents, inhibit bacterial protein synthesis by structure-based mechanisms: 14-membered macrolides such as erythromycin block translocation of peptidyl-tRNA, whereas 16-membered macrolides inhibit peptidyl-transfer reactions (64). There are apparently additional interactions of macrolides of both groups with ribosomes, but even after 30 years of research, these are not fully understood (64). Resistance to macrolides is most often via the MLS resistance mechanism, which is based on mechanism of action. MLS resistance, conferred by several different genes, results in methylation of adenosine-2058 of 23S rRNA. MLS-I resistance, exemplified by the *tlrD* gene, is result of monomethylation, whereas MLS-II (*ermE* -type) is the result of dimethylation (65).

Erythromycin itself is a poorly soluble, acid-labile molecule that nevertheless has enjoyed a long history of successful use. Erythromycin has been an important orally active antibiotic for many years. The efforts for years in the macrolide field have been aimed at developing new drugs that are acid-stable, possess a longer half-life, have broader ranges of activity, and are not subject to MLS resistance mechanisms. Recently, several semisynthetic macrolides arising from modifications of erythromycin, including clarithromycin (Biaxin) the 15-membered azalide, azithromycin (Zithromax), and most recently, dirithromycin (Dynabac), have made significant contributions to the antibacterials market. Among them, these macrolides represent more than $2 billion in market share.

The most exciting research area in antibacterials today may be the discovery of ketolides, erythromycin-analogue macrolides in which the L-cladinose at the 3-position of

Table 5. Clinical and Experimental Macrolides of Interest as Sorted by Ring Size

Compounds	Ring size	Comments
Spinosad (spinosyns)	12	New insecticidal (esp. against lepidopterans) drugs from Dow AgroSciences
Methymycin	12	Antibacterial; produced by same PKS and gene cluster as picromycin (63)
Clarithromycin	14	Biaxin; 6-*O*-methylerythromycin; acid stable form available in 1-day dosing
Dirithromycin	14	Dynabac; approved in 1995 as orally available, single dose/day formulation
Erythromycin A	14	Original 14-membered macrolide antibacterial agent
Flurithramycin	14	8-Fluoro-erythromycin
Galbonolides	14	Group of 14-membered antifungal macrolides
Ketolides	14	HMR3004 and HMR3647 are new, experimental, 3-oxo-semisynthetic macrolides
Oleandomycin	14	Antibacterial compound used as agricultural feed additive
Picromycin	14	Synthesized from same PKS and gene cluster as methymycin (63)
Roxithramycin	14	9-[*O*]-(2-Methoxyethoxy)methyloxime] derivative of erythromycin
Rustmicin	14	Experimental 14-membered ring antifungal agent of the galbonolide group
Troleandomycin	14	Acid-stable acetylated ester of oleandomycin; antibacterial
Azithramycin	15	Zithromax; Orally active azalide; semisynthetic analogue of erythromycin
Avermectin	16	Veterinary use; also use to treat human onchocerciasis (river blindness)
Doramectin	16	Mutational biosynthetic analogue of avermectin; antiparasitic
Epothilone	16	Experimental myxobacterial antitumor drug; targets microtubule formation
Ivermectin	16	Semisynthetic analogue of natural product avermectin; antiparasitic
Josamycin	16	Used in both human and veterinary treatments
Midecamycin A_1	16	Antibacterial agent
Milbemycin	16	Used in Europe as antiparasitic agent
Miocamycin	16	Acetylated ester of midecamycin gives oral bioavailability
Rokitamycin	16	Propionyl ester of leucomycin A5; active against macrolide-resistant Gram-positives
Spiramycin	16	Used in France as antibacterial agent
Tylosin	16	Veterinary use as antibacterial feed additive
Soraphen	18	Myxobacterial macrolide patented as an agricultural antifungal agent
Ascomycin	23	Also called FK-520 and immunomycin; immunosuppressive compound
Tacrolimus (FK-506)	23	Immunosuppressive agent
Rapamycin	31	Immunosuppressive agent

erythromycin has been replaced by a keto moiety. The ketolides are particularly interesting because they do not induce MLS-resistance mechanisms in pathogenic organisms. This is critical because MLS resistance is typically not constitutive. The two leading ketolide drug candidates, RU64004 (also known as HMR 3004) and RU66647 (also known as HMR 3647), have been shown to be potent inhibitors of Gram-positive bacteria, particularly the pneumococci, including those that show marked resistance to other macrolides such as erythromycin and azithromycin (66). The ketolides, however, show poor activity against Gram-negative bacteria.

Recently, a series of new, 14-membered macrolide antifungal compounds known collectively as galbonolides was discovered (67), of which one, rustmicin, has been investigated in more detail (68). The data indicate that rustmicin and, by extension, possibly the other galbonolides appear to interact with the α-mannan of the cell wall by a potentially novel mode of action (68). Several years ago, interest was sparked when a macrolide known as FK506 (Tacrolimus) was found to possess significant immunosuppressive action (69). Additional compounds of interest in this area included rapamycin, which is expected to be approved soon, and immunomycin, which was dropped because of a narrow efficacy-to-toxicity window (69). Both the milbemycins and avermectins have been used for several years as antiparasitic drugs (70). Ivermectin has been used and approved to treat human diseases such as river blindness and intestinal threadworm. Additionally, macrolides such as tylosin have long been important as animal-feed additive growth promotants in the agricultural area (71).

During the past few years, interest in macrolides as bioactive natural products has grown significantly in areas other than the anti-infective area. As shown in Table 6, there currently is active experimental interest in macrolides in multiple therapeutic areas, including antibacterial, antifungal, antiparasitic, anticancer, immunosuppressive, neuroimmunophilin, gastrointestinal, respiratory, and agricultural areas.

There are five human therapeutic areas in which macrolides have not been used clinically, but in which there is currently considerable interest. An acid-catalyzed breakdown product of erythromycin, called motilide, has been shown to be a potent agonist of the motilin receptor, but it is unstable. Abbott has the lead drug candidate in this area, ABT-229, a 12-deoxyerythromycin homologue possessing modification in the sugars, which is stable and has reasonably high potency (74). It has recently been demonstrated that tacrolimus (formerly FK506) and rapamycin possess additional activities other than those related to immunosuppression. Both tacrolimus and rapamycin have been demonstrated to promote neuron cell growth under conditions in which they can still bind the FK-binding protein in the absence of the second ligand (calcineurin or RAFT, respectively) (73). The macrolide, roxithramycin, currently an investigational drug for treatment of asthma (75), potentially has multiple modes of action for this indication, including inhibition of glycoconjugate secretion to

Table 6. Therapeutic Areas in which Macrolides Are Being Studied as Potential Bioactive Molecules

Therapeutic area	Example macrolides	Comments
Antibacterial	Erythromycin analogues	Traditional area of strength based on erythromycin (64)
Antifungal	Rustmicin, galbonolides	New activities of interest; interaction with mannans (67,68)
Antiparasitic	Avermectin, doramectin	Avermectin recently approved for human use in U.S. (70)
Anticancer	Epothilone	Tubulin stabilizer activity similar to Taxol (72)
Immunosuppressive	Tacrolimus, rapamycin	Substitutes for cyclosporin (69)
Neuroimmunophilin	Rapamycin and analogues	Stimulation of neuronal growth (73)
Motilin receptor	Motilide, ABT-229	Treatment of gastroesophageal reflux disorder (74)
Respiratory	Roxithramycin	Antiasthmatic by multiple potential mechanisms (75)
Agricultural	Tylosin	Antibacterial and growth promoting feed additive (71)

increase mucous clearance; clearance of *Chlamydia pneumoniae*, a possible causative or implicated agent in asthma; suppression of T-cell cytokines interleukins 2, 3, and 4; and inhibition of cholinergic neuroeffector transmission in airway smooth muscle cells. Finally, a macrolide produced by the myxobacterium, *Sorangium cellulosum*, was found to possess tubulin-stabilizing activity similar to Taxol (72), a $1 billion per year anticancer drug. Several companies and research laboratories have shown strong interest in trying to develop epothilone analogues as potential antitumor drugs.

Streptogramins

For more than 25 years, pristinamycin, a combination of two antibiotics, pristinamycin I_A (a peptidic macrolactone belonging to the streptogramin B group) and pristinamycin II_A (a polyunsaturated macrolactone belonging to the streptogramin group A) (76), has been used in Europe as a drug to combat severe infections caused by both Gram-positive and Gram-negative bacteria, without development of resistance against it (76). Each component of pristinamycin is bacteriostatic with poor efficacy, but the combination is bacteriocidal with good potency against a broad range of bacteria (76). The usefulness of pristinamycin, however, has been limited because of its poor solubility (77). In February 1996, however, Rhône-Poulenc Rhorer introduced Synercid, a new variation on pristinamycin. A mixture of the two water-soluble, semisynthetic streptogramins, quinupristin (RP57669) and dalfopristin (RP54476) (78), Synercid (RP 59500) is currently in phase III clinical trials worldwide and is currently available in the United States and Europe in an emergency-use program. More than 700 patients have received Synercid in this program. In one study, 70% of 115 patients with VRE infections were successfully treated (i.e., infection cleared) with Synercid (5). Synercid has been found to be bactericidal for most Gram-positive bacteria, but bacteriostatic for *Listeria monocytogenes* (78). In one study, 96% of all staphylococci tested were sensitive to Synercid in vitro (76), suggesting that this drug may become the "vancomycin of the future," a drug of last resort for use against severe bacterial infections. One hopeful aspect concerning Synercid is that it is unaffected by classical rRNA-methylase-conferred MLS resistance mechanisms, the most common resistance mechanisms found in staphylococci (79). Other MLS resistance mechanisms such as acetylation or hydrolysis of streptogramins still affect these compounds, however, so they are not completely impervious to bacterial resistance mechanisms (79). A recent study showed that Synercid at 2 μg/mL inhibited nearly 99% of vancomycin-resistant *E. faecium* clinical isolates tested in the United States (80), giving rise to hopes that Synercid will help to contain the growing VRE problem.

Tetracyclines

The tetracyclines are among the earliest of antibacterial antibiotics to be discovered, with chlortetracycline being discovered in 1948. The natural tetracyclines are produced by various species of actinomycetes; *Streptomyces aureofaciens* produces both chlortetracycline and tetracycline, the production of each which can be optimized by specific fermentation conditions (81), *Streptomyces rimosus* produces oxytetracyclines, and dactylocyclines are produced by *Dactylosporangium* spp. and *Actinomadura brunnea* (81).

Tetracyclines function as antibiotics by inhibiting protein translation, specifically by disrupting codon-anticodon interactions of tRNA and mRNA at the acceptor site of the ribosome. Because they inhibit protein synthesis, tetracyclines, like macrolides, are bacteriostatic. Tetracyclines are widely used not only in human health, but also in animal health programs; perhaps the widest current use of chlortetracycline and oxytetracycline is as prophylactic drugs in the fish husbandry industry (81).

Bacteria typically develop resistance to tetracyclines by acquiring a plasmid- and transposon-borne resistance gene (*tetA*) encoding an efflux pump (81). Thus, tetracycline resistance is one of the primary resistances observed as being transmitted horizontally among different groups of bacteria. Additionally, a second mechanism, mediated by TetM, gives resistance by inhibition of the interaction of the tetracyclines with the ribosome. This, unfortunately, has led to the diminution of use of the classical tetracyclines for most human health indications. Some of the more modern tetracyclines, such as minocycline, are important for treating MRSAs (81).

The most exciting advance in the tetracycline area has been the development of the glycylglycine tetracyclines (81). Two new semisynthetic glycylglycines, *N,N*-dimethylglycylamido-9-aminominocycline (DMG-MINNO) and 9-amino-6-demethyl-6-deoxytetracycline (DMG-DMDOT), have been developed by Lederle for their very broad spectrum of activity against Gram-positive, Gram-negative, aerobic, and anaerobic bacteria (54). Their spectrum includes, perhaps surprisingly, bacteria as widely disparate as mycoplasmas, MRSAs, VREs, *Neisseria* spp., *Haemophilus* spp., and *Moraxella* spp. Most importantly, glycylglycine tetracyclines bind the ribosome more tightly than traditional tetracyclines so that TetM, which mediates ribosomal resistance mechanism, cannot interfere with their activity, and they are not substrates for the TetA-mediated efflux mechanisms (81). Thus, there currently is no effective resistance mechanism yet known for these modified tetracyclines. For more information, Hunter and Hill (81) have recently described the tetracyclines in detail.

Glycopeptides

Vancomycin was first isolated in the early 1950s, making it another of the group of long-standing, well-known antibacterial drugs. For a long time, vancomycin was not used to a significant degree, but as the incidence of infections with drug-resistant organisms increased, the use for vancomycin became significantly greater. Because vancomycin is absorbed poorly by the gastrointestinal tract, it is rarely used as an oral drug. Instead, its primary use today comes as a parenteral drug as the last line of defense against many potentially lethal pathogenic bacteria, including MRSAs and antibiotic-resistant *Enterococcus faecium*. In 1987, however, after 30 years of vancomycin being on the market, resistant strains of *E. faecium* (VREs) were finally observed (39). The increased incidence of vancomycin resistance, especially in *S. aureus*, is of enormous concern (44).

Little more than a decade ago, resistance to vancomycin was unheard of and totally unsuspected, because its mechanism of action was so different from that of any other antibiotic class. Nevertheless, resistance to vancomycin and other glycopeptides was discovered and is on the rise. Vancomycin and other glycopeptides exert their activity by inhibiting peptidoglycan biosynthesis, making them potent bacteriocidal drugs. Peptidoglycan consists of a sugar backbone to which pentapeptide chains are attached. The terminal residues of the pentapeptide chain are D-alanine residues, which are critical to processing the peptidoglycan matrix as it is built. Vancomycin and other glycopeptides bind, through multiple hydrogen bonding, to the terminal D-alanine residue of the pentapeptide chain and prevent both transglycosylation and transpeptidation (82). Resistance occurs when a bacterium contains special genes (*vanAHXY*) that allow for incorporation of a D-lactic acid residue in place of the terminal D-alanine residue of the peptidoglycan pentapeptide chains, rendering the peptidoglycan unrecognizable to vancomycin binding. Two genetically distinct forms of resistance have emerged, although both operate by the same mechanism. VanA-mediated resistance gives high-level resistance to both vancomycin and teichoplanan, whereas VanB-mediated resistance is restricted to vancomycin (82). A recent study of clinical VRE isolates in the United States showed that 72% possessed the VanA phenotype, whereas only 28% possessed the VanB phenotype (80).

There are several relatively new analogues of vancomycin on the market to combat vancomycin resistance. Teichoplanan has been around for several years, but its use is restricted to Europe so far. Ramoplanan, another glycopeptide that has been around for a few years, is currently in phase II clinical trials at BioSearch Italia. Eli Lilly has developed a new chlorobiphenyl analogue of vancomycin, LY333328, that is considerably more potent than vancomycin (83). Moreover, LY333328 appears not only to inhibit bacterial growth, but to kill existing bacteria, which may give it a distinct clinical advantage over other existing antibiotics. LY333328 is very potent against VRE in vitro (84), giving hope for treatment of vancomycin-resistant pathogens.

Aminoglycosides

Aminoglycosides are not widely used in clinical therapy in the United States because of their relative toxicity profile as compared to other available antibacterials and to the rapid development of plasmid-borne resistance mechanisms through pathogen populations. Streptomycin and neomycin were first described in 1944 and 1949, respectively, so this group of antibacterials has a rich history. An incredible amount of knowledge about the biosynthesis, the genes, and the molecular mechanisms for regulation of aminoglycoside production is now known, and Piepersberg, the architect of much of this effort, has coalesced this enormous wealth of information in a recent review (85). One new aminoglycoside worth mentioning is the development in Japan of arbekacin, a new aminoglycoside that has potent activity against methicillin-, gentamycin- and tobramycin-resistant *S. aureus* strains (86). The incidence of methicillin-resistance in Japan has risen to nearly 60%, and only vancomycin and the newly developed arbekacin are approved for treatment of these strains. The new aminoglycoside is apparently effective because the widely distributed resistance mechanism, a bifunctional aminoglycoside phosphotransferase (APH[2″])-6′-acetyltransferase (AAC[6′]) which confers resistance to other aminoglycosides such as gentamycin and tobramycin, is relatively ineffective at modifying the new drug (86). It is probable that within a few years, APH(2″)-AAC(6′) will evolve so that it will be able to modify arbekacin.

Other Antibacterial Compounds and New Activities of Interest

Several new antibacterial drugs recently have been introduced to the market or in clinical trials that hold hope for containing the problems of newly emergent pathogens and increased resistance to existing drugs. Several of these are listed in Table 7. A comparison of the minimal inhibitory concentrations (MICs) to kill 90% of *S. pneumoniae* treated in vivo (MIC_{90}s, in μg/mL) was made of several of the new antibacterials with the following results (in order of efficacy) (87): ketolides (new macrolides containing keto groups at C-3), 0.016 to 0.25; everninomicins (novel oligoglycoside from marine organism), 0.06 to 0.25; glycylglycines (modified tetracyclines), 0.12 to 0.5; new fluoroquinolones (e.g., sparfloxacin and trovafloxacin), 0.12 to 0.5; oxazolidinones (new class of chemosynthesized antibacterials), 0.5 to 1.0; and new streptogramins (e.g., Synercid), 0.25 to 2.0. These can be compared to MIC_{90}s for the ex-

Table 7. Newly Approved and Experimental Antibacterial Agents

Compound or drug	Class[a]	Comments
Azithromycin	Azalide	New 15-membered macrolide containing a nitrogen in the ring; approved for pediatric use against otitis media
Trovafloxacin	Quinolone	New very broad-spectrum, high potency quinolone (Pfizer)
Mupirocin	Pseudomonic acid	The pseudomonad product, which inhibits bacterial isoleucyl tRNA synthetase, was recently approved in the U.S. as a topical drug
Cefepime	Cephalosporin	Fourth generation very broad spectrum cephalosporin; has potency of cefotaxime against Gram-positive bacteria and of ceftazidime against Gram-negative bacteria
Oxazolidinones	Chemically made	Two new oxazolidinones are being tested as potential broad-spectrum antibiotics for use against VRE and staphylococci (Upjohn)
Synercid	Streptogramins	Now in phase III clinical trials; combination of quinupristin (RP57669) and dalfopristin (RP54476), both semisynthetic streptogramins
Ziracin (SCH27899)	Everninomycin	Although discovered in 1979, everninomycins were not pursued because of their narrow spectrum; now Ziracin is a leading clinical candidate for treatment of VRE and staphylococci
PA 1648	Ansamycin	Rifampin-derivative being developed that has excellent activity against *Mycobacterium tuberculosis* and *M. avium–M. intracellulare* complex, the most common opportunistic bacteria infection of AIDS patients (PathoGenesis)
Meropenem	Carbapenem	New, safer penem with better activity against Gram-negative bacteria
Rifapentine	Ansamycin	FDA approved in 1998 for treatment of pulmonary tuberculosis
MK-826	Carbapenem	Long-acting, stable, broad-spectrum carbapenem now in phase II clinical trials (Merck)
Trinem	Novel β-lactam	New, broad-spectrum β-lactam being tested that is resistant to β-lactamases (Glaxo-Wellcome)
LY333328	Glycopeptide	New chlorobiphenyl vancomycin analogue with improved potency over vancomycin; currently in phase II clinical trials (Eli Lilly)

[a]Natural product class, except where noted.

isting drugs on various (and variably resistant) strains of *S. pneumoniae*: amoxicillin-clavulanate (Augmentin), 0.06 to 1.0; cefuroxime (semisynthetic cephalosporin), 0.12 to 2.0; cefpodoxime (semisynthetic cephalosporin), 0.25 to 2.0; and cefixime (semisynthetic cephalosporin), 0.5 to 16 (87). As important as MICs are a low MIC does not alone make a good drug. The other critical factor is the development of antibacterials that can be taken orally in once-per-day doses to increase patient compliance. Thus, the time for a particular drug dose to maintain in vivo efficacy above MIC (T > MIC) value is also critical to dial into the structure of newly developed drugs (87).

Most of the new antibacterials on the market are modified analogues of existing, well-known classes of antibacterials: azithromycin and clarithromycin, as well as the new experimental ketolides, are based on modifications to the erythromycin core structure; trovafloxacin and sparfloxacin are new fluoroquinolones; cefepime is a new cephalosporin; meropenem and MK-826 are new carbapenems; trinem is a novel β-lactam; LY33328 is a new glycopeptide; and Synercid is comprised of two new streptogramins. Only the oxazolidinones, a chemical synthesis class, represents a new, novel class of antibacterials.

Pharmacia-Upjohn is testing two new oxazolidinones for their ability to treat multiple-drug-resistant strains of *Mycobacterium tuberculosis* (23). These new chemically synthesized antibiotics inhibit protein synthesis in a manner different from natural product protein synthesis-inhibitory antibiotics, have novel structures unrelated to structures of other antibiotics, and have a wide spectrum of activity (23). Experiments have shown that, at least in in vitro tests, pathogens do not form resistance against these antibiotics at significant frequencies.

Mupirocin (Bactroban) was approved in 1997 by the U.S. FDA for use as a topical agent, especially against nasal MRSA. Mupirocin has been used in the U.K. since 1985 and is currently approved in 90 countries. Produced as pseudomonic acid by *Pseudomonas fluorescens*, mupirocin is interesting because of its novel mode of action; it inhibits bacterial protein synthesis specifically through its binding to isoleucyl tRNA synthetase (88). Unfortunately, mupirocin-resistant MRSA have emerged, which is not surprising since the target is a single enzyme.

Ziracin is an interesting, if still-evolving story. The everninomycins have been known since the early 1960s as bacterial protein synthesis inhibitors, but they have attracted little attention for many years (89). Ziracin, an aromatic moiety-containing oligosaccharide, was first isolated in 1979 from a marine actinomycete, *Micromonospora cabonacea* (89). At that time, the everninomycin was not pursued because of its relatively narrow spectrum of being potent against only Gram-positive bacteria. With the increased resistance of Gram-positive bacteria to existing drugs, interest in the everninomycin, Sch27899, was revived. Now, Sch27899 is in clinical development as a novel antibacterial. The problem is that a similar everninomycin, avilamycin (MaxusG), has been used for several years in the European Union as a feed additive in the agricultural market. Recently, avilamycin-resistant isolates of *E. faecium* have been found in broiler chickens that are cross-resistant to Ziracin (90). This discovery is of great concern, potentially cutting short the effectiveness, and certainly the life span, of Ziracin if it is ever marketed (90).

ANTIFUNGAL AGENTS

The need for new antifungal drugs has never been greater for a variety of reasons. Fungal infections fall into two major categories, those that are localized, dermal, and mostly annoyances, such as ringworm, athlete's foot, and nail infections (onychomycoses), and those that are disseminated and life threatening. Life-threatening, disseminated fungal diseases have risen dramatically during the past few decades, making this an important area for new drug development. Typically, life-threatening fungal infections occur in patients with impaired defenses, such as patients with AIDS; with other debilitating diseases; who are under treatment with immunosuppressive drugs (usually after organ transplants, to prevent tissue rejection); who are receiving chemotherapy; or with severe burns. Additionally, with the potentially broader use of immunosuppressive drugs for other indications such as arthritis, the potential for fungal infections may be expanded.

The most important pathogenic fungi are *Candida albicans* (disseminated candidiasis), *Candida* spp., *Aspergillus fumigatus* (systemic aspergillosis), *Cryptococcus neoformans* (disseminated cryptococcosis leading to meningoencephalitis), *Pneumocystis carinii* (*P. carinii* pneumonia [PCP]), *Histoplasma capsulatum* (histoplasmosis), and *Coccidioides immitis*, the final example being a fungal pathogen very rich in chitin (91). The most important fungal pathogens associated with immunocompromised patients, such as those with AIDS or receiving immunosuppressive therapy, are *P. carinii*, *C. neoformans*, and *Candida* spp.

The current antifungal drugs and leading clinical candidates are listed in Table 8. Current potent antifungal drugs have problems that make them less than optimal for clinical use. For example, amphotericin B, the antifungal drug of choice in life-threatening cases, causes acute toxicity, resulting in it not being well tolerated in a large percentage of patients, although it is efficacious against PCP (92). Azoles (e.g., clotrimazole, fluconazole, ketoconazole), which target membrane ergosterol synthesis, are static in vitro; they stop fungi from growing but do not kill them. Such agents, which depend on the immune system to clear the existing fungal infection, are used in immunocompromised patients but, because they do not clear the infections, require lifelong therapy to prevent recurrence. Additionally, *Aspergillus* spp. are naturally resistant to azoles, preventing their use in typically aggressive systemic aspergillosis infections that result in high mortality rates. Finally, the incidence of azole-resistant fungi is increasing, particularly among *Candida* spp. such as *C. glabrata* and *C. krusei*.

Because most of the current antifungal drugs suffer from either toxicity (amphotericin B) or the limitations of being static in vitro and increasing resisted by target populations (azoles), there is a critical need for new antifungal agents that act against new targets. Azoles target lanosterol demethylase, a P450 enzyme involved in the synthesis of ergosterol (93), a sterol found in fungal cell membranes but not mammalian cell membranes. Antifungal polyenes, including amphotericin B, nystatin, candicidin, and faerifungin, also target the cell membrane of fungi, but in this case by the formation of transmembrane pores and other membrane-associated effects (94). Although polyene antifungal agents are typically fungicidal, they also are not generally well tolerated by a large majority of patients, making them drugs used as last resorts.

Drug resistance in fungi, which more closely resembles drug resistance in tumor cells, is inherently different from drug resistance in bacteria. First, there is no evidence for plasmid-borne drug resistance mechanisms in fungi, so resistance is limited to chromosomal and nontransferable resistance genotypes. Most fungal drug resistance is effected

Table 8. Clinical and Lead Experimental Antifungal Drugs

Antifungal drug	Type	Comments
Amphotericin B	Polyene (NP)	Clinical; broad-spectrum antifungal agent
Atovaquone	Substituted naphthoquinone (CS)	Clinical; used for *P. carinii* infections
Clotrimazole	Oral fluconazole (CS)	Clinical; for vaginal candidiasis and dermal use
Fluconazole	Bis-triazole (CS)	Clinical; Orally active, relatively broad spectrum
Flucytosine	Flurocytosine	Clinical; for cryptococcal meningitis and candidiasis
Itraconazole	Triazole (CS)	Clinical; used for histoplasmosis and blastomycosis
Ketoconazole	Triazole (CS)	Clinical; broad spectrum
LY303366	Lipopeptide echinocandin (NP)	Phase II clinical trials (Eli Lilly)
Miconazole	Azole (CS)	Clinical; oral, parenteral, and topical; broad spectrum
MK-0991	Lipopeptide pneumocandin (NP)	Phase III clinical trials (Merck); broad spectrum
Natamycin	Polyene (NP)	Clinical; for aphthalmic treatment
Nikkomycin Z	Nucleoside dipeptide (NP)	Phase I clinical trials; chitin synthesis inhibitor
Nystatin	Polyene (NP)	Clinical; orally active and broad spectrum
Pentamidine isethionate	Aromatic diamidine (NP)	Clinical; for *P. carinii* infections
Pradamycins	Benzonaphthacene quinone (NP)	Phase I clinical trials
Rustmicin	Macrolide (NP)	Unusual target; plasma protein interactions
Sordarin analogue	Unusual natural product (NP)	Experimental compound
Sulconazole nitrate	Imidazole (CS)	Clinical; for topical treatment
Sulfanilamide	Sulfonamide (CS)	Clinical; for topical vulvovaginitis
Terbinafine	Allylamine (CS)	Clinical (Lamisil); orally active; for nail fungal infections
Thiabendazole	Benzimidazole (CS)	Clinical; for treatment of worms and fungi
Trimetrexate	Diaminoquinazoline (CS)	Clinical; for treatment of *P. carinii*

Notes: This list may not represent all clinical or experimental antifungal drugs. Generic names are used. NP = natural product; CS = chemically synthesized.

by changes in membrane permeases, leading to exclusion of the drugs, induction of membrane-bound efflux mechanisms, modifications of activases required for the activation of some antifungal drugs into their active forms, and site-specific mutations of target enzymes. Degradation of antifungal drugs by target organisms (e.g., similar to the action of β-lactamases) or modification of antifungal drugs to inert forms (e.g., aminoglycoside phosphotransferases) has not been observed thus far.

With the sequencing of the entire *Saccharomyces cerevisiae* genome, a study was conducted to search for novel targets, unique to fungi or different enough in fungi, which might be used to develop new antifungal agents with novel mechanisms of action (95). Of 6,000 genes analyzed, only a dozen or so potential targets were identified, including chitin synthetase, β-1,4-glucan synthetase, tubulin, elongation factor 2,*N*-myristoyl transferase, acetyl-CoA carboxylase, inositol phosphoryl ceramide synthase, membrane ATPase, mannosyl transferase, tRNA synthetases, lanosterol dehydrogenase, lanosterol synthase, and squalene epoxidase.

The most widely considered new targets for fungicidal activity are all involved with cell wall biosynthesis or structure. Yeast cell walls, which make up 15 to 25% of the cell dry weight, are typically comprised of β-1,3- and β-1,6 glucan polymers (about 50%), α-mannan (highly glycosylated mannoproteins; about 50%), and chitin (about 2%). Although different species of fungi contain different percentages of these components in their cell walls, the major constituents are typically present (93). Thus, the biosynthesis or structural integrity of β-1,3-glucans, β-1,6-glucans, and α-mannans represent potential excellent targets. The pneumocandins and echinocandins, novel lipopeptides produced by *Glarea lozoyensis* and similar fungi, inhibit the synthesis of β-1,3-glucans (96). Two new drug candidates, Merck's MK-0991 and Eli Lilly's LY303366, are semisynthetic lipopeptide echinocandin derivatives that specifically target the Rholp dissociable subunit of the heterodimeric transmembrane 1,3-β-(D)-glucan synthase (92,93,96). Because there is apparently only a single gene that is responsible for the synthesis of this subunit, this is considered an excellent target. The pneumocandin experimental drug, MK-0991, has been shown to possess efficacy against *Aspergillus fumigatus, C. albicans, H. capsulatum*, and *P. carinii* (but not against *Cryptococcus neoformans*), suggesting that this would make an outstanding potential broad-spectrum candidate for disseminated, life-threatening fungal diseases, particularly in immunocompromised patients (92).

Other targets of interest that have been pursued recently include chitin and sphingolipid biosynthesis. Fungi contain three chitin synthetases (CSI, CSII, CSIII) that catalyze the synthesis of the polymer from individual units of UDP-*N*-acetylglucosamine. The leading antifungal chitin biosynthesis-inhibitor drug candidate is nikkomycin Z, a nucleoside dipeptide substrate analog of UDP-*N*-acetylglucosamine which competitively inhibits chitin synthases (91). Although nikkomycin Z is effective in vitro against *C. albicans, C. immitis*, and other fungi with relatively high chitin contents, it does not kill *Saccharomyces cerevisiae* (92,94). The polyoxins, also peptidyl-nucleoside natural products, also inhibit chitin synthase, but none of these has been developed yet as a clinical candidate. Rustmicin (also called galbonolide A), a 14-membered macrolide produced by *Micromonospora chalcea*, inhibits inositol phosphoceramide synthase at pM levels, resulting in the loss of all complex sphingolipids and the accumulation of ceramide (68). Although rustmicin was reasonably efficacious in a murine cryptococcosis model, it suffers from severe in vivo instability, brought about by serum protein binding, and by being an excellent substrate for *S. cerevisiae* efflux pump Pdr5 (68). The pradamycins are D-amino-acid-substituted benzonaphthacene quinone glycosides that appear to act by binding to the saccharide moieties of the mannoproteins in fungal cell walls, leading to a loss of intracellular potassium that results in gross morphological changes (6). Another unusual antifungal compound is sordarin, which exerts antifungal activity via inhibition of the elongation step in protein synthesis (6,97). Glaxo-Wellcome has developed a semisynthetic sordarin, GM237354, that possesses a broad spectrum of activity and high in vivo potency (6).

ANTICANCER AGENTS

The search for natural product anticancer agents began in the late 1950s and became a reality with the discovery in 1964 of daunorubicin, a cytotoxic, antitumor drug of the anthracycline class with efficacy against a wide variety of solid tumors and leukemias (15). Five years later, doxorubicin, the 14-hydroxy analogue of daunorubicin, was discovered at Farmitalia (15). Doxorubicin, which showed a wider range of efficacy for treatment of various cancers, has been a mainstay in chemotherapy for more than 25 years. It was shown recently that doxorubicin as well as many other cytotoxic antitumor drugs (e.g., etoposide) act by trapping topoisomerase II with DNA in what has been referred to as the cleavable complex, resulting in multiple DNA strand breaks (15). Another class of anthracyclines, exemplified by aclarubicin (aclacinomycin A), were recently shown to inhibit by a different mechanism of action, by inhibition of topoisomerae II–DNA interaction (15).

Table 9 shows a list of the major antitumor drugs derived from natural products. Other natural product antitumor chemotherapy agents used for several years, but more restrictively than doxorubicin, include mithramycin, discovered in the early 1950s; the mitomycins, first described in 1956; bleomycin, described by Umezawa in 1965; and the vinca alkaloids, vinblastine and vincristine, which were first described in the period of 1958–1962. In all these cases, the mode of action is cytotoxicity through drug-DNA interactions. With the ultimate desire to get away from cytotoxic drugs that are detrimental to all cells and not just cancerous cells, these cytotoxic natural product antitumor drugs are considered to be drugs of the past and not of the future.

Interest in natural product antitumor drugs was renewed, however, with the discovery of taxanes from the pacific yew tree. Paclitaxel (Taxol), a complex diterpenoid alkaloid, was discovered as a potential antitumor agent in 1971 (98), and it has taken nearly three decades to bring

Table 9. Clinical and Lead Experimental Natural Product-Derived Anticancer Drugs

Anticancer drug	Type	Comments
Aclacinomycin A	Anthracycline	Produced by *Streptomyces galilaeus*; used in Japan and France
Bleomycin	Glycopeptide	Produced by *Streptomyces verticillus*
Carminomycin	Anthracycline	Produced by *Actinomadura carminata*; used in Russia
Dactinomycin	Acylpeptidolactone	Produced by *Streptomyces parvulus*
Daunorubicin	Anthracycline	Produced by *Streptomyces peucetius*
Docetaxel	Taxane	Taxotere; recently approved for treatment of ovarian cancer
Doxorubicin	Anthracycline	Adriamycin; Produced by *S. peucetius* subsp. *caesius*
Epothilone	Substituted macrolide	Produced by *Sorangium cellulosum*; experimental only
Idarubicin	Anthracycline	Synthetic (mimetic), orally available analogue of doxorubicin
Irinotecan	Camptothecin	Approved in 1996 for metastatic colon and rectal cancers
Mitomycin C	Cytotoxic drug	Produced by *Streptomyces caespitosus*
Mitoxantrone	Anthracenedione	Biomimetic compound
Paclitaxel	Taxane	Produced by *Taxus* spp. (yew trees); approved for use in 1994
Pentostatin	Nucleoside analogue	Produced by *Streptomyces antibioticus*
Plicamycin	Polyketide	Mithramycin; Produced by *Streptomyces argillaceus*
Tenipocide	Podophyllotoxin	Cytotoxic antitumor drug
Topotecan	Camptothecin	Semisynthetic; from *Camptotheca acuminata* (Chinese tree)
Vinblastine sulfate	Vinca alkaloid	Produced by *Catharanthus roseus* (periwinkle)
Vincristine sulfate	Vinca alkaloid	Produced by *Catharanthus roseus* (periwinkle)
Vinorelbine tartrate	Vinca alkaloid	Semi-synthetic vinblastine analogue

it to the market. One of the unique features that helped move the taxanes from experimental drugs to clinical candidates was the finding that they acted as tubulin stabilizers, a novel mode of action (99). Docetaxel (Taxotere) and paclitaxel (Taxol) have excellent activity against ovarian and breast cancers. Additionally, Taxol and semisynthetic taxanes may be efficacious against a wide range of tumors. Thus far, the greatest impediment to the development of these drugs on a large scale was that they are only found at levels of about 0.01% (w/w) in the bark of several related yew trees, *Taxus brevifolia* (Pacific yew, an endangered species), *Taxus baccata, Taxus cuspidata* (Japanese yew), and others (100). Moreover, the chemical synthesis of paclitaxel is very difficult, requiring more than 30 steps. Thus, paclitaxel is currently made semisynthetically from 10-deacetylbaccatin III, produced by the needles of the Himalayan yew (100). Other novel antitumor drugs developed in recent years include irinotecan (Camptosa) and topotecan (Hycamptin), both which are inhibitors of topoisomerase I.

FUTURE DIRECTIONS IN NATURAL PRODUCTS DRUG DISCOVERY

Several approaches are being taken to address the issues of MRSA, VRE, other antibiotic-resistant bacteria, newly emerging infectious diseases and related problems. One approach is the use of genomics to help to identify new targets amongst bacteria and fungi. As mentioned previously, sequencing of the yeast genome helped to identify about a dozen unique targets for new antifungal drugs. It appears that the incidence of fungal resistance to new drugs targeted against single, critical enzymes, such as glucan synthase, is very low. Similarly, it is hoped that genomic analysis of bacteria is as useful in identifying new targets for novel antibacterials. Unfortunately, there is a significant difference between bacteria and fungi and between bacterial infections and fungal infections that must be considered. First, any single-site bacterial target for which single mutations can render resistance cannot be seriously considered because of the high frequency of mutations (10^{-8}) that occur in bacteria (77). Second, many resistance mechanisms are passed horizontally among bacterial populations by plasmid- and transposon-mediated transfer, which renders many antibiotics useless. Even the fluoroquinolones, which for years were thought to target only DNA gyrase, are now known to target at least two separate enzymes (101). This is critical, too, because single mutations in bacterial gyrases have been shown recently to yield resistance to several fluoroquinolones (102). What is the best answer? Although there is probably no single best answer for all resistance problems, the clear-cut best approach is still to inhibit cell wall biosynthesis, preferably with a novel mechanism of action or at several sites. Alternatively, inhibition of systems or processes, such as protein or nucleic acid synthesis, again at multiple sites, appears to be back in vogue. The conundrum, however, is that protein synthesis inhibitors are bacteriostatic agents, which in recent years have been thought not to be desired. Nevertheless, the newer macrolides, dirithromycin, clarithromycin, and azithromycin, all protein synthesis-inhibiting bacteriostats, are effective drugs.

The Pharmaceutical Research and Manufacturers of America organization announced that as of September 1998, 136 new anti-infective drugs were currently in advanced stages of development by 78 different companies. Of these, 27 compounds are antibacterials, 31 are antivirals, 12 are antifungals, 42 are vaccines, and 24 are in the "other" category, but still relating to anti-infection indications. As described throughout this review, there are unmet needs in all fields of antibiotics. The wealth of new drugs on the horizon bodes well for the future of antibiotics of all classes.

BIBLIOGRAPHY

1. S.B. Levy, *The Antibiotic Paradox: How Miracle Drugs Are Destroying the Miracle*, Plenum, New York, 1992.
2. E.J. Vandamme, in E.J. Vandamme ed., *Biotechnology of Industrial Antibiotics*, Dekker, New York, 1984, pp. 3–31.
3. A.L. Demain, in C.L. Hershberger, S.W. Queener, and G. Hegeman eds., *Genetics and Molecular Biology of Industrial Microorganisms*, American Society for Microbiology, Washington, D.C., 1989, pp. 1–11.
4. K.A. Reynolds and A.L. Demain, in W.R. Strohl ed., *Biotechnology of Antibiotics*, Dekker, New York, 1997, pp. 497–520.
5. W.R. Strohl, in W.R. Strohl ed., *Biotechnology of Antibiotics*, Dekker, New York, 1997, pp. 1–47.
6. Y.-Z. Shu, *J. Natl. Prod.* **61**, 1053–1071 (1998).
7. G.M. Cragg, D.J. Newman, and K.M. Snader, *J. Natl. Prod.* **60**, 52–60 (1997).
8. J. Bérdy, in V.G. Debabov, Y.V. Dudnik, V.N. Danilenko eds., *Proceedings of the Ninth International Symposium on the Biology of the Actinomycetes*, All-Russia Scientific Research Institute for Genetics and Selection of Industrial Microorganisms, Moscow, 1995, pp. 13–34.
9. E. Culotta, *Science* **264**, 362–363 (1994).
10. H. Umezawa, *Annu. Rev. Microbiol.* **36**, 75–99 (1982).
11. A.W. Alberts, J. Chen, G. Kuron, V. Hunt, J. Huff, C. Hoffman, J. Rothrock, M. Lopez, H. Joshua, E. Harris, A. Patchett, R. Monaghan, S. Currie, E. Stapley, G. Albers-Schonberg, O. Hensens, J. Hirschfield, K. Hoogsteen, J. Liesch, and J. Springer, *Proc. Natl. Acad. Sci. U.S.A.* **77**, 3957–3961 (1980).
12. American Druggist, *http://www.rxlist.com/top200a.htm*, Feb. 1998.
13. N. Pfeiffer, *Gen. Eng. News* **16**, 18–19 (1996).
14. SCRIP's 1995 Yearbook, PharmaBooks Ltd., New York.
15. W.R. Strohl, M.L. Dickens, V.B. Rajgarhia, A.J. Woo, and N.D. Priestley, in W.R. Strohl ed., *Biotechnology of Antibiotics*, Dekker, New York, 1997, pp. 577–657.
16. T.J. Walsh and N. Giri, *Eur. J. Clin. Microbiol. Infect. Dis.* **16**, 93–97 (1997).
17. P.R. August, L. Tang, Y.J. Yoon, S. Ning, R. Muller, T.W. Yu, M. Taylor, D. Hoffman, C.G. Kim, X. Zhang, C.R. Hutchinson, and H.G. Floss, *Chem. Biol.* **5**, 69–79 (1998).
18. L.C. Vining and C. Stuttard eds., *Genetics and Biochemistry of Antibiotic Production*, Butterworth-Heinemann, Boston, 1995.
19. W.R. Strohl ed., *Biotechnology of Antibiotics*, Dekker, New York, 1997.
20. L.C. Vining, in L.C. Vining and C. Stuttard eds., *Genetics and Biochemistry of Antibiotic Production*, Butterworth-Heinemann, Boston, 1995, pp. 499–504.
21. J. Mann, R.S. Davidson, J.B. Hobbs, D.V. Banthorpe, and J.B. Harborne, *Natural Products. Their Chemistry and Biological Significance*, Addison-Wesley Longman, Essex, U.K., 1996.
22. S. Budavari ed., *The Merck Index: An Encyclopedia of Chemicals, Drugs, and Biologicals*, 12th ed., Merck Research Laboratories, Merck and Co., Whitehouse Station, 1996.
23. C.W. Ford, J.C. Hamel, D. Stapert, J.K. Moerman, D.K. Hutchinson, M.R. Barbachyn, and G.E. Zurenko, *Trends Microbiol.* **5**, 196–200 (1997).
24. J.E. McGowan, *Rev. Infect. Dis.* **5**, 1033–1048 (1983).
25. J. Davies, *Science* **264**, 375–382 (1994).
26. M.L. Cohen, *Science* **257**, 1050–1055 (1992).
27. J. Travis, *Science* **264**, 360–362 (1994).
28. S. Falkow, *Infectious Multiple Drug Resistance*, Pion Ltd., London, 1975.
29. J. Lederberg, R.E. Shope, and S.C. Oaks eds., *Emerging Infections: Microbial Threats to Health in the United States*, Institute of Medicine, National Academy of Sciences Press, Washington, D.C., 1992.
30. R.A. Weinstein, *Emerg. Infect. Dis.* **4**, 416–420 (1998).
31. Centers for Disease Control and Prevention, *Morb. Mortal. Wkly. Rep.* **42**, 597–599 (1993).
32. E. Rubinstein and F. Bompart, *J. Antimicrob. Chemother.* **39**, 139–143 (1997).
33. *World Health Report 1996*, Office of World Health Reporting, World Health Organization, Geneva, 1996.
34. J. May, K. Shannon, A. King, and G. French, *J. Antimicrob. Chemother.* **42**, 189–197 (1998).
35. K. Sunakawa, H. Akita, S. Iwata, Y. Sato, and R. Fujii, *Infection* **23**, S74–S78 (1995).
36. B.R. Bloom and C.J.L. Murray, *Science* **257**, 1055–1064 (1992).
37. E. Collatz, R. Labia, and L. Gutman, *Mol. Microbiol.* **4**, 1615–1620 (1990).
38. P.M. Rowe, *Lancet* **347**, 252 (1996).
39. A.H.C. Uttley, C.H. Collins, J. Naidoo, and R.C. George, *Lancet* **1**, 57–58 (1988).
40. G.L. French, *Clin. Infect. Dis.* **27**, S75–S83 (1998).
41. P. Moreillon and A. Wenger, *Schweiz. Mediz. Wochensch.* **126**, 255–263 (1996).
42. W. Brumfitt and J.M.T. Hamilton-Miller, *Drugs Exptl. Clin. Res.* **20**, 215–224 (1994).
43. W.C. Noble, Z. Virani, and R.G.A. Cree, *FEMS Microbiol. Lett.* **93**, 195–198 (1992).
44. H.C. Neu, *Scand. J. Infect. Dis.* **91** 7–13 (1993).
45. K.J. Christensen and P.O. Gubbins, *Ann. Pharmacother.* **30**, 288–290 (1996).
46. Y. Sumita and M. Fukusawa, *Chemotherapy* **42**, 47–56 (1996).
47. G. Cohen and Y. Aharonowitz, in P.A. Hunter, G.K. Darby, and N.J. Russell eds., *Fifty Years of Antimicrobials: Past Perspectives and Future Trends*, Soc. Gen. Microbiol. Ltd., Cambridge, 1995, pp. 139–163.
48. H. Vanderhaeghe, *Verh. K. Acad. Geneeskd. Belg.* **53**, 39–58 (1991).
49. A.S. Paradkar, S.E. Jensen, and R.H. Mosher, in W.R. Strohl ed., *Biotechnology of Antibiotics*, Dekker, New York, 1997, pp. 241–277.
50. P. Zuber and M. Marahiel, in W.R. Strohl ed., *Biotechnology of Antibiotics*, Dekker, New York, 1997, pp. 187–216.
51. B.O. Bachmann, R. Li, and C.A. Townsend, *Proc. Natl. Acad. Sci. U.S.A.* **95**, 9082–9086 (1998).

52. H. Palissa, H. von Dohren, H. Kleinkauf, H.H. Ting, and J.E. Baldwin, *J. Bacteriol.* **171**, 5720–5728 (1989).
53. D.M. Allen and M.L. Ling, *Singapore Med. J.* **35**, 626–630 (1994).
54. P.M. Southern, *Compr. Ther.* **23**, 575–582 (1997).
55. M.A. Wynd and J.A. Paladino, *Ann. Pharmacother.* **30**, 1414–1424 (1996).
56. R.C. Moellering Jr., G.M. Eliopoulos, and D.E. Sentochnik, *J. Antimicrob. Chemother.* **24**, 1–7 (1989).
57. D.N. Fish and T.J. Singletary, *Pharmacotherapy* **17**, 644–669 (1997).
58. J. Shimada and Y. Kawahara, *Drugs Exptl. Clin. Res.* **20**, 241–245 (1994).
59. C.J. Gill, J.J. Jackson, L.S. Gerckens, B.A. Pelak, R.K. Thompson, J.G. Sundelof, H. Kropp, and H. Rosen, *Antimicrob. Agents Chemother.* **42**, 1996–2001 (1998).
60. D.M. Livermore, *J. Antimicrob. Chemother.* **39**, 673–676 (1997).
61. E. Di Modugno, R. Broggio, I. Erbetti, and L. Lowther, *Antimicrob. Agents Chemother.* **41**, 2742–2748 (1997).
62. L. Katz and S. Donadio, in L.C. Vining and C. Stuttard eds., *Genetics and Biochemistry of Antibiotic Production*, Butterworth-Heinemann, Boston, 1995, pp. 385–420.
63. Y. Xue, D. Wilson, L. Zhao, H.-W. Liu, and D.H. Sherman, *Abstr. 8th Intl. Symp. Genet. Industr. Microorg.*, Jerusalem, 1998, p. 39.
64. T. Mazzei, E. Mini, A. Novelli, and P. Periti, *J. Antimicrob. Chemother.* **31**, 1–9 (1993).
65. J.L. Pernodet, S. Fish, M.H. Rouault-Blondelet, and E. Cundliffe, *Antimicrob. Agents Chemother.* **40**, 581–585 (1996).
66. R.R. Reinert, A. Bryskier, and R. Luetticken, *Antimicrob. Agents Chemother.* **42**, 1509–1511 (1998).
67. H. Achenbach, A. Muhlenfeld, U. Fauth, and H. Zahner, *Ann. N.Y. Acad. Sci.* **544**, 128–140 (1988).
68. S.M. Mandala, R.A. Thornton, J. Milligan, M. Rosenbach, M. Garcia-Calvo, H.G. Bull, G. Harris, G.K. Abruzzo, A.M. Flattery, C.J. Gill, K. Bartizal, S. Dreikorn, and M.B. Kurtz, *J. Biol. Chem.* **273**, 14942–14949 (1998).
69. K.W. Mollison, T.A. Fey, R.A. Krause, V.A. Thomas, A.P. Mehta, and J.R. Luly, *Ann. N.Y. Acad. Sci.* **685**, 55–57 (1993).
70. D.J. MacNeil, in L.C. Vining and C. Stuttard eds., *Genetics and Biochemistry of Antibiotic Production*, Butterworth-Heinemann, Boston, 1995, pp. 421–442.
71. F.G. Day, *J. Am. Vet. Med. Assoc.* **208**, 655–656 (1996).
72. D.M. Bollag, P.A. McQueney, J. Zhu, O. Hensens, L. Koupal, J. Liesch, M. Goetz, E. Lazarides, and C.M. Woods, *Cancer Res.* **55**, 2325–2333 (1995).
73. J.P. Steiner, M.A. Connolly, H.L. Valentine, G.S. Hamilton, T.M. Dawson, L. Hester, and S.H. Snyder, *Nature Med.* **3**, 421–428 (1997).
74. P.A. Lartey, H.N. Nellans, R. Faghih, A. Petersen, C.M. Edwards, L. Freiberg, S. Quigley, K. Marsh, L.L. Klein, and J.J. Plattner, *J. Med. Chem.* **38**, 1793–1798 (1995).
75. S. Konno, K. Asano, M. Kurokawa, K. Ikeda, K. Okamoto, and M. Adachi, *Int. Arch. Allergy Immunol.* **105**, 308–316 (1994).
76. J.C. Barrière, D.H. Bouanchaud, J.M. Paris, O. Rolin, N.V. Harris, and C. Smith, *J. Antimicrob. Chemother.* **30** 1–8 (1992).
77. L.L. Silver and K.A. Bostian, *Antimicrob. Agents Chemother.* **37**, 377–383 (1993).
78. T. Nichterlein, M. Kretschmar, and H. Hof, *J. Chemother.* **8**, 107–112 (1996).
79. R. Leclercq, L. Nantas, S.-J. Soussy, and J. Duval, *J. Antimicrob. Chemother.* **30**, 67–75 (1992).
80. G.M. Eliopoulos, C.B. Wennersten, H.S. Gold, T. Schuelin, M. Souli, M.G. Farris, S. Cerwinka, H.L. Nadler, M. Dowzicky, G.H. Talbot, and R.C. Moellering Jr., *Antimicrob. Agents Chemother.* **42**, 1088–1092 (1998).
81. I.S. Hunter and R.A. Hill, in W.R. Strohl ed., *Biotechnology of Antibiotics*, Dekker, New York, 1997, pp. 659–682.
82. T.I. Nicas and R.D.G. Cooper, in W.R. Strohl ed., *Biotechnology of Antibiotics*, Dekker, New York, 1997, pp. 363–392.
83. T.I. Nicas, D.L. Mullem, J.E. Flokowitsch, D.A. Preston, N.J. Snyder, M.J. Zweifel, S.C. Wilkie, M.J. Rodriguez, R.C. Thompson, and R.D.G. Cooper, *Antimicrob. Agents Chemother.* **40**, 2194–2199 (1996).
84. R.S. Schwalbe, A.C. McIntosh, S. Qaiyum, J.A. Johnson, R.J. Johnson, K.M. Furness, W.J. Holloway, and L. Steele-Moore, *Antimicrob. Agents Chemother.* **40**, 2416–2419 (1996).
85. W. Piepersberg, in W.R. Strohl ed., *Biotechnology of Antibiotics*, Dekker, New York, 1997, pp. 81–163.
86. M. Inoue, M. Nonoyama, R. Okamoto, and T. Ida, *Drugs Exptl. Clin. Res.* **30**, 233–240 (1994).
87. W.A. Craig, *Diagn. Microbiol. Infect. Dis.* **27**, 49–53 (1997).
88. J.S. Bertino Jr., *Am. J. Health Syst. Pharm.* **54**, 2185–2191 (1997).
89. A.L. Demain, *Nature Biotechnol.* **16**, 3–4 (1998).
90. F.M. Aarestrup, *Microb. Drug Resist.* **4**, 137–141 (1998).
91. J.R. Graybill, L.K. Najvar, R. Bocanegra, R.F. Hector, and M.F. Luther, *Antimicrob. Agents Chemother.* **42**, 2371–2374 (1998).
92. M.A. Powles, P. Liberator, J. Anderson, Y. Karkhanis, J.F. Dropinski, F.A. Bouffard, J.M. Balkovec, H. Fujioka, M. Aikawa, D. McFadden, and D. Schmatz, *Antimicrob. Agents Chemother.* **42**, 1985–1989 (1998).
93. M.B. Kurtz, *ASM News* **64**, 31–39 (1998).
94. J.A. Gil and J.F. Martin, in W.R. Strohl ed., *Biotechnology of Antibiotics*, Dekker, New York, 1997, pp. 551–575.
95. Y. Koltin, *Abstr. 8th Intl. Symp. Genet. Industr. Microorg.*, Jerusalem, 1998, p. 5.
96. W.W. Turner and W.L. Current, in W.R. Strohl ed., *Biotechnology of Antibiotics*, Dekker, New York, 1997, pp. 315–334.
97. M.C. Justice, M.-J. Hsu, B. Tse, T. Ku, J. Balkovec, D. Schmatz, and J. Nielsen, *J. Biol. Chem.* **273**, 3148–3151 (1998).
98. M.C. Wani, H.L. Taylor, M.E. Wall, P. Coggen, and A.T. McPhail, *J. Am. Chem. Soc.* **93**, 2325–2327 (1971).
99. P.B. Schiff, J. Fant, and S.B. Horowitz, *Nature* **277**, 665–667 (1979).
100. Y. Yukimune, H. Tabata, Y. Higashi, and Y. Hara, *Nature Biotechnol.* **14**, 1129–1132 (1996).
101. X.-S. Pan and L. Fisher, *Antimicrob. Agents Chemother.* **41**, 471–474 (1997).
102. B. Waters and J. Davies, *Antimicrob. Agents Chemother.* **41**, 2766–2769 (1997).

SECRETION FROM ANIMAL CELLS

RANDAL J. KAUFMAN
University of Michigan Medical School
Ann Arbor, Michigan

KEY WORDS

Disulfide bond formation
Endoplasmic reticulum
Glycosylation
Golgi apparatus
Posttranslational modification
Protein chaperones
Protein folding
Protein trafficking
Proteolytic processing
Tyrosine sulfation

OUTLINE

INTRODUCTION

The specialized process of secretion in mammalian cells is directed through membrane-enclosed compartments composed of vesicles. Each compartment is unique to ensure specificity and directionality for the secretion process. The compartments are maintained by mechanisms of selective targeting of newly made proteins to their correct vesicle. As proteins destined for the cell surface transit the different compartments, they attain their final conformation through a maturation process that involves protein folding facilitated by molecular chaperones, covalent modification of the polypeptide backbone, and assembly into higher-order structures.

PROTEIN TRANSLOCATION INTO THE ENDOPLASMIC RETICULUM

Targeting to the Endoplasmic Reticulum

All proteins destined for the cell surface, the extracellular space, or intracellular organelles such as lysosomes or the Golgi apparatus are first translocated across the membrane of the endoplasmic reticulum (ER). In mammalian cells the process of protein transport across the ER membrane occurs in the majority of cases while the polypeptide is being translated, which is called cotranslational translocation. In contrast, in yeast, translocation can occur either cotranslationally or after translation of the polypeptide is complete, that is, posttranslational translocation. Proteins that are targeted for transport across the ER must contain a signal that directs the process. Most secretory proteins contain an amino-terminal set of amino acids that compose the signal that directs cotranslational translocation of the protein across the ER membrane. The signal peptide is exposed as the polypeptide emerges from the 60S ribosomal subunit and mediates association of the nascent polypeptide with the cytosolic face of the ER. Although the exact amino acid sequence of signal peptides is not conserved for a given protein across species, its general character is very conserved. The signal sequence may be approximately 30–60 amino acids in length and is composed of three regions: (*1*) an amino terminal segment of variable length that has a net positive charge, (*2*) a central hydrophobic core of 6–15 residues, and (*3*) a C-terminal region that often has a helix-breaking amino acid residue such as glycine proline or serine (1). In yeast, the more hydrophobic the character of the signal peptide, the more likely the transport process occurs by a cotranslational mechanism.

The signal recognition particle (SRP) is a complex of six polypeptides and one 7S RNA molecule that targets substrates for cotranslational translocation (for review see Ref. 2). The 300-nucleotide RNA component is very conserved across all species and is homologous with the highly repetitive Alu DNA family. The mechanism by which cotranslational translocation occurs was elucidated using in vitro studies of protein import into canine pancreatic microsomes, because they are efficient at directing SRP-dependent translocation. Analysis of the SRP has identified subcomplexes that perform different functions. SRP54 (the number refers to the molecular weight of the protein) binds the signal sequence and targets it to the ER membrane. SRP54 contains a stretch of methionine residues that provide a flexible surface, much like bristles of a brush, for interaction with the variable hydrophobic stretches of amino acids in different signal peptides. After interaction with SRP54, translation elongation pauses by a subcomplex of SRP9 and SRP14. This pause helps ensure proper targeting to the ER membrane before significant portions of the polypeptide are synthesized and begin to fold.

After SRP binds the signal sequence of the nascent chain–ribosome complex, it targets the complex to the ER translocation machinery (Fig. 1). The SRP receptor (SRPR) on the ER membrane is composed of two subunits SRα and SRβ and receives the nascent polypeptide after release from SRP. SRP then recycles back to the cytosol. Both SRP54 of the SRP and each of the SRP receptor subunits are GTPases. SRP and SRPR reciprocally stimulate each others' GTPase activity. SRP in the GTP-bound form promotes high-affinity association with SRPR (4). Subsequent transfer of the signal to the translocation apparatus depends on GTP binding by SRα (5). For recycling, SRP must be released from SRPR, a process that requires GTP hydrolysis by SRP54 (6). The nascent chain–associated complex (NAC), an ER-localized protein complex, ensures fidelity of cotranslational targeting to the translocon in the ER membrane by binding nascent polypeptides that lack signal sequences, preventing their targeting to the ER membrane, and preventing free ribosomes from interacting with the ER membrane (7–9). In addition, some polypeptides also require the translocating chain–associated membrane protein (TRAM) for translocation. TRAM can be cross-linked to the signal peptide of some secretory proteins (10). The signal-containing polypeptide inserts into the membrane, and the polypeptide threads through the protein channel, composed of Sec61α, into the lumen of the ER as an extending loop.

Translocation into the ER

The proteins that are required for translocation in vitro and in vivo have been identified through genetic studies in yeast. The most important protein is Sec61p (40 kDa), a protein predicted to span the ER membrane a number of times; it is the primary subunit of the protein translocation channel (11). Sec61p is part of a heterotrimeric complex with Sbh1p and Sss1p, which are the yeast homologues of mammalian Sec62β and Sec61γ (12). In addition to the Sec61p complex, another complex, Sec62p, Sec63p, Sec71p, and Sec72p, is required for posttranslocational translocation in yeast. Sec63p has a luminal region homologous to the J-domain that is highly conserved in the DnaJ protein family, a protein family that stimulates the ATPase activity of Hsp70 family members. The yeast homologue of BiP, Kar2p, is required for posttranslational translocation, probably through its ATPase activity, which is regulated by the J-domain of Sec63p. In mammalian cells, purified SRP, SRPR, and Sec61p are the only proteins required for cotranslocational translocation (12). TRAM is required for only a subset of precursors and stimulates translocation of others.

The role of resident proteins in the lumen of the ER was studied by selectively solubilizing and reconstituting microsomal translocation activity into proteoliposomes (13). Alkaline extraction at pH 9.5 releases a class of luminal proteins that is essential for completion of polypeptide translocation (14). Among the proteins released are the stress proteins BiP/GRP78, GRP94, protein disulfide isomerase, calreticulin, and peptidyl prolyl *cis-trans*-isomerase enzymes (cyclophilin and FKBP). These studies demonstrated that BiP/GRP78 was required for translocation in these yeast-derived proteoliposomes. However, experiments in mammalian systems have not provided compelling evidence that BiP/GRP78 is required for translocation. In addition, expression of an ATPase-defective BiP only prevented the secretion of selective proteins, supporting the view that the BiP ATPase activity is not absolutely required for the secretion of all proteins (15).

Proteins that become integral membrane proteins pose a unique problem for the translocation machinery. Type I transmembrane proteins contain a cleavable N-terminal signal peptide and an internal transmembrane domain. They are oriented such that their amino terminus is in the extracellular space. For these proteins, a span of hydro-

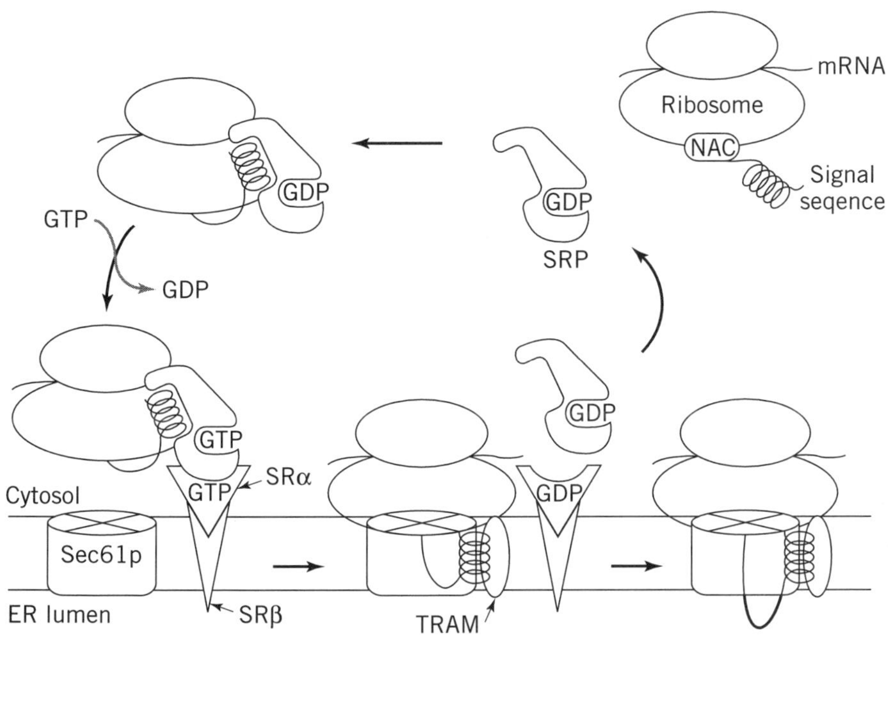

Figure 1. SRP-dependent cotranslocational translocation into the ER. SRP binds to the signal sequence of a nascent polypeptide emerging from the 60S ribosome to form a targeting complex in which elongation is arrested. SRP in the targeting complex then binds SRP receptor (composed of SRα and Srβ subunits) in the ER membrane. After GTP binding, SRP interacts with SRP receptor to result in formation of the ribosome-translocon junction, and translation elongation resumes. GTP hydrolysis stimulates the dissociation of SRP from SRP receptor so they can be recycled. As the nascent chain grows, the translocon pore allows passage of the nascent protein across the membrane. In most cases, the signal sequence is removed on the luminal side of the ER membrane by signal peptidase (not shown). *Source*: Reprinted from Ref. 3, with permission.

phobic amino acids (called the stop-transfer sequence) anchors the polypeptide in the lumen of the ER. In type II transmembrane proteins the carboxy terminus of the protein is in the extracellular space. The topology of type I and type II membrane proteins is determined by the charge distribution of polar amino acids flanking the signal-anchor domain (16). The side of the signal anchor that has more positively charged flanking residues remains intracellular, likely reflecting interactions with the negatively charged phospholipid bilayer. For multiple-spanning transmembrane proteins, each transmembrane domain displaces the prior one from the protein conducting channel where it moves laterally into the lipid bilayer. During the process of translocation, the channel is always gated, either by the ribosome on the cytoplasmic side or by BiP/GRP78 on the luminal side (17).

The cleavage site for signal peptidase is marked by small amino acids (alanine or glycine) in the 3- and 1-positions, relative to the cleavage site. Insertion into the ER membrane involves the formation of a loop structure with the amino terminus remaining in the cytoplasm and the growing carboxy terminus being continuously translocated across the membrane through Sec61p, a protein-conducting channel (18). Cleavage of the signal sequence by the signal peptidase (19,20) then releases the mature amino terminus into the lumen of the ER and is required for translocation of proteins into the secretory pathway.

The transport of a polypeptide across the ER membrane requires energy. GTP is used in the cycling of SRP and SRPR. The energy in cotranslational translocation is derived from the process of translational elongation. For posttranslational translocation, ATP is required for both function of Hsp70 in the cytosol and BiP function in the lumen of the ER. Depletion of intracellular ATP can block the secretion of some proteins, presumably by preventing dissociation of proteins from BiP (21). An ER ATP transporter has been characterized (22), and studies have directly detected ATP in the lumen of the ER by directing firefly luciferase expression to be localized in the ER (23). It is possible that increases in the intraluminal concentration of ATP will increase the secretion capacity in mammalian cells.

POSTTRANSLATIONAL MODIFICATIONS WITHIN THE SECRETORY PATHWAY

Protein translocation into and transport through the secretory pathway in eukaryotic cells is accompanied by a multitude of covalent modifications that occur on the polypeptide backbone. The sequence of posttranslational modifications that occur on a polypeptide is carefully regulated both temporally and spatially; specific modifications occur only within vesicles of the ER or Golgi compartment (Fig. 2). In many cases these modifications are required for proper polypeptide folding and secretion. In addition, posttranslational modifications can affect the half-life of the protein in the plasma and may be required for functional activity of the polypeptide. The mechanisms that direct posttranslational modifications recognize specific structural determinants within the polypeptide backbone. The efficiency of these modifications is determined by the specific host cell enzymatic machinery, the availability of substrates and cofactors, as well as structural properties of the polypeptide. To evaluate the role of posttranslational modification in protein function, investigators have studied proteins treated with chemicals or enzymes to remove modifications, proteins synthesized in the presence of inhibitors of specific modification reactions, proteins expressed in either different cell types or in cell mutants with defects in specific enzymatic machinery required to perform modifications, or proteins engineered through recombinant DNA technology that contain mutations that prevent modifications. The interpretation of results utilizing different strategies needs to be qualified to consider secondary effects due to the approach utilized. As a consequence, it is desirable to utilize several independent approaches to confirm the importance of any specific modification.

Protein Folding and Catalysts of Protein Folding in the ER

Analysis of protein folding in vitro has identified several key steps that occur during spontaneous refolding of a protein upon dilution from a denaturant. First, the polypeptide chain collapses into a compact shape to bury hydrophobic structures; this coincides with formation of secondary structures. Then native tertiary structures form as the protein passes through kinetically defined intermediates; this process may take several minutes. The partially folded intermediate is called the molten globule. Finally, the rate-limiting step lies close to the native folded state and involves formation of hydrogen bonds and disulfide bonds. Although proteins can fold into correct tertiary conformations in vitro (25), additional factors such as protein chaperones assist protein folding in vivo. Molecular chaperones compose a group of unrelated proteins that mediate the correct assembly of other proteins but are not themselves components of the final folded structure. Some molecular chaperones actually catalyze protein folding, whereas others maintain proteins in a folding-competent state and prevent protein aggregation.

A number of protein chaperones bind and stabilize partially folded polypeptides. One of the most well studied members of this family is the immunoglobulin binding protein (BiP) that is the same as the glucose-regulated protein of 78 kDa (GRP78) within the lumen of the ER (26). BiP is a member of the heat shock protein family, which exhibits a peptide-dependent ATPase activity (27), and for which expression is induced by the presence of abberently folded protein or unassembled protein subunits within the ER (28,29). Polypeptide release from BiP and transport out of the ER requires high levels of intracellular ATP (21); however, not all proteins require functional BiP for secretion (15). Intensive investigation into the role of BiP in protein folding has led to two different, but not mutually exclusive, functions for BiP. One hypothesis is that BiP assists protein folding by maintaining proteins in a conformation where they are folding competent, (30,31). This model is supported by the transient association with polypeptides destined for secretion (32–35) and an ATP dependence for proper folding and disulfide bond formation (36). In

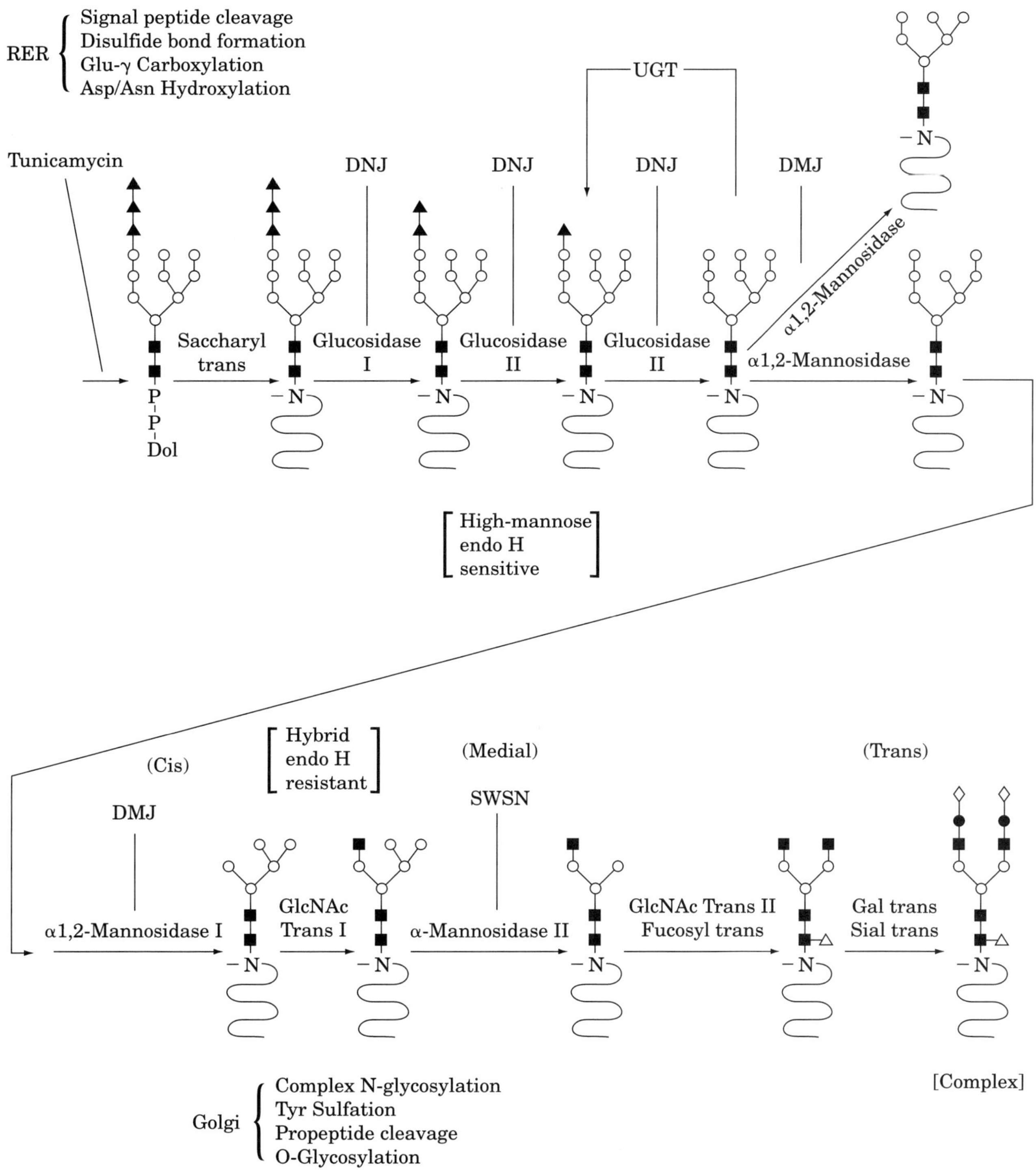

Figure 2. Posttranslational modifications within the secretory pathway. The posttranslational modifications that occur within the rough endoplasmic reticulum (RER, top) and the Golgi compartment (below) are shown. In addition, the sequence of oligosaccharide processing of N-linked oligosaccharides is shown, including the enzymes responsible and the known inhibitors (DNJ, deoxynojirimycin; DMJ, deoxymannojirimycin; SWSN, swainsonine). The point at which N-linked oligosaccharides become resistant to endoglycosidase H is shown. Sugars: ▲, glucose; ○, mannose; ■, *N*-acetylglucosamine (GlcNAc); △, fucose; ●, galactose (Gal); ◇, sialic acid (Sial) trans, transferase. UGT, UDP-glucose:glycoprotein glucosyltransferase. *Source*: Adapted from Ref. 24, with permission.

contrast, other studies (37–40) suggest that these proteins act as a retention mechanism for quality control to prevent aberrantly folded proteins from exiting the secretory pathway. However, despite many efforts to date, there is no direct demonstration that BiP binding to protein substrates actually catalyzes protein folding (41).

Disulfide bond formation occurs in the oxidizing environment of the ER. The tripeptide glutathione (γ-glutamyl, cyteinyl, glycine) is the major thiol-containing molecule in eukaryotic cells and prevents disulfide bond formation in the cytosol and catalyzes their formation in the ER. Glutathione is in equilibrium between a reduced form (GSH) and an oxidized form (GSSG). In the ER the ratio of GSH to GSSG is 5:1, an optimum for formation of disulfide bonds. In contrast, the ratio is 50:1 in the cytosol (42). Oxidized glutathione is reduced by glutathione reductase in a reaction that uses NADPH as a cofactor and occurs in the cytosol. One identified catalyst of protein folding is protein disulfide isomerase (PDI), which exchanges disulfide bonds on substrates and ensures proper disulfide bond formation, and exchange occurs prior to exit from the ER (43). PDI is also the β-subunit of prolyl hydroxylase, which catalyzes posttranslational hydroxylation of proline residue in pro-α-collagen. PDI is a soluble protein of the ER and is the most effective catalyst of protein folding known. PDI does not change the intermediates in the folding reaction but increases the kinetics of the overall process. Proper pairing of disulfide bonds is essential for transport through the ER, and pairing does not always occur in a sequential order.

Another rate-limiting step in protein folding is proline isomerization. Proteins that catalyze isomerization around peptidyl-prolyl bonds are peptidyl-prolyl isomerases (PPIases). There are two classes of these enzymes, based on their different substrate specificities and sensitivities to cyclophilin or FK506. Cyclophilin binds to cyclosporin A, a PPIase that has broad specificity. FK506 binds to FKBP, a PPIase that has narrower specificity with preference for hydrophobic residues in the P1 position. There are some very specific PPIases, one of the best studied is NinaA, a PPIase of the ER that specifically folds rhodopsins 1 and 2. *Drosophila* that have mutations in NinaA lack proper visual responses.

Asparagines and Serine/Threonine-Linked Glycosylation

High-mannose-containing oligosaccharide structures are added to asparagine residues of glycoproteins as an obligatory event for the folding and assembly of many newly synthesized polypeptides (44). This is an essential process in eukaryotic cells (45). The presence of oligosaccharides may be required for the efficient transport of individual glycoproteins through the secretory pathway (33,46), to increase the plasma half-life, or for the biological activity of glycoproteins. The luminal enzyme oligosaccharyltransferase catalyzes the transfer of a preassembled high-mannose-containing oligosaccharide core structure [glucose $(Glc)_3$mannose $(Man)_9$$N$-acetylglucosamine $(GlcNAc)_2$] from a dolichol pyrophosphate precursor onto asparagine acceptor sites within the consensus sequence Asn-X-Ser/Thr, where X can be any amino acid except proline. The mammalian oligosaccharyltransferase is a heterotrimeric transmembrane complex of ribophorin I, ribophorin II, and the oligosaccharyltransferase, which has been purified and molecularly cloned (47). The utilization of a particular consensus site for N-linked oligosaccharide attachment is determined by the structure of the growing polypeptide. As a consequence, proteins expressed in heterologous cells most frequently exhibit occupancy of N-linked sites very similar to that of the native polypeptide (48). To date there is no evidence that the N-linked oligosaccharide addition machinery can be saturated at high expression levels of glycoproteins.

After addition of the high-mannose-containing oligosaccharide core structure, trimming begins with the removal of the three terminal glucose residues, which is mediated by the action of glucosidases I and II (Fig. 2). Glucosidase I removes the terminal α1-3-glucose, and glucosidase II subsequently removes the two α1-2-glucose residues. Glucose trimming is required for binding to the protein chaperones calnexin and calreticulin within the lumen of the ER (49) (Fig. 3). Calnexin (IP90, p88, or CNX; an integral transmembrane protein) (50) and calreticulin (CRT; an ER luminal protein) (51) are homologous lectin-binding protein chaperones that transiently and selectively bind to overlapping sets of newly synthesized glycoprotein folding intermediates, thereby preventing their transit through the secretory pathway. Prolonged association with CNX and/or CRT is observed when proteins are unfolded, misfolded, or unable to oligomerize. CNX and CRT bind most avidly to monoglucosylated forms of the N-linked core structure and promote folding, delay oligomerization, and prevent degradation of some glycoproteins (52). In contrast, interaction with CNX correlates with intracellular degradation of some glycoproteins (53). Removal of the third glucose from the oligosaccharide core structure correlates with release from CNX and CRT and transport to the Golgi apparatus. It is proposed that the selective binding of unfolded glycoproteins to CRT and CNX is mediated by reglucosylation of the deglucosylated N-linked oligosaccharide. This reglucosylation activity is performed by a UDP-glucose:glycoprotein glucosyltransferase (UGT). The activity of the UGT is activated by unfolded protein (54). Thereby, only unfolded, mutant, or unassembled proteins are subject to reglucosylation. Reglucosylated proteins can rebind CNX and/or CRT and in this manner; unfolded proteins are retained in the ER through a cycle of CNX/CRT interaction, glucosidase II activity, and UGT activity. Inhibition of glucose trimming by inhibitors of glucosidase I and II, such as deoxynojirimycin or castanospermine, can inhibit this cycle and the secretion of some proteins. Because glucosidase inhibitors are fairly nontoxic and can inhibit the secretion of HIV envelope glycoprotein gp160, they have been used in the treatment of AIDS (55). Subsequent to glucose trimming in the ER, at least one α1-2-linked mannose is removed by an ER α1-2-mannosidase prior to transport out of the ER. Transit out of the ER is the rate-limiting step in secretion for the majority of proteins and may vary from 15 min to days, depending upon the rate at which a polypeptide attains a properly folded conformation.

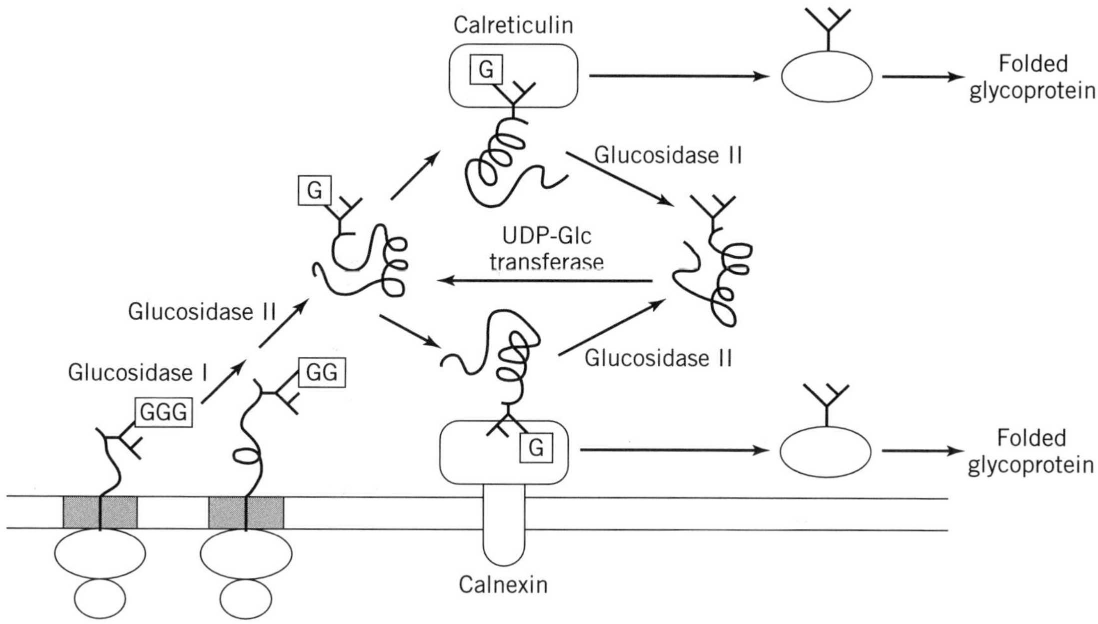

Figure 3. Role of calnexin/calreticulin and UDP-glucose:glycoprotein glucosyltransferase in the recognition and retention of unfolded glycoproteins in the ER. As a protein is cotranslationally translocated into the lumen of the ER, high-mannose core oligosaccharides are attached to asparargine residues. Subsequently, the glucose residues on the core-N-linked oligosaccharides are trimmed by the action of glucosidases I and II to expose a monoglucosylated core oligosaccharide that is a ligand for calnexin (a transmembrane protein) and/or calreticulin (a luminal ER protein) binding. After further action by glucosidase II, the glycoprotein is released from calnexin and calreticulin interaction. Unfolded glycoproteins are recognized as a substrate for the UDP-glucose:glycoprotein glucosyltransferase to form a monoglucosylated ligand to promote another round of calnexin and/or calreticulin binding. Folded glycoproteins bypass reglucosylation and are subsequently transported to the Golgi compartment. *Source*: Adapted from Ref. 44, with permission.

A sequential series of additional carbohydrate modifications occur that are separated spatially and temporarily as the protein transits through the *cis*, *medial*, and *trans* Golgi compartments. Mannose residues are removed by Golgi mannosidases I and II, and then *N*-acetylglucosamine, fucose, galactose, and sialic acid residues are added. These reactions are catalyzed by specific glycosyltransferases that modify the high-mannose carbohydrate to complex forms. Also within the Golgi apparatus, O-linked oligosaccharides are attached to the hydroxyl of serine or threonine residues through an O-glycosidic bond to *N*-acetylgalactosamine. Serine and threonine residues subject to glycosylation are frequently clustered together and contain an increased frequency of proline residues in the region, especially at positions -1 and $+3$, relative to the glycosylated residue (56). Galactose, fucose, and sialic acid are frequently attached to the serine/threonine-linked *N*-acetylgalactosamine. O-glycosylation occurs in the Golgi complex concomitant with complex processing of N-linked oligosaccharides.

The role of N- and O-linked oligosaccharide addition in glycoprotein function can be studied by using specific enzymatic inhibitors, glycosidases, and mutant cell lines (24). N-linked glycosylation is most easily evaluated by using specific glycosidases. The acquisition of resistance to endo-β-*N*-acetylglucosaminidase (Endo-H), which cleaves high-mannose and some hybrid-type oligosaccharides at the GlcNAcβ1-4GlcNAc linkage to leave a single GlcNAc residue attached to the asparagine (57), is frequently used to monitor movement of the protein from the ER to the medial Golgi apparatus. Resistance to Endo-H occurs in the medial Golgi following action of GlcNAc transferase I and mannosidase II. Peptide-*N*4-(*N*-acetyl-β-glucosaminyl)-asparagine amidase (N-gly) removes all N-linked oligosaccharides, regardless of the complexity of their structure, to leave a free aspartic acid residue (58). Also of use are a number of inhibitors that are specific for selective steps in the N-linked oligosaccharide processing pathway (see Fig. 2). In addition, a number of cell lines have been isolated that are defective in specific steps in oligosaccharide processing (59).

There are no specific inhibitors that can be used to study the requirement for O-linked glycosylation. It is possible to inhibit O-glycosylation, as well as complex modification of N-linked oligosaccharides, by depletion of the divalent metal ion manganese from the secretory pathway, and this may be used as an indication of whether O-linked sugars are present on the polypeptide (60). Of greater utility is a mutant cell line, ldlD, that is deficient in the UDP-galactose and UDP-*N*-acetylgalactosamine-4-epimerase and cannot synthesize Gal or GalNAc under normal growth conditions in the presence of glucose (61,62).

Finally, it is possible to enzymatically remove O-linked glycans with endo-α-*N*-acetyl-D-galactosaminidase (*O*-glycanase) that cleaves the Galβ1-3GalNAc disaccharide unit linked to serine or threonine residues (63). Sialic acid residues on the Gal or GalNAc will inhibit *O*-glycanase cleavage and therefore need to be removed by prior digestion with neuraminidase.

Glycosylphosphatidylinositol Anchor Addition

The addition of glycosylphosphatidylinositol (GPI) to many cell-surface-associated proteins occurs in order to anchor the protein in the lipid bilayer. The addition of GPI anchors occurs in the ER. A signal in the carboxy terminus of the polypeptide substrate contains a hydrophobic sequence of 17–30 amino acids followed by a spacer of 5–10 amino acids and a cleavage/attachment site. The attachment site must be an amino acid with a small side chain. The reaction is catalyzed by GPI:protein transamidase. This activity is defective in individuals with paraxysmal nocturnal hemoglobinuria. This disease results from a defect in the cell surface expression of inhibitors of complement-mediated cell lysis (such as decay accelerating factor). As a consequence, hemolytic anemia occurs because blood cells are unprotected from complement-mediated lysis.

Gamma-Carboxylation of Glutamic Acid Residues

The precursor of the vitamin K–dependent coagulation factors and a number of bone proteins and growth factors contain a propeptide that directs γ-carboxylation of up to 12 glutamic acid residues at the amino terminus of the mature protein (64). The propeptides of these proteins exhibit amino acid conservation within the γ-carboxylase recognition site and the site for cleavage of the propeptide. NMR structural analysis identified that the propeptide is an amphipathic α-helix with the carboxylase recognition site N-terminal to the helix (65).

The vitamin K–dependent γ-glutamyl carboxylase enzyme converts glutamate residues to γ-carboxyglutamate Gla residues. In the presence of CO_2, O_2, and vitamin K hydroquinone (KH_2) the enzyme is able to carboxylate a peptide containing glutamic acid residues, yielding a Gla-containing peptide, vitamin K epoxide, and H_2O (66). The vitamin K epoxide formed is subsequently reduced to regenerate KH_2 by either a thiol or the enzyme vitamin K epoxide reductase. The bovine (67) and human (68) cDNAs encoding the vitamin K–dependent γ-glutamyl carboxylase were isolated and demonstrate that the protein is a single-chain polypeptide of 94,000 kDa that spans the membrane 3 to 5 times. The overexpressed protein product directed increased carboxylation activity in vitro using isolated microsomes from transfected mammalian or insect cells and a synthetic peptide substrate (67,69). The ability to express γ-carboxylase activity from the cloned cDNA has permitted identification of the propeptide binding site between residues 50 and 225 (70) and the γ-carboxylase active site for glutamate binding within 218 residues of the amino terminal (71), whereas the vitamin K–reactive site is in the carboxy terminus of the protein (72). Vitamin K epoxidase activity is also catalyzed by the enzyme, and the carboxy terminus of the enzyme is required for this activity.

High-level expression of the vitamin K–dependent plasma proteins in transfected mammalian cells is limited by the ability of the mammalian host cell to efficiently perform γ-carboxylation of amino-terminal glutamic acid residues and also to efficiently cleave the propeptide (73,74). Analysis of factor IX expressed in CHO cells revealed that the protein had a much lower specific activity compared with the natural human plasma–derived protein. The reduced specific activity was attributed to both the limited ability of CHO cells to cleave the propeptide of factor IX and its ability to efficiently perform γ-carboxylation (67,73). Overexpression of the γ-carboxylase cDNA yielded increased activity in microsomal fractions when measured in vitro using small substrates (67). However, overexpression of the γ-carboxylase did not improve γ-carboxylation of factor IX when coexpressed in transfected mammalian cells (67). These results suggest that the amount of carboxylase protein is not a limiting factor to direct vitamin K–dependent γ-carboxylation in vivo. Several possibilities exist for the inability of the overexpressed γ-carboxylase to improve γ-carboxylation in vivo. First, the overexpressed γ-carboxylase may be mislocalized within the secretory pathway. It is possible that another protein, such as a protein chaperone, may be required to utilize a more complex protein substrate as opposed to a small peptide substrate. It is possible that another cofactor, possibly reduced vitamin K, is limiting for factor IX carboxylation in vivo. Further information on the mechanism of γ-carboxylation reaction in vivo is required in order to elucidate the rate-limiting step for γ-carboxylation in vivo.

β-Hydroxylation of Amino Acids: Aspartic acid, Asparagine, Lysine, and Proline

Proline and lysine residues are hydroxylated in procollagen by prolyl-3-hydroxylase and prolyl-4-hydroxylase. Prolyl-4-hydroxylase acts on prolines if the amino acid sequence is G-X-P, where X is any amino acid. This reaction requires O_2, Fe^{2+}, ascorbic acid, and α-ketoglutarate. This modification is important for stability of the collagen triple helix. Vitamin C deficiency results in scurvey because of reduced proline hydroxylation to form a stable collagen triplex.

A number of proteins that contain epidermal growth factor (EGF) domains, such as blood coagulation factor IX, have an aspartic acid or asparagine that is β-hydroxylated (75). This modification occurs by posttranslational hydroxylation of aspartic acid and/or asparagine within the ER. The β-hydroxylase has been molecularly cloned and characterized to some extent in recent years (76,77). β-Hydroxylation does not require the propeptide, vitamin K, or concomitant γ-carboxylation. A consensus β-hydroxylation site within EGF domains (Cys-X-Asp/Asn-X-X-X-X-Phe/Tyr-X-Cys-X-Cys) was proposed (75). Hydroxylation of both aspartic acid and asparagine is catalyzed by aspartyl β-hydroxylase, requires 2-ketoglutarate and Fe^{2+} (78,79), and is inhibited by agents that inhibit 2-ketoglutarate-dependent dioxygenases (80). It is interesting that only 0.3 mol/mol of plasma factor IX is modified by β-hydroxylation at Asp64, and this same amount of β-hydroxylation occurs in recombinant factor IX expressed at high levels in CHO cells (80). Thus, the low efficiency of β-hydroxylation appears to be a consequence of the factor IX polypeptide backbone and independent of the cell type used for expression.

Sulfation of Tyrosine Residues

Sulfate addition to tyrosine as an *O*4-sulfate ester is a common postranslational modification of secretory proteins that occurs in the trans Golgi apparatus (81) and is mediated by tyrosylprotein sulfotransferase that utilizes the activated sulfate donor 3′-phosphoadenosine 5′-phosphosulfate (PAPS) (82,83). Although the enzyme has been biochemically characterized (83), to date the enzyme has not been molecularly cloned. This modification occurs on many secretory proteins including a number of proteins that interact with thrombin, such as hirudin (84), fibrinogen (85), heparin cofactor II (86), α2-antiplasmin (87), vitronectin (88), bovine factor X (89), factor V, and factor VIII (90–92). Comparison of all known tyrosine sulfation sites yielded a consensus sequence that is primarily characterized by a large number of acidic amino acid residues. The consensus tyrosine sulfation site has three aspartic acid and/or glutamic acid residues near the sulfated tyrosine (± 5 amino acids), a turn-inducing residue within $+7/-7$, and the absence of cysteine or N-linked glycosylation sites within 7 residues of the sulfated tyrosine (81). Although tyrosine sulfation modification is found on many proteins that transit the secretory apparatus, there are few examples where this modification is required for secretion or functional activity of the molecule; for example, tyrosine sulfation in the hormone cholecystokinin is required for its biological activity (93). However, tyrosine sulfation can modulate the biological activity, binding affinities, and secretion of specific proteins (81). For example, tyrosine sulfation at the carboxy terminus of hirudin increases its binding affinity to the anion binding exosite of thrombin (94,95).

The most direct method to measure sulfation of a particular protein is to measure incorporation of [^{35}S]-sulfuric acid into protein as it is synthesized in cultured cells as described in Ref. 24. [^{35}S]-Sulfuric acid will not be incorporated in cysteine. Sodium chlorate is an inhibitor of ATP sulfurylase, the first of two enzymes involved in the synthesis of 3′-phosphoadenosine 5′-phosphosulfate (PAPS). PAPS is the donor for sulfation of both tyrosine and carbohydrate residues in intact cells (96). Analysis of proteins synthesized in the presence of sodium chlorate can provide a useful means to study the role of tyrosine sulfation.

Proteolytic Processing of Precursor Polypeptides

Early studies demonstrated that the budding yeast *Saccharomyces cerevisiae* gene product Kex2 could cleave mammalian precursor polypeptides after paired basic residues (97), suggesting that a human homologue of Kex2 may be the protease required for propeptide processing. Kex2 is a membrane-bound Ca^{2+}-dependent subtilisin-like serine protease that cleaves substrates, such as alpha mating factor, within the *trans*-Golgi compartment (98,99). A computer search identified a human homologue of Kex2 upstream of the c-*fes*/*fps* protooncogene and was subsequently named furin (fes/feps upstream coding region) or PACE, an acronym for "paired basic amino acid cleaving enzyme." The cDNA was subsequently cloned and shown to encode a protein that could cleave pro-vWF (100,101), pro–nerve growth factor (102), proalbumin (103), complement pro-C3 (104), and pro–factor IX (73) after pairs of basic amino acids. PCR-based cloning strategies subsequently identified over a half dozen members of this subtilisin-like serine protease family (105). Expression of some of these enzymes (PC1/PC3, PC2, and PC4) is restricted to neuroendocrine tissues, and the activities are likely responsible for processing of neuropeptides and endocrine hormones (106,107). In contrast, furin/PACE and PACE4 are ubiquitously expressed, but to a greater extent in the hepatocyte (108,109), and are likely candidates for processing of many other proteins including coagulation factors.

Propeptide cleavage occurs in the trans Golgi compartment just prior to secretion from the cell. The activity of the propeptide processing enzymes is uniquely regulated where a calcium-dependent induced autocatalytic activation occurs within the trans Golgi compartment to activate these proteases to their active form (110–113). In the process of activation, a propeptide is cleaved away from the zymogen. The specific localization of propeptide processing to the trans Golgi compartment ensures that the propeptide is associated with the mature polypeptide as proteins transit the secretory compartment.

Characterization of the amino acid requirements around the propeptide cleavage site has identified that both the P1 and P4 arginine are important for efficient processing mediated by furin/PACE and PACE4 (114,115). Cotransfection experiments were performed to test whether these enzymes could process pro–factor IX and or pro-vWF. Whereas both furin/PACE and PACE4 were effective at enhancing pro-vWF processing, only furin/PACE was capable of improving pro–factor IX processing. Thus, it appears that different members of this class of enzymes recognize and process overlapping sets of substrates. Overexpression of furin/PACE in transfected cells (103) as well as in transgenic animals (116) improves the ability to yield fully processed proteins. Recombinant factor IX is produced by coexpression with furin/PACE to ensure complete processing of the propeptide.

VESICULAR PROTEIN TRANSPORT

Transport vesicles are the intermediates for the trafficking of proteins between compartments of the secretory pathway (117). The principal mechanism of transport vesicle formation is through cytosolic coat proteins that bind to specific sites on the membrane (118). Interactions between these coat proteins and the cytosolic tails of membrane proteins play a fundamental role in sorting cargo into vesicles (119,120). Binding of coat proteins to the membrane induces the configuration of the membrane into a coated bud. Membrane fission at the neck of this bud releases the vesicle. The most well-characterized coat protein is the clathrin coat that is responsible for endocytic vesicle formation at the plasma membrane. Two additional types of coat components mediate vesicle formation and fusion within the secretory pathway. COP II (*co*at *p*rotein) components are required for the formation of vesicles that export cargo from the ER. COP I components were identified as coat proteins that form Golgi-derived transport vesicles (121,122) (Fig. 4).

Protein transport through the secretory pathway begins at the ER. Secretory proteins leave the ER in vesicles that bud from a specialized region of the ER. Protein transport between the ER and the intermediate compartment occurs in both directions. Retrograde transport from the intermediate compartment to the ER was first demonstrated for ER-resident proteins. If such proteins are erroneously incorporated into transport vesicles and leave the ER, they are efficiently recycled by retrograde transport. COP I–coated vesicles mediate the recycling of ER-resident proteins from intermediate compartments back to the ER, and COP II–coated vesicles mediate the export from the ER. The interaction of COP I and COP II components with the cytosolic tail domains of secretory transmembrane and ER-resident transmembrane proteins is likely to play an important role in sorting cargo into transport vesicles (119,120). Recently studies have used green fluorescent protein fused to vesicular stomatitis virus glycoprotein to visualize protein trafficking between the ER and Golgi compartments in living cells. The results demonstrate that the COP II–coated vesicles that are derived from the ER membrane have their coat proteins replaced by COP I and then move in a manner that depends on microtubules to the Golgi complex (Fig. 4). In addition, the vesicles are more elongated (appearing more like tubules) and are juxtaposed to microtubules that probably guide their movement (122,123).

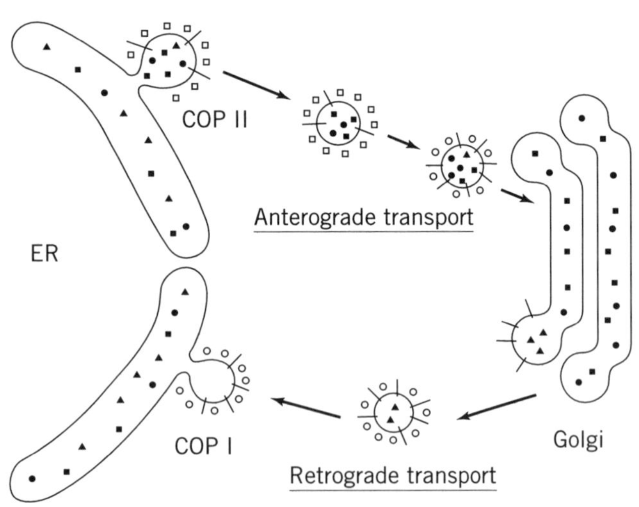

Figure 4. Proposed roles of COP I and COP II in vesicle traffic between the ER and cis Golgi. COP II vesicles transport normal cargo, some escaped ER resident proteins, and some anterograde targeting membrane proteins from the ER to the intermediate compartment where the COP II coat is exchanged for a COP I coat. The COP I–coated vesicle then fuses with the cis Golgi compartment. COP I vesicles are also shown mediating retrograde transport to the ER for escaped ER proteins.

Secretory proteins pass through the Golgi in the cis to trans direction. As proteins pass through the Golgi compartments they are modified by sequence of glycosylation and deglycosylation steps (124). Best understood is the modification of N-linked glycans in the Golgi. Initial trimming of mannose residues is followed by attachment of *N*-acetylglucosamine (GlcNAc), which is followed by more trimming of mannose residues and attachment of more GlcNAc. Mannose trimming and GlcNAc addition occur in cis and medial Golgi compartments. Although each Golgi subcompartment differs from its neighboring compartments in the specific set of enzymes contained within it, this separation is not complete. The same enzymes may even have different localization in different cell types. Golgi enzymes are not restricted to just one cistern. When the distributions of two different Golgi enzymes were determined relative to each other in the same samples, both were found spread out over several cisternae. The maximum concentration of each enzyme was in different in different compartments, but their distribution showed considerable overlap. Furthermore, Golgi enzymes seem to be in flux. Experiments with yeast demonstrated that a Golgi enzyme that is predominantly localized in the early Golgi was modified by late-acting Golgi enzymes (125). In a similar study using mammalian cells it was shown that an enzyme that was localized in the medial Golgi could be modified by an enzyme in the cis Golgi, even after the medial Golgi enzyme had reached its localization (126). These results indicate that Golgi enzymes must be efficiently recycled from late to early Golgi compartments and move again forward with the secretory flow. Golgi enzymes are not immobile in Golgi cisternae, but rather are freely diffusable.

Two models are proposed to explain protein transport through the Golgi: the vesicular transport model and the cisternal maturation model (127–129). The vesicular transport model assumes several distinct subcompartments that exchange cargo by vesicular transport in the anterograde and retrograde direction. The cisternal maturation model proposes that cisternae form de novo by fusion of ER-derived transport vesicles. These ER-derived transport intermediates would acquire Golgi enzymes by fusion with Golgi-derived retrograde transport vesicles from the cis-most compartment of the Golgi stack. At the same time, resident Golgi enzymes would move by retrograde transport in the trans to cis direction, from cisternae that formed earlier to more recently formed cisternae. Ultimately, cisternae would emerge on the trans face of the Golgi stack and disperse into vesicles of the constitutive and regulated secretory pathways and into transport vesicles to endosomes and lysosomes. Both models were first formulated shortly after the Golgi apparatus was characterized by electron microscopy, and which one best describes the Golgi has been controversial ever since.

Fusion of the vesicle membrane with its target membrane requires the cytosolic protein NSF (NEM-sensitive factor for fusion). NSF interacts with another cytosolic protein, SNAP (soluble NSF attachment protein), and SNARE (SNAP receptor) membrane proteins (130). It is the interaction of vesicle SNARE (v-SNARE) with target SNARE (t-SNARE) that is thought to play an important role in

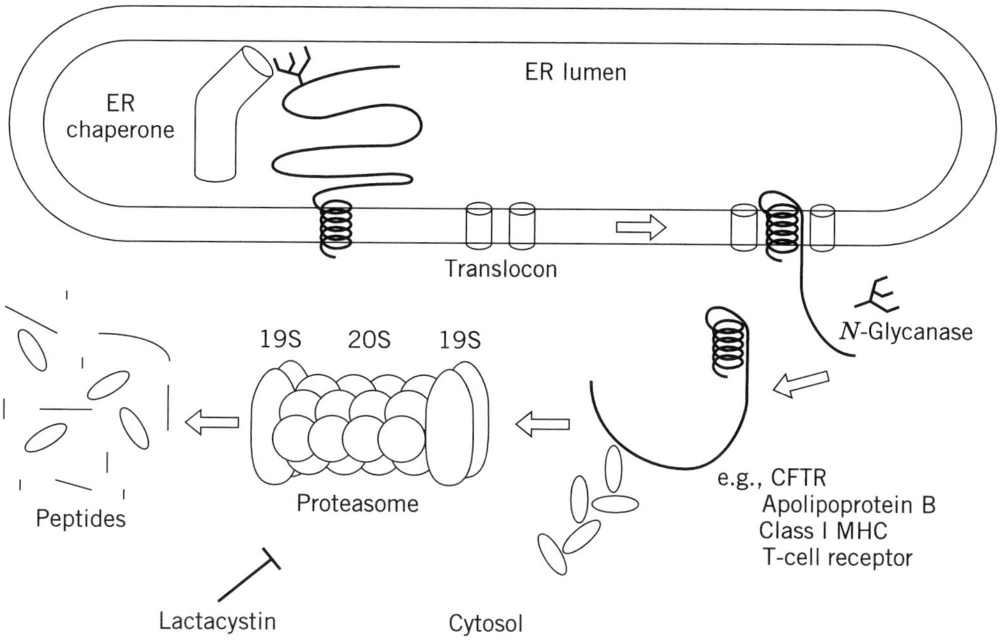

Figure 5. Degradation of ER-associated proteins. A putative ER chaperone facilitates association of a protein, with the Sec61 translocon. Reverse translocation from the lumen to the cytosol occurs with subsequent deglycosylation by *N*-glycanase. The protein may be subject to modification by ubiquitination and eventual targeting to the proteasome. Soluble proteins of the ER may also be degraded by reverse translocation through the Sec61 translocon, or alternatively they may associate with transmembrane proteins, such as CNX, that mediate their degradation. *Source*: Adapted from Ref. 141, with permission.

controlling the specificity of the vesicle fusion reaction. The precise role of NSF in this reaction is unclear. NSF action disrupts a v-SNARE/t-SNARE complex (131). This disruption could be what mediates the rearrangement of SNAREs that is required during fusion (132). These results therefore support the hypothesis that NSF directly mediates vesicle fusion. However, in yeast NSF mutants, vesicles accumulate in the cytosol, and this observation lends support to a model in which NSF is required for vesicle docking (133). Consistent with this hypothesis is that although yeast vacuole fusion is NSF dependent, the NSF action in this process precedes the docking of the membranes (134). These results indicate that NSF acts by activating the SNAREs for vesicle docking, rather than being involved in fusion itself. Small GTP-binding proteins of the rab family, as well as other proteins such as p115/Uso1p, are likely to be involved in further control of vesicle docking and fusion (135–138).

Mammalian cells constitutively transport proteins within vesicles from the ER to the plasma membrane, where they fuse through exocytosis. There are also specialized secretory cells, for example neuroendocrine cells, that store granules and release them upon appropriate stimulation through a regulated secretory pathway. Different proteins are sorted to the regulated secretory granules in different cell types through a common mechanism. The sorting process involves selective protein aggregation into clathrin-coated vesicles at the trans Golgi network. The aggregation occurs at pH 6.5 with 1 mM Ca^{2+}.

PROTEIN RETENTION AND DEGRADATION IN THE ER

A protein within the ER has several possible destinations that include transport to the Golgi compartment, degradation, and retention in the ER. What determines the fate of any particular polypeptide? For resident proteins of the ER, there are two primary mechanisms that ensure that proteins do not exit the ER compartment (139). Soluble luminal proteins of the ER, such as BiP and PDI, contain a carboxy-terminal tetrapeptide, with the consensus sequence KDEL, that is necessary and sufficient for preventing secretion of the polypeptide. A KDEL receptor, that is, a transmembrane-spanning protein, binds KDEL-containing proteins that leave the ER and retrieves them to the ER from the intermediate compartment. For transmembrane-resident proteins of the ER, there is a signal within the cytosolic domain that mediates retrieval from the intermediate compartment by COP proteins. Type I transmembrane-resident ER proteins contain a dilysine motif in the carboxy terminus (i.e., cytosolic tail). For type II transmembrane proteins, a diarginine motif in the amino terminus mediates the interaction with COP proteins. COP I vesicles then retrieve these proteins to the ER from the intermediate compartment.

As mentioned earlier, for many years investigators searched for signals that direct protein transport out of the ER. Although it is now known that specific short sequences exist that facilitate export from the ER, it appears now that the majority of polypeptides have retention signals that prevent their export. As a polypeptide folds, these retention signals become buried within the polypeptide, and thus, the polypeptide can transit to the Golgi compartment. A variety of polypeptides that recognize and bind unfolded proteins have been identified, including BiP/GRP78, CNX, and CRT, which were discussed earlier. These protein chaperones play a fundamental role in preventing unfolded proteins from exiting the ER compartment.

Proteins that are unable to attain their native conformation are targeted for degradation. A nonlysosomal pathway has been identified in the degradation of these proteins (140,141). The discovery that lactacystin, a fungal metabolite that specifically binds and inhibits the cytosolic 26S proteasome, inhibits the degradation of many ER-associated proteins pointed to the involvement of the cytosolic proteasome in the degradation of ER-localized proteins. There are now a number of examples of ER-associated proteins, such as the cystic fibrosis transmembrane conductance regulator, the T-cell receptor, factor VIII, class I MHC molecules, and so forth, that are degraded by this pathway (53,142–146). Insight into the mechanism of degradation was obtained by studying how human cytomegalovirus (HCMV) evades the immune system (Fig. 5). HCMV encodes two gene products, US2 and US11, that are able to dislocate ER MHC class I proteins to the cytosol (146,147). This dislocation occurs through the Sec61 protein conducting channel (148). Once in the cytosol, the class I molecules are deglycosylated by an *N*-glycanase and are then polyubiquitinated. The ubiquitinated protein is then degraded by the cytosolic 26S proteasome. How are unfolded proteins identified to be degraded? The answer to this question is unknown, but present data suggest that CNX may be involved in this quality control decision. Proteins that are destined to be degraded are bound to CNX but not to CRT (53,143,149,150). CNX is a transmembrane protein that is in close proximity to the Sec61p translocon. It is speculated that CNX may mediate the transfer of unfolded proteins in the lumen of the ER to the proteasome.

ACKNOWLEDGMENTS

I thank Joe Nowak and Renee Strauss for assistance in preparation of this article.

BIBLIOGRAPHY

1. H. Nielsen, J. Engelbrecht, S. Brunak, and G. von Heijne, *Protein Eng.* **10**, 1–6 (1997).
2. P. Walter and A.E. Johnson, *Annu. Rev. Cell Biol.* **10**, 87–119 (1994).
3. A.K. Corsi and R. Schekman, *J. Biol. Chem.* **271**, 30299–39302 (1996).
4. P.J. Rapiejko and R. Gilmore, *J. Cell Biol.* **117**, 493–503 (1992).
5. P.J. Rapiejko and R. Gilmore, *Mol. Biol. Cell* **5**, 887–897 (1994).
6. J.D. Miller, H. Wilhelm, L. Gierasch, R. Gilmore, and P. Walter, *Nature* **366**, 351–354 (1993).
7. B. Wiedmann, H. Sakai, T.A. Davis, and M. Wiedmann, *Nature* **370**, 434–440 (1994).
8. B. Lauring, G. Kreibich, and M. Weidmann, *Proc. Natl. Acad. Sci. U.S.A.* **92**, 9435–9439 (1995).
9. B. Lauring, H. Sakai, G. Kreibich, and M. Wiedmann, *Proc. Natl. Acad. Sci. U.S.A.* **92**, 5411–5415 (1995).
10. S. Voigt, B. Jungnickel, E. Hartmann, and T.A. Rapoport, *J. Cell Biol.* **134**, 25–35 (1996).
11. C.J. Stirling, J. Rothblatt, M. Hosobuchi, R. Deshaies, and R. Schekman, *Mol. Biol. Cell* **3**, 129–142 (1992).
12. D. Gorlich and T.A. Rapoport, *Cell* **75**, 615–630 (1993).
13. T.A. Rapoport, *Science* **258**, 931–936 (1992).
14. C.V. Nicchitta and G. Blobel, *Cell* **73**, 989–998 (1993).
15. J.A. Morris, A.J. Dorner, C.A. Edwards, L.M. Hendershot, and R.J. Kaufman, *J. Biol. Chem.* **272**, 4327–4334 (1997).
16. S. High and B. Dobberstein, *Curr. Opin. Cell Biol.* **4**, 581–586 (1992).
17. B.D. Hamman, L.M. Hendershot, and A.E. Johnson, *Cell* **92**, 747–758 (1998).
18. B. Martoglio, M.W. Hofmann, J. Brunner, and B. Dobberstein, *Cell* **81**, 207–214 (1995).
19. R.E. Dalbey, M.O. Lively, S. Bron, and J.M. van Dijl, *Protein Sci.* **6**, 1129–1138 (1997).
20. K.U. Kalies, T.A. Rapoport, and E. Hartmann, *J. Cell Biol.* **141**, 887–894 (1998).
21. A.J. Dorner, L.C. Wasley, and R.J. Kaufman, *Proc. Natl. Acad. Sci. U.S.A.* **87**, 7429–7432 (1990).
22. P. Mayinger, V.A. Bankaitis, and D.I. Meyer, *J. Cell Biol.* **131**, 1377–1386 (1995).
23. A.J. Dorner and R.J. Kaufman, *Biologicals* **22**, 103–112 (1994).
24. A.J. Dorner and R.J. Kaufman, *Methods Enzymol.* **185**, 577–596 (1990).
25. C.B. Anfinsen, E. Haber, M. Sela, and F.H. White, *Proc. Natl. Acad. Sci. U.S.A.* **47**, 1309–1314 (1961).
26. S. Munro and H.R.B. Pelham, *Cell* **46**, 291–300 (1986).
27. G.C. Flynn, T.G. Chappell, and J.E. Rothman, *Science* **245**, 385–390 (1989).
28. Y. Kozutsumi, M. Segal, K. Normington, M.J. Gething, and J. Sambrook, *Nature* **332**, 462–464 (1988).
29. A.S. Lee, *Trends Biochem. Sci.* **12**, 20–24 (1987).
30. M.-J. Gething and J. Sambrook, *Nature* **355**, 33–45 (1992).
31. J.E. Rothman, *Cell* **59**, 591–601 (1989).
32. D.T.W. Ng, R.E. Randall, and R.A. Lamb, *J. Cell Biol.* **109**, 3273–3289 (1989).
33. A.J. Dorner, D.G. Bole, and R.J. Kaufman, *J. Cell Biol.* **105**, 2665–2674 (1987).
34. T. Hai, F. Liu, E.A. Allegretto, M. Karin, and M.R. Green, *Genes Dev.* **2**, 1216–1226 (1988).
35. P. Blount and J.P. Merlie, *J. Cell Biol.* **113**, 1125–1132 (1991).
36. I. Braakman, J. Helenius, and A. Helenius, *Nature* **356**, 260–262 (1992).
37. L. Hendershot, D. Bole, G. Kohler, and J.F. Kearney, *J. Cell Biol.* **104**, 761–767 (1987).
38. A.J. Dorner, M.G. Krane, and R.J. Kaufman, *Mol. Cell. Biol.* **8**, 4063–4070 (1988).
39. A.J. Dorner, L.C. Wasley, and R.J. Kaufman, *EMBO J.* **11**, 1563–1571 (1992).

40. B.A. Pollok, R. Anker, P. Eldridge, L. Hendershot, and D. Levitt, *Proc. Natl. Acad. Sci. U.S.A.* **84**, 9199–9203 (1987).
41. A. Zapun, T.E. Creighton, P.J.E. Rowling, and R.B. Freedman, *Proteins* **14**, 10–15 (1992).
42. C. Hwang, A.J. Sinskey, and H.F. Lodish, *Science* **257**, 1496–1502 (1992).
43. R. Noiva and W.J. Lennarz, *J. Biol. Chem.* **267**, 3553–3556 (1992).
44. A. Helenius, *Mol. Biol. Cell* **5**, 253–265 (1994).
45. S. Silberstein, P.G. Collins, D.J. Kelleher, P.J. Rapiejko, and R. Gilmore, *J. Cell Biol.* **128**, 525–536 (1995).
46. J.L. Guan, C.E. Machamer, and J.K. Rose, *Cell* **42**, 489–496 (1985).
47. D.J. Kelleher, G. Kreibich, and R. Gilmore, *Cell* **69**, 55–65 (1992).
48. Y. Gavel and G. von Heijne, *Protein Eng.* **3**, 433–442 (1990).
49. D.N. Hebert, B. Foelimer, and A. Helenius, *Cell* **81**, 425–433 (1995).
50. J.J. Bergeron, M.B. Brenner, D.Y. Thomas, and D.B. Williams, *Trends Biochem. Sci.* **19**, 124–128 (1994).
51. K.H. Krause and M. Michalak, *Cell* **88**, 439–443 (1997).
52. D.N. Hebert, B. Foellmer, and A. Helenius, *EMBO J.* **15**(12), 2961–2968 (1996).
53. S.W. Pipe, J.A. Morris, J. Shah, and R.J. Kaufman, *J. Biol. Chem.* **273**, 8537–8544 (1998).
54. S.E. Trombetta and A.J. Parodi, *J. Biol. Chem.* **267**, 9236–9240 (1992).
55. A.D. Elbein, *Semin. Cell Biol.* **2**, 309–317 (1991).
56. I.B. Wilson, Y. Gavel, and G. von Heijne, *Biochem. J.* **275**, 529–534 (1991).
57. A.L. Tarentino and F. Maley, *J. Biol. Chem.* **249**, 811–817 (1974).
58. A.L. Tarentino, C.M. Gomez, and T.H. Plummer, Jr., *Biochemistry* **24**, 4665–4671 (1985).
59. P. Stanley, in M.M. Gottesman ed., *Molecular Cell Genetics*, Wiley, New York, 1985, pp. 745–772.
60. R.J. Kaufman, M. Swaroop, and P. Murtha-Riel, *Biochemistry* **33**, 9813–9819 (1994).
61. C.L. Wasley, P. Horgan, G. Timony, J. Stoudemier, M. Krieger, and R.J. Kaufman, *Blood* **77**, 2624–2632 (1991).
62. D.M. Kingsley, K.F. Kozarsky, L. Hobbie, and M. Krieger, *Cell* **44**, 749–759 (1986).
63. J. Umemoto, V.P. Bhavanandan, and E.A. Davidson, *J. Biol. Chem.* **252**, 8609–8614 (1977).
64. M.J. Jorgensen, A.B. Cantor, B.C. Furie, C.L. Brown, C.B. Shoemaker, and B. Furie, *Cell* **48**, 185–191 (1987).
65. D.G. Sanford, C. Kanagy, J.L. Sudmeier, B.C. Furie, B. Furie, and W.W. Bachovchin, *Biochemistry* **30**, 9835–9841 (1991).
66. J.W. Suttie, *FASEB J.* **7**, 445–452 (1993).
67. A. Rehemtulla, D.A. Roth, L.C. Wasley, A. Kuliopulos, C.T. Walsh, B. Furie, B.C. Furie, and R.J. Kaufman, *Proc. Natl. Acad. Sci. U.S.A.* **90**, 4611–4615 (1993).
68. S.M. Wu, W.F. Cheung, D. Frazier, and D.W. Stafford, *Science* **254**, 1634–1636 (1991).
69. D.A. Roth, A. Rehemtulla, R.J. Kaufman, C.T. Walsh, B. Furie, and B.C. Furie, *Proc. Natl. Acad. Sci. U.S.A.* **90**, 8372–8376 (1993).
70. M. Yamada, A. Kuliopulos, N.P. Nelson, D.A. Roth, B. Furie, B.C. Furie, and C.T. Walsh, *Biochemistry* **34**, 481–489 (1995).
71. A. Kuliopulos, N.P. Nelson, M. Yamada, C.T. Walsh, B. Furie, B.C. Furie, and D.A. Roth, *J. Biol. Chem.* **269**, 21364–21370 (1994).
72. D.A. Roth, M.L. Whirl, L.J. Velazquez-Estades, C.T. Walsh, B. Furie, and B.C. Furie, *J. Biol. Chem.* **270**, 5305–5311 (1995).
73. L.C. Wasley, A. Rehemtulla, J.A. Bristol, and R.J. Kaufman, *J. Biol. Chem.* **268**, 8458–8465 (1993).
74. R.J. Kaufman, L.C. Wasley, B.C. Furie, B. Furie, and C.B. Shoemaker, *J. Biol. Chem.* **261**, 9622–9628 (1986).
75. J. Stenflo, A. Lundwall, and B. Dahlback, *Proc. Natl. Acad. Sci. U.S.A.* **84**, 368–372 (1987).
76. K. McGinnis, G.M. Ku, W.J. VanDusen, J. Fu, V. Garsky, A.M. Stern, and P.A. Friedman, *Biochemistry* **35**, 3957–3962 (1996).
77. S. Jia, W.J. VanDusen, R.E. Diehl, N.E. Kohl, R.A. Dixon, K.O. Elliston, A.M. Stern, and P.A. Friedman, *J. Biol. Chem.* **267**, 14322–14327 (1992).
78. R.S. Gronke, W.J. VanDusen, V.M. Garsky, J.W. Jacobs, M.K. Sardana, A.M. Stern, and P.A. Friedman, *Proc. Natl. Acad. Sci.* **86**, 3609–3613 (1989).
79. J. Stenflo, E. Holme, S. Lindstedt, N. Chandramouli, L.H. Huang, J.P. Tam, and R.B. Merrifield, *Proc. Natl. Acad. Sci.* **86**, 4440–4447 (1989).
80. C.K. Derian, W. VanDusen, C.T. Przysiecki, P.N. Walsh, K.L. Berkner, R.J. Kaufman, and P.A. Friedman, *J. Biol. Chem.* **264**, 6615–6618 (1989).
81. W.B. Huttner, *Annu. Rev. Physiol.* **50**, 363–376 (1988).
82. P.A. Baeuerle and W.B. Huttner, *J. Cell Biol.* **105**, 2655–2664 (1987).
83. C. Niehrs and W.B. Huttner, *EMBO J.* **9**, 35–42 (1990).
84. P.J. Braun, S. Dennis, J. Hofsteenge, and S.R. Stone, *Biochemistry* **27**, 6517–6522 (1988).
85. D.H. Farrell, E.R. Mulvihill, S.M. Huang, D.W. Chung, and E.W. Davie, *Biochemistry* **30**, 9414–9420 (1991).
86. G. Hortin, D.M. Tollefsen, and A.W. Strauss, *J. Biol. Chem.* **261**, 15827–15830 (1986).
87. G. Hortin, K.F. Fok, P.C. Toren, and A.W. Strauss, *J. Biol. Chem.* **262**, 3082–3085 (1987).
88. D. Jenne, A. Hille, K.K. Stanley, and W.B. Huttner, *Eur. J. Biochem.* **185**, 391–395 (1989).
89. T. Morita and C.M. Jackson, *J. Biol. Chem.* **261**, 4008–4014 (1986).
90. D.D. Pittman, J.H. Wang, and R.J. Kaufman, *Biochemistry* **31**, 3315–3323 (1992).
91. D.D. Pittman, K.N. Tomkinson, D. Michnick, U. Selighsohn, and R.J. Kaufman, *Biochemistry* **33**, 6952–6959 (1994).
92. D.A. Michnick, D.D. Pittman, R.J. Wise, and R.J. Kaufman, *J. Biol. Chem.* **269**, 20095–20102 (1994).
93. V. Mutt, in G.B.J. Glass ed., *Gastrointestinal Hormones* Raven Press, New York, 1980, pp. 169–221.
94. C. Niehrs, W.B. Huttner, D. Carvallo, and E. Degryse, *J. Biol. Chem.* **265**, 9314–9318 (1990).
95. T.J. Rydel, K.G. Ravichandran, A. Tulinsky, W. Bode, R. Huber, C. Roitsch, and J.W. Fenton, *Science* **249**, 277–280 (1990).
96. E. Friederich, H.J. Gritz, and W.B. Huttner, *J. Cell Biol.* **107**, 1655–1667 (1988).

97. D.C. Foster, R.D. Holly, C.A. Sprecher, K.M. Walker, and A.A. Kumar, *Biochemistry* **30**, 367–372 (1991).
98. I.C. Bathurst, S.O. Brennan, R.W. Carrell, L.S. Cousens, A.J. Brake, and P.J. Barr, *Science* **235**, 348–350 (1987).
99. D. Julius, A. Brake, L. Blair, R. Kunisawa, and J. Thorner, *Cell* **37**, 1075–1089 (1984).
100. R.J. Wise, P.J. Barr, P.A. Wong, M.C. Kiefer, A.J. Brake, and R.J. Kaufman, *Proc. Natl. Acad. Sci. U.S.A.* **87**, 9378–9382 (1990).
101. W.J. van de Ven, J. Voorberg, R. Fontijn, H. Pannekoek, A.M. van den Ouweland, H.L. van Duijnhoven, A.J. Roebroek, and R.J. Siezen, *Mol. Biol. Rep.* **14**, 265–275 (1990).
102. P.A. Bresnahan, R. Leduc, L. Thomas, J. Thorner, H.L. Gibson, A.J. Brake, P.J. Barr, and G. Thomas, *J. Cell Biol.* **111**, 2851–2859 (1990).
103. S.O. Brennan and R.J. Peach, *J. Biol. Chem.* **266**, 21504–21508 (1991).
104. Y. Misumi, K. Ohkubo, M. Sohda, N. Takami, K. Oda, and Y. Ikehara, *Biochem. Biophys. Res. Commun.* **171**, 236–242 (1990).
105. P.A. Halban and J.C. Irminger, *Biochem. J.* **299**, 1–18 (1994).
106. S. Benjannet, N. Rondeau, R. Day, M. Chretien, and N.G. Seidah, *Proc. Natl. Acad. Sci. U.S.A.* **88**, 3564–3568 (1991).
107. L. Thomas, R. Leduc, B.A. Thorne, S.P. Smeekens, D.F. Steiner, and G. Thomas, *Proc. Natl. Acad. Sci. U.S.A.* **88**, 5297–5301 (1991).
108. A.J. Roebroek, J.A. Schalken, J.A. Leunissen, C. Onnekink, H.P. Bloemers, and W.J. van de Ven, *EMBO J.* **5**, 2197–2202 (1986).
109. M.C. Kiefer, J.E. Tucker, R. Joh, K.E. Landsberg, D. Saltman, and P.J. Barr, *DNA Cell Biol.* **10**, 757–769 (1991).
110. A. Rehemtulla, A.J. Dorner, and R.J. Kaufman, *Proc. Natl. Acad. Sci. U.S.A.* **89**, 8235–8239 (1992).
111. R. Leduc, S.S. Molloy, B.A. Thorne, and G. Thomas, *J. Biol. Chem.* **267**, 14304–14308 (1992).
112. E.D. Anderson, J.K. VanSlyke, C.D. Thulin, F. Jean, and G. Thomas, *EMBO J.* **16**, 1508–1518 (1997).
113. S.S. Molloy, L. Thomas, J.K. VanSlyke, P.E. Stenberg, and G. Thomas, *EMBO J.* **13**, 18–33 (1994).
114. A. Rehemtulla and R.J. Kaufman, *Blood* **9**, 2349–2355 (1992).
115. A. Rehemtulla, P.J. Barr, C.J. Rhodes, and R.J. Kaufman, *Biochemistry* **32**, 11586–11590 (1993).
116. R. Drews, R.K. Paleyanda, T.K. Lee, R.R. Chang, W.N. Drohan, A. Rehemtulla, and R.J. Kaufman, *Proc. Natl. Acad. Sci. U.S.A.* **92**, 10462–10466 (1995).
117. J.E. Rothman, *Protein Sci.* **5**, 185–194 (1996).
118. R. Schekman and L. Orci, *Science* **271**, 1526–1533 (1996).
119. M.J. Kuehn, J.M. Herrmann, and R. Schekman, *Nature* **391**, 187–190 (1998).
120. P. Cosson and F. Letoumeur, *Science* **263**, 1629–1631 (1994).
121. R. Schekman and I. Mellman, *Cell* **90**, 197–200 (1997).
122. S.J. Scales, R. Pepperkok, and T.E. Kreis, *Cell* **90**, 1137–1148 (1997).
123. J.F. Presley, N.B. Cole, T.A. Schroer, K.J. Hirschberg, K.J. Zaal, and J. Lippencott-Schwartz, *Nature* **389**, 81–85 (1997).
124. R. Kornfeld and S. Kornfeld, *Annu. Rev. Biochem.* **54**, 631–664 (1985).
125. J. Ting, S.K. Wooden, K.S. Kelleher, R. Kritz, R.J. Kaufman, and A.S. Lee, *Gene* **55**, 147–152 (1987).
126. R.J. Jenny, D.D. Pittman, J.J. Toole, R.W. Kriz, R.A. Aldape, R.M. Hewick, R.J. Kaufman, and K.G. Mann, *Proc. Natl. Acad. Sci. U.S.A.* **84**, 4846–4850 (1987).
127. S.I. Bannykh and W.E. Balch, *J. Cell. Biol.* **138**, 1–4 (1997).
128. B.S. Glick, T. Elston, and G. Oster, *FEBS* **414**, 177–181 (1998).
129. H.D. Love, C.C. Lin, C.S. Short, and J. Ostermann, *J. Cell Biol.* **140**, 541–551 (1998).
130. T. Sollner, S.W. Whiteheart, M. Brunner, H. Erdjument-Bromage, S. Geromanos, P. Tempst, and J.E. Rothman, *Nature* **362**, 318–324 (1993).
131. J.E. Rothman, and G. Warren, *Curr. Biol.* **4**, 220–233 (1994).
132. T. Sollner, M.K. Bennett, S.W. Whiteheart, R.H. Scheller, and J.E. Rothman, *Cell* **75**, 409–418 (1993).
133. C.A. Kaiser, and R. Schekman, *Cell* **61**, 723–733 (1990).
134. A. Mayer, W. Wickner, and A. Haas, *Cell* **85**, 83–94 (1996).
135. V. Rybin, O. Ullrich, M. Rubino, K. Alexandrov, I. Simon, C. Seabra, R. Goody, and M. Zerial, *Nature* **383**, 266–269 (1996).
136. V.V. Lupashin, and M.G. Waters, *Science* **276**, 1255–1258 (1997).
137. S.K. Sapperstein, V.V. Lupashin, H.D. Schmitt, and M.G. Waters, *J. Cell Biol.* **132**, 755–767 (1996).
138. N. Nakamura, M. Lowe, T.P. Levine, C. Rabouille, and G. Warren, *Cell* **89**, 445–455 (1997).
139. H.R. Pelham, *Curr. Opin. Cell Biol.* **7**, 530–535 (1995).
140. R.D. Klausner and R. Sitia, *Cell* **62**, 611–614 (1990).
141. R.R. Kopito, *Cell* **88**, 427–430 (1997).
142. C.L. Ward, S. Omura, and R.R. Kopito, *Cell* **83**, 121–127 (1995).
143. R. Halaban, E. Cheng, Y. Zhang, G. Moellmann, D. Hanlon, M. Michalak, V. Setaluri, and D.N. Hebert, *Proc. Natl. Acad. Sci. U.S.A.* **94**, 6210–6215 (1997).
144. D. Qu, J.H. Teckman, S. Omura, and D.H. Perlmutter, *J. Biol. Chem.* **271**, 22791–22795 (1996).
145. M.M. Hiller, A. Finger, M. Schweiger, and D.H. Wolf, *Science* **273**, 1725–1728 (1996).
146. E.J. Wiertz, T.R. Jones, L. Sun, M. Bogyo, H.J. Geuze, and H.L. Ploegh, *Cell* **84**, 769–779 (1996).
147. E.J. Wiertz, D. Tortorella, M. Bogyo, J. Yu, W. Mothes, T.R. Jones, T.A. Rapoport, and H.L. Ploegh, *Nature* **384**, 432–438 (1996).
148. M. Pilon, R. Schekman, and K. Romisch, *EMBO J.* **16**, 4540–4548 (1997).
149. K.S. Cannon, D.N. Hebert, and A. Helenius, *J. Biol. Chem.* **271**, 14280–14284 (1996).
150. W. Chen, J. Helenius, I. Braakman, and A. Helenius, *Proc. Natl. Acad. Sci. U.S.A.* **92**, 6229–6233 (1995).

See also PROTEIN SECRETION, *SACCHAROMYCES CEREVISIAE*.

SHEAR SENSITIVITY

Yusuf Chisti
University of Almería
Almería, Spain

KEY WORDS

Air-lift bioreactors
Animal cell culture
Bioreactor hydrodynamics
Bubble columns
Cell damage
Shear damage
Shear protection
Shear rate
Shear stress

OUTLINE

INTRODUCTION

Susceptibility to hydrodynamic and mechanical shear forces affects performance of cultured cells of animals, plants, microalgae, and cyanobacteria. In addition, hydromechanical forces affect or otherwise damage some commercially relevant mycelial fungi, filamentous bacteria, microbial flocs, and biofilms. In specific cases, intense shear fields may also damage the larger bioactive molecules such as enzymes. This article details the shear sensitivity of the major types of biocatalysts and the approaches available for mitigating the damaging effects. Methods for estimating shear rate and shear stress are discussed for various configurations of bioreactors operated in process-relevant conditions.

Depending on its intensity, a shear field may be stimulatory, inhibitory, or outright destructive. Intense shear fields will disintegrate even the most robust of microorganisms (1). Lesser levels of turbulence impact upon cellular morphology (2,3), biochemistry and physiology (4), floc size (5), and attachment and detachment from surfaces (6,7). These influences of shear stresses are reflected in the industrially important aspects of production rate, productivity, product spectrum, and characteristics. Among the most susceptible to shear effects are the animal cells, plant cells, and cells of some microalgae and cyanobacteria. Shear-associated problems have the greatest economic impact in the animal cell culture industry because it is among the largest ones based on shear-sensitive cells and because of the high-value nature of most cell-culture-derived products (8). Damaging consequences of intense shear fields are also well documented for other microorganisms. For example, shear effects may alter characteristics of extracellular microbial polysaccharides by affecting the secretery process in some bacteria (9,10). Considerations of shear stress levels are also important in design and operation of biomedical devices such as dialysis machines, heart-lung machines, and transfusion filters. Shear stresses associated with implanted devices such as artificial heart valves can have medically significant implications. Increasingly, shear-sensitive human cells are being used to regenerate damaged tissue in vitro for later implanting or grafting. In other instances, susceptibility to shear may be beneficial, for example, during recovery of intracellular material by mechanical disruption (1,11,12). Some of the factors that determine shear sensitivity are summarized in Table 1.

SHEAR FORCES IN BIOREACTORS

Substantial information exists on the effects of hydrodynamic forces on cells in defined flow geometries such as viscometers and capillaries (5,13–18), but little is known about shear fields in bioreactors (19–21). Whereas selection of more shear-tolerant cell lines can be helpful, successful culture of shear-sensitive biocatalysts requires attention to bioreactor design and operation.

Table 1. Factors Determining Shear Sensitivity

1. Type of cell and species
2. Composition and thickness of cell wall when present
3. Size and morphology of cell
4. The intensity and nature of shear stress, whether turbulent or laminar, or associated with interfaces (e.g., during bubble rise and rupture)
5. Growth history, both short-term (e.g., starvation) and long-term adaptation
6. Growth environment (pH, temperature, agitation intensity, light in photosynthetic cultures)
7. Growth medium (trace elements, vitamins, carbon and nitrogen sources)
8. Growth rate
9. Growth stage
10. Type and concentration of shear-protective agents if present

The magnitude of the fluid mechanical forces is often expressed as shear stress, τ, or shear rate, γ. These quantities are related; thus, in laminar newtonian flow,

$$\tau = \gamma \mu_L \qquad (1)$$

where μ_L is the viscosity of the fluid. Shear rate is a measure of spatial variation in local velocities in a fluid. Cell damage in a moving fluid is sometimes associated with the magnitude of the prevailing shear rate or the associated shear stress, but these quantities are neither easily defined nor easily measured in the relatively turbulent environment of most bioreactors. Moreover, shear rate varies with location within a vessel. Attempts have been made to characterize an average shear rate or a maximum shear rate in various types of bioreactors and process flow devices, as discussed later for the more common cases.

Bubble Columns and Airlift Bioreactors

Expressions for mean shear rate in bubble columns have been summarized by Chisti (21), Chisti and Moo-Young (19), and Shi et al. (22). These expression generally correlate the average shear rate with the superficial gas velocity; thus

$$\gamma = kU_G^a \qquad (2)$$

where the parameter a equals 1.0 in most cases, but the k value has been reported variously as 1,000, 2,800, 5,000 m^{-1}, and so forth (20). The enormous disagreement among various authorities is obvious in Figure 1, where the shear rate calculated according to several available equations is shown as a function of the superficial aeration velocity in air–water system in a bubble column.

Equation 2 has also been applied to airlift bioreactors, using the superficial gas velocity in the riser zone as a correlating parameter; however, that usage is incorrect (19). A more suitable form of equation 2 for airlift reactors is

$$\gamma = \frac{kU_{Gr}}{1 + \dfrac{A_d}{A_r}} \qquad (3)$$

where U_{Gr} is the superficial gas velocity in the riser, A_r is the cross-sectional area of the riser, and A_d is the cross-sectional area of the downcomer. In addition to the already-noted discrepancies (Figure 1), equations 2 and 3 have other significant flaws. The shear rate should depend also on the momentum transfer capability of a fluid, that is, on the density and the viscosity of the fluid, but equations 2 and 3 show no such dependence. Indeed, it is well known that the bubble size in a turbulent field depends on the viscosity and the density of the fluid as well as on the specific energy input rate. It is therefore reasonable to assume that correlations that express the shear rate as a function of U_G (or U_{Gr}) alone are incomplete (19). Furthermore, correlations such as equation 2 have generally been based on observations of phenomena at solid–liquid interfaces (e.g., heat transfer from coils or jackets), and their extension to phenomena at the gas–liquid interface or the bulk fluid is absurd at best. Shear stress and, hence, shear rate at walls of riser and downcomer zones of an airlift device are readily calculated using methods developed for pipes and channels, as discussed in a later section.

Another equation for estimation of an "effective" shear rate in airlift reactors is

$$\gamma = 3.26 - 3.51 \times 10^2 U_{Gr} + 1.48 \times 10^4 U_{Gr}^2 \qquad (4)$$

which was developed for $0.004 < U_{Gr}\ (m\ s^{-1}) < 0.06$ (22); the shear rate range covered was 2–35 s^{-1}. Equation 4 was developed in an external-loop airlift reactor. First, the effect of viscosity on the induced liquid circulation velocity in the downcomer was established using Newtonian glycerol solutions at various gas flow rates. The maximum Reynolds number in the downcomer was about 3,200, or barely in the turbulent regime. In a second step, pseudoplastic media were used in the reactor, and the effective viscosity (μ_{ap}) of these fluids in the circulation loop was determined as being equal to the viscosity of the Newtonian glycerol solutions, when the two systems were at identical aeration rates and liquid circulation rates. The effective viscosity and the known values of K and n were used in the power law equation to calculate the prevailing shear rate:

$$\gamma = \left(\frac{\mu_{ap}}{K}\right)^{1/n-1} \qquad (5)$$

The calculated shear rates were correlated with the superficial gas velocity in the riser, as noted in equation 4 (22). Although written in terms of the superficial gas velocity in the riser, equation 4 may usefully be expressed in terms of the specific power input in the reactor, as recommended elsewhere (19).

Because the procedure employed by Shi et al. (22) equated the viscosity-associated reduction in the liquid cir-

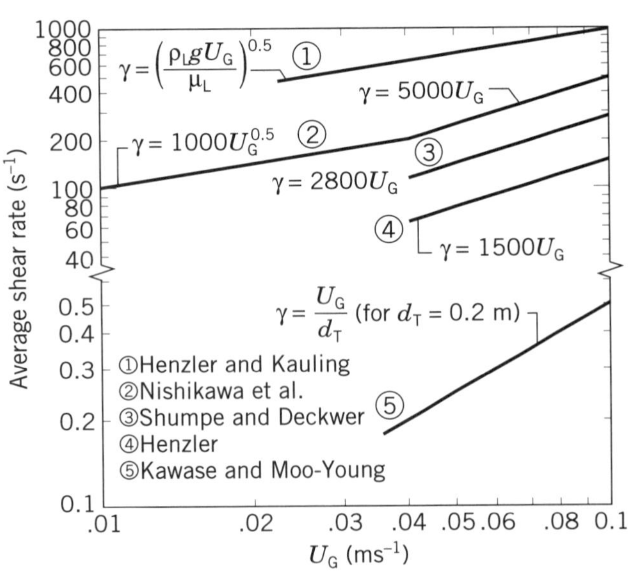

Figure 1. Average shear rate in air–water system in a bubble column according to various sources. *Source*: Based on Ref. 21.

culation velocity in different fluids, it gave in some sense a shear rate in the vicinity of the interface between the fluid and the walls of the reactor; shear rate in the bulk flow, which is the quantity of interest in most cases, was not quantified. Furthermore, the method of analysis used (22) applies strictly to a laminar flow regime, quite unlike the flow situations encountered in most practical operations.

Compared with bubble columns, equation 4 yields quite low values for shear rates in airlift reactors, as noted by Shi et al. (22). In such comparisons, care needs to be taken to ensure that the devices are being compared at identical values of specific power inputs (19,21). Although Shi et al. (22) did not adhere to this criterion, the specific geometry of the reactor they used was such that the error was small. Unlike what Shi et al. (22) concluded, equation 4 is not suitable for correlating mass transfer from gas bubbles or suspended solids because it does not give shear rates at gas–liquid or particle–liquid interfaces. Similarly, the usefulness of the shear rate calculated using equation 4, for correlating survival of fragile biocatalysts, remains questionable. An alternative, mechanistic approach to quantifying the bulk shear rate in various zones of airlift bioreactors has been advanced by Molina Grima et al. (14). Because the hydrodynamic environment in various zones of airlift reactors tends to be quite different, characterization of shear rate by a single global value is not sensible. The overall shear rates can be deceptively low even though damaging levels may be experienced in the high-shear zones (14); hence, the approach of Molina Grima et al. (14) is preferred.

Yet other attempts at characterizing the hydrodynamic forces relied on measurement of noises associated with events such as bubble formation at the sparger, bubble disengagement at the surface, and liquid flow over the baffle into the downcomer (23). Spectra of acoustic signals were measured in water, a salt solution, and aqueous glycerol media in a bubble column and split-cylinder airlift reactor (23). Because several events occur simultaneously in bubbling reactors, there were substantial ambiguities in assignment of the noise signals to specific events. The practical significance of the results was unclear even though the intended aim was to somehow relate the data to damage of fragile cells that has been reported by others in such reactors. Yet other work (24) examined the structure of turbulence in water and power law solutions ($K = 0.0194$ Pa $s^{0.973}$ and 0.0596 Pa $s^{0.958}$) in an external-loop airlift reactor. The reactor achieved complete gas–liquid separation and there was no gas in the downcomer. Measurements of local root mean square velocity fluctuations as an indicator of turbulence intensity showed a slight decrease from the center of the riser to the wall. These measurements were at a constant gas velocity of 2.05×10^{-2} m s^{-1}. The magnitudes of the velocity fluctuations were similar for all media; however, the velocity fluctuations were much lower in the gas-free downcomer than in the riser despite similar values of Reynolds numbers in the two zones. Based on measurements of one-dimensional energy spectra in the center of the riser (24), turbulence could not be considered isotropic, particularly in power law fluids. Other substantial evidence for the absence of isotropic turbulence in airlift and bubble column reactors has been documented (25,26).

Stirred-Tank Fermenters

The local velocity at a fixed position in any vessel fluctuates around a mean value; hence, the shear rate fluctuates. The extent of fluctuations is position dependent. In stirred tanks with radial-flow impellers such as Rushton turbines, the velocity fluctuations are greatest near the impeller tip and decline rapidly as one moves radially outward from the tip. Elsewhere in the vessel, the velocity fluctuations are reduced yet further. The time-averaged mean shear rate generated by a six-bladed Rushton turbine agitating a Newtonian fluid in a baffled vessels of diameter d_T is given as (27)

$$\gamma_{av} = 4.2N\left(\frac{d_i}{d_T}\right)^{0.3}\frac{d_i}{W} \tag{6}$$

where W is the width of the impeller blade, d_i is the impeller diameter, and N is it's rotational speed. The time averaged maximum shear rate value is 2.3-times γ_{av} (27). Another expression for the maximum shear rate at Rushton turbine blades is (28)

$$\gamma_{max} = 3.3N^{1.5}d_i\left(\frac{\rho_L}{\mu_L}\right)^{1/2} \tag{7}$$

which applies to Newtonian and non-Newtonian liquids when $100 \le (Nd_i^2\,\rho_L/\mu_L) \le 29{,}000$. The uncertainty in the coefficient (equation 7) is said to be $\pm 20\%$. For non-Newtonian fluids, μ_L in equation 7 is the zero shear viscosity. A different equation has been proposed by Wichterle et al. (28):

$$\gamma_{max} = N(1 + 5.3n)^{1/n}\left(\frac{N^{2-n}d_i^2\rho_L}{K}\right)^{1/1+n} \tag{8}$$

Equation 8 provides the maximum shear rate on a Rushton turbine blade. An average shear rate expression applicable to a broader range of impellers and media is (29):

$$\gamma_{av} = k_i\left(\frac{4n}{3n + 1}\right)^{n/n-1}N \tag{9}$$

where n, the flow index of a fluid, equals 1.0 for a Newtonian liquid. In equation 9, k_i is an impeller-dependent constant. Some typical k_i values are noted in Table 2. The various shear rate correlations discussed here are compared in Figure 2 for water in a standard stirred tank (30) agitated

Table 2. The k_i Values for Use in Equation 9

Impeller	k_i
Six-bladed disc turbines	11–13
Paddle impellers	10–13
Propellers	~10
Helical ribbon impellers	~30

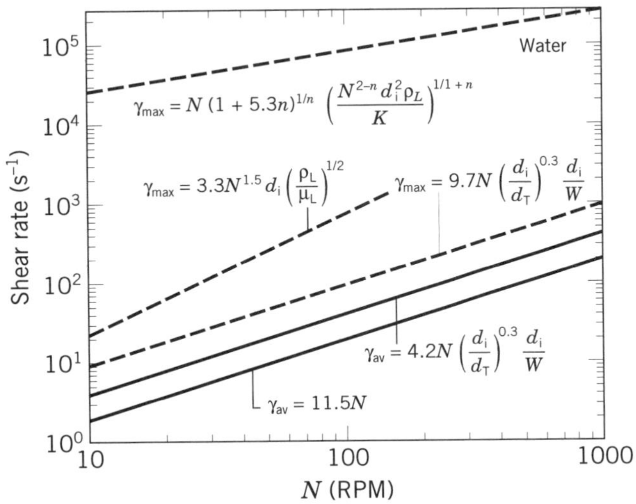

Figure 2. Comparison of the various shear rate correlations for water in a standard stirred-tank agitated with a 0.1-m-diameter six-bladed Rushton turbine. The dashed lines are for the maximum shear rate; the solid ones are for the average shear rate. See text for additional details.

with a 0.1-m diameter six-bladed Rushton turbine. For some contexts, the shear rate around a rising bubble may be approximated as the ratio of the terminal rise velocity to the bubble diameter (or radius); hence,

$$\gamma_{av} = \frac{2U_B}{d_B} \tag{10}$$

In air–water, under conditions typical of bubble columns and airlift reactors, the bubble rise velocity is about 0.2 m s^{-1}, and the bubble diameter is about 0.006 m. Thus, the interfacial shear rate approximates to 67 s^{-1} if the interface is nonmobile. Lower shear rates are expected at circulating interfaces.

Spinfilter

Spinfilters—rotating cylinders made of wire screen with openings that are significantly bigger (e.g., 25 μm) than the cells—are used to retain suspended animal cells in perfusion culture continuous bioreactors. Most of the cells are kept on the upstream side of the wire mesh by a hydrodynamic mechanism that depends on rapid rotation (e.g., 500 rpm) of the cylindrical screen (Fig. 3). The spent, largely cell-free medium is withdrawn from the zone within the rotating screen. Shear stresses associated with rotation of spinfilters do not generally damage animal cells (31). Sometimes the cell-free zone within the confines of the screen is also used for submerged aeration.

Other Flow Devices

Bioreactors and process vessels are where much of the shear-sensitive material is processed; however, at least part of the processing usually requires transfer of material between vessels. Transfer is accomplished through pipework, and centrifugal pumps are sometimes used. Methods

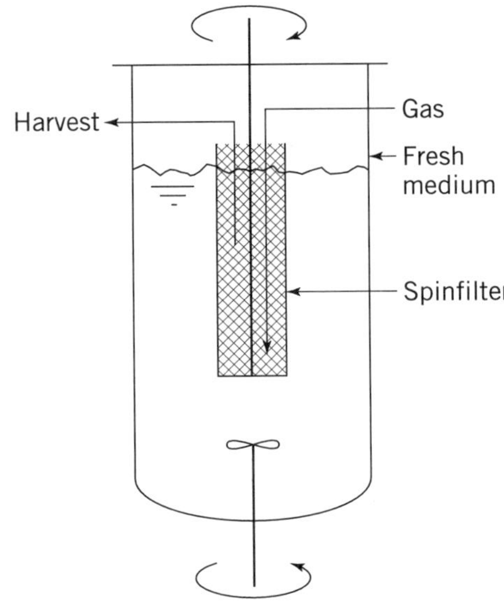

Figure 3. Perfusion culture with spinfilter for hydrodynamics-based cell retention in bioreactors.

of estimating shear stress levels in these devices are briefly summarized next.

Pipes, Tubings, and Flow Channels. ***Newtonian Fluids.*** In developed laminar flow of a Newtonian fluid through a straight tube of diameter d, the wall shear rate is given as

$$\gamma_w = \frac{8U_L}{d} \tag{11}$$

where U_L is the mean flow velocity. For rectangular channels of height h, the maximum or wall shear rate in developed laminar flow is

$$\gamma_w = \frac{6U_L}{h} \tag{12}$$

where U_L is again the mean flow velocity.

The wall shear stress (i.e., the maximum value) in a flow channel such as the riser of an airlift reactor is related to the pressure drop (ΔP), the length L of channel, and the hydraulic diameter (20); thus,

$$\tau_w = \frac{d}{4L}\Delta P \tag{13}$$

Consequently, in turbulent flow, the wall shear stress is

$$\tau_w = \frac{1}{2}C_f\rho_L U_L^2 \tag{14}$$

where ρ_L is the liquid density and C_f is the Fanning friction factor. The latter is related with the Reynolds number as follows:

$$C_f = 0.0792\left(\frac{\rho_L U_L d}{\mu_L}\right)^{-0.25} \quad (15)$$

where d is the hydraulic diameter of the flow channel or pipe. From equations 14 and 15, the wall shear stress can be shown to depend on Newtonian viscosity of the fluid: $\tau_w \propto \mu_L^{-0.75}$ (14). Use of equations 11 and 12 and others given later that are similar presupposes that the liquid velocity is known. This is normally the case in pipes and channels; however, with airlift vessels the induced liquid circulation rate will often need to be estimated using available methods (20,21,32).

Non-Newtonian Power Law Fluids. Methods for calculation of wall shear stress in laminar and turbulent flow of power law fluids with and without dispersed gas, have been discussed by Chisti (21) and Chisti and Moo-Young (33). For a non-Newtonian pseudoplastic medium flowing through a tube such as the riser or the downcomer of an airlift bioreactor, the flow is laminar when (34)

$$Re \leq 2300\,\frac{3n + 1}{4n} \quad (16)$$

where the Reynolds number is calculated using the equation

$$Re = \frac{8U_L^{2-n} d^n \rho_L}{K\left(\frac{6n + 2}{n}\right)^n} \quad (17)$$

In the laminar regime, the wall shear stress for flow of a gas-free liquid has been expressed as (35)

$$\tau_w = \frac{8\rho_L U_L^2}{Re} \quad (18)$$

When the flow is turbulent (i.e., the Reynolds number exceeds the value calculated with equation 16), the wall shear stress in a gas-free liquid is (34,35)

$$\tau_w = \frac{0.316 n^{0.121} \rho_L U_L^2}{8Re^{2/n(x-1)+2}} \quad (19)$$

In presence of a gas, the shear stress is a complex function of τ_w, the gas holdup, the properties of the liquid (ρ_L, K, n), and the superficial velocities of the gas and the liquid phases, as discussed elsewhere (21,33). In equations 16–19, d is the channel diameter, U_L is the superficial liquid velocity, K is the consistency index, and n is the flow behavior index. The parameter x in equation 19 is given as

$$x = 0.70n^{-0.591} \quad (20)$$

as reported by Chisti and Moo-Young (33) based on data of Sokolov and Metkin (34).

Submerged Jets. Submerged jets are encountered wherever a pipe or nozzle discharges a fluid beneath the surface of the same fluid in a larger reservoir. For example, culture broth recirculating from a tank to an ultrafiltration module and back to the tank could form a submerged jet. If the cross-sectional area of the discharge nozzle is less than about 25% of that of the reservoir, the wall effects can be neglected and the jet is said to be "free." In a stable submerged turbulent jet, the maximum shear stress occurs in the direction of discharge, six to seven nozzle diameters downstream from the orifice. This shear stress is given as

$$\tau_{max} = 0.018\rho_L u_o^2 \quad (21)$$

where u_o is the velocity at the orifice.

Centrifugal Pumps. Centrifugal pumps are commonly employed in bioprocessing; hence, a knowledge of the shear rate in these devices is essential to assessing their suitability for a given use. The shear rate in a Newtonian fluid adjacent to the rotating impeller of a centrifugal pump depends on the radial position. The local shear rate at any position has been expressed as

$$\gamma = 6.30 N Re_L^{0.5} \quad (22)$$

where the local Reynolds number at the local diameter d_L is

$$Re_L = \frac{N d_L^2 \rho_L}{\mu_L} \quad (23)$$

Equation 22 is based on theoretical considerations, but it has closely correlated measured shear rate data over the range $10^2 \leq Re_L \leq 2 \times 10^6$ in a centrifugal pump with the impeller rotating in a standard volute casing (36). Over the approximate impeller rotational speed range of 0.01–100 s^{-1}, the shear rate values ranged over 1–10^5 s^{-1} in aqueous electrolyte solutions (36).

As a general guideline, centrifugal pumps may be used to transfer most low-viscosity microbial broths and enzyme solutions so long as gas–liquid interfaces are rigorously excluded. Centrifugal pumps should not be used for animal cells, biofilm supporting particles, soft immobilized enzyme supports, suspensions of nematodes and protozoa, protein precipitates, and plant cells. These pumps will be satisfactory for some microalgae but not for others.

Shear Rate in Isotropic Turbulence

Irrespective of the bioreactor configuration, a turbulence field is deemed isotropic when the size of the primary eddies generated by the turbulence-producing mechanism is a thousandfold or more compared with the size of the energy-dissipating microeddies. The length scale of the primary eddies is often approximated as the width of the impeller blade or the diameter of the impeller in a stirred tank. In bubble columns and airlift bioreactors, the length scale of primary eddies is approximated as the diameter of the column (or the riser tube), or the diameter of the bubble issuing from the gas sparger.

Shear stress, shear rate, the dimensions of microeddies, and other characteristics of flow are linked ultimately to the energy input and dissipation rates in the fluid. The local shear rate in the vicinity of an eddy in an isotropically turbulent field may be estimated as the ratio of the velocity and the length of the eddy; hence,

$$\gamma_i = \frac{u}{l} = \frac{\mu_L}{\rho_L l^2} \tag{24}$$

where ρ_L and μ_L are, respectively, the density and the viscosity of the fluid. The mean length, l, and the velocity, u, of the microeddies are related with the energy dissipation in the turbulence field; thus,

$$l = \left(\frac{\mu_L}{\rho_L}\right)^{3/4} E^{-1/4} \tag{25}$$

and

$$u = \left(\frac{\mu_L E}{\rho_L}\right)^{1/4} \tag{26}$$

where E is the rate of energy dissipation per unit mass of fluid. In most cases, all the energy input to the fluid is dissipated in fluid eddies, and E equals the rate of energy input. Equations 24–26 apply when local isotropic turbulence prevails.

Energy dissipation rate in bubble columns and airlift vessels is a function of the superficial aeration velocity (20,21); thus

$$E = gU_G \text{ (bubble columns)} \tag{27}$$

and

$$E = g\frac{U_{Gr}}{1 + \dfrac{A_d}{A_r}} \text{ (airlift bioreactors)} \tag{28}$$

Methods for calculating the specific energy dissipation rate in stirred vessels have been discussed elsewhere (30,37). The energy dissipation varies greatly from the mean value in a stirred tank. The maximum energy dissipation occurs in the vicinity of the impellers. This maximum value can be calculated using the equation

$$E = \frac{\text{Po}N^3 d_i^2}{\rho_L} \tag{29}$$

where Po is the power number, d_i is the diameter of the impeller, and N is the rotation speed (s^{-1}). In developed turbulent flow in stirred vessels, the power number is generally constant for a given type of impeller and tank geometry. For single-phase pipe flow, the specific energy dissipation rate is

$$E = \frac{U_L \Delta P}{\rho_L} \tag{30}$$

where ΔP is the pressure drop per unit length, and U_L is the mean flow velocity.

Generally, if the dimensions of the biocatalyst particle are much smaller than the calculated length, l, of the microeddies, the particle is simply carried around by the fluid eddy; the particle does not experience any disruptive force. On the other hand, a particle that is larger than the length scale of the eddy will experience pressure differentials on its surface, and if the particle is not strong enough it could be broken by the resulting forces.

In addition to turbulence within the fluid, other damage-causing phenomena in a bioreactor include interparticle collisions; collisions with walls, other stationary surfaces, and the impeller; shear forces associated with bubble rupture at the surface of the fluid; phenomena linked with bubble coalescence and breakup; and bubble formation at the gas sparger. Some of these aspects are discussed elsewhere in this article. Summarizing, several possible shear rate values may be calculated for a given situation in a bioprocess device. Not every calculated value is appropriate or relevant to the problem at hand. For pneumatically agitated bioreactors, when the turbulence characteristics in the bulk fluid are the relevant ones, the preferred approach is to use equation 24 for shear rate in the vicinity of eddies. The same applies to mechanically stirred tanks. In some cases, the relevant shear rate may be that at the interface of a rising bubble unless turbulence is so intense that bubbles do not rise freely. Other situations would be controlled by the bubble rupture events, as discussed later. In yet other cases, the fluid eddy shear rate and the maximum shear rate at the impeller will need to be taken into account. Factors such as the frequency of passage of a sensitive biocatalyst through high-shear regions may need consideration.

RESPONSE TO SHEAR

Response to shear has been characterized using various indices, including growth rate (38–40); cell viability (41); regrowth potential (42); release of intracellular material (38,43,44); changes in oxygen uptake rate, ATP, productivity of metabolites, and biochemical composition of the cells; and alterations in morphology of cells and cell aggregates (2,42,44).

Most studies of shear response of cells have utilized batch culture with or without aeration. Such cultures contain cells at different stages of growth; hence, the size and metabolic state of cells are nonuniform. In contract, continuous culture is time-consuming (38), differs substantially from the culture methodologies that are typically employed commercially, and, through natural selection, the cells in continuous culture adapt to shear fields, hence, masking the effect of shear. For example, Wu et al. (45) noted for insect cells that the beneficial effect of shear-protective agents (serum, methyl cellulose, Pluronic F68) were not as important as the effect of passage number. Similar results have been alluded to for plant cells (38). To overcome problems associated with cell growth and adaptation to shear, Zhang et al. (43) recommended the use of erythrocytes for comparative evaluation of the hydrodynamic environment in bioreactors. A nongrowing, well-defined, and shear-susceptible cell population such as that provided by erythrocytes can help identify suitable regimes of operation of a given reactor. In addition, shear-damaging potential of different bioreactor configurations can be assessed on a common basis using cells such as erythrocytes. Specific information on shear susceptibility of animal, plant, and various microbial cells is discussed in the following sections.

Bacteria, Yeasts, and Mycelial Microfungi

Yeasts, bacteria, and mycelial microfungi are generally quite robust and not easily damaged under typical processing conditions; however, even the most robust cells are broken when subjected to suitable combinations of high stress and exposure time (1). Often, genetically modified strains of an otherwise robust microorganism are less shear tolerant (1,10,46–48). With microorganisms, the composition of the growth medium and the specific growth rate used for the culture are known to influence the robustness of the cell (1). Cells grown in fully defined media that may lack one or more components essential to synthesis of a normal cell wall are often less robust (1). Growth at a higher specific rate will often produce weaker cells (1). When the shear field is insufficiently intense to damage cells, it may still affect product formation and characteristics (9,10). In one case, properties of a shear-sensitive polymer were affected, but not the producing bacterial cells (10). Cells experiencing physiological stress (e.g., starvation) are more susceptible to shear-stress-related damage (44).

Reports cited by Märkl et al. (44) document increasing release of intracellular material from mycelium of *Rhizopus javanicus* and *Mucor javanicus* as the stirrer speed increased. Increasing release of nucleotides was noted with increasing speed of agitation, or with increasing agitation time at a constant agitation speed. Effect of Rushton turbine tip speed on performance of a process for converting cellulose to protein-enriched microbial biomass were clearly demonstrated by Moo-Young et al. (39). The process utilized the filamentous microfungus *Neurospora sitophila* for the conversion. Increasing impeller tip speed from 2.35 m s^{-1} to 3.29 m s^{-1} reduced the specific protein production rate and the final yield (Fig. 4). In addition, a distinct lag phase was observed at the higher agitation rate (Fig. 4) when the inoculum grown in a more quiescent environment was transferred to the fermenter. At the highest tip speed shown in Figure 4, the specific protein production rate was only 55% of that at the lowest tip speed. Similarly, the yield of crude microbial protein was reduced to ~67% of the value obtained at the lowest agitation speed used (39). Further work revealed that agitation-associated damage could be reduced substantially by replacing the Rushton turbines with axial flow hydrofoil impellers (40); thus confirming that different types of impellers generate very different shear fields at identical tip speeds. For equal or lower power draw, hydrofoil impellers are typically better bulk mixers than Rushton turbines (30,37).

In *Penicillium chrysogenum* fermentations performed in turbine-impeller-agitated vessels with working volumes of 0.005–1 m^3, operated at impeller tip speeds of 2.5–6.3 m s^{-1}, the specific penicillin production rate and the mean length of the main hyphae were lower at high agitation intensities (2). High agitation speeds affected the fungal

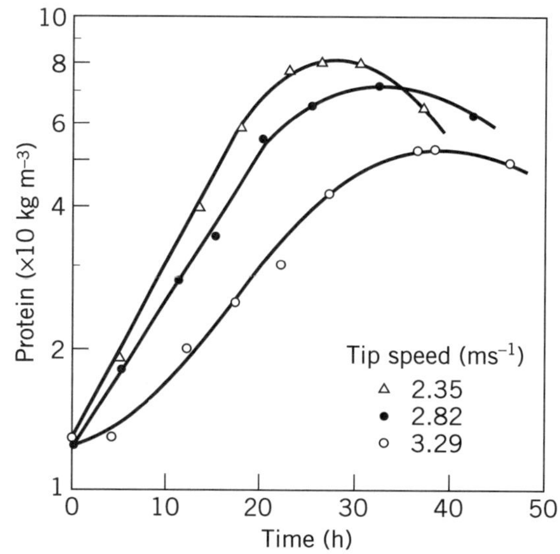

Figure 4. Effect of Rushton turbine tip speed on protein production by the mold *N. sitophila*. *Source*: From Refs. 39 and 40.

morphology, promoting mycelial fragmentation and a greater frequency of branching (2). The behavior of the fermentation at various scales correlated poorly with the impeller tip speed, but correlation was superior in terms of a parameter that took into account the frequency of passage of the culture through the intensely agitated zones around the impellers (2). This behavior suggests that the degree of hyphal damage in stirred bioreactors depends on a combination of the shear intensity in the impeller zone and the cumulative time of exposure of the microbial solids to the shear field.

Sometimes, especially with certain mycelial microfungi, the biomass yield may not be affected by agitation rate, but the cell morphology and production of metabolites may be influenced greatly. Using different strains of citric acid–producing *Aspergillus niger*, Ujcova et al. (3) showed that the yield of citric acid first increased with increasing agitation rate, presumably because of improved oxygen supply, but the yield declined because of cell damage if the agitation speed was increased further. The yield maxima occurred at different agitation rates for different strains, and the sensitivity to agitation differed among strains. Other morphological effects of nondamaging shear levels include changes in growth form, whether pelletlike or pulplike; changes in the degree of branching; the extent of entanglement; and whether growth occurs in filamentous or spherical cell form in certain dimorphic species.

Evidence suggests that even a supposedly robust cell such as *Saccharomyces cerevisiae* (1,11) is susceptible to mechanical shear damage in fermenters (49,50). In one study, *S. cerevisiae* suspended in isotonic saline lost viability at increasing rate over a 25-h period as the agitation speed of the paddle impeller was increased (50). At 1.43 m s^{-1} impeller tip speed, >90% of the cells were dead over the course of 25 h. In contrast, at 0.66 m s^{-1} tip speed, only about 15% of the cells died over the same period (50). Excessive shear has been alleged to also influence the productivity of ethanol in fermentations employing *S. cerevisiae* (49).

Among methane bacteria, *Methanobacterium thermoautotrophicum* with rodlike cells that exceeded 2 μm in length was highly resistant to shear stress (44). In contrast, and surprisingly, the relatively smaller (<2 μm) spheroidal cells of *Methanococcus vanielii* were quite sensitive to shear stresses (44). Supplementing the medium with carboxymethyl cellulose viscosity enhancer increased the extent of the cell damage (44). Failure to provide an essential nutrient, selenium is this case, also enhanced shear susceptibility of the culture (44). Although many authors have successfully correlated cell damage with the tip speed of the impeller, Märkl et al. (44) believe that power input per unit culture volume is a superior correlating parameter that is also suitable for comparing the cell damage effects of mechanically stirred and other nonmechanically agitated bioreactors. A constant power input per unit culture volume has also been suggested as a bioreactor scale-up criterion for fragile biocatalysts (44). This recommendation notwithstanding, average power input alone cannot be a universally satisfactory determinant of cell damage; as noted in an earlier section, the maximum shear rate and the frequency of passage (or cumulative exposure) through the high-shear zone need to be taken into account, especially when the shear rate in a given device is strongly position dependent. Several studies confirm that frequency of passes through a high-shear environment have a clear impact on biocatalyst survival (1,2,11,12,16,20,51). Effects of variously intense shear fields on microbial cells are further discussed by Thomas (5,52), Prokop and Bajpai (53), and Chisti and Moo-Young (1). Shear effects on microbial flocs and pellets are treated in a separate section of this article.

Microalgae and Cyanobacteria

Cell fragility has been referred to as the key problem of culture of microalgae in photobioreactors (54); however, the real situation is not quite that bad. Among microalgae, the sensitivity to shear varies greatly with the species. Algae such as *Dunaliella* are indeed extremely fragile, but many other species will tolerate high levels of turbulence and sparging with gas. Thus, gas-sparged airlift devices are commonly used to circulate algal cultures through the solar receivers of photobioreactors (20,55–57). Similarly, relatively low-shear positive displacement rotary pumps have been successfully employed. High-shear centrifugal pumps have also been used; however, at least in some cases, the centrifugal pumps are known to be more damaging than the other fluid-moving systems. In one case, for otherwise equivalent conditions, replacement of the centrifugal pump with a positive displacement device doubled the productivity of *Porphyridium cruentum* (54), a producer of phycoerythrin and polyunsaturated fatty acids. Irrespective of the method used to circulate the culture, the culture flow velocities through tubular solar receivers typically range over 0.3 to 0.5 m s^{-1} in tubes up to 0.06 m in diameter. These conditions have been used with a variety of algae.

In cultures of *Haematococcus pluvialis*, a producer of astaxanthin, early onset of the stationary phase has been associated with turbulence-induced deflagelation of the cells (54). Fragility of species such as *H. pluvialis* depends also on the phase of growth; thus, tailoring the bioreactor operational regime to the growth phase can improve overall productivity (54). The stationary phase red cysts of *H. pluvialis* are significantly more robust than the green vegetative cells (54). The cyanobacterium *Spirulina*—a healthfood and source of the food colorant phycocyanin—is another species that is susceptible to damage by excessive shear forces. The marine alga *Dunaliella* is particularly fragile: the cell lacks a ridged wall; only a thin cytoplasmic membrane encloses it. *Dunaliella* is a good source of natural β-carotene; hence, it is commercially useful. The cell also produces glycerol. Silva et al. (58) showed that *Dunaliella* was susceptible to damage by gas bubbles as well as by turbulence in the liquid. Supplementing the culture with carboxymethyl cellulose or agar improved cell survival.

In one study with *Dunaliella tertiolecta*, culture in a bubble column was quite successful, but when the bubble column was converted to an airlift device by inserting a vertical baffle, the productivity declined (59). Under conditions that were earlier identified as optimal, no growth was observed in the airlift reactor, whereas good growth occurred in the bubble column. Microscopic examination showed significant disruption of the cells in the airlift device (59). This was associated with the hydrodynamic stresses generated as the culture flowed over the upper edge of the baffle into the downcomer (59). This effect could have been avoided, or at least minimized, by hydrodynamic smoothing of the upper and lower parts of the baffle to prevent flow separation (20). In the bubble column, the growth was sensitive to aeration rate: growth rate increased with increasing superficial gas velocity until a velocity of about 0.6 m min^{-1}. Further increase in aeration rate reduced growth, apparently because of hydrodynamic stresses in the fluid (59). Under nongrowth conditions (no light), the specific death rate in the bubble column was shown to increase with superficial gas velocity for velocities exceeding 0.6 m min^{-1} (59). At a fixed aeration velocity ($U_G = 1$ m min^{-1}), the specific death rate decreased with increasing height of the culture fluid in the column (59), probably because the specific power input and, hence, the turbulence intensity declined (20). Similar behavior has been reported with animal cells in bubble columns (60–64).

For otherwise fixed conditions, increasing the number of aeration nozzles (i.e., decreasing the gas jet velocity) caused a marginal reduction in the specific death rate of *D. tertiolecta* (59). This led to the conclusion that the events at the gas sparger did contribute to cell death (59), although the effect was minimal. Unlike Suzuki et al. (59), others (56,57) have quite successfully cultured various microalgae in pilot-scale airlift devices. Continuous culture of the green alga *Chlorella pyrenoidosa* in a rectangular airlift device with multiple light-emitting diodes was reported by Matthijs et al. (65). With a relatively robust photosynthetic organism such as *Chlorella*, increasing turbulence within limits improves productivity in photobioreactors because of enhanced fluid interchange between the dimly lit interior and the better-illuminated exterior parts of a photobioreactor.

In addition to an organism's morphology, the biochemical composition of the cell wall has been associated at least qualitatively with the cell's susceptibility to hydrodynamic stresses. The microalga *Chlorella vulgaris*, which is shear tolerant, has a thick (100–200 nm) cell wall composed of cellulose, pectin, and sporopollenin (44). The somewhat more sensitive alga *Chlorella reinhardii* has only glycoproteins in its relatively thin (40 nm) cell wall (44), whereas a highly sensitive cell wall defective mutant of *C. reinhardii* is devoid of cell wall (44). Similarly, the cyanobacterium *Synechococcus*, with a murein-containing cell wall, and the methane bacterium *M. thermoautotrophicum*, with a pseudomurine-containing wall, are noticeably more resistant than methane bacteria such as *Methanococcus vanielli*, which has a protein-based wall.

Although earlier work with the mold *M. javanicus* showed a reduced extent of damage to mycelia as the viscosity was increased at constant agitation rate, and similar reports exist with animal cells (66,67) and the halotolerant alga *Dunaliella* (58), Märkl et al. (44) observed an opposite effect: with a highly shear-sensitive strain of *C. reinhardii*, even slight increase in viscosity enhanced cell damage (44). The viscosity was increased by adding carboxymethyl cellulose to the culture broth. With a relatively less shear sensitive strain of *C. reinhardii*, increasing the viscosity reduced the impeller agitation speed at which the damaging effects were first discerned (44). Apparently, at least with the less sensitive strain, supplementing the medium with the viscosity enhancer caused the cells to grow in clumps, and aggregation supposedly was an effective shield against cell damage (44). Increasing shear damage with increasing culture viscosity has been seen before with a protozoan (68). These apparently contradictory results are easily explained: whether changes in viscosity increase or reduce shear damage depends on the peculiarities of the mechanism causing the damage. For a given impeller tip speed, increasing viscosity will not affect the tip speed–associated shear rate, but the turbulence will be reduced; hence, the shear rate in the fluid should be lower. However, the shear stress around a particle may increase if the fluid eddy length scale remains comparable to the dimensions of the particle. In other cases, increased viscosity may reduce both shear rate and shear stress: only eddies that are larger than the particle may survive and there may be little or no relative motion between the fluid and the particle. In yet other cases, the frequency of passage through high-shear zones may decline with increasing viscosity.

Plant Cells

Cultured plant cells are a potential source of numerous high-value compounds, although few production processes have been commercialized (13,69). Low culture productivities continue to be the primary limitation. Investigational culture in large bioreactors has been carried out since at least 1959. A variety of reactors have been used, including bubble columns, airlift devices, stirred vessels, rotary-drum reactors, and immobilized cell systems (13). Airlift bioreactors have proven especially effective for large-scale culture of heterotrophic, photomixotrophic, and photosynthetic plant cell suspensions (20,70). Relatively low values of pneumatic power inputs have given satisfactory results during cultivation of plant cells. For example, Fischer and Alfermann (70) obtained good results with about 13 W m^{-3} specific power input during culture of *Chenopodium rubrum* cells in a concentric draft-tube airlift device. Similarly, Ballica and Ryu (71) obtained maximum biomass and product yield at an approximate power input value of 35 W m^{-3}, also in a draft-tube airlift reactor, while culturing *Datura stramonium* cells. In the latter case, power inputs above ~60 W m^{-3} caused large declines in biomass and product yields. The size of the cell aggregates declined with increasing specific power input (71). Reinhard et al. (72) described culture of *Digitalis lanata* in a 300-L airlift bioreactor for producing β-methyldigoxin.

Plant cells in suspension tend to be large: 20–50 μm in diameter and 100–500 μm in length. The cells have a cellulose-based wall. Most of the cell's volume is taken up by a vacuole that may occupy >95% of the available space. The fluid in the vacuole is much less viscous than the cytoplasm; hence, a cell with a large vacuole is likely to be less resistant to certain types of deformations than one with a smaller vacuole. Cells frequently occur as aggregates of up to a hundred cells. The aggregates settle easily. Plant cells grow slowly, with doubling times of the order of 20–100 h. Cells may secrete extracellular polysaccharides, producing non-Newtonian broths. Also, the biomass concentration may be high, up to 70 kg m^{-3}. Heterotrophically growing plant cells require oxygen, but demands are low, of the order of 10^{-6} kg O_2 kg^{-1} cell s^{-1}. Critical dissolved oxygen concentration generally tends to be ≤20% of air saturation. The polysaccharide-containing cultures tend to foam a lot when aerated. Also, excessive aeration may reduce productivity by stripping carbon dioxide and other essential volatiles produced by the cells themselves (5,69).

As has been demonstrated with some microorganisms (1), the composition of the culture medium, the culture conditions, and the specific growth rate of a cell may affect the wall strength of plant cells also. For otherwise identical conditions, agitation intensity is reported to influence the composition of plant cell walls. Thus, the relative amounts of pectin, cellulose, and hemicellulose in walls of *Catharanthus roseus* were affected by the intensity of agitation in one case (73). In another case cited by Doran (13), the mass ratio of cell wall relative to the whole cell increased with increasing intensity of shear, suggesting adaptation of the cell to the prevailing environment even over short periods. Even though the turbulence level may be insufficient to affect individual cells, aggregates may be affected; hence, a shear-related metabolic response may occur due to altered cell–cell interactions (69).

Based on laminar flow viscometric measurements, a critical shear stress level of 80–200 N m^{-2} has been suggested for *Morinda citrifolia* cells. Whereas for *Daucus carota*, a shear stress level of 50 N m^{-2} has been associated with cell damage. In another study, carrot cells in a laminar flow Couette viscometer lost the ability to grow and divide in the shear stress range of 0.5–100 N m^{-2} (53). The intracellular enzyme activity was impaired at shear stress levels above 3,000 N m^{-2}, but significant lysis did not occur until a shear stress level of 10,000 N m^{-2} was applied over a prolonged period (≥1 h) (53). In contrast to behavior in laminar flow, the cells were quite sensitive to turbulent impeller agitation. Impeller tip speeds of ~1.1 m s^{-1} lysed a significant proportion of the cells within 40 min (53).

In studies with *Perilla frutescens* cell suspensions, Zhong et al. (74) noted that the cell viability declined with increasing shear rate (constant exposure time) or increasing shear time at constant rate of shear in a concentric cylinder shearing device. A constant shear rate of about 140 s^{-1} produced a noticeable loss of viability within 20 min. At higher shear rate of ~860 s^{-1}, less than 75% of the cells were viable after 20 min (74). In marine propeller–agitated fermenters, the impeller tip speed affected the specific growth rate and the yield of the cells (74). Tip speeds >0.6 m s^{-1} reduced the specific growth rate and the cell biomass yield (74). In other cases, impeller tip speeds as low as 0.35 m s^{-1} have extensively lysed certain plant cells.

With *M. citrifolia* cells subjected to 25–350 N m^{-2} turbulent shear stress in a capillary, the viability decline followed first-order kinetics in which the rate constant increased with the applied shear stress (75). The culture was circulated through the capillary by a peristaltic pump, which also contributed to cell damage (75). The specific death rate constant was linearly related with the cumulative energy dissipated (75). Cell aggregates subjected to shear were reduced in size. For the same species, damage was reported when cells grown in a quiescent environment were subjected to turbulent jet flow through 1- and 2-mm nozzles (51). Cells lost viability with increasing number of passes through the nozzles. The viability loss was first order in the number of passes (51). For a fixed number of nozzle passes, increased nozzle flow velocity enhanced the death rate constant. Similarly, the maximum and the mean aggregate size were reduced when the suspension was subjected to jet flows (51). The average aggregate size was lower for higher jet velocities. The average energy dissipation rates in the jet ranged over 10^3–10^5 W kg^{-1} (51).

Meijer et al. (38) assessed the hydrodynamic stress sensitivity of *C. roseus, Nicotiana tabacum, Tabernaemontana divaricata*, and *Cinchona robusta* in a 3-L Rushton turbine-stirred vessel. Effects of stress were quantified in terms of culture growth rate, respiration rate, and release of organic compounds from the cells. Cultures of *C. roseus* carried out at 0.59 m s^{-1} and 2.36 m s^{-1} impeller tip speeds showed identical behavior in terms of biomass growth rate and glucose consumption rate; hence, hydrodynamics-associated stress was concluded to not affect the cells over the ranges examined (38). Similarly, *N. tabacum* was little affected by differences in agitation rate, although the growth rate and the final yield were noticeably greater at the lower agitation speed. Cells of *C. robusta* responded differently: at 0.59 m s^{-1} impeller tip speed the cells grew and there was little release of intracellular protein; however, at 2.36 m s^{-1}, there was no growth and the specific protein concentration increased with culture time, indicating leakage from damaged cells (38). The *T. divaricata* cells also were damaged at 2.36 m s^{-1}, but less than the cells of *C. robusta*. The damage to *T. divaricata* was reduced at a lower impeller tip speed of 0.59 m s^{-1}. In separate experiments, cells of *N. tabacum* that had earlier withstood higher shear stress levels were shown to be more susceptible to damage after a prolonged period (10 days) of starvation (38). This effect is similar to that noted with other cells (44).

Table 3 is a summary of the damage thresholds levels—either impeller tip speed or the specific power input due to aeration—noted for various plant cell suspensions in stirred fermenters and airlift bioreactors. The studies just referenced clearly affirm that the susceptibility to shearing forces varies a great deal with cell line. Moreover, for the same cell line, cultivation history (e.g., prolonged starvation) and age of cell affect the cell's response to shear fields (38,44,69), which is consistent with similar findings for microorganisms (1). That plant cells adapt to shear fields is also well known: cells cultured in high-shear environments better withstand hydrodynamic forces after many generations.

Protozoa

Midler and Finn (68) provided a clear demonstration of shear-related damage to cells of the large (~80-μm cell diameter) ciliated protozoan *Tetrahymena pyriformis*. Lysis correlated with impeller tip speed. Agitation at impeller tip speeds approximately ≥0.5 m s^{-1} destroyed most of the cells. In a Couette viscometer, the critical shear stress level when lysis first occurred was ~24 N m^{-2} (68). Eighty percent of the cells lysed within the first minute when exposed to shear stress levels of 10^3 N m^{-2}. The extent of damage increased with shear stress level and exposure time. At a given shear rate, increasing the broth viscosity made the protozoan more susceptible to shear damage, implying that the damage was associated with shear stress rather than shear rate (68). The cells were more sensitive to turbulent shear stress: an average turbulent shear stress of ~10 N m^{-2} lysed over half the population within 10 min. In contrast, in laminar flow, similar levels of damage occurred at shear stress levels of almost 300 N m^{-2}. Note that *T. pyriformis* cells are larger than most animal cells, but smaller than most cultured plant cells.

Animal Cells

Animal cells are the most shear-susceptible of biocatalysts. Animal cells are relatively large—typically 10–20 μm in diameter—and they lack a protective cell wall. Osmotic

Table 3. Damaging Threshold Values of Impeller Tip Speed or Specific Power Input for Some Plant Cells

Stirred fermenters	Impeller tip speed (m s^{-1})
Catharanthus roseus	≥2.36
Nicotiana tabacum	~2.36
Cinchona robusta	<2.36
Daucus carota	<1.10
Tabernaemontana divaricata	~0.59
Perilla frutescens	0.60
Others	≤0.35
Airlift bioreactors	**Power input (W m^{-3})**
Chenopodium rubrum	≥13
Datura stramonium	~60

shock is often sufficient to lyse animal cells. Other than blood cells, most normal animal cells are anchorage dependent. Cancerous cells, hybridomas, and established lines will normally grow in suspension. Because of these distinct growth characteristics, two types of culture methods are employed in industrial practice: free suspension culture, in which cells are freely suspended in a nutrient broth; and microcarrier culture, in which anchorage-dependent cells grow on surfaces of solid carriers that are suspended in a culture fluid. Irrespective of the culture method, the suspending fluid is invariably directly sparged with air or other gas mixture to supply the cells with oxygen. Although other aeration alternatives are available (see MASS TRANSFER article), direct sparging with a gas is likely to remain the dominant method of providing oxygen to animal cells (76). Both stirred tanks and airlift devices have been successfully employed for commercial processing of animal cells (77–79). Anchorage-dependent cells may also be cultured on surfaces of large static particles held in packed beds (80,81), but this method is uncommon. Devices used in generation of implantable tissue typically grow normal cells supported on stationary surfaces or within a static matrix of gel or woven fabric (82). The supporting surface may be flat or variously shaped.

Cells in bioreactors are subject to two main kinds of fluid mechanical forces: (*1*) those due to turbulence within the liquid, and (*2*) forces associated with the gas–liquid interface. The latter category includes effects of bubble formation, rise, and rupture. In addition in microcarrier culture, collisions among carriers and carrier–impeller (or carrier–wall) interactions may play a role. Furthermore, in tall bioreactors cells may be subjected to substantial changes in hydrostatic pressure as they circulate with the fluid. Cells in tissue regeneration bioreactors experience the same kind of forces as occur in flow past a flat plate, wall, or other submerged object. The specific considerations relevant to the various culture methodologies are treated in the following sections.

Resistance to shear stress is determined at least partly by the cytoskeletal structure of the cell. The cytoskeleton determines a cell's mechanical properties, shape, and movements such as occur during phagocytosis. The cytoskeleton is made of protein filaments, mainly of actin. The filaments interact closely with the cell membrane by anchoring to specific membrane proteins. A cell exposed to mechanical forces undergoes cytoskeletal reorganization. Cytoskeleton-mediated stress signal transmission to intracellular organelles (Fig. 5) has been associated with various biochemical, physiological, and gene-level responses in vascular cells (4). Some cells are known to have surface stress receptors (e.g., Ca^{2+} ion channels) that have been implicated in biochemical responses (4,83,84) independent of cytoskeletal mediation (Fig. 5). Cytoskeletal properties are known to affect the shear tolerance of hybridomas. Effects of turbulence on the cytoskeleton and other biological responses have been treated by several authors (4,53,85). Work cited by Al-Rubeai et al. (86) suggests that inhibitors of actin polymerization decrease a cell's resistance to shear forces, whereas inducers of polymerization make the cells more robust. Thus, supplementing the culture medium with actin polymerization inducers may be one way of improving performance of bioreactor-cultured animal cells, so long as the requisite production biochemistry is not adversely affected.

Plasma membrane fluidity (PMF), which is a measure of the degree of packing of the membrane and its constituent's freedom to move about laterally through the bilayer membrane, to rotate about their major axes, and to oscillate in the plane of the membrane, is another factor that appears to influence shear susceptibility of cells (41). Increasing membrane fluidity correlates with increasing shear sensitivity (41). PMF increases with increasing temperature and can be manipulated in both directions using various additives. In studies with a freely suspended mouse–mouse hybridoma in a laminar shear stress Couette flow device, conditions that yielded higher membrane fluidity produced more fragile cells (41). For example, cells sheared at elevated temperatures showed greater loss of viability than unsheard controls at the same temperature (41).

Shear Effects in Bubble-Free Environments ***Suspended Cells.*** Freely suspended cells in bubble-free bioreactors are not damaged by mechanical agitation even at intensities much greater than the ones used in typical processing. Exceptions occur in extensional or elongational flow in certain high-shear devices even when the flow is laminar. Extensional or elongational flow occurs whenever the cross-sectional area of the flow channel reduces (e.g., at an orifice on the wall of a tank or at the entrance of a capillary connected to a larger reservoir). The fluid elements undergoing extensional flow stretch and thin. Suspended particles also experience elongational forces in the direction of flow, and compression perpendicular to the flow streamlines. Drops subjected to extension flow can rupture. Similarly, extensional flow through orifices of high-pressure homogenizers contributes to rupture of even the very robust microbial cells (1). Rupture of erythrocytes at entrances to capillaries is well known (87). Shear effects on suspended erythrocytes are discussed in detail in a later section.

For a hybridoma line, Born et al. (88) reported that exposure to laminar shear stress (208 N m^{-2}) in unaerated flow in a cone-and-plate viscometer led to substantial loss in cell count and viability within 20 min. At a constant 180-s exposure, increasing shear stress over 100–350 N m^{-2} linearly enhanced cell disruption, with >90% of the cells being destroyed at 350 N m^{-2} stress level (88). Shear stress levels associated with bubble rupture at the surface of a bioreactor may range over 100–300 N m^{-2} (13). These values are remarkably consistent with shear rates that damaged hybridomas in *unaerated* laminar flow experiments (88).

For a suspended mouse myeloma line in turbulent capillary flow, McQueen et al. (18) noted a threshold average wall shear stress value of 180 N m^{-2} when lysis first commenced. Although the flow caused lysis, it had no effect on viability (18), suggesting that cells at various growth stages were equally affected. The sudden flow contraction at the entrance to the capillary may have contributed to cell lysis, but the residence time in the capillary also had an effect at otherwise constant average wall shear stress level. The rate of lysis was first order in cell number. Above

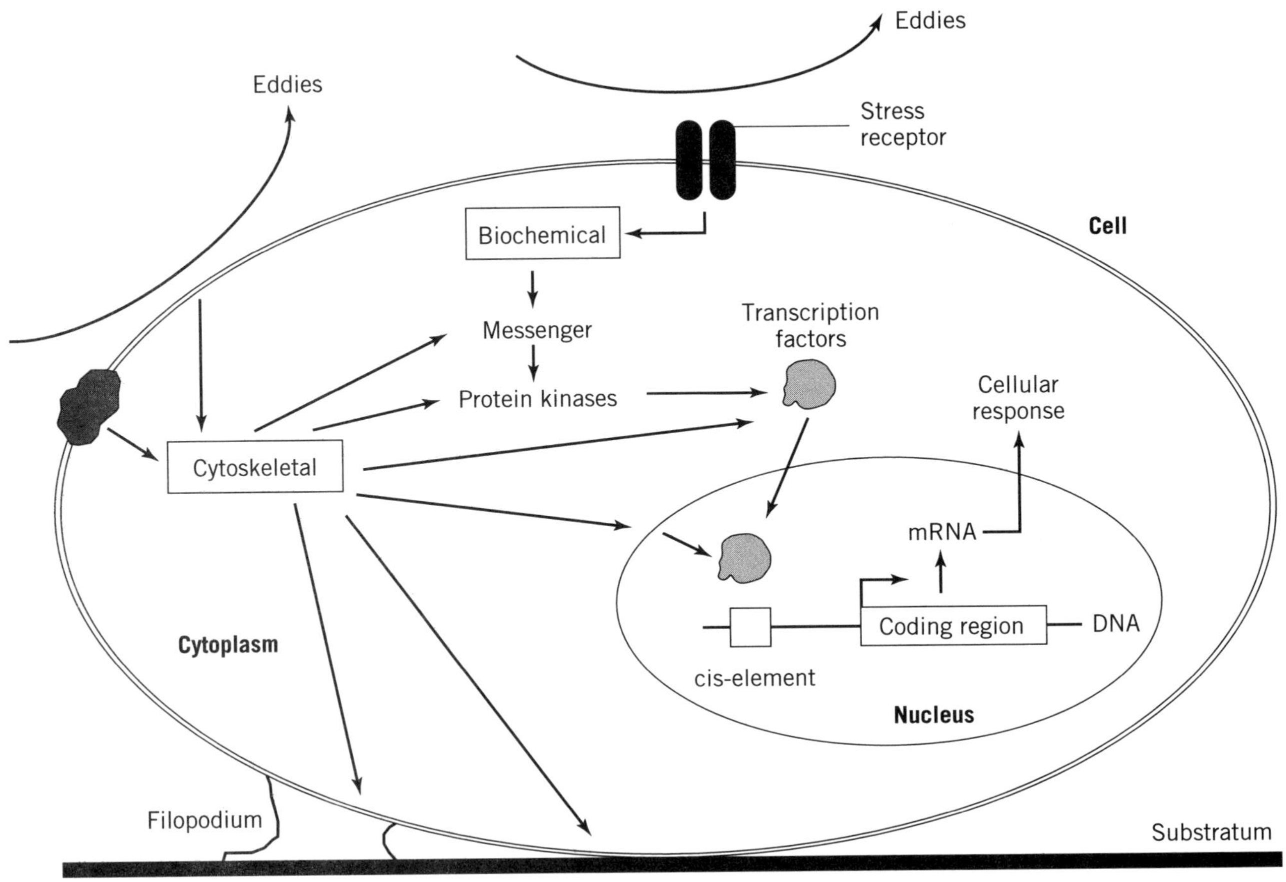

Figure 5. Stress signal transmission and transduction via cytoskeletal and biochemical routs in an anchored cell. *Source*: Adapted from Ref. 4.

the threshold shear stress value, the specific lysis rate increased with increasing level of shear stress (18). The growth rate and the DNA synthesis rate of the cells exposed to the shearing environment were unaffected when the surviving cells were returned to a normal quiescent growth environment (18). In other studies cited by McQueen et al. (18), the shear stress threshold for damage has been reported as 0.87 N m^{-2} for a mouse hybridoma and 1.5 N m^{-2} for insect cells. Higher shear sensitivity of another mouse cell line relative to a human carcinoma has been reported (16).

For human cervical carcinoma HeLa S3 and mouse abdominal fibroblast L929, Augenstein et al. (16) observed lysis of suspended cells in turbulent flow through stainless steel capillaries. Cell death could be correlated with the average wall shear stress level or the power dissipation within the capillaries. The L929 line was more sensitive than the human cell. Control experiments showed that the positive displacement pumps used to circulate the cells through capillaries contributed little to cell lysis (16). Average wall shear stress levels of $0.1–2.0 \times 10^3$ N m^{-2} were sufficient to induce cell inactivation for the two lines. According to Shiragami (17), the mean shear stress acting on cells suspended in capillary flow is 4/3 the shear stress at the capillary walls, so long as the ratio of the cell's diameter to that of the capillary is <0.08.

For a hybridoma examined by Shiragami (89), the specific rate of monoclonal antibody production in a surface-aerated spinner flask depended on the agitation speed. In a 250-mL spinner vessel an agitation rate of ~180 rpm gave the highest specific antibody production rate. The specific productivities were reduced at higher or lower values of agitation speed. The increased antibody production with increasing agitation was associated supposedly with enhanced secretion in a more turbulent environment (89). Oxygen transfer effects may have better explained the observations (see MASS TRANSFER), but no data were reported on this aspect.

Damage to murine hybridomas was observed by Jan et al. (31) in stirred tanks equipped with marine impellers agitated at sufficiently high speeds that vortexing occurred and gas entrained into the medium. Even at these high speeds, damage could be prevented by baffling the tank, which suppressed vortex formation. Usually though, vortexing is not a problem in large-scale cell culture. Unbaffled, marine impeller–stirred tanks were successfully used by Chisti (78) in industrial culture of several hybridoma lines. Effects of agitation on hybridoma culture in the absence of sparging or surface entrainment were further examined by Smith and Greenfield (90). Culture growth was unaffected by agitation intensity (100 or 600 rpm) in the RPMI medium supplemented with fetal bovine serum

(10% vol/vol). However, in PFHM II medium supplemented with either Pluronic F68, fetal bovine serum, or bovine serum albumin, and agitated at 600 rpm (impeller tip speed = 1.6 m s^{-1}, power input = 1 kW m^{-3}) the results were different: the agitation intensity did not affect the exponential growth rate, but once growth had ceased, the decline phase was substantially faster than in control experiments (90).

A certain level of hydrodynamic shear is generally beneficial to culture processes, especially in intraparticle immobilized culture where mass transfer limitations can be severe. In producing engineered tissue of smooth muscle cells/polymer matrices (biodegradable 2-mm-diameter fibers of polyglycolic acid assembled into a nonwoven matrix), Kim et al. (82) noted that seeding of the matrix under agitation led to significantly higher intramatrix cell densities than when cells were seeded under static conditions. Moreover, the higher cell densities were attained more rapidly than the lower densities of static culture. In addition, the relative rates of synthesis of elastin and collagen were significantly greater in seeded matrices cultured with agitation than in ones grown statically (82). A lower possible supply of oxygen or other nutrient may have reduced the performance of static seeding methodology. In view of the above referenced studies and similar others, sufficiently intense fluid mechanical forces other than those associated with aeration do affect cells.

Adherent Cells. Quite low shear stress levels, for example, between 0.25 and 0.6 N m^{-2} in laminar flow (91), can interfere with the process of cell attachment to surfaces; however, once the cells are attached and spread out, they may tolerate higher stresses. Laminar shear stress of the order of 0.5–10 N m^{-2} may remove adherent cells from surfaces (92), but even lower values (e.g., 0.1–1.0 N m^{-2}) are known to affect cellular morphology, permeability, and gene expression (92). Sublethal shear stress levels cause no obvious physical damage but are known to produce various biochemical and physiological responses (4,83,84). In studies with rat aortic endothelial cells anchored on the internal walls of glass capillaries, laminar shear stress was shown to affect the cytosolic pH because of preferential leakage of certain ions out of the cells into the buffer saline (84). This reversible permeability enhancement occurred even at stress levels as low as 0.05 N m^{-2} applied over short durations (~2 min). Similar effects were noted with human aortic endothelial cells but not with human skin fibroblasts or rat intestinal epithelial cells exposed to shear stress levels of up to 1.34 N m^{-2} (84). With rat aortic cells, the flow-induced cytosolic acidification could be maintained for at least 30 min at 1.34 N m^{-2} shear stress level, but the cytosolic pH returned to unstressed levels within about 15 min when the sustained shear stress was ≤0.027 N m^{-2} (84).

In studies with anchorage-dependent cells attached to the flat glass walls of a rectangular flow channel, Shiragami and Unno (93) observed increased activity of lactate dehydrogenase (LDH) in cells that had been exposed to a steady-state shear stress of 0.5 N m^{-2} for 12 h: the activity was fourfold greater relative to controls. The LDH activity correlated with the transmission of energy from the fluid to the attached cells (93).

In vivo hemodynamic forces have been implicated in various physiological and pathophysiological processes (4). Atherosclerotic lesions in humans tend to develop in zones of flow separation (4) such as regions of arterial branching and sharp curvature. Arteries adapt to chronic changes in blood flow, increasing in circumference under high flow and declining under reduced flow (4). Shear stress signals transmitted throughout the vascular cell via cytoskeletal and biochemical elements result in changes to structure, metabolism, and gene expression (4).

Shear effects need to be considered also in perfusion devices that retain cells over long periods in continuous culture (94). Shear effects on adhering erythrocytes and leukocytes are treated separately.

Erythrocytes and Leukocytes. Mammalian erythrocytes, or red blood cells, are among the best studied of animal cells. Properties of erythrocytes and other blood cells are relevant to various pathologies and biomedical processes. Erythrocytes and leukocytes, being suspended cells in vivo, likely experience the kind of stresses encountered in bioreactors and various other industrial processing devices; hence, these cells may provide a broad general insight into mechanical behavior of other cells of mammalian origin. The normal lifetime of a human erythrocyte in circulation is 121 days. The normal daily destruction rate is about 2 × 10^{11} cells (87). Some cells lyse during circulation; most are destroyed in the spleen. Aging cells resist deformation more than the younger ones. All cells unable to deform in the 3-μm openings of the splenic circulation are delayed and ultimately phagocytized (87).

Erythrocytes are highly deformable cells that orient in laminar flow so that least possible surface area of the disc-shaped cell is perpendicular to the flow. The cytoplasm of erythrocytes is a viscous Newtonian fluid (95). The cell membrane behaves as an elastic solid: the cell deforms but almost instantly recovers its shape when the deforming force is removed. The membrane has little resistance to bending but substantially resists increase in area (95). Erythrocytes suspended in turbulent isotonic saline (viscosity ~1 × 10^{-3} Pa s) have been observed to undergo elongation and deformation; however, the cell appears to be less vulnerable to turbulent shear stress than a cell at the same stress level in a viscous suspending medium (87). Based on measurements in turbulent jets, a critical lytic shear stress level of 4,000 N m^{-2} has been reported for very brief exposures (~10^{-5} s) (87). Measurements on erythrocytes of different mammals reveal that the critical shear stress increases dramatically as the cell volume declines.

According to Blackshear and Blackshear (87), "Areal change occurs when the cell membrane is subjected to stress; hemolysis occurs when the area increases by approximately 6.4 percent." Hemolysis is associated at least in part with physical factors and flow, which produce the hemolysis threshold strain in membranes of erythrocytes (87). Once the threshold strain is exceeded, membrane pores open and the membrane eventually tears (87). A briefly (e.g., 10 ms) imposed uniaxial tension of 0.058 N m^{-1} is a sufficient criterion for lysis (87). When the cell is subjected to biaxial stress, a tension of about 0.029 N m^{-1}

may produce lysis. In viscometric stress for prescribed periods, time to lysis declines as the imposed stress is increased. However, it has been conclusively shown that shear stress alone is not a sufficient predictor of hemolysis rate or thresholds (87); cell shape and tumbling also play a role. Membrane shear appears to be far lower when the cells are suspended in higher-viscosity media in viscometric studies.

Erythrocytes allowed to adhere to a glass surface and then subjected to a fluid shear commence movement when the fluid shear force exceeds about 10^{-11} N (95). During this process the cell gradually moves downstream, but the membrane may remain attached to the surface (95). The membrane can be deformed permanently when the deforming force persists for more than a few minutes (95). Shear elasticities of nonnucleated mammalian red cells are generally similar, but elasticities are about an order of magnitude greater for cells of nucleated species. Small amounts of thiol reagents are known to decrease the elongation of human red cells suspended in a shear field (95), presumably by producing some sort of cross-linking. Potentially, this methodology may improve survival of any fragile cell with a significant amount of cross-linkable protein in the membrane. Other additives may render the cell more susceptible to shear damage. Certain chemical lysins and some antigen–antibody reactions cause perforation of the cell membrane and leakage of intracellular material (87). Cholesterol enrichment or depletion of the human erythrocyte membrane does not affect its viscosity.

Hemolysis of red cells is known to occur intravascularly in vivo as well as in various in vitro flow systems (87). A number of studies have correlated hemolysis to flow in pumps, valves, heart–lung machines, blood dialyzers, and transfusion filters. In flow through tubes, wall roughness of the scale of erythrocyte correlates with hemolysis (87). Bubbles trapped in surface imperfections appear to aid lysis. In flow through tubes, hemolysis correlates with the shear rate and the surface-to-volume ratio (87). This type of lysis occurs at shear stress thresholds lower than the ones required to produce lysis in a fluid shear field. Wall contact–associated lysis has been observed to depend on the chemical nature of the wall material. Lysis may decline with time as surfaces become passivated by prolonged contact with plasma proteins (87).

In capillaries of ~1 mm in diameter, an upper limit on the mean tube velocity of 6 m s^{-1} has been suggested for capillaries with sharp-edged entrances, and blood with a viscosity of 4×10^{-3} Pa s (87). This corresponds to a Reynolds number of 1,500 inside the capillary and an average wall shear rate of about 4,800 s^{-1}. Velocities as high as 17 m s^{-1} may be employed inside capillaries with carefully flared entrances (87). As with other animal cells, erythrocytes subjected to bubbling are susceptible to bubble rupture–associated damage, as shown in Figure 6 for porcine erythrocytes. The normalized cell number (Fig. 6) declined more rapidly in a sparged airlift bioreactor in comparison with cells in a surface-aerated shake flask (43). In surface-aerated shake flasks, increasing shaker platform speed over 100–400 rpm increased the specific cell lysis rate (Fig. 7). The slight decline in lysis rate at 100 rpm (Fig. 7) was associated with improved surface aeration

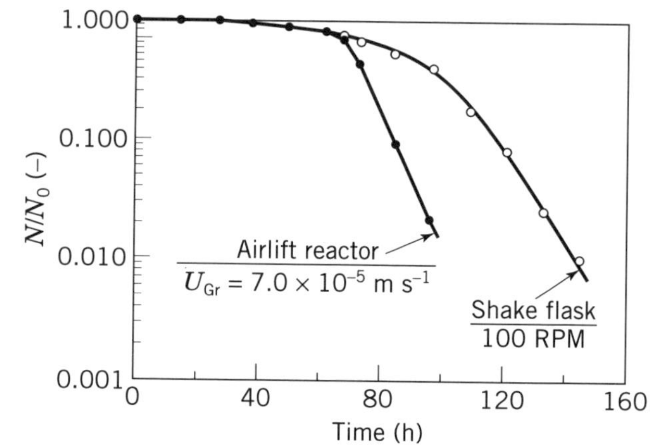

Figure 6. Survival of suspended porcine erythrocytes in a sparged airlift bioreactor and a surface aerated shake flask. Normalized cell count (N/N_0) is plotted as a function of time. *Source*: From Ref. 43.

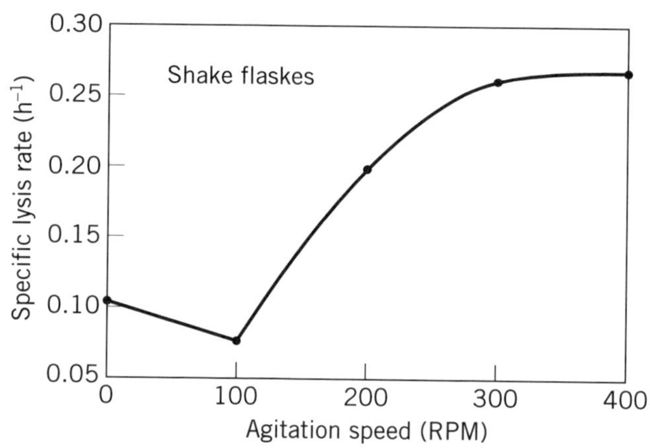

Figure 7. Effect of shaker platform agitation speed on specific lysis rate of suspended porcine erythrocytes in surface aerated shake flasks. *Source*: From Ref. 43.

relative to static flask. Because erythrocytes do not multiply in vitro, the effects of cell-damaging forces are not masked by growth-associated adaptation, changes in cell size, and stages of growth; hence, suitably chosen erythrocytes provide a well-defined cell population for comparative assessment of the damaging potential of various hydrodynamic environments (43).

In comparison with erythrocytes, the cytoplasm of leukocytes has markedly different rheological properties (96), but the properties of the leukocyte membrane are similar to those of the red cell membrane. Leukocytes adhering to vascular endothelium detach when the shear stress is between 26.5 and 106.0 N m^{-2} (96). According to work cited by Prokop and Bajpai (53), a shear stress level of 60 N m^{-2} applied over 10 min should lyse about one-fourth of a leukocyte population. In another study, sublethal shear stresses of 10 and 20 N m^{-2} applied over 10 min in a Couette viscometer affected the biochemical response of human T cells relative to unsheared controls (83). It seems, therefore, that cells in vivo in circulation are apparently more shear resistant than ones studied in vitro.

Microcarrier Culture. Both stirred-tank and airlift bioreactors (97) may be used for suspended microcarrier culture. Microcarrier-supported cells likely experience hydrodynamic forces of greater magnitude than do freely suspended cells. This is because in highly agitated or aerated systems, the length scale of fluid eddies can be of the same order as the dimensions of microcarriers, resulting in high local relative velocities between the solid and the liquid phases (15,78,98). Furthermore, the carriers have greater inertia than free cells; hence, collisions among microcarriers and between the impeller and microcarriers likely damage attached cells. Similarly, fluid eddy impact and shear stress forces on a high-inertia particle are greater than on free cells.

Resistance of a cell to rupture by impact (i.e., burst resistance) is a possible measure of the cell's survivability in the culture environment. Measurements of burst force are potentially useful also for comparative assessment of cell strength and in establishing culture conditions that give rise to more robust cells. It may eventually be possible to correlate the sensitivity to shear of a cell to its resistance to mechanical rupture (99). Burst strength may be directly relevant in microcarrier culture where impact-associated damage is likely; however, in view of the cytoskeleton-mediated pressure signal transmission to internal parts of a cell, a more likely scenario is that the force required to cause impact-associated damage is far lower than the rupture threshold. This issue notwithstanding, resistance to rupture of a mouse hybridoma grown in serum-containing continuous culture has been measured by squeezing single cells between flat surfaces of micromanipulator arms (100). The bursting strength increased with cell size (101). The bursting force correlated with the initial cell diameter as follows:

$$S = 0.2d_c - 0.1 \times 10^{-6} \quad (31)$$

where the intercept was not significantly different from zero (100). The cell diameter ranged over ~(10–17) × 10^{-6} m. Other similar data (101) have been correlated (99) as follows:

$$S = 0.27d_c - 0.86 \times 10^{-6} \quad (32)$$

where the burst strength S is in newtons, and the hybridoma cell diameter, d_c, is in micrometers. The calculated bursting tension had a mean value of $(1.8 \pm 0.5) \times 10^{-3}$ N m^{-1}, which was essentially independent of cell size (100). Because the bursting tension was size independent, the bursting pressure fell with increasing cell diameter. At $(0.8 \pm 0.3) \times 10^{-3}$ N m^{-1}, the calculated mean compressibility modulus of the cells was roughly independent of cell diameter (100).

Sinskey et al. (102) advanced the concept of an integrated shear factor (ISF)—a measure of the strength of the shear field between the impeller and the spinner vessel walls—to correlate shear damage to mammalian cells. The ISF was defined as

$$\text{ISF} = \frac{2\pi N d_i}{d_T - d_i} \quad (33)$$

or, effectively, the laminar shear rate between the impeller tip and the walls. For a range of stirred vessels (0.25–2.0 L, $0.032 \le d_i\,(\text{m}) \le 0.085$), cell damage occurred once the ISF value exceeded about 18 s^{-1} during culture of microcarrier-supported human fibroblasts (103). Damage could be correlated also with the impeller tip speed, but unlike the ISF, the damaging value of the tip speed depended on the size of the culture vessel (103).

Relying on the earlier work of Nagata, Croughan et al. (103) arrived at the following expression for a time-averaged shear rate:

$$\gamma_{\text{av}\cdot\text{T}} = \frac{112.8 r_i^{1.8}(r_T^{0.2} - r_i^{0.2})(r_c/r_i)^{1.8}}{r_T^2 - r_i^2} \quad (34)$$

where r_i and r_T are the impeller and the tank radii, respectively. The radius r_c of the formed vortex zone would need to be estimated using Nagata's expression:

$$\frac{r_c}{r_i} = \frac{Re_i}{1{,}000 + 1.6Re_i} \quad (35)$$

The impeller Reynolds number in equation 35 is calculated as follows:

$$Re_i = \frac{N d_i^2 \rho_L}{\mu_L} \quad (36)$$

Equation 34 was developed for unbaffled stirred tanks operated in the turbulent regime (103). For human fibroblasts on microcarriers in various stirred vessels, cell damage occurred when the time-averaged shear rate exceeded about 2.5 s^{-1} (103). For chicken embryo fibroblasts, also on microcarriers, the damage threshold was 6 s^{-1} time-averaged shear rate (103). Cell damage correlated also with the Kolmogoroff eddy length scale (l): for human fibroblasts the cell damage occurred when the length scale declined to approximately below 125 μm, whereas for the chicken embryo cells damage was observed when the l-value declined to below 100 μm (103).

Analyzing data from several sources, Croughan et al. (103) further showed that the specific death rate correlated with the average energy dissipation rate per unit mass; thus,

$$k_d \propto E^m \quad (37)$$

where m was 0.72, 0.76, and 0.82, respectively, for microcarrier-supported Vero cells, similarly supported human fibroblasts, and a freely suspended protozoan. For attaining a more homogeneous shear field in stirred culture vessels (i.e., for the minimum value of the maximum-to-average shear rate ratio) Croughan et al. (103) noted an optimal vessel geometry corresponding to $r_i/r_T = 0.74$.

In microcarrier culture, collisions between microcarriers and interactions between carriers and internals of a reactor are other possible causes of cell damage, particularly in stirred bioreactors (15,104,105). Cherry and Papoutsakis (98,104,105) advanced the concept of "severity of collision" to account for at least some of the damage to cells in suspended microcarrier culture in stirred vessels. Severity of collision combined collision frequency and energy of the impact. Two collision severities were defined: a turbulent collision severity (TCS) for turbulence-associated particle-to-particle impacts, and an impeller collision severity (ICS) for particle-to-impeller collisions; thus,

$$\mathrm{TCS} = \left(\frac{E\rho_L}{\mu_L}\right)^{3/2}\left(\frac{\pi^2\rho_S d_p^5 \varepsilon_S}{72}\right) \quad (38)$$

and

$$\mathrm{ICS} = \frac{9\pi^4\rho_S n_B N^3 d_t^4 d_p^4}{512V_L} \quad (39)$$

where E is the energy dissipation rate per unit liquid mass, ρ_S and d_p are the density and diameter of the microcarriers, ε_S is the volume fraction of the carriers, n_B is the number of impeller blades, N and d_i are the impeller rotational speed and diameter, and V_L is the volume of the liquid in the vessel. Improved cell growth was observed with smaller microcarriers, as predicted by equations 38 and 39. The specific cell death rate increased with increasing values of TCS and ICS (104,105); however, the influence of hydrodynamic forces on culture performance correlated also in terms of the ratio of Kolmogoroff eddy scale to bead diameter: the specific death rate declined as l/d_p increased. As culture viscosity was raised, damage to cells declined in conformance with equation 38 and in agreement with interpretations based on the l/d_p ratio (105).

Unlike in stirred tanks, animal cell microcarriers suspended in airlift bioreactors under typical operating conditions do not significantly interact with each other or with the walls of the vessel (106); instead, the microcarrier particles follow the laminar streamlines of the fluid (106). Consequently, in airlift reactors at least, the effects of particle–particle or particle–wall collisions on monolayers of cells may be disregarded. Observations of Ganzeveld et al. (106) spanned microcarrier loadings of up to 30 kg m^{-3}, with carriers of 150–300 μm diameter, and 1,030–1,050 kg m^{-3} density. The observations covered a power input value of up to 33 W m^{-3}, which is about the upper limit for cell culture in pneumatically agitated bioreactors. These results applied to a split-cylinder airlift bioreactor with an aspect ratio of 7.6, which would not normally be exceeded in large-scale cell culture systems. As an additional design constraint, the Reynolds number in the riser and the downcomer should not exceed about 3,000, or the flow will be more chaotic (106).

Numerical analysis of forces exerted by laminar flow on anchorage-dependent cells attached to flat surfaces suggests that shear stress between 0.25 and 0.6 N m^{-2} is sufficient to detach round cells, but much higher values are needed to dislodge spread out cells (91). These observations are relevant to microcarrier culture where, during inoculation, the round cells must first attach to microcarriers before spreading on the solid surface. During proliferation, bead-to-bead transfer of cells may also require a level of turbulence that is not so high as to hinder the reattachment process, yet not so low that bead-to-cell encounters are few.

Typically, the shear rate values in airlift bioreactors range over 250–4,000 s^{-1} for operational conditions that are relevant to animal cell culture (14). These shear rates are substantially lower than the ~10^5 s^{-1} that would be needed to damage cells if a 100-N m^{-2} shear stress value (13,88) is taken as the threshold of mechanical damage. However, based on the shear stress data of Olivier and Truskey (91), during the process of attachment of cells to microcarriers, shear rate levels of 250–600 s^{-1} may well be detrimental. Thus, during initial attachment of cells, the reactor would need to be operated at reduced aeration rates—a practice that is well established through empirical experience, but previously explained only intuitively (14).

Under conditions typical of microcarrier culture in airlift bioreactors, the specific energy dissipation rates are different in different zones of the vessel. The specific energy dissipation rates increase in the following order: downcomer < riser < bottom, for any fixed value of the aeration rate (14). The dissipation rates in all zones decline with increasing loading of microcarriers; however, the prevailing shear rates are not particularly sensitive to the density or the diameter of microcarriers within the ranges that are relevant to anchorage-dependent cell culture. In one case, typical shear rates ranged over 250–4,000 s^{-1} in microcarrier-containing systems, but much higher values, up to 12,000 s^{-1} could occur in solids-free media (14). These values compare favorably with shear rates of ~10^5 s^{-1} that have been reported as the threshold of damage to cells. Cells in airlift bioreactors would experience a substantial increase in the riser shear rate only when the fluid eddy length–to–microcarrier diameter ratio declines to ~1 (14).

Hydrostatic Pressure As the scale of operation increases, suspended animal cells will inevitably experience substantial hydrostatic pressure changes as they move with fluid elements in a large bioreactor. At least for hybridomas, artificially imposed pressure cycling of up to 1.2 atm (18 psig) has shown no harmful effect (79). This level of pressure fluctuation will be encountered in bioreactors that are about 12 m tall; hence, for a 2:1 aspect ratio, an animal cell culture vessel could be scaled up to about 340 m^3. If, as recommended for airlift and bubble column bioreactors (20,21), an aspect ratio of 5:1 is employed, a maximum vessel volume of about 50 m^3 should be feasible. Suspension cell culture vessels in current commercial use do not generally exceed 10 m^3, or about 20% of the feasible volume based on hydrostatic pressure considerations.

Bubble-Associated Damage That sparging damaged cells has been known since the 1960s; however, the specific events responsible for the damage—whether bubble formation at the sparger, bubble detachment, rise through the fluid, or events at the disengagement surface—and the mechanism(s) of damage remained unclear until recently.

The growth rate, viability, and antibody productivity of a mouse hybridoma in a surface-aerated stirred vessel were not affected by the agitation rate, but damage occurred when gas was sparged through the culture (107). In surface-aerated operation, the average and the maximum shear stress values varied over the respective ranges of 0.001–0.4 N m^{-2} and 0.06–6.4 N m^{-2}. Subsequent work showed that bubbles larger than 5 mm were significantly less damaging to cells than smaller bubbles (108). Furthermore, placing the sparger below the impeller enhanced cell damage, presumably because the impeller-associated bubble breakup gave rise to the more damaging smaller bubbles (108).

Cell lines differ tremendously in sensitivity to aeration as was demonstrated by Handa et al. (109). At constant aeration rates in geometrically identical bubble columns, smaller diameter bubbles (0.2 mm in diameter) were shown to be significantly more damaging to a hybridoma than larger bubbles (1.62 mm in diameter). In industrial-scale culture of several hybridomas, Chisti (78) preferred yet larger bubbles (10–20 mm diameter) for aeration. Similarly, Tramper et al. (63) observed that smaller bubbles were more damaging to insect cells than larger ones. However, the effect of bubble size may not be prevalent across cell types. Thus, in another study, baby hamster kidney (BHK-21) cells were not sensitive to the size of gas bubbles (110). Over a superficial gas velocity range of $(0.42–8.5) \times 10^{-4}$ m s^{-1}, higher velocities led to faster decline in cell viability of a hybridoma (109), which is consistent with other similar observations (61).

In commercial stirred-tank culture of several hybridoma lines at up to 0.3 m^3, Chisti (78) employed submerged aeration with bubbles of 0.01–0.02 m in diameter. Because of high rise velocities and mobile interfaces, such bubbles did not carry many attached cells to the surface. In contrast, during aeration with sintered metal spargers that produced slower rising bubbles of ~0.002–0.003 m in diameter, the cells were rapidly carried into a persistent foam layer at the top by a froth flotation mechanism (78). The foam produced by larger bubbles collapsed soon after the bubble rose to the surface. The reactors employed were agitated by magnetically coupled agitators with support bearings submerged in the culture. Reactors with submerged mechanical seals were also used. No loss in culture performance occurred relative to roller bottles. The impeller tip speeds >1 m s^{-1} produced no ill effects. Under typical operating conditions, the length scale of the microeddies was about 130 μm, or about 10-fold greater than the dimensions of the cells (78).

Based on energetics of bubble-associated damage to freely suspended cells, Wang et al. (111) concluded that the principal determinants of cell damage were the cell–bubble encounter rate, the rate of bubble breakup within the fluid, and the bursting rate at the surface. Cell death correlated linearly with specific gas–liquid interfacial area, with the proportionality constant being 0.0125 m h^{-1} for a murine hybridoma (111).

Using an elegant mechanistic analysis, Tramper et al. (62,63) related the first-order rate constant for cell death due to all causes to aeration rate in a bubble column; thus,

$$k_d = \frac{24QV_k}{\pi^2 d_B^3 d_T^2 h_L} \tag{40}$$

where k_d is the first-order death rate constant, Q is the volumetric aeration rate, h_L is the height of fluid, d_T is the column diameter, and V_k is a hypothetical killing volume associated with the bubbles. The latter is actually the total volume of fluid associated mainly with the bubble wakes; the cells entrapped in the circulating wake are carried with the bubble to the surface where most of the damaging events take place. According to equation 40, the specific cell death rate is proportional to the frequency of bubble generation, or $6Q/\pi d_B^3$. Alternatively, because at hydrodynamic steady state the bubble generation frequency equals the frequency of rupture at the surface, the death rate may be interpreted as being dependent on the frequency of bubble rupture and the killing volume associated with the bubbles.

Based on equation 40, the death rate constant should decline with increasing height of fluid, as experimentally confirmed: for otherwise fixed conditions, increasing liquid height in bubble columns improved survival of insect cells (62,63), myelomas, and hybridomas (60,64,110), supporting the view that cell death occurred predominantly in the bubble disengagement zone at the surface (109,110). For a myeloma grown in serum-supplemented medium, the minimum culture height that provided a performance equivalent to that of surface aerated flasks, was ~0.7 m, corresponding to a bubble column aspect ratio of ~14 (64). The 0.05-m diameter column was aerated through a sintered disc sparger (150- to 250-μm pore size). The aeration rate was 5 mL min^{-1} (64), corresponding to a specific power input of about 0.42 W m^{-3}.

If bubble rise were to predominantly contribute to cell death, then the specific cell death rate should be independent of the height of fluid (112). Because this is contrary to experimental observations, the rise events may be disregarded as principal contributors to cell damage (112). Whereas, if gas entry at the sparger contributed predominantly to cell damage, then at a given gas flow rate in the reactor, decreasing the number of sparger holes should enhance the specific cell death rate constant (112). In bubble columns, Jöbses et al. (112) noted that increasing the sparger hole gas velocity over 0.6–2.5 m s^{-1} at a constant volumetric gas flow in the reactors had no effect on the first-order rate constant for cell death; hence, events at the sparger could be disregarded as contributing to cell damage. This process of elimination identified bubble breakup at the culture surface as the principal cause of damage to suspended cells in sparged bioreactors such as bubble columns (112). Equation 40 applies also to airlift devices (61), especially ones in which gas is not carried into the downcomer. Because most agitated tanks used in animal cell culture operate at quite low mechanical power inputs and the impeller does not usually serve as a bubble breaking device, equation 40 should provide a lower limit on k_d also in such tanks.

Using a hybridoma culture van der Pol et al. (60) showed that the first-order death rate constant increased with increasing $1/d_T^2$ in bubble columns, as expected from equation 40. Equation 40 suggests that cell damage increases in direct proportion to the superficial gas velocity in the column, or in direct proportion to the specific power input. Note that increasing liquid level at constant aeration rate in a column of fixed diameter decreases the specific power input (14,20,21). Equation 40 further suggests that for a given V_k, larger bubbles should be less damaging than smaller ones. This agrees with substantial empirical evidence (62,63,78,108,110,113).

That the death rate increases linearly with increasing aeration rate has been proven (62,63,113). Also, the killing volume, V_k, has been shown to be independent of the aeration rate and the height of fluid, so long as the bubble diameter is not affected (62,63). The killing volume apparently depends on bubble diameter, increasing with diameter but in such a way that the ratio V_k/d_B^3 declines as the bubble diameter increases. This has been confirmed by Wu and Goosen (113) using *Spodoptera frugiperda* insect cells in bubble columns that were sparged with bubbles of different mean diameters while maintaining a constant volumetric gas flow rate. Increasing bubble diameters reduced the specific death rate, as shown in Figure 8. In view of the numerous contrary observations, analyses suggesting lesser cell damage with smaller bubbles are clearly flawed.

The dimensionless specific killing volume, or $6V_k/\pi d_B^3$, in equation 40 depends on the concentration of serum in the culture medium. Van der Pol et al. (60) showed that the dimensionless killing volume declined as the serum concentration was raised over the range 0–2.5% (vol/vol); the dimensionless killing volume in the absence of serum was ~2.7-fold that at 2.5% serum in a hybridoma culture (60). This and other similar observations (61) at least partly explain the well-known protective effect of serum. Cell-protective effects of other proteins may be similarly explained. Serum-free media with 1 g protein L^{-1} (equivalent to supplementing the medium with 2% vol/vol serum) are commercially used in sparged airlift culture vessels (79).

Why rupture of larger bubbles is less damaging to cells than breakup of smaller ones has become clear through

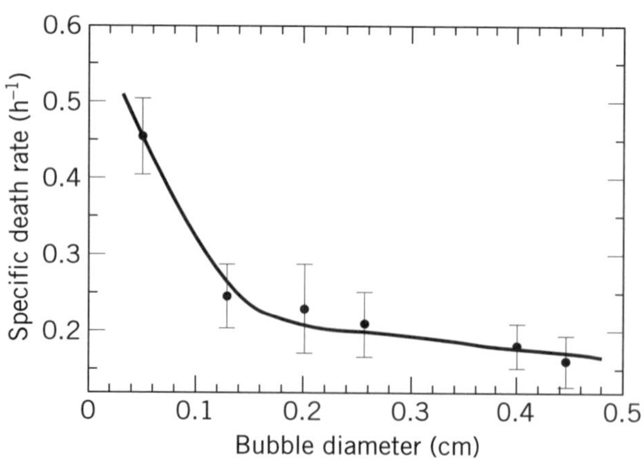

Figure 8. Effect of bubble diameter on the specific death rate of *S. frugiperda* insect cells in bubble columns aerated at a constant volumetric gas flow rate of 10 mL min^{-1}. *Source*: Adapted from Ref. 113.

analyses of mechanics of rupture (114,115). Numerical simulations and experimental observations of bubble rupture at an air–water interface confirm that smaller bubbles rupture much more violently than larger ones (114,115). A bubble reaching the surface rises to different heights above the flat liquid surface before rupturing. For a bubble that is stationary at the surface, a part is always submerged below the flat surface of the liquid. The degree of submergence is generally greater for smaller bubbles. During rupture, the liquid drains from the film between the raised liquid surface and the bubble (116,117). As the film thins, a hole develops in the center of the dome, and the film rapidly retreats into the bulk liquid. The rupture of the film and consequent elongational and accelerational flow have been postulated to damage cells in the film and those adhering to the bubble cavity (116). A cell on the surface of a rupturing film would apparently not experience any motion until struck by the torroidal ring of fluid at the receding edge of the film (116). The force experienced by the cell will depend on the film thickness at the instance of rupture (116). Rupture of thinner films is likely to be less damaging than the breakage of a thicker film. Once the bubble ruptures, fluid rapidly accelerates and moves down the interior walls of the bubble cavity. The flow from around the walls impacts at a stagnation point located immediately below where the bubble had been. The high pressure produced by the impacting streams forces a jet of fluid upward into the gas phase above the liquid surface. A second opposing jet is forced into the fluid below the stagnation point. Intense accelerational flows around the bubble cavity and formation of liquid jets occur mostly during rupture of small bubbles. Bubbles larger than about 6 mm rupture much less violently.

According to Boulton-Stone and Blake (114), "one of the most important factors determining the motion following film rupture, in terms of the energy released, is the height of the top of the bubble above the equilibrium free-surface" of the liquid. Bubbles rupturing from a lower position beneath the equilibrium surface of the liquid release more energy because of their higher internal pressure. The energy is released as high-speed liquid jets. Bubbles larger than about 6 mm bursting at the surface do not form significant jets (114) which is consistent with numerous observations dating from 1950s and earlier. When a jet is produced, the speed of the jet declines as the size of the rupturing bubble increases: for example, for air bubbles rupturing on water, a maximum jet speed of 6.4 m s^{-1} has been calculated for rupture of a 1-mm bubble; for rupture of a 6-mm bubble the maximum jet speed declines to 0.94 m s^{-1} (114). In addition to a liquid jet rising into the atmosphere above the liquid, a second liquid jet moves down into the fluid from the region that had been below the base of the bubble. The maximum pressure produced during bursting of bubbles of various sizes declines with increasing bubble diameter, as illustrated in Figure 9. The maximum energy dissipation rate occurs just beneath the bubble immediately before the jet forms (114). The calculated maximum energy dissipation rates for the smallest bubbles are equivalent to a stress of about 10.4 N m^{-2} (114). The maximum energy dissipation rates decline with increasing bubble diameter, as depicted in Figure 10.

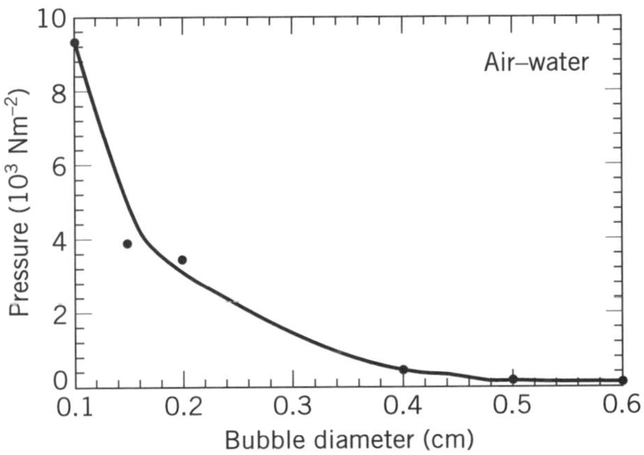

Figure 9. The maximum pressure produced during bursting of air bubbles of various sizes on the surface of water. *Source*: Adapted from Ref. 114.

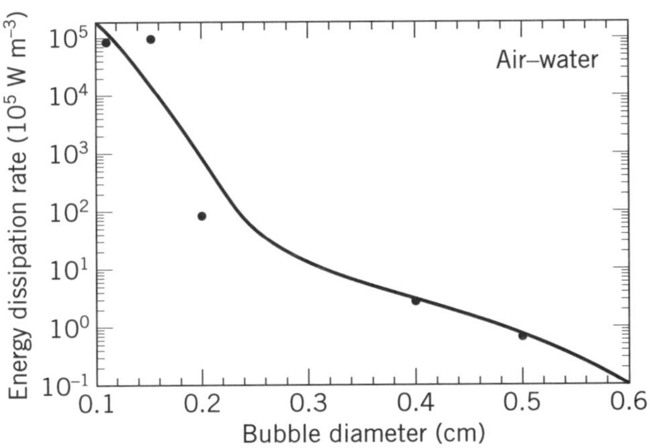

Figure 10. The maximum energy dissipation rates produced during bursting of air bubbles of various sizes on the surface of water. *Source*: Adapted from Ref. 114.

Even a violent bubble rupture event would be inconsequential if there were no cells in the vicinity. Unfortunately, cells attach to rising bubbles, and the fluid behind bubbles is especially rich in cells. That some animal cells will attach to rising bubbles has been shown with *Trichoplusia ni* and *S. frugiperda* insect cells (116,118). The cells adhered to rising microbubbles (diameter ≤3 mm) of the type used commonly in froth flotation. Hydrophobic interactions are apparently responsible for cell attachment—a mechanism observed also with ore particles attaching to bubbles during foam flotation. Although, very small bubbles are generally not suited to aeration of cell cultures (78), larger bubbles also transport cells to the surface, but not predominantly as adhering cells. Instead, the cells are trapped in the circulating wake carried behind large bubbles. The fluid associated with the wakes is known to be richer in particles than the bulk fluid.

More than 95% of the cells in the liquid ejected as jet during rupture of bubbles have been shown to be nonviable even when the cells in the bulk culture contained fewer than 10% dead cells (115). Clearly, the cells carried in the bubble wake are destroyed by the rupture event. Based on simulations of rupture of bubbles with diameters of 0.77 mm, 1.70 mm, and 6.32 mm, Garcia-Briones et al. (115) calculated maximum energy dissipation rate values (W m^{-3}) of 9.52×10^7, 1.66×10^7, and 9.40×10^4, respectively. Comparing these values with energy dissipation rates that have damaged cells in various unaerated flow devices (capillaries and viscometers), Garcia-Briones et al. (115) noted that values of the order of 5.81×10^2 W m^{-3} did not damage the cells significantly, whereas appreciable damage occurred when the energy dissipation rate exceeded about 2.25×10^4 W m^{-3}. Clearly, therefore, a 6.32-mm bubble still has a significant potential to damage cells, and a smaller bubble would produce instantaneous damage. For example, in gas-free capillary flow in experiments reported by Augenstein et al. (16), more than 80% of the cells died within 0.3 s of exposure to energy dissipation rates of about 4.80×10^7 W m^{-3} (115). Interpreting data from several sources, Garcia-Briones et al. (115) further note that average energy dissipation rates of the order of 2,990–29,000 W m^{-3} have generally produced a significant decline in growth rate in impeller-agitated bioreactors. These dissipation rates are average values; the cells probably experienced much higher local energy dissipation rates in the vicinity of the impellers.

If stable foam is not allowed to form, and relatively large bubbles (diameter ≈ 10–20 mm) are used for aeration so that the energy dissipated during rupture remains below the cell damaging threshold, then the cells carried in the wake will not be damaged or removed into a foam layer. For bubbles of a given size, the volume of the circulating wake can be reduced by increasing the viscosity of the culture broth. Small increases will reduce the rise velocity of the bubbles, and the circulation in the wakes will be slowed down. Consequently, slight enhancements in viscosity could reduce the transport of particles in the wake. This may indeed be one explanation of why viscosifying additives have been frequently observed to reduce cell damage. Also, for a given volumetric aeration rate, use of larger bubbles reduces bubbling frequency; hence, the frequency of bubble rupture is reduced.

Based on the studies reviewed, several important observations can be made regarding freely suspended animal cells in sparged and agitated bioreactors: (*1*) the aeration rate should be kept as low as possible; (*2*) the mean bubble size should larger than 7 mm, preferably about 10–20 mm; (*3*) the location of the sparger should be such that the rising bubbles do not interact with any impellers; (*4*) with small airlift and bubble column reactors the aspect ratio should be about 14; however, in larger columns where the sparged portion of the cross section can be a small fraction of the total cross-sectional area, more realistic aspect ratios of about 6 or 7 are satisfactory; and (*5*) when an impeller is used, the average energy dissipation (input) rate should remain below about 1.0×10^3 W m^{-3}. Observations (*1–5*) are consistent with industrial practice (78). The sole purpose of the impeller should be to suspend the cells and mix the fluid gently so that the oxygen transferred from the bubbles is distributed throughout the vessel. Consistent with this purpose, the impeller should be of a kind that does not produce excessively high local rates of energy dissipation. For example, relatively large axial flow hydrofoils are the preferred impellers (30,78). In addition, a suitably selected additive such as Pluronic F68 should be used whenever feasible.

Mitigation of Shear Damage Various approaches have been used to reduce bubble rupture–associated damage to cells. Some common approaches are the use of cell protective additives as noted here; compartmentalization of the culture into cell-free and cell-containing zones, with sparging confined to cell-free regions (80,81); use of bubble-free oxygenation through microporous or diffusion tubing made of silicone rubber or other polymer; use of liquid oxygen vectors such as perfluorocarbons; and surface aeration (see MASS TRANSFER). Alternatives to bubbling are suited typically only to specific cases; direct sparging remains the predominant method of aeration of cell cultures, and it is not likely to be displaced in the foreseeable future (76).

One adaptation of submerged aeration is the "bubble bed" bioreactor, which can substantially reduce surface rupture of bubbles in short-term batch culture (119). The bubble bed device (Fig. 11) is similar to a concentric draft-tube airlift bioreactor, but a downward pumping axial flow impeller located in the draft tube circulates the fluid. The downward liquid flow in the draft tube is sufficiently fast that gas bubbles injected into the draft tube cannot escape; hence, the bubbles do not rupture at the surface of the culture. Pure oxygen or oxygen-enriched gas is used to reduce the needed gas injection rate to very low values. The gas holdup in the downcomer increases during batch operation, and a noncirculating bed of bubbles develops. The cross-sectional area of the draft tube increases downward (Fig. 11) so that the superficial velocity of the liquid declines as it moves down; consequently, the downflowing liquid cannot drag the bubbles out of the draft tube and into the bubble-free annular zone. According to Sucker et al. (119), the cell growth and antibody production in bubble bed devices is comparable to values obtained in surface-aerated spinner flasks even when the sparged culture in the bubble bed units is not supplemented with serum, Pluronic F68, or other protective additives.

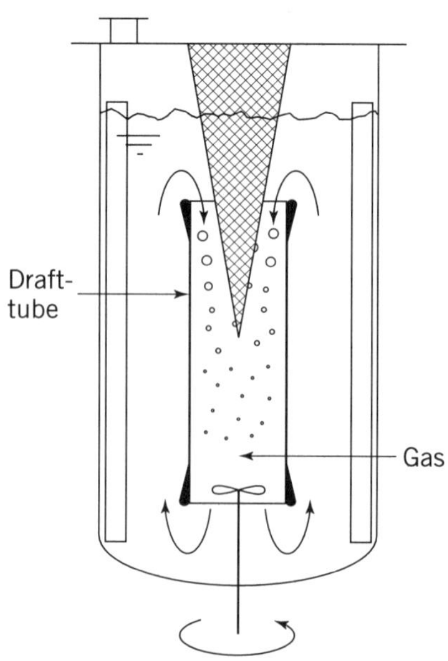

Figure 11. A bubble bed reactor.

Shear Protectants. Bubble-associated damage can be significantly reduced by adding shear-protective substances to the culture medium. Many additives have been examined for shear-protective function (120). Some of the protective additives are noted in Table 4. The protective effect depends on the properties of the specific additives, the concentration used, the type of cell, and the mechanism(s) of protection. Some common protective agents and their mechanisms of protection are treated next.

Serum. A concentration-dependent protective effect of fetal calf serum, horse serum, and others, has been widely observed in aerated and bubble-free suspension cultures of many types of animal cells (41,60,61,66,110,120–122). Typically, protection increases with increasing serum concentration until about 10% (vol/vol) serum. In one study reduction in fetal calf serum level from 2.5% to 0% substantially increased the death rate constant of a hybridoma in aerated culture (60). Investigations revealed that serum had no long-term biological effect, but only a direct nonspecific physical protective effect (60). Thus, cells growing in low-serum medium were immediately protected upon transfer to a medium with 2.5% serum. A physical protective effect of serum has been claimed to originate in its ability to reduce plasma membrane fluidity (41,121). According to Martens et al. (61), serum has an immediate physical protective effect, but also a physiological protective effect that takes longer to become effective. The latter finding concurs with the results of Michaels et al. (122). In bubble-free experiments in a Couette viscometer (shear rate $= 5{,}000\ s^{-1}$, shear stress $= 5\ N\ m^{-2}$), prolonged exposure of cells to fetal bovine serum (FBS) reduced their shear sensitivity in laminar flow studies (122); however, shorter (30–120 min) exposure to FBS did not affect shear tolerance of cells. Based on these results, the protective effect of FBS was claimed to have both physical and biological (biochemical) origins (122).

The precise nature of the physical protective effect of serum is not clear. Supplementation with serum enhances the viscosity of the culture medium, but the effect is small (92,123). Nevertheless, Croughan et al. (66) and others (67,104) have associated the observed protection to the turbulence-dampening effect of serum. In contrast, other evidence suggests a lack of relationship between the amount of protection afforded by the serum and enhance-

Table 4. Media Additives Associated with a Cell-Protective Effect

Serum	Pluronics
Proteins (e.g., serum albumin)	Polyvinyl alcohols
Derivatized celluloses (e.g., methyl cellulose)	Polyvinyl pyrrolidone
Derivatized starches (e.g., hydroxyethyl starch)	Polyethylene glycols
Dextrans	

ment of viscosity (124,125). In addition to dampening turbulence, the serum proteins may coat the cells so that the fluid eddies no longer penetrate to the vicinity of the cell membrane (66). In one study, increasing the concentration of fetal bovine serum over 0–10% (vol/vol) reduced plasma membrane fluidity of a hybridoma line, making the cells more resistant to shear damage (41,121).

Pluronic® F68. Another well-known shear protectant is the nonionic surfactant polyol Pluronic® F68, a block copolymer of poly(oxyethylene) and poly(oxypropylene). Highly pure Pluronic F68 is added to cell culture media, typically at 0.5–3 kg m^{-3} (92,109,110,112,120,122). The protective effect of the surfactant is concentration dependent, increasing with concentration but leveling off at ~0.5 g L^{-1} (Fig. 12).

The protective effect of Pluronic F68 is commonly associated with its ability to suppress attachment of cells to bubbles (110): in the presence of the surfactant, fewer cells are carried to the surface where much of the cell damage occurs in aerated culture. However, evidence exists for other protective mechanisms. For example, a cell-protective effect of Pluronic F68 has been observed even in unaerated culture (41,112). In one case this effect was observed in a strongly agitated culture sample (112). Although no bubbles were present, a deep vortex was drawn into the fluid by the marine impeller used. Extensive cell damage occurred, but the damage was significantly reduced when the culture was supplemented with Pluronic F68. Based on these observations, Jöbses et al. (112) concluded that the protective effect of Pluronic F68 was due to its direct influence on cells and to some effect of the surfactant on the gas–liquid interface. One possible explanation for the protection observed by Jöbses et al. (112) in the absence of the bubbles follows: the damage may have been occurring at the rapidly moving surface of the vortex (126), and supplementation of the culture with Pluronic F68 may have reduced the surface adherent tendency of the cells; hence, fewer cells may have attached to the surface during its formation at the periphery of the vessel.

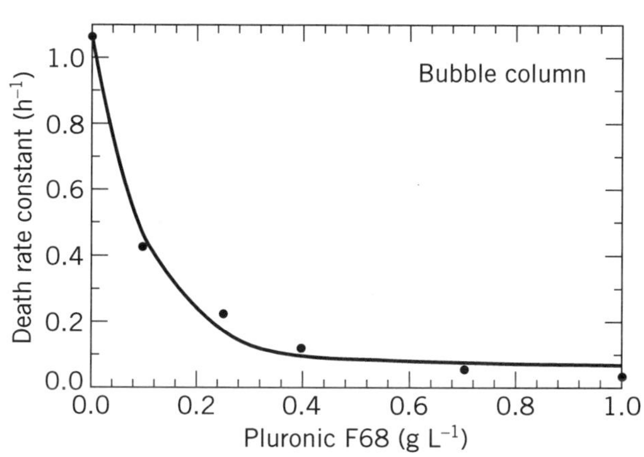

Figure 12. Cell protective effect of Pluronic® F68 in a bubble column. The specific first-order cell death rate constant is plotted as a function of the concentration of Pluronic F68 in the medium. All data were obtained at a constant specific power input of ~40 W m^{-3}. *Source*: Adapted from Ref. 112.

Using flow cytometry, Al-Rubeai et al. (86) showed that leakage of a positively charged dye, fluorescein diacetate, from murine hybridomas increased with rising intensity of agitation, suggesting a possible link between the permeability of the cell membrane and the shear rate. However, because intense agitation was accompanied by entrainment of gas bubbles into the culture (86), the observed leakage could not be conclusively associated with fluid turbulence. The authors noted that for otherwise fixed agitation conditions, the leakage of the dye reduced when the culture medium was formulated with 1% Pluronic F68. The protective effect of the surfactant was apparently linked at least partly with its biochemical association with the cell membrane. The agitation- and vortex-associated hydrodynamic forces were reported to also affect the antigen receptors embedded in the membrane of the hybridoma (86). The number of active surface receptors declined with increasing intensity of agitation and the accompanying increase in the size of the vortex; however, the viability of the cells was unaffected. Pluronic F68 had a protective effect on the receptors.

Pluronic F68 is known to reduce the plasma membrane fluidity (PMF) of cells, and this has been suggested as a possible mechanism of protection (41,121). Other additives that affect PMF also affect a cell's shear tolerance. For example, in one case cholesterol enrichment of the culture medium reduced membrane fluidity and enhanced the shear resistance of hybridomas, whereas supplementing the culture with benzyl alcohol increased PMF and made cells more shear sensitive relative to controls (41,121).

Other protective mechanisms speculated for Pluronic F68 include its stabilizing action on the gas–liquid interface and consequent slower film drainage during bubble rupture (110), improved nutritional transport because of a reduced cell–fluid interfacial tension, and incorporation of the surfactant into the cell membrane accompanied with membrane stabilization (127,128). Although the protective effect of Pluronic F68 is said to be solely due to physical factors (122), the surfactant is known to produce physiological and biochemical effects in at least some cases. Growth suppressing and stimulatory effects of Pluronic F68 have been seen even in some static cultures. Such effects are sometimes associated with impurities in Pluronic F68 samples. Pluronic F68 affects the permeability of some cells. Exposure of *Saccharomyces cerevisiae* to 0.2% (wt/vol) Pluronic F68 for 30 min at 20 °C made the cells more permeable, and they became susceptible to otherwise nonlethal doses of certain antibiotics (129). The mean activity of alcohol dehydrogenase in permeablized cells was more than 3.5-fold of that in normal cells. Although Pluronic F68 could permeablize the yeast cells, it did not inhibit growth or affect cell viability relative to controls when added to culture media at up to 1% (wt/vol), which was the highest concentration tested (129).

Other data suggest that the cell-protective effect of Pluronic F68 may not be quite general (43). For porcine erythrocytes suspended in isotonic buffer, Zhang et al. (43) noted that addition of Pluronic F68 actually increased cell lysis in agitated environments relative to results in the surfactant-free medium. Whether Pluronic F68 incorporates into an erythrocyte's membrane is not known; however, cholesterol enrichment or depletion of erythrocyte membrane is reported to not affect membrane viscosity and presumably, therefore, fluidity. In addition to Pluronic F68, several other pluronics are known to have a shear-protective capability.

Poly(ethylene glycol). Poly(ethylene glycol) (PEG) supplementation of the medium has a protective effect on some cells but not on others. In one study (129) that employed different PEGs with molecular weights of 400–35,000 Da in the concentration range of 0–2 kg m^{-3}, the polymer had a protective effect in sparged cultures when the molecular weight and the concentration exceeded 1,000 Da and 0.25 kg m^{-3}, respectively. The extent of protection afforded depended on the molecular weight and the concentration used. Under the best conditions, the PEG supplementation reduced the first-order cell death rate constant to 0.5% of the value in the PEG-free medium (130). The PEG additives affected the surface tension of the culture medium; however, in extreme cases the surface tension varied only between $\sim 60 \times 10^{-3}$ and 70×10^{-3} N m^{-1} (130). The pattern (but not the magnitude) of variation in surface tension with addition of PEG was similar to the pattern of change in the death rate constant. This led the authors to suggest that the presence of PEG at the gas–liquid interface was somehow related to the protective mechanism.

Michaels et al. (122) showed that, in sparged bioreactors, supplementing the culture medium with Pluronic F68, fetal bovine serum (FBS), or PEG (8,000 Da, 1 kg m^{-3}) protected cells irrespective of whether the CRL-8018 hybridomas were grown in the presence of the additive, or the additive was added after the damaging levels of agitation were imposed.

Composite Particles and Cell Aggregates

Microbial Biofilms on Suspended Particles. Gas–liquid–solid three-phase suspension reactors are increasingly being used in wastewater treatment (20,131). The suspended solid phase (usually sand, glass beads, or carbon particles) supports a biofilm of waste-degrading microorganisms (132). The hydrodynamic environment of such systems must be tailored to control the biofilm thickness without dislodging it altogether (20). Biofilm slurry reactors are conceptually similar to microcarrier-supported suspension culture of animal cells, but there the similarity ends. Unlike animal cell microcarrier suspensions, the biofilm reactors are much more turbulent, the solids loading is generally higher, the density difference between the solid and the liquid phase is much greater, and the biofilm is typically four or five times thicker than the monolayer of cells on microcarriers.

In studies of degradation of naphthalene-2-sulfonate by *Pseudomonads* immobilized as biofilms on sand particles suspended in an external-loop airlift bioreactor, Wagner and Hempel (133) noted that at solids loadings of 20–30 kg m^{-3}, only about 15–20% of the particles were covered with biofilm, but high solids loadings were necessary to control the film thickness at 75 ± 10 μm by providing a high-attrition environment (133). These observations were for particles about 200 μm in diameter, covered with a biofilm ~ 75 μm thick, suspended in the reactor under fairly turbulent conditions (230–440 W m^{-3} specific power inputs). Even when oxygen limitation is not an issue, diffusion of other substrates limits the useful thickness of microbial biofilm to less than 400 μm in most wastewater treatment operations (131).

The development of microbial biofilms on static surfaces and carriers suspended in airlift bioreactors has been investigated quite extensively (6,7,132,134–140). In one case, 21 carriers were tested at a constant carrier loading of 125 kg m^{-3} (132). The superficial gas velocity was constant at 0.038 m s^{-1}. The hydraulic retention time of wastewater was one or two hours. Based on the tests, carrier diameter and surface roughness were the major factors that influenced biofilm formation (132). Smaller particles, 0.1–0.3 mm in diameter, tended to show better biofilm development. This was said to be due to more frequent collisions among larger particles, causing biofilm detachment (132); however, with larger particles, a greater *severity* of collisions would have been a better explanation (20). For equal mass loadings of solids in two systems, the system with larger particles should have fewer particles per unit volume and, therefore, a lower frequency of particle–particle collisions. Cell disruption experiments in agitated slurries of glass beads confirm this (11). The explanation for the impact of surface roughness on biofilm formation lay in the fact that a rough surface provided many relatively protected, low-shear sites where microbial colonies could take hold in the process of gradually spreading over the whole surface. Based on the criteria of surface roughness, particle size, cost, and the requirement of a high-density carrier for ease of separation, basalt (ρ_S = 2,900 kg m^{-3}) was the best carrier (132).

Irrespective of carrier type, biofilm formation is favored at relatively high dilution rates when any detached cells are rapidly washed out (132,136). Tijhuis et al. (136) have shown that during start-up at high dilution rates, the concentration of bare carriers declines as an increasing number are covered with the biofilm. The total amount of the biomass in the reactor increases, but the amount per carrier remains constant until most of the carriers are covered. With few bare carriers remaining, the extent of abrasive loss of film declines, and the film thickness begins to increase until a stable thickness is reached (136). These observations suggest that varying the reactor operation to suit different stages of film development may be worthwhile (20).

In other similar work, detachment of biomass from non-growing biofilms supported on carriers suspended in four configurations of draft-tube sparged internal-loop airlift reactors was examined (139). The superficial aeration velocity in the riser varied over the approximate range 0.03–0.09 m s^{-1}. The detachment of biomass from biofilm-covered carriers in the presence of bare solids was dominated by collisions between bare carriers and the covered particles (139,140). The concentration of bare solids and the particle roughness strongly influenced the detachment rate (139). For otherwise identical conditions, biofilm detachment due to bare carriers increased in the following order: glass, sand, basalt. The glass particles were the most round and smooth; the basalt carriers were the roughest. Whereas biofilms develop more easily on rough surfaces (132), once developed, the film is readily damaged by biofilm-free rough carriers. Gjaltema et al. (139) observed that a "change in flow regime from bubbling to slug flow considerably increased the detachment rate." This change apparently occurred at a U_{Gr} value of ~0.06 m s^{-1}. The "slug flow" regime noted by Gjaltema et al. (139) was more likely the highly chaotic churn turbulent regime in which particle-to-particle collisions are more frequent and severe than in the bubble flow regime. The bottom clearance of the draft tube had no effect on detachment rate (139), but only a small range of the A_d/A_b ratios was investigated. The volume fraction of carriers was 3.3–3.8%, and other properties were: basalt, d_p = 0.32 mm, ρ_S = 3,010 kg m^{-3}; sand, d_p = 0.31 mm, ρ_S = 2,606 kg m^{-3}; and glass, d_p = 0.28 mm, ρ_S = 2,887 kg m^{-3} (139). Another similar study focused on the effects of particle shape, structure, and concentration on biofilm detachment (140).

Strategies for development of biofilms with nitrification activity on carriers suspended in draft-tube sparged internal-loop airlift reactors were investigated by Tijhuis et al. (141) and van Benthum et al. (138). The development of the nitrifying films was similar to what had been reported for heterotrophic biofilms (134–137); however, the nitrifying films were more robust (141). This suggests that biofilms formed by different microorganisms can have markedly different physical properties and strength of attachment to substrate (20); hence, the operating regime of the reactor and, in extreme cases, the design peculiarities are likely to be affected.

In further studies of the dynamics of biofilm detachment from carriers suspended in airlift reactors, Tijhuis et al. (135,137) noted that the common assumptions of uniform growth and detachment of biomass over the surface of the film under fixed hydrodynamic conditions were inconsistent with observations. Instead, nonuniform local detachment of biofilm with formation of cracks and fissures on the surface and rapid filling of the spaces formed with new growing biomass was proposed as a more realistic picture of film dynamics (135).

During development of mixed bacterial biofilm on a static glass surface in a rectangular turbulent flow cell, Stoodley et al. (7) noted that parts of the biofilm sloughed off at a wall shear stress level of 1.32 N m^{-2}. In the region of detachment, pieces of the biofilm were still attached to the surface, suggesting that the glass-to-biofilm bond was stronger than the biofilm structure (7). Relative to the clean flow cell, the pressure drop through the channel was higher in presence of the biofilm (7), presumably because a patchy development of the film increased the effective surface roughness.

Flocs and Agglomerates. Microbial flocs are encountered in several commercially relevant processes, especially in wastewater treatment and beer fermentation with autoflocculating yeasts. Similarly, agglomerates of protein precipitates form during certain recovery operations. The size of flocs and agglomerates often needs to be preserved or increased during processing. Stable size of aggregates and flocs, and whether the flocculated mass further increases in size or breaks up, depends on the level of turbulence in the suspending fluid. Turbulent stress levels below about 0.75 N m^{-2} are generally too mild to break aggregates (142). Physical disruption of aggregates commences at stress levels of ~2.6 N m^{-2}, and especially when the stress level exceeds about 5 N m^{-2} (142). In a turbulent field, aggregates may deform or breakup because of the influence of instantaneous velocity differences acting on different parts of the floc (143). Also, turbulent drag forces originating from the local motion of the fluid relative to the aggregate may contribute to fragmentation (142). The shear force across the aggregate due to instantaneous velocity difference is related to the energy dissipation rate in the fluid and the size of the aggregate (142); thus,

$$F_s = \frac{\pi}{4}(\rho_L \mu_L E)^{1/2} d_F^2 \tag{41}$$

The dynamic pressure fluctuations acting on opposite sides of the floc create a normal stress. Assuming that the eddy size is in the Kolmogoroff microscale range and that the particle dimensions are similar to those of the eddies, the normal force can be expressed as follows (142):

$$F_n \propto \left(\frac{E}{\mu_L}\right)\rho_L^2 d_F^2 \tag{42}$$

The turbulent drag force acting on an aggregate may be calculated using the expression (142)

$$F_D \propto \frac{C_D}{2}\left(\frac{d_F}{2}\right)^3 (\Delta u d_F)^2 \rho_L \tag{43}$$

where Δu is the difference in velocities of the particle and the fluid, and C_D is the drag coefficient. For low-density particles in the Stokes regime ($Re < 1$), the drag coefficient equals $24/Re$, where Re is the Reynolds number, or $(\rho_L \Delta u d_F)/\mu_L$. Reynolds shear stresses (i.e., turbulent stresses associated with the fluctuating component of the mean velocity) may be calculated using the equation (144)

$$\tau = 0.37\rho_L\left(\frac{E\rho_L}{\mu_L}\right)d_p^2 \tag{44}$$

which applies to stresses around a spherical particle. Reynolds stresses are independent of viscosity.

The maximum stable size of the aggregate in a turbulent field can be established by a force balance on a particle: at stable size the instantaneous turbulence force accelerating the particle just equals its mechanical strength (142). In the inertial subrange of turbulence when the eddy size is greater than the Kolmogoroff scale (l),

$$d_{F\,max} \propto \left(\frac{\rho_L \sigma_M}{\mu_L}\right)^{1/3} E^{-1} \tag{45}$$

whereas in the viscous subrange of turbulence when the eddy size is $\leq l$, we have

$$d_{\mathrm{F\ max}} \propto \left(\frac{\sigma_{\mathrm{M}}\mu_{\mathrm{L}}}{\rho_{\mathrm{L}}^{2}E}\right)^{1/2} \quad (46)$$

where σ_{M} is the mechanical strength of the aggregate. In practice, the maximum stable floc size has correlated with the energy dissipation per unit mass of fluid,

$$d_{\mathrm{F\ max}} \propto \Gamma E^{-m} \quad (47)$$

where the coefficient Γ depends on the nature of the particle, and m is usually between 0.2 and 0.4 (145).

In stirred tanks, the maximum floc size for particles of uniform composition has been theoretically and experimentally correlated with the rotational speed of the impeller (143); thus,

$$d_{\mathrm{F\ max}} \propto N^{-1.5} \quad (48)$$

which applies to particles smaller than the microscale of turbulence. The dependence shown in equation 48 can be expressed in terms of the energy dissipation per unit mass; hence,

$$d_{\mathrm{F\ max}} \propto E^{-0.5} \quad (49)$$

Expression 49 is consistent with that reported by Thomas (143), but the exponent on E is somewhat stronger than the values noted by Gregory (145) (see equation 47). In precipitation vessels, shear rate values of $\sim 80\ \mathrm{s}^{-1}$ appear to be a damaging threshold for soya protein precipitate particles of 20–30 μm initial size (5). Particles exposed to approximate shear rates of $\geq 80\ \mathrm{s}^{-1}$ over 10–12 min would be reduced to 5–8 μm. The shear rate numbers were calculated from the specific energy dissipation rates in the vessels, as follows:

$$\gamma = \left(\frac{\rho_{\mathrm{L}}E}{\mu_{\mathrm{L}}}\right)^{0.5} \quad (50)$$

Protein precipitates are easily damaged by pumping. Moyno screw pumps, centrifugal pumps, and gear pumps all damage precipitates to similar levels (5). Peristaltic pumps are apparently the only suitable ones.

Fragmentation of mycelial pellets of the molds *Lentinus edodes* and *Aspergillus niger* in a standard Rushton turbine stirred tank was examined by Taguchi et al. (145). Two types of pellet fragmentation patterns were noted: (*1*) decrease in diameter by erosion of surface, and (*2*) direct rupture of the spherical pellet. The rates of these processes were expressed as

$$-\frac{dd_{\mathrm{p}}}{dt} = k_{\mathrm{j}}(Nd_{\mathrm{i}})^{5.5}d_{\mathrm{p}}^{5.7} \quad (51)$$

for pellet surface erosion, and

$$-\frac{dN_{\mathrm{p}}}{dt} = k'N_{\mathrm{p}} = \alpha(d_{\mathrm{p}}^{3.2}N^{6.65}d_{\mathrm{i}}^{8.72})^{5.5}N_{\mathrm{p}} \quad (52)$$

for rupture of the particle. In equations 51 and 52, d_{p} is the pellet diameter at time t, N is the rotational speed of the impeller, d_{i} is the impeller diameter, N_{p} is the concentration of unbroken pellets at time t, and k_{j}, k', and α are constants. Clearly, the rates of pellet erosion and rupture depend strongly the agitation intensity–associated factors (i.e., on N and d_{i}) and the dimensions of the pellet.

In submerged stirred-tank culture of the filamentous fungus *Aspergillus awamori*, Cui et al. (147) correlated the pellet size with the specific energy dissipation rate; thus,

$$d_{\mathrm{p}} \propto E^{-0.16} \quad (53)$$

The E values varied over 0.2–5.7 W kg^{-1} (146), whereas the pellet diameter ranged roughly over 1–2 mm. The length of the "hairy" hyphal extensions from the well-defined surface of the denser pellet also correlated with the specific energy dissipation rate (147); thus,

$$L_{\mathrm{h}} = 1.17 \times 10^{-4}E^{-0.25} \quad (54)$$

where L_{h} (m) is the length of the hairy extensions.

Bioactive Proteins

Enzymes are not generally susceptible to the shear rates that are commonly encountered during processing including high-shear operations such as cell disruption (1). Exceptions occur when shear stresses are applied in the presence of gas–liquid interfaces, at which proteins tend to unfold (1,8,148). Also, membrane-associated proteins are more prone to shear damage (1). In addition to processes such as high-pressure homogenization, high-shear processing of proteins occurs during liquid–liquid emulsification, which is used to produce microencapsulated proteins for drug delivery systems (149). Effects of extremely high shear rate levels ($>10^{5}\ \mathrm{s}^{-1}$) on recombinant human growth hormone (rhGH) and recombinant human deoxyribonuclease (rhDNase) were examined by Maa and Hsu (149). The rhGH and rhDNase had respective molecular masses of 22.13 kDa and 32.74 kDa. The shearing was done at a controlled temperature of 20 °C. The aggregation and the bioactivity of the two proteins were not significantly affected at shear rates greater than $10^{5}\ \mathrm{s}^{-1}$ applied over 60 min in the absence of gas–liquid interfaces (148,149). Thermodynamic properties of the proteins were determined before and after shearing using scanning microcalorimetery (SM). Alterations in secondary and tertiary structure of a protein are manifested in its scanning calorimetery data. The SM thermograms of rhDNase were not significantly affected by shearing; however, for rhGH, the SM indicated possible conformational changes (149). The silver-stained SDS-PAGE gels of rhGH showed no significant differences in the high molecular weight region, but low molecular weight fragments were observed in the sheared sample, which suggested some clipping of rhGH in the intense shear field. No significant changes were noted in the SDS-PAGE gels of sheared and unsheared rhDNase (148). If a globular protein's molecular weight is assumed to indicate it's size, the rhDNase molecule should be larger by a factor of approximately $(32.74/22.13)^{1/3}$, or 1.14, relative to the rhGH. It seems therefore that the molecular size is not the sole determinant of a protein's susceptibility to damage; the specific structure or morphology also appear to play a role.

The damaging effect of shear in presence of gas–liquid and solid–liquid interfaces may be reduced by supplementing the protein solution with a surfactant. Added surfactants tend to reduce gas–liquid interface–induced aggregation and precipitation of rhGH (148). Proteins also adsorb on hydrophobic solid surfaces. Adsorption of insulin at hydrophobic surfaces, accompanied by denaturation, is well known. Similarly, tPA is known to adhere to solid surfaces (150), but the surface adherent tendency is reduced in presence of surfactants such as Tween-80 (polyoxyethylene sorbitane monooleate) added at 0.01% typical concentration (150). Shear susceptibility of proteins is further discussed elsewhere (5,151).

CONCLUDING REMARKS

Biocatalysts, especially molds, filamentous bacteria, certain microalgae, and plant and animal cells are susceptible to damage by hydrodynamic and mechanical forces encountered in bioprocessing. Typically, recombinant strains are more prone to shear damage than equivalent wild types. Extent of damage depends on the cell and the specific damaging mechanism operating. Often, the damage may be reduced by one or more of several approaches: adaptation of cells, supplementing culture media with shear-protectants, modification of processing operations and equipment, and other methods. In many cases, the mechanisms of protection and damage are not entirely clear, but multiple mechanisms seem to operate. As noted throughout this monograph, some sound guidance is available for engineering and operation of fragile culture processes; nevertheless, a priori quantitative prediction of culture behavior is not reliably possible. Empirical approaches must continue to be relied upon in engineering practice.

NOMENCLATURE

A_b	Cross-sectional area under the baffle or draft tube (m^2)
A_d	Cross-sectional area of the downcomer (m^2)
A_r	Cross-sectional area of the riser (m^2)
a	Parameter in equation 2
C_D	Drag coefficient
C_f	Fanning friction factor
d	Diameter or hydraulic diameter (m)
d_B	Bubble diameter (m)
d_c	Cell diameter (μm)
d_F	Floc diameter (m)
d_{Fmax}	Maximum stable diameter (m)
d_i	Impeller diameter (m)
d_L	Local diameter (m)
d_p	Microcarrier or particle diameter (m)
d_T	Tank or column diameter (m)
E	Energy dissipation rate per unit mass ($W\ kg^{-1}$)
F_D	Turbulent drag force on a particle (N)
F_n	Normal force on a particle (N)
F_s	Shear force across the aggregate due to instantaneous velocity difference (N)
g	Gravitational acceleration ($m\ s^{-2}$)
h	Channel height (m)
h_L	Height of liquid (m)
ICS	Impeller collision severity defined by equation 39 ($kg\ m^{-2}\ s^{-3}$)
ISF	Integrated shear factor defined by equation 33 (s^{-1})
K	Consistency index (Pa sn)
k	Parameter in equation 2 (m^{-1})
k'	Constant in equation 52 (s^{-1})
k_d	Death rate constant (s^{-1})
k_i	Constant in equation 9
k_j	Constant in equation 51 ($s^{4.5}\ m^{-10.2}$)
L	Channel length (m)
L_h	Length of hairy extensions from pellet surface (m)
l	Mean microeddy length (m)
m	General exponent
N	Rotational speed (s^{-1}) or cell concentration (m^{-3})
N_o	Initial cell concentration (m^{-3})
N_p	Concentration of unbroken pellets at time t (m^{-3})
n	Flow behavior index
n_B	Number of impeller blades
Po	Power number
ΔP	Pressure drop (Pa)
Q	Volume flow rate of gas ($m^3\ s^{-1}$)
Re	Reynolds number
Re_i	Impeller Reynolds number defined by equation 36
Re_L	Local Reynolds number defined by equation 23
r_c	Radius of the vortex zone defined by equation 35 (m)
r_i	Impeller radius (m)
r_T	Tank radius (m)
S	Burst strength (N)
TCS	Turbulent collision severity for particle-to-particle collisions ($kg\ m^{-2}\ s^{-3}$)
t	Time (s)
U_B	Bubble rise velocity ($m\ s^{-1}$)
U_G	Superficial gas velocity ($m\ s^{-1}$)
U_{Gr}	Superficial gas velocity in riser ($m\ s^{-1}$)
U_L	Average liquid velocity ($m\ s^{-1}$)
u	Mean velocity of the microeddies ($m\ s^{-1}$)
Δu	Difference in velocities of the particle and the fluid ($m\ s^{-1}$)
u_o	Jet velocity at orifice ($m\ s^{-1}$)
V_k	Dimensionless hypothetical specific killing volume

V_L	Volume of liquid (m^3)
W	Width of the impeller blade (m)
x	Parameter in equation 19

Greek Symbols

α	Constant in equation 52 ($s^{35.575}$ $m^{-65.56}$)
Γ	Coefficient in equation 47
γ	Shear rate (s^{-1})
γ_{av}	Average shear rate (s^{-1})
$\gamma_{av \cdot T}$	Time averaged shear rate (s^{-1})
γ_i	Isotropic turbulence shear rate defined by equation 24 (s^{-1})
γ_{max}	Time averaged maximum shear rate (s^{-1})
γ_w	Wall shear rate (s^{-1})
ϵ_S	Volume fraction of microcarriers or solids (−)
μ_{ap}	Effective or apparent viscosity (Pa s)
μ_L	Viscosity of liquid (Pa s)
π	Pi
ρ_L	Density of liquid (kg m^{-3})
ρ_S	Density of microcarriers or solid (kg m^{-3})
σ_M	Mechanical strength of aggregate (N m^{-2})
τ	Shear stress (N m^{-2})
τ_w	Wall shear stress (N m^{-2})

BIBLIOGRAPHY

1. Y. Chisti and M. Moo-Young, *Enzyme Microb. Technol.* **8**, 194–204 (1986).
2. H.Y. Makagiansar, P.A. Shamlou, C.R. Thomas, and M.D. Lilly, *Bioprocess Eng.* **9**, 83–90 (1993).
3. E. Ujcová, Z. Fencl, M. Musi'lková, and L. Seichert, *Biotechnol. Bioeng.* **22**, 237–241 (1980).
4. M. Papadaki and S.G. Eskin, *Biotechnol. Prog.* **13**, 209–221 (1997).
5. C.R. Thomas, in M.R. Winkler ed., *Chemical Engineering Problems in Biotechnology*, Elsevier, London, 1990, pp. 23–93.
6. W.K. Kwok, C. Picioreanu, S.L. Ong, M.C.M. van Loosdrecht, W.J. Ng, and J.J. Heijnen, *Biotechnol. Bioeng.* **58**, 400–407 (1998).
7. P. Stoodley, Z. Lewandowski, J.D. Boyle, and H.M. Lappin-Scott, *Biotechnol. Bioeng.* **57**, 536–544 (1998).
8. Y. Chisti, in G. Subramanian ed., *Bioseparations and Bioprocessing: A Handbook*, vol. 2, Wiley-VCH, New York, 1998, pp. 3–30.
9. E. Dreveton, F. Monot, J. Lecourtier, D. Ballerini, and L. Choplin, *J. Ferment. Bioeng.* **82**, 272–276 (1996).
10. M. Moo-Young and Y. Chisti, *Biotechnology*, **6**(11), 1291–1296 (1988).
11. F. Garrido, U.C. Banerjee, Y. Chisti, and M. Moo-Young, *Bioseparation*, **4**, 319–328 (1994).
12. I.M. Tamer, M. Moo-Young, and Y. Chisti, *Ind. Eng. Chem. Res.* **37**, 1807–1814 (1998).
13. P.M. Doran, *Adv. Biochem. Eng. Biotechnol.* **48**, 115–168 (1993).
14. E. Molina Grima, Y. Chisti, and M. Moo-Young, *J. Biotechnol.* **54**, 195–210 (1997).
15. E.T. Papoutsakis, *Trends Biotechnol.* **9**, 427–437 (1991).
16. D.C. Augenstein, A.J. Sinskey, and D.I.C. Wang, *Biotechnol. Bioeng.* **13**, 409–418 (1971).
17. N. Shiragami, *Bioprocess Eng.* **10**, 47–51 (1994).
18. A. McQueen, E. Meilhoc, and J.E. Bailey, *Biotechnol. Lett.* **9**, 831–836 (1987).
19. Y. Chisti and M. Moo-Young, *Biotechnol. Bioeng.* **34**, 1391–1392 (1989).
20. Y. Chisti, *Appl. Mech. Rev.* **51**, 33–112 (1998).
21. Y. Chisti, *Airlift Bioreactors*, Elsevier, London, 1989.
22. L.K. Shi, J.P. Riba, and H. Angelino, *Chem. Eng. Commun.* **89**, 25–35 (1990).
23. L.A. Glasgow, J. Hua, T.-Y. Yiin, and L.E. Erickson, *Chem. Eng. Commun.* **113**, 155–181 (1992).
24. K. Okada, S. Shibano, and Y. Akagi, *J. Chem. Eng. Jpn.* **26**, 637–643 (1993).
25. A. Lübbert and B. Larson, *Chem. Eng. Sci.* **45**, 3047–3053 (1990).
26. A. Lübbert, B. Larson, L.W. Wan, and S. Bröring, *Inst. Chem. Eng. Symp. Ser.* **121**, 203–213 (1990).
27. R. Bowen, *Chem. Eng.* June 9, 55–63 (1986).
28. B. Robertson and J.J. Ulbrecht, in C.S. Ho and J.Y. Oldshue eds., *Biotechnology Processes: Scale-Up and Mixing*, American Institute of Chemical Engineers, New York, 1987, pp. 31–35.
29. P.H. Calderbank and M.B. Moo-Young, *Trans. Inst. Chem. Eng.* **37**, 26 (1959).
30. Y. Chisti and M. Moo-Young, in V. Moses and R.E. Cape eds., *Biotechnology: The Science and the Business*, Harwood Academic, New York, 1991, pp. 167–209.
31. D.C.-H. Jan, A.N. Emery, and M. Al-Rubeai, *Biotechnol. Techn.* **7**, 351–356 (1993).
32. Y. Chisti, B. Halard, and M. Moo-Young, *Chem. Eng. Sci.* **43**, 451–457 (1988).
33. Y. Chisti and M. Moo-Young, *J. Chem. Technol. Biotechnol.* **42**, 211–219 (1988).
34. V.N. Sokolov and V.P. Metkin, *Zh. Prikl. Khim.* **54**, 1321–1326 (1981): *J. Appl. Chem. USSR* **54**, 1103–1107 (1981).
35. V.P. Metkin and V.N. Sokolov, *J. Appl. Chem. USSR* **55**, 558–563 (1982).
36. K. Wichterle, V. Sobolík, M. Lutz, and V. Denk, *Chem. Eng. Sci.* **51**, 5227–5228 (1996).
37. A.W. Nienow, *Appl. Mech. Rev.* **51**, 3–32 (1998).
38. J.J. Meijer, H.J.G. ten Hoopen, Y.M. van Gameren, K.Ch.A.M. Luyben, and K.R. Libbenga, *Enzyme Microb. Technol.* **16**, 467–477 (1994).
39. M. Moo-Young, Y. Chisti, and D. Vlach, *Biotechnol. Lett.* **14**, 863–868 (1992).
40. M. Moo-Young, Y. Chisti, and D. Vlach, *Biotechnol. Adv.* **11**, 469–479 (1993).
41. O.T. Ramírez and R. Mutharasan, *Biotechnol. Bioeng.* **36**, 911–920 (1990).
42. E.H. Dunlop, P.K. Namdev, and M.Z. Rosenberg, *Chem. Eng. Sci.* **49**, 2263–2276 (1994).

43. Z. Zhang, Y. Chisti, and M. Moo-Young, *J. Biotechnol.* **43**, 33–40 (1995).

44. H. Märkl, R. Bronnenmeier, and B. Wittek, *Int. Chem. Eng.* **31**, 185–197 (1991).

45. J. Wu, G. King, A.J. Daugulis, P. Faulkner, D.H. Bone, and M.F.A. Goosen, *J. Ferment. Bioeng.* **70**, 90–93 (1990).

46. Y. Chisti and M. Moo-Young, *Trans. Inst. Chem. Eng.* **74**A, 575–583 (1996).

47. M. Moo-Young and Y. Chisti, in G. Durand, L. Bobichon, and J. Florent eds., *Proceedings, 8th International Biotechnology Symposium*, vol. 1, Société Française de Microbiologie, Paris, 1988, pp. 454–466.

48. M. Moo-Young, Y. Chisti, Z. Zhang, F. Garrido, U. Banerjee, and D. Vlach, *Ann. N. Y. Acad. Sci.* **782**, 391–401 (1996).

49. A. Converti, C. Sommariva, M.D. Borghi, and G. Ferraiolo, *Bioprocess Eng.* **9**, 183–189 (1993).

50. A. Converti, M.D. Borghi, G. Ferraiolo, and C. Sommariva, *Chem. Eng. J.* **62**, 155–167 (1996).

51. P.F. MacLoughlin, D.M. Malone, J.T. Murtagh, and P.M. Kieran, *Biotechnol. Bioeng.* **58**, 595–604 (1998).

52. C.R. Thomas, in P.A. Shamlou ed., *Processing of Solid–Liquid Suspensions*, Butterworth-Heinemann, Oxford, 1993, pp. 158–191.

53. A. Prokop and R.K. Bajpai, *Adv. Appl. Microbiol.* **37**, 165–232 (1992).

54. C. Gudin and D. Chaumont, *Bioresource Technol.* **38**, 145–151 (1991).

55. Y. Watanabe, J. de la Noüe, and D.O. Hall, *Biotechnol. Bioeng.* **47**, 261–269 (1995).

56. E. Molina Grima, J.A. Sánchez Pérez, F. García Camacho, J.L. García Sánchez, F.G. Acién Fernández, and D. López Alonso, *J. Biotechnol.* **37**, 159–166 (1994).

57. E. Molina Grima, J.A. Sánchez Pérez, F. García Camacho, J.M. Fernández Sevilla, F.G. Acién Fernández, and J. Urda Cardona, *Appl. Microbiol. Biotechnol.* **42**, 658–663 (1995).

58. H.J. Silva, T. Cortiñas, and R.J. Ertola, *J. Chem. Technol. Biotechnol.* **40**, 41–49 (1987).

59. T. Suzuki, T. Matsuo, K. Ohtaguchi, and K. Koide, *J. Chem. Technol. Biotechnol.* **62**, 351–358 (1995).

60. L. van der Pol, W.A.M. Bakker, and J. Tramper, *Biotechnol. Bioeng.* **40**, 179–182 (1992).

61. D.E. Martens, C.D. de Gooijer, E.C. Beuvery, and J. Tramper, *Biotechnol. Bioeng.* **39**, 891–897 (1992).

62. J. Tramper, D. Joustra, and J.M. Vlak, in C. Webb and F. Mavituna, eds., *Plant and Animal Cells: Process Possibilities*, Ellis Horwood, Chichester, 1987, pp. 125–136.

63. J. Tramper, D. Smit, J. Straatman, and J.M. Vlak, *Bioprocess Eng.* **2**, 37–41 (1987).

64. A.N. Emery, M. Lavery, B. Williams, and A. Handa, in C. Webb and F. Mavituna, eds., *Plant and Animal Cells: Process Possibilities*, Ellis Horwood, Chichester, 1987, pp. 137–146.

65. H.C.P. Matthijs, H. Balke, U.M. Van Hes, B.M.A. Kroon, L.R. Mur, and R.A. Binot, *Biotechnol. Bioeng.* **50**, 98–107 (1996).

66. M.S. Croughan, E.S. Sayre, and D.I.C. Wang, *Biotechnol. Bioeng.* **33**, 862–872 (1989).

67. A. McQueen and J.E. Bailey, *Biotechnol. Lett.* **11**, 531–536 (1989).

68. M. Midler and R.K. Finn, *Biotechnol. Bioeng.* **8**, 71–84 (1966).

69. P.M. Kieran, P.F. MacLoughlin, and D.M. Malone, *J. Biotechnol.* **59**, 39–52 (1997).

70. U. Fischer and A.W. Alfermann, *J. Biotechnol.* **41**, 19–28 (1995).

71. R. Ballica and D.D.Y. Ryu, *Biotechnol. Bioeng.* **42**, 1181–1189 (1993).

72. E. Reinhard, W. Kreis, U. Barthlen, and U. Helmbold, *Biotechnol. Bioeng.* **34**, 502–508 (1989).

73. H. Tanaka, *Biotechnol. Bioeng.* **23**, 1203–1218 (1981).

74. J.-J. Zhong, K. Fujiyama, T. Seki, and T. Yoshida, *Biotechnol. Bioeng.* **44**, 649–654 (1994).

75. P.M. Kieran, H.J. O'Donnell, D.M. Malone, and P.F. MacLoughlin, *Biotechnol. Bioeng.* **45**, 415–425 (1995).

76. J.J. Chalmers, *Appl. Mech. Rev.* **51**, 113–120 (1998).

77. W.R. Arathoon and J.R. Birch, *Science*, **232**, 1390–1395 (1986).

78. Y. Chisti, *Bioprocess Eng.* **9**, 191–196 (1993).

79. J.R. Birch, K. Lambert, P.W. Thompson, A.C. Kenney, and L.A. Wood, in B.K. Lydersen ed., *Large Scale Cell Culture Technology*, Hanser, New York, 1987, pp. 1–20.

80. Y. Chisti and M. Moo-Young, *Trans. Inst. Chem. Eng.* **71C**, 209–214 (1993).

81. Y. Chisti and M. Moo-Young, *Trans. Inst. Chem. Eng.* **72C**, 92–94 (1994).

82. B.-Y. Kim, A.J. Putnam, T.J. Kulik, and D.J. Mooney, *Biotechnol. Bioeng.* **57**, 46–54 (1998).

83. K.K. Chittur, L.V. McIntire, and R.R. Rich, *Biotechnol. Prog.* **4**(2), 89–96 (1988).

84. R.C. Ziegelstein, L. Cheng, and M.C. Capogrossi, *Science*, **258**, 656–659 (1992).

85. R.S. Cherry, *Biotechnol. Adv.* **11**, 279–299 (1993).

86. M. Al-Rubeai, A.N. Emery, S. Chalder, and M.H. Goldman, *J. Biotechnol.* **31**, 161–177 (1993).

87. P.L. Blackshear and G.L. Blackshear, in R. Skalak and S. Chien eds., *Handbook of Bioengineering*, McGraw-Hill, New York, 1987, pp. 15.1–15.19.

88. C. Born, Z. Zhang, M. Al-Rubeai, and C.R. Thomas, *Biotechnol. Bioeng.* **40**, 1004–1010 (1992).

89. N. Shiragami, *Bioprocess Eng.* **16**, 345–347 (1997).

90. C.G. Smith and P.F. Greenfield, *Biotechnol. Bioeng.* **40**, 1045–1055 (1992).

91. L.A. Olivier and G.A. Truskey, *Biotechnol. Bioeng.* **42**, 963–973 (1993).

92. J.G. Aunins and H.-J. Henzler, in H.-J. Rehm and G. Reed eds., *Biotechnology*, vol. 3, 2nd ed., VCH Weinheim, 1993, pp. 219–281.

93. N. Shiragami and H. Unno, *Bioprocess Eng.* **10**, 43–45 (1994).

94. T.L. LaPorte, J. Shevitz, Y. Kim, and S.S. Wang, *Bioprocess Eng.* **15**, 1–7 (1996).

95. R.M. Hochmuth, in R. Skalak and S. Chien eds., *Handbook of Bioengineering*, McGraw-Hill, New York, 1987, pp. 12.1–12.17.

96. G.W. Schmid-Schönbein, in R. Skalak and S. Chien eds., *Handbook of Bioengineering*, McGraw-Hill, New York, 1987, pp. 13.1–13.25.

97. X. Dai, F. Ouyang, and Y. Qi, in W.K. Teo, M.G.S. Yap, and S.K.W. Oh, eds., *Better Living through Biochemical Engineering*, University of Singapore, Singapore, 1994, pp. 408–410.
98. R.S. Cherry and E.T. Papoutsakis, *Bioprocess Eng.* **1**, 29–41 (1986).
99. Y. Chisti and M. Moo-Young, *Chimica Oggi*, **11**(3–4), 25–27 (1993).
100. Z. Zhang, M.A. Ferenczi, and C.R. Thomas, *Chem. Eng. Sci.* **47**, 1347–1354 (1992).
101. Z. Zhang, M.F. Ferenczi, A.C. Lush, and C.R. Thomas, *Appl. Microbiol. Biotechnol.* **36**, 208–210 (1991).
102. A.J. Sinskey, R.J. Fleishaker, M.A. Tyo, D.-J. Giard, and D.I.C. Wang, *Ann. N. Y. Acad. Sci.* **369**, 47–59 (1981).
103. M.S. Croughan, J.-F. Hamel, and D.I.C. Wang, *Biotechnol. Bioeng.* **29**, 130–141 (1987).
104. R.S. Cherry and E.T. Papoutsakis, *Biotechnol. Bioeng.* **32**, 1001–1014 (1988).
105. R.S. Cherry and E.T. Papoutsakis, *Bioprocess Eng.* **4**, 81–89 (1989).
106. K.J. Ganzeveld, Y. Chisti, and M. Moo-Young, *Bioprocess Eng.* **12**, 239–247 (1995).
107. S.K.W. Oh, A.W. Nienow, M. Al-Rubeai, and A.N. Emery, *J. Biotechnol.* **12**, 45–62 (1989).
108. S.K.W. Oh, A.W. Nienow, M. Al-Rubeai, and A.N. Emery, *J. Biotechnol.* **22**, 245–270 (1992).
109. A. Handa, A.N. Emery, and R.E. Spier, *Dev. Biol. Stand.* **66**, 241–253 (1987).
110. A. Handa-Corrigan, A.N. Emery, and R.E. Spier, *Enzyme Microb. Technol.* **11**, 230–235 (1989).
111. N.S. Wang, J.-D. Yang, R.V. Calabrese, and K.-C. Chang, *J. Biotechnol.* **33**, 107–122 (1994).
112. I. Jöbses, D. Martens, and J. Tramper, *Biotechnol. Bioeng.* **37**, 484–490 (1991).
113. J. Wu and F.A. Goosen, *Enzyme Microb. Technol.* **17**, 1036–1042 (1995).
114. J.M. Boulton-Stone and J.R. Blake, *J. Fluid Mech.* **254**, 473–466 (1993).
115. M.A. Garcia-Briones, R.S. Brodkey, and J.J. Chalmers, *Chem. Eng. Sci.* **49**, 2301–2320 (1994).
116. J.J. Chalmers and F. Bavarian, *Biotechnol. Prog.* **7**, 151–158 (1991).
117. R.S. Cherry and C.T. Hulle, *Biotechnol. Prog.* **8**, 11–18 (1992).
118. F. Bavarian, L. S. Fan, and J.J. Chalmers, *Biotechnol. Prog.* **7**, 140–150 (1991).
119. H.G. Sucker, M. Jordan, H.M. Eppenberger, and F. Widmer, *Biotechnol. Bioeng.* **44**, 1246–1254 (1994).
120. E.T. Papoutsakis, *Trends Biotechnol.* **9**, 316–324 (1991).
121. O.T. Ramirez and R. Mutharasan, *Biotechnol. Prog.* **8**, 40–50 (1992).
122. J.D. Michaels, J.F. Petersen, L.V. McIntire, and E.T. Papoutsakis, *Biotechnol. Bioeng.* **38**, 169–180 (1991).
123. M. Lavery and A.W. Nienow, *Biotechnol. Bioeng.* **30**, 368–373 (1987).
124. K.T. Kunas and E.T. Papoutsakis, *J. Biotechnol.* **15**, 57–70 (1990).
125. S. Goldblum, Y.-K. Bae, W.F. Hink, and J.J. Chalmers, *Biotechnol. Prog.* **6**, 383–390 (1990).
126. Y. Chisti and M. Moo-Young, *J. Chem. Technol. Biotechnol.* **58**, 331–336 (1993).
127. D.W. Murhammer and C.F. Goochee, *Biotechnology*, **6**, 1411–1418 (1988).
128. D.W. Murhammer and C.F. Goochee, *Biotechnol. Prog.* **6**, 142–148 (1990).
129. L. Laouar, K.C. Lowe, and B.J. Mulligan, *Enzyme Microb. Technol.* **18**, 433–438 (1996).
130. L.A. van der Pol, I. Beeksma, and J. Tramper, *Enzyme Microb. Technol.* **17**, 401–407 (1995).
131. M.C.M. van Loosdrecht and S.J. Heijnen, *Trends Biotechnol.* **11**, 117–121 (1993).
132. J.J. Heijnen, M.C.M. van Loosdrecht, A. Mulder, and L. Tijhuis, *Water Sci. Technol.* **26**, 647–654 (1992).
133. K. Wagner and D.C. Hempel, *Biotechnol. Bioeng.* **31**, 559–566 (1988).
134. L. Tijhuis, H. Zwols, K.-H. van Meekeren, M.C.M. van Loosdrecht, and J.J. Heijnen, L. Alberghina, L. Frontali, and P. Sensi eds., *Progress in Biotechnology*, vol. 9, *Proceedings of the 6th European Congress on Biotechnology*, Elsevier, Amsterdam, 1994, pp. 953–956.
135. L. Tijhuis, W.A.J. van Benthum, M.C.M. van Loosdrecht, and J.J. Heijnen, *Biotechnol. Bioeng.* **44**, 867–879 (1994).
136. L. Tijhuis, M.C.M. van Loosdrecht, and J.J. Heijnen, *Biotechnol. Bioeng.* **44**, 595–608 (1994).
137. L. Tijhuis, M.C.M. van Loosdrecht, and J.J. Heijnen, *Biotechnol. Bioeng.* **45**, 481–487 (1995).
138. W.A.J. van Benthum, J.M. Garrido-Fernández, L. Tijhuis, M.C.M. van Loosdrecht, and J.J. Heijnen, *Biotechnol. Prog.* **12**, 764–772 (1996).
139. A. Gjaltema, L. Tijhuis, M.C.M. van Loosdrecht, and J.J. Heijnen, *Biotechnol. Bioeng.* **46**, 258–269 (1995).
140. A. Gjaltema, J.L. Vinke, M.C.M. van Loosdrecht, and J.J. Heijnen, *Biotechnol. Bioeng.* **53**, 88–99 (1997).
141. L. Tijhuis, J.L. Huisman, H.D. Hekkelman, M.C.M. van Loosdrecht, and J.J. Heijnen, *Biotechnol. Bioeng.* **47**, 585–595 (1995).
142. P.A. Shamlou and N. Titchener-Hooker, in P.A. Shamlou ed., *Processing of Solid–Liquid Suspensions*, Butterworth-Heinemann, Oxford, 1993, pp. 1–25.
143. D.F. Bagster, in P.A. Shamlou ed., *Processing of Solid–Liquid Suspensions*, Butterworth-Heinemann, Oxford, 1993, pp. 26–58.
144. T. Matsuo and H. Unno, *J. Environ. Eng. Div. Am. Soc. Civ. Eng.* **107**, 527–545 (1981).
145. J. Gregory, in P.A. Shamlou ed., *Processing of Solid–Liquid Suspensions*, Butterworth-Heinemann, Oxford, 1993, pp. 59–92.
146. H. Taguchi, T. Yoshida, Y. Tomita, and S. Teramoto, *J. Ferment. Technol.* **46**, 814–822 (1968).
147. Y.Q. Cui, R.G.J.M. van der Lans, and K.Ch.A.M. Luyben, *Biotechnol. Bioeng.* **57**, 409–419 (1998).
148. Y.-F. Maa and C.C. Hsu, *Biotechnol. Bioeng.* **54**, 503–512 (1997).
149. Y.-F. Maa and C.C. Hsu, *Biotechnol. Bioeng.* **51**, 458–465 (1996).
150. S.A. Rouf, M. Moo-Young, and Y. Chisti, *Biotechnol. Adv.* **14**, 239–266 (1996).
151. D.J. Bell, M. Hoare, and P. Dunnill, *Adv. Biochem. Eng.* **26**, 1–72 (1983).

SOLID STATE FERMENTATION, MICROBIAL GROWTH KINETICS

DAVID A. MITCHELL
Universidade Federal do Parana
Curitiba, Brazil
DEIDRE M. STUART
Queensland University of Technology
Brisbane, Australia
ROBERT D. TANNER
Vanderbilt University
Nashville, Tennessee

KEY WORDS

Growth kinetics
Mathematical modeling
Solid-state fermentation
Transport phenomena

OUTLINE

INTRODUCTION

Solid-state fermentation (SSF) involves the growth of microorganisms on moist solid substrates in the absence of free water. The absence of free water makes the system quite different from submerged liquid fermentations and makes SSF superior for the production of some products. However, compared with submerged liquid fermentation, relatively little is known about how to design and operate bioreactors for large-scale SSF processes.

Despite this, there are commercially successful large-scale SSF processes for the production of soy sauce koji and other traditional fermented foods, citric and gluconic acids, and fungal enzymes such as cellulases, amylases, lipases, and pectinases (1,2). In addition, fungal spores have been produced by SSF for use in steroid transformations and as inocula for production of blue-vein cheeses (2). Composting is another SSF application that is widely practiced, although composting is different from other SSF processes, because self-heating of the fermenting substrate is desired in composting processes, whereas avoidance of overheating is a key problem in most SSF processes.

In addition to these commercial processes, over the last 20 years there has been a growing interest in other SSF applications (2), including the following:

- Upgrading of solid wastes from agriculture and food processing for use as fermented animal feeds. In some cases the microorganism is used to enrich the protein content; in other cases it is used to degrade toxic substances in the waste.
- Production of fungal spores for use as mycopesticides. SSF is superior for the production of fungal spores because most fungi do not sporulate well in liquid culture, and even when they do, spores produced in SSF tend to be more robust.
- Production of other enzymes such as dehairing enzymes, fungal rennets, beta-glucanases and xylanases.
- Biopulping of woodchips during the production of paper.
- Production of various other products such as gibberellic acid, aroma and flavor compounds, antibiotics, and ethanol.

Unlike the situation with submerged liquid fermentations, there is no systematic framework guiding the design and operation of large-scale bioreactors for SSF. As a result, there has been relatively little success in developing successful large-scale processes for these newer SSF products, although large-scale bioreactors have been used to make soy sauce koji. Intermediate-scale processes can be carried out using tray fermentations, but such operations are quite labor intensive. Mathematical models of bioreactor operation will be key tools in the development of strategies for the design and operation of large-scale SSF bioreactors. This article summarizes the current state of development of modeling of SSF processes.

AIMS OF THIS ARTICLE

This article addresses the importance, to the development of large-scale SSF technology, of understanding and modeling microbial growth kinetics and transport phenomena, which combine to determine the overall performance of a solid-state bioreactor. Transport phenomena are of special importance in SSF due to the unavoidable heterogeneity of SSF systems. Good reactor performance will be achieved only through an understanding of these transport phenomena and how they interact with the microbial growth kinetics. Mathematical models represent a convenient, concise, and powerful way of describing these phenomena and their interactions and provide a sound foundation for process development, control, and optimization. They can also guide us in learning where the problems are and how to attack them.

The main message of this article is that modeling of phenomena in SSF and of the performance of SSF bioreactors has not yet progressed to a stage where it can be usefully applied to industrial applications. In fact, it was only in 1990 that the first model was proposed for SSF bioreactor operation (3). Although advances have been made since then, currently we still do not have mechanistic models of SSF that are able to make accurate predictions based fully on independently determined parameter values. To date the focus has been on appropriate model structures. Relatively little attention has been given to parameter estimation, with many parameters being borrowed from other systems, systems that often involve different substrates, organisms, and cultivation systems. As a result, none of the models has been properly validated against experimental data, and although the models are now being used to predict improved operating conditions, they have not yet been used to improve operation of actual bioreactors. In most cases some of the model parameters have been estimated by adjusting them to enable the model to fit the experimental data. Furthermore, although the ultimate aim of SSF processes is to produce useful products industrially, to date models have focused on growth. In fact, in some cases the main aim has been to predict temperatures rather than growth itself.

The aim of this article is to draw together the basic principles from the work that has been done so far, to show the general concepts that underpin all models, to identify the features that will make models most useful for application in industrial processes, and thereby to provide a foundation for continued progress in the modeling of SSF.

ORGANIZATION AND SCOPE

Filamentous fungi are the most important group of microorganisms used in SSF, and most processes are aerobic. Most of the work discussed in this article addresses this type of system, although other systems are mentioned as appropriate.

The article begins by describing the biological and transport phenomena occurring in an SSF bioreactor and identifying the various phenomena that can potentially limit the kinetic performance of the bioreactor:

- Microbial growth phenomena
- Local transport phenomena (occurring at the level of a single particle)
- Bulk transport phenomena (occurring at the level of the whole bioreactor)

Current models of SSF address the influence of either local or bulk transport phenomena on growth and are therefore labeled as microscopic or macroscopic models. The general aims and usefulness of these two types of models are discussed. The mathematical expressions appropriate for describing the three types of phenomena that may be incorporated into these models are discussed in turn.

Readers interested in modeling of SSF processes are urged to consult the original references because no attempt has been made to present the complete models of individual workers. The original articles do that better than we could hope to do. Rather we have organized this article around the ways in which the various kinetic and transport phenomena have been handled. Therefore, where we present equations, these have been separated from the author's discussion of their underlying assumptions and the notation and units they have used.

DESCRIPTION OF KEY FEATURES OF SOLID-STATE FERMENTATION

Growth in an SSF bioreactor results from and is influenced by processes operating at different scales (Fig. 1): microbial processes and local and bulk transport processes. Figure 1 also highlights some of the complex interactions between these phenomena during SSF. This article addresses the approaches that have been made to modeling the various phenomena shown in Figure 1, and the interactions between the various phenomena. Thermodynamic phenomena, such as the ability of solid substrates to absorb heat and the ability of gases to carry water vapor, also affect the overall bioreactor performance; however, these phenomena are not addressed in this article.

In developing a mathematical model, it is important to identify the processes occurring before attempting to describe them mathematically. This focuses attention on the processes that are potentially rate limiting. An understanding of the rate-limiting processes facilitates development of strategies to overcome the limitation, although SSF processes can be difficult to optimize because the limiting step might be different under different conditions, and at different stages of an SSF process the growth rate may be limited by different phenomena. Therefore this section provides a qualitative description of our current understanding of the phenomena occurring in SSF.

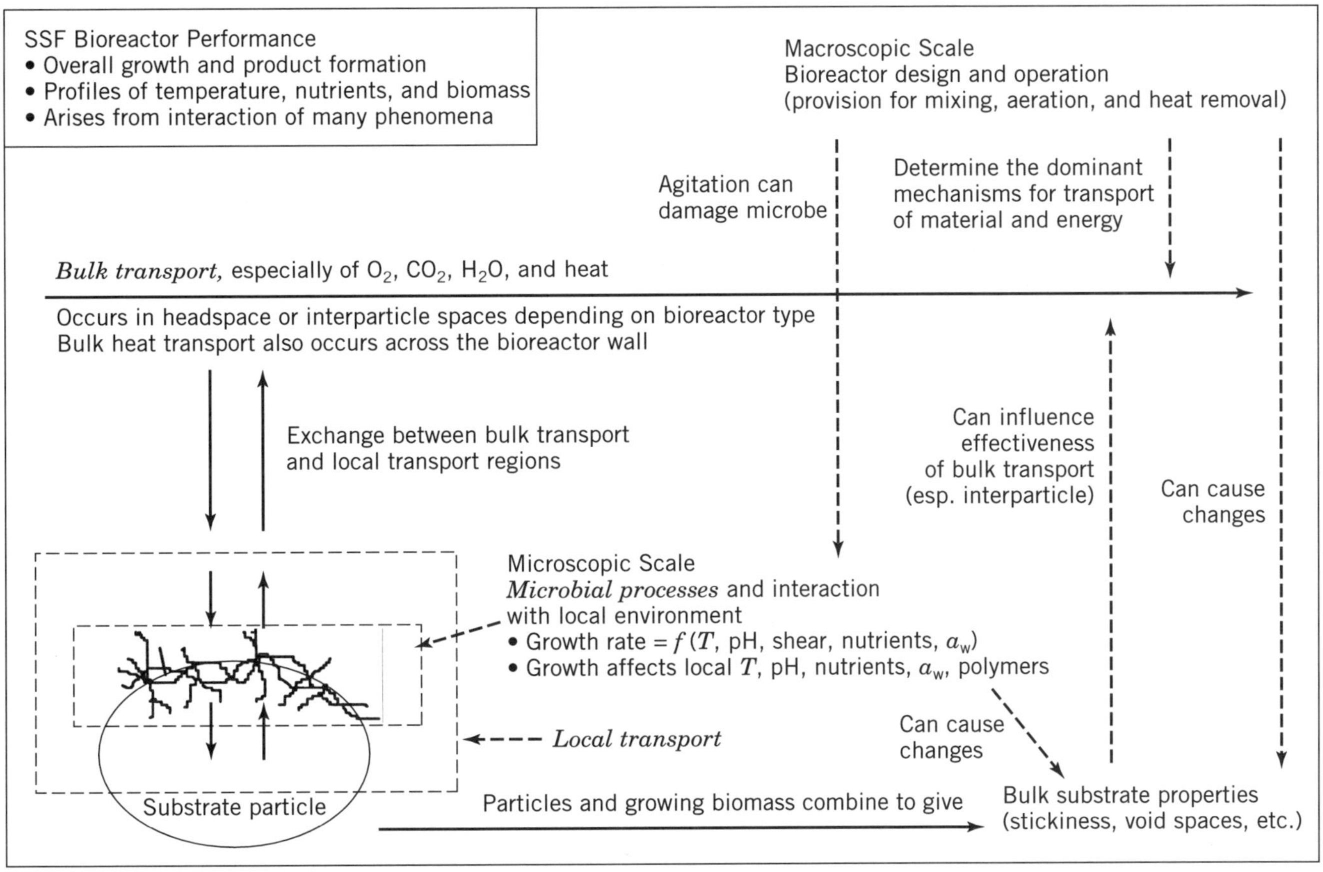

Figure 1. An overview of the microscopic and macroscopic processes that combine to determine the overall performance of an SSF bioreactor. This diagram describes transport processes in general terms and does not identify all the individual mass and heat transfer processes occurring. Although the diagram represents an SSF bioreactor, it is not drawn to scale. The nature and extent of the local and bulk transport regions depend on the particular bioreactor and how it is operated.

The Physical Nature Of Solid-State Fermentation Systems

From a macroscopic viewpoint, up to four subsystems are apparent (Fig. 2), depending on the bioreactor type: the material from which the bioreactor is constructed, a headspace containing gases, the solid particles themselves, and the interparticle gas phase entrapped between the solid particles. It is often appropriate to think of the solid particles and interparticle gas as a pseudohomogeneous "substrate bed."

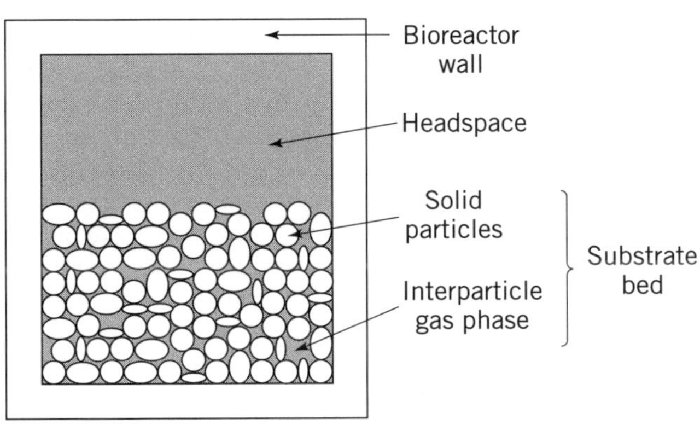

Figure 2. Subsystems within a solid-state fermentation bioreactor.

From a microscopic perspective the system is considered in greater detail. The description focuses on the region near an individual particle and can include (Fig. 3) the following:

- The biomass, the majority of which is concentrated as a layer at the particle surface. Part of the fungal mycelium is immersed in the liquid film at the surface, but many hyphae extend into the gaseous region. These aerial hyphae may support films or droplets of water. Other hyphae may penetrate into the substrate particle. In contrast, yeasts and bacteria form a compact moist layer on the particle surface.
- The solid matrix, which provides the physical support for the liquid and biomass phases, and which may or may not be degraded during the fermentation.
- A liquid phase consisting of a stagnant liquid film at the substrate surface and the moisture inside the particle. This phase may be continuous, depending on the internal structure of the substrate particle.

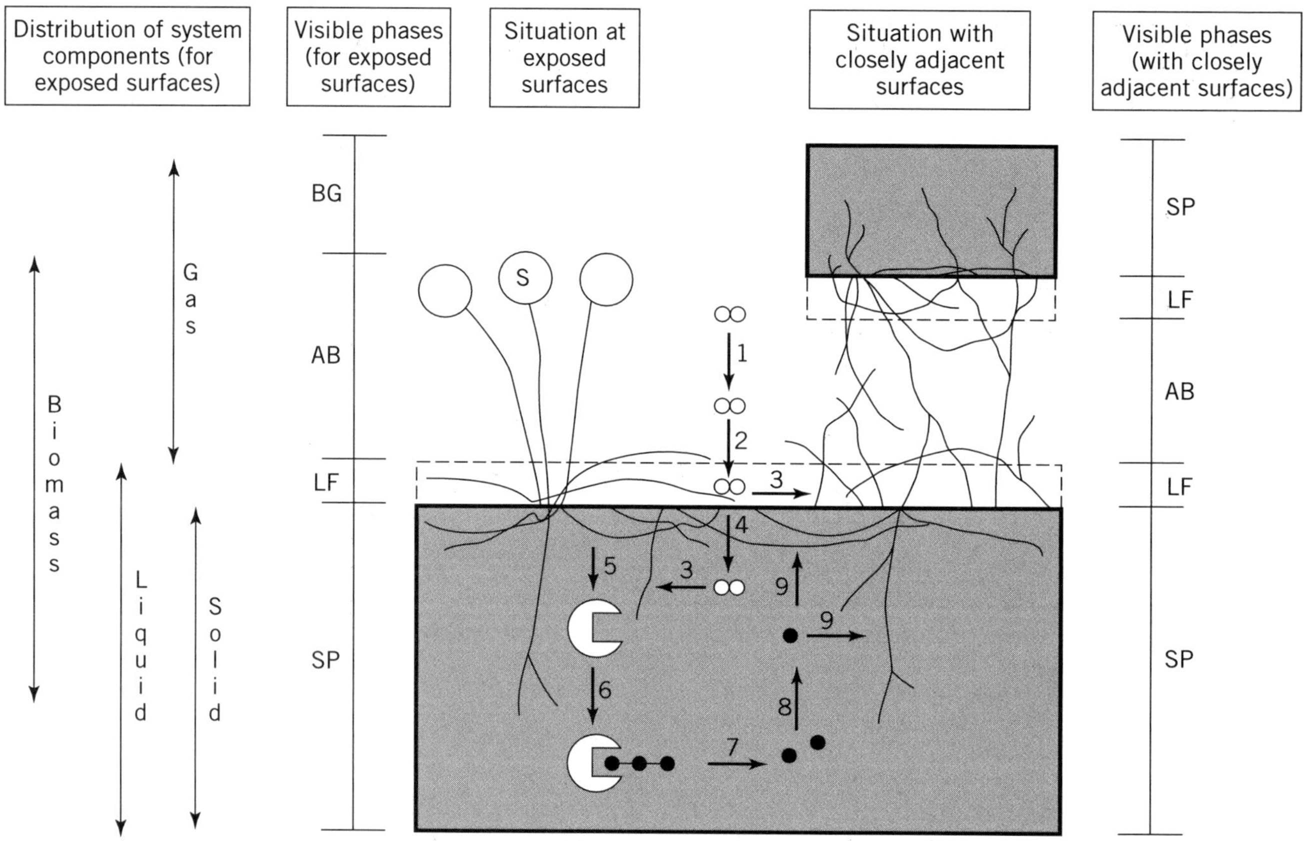

Figure 3. Structure of SSF systems at the microscopic scale and local mass transport processes. The example is for growth of a filamentous fungus, such as *Rhizopus* sp., on a starchy substrate. The diagram represents a cross-sectional view near the surfaces of the substrate particles and indicates the difference between substrate particles exposed to the bulk gas phase and substrate particles that are packed closely together. Simplifications, such as the absence of penetrative hyphae, and the liquid film as the only site of O_2 uptake, have been used in modeling studies. ∞, oxygen; ●, glucose; ●—●—●, starch; S, sporangium; SP, substrate particle; LF, liquid film; AB, aerial biomass; BG, bulk gas. The processes occurring are 1, bulk transport of oxygen; 2, diffusion of O_2 through stagnant gas regions and diffusion across the gas–liquid interface; 3, uptake of oxygen by hyphae; 4, diffusion of oxygen through the liquid phase of the substrate; 5, release of glucoamylase from the hyphae; 6, diffusion of glucoamylase through the solid substrate particle; 7, hydrolysis of starch by the glucoamylase; 8, diffusion of the released glucose through the solid substrate particle; 9, uptake of glucose by the hyphae.

- A gaseous phase. Although this gas phase is continuous, it may be conceived as comprising up to three regions. First, the aerial hyphae of the biomass are surrounded by regions of stagnant gas, and there will be a stagnant gas film adjacent to the liquid film. Second, there may be a stagnant gas region above the biomass region (i.e., part of the interparticle space). Third, there may be a mobile region in which there is bulk flow. The locations and extents of the stagnant and mobile gas regions will depend on whether the substrate bed is agitated and how it is aerated. For example, in static packed beds the interparticle spaces are filled with a network of hyphae, but there is still bulk gas flow through parts of this region. On the other hand, in tray bioreactors, the gas phase within the substrate bed is relatively stagnant, although there may be some free convection due to thermal gradients.

The Processes Occurring During Solid-State Fermentation

Aufeuvre and Raimbault (4) described the changes in morphology during growth of the filamentous fungus *Aspergillus niger* in a packed bed bioreactor, on a granular substrate prepared from cassava flour. Initially, the spores used as inoculum are spread across the substrate surface. During the germination period of about 8 h, the spores swell and then extend germ tubes across the substrate surface. These germ tubes extend for up to 50 μm before branching begins. Branching of these germ tubes then rapidly leads to a loose covering of hyphae across the whole surface. As branching continues, the hyphal density increases, and it is soon impossible to distinguish individual mycelia. The network of hyphae increases in thickness, so that many hyphae are not in contact with the substrate surface. Some hyphae form together into multihyphal filaments, which can extend up to 80 μm above the substrate surface. These filaments form bridges between the substrate granules and bind the whole bed of substrate particles together into a single compact mass.

This description addresses only the features of the growth process that can be seen with a microscope. Many less obvious processes are occurring and influencing growth (Fig. 3). The growth activities of the microbe are determined by the local environment experienced by the microbe, especially the local temperature, pH, water activity (a_w), oxygen concentration, and nutrient concentrations. In turn, growth processes also affect the local environment, for example, hydrolytic enzymes are secreted, carbon and nitrogen substrates are absorbed into the mycelium, oxygen is consumed, and various products are excreted. In addition, waste metabolic heat is released by the microorganism. Therefore the temperature, pH, water activity, and oxygen and nutrient concentrations in the local environment are affected by the microorganism. This induces the transport processes, which tend to equilibrate chemical potential and temperature throughout a material. Within solid particles and within stagnant gas regions, mass transfer is limited to diffusion and the exchange of gases (O_2, CO_2, and water vapor) at gas–liquid interfaces, and heat transfer is limited to conduction, free-convection, and evaporation of water at the gas–liquid interface. These dissipative processes can be relatively slow compared with the rate of the microbial processes, leading to the establishment of local concentration and temperature gradients around the microbe.

The purpose of a bioreactor is to maintain an optimal local environment for the desired microbial activity, within the limits of physical and economical constraints. In SSF the ability to control the local environment is largely limited to manipulating either the bulk gas flow or agitation of the substrate, or both, providing oxygen for aerobic processes, removing carbon dioxide and evaporated water, and removing heat. Almost nothing can be done during the fermentation to prevent nutrient concentration gradients within individual substrate particles, because it is not possible to have bulk mixing below the scale of the substrate particles. This is an intrinsic consequence of the physical form of SSF systems.

Given that control of the environment is achieved by manipulating the bulk phases, the rate of removal of heat and the rate of supply of oxygen can be maximized by minimizing gradients within the bulk gas phase and minimizing the thickness of the stagnant gas film that separates the bulk gas phase from the local environment. Different reactor types enable this to be achieved to different extents. In bioreactors with bulk flow of air past the solids, due either to forced airflow or to mixing of the substrate bed, gases diffuse from the stagnant gas regions into the bulk flow and are carried away. Heat can be removed by convection and by evaporation of water into this air. Evaporative heat removal can be maximized by aerating with dry air, but this will tend to dry the substrate. Depending on the bioreactor operation there might be significant gas concentration gradients and temperature gradients in the bulk air phase across the bioreactor. Plug flow of gas (and maybe substrate) will tend to favor macroscopic gradients, whereas mixing of the gas phase and of the substrate bed will tend to minimize such gradients. In bioreactors without bulk flow of air past the solids (i.e., trays), mass transfer and heat transfer occur mainly by diffusion and conduction, respectively. In addition to heat and mass transfer within the bioreactor, there are also heat transfers between the substrate bed and the bioreactor wall, the headspace gases and the bioreactor wall, and the bioreactor walls and the surroundings (Fig. 1).

During microbial growth in SSF, growth activities and the aeration and agitation strategies can affect the bulk substrate properties, which in turn can influence the bulk transport processes, as indicated in Figure 1. For example, in an unstirred packed bed, fungal mycelia grow in the interparticle spaces and can significantly increase the pressure drop through the bed, affecting airflow patterns within the bed (5). Microbial activities on the substrate may decrease the strength or increase the stickiness of the particles, leading to compaction or cohesion of particles when the substrate is agitated, thereby eliminating interparticle spaces and preventing bulk transport to or from the microbes within the bed.

In conclusion, there can be significant profiles of temperature, nutrients, and biomass at both the microscopic scale (looking at a single particle with a microbe growing on it, as shown in Fig. 3) and the macroscopic scale (looking across the bioreactor as a whole, as shown in Fig. 1). The success of transport processes at both these scales in dissipating thermal and concentration gradients determines how readily the local conditions can be maintained near the optimal state for the desired microbial activity. Therefore, the overall performance of an SSF bioreactor arises from the interaction of biological growth kinetics, local transport processes, and bulk transport processes. The next section discusses the nature of models of SSF in general terms, whereas the remaining sections address how the underlying phenomena have been described in mathematical models of SSF processes.

Approaches to Modeling Solid-State Fermentation

A fully mechanistic description of the bulk and local transport processes occurring within SSF (Figs. 1 and 3), when combined with the growth kinetics, would lead to a very complex model of the system. Models of SSF to date have focused on either local or bulk transport phenomena, and therefore it is convenient to classify the models as being either microscopic models or macroscopic models. Microscopic models concern themselves with growth at the level of individual particles, whereas macroscopic models describe overall bioreactor performance. Both types of models have a role to play in the development of SSF technology. However, in the future, models should be developed that integrate mechanistic descriptions of both local and bulk transport phenomena, although this will depend on further advances in computing power and methods for constructing and solving complex dynamic models. Such models will describe performance under a wide range of operating conditions. This trade-off between the power of the model and the ease of model solution should be kept in mind. As well, the purpose of the model and the ability to estimate parameters should be considered when developing models that incorporate both microscopic and macroscopic phenomena. Because microscopic and macroscopic models are currently developed separately, the modeling work usually has related but different aims.

Microscopic models combine local transport phenomena (Fig. 3) and the effect of the local environment on growth, in an attempt to identify the mechanisms by which these may control the growth rate and the total amount of growth that occurs during SSF. For example, due to the absence of mixing at the microscopic scale, diffusive phenomena within the substrate particle can limit the growth rate (6). Microscopic models tend to consider the biomass in more detail than do macroscopic models, and it may be appropriate to describe variations in hyphal density in space, or the presence of differentiated forms of biomass. However, for the modeling of bioreactor performance, it is not appropriate to describe the extension and branching of individual fungal hyphae. Models incorporating these growth phenomena are powerful tools for fundamental studies aimed at understanding the mechanisms of fungal growth but are not discussed in this article.

Macroscopic models attempt to describe the performance of the whole bioreactor (Fig. 1). Such models may be developed to contribute to various aspects of bioreactor development such as design, operation, scale-up, control, and optimization. The main aim is to describe growth, although this may require prediction of temperature or nutrient or oxygen concentrations, or properties of the substrate bed. Their level of sophistication is in keeping with the ability or lack of ability to measure fermentation parameters such as biomass, nutrient concentrations, gas concentrations, temperature, and pH. Macroscopic models should describe the phenomena controlling the bioreactor performance, but current macroscopic models simply combine bulk transport phenomena with empirical descriptions of microbial growth. They do not address local transport phenomena, because of the complexity that these introduce. Simple empirical descriptions of growth may subsume biological and local transport limitations on growth but conceal the mechanisms of the limitation. Therefore, current macroscopic models are most useful for determining whether bioreactor performance is limited by gas or heat exchange, and guiding bioreactor design and operation to overcome any such limitations.

Several other characteristics of SSF lead to common features of models:

- The vast majority of SSF processes are batch processes, and the few continuous processes involve plug flow systems rather than well-stirred vessels. Therefore dynamic models are most useful for describing SSF. Steady-state models are usually not appropriate.
- The presence of mass and thermal gradients means that in most cases it is desirable to include spatial coordinates in addition to the temporal coordinate. Unfortunately this makes the equation sets more complicated to set up and solve.
- Balances are commonly set up over more than one of the subsystems shown in Figure 2.
- Depending on what is expected to limit growth, either heat or mass transfer may be described. Models incorporating both of these transport phenomena will be most flexible in describing what controls the rate of growth under a wide range of operating conditions.
- Due to the complexity that transport phenomena introduce into problems, plus the difficulties of measuring biomass and metabolic states, the majority of models for SSF are unstructured and nonsegregated, although some do have very simple structure or segregation.

The remaining sections discuss the various expressions that have been used within models of SSF to describe microbial, local transport, and bulk transport phenomena in SSF.

MODELING OF MICROBIAL PHENOMENA

This section addresses microbial phenomena and how models of SSF have treated these phenomena. It focuses on studies of either SSF itself or similar systems involving overculture on solid nutrient media. Overculture occurs when a fungus is inoculated across the whole surface, and therefore the mycelium develops simultaneously from many points. This contrasts with colony growth, where the organism is inoculated at one point, and a colony spreads outward across the surface. The kinetics of colony growth on agar plates are not considered in this article. Note also that the word biomass is used here in the restrictive sense of microbial biomass, and not in the broad sense of agricultural biomass. This distinction is important because many substrates are of agricultural origin.

A word of warning is appropriate about modeling growth kinetics. The basis for expression of the microbial biomass must be clear (e.g., total biomass in the bioreactor, or biomass per kilogram of initial dry substrate, or biomass per kilogram of wet substrate present, etc.), and conservation equations must originally be formulated on a conserved quantity (i.e., absolute amounts of biomass rather than biomass concentrations). This is crucial in SSF because there is a significant decrease in substrate mass due to conversion of dry matter to CO_2, and therefore increases in biomass concentration result from two effects: the increase in the overall amount of biomass, plus the decrease in the overall amount of dry matter. Water transfer between the substrate and gas phases causes further complications if concentrations are expressed on a wet substrate basis. Due to the wide range of ways that different workers have expressed biomass, we have used general equations in which the variables X and S represent absolute amounts of biomass and substrate respectively, whereas the variables C_X and C_S represent concentrations. Rates are expressed as mass or energy per unit time and are represented by the symbol r with a subscript. Readers should refer to original articles to find the units and nomenclature used by the authors.

Substrate-Independent Growth Kinetic Equations

Sangsurasak et al. (7) note that four general shapes of growth kinetic profiles in SSF have been reported. In SSF models the simplest kinetic expressions describing these profiles relate the growth rate (r_g) directly to the biomass:

- Linear growth kinetics, described by

$$r_g = \text{constant} \tag{1}$$

where r_g is the rate of growth.

- Exponential growth kinetics, described by

$$r_g = \mu_m X \tag{2}$$

where μ_m is the observed maximal specific growth rate.

- Early acceleration followed by deceleration, with the rates of acceleration and deceleration being nearly equal, such that the biomass profile is symmetrical by rotation around the transition point. These kinetics are described by the logistic equation

$$r_g = \mu_m X\left(1 - \frac{X}{X_m}\right) \tag{3}$$

where X_m is the maximum possible amount of biomass.

- Early rapid acceleration, followed by an extended period during which growth decelerates slowly. This profile can be approximated by raising the ratio X/X_m to an exponent n to give the power-law logistic model (8,9)

$$r_g = \mu_m X\left(1 - \left(\frac{X}{X_m}\right)^n\right) \tag{4}$$

This allows the logistic curve to be skewed (i.e., for the rates of acceleration and deceleration to be different). The value of n is a measure of the relative sensitivity of the culture to self-inhibition at high biomass densities (8). For values of n less than 1, the organism is relatively sensitive to self-inhibition, and self-inhibition occurs for quite low values of X. For $n = 1$ the logistic equation is obtained. Finally, for values of n greater than 1 the organism is relatively resistant to self-inhibition, and significant self-inhibition occurs only at values of X close to X_m. The exponent n may have biological significance. The mycelium arising from a spore undergoes a maximum number (k_m) of branching events before growth ceases, and mycelia arising from a large number of different spores have a distribution of k_m values. If n is less than 1 then many of the mycelia have low k_m values, whereas if n is greater than 1 then many of the mycelia have high k_m values (8).

The expressions for both the linear and exponential growth kinetics do not describe any limitation on growth and are of little use for incorporating into models of growth in SSF. On the other hand, the logistic equation (and by extension, its power-law modification) describes a limitation on growth, and for this reason has quite commonly been incorporated into models of growth in SSF (10–12), especially macroscopic models. However, the significance of X_m is not clear. Growth might potentially be limited by the amount of substrate or diffusional limitations, but it can also be limited sterically.

Assumption of symmetric branching enables a simplified description of the mycelial mode of growth and enables the macroscopic parameter μ_m in the logistic equation to be related to the specific growth rate of a segment of biomass just before it is about to divide (μ'_m). The value of μ'_m can be determined from microscopic observation of extending hyphae. After a few branching events the following equation holds (8,9):

$$\mu_m = \mu'_m \frac{(X + X_c)}{2X} \approx \frac{\mu'_m}{2} \tag{5}$$

where X_c is the biomass in one hyphal segment just prior to a branching event. The approximation arises because X_c is small compared with the total biomass. This relationship between the two specific growth rates occurs because, after a few symmetric branchings, the biomass in the actively extending segments of a mycelium comprises approximately half the total biomass. The remaining biomass is in the older section of the mycelium where there are no actively extending hyphal tips. The microscopic growth parameters (X_c and μ'_m) can be determined by image analysis, which might even be used to monitor growth during SSF itself (9). If the effect of environmental parameters on growth and branching is known, some prediction can be made about how the environment will influence macroscopic growth kinetics.

Substrate-Dependent Growth Kinetic Equations

Kinetic expressions that rely only on biomass are simple but lack flexibility in describing different systems because new values for the parameters must be determined for each new system. Another approach is to assume that growth is related to the concentration of a growth-limiting substrate (such as a soluble sugar or oxygen) by the Monod equation (6,13,14),

$$r_g = \mu_m X\left(\frac{C_S}{K_S + C_S}\right) \tag{6}$$

where K_S is the saturation constant for the substrate. The models of Muck et al. (13) and Kaiser (14), which are based on the Monod equation, are relatively complex because they describe composting processes involving several organisms and several substrates. Muck et al. have four microbial groups (fungi and yeasts, each of which is divided into thermophilic and mesophilic subgroups), each with different Monod growth parameters, and there are several substrates that are assumed to be used sequentially.

The substrate concentration in the Monod equation must be the actual concentration that the microorganism experiences, which may not be simple to define due to the presence of both solids and water. Due to the concentration gradients that arise during growth, substrate concentrations at the surface of a substrate where the microorganism is located can be quite low compared with the average concentration (6). Therefore the effect of substrate concentration should be incorporated into kinetic models only if the growth equations are coupled with local transport equations. Neither Muck et al. (13) nor Kaiser (14) modeled local transport, which limits the usefulness of their models.

Various modifications of simple Monod kinetics have been used by different workers:

- Simultaneous limitation of growth by oxygen according to Monod kinetics and self-inhibition by the biomass according to the logistic equation (15)

$$r_g = \mu_m\left(\frac{C_{O_2}}{K_{O_2} + C_{O_2}}\right)X\left(1 - \frac{X}{X_m}\right) \tag{7}$$

- Simultaneous limitation by oxygen and glucose (16)

$$r_g = \mu_m X\left(\frac{C_{O_2}}{K_{O_2} + C_{O_2}}\right)\left(\frac{C_G}{K_G + C_G}\right) \tag{8}$$

- Self-inhibition of growth by biomass late in growth (according to logistic kinetics) combined with substrate inhibition kinetics (17)

$$r_g = \mu_m\left(\frac{C_S}{K_S + C_S + C_S^2/K_i}\right)X\left(1 - \frac{X}{X_m}\right) \tag{9}$$

where K_i is the inhibition constant for the substrate.

Modeling Effects of Microbial Activities on the Environment

Microbial activities affect the environment in SSF, and the changes in the local environment in turn affect growth. The effect of microbial activities on the local pH and water activity in SSF have received almost no attention, with the exception of Muck et al. (13) who related pH changes to acetic and lactic acid production rates and the buffer index of the solids. This section therefore is restricted to describing how workers have modeled consumption of nutrients (either the carbon source or oxygen), the generation of metabolic heat, and effects of growth on porosity in the substrate bed.

In models that describe the dependence of the growth rate on substrate concentration, it is also important to describe the effect of the growth activities on the substrate (i.e., substrate consumption). Sometimes a constant growth yield coefficient has been assumed,

$$r_s = -\frac{1}{Y_{XS}} r_g \tag{10}$$

where Y_{XS} is the observed yield coefficient. The negative sign indicates that substrate is consumed during the growth reaction. However, many workers have assumed that substrate is used for both growth and maintenance and have assumed the relationship

$$r_s = -\frac{1}{Y'_{XS}} r_g - m_s X \tag{11}$$

where Y'_{XS} is the true growth yield on the substrate, and m_s is the maintenance coefficient. This equation has been applied both to nutrients within the substrate and to oxygen.

If the particle structure is provided by a macromolecule that is degraded by the microorganism, such as during the utilization of predominantly starchy materials by amylolytic microbes, or the degradation of cellulosic substrates by cellulolytic organisms, then the particle will shrink during the fermentation. This can affect bioreactor performance through effects on bulk transport. For example, a decrease in particle size caused the substrate bed in a packed-bed bioreactor to pull away from the walls, allowing the air to bypass the bed and reducing the availability of oxygen within the bed (18). A simple model for particle size reduction assumes that oxygen is the limiting substrate and that it is consumed at the substrate–biomass interface (19,20). As the fermentation proceeds, the oxygen must diffuse through a biomass layer that increases in thickness, lowering the rate of diffusion and therefore the rate of reaction. Further developments of shrinking-substrate models are necessary because most substrates have complex compositions, and microbes can produce a range of enzymes, often in sequential order according to preferences for different macromolecules.

The generation of large amounts of metabolic waste heat can cause overheating of the substrate, and the kinetics of waste heat generation must be described in models of heat transfer in SSF. The simplest approach is to assume a constant relationship between the rate of production of new biomass and the rate of generation of metabolic heat, r_Q (11,15):

$$r_Q = Y_{QX} r_g \tag{12}$$

where Y_{QX} is the differential yield coefficient for waste heat production during growth. However, if maintenance metabolism is significant, then a growth and maintenance model is appropriate (21),

$$r_Q = Y'_{QX} r_g + m_Q X \tag{13}$$

where Y'_{QX} is the differential yield of heat from the metabolism associated with growth, and m_Q is the maintenance coefficient for heat production.

The rate of metabolic heat production does not have to be directly related to the growth rate. Saucedo-Castaneda et al. (3) related heat generation to the carbon dioxide evolution rate, which in turn was related to the substrate utilization rate by a constant yield,

$$r_Q = Y_{QCO_2} r_{CO_2} = Y_{QCO_2}[Y_{CO_2S}(-r_s)] \tag{14}$$

where Y_{QCO_2} is the differential relationship between heat released and CO_2 evolved (J/g), and Y_{CO_2S} is the yield of carbon dioxide from the substrate (g/g). This equation was indirectly related to the growth equation because logistic growth kinetics were assumed, and substrate consumption was modeled by a growth and maintenance model.

Instead of simply using experimentally determined heat yield coefficients (e.g., Y_{QX}) several workers have applied a bioenergetic approach. Gutierrez-Rojas et al. (17)

proposed a stoichiometric equation for microbial growth on sucrose and estimated the enthalpy change associated with the growth reaction from the enthalpies of formation of the reactants and products. Larroche and Gros (22) considered the energetics of growth in SSF in even greater detail, estimating ATP generation and expenditure during growth. Muck et al. (13) related the rate of energy release to the rate of consumption of each of four substrates (water-soluble carbohydrate, ethanol, lactic acid, and acetic acid), estimating the fraction of the energy of combustion of the substrates that is actually released as waste metabolic heat. Finally, Rodriguez Leon et al. (23) estimated the heat yield coefficient from considerations of the degree of reductance of the substrate and biomass.

The growth of the microorganism into the interparticle voids affects the porosity of the substrate bed according to the relationship (24)

$$E = E_0 - X/(\rho_x W_s V_r) \tag{15}$$

where ρ_x is the wet biomass density (g/cm^3); W_s is the biomass dry matter content (g/g); V_r is the reactor volume, and E_0 is the initial porosity. This reduction in void fraction affects the pressure drop through packed beds (5) and reduces the effective diffusion coefficient within the substrate bed in tray bioreactors (24).

Death Kinetics

Overheating is a common problem in SSF, and several models have attempted to describe both growth and death of microorganisms. To date, death kinetics have largely been assumed to be first order (11,12),

$$r_d = k_d X_v \tag{16}$$

where the specific death rate, k_d, is a function of temperature. The effect of temperature on the parameters in growth and death kinetic expressions will be discussed later. The incorporation of death kinetics into a model divides the population into viable (X_v) and dead (X_d) subpopulations, representing a very simple level of segregation. On the other hand, Sargantanis et al. (21) assumed that their organism was killed instantly when the temperature reached 43 °C. Not all models considering the effect of temperature actually have death kinetics in them; some describe only a decrease in net growth rate with temperature (3).

Effects of Environmental Variables on Growth and Death Kinetics

The effect of the environment on kinetic parameters is quite important because it is difficult to control environmental variables; some may change quite significantly during growth, and some might limit growth. The effect of nutrient concentrations (including O_2) have already been considered. Of most interest here are temperature, water activity, and pH. Of these, temperature has received most attention due to the overheating problems that are commonly encountered in SSF. For bioreactors with large temperature gradients, heat transport equations must be built into the model if growth is to be adequately described.

Regarding the effect of temperature, most attention has been given to the effect of the temperature on the rate constants μ_m and k_d. Simple empirical equations are normally fitted by least squares regression to growth rate data collected between the minimum and maximum growth temperatures. Polynomial expressions (11,14,15) or double exponential expressions (3,17) have been used to describe the variation in specific growth rate with temperature. Arrhenius-type relationships have also been used to describe the effect of temperature on both the specific growth rate and the specific death rate (12),

$$\mu_m = \mu_{mo} \exp\left(\frac{-E_g}{RT}\right) - k_{do} \exp\left(\frac{-E_d}{RT}\right) \tag{17}$$

where R is the universal gas constant, T is the absolute temperature, E_g and E_d are the activation energies for growth and death, and μ_{mo} and k_{do} are rate constants. Least squares regression usually allows an adequate fit of such equations, although the biological significance of the parameters is not clear because the equations are not formulated on any theoretical basis.

Other growth parameters may be expressed as functions of the environmental conditions. Sargantanis et al. (21) expressed not only μ_m but also X_m in the logistic equation as empirically fitted functions of temperature and moisture content. Saucedo-Castaneda et al. (3) expressed the maximum biomass, X_m, in the logistic equation as an empirically fitted fourth-order polynomial function of temperature.

The model of Muck et al. (13) has described the effect of the environment on growth in most detail. In addition to using Arrhenius-type equations to describe the specific growth and death rates, with different parameter values for the different microbial groups, they also described the effect of pH and water activity on growth. However, for both these environmental variables the effects on the specific growth rates of the various microbial groups were incorporated simply by using empirical equations, fitted to experimental kinetic data, to modify the maximum specific growth rate. Gervais and Simatos (25) have addressed the effect of water activity on growth on solid surfaces in greater depth, although they consider the radial extension rates of fungal colonies. The effect of water activity on the overculture situation in SSF has not been addressed.

Modeling Branching, Penetration, and Differentiation

Fundamental phenomena in the growth of filamentous fungi on solid surfaces have received some attention. Some models address extension of the hyphal tip at a very fundamental level, but these models have not been extended to describe morphological development. Others can describe branching and certain aspects of hyphal morphogenesis but have not been extended to describe overall growth kinetics (8). In any case, models describing the extension and branching of individual hyphae describe the system in far more detail than is required in models of bioreactor performance, where only overall mycelial concentrations are of interest. However, models describing hyphal phenomena might be useful if they concern themselves with the overall population of hyphae or hyphal tips, and this section addresses such models briefly. The equations are not presented but can be found in the original articles.

The symmetric branching model of Viniegra-Gonzalez et al. (8,9), described earlier in conjunction with the logistic equation, makes many simplifying assumptions about the branching process, essentially that all hyphae branch after reaching a certain length, and the two branches arising are identical. These do not happen in practice but may be approximated on average in the large population of hyphae within a mycelium. As shown in equation 5, on this basis the relationship between some microscopic and macroscopic growth parameters can be derived.

Edelstein and Segel (26) used a modeling approach based on average tip and hyphal concentrations per unit area of substrate surface and describing the generation of tips by branching and the movement of hyphal tips across the surface due to hyphal extension. Growth of tips depended on the local nutrient concentration. This approach avoids the great complexity that would be involved in constructing a model on the basis of intracellular events, tip extension, and branching events, and some of the parameters such as tip densities can be estimated experimentally (27). Laszlo and Silman (28) used a cellular automata approach to model mycelial densities during fungal colony expansion over solid surfaces. In this approach a matrix of small squares is superimposed on a flat substrate surface. Initially, one or more squares are designated as living, to represent an inoculum. The fungus can then grow into adjacent squares, with simple rules based on the number of occupied squares adjacent to a vacant square being used to determine whether that vacant square will be colonized. This approach can give insights into density patterns within fungal colonies (28) but otherwise is of limited usefulness for modeling growth kinetics in bioreactors. Edelstein (29) developed a model describing colony expansion, assuming no nutrient limitations, but based on processes of branching, anastomosis, tip death, and hyphal death. Only the average frequencies of these processes need to be known. Different frequencies for the various processes can lead to very different colony morphologies. These approaches could be adapted to describe growth during overculture in SSF.

Penetration into the substrate is an interesting aspect that has not received much attention, although Jurus and Sundberg (30) gave a qualitative account of the penetration of *Rhizopus oligosporus* into soybeans. More recently Ito et al. (31) observed an exponential decrease in the concentration of hyphae below the surface of the rice in koji. Experimentally they observed that the quality of the koji was strongly correlated with the degree of mycelial penetration: at lower degrees of penetration, the activities of undesirable enzymes was high, leading to poorer quality koji. Although they proposed a mathematical equation to describe the decrease in hyphal concentration with depth, it was simply fitted to the data collected at each time point. A dynamic model describing the increase in concentration of penetrative hyphae as a function of time and distance was not proposed. Such dynamic models might be important because penetration into the substrate has the potential to increase substrate degradation.

Differentiation can be quite important in SSF processes as an essential step in the production of spores or secondary metabolites (32). Georgiou and Shuler (33) modeled growth and differentiation of *Aspergillus nidulans* on a solid medium containing glucose. The model could predict the increases in density of the various types of biomass (vegetative, competent, differentiated, and conidial) during overculture.

MODELING OF LOCAL TRANSPORT PHENOMENA

Local mass transfer processes have the potential to limit the rate of growth in solid-state fermentation. Most of the local transport phenomena occurring in SSF, such as diffusion through porous media and stagnant films, have been well studied in other systems. However, the translocation of nutrients within the fungal mycelium itself is poorly understood. The work done on nutrient diffusion during radial expansion of colonies on agar surfaces will not be described here because SSF usually involves overculture of the organism across the whole surface, for which nutrient diffusion horizontally toward the organism is not important. In overculture the key spatial dimension is the one passing vertically through the substrate (Fig. 3).

This section indicates how modeling of these local transport phenomena has improved our understanding of what controls growth in SSF. Emphasis is placed on the mathematical descriptions of the phenomena. Model predictions about system behavior are mentioned but not discussed in detail. This section concentrates on mass transport phenomena because local heat transfer has not been considered in any mathematical model of SSF to date. Indeed, local heat transport is less likely than local mass transfer to limit SSF performance, although it might have unknown significance, for example, through indirect effects on mass transfer coefficients.

Diffusion and Reaction

Early modeling studies relevant to SSF considered the overculture of microorganisms on solid agar media, with glucose as the substrate (33,34). Two growth periods are predicted: an early period during which the glucose concentration at the substrate surface is relatively high and consequently the microbe grows at its maximum specific growth rate (i.e., exponential growth), followed by diffusion-limited growth during which the glucose concentration at the substrate surface is very low and the growth rate depends on the flux of glucose to the surface. The duration of the exponential growth phase depends strongly on the initial concentration of glucose because this determines how long it takes for the surface glucose concentration to fall to limiting levels. Analytical solutions to the diffusion equation can be obtained for short time periods but cannot describe the whole growth period. In any case, the usefulness of these models is limited because relatively few SSF processes contain a soluble sugar as the major carbon source.

Mitchell et al. (6) proposed and solved numerically a model for growth of *R. oligosporus* in overculture with starch as the carbon source. This allows some insight as to how mass transfer within the substrate itself can limit growth. In the experimental system a membrane filter prevented penetration of hyphae into the substrate. Five steps in the growth process were modeled (these correspond to processes 5 to 9 in Fig. 3):

- Glucoamylase release into the substrate by the fungus located in a plane at the substrate surface; an empirical equation was used
- Diffusion of glucoamylase through the substrate according to Fick's law
- Hydrolysis of starch by the glucoamylase according to Michaelis–Menten kinetics
- Generation of glucose by the glucoamylase according to Michaelis–Menten kinetics and the diffusion of the glucose through the substrate; this is a process of simultaneous reaction and diffusion
- The uptake of glucose by the fungus at the substrate surface according to Monod kinetics based on the surface glucose concentration.

The expression of interest here is that for simultaneous reaction and diffusion. For diffusion in an overculture system in which net diffusion occurs only in the vertical direction, as shown in Figure 3, the balance equation for compound A has the general form

$$\frac{\partial A}{\partial t} = r_a + D_a \frac{\partial^2 A}{\partial z^2} \tag{18}$$

where r_a is the rate of generation of compound A, D_a is the diffusivity of A, and z is the vertical spatial coordinate. This equation, with the appropriate sign on the reaction term, can refer to the generation and diffusion of hydrolysis products within the substrate, or to the diffusion and consumption of nutrients within a biomass layer.

Microscopic models incorporating local transport equations are useful tools for understanding the role of local transport phenomena in limiting growth in SSF. The model of Mitchell et al. (6) predicts a short phase of exponential growth during the early period when the glucoamylase is still near the surface and the surface glucose concentration is much higher than the saturation constant for growth. As the biomass at the surface increases, the increased glucose uptake rate causes the surface glucose concentration to fall below the saturation constant. This is followed by a period of slow deceleration of growth. This kinetic profile has been observed in SSF by several workers. The model is able to explain how such kinetics might arise from interactions among the processes of glucoamylase diffusion, starch hydrolysis, and glucose diffusion.

Recently, Rajagopalan and coworkers have made two extensions to this model. Rajagopalan and Modak (16) incorporated an expanding biomass film of constant density, with simultaneous reaction and diffusion of oxygen from the bulk air above the biomass layer and of glucose from the substrate below the biomass layer. This model suggests that at different times during growth either glucose or oxygen may be limiting, but that overall oxygen limitation might be more important than glucose limitation. They ran simulations under different bulk oxygen concentrations, such as might occur in different positions in a bioreactor, foreshadowing the linking of both local and bulk transport equations in a single model. Rajagopalan et al. (35) then further modified the model to describe a decrease in particle size as glucose is used up. Their predictions agreed reasonably well with the experimental data of Nandakumar et al. (20), who measured decreases in particle size during growth of *Bacillus coagulans* on wheat bran particles. The predictions also confirm the observations of Mitchell et al. (6) that glucoamylase diffusion plays a crucial role. At low diffusivities most of the glucoamylase remains near the surface. This quickly depletes the starch at the surface, leaving much of the enzyme without substrate to act upon. As a consequence, the rate of glucose production within the substrate particle falls, and growth becomes limited by the rate of glucose production in the deeper regions of the particles.

These modeling studies have used homogeneous gel substrates with very high water contents in which diffusion is the major mass transport phenomenon. However, SSF substrates may have high solute concentrations and contain cellular structures. Sorption phenomena may restrict diffusion (25), capillary action may occur, and cell walls may present an impermeable barrier. The mechanisms of mass transport within real solid substrates deserve greater attention.

Film Transfer of Oxygen

Another important local transport process is that of diffusion of oxygen across the stagnant gas film into the thin liquid film at the substrate surface (Fig. 3). If the biomass is assumed to be surrounded by a thin liquid film that offers the major resistance to mass transfer, then the oxygen transfer rate, N, can be expressed as (36)

$$N = K_L a(C^* - C) \tag{19}$$

where C^* is the concentration of oxygen in the bulk gas phase, C is the concentration of oxygen at the biomass–liquid interface, a is the area over which mass transfer occurs, and K_L is the mass transfer coefficient. This equation was used to estimate the overall mass transfer coefficient ($K_L a$) for a packed-bed bioreactor. It can be applied also at the microscopic scale, although it can be difficult to measure local oxygen consumption rates and oxygen concentrations (36).

Alternatively, the biomass film can be assumed to be directly in contact with the gas phase, with the transport resistance residing in a stagnant gas film. In this case the variable C in the above equation is replaced by $HC_{O_2,F}$, where H is Henry's constant, and $C_{O_2,F}$ is the concentration of oxygen in the biomass film. This expression has been used to describe oxygen transfer to the biomass in a tray bioreactor (15). Film transfer of oxygen in SSF needs more attention because significant amounts of fungal mycelium are in direct contact with the air phase and are not immersed in the liquid film at the substrate surface (37).

MODELING OF BULK TRANSPORT PHENOMENA

Recognizing Subsystems within SSF Bioreactors

In developing a model of bulk transport phenomena occurring during SSF, we must consider the level of detail to be modeled and identify simplifications of the model structure. As pointed out earlier, at the macroscopic scale, up to four subsystems can be identified in SSF bioreactors: the bioreactor wall, the headspace gases, the solid particles themselves, and the interparticle gases (Fig. 2). The presence and arrangement of these various subsystems and whether the subsystems themselves are homogeneous or have concentration or temperature gradients depend on the particular bioreactor and how it is operated. Macroscopic models contain transport equations to describe the heat and mass transfer processes within and between the subsystems recognized by the model, and between the subsystems and the surroundings.

It is often convenient to ignore the bioreactor wall in order to simplify the model. This may be appropriate if the wall is maintained at a constant temperature by a cooling system, making the bioreactor wall part of the surroundings. Alternatively, if the resistance to heat transfer between the bioreactor wall and the outside air is much greater than the resistance to heat transfer from the substrate material to the bioreactor, the overall heat transfer coefficient can be assumed to be similar to the outside heat transfer coefficient. This also has the implicit assumption that the energy stored in the bioreactor wall is negligible.

It is usually convenient and appropriate in macroscopic models to lump two or more of the phases together as a pseudohomogeneous subsystem having the mass or volume-averaged properties of the individual phases (Table 1). Most commonly, a substrate bed consisting of the moist solid particles and the interparticle gases is defined as a single subsystem. In the pseudohomogeneous matrix that comprises this system, the temperatures and water potentials of the gas and solid at the same location are assumed to be the same. The validity of this assumption depends on whether the bioreactor operation allows the gas and solid to approach equilibrium. In addition, note that the temperature and water potential of this pseudohomogeneous matrix can vary from position to position.

Depending on how well subsystems are mixed, macroscopic models may only need to describe transfer between subsystems, or they may also need to describe transfer within subsystems. If a subsystem is well mixed, then there will be no thermal and concentration gradients across the subsystem, and there will only be transfer equations to describe the transport between subsystems. Therefore the spatial coordinate does not appear in the transfer equations for such subsystems. For subsystems that are static or poorly mixed, there will also be transfer equations to describe the transfer processes within subsystems. For example, equations incorporating the spatial coordinate may be required to describe solid or gas flow patterns in the bioreactor, and to describe temperature and concentration gradients within the substrate bed. Such models can be used to explore bioreactor design and operation strategies to minimize these gradients.

The Bulk Transport Phenomena Occurring in SSF Bioreactors

Through their interactions with local transport processes, bulk transport processes determine the local environment experienced by the microorganism and therefore influence the performance of SSF bioreactors. Macroscopic models predict bioreactor performance by combining empirical descriptions of growth kinetics with expressions describing bulk transport between and within the subsystems (solid, gas, and bioreactor wall), and between these subsystems and the surroundings. Local transfer mechanisms are currently ignored in these models. Momentum transfer, heat transfer, and mass transfer may be involved, and any of these may potentially limit the performance of an SSF bioreactor. Table 2 indicates the mechanisms for these phenomena in various SSF bioreactor types. Of these phenomena, heat transfer has received the most attention.

Conversely, there is very little information regarding momentum transfer processes, that is, those processes involving material flow and that affect mixing and shear phenomena. Within bioreactors, both mixing of gas molecules within the gas phases in the bioreactor and mixing of solid substrate particles may be important. To date, the mixing of solids in agitated SSF systems has not been studied, and there are only two investigations of gas mixing and flow behavior (40,42). Residence time distributions for gas molecules indicate that operational parameters such as the gas velocity and solids loading affect the presence and extent of stagnant gas regions within the substrate bed and the homogeneity of the headspace gas phase (40,42). Further, such investigations are crucial because these flows underlie the other transport phenomena, such as heat and mass transfer. Due to the paucity of information, current macroscopic models assume ideal flow behavior such as plug flow of gases in packed beds or perfect mixing of the solids and gases in rotating drums.

The flow of solids leads to shear forces due to interactions of the solid substrate particles with each other and with the bioreactor wall. These shear forces are complex and therefore difficult to characterize. Thus, despite their potential to affect substrate characteristics and adversely affect microbial growth, and significant speculation regarding their importance in SSF systems, shear forces are not described in any macroscopic model of SSF.

Heat transfer processes have received attention because overheating due to release of metabolic waste heat is a very significant problem in SSF even at small scales. Rates of metabolic heat production reported in SSF range from about 3 to 330 kJ (kg dry matter)$^{-1}$ h^{-1}. Removal of metabolic heat is more difficult in SSF than in liquid fermentation processes for a number of reasons. First, compared to liquid media, solid substrates have low heat capacities and low thermal conductivities, which means that although less heat is required to raise the temperature of the solid substrate by the same increment, the solid substrate loses heat less readily than liquid media. As well, due to the concentrated nature of SSF systems, biomass concentrations are higher and heat production on a volumetric basis is therefore also correspondingly higher. Heat removal in SSF system occurs by convection, conduction,

Table 1. Subsystems in SSF Bioreactors and Their Simplification

Subsystem and references	Detailed view of the system (macroscopic viewpoint)	Simplification of the system appropriate for modeling	Mixing
Trays: may be stirred (the whole room is the bioreactor) (10,15,24)	Headspace gas (flowing around tray)	Headspace gas	Assume headspace infinite and well mixed, such that one tray represents all trays
	Solid particles (static) Interparticle gas (no forced flow)	Assume the bed consists of a pseudohomogeneous matrix	May be intermittently mixed, but usually not appropriate to assume that the bed is well mixed; gradients will exist
Packed beds and similar stirred reactors (3, 38, 39)	Solid particles (mostly static) Interparticle gas (flowing due to movement within bed)	Assume the bed consists of a pseudohomogeneous matrix	If mixed, usually intermittently; therefore will still have end-to-end gradients within the bed
Rotating drums and similar stirred reactors (21, 40)	Headspace gas (flowing) Solid particles (flowing) Interparticle gas (flow depends on movement within bed)	Headspace gas Assume the bed consists of a pseudohomogeneous matrix	Depending on operation may assume these two subsystems are each well mixed, or may allow for axial variation; if well mixed, may assume that headspace and bed are in equilibrium (i.e., 1 subsystem)
Air–solid fluidized beds (growth in this bioreactor has not yet been modeled)	Solid particles (flowing) Interparticle gas (flowing)	Assume a pseudohomogeneous matrix, with mixing keeping the whole bed homogeneous, and that aeration air is moist so that system is close to moisture equilibrium	Typically well mixed

Note: Subsystems are in addition to the bioreactor wall itself.

and evaporative cooling, with the contributions of each mechanism depending on the bioreactor design and operation.

The important bulk mass transfers in SSF bioreactors involve the gas phase: oxygen supply and carbon dioxide removal for aerobic processes, and the removal of water vapor during evaporative cooling. The transfer processes described by macroscopic models may include transfer across interfaces, diffusion within stagnant gas regions, and convection by bulk gas flow. As noted in Table 1, it may be appropriate to divide the gas phase into a headspace gas phase and an interparticle gas phase. The degree of continuity between these phases depends on the type of bioreactor and how it is operated.

Solid-state bioreactors differ in the types of mixing that they provide and in the relative importance of flow and diffusion processes in the bulk transport phenomena that occur. This section discusses mathematical modeling of bulk transport in tray, packed-bed, and rotating drum bioreactors. Models have not been developed to describe transport phenomena within other bioreactor types. The following sections discuss, for each bioreactor, the approaches to modeling the transport phenomena identified in Table 2. Although these transport phenomena have also been studied in non-SSF systems, the discussion is limited to SSF itself.

Macroscopic Transport Phenomena in Tray Bioreactors

Tray bioreactors often consist of a large number of individual trays incubated within a room in which the temperature and the humidity of the air may be controlled. However, mathematical models of trays usually describe only a single tray, assuming homogeneous conditions within the headspace surrounding the multiple trays in the bioreactor. This single tray comprises either an intact or a perforated tray on which rests a thin layer of static solid substrate. Two distinct gas phases exist in tray SSF systems: the first is the interparticle voids within the substrate itself; the second is the bulk gas phase, which consists of the headspaces above and below the tray.

Heat Transfer in Tray Bioreactors. Only Rajagopalan and Modak (15) have described heat transfer during tray SSFs. In their model, sensible heat transfer occurs in the vertical direction by conduction through the substrate,

$$\frac{\partial T}{\partial t} = \frac{k}{\rho_s C_p}\frac{\partial^2 T}{\partial y^2} + \frac{\Delta H r_g}{C_p} \tag{20}$$

and by convection from the substrate surfaces to the headspaces above and below the substrate bed (the base of the tray was assumed to be open),

$$-k\frac{\partial T}{\partial y} = h(T_0 - T) \tag{21}$$

where T is the temperature of the bed; t is time; k, C_p and ρ_s are the thermal conductivity, heat capacity, and density of the bed, respectively; ΔH is the enthalpy of the growth reaction; y is the spatial coordinate; T_0 is the temperature

Table 2. Important Transport Phenomena within SSF Bioreactors

Bioreactor and subsystem	Heat transport phenomena	Mass transport phenomena	Momentum transport phenomena
All bioreactor systems: the microscopic phenomena within the substrate particle are usually greatly simplified in macroscopic models describing bioreactor performance. Also, as noted in Table 1, the solid and gas subsystems are often assumed to form a single pseudohomogeneous matrix.			
Within the solid particle	This is the site of heat generation by the microbe growing at the surface of the solid particles.	Diffusion of enzymes and nutrients within the solid particle (6, 11, 33, 34)	
Between the solid and gas subsystems within the substrate bed	Conduction, evaporation of water, and free or forced convection (depending on bioreactor operation) from solids to interparticle air	Evaporation of water, and transfer of O_2 and CO_2 between the solid particles and the interparticle air (15)	
After assumption of a pseudohomogeneous matrix	Solid and gas in thermal and moisture equilibrium	Transfer of O_2 and CO_2 between solid particles and interparticle air Transfer of water to maintain saturated air (38,39)	
Trays: the solid particles and interparticle air may or may not be considered as a single pseudohomogeneous substrate bed. The whole room is taken as the bioreactor. Individual trays are assumed to be open at the top and to have perforated bases.			
Transfer across the substrate bed	Conduction (15) and natural convection within the bed, leading to spatial gradients	Diffusion (10, 15) and free convection of gases and water vapor within the bed, leading to spatial gradients	
Involving the tray wall or the headspace gases	Conduction to the tray walls, conduction within the tray wall, and then convection between the tray wall and the headspaces Evaporation and free or forced convection (depending on airflow), between the exposed bed surfaces and the headspace gases	Convection of gases (15) from the exposed bed surfaces into the headspace Evaporation of water from the substrate into the headspace at the exposed bed surfaces	Flow of headspace gases around the trays (15)
At inner surface of the room walls	Convection to the tray room walls Condensation on room walls		
	Packed beds and similar stirred reactors, assumed to be vertical cylinders		
Transport through the pseudohomogeneous matrix	Conduction, convection and evaporation axially along the substrate bed (38,39) Conduction radially within the substrate bed (3,38) Both axial and radial gradients may exist (38)	Convection of air in the axial direction due to the forced aeration of the bed. Although axial O_2 gradients occur, they are usually not as important as the temperature gradients Radial mass transfer should be negligible	Gas flow patterns through the bed (plug flow without dispersion is often assumed) (3,38,39) Pressure drop between the bottom and top of the substrate bed (40)
At the inner surface of the bioreactor wall	Conduction radially between the substrate and the inside surface of the bioreactor wall (3,38)		
	Rotating drums and similar stirred reactors, assumed to be horizontal cylindrical drums		
Transport through the pseudohomogeneous matrix	The role of conduction depends on the degree of mixing; may assume that bed is well mixed (21,41) Convective flow within the bed due to mixing Usually well mixed radially, but axial gradients may exist	The role of diffusion depends on the degree of mixing Convection of gases within substrate voids due to mixing Axial O_2 gradients may occur but are usually not important	Mixing and flow of the solid substrate Shear forces; impact and abrasion due to solids interacting with each other

Table 2. Important Transport Phenomena within SSF Bioreactors *(continued)*

Bioreactor and subsystem	Heat transport phenomena	Mass transport phenomena	Momentum transport phenomena
Between the bed and the headspace	Convection between the substrate bed surface and the headspace (41) Evaporative cooling at the surface of the substrate bed (41) Bulk flows of gases between the interparticle voids and the headspace carry sensible energy	Gas and water vapor exchange between the substrate bed and headspace gases (41). This may involve bulk flows exchanging gas between the headspace and interparticle voids.	Bulk gas flows between the interparticle voids and the headspace
Within the headspace	Convective flow of the headspace gases, carrying sensible energy (41)	Convective flow of the headspace gases (41)	Flow patterns of the headspace gases (41)
At the inner surface of the drum wall	Conduction from the substrate to the drum wall and convection between the headspace gases and the inside surface of the drum wall (41)		Shear forces, impact, and abrasion due to solids interacting with the drum wall
Air–solid fluidized beds, assumed to be vertical cylinders. The aeration air is assumed to be moist so that the bed is close to moisture equilibrium. Growth in these bioreactors has not yet been modeled.			
Within the well-mixed substrate bed	Whole bed well mixed such that there are no thermal or moisture gradients	Whole bed well mixed, such that there are no water vapor, CO_2, or O_2 gradients	Pressure drop across the bed itself, fluidization phenomena Shear, impact, and abrasion due to interactions between particles in the bed
At the inner surface of the bioreactor wall	Convection from the bed to the bioreactor wall		Shear, impact, and abrasion due to interactions between the solid particles and the wall
All bioreactors			
Involving the entry and exit of the airstream	Energy enters and leaves the system in the airstream blown through the bioreactor (3,21,38,41).	Gases and water vapor enter and leave the system in the airstream blown through the bioreactor (21,41)	Pressure drop between air pump and air exhaust
Within the bioreactor wall and between the wall and the surroundings	Conduction through the wall (the room wall for the tray system) and loss to the surroundings, often by free convection to air (21,41) or forced convection to cooling water (3,38)		

Note: These are phenomena that should be considered in the construction of mathematical models of bioreactor performance. As discussed in the text, most models greatly simplify the system by including only the most important transport phenomena or lumping phenomena together. In this table not all phenomena have been mentioned. In some places where transport processes occur in parallel, minor contributors have been ignored. In addition, the importance of the transport phenomena mentioned will depend on how the particular bioreactor is designed and operated. After identifying the phenomena to be modeled, the next task in modeling is to find appropriate expressions to include in the mass and energy balances that describe the system. Expressions that have been used to date in macroscopic models of SSF bioreactors are discussed in the text. Numbers in parentheses indicate references.

of the surrounding air, and h is the heat transfer coefficient between the bed and the surroundings. The model predicted that substrate temperatures would be highest in the center of the tray, and that the incubation temperature and substrate depth were critical to the success of tray SSFs. For a tray bed 6.35 cm high, an incubation temperature of 30 °C produced the greatest total biomass, although the maximum specific growth rate occurred at 40 °C. Likewise, substrate depths of 3 cm and under gave significantly better growth when incubated at 38 °C than when incubated at 40 °C. Simulations involving simultaneous changes of incubation temperature and substrate depth indicate that, as the bed height decreases, the incubation temperature supporting optimal growth increases and overall growth is optimal in trays less than 1 cm high. Clearly, such shallow trays would not be practical at large scales. Gas flow velocity around the trays had little effect on heat transfer between the gas phase and the bed. This model did not describe evaporative cooling, convective heat transfer through the substrate bed due to free convection, or heat removal through the tray material.

Mass Transfer in Tray Bioreactors. A number of workers (10,15,24) have modeled mass transfer in trays. All of these models describe mass transfer within the interparticle voids; however, only the model of Rajagopalan and Modak (15) describes mass exchange between the interparticle voids and the bulk headspace.

In describing mass transfer within the interparticle voids, all three models consider transfer by diffusion in the vertical direction only. The models consider transfer of either oxygen only (10) or both oxygen and carbon dioxide (15,24). None of the models describes the mass transfer of water. All three models neglect the possibility of convective mass transfer due to thermal gradients within the substrate. The equations used in these models for mass transfer within the substrate bed take the general form of a balance of compound A (either O_2 or CO_2) in the interparticle gas phase,

$$\frac{\partial \varepsilon C_a}{\partial t} = D_a \frac{\partial^2 C_a}{\partial y^2} - R_a \quad (22)$$

where t is time, y is vertical distance within the bed, ε is the porosity, and D_a and C_a are the diffusivity and the concentration, respectively, of A in the bed. In the case where A is oxygen, R_a is the rate of oxygen consumption (10,24) or the rate of oxygen dissolution into a liquid film layer on the substrate particles (15). For film transfer, $R_a = K_a(C_{O_2} - HC_{O_2,F})$, where K_a is an overall gas side mass transfer coefficient, H is Henry's constant for oxygen and $C_{O_2,F}$ is the oxygen concentration in the film liquid. In the case where A is carbon dioxide, the minus sign in front of R_a in the previous equation is changed to a plus sign, and R_a is the rate of carbon dioxide production (15,24).

The porosity in the previous equation is an important term and received further attention within the models of Rajagopalan and Modak (15) and Auria et al. (24). Rajagopalan and Modak (15) calculated D for the oxygen and carbon dioxide gases as

$$D_i^b = \frac{\varepsilon}{\tau} D_i' \quad (23)$$

where ε is the porosity of the substrate bed, τ is the tortuosity factor (taken to equal 1.7), i is the gas (either carbon dioxide or oxygen), and D^b and D' are the diffusivities in the bed and in air. However, the porosity was held constant for any one simulation; therefore porosity changes caused by growth were neglected. Auria et al. (24) used their model to estimate the diffusivity within static beds consisting of *A. niger* growing on beads of ion exchange resin impregnated with a nutrient medium. The diffusivity decreased with increasing biomass concentration, and at high biomass concentrations of 27 mg g^{-1} dry matter, the diffusivities of oxygen and carbon dioxide were less than 5% of their bulk diffusivities in air. The effect of growth on porosity and gas diffusivity deserves further attention.

The tray models of Auria and Rajagopalan are dynamic models. The model developed by Ragheva Rao et al. (10) is different in that it consists of only a single pseudo-steady-state equation. From a balance for oxygen within the bed, they developed an equation for the maximum thickness, H_c, that a tray could be without becoming depleted of oxygen at any depth within the substrate layer during the fermentation,

$$H_c = \sqrt{\frac{2D_e C_{O_2} Y}{\mu_m X_m}} \quad (24)$$

where μ_m and X_m are the maximum specific growth rate (h^{-1}) and maximum biomass concentration (mg cm^{-3}), respectively. Use of μ_m and X_m represents the worst case scenario for which the rate of oxygen demand would be greatest. This model has a number of implications for the design and operation of tray SSFs. First, with the use of perforated rather than unperforated trays, the critical bed thickness, H_c, could be increased by a factor of up to two, depending on the fraction of the area of the underside of the bed exposed to air. Second, greater substrate depths can be used if the air above the bed is oxygen enriched or if the headspace is continually replenished during operation (thereby keeping C_{O_2} as high as possible). This model is one that could be readily applied in practice, requiring only values of the D and kinetic parameters μ_m and X_m, although it has not yet been tested experimentally.

The model of Rajagopalan and Modak (15) goes further than the other models in not only describing diffusion within the bed, but also the convection of gases from the upper and lower bed surfaces into the headspaces. Headspace air is assumed to move horizontally across the substrate surface, being well mixed normal to the flow direction but with no axial dispersion. At the plane of contact with the substrate, gases diffuse into or out of the bed. The balance of oxygen or carbon dioxide at the bed surface is given by

$$\frac{\partial C_a^h}{\partial t} = -v_x \frac{\partial C_a^h}{\partial x} - \frac{D_a}{z} \frac{\partial C}{\partial y}\bigg|_{y=L} \quad (25)$$

where A is either carbon dioxide or oxygen, x is the horizontal coordinate, y is the vertical coordinate, L is the thickness of the substrate, z is the thickness of the headspace above the bed, C_a^h is the headspace concentration, and v_x is the headspace gas flow velocity. Model simulations showed that substrate depth and porosity greatly affected mass transfer within the bed; however, the gas flow velocity around the trays had little effect on mass transfer between the headspaces and the bed. These predictions can guide design and operation of tray bioreactors.

Combined Heat and Mass Transfer in Tray Bioreactors. Finally, both heat and mass transfer processes may be important at different times during SSFs in trays. Therefore, mathematical models describing these systems will need to describe both transport mechanisms. Currently, only the model by Rajagopalan and Modak (15) does this. The phenomena described by their model are shown in Figure 4.

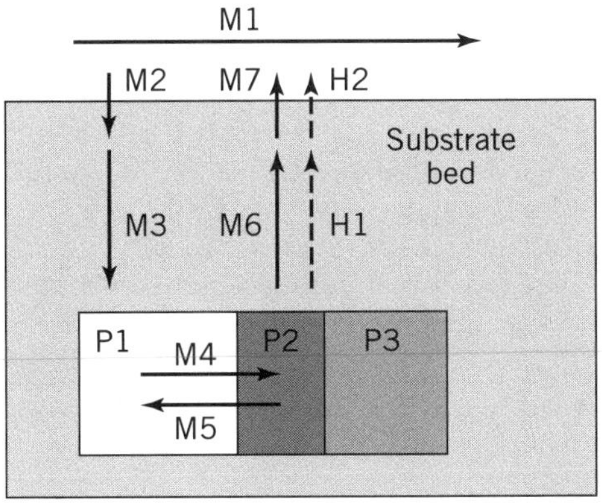

Figure 4. Diagrammatic representation of the tray bioreactor as modeled by Rajagopalan and Modak (15), indicating the three phases within the substrate bed. P1, Interparticle gas phase; P2, thin water film on the surface of substrate particles, which is the location of oxygen consumption, carbon dioxide production, and generation of waste metabolic heat; P3, solid substrate particle. The heat and mass transfer processes occurring are M1, bulk flow of air above the tray surface; M2, transfer of oxygen into the bed; M3, diffusion of oxygen within the bed; M4, transfer of oxygen from the interparticle air phase to the liquid film; M5, transfer of carbon dioxide from the liquid film to the interparticle air phase; M6, diffusion of carbon dioxide within the bed; M7, transfer of carbon dioxide from the bed to the surrounding air; H1, conduction of heat through the bed; H2, convective removal of heat at the bed surface.

The predictions of this model for substrate temperature, biomass, and gas phase concentrations of oxygen and carbon dioxide showed trends similar to those noted experimentally by Rathbun and Shuler (43) during the tempe fermentation of soybeans by *R. oligosporus*. The model predicts that the substrate temperature is the dominant influence on growth in the initial and final stages of the fermentation. During intermediate stages of the fermentation, oxygen has a large effect at the bottom of the bed (in an unperforated tray), whereas temperature effects are more important in the center of the bed.

Macroscopic Transport Phenomena in Packed-Bed Bioreactors

Heat Transfer in Packed-Bed Bioreactors. Heat transfer in packed beds has received most modeling attention because the temperature gradients obtained are more important than the gas concentration gradients, and if aeration satisfies the heat removal requirements it will also satisfy aeration requirements (44,45). In bioreactors of different designs and operated under different conditions, both axial and radial temperature gradients have been observed (45–47). Under some operating conditions, temperatures in the middle of the bed can be over 20 °C higher than the temperature of the inlet air, leading to large variations in growth and product formation across the bioreactor (46).

Saucedo-Castaneda et al. in 1990 (3) modeled radial heat transfer in an experimental packed bed in which *A. niger* was grown on cassava meal. This model was later extended by Sangsurasak to describe both axial and radial heat transfer (38). A general expression for heat transfer in a pseudohomogeneous substrate bed within a cylindrical packed-bed bioreactor, which takes into account the heat transfer processes shown in Figure 5, can be expressed as (38)

$$\rho_b C_{pb} \frac{\partial T}{\partial t} + (\rho_{ma} C_{pma} + \rho_a f \Delta H_v) V_z \frac{\partial T}{\partial z} = \frac{k_b}{r}\frac{\partial T}{\partial r} + k_b \frac{\partial^2 T}{\partial r^2} + k_b \frac{\partial^2 T}{\partial z^2} + r_Q \qquad (26)$$

where ρ_a, ρ_{ma}, and ρ_b are the densities of air, moist air, and the bed, respectively; C_{pma} is the heat capacity of moist air; C_{pb} the heat capacity of the bed; f describes the relationship between saturation humidity and temperature; ΔH_v is the enthalpy of vaporization of water; k_b is the thermal conductivity of the bed; V_z is the superficial air velocity; z and r are the axial and radial spatial coordinates; and r_Q is the rate of heat generation from the growth reaction. The factor $\rho_a f \lambda$ arises because the equilibrium assumed between the substrate and air means that the evaporation of water to keep the air saturated as the air temperature increases gives the air a higher apparent heat capacity. This is a simplified approach to handling evaporative cooling,

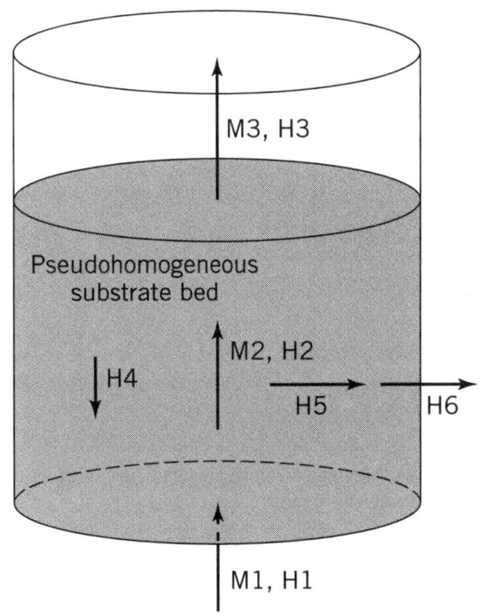

Figure 5. Diagrammatic representation of the packed-bed bioreactor as modeled by Sangsurasak (38). The substrate bed is treated as a single pseudohomogeneous phase. The microbe grows in the bed and releases waste metabolic heat. The heat and mass transfer processes occurring are M1, flow of moist air into the pseudohomogeneous bed; M2, flow of moist air within the bed, with the air being saturated at the bed temperature; M3, flow of moist air out of the pseudohomogeneous bed; H1, bulk transfer of energy into bioreactor with airflow; H2, convective heat transfer within the bed; H3, bulk transfer of energy out of bioreactor with airflow; H4, axial conduction within the bed (opposes the temperature gradient caused by convection); H5, radial conduction within the bed; H6, conduction across the bioreactor wall and convective removal by cooling water.

which can contribute as much as 65% of the heat removal from a packed bed (47).

For thin packed-bed bioreactors of about 6 cm diameter, operated with low superficial air velocities (e.g., 0.01 m s^{-1}), radial conduction is the major heat removal mechanism (3). In this particular case the terms for convection and evaporation and axial conduction can be omitted.

In addition to transfer phenomena within the bed, the other important transfer process is from the outer edge of the substrate bed to the surroundings (the bioreactor wall itself has been assumed to offer negligible resistance to heat transfer). Convective cooling at this boundary has been assumed (3,38):

$$\left.\frac{\partial T}{\partial r}\right|_{\text{wall}} = \frac{ha}{k_{\text{b}}}(T_{\text{surr}} - T_{\text{wall}}) \quad (27)$$

Overheating problems occur at the top of the column when it is aerated from the bottom because the air reaching higher regions has been prewarmed in the lower regions and therefore does not cool the top as effectively. Even for quite small columns, when they are operated with the cross-sectional airflow rates of 0.01 to 0.025 m s^{-1} that are commonly used in packed-bed bioreactors, temperatures that can cause significant microbial death are predicted. The model of Sangsurasak (38) can be used as a tool in bioreactor design, operation, and scale-up. Sangsurasak used the model to explore strategies for minimizing the overheating problem at large scale. Higher aeration rates and lower height-to-diameter ratios appear to be the best strategies. However, the model has not yet been used in the actual construction of a larger-scale packed bed.

Gutierrez-Rojas et al. (17) used a different approach to modeling heat transfer during growth of *A. niger* in a packed bed on beads of ion exchange resin impregnated with nutrient medium. They defined an "elementary representative volume" of 20 mL, over which scale variations in the process variables were assumed to be negligible. However, they modeled only an individual elementary representative volume, and although their model was able to describe the experimental results obtained in a 20-mL packed bed, it is not able to describe the gas concentration and temperature profiles that occur across larger-scale packed-bed bioreactors (46).

Heat and Mass Transfer in Packed-Bed Bioreactors. The reliance on evaporative cooling means that water balances over the bed are important (45). A combined heat and mass transfer model is required to take these into account. Such a model, considering only gradients in the vertical direction, was proposed by Van Lier et al. (39), who modeled the composting of a mixture of wheat straw and horse manure. The energy balance is similar to that used by Sangsurasak (38), but because a material balance was also done for water, a simplified description of evaporation was not required. The equation was of the form

$$\rho_{\text{b}}C_{\text{pb}}\frac{\partial T}{\partial t} + \rho_{\text{ma}}C_{\text{pma}}V_{\text{z}}\frac{\partial T}{\partial z} = k_{\text{b}}\frac{\partial^2 T}{\partial z^2} - h(T - T_0) - \Delta H_{\text{v}}r_{\text{w}} + r_{\text{Q}} \quad (28)$$

where h is the heat transfer coefficient for conduction through the side wall to the surroundings at temperature T_0, and r_{W} is the rate of evaporation of water from the solid phase into the gas phase.

A general mass balance equation for compound A in the gas phase, for example, oxygen or water, can be written in the following form with terms for axial convection, axial diffusion, and exchange with the metabolizing organism attached to the solid phase:

$$\frac{\partial(\varepsilon C_{\text{A}})}{\partial t} = -V_{\text{Z}}\frac{\partial C_{\text{A}}}{\partial z} + \frac{\varepsilon}{\beta^2}D_{\text{A}}\frac{\partial^2 C_{\text{A}}}{\partial z^2} + r_{\text{A}} \quad (29)$$

where C_{A} is the concentration of A at a location, D_{A} is the diffusivity of A in air, ϵ is the bed void fraction, β is the tortuosity and r_{A} is the rate of exchange of A with the solid phase. In the water balance Van Lier et al. assumed that the vapor phase remained saturated with water at the local temperature in the bed, and used this equation to calculate the evaporation rate (r_{W}). Film transfer of oxygen was not incorporated into the equations.

Solution of the equation set was complicated because the porosity and height of the bed changed during the composting process. However, good agreement was achieved between the model predictions and measured temperatures and bed heights. In composting, unlike other SSF processes, some self-heating is desired, and this reduces problems with overheating. Despite this, the aeration rate required to remove excess heat is still about 10-fold greater than that required to supply oxygen (39).

Other Phenomena in Packed-Bed Bioreactors. As air flows through the porous bed in a packed-bed bioreactor, there are viscous and kinetic phenomena (5). Substantial pressure drops can occur in packed-bed bioreactors (5,18,41). To date characterization of this phenomenon has been limited to demonstrating the relationship between pressure drop and the bed void fraction, which decreases as the microbe fills in the interparticle spaces. At high biomass densities, the microorganism can occupy up to 34% of the free interparticular space, reducing the relative permeability of the bed to around 1% of the original value (40). In the future, such considerations could be built into momentum balances to describe how the flow through the bed changes during the fermentation.

Macroscopic Transport Phenomena in Rotating-Drum Bioreactors

Only heat transfer phenomena (and the related mass transfer of water) have been modeled in rotating-drum bioreactors and the related rocking-drum bioreactor (RDBs). As in packed-bed systems, it is likely that in order to achieve adequate heat removal in RDBs, aeration requirements will already be met. Neither of the models developed so far (21,40) describes all of the heat transfer processes mentioned in Table 2.

Sargantanis et al. (21) modeled heat transfer during growth of *R. oligosporus* on corn grit in a rocking-drum bioreactor with a 1.2-L working volume. They described heat removal very simplistically by assuming equilibrium between the headspace and the substrate bed; in essence, this was a single homogeneous subsystem. The model was useful for determining better operation of their rocking-drum bioreactor system; however, it is not very versatile and can not readily be adapted to describe larger-scale SSFs in RDBs, where the assumption of thermal and moisture equilibrium between the solid substrate and the headspace could be unreasonable, particularly if high aeration rates are used. For such cases a model predicting heat removal in terms of heat transfer rather than equilibrium relationships is likely to be more useful.

Stuart (41) modeled heat transfer in a 20-L rotating-drum bioreactor in which *A. oryzae* was grown on both an artificial gel substrate and on wheat bran. The model is more mechanistic than that of Sargantanis et al. (21) and considers several of the heat transfer processes listed in Table 2. As shown in Figure 6, it simplifies the bioreactor system into three subsystems: the two internal subsystems, the fermenting substrate and the headspace gas, which are contained by the third subsystem, the fermenter wall. Each subsystem is assumed to be homogeneous with respect to composition and temperature in both the radial and axial directions. The model, therefore, does not describe heat transfer processes within any of the subsystems, and the temperature of each subsystem is represented by a single time-dependent value. Stuart (41) neglected water mass transfer within the substrate bed by assuming that water at 25 °C is added to maintain the water activity within this pseudohomogeneous subsystem close to unity. Mass transfer from this bed to the headspace gas is assumed to be limited by the rate at which water vapor in the gas sublayer of the substrate bed diffuses into the bulk headspace gas.

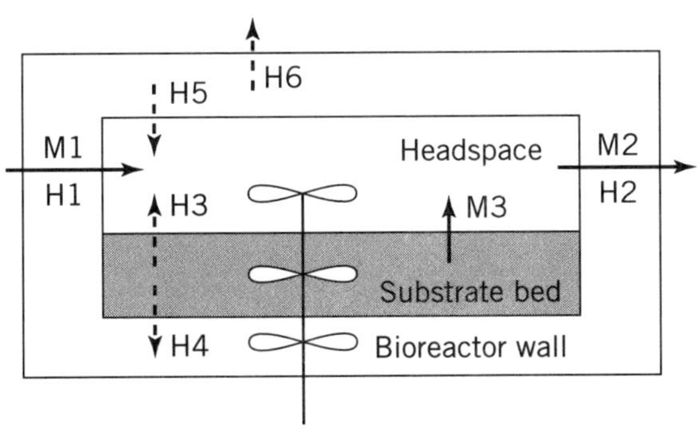

Figure 6. Diagrammatic representation of the rotating-drum bioreactor as modeled by Stuart (41), showing the three subsystems that were assumed. The pseudohomogeneous substrate bed is the site of generation of water and waste metabolic heat during growth. The mixer indicates that each subsystem is assumed to be at the same temperature at all locations within the subsystem. The heat and mass transfer processes occurring are M1, flow of moist air into the headspace; M2, flow of moist air out of the headspace; M3, transfer of water from the substrate bed to the bioreactor headspace; H1, bulk transfer of energy into bioreactor with airflow; H2, bulk transfer of energy out of bioreactor with airflow; H3, convection from the substrate bed to the headspace; H4, conduction from the substrate bed to the bioreactor wall; H5, convection from the bioreactor wall to the headspace; H6, convection from the bioreactor wall to the surroundings.

Three energy balance equations, one for each subsystem, describe heat transfer phenomena within the RDB system, and heat exchange between the RDB system and the surroundings (41). Metabolic heat is generated within the substrate and then dissipated by conduction to the bioreactor wall and by convection and evaporative cooling to the headspace gases. The overall energy balance for the substrate is

$$\begin{aligned}\frac{d[T_sM(C_{pm} + C_{pw}W)]}{dt} = {} & r_Q - h_{sf}A_{sf}(T_s - T_f) \\ & - h_{sa}A_{sa}(T_s - T_a) \\ & - (1.2 \times 25)C_{pw}(W_i - W)M \\ & - kA_{sa}(C_I - C_B)[T_sC_{pw} + \Delta H_v \\ & - (T_s - T_a)C_{ph}] \qquad (30)\end{aligned}$$

Energy enters the headspace gas with the evaporated water from the substrate and with incoming aeration gases, and through convection from the substrate and from the bioreactor wall:

$$\begin{aligned}\frac{d[T_aG(C_{pg} + C_{ph}H)]}{dt} = {} & T_iF_i(C_{pg} + C_{ph}H_i) \\ & + kA_{sa}(C_I - C_B)T_aC_{ph} \\ & - T_aF_o(C_{pg} + C_{ph}H) \\ & + h_{sa}A_{sa}(T_s - T_a) \\ & + h_{fa}A_{fa}(T_f - T_a) \qquad (31)\end{aligned}$$

The exiting air flow carries energy out of the headspace and thereby away from the RDB system. During fermentation, the bioreactor wall heats up due to its contact with the substrate. Energy, when it leaves the bioreactor wall, passes either to the surroundings or to the headspace gas inside the RDB:

$$\begin{aligned}\frac{d(T_fV_f\rho_fC_{pf})}{dt} = {} & h_{sf}A_{sf}(T_s - T_f) - h_{fa}A_{fa}(T_f - T_a) \\ & - h_{fe}A_{fe}(T_f - T_e) \qquad (32)\end{aligned}$$

Thus there are two overall routes by which metabolic heat is removed from the substrate subsystem to the surroundings: with the exiting air stream or as heat lost through the bioreactor wall.

In these equations, T is temperature, M is the mass of dry solid, W is the mass of water per mass of dry solid, G is the mass of dry gas, H is the mass of water per mass of dry gas, V_f and ρ_f are the volume and density of the bioreactor material, respectively, C_P is a heat capacity, r_Q is the rate of production of metabolic heat, h is a heat transfer coefficient, A is an area of heat or mass transfer, k is a mass transfer coefficient, C is a concentration of water at either the interface (subscript I) or within the headspace bulk (subscript B), ΔH_v is the heat of vaporization of water, F is a flow of dry gas either entering (i) or leaving (o) the bioreactor system during aeration. The subscripts S, A, and F denote the three subsystems: the substrate, headspace gas, and fermenter wall, respectively. Combinations of these subsystem subscripts indicate heat or mass transfer between these materials, for example, the subscript combination "SF" represents heat transfer from the substrate to the bioreactor wall. The other subscript terms denote dry solid matter (m), dry gases (g), liquid water (w), water vapor (h), and the external surroundings (e).

Estimation of the coefficients and areas involved in the heat and mass transfer processes during SSF in RDBs is difficult, Stuart (41) used empirical equations to estimate transfer coefficients and assumed that the substrate and headspace gas subsystems occupy constant volumes throughout the fermentation. Likewise, transfer areas were assumed to be constant and were calculated based on the interfacial areas between the substrate, headspace gas, and the bioreactor wall, when the surface of the substrate bed forms a flat plane. Heat and mass transfer coefficients and interfacial areas of contact urgently require experimental attention in SSF systems.

In order to incorporate evaporative cooling into a model of heat transfer during SSF, water balances must be determined over the substrate and headspace gas subsystems. Stuart (41) assumed that the substrate and headspace gas subsystems each consist of a water component and a dry component. Therefore four material balance equations were written, two equations to describe each subsystem.

- Balance of dry matter in the substrate:

$$\frac{dM}{dt} = r_m \tag{33}$$

- Balance of liquid water in the substrate:

$$\frac{d(MW)}{dt} = -kA_{sa}(C_I - C_B) + r_w + 1.2(W_i - W)M \tag{34}$$

- Balance of dry gas in the headspace gas subsystem:

$$\frac{dG}{dt} = F_i - F_o + r_{dg} \tag{35}$$

- Balance of water vapor in the headspace gas subsystem:

$$\frac{d(GH)}{dt} = F_iH_i - F_oH_o + kA_{sa}(C_I - C_B) \tag{36}$$

In these equations, r_m, r_w, and r_{dg} are the net rates due to microbial metabolism of substrate dry matter consumption, water production, and dry gas production, respectively.

Due to consumption of substrate dry matter during microbial growth in SSF and loss of water from the substrate to the headspace gas, the flow of gases out of the RDB ($F_o[1 + H]$) will often be greater than the air flow into the RDB ($F_i[1 + H_i]$). Because RDBs are open systems and large pressure variations in the headspace are unlikely, ideal gas behavior and a constant pressure within the headspace gas are assumed. This gives an equation (not shown) for the flow of gas out of the bioreactor (F_o) and enables the calculation of C_B (the headspace concentration of water vapor).

This model was used to explore SSF performance in RDBs over a range of scales, operations, and designs. The model predictions of temperature profiles for the substrate, bioreactor wall, and headspace gas agreed reasonably well with experimental results for SSFs in real small-scale RDB systems (41). Model simulations indicate that, whereas at small scales heat removal from the substrate bed through the bioreactor wall is most important, the required contribution of evaporative cooling increases with scale. This requires very high aeration rates with relatively dry air: simulations showed that the aeration rate in air volumes per bioreactor volume per minute (vvm) required to achieve adequate heat removal from the substrate might increase by a factor of 25 over a scale change of 1,000. Such high aeration rates would be costly. Therefore heat transfer through the bioreactor wall should be maximized to minimize reliance on evaporative cooling and to limit the amount of water that needs to be added to the system during the fermentation. At all scales, heat removal from the substrate to the headspace gases by sensible heat transfer was insignificant compared with heat transport from the substrate subsystem by conduction and evaporative cooling. However, sensible heat transfer to the headspace gases was always indirectly important in the model predictions because it affected the temperature of the headspace gases and thereby determined the amount of moisture that the headspace could hold and the driving force for evaporative cooling. This investigation of RDB performance demonstrates how models can be used as tools to explore interactions between important processes occurring during SSFs as well as operation and control options for the design of larger-scale bioreactors. This model, with its mechanistic basis, represents a first step toward developing semifundamental scale-up criteria for SSF in RDBs.

PARAMETER ESTIMATION

As highlighted in the introduction to this article, much more attention needs to be given to the estimation of the parameters involved in models of SSF systems. Models involve four main types of parameters: operational and design parameters, microbial parameters, transport parameters, and thermodynamic parameters. In the modeling work to date, many of these parameters have not been independently determined. In some cases parameters have been allowed to vary to enable the model to fit the experimental data. In other cases parameters have been borrowed from other systems involving different substrates and microorganisms, and even from work done in submerged liquid culture. Even physical parameters may depend on the substrate and its preparation and the bioreactor and how it is operated, making it important for parameters to be measured in the actual system studied. This section describes the work that has been done to determine model parameters in SSF systems.

Operational and Design Parameters

Operational and design parameters come directly from the experimental system being modeled. They include bioreactor geometry, air temperatures and flow rates, and rotation or mixing rates. These parameters are important in optimizing bioreactor performance, but these considerations are beyond the scope of this article. For modeling purposes, these parameters are easy to obtain because they are determined by the operator.

Microbial Parameters

Microbial parameters include specific growth and death rates, saturation constants, yields, activation energies, lag periods, and biomass composition. Ideally these parameters should be measured in SSF systems, although this is difficult for those parameters requiring direct measurement of biomass because it is usually impossible to remove the biomass from the substrate particles. Parameters such as specific growth rates and yields are often based on indirect indicators of growth, such as oxygen uptake or protein content, but the relationship of these indicators to growth can often vary during the growth phase. In other cases, parameter values determined in liquid culture have been used, but due to the different physical environment in SSF, these values may not be appropriate for SSF.

Specific growth rates depend on the particular system used. A wide range of values have been reported, from very low values of 0.05 h^{-1} to values as high as 0.5 h^{-1} (45). In some modeling work the specific growth rate has been used as one of the fitting parameters (3,6). The stoichiometry of growth will also be highly variable, and most models consider overall yields and do not balance elements. Only Larroche and Gros (22) have done detailed stoichiometric analyses (for growth and sporulation of *Penicillium roqueforti* on buckwheat seeds). For the anabolic reactions during biomass production from starch they obtained

$$1.104CH_{1.667}O_{0.833} + 0.087NH_3 \rightarrow CH_{1.882}O_{0.603}N_{0.087} + 0.014CO_2 + 0.109H_2O \quad (37)$$

Because a range of different empirical equations have been used to describe the effect of temperature on growth, the values of the parameters do not have general significance. In any case, these parameters are typically determined by incubating cultures at different temperatures, but with any one culture being held isothermally throughout the whole growth phase. The models of Stuart (41) and Saucedo-Castaneda et al. (3) suggest that the relationships obtained from such experiments may not describe the situation in SSF where the organism is subjected to high temperatures after an initial period at the optimum temperature for growth, because their models could describe their data only if the specific growth rate was maintained unchanged at the value for the optimum temperature as the temperature increased.

Transport Parameters

Transport parameters include heat and mass transfer coefficients, diffusivities, and thermal conductivities. Relatively few efforts have been made to determine these in SSF systems. With respect to mass transfer, Auria et al. (24) estimated the diffusivities of oxygen and carbon dioxide within substrate beds as 2 to 25% of their diffusivities in air, with the values decreasing with increasing biomass content. For transfer of oxygen into the moisture films on substrate surfaces, values of K_La have been estimated by a sulfite oxidation method to range from 1,140 to 3,050 h^{-1}, with the value increasing with increasing airflow rate (48). For nutrient diffusion within particles, Mitchell et al. (6) estimated the diffusivity of glucose at 37 °C as 8.2×10^{-3} $cm^2\ h^{-1}$ compared with a bulk diffusivity in water of 2.5×10^{-2} h^{-1}. However, because they used an artificial gel substrate, this value is not of general applicability. They were not able to measure glucoamylase diffusivity using their experimental system. Diffusion has not been investigated for real SSF substrates containing structures such as fibers and cell walls.

With respect to heat transfer, coefficients for heat transfer between subsystems in SSF bioreactors have typically been estimated from literature values for non-SSF systems. However, because these coefficients can depend strongly on bioreactor design and operation, they need to be determined experimentally for the particular system used. Thermal conductivities of composts varied from 0.25 $W\ m^{-1}\ K^{-1}$ at 20% moisture content to 1.0 $W\ m^{-1}\ K^{-1}$ at 70% moisture content (39).

Thermodynamic and Physical Parameters

Thermodynamic parameters include enthalpies of vaporization and saturation humidities that are usually straightforward to obtain from thermodynamic databases. However, it may be difficult to estimate other parameters such as the heat capacities of substrates with complex compositions, and therefore experimental measurements should be made. Physical parameters such as substrate densities and bulk packing densities are readily measured experimentally. They vary widely between different systems.

SUMMARY AND NEEDS FOR THE FUTURE

Growth in SSF results from the interaction of three types of phenomena: microbial growth, local transport, and bulk transport. This article has surveyed the ways in which these phenomena have been described in mathematical models of SSF processes. In these models, many simplifications are made, reducing both the complexity of the system and the complexity of the processes occurring within the system. Despite these simplifications and assumptions, the usefulness of the models has been identified. Microscopic models give insights into how diffusion of enzymes, their hydrolysis products, and oxygen can limit growth. Macroscopic models give insights into how interparticle oxygen transport, and more importantly, how heat transport can limit growth within bioreactors.

SSF and traditional liquid culture systems are quite different: compared with liquid culture systems, in SSF systems water availability is more restricted, and diffusional limitations play a greater role. It is these differences that can give SSF advantages over liquid culture in certain cases; the unique SSF environment can stimulate certain products, and the low water content can lead to higher concentrations and therefore higher volumetric productivities and reduced recovery costs. These differences also lead to some differences in approaches to modelling SSF and liquid culture. Due to the heterogeneity of SSF systems, differential equations must often be formulated and solved for each of several subsystems, and due to the importance of transport processes, many of the equations involve partial differential terms. Also, due to diffusional limitations within substrate particles, it is not simple to measure the substrate concentration experienced by the microbe. For this reason, growth kinetic expressions for the majority of models in SSF will continue to be substrate independent. This is in contrast to liquid culture, where even simple models usually take substrate concentration into account, often assuming a Monod-type relationship. However, in the future, some models will be developed that integrate the microscopic- and macroscopic-scale processes occurring within SSF. These models will give interesting insights into how the interactions among the various phenomena control growth at various stages during SSF.

More work needs to be done in establishing the validity and flexibility of current models. Due to the difficulties of measuring key process parameters in SSF, many of the parameters used in these models have been estimated, or borrowed from other systems. Therefore, more effort is required to measure model parameters in SSF systems. More attention must be paid to microbial physiology. Current expressions describing the effect of elevated temperatures on microbial growth are based on experiments where a number of cultures have been incubated isothermally at a range of temperatures. Better expressions will be developed from experiments that mimic the temporal temperature profiles characteristic of SSF systems. In addition, product formation is usually the main objective of an SSF process. Accurate description of product formation kinetics will require an understanding of how the SSF environment influences microbial product formation.

Our current ability to apply SSF technology at large scale is limited by our relatively poor knowledge of the basic engineering principles underlying SSF bioreactor performance. Rational rules for the design and operation of large-scale SSF bioreactors are crucial to the future commercial success of SSF technology, and mathematical models of bioreactor performance will be essential tools in the development of such rules. The power of the modeling approach is obvious. Modeling is a more cost-effective way of searching for optimal bioreactor designs than building large bioreactors and encountering problems experimentally. It does not replace the need to experiment, but can reduce the experimental program by avoiding poor designs and can guide the program by raising questions about the dominant mechanisms and the key process variables. However, because the models to date have been based on laboratory-scale systems, work needs to be done to confirm that the mechanisms incorporated into the models remain dominant at large scale. Several of the bioreactor designs, such as stirred bioreactors and air solid fluidized beds, have not received any modeling attention. Air solid fluidized beds have the potential to overcome many of the problems faced in other SSF bioreactors, and therefore urgently need modeling attention.

BIBLIOGRAPHY

1. R.E. Mudgett, in A.L. Demain and N.A. Solomon eds., *Manual of Industrial Microbiology and Biotechnology*, American Society for Microbiology, Washington, D.C., 1986, pp. 66–83.
2. A. Pandey, *Process Biochem.* **27**, 109–117 (1992).
3. G. Saucedo-Casteneda, M. Gutierrez-Rojas, G. Bacquet, M. Raimbault, and G. Viniegra-Gonzalez, *Biotechnol. Bioeng.* **35**, 802–808 (1990).
4. M.A. Aufeuvre and M. Raimbault, *Comptes Rendus des Seances de L'Academie des Sciences, Serie III Sciences de la Vie* **294**, 949–956 (1982).
5. R. Auria, M. Morales, M., E. Villegas, and S. Revah, *Biotechnol. Bioeng.* **41**, 1007–1013 (1993).
6. D.A. Mitchell, D.D. Do, P.F. Greenfield, and H.W. Doelle, *Biotechnol. Bioeng.* **38**, 353–362 (1991).
7. Sangsurasak, M. Nopharatana, and D.A. Mitchell, *J. Sci. Ind. Res.* **55**, 333–342 (1996).
8. G. Viniegra-Gonzalez, G. Saucedo-Castaneda, F. Lopez-Isunza, and E. Favela-Torres, *Biotechnol. Bioeng.* **42**, 1–10 (1993).
9. G. Viniegra-Gonzalez, C.P. Larralde-Corona, and F. Lopez-Isunza, *Adv. Bioprocess Eng.* **11**, 183–189 (1994).
10. K.S.M.S. Ragheva Rao, M.K. Gowthaman, N.P. Ghildyal, and N.G. Karanth, *Bioprocess Eng.* **8**, 255–262 (1993).
11. P. Sangsurasak and D.A. Mitchell, *J. Chem. Technol. Biotechnol.* **64**, 253–260 (1995).
12. K.W. Szewczyk and L. Myszka, *Bioprocess Eng.* **10**, 123–126 (1994).
13. R.E. Muck, R.E. Pitt, and R.Y. Leibensperger, *Grass Forage Sci.* **46**, 283–299 (1991).
14. J. Kaiser, *Ecol. Model.* **91**, 25–37 (1996).
15. S. Rajagopalan and J.M. Modak, *Chem. Eng. Sci.* **49**, 2187–2193 (1994).
16. S. Rajagopalan and J.M. Modak, *Chem. Eng. Sci.* **50**, 803–811 (1995).
17. M. Gutierrez-Rojas, R. Auria, J.C. Benet, and S. Revah, *Chem. Eng. J.* **60**, 189–198 (1995).
18. E. Gumbira-Sa'id, P.F. Greenfield, D.A. Mitchell, and H.W. Doelle, *Biotechnol. Adv.* **11**, 599–610 (1993).
19. M.P. Nandakumar, M.S. Thakur, K.S.M.S. Raghavarao, and N.P. Ghildyal, *Process Biochem.* **29**, 545–551 (1994).
20. M.P. Nandakumar, M.S. Thakur, K.S.M.S. Raghavarao, and N.P. Ghildyal, *Enzyme Microb. Technol.* **18**, 121–125 (1996).
21. J. Sargantanis, M.N. Karim, V.G. Murphy, D. Ryoo, and R.P. Tengerdy, *Biotechnol. Bioeng.* **42**, 149–158 (1993).
22. C. Larroche and J.B. Gros, *Biotechnol. Bioeng.* **39**, 815–827 (1992).
23. J.A. Rodriguez Leon, A. Torres, J. Echevarria, and G. Saura, *Acta Biotechnol.* **11**, 9–14 (1991).
24. R. Auria, J. Palacios, and S. Revah, *Biotechnol. Bioeng.* **39**, 898–902 (1992).
25. P. Gervais and D. Simatos, in S. Thorne ed., *Mathematical Modelling of Food Processing Operations,* Elsevier, London, 1992, pp. 137–183.
26. L. Edelstein and L.A. Segel, *J. Theor. Biol.* **104**, 187–210 (1983).
27. L. Edelstein, Y. Hadar, I. Chet, Y. Henis, and L.A. Segel, *J. Gen. Microbiol.* **129**, 1873–1881 (1983).
28. J.A. Laszlo and R.W. Silman, *Biotechnol. Adv.* **11**, 621–633 (1993).

29. L. Edelstein, *J. Theor. Biol.* **98**, 679–701 (1982).
30. A.M. Jurus and W.J. Sundberg, *Appl. Environ. Microbiol.* **32**, 284–287 (1976).
31. K. Ito, A. Kimizuka, N. Okazaki, and S. Kobayashi, *J. Ferment. Bioeng.* **68**, 7–13 (1989).
32. C. Larroche, *J. Sci. Ind. Res.* **55**, 408–423 (1996).
33. G. Georgiou and M.L. Shuler, *Biotechnol. Bioeng.* **28**, 405–416 (1986).
34. G. Lazarova, M. Kancheva, and I. Panayotov, *Acta Microbiol. Bulgarica* **15**, 63–71 (1984).
35. S. Rajagopalan, D.A. Rockstraw, and S.H. Munson-McGee, *Bioresource Technol.* **61**, 175–183 (1997).
36. M.K. Gowthaman, K.S.M.S. Ragheva Rao, N.P. Ghildyal, and N.G. Karanth, *Process Biochem.* **30**, 9–15 (1995).
37. D.A. Mitchell, P.F. Greenfield, and H.W. Doelle, *World J. Microbiol. Biotechnol.* **6**, 201–208 (1990).
38. P. Sangsurasak, Ph.D. Thesis, The University of Queensland, Brisbane, Australia, 1996.
39. J.J.C. Van Lier, J.T. Van Ginkel, G. Straatsma, J.P.G. Gerrits, and L.J.L.D. Van Griensven, *Netherlands J. Agric. Sci.* **42**, 271–292 (1994).
40. R. Auria, I. Ortiz, E. Villegas, and S. Revah, *Process Biochem.* **30**, 751–756 (1995).
41. D.M. Stuart, Ph.D. Thesis, The University of Queensland, Brisbane, Australia, 1996.
42. P. Gervais, C. Bazelin, and A. Voilley, *Biotechnol. Bioeng.* **28**, 1540–1543 (1986).
43. B.L. Rathbun and M.L. Shuler, *Biotechnol. Bioeng.* **25**, 929–937 (1983).
44. M.K. Gowthaman, N.P. Ghildyal, K.S.M.S. Raghava Rao, and N.G. Karanth, *J. Chem. Technol. Biotechnol.* **56**, 233–239 (1993).
45. G. Saucedo-Castaneda, B.K. Lonsane, M.M. Krishnaiah, J.M. Navarro, S. Roussos, and M. Raimbault, *Process Biochem.* **27**, 97–107 (1992).
46. N.P. Ghildyal, M.K. Gowthaman, K.S.M.S. Raghava Rao, and N.G. Karanth, *Enzyme Microb. Technol.* **16**, 253–257 (1994).
47. M. Gutierrez-Rojas, S. Amar Aboul Hosn, R. Auria, S. Revah, and E. Favela-Torres, *Process Biochem.* **31**, 363–369 (1996).
48. A. Durand, P. Pichon, and C. Desgranges, *Biotechnol. Tech.* **2**, 11–16 (1988).

SOLID SUBSTRATE FERMENTATION, AUTOMATION

J. Ricardo Pérez-Correa
Eduardo Agosin
Pontificia Universidad Católica de Chile
Santiago, Chile

KEY WORDS

Bacterial α-amylase
Differential pressure measurement
Evaporative cooling
Fermented foods
Fixed-bed bioreactor
Fluidized-bed bioreactor
Gibberellic acid
Packed-bed bioreactor
Stirred bioreactor
Thermocouples

OUTLINE

INTRODUCTION

Solid substrate fermentation (SSF) processes may be defined as four-phase systems. The first, an air phase, is a continuous system that usually flows through a solid bed. The solid second phase consists of a water-insoluble support containing the third phase, a nutrient-rich aqueous solution. This solution is tightly absorbed within the matrix of the insoluble support; some water, however, may exist in a free state in capillaries, although this can never exceed the water-holding capacity of the matrix (1). The fourth phase is the microorganism itself, which grows in or around the support and may grow in the interparticular free space.

SSF has been used by humans for centuries. Traditional uses of SSF systems include the production of fermented foods, pigments, and koji in the Far East. Within the past decade, the production of other, higher-value microbial metabolites such as antibiotics (2,3), biopesticides (4), aromas (5), gibberellic acid (6–9), and bacterial α-amylase (10), to name a few, have been evaluated with highly promising results. Polymeric and insoluble humid agricultural products or by-products are usually employed in SSF processes. These serve a dual functions (*1*) as a source of carbon, energy, nitrogen, and other nutrients and (*2*) as the support for microorganism growth (11). A novel system has been recently developed (12) whereby a solution of nutrients is adsorbed on an inert support.

This type of process has several features that make SSF particularly attractive to industry. One of the main advantages is the use of agroindustrial by-products as raw materials, because these are not only low in cost but are usually very difficult to dispose of properly, given the organic nature of the by-products and the large volumes generated. Differential expression of certain microbial metabolites, which are not expressed in liquid medium (13), is another significant attribute of this culture system, as, for example, in the production of active *Trichoderma* propagules for biological control programs as well as in bioremediation applications.

Despite certain drawbacks of SSF, higher yields, measured in terms of product per volume, make the process economically viable. Reports of SSF metabolite concentrations 100-fold higher than those obtained from submerged fermentation (14) and biomass concentrations up to 200 g/L have been published. SSF does not need large bioreactors, and downstream processing costs might also be lower. The low energy consumption of SSF bioreactors, which require only occasional agitation of the solid bed, is a further asset from the standpoint of operating costs. Last, but not least, SSF generates less effluent, thereby reducing the environmental impact of the process as well as treatment costs.

In spite of these promising features, few applications of SSF processes are currently utilized in industry. Most SSF research has been carried out only at the laboratory level, where its interest as a fermentation system has been highlighted. Unfortunately, only a few of these processes have been evaluated at pilot or preindustrial levels because of problems inherent to solid substrate cultivation technology:

- Relatively slow rates of production, due to the additional barrier to growth imposed by the solid substrate
- Limited capacity of dissipating metabolic heat generated during cultivation
- Mass transfer problems, in particular, slow diffusion of nutrients from substrate to microorganism
- Lack of readily available, appropriate sensors for monitoring
- Limited research on SSF reactor models, control strategies, and scaling-up procedures

The latter topics are considered in this article in an effort to highlight areas that call for substantial development to bring about widespread industrial acceptance of this unique process.

The production of gibberellic acid by the filamentous fungus *Gibberella fujikuroi* is used throughout to illustrate the technical difficulties in the automation and scaling-up of SSF technology. Gibberellic acid is a biologically active compound, naturally present in plants, which acts as an hormone and regulates many plant growth and development processes. Gibberellic acid is a typical secondary metabolite, which in *G. fujikuroi* is triggered on exhaustion of assimilative nitrogen sources.

This production process presents a real challenge to those involved with automation and scaling. *G. fujikuroi* is a slow-growing filamentous fungus, generates significant amounts of metabolic heat during its exponential growth phase, and is highly susceptible to microbial contamination. Hence, successful advances in this process will boost the industrialization of SSF technology.

AUTOMATION OF SSF BIOREACTORS

Design

This section covers the design and operation of current SSF bioreactors, many of which are still the subject of research and development.

Bioreactor Types. SSF bioreactors may be of static (fixed-bed) or agitated (stirred) design. Tray and packed bioreactors are most commonly used among the former, whereas rotating drums, periodically agitated tanks, and rocking and fluidized-bed bioreactors are examples of the latter. Most SSF bioreactors operate in batch form, although water and nutrients are added periodically in many systems. Very few SSF bioreactors have been designed to operate continuously (15–17), and no industrial continuous processes have been reported.

Tray Bioreactors. Trays, one of the simplest types of SSF bioreactor, have been employed in the Far East for many years. They are readily available commercially (18) from laboratory to industrial scale. The main characteristics of this type of bioreactor follow (11):

1. A set of trays are covered with a thin layer of solid substrate.
2. Trays are kept inside a chamber, normally under controlled environment.
3. Forced air is not commonplace, although trays can be perforated allowing air to circulate smoothly through their bases.
4. In some cases the solid bed is agitated gently, either manually or automatically, using simple mechanical devices.

The main drawback of this type of bioreactor is that it is highly labor intensive, so that automation of large-scale processes is very difficult. Also, because the solid layer in the tray must be thin and the trays cannot be too close to one another, very large chambers are required to achieve industrial levels of production. Industrial applications of this type of reactor include production of mushrooms (*Agaricus bisporus*) and tempeh as well as aseptic production of soy sauce koji mold spore inoculum.

Packed-Bed Bioreactors. Packed-bed bioreactors are normally cylindrical tanks, filled with a solid substrate suspended on a perforated surface through which air is forced. The inlet air conditions (temperature, humidity, and flow rate) are generally regulated. At laboratory scale, the most common types of packed bioreactors are known as Raimbault columns. The latter are submerged in a thermoregulated bath and supplied with saturated air. In larger systems, cooling fluid is forced through double walls, coils, or plates (19) immersed in the solid bed to overcome the difficulty of removing metabolic heat. The main disadvantage of this type of bioreactors is its propensity for heterogeneous growth that hinders scaling up to industrial size.

Plastic bags filled with solid substrate provide a typical example of commercial application of packed bioreactors. The latter includes mushroom cultivation (*Agaricus bisporus, Pleurotus ostreatus*, etc.), production of biopesticides (*Trichoderma harzianum*), tempeh, and koji mold spore inoculum. In addition, biofilters (microbial flora adsorbed to a porous solid support) have recently been developed and commercialized for treating toxic gaseous and liquid streams.

Rotating Drums. Features of rotating drums used as bioreactors (11) include the following:

1. A partially inclined cylinder is filled with substrate up to a predefined level.
2. Smooth cylinder rotation mixes the solid bed; this process is usually enhanced with baffles.
3. Aeration is occasionally provided by means of tubes placed in the free space of the cylinder or in its shaft.

Rotation must be controlled to avoid damaging the microorganisms or causing substrate agglomeration or particle breakup. The control of temperature in this type of bioreactor is very difficult and becomes more difficult when increasing the scale of the reactor. Rotating drums are particularly appropriate for bioremediation of contaminated soils. Klein (16) has described several industrial applications of this technology to kerosene-contaminated soils and to degradation of mineral oil hydrocarbons, chlororganics, and polycyclic compounds. Bioreactor sizes reach 40 m in length and 3 m in diameter, with a treatment capacity of 0.5 to 1 ton/h.

Rocking Bioreactor. Although similar to the rotating drum, the rocking bioreactor cylinder rocks gently instead of performing a complete rotation. It achieves acceptable agitation and mixing without damaging either the microorganism or the substrate. This bioreactor has only been built and tested at a laboratory scale (20).

Fluidized-Bed Bioreactors. This bioreactor is a vertical cylinder inside which the solid substrate rests on a perforated plate. Air forced through the plate at high flow rates suspends the solid in the air, causing the solid–air mixture to behave like a fluid. The main advantage of this type of bioreactors is (11) the high rate of heat transfer and aeration resulting from the enormous solid–gas interface area. Additionally, gases and volatile metabolites are easily removed from the bioreactor, a very good degree of mixture of the solid is achieved, and the absence of mechanical components simplifies the automatic control of the process. These factors permit high yields, reducing the needs for large bioreactors.

Most investigations with this technology have included yeasts, either for cell biomass or ethanol production. Other microorganisms such as filamentous fungi have also been assayed. Even though these cells could be damaged by the vigorous agitation, this system causes less harm than conventional, stirred bioreactors. To our knowledge, this technology has not yet been applied at an industrial scale.

Stirred Bioreactors. Stirred bioreactors are probably the type of SSF bioreactor most widely used at the industrial scale. Here the solid bed is agitated by screws, palettes, or other mechanical devices. Agitation is intermittent and is varied during the process. These bioreactors may be either vertical or horizontal.

- Among vertical bioreactors, the most popular are the Japanese Fujiwara bioreactors (21), which are expensive. These are specifically designed and mostly used for koji production in the Far East and in the United States.
- Among the horizontal type, the best known is the INRA-Dijon bioreactor (22), which is used by some French companies in the production of enzymes (Lyven) and biopesticides (Calliope). This open, rectangular reactor is based on a design for a barley malting process. The solid bed rests on a perforated plate through which air is forced. Agitation is performed by vertical rotating screws mounted on a conveyor that moves them back and forth across the main axis of the tank. The shape and size of the screws must be designed specifically for each substrate and microorganism to minimize abrasion and breakup damage.

In both types, water, nutrients, and preservatives can be added by aspersion over the solid mass. Air can be conditioned at the inlet, and the temperature and pH of the solid may be measured at several places within the tank.

Neither the Fujiwara bioreactors nor the INRA-Dijon bioreactor is designed for a completely aseptic operation. However, results obtained using pilot stirred bioreactors with aseptic operation have been published (23–26). These required a high degree of automation and are described extensively next.

Continuous Countercurrent Bioreactor. The continuous countercurrent bioreactor consists of a ribbon-flighted screw inside a horizontal cylindrical vessel. The screw transports the solid matter backward and forward. Roussos and Pyle (17) evaluated this design for the continuous cultivation of *Aspergillus niger* at a substrate feed rate of 0.5 kg/h in the production of glucoamylase. Regulation of aeration, cooling, and water content was achieved through natural convection. It was shown that the mycelium did not suffer any damage. The general performance of this bioreactor was comparable to that obtained in laboratory-scale batch cultivation. Currently, this kind of bioreactor is used for large-scale soil bioremediation in a manner similar to rotating drums (16).

Critical Operating Conditions. The most critical operating variables in controlling an SSF bioreactor are the bed temperature, pH, and water content of the solid support, because microorganisms can only grow and produce within a relatively small range of these operating conditions. Agitation and aeration of the solid bed are further significant factors. Finally, the concentration of nutrients and metabolic gases affect biomass and metabolite production rates, although these variables are very difficult to control.

Temperature. Significant temperature gradients, to 3 °C/cm, can be observed with almost every type of static SSF bioreactor. Any microorganism generates energy (metabolic heat) during its growth. This heat must be dissipated to avert undesirable solid bed temperature conditions. In certain types of reactor, this requirement may become critical. Keeping laboratory-scale reactors, such as Raimbault columns, in a thermostatic bath overcomes the problem. Controlling the chamber temperature of tray reactors suffices in regulating the solid bed temperature; but only if the solid bed is thin enough. Larger pilot- or industrial-scale reactors require additional cooling techniques. Because of simplicity, the most commonly used cooling method is to force air through the solid bed (27). In extreme cases, very cold air or very high flow rates are required to ensure proper regulation of the solid bed temperature. In these situations, good temperature regulation may be achieved using refrigerated jackets, cooled coils, or plates (19), although these may diminish the useful volume of the reactor. Evaporative cooling is likely one of the most effective methods; the relative humidity of the inlet air is regulated to control the rate of evaporation from the solid bed, thereby governing the amount of heat removed by the air stream. At some point in the process, however, very dry inlet air may be required, which points out the main drawback of this method; excessive drying of the solid bed requires efficient water content regulation in the control system. Ryoo et al. (20) used the technique successfully in controlling the solid temperature of a rocking laboratory-scale reactor. Fernández et al. (25) reported that evaporative cooling was capable of regulating the solid temperature in a 50-kg pilot-scale reactor. Recent simulation studies (28,29) have established the convenience of regulating the temperature of the solid bed through simultaneously manipulating the relative humidity, temperature, and flow rate of the inlet air.

Water Content of the Solid Bed. Regulation of water activity in the solid bed is crucial because water content directly defines microbial growth and metabolite production rates. Laboratory studies should thus determine the optimal water activity for the rates of both growth and production. The manner in which water activity of the solid bed varies with water content as the result of substrate consumption and subsequent biomass accumulation also needs to be assessed.

Control of water content is difficult because on-line sensors are expensive and unreliable. However, this limitation can be overcome using soft sensors based on secondary online measurements such as solid bed weight (22). Periodic samples could also be taken manually and analyzed for water content in the laboratory (25). There are several ways to regulate bed humidity, although the easiest is the direct addition of freshwater to the solid bed. Water may also be supplied through the air stream as fog or vapor. Sato et al. (30) were able to control water content by incorporating particles with high water retention capacity in the solid bed. The method ultimately adopted depends on the type of reactor selected and the way in which control of the solid temperature is managed.

pH. The high buffering capacity of the fermenting bed and the lack of appropriate sensors for solid substrate conditions makes pH control difficult to implement. However, pH regulation could be achieved by selecting supplementary nutrients that can be fed periodically during the cultivation; this is particularly true for the nitrogen source (22).

Agitation. Agitation in SSF reactors accomplishes several functions:

- Maintains homogeneous operating conditions
- Permits uniform distribution of water, nutrients, and preservatives supplied during cultivation
- Renews the air occluded between the particles
- Avoids particle agglomeration, crucial when a filamentous fungus is cultivated

The choice of agitation policy is determined by its main purpose. Normally, several objectives must be considered simultaneously, and it is very difficult to establish a priori how agitation could influence each objective independently. The process engineer also must ensure that the agitation neither harms the active biomass (e.g., mushroom culture) nor the substrate. Consequently, the agitation policy, at this point, is fundamentally empirical. Parameters associated with agitation are speed, sensing, and, for reactors employing intermittent agitation, frequency and duration. In fixed-bed reactors, for example, intermittent agitation could assist in the removal of heat upon the saturation of the other temperature loop control variables. Additionally, stirring could be used when air pressure loss exceeds a certain value. When using these types of reactors, however, it is essential to agitate every time water or nutrients are added.

Aeration. Because most SSF processes involve aerobic microorganisms, oxygen transfer toward the biomass is fundamental (11). Aeration also plays a key role in the removal of metabolic heat, CO_2, and volatile metabolites occluded within interparticular spaces. The following mechanisms are potentially important as relates to mass transfer efficiency between the gas phase and the microorganisms in SSF bioreactors:

- Convection from the bulk of the gas phase toward the interparticular spaces.
- Diffusion through the stagnant gas layer that surrounds the particles.
- Diffusion through the liquid layer that covers the surface of the particles. The high solute concentration of the liquid layer characteristic of SSF processes may significantly affect gaseous diffusion through this layer. To our knowledge, no research has directly targeted this important factor in SSF reactor technology.
- Diffusion inside the porous particles.
- Microbial oxygen consumption and CO_2 production.

Only the first two of these mechanisms could be influenced by either reactor design or operating conditions. Convection in particular is important in packed reactors or in those in which agitation is very slow or infrequent or aeration is very weak. Interparticular oxygen transfer is also retarded in substrates formed by a mixture of fine and thick particles, above all when filamentous microorganisms are cultivated. This situation creates an additional barrier to air circulation. The other mechanisms are fundamentally determined by the characteristics of the substrate (particle size, form, porosity, texture, etc.) and of the microorganism.

Case Study: An Aseptic Pilot Bioreactor for SSF Processes. An SSF general-purpose aseptic pilot reactor of 50-kg nominal capacity that has been built and evaluated for the production of gibberellic acid (GA_3) is now described.

General Features. The structure of the stainless steel reactor encompasses the following parts:

1. A hermetic lid with perforations (for air sampling and nutrient feeding) and an observation window;

the agitation system (motor, double arm, and blades) and the feeding system are mounted on the lid.

2. The principal body has a double-jacket system for sterilization.
3. An air chamber, located in the lower part of the reactor, is used for inlet air homogenization. Air temperature and relative humidity are measured here.
4. A rotating basket 1.15 m in diameter and 0.28 m high has a nominal capacity of 50 kg of wet solid. The basket base contains perforations of 2 mm in diameter with a separation of 3.5 mm.

During reactor operation, the following main steps can be distinguished: (*1*) sterilization, (*2*) substrate loading, (*3*) inoculation, (*4*) cultivation, (*5*) substrate unloading, and (*6*) cleaning. All these steps except cultivation are carried out manually.

Air Conditioning. As shown in Figure 1, the air is forced into the reactor by a fan (a) with a prefilter (b). The speed of the air can be regulated manually between 0.25 and 7 m/s through a purge system (c). An absolute filter (d) ensures the purity of the air, it is able to retain particles larger than 0.3 μm with a minimal efficiency of 99.99%. A heating system (e), with 6-kW electric resistance, can heat the air for reactor sterilization and process needs. In addition, the air can be cooled to 0 °C with fins (f) and a refrigeration system (g). The relative humidity is controlled by vapor addition through a solenoid valve (h) connected to a steam boiler (i) with a working pressure of 5–10 psi.

Agitation. The bed agitation system includes two arms: one contains blades that operate like a plough, and the other has curved paddles that scrape the basket base. Both arms can be manually regulated with cranks, according to process needs. The basket movement is provided by a reducing motor of 1.5 kW that is connected to the basket shaft. An inverter driver controls the rotation speed, achieving a minimum of 5 rpm with a torque of 5,500 Nm. An aseptic water seal ensures that all the inlet air passes through the solid bed. The basket is supported on roller bearings for a uniform and soft movement.

Water and Nutrient Feeding. A set of three sprinklers, mounted on the lid, are used to add water and dissolved nutrients periodically. The solution is based forced into the reactor with a peristaltic pump with a maximum flow of 10 L/min. To distribute the solution over the entire solid bed, the agitation system is turned on each time the solution is fed. The monitoring and control system of this reactor is described later.

Instrumentation and Control

Automation of SSF bioreactors is a complex problem that has yet to be resolved satisfactorily before scaling up the process. In fact, this is why much laboratory-scale research has not yet been scaled up to industrial level (1,6,8). Many of the difficulties complicating the operation and control of SSF bioreactors concern the regulation of temperature and water content in the solid bed. If the high rate of heat generation during the growth period is poorly controlled, overheating of the bed can result. The control of water content is particularly cumbersome. Part of the water is strongly bound to the solid, while the rest may remain free in the capillary areas of the material. Care must be taken to

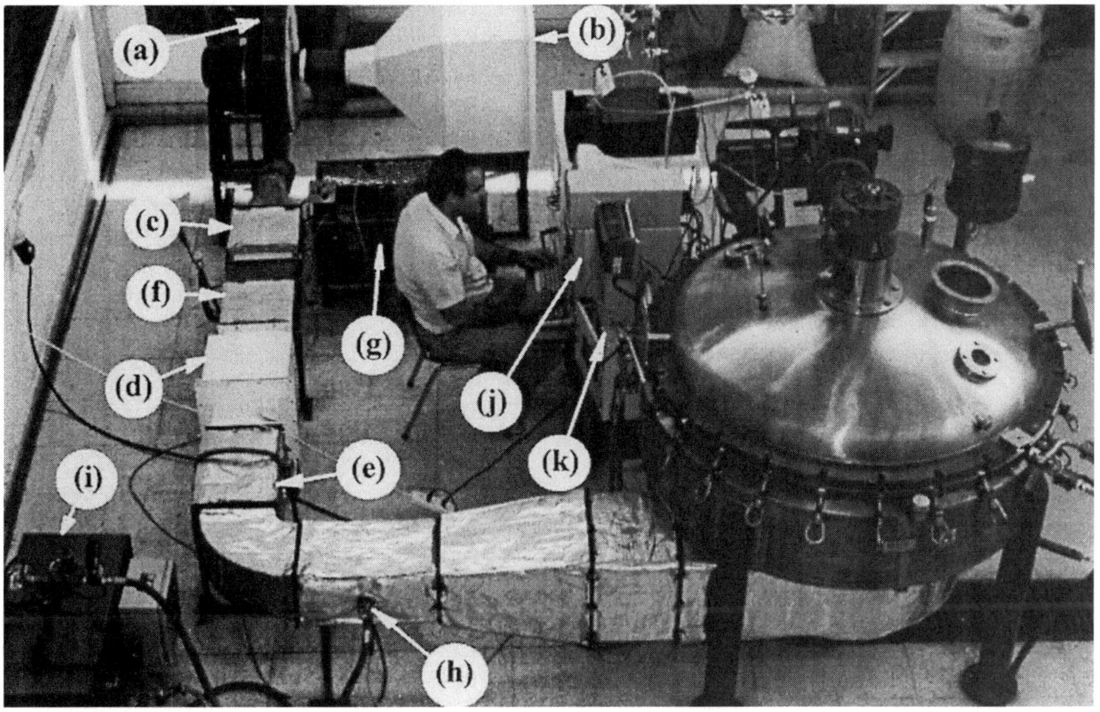

Figure 1. Aseptic pilot solid substrate fermentation (SSF) reactor with its air conditioning system: (**a**) fan, (**b**) prefilter, (**c**) purge system, (**d**) absolute filter, (**e**) heating system, (**f**) fins, (**g**) refrigeration system, (**h**) solenoid valve, (**i**) steam boiler, (**j**) control system, and (**k**) programmable logic controller (PLC).

avoid saturation of the solid matrix, which will lead to anaerobic conditions. The lack of appropriate and inexpensive sensors for monitoring water content on-line results in difficult, slow, and imprecise regulation of water content. The development of new instruments and control strategies specific to SSF processes would unquestionably simplify the development of scalable and economically competitive SSF bioreactor technology.

Instrumentation One of the main limitations of SSF as opposed to submerged fermentation (SMF) bioreactor development is the lack of specific and inexpensive instrumentation enabling proper monitoring and control of the processes. Much of the instrumentation used in SSF bioreactors is associated with measurements of the gaseous phase only. The following section discusses instrumentation commonly used in the monitoring and control of solid substrate bioreactors.

Temperature. The solid bed temperature is a primary control objective because overheating can seriously impair process performance. The temperature of the air stream is an important manipulated variable in many SSF bioreactor control strategies. Several relatively inexpensive and efficient temperature sensors are available, with thermocouples being the most widely used. When two wires of different metals are welded at two junctions and kept at different temperatures, an electric current will flow in the circuit, proportional to the temperature difference. This is the principle behind thermocouples, of which there are of several types, differing with the kind of metals used: J, iron-constantan; T, copper-constantan; K, chromel-alumel; etc. These types are appropriate for different temperature ranges and ambient conditions (reducing or oxidizing). Error limits range from ± 0.5 °C to ± 2.2 °C, depending on thermocouple type and quality. Their main advantages are fast response, shock resistance, small size, probe flexibility, heavy use capability, low cost, and suitability for both high and low temperatures. They must be calibrated periodically.

The electrical resistance of metals changes with temperature. This principle is applied in resistance temperature detectors (RTDs), which are generally more expensive than thermocouples and are suitable over a narrower temperature range; they are however more accurate and do not need periodic calibration. Platinum RTDs are most commonly used.

Air Flow Rate. Differential pressure measurement, with accuracy ranging from $\pm 0.8\%$ to $\pm 5\%$, relies on the pressure drop created as a fluid posses an obstruction (such as a concentric orifice, venturi, or pitot tube), and is therefore proportional to flow rate. The main limitation of these devices is that the energy cost from the loss of pressure can be considerable. A differential pressure transducer is needed to include this measurement in a digital control system.

Another air velocity measurement instrument commonly used is the turbine anemometer. A rotating device is placed in the path of the fluid, where its rotational speed is proportional to the fluid velocity and provides accurate flow measurement over wide ranges. Thermal anemometers, in which a thin wire immersed in the fluid is heated by an electric current, can also be used to measure air velocity. The fluid velocity is related to the heat dissipated from the wire. Full calibration is required for anemometers used for flow rate measurements because velocity depends on the anemometer's transversal position inside the duct.

Relative Air Humidity. Electrochemical sensors are most commonly used for determining relative humidity. Two widely used electronic relative humidity (RH) sensors are the Dunmore element and the Pope cell. The former employs a solid substrate coated with a LiCl solution. The hygroscopic nature of this salt causes it to take up water vapor from the surrounding atmosphere. Dunmore cells, while excellent RH sensors, are normally designed to cover only a narrow range (from 40% to 60% RH). The Pope cell is similar to the Dunmore cell with the substrate being made from sulfuric acid-treated polystyrene. The main advantage of Pope cell is its wide range of response covering from 15% to 99% RH.

Pressure and Differential Pressure. When certain asymmetric crystals are deformed along specific axes, an electrical potential created within the crystals produces an electric current in external circuits. Such devices are called piezoelectric sensors. Cultured quartz is both readily available and reasonably priced for this purpose. In addition, it presents near-perfect elasticity and stability and is temperature insensitive. Instrument response times are of the order of submicroseconds. Differential pressure transducers consist of two chambers connected internally to a rod and externally to high and low pressure measuring diaphragms. When a difference of pressure is encountered the rod is displaced in the oil-filled chambers, producing a signal.

Air Composition. Analysis of exhaust gases is very useful in the monitoring and control of SSF bioreactors (31). Information on the physiological state and respiration rate of the culture can be obtained through continuous measurement of CO_2 and O_2 concentrations in the gas outlet stream. The specific growth rate (μ), biomass concentration, and respiratory quotient (RQ) can also be resolved online. Other volatile compounds associated with the metabolite of interest can be measured on-line, facilitating production control and optimization. On-line measurement of CO_2 and O_2 can be performed by gas chromatography or special-purpose gas analyzers. The main advantages of the first method are that many compounds can be monitored with the same instrument and over a wide range of values. Other gas analyzers may be more precise and provide faster response. The several kinds of instruments include paramagnetic analyzers for CO_2, O_2, and ammonia, infrared instruments for CO_2, and electrochemical probes for O_2. Care must be taken to dry the sampling air beforehand and to provide a regulated flow rate entering the instrument to obtain meaningful results.

Solid Bed Water Content. Water content of the reactor bed is very difficult to measure on line. There are no instruments commercially available that can measure water content in solid substrates at levels greater than 80% with good performance and a small error. Conduction and capacitance principles are applied, to good effect, in measuring water content up to 50%. The main problem in measuring high water content in solid substrates, with instruments based on conductance/capacitance, is that the apparent density varies unpredictably during cultivation, either because of agitation or growth of the biomass. Devices based on infrared (IR) measurements of solid samples yield very good results, but, they are expensive and their application for on-line measurement in aseptic SSF bioreactors is not well developed. A probe that can measure the water activity of a solid medium consists of a chamber which is separated from the medium by a hydrophobic membrane and contains a capacitive device to measure the relative air humidity within it (32).

Control Strategies. Proper regulation of the temperature and water content of the solid bed over several days has long been recognized as crucial to optimum performance (33). The major challenge in temperature control is to overcome the limited dissipation capacity of the metabolic heat generated during the cultivation. Temperature gradients across a bed can be as much as 20 °C, especially during the period of high heat generation. Although agitation of the bed can help in reducing the temperature gradient, in practice this heat cannot be removed by vigorous or frequent agitation, as this often has a negative impact on microbial growth and metabolism. Several methods, linked to reactor type, have been proposed and experimentally tested in attempting to control the temperature of the solid bed. Forced air and the addition of freshwater have been successfully used (27) in controlling the temperature of a packed column (35 cm height × 6 cm diameter). Temperature control has also been achieved with intermittent agitation in a laboratory rotating drum reactor (34) that homogenized the bed and also decreased its compactness. However, agitation can damage the hyphae of filamentous fungi and may alter structure morphology.

The most effective method of controlling bed temperature may be evaporative cooling. In this strategy, the rate of water evaporation from the bed can be modified by manipulating the relative humidity of the inlet air, considerably increasing the rate of heat removal. Applying this principle, good temperature control has been attained at laboratory scale in a rotary drum bioreactor (31) and a rocking bioreactor (20). A further consideration, however, is that freshwater must be supplied to avoid drying of the bed. In a packed bed, the rotation or rocking motion has to be replaced with discontinuous agitation. Our experience with 50-kg and 200-kg aseptic pilot reactors indicates that none of these strategies is sufficient to maintain proper control of the solid temperature (25); periodic agitation of the bed and manipulation of the inlet air temperature are also required. Simulation studies with SSF pilot-scale reactor models have shown that a multivariable algorithm can control solid temperature efficiently by simultaneous manipulation of relative humidity, flow rate, and inlet air temperature (28,29).

Few publications have considered the regulation of solid water content. For example, wet wood pulp particles have been placed in the solid substrate during cultivation of *Aspergillus oryzae* to provide a reservoir of free water (35). However, this practice reduces the effective bioreactor volume. In other works, the water content of the bed has been controlled by manipulating the ambient air temperature (36); the algorithm was based on energy and water balances. While this method was successfully applied in a laboratory bioreactor (20 × 16 cm), it is difficult to scale up to pilot level. Fernández et al. (25) controlled the water content to keep the water activity constant in the solid medium in a 50-kg pilot reactor. As no on-line measurement of water content was available, samples were taken manually to define the addition of freshwater based on a water balance and heuristics.

Carbon dioxide and oxygen in the outlet gas can be controlled by manipulating the inlet air flow rate (31) and pH can be controlled by spraying an acid or base solution over the solid bed (22). Nutrient concentration, can be maintained by spraying a solution directly over the solid bed (37) or into the inlet air (38).

Control Algorithms. Discrete PID and on–off control algorithms are currently used in both pilot and industrial SSF bioreactors. In light of the established process difficulties, these controllers prove difficult to tune and demand manual supervision to ensure effective operation (25).

There are remarkably few reports concerning applications of advanced techniques for controlling SSF bioreactors. The work of Ryoo et al. (20) can be mentioned as an isolated example in which an adaptive algorithm was used to control the temperature of the solid in a rocking reactor by evaporative cooling. These authors also estimated biomass concentration and total dry mass on-line using an extended Kalman filter. Optimizing the addition of nutrients to maximize the production of secondary metabolites in fed-batch SSF reactors remains a challenge, although some attempts have been made to solve the problem empirically (37,38). A short description of a number of advanced control methodologies that have been applied successfully in SMF bioreactors follows. We believe that these techniques can also be useful in the automation of SSF reactors.

Adaptive Linear Control. In essence, an adaptive controller is an on-line algorithm that estimates parameters that are, directly or indirectly, associated with the dynamics of the process. These parameters are then used to "tune" a control law on-line. Several adaptive control algorithms have been described (39,40). These controllers are particularly appropriate for processes that have variable time delays and that exhibit nonlinear dynamic behavior. Many successful applications of adaptive control in biotechnology have been published (41–44). As mentioned, application of this technique in SSF automation has been experimentally assessed by Ryoo et al. (20).

Nonlinear Control. Because real processes present nonlinear dynamics, nonlinear control techniques have attracted the attention of many researchers, both as to the theory and in the development of real-time applications. The employment of adaptive techniques in estimating uncertain or varying model parameters greatly improves the likelihood of using nonlinear control successfully in industrial processes. Several applications of this technique in biotechnology have been reported. Bastin and Dochain (45) have even developed a general methodology for designing nonlinear bioreactor controllers. Successful industrial applications have been reported by both Pomerleau and Viel (46) and Chen et al. (47).

Predictive Controllers. Paján et al. (28) compared the performance of a dynamic matrix control (DMC) predictive algorithm (48) with that of PID in the control of a simulated fed-batch SSF process and established that employing DMC greatly simplified the design and start-up of the control system. DMC possesses a flexible structure that can be adapted to the process facilities and limitations. In this particular application, the effort devoted to tuning the temperature control loop was significantly reduced and a better closed-loop response was achieved with a DMC algorithm than with a PID one. Real-time application has not yet been described, but the authors believe that this could be on optimal control strategy.

Case Study. To illustrate how the foregoing ideas can be applied to a real SSF process, a control system is now fully described for pilot SSF bioreactors (24,25).

The Process. Two aseptic packed-bed bioreactors (50 kg and 200 kg of total solid nominal capacity) using periodic agitation and forced air were built to scale-up the production of gibberellic acid by the filamentous fungus *Gibberella fujikuroi*. This process is demanding because *G. fujikuroi* is sensitive to high temperatures (above 36 °C) and to contamination by other microorganisms. Because it is filamentous, this fungus is sensitive to mechanical stresses when the solid bed is agitated, and its growth increases bed compactness, which creates a reduction in heat and mass transfer rates. In addition, bed overheating can easily occur as the result of the high rate of metabolic heat generated during the growth period. Finally, the processing time is long, demanding a robust control system.

Monitoring. Variables measured on-line are as follows (Fig. 2):

- Temperatures: in the solid bed and in the inlet and outlet air (K-type thermocouples)
- Relative humidity of the inlet air
- CO_2 concentration in the outlet air
- Air pressure loss throughout the bed
- Inlet air speed

Bed water content was measured off-line every hour.

The following remote commands are available for manipulation:

- Inlet air electric heaters (on–off; 6 kW maximum for sterilization and 3 kW maximum for operation)
- Inlet air cooler (on–off; 6.6 kW capacity)
- Bed agitation speed (regulated between 3 and 15 rpm) for both reactors
- Local screw stirrers (inverted drive) for 200-kg reactor only
- Water addition flow rate (on–off; peristaltic pump, 600 rpm)
- Steam addition (on–off; solenoid valve)

Finally, the inlet air flow rate can be set manually between 50 and 500 m^3/h.

A programmable logic controller (PLC) is used for data acquisition and control of primary actuators (Fig. 2). Operator and control calculation interaction are carried out via an IBM-compatible PC computer linked to the PLC. Project-specific software was developed for the PC with a graphic interface to handle the control systems in either automatic or manual mode (Fig. 3).

Control System. The primary objectives of this control system are to regulate solid bed temperature and to control bed water content according to a varying reference. The secondary objectives are the minimization of temperature gradients within the bed, as well as bed compactness.

Control of the bed temperature is based on evaporative cooling, by manipulating the relative humidity of the inlet air and maintaining its temperature at a low value. During the period of high heat generation, manual operation of inlet air flow rate, inlet air temperature, and agitation is also necessary to avoid bed overheating. The average bed temperature is fed into a digital PID control algorithm to drive the setpoint of the inlet air relative humidity. This, in turn, is controlled with an on–off algorithm with a dead band and hysteresis that commands a solenoid valve adding steam. The inlet air temperature is further controlled with another on–off algorithm with a dead band and hysteresis that manipulates the heaters or coolers, according to process needs.

The water content of the bed is controlled through the periodic addition of freshwater. The reference trajectory of the water content was computed in the laboratory based on water activity studies. A solid sample was taken each hour from the reactor for the laboratory measurement of the water content. The amount of water to be added is determined by the operator using an approximated water balance and their own experience. The bed is agitated at each addition of water.

To keep the bed as homogeneous as possible and to avoid excessive interparticular aerial growth, which would reduce porosity, a periodic agitation policy was established. The degree of homogeneity is defined by the temperature gradient inside the bed, while interparticular aerial growth is estimated by the loss in air pressure through the bed, as suggested by Auria et al. (49); this is a semiautomatic loop that employs heuristic logic. The operator may establish the agitation speed, its duration, and, as for the 200-kg reactor, also its path (left or right, up or down for individual stirrers).

Control Performance. The control strategy adopted here provides efficient operation of an aseptic pilot SSF bioreactor. When the reactor was run manually, it required the constant attention of at least two operators. The current control system, however, requires only one operator, who has relatively little interaction with the process; the system even allows bioreactor operation without direct supervision at certain times. Moreover, the control of bed temperature and water content is more precise than was possible without automation. Although the inlet air conditioning control system exhibits limitations, such as oscillations and offsets (25), these do not influence bed temperature control. The latter loop is critical to the achievement of high biomass and product concentrations.

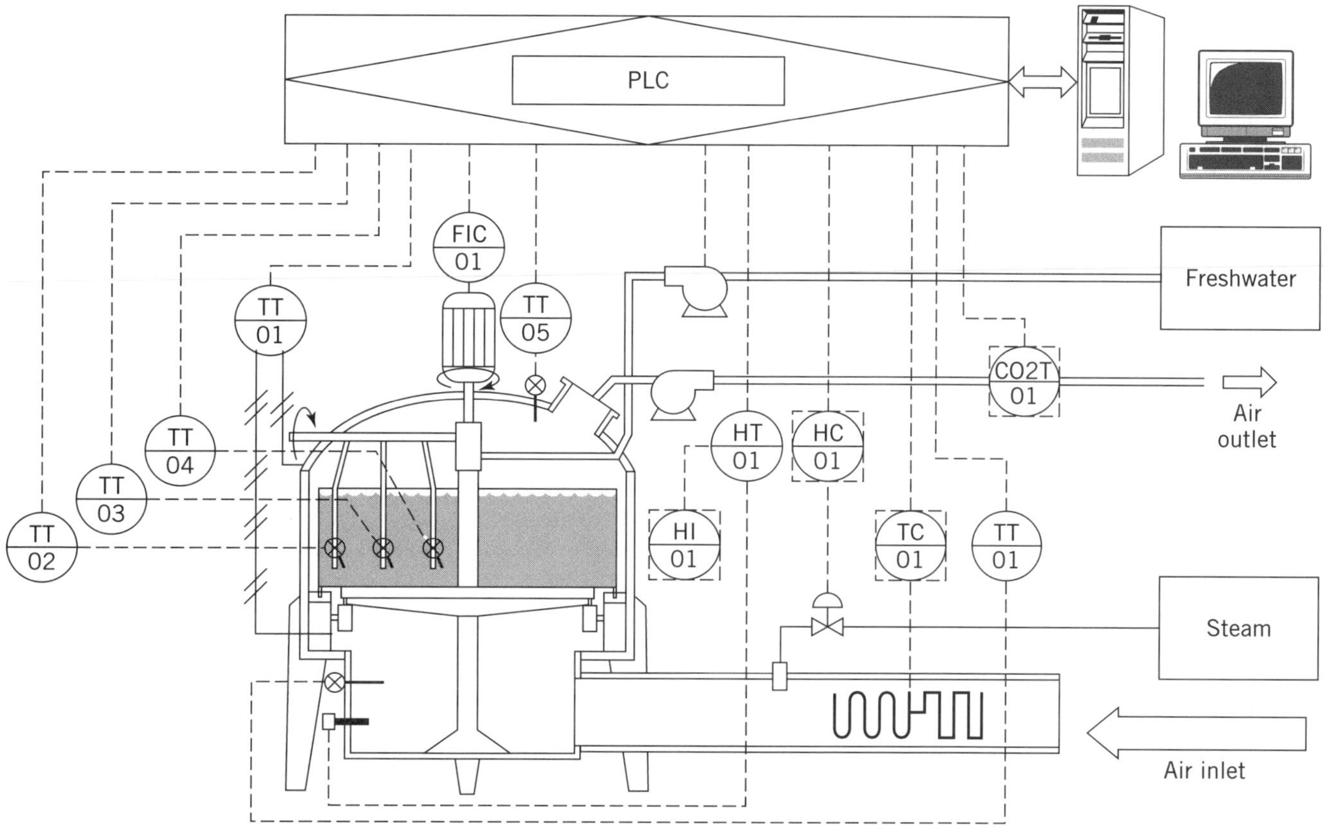

Figure 2. Process and instrumentation diagram (P&ID) of a pilot SSF reactor.

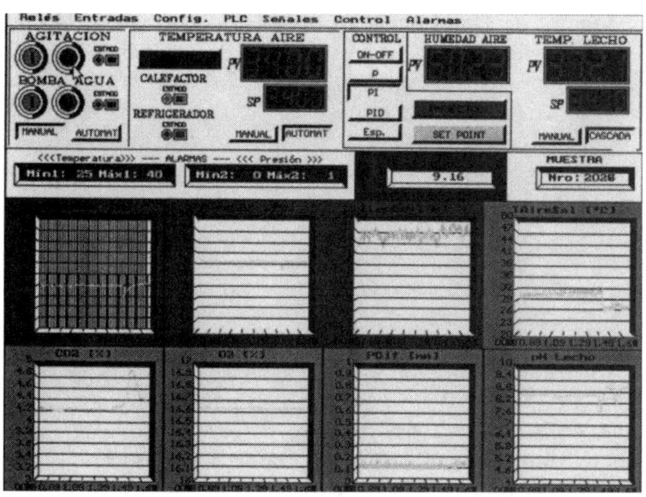

Figure 3. Example of graphics interface for SSF reactor operation.

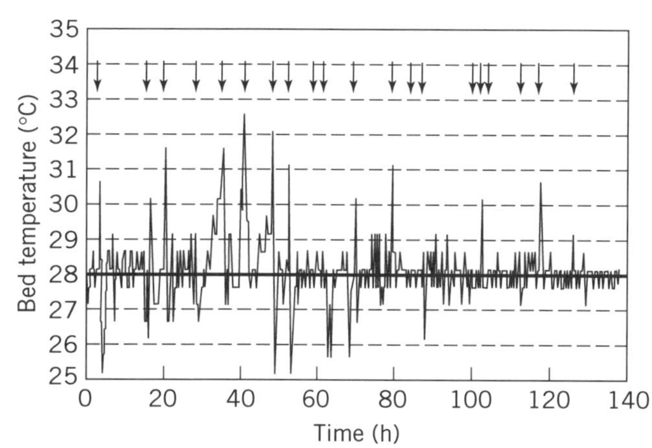

Figure 4. Bed temperature control of a SSF reactor. Agitation time is shown by arrows.

Adequate average bed temperature (Fig. 4) and water content control (Fig. 5) are achieved, although they are not fully automatic. For temperature control, evaporative cooling is insufficient during high heat generation and must be complemented with agitation and direct manual operation of the inlet air temperature and flow rate. Lacking appropriate sensors, control of bed water content is performed manually. Another significant limitation of the current system is that it is difficult to control the temperature gradient of the bed during high heat generation (not shown). However, this is undoubtedly more a limitation of the current agitation system than of the controller. It is also noteworthy that the control strategy was successfully scaled up from a 50-kg reactor to a 200-kg design with only minor adjustments.

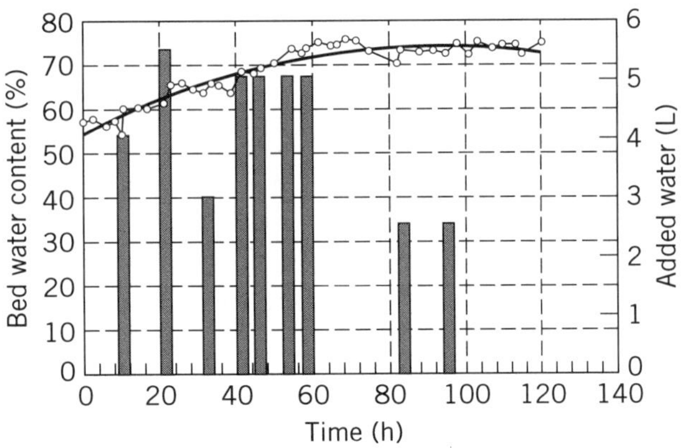

Figure 5. Water content control of a SSF reactor. *Bars* indicate addition of freshwater.

Future Developments. Before scaling the bioreactor up to industrial capacities (2 or 3 tons), a multivariable model-based control algorithm that has already been tested using simulations (29) will be installed in the 200-kg bioreactor to achieve control of the bed temperature. In this strategy, the inlet air temperature, flow rate, and relative humidity will be manipulated automatically considering prespecified priorities.

Research is under way to develop on-line estimators of water content, biomass, and product. Thus, through the addition of freshwater and nutrient solutions, these variables will be controlled automatically in the near future.

Modeling. The first step in modeling is to clearly establish the purpose of the model (50,51). This section mainly covers physical models that may be used in scaling-up, optimization, and control of solid substrate cultivation of filamentous microorganisms; black-box (fuzzy, neural networks, etc.) and microscopic models are purposely excluded. Because industrial SSF processes are carried out in batch or semibatch mode, only dynamic models are discussed.

To simplify matters, it is convenient to divide the modelling of solid state bioreactors into three stages. First, it is necessary to describe *biomass growth* and its relationship with respiratory activity (CO_2 and O_2 evolution). The effect of culture conditions such as temperature, water content, pH, CO_2, O_2, and limiting nutrients on biomass growth must be defined. Then, the *production* and *degradation* rates of the metabolite of interest are studied. Finally, *mass* and *heat transfer* phenomena must be included using appropriate energy and mass balances. This step is critical when it comes to scaling-up from laboratory to pilot and on to industrial reactors.

Biomass Growth. Almost all published work on solid substrate fermentation models considers logistic growth kinetics (27,52,53), which assumes that steric effect is the only limitation for microbial growth:

$$\frac{dx}{dt} = \mu_{\max} \cdot \left(1 - \frac{x}{x_{\max}}\right) \cdot x$$

This model may be useful in describing the evolution of the biomass during the growth phase, although it cannot be applied directly during stationary or lytic phases. If the aim of the fermentation is to produce a secondary metabolite, substrate limitation must be included in the kinetic growth model. The most widely used expression for this is the Monod equation, in which the growth rate decreases steadily after the limiting substrate begins to be depleted. This equation may be derived theoretically assuming that the growth rate is controlled by a single enzymatic reaction. Although real systems are much more complex, the Monod equation fits experimental results well. Many other kinetic equations also can be used to fit experimental results, for example, the Contois, Tessier, and Moser (54). Each of these equations considers a single limiting substrate; multiple substrate models are discussed by Blanch (55). Few SSF models consider growth kinetics other than the logistic equation. For example, Mitchell et al. (56) modeled the growth of *Rhizopus oligosporus* on a solid substrate using the Monod equation with glucose as the limiting substrate, and Gutiérrez-Rojas et al. (57) presented a growing kinetic expression that simultaneously included a logistic model and a substrate-dependent model with substrate inhibition.

Temperature and water content strongly influence biomass growth. Few models consider the effect of water content on the growth kinetic in SSF processes. Sargantanis et al. (52) modeled the dependence of the logistic kinetic parameters on temperature and water content using polynomials. Gervais et al. (58) considered the effect of water activity on the biomass growth rate; this variable better represents the physicochemical state of the water in the fermenting solid substrate. The effect of temperature on biomass growth is modeled in several works using different mathematical expressions, such as polynomials (53,59), the Ratkowsky equation (60,61), the Esener equation (27), and the Arrhenius equation (62).

Other important aspects that have received little research attention include dependence on other environmental variables (pH, CO_2, O_2) and the description of the reduction of biomass activity during the lytic phase. Rajagopalan and Modak (59,63,64) attempted modeling the effect of O_2 concentration on biomass growth using the Monod equation. Sangsurasak and Mitchell (53) and Smits (61) included a death rate in their growth models to describe the reduction of biomass activity. Finally, almost all authors use the Pirt linear growth model (65), which relates biomass with respiration activity. This model assumes that respiration activity is proportional to the biomass growth rate, and that a little respiration activity is still present (maintenance) even after the growth has ceased.

Metabolite Production. Although many authors have presented experimental results on metabolite production in SSF processes (66–68), modeling is in its infancy (54,69).

Heat and Mass Transfer. Many of the experimental systems used in the study of fungal growth on solid substrates are small enough (shake flasks, Raimbault columns, thin trays) to ensure homogeneity of the medium; that is, biomass concentration, temperature, water content, O_2 concentration, pH, and nutrient concentration are approximately equal throughout the solid bed. Because these systems are small, it is relatively easy to keep environmental variables (temperature, water content, O_2) under control. In larger bioreactors (pilot and industrial scale), the situation is very different; in these, it is extremely difficult to regulate the environmental variables and almost impossible to ensure homogeneity of the solid medium.

Few researchers have developed models aimed at predicting or explaining temporal and spatial dependence of temperature, water content, and O_2. These models can be classified as lumped parameter or distributed parameter and as pseudo-homogeneous or heterogeneous. Lumped parameter models are those that do not consider space distribution. In this scheme, process variables depend only on time and are mathematically described by ordinary differential equations. When space distribution is considered, process variables are described by partial differential equations. Lumped parameter models are used to describe very small bioreactors or stirred bioreactors. Larger static bioreactors should be modeled as a distributed parameter system. Models not explicitly considering distribution inside the solid particles are termed pseudo-homogeneous (70), while models that describe gradients inside the solid particles are called heterogeneous.

An early complete model describing a SSF process that was developed by Raimbault (71) purports that the growth of the filamentous fungus *Aspergillus niger* on amylaceous substrates is limited by the lack of surface and free water. These limitations are specific to the fermentation on solid substrates and hence are not applicable to liquid fermentation. Thus, a physiological base is established to explain the logistic model of growth. Raimbault obtained a global model, based on these principles, to predict the evolution of relevant process variables, such as sugars, biomass, CO_2, O_2, water, and metabolic heat. The determination of the maintenance coefficients for sugars, O_2, and CO_2, and the establishment of a global stoichiometric equation of growth, are significant contributions of this work. This author used small laboratory columns (known as Raimbault columns) specifically designed to minimize temperature and humidity gradients.

Another significant contribution to SSF modeling has been presented by Ryoo (72). The main feature of this lumped parameter model is the incorporation of the latent heat of evaporation in the energy balance. This model also considers the degradation kinetics of the substrate (extruded corn) by the filamentous fungus *Rhizopus oligosporus*. Like Raimbault, Ryoo used a logistic model to describe the fungal growth. The model also includes the evolution of CO_2, water content, and temperature. The growth kinetic parameters are calculated in terms of the water content and temperature of the solid bed. This work was carried out in a 20- × 25-cm rocking reactor.

More recently, Gutiérrez-Rojas et al. (57) proposed a lumped parameter model for the growth of *Aspergillus niger* on inert support in a Raimbault column. The model describes the evolution of biomass, sugar, water, CO_2, O_2, and temperature. It incorporates several new features:

- Growth kinetic with simultaneous steric and substrate limitations
- A temperature-dependent μ_{max}
- An energy balance that includes conduction, convection, and metabolic heat

Using this model, which did not consider evaporative cooling, the authors concluded that the main growth rate-limiting factor was sugar depletion; water and void spaces remained largely available.

Saucedo-Castañeda et al. (27) applied a distributed pseudo-homogeneous model in describing the evolution of the radial temperature gradient of a cylindrical static packed-bed SSF bioreactor with forced air in laminar regime. The energy balance included metabolic heat generation and heat transfer by two mechanisms, conduction inside the solid bed and convection through the reactor wall. The model was calibrated and validated with experimental data obtained in a reactor of 6 cm radius × 35 cm height during the growth of *Aspergillus niger* on cassava meal. In this example, conduction presented the largest heat transfer resistance.

A similar model was developed by Rajagopalan and Modak (59) to describe a 6-cm-deep tray reactor without forced air. The model was able to reproduce the trend of experimental observations, such as maximum temperatures at the center of the bed, decrease of oxygen concentration, and increase of carbon dioxide from the top to the bottom of the tray. However, it failed to quantitatively fit the experimental data. The authors argued that this is because the model neglects some observed phenomena, for example, evaporative cooling, loss of moisture, and solid shrinkage. Smits (61) included evaporative cooling in a tray bioreactor model, concluding that evaporation was not significant because the differences in water vapor concentrations between the tray and the environment were small.

A heterogeneous form of their original model was developed by Rajagopalan and Modak (64); that is, oxygen diffusion inside a solid particle is also considered. Assuming a zero-order oxygen consumption rate, they found an upper limit for cell density. In addition, the model predicted that, despite aerobic conditions in the gas phase, anaerobic conditions may exist within the biofilm. The same authors (63) also developed a model for fungal growth over an spherical particle of wheat bran, where the growth kinetics included simultaneous O_2 and glucose limitations, using the Monod equation in both cases. Simulations with this model showed that for a typical tray fermentation, oxygen depletion inside the biofilm is the primary cause of growth rate decline. However, as fermentation proceeds, the depletion of glucose causes growth to halt completely. In contrast, for a highly aerated solid bed of spherical particles, J. Thibault, K. Pouliot, E. Agosin, and R. Pérez-Correa (unpublished), using an oxygen transfer model calibrated with published data, concluded that in this case diffusion of oxygen through the biofilm is the main limiting step for oxygen transfer. Therefore, a direct consequence of this result is that for large bioreactors, optimizing operating conditions (air flow rate, oxygen concentration in the gas phase, bed agitation) is not enough to maximize oxygen availability to the microorganism because this also significantly depends on the thickness and the diffusion coefficient of the biofilm. In our opinion, data resulting from this exciting research subject can significantly improve the overall performance of SSF technology.

Case Study: Modeling of Fungal Growth and Metabolite Production by *Gibberella fujikuroi*. A homogeneous, lumped parameter model that described the evolution of biomass of *G. fujikuroi* growing on an inert support in a Raimbault column was developed. The lag phase was not incorporated either in the model or with the data used for calibration. The main assumptions are as follows:

- Oxygen transfer resistance is negligible.
- Available nitrogen is the limiting substrate.
- Carbon source is not limiting.
- Temperature and water activity are constant throughout cultivation.

The major novelties of this model are the inclusion of the lytic phase of growth and the production of a secondary metabolite in SSF. A more detailed description of the model is under preparation.

Fungal Growth. The Contois equation is used to model the growth kinetics. Total biomass is determined by the glucosamine method (73). To be consistent with the respiration activity, the model also considers active and inactive biomass. Moreover, it was assumed that biomass lysis occurs.

The active biomass concentration is described by

$$\frac{dX}{dt} = \mu \cdot X - k_d \cdot X$$

where μ is the specific growth rate: and k_d is the death rate. The measured biomass, X_m, includes both active and inactive biomass:

$$\frac{dX_m}{dt} = \mu \cdot X - k_m$$

k_m represents a constant lysis rate. The consumption rate of the nitrogen source, urea, is given by

$$\frac{dU}{dt} = -k \cdot U$$

where k is the conversion rate from urea to available nitrogen for the microorganism, N_l, which can be directly metabolized into active biomass. The following expression represents its evolution during cultivation:

$$\frac{dN_I}{dt} = 0.47 \cdot k \cdot U - \mu \cdot \frac{X}{Y_{X/N_I}}$$

where 0.47 corresponds to the nitrogen content of the urea.

The specific growth rate is then

$$\mu = \mu_m \cdot \left(\frac{N_I}{k_N \cdot X + N_I}\right)$$

The evolution of the carbon source, soluble starch, is given by

$$\frac{dG}{dt} = -\mu \cdot \frac{X}{Y_{X/G}} - m_G \cdot X$$

where m_G is the maintenance coefficient. Finally, the respiration activity is described by the accumulated CO_2 using a standard linear growth model:

$$\frac{dCO_2}{dt} = \mu \cdot \frac{X}{Y_{X/CO_2}} + m_{CO_2} \cdot X$$

The following figures compare experimental data, obtained at 31 °C, with simulations using this model (the parameters are given in Table 1).

Figure 6 shows the expected three growth phases: exponential, deceleration, and lysis. The model reproduces accurately the first and second phases, as well as the trend of the lytic phase. The data of the latter show high dispersion, which makes it difficult to fit it adequately. Urea concentration disappeared before 20 h of cultivation, much earlier than the maximum biomass concentration was reached (Fig. 7). This result supports the assumption that urea is not directly used by the microorganism for growth.

Table 1. Model Parameters

Parameter	31 °C
μ_m (h^{-1})	0.20
k_N (g N_l/g X)	0.55
k_d (h^{-1})	0.022
k_m (h^{-1})	5.0×10^{-5}
$Y_{X/G}$ (g X/g G)	0.35
m_G (g G/g X h)	0.028
$Y_{X/N\,l}$ (g X/g N_l)	4.4
m_{CO_2} (g CO_2/g X h)	0.12
Y_{X/CO_2} (g X/g CO_2)	0.19
β_m (h^{-1})	1.22×10^{-3}
γ (g/g i.s.)	0
K_l (g i.s./g)	996
k_p (h^{-1})	1.0×10^{-6}
k (h^{-1})	0.1330

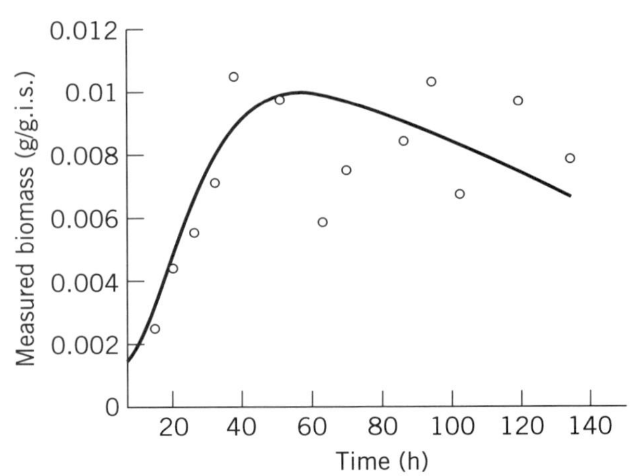

Figure 6. Biomass: measured (*circles*) and simulated (*line*).

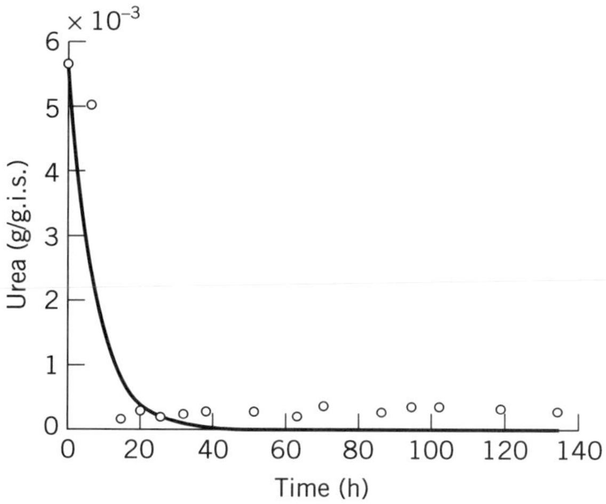

Figure 7. Urea concentration: measured (*circles*) and simulated (*line*).

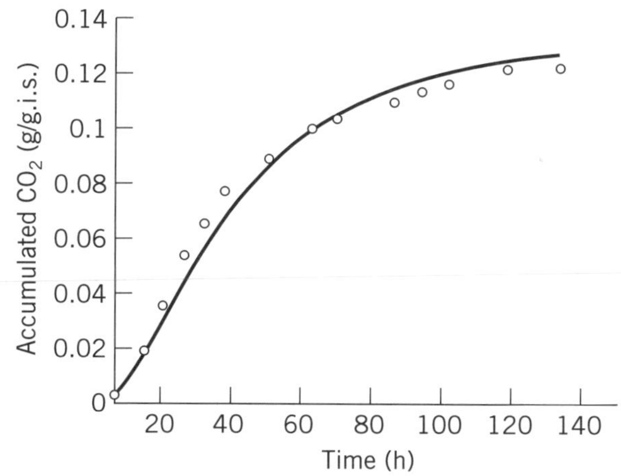

Figure 8. Accumulated CO_2: measured (*circles*) and simulated (*line*).

The model is able to reproduce well the observed decline in biomass respiration activity (Fig. 8). The carbon source is not limiting for growth as seen in Figure 9, because enough starch remains even after growth has stopped.

These figures show that the model is able to describe the complex interaction between active biomass concentration, nutrients, and respiration activity. This model can serve several purposes, such as establishing a fed-batch policy to maximize biomass concentration or developing on-line biomass estimators (74).

Production. The product of interest is a growth hormone called gibberellic acid (GA_3), which is a secondary metabolite; its production starts after nitrogen depletion. GA_3 production in SSF is difficult because it decomposes in water (75), so that if cultivation continues too long, most of the product will be lost. The following model can reproduce this observed behavior:

$$\frac{dP}{dt} = \beta \cdot X - k_P \cdot P$$

where k_p is the degradation rate and $\boldsymbol{\beta}$ is the specific production rates, which depends on the available nitrogen:

$$\beta = \frac{N_I \cdot \beta_m}{\gamma + N_I + K_l \cdot N_I^2}$$

Figure 10 shows the evolution of product concentration. The model fits the experimental data well, so it can be used to developed an on-line GA_3 estimator or a fed-batch policy to maximize product concentration at a given time.

INDUSTRIAL IMPLEMENTATION

To develop a complete SSF process that can be scaled up to industrial level, production must be integrated with the following two critical stages.

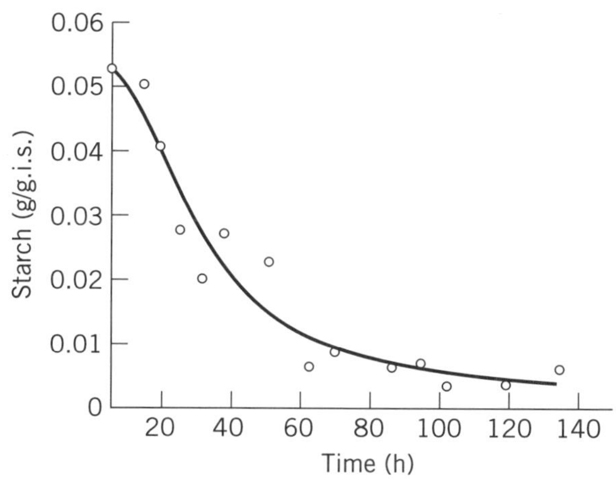

Figure 9. Carbon source evolution: measured (*circles*) and simulated (*line*).

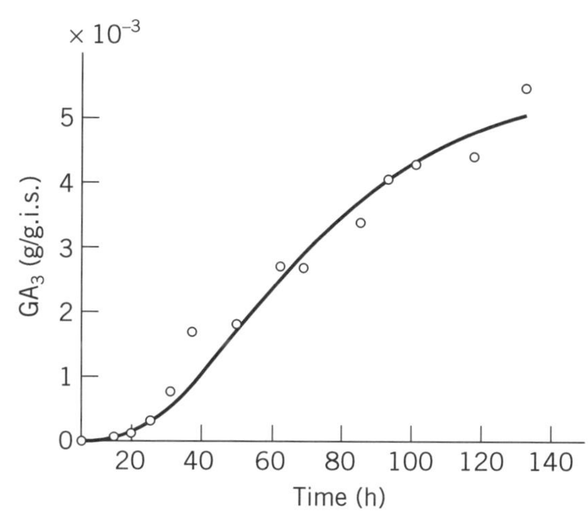

Figure 10. GA_3 concentration: measured (*circles*) and simulated (*line*).

Upstream Processing

In SSF processes, supports are normally made from either agricultural products (soybean, rice, wheat, corn) or by-products (wheat bran, sugarcane bagasse, cassava bagasse). These materials are generally supplemented with easily metabolized carbon and energy sources (starch, maltodextrin, molasses, etc.) and other nutrients (urea, calcium, phosphates, etc.). The preprocessing of these substrates, mainly thermal treatment, has a double objective:

- Sterilize the support
- Change substrate physicochemical properties (starch gelatinization, increase porosity, etc.) to favor the support's structural strength and also the degradability and accessibility of its polymeric components (polysaccharides and proteins)

Minimizing the endogenous microbial load present in these substrates to levels compatible with the growth of the microorganism of interest is fundamental to attaining proper cultivation. It is easy to understand that this operation shares a significant part of the production costs when considering the high volumes of solids that are handled in industrial reactors. Lonsane et al. (76) have proposed the use of batch sterilization for 3–6 h at 121 °C, emphasizing that each solid particle must be subjected to 121 °C for at least 60 min. This, however, does not seem feasible for industrial-scale processes.

Traditionally, this stage is performed through prolonged cooking of the support, that is, rice and soybean in the particular case of koji production. For the latter, toasting has been recently been introduced as an alternative to cooking, with encouraging results in terms of productivity (Tomomori, personal communication, 1992). In 1995 Chou and Rwan (77) evaluated the extrusion cooking of rice for koji production. Here, *Aspergillus oryzae* was cultivated on rice extruded at different temperatures and water content levels, and achieved enzymatic yields superior to those obtained with traditional cooking techniques. Extrusion is a technology widely applied in the formulation of foods for human and animal consumption. However, its use in the preprocessing of substrates for SSF has been limited (78,79). Extrusion is essentially a fast process at high temperature in which proteic nutrients, rich in starch, are subjected to considerable heat and mechanical stresses, which modify the structure of the original material (80). Extrusion provides cooked substrates with the following characteristics:

- Elimination or at least significant reduction of the initial microbial load
- Liquefaction and gelatinization of the starch
- Denaturation of proteins and inactivation of antinutritional factors
- Solid form and porosity, which improve microbial growth and resistance to mechanical stresses
- Homogeneous substrate granulometry, increasing interparticular gas mass transfer

Special substrates made from products rich in fiber, such as sugar beet (81) or wheat bran (82), and amylaceous products, have also been developed. Recent results obtained in our laboratory in the production of gibberellic acid by *G. fujikuroi* have demonstrated the benefits of using this treatment. A mixture of wheat bran and starch (80%: 20%) extruded at 160 °C and at 22% humidity achieved a volumetric productivity 25% higher than that obtained when using nonextruded material.

In conclusion, this technique appears particularly appropriate for industrial-level SSF preprocessing. An aseptic support is produced at shorter processing times, with higher productivities and reduced costs. However, the process requires considerable investment. Other promising techniques that are currently being evaluated are sterilization by microwave (83), high voltages, and magnetic fields.

Downstream Processing

Downstream processing is one of the most neglected aspects of SSF processes in spite of its importance from both technical and economical standpoints. Traditional SSF-derived products generally encompass the whole fermented mass (koji, ang-kak, tempeh, etc.), which is, at most, dried (tane koji, for example) to improve keeping quality. However, there is a growing number of products that now require purification and formulation before being marketed. It is worth noting that the usual operations tied to these stages, although well developed for liquid cultivation, cannot be directly applied to SSF. Indeed, it is necessary first to leach the product from the fermented solids.

Following solid substrate cultivation, the fermented mass consists of unused substrate, biomass, spores (when formed), the product of interest, and a series of other cometabolites (76,84). The product in question must be extracted from this mass with an appropriate solvent. Efficiency of leaching depends on a series of factors:

- Effectiveness of the selected solvent
- Solid-solvent proportion
- Product and solvent diffusion
- Solvent retention by the solids
- Time, degree of mixing, temperature, and pH
- Techniques used to separate the leached solution

Among the available separation techniques, the most commonly used are percolation, pulsed extraction in plug flow, multistage countercurrent extraction, hydraulic press, and supercritical extraction. It is imperative to the economic viability of the SSF process that the selected extraction technique maintains the high concentration of the product to the greatest extent possible. Indeed, if the leaching process resulted in the dilution of the product, even at concentrations similar to those obtained by submerged fermentation, one of the main advantages of SSF would be lost. As a result, while common, percolation is not an appropriate technique because the high dilution of the resulting product (1:10 on the average) demands an additional concentration stage (84).

Plug flow pulsed extraction is a variation of percolation (85) in which the solvent is added by pulses through aspersion of the substrate. With such a flow, the solvent completely wets the solid, achieving optimum solid to solvent proportions (1:2 to 1:3). A further advantage is that this is a simple, low-cost technique that requires minimum supervision. On the negative side, the extraction period is long, which may result in product modification or degradation.

Countercurrent extraction (86) allows adequate product concentration (1:1 solid:solvent proportion). In this system, the solids and the solvent move in opposite direction. Initially, the fresh solvent makes contact with the leached solids then with partially leached solids, contacting increasing concentrations of product. Finally, the enriched solvent contacts fresh solid. The required number of contact stages varies between 2 and 5 (86,87). In spite of the high concentrations achieved, the method presents some limitations in that the operation needs constant attention from trained operators. In addition, the extract obtained can be highly viscous, resulting from the extraction of exopolysaccharides. Extraction by hydraulic press is an efficient operation, although it requires high pressures and a second press to improve efficiency. Moreover, the operation is slow and tedious.

Product extraction with supercritical fluids has been studied for solid-state fermentation; in particular with CO_2 (88). Solvents have greater extraction ability at near-critical temperatures and pressures. This property is being used increasingly in the chemical industry as an alternative to conventional distillation and to other extraction processes. Here, the separation stages are simplified because the products are volatilized or precipitated from the supercritical fluid. However, the capital and operational costs of this process are high, restricting its applicability to high-value products. Once the product is leached from the solids, the extract is similar to a cultivation broth obtained following submerged fermentation. Conventional bioseparation (concentration, purification) and product formulation methods normally used for liquid cultivation processes can then be applied.

Case Study: Production of Gibberellins by Solid Substrate Cultivation of *Gibberella fujikuroi*

All commercially available gibberellins, which are typical secondary metabolites, are produced by liquid cultivation (LC) of the ascomycetous fungus *G. fujikuroi* (88). On exhaustion of the nitrogen source, exponential fungal growth ceases. At this point, secondary metabolism is triggered, with the concomitant biosynthesis of gibberellins. Commercially, the most important resultant product is gibberellic acid (GA_3). Several authors have reported on gibberellic acid production in solid substrate fermentation with higher yields than those obtained from liquid fermentation (6,9,37,89). A flowsheet of a pilot plant (Fig. 11) used for the production of gibberellic acid is presented and discussed.

This process, which has been scaled up from laboratory (12 g) to prepilot level (50 kg) and, recently, on up to a 200-kg reactor (Fig. 12), is carried out in three stages. The first, *upstream*, includes fungal propagation and substrate

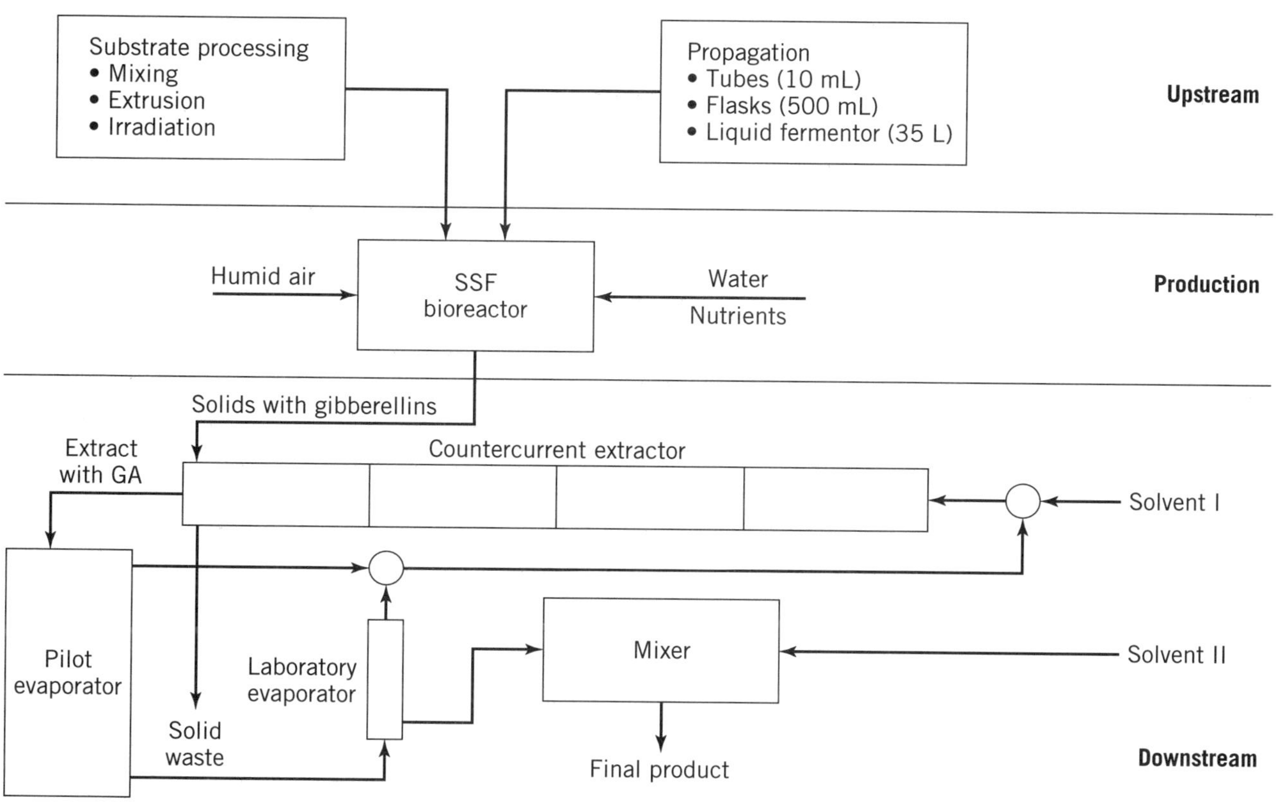

Figure 11. Flowsheet of a SSF pilot plant for gibberellin production.

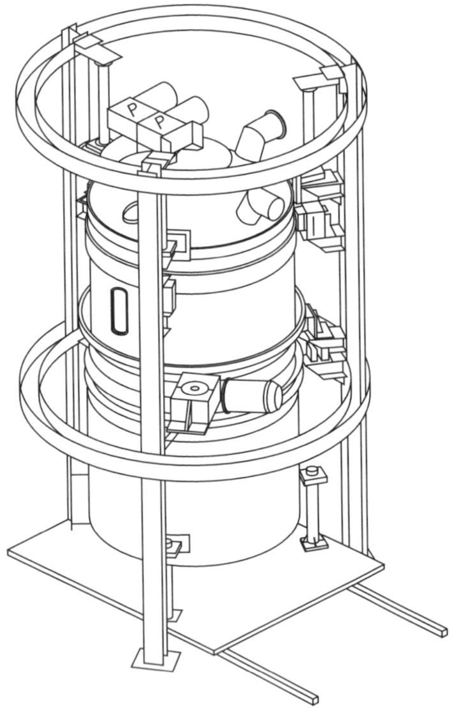

Figure 12. A 200-kg SSF pilot bioreactor.

preparation; second, the *production phase*; and finally, *downstream*, which consists of extraction and formulation. Substrate preparation involves mixing wheat bran with starch, extrusion of the mixture at high temperatures, and gamma ray irradiation. Mycelial propagation of the fungus was achieved in 72 h, from 10 mL up to 35 L. Production is carried out in batch form following a defined sequence: reactor sterilization (3 h), substrate loading (1 h), and inoculation (0.5 h), water content adjustment (0.5 h), cultivation (5 days), unloading (2 h), and bioreactor cleaning (4 h).

Extraction is done in batch form in four countercurrent stages. A primary concentration is then performed in a 15-L batch evaporator followed by final concentration in a laboratory evaporator. The last step is the formulation of the product according to market needs. The results obtained are easily reproducible (Fig. 13), and the economic viability and technical feasibility of the whole process have been demonstrated.

On field evaluation in production-scale vineyards, the product obtained (FITOPLUS) via this process exhibited considerable benefits over commercially available gibberellic acid products. Thompson seedless grapes of superior quality (berry size, reduction in berry loss, berry stalk color, etc.) were consistently obtained when treated with the SSF-derived product. This product has been placed under patent protection for agricultural applications (90).

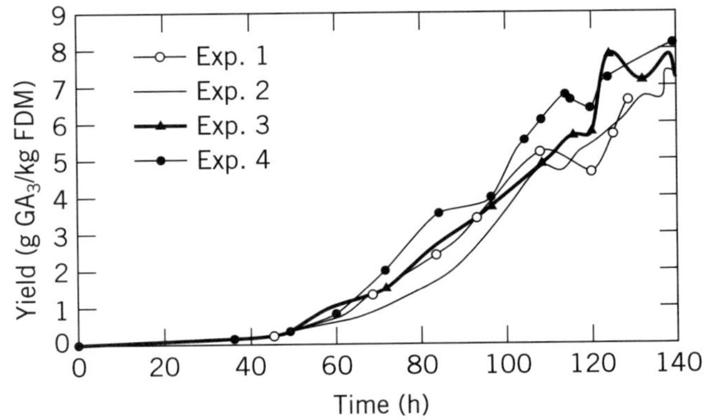

Figure 13. Gibberellic acid production in a SSF pilot bioreactor.

NOMENCLATURE

G	Starch concentration (g/g i.s.)
k	Urea transformation constant (L/h)
k_d	Biomass death constant (L/h)
K_i	Inhibition constant (g i.s./g)
k_m	Lysis constant (L/h)
k_N	Contois model parameter (g N_I/g X)
k_p	GA_3 degradation constant (L/h)
m_G	Starch maintenance constant (g G/g X h)
m_{CO_2}	CO_2 maintenance constant (g CO_2/g X h)
N_1	Available nitrogen (g/g i.s.)
P	Gibberellic acid (GA_3) (g/g i.s.)
U	Urea (g/g i.s.)
X	Active biomass (g/g. i.s.)
X_m	Measured biomass (g/g. i.s.)
$Y_{X/G}$	Biomass/starch yield coefficient (g X/g G)
$Y_{X/NI}$	Biomass/nitrogen yield coefficient (g X/g N_I)
Y_{X/CO_2}	Biomass/CO_2 yield coefficient (g X/g CO_2)
β	Specific production rate (L/h)
β_m	Maximum specific production rate (L/h)
γ	Parameter in the production rate (g/g i.s.)
μ	Biomass specific growth rate (L/h)
μ_m	Maximum specific growth rate (L/h)

ACKNOWLEDGMENTS

Financial support through the projects Fondecyt 1960360 and FONDEF 2-50, and from DICTUC S.A. and BASF-Chile S.A., is greatly appreciated. The work accomplished would not have been possible without the fruitful and continual discussions with our colleagues Iván Solar, Luciano Chiang, and Héctor Jorquera (Pontificia Universidad Católica de Chile), Gonzalo Acuña (Universidad de Santiago de Chile), Georges Corrieu and Erik Latrille (INRA, France), Paul Molin (ENSBANA, France), Maurice Raimbault (ORSTOM, France), and Jules Thibault (Université de Laval, Canada). Our dedicated postgraduate students undertook most of the experimental and computational work and also contributed several original ideas; our thanks go to Andrés Corona, Armin Ebner, Mario Fernández, Claudio Gelmi, Mauricio González, Harold Paján, Marcial Peña y Lillo, Ariel Rossenblitt, and Daniel Volpe. The skillful technical assistance of Andrea Belancic, Lenka Torres, and Cristián Valenzuela is also acknowledged. The English translation was carefully revised by Alex Crawford.

BIBLIOGRAPHY

1. R.E. Mudgett, in A. Demain and N. Solomon eds., *Manual of Industrial Microbiology and Biotechnology*, American Soc. for Microbiology, Washington, D.C., 1986, pp. 66–83.
2. A. Barrios-González, A.G. Tomasini, G. Viniegra-González, and J. Lopez, *Biotechnol. Lett.* **10**, 793–798 (1988).
3. M. Jermini and A.L. Demain, *Experientia (Basel)* **45**, 1061–1065 (1989).
4. S. Roussos, A. Olmos, M. Raimbault, G. Saucedo-Castañeda, and B.K. Lonsane, *Biotechnol. Tech.* **5**, 415–420 (1991).
5. P. Christen, E. Villegas, and S. Revah, *Biotechnol. Lett.* **16**, 1183–1188 (1994).
6. P.K. Kumar and B.K. Lonsane, *Biotechnol. Bioeng.* **30**, 267–271 (1987).
7. P.K. Kumar and B.K. Lonsane, *Appl. Microbiol. Biotechnol.* **34**, 145–148 (1990).
8. X. Qian, J. du Preez, and S. Kilian, *World J. Microbiol. Biotechnol.* **10**, 93–99 (1994).
9. E. Agosin, M. Maureira, V. Biffani, and F. Pérez, in S. Roussos, B.K. Lonsane, M. Raimbault, and G. Viniegra-González eds., *Advances in Solid State Fermentation*, Kluwer, Paris, 1997, pp. 355–366.
10. M.V. Ramesh and B.K. Lonsane, *Appl. Microbiol. Biotechnol.* **33**, 501–505 (1990).
11. D.A. Mitchell and B.K. Lonsane, in H.W. Doelle, D.A. Mitchell, and C.E. Rolz eds., *Solid Substrate Cultivation*, Elsevier, London, 1992, pp. 1–13.
12. R. Auria, S. Hernández, M. Raimbault, and S. Revah, *Biotechnol. Tech.* **4**, 391–396 (1990).
13. G. Muñoz, E. Agosin, M. Cotoras, R. San Martin, and D. Volpe, *FEMS Microbiol. Lett.* **125**, 63–70 (1995).
14. E. Agosin and J.M. Aguilera, in G.E. Harman and C.P. Kubicek eds., *Trichoderma and Gliocladium Vol. 2*, Taylor and Francis, London, 1998, pp. 205–228.
15. W.R. Gibbons, C.A. Westy, and T.L. Dobbs, *Biotechnol. Bioeng.* **26**, 1098–1107 (1984).
16. J. Klein, in D.L. Wise ed., *Global Environmental Biotechnology*, Kluwer, Dordrecht, 1997, pp. 489–504.
17. S. Roussos and D.L. Pyle, in M. Raimbault, R. Soccol, and G. Chuzel eds., *International Training Course on Solid State Fermentation*, ORSTOM, Montpellier, 1998, pp. 21–35.
18. M.G. Byndoor, N.G. Karanth, and G.V. Rao, in S. Roussos, B.K. Lonsane, M. Raimbault, and G. Viniegra-González eds., *Advances in Solid State Fermentation*, Kluwer, Paris, 1997, pp. 113–119.
19. A.A. De Araujo, C. Lepilleur, S. Delcourt, P. Colavitti, and S. Roussos, in S. Roussos, B.K. Lonsane, M. Raimbault, and G. Viniegra-González eds., *Advances in Solid State Fermentation*, Kluwer, Paris, 1997, pp. 93–111.
20. D. Ryoo, V.G. Murphy, M.N. Karim, and R.P. Tengerdy, *Biotechnol. Tech.* **5**, 19–24 (1991).
21. E. Cannel and M. Moo-Young, *Process Biochem.* **15**, 24–28 (1980).
22. A. Durand and D. Chereau, *Biotechnol. Bioeng.* **31**, 476–486 (1988).
23. Y. Chameliec, R. Renaud, J. Maratray, S. Almanza, M. Diez, and A. Durand, *Biotechnol. Tech.* **8**, 245–248 (1994).
24. M. Fernández, J. Ananias, I. Solar, R. Pérez, L. Chang, and E. Agosin, in S. Roussos, B.K. Lonsane, M. Raimbault, and G. Viniegra-González eds., *Advances in Solid State Fermentation*, Kluwer, Paris, 1997, pp. 155–168.
25. M. Fernández, R. Pérez-Correa, I. Solar, and E. Agosin, *Bioprocess Eng.* **16**, 1–4 (1996).
26. E. Agosin, R. Pérez-Correa, M. Fernández, I. Solar, and L. Chiang, in D.L. Wise ed., *Global Environmental Biotechnology*, Kluwer, Dordrecht, 1997, pp. 233–243.
27. G. Saucedo-Castañeda, M. Gutierrez-Rojas, G. Bacquet, M. Raimbault, and G. Viniegra-González, *Biotechnol. Bioeng.* **35**, 802–808 (1990).
28. H. Paján, R. Pérez-Correa, I. Solar, and E. Agosin, in D.L. Wise ed., *Global Environmental Biotechnology*, Kluwer, Dordrecht, 1997, pp. 221–232.
29. M. Fernández, R. Pérez-Correa, and M. Peña y Lillo, in G. Lefranc, M. Duarte, and R. Pérez-Correa eds., *Anales del VIII Congreso Latinoaméricano de Control Automático*, ACCA, Santiago, 1998.
30. K. Sato, M. Nagatani, and S. Sato, *J. Ferment. Technol.* **60**, 607–610 (1982).
31. G. Saucedo-Castañeda, M. Trejo-Hernández, B.K. Lonsane, J.M. Navarro, S. Roussos, D. Dufour, and M. Raimbault, *Process Biochem.* **29**, 13–24 (1994).
32. A. Durand, R. Renaud, J. Maratry, and S. Almanza, in S. Roussos, B.K. Lonsane, M. Raimbault, and G. Viniegra-González eds., *Advances in Solid State Fermentation*, Kluwer, Paris, 1997, pp. 71–111.
33. M.V. Ramana, N.G. Karanth, and K.S.M.S. Raghava Rao, *Adv. Appl. Microbiol.* **39**, 214–265 (1993).
34. J. de Reu, M.H. Zwietering, F.M. Rombouts, and M.J.R. Nout, *Appl. Microbiol. Biotechnol.* **40**, 261–265 (1993).
35. L.M. Barstow, B.E. Dale, and R.P. Tengerdy, *Biotechnol. Tech.* **2**, 237–242 (1988).
36. H. Narahara, Y. Koyama, T. Yoshida, P. Atthasampunna, and H. Taguchi, *J. Ferment. Technol.* **62**, 453–459 (1984).
37. S. Bandelier, R. Renaud, and A. Durand, *Process Biochem.* **32**, 141–145 (1997).
38. S. Tao, L. Beihui, and L. Zuohu, *J. Chem. Technol. Biotechnol.* **69**, 429–432 (1997).
39. C.J. Harris and S.A. Billings, *Self-Tuning and Adaptive Control: Theory and Applications*, IEEE Control Engineering Series, Peregrinus, London, 1981.
40. G.C. Goodwin and K.S. Sin, *Adaptive Filtering Prediction and Control*, Prentice Hall, Englewood Cliffs, N.J., 1984.
41. K. Schügerl, J. Müller, J. Niehoff, R. Hiddessen, K. Früh, and A. Lübbert, in N.M. Fish, R.I. Fox, and N.F. Thornhill eds., *Computer Applications in Fermentation Technology*, Elsevier, London, 1989.
42. S. Freyer, R. Eberhardt, D. Schurbuscher, and C. Wandrey, in N.M. Fish, R.I. Fox, and N.F. Thornhill eds., *Computer Applications in Fermentation Technology*, Elsevier, London, 1989.
43. J.M. Flaus, A. Cheruy, and J.M. Engasser, *J. Process Control* **1**, 271–281, 1991.
44. G. Montague, A. Morris, A. Wright, M. Aynsley, and A. Ward, *Can. J. Chem. Eng.* **64**, 567–580 (1986).
45. G. Bastin and D. Dochain, *On-Line Estimation and Adaptive Control of Bioreactors*, Elsevier, Amsterdam, 1990.
46. Y. Pomerleau and G. Viel, in *Proc. IFAC Symp. on Modelling and Control of Biotechnical Processes*, Keystone, Colo., 1992, pp. 315–318.

47. L. Chen, G. Bastin, and V. Van Breusegem, *Automatica* **31**, 55–65 (1995).
48. C.R. Cutler and B.L. Ramaker, *Proc. Joint Auto. Control Conf.*, paper WP5-B, San Francisco, 1980.
49. R. Auria, M. Morales, E. Villegas, and S. Revah, *Biotechnol. Bioeng.* **41**, 1007–1013, (1993).
50. J.M. Brass, F.W.J.M.M. Hoeks, and M. Rohner, *J. Biotechnol.* **59**, 63–72 (1997).
51. R. Simutis, R. Oliveira, M. Manikowski, S. Feyo de Azevedo, and A. Lübbert, *J. Biotechnol.* **59**, 73–89 (1997).
52. J. Sargantanis, M. Karim, V. Murphy, D. Ryoo, and R.P. Tengerdy, *Biotechnol. Bioeng.* **42**, 149–158 (1993).
53. P. Sangsurasak and D.A. Mitchell, *Chem. Eng. J.* **60**, 199–204 (1995).
54. J. Bailey and D. Ollis, *Biochemical Engineering Fundamentals*, McGraw-Hill, New York, 1986.
55. H.W. Blanch, *Chem. Eng. Commun.* **8**, 181–211 (1981).
56. D. Mitchell, D. Duong, P. Greenfield, and H. Doelle, *Biotechnol. Bioeng.* **38**, 353–362 (1991).
57. M. Gutiérrez-Rojas, R. Auria, J.C. Benet, and S. Revah, *Chem. Eng. J.* **60**, 189–198 (1995).
58. P. Gervais, P. Molin, W. Grajek, and M. Bensoussan, *Biotechnol. Bioeng.* **31**, 457–463 (1988).
59. S. Rajagopalan and J.M. Modak, *Chem. Eng. Sci.* **49**, 2187–2193 (1994).
60. J.C. De Reu, Ph.D. Thesis, Agricultural University of Wageningen, 1995.
61. J.P. Smits, *Solid-State Fermentation: Modelling Fungal Growth and Activity*, Ph.D. Thesis, Agricultural University of Wageningen, 1998.
62. K.W. Szewczyk and L. Myszka, *Bioprocess Eng.* **10**, 123–126 (1994).
63. S. Rajagopalan and J.M. Modak, *Chem. Eng. Sci.* **50**, 803–811 (1995).
64. S. Rajagopalan and J.M. Modak, *Bioprocess Eng.* **13**, 161–169 (1995).
65. S.J. Pirt, *Proc. R. Soc. London* **163**, 224–231 (1965).
66. P.J. Blanc, M.O. Loret, and G. Goma, in S. Roussos, B.K. Lonsane, M. Raimbault, and G. Viniegra-González eds., *Advances in Solid State Fermentation*, Kluwer, Paris, 1997, pp. 379–391.
67. S. Kumaran, C.A. Sastry, and S. Vikineswary, in D.L. Wise ed., *Global Environmental Biotechnology*, Elsevier, Amsterdam, 1997, pp. 239–248.
68. S. Vikineswary, K.S. Kumaran, S.K. Ling, N. Dinesh, and Y.L. Shim, in D.L. Wise ed., *Global Environmental Biotechnology*, Kluwer, Dordrecht, 1997, pp. 301–305.
69. E.L. Gaden, Jr., *J. Biochem. Microbiol. Technol. Eng.* **1**, 413 (1959).
70. G.F. Froment and K.B. Bischoff, *Chemical Reactor Design and Analysis*, Wiley, New York, 1979.
71. M. Raimbault, *Fermentation en Milieu Solide: Croissance de Champignons Filamenteux sur Substrat Amylacé*, ORSTOM, Paris, 1981.
72. D. Ryoo, Ph.D. Thesis, Colorado State University, 1990.
73. P.D. Sharma, P.J. Fisher, and J. Webster, *Trans. Br. Mycol. Soc.* **69**, 479–483 (1977).
74. A. Ebner, I. Solar, G. Acuña, R. Pérez-Correa, and E. Agosin, in D.L. Wise ed., *Global Environmental Biotechnology*, Kluwer, Dordrecth, 1997, pp. 211–220.
75. F. Pérez, A. Vecchiola, M. Pinto, and E. Agosin, *Phytochemistry* **41**, 675–679 (1996).
76. B.K. Lonsane, G. Saucedo-Castañeda, M. Raimbault, S. Roussos, G. Viniegra-Gonzalez, N. Ghildyal, M. Ramakrishna, and M. Krishnaiah, *Process Biochem.* **27**, 259–273 (1992).
77. C.C. Chou and J.H. Rwan, *J. Ferment. Bioeng.* **79**, 509–512 (1995).
78. A. Fujita, Y. Watanabe, K. Kishi, G. Ogawa, and T. Yoshizaki, *Nippon Shokuhin Kogyo Gakkaishi* **27**, 483–488 (1980).
79. Japanese Pat. Showa 61-205479 (1986).
80. J.M. Harper, *CRC Crit. Rev. Food Sci. Nutr.* **11**, 155–215 (1981).
81. S. Lue, F. Hsieh, and H.E. Huff, *Cereal Chem.* **68**, 227–234 (1991).
82. W.M. Wang, F. Klopfenstein, and G. Ponte Jr., *Cereal Chem.* **70**, 707–711 (1993).
83. M. Torrejón, Tesis de Ingeniero Civil de Industrias, Pontificia Universidad Católica de Chile, 1996.
84. B.K. Lonsane and M.M. Kriahnaiah, in H.W. Doelle, D.A. Mitchell, and C.E. Rolz eds., *Solid Substrate Cultivation*, Elsevier, London, 1992, pp. 147–171.
85. N.P. Ghildyal, M. Ramakrishna, B.K. Lonsane, and N.G. Karanth, *Process Biochem. Int.* **25** (1990).
86. R.W. Treybal, *Mass Transfer Operations*, Kin Keong, Singapore, 1981, pp. 737–765.
87. P.K. Kumar and B.K. Lonsane, *Process Biochem.* **22**, 139–143 (1987).
88. A. Borrow, E.G. Jefferys, R.H.J. Kessels, E.C. Lloyd, P.B. Lloyd, and I.S. Nixon, *Can. J. Microbiol.* **7**, 227–255 (1964).
89. P.K. Kumar and B.K. Lonsane, *Biotechnol. Lett.* **9**, 179–182 (1987).
90. U.S. Pat. Pending (1998), M. Pinto, E. Agosin, J.R. Pérez, F. Pérez, and A. Vecchiola.

SOLID SUBSTRATE FERMENTATIONS, ENZYME PRODUCTION, FOOD ENRICHMENT

YUSUF CHISTI
University of Almería
Almería, Spain

KEY WORDS

Bioreactors
Enzymes
Fermentors
Fermented foods
Indigenous fermented foods
Leaching
pH control
Product recovery
Scale-up
Temperature control
Water activity

OUTLINE

INTRODUCTION

Solid substrate fermentations involve microbial modification of a solid, undissolved substrate. This definition includes solid-state fermentations in which microbial cultures grow on a moist solid with little or no free water, although capillary water may be present (1). Examples of this type are seen in mushroom cultivation (2–4), bread making (5,6), and production of cheeses (7). Examples in which the solid substrate is submerged in large amounts of free water are pickling of vegetables (8) and brewing of soy sauce and sake (9,10), as well as bioconversion of cellulosic materials (11,12).

Solid substrate fermentations have been practiced worldwide since ancient times (9,13,14). Bread, sausages, and soy sauce are a few familiar products of solid substrate fermentation. Such fermentations are used in processing coffee and cocoa, as well as in environmentally significant operations: composting, silage production, soil bioremediation, biodesulfurization of coal, and metal recovery from low-grade ores (15). Solid substrate fermentations are employed also in producing enzymes and chemicals and in biotransformations.

Despite long history and widespread use, most solid substrate fermentations are not well understood. Processing is often an art, commonly employing poorly defined, naturally occurring, mixed microbial populations. Mixed cultures are sometimes essential to development of the desired product characteristics, including appearance, aroma, texture, and taste. With few exceptions, fermentations are labor-intensive batch operations. The level of technical sophistication and capital outlay vary a great deal. Solids concentration is generally high, whereas the water content is usually low (16). The product may be highly concentrated in comparison with submerged fermentations. Similarly, water consumption is comparatively lower, but space requirements can be substantial. Solid substrate culture generally requires less power because agitation is not excessive. A low-pressure blower is sufficient for air supply, unlike the higher-pressure compressors required in submerged fermentations (17). Unlike submerged culture, solid-state processes require extensive solids handling, which is difficult; nevertheless, highly mechanized, automated, and continuous processing of solids is being practiced in large *koji* (or molded grain) factories (9,10,18), as well as for other processes. Processing machinery tends to be complex. Hygienic processing practices are followed, but sterility standards that are common in submerged culture production of pharmaceuticals are not attained. Because of fungal spores and product- or microbe-contaminated dust from substrate solids, solid substrate fermentations often pose a greater health risk to operators than do submerged fermentations.

FERMENTATION EQUIPMENT

Solid substrate fermentation devices vary in technical sophistication from the very primitive banana leaf wrappings to highly automated machines used mainly in Japan. Simple fermentation systems (e.g., fermentation of cocoa beans in heaps) are quite effective in some large-scale processes. Commonly used fermentation devices are detailed in the following sections.

Tray Fermentors

One of the simplest and widely used fermentors is a wooden, metal, or plastic tray, often with a perforated or wire mesh bottom to improve air circulation (1,19,20). A shallow layer, usually less than 0.15 m deep, of pretreated (e.g., steamed) substrate is placed on the tray for fermentation. Individual trays or stacks may be located in temperature- and humidity-controlled chambers (Fig. 1) or simply in ventilated areas. A spacing of at least one tray height is usually allowed between stacked trays. Trays may be covered with cheesecloth to reduce contamination (10), but strict monosepticity is not attempted. Inoculation and occasional mixing are done manually, often by hand

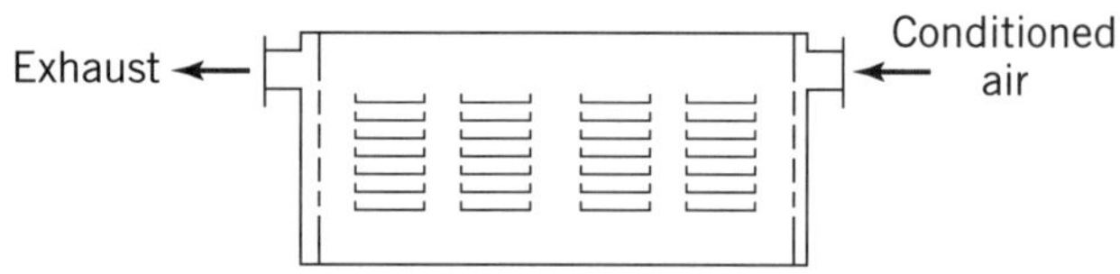

Figure 1. Tray fermentation chamber.

(10). This type of technology is quite common in small- and medium-scale koji operations in Asia. In some cases, filling, emptying, and other aspects of handling (4) of trays may be automated. For example, hinged trays with open ends may be dropped for emptying (19). Despite some automation, tray fermentors are labor intensive, require a large area, and difficulties with processing hundreds of trays limit their scalability (19). In parts of Asia, the bamboo baskets historically used instead of trays (16) continue to be employed.

Static Bed and Tunnel Fermentors

A commercially used modification of the tray fermentor employs a single, larger and deeper, static bed of substrate with forced aeration through the bed (Figs. 2 and 3). The substrate is located in an insulated chamber. In one version of this device, the tunnel fermentor (Fig. 3), the bed of solids may be quite long (1,20), but is usually no deeper than 0.5 m. Tunnel fermentors may be highly automated with mechanisms for mixing (Fig. 3), inoculation, continuous feeding, and harvest of substrate.

Rotary Disk Fermentors

Some large-scale commercial koji-making operations in Japan use a rotary disk fermenter configuration shown schematically in Figure 4 (18). The fermentor consists of upper and lower chambers, each with a circular perforated disk

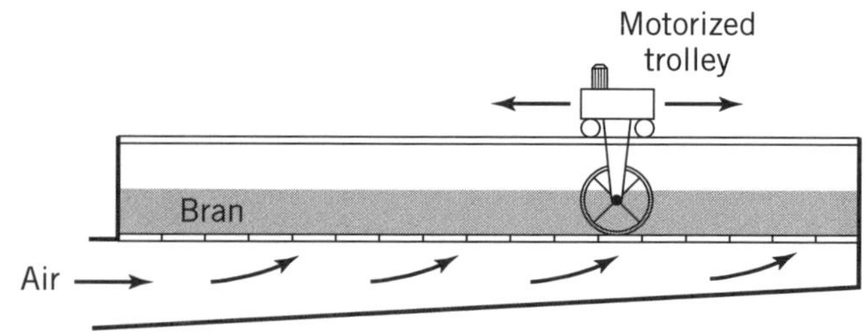

Figure 3. Mechanical mixing of fermenting substrate in a tunnel fermentor.

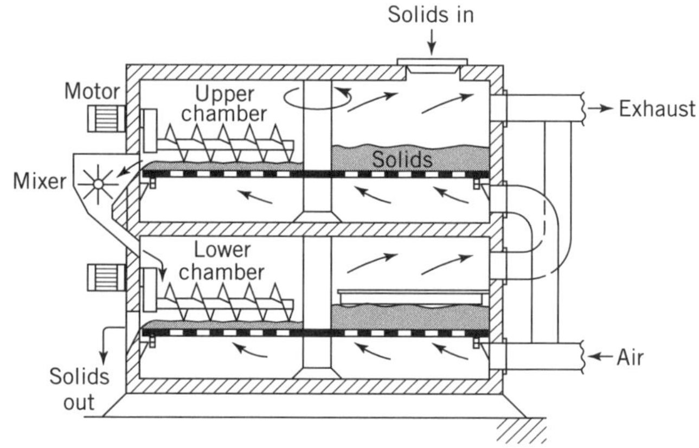

Figure 4. Automatic rotary koji fermentor.

to support the substrate. A common central shaft rotates the disks. Inoculated substrate is introduced in the upper chamber and slowly moved to the transfer screw. The upper screw transfers the partly fermented solids through a mixer to the lower chamber where further fermentation occurs. The mixer breaks up the partly fermented

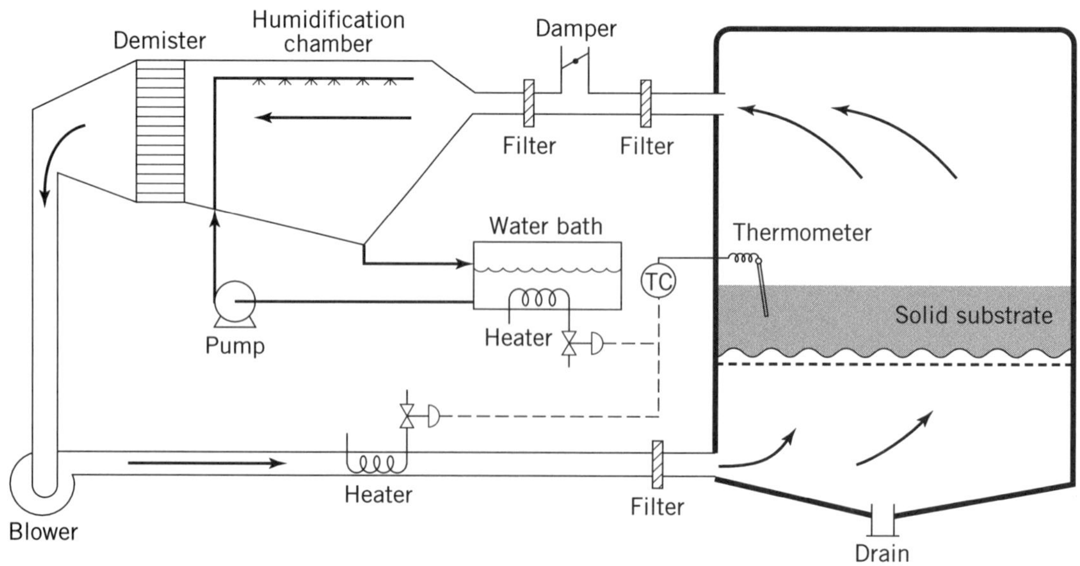

Figure 2. Static bed-type koji fermentor.

substrate–mycelium aggregates halfway through the fermentation process. Fermented substrate is eventually harvested using the lower transfer screw (Fig. 4). Both chambers are aerated with humidified, temperature-controlled air.

A tower fermentor similar in concept to the rotary koji fermentor is used in certain composting operations. The solids are supported on plates or trays that may be perforated. A stack of several tray chambers forms the tower. Mechanical devices mix and move solids down the tower (20–22) in a manner similar to that described for the rotary koji device. The gradually moving bed of solids is aerated either by forced aeration through the solid mass or solely from the surface exposed to the air above the bed.

Rotary Drum Fermentors

Rotary drum fermentors consist of a cylindrical vessel mounted horizontally on rollers and rotated around the long axis (Fig. 5). The rotational speed is commonly 1–5 rpm, but speeds up to 15 rpm are used occasionally. Rotation may be intermittent, and the speed may vary with the fermentation stage. The vessel may have straight or curved internal baffles that aid aeration and temperature control by imparting a tumbling motion to the substrate (23). Sometimes the drum may be inclined, causing the substrate to move from the higher inlet end to the lower outlet during rotation. The amount of substrate that can be processed in a batch operation depends directly on the length of the vessel and the square of its diameter. Aeration is through coaxial inlet and exhaust nozzles (Fig. 5). The air inlet pipe may extend into the vessel and may branch into several arms (19). Air is either supplied by a compressor or may be sucked in by an exhaust fan located at the air outlet (19).

Fluidized Beds

Fluidized-bed fermentors (Fig. 6) are relatively uncommon, but sufficiently promising that large-scale devices are being tested (9). A fluidized-bed fermentor consists of a relatively shallow bed of substrate supported on a perforated plate. Compressed conditioned air is normally used for fluidization, but carbon dioxide or nitrogen may be employed for anaerobic processes. Superficial gas velocity for fluidization is typically 0.24–1 m s^{-1}; the specific value depends on the density and the particle size of the substrate. In the upper regions of the fermentor, the cross section is expanded (Fig. 6) to reduce the gas velocity to values that are insufficient to suspend the larger particles. Loss of fines in the exhaust gas is prevented by screens, filters, or cyclonic separators located at the exhaust nozzle. Fluidized beds achieve good aeration, mixing, and temperature control. Intermittent water spray may be necessary to prevent dehydration. A mechanical agitator as in Figure 6 is useful for periodic loosening of clumped solids.

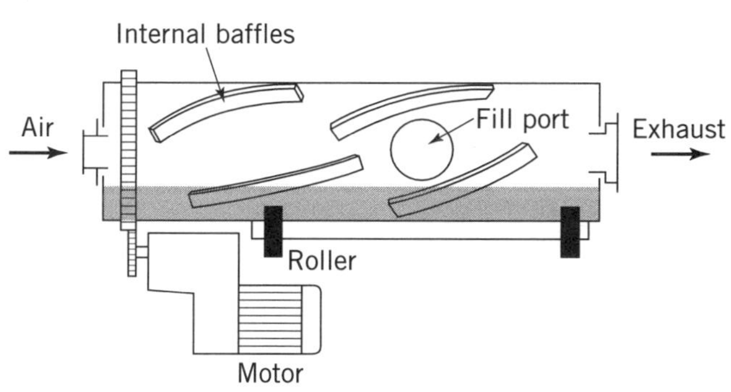

Figure 5. Rotary drum fermentor.

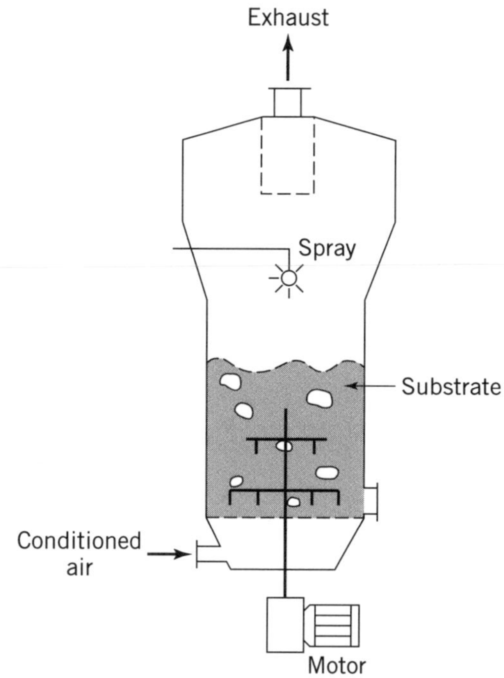

Figure 6. Fluidized-bed fermentor.

Agitated Tank Fermentors

Helical ribbon-stirred tank fermentors have been employed for solid-state culture of fungi such as *Chaetomium cellulolyticum* on wheat straw (24). Other similar designs have utilized multiple helical screws for agitation of large rectangular tanks. The screws extend into tanks from mobile trolleys that ride horizontal rails located above the tanks (25). Yet another stirred tank configuration is the paddle fermentor. This design is similar to the rotary drum device (Fig. 5), except that the drum is stationary and motor-driven paddles on a concentric shaft provide periodic mixing of the substrate (20). As with rotary drums, the quantity of the substrate in paddle fermentors is generally restricted to less than 50% of the volume of the vessel. Although simpler than the rotary drum, the paddle fermentor is unlikely to be as effective in moving the entire mass of solids.

Continuous Screw Fermentors

A screw fermentor suitable for continuous fermentations is shown in Figure 7. Sterilized, cooled, and inoculated sub-

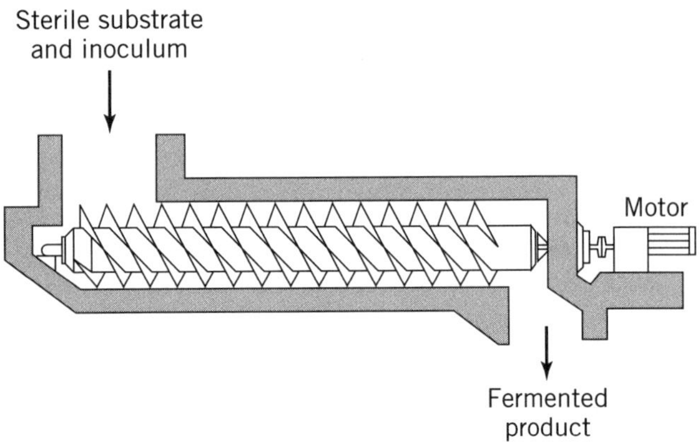

Figure 7. Screw fermentor for continuous fermentation.

strate is fed at the inlet. The screw moves the fermenting solids toward the harvest port. The fermentation time depends on the rotational speed and the length of the screw. Because the device is not aerated, only anaerobic or microaerophilic fermentations may be carried out.

Other Systems

Other miscellaneous types of fermentation devices that are used mostly in small-scale studies have been described by Lonsane et al. (19). Equipment for fermenting undissolved solid substrate slurries is identical to that used in submerged culture processes. Exceptions may be in devices used to feed slurried solids to the fermentor and remove the fermented product. Commonly used slurry fermentors include stirred tanks, bubble columns, and air-lift vessels (26–28). Simple vats are frequently employed in anaerobic slurry fermentations. Fermentors used in various types of composting have been described (14,21,22).

Construction and Cleaning

Modern commercial solid-state fermentors are constructed predominantly of stainless steel, but pressure vessel construction that is the norm for submerged culture processes (29) is not employed. Large concrete or brick fermentation chambers, or koji rooms (Fig. 1), may be lined with steel. Type 304 steel is used typically, but more corrosion-resistant grades such as Typed 304L and 316L are also utilized. Materials of construction have been discussed further by Chisti (29).

Because pressure vessel construction is not used, the fermentors cannot be steam sterilized; however, the design usually allows for atmospheric steaming. Typically, the fermentation chamber and process vessels are washed with alkaline detergent, sanitized with sodium hypochlorite, formaldehyde, or quaternary ammonium salts, and steamed between successive batches (17). Automated clean-in-place methods similar to those noted for submerged culture processes (30) are increasingly employed.

Design and Scale-Up

Design and scale-up of solid-state fermentation equipment relies heavily on empirical experience because the process kinetics and transport phenomena, such as heat transfer, oxygen transfer, and mixing, in these systems are insufficiently understood for more formal approaches to be usefully applied (20). Potentially useful strategies for scale-up have been discussed by Lonsane et al. (31), and Moo-Young et al. (22) have considered aspects of transport phenomena.

ENVIRONMENTAL FACTORS IN SOLID SUBSTRATE FERMENTATIONS

Moisture

Moisture content of the substrate is an important consideration in solid substrate processes. Free water indicates a saturated substrate. The moisture content at which free water is apparent depends on the nature of the solid (Table 1). Many solid substrate fermentations are carried out with little or no free water; typically, the substrate moisture content is 30–80% w/w (1,32). High moisture levels lead to aggregation of substrate particles, poor aeration, and possible anaerobic conditions (24). Steamed rice, a common substrate, becomes sticky when the moisture level exceeds 30–35% w/w. Optimal moisture content depends on the microorganism and the substrate. For the commonly used fungi, the optimal moisture requirement varies between 40% and 80% w/w. For the same organism growing on different substrates, the optimal moisture levels may differ widely; hence, percent moisture by itself is unreliable for predicting growth (32), and water activity is preferred instead. The water activity or relative humidity of the substrate is the ratio of the vapor pressure of water in the substrate to the vapor pressure of pure water at the temperature of the substrate. Water activities below 0.9 do not support most bacterial growth, but yeasts and fungi can grow at water activities of 0.7 and greater. Thus, the low moisture environment of many solid substrate fermentations favors yeasts and fungi. The water activity in a substrate may be determined by hygrometric measurement of relative humidity of a small air space in equilibrium with the substrate.

In addition to affecting growth, water activity affects product formation. The spectrum of products produced and characteristics such as aroma are affected. In some cases, the optimal water activities for growth and product formation differ (32). The optimal water activity depends also on factors such as agitation rate and cultivation temperature (32). Because the water activity depends on the concentration of dissolved solutes, sometimes salts, sugar, or other solutes are added to alter the activity. Different additives may influence the fermentation differently even though the water activity produced may be the same. Furthermore, the fermentation process itself leads to changes

Table 1. Saturation Moisture Content of Some Substrates

Substrate	Percent moisture (w/w)
Maple bark	>40
Rice and cassava	>50–55
Straw	75

in water activity as products are formed and the substrate is hydrolyzed. Oxidation of carbohydrates produces water.

During fermentation the water activity is controlled by aeration with humidified air and, sometimes, by intermittent water spray. Aeration with water-saturated air has commonly been found to increase the moisture content of the substrate. Relative humidity of the aeration gas is typically 60–80%. Ideally, prevention of water loss or gain by the uninoculated substrate would require that the water activity of the aeration gas and the substrate be identical. In practice, water generated during fermentation, and the additional requirement of evaporative cooling, demand aeration gas that is somewhat drier than the substrate. The temperature and humidity of the aeration gas are controlled in the same way as in air conditioning systems: temperature and humidity sensors (hygrometers) in the fermentation chamber provide the necessary data, and cooler air is humidified by water spray or injection of clean steam, and warmed to reduce the humidity to the required value (Fig. 2). Humidity–temperature relationships available in standard psychrometric charts are the basis of control.

Temperature

At 10–30 kg m^{-3}, the maximum microbial biomass concentration in solid-state fermentations is lower than 40–50 kg m^{-3} typically seen in submerged culture (24); but, because there is little water, the heat generation per unit fermenting mass tends to be much greater in solid-state fermentations. Temperature can rise rapidly, again, because there is little water to absorb the heat (i.e., the mean specific heat capacity of the fermenting mass is much lower than that of water). Consequently, temperature control in large-scale fermentations can be particularly difficult. Removal of the metabolic heat sometimes becomes the major limitation, especially because of the poor thermal conductivity of the porous fermenting mass. Cumulative metabolic heat generation in koji fermentations for a variety of products has been noted at 419–2,387 kJ kg^{-1} solids (14). Higher values, up to 13,398 kJ kg^{-1}, have been observed during composting (32). Peak heat generation rates in koji processes range over 71–159 kJ kg^{-1} h^{-1}, but the average rates are more moderate, at 25–67 kJ kg^{-1} h^{-1} (32). Peak metabolic heat production rate during fermentation of readily oxidized substrates such as starch can be much greater than in typical koji processes. For example, in one case, a generation rate of 330 kJ kg^{-1} h^{-1} was observed for *Aspergillus niger* growing on cassava starch. Heat generation during composting has been treated in depth by Stentiford and Dodds (33).

Temperature control during fermentation is obtained mostly through evaporative cooling (34); hence, drier air provides a better cooling effect. Air temperature is controlled. Intermittent spray of cool water is sometimes necessary to prevent dehydration of the substrate. Occasionally, the substrate-containing metal trays are also cooled by a circulating coolant, even though most relatively dry and porous substrates are poor conductors. Intermittent agitation further aids heat removal, particularly in large heaps and piles. Despite much effort, temperature gradients in the substrate do occur, particularly during peak growth. Gradients as steep as 3 °C cm^{-1} have been recorded during rapid growth of *Rhizopus oligosporus* on soybeans in *tempe* fermentations, even though the substrate layer was less than 7 cm deep (35). Similar values were noted in solid-state citric acid production by *Aspergillus niger* (34).

pH

Solid-state fermentations are practiced without pH control (19) other than any adjustments made during substrate preparation. In some nonsterile processes the initial pH may be adjusted to pH 4 or less to suppress bacterial contamination. Unlike most bacteria, yeasts and fungi are generally tolerant of more acidic conditions. Either mineral acids (e.g., 0.2 M hydrochloric acid) or organic acids (e.g., 0.5% w/w acetic acid) may be used for pH adjustments. Other than initial adjustments, the buffering capacity of substrates is relied on to check large changes in pH during fermentation (19). Many substrates are effective buffers. This is particularly true of protein-rich substrates, especially if deamination of protein is minimal. Some pH stability can be obtained by using a combination of urea and ammonium sulfate as the nitrogen source in the substrate. Decomposition of urea produces ammonia that takes up a proton to become $NH^+{}_4$, causing an increase in pH. This effect is countered by ejection of a proton from cells that take up the ammonium ion but incorporate it into proteins as $—NH^+{}_3$. Uptake of ammonium causes rapid acidification. In the absence of other contributing nitrogen sources, an equimolar combination of ammonium sulfate and urea is expected to yield the greatest pH stability. In media that are rich in protein or amino acids, deamination may contribute to pH rise. Attention to stabilizing the pH during fermentation can be particularly important for certain processes. Stability of enzymes and some secondary metabolites is pH dependent, and even though the rate of production may not be affected by changes in pH, the overall process productivity may decline because of destruction of the product. This behavior has been observed in producing pectinases by *Aspergillus niger* cultured on glucose-enriched sugar cane bagasse (36).

PRODUCTION OF ENZYMES

Process Technology

Only extracellular enzymes, those secreted by the microbial cell into the extracellular medium, can be produced by solid-state fermentation technology (37). Predominantly, these are hydrolytic enzymes (38,39), as noted in Table 2. Commercial production of many of those enzymes in the West utilizes submerged culture, but solid-state fermentation persists particularly in Asia where koji-type processing is commonly used. Koji is molded grain that has been used in oriental food preparations for thousands of years (40,41). Koji is a source of fungal enzymes that digest proteins, carbohydrates, and lipids into nutrients used by other microorganisms in subsequent fermentation. Koji

Table 2. Enzymes That May Be Produced by Solid-State Fermentation

Enzyme	Producing organisms
Amylase	*Aspergillus oryzae, Aspergillus niger, Bacillus licheniformis, Bacillus subtilis, Bacillus megaterium*
Catalase	*A. oryzae, Rhizopus niveus*
Cellulase	*Trichoderma reesei, Trichoderma viride, Trichoderma koningi, Trichoderma harzianum, Aspergillus ustus, A. oryzae, Sporotrichum pulverulentum, Penicillium spinulosum, Penicillium capsulatum, Pestalotiopsis versicolor*
Chitinase	*A. niger*
Glucoamylase	*A. niger, A. oryzae, Rhizopus* sp.
Invertase	*Aspergillus awamori, A. niger, A. oryzae*
Lactase	*A. oryzae*
Linamarase	*Aspergillus sydowi, Penicillium steckii*
Lipase	*A. niger, Aspergillus luchuensis, Rhizopus delemar, Penicillium candidum*
Pectinase	*A. niger, Aspergillus sojae, Aspergillus fumigatus, Bassochlumys fulva, Penicillium expansum*
Phytase	*Aspergillus ficuum, Rhizopus oligosporus*
Protease	*A. oryzae, A. niger, Aspergillus flavus, Mucor dispersus*
Rennet	*Mucor pusillus, Mucor meihei, R. oligosporus*
Ribonuclease	*Aspergillus candidus*
Xylanase	*A. niger, A. fumigatus, Aspergillus terreus, Gibberella fujikuroi*

comes in many varieties depending on the mold, substrate, method of preparation, and stage of harvest. Traditional and mechanized koji production has been described by several sources (9,10,18), and concise outlines of food koji processes are given later in this article. In typical processing for enzymes, a suitable solid substrate is mixed with water and mineral salts, heat sterilized/cooked, cooled, and inoculated. Incubation at controlled temperature (20–45 °C) and humidity follows. Subsequently, the fermented substrate is extracted with buffer to yield an enzyme-containing liquor. Further purification and recovery methods are identical to those that have been described for submerged culture processes (17,27,42,43). The crude liquor is filtered or centrifuged to remove suspended solids and concentrated by ultrafiltration. The concentrate may be stabilized and packaged. Extensive purification is uncommon for many bulk enzymes. In some cases, the fermented substrate may be dried at 35–40 °C, and ground before extraction (44).

The commonly used solid substrates include wheat bran, wheat, soybean, rice, barley, oats, and other cereals, but many more substrates have been used occasionally (Table 3). Studies with nutritionally inert substrates (e.g., polyurethane foam) wetted with dissolved nutrients have been reported (45). Two or more substrates are sometimes used in combination. Cereal grain may be used whole or cracked into pieces. Substrate particle size affects the extent and the rate of microbial colonization, air penetration, and carbon dioxide removal as well as the downstream extraction and handling characteristics (19,46). Relatively small particles are preferred because they present a larger surface for microbial action. However, particles that are too small pack together to reduce the interparticle voids that are essential for aeration (46). Similarly, too many fines in a batch of larger particles will fill up the voids. Particle shapes that pack together tightly (e.g., flat flakes, cubes) are undesirable. Lignocellulosic substrates such as straw, sugar cane bagasse, and wood chips are sometimes extensively pretreated to ease microbial colonization and substrate utilization. Pretreatments include size reduction, steaming, steam explosion, solubilization of lignin with organic solvents, acid and alkaline hydrolysis, and irradiation (12,47,48). Most of those pretreatments have proven too expensive for commercial utilization. For cereal substrates, common pretreatments include dehulling, pearling, size reduction, soaking, and cooking or steaming (9,10,18).

Table 3. Solid-State Fermentation Substrates

Animal	Milk solids
Apple pulp	Millet
Bagasse	Newspaper
Banana meal	Nutrient wetted inerts
Barley	Oats
Beet pulp	Orange peel
Buckwheat	Paper pulp
Canola meal	Peanut press cake
Cardboard	Potato starch
Cassava starch	Rice
Coffee pulp	Rice bran
Corn cobs	Rye meal
Corn grits	Sawdust
Cotton seed meal	Sewage sludge
Fish	Soybean
Fruit pulp	Straw
Jerusalem artichoke meal	Sweet potato meal
Lignocellulosic waste	Wheat bran
Logs	Wood chips
Meat	Vegetables

The selection of substrate depends to some extent on the enzyme of interest. Thus, pectinases are often produced using fungi grown on fruit pulp (e.g., apple, avocado), or on a combination of pulp and bran, or bran and fruit juice. Similarly, cellulose-rich substrates such as rice and wheat straw are commonly employed in producing cellulases.

Enzyme production processes rely overwhelmingly on filamentous microfungi (43), the main producer genera being *Aspergillus*, *Mucor*, *Penicillium*, *Trichoderma*, and *Rhizopus*. *Bacillus* species predominate among bacteria used in enzyme production (38,43,44). Examples of production from yeasts are few. Some specific enzyme producers are listed in Table 2. Production of β-galactosidase by the yeast *Kluveromyces lactis* grown on corn grits and wheat bran moistened with deproteinated milk whey has been reported (49). Although β-galactosidase is an intracellular enzyme in yeasts and cultivation on solid substrate does not promote secretion, solid-state fermentation has been claimed to be superior to liquid culture for this enzyme (49); however, recovery questions have not been addressed.

Microbial inocula for commercial enzyme production are maintained as pure cultures (17). Routine checks on culture purity, freedom from variant forms, physiological characteristics, and enzyme productivity are used to ensure process consistency (17). The sterile solid substrate is inoculated in one of several ways: by mixing with a liquid suspension culture inoculum; by mixing with solid substrate-grown inoculant; by using a suspension of solid substrate-produced spores in a sterile liquid; or by blowing the solid substrate-produced spores with sterile air going into the fermentation chamber. Aseptic processing practice is observed (17), but fermentations are never entirely contamination free. Fermentations last 1–7 days, and a significant amount of the substrate is oxidized to carbon dioxide and water (17). Thin uncompacted and porous layers of nonadhering solid substrate particles generally ensure good oxygen penetration from the exposed surface. Excessive agitation is not wanted. Occasional turning and mixing improve oxygen transfer and reduce compaction and mycelial binding of substrate particles. Because agitation continually damages the surface hyphae, mixing suppresses often-unwanted sporulation (46). The frequency of agitation may be purely experience based, as in occasional turning of a fermenting cocoa heap, or it may be determined by a temperature controller.

Production of spore inocula requires static conditions. Deep substrate layers and heaps may require forced aeration and agitation. Oxygen supply affects product formation. Depending on the microorganism and the product, both enhanced and reduced productivities have been observed to accompany improved aeration or agitation. Aeration plays an important role in removing carbon dioxide and in controlling temperature and moisture. In some cases, increased concentrations of carbon dioxide have been severely inhibitory, whereas enhanced oxygen partial pressures have improved productivity. If used, forced aeration rates may vary widely; a typical range being 0.05–0.2×10^{-3} m^3 kg^{-1} min^{-1} (1).

Regulatory Considerations

Production of food, medicinal, and medical–diagnostic enzymes must comply with standards of "good manufacturing practices" as established by the U.S. Food and Drug Administration (FDA) or other regulatory agency appropriate to the particular jurisdiction. Safety of enzyme preparations, especially those derived from non-GRAS species (GRAS, generally recognized as safe; i.e., microorganisms such as baker's yeast that have proven to be safe through extended use and have been recognized as such by placement on the GRAS list of the FDA), must be demonstrated to gain marketing approval. Specific requirements vary with the intended use of the product. Freedom from toxins, antibiotics, other contaminants, and unwanted effects may have to be ascertained. Multigeneration feeding trials and studies of teratogenicity may be necessary. Requisite purity, potency, and stability may have to be demonstrated. The quality assurance requirements are similar to those for therapeutic proteins produced by other methods (42,50–52). Enzymes for industrial and household use (e.g., detergent enzymes) also must comply with product safety criteria, although the requirements are less stringent than in food, feed, veterinary, and medical applications.

OTHER PRODUCTS AND APPLICATIONS

Secondary Metabolites

Although solid-state surface culture technology was employed in the earliest attempts at large-scale production of penicillin, submerged culture is now the only method used in commercial production of antibiotics. Nevertheless, promising research on solid-state production methods continues, albeit at a slow pace. Solid-state production of oxytetracycline by *Streptomyces rimosus* grown on corncob has been described (53). Cephalosporins have been produced using *Cephalosporium acremonium*, *Streptomyces clavulgerus*, and *Aspergillus chrysogenum*. *Streptomyces cinnamonensis* has been used to produce monensin, an anticoccoidal agent for poultry. Other potentially antibiotic-producing *Streptomyces* that have been cultured in solid state include *S. badius*, *S. aureofaciens*, and *S. flavovirens* (1). Studies of penicillin production by several strains of *Penicillium chrysogenum* have shown that the best producers in submerged culture are not necessarily the best in solid-state culture (54). Improving productivity demands selection of strains suited specifically to solid-state fermentation.

Solid substrate fermentation for producing gallic acid used in tanning, printing, and other applications has been described (13,14). Koji technology is employed in citric acid production in Japan (16). In producing citric acid by *Aspergillus niger*, Pallares et al. (45) noted that solid-state culture had doubled the productivity (g L^{-1} d^{-1}) of submerged culture, reducing the fermentation time from 14 to 6 days. Medium-impregnated polyurethane foam cubes were used as substrate (45). Productivity was sensitive to substrate particle size; 0.3-cm cubes gave the best results (45). Other products that have been produced by solid-state culture include several mycotoxins (14,46,55) and ergot alkaloids from various fungi, gibberellic acid from *Gibberella fujikuroi* (56,57), and kojic acid (14). In one case involving production of gibberellic acid and proteases, the yield of products was improved by fed-batch strategies (57), but nutrient feeding of solid-state batches other than at start-up remains a rare practice.

Ethanol has been commercially produced by aerobic–anaerobic solid-state fermentation of sweet sorghum grains in Taiwan (24). Production of γ-linolenic acid, a polyunsaturated fatty acid, by *Cunninghamella japonica* grown on rice and millet has been reported (58). The fungus *Monascus purpureus* or *Monascus anka* is used to commercially produce a bright red polyketide pigment for food coloring and flavoring in Asia (41). An additional and developing application of solid-state fermentations is in producing fungal spores for use in a variety of biotransformations (1,14). Use of phytopathogenic and entomopathogenic fungi as herbicidal and insecticidal biocontrol agents promises further applications of solid-state fermentation technology for spore production (14,59–61).

Environmental Remediation

Because of their ability to degrade otherwise persistent toxic organics, white rot fungi—for example, *Phanerochaete chrysosporium*, *Phanerochaete sordida*, and *Trametes versicolor*—are being tested for bioremediation of soils contaminated with polychlorinated biphenyls, pentachlorophenol, and other polycyclic aromatics (62). In the preferred mode of implementation, solid-state cultured fungal carriers such as corncobs, wood chips, and straw are mixed with the soil in situ. Commercial implementation requires, among other factors, the ability to economically and consistently produce large inocula. Solid substrates enriched with nutrients and concentrates of fugal spores, and formulated into pellets, are being developed to improve consistency of performance (62). Other pollution remediation applications of solid-state fermentations are composting (22,33,63), anaerobic digestion for producing methane in slurry fermentations and landfills (64–66), and treatment of gaseous pollutants and vapors in biofilters (65).

Animal Feeds

Solid substrate fermentation methods are used to enrich many wastes and low-grade substrates for animal feed (1,11,12,22,25,46,65,67). One such widely used process is ensiling. Ensiling is used to stabilize and preserve forage crops; any nutrient enrichment is incidental. Typically, crops are ensiled immediately on harvest. An uncontrolled anaerobic primary fermentation by naturally occurring or inoculated (68,69) lactic acid bacteria takes place. Sugars present in the fodder are converted to lactic and acetic acids, causing a drop in pH. A low pH suppresses secondary fermentation by *Clostridia* that leads to butyric acid and spoilage. A good silage should generally have a pH of 4.0 to 4.2 and a butyric acid content of less than 5% of dry weight. A high dry matter content, for example, 20–30% w/w, of biomass at ensiling is preferred because secondary fermentation is difficult to suppress when dry solids are less than 15% w/w. Unlike *Clostridia*, the lactic acid bacteria remain active up to 70% w/w dry matter. Microbiology of silage has been detailed by Whittenbury (70) and Fenlon et al. (68). Silage inoculants have been described by Seale (69). Certain silages may contain animal and food processing waste (e.g., cattle and poultry manure, fish waste, or offal) and grain (e.g., corn), in addition to grass and other fodder crops. Fish silage has been described by Wignall and Tatterson (71); in-depth treatments of silage fermentations have been reported by McDonald (72) and Woolford (73).

PRODUCT RECOVERY FROM FERMENTED SOLIDS

Unless the fermented solids are the desired product, the solids must be further processed to recover the enzyme or other metabolite. Volatile products such as ethanol may be recovered by direct distillation of the fermented mash (24). Most other products require extraction from fermented solids. The solids may be extracted moist or dry (44); moisture content can affect the extraction behavior. Because drying generally improves keeping quality, dried material from several batches can be accumulated for later extraction. The solid cake should be crumbled or otherwise size reduced for improved extraction. Either batch or continuous extraction may be employed. When feasible, continuous extraction with countercurrent flow of solids and solvent allows greater product recovery in a more concentrated solution. A suitable continuous extraction device is the Chisti contactor shown in Figure 8. Leaching machinery developed for the chemical industry can also be potentially adapted for use with fermented solids. A series of mixer-settlers has been used in small-scale extractions (74), but that arrangement is too cumbersome for commercial practice. Between two and five contacting stages are generally sufficient, but a larger number may be practicable with high-value products.

Cumulative contact periods typically do not exceed 50 min. Water, aqueous buffers, or dilute solutions of salts (e.g., 1% sodium chloride) or glycerol (e.g., 1% glycerol) are used to leach enzymes (74). Aqueous solutions of ethyl acetate, acetone, ethanol, or chloroform may be needed for nonprotein products such as most antibiotics and mycotoxins. In one case, aqueous ethanol (10%) was the best solvent for extracting gibberellic acid from fermented wheat bran (74). Pure methylene chloride has been used to extract mycotoxins (1). A suitable solvent must be selected empirically for each application. Factors that affect solvent choice are cost, toxicity, flammability, corrosion effects, disposal issues, solubility and stability of product, solubilities of other solutes, and compatibility with end use. The

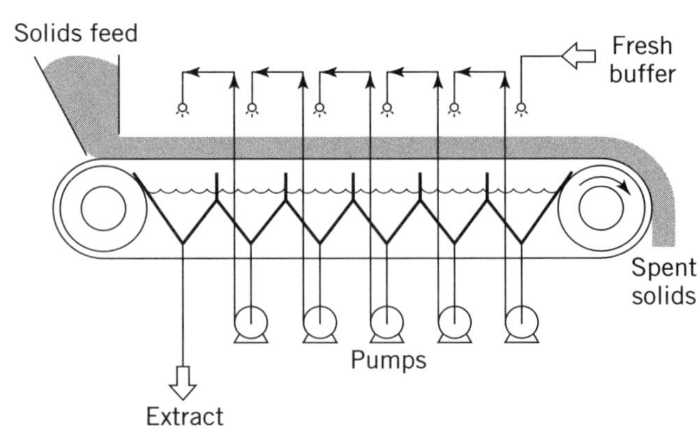

Figure 8. Continuous countercurrent leaching of solids in an extractor. Solids are placed continuously on the stainless steel conveyer mesh (0.6 m wide, 6 m long) that moves at 0.1–0.2 m min^{-1}. Fresh buffer is sprayed at the downstream end of the conveyer; the percolated extract is collected for extracting solids further upstream.

solvent-to-solid ratio generally ranges over 1 to 8; higher values increase the fraction of the product leached at the expense of producing a more dilute solution. Extraction temperature does not normally exceed 30 °C, but enzymes are commonly extracted at lower temperatures (e.g., 4–10 °C) to reduce denaturation, proteolysis, and microbial growth. Extraction performance and stability of the product may depend critically on pH. The pH of greatest stability may not be the same as that required for optimal leaching (75). Enzymes are generally extracted at pH values close to 7; however, to prevent precipitation, the extraction pH must not be the same as the isoelectric pH. Once the product has been extracted into solution, further purification employs the same well-known methods that are used in recovering extracellular products from submerged fermentations (27,42,43).

FERMENTATION-ENRICHED AND FERMENTATION-MODIFIED FOODS

Solid-state fermented foods occur worldwide. In some regions, such foods are the primary source of calories, proteins, and other nutrients for the consumers. Historically, fermented foods arose more by accident than by guided effort; nonetheless, recognition of the numerous advantages of the fermented products led to their established use. In comparison with the unfermented substrate, the fermented material may have one or more of the following characteristics: enhanced flavor, texture, and digestibility; reduced or eliminated undesirable compounds or flavors; protein enrichment or otherwise improved nutritional value; changed physical character; improved keeping quality; fast food quality of easier or quicker preparation; and other economic advantages such as reduced cooking time and hence lower energy demand. Table 4 notes some of the world's significant fermented products. Further details appear in the *Handbook of Indigenous Fermented Foods* (76), as well as other sources (6,8,40,41,77). The contribution of fermented products to the human food supply is likely to remain strong (78), and increasing cross-cultural spread is expected considering the popularity of tempe, tofu, and soy sauce in nonindigenous regions. The following sections discuss a small selection of the important foods as representative illustrations of the solid-state and solid substrate fermentation technologies.

Miso

Miso is a semisolid fermented soybean product that is widely consumed in Japan (18,40,41,79). Similar foods are produced in other parts of Asia (Table 4). In Japan miso is made into soup with added ingredients such as vegetables, meat, and seafood (18). Modern miso is generally produced from protein-rich yellow soybeans, but other varieties can be used. Cleaned soybeans are continuously washed, soaked overnight, drained, and cooked (Fig. 9). Cooking conditions vary. Large plants employ continuous pressure cookers with cooking done at 1.3–2.0 kg cm^{-2} for 2–7 min (18). Cooked beans are ground using a meat grinder with 5-mm-diameter holes (18) before being mixed with separately produced koji for fermentation. Barley koji is used in making "rural miso"; commercial processes employ rice koji (18). Miso koji is made from milled, cleaned, and usually cracked rice that is washed and soaked overnight in water at approximately 15 °C (18). Soaked rice is drained (1 h) and steamed. A horizontal belt steamer (Fig. 10) is commonly used. Steamed rice is unloaded onto a cooling conveyer, where a stream of air is blown through the rice layer to lower the temperature to 35 °C (18). Rice is then inoculated with the starter culture (called *tane-koji* in Japanese) of *Aspergillus oryzae*. Tane-koji is an olive-green, sporulated culture that has been grown on cooked rice mixed with a small amount of wood ash (18). Tane-koji is produced under aseptic and strictly controlled conditions. One kilogram of tane-koji is sufficient to inoculate 1,000 kg of rice (18). Many varieties of tane-koji are available with different capabilities for hydrolyzing proteins, carbohydrates, and lipids (18). An appropriate variety should be selected with regard to the amounts of proteins, oils, and carbohydrates in the mix of miso substrates (18).

Inoculated rice is transferred to a koji fermentor. Either a tunnel fermentor (Fig. 3) or, in larger plants, a rotary fermentor (Fig. 4) is used. The tunnel fermentor is mixed twice during the fermentation, but the rotary machine is mixed only once. Fermentation occurs at 30 °C for 40 h (18). Rice koji is then harvested and mixed with ground beans, salt, and a miso starter. The starter consists of halophilic yeasts, *Zygosaccharomyces rouxii* and *Candida versatilis*, and a halophilic lactic acid bacterium, *Pediococcus halophilus* (18). Fermentation is carried out in wooden vats, or tanks made of stainless steel or fiberglass-reinforced plastic. During fermentation miso is occasionally transferred from one vat to another to maintain homogeneity (18). The fermentation temperature is 25–30 °C (18). Ripened miso is taken out of vats, blended, pasteurized, and packaged. Process variations occur depending on the type of product desired (41).

The miso koji mold, *A. oryzae*, grows well around 30 °C (18), but optimal conditions for spore germination have been found to be 36 °C, ≥97% relative humidity, 0.1% carbon dioxide, and 20% oxygen (18). Optimal conditions for mycelial growth are 35–37 °C, ≥75% relative humidity, <1% carbon dioxide, and 20% oxygen (18). In practice, a lower than optimum temperature (28–30 °C) is used for germination, and 30–35 °C for cultivation in the koji chamber (18), apparently to avoid overheating. Desirable temperatures for production of enzymes are as follows: glucoamylase, 30 °C; α-amylase, 35 °C; protease, 25–30 °C; and carboxypeptidase, 35 °C (18). The optimal moisture level of cooked rice is 36–38%, and the optimal pH is 6.0–6.4 (18). Fermentation pH can be effectively maintained at 6–7 by supplementing the substrate with sodium glutamate or sodium succinate. Controlled pH favors higher-activity proteases (18). The microbiology and biochemistry of miso fermentation are further discussed by Abiose et al. (82).

Sake

Sake is a Japanese alcoholic beverage made from rice (10,40). Rice koji preparations for sake brewing are obtained by methods similar to those described for miso. The amount of tane-koji inoculum is 0.6–1 kg per kilogram of

Table 4. Solid Substrate Fermented Foods of the World

Sufu. A Chinese cheese made from soybean curd (*tofu*) by fermentation with molds such as *Actinomucor elegans* and *Mucor hiemalis* (41).

Idli. Small, white, steamed rice cake made by lactic acid fermentation of a thick batter of coarse-ground rice and fine-ground black gram in India. Idli cakes are moist, soft, and spongy, with a slightly sour flavor. Several varieties are produced. Other similar products of India are *dhokla* and *khaman*.

Dosa. Pancake made with thin pourable fermented batter of fine-ground rice and gram. Important part of diet in South India and Sri Lanka.

Ambali. A cooked product of fermented millet flour batter and rice. India.

Puto. Steamed cake of milled, acid-fermented rice. Similar to idli, but without legume. Important food in Philippines. Several varieties.

Enjera. Very flexible, soft, sour leavened pancake containing many small uniformly distributed gas bubbles. Made from thin, pourable fermented batter of *tef* (*Eragrostis abyssinica*) flour, or other ground grain (corn, sorghum, millet, barley, wheat). Consumed as staple in Ethiopia.

Hopper or ***appa***. Steam-baked product made from acid-fermented dough (rice or wheat flour) formulated into liquidy batter with coconut water. Sri Lanka.

Kisra. Flat bread made from fermented sorghum flour batter in Sudan.

Ogi (also *furah* and other names). Fermented cereal flour (corn, sorghum, or millet) made into a slightly sour gruel. Important source of calories in Nigeria.

Uji (and other names). Suspension of cereal (corn, millet, sorghum) or cassava flour (relatively large particles) in water. Sour-acid fermented either before or after cooking. East Africa.

Mahewu (also *magou* or *mageu*). Suspension of acid-fermented corn flour used as high-calorie beverage/food. More dilute than uji or ogi. Cooked cornmeal in water is fermented after adding a small amount of wheat flour. Product is very slightly alcoholic. Produced in large commercial operations in South Africa. Similar cereal-derived alcoholic beverages/foods occur throughout sub-Saharan Africa.

Sake. Rice wine of Japan, and other similar products: *brem bali* (Indonesia); *mie-chiu* (China); *tapoi* (Philippines); *sato* (Thailand); *yakju* (Korea); and *sonti* (India). See text.

Gari. A dry granular starchy product of West Africa (80,81). Made into staple food by adding hot or cold water. Made from peeled, grated cassava by fermenting in cloth bags while squeezing out the water over 12–96 h. The anaerobic acid fermentation produces a sour taste. Fermented solids are sun dried and lightly toasted in iron pans. A small amount of palm oil may be added during toasting to impart a pale-yellow color. Solids are sieved and sold in markets. Very poisonous varieties of cassava may be used in making gari. Crushing of cassava releases naturally present linamarase, which breaks down the cyanogenic linamarin to reduce the cyanide content. Cyanide produced boils off as hydrogen cyanide gas, indicating the need for good ventilation during processing. Acid pH produced during fermentation enhances the action of linamarase (pH optimum of 5.5). In addition, microbial action may produce enzymes that enhance the breakdown of cyanoglycosides.

Miso. Fermented soybean paste of Japan. Other similar products: *jang* (China); *doenjang* and *kochujang* (Korea); *taucho* or *tauco* (Indonesia); *tao-tsi* (Philippines) (18,40,41,79,82). See text.

Soy sauce. Japanese *shoyu*, and similar products throughout Asia: *chiang-yu* (China); *kan jang* (Korea); *kecap* (Indonesia) (9,40,41,76,79). See text.

Natto. Soaked and cooked soybeans fermented by bacteria, principally *Bacillus natto*, consumed in Japan. Other similar products: *thua-nao* of Thailand; *kinema* of Nepal, Bhutan, Sikkim, and parts of northeastern hills of India; *dagé*, an Indonesian produce made of bacterial fermented peanuts and other oilseeds.

Tempe (or *tempeh*) and similar products: *tempe kedele*, *tempe gembus*, *tempe benguk*, *tempe bongkrek*, *oncom*, or *ontjom*. See text. A Japanese solid food, *hamanatto*, is also made from soy beans fermented by fungi (41).

Dawadawa. Fermented African locust bean (*Parkia filicoidea*) product used as flavoring agent in West Africa (14).

Kenkey (and other similar products). Fermented corn dough used as staple in Ghana (8).

Mawé. Acid-fermented corn dough used in Benin and Togo (8).

Pozol. Fermented corn dough used as staple in parts of Mexico.

Kocho. Unleavened flat bread made from fermented pulp of ensete (*Ensete ventricosum*) or 'false banana' plant (a banana-like plant that does not bear bananas). Used as staple in certain regions of Ethiopia.

Tapai. Malaysian fermented glutinous rice or cassava gratings. Sweet and mildly alcoholic. Used as dessert. Similar products occur elsewhere: *lao chao* in China (41); *tapé ketan* and *tapé ketella* of Indonesia (40); and *peujeum* in Sudan.

Nham. Fermented minced raw pork similar to sausages. Thailand.

Fermented seafood. Usually fish or shrimp pastes, or sauces: *bagoong, patis* (Philippines); *ngapi* (Burma); *trassi, kccap ikan* (Indonesia); *belachan, budu* (Malaysia); *mam, nuoc-mam* (Vietnam); *jeotkal* (Korea); *prahoc* (Cambodia); *kapi* (Thailand); *padec* (Laos); *shottsuru, shiokara, katsuobushi* (Japan); *faseich* (Sudan) (14); *kobi* (Ghana) (14); and fermented fish of Northern Europe.

Balao balao. Fermented rice–shrimp mixture eaten in Philippines. Cooked, somewhat pasty rice is blended with fresh, unshelled, salted shrimps, and fermented for 7–10 days. Shrimp shell becomes quite soft. Product cooked before consumption.

Burong dalag. A fermented fish and rice product of the Philippines.

Fermented milk products. Yogurt; sour cream; cheeses; *ergo* and others (Ethiopia); *kefir, koumiss* (Russia); *dahi* (India); *tairu* (Malaysia); *laban rayeb, laban zeer, laban zabadi* (Egypt); *liban argeel, liban khather, mast, mass taw, shenina, dabbo* (Iraq).

Kishk. A fermented milk–wheat mixture of Egypt. Stored as dry balls.

Trahanas (and other names). Made from crushed wheat and fermented sheep milk in Greece, Turkey, and Cyprus. Stored as dry biscuits or dry crumbled solids.

Fermented vegetables (8). Sauerkraut (Europe, North America); Chinese sauerkraut *hum choy*; acid-pickled olives and cucumbers; various Indian and Malaysian pickles; various varieties of *kimchi* (Korea); kimchilike products of Thailand and elsewhere.

Source: Refs. 6, 76, and 77.

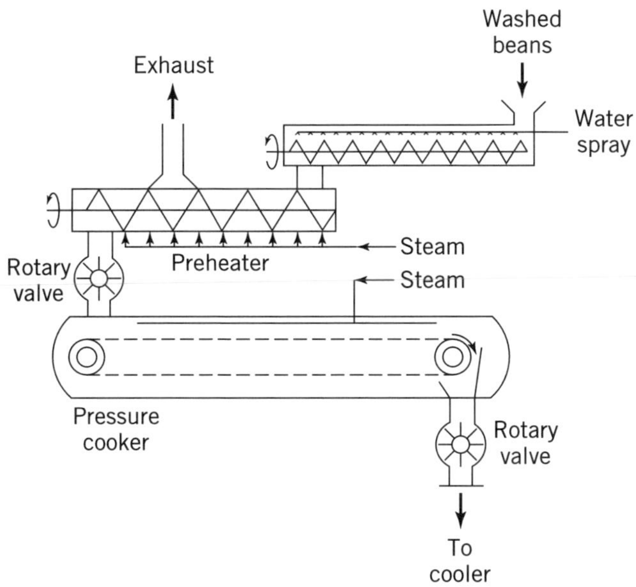

Figure 9. Continuous washing, soaking, preheating, and pressure cooking of soybeans.

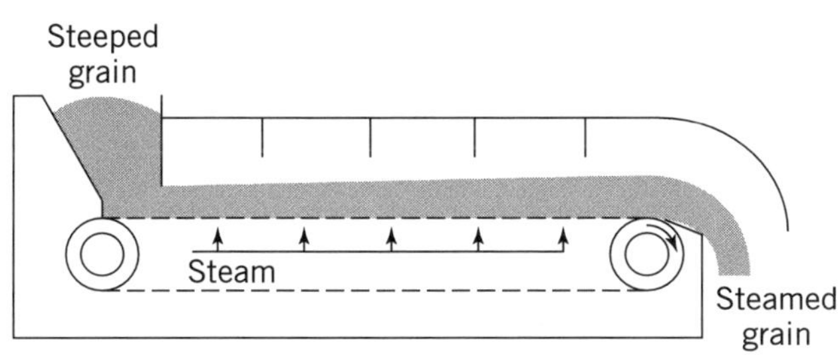

Figure 10. Continuous grain steamer.

the original rice, corresponding to about 2,000 fungal spores for each rice grain (10). The koji used is white because mold growth is arrested before sporulation (10). The unsporulated koji is relatively dry; the rice kernels are relatively hard and unclumped (10). The koji is then used in preparing the sake yeast starter mash, or *moto*. The traditional method of moto making required about a month-long fermentation of koji (300 kg) and water (1 m^3) (10). A succession of microbial populations developed, including lactic acid bacteria that reduced the pH and created an environment in which only the sake yeast, *Saccharomyces cerevisiae*, thrived (10). In modern moto production, fermentation time is reduced by adding lactic acid to water (pH 3.5–3.8) before mixing with koji. A large inoculum of sake yeast and steamed rice is also added. In a few days the starch is hydrolyzed, and fermentation begins. The next step is a *moromi* fermentation; it involves further addition of steamed rice, koji, and water to moto. Three additions are made at 1-day intervals (10). Temperature is lowered at each addition to control contamination because the added materials dilute the acids and alcohol. The temperature of moromi is controlled carefully, and after the final addition the temperature is 7–8 °C (10). Saccharification of starch and fermentation occur simultaneously in moromi (unlike in moto) until the alcohol content reaches nearly 20% v/v (10). Some steamed rice may be added during the final stages of fermentation to produce a sweeter product (10). Once the fermentation has almost ceased, alcohol is added to the mash to bring the concentration to 20–22% by volume (10). Solids are filtered out; the liquid product is further clarified, pasteurized, aged, blended, diluted to 15.0–16.5% v/v alcohol, and packaged (10). Unlike Western beer fermentations, sake mash is very high in solids that help to keep a large amount of yeast in suspension (10).

Beverages similar to sake are also found in China and many other Asian countries (Table 4). Other quite different indigenous alcoholic beverages, some made through solid substrate fermentation, are found in Africa and elsewhere (76). Some of these products are now produced in large industrial operations (77).

Soy Sauce

Soy sauce originated in China, but it is widely produced and consumed throughout Southeast Asia, Japan, and, increasingly, Europe and North America. The history, variety, and production technology of soy sauce have been detailed by Fukushima (9) and others (40,41,76,79). Here, a short description of only the latest, large-scale Japanese production methods is given.

Manufacture of soy sauce (*shoyu* in Japanese) involves a solid substrate koji fermentation, a brine or *moromi* fermentation, and refining steps (40,41,79). In koji making, soybeans or, more usually, defatted soybean flakes or grits are moistened and cooked in continuous pressure cookers (Fig. 9). The latest cooking methods employ a cooking time of 0.25 min or less, at approximately 7 kg cm^{-2} gauge pressure, equivalent to 170 °C (9). Rapid high-temperature cooking reduces loss of proteins and improves the yield of hydrolyzed amino acids in the final product. Cooked beans are mixed with roasted wheat that has been cracked into four or five pieces (9). Wheat roasters operate in a continuous mode at 170–180 °C (9). The ratio of wheat to beans is commonly 1:1, but it varies with the variety of shoyu (9). The mixed substrate is inoculated with a pure culture of *Aspergillus oryzae* (or *A. sojae*). The fungal spore density at inoculation is about 2.5×10^8/kg of wet solids (1). After a 3-day fermentation the substrate mass becomes green-yellow because of sporulation (9). Too long an incubation causes excessive, unwanted sporulation, production of ammonia from deamination of amino acids, and off-flavors in the final product (41). Koji is now harvested for use in the moromi step. Koji production is highly automated and continuous (9,18). Among the largest of koji-making processes is that of Kikkoman Corporation; it produces up to 4,150 kg h^{-1} koji, which is equivalent to processing 1,650 kg h^{-1} raw soybeans and 1,775 kg h^{-1} wheat (9).

Harvested koji is mixed with brine containing at least 20% salt for the moromi fermentation step (9,41). Large outdoor tanks are used for moromi fermentation. A specially selected culture of lactic acid bacteria is added to moromi mash, and the temperature is kept relatively low (15–20 °C) for the first month or so (9). Low temperature prevents a too rapid growth and attendant rapid decline in pH. A pH drop that is too rapid causes undesirable early arrest of the proteolytic activity of koji enzymes; action of koji glutaminase is also obstructed (9). When the pH of moromi has declined to pH 5, a separately grown culture of the yeast *Zygosaccharomyces rouxii* is added (9). The temperature is then gradually raised to nearly 30 °C until a vigorous alcoholic fermentation occurs (9). On cessation of the alcoholic fermentation, the temperature is again reduced to about 25 °C and maintained there for the last 2 months (9). Aged moromi is filtered using automatic filter presses; the liquor is pasteurized in continuous flow plate heat exchangers and packaged in plastic bottles (9). The production schemes vary a great deal within Japan, and many quite primitive methods are encountered elsewhere in Asia. Chinese practices have been outlined by Wood and Min (41).

Tempe

Tempe (or tempeh) kedele is a fermented soybean product consumed in Indonesia, Malaysia, Singapore, the Netherlands, the West Indies, and, increasingly, the United States and Canada (8,40,41,76). Tempe kedele is made from partly cooked soybeans that are fermented with the mold *Rhizopus oligosporus*. The mold mycelium completely covers, penetrates, and binds the beans into a cake. In addition to the beans, lower-quality tempe varieties may contain solids such as cassava grits, soybean coats, or tofu that are added before fermentation. Freshly fermented tempe is creamy white; it has a mushroomlike flavor that becomes nutty and peppery on deep frying (8,76). Raw tempe does not keep well. Unless consumed within 24–48 h, it should be deep fried, dried, or blanched and refrigerated (76).

Production processes vary a great deal. In typical cottage-level processing in Indonesia, cleaned dry soybeans are washed and partially cooked by boiling for 2.5–3 h. An overnight soaking in clean water follows (8). A bacterial acid fermentation occurs and the pH drops to 4.5–5.3. The beans are now dehulled by rubbing with hands or feet. Hulls float on top and are washed away. Well-drained beans are inoculated using a previous batch of sporulated tempe that may have been sun dried. Alternatively, mold grown and dried on large leaves of *Hibiscus tiliaceus* may be used (8,76). Inoculum densities of 10^7 spores (or 10^7 cfu) per kilogram of wet solids have been reported (1,8). Overinoculation may produce excessive heat that inactivates fungal mycelium; *Bacillus* species proliferate, causing spoilage (8). Similarly, suboptimal inoculum density may cause outgrowth of yeasts and bacteria and again spoilage (8). Inoculated beans are lightly packed in relatively flat packets of *Hibiscus*, banana, or other broadleaves for 30–40 h of fermentation at ambient conditions (76). Similar general schemes are followed in large-scale production processes operated in the United States, Asia, and other countries; exceptions are mechanized processing, better-defined inoculum, and fermentation in perforated stainless steel trays (2- to 3-cm-deep bean layer) or perforated plastic bags (76). Fermentation is nonsterile; the bacterium *Klebsiella pneumoniae* that is invariably present in commercial preparations leads to vitamin B_{12} enrichment of the product (76).

Studies on production of *R. oligosporus* spores (pure culture) for commercial tempe inoculum indicate that viable spore counts of 10^9–10^{10}/kg dry substrate can be obtained on a variety of grains (1). Spores may be freeze-dried, but typically about one-log reduction in viability accompanies drying (1). Freeze-dried spores generally keep for 6 months (1).

Other tempelike products are consumed in Indonesia, including *tempe gembus*, made from waste produced by soybean curd factories; *tempe benguk*, produced using the legume *Mucuna pruriens* instead of soybeans; *tempe mata kedele*, made from soybean hulls; and *tempe bongkrek*, made from coconut press cake. The latter type can be potentially contaminated with the bacterium *Pseudomonas cocovenenans*, which produces lethal toxins (76). *P. cocovenenans* suppresses the growth of the tempe mold; hence, tempe bongkrek must never be consumed if a luxuriant growth of the mold has not developed (76). Formation of bacterial toxins is suppressed below pH 6, and an incubation temperature of 37 °C or higher prevents development of *P. cocovenenans* (76). The mold that is primarily responsible for most tempe fermentations is *Rhizopus oligosporus*, but other *Rhizopus* species, such as *R. oryzae, R. achlamydosporus*, and *R. arrhizus*, as well as non-*Rhizopus* fungi, may be involved (8).

Oncom

Oncom (or ontjom) is another tempelike product made from peanut press cake in Indonesia (41,76). *Rhizopus* species are used to produce the black oncom (*oncom hitam*), whereas the preferred red oncom (*oncom merah*) is produced using *Neurospora* species (76). *Neurospora sitophila, N. crassa*, and *N. intermedia* are typical oncom molds (76). In quite different processes, *N. sitophila* has been investigated also for protein enrichment of low-grade substrates such as sugar cane bagasse for animal feed (11).

Sausages

Fermented meat products such as bologna, summer sausage, thuringer, teewurst, dry sausage, pepperoni, and salami are produced by solid-state anaerobic fermentations utilizing acid-forming bacteria, particularly *Lactobacillus, Pediococcus*, and *Micrococcus* species. Typically, ground meat is mixed with curing agents and a starter culture at low temperature (−1 °C or less), and placed in cloth or other casings. Fermentation occurs at 20–40 °C. In dry sausage production in the United States, for example, a fast acid fermentation is employed to prevent growth of spoilage microorganisms. The curing mixture is predominantly salt and nitrite. Nitrite reacts with myoglobin in meat to produce the characteristic dark-red color of the product. Either *Pediococcus acidilactici* or *P. pentosaceus* is used for starter, depending on the selected fermentation temperature. Slower fermentation at 24 °C is commonly used in Europe. The curing mixture consists of salt, nitrate, and other spices. Naturally present micrococci and staphylococci convert the nitrate to nitrite for the desired coloring effect. After the lactic acid fermentation, a further incubation period of up to 7 days at low temperature is practiced, with the meat held in 0.2-m-deep layers in metal pans.

Coffee

Coffee, after oil, is the second largest traded commodity, and more than 1.2 billion cups are consumed daily (83). Coffee processing employs a solid substrate fermentation for removing adhering mucilage (mesocarp) from ripe, mechanically pulped coffee cherry. This method is used predominantly only for *Coffea arabica* (84), which supplies the high-quality end of the market (83). Apparently, fermentation is not essential to flavor development, serving merely to remove the mesocarp from the parchment that encloses the bean. Nevertheless, the fermentation needs to be carefully conducted, or off-flavors develop that affect the final cup quality (84). In the preferred mode of fermentation, pulped cherry is placed in concrete or wooden tanks arranged for continuous drainage of the liquors produced (84). An alternative fermentation scheme utilizes under water fermentation with or without continuous water flow (84). The fermentation lasts 20–100 h, depending on factors such as pH, the coffee variety and stage of ripeness, aeration, and concentration of ions (for example, Ca^{2+} enhances the activity of certain enzymes) (84). Once the mucilage has degraded, the solids are washed and graded by water in concrete channels, and dried to produce parchment coffee (84). Milling (dehulling) removes the parchment to yield green beans used in commerce. The well-known coffee grounds are obtained by crushing blended and roasted green beans. Further processing is required to produce the instant varieties of coffee (83).

Cocoa

Cocoa beverages and chocolate are made from fermented cocoa (*Theobroma cacao*) seed or beans (85). Mature cocoa fruits or pods are harvested and split open to release the pulp-enveloped seeds. The pulp ferments as the seeds are piled in 2- to 4-ft-diameter heaps placed on perforated banana leaves that allow drainage of the liquids released (14,85). Alternative fermentation schemes employ perforated wooden boxes and other methods (85,86). The fermentation usually lasts 2–8 days, depending on the cocoa variety. The cocoa mass is turned several times during fermentation to ensure aeration and homogeneity and to avoid clumping (85,86). Turning also discourages mold growth on the better-aerated, exposed parts (86). A succession of microorganisms develop, including yeasts that are followed in turn by lactic acid producers, acetic acid bacteria, and, finally, aerobic spore-forming bacteria (86). The heat developed during the two-stage anaerobic–aerobic fermentation kills the cocoa embryo. Fermentation degrades the pulp, making removal easier, and products of fermentation penetrate the beans, stimulating complex biochemical reactions that are essential to flavor development (85,86). Fermented beans are dried, cleaned, and sold to chocolate processors or marketing organizations.

Mushrooms

The variety of edible mushrooms is tremendous, but only a few types are commercially cultivated. Production of mushrooms, mushroom compost and other substrates, and mushroom spawn utilize solid substrate fermentation methods that are comprehensively described elsewhere (2–4).

SAFETY CONSIDERATIONS

Microorganisms used in solid substrate fermentations have the potential to produce harmful effects in humans and animals. The history of occupational disease in solid-state cultivation is reflected in names such as cheese-maker's lung and mushroom worker's lung (87). Allergenic reactions and immunological hypersensitivity have been associated with exposure to fungal spores, including those of *Aspergillus niger, Aspergillus oryzae, Aspergillus awamori*, and *Aspergillus batatae* in industrial settings (88,89). Conditions such as dermatitis, asthma, vasomotor rhinitis, and itchy skin have been observed among exposed workers (87,88). Exposure to *Bacillus thuringiensis* has produced similar effects in humans (88). *Aspergillus* and *Pseudomonas* species have caused fatal infection in immunocompromised individuals and serious lung infection in cystic fibrosis patients (89). Fatal infections with *Aspergillus fumigatus* in previously healthy adults have been observed (89).

Inhalation of spores of edible mushrooms has long been associated with disease among mushroom growers (87). Allergenic reactions to spores of *Pleurotus* species, *Lentinus edodes* (3), and others have been documented. *Pleurotus* spores survive for several days after inhalation in the human respiratory tract, suggesting the potential for germination in the body (3). Allergenic reactions encountered in cultivation of *Agricus* species are linked to other molds that inhabit the substrate (3). In best-practice cultivation, attention to handling and processing of mushroom substrate and product should accompany ventilation of work areas and the use of personnel protective clothing.

In addition to microbes, fermentation products such as antibiotics and enzymes have been implicated in occupational asthma (87,89). Severe allergenic reactions to *Bacillus subtilis* proteases are well known (89). Under certain environmental conditions, organisms such as *Aspergillus flavus* and *Aspergillus oryzae* are known to produce lethal toxins (89). Specific strains of the blue-veined cheese mold, *Penicillium roqueforti*, also produce mycotoxins under narrowly defined environmental conditions (7). Species of the genus *Claviceps* and some members of other genera produce toxic ergot alkaloids. Some other toxin-producing fungi are noted in Table 5. Potentially, most physiologically active fermentation products can be disruptive to health, and certain products are highly toxic. Aflatoxins are potent carcinogens. For a given fungus, the product spectrum often depends on the fermentation conditions (temperature; pH; moisture content; type of substrate; agitation; metabolic energy source; amounts of oxygen and carbon dioxide; and nature and concentration of micronutrients, metal ions, and other chemicals); hence, evaluations under relevant conditions are essential. In particular, information from submerged culture may be an unsatisfactory indicator of behavior in solid-state fermentation.

Table 5. Toxin-Producing Fungi

Aspergillus alliaceus
Aspergillus clavatus
Aspergillus melleus
Aspergillus niger
Aspergillus ochraceus
Aspergillus parasiticus
Aspergillus sclerotiorum
Aspergillus sulphureus
Chaetomium cellulolyticum
Fusarium moniliforme
Fusarium graminearum
Penicillium islandicum
Penicillium roqueforti
Penicillium viridicatum

Handling of the substrate used in solid substrate fermentation requires care. A seemingly innocuous substrate may be contaminated with microbial spores or may generate allergenic dust. Raw material for composting may be heavily contaminated with human and animal pathogens, including eggs of parasites. Hence, certain high temperatures and holding time combinations are essential in composting to ensure destruction of pathogens and parasites. Depending on the composting process and the end use of the product, a uniform maximum temperature of 55 °C maintained for 3–15 days may be considered acceptable (33). Special care in process design and operation may be necessary to prevent recontamination of the final product. Cases of such contamination do occur perennially (e.g., infection from vegetables fertilized with manure; *Listeria* contamination in cheese). Sometimes inadequacies in processing may lead to a dangerous product: cases of cyanide poisoning from consumption of insufficiently processed *gari* occur from time to time in West Africa. Normally, fermentation and other process steps in gari making should substantially eliminate the cyanogenic glycosides (mostly linamarin and lotaustralin) that occur naturally in cassava tubers. Cases of fatal poisonings from consumption of poorly prepared tempe bongkerk have occurred in Indonesia.

Poor operational practice and failings in process and plant design can exacerbate the hazard. Experience suggests the following guidelines for minimizing process risks: (*1*) the number of operators should be kept to a minimum through mechanization and automation; (*2*) personnel–product contact should be minimal to reduce contamination of the product and ensure safety of operators; (*3*) operations should be enclosed and should be designed to minimize generation of aerosols and particulates; (*4*) the applicable GMP standards for food and pharmaceutical products should be followed; (*5*) the environment in and around the process plant must meet acceptable workplace air quality standards, especially for particulates, microorganisms, spores, humidity, and temperature; (*6*) the process equipment and the facility (including the air conditioning system) must be frequently sanitized; (*7*) facility design and operation must eliminate the risk of recontaminating the processed product; (*8*) there should be checks on the quality of the in-process material and the final product; and (*9*) satisfactory disposal of solid, liquid, and gaseous wastes must be ensured.

Operators should be fully informed of risks, and they should be given the training necessary for safe processing. Where required, personnel protective equipment should be provided. There should be adequate supervision, including routine health surveillance. Personnel who are ill, and those with open sores and cuts, should be sent home or moved to other, noncritical work. Attention to protection of peripheral support staff, for example, maintenance and cleaning personnel, is especially important because they may not have the knowledge or training for the potential risks. Thus, for example, trained, regular process operators should clean and sterilize a bioreactor before maintenance work.

The solid substrate fermentation facility should be designed and operated to prevent cross-contamination between dirty and clean areas. This goal requires attention to plant layout; movement of personnel, in-process material, process machinery; and flow of air in the facility. A minimum 12-Pa (0.05-in water gauge) pressure differential is recommended between adjacent clean and dirty areas. Negative pressure with respect to surroundings should be maintained in most solid-state fermentation chambers. Alternatively, the fermentation chamber may be under positive pressure with respect to the adjacent area, but that area should be under negative pressure with respect to other surrounding areas. Most work areas should comply with a relative humidity standard of 40–50%. Relative humidities lower than 40% cause problems with static electricity, whereas values greater than 50% promote corrosion. Between 10 and 15 air changes per hour are recommended for most plant areas. Workspaces for substrate preparation, downstream processing, and fermentation should attain 15–20 air changes per hour. Generally, conditioned air should be supplied at the ceiling level and return at lower wall locations. Preferably, particle counts should not exceed $3.5 \times 10^6\ m^{-3}$ (10^5 per cubic foot) for particles 0.5 μm or larger, and $2.5 \times 10^4\ m^{-3}$ (700 per cubic foot) for particles 5.0 μm or larger. There should be no more than 25 cfu/10 ft^3 of air. Air entering and exiting the fermentation chamber should be filter sterilized with 0.45-μm (or better) absolute filters that are also capable of in situ steam sterilization. A filter maintenance and replacement program should be in place. The practice of sterilizing solid substrate aeration gas by bubbling it through caustic or acid baths is misguided. Admittedly, the standards recommended here are not intended for processes such as composting and mushroom cultivation, but for plants producing foods, food additives, bulk pharmaceuticals, and secondary metabolites. The recommended practices are based on the author's experience. Note that the standards for packaging, sterile processing, and final dosage pharmaceuticals processing areas are significantly higher than the ones given here (90,91).

Much of the indigenous fermented food production in Asia, Africa, and South America is carried out at the cottage- or village-scale level with little consideration for hygienic processing. With few exceptions, even the larger plants do not come close to meeting the minimum standards expected in the developed world. The practices recommended in this review are for best-case production processes in the developed world where attention to hygiene and safety have transformed the traditional processing methods. Finally, as examples of toxic gari and tempe bongkerk prove, a thorough understanding of the fermentation microbiology and biochemistry is essential if disasters are to be prevented (78). Such knowledge is indispensable to developing large-scale production processes and in gaining increased acceptance of products beyond the regions of origin.

BIBLIOGRAPHY

1. R.E. Mudgett, in A.L. Demain and N.A. Solomon eds., *Manual of Industrial Microbiology and Biotechnology*, American Soc. for Microbiology, Washington, D.C., 1986, pp. 66–83.
2. T.E. Tautorus, *Adv. Biotechnol. Processes* **5**, 227–273 (1985).
3. F. Zadrazil, D. Ostermann, and G.D. Compare, in H.W. Doelle, D.A. Mitchell, and C.E. Rolz eds., *Solid Substrate Cultivation*, Elsevier, London, 1992, pp. 283–319.
4. W.A. Hayes and N.G. Nair, in J.E. Smith and D.R. Berry eds., *The Filamentous Fungi*, vol. 1, E. Arnold, London, 1975, pp. 212–248.
5. W. Röcken and P.A. Voysey, *J. Appl. Bacteriol. Symp. Suppl.* **79**, 38S–48S (1995).
6. A.H. Rose, ed., *Fermented Foods*, Academic Press, London, 1982.
7. B.A. Law, in A.H. Rose ed., *Fermented Foods*, Academic Press, London, 1982, pp. 147–198.
8. F.M. Rombouts and M.J.R. Nout, *J. Appl. Bacteriol. Symp. Suppl.* **79**, 108S–117S (1995).
9. D. Fukushima, in K.H. Steinkraus ed., *Industrialization of Indigenous Fermented Foods*, Dekker, New York, 1989, pp. 1–88.
10. K. Yoshizawa and T. Ishikawa, in K.H. Steinkraus ed., *Industrialization of Indigenous Fermented Foods*, Dekker, New York, 1989, pp. 127–168.
11. M. Moo-Young, Y. Chisti, and D. Vlach, *Biotechnol. Adv.* **11**, 469–479 (1993).
12. U.C. Banerjee, Y. Chisti, and M. Moo-Young, *Resour. Conserv. Recycl.* **13**, 139–146 (1995).
13. L.M. Miall, in J.E. Smith and D.R. Berry eds., *The Filamentous Fungi*, vol. 1, E. Arnold, London, 1975, pp. 104–121.
14. K.E. Aidoo, R. Hendry, and B.J.B. Wood, *Adv. Appl. Microbiol.* **28**, 201–237 (1982).
15. K. Bosecker and M. Kürsten, *Process Biochem.* **13**, 2–4 (1978).
16. E. Cannel and M. Moo-Young, *Process Biochem.* **15**, 2–7 (1980).
17. L.A. Underkofler, *Chem. Eng. Prog. Symp. Ser. No. 69* **62**, 11–20 (1966).
18. H. Ebine, in K.H. Steinkraus ed., *Industrialization of Indigenous Fermented Foods*, Dekker, New York, 1989, pp. 89–126.
19. B.K. Lonsane, N.P. Ghildyal, S. Budiatman, and S.V. Ramakrishna, *Enzyme Microb. Technol.* **7**, 258–265 (1985).
20. P. Weiland, in F. Zadrazil and P. Reiniger eds., *Treatment of Lignocellulosics with White Rot Fungi*, Elsevier, London, 1988, pp. 64–76.
21. E. Cannel and M. Moo-Young, *Process Biochem.* **15**, 24–28 (1980).
22. M. Moo-Young, A.R. Moreira, and R.P. Tengerdy, in J.E. Smith, D.R. Berry, and B. Kristiansen eds., *The Filamentous Fungi*, vol. 4, E. Arnold, London, 1983, pp. 117–144.
23. C.W. Hesseltine, *Process Biochem.* **12**, 29–32 (1977).
24. R.P. Tengerdy, *Trends Biotechnol.* **3**, 96–99 (1985).
25. A. Durand and D. Chereau, *Biotechnol. Bioeng.* **31**, 476–486 (1988).
26. Y. Chisti, *Airlift Bioreactors*, Elsevier, London, 1989.
27. Y. Chisti and M. Moo-Young, in V. Moses and R.E. Cape eds., *Biotechnology: The Science and the Business*, Harwood, New York, 1991, pp. 167–209.
28. Y. Chisti and M. Moo-Young, *Trans. I. Chem. E.* **74A**, 575–583 (1996).
29. Y. Chisti, *Chem. Eng. Prog.* **88**, 55–58 (1992).
30. Y. Chisti and M. Moo-Young, *J. Ind. Microbiol.* **13**, 201–207 (1994).
31. B.K. Lonsane, G. Saucedo-Castaneda, M. Raimbault, S. Roussos, G. Viniegra-Gonzalez, N.P. Ghildyal, M. Ramakrishna, and M.M. Krishnaiah, *Process Biochem.* **27**, 259–273 (1992).
32. B.A. Prior, J.C. Du Preez, and P.W. Rein, in H.W. Doelle, D.A. Mitchell, and C.E. Rolz eds., *Solid Substrate Cultivation*, Elsevier, London, 1992, pp. 65–85.
33. E.I. Stentiford and C.M. Dodds, in H.W. Doelle, D.A. Mitchell, and C.E. Rolz eds., *Solid Substrate Cultivation*, Elsevier, London, 1992, pp. 211–246.
34. M. Gutiérrez-Rojas, S.A.H. Hosn, R. Auria, S. Revah, and E. Favela-Torres, *Process Biochem.* **31**, 363–369 (1996).
35. B.L. Rathbun and M.L. Shuler, *Biotechnol. Bioeng.* **25**, 929–938 (1983).
36. S. Solis-Pereyra, E. Favela-Torres, M. Gutiérrez-Rojas, S. Roussos, G. Saucedo-Castañeda, P. Gunasekaran, and G. Viniegra-González, *World J. Microbiol. Biotechnol.* **12**, 257–260 (1996).
37. Y. Chisti, *Biotechnol. Adv.* **11**, 385 (1993).
38. A. Pandey, *Process Biochem.* **27**, 109–117 (1992).
39. P.W. Lambert, in J.E. Smith, D.R. Berry, and B. Kristiansen eds., *The Filamentous Fungi*, vol. 4, E. Arnold, London, 1983, pp. 210–237.
40. K.H. Steinkraus, in J.E. Smith, D.R. Berry, and B. Kristiansen eds., *The Filamentous Fungi*, vol. 4, E. Arnold, London, 1983, pp. 171–189.
41. B.J.B. Wood and Y.F. Min, in J.E. Smith, and D.R. Berry eds., *The Filamentous Fungi*, vol. 1, E. Arnold, London, 1975, pp. 265–280.
42. Y. Chisti and M. Moo-Young, *I. Chem. E. Symp. Ser.* **137**, 135–146 (1994).
43. K. Aunstrup, in L.B. Wingard, Jr., E. Katchalski-Katzir, and L. Goldstein eds., *Applied Biochemistry and Bioengineering*, vol. 2, Academic Press, New York, 1979, pp. 27–69.
44. B.K. Lonsane and N.P. Ghildyal, in H.W. Doelle, D.A. Mitchell, and C.E. Rolz eds., *Solid Substrate Cultivation*, Elsevier, London, 1992, pp. 191–209.
45. J. Pallares, S. Rodríguez, and A. Sanromán, *Bioprocess Eng.* **15**, 31–33 (1996).
46. C.W. Hesseltine, *Biotechnol. Bioeng.* **14**, 517–532 (1972).
47. A. Singh and P. Mishra, *Microbial Pentose Utilization: Current Applications in Biotechnology*, Elsevier, Amsterdam, 1995, pp. 71–98.
48. M.A. Millett, A.J. Baker, and L.D. Satter, *Biotechnol. Bioeng. Symp.* **5**, 193–219 (1975).
49. M. Becerra and M.I. González Siso, *Enzyme Microb. Technol.* **19**, 39–44 (1996).
50. V.K. Garg, M.A.C. Costello, and B.A. Czuba, in S. Seetharam and S.K. Sharma eds., *Purification and Analysis of Recombinant Proteins*, Dekker, New York, 1991, pp. 29–54.

51. V.R. Anicetti, B.A. Keyt, and W.S. Hancock, *Trends Biotechnol.* **7**, 342–349 (1989).
52. S.A. Rouf, M. Moo-Young, and Y. Chisti, *Biotechnol. Adv.* **14**, 239–266 (1996).
53. S.S. Yang and W.J. Swei, *World J. Microbiol. Biotechnol.* **12**, 43–46 (1996).
54. J. Barrios-González, T.E. Castillo, and A. Mejía, *Biotechnol. Adv.* **11**, 525–537 (1993).
55. C.W. Hesseltine, *Process Biochem.* **12**, 24–27 (1977).
56. P.K.R. Kumar and B.K. Lonsane, *Process Biochem.* **22**, 139–143 (1987).
57. P.K.R. Kumar and B.K. Lonsane, *Process Biochem.* **23**, 43–47 (1988).
58. E.V. Emelyanova, *Process Biochem.* **31**, 431–434 (1996).
59. R.W. Silman, *Biotechnol. Adv.* **11**, 559–560 (1993).
60. R.W. Silman, R.J. Bothast, and D.A. Schisler, *Biotechnol. Adv.* **11**, 561–575 (1993).
61. C. Desgranges, C. Vergoignan, A. Léréec, G. Riba, and A. Durand, *Biotechnol. Adv.* **11**, 577–587 (1993).
62. D. Leštan, M. Leštan, J.A. Chapelle, and R.T. Lamar, *J. Ind. Microbiol.* **16**, 286–294 (1996).
63. J.H. Crawford, *Process Biochem.* **18**, 14–18 (1983).
64. D. Augenstein, D.L. Wise, N.X. Dat, and N.D. Khien, *Resour. Conserv. Recycl.* **16**, 265–279 (1996).
65. M. Moo-Young and Y. Chisti, *Resour. Conserv. Recycl.* **11**, 13–24 (1994).
66. J.F. Rees and J.M. Grainger, *Process Biochem.* **17**, 41–44 (1982).
67. H.W. Doelle, D.A. Mitchell, and C.E. Rolz eds., *Solid Substrate Cultivation*, Elsevier, London, 1992.
68. D.R. Fenlon, A.R. Henderson, and J.A. Rooke, *J. Appl. Bacteriol. Symp. Suppl.* **79**, 118S–131S (1995).
69. D.R. Seale, *J. Appl. Bacteriol. Symp. Suppl.* **61**, 9S–26S (1986).
70. R. Whittenbury, *Process Biochem.*, 27–31 (February 1968).
71. J. Wignall and J. Tatterson, *Process Biochem.* **11**, 17–19 (1976).
72. P. McDonald, *The Biochemistry of Silage*, Wiley, Chichester, 1981.
73. M.K. Woolford, *The Silage Fermentation*, Dekker, New York, 1984.
74. B.K. Lonsane and M.M. Kriahnaiah, in H.W. Doelle, D.A. Mitchell, and C.E. Rolz eds., *Solid Substrate Cultivation*, Elsevier, London, 1992, pp. 147–171.
75. L. Ikasari and D.A. Mitchell, *Enzyme Microb. Technol.* **19**, 171–175 (1996).
76. K.H. Steinkraus ed., *Handbook of Indigenous Fermented Foods*, Dekker, New York, 1983.
77. K.H. Steinkraus ed., *Industrialization of Indigenous Fermented Foods*, Dekker, New York, 1989.
78. C.W. Hesseltine, *Process Biochem.* **16**, 2–6, 13 (1981).
79. B.J.B. Wood, in A.H. Rose ed., *Fermented Foods*, Academic Press, London, 1982, pp. 39–86.
80. N. Okafor and A.O. Ejiofor, *Process Biochem.* **25**, 82–86 (1990).
81. C.O. Ikediobi and E. Onyike, *Process Biochem.* **17**, 2–5 (1982).
82. S.H. Abiose, M.C. Allan, and B.J.B. Wood, *Adv. Appl. Microbiol.* **28**, 239–265 (1982).
83. M. Lockie, *Chem. Eng. (Lond.)* **615/616**, 22–25 (1996).
84. R.O. Arunga, in A.H. Rose ed., *Fermented Foods*, Academic Press, London, 1982, pp. 259–274.
85. J.G. Carr, in A.H. Rose ed., *Fermented Foods*, Academic Press, London, 1982, pp. 275–292.
86. R.F. Schwan, A.H. Rose, and R.G. Board, *J. Appl. Bacteriol. Symp. Suppl.* **79**, 96S–107S (1995).
87. B.K. Lonsane, N.P. Ghildyal, M. Ramakrishna, and F. Stutzenberger, in H.W. Doelle, D.A. Mitchell, and C.E. Rolz eds., *Solid Substrate Cultivation*, Elsevier, London, 1992, pp. 385–401.
88. A. Rimmington, in P. Hambleton, J. Melling, and T.T. Salusbury eds., *Biosafety in Industrial Biotechnology*, Chapman & Hall, London, 1994, pp. 67–89.
89. A.M. Bennett, in P. Hambleton, J. Melling, and T.T. Salusbury eds., *Biosafety in Industrial Biotechnology*, Chapman & Hall, London, 1994, pp. 109–128.
90. Y. Chisti, in G. Subramanian ed., *Bioseparations and Bioprocessing: A Handbook*, vol. 2, VCH, New York, 1998, pp. 3–300.
91. Y. Chisti, in G. Subramanian ed., *Bioseparations and Bioprocessing: A Handbook*, vol. 2, VCH, New York, 1998, pp. 379–415.

See also FOOD PROCESS ENGINEERING; MEMBRANE SURFACE LIQUID CULTURE, MICROORGANISMS, FUNGI.

SOLVENT FERMENTATION. See ANAEROBES; ANAEROBES, INDUSTRIAL USES; CLOSTRIDIA, SOLVENT FERMENTATION.

SOYBEAN (FERMENTATION, MEAL OIL)

ZIVKO L. NIKOLOV
ROGER FUENTES-GRANADOS
Iowa State University
Ames, Iowa

KEY WORDS

Concentrates
Extraction
Fermented products
Flour
Isolates
Meal
Non-fermented products
Processing
Refining
Soybean oil

OUTLINE

INTRODUCTION

Soybean, *Glycine max*, is a crop native to Asia. It is not known when soybeans were introduced to the United States. Morse (1) reports that the first written mention of soybeans in the United States occurred in 1804, although there are references dating back to 1770, when Benjamin Franklin sent soybeans home from England (2). It is certain, however, that in the United States soybeans as a crop did not begin to gain popularity until the early 1920s. The primary reason for soybean's increased popularity was a newly developed ability to extract oil from it. Technological advances allowing utilization of soybeans for oil production led to the development of a large soybean industry. Today, the annual soybean crop value is about $14.5 billion. The advent of crushing facilities (continuous expellers) in the 1920s encouraged domestic demand for soybeans, and so for several decades since the 1920s, steady increases in the production and consumption of soybeans have occurred (Table 1) (3).

Economically, meal production today is more important than oil production. One metric ton (MT) of soybeans yields about 180 kg of oil and 800 kg of defatted meal. The unit price of oil is more than twice that of meal; however, a metric ton of soybeans produces meal and oil worth $160 and $80, respectively. The meal usually is the driving force behind soybean price because, on the average, it accounts for approximately two-thirds of the value of the beans (4). Soybean oil, the other direct product of soybean processing, is also important especially in the food industry. Soybean oil has 77% of the edible fats and oils market in the United States (4).

Most soy protein in the form of defatted meal is used as animal feed. Because of an ever-increasing population, however, demand for food protein has increased tremendously in recent years. Many types of soy food products have been developed, mostly from defatted soybean meal. These include soy flour and grits, soy protein concentrate, soy protein isolates, texturized soy proteins, defatted soy flakes, soy meal, full-fat soy flour, and enzyme-active soy flour. Advances in processing technology have resulted in soy protein products that can perform many functions in food while retaining their nutritive value. The development of technologies and processes has pioneered the way for soybean production and utilization. Improving existing and developing new ways to process and utilize soybeans will continue to be necessary if increasing demand for soybean products is to be met. This article reviews major soybean products and commercial processes currently in use.

Table 1. U.S. Production of Soybeans, Soybean Oil, and Total Crop Value from 1925 to 1995

Year	Total soybean production (million Bu.)	Total oil production (million lb)	Total crop value ($ million)
1925	4.9	2.5	11.4
1930	14	14	19
1935	49	105	36
1940	78	533	70
1945	193	1,392	402
1950	299	2,075	738
1955	374	2,827	831
1960	555	4,420	1,185
1965	846	5,800	2,151
1970	1,127	8,265	3,205
1975	1,548	9,630	7,618
1980	1,798	11,270	13,560
1985	2,099	11,617	10,571
1990	1,926	13,408	11,042
1995	2,152	15,155	14,564

Source: Soya Bluebook Plus, 1997 (7).

VARIETIES AND COMPOSITION

A great number of soybean varieties are grown in the United States. Most differ in agronomic characteristics, so maximum yield under diverse growing conditions can be achieved. Because chemical composition is not considered in the grading standards for trading soybeans, plant breeders have focused plant improvement programs on increasing yield. But in the last few years, new market opportunities for specialty soybeans have led to an increased release of soybean varieties with improved quality traits. Some quality characteristics recently developed in soybeans include lower amounts of linolenic and palmitic acids and higher levels of oleic acid in the oil and reduced or complete elimination of lipoxygenase and lower concentration of trypsin inhibitor in the meal.

Processing and utilization of soybeans have led to the creation of soybean varieties of two types: those processed for production of oil and meal and those processed for direct food use (5). Soybeans used directly for food are generally lighter in color, have clear hilum, and contain more protein and less oil than soybeans used for oil and meal do (2). The most desired characteristics of soybeans for food include a large seed, high protein content, and high quality at a reasonable price (6). Typical dry weight composition of soybeans is 12 to 25% oil, 35 to 50% protein, 33 to 35% carbohydrate, and 5% ash (6).

PROCESSING SOYBEANS FOR OIL AND MEAL

Soybeans are the dominant oilseed crop among the seven major oilseeds grown in the United States. More than half of the U.S. soybean crop is used by the domestic crushing industry to produce meal and oil. More than 15 billion pounds of soybean oil were produced in 1996 (7). Improved understanding of lipid chemistry and enhanced oil processing technologies were the most vital factors in the acceptance of soybean oil as a food-grade product. Refined, bleached, and deodorized soybean oil is used by a variety of industries, including food processors and industrial concerns, by food service organizations, and by consumers (8). Modern soybean crushing for oil and meal production involves these unit operations: preparation and handling, cleaning, drying, cracking and dehulling, conditioning, flaking, extraction, and product desolventizing.

Preparation and Handling

Conventional preparation and handling of soybeans before extraction consists of cleaning, drying, cracking, dehulling, conditioning, and flaking (9).

Cleaning. Sand, stones, dust, and other foreign material usually are removed by conventional seed conditioners, which consist of a two-deck vibrating screen (5). The upper screen retains the stones and coarser materials but allows the passage of soybeans, and the lower screen retains the soybeans and filters out finer particles. Light trash and hull particles are removed by aspiration, and magnetic separators are used to remove tramp iron (5).

Drying. Drying is conducted to efficiently remove the hulls from cotyledons. Beans are dried to a moisture content of 10% or less. Soybeans are usually dried in countercurrent open-flame grain dryers heated with natural gas or fuel oil (10). Soybean temperature must be raised sufficiently to achieve the desired final moisture of 10%, but it must not exceed 76 °C because discoloration and protein denaturation will occur (10). Dried soybeans are stored in bins for 2 to 5 days of tempering to allow for moisture equilibration by diffusion. The tempering bins also serve as working bins for uninterrupted feeding of processing plants (5). Improvements have been made in reducing fuel and energy costs. Some relatively new drying concepts include microwave vacuum drying and solar drying (10).

Cracking and Dehulling. To break the soybeans into small pieces for dehulling and flaking, cracking is achieved by counter-rotating corrugated rollers (9). To save power and to obtain more uniform granulation with fewer fines in the cracked beans, the corrugation of the cracking rolls was changed some years ago by widening the pitch between the teeth of the corrugated rolls (11). This alteration resulted in cleaner cuts or shearing with less power consumption (11). Cracking rollers usually are 25 cm in diameter and at least 107 cm long, having a processing capacity as great as 500 to 600 MT/day (9). The cracking step is very important in traditional dehulling. The cotyledons of properly dried and cracked soybeans should separate easily from the hulls (12). If the soybeans are too dry, they will pulverize in the cracking rolls and produce excessive fines, which could cause problems in the extractor. If beans are too moist, they tend to mash and not to release the hulls (12).

Conditioning. Cotyledon fragments (meats) are conditioned by heat and moisture to obtain the plasticity required for flaking. The temperature of the meats at the end of conditioning is between 65 and 70 °C (9). Conditioners may be vertical stack cookers or rotary horizontal cookers in which steam or water spray is added to adjust moisture to approximately 11% (12). Effective preheating before flaking decreases power consumption of the flaker (11). Flaking cold beans (25–30 °C) requires a power of about 5 to 6 kWh/ton of soybeans; however, flaking at 55 to 60 °C requires only 3 to 4 kW h/ton (11).

On-line dryers have been developed that combine the drying, dehulling, and conditioning steps into a continuous operation. Soybeans are cracked hot, which decreases the amount of fines produced. Different systems have been designed for this combined operation. The Escher-Wyss system uses a fluid bed dryer-heater to heat and dry soybeans so that the hulls detach from the meat. This is followed by cracking rolls and impactors to break up the soybeans and loosen the hulls. Hulls then are separated by air aspiration and meats are sent to a fluidized-bed conditioner-cooler for final moisture and temperature adjustment (13).

Flaking. Flaking is achieved by passing the conditioned meats between two horizontal smooth rolls. Flaking mills consist of a pair of smooth surfaced rolls with a minimum diameter of 50 cm (12). Rolls usually are 120 cm long and 70 cm in diameter. Pressure between the rolls is generated by mechanical or hydraulic systems, and flake thickness is controlled by the pressure (12). Flaking before solvent extraction is important because (*1*) solvent can flow easier through a bed of flaked particles than through a bed of cotyledon fragments or fine particles and (*2*) pressure generated between the rolls ruptures oil cells and facilitates subsequent solvent extraction (9).

Thickness of the flakes is crucial because solvent and oil diffusion are the rate-limiting steps during hexane extraction. Optimal flake thickness is 0.025 to 0.037 cm; thickness outside this range does not yield satisfactory results. Thicker flakes offer greater diffusion resistance; thinner ones easily crumble, producing fines that hinder hexane penetration into the soy-flakes bed (9).

A common problem in flaking is greater wear at the center than at the ends of the rolls. As the wear progresses, soy flakes at the center of the rolls become thicker than those at the ends. Continued pressure adjustment to correct flake thickness near the center may cause excessive pressure at the roll ends and chipping or pitting of the metal (12).

Improvements in Pre-Extraction Technology. The use of expanders in preparing oilseeds for extraction has been one of the most significant developments in the oilseed industry (13). Expanders produce collets that improve drainage and reduce solvent carryover to the desolventizer toaster. The three main benefits of the expander in oil extraction are (*1*) as great as a 40% increase in bulk density, potentially as great as 40% increase in contact time; (*2*) increase of bed porosity resulting in more rapid drainage rates; and (*3*) additional rupture of oil cells, which substantially increases the oil diffusion rate (14).

Solvent Extraction

The most prevalent solvent extraction method in modern soybean processing facilities is hexane extraction (15). The oil extraction process is basically a process of solid–liquid extraction (5). The transfer of oil from solids to the surrounding oil–solvent solution may be described in three steps: (*1*) diffusion of solvent into the solids, (*2*) oil solubilization, and (*3*) diffusion of oil from the solid particles into the surrounding liquid phase.

The driving force of diffusion is the oil concentration gradient. Although hexane has been the solvent of choice in most commercial extraction plants, other solvents also have been considered (16,17). The properties of an ideal solvent have been discussed in detail by Johnson (17); the most important of these include high-solvent power, nontoxicity, triglyceride selectivity, nonflamability, low specific heat and heat of vaporization, and low cost. The two streams exiting from the solvent extractor are the miscella, or oil-hexane mixture, and the marc, or defatted flakes containing hexane (Fig. 1).

Throughout the world, mechanical extraction often is preferred by small extraction plants, particularly those processing a variety of oilseeds. Screw presses consist of a shaft with an interrupted worm gear rotating in a cage of metal bars separated by small spaces. High pressure forces the oil out between the cage bars as the press cake moves along the shaft. Mechanical extraction can be used for the cold pressing of oils and the production of meal with a high oil content. Advantages are low initial capital cost and no solvent requirement; the efficiency of oil extraction, however, is less than with other methods.

Extraction Equipment. The basic types of commercial oil extractors in use today fall into two main categories: percolation and immersion (15,18). Percolation types are used worldwide and are considered more efficient. Immersion extractors are not commercially used for soybean oil extraction (15).

Four types of percolation extractors are actively marketed at present: the rotary basket extractor (Rotocel), the stationary basket extractor, the horizontal belt extractor, and the continuous loop extractor (12). A description of each extractor is beyond the scope of this article, and an excellent review of commercial extractors has been given by Milligan (19). In percolation extractors, hexane or miscella is pumped over the bed of soy flakes, percolates through the bed, and leaves at the bottom through a perforated plate or mesh screen. Extractors can be classified as deep-bed (> 5 m deep) and shallow-bed (< 5 m deep) extractors. Deep-bed extractors (rotary or stationary basket extractors) have a material bed whose overall length is approximately five times greater than its depth (14). The bed surface area per miscella stage is relatively small, allowing the required miscella flow to permeate the bed completely. The solvent-flakes contact time is maximized, and these extractors can operate with relatively thick flakes. Shallow-bed extractors (horizontal belt and continuous loop extractors) are designed so that overall length of the bed material is 50 times greater than its depth. Bed surface area per miscella stage is large, and contact time between solvent and flakes is not maximized. Shallow-bed extractors operate effectively by using thin flakes, thereby increasing the diffusion rate (14).

Miscella Stripping. Full miscella contains about 30% oil; for every ton of crude oil, therefore, about 2.5 tons of solvent must be removed by distillation (5). The hexane must be efficiently removed and recycled to ensure the oil safety and to conform to environmental regulations (9).

A solvent recovery system consists of a two-stage stripping evaporator and an oil stripping column. Solvent is removed from the miscella by double-effect evaporators and steam stripping. The first stage usually consists of a long-tube vertical evaporator, where the miscella is heated and solvent evaporated using vapors from the desolventizer-toaster (DT) unit at below atmospheric pressure, concentrating the miscella to about 70% oil (12,15). The second-stage evaporator is a rising-film evaporator heated with steam at 10 to 20 psi in which the miscella is further concentrated to 90 to 95%. Vapors generated go to the condenser, and oil goes to the oil stripper, a disk-donut type operating under vacuum provided by steam injection (12,15). Oil flows downward in contact with stripping steam introduced at the bottom. At the end of this operation, the amount of residual solvent is sufficiently low to achieve a flash point of 150 °C. A flash point of 121 °C corresponds to 1,000 ppm of hexane in the crude oil, which is the required level for transfer to the refinery (9).

Desolventizing and Toasting Soybean Meal for Feed. Desolventizing is a key operation in solvent extraction (13). Defatted flakes from the extractor contain about 30% (w/w) hexane, and before they can be processed further for feed they need to be processed through a desolventizer. The meal, which is used usually as livestock feed, must be heated sufficiently to inactivate antinutritional factors such as trypsin inhibitors (9). This process is known as *toasting*. It is carried out commercially in conjunction with desolventizing in a DT unit (9). Hexane-wet flakes leave the extractor at temperatures of about 57 °C. Live steam is injected at the bottom of the DT unit, and the steam that condenses provides the latent heat required for hexane evaporation. The condensed steam raises the moisture level of the flakes to 16 to 24%, facilitating the toasting operation at 100 to 105 °C (20). The residual hexane content of the flakes is less than 500 ppm. The flakes then are hot-air dried to about 10% moisture, cooled to a temperature below 32 °C, and ground and screened to a desired grit size.

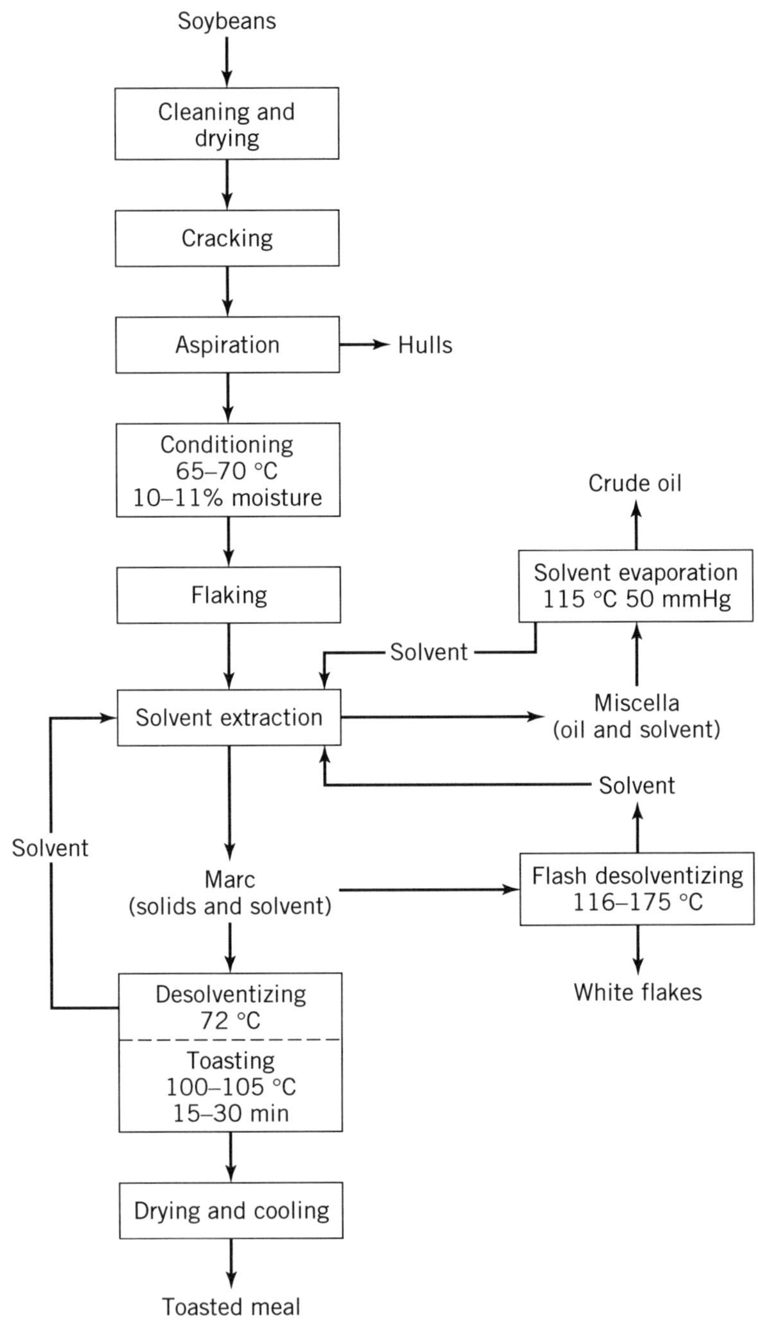

Figure 1. Principal unit operations in soybean oil extraction facilities.

Soybean meal with greater than 50% (w/w) protein, which is produced by grinding toasted flakes, provides greater than 60% of the value of processed products from soybeans. In the United States, meal is used primarily in swine and poultry rations (8). The other components of meal are carbohydrates (30 to 35%), lipids (1%), ash, and moisture.

A predesolventizing step and the use of expanders in flake preparation have marked effects on the desolventizing operation (20). Predesolventizing reduces the amount of hexane going into the DT and the amount of water condensed on the flakes and, consequently, decreases the steam requirements (20). Moisture reduction of the flakes, however, increases the possibility of overheating and decreases the nutritive value of the meal (20). Expanded flakes have higher bulk density and lower residual solvent content than unexpanded flakes do (9).

Desolventizing and Deodorizing Soy Flakes for Food. For food and industrial applications of meal, it is necessary to minimize the heat denaturation of soy proteins and to maintain their water solubility (9). Desolventizing to produce white flakes can be performed (*1*) in steam-jacketed conveyor tubes arranged in a vertical stack; (*2*) by using superheated circulating solvent vapor in a horizontal drum equipped with an agitator/conveyor pipe to facilitate contact between the flakes and the superheated vapor; and (*3*) by using superheated solvent vapor to contact the flakes in a conveyor pipe, where the conveying action results from the velocity of the superheated vapor (20). The last process, also called *flash desolventizing* because of the relatively short residence time of flakes in the apparatus, is the best method for desolventizing white flakes (5). The superheated vapor provides energy for the evaporation of solvent from the flakes. The turbulent nature of the flake-vapor flow permits extremely rapid heat and mass transfer. In seconds, the superheated hexane will flash off the residual liquid hexane, leaving desolventized flakes to be collected in the cyclone (21). Protein denaturation is minimized primarily because of the short heating time. Hexane losses in white flake desolventizing systems are generally higher than those in DT systems (20). After desolventizing, white flakes are deodorized with live steam to remove the remaining hexane. Depending on end use, deodorization conditions can be adjusted to produce flakes with varying protein solubilities, such as with protein dispersibility indices (PDIs) of 10 to 90%.

One negative impact on the environment from the oilseed processing industry is the unrecovered solvent from the oil extraction (14). By November 15, 2000, maximum achievable control technology standards for hexane emissions will be established (23). What the standards will be is unknown; it will not, however, be easy to meet them, even by the most efficient plants (24). The most significant hexane loss points in the extraction plants include desolventizing, meal drying and cooling, vent systems, water and oil carrying solvents, fugitive emissions, and plant shutdowns (24). In most plants, hexane loss is less than half that of 10 years ago. This decrease was accomplished by emphasizing the basic engineering and processing principles upon which most plants were designed and built (24).

Storing and Handling Crude Soybean Oil. Crude soybean oil contains a significant amount of nontriglyceride components that should be removed before bulk storage, handling, and transport (25). Deterioration of oil in storage depends not only on external factors but also on the initial conditions of oil, including moisture content, free fatty acids content, organic impurities, and contamination with molds (25). Crude oil must be stored under specific conditions to avoid oxidation caused by moisture, temperature, air, and metals promoting oxidation (9). To avoid the settling of fines and gums, oil with a moisture content below 0.2% is cooled to 40 °C and stored in stainless steel tanks with agitation. Oil stability can be improved if impurities are removed before storage (9).

REFINING SOYBEAN OIL

Refining freshly extracted oil means less product loss and improved product quality (13). Crude oil contains both insoluble and oil-soluble impurities. Insoluble impurities include seed fragments, excess moisture, and sometimes a waxy fraction appearing in refrigerated oils (9). Oil-soluble impurities include phosphatides, free fatty acids, gummy or mucilaginous material, color bodies, tocopherols, sterols, hydrocarbons, ketones, and aldehydes. The refining process removes those impurities, converting the crude oil into food grade soybean oil (9). Oil refining consists of several major unit operations: degumming, neutralization, bleaching, hydrogenation, and deodorization, each of which is designed to remove one or more specific compounds (Fig. 2).

Crude Oil Clarification

The removal of fines from crude oil is done traditionally by filtration. In integrated facilities, filters are being replaced by centrifugal decanters (13). In facilities with acid and water degumming and when lecithin is not produced, the clarification step is combined with the degumming step by removing both solids and gums in a high-speed decanter (13).

Degumming. The degumming step removes the phospholipids, also referred to as *phosphatides*, *lecithin*, or *gums*, which are the major source of oil phosphorus. Removal of phospholipids is essential in the production of oil with optimal color, flavor, and oxidative stability. Direct mixing of warm oil (70 °C) and soft water (1 to 3%) is the most commonly used method for degumming. Hydrated and insolubilized phospholipids partition to the water phase, and the oil temperature disrupts the possible formation of emulsion (9). Degumming of the oil from undamaged soybeans decreases phosphorus content in crude oil from 500 to 900 ppm to 12 to 170 ppm (9). Soy phospholipids also are extracted, because they are used in the food industry as soy lecithin (a mixture of phosphatidylcholine, phosphatidylinositol, and phosphatidylethanolamine). Because of the limited market for lecithin, a common practice in the United States is to alkali refine crude soybean oil and to dispose of lecithin in soapstock (26). The next important development in the degumming process probably will be the implementation of membrane filtration (13).

Neutralization

Neutralization is the removal of free fatty acids from the oil. Free fatty acids must be removed because they reduce the smoke point of oil, increase foaming, and may impart soapy flavor in foods (9). There are several methods of neutralizing crude soybean oil, alkali refining, Zenith refining, and physical refining are commercially used.

Alkali Refining. Alkali refining, also called *caustic refining*, is the most common method of neutralization and can be performed on crude or degummed oil. It involves adding alkali solution, in-line mixing with oil, centrifugation, in-line washing with water, and a second centrifugation (27). The sodium salt of the fatty acid (soaps), which are produced during neutralization, are removed during centrifugation. Some loss of triglycerides may occur with soaps. A typical level of free fatty acids before refining is 0.3 to 0.7%, and refining lowers this to <0.05% (9).

Zenith Refining. The Zenith refining process was developed in Sweden in 1960 and consists of degumming with

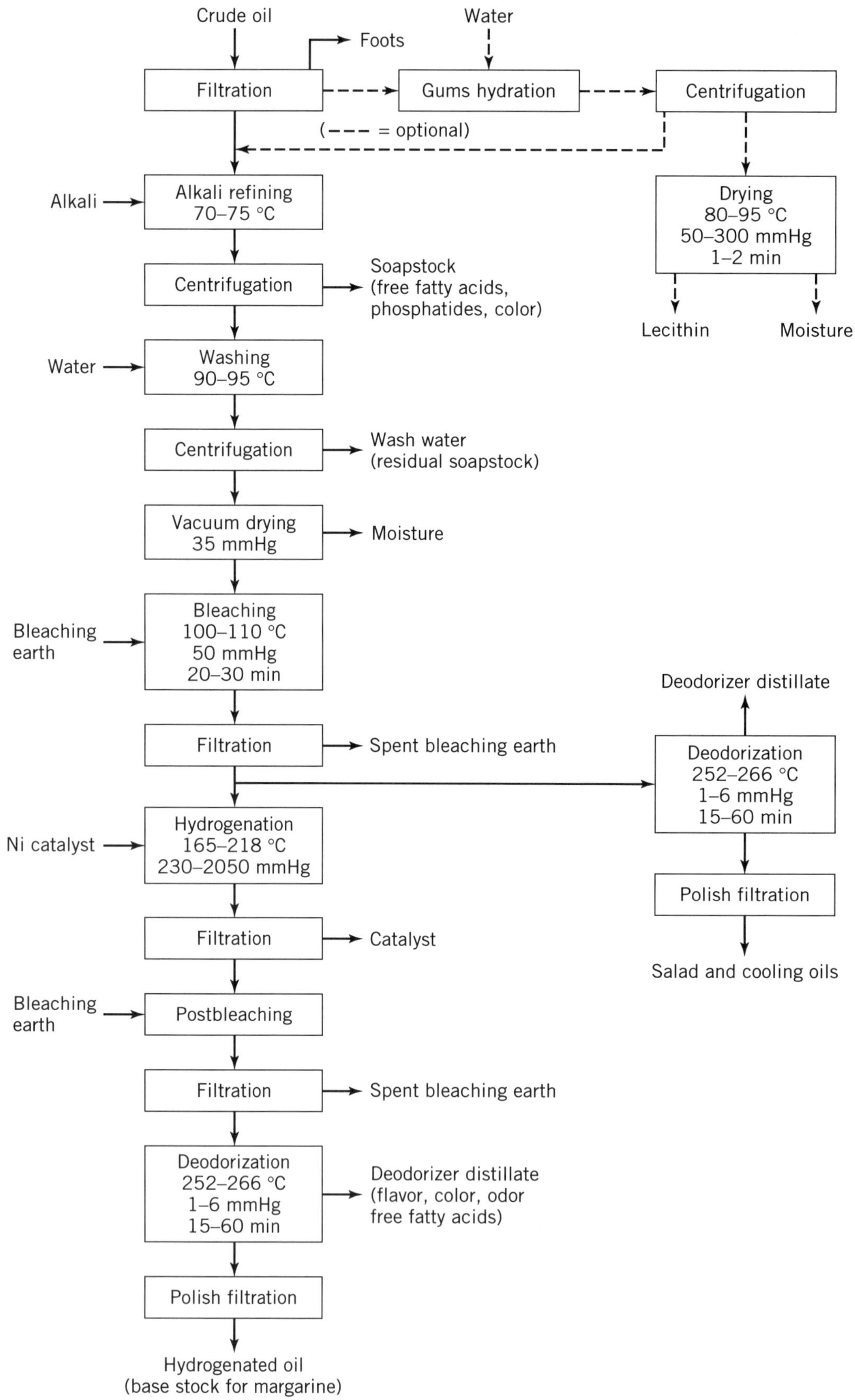

Figure 2. Unit operations involved in soybean oil refining for the production of hydrogenated oil, salad and cooking oils, and optional lecithin production.

0.1 to 0.4% (w/w) of phosphoric acid and alkali refining with a dilute sodium hydroxide solution (0.35 M). The use of dilute alkali obviates washing the oil with water. The neutralized oil contains about 1% water and 50 ppm soap and requires bleaching with citric acid and bleaching earth (27).

Physical/Steam Refining. This process is essentially a deodorization of lecithin-free oil under high vacuum at high temperatures. The process conditions are the same as those used in the deodorization. Physical refining reduces water usage and generates less wastewater than alkali refining does. Because of flavor instability, physically refined oils are not acceptable in the United States (27).

Bleaching

Treatment of refined oil with a bleaching adsorbent removes pigments, metal compounds, residual soap, and trace amounts of prooxidants that were not removed by alkali refining (15). Neutral earth, acid-activated earth, activated carbons, and silicates are used as adsorbents for soybean oil bleaching (28). This process has evolved basically from an open batch system to a continuous process under vacuum with alternating filters for clay removal (13). Typically, neutralized oil is mixed at about 70 °C with an appropriate dosage of earth (0.3 to 0.6%), heated to bleaching temperature (100 to 110 °C), mixed for 20 to 30 min, cooled to 70 to 80 °C, and filtered (28). Bleaching is an important step in enhancing oil appearance, flavor, and stability and in obtaining a light color appealing to the consumer (9).

Hydrogenation

The goal of hydrogenation is to decrease the amount of unsaturated fatty acids and to convert liquid oils to fluid, semisolids, or plastic fats (9,29). Hydrogenation is performed with hydrogen gas in the presence of a nickel catalyst. Reaction conditions differ with the desired product, and the ranges of process temperatures, pressures, and catalyst concentrations are 165 to 218 °C, 0.3 to 2.7 atm, and 0.02 to 0.05%, respectively (29). Hydrogenation increases the oxidative stability by reducing the number of double bonds in fatty acids, which are the sites of peroxide formation and subsequent off-flavor development (9).

Deodorization

The main purpose of deodorization is to remove free fatty acids, odors, and flavor compounds from the oil (13). Deodorization is a steam-stripping process in which good quality water is injected into soybean oil under a strong vacuum (1 to 6 mm Hg) and a temperature sufficiently high (252 to 266 °C) to vaporize undesirable volatile compounds (30). Deodorization is the last step in refining edible oils to be used as ingredients in salad dressings, cooking oil, margarines, shortenings, hard butters for the confectionery industry, and other food products (9).

Winterization

Winterization is the removal of solids that settle out from the oil at temperatures between 4 and 10 °C. The solid fraction (stearine) is separated so that it does not precipitate during refrigeration of salad oils (9). Winterization begins with dry, hydrogenated oil being pumped into a large insulated tank fitted with cooling coils. The temperature is lowered slowly by controlling the amount of propylene glycol circulating through the cooling coils. The rate of cooling must be controlled to promote the formation of larger crystals for easy separation. Crystal formation and growth is determined by nucleation temperature, triglyceride composition, and cooling rate. Commercial winterization is done by cooling the oil from 30 to 35 °C to about 6 °C. The temperature difference between oil and coolant is maintained at about 13 °C to ensure slow cooling. As the final temperature is approached, the cooling rate is decreased further. The cooling process lasts between 12 and 24 h (9). At the end of the cooling period, the external cooling is stopped and the winterized oil is allowed to stand for 6 to 8 h at 6 °C before filtration (15). Typical oil recovery is 75 to 80%.

Industrial Uses of Soybeans

Industrial (nonfood, nonfeed) uses of the two major soybean products, protein and oil, comprise less than 0.5% of total soy protein and about 2.6% of the total soy oil produced in the United States (31). Economically viable industrial uses today of soy protein and oil include paper coatings (protein), wood adhesives (protein), alkyd resins (oil), printing inks (oil), and oleochemicals. Potential uses include plastics and textile fibers from soy protein and soybean-based fuel. The most important current and potential applications of major soybean products are summarized in Table 2. For in-depth review of the industrial uses of soybeans, interested readers are referred to the work of Johnson and Myers (31).

SOYBEAN PROTEIN PRODUCTS

The term *soy protein* is given to processed edible dry soybean products with protein contents greater than 50% (w/w). Typically, food-grade soy proteins are produced from white flakes, which are processed on different lines than those used for producing oil and feed meal (32).

Defatted Soybean Grits and Flour

Grits are obtained by grinding defatted flakes coarsely followed by screening. Grits are sold as coarse (10 to 20 mesh), medium (20 to 40 mesh), or fine (40 to 80 mesh) grade. Soy flour is produced by grinding soy flakes to very fine particles to achieve 97% product pass through a 100-mesh screen. Soybean flour and grits are used in a variety of food products, including soups, sweets, beverages, desserts, bakery goods, breakfast cereals, and meat products.

Soy Protein Concentrates

Soy protein concentrates are prepared by removing soluble carbohydrates and volatile compounds from defatted meal or flour to obtain a product with a protein content of 65 to 75% (w/w). Nondenatured soybean protein is most soluble at pH values between 1.5 and 2.5 and between 7 and 12

Table 2. Current and Potential Industrial Uses of Soybean Products

Product	Main processes	Uses
Adhesives	Alkali dispersion of soy flour, mixing and treatment with other ingredients such as sodium silicate, antifoam, and preservatives	Wood adhesives
Plastics	Mixing of soybean meal with formaldehyde and phenol followed by heating	Potential use for foam cups, food packaging, and other fast-food industry items
Textile fibers	Treatment of soybean meal with alkaline solution and with acid to precipitate protein; dissolving, wet-spinning, recoagulating in reel, and stretching	Blending with other fibers such as cotton, rayon, nylon, wool
Paper coating	Alkali dispersion, mixing with additives	Decorative papers
Paints, varnishes, and coatings	Heating of soybean oil, and mixing with maleic anhydride; esterification of maleinized oil to a polyol	Alkyd resins
Plasticizers	Oxidation of soybean oil with performic acid, peracetic acid, or with hydrogen peroxide in the presence of a strong ion-exchange resin	To improve heat and light stability of plastics
Diesel fuel	Dissolving of 0.1–0.5% (based on oil) sodium methoxide in methanol, and mixing with heated soybean oil (80 °C); vacuum distilling, drying, and blending with diesel fuel, of the ester layer	Interest in biodiesel in cities with air pollution problems
Printing ink	Polymerization of oil by heat, blending the polymerized oil with 1% butilated hydroxitoluene and 6–20% pigments	Lithographic and letterpress newsprint applications

Source: Adapted from Johnson and Myers, 1995 (31).

and least soluble at its isoelectric region of pH 4.2–4.6 (32). Three basic processes have been used to remove carbohydrates (Fig. 3): acid extraction at pH 4.5, 60 to 80% aqueous ethanol extraction, and hot water extraction (33). In all three processes, protein remains insoluble while solubilized carbohydrates are removed by centrifugation. Solids from the acid extraction process are resuspended in water and neutralized to pH 6.8, whereas those from alcohol extraction are flash desolventized before spray drying. The most obvious difference between soybean protein concentrates is the lower ash content of concentrates prepared by acid or hot water extraction, indicating more thorough removal of minerals (32). Figure 3 summarizes basic unit operations and process conditions required to make soy protein concentrate by each of the three methods.

Soy Protein Isolates

Many processing options exist for making soy protein isolates. The preferred approach is to tailor the extraction process so that the isolate is made optimally compatible with the intended consumer product (32). Soy protein isolates are prepared traditionally from defatted soybean meal using aqueous or mild alkali extraction (pH 7 to 10) of proteins and soluble carbohydrates (33). The insoluble residue, mostly carbohydrates, is removed by centrifugation, and soluble protein is then precipitated at the isoelectric point (4.2 to 4.5) of soy proteins (Fig. 4). The precipitated protein is separated by mechanical decanting, washed, neutralized to about pH 6.8, and spray dried to obtain a product with protein content greater than 90% (w/w).

Textured Soy Protein

Textured soy protein products are obtained by thermoplastic extrusion of soy flour or concentrates to obtain a meatlike texture. Flours and concentrates are mixed with water and additives to form a dough and extruded under high pressure and temperature to obtain fibrous texture (33). Processes for textured proteins include thermoplastic extrusion, fiber spinning, direct steam texturization, shaping and heating, enzymatic texturization, and coagulation (34). Texturizing processes differ based on the type of raw protein material used, which can include defatted soy flour, soy protein concentrate, soy protein isolate, or a blend of several products (34). The most popular raw material has been slightly toasted, defatted soy flour.

Textured soy products have been produced with raw materials ranging in PDI from 20 to 70 (35). The PDI is the percentage of total protein dispersible in water under controlled conditions of extraction. During extrusion, the raw material is placed in the feeder and moved to a conditioning section where the food is mixed with 80 °C steam. In the conditioning section, the moisture content is raised to between 10 and 25%. The mixture then is sent to the extrusion cooker, where it is processed under conditions depending on type of raw material, equipment, and desired final product. Processing conditions include pressures of 400 to 1,500 psi and temperatures of 105 to 113 °C (35).

PROCESSING OF WHOLE SOYBEANS FOR FOOD

In the United States, only 3 to 5% of the soybean crop is used for food. This proportion is much larger in other countries, especially those in Asia, where soybeans are a staple of the human diet (8). In Japan alone, about 1 million tons

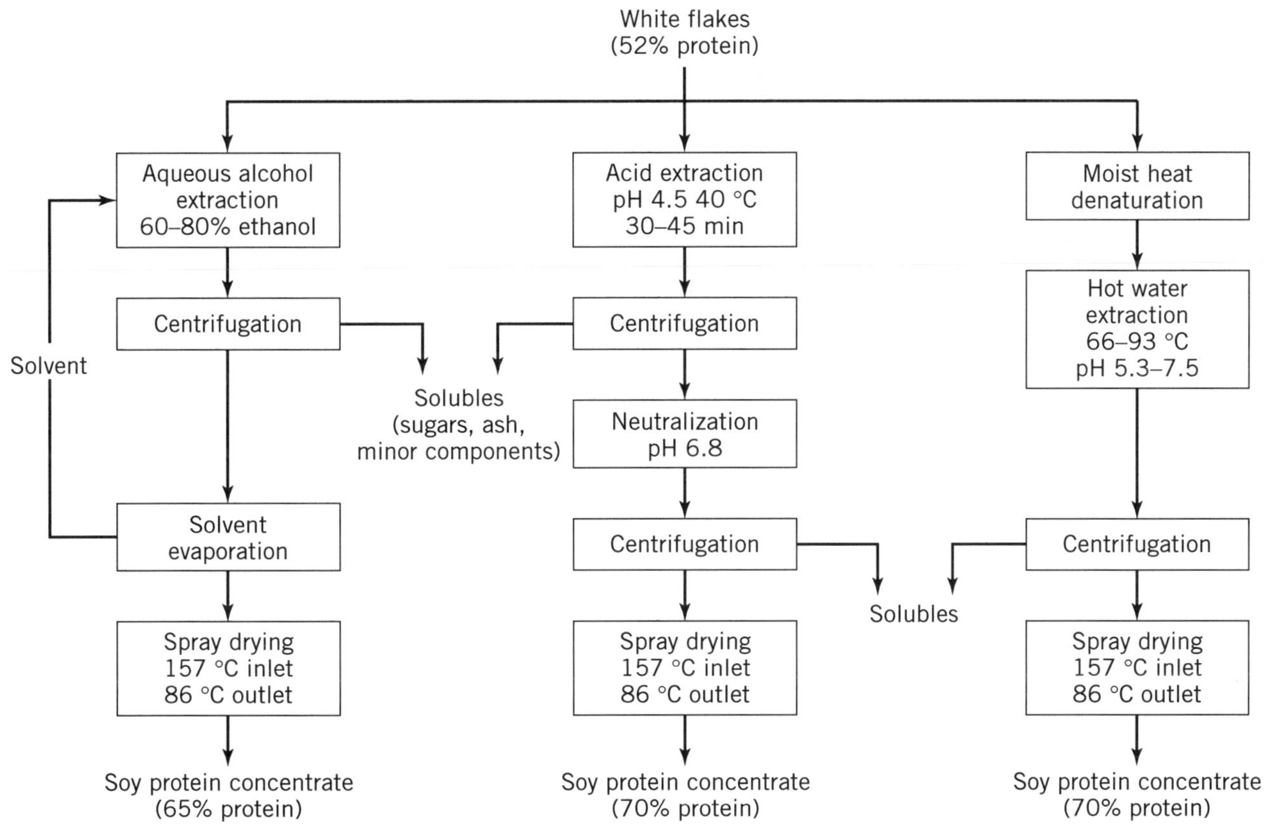

Figure 3. Alternative processes for the production of soy protein concentrates.

of soybeans are used to produce traditional Japanese foods such as tofu, shoyu, and miso (36). Both fermented and nonfermented foods are made from soybeans, and a brief discussion of most important products will be presented next.

Nonfermented Foods

A variety of nonfermented food products are made from whole soybeans. Raw soybeans have an undesirable bitterness and beany off-flavor and are rarely consumed without some sort of processing. In addition, raw soybeans contain several enzymes and antinutritional factors that need to be eliminated or inactivated if soybeans are to be palatable and digestible for humans (5,37). Growing interest in soybean foods in western countries has led to modernization of the manufacturing of traditional products and to the development of totally new products such as tofu-based ice cream and soybean yogurt (5). Soy milk, tofu, toasted full-fat soy flour, and enzyme-active soy flour are among the most important and best known commercial nonfermented food products.

Soy Milk. Soy milk is basically a water extract of soybeans obtained by wet grinding of soaked soybeans and subsequent filtering of solids (Fig. 5). Soy milk quality is related directly to raw soybean quality. Differences among soybean varieties affect not only yield but also important properties such as aroma and taste of soy milk. Soybean standards for soy milk production have been developed in Asian countries, and their use in the selection the raw soybeans have contributed to the consistently good quality of soy milk.

There are many different methods for producing soy milk, and the majority are variations of those described by Nelson et al. (38) (the Illinois method) and by Wilkens et al. (39) (the Cornell method). Both methods were developed primarily to improve soy milk flavor and stability. Regardless of process variations, the basic steps and principles of soy milk preparation are very similar and consist of soaking and grinding the beans and heating the extracted soy milk (Fig. 5). Modern large-scale soy milk plants use several additional steps and more sophisticated equipment. Several new processes have been described by Chen (40), Wilson (41), Kwok and Niranjan (42), and Liu (43).

Cleaning. Cleaning and washing, which remove dirt, dust, and microorganisms from the surface of soybeans, is crucial if contamination of the final product is to be avoided. The precleaning steps also include the removal of foreign material such as straws, stones, metal, weeds and their seeds, and damaged soybeans.

Dehulling. There are arguments in favor of and against dehulling. Most soy milk manufacturers use whole beans because dehulling disrupts the cotyledon tissue and releases the lipoxygenase enzymes, which diminishes the quality of soy milk. In addition, the presence of hulls helps filtration. Dehulling supporters believe that dehulling im-

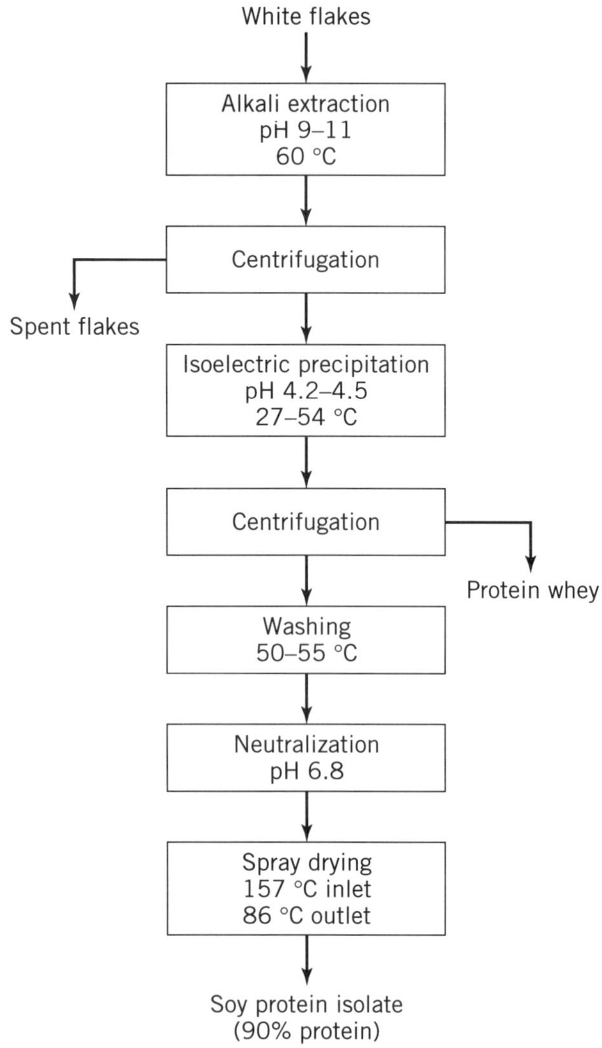

Figure 4. Production of soy protein isolate from white flakes.

proves the flavor and color of soy milk. Dehulling of soybeans is a process that consists of three steps, a heat treatment (93 °C for 15 min) to break the bonds between the hulls and the cotyledon, soybean cracking, and hull aspiration. A final dehulling efficiency of 88% is required for good quality soy milk (40). Most dehulling is done dry, but wet dehulling is also possible.

Soaking. Soybeans are soaked in water at a 1:3 solids-to-liquid ratio. The length of soaking depends on water temperature; the cooler the water, the longer the soaking. Typically, soaking time is 8 to 10 h at 20 °C in the summer and 14 to 20 hours at 10 °C in winter (40). Soybeans can be soaked in circulating water to be rinsed continuously throughout the soaking period. Most commercial soy milk production calls for soaking of soybeans to decrease the energy required for grinding, to achieve better dispersion and suspension of the solids, to increase yield, and to decrease cooking time (40,43).

Increased soaking time increases loss of water-soluble solids. Soaking of dehulled soybeans for 24 h results in a 5% solids loss as compared with a 1.5 % loss for whole (not dehulled) soybeans. Dehulled soybeans reach full hydration much faster than whole soybeans do and require only 2 to 3 h of soaking at 30 °C, or 1 h at 50 °C; whole soybeans require 7 h at 30 °C or 3 h at 50 °C (40). The addition of alkaline salts such as sodium bicarbonate, sodium citrate, or sodium hydroxide to the soaking water has been shown (*1*) to increase the protein and solids recovery, (*2*) to decrease beany off-flavor, (*3*) to inactivate trypsin inhibitors, (*4*) to tenderize soybeans, and (*5*) to affect soy milk quality by increasing soy protein solubility (43).

Fine Grinding. The purpose of fine grinding is to disrupt soybean tissue at both the cellular and the subcellular level and to suspend or to solubilize proteins, lipids, and other solids (43). Soaked soybeans may be ground with hot or cold water by a stainless steel disintegrator (Rietz), hammermill (Fiztpatrick), pin mill, or large blender (Waring) to produce coloid solution that will pass through a 150-mesh screen (40). Grinding in hot water (above 80 °C) inactivates lipoxygenases and prevents the formation of the beany off-flavor. Both very fine and rather coarse grinding are undesirable. Finely ground slurry contains maximum amount of soy protein but makes filtration difficult, thereby lowering soy milk yield. Coarse grinding may not result in the highest solubilization of the soy protein and other soluble solids and will be retained in the okara after the filtration (43). To improve flavor and mouthfeel, insoluble soybean residue is removed from the soy slurry by filtration or centrifugation.

Thermal Treatments. One or more heat treatments are used in soy milk processing. For example, in addition to cooking the slurry, the Illinois method calls for blanching (steaming), whereas the Cornell method calls for grinding the beans in hot water. Commercial processes also require pasteurization or sterilization of milk before packaging. Thermal treatment is one of the most important variables in the processing of soy milk (42). The positive effects of heat on soy milk include inactivation of antinutritional factors and lipoxygenases, increased extraction efficiency, denaturation of soy protein, and increased product shelf life. It is recommended that soy slurry or soy milk be heated for about 30 min at 100 °C to obtain optimal nutritional value and flavor. Heating also is applied to lower viscosity of the slurry to increase recovery of proteins and solids (40–43). Different soy milk processes use different temperature-time regimes during thermal processing. Effects of thermal processing on soy milk have been extensively reviewed by Kwok and Niranjan (42).

Homogenization. Homogenization is an optional step that is used to stabilize the soy milk emulsion and to disperse solids that otherwise may settle on the bottom of the container. Homogenization makes soy milk creamier and more uniform in consistency. For most soy milk products, a single pass at 2000 to 3500 psi and 90 °C through a dairy-type homogenizer is sufficient to obtain a good product. Some soy milk products exhibit a chalky taste caused by relatively large particles (>150 mesh) suspended in the milk (40). Chalkiness may be reduced by increasing the alkalinity of the blanching solution, by adjusting the final pH of soy milk to about 7.5, by decreasing solids content, or by homogenizing at high temperatures and pressures.

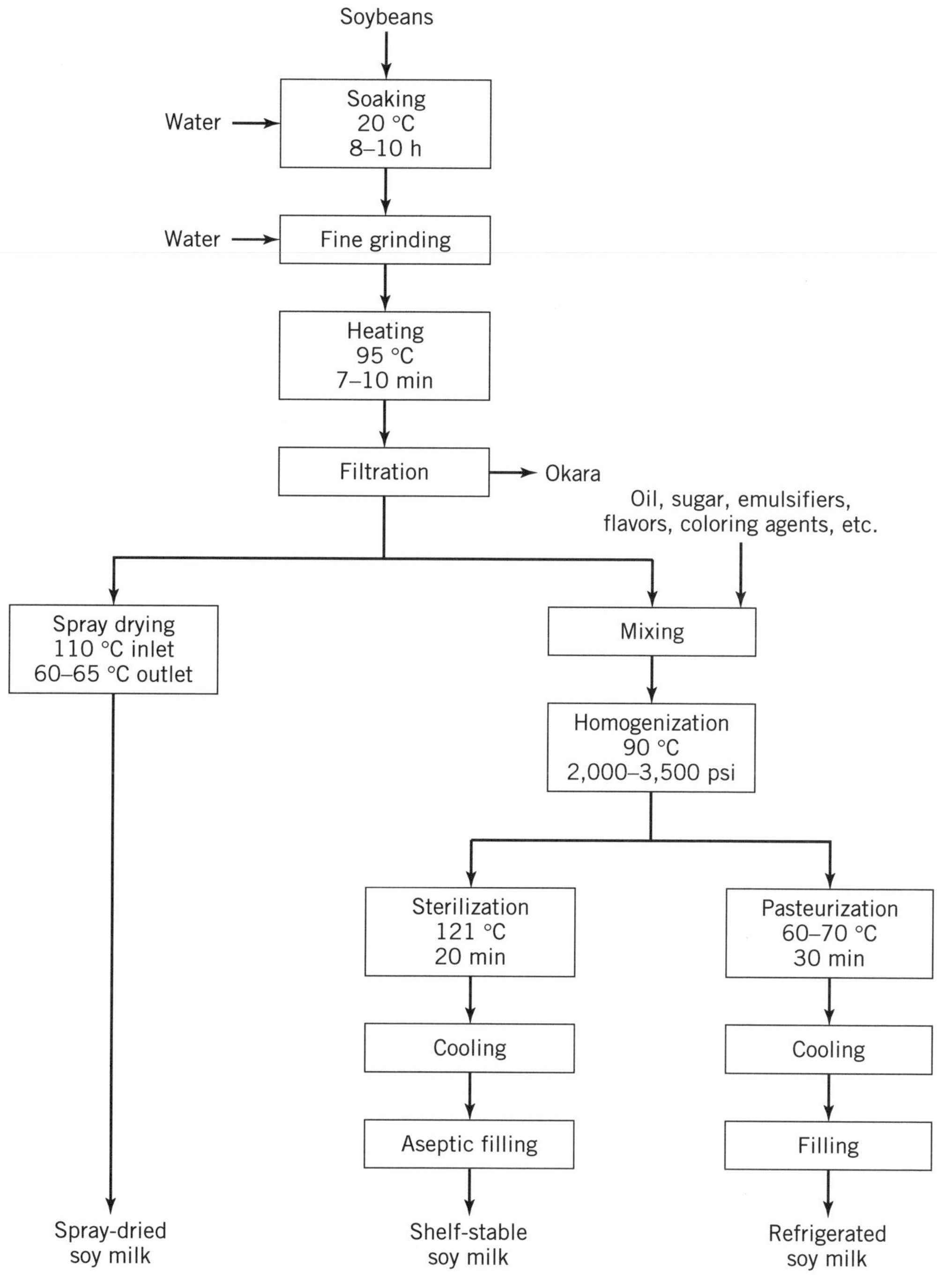

Figure 5. Alternative processing methods for soy milk production.

Pasteurization/Sterilization. After homogenization, soy milk can be pasteurized (75 °C for 15 min) or sterilized (12 °C for 15 to 20 min) and packaged for distribution. Pasteurized soy milk requires refrigeration at 4 °C to maintain a typical shelf life of about 1 week, whereas sterilized milk is shelf stable (Fig. 5).

Tofu. Tofu is a curd that is made directly from soybeans and resembles a soft white cheese. Basically, tofu is water-extracted and salt- or acid-coagulated soy protein gel with water, lipids, and other constituents trapped in its network (43). Typical pressed tofu has a moisture content of about 85% and contains about 7.8% protein, 4.2% lipids, and 2 mg/g calcium. On a dry basis, it contains about 50% protein and 27% oil.

In the United States, the tofu industry has shown steady growth during the past 10 years, partly because of the demand for healthy and low-fat ingredients (44). The quality and yield of produced tofu depend on several factors including soybean variety. For example, the ratio of the 7S and 11S proteins in soybeans contribute to the textural characteristics of tofu (45). Tofu made from crude 11S protein is significantly harder than that made from 7S protein, and the crude 11S fraction adds to the springiness, chewiness, and gumminess of tofu (43,45). During the preparation of soy extract (milk), water-to-bean ratio, temperature and hardness of water, degree of maceration, extraction time, and grinding equipment affect amount of extracted protein (46).

There are three main steps involved in the making of tofu: preparation of soy milk, coagulation of protein, and pressing and molding (Fig. 6) (43,45,46). The processing steps for preparation of soybean extract are similar to those discussed in the previous section. A greater yield and better quality of tofu was obtained when the bean slurry was filtered before heating (45). The bean slurry is filtered through a cloth sack to separate the undispersible fiber (okara) from the soy milk. The okara normally is washed with hot water, stirred, and pressed again to increase dispersible solids recovery. Previous studies have shown that tofu yield and texture depend on the heating temperature. The texture of tofu produced from soy milk heated below 70 °C is soft and watery, between 70 and 80 °C is good and above 80 °C is hard and uneven (48).

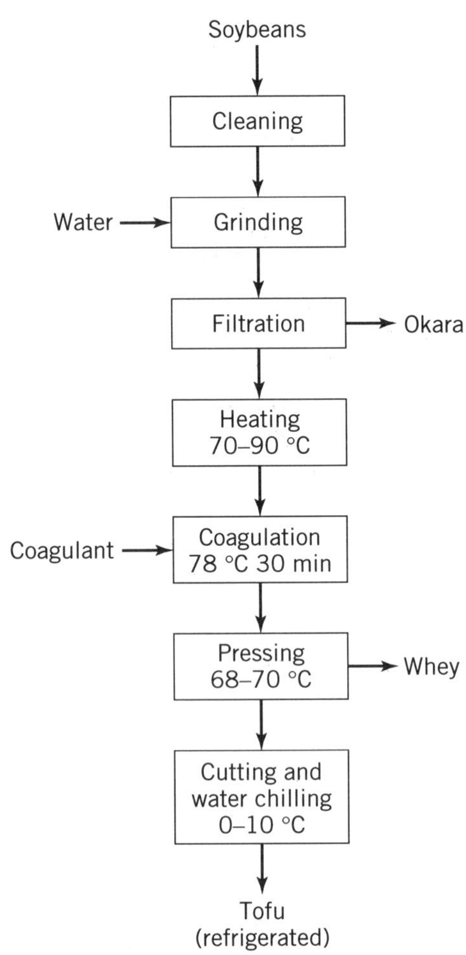

Figure 6. Basic unit operations involved in processing of soy milk into tofu.

Coagulation. Coagulation of soy milk is the most important step in tofu making and the most difficult to control because it depends on the complex interaction of many variables (49,50). Temperature of the soy milk, coagulant amount and concentration, and mode and rate of stirring all affect yield and texture of tofu. The ideal pH for soy milk coagulation is at or near 6. Soy proteins start coagulating near pH 6, and the addition of coagulant should be stopped at that pH (45). The amount and type of coagulant are important in producing reproducible tofu. Insufficient amount of coagulant would fail to coagulate the soy milk, but too much would result in low yield of small, hard aggregates with an uneven texture and a bitter (magnesium chloride) or chalky (calcium sulfate) taste (41). Data indicate that a coagulant concentration of 0.02 to 0.04 M is likely to produce a firm product (45). The characteristics of the five most common coagulants (magnesium sulfate, magnesium chloride, calcium sulfate, calcium chloride, and glucono-*d*-lactone) have been described by Karta (45). Stirring is necessary to keep the coagulant suspended, and the speed of stirring must be sufficient to maintain uniform distribution of the coagulant. A short mixing time is preferred because prolonged stirring may result in breakage of protein aggregates (50). In general, higher speed and shorter mixing time increase tofu yield.

Pressing. The soy protein coagulate (curd) is stirred and transferred to a shallow box for pressing. The purpose of breaking the gel is to disrupt the protein network to release the whey. Part of the whey is separated from the broken curd before transferal to the press box, and the rest is separated during pressing. Degree of curd breakage affects tofu yield and texture. Greater curd breakage generally leads to firmer texture and lower bulk yield. Variations in the pressing step include the amount of pressure and time applied (43). A study by Gandhi and Bourne (51) indicated that increasing the pressure from 0.19 Pa to 0.74 Pa reduced decrease moisture content from 82 to 60% and yield from 2 to 1.2 kg fresh tofu/kg soybeans. It also increased the hardness, chewiness, and gumminess of the final product. The ideal pressing temperature is between 68 and 70 °C; and, from a microbiological quality standpoint, tofu temperature should not fall below 60 °C (41).

Toasted Full-Fat Soy Flour. Full-fat soy flour, with different degrees of heat treatment, may be considered as modern industrial version of oriental soybean powder (5). It is made from dehulled soybeans and used mainly in bakery and dietetic foods. In the preparation of full-fat soy flours, soybeans first are steamed under light pressure for 20 to 30 min to inactivate lipoxygenases (52). Beans then are dried, cracked, dehulled, and ground finely to obtain full-fat soy flour.

Enzyme-Active Full-Fat Soy Flour. Soybeans used for producing enzyme-active full-fat soy flour are dried at low temperatures to minimize protein denaturation (52). Enzyme-active flour is produced by a process similar to that described for full-fat soy flour except that the initial steaming step is omitted. The resulting product is used widely in the food industry for bleaching wheat flour and for conditioning doughs in western-type breads. Enzyme-active full-fat flour contains several lipoxygenases that bleach the carotenoid content of wheat flour to produce a whiter bread (5). The enzyme-active flour also possesses β-amylase activity, which improves the textural properties of dough. The U. S. Food and Drug Administration permits a maximum of 0.5% (w/w) of enzyme-active soy flour in wheat flour.

Fermented Foods

Fermented soybean food products are consumed commonly in Asian countries and are becoming increasingly popular in human diets in the United States and other western countries. Different procedures are used to produce somewhat similar and yet unique fermented soybean products, whose names differ between countries. This section will focus on manufacturing soy sauce, soy paste, and tempeh because they are representative and the commercially most important fermented products and because their process operations are similar to those of other fermented foods.

Soy Sauce. Soy sauce is a dark-brown liquid extracted from a fermented mixture of soybeans and wheat. Among the fermented soy foods, soy sauce is the most widely accepted, not only in the Far East but also in western countries (47). Depending on the geographic region, there are great variations in processing method and variety of soy sauce. Household production of soy sauce is still popular in some regions, but most soy sauce is made in commercial plants. Regardless of production level and method used, the basic steps and principles involved in soy sauce manufacture are very similar. Based on the method of preparation, soy sauce is divided into three types: fermented, chemical, and semichemical (47,53,54). In fermented soy sauce, proteins and carbohydrates in the raw material are enzyme hydrolyzed very slowly under mild conditions, below 30 °C for longer than 6 months. In chemical soy sauce, protein and carbohydrates are hydrolyzed quickly with hydrochloric acid at greater than 80 °C for 8 to 10 h. Semichemical soy sauce usually involves chemical hydrolysis of the raw material followed by lactic acid or yeast fermentation (47,53,54).

Fermented Soy Sauce. A typical process for the manufacture of a representative Japanese soy sauce (shoyu) generally involves treatment of raw material, koji preparation, brine fermentation, pressing, and refining.

Cleaned soybeans are soaked overnight in water at ambient temperature and then cooked for several hours under pressure. Soybeans are cooled quickly to 40 °C to prevent growth of undesirable microorganisms before fermentation (Fig. 7). Wheat kernels are roasted at 170 to 180 °C for a few minutes and then cracked into several pieces to be able to adsorb moisture from soybeans upon mixing (47,54).

Two fermentations are used in making soy sauce (Fig. 7). The first fermentation is a solid state and occurs during koji preparation, during which various enzymes are produced under aerobic conditions. The second fermentation, known as *brine fermentation*, is mainly anaerobic and begins after the addition of brine to the koji (47,54). The koji is the source of hydrolytic enzymes for converting polysaccharides and proteins into simple sugars, peptides, and amino acids, all of which are utilized by lactic acid bacteria and yeast in the subsequent brine fermentation (53). The soybean-wheat mixture is inoculated with a 0.1 to 0.2% (w/w) culture of *Aspergillus oryzae* or *A. sojae* and allowed to ferment in shallow, perforated vats for 3 to 4 days at 30 °C. Before brine fermentation, mature koji is mixed with 17 to 23% NaCl solution to achieve a final salt concentration of 17 to 19%. Lower salt concentrations allow growth of undesirable bacteria, whereas higher concentrations (greater than 23%) may retard growth of desirable halophilic bacteria and osmophilic yeast. The mash (*moromi*) is kept in large wooden containers or concrete vats fitted with aeration devices.

Temperature is an important factor during brine fermentation. In general, the higher the temperature, the shorter the fermentation time. Fermentation at lower temperatures results in a better product because enzyme inactivation rate is lower. A good quality soy sauce can be produced in 6 months by maintaining the mash temperature at 15 °C for the first month, at 28 °C the next 4 months, and at 15 °C the last month. Occasional stirring is necessary to provide aeration for yeast growth, to prevent growth of undesirable microorganisms, to maintain uniform temperature, and to facilitate CO_2 removal (47,54).

At the end of the fermentation and aging period, matured mash should have a bright reddish brown color, a pleasant aroma, and a salty but palatable taste. Raw soy sauce is commonly removed from the mash by pressing. Filtrate is stored in tanks to separate sediment (on the bottom) and floating oil (47,54). Refining of raw soy sauce consists of pasteurization (60 to 70 °C) and clarification by either sedimentation or filtration. Kaolin, diatomite, or alum may be added to enhance clarification. Heating during pasteurization inactivates enzymes and microorganisms, darkens the sauce, enhances the aroma, and induces formation of flocculants, which also facilitate clarification (47,54).

Chemical Soy Sauce. The chemical process, much simpler, faster, and cheaper than the fermentation process, results in a product lacking some of the fermented soy sauce flavor. Chemically processed soy sauce has a very limited market in Asia. In the western countries, where consumers are unaccustomed to the fermented product, acceptance of chemical soy sauce has been more rapid (54). In the chemical process, raw materials (soybeans and wheat) are hydrolyzed by boiling with 18 to 20% hydrochloric acid for 8 to 16 h (47,54). Hydrolysate is neutralized with sodium hydroxide or sodium carbonate to pH 4.5, pasteurized, and filtered. The resulting product is a clear dark-brown liquid with an approximate concentration of 18% (w/v) NaCl (47,54).

Semichemical Soy Sauce. Chemical soy sauce does not have the taste and flavor of fermented soy sauce. To improve its quality, chemical soy sauce is often blended with fermented soy sauce. A semichemical process has been developed whereby chemical hydrolysis is followed by lactic acid or yeast fermentation. Dilute hydrochloric acid (7 to 8%) is used to partly hydrolyze defatted soybeans, koji is added to the neutralized hydrolysate, and the mixture is inoculated with yeast. Fermentation can be shortened by

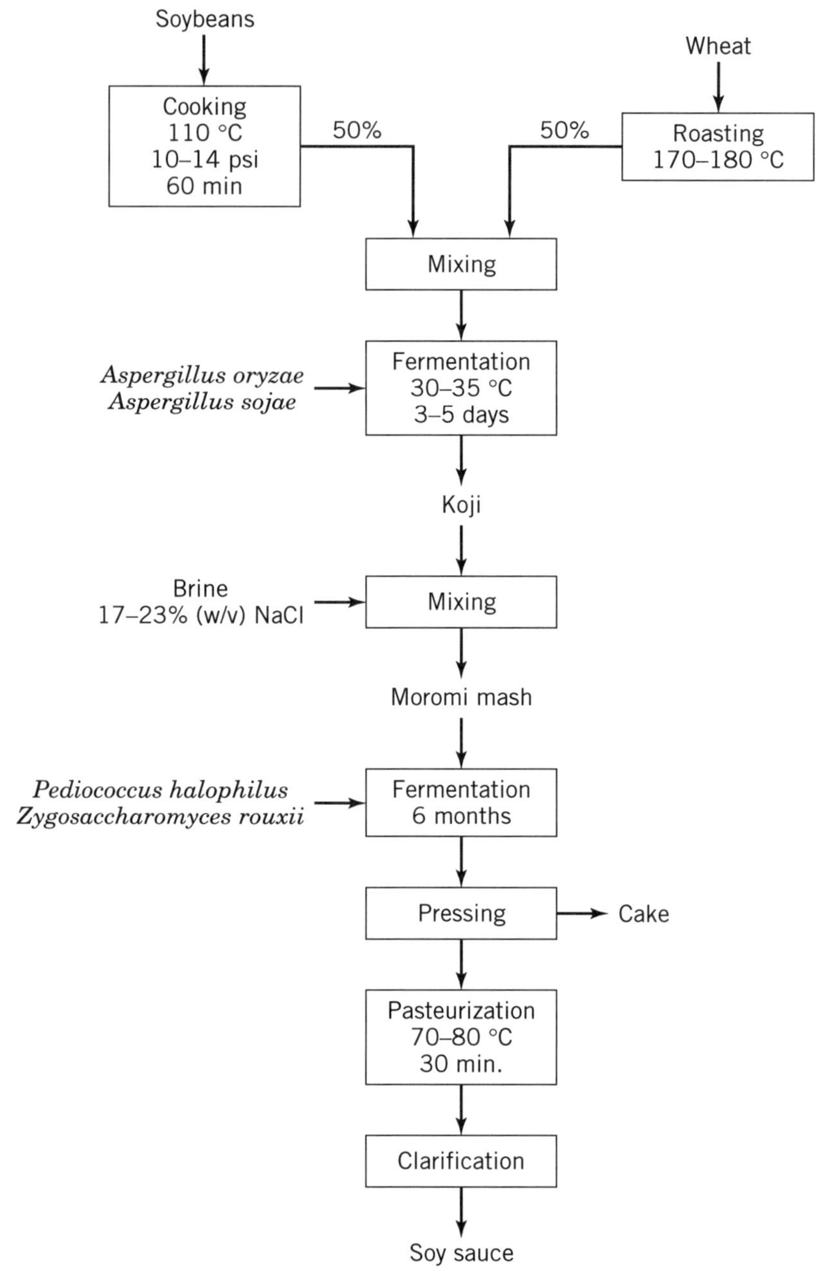

Figure 7. Flow diagram of the unit operations involved in the production of fermented soy sauce.

2 to 3 months, but the product retains some odors characteristic of chemically hydrolyzed beans (47,54).

Soy Paste. Soy paste is one of the most important fermented oriental soy foods. It is known commonly as *jiang* in China, *miso* in Japan, *jang* in Korea, *taucho* in Indonesia, and *taotsi* in Philippines. The process of making soy paste depends on the region, but the basic processes are essentially similar. The production of Japanese rice miso involves five steps (rice koji preparation, soybean treatment, ingredient mixing, fermentation, pasteurization, and packaging) that are very similar to those used in producing soy sauce (47,54). The major differences are that koji for miso is grown only on rice and that, because the whole fermented mixture is pasteurized and packaged for distribution, no pressing and filtering is required, as it is in the soy production. In this section, pertinent processing steps and conditions for making Japanese rice miso will be briefly discussed (Fig. 8).

Rice koji is prepared by soaking polished rice in water at 15 °C overnight or until the grain has a moisture content of 35%. After draining, the rice is cooked in an open cooker for 40 min, cooled to 35 °C, inoculated with a 0.1% (w/w) koji starter with *A. oryzae*, and incubated at 30 °C and 96% relative humidity (RH) for 15 h. After 15 h incubation, rice is spread on trays and maintained at 35 °C for an additional 25 to 30 h (47,54).

Soaked soybeans are steam cooked under pressure (10 to 15 psi) for 20 to 30 min. For white or light colored miso, soybeans are boiled with water at a 1:4 solids-to-water ratio to prevent browning. The advantage of boiling over steam cooking is the retention of soybean color, but the disadvantage is a 10 to 20% solids loss. Batch-type cookers are used widely, but continuous cookers with rotary bulbs and belt conveyers have been used recently in larger factories (47,54).

Pretreated soybeans are mixed with the rice koji and salt and blended in a closed blending container or ground with a sausage-type grinder. The mixture is packed tightly into wood, concrete or steel vats, covered with resin sheets or plates, and allowed to ferment at 30 to 38 °C. Fermentation for sweet miso production lasts between 10 and 15 days and for salty miso—2 and 12 months.

After ripening, miso is blended and, if needed, mashed again through a chopper with a plate cutter with perforations of 1 to 2 mm. The mashed miso is packaged in resin bags or cubic containers. For preservation, miso can be pasteurized or mixed with preservatives such as 2% ethyl alcohol or less than 0.1% sorbic acid (47,54).

Tempeh. Tempeh is a compact, cakelike product made by fermentation of dehulled and cooked soybeans with *Rhizopus oligosporous*. Tempeh is one of the most popular traditional foods in Indonesia and is becoming a popular vegetarian dish in the United States because it serves as a major source of protein, calories, and vitamins (55). A broad range of tempeh preparation methods have been reported; principles are the same, however, regardless of the method. Steinkraus et al. (54) have described a large-scale tempeh production method consisting of size grading, dry dehulling, hydration with optional addition of lactic acid, boiling, and fermentation (Fig. 9).

First, soybeans are cleaned to remove dirt, stones, weed seed, damaged and decomposed beans, and any other matter. Cleaned and graded soybeans are shrivelled slightly by dry heating for 10 min at 93 °C, cooled, and cracked to facilitate hull removal (47). Dehulled soybeans are soaked

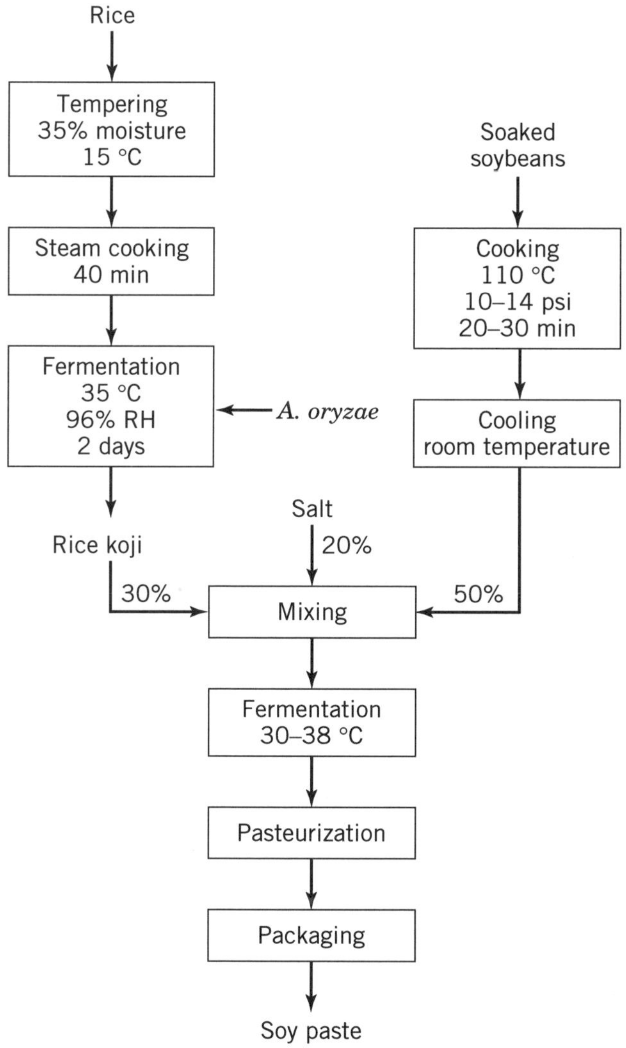

Figure 8. Processing steps in the production of Japanese rice miso.

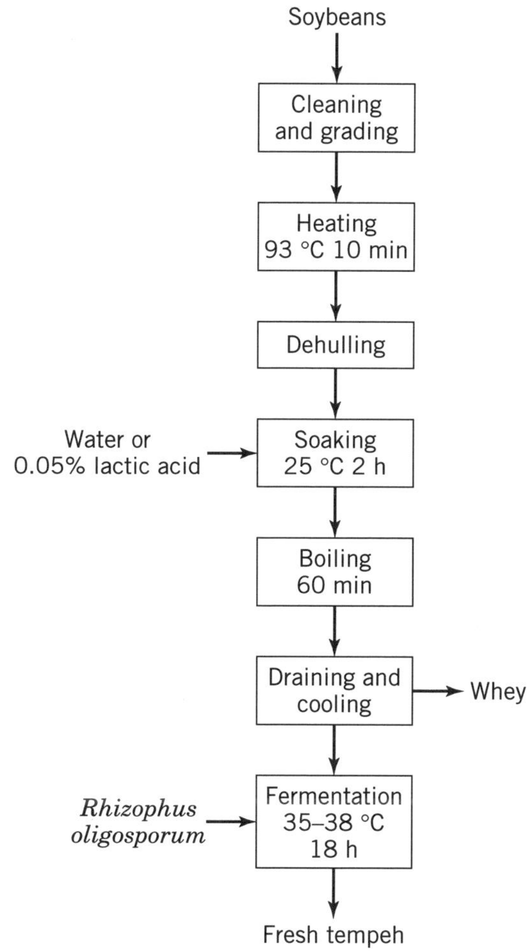

Figure 9. Basic unit operations in the processing of soybeans into tempeh.

Table 3. Principal Processing Steps and Uses of Fermented Soy Food Products

Food product (product description)	Main processing steps	Uses
Natto (cakelike product)	Soaking and cooking, fermenting with *Bacillus natto*	Side dish
Tao tio (soy paste type)	Cooking, fermenting with *Aspergillus* sp., sun-drying	Side dish, flavoring
Kochu chang (soy paste)	Boiling, mashing, fermenting, and drying	Flavoring
Ketiap (soy sauce)	Soaking, cooking, fermenting with *Aspergillus oryzae*, filtrating and washing, concentrating, and bottling	Flavoring
Sufu (soft cheese type)	Cutting, sterilizing, fermenting with *Mucor* and *Antinomucor* sp., brine aging	Side dish

Source: Data adapted from Johnson (1989) and Lui (1997) (47,52).

in water or 0.85% lactic acid solution for 2 h at 25 °C or for 30 min at 100 °C. Addition of lactic acid results in a pH drop to 4.3 to 5.3 and controls contaminating bacterial growth.

Hydrated soybeans are boiled for 60 min, drained, and cooled before inoculation with a pulverized pure *R. oligosporus* starter containing both mycelia and spores. Inoculated beans are spread out on perforated trays and covered with waxed paper to prevent dehydration and excessive aeration. Trays are incubated at 35 to 37 °C and 75 to 78% RH for about 18 h. At the end of the fermentation period, fresh tempeh is harvested, dried, and packaged before storage and shipping (47).

There are other traditional fermented products made from soybeans. Table 3 summarizes some of the best known, their principal processing steps, and their uses. Until now, most have been prepared on a small scale for family or regional consumption. Globalization of the world trade and economy is expected to increase demand for some of those products, whose production, in turn, will require further scale-up and modernization.

CONCLUDING REMARKS

Processing of soybeans into food, feed, and industrial products has experienced significant growth in the past, but new developments and challenges are foreseen. Continuous efforts must be made to further reduce energy consumption and to address safety issues and various environmental concerns such as air pollution, odor control, and waste treatment. New products and associated processing issues will emerge as a result of innovations in agricultural biotechnology. Plant biotechnology progress will not bypass the soybean industry; on the contrary, soy growers and processors will play an important role in capturing added value from new plant traits. For example, the precise alteration of metabolic processes in soybean plant will allow modification of quality, functional properties, and quantity of the protein, oil, and phytochemicals to fit end application. The new traits imparted to soybeans will require better understanding and control of soybean handling and processing to achieve the full impact and capture the added value of improved food and feed products. The ability to transfer new genes into soybeans (transgenic soybeans) will enable their use not only as a source for modified oil and protein but also as bioreactors to produce pharmaceutical proteins and drugs, industrial enzymes, biodegradable polymers, and nutritional supplements. Transgenic soybeans producing novel biomolecules will require significant processing adjustments and, in many instances, dedicated processing facilities. The success of newly derived products from genetically modified soybeans will probably depend on the close interaction or integration of the production, processing, and end application. The developments in processing of genetically engineered soybeans will ultimately affect processing and utilization of commodity soybeans.

ACKNOWLEDGMENTS

The authors thank Drs. L. A. Johnson and L. A. Wilson for reviewing the manuscript and for providing useful suggestions throughout its preparation. This work has been supported in part by the Center for Crops Utilization Research of Iowa State University. This is Journal Paper no. J-17843 of the Iowa Agriculture and Home Economics Experiment Station, Ames, Iowa, Project no. 3331.

BIBLIOGRAPHY

1. M.J. Morse, in K.S. Markley ed., *Soybeans and Soybean Products*, Interscience Publishers, New York, 1950, pp. 3–60.
2. K.S. Liu, in K. Liu ed., *Soybeans: Chemistry, Technology, and Utilization*, Chapman & Hall, New York, 1997, pp. 1–24.
3. R.E. Fiedler, *JAOCS* **48**, 43–46 (1971).
4. D.D. Asbridge, in D.R. Erickson ed., *Practical Handbook of Soybean Processing and Utilization*, AOCS Press, Champaign, Ill., 1995, pp. 1–8.
5. Z. Berk, *Technology of Production of Edible Flours and Protein Products from Soybeans*, FAO Agricultural Services, Bulletin 97, Rome, Italy, 1992, p. 178.
6. K.J. Smith, in E.W. Lusas, D.R. Erickson, and W.K. Nip eds., *Food Uses of Whole Oil and Protein Seed*, AOCS Press, Champaign, Ill., 1989, pp. 1–11.
7. *Soya Bluebook Plus*, Soyatech, Inc., Bar Harbor, Maine, 1997.
8. T.H. Applewhite, in Council for Agricultural Science and Technology ed., *Quality of U.S. Agricultural Products*, Report no. 126, Council for Agricultural Science and Technology, Ames, Iowa, 1996, pp. 86–110.

9. A. Proctor, in K. Liu ed., *Soybeans: Chemistry, Technology, and Utilization*, Chapman & Hall, New York, 1997, pp. 297–346.
10. J.B. Woerfel, in D.R. Erickson ed., *Practical Handbook of Soybean Processing and Utilization*, AOCS Press, Champaign, Ill., 1995, pp. 39–55.
11. H. Schumacher, in T.H. Applewhite ed., *Vegetable Protein Utilization in Human Foods and Animal Feedstuffs*, AOCS Press, Champaign, Ill., 1989, pp. 37–40.
12. J.B. Woerfel, in D.R. Erickson ed., *Practical Handbook of Soybean Processing and Utilization*, AOCS Press, Champaign, Ill., 1995, pp. 65–91.
13. K.F. Carlson and J.D. Scott, *Inform* **2**, 1034–1060 (1991).
14. T.G. Kemper, *Inform* **6**, 1231–1236 (1995).
15. E.F. Sipos and B.F. Szuhaj, in Y.H. Hui ed., *Bailey's Industrial Oil and Fat Products*, 5th ed., Wiley, New York, 1996, pp. 497–601.
16. L.A. Johnson and E.W. Lusas, *JAOCS* **60**, 229–242 (1983).
17. L.A. Johnson, in P.J. Wan and P.J. Wakelyn eds., *Technology and Solvents for Extracting Oilseeds and Nonpetroleum Oils*, AOCS Press, Champaign, Ill., 1997, pp. 4–47.
18. G.C. Mustakas, in D.R. Erickson, E.H. Pryde, O.L. Brekke, T.L. Mounts, and R.A. Falb eds., *Handbook of Soy Oil Processing and Utilization*, AOCS, Champaign, Ill., 1980, pp. 49–66.
19. E.D. Milligan, *JAOCS* **53**, 286–290 (1976).
20. N.H. Witte, in D.R. Erickson ed., *Practical Handbook of Soybean Processing and Utilization*, AOCS Press, Champaign, Ill., 1995, pp. 93–116.
21. R.W. Fulmer, in T.H. Applewhite ed., *Vegetable Protein Utilization in Human Foods and Animal Feedstuffs*, AOCS Press, Champaign, Ill., 1989, pp. 55–61.
22. T.G. Kemper, *Inform* **5**, 898–901 (1994).
23. P.J. Wakelyn, in P.J. Wan and P.J. Wakelyn eds., *Technology and Solvents for Extracting Oilseeds and Nonpetroleum Oils*, AOCS Press, Champaign, Ill., 1997, pp. 48–74.
24. T. Gum, *Inform* **5**, 902–905 (1994).
25. G.R. List and D.R. Erickson, in D.R. Erickson, E.H. Pryde, O.L. Brekke, T.L. Mounts, and R.A. Falb eds., *Handbook of Soy Oil Processing and Utilization*, AOCS Press, Champaign, Ill., 1980, pp. 267–354.
26. D.R. Erickson, in D.R. Erickson ed., *Practical Handbook of Soybean Processing and Utilization*, AOCS Press, Champaign, Ill., 1995, pp. 174–183.
27. D.R. Erickson, in D.R. Erickson ed., *Practical Handbook of Soybean Processing and Utilization*, AOCS Press, Champaign, Ill., 1995, pp. 184–202.
28. D.R. Erickson, in D.R. Erickson ed., *Practical Handbook of Soybean Processing and Utilization*, AOCS Press, Champaign, Ill., 1995, pp. 203–217.
29. D.R. Erickson and M.D. Erickson, in D.R. Erickson ed., *Practical Handbook of Soybean Processing and Utilization*, AOCS Press, Champaign, Ill., 1995, pp. 218–238.
30. C.T. Zehnder, in D.R. Erickson ed., *Practical Handbook of Soybean Processing and Utilization*, AOCS Press, Champaign, Ill., 1995, pp. 239–257.
31. L.A. Johnson and D.J. Myers, in D.R. Erickson ed., *Practical Handbook of Soybean Processing and Utilization*, AOCS Press, Champaign, Ill., 1995, pp. 380–427.
32. E.W. Lusas and K.C. Rhee, in D.R. Erickson ed., *Practical Handbook of Soybean Processing and Utilization*, AOCS Press, Champaign, Ill., 1995, pp. 117–160.
33. N. Hettiarachchy and U. Kalapathy, in K. Liu ed., *Soybeans: Chemistry, Technology, and Utilization*, Chapman & Hall, New York, 1997, pp. 379–411.
34. K.S. Liu, in K.S. Liu ed., *Soybeans: Chemistry, Technology, and Utilization*, Chapman & Hall, New York, 1997, pp. 412–441.
35. J.P. Kearns, G.J. Rokey, and G.R. Huber, in T.H. Applewhite ed., *Vegetable Protein Utilization in Human Foods and Animal Feedstuffs*, AOCS Press, Champaign, Ill., 1989, pp. 353–362.
36. K. Seguro and M. Motoki, *Inform* **5**, 308–313 (1994).
37. L.P. Hanson, *Vegetable Protein Processing*, Noyes Data Corporation, New Jersey, 1974.
38. A.I. Nelson, M.P. Steinberg, and L.S. Wei, *J. Food Sci.* **41**, 57–61 (1976).
39. W.F. Wilkens, L.R. Mattick, and D.B. Hand, *Food Technol.* **21**, 86–89 (1967).
40. S. Chen, in T.H. Applewhite ed., *Vegetable Protein Utilization in Human Foods and Animal Feedstuffs*, AOCS Press, Champaign, Ill., 1989, pp. 341–352.
41. L.A. Wilson, in D.R. Erickson ed., *Practical Handbook of Soybean Processing and Utilization*, AOCS Press, Champaign, Ill., 1995, pp. 428–459.
42. K.C. Kwok and K. Niranjan, *Int. J. Food Sci. Technol.* **30**, 263–295 (1995).
43. K.S. Liu, in K.S. Liu ed., *Soybeans: Chemistry, Technology, and Utilization*, Chapman & Hall, New York, 1997, pp. 137–217.
44. K. Kevin, *Food Process. U.S.A.* **55**, 81–82 (1994).
45. S.S. Karta, in T.H. Applewhite ed., *Vegetable Protein Utilization in Human Foods and Animal Feedstuffs*, AOCS Press, Champaign, Ill., 1989, pp. 382–387.
46. C.G. Beddows and J. Wong, *Int. J. Food Sci. Technol.* **22**, 15–21 (1987).
47. K.S. Liu, in K.S. Liu ed., *Soybeans: Chemistry, Technology, and Utilization*, Chapman & Hall, New York, 1997, pp. 218–296.
48. C.G. Beddows and J. Wong, *Int. J. Food Sci. Technol.* **22**, 23–27 (1987).
49. C.G. Beddows and J. Wong, *Int. J. Food Sci. Technol.* **22**, 15–21 (1987).
50. H.J. Hou, K.C. Chang, and M.C. Shih, *J. Food Sci.* **62**, 824–827 (1997).
51. A.P. Gandhi and M.C. Bourne, *J. Textural Studies* **19**, 137–142 (1988).
52. D.W. Johnson, in E.W. Lusas, D.R. Erickson, and W.K. Nip eds., *Food Uses of Whole Oil and Protein Seeds*, AOCS Press, Champaign, Ill., 1989.
53. D. Fukushima, in C.W. Hesseltine and H.L. Wang eds., *Indigenous Fermented Foods of Non-Western Origin*, J. Cramer, Berlin, 1986, pp. 121–150.
54. K.H. Steinkraus, R.E. Cullen, C.S. Pederson, L.F. Nellis, and B.K. Gavitt, *Handbook of Indigenous Fermented Foods*, Dekker, New York, 1983.
55. F.G. Winarno, in T.H. Applewhite ed., *Vegetable Protein Utilization in Human Foods and Animal Feedstuffs*, AOCS Press, Champaign, Ill., 1989, pp. 363–368.

STAINLESS STEELS

C.P. Dillon
C.P. Dillon & Associates
St. Albans, West Virginia

KEY WORDS

Cleaning
Corrosion
Design
Materials selection
Passivation
Passivity
Pickling
Stainless steel
Superaustenitic
Welding

OUTLINE

Stainless steels offer an unusual combination of strength, ductility, formability, weldability, corrosion resistance, and amenability to cleaning that makes them uniquely suitable for pharmaceutical and bioprocessing operations. This article discusses the origin and nature of stainless steels, the several groups and many grades of stainless steels, together with the fundamentals of the metallurgical differences. Specific corrosion phenomena that pose potential problems, together with preventive and remedial measures, are presented. I explain the basic nominal workhorse compositions, together with their potential weaknesses, and discuss the possible utilization of other more highly alloyed substitute alloys. Particular emphasis is laid upon measures for quality assurance, postfabrication cleanup procedures, and passivation treatments to ensure improved resistance toward specific types of localized corrosion.

THE NATURE OF STAINLESS STEELS

As a whole, stainless steels and similar chromium-rich alloys are characterized by their *passivity*. Passivity is a condition in which a base metal, such as iron, exhibits the corrosion behavior of a more noble metal or alloy. As described further here, passivity cannot be maintained under some specific conditions.

Steel is an alloy of iron and carbon plus other minor elements and is notorious for susceptibility to rusting. With even minimal corrosion, it is a source of iron contamination and is therefore objectionable in many services. However, when a minimum of 12% chromium (Cr) is alloyed with steel, the resulting metal is generally not susceptible to rusting under ordinary circumstances of exposure to the atmosphere or in substantially neutral aqueous solutions. (Certain exceptions, however, are described further here.) Actually, several types of stainless steels, which are characterized by specific properties, have been developed.

TYPES OF STAINLESS STEELS

The 12% Cr grades of stainless steels are predominantly "martensitic." Martensite is a needlelike structure resulting from heating and quenching iron-base alloys, which gives increased hardness and strength with a concurrent diminution of ductility. (A certain amount of ductility is restored by tempering, i.e., heating to some intermediate temperature.) The martensitic grades are exemplified by Type 410 (UNS S41000) (1) and are characteristically magnetic.

When the chromium is increased to about 17%, the material is magnetic but not amenable to hardening by heat treatment. Such alloys, exemplified by Type 430 (S43000), are classified as "ferritic" and are more corrosion resistant than the martensitic grades by virtue of their higher chromium content, although they have lower strength than heat-treated martensitic grades.

The "workhorse" materials for the process industries, however, are the "austenitic" grades, often loosely called "18-8" stainless steels. When about 8% nickel (Ni) is added to an 18% Cr steel, a face-centered crystal structure called austenite is formed that is characteristically nonmagnetic. (This structure exists only at elevated temperatures with steels and the straight chromium grades.) Austenitic stainless steels, exemplified by Type 304 (S30400) and Type 316 (S31600), have in general a remarkable combination of properties that occasion widespread use in the process industries as compared with the martensitic and ferritic grades, which are used much less frequently.

Stainless steels that contain 18% Cr or more are also alloyed with a lesser amount of nickel (about 5%) to form "duplex" grades, which are about 50:50 austenite and ferrite. This method results in alloys of higher strength than Types 304 (S30400), 316 (S31600), and so forth, with an improved resistance in aqueous environments of high chloride content. The newer duplex grades are exemplified by Alloys 2205 (S31803) and 255 (S32550).

Finally, there are "superferritic" grades such as Alloy 26-1 (S44626) and several "superaustenitic" alloys, exemplified by Alloys 254SMO (S31254) and 6XN (N08367). The superaustenitic grades have superior resistance to localized corrosion such as pitting, crevice corrosion, and stress corrosion cracking (SCC) in chloride-bearing environments. Also, they are as strong as duplex grades by virtue of nitrogen addition.

The compositions of some typical specific grades in these basic categories are given in Table 1, wherein the midrange of major elements is given with the maximum allowable concentration for minor elements. More detailed listings of stainless steel types and compositions are available (2).

CORROSION MECHANISMS

Corrosion of metals is electrochemical in nature, metal ions being formed at anodic sites by oxidation, accompanied by a discharge of electrons through the external (metallic) circuit. At cathodic sites, electrons provide reduction of specific species (e.g., dissolved oxygen, cations) in the electrolyte to complete the reaction equilibrium. This change is most apparent in galvanic (i.e., bimetallic) corrosion in which the anodes and cathodes are different materials (e.g., the galvanic corrosion of steel by contact with copper, or of zinc by contact with steel). However, it also occurs in monometallic corrosion, such as steel in water, wherein discrete local anodes and cathodes of a transient nature provide the reactions.

$$Fe^{\circ} \rightarrow Fe^{2+} + 2e \text{ (anodic)}$$

$$2H^{+} + 1/2O_2 \rightarrow H_2O \text{ (cathodic)}$$

The formation and dissipation of discrete anodes and cathodes cause the general, uniform corrosion typical of base metals.

When a metal or alloy is protected by a superficial layer of a composition different from the substrate (mill-scale on steel; oxide films on copper-, nickel-, or chromium-bearing alloys), anodic sites may be relatively stable at defects in the otherwise protective film, causing a more localized type of attack. This reaction is somewhat analogous to galvanic corrosion, the film acting as a cathode to the more anodic substrate.

CORROSION SUSCEPTIBILITY OF STAINLESS STEELS

The stainless steels in general resist corrosion because of the formation of an invisible oxide film that renders them

Table 1. Composition of Typical Stainless Steels

Grade	UNS #	C	Cr	Ni	Mo	Other
			Martensitic grades			
410	S41000	0.15	12.5	—	—	1 Si
420	S42000	>0.15	13.0	—	—	1 Si
			Ferritic grades			
409	S40900	0.08	11.0	—	—	1 Si, Ti 6 × C
430	S43000	0.12	17	—	—	1 Si
444	S44000	0.025	18.5	1.0	2.0	Ti, Nb, N
			Austenitic grades			
304	S30400	0.08	19	9	—	1 Si
304L	S30403	0.03	19	10	—	1 Si
316	S31600	0.08	17	12	2.5	1 Si
316L	S31603	0.03	17	12	2.5	1 Si
317L	S31703	0.03	19	13	3.5	1 Si
Duplex grades						
329	*S32900*	*0.08*	*25*	*3.5*	*1.5*	*0.75 Si*
2205	*S31803*	*0.03*	*22*	*5.5*	*3.0*	*0.08–0.20 N*
255	*S32550*	*0.04*	*25*	*5.5*	*3.0*	*1.5–2.5 Cu*
			Superaustenitic grades			
254SMO	S31254	0.02	20	18	6.0	0.8 Cu, 0.20 N
367XN	N08367	0.03	21	24.5	6.5	1 Si
28	N08028	0.03	27	31	3.5	1.0 Cu
20Mo6	N08026	0.03	26	35	5.5	0.5 Si

passive in oxidizing environments. The semipermanent surface passive film is very thin (about 0.1 nm), analogous to ice on a pond. This passive film of chromic and silicon oxides plus adsorbed oxygen can be increased in thickness by exposing it to oxidizing solutions or by anodic passivation using an electrochemical technique. (Note: Electropolishing is followed by another electrochemical treatment to reform and enhance the passive film following anodic dissolution of the alloy. If the film is solubilized, as by reducing acids such as hydrochloric or dilute sulfuric, stainless steels change from a passive to an active state and are actually less resistant than ordinary iron and steel.) Most applications of stainless steels are in air, aerated water, and other oxidizing media wherein the passive film is retained. It should be noted that stainless steels are *autopassivating*; the film reforms immediately if removed by mechanical means (e.g., abrasion) when the alloy is subsequently exposed to air, water, or other oxidizing environments. They do not demonstrate the active state except under continuous abrasion or conditions of immersion in reducing environments.

Stainless steels are resistant to general corrosion in most applications, with corrosion rates <4–5 mpy. When data are reported showing general corrosion rates of 10, 20, or 50 mpy, this indicates that the alloy is fluctuating between an active and a passive state and should not be used in that specific environment. Unfortunately, despite their passive nature, stainless steels are subject to several forms of localized attack. These are described further for the austenitic grades, which are the primary alloys for bioprocessing applications (3).

FORMS OF LOCALIZED CORROSION

Pitting

Pitting can occur on boldly exposed surfaces in high-chloride environments, such as brackish cooling waters or dilute hypochlorite solutions (e.g., during sanitizing operations). The molybdenum-free alloys such as Type 304L (S30403) are more susceptible to pitting than molybdenum-bearing alloys in chloride-bearing environments, and the problem is combated by specifying successively higher grades, that is, Type 316L, Type 317L (S31703), and the 6% Mo grades as required by the severity of the environment.

Crevice Corrosion

Crevice corrosion is a special form of pitting occurring in physical or mechanical crevices (e.g., flanges, threaded fasteners) or under surface films. When it occurs under calcareous deposits, biomasses, or other foreign material, it is known as "under-deposit corrosion" or UDC. Typically, crevice corrosion is driven by a dissolved oxygen (DO) cell, the oxygen-starved crevice functioning as an anode to the boldly exposed cathodic area where the DO constitutes a cathodic depolarizer. However, the cell can also be driven by dissolved sulfur (DS) or by differences in metal ion concentrations. Crevice corrosion is aggravated by chloride contamination, the chlorides being attracted to the anodic sites and producing a locally low pH (e.g., pH 2) from acid chlorides within the crevice.

Sensitization

When the regular carbon grades such as Types 304 (S30400; 18 Cr–8 Ni–0.08 C maximum) or 316 (S31600) are heated in the temperature range from about 800° to 1,500 °F (as by slow cooling from the annealing process or from thermal stress-relieving treatment, or in the heat-affected zones of welds), carbon migrates to the grain boundaries and precipitates as chromium carbides. This leaves a chromium-depleted zone around each grain and causes a susceptibility to intergranular corrosion (IGA) in many environments. When IGA is observed in the heat-affected zone of welds, it is commonly called by the misnomer "weld decay," although it is not the weld itself that suffers corrosion. Sensitized areas are also more prone to chloride pitting than are unaffected areas such as those at some distance from welds.

Conventional sensitization by chromium carbide precipitation was originally combatted by addition of stabilizing elements, that is, columbium (niobium) or titanium, to an 18-8 alloy such as Type 304 (S30400). When stabilizing elements are present in concentrations several times the carbon content, it is their carbides that precipitate, thereby avoiding chromium depletion. The stabilized grades such as Types 321 (S32100) and 347 (S34700) are therefore derivatives of Type 304 containing Ti and Cb, respectively, and are still used in some special applications. It should be noted that titanium-stabilized grades such as S32100 require a stabilizing anneal at about 1,800 °F to tie up the carbon in the form of titanium carbides before welding. IGA may occur even in stabilized or low-carbon grades if weldments are carbon-contaminated by oil or grease.

The stabilized grades can themselves become sensitized by multipass welding or cross-welding, causing knife-line attack (KLA); this is caused by a resolubilizing of the columbium or titanium carbides by the heat of welding, followed by slow cooling. This step requires about 1,250 °C (2,300 °F) reheating in 18-8 grades, moving the susceptible area to the fusion zone, and permitting a very narrow band of corrosion immediately adjacent to the weld (KLA). With higher nickel content, the resolubilization temperature drops; Alloy 20Cb3 (N08020) and Alloy 825 (N08825) with nickel contents in the 35% range are more easily sensitized to KLA (about 980 °C or 1,800 °F) than is type 347.

In modern practice, the problems of sensitization have been largely alleviated by use of low-carbon grades (Type 316L, S31603; 17 Cr–12 Ni–2.5 Mo–0.030 C maximum), although sensitization will occur with very prolonged exposure in the sensitizing range (e.g., in long-time service >800 °F or when stainless steel–clad vessels are thermally stress relieved). Modern-day steel-making practice using argon–oxygen decarbonation (AOD) makes very low carbon grades of stainless steels routinely available.

IGA may also occur due to mechanisms other than carbide precipitation. A ferrite phase, if present for any reason, may be selectively attacked by reducing acids such as hydrochloric or sulfuric; its thermal conversion product, sigma phase, is selectively attacked by oxidizing acids (e.g., nitric). Chi phase can form in molybdenum-bearing grades (S31603) as the result of prolonged heating in the sensitizing range (e.g., thermal stress relief of clad-steel vessels), producing IGA in alloys of <0.030% carbon in certain aggressive environments.

Selective Weld Corrosion

Selective or preferential attack is sometimes encountered in weldments, particularly in the molybdenum-bearing grades. This can occur either in conventional welds or autogenous welds (i.e., welds made without the addition of filler metal) in some environments. This type of attack is ascribed to microsegregation, which causes a maldistribution of the molybdenum component of the alloy.

Selective weld corrosion is the reason for the cold-work requirement for welded stainless steel condenser tubes (see ASTM A249). Cold-working to minimum requirements before the final anneal allows the weld to recrystallize and homogenize, alleviating the microsegregation.

For butt welds in piping and fabrication welds in tanks and vessels, no such cold-work is feasible. Instead, an overmatching rod is used to confer higher resistance to the weld itself, for example, alloy 254SMO (S31254) to weld S31603 or S31703 or alloy 625 (N06625) to weld S31254.

Stress Corrosion Cracking

The predominant form of environmental stress cracking (ESC) in the 18-8 varieties of stainless steels is anodic stress corrosion cracking (SCC) under either applied or residual tensile stress. Compressive stress will not cause SCC, and indeed application of *compressive* surface stress, as by shot-peening, is sometimes used as a palliative surface treatment against SCC. Only the martensitic grades are subject to hydrogen-assisted cracking (HAC), a form of ESC found, for example, in "sour" service due to hydrogen sulfide effects.

The most common form of SCC of 18-8 austenitic grades is caused by chloride contamination. The ubiquitous chloride is sodium chloride (NaCl, table salt), as found in water and atmospheric exposures. NaCl usually causes SCC only between about 50 °C (120 °F) and 200 °C (390 °F). However, SCC can occur at room temperature and even at cryogenic temperatures in the presence of other chlorides, such as ferric chloride, sulfuric–NaCl mixtures, aqueous solutions of hydrogen sulfide–NaCl, and hydrogen chloride (HCl). For boldly exposed surfaces, there are rough correlations between pH and chloride concentration at which SCC may be anticipated. There is no minimum chloride concentration below which SCC will not occur if there is any possibility of concentration, either by evaporation or by occlusion or adsorption in films adhering to the stainless surface (e.g., mill scale, heat tints, welding slag, rust deposits, calcareous deposits, biomasses).

External stress corrosion cracking (ESCC) occurs when chlorides from the atmosphere, water leaks, or insulation concentrate on the surface of an 18-8 type stainless steel pipe or vessel. This is best combatted by coating vessels and piping that are more than 4 in in diameter and that operate between 60 and 200 °C, using a zinc-free and chloride-free paint system (4,5).

Hot caustic can also cause SCC of 18-8 stainless steels, as frequently happens when high-pressure steam is contaminated by carryover of alkaline boiler-treating chemicals. This is a potential problem in steam lines and during sterilization procedures. Use of nickel-base alloys such as N06625 or better (but not N08800 or N08825) is effective against caustic SCC.

Rouging in High-Purity Water

Austenitic stainless steels are subject to superficial rusting under some circumstances, either because of tramp iron embedded in the surface or from chloride contamination by corrosive atmospheres or water environments. The former problem is addressed through standard cleaning and passivation techniques described further next.

An apparently separate problem, "rouging," is sometimes observed with clean stainless steels in high-purity water at about 80 °C (175 °F) (6). Strictly speaking, rouge consists of a thin red deposit of α-Fe_2O_3 which is usually loosely adherent, wiping off easily. Generally, very little corrosion of the surface is evident. Typically, rouging is experienced in high-purity deionized water such as WFI ("water for injection") in the pharmaceutical industry, and UHPW ("ultrahigh-purity water") in the electronics industry and in nuclear reactor systems. When the U.S. Food and Drug Administration and solutions for human injection are involved, the presence of red-colored oxide on the surfaces of stainless steel equipment is not acceptable. Also, the standards for ultrahigh-purity water in the electronics industry do not permit this sort of condition.

One explanation that has been advanced to explain the fairly recent development of this phenomenon relates to modern steel-making processes. The AOD process itself gives a precise control that permits the manufacturer to work at the lowest ranges permissible for alloying elements, for example, chromium, within the nominal range specified. During long-time hot-working and annealing, the stainless steels form a chromium oxide scale that depletes the metal substrate of Cr. Modern short-time "spray pickling" may not be adequate to dissolve the depleted layer, leaving a surface of somewhat lesser corrosion resistance than the nominal analysis. However, even minor mechanical abrasion restores the proper surface chemistry, and electropolishing, although not simple nitric acid "passivation," which does not attack alloys with as little as 11% Cr, also does so.

A rouge of loose ferric oxide has been developed on Type 304L in laboratory tests in deionized water at 250 °C, simulating a reactor loop. When the water was cooled for sampling and filtered through a 0.1-μm PTFE filter supported on a 0.2-μm silver filter, no deposits were found on the PTFE but large crystals of spinel ferric oxide (with traces of Cr and Ni) had developed on the silver filter and grown back into the PTFE from the backside. This reaction suggests either soluble or colloidal iron in the nearly supercritical (288 °C or higher) water. Loose rouge has also been observed in hot aggressive waters (e.g., WFI) as low as 80 °C (175 °F) from sterilizing procedures with stainless steels in the pharmaceutical industry. Some operators have used solutions of sodium hydrosulfite ($NaHSO_3$) to reduce and remove rouge from their systems.

Some authorities state that the term *rouging* is a misnomer because the films may be either the powdery red iron oxide or adherent films of various colors including gray, black, purple, blue, and even yellowish green. In practice, it is often difficult to distinguish between superficial corrosion (involving a black substrate under a reddish-brown crust, associated with craters and white debris) and extraneous deposits (7). Heat tints and similar films exhibit colors related to their thickness and, although they can occlude and concentrated corrosive species such as chlorides, should not be considered rouge.

In one case, an electropolished Type 316L filter plate developed a blue-black tint, with no apparent corrosion of the parent metal, in ultrahigh-purity water of reportedly ambient temperature or at least not more than about 55 °C (130 °F). Analysis determined that the film contained more than 85% iron as a black oxide, probably derived from equipment upstream and entering the system, perhaps during regeneration of the mixed ion exchange beds.

FACTORS CONTRIBUTING TO LOCALIZED CORROSION

Some factors that contribute to the occurrence of localized corrosion in the stainless steels are surface defects or foreign matter. Surface defects are undercut welds, scratches, and grinding marks, which can initiate localized attack.

Foreign matter constitutes embedded iron or iron oxides (usually from mill or shop contamination of the stainless steel surface), the bluish heat tints adjacent to welds, weld splatter or arc strikes, and paint or crayon markings. Iron oxides and heat tints can occlude and concentrate chlorides and are themselves susceptible to corrosive attack. Weld splatter and arc strikes weaken the surface film and may constitute crevices for the initiation of attack. Paints and crayon markings may contain organic chlorides, which can hydrolyze or thermally degrade to release traces of hydrochloric acid and promote localized attack.

Bacterial effects in unsterile water, for example, residual hydrotest water, can also cause localized attack. The production of traces of hydrogen sulfide by sulfate-reducing bacteria (SRBs) and the excretion of iron or manganese oxides by aerobic strains can cause serious pitting in a very short time. Sulfides compete with oxygen and weaken the passive film, contributing to pitting and SCC. Microbiologically induced corrosion (MIC) has attracted increasing interest in the process industries.

PREVENTION OF LOCALIZED ATTACK

To prevent or minimize the possibility of localized corrosion, a specific sequence of events must be established. The methodology for this starts with proper materials and design and the establishment of certain procurement details in the materials selection, specification, and inspection phases, plus postfabrication cleanup and preservice sanitizing or other preparatory steps.

Materials

Materials must be properly chosen for the expected conditions of fabrication and service. The alloys chosen should be indicated by the proper alloy designation or corresponding UNS number, and these should be allied with an appropriate ASTM or ASME specification and any necessary supplements.

For best results, cast components, such as pumps and valves (which are predominantly CF8M and CF3M, the cast equivalents of types 316 and 316L), should be purchased to ASTM A743 with a special annealing temperature requirement. Annealing at about 1,120 °C reduces the tendency toward molybdenum gradients within the casting, ensuring a more uniform corrosion resistance.

To avoid materials mixups, such as may occur either at the manufacturer's plant or in warehousing, the composition of the alloy should be verified not only by checking the mill analysis but also by spot-test or nondestructive metal identification procedures; this is particularly necessary when a molybdenum-bearing alloy has been chosen. Inadvertent substitution of Type 304L for 316L has caused many problems in process applications. The fact that a piece of stainless steel is clearly marked as Type 316L, Heat No. *X*, is not a guarantee.

Finally, special finishes may be required. The several commercial finishes (in decreasing order of surface roughness) are numbered 1 (hot-rolled, annealed, and pickled) through 8 (a mechanically polished and buffed specular finish, with no visible abrasive scratches). For bioprocessing applications, a number 2B finish (cold-rolled, annealed, pickled, and temper-rolled) or a number 4 (which adds a 150-grit abrasive belt grinding to the 2 B finish) is usually specified.

For extra smoothness in special nonstick applications or to facilitate removal of bacterial contaminants, stainless steel may be electropolished to a mirror finish in lieu of a number 8 finish. Electropolishing aids surface homogenization and autopassivation, and an actual anodizing can be added as the final step to enhance the passive film.

Design

Because localized corrosion causes about 90% of the failures experienced with stainless steels in process applications, vessels, piping, and ancillary equipment must be designed, purchased, and maintained to minimize the risk of such attack. This requires that crevices, pockets, and traps should be avoided, thereby minimizing the probability of localized corrosion.

All fabrication steps, such as bending, forming, and welding, should conform to standard procedures. The welds must be full penetration to prevent crevice corrosion in incompletely fused joints. Fillet welds should be smooth and continuous, with no undercutting. Butt welds should be smooth and corner welds should be avoided if possible; side-wall welds in tanks are preferable. Lines and vessels should be sloped and self-draining, and top nozzles, side ports, and bottom nozzles must be accessible for cleaning.

A postweld chemical cleaning with an ammoniated citric acid solution, as described next, will remove the heat tints from the welded heat-affected zones (HAZ).

REMEDIAL MEASURES

Pickling

Because stainless steels are chosen for appearance, corrosion resistance, and freedom from product contamination in storage and shipment, pickling (to remove contaminants) or passivation (to augment the naturally occurring surface film) is often required. A surface of assured cleanliness, uniformity, and corrosion resistance is desirable and, in some services, absolutely required.

During fabrication, stainless steels are subject to contamination by embedded steel particles or iron oxides, as well as heat tints caused by welding. The alloy surface may be cleaned with acid solutions or pastes, mechanically cleaned with a stainless steel wire brush or a clean abrasive disk, or hydroblasted with high-pressure water of potable water quality (i.e., <250 ppm Cl^-). ASTM A 380, "Cleaning and Descaling Stainless Steel Parts, Equipment, and Systems," provides detailed procedures. Demineralized water is usually specified for critical services, but a 25-ppm maximum chloride limit should be adequate.

Embedded iron is subject to air oxidation to hydrated ferric oxide ($Fe_2O_3 \cdot H_2O$). Foreign rust particles or mill scale (Fe_3O_4) may also be present. "Pickling" (see following) is intended to effect the desired cleanliness while simultaneously homogenizing the surface. A solution of 10–15% nitric acid and 1–3% hydrofluoric (HF) acid has been used as a "bright pickle" to remove visible oxides, while pastes of similar composition are available for localized application. Such solutions and pastes constitute somewhat of a safety hazard to personnel, as well as potential disposal problems, but are very effective. (Note: There is increasing resistance to using HF mixtures today because of the safety concerns and disposal problems related to environmental considerations.)

A HNO_3–HF mixture, unlike nitric acid itself, is a reducing acid intended to remove iron oxides and other metallic oxides by chemical reduction. This mixture is not a passivating medium, and corrosion rates are high during the actual exposure. However, a uniform passive oxide film will reform over the freshly cleaned surface after removal of the pickling environment by autopassivation.

Heat tints from welding tend to adsorb chlorides, with attendant pitting or stress corrosion cracking. They may be removed by abrasive disks or with a nitric–HF paste. Citric acid solutions, such as a 5% solution adjusted to pH 3.5 with ammonia and used at about 80 °C, are very effective and are to be preferred for removing heat tint from bioprocessing equipment. This solution is a chelating agent as well as an acid cleaning medium. Unlike the nitric–HF materials, citric acid solutions are benign and quite readily disposable from the environmental standpoint. However, chelants incorporate the metal ions within a clawlike ringed structure that may pass through conventional waste treatment systems. When heavy metals (i.e., chromium, copper, nickel, and zinc) are involved, special waste treatment systems, rather than simple sanitary drain systems, may be employed.

Antisplatter pastes are sometimes used to minimize weld-splatter sticking. If such splatter has adhered to the adjacent areas, it should be removed by a flapper wheel, wire-brushing, or hydroblasting. A cleaning operation with organic reducing acids may be followed with an alkaline sodium nitrite neutralizing–oxidizing rinse. (Note: Adequate provision must be made for proper disposal of the chemical solutions in accordance with environmental regulations.)

Electropolishing, in which the work serves as the anode during electrochemical dissolution to produce a highly polished nonstick and uncontaminated surface, is a specialized form of chemical cleaning in which the surface is actively corroded during the operation. Although the stainless steel surface will passivate on rinsing and exposure to air, many users specify a final anodizing step, which is tantamount to chemical passivation.

Passivation

Cleaning and pickling, as described, and passivation of stainless steels, particularly in the widely used austenitic 300 series, for example, Types 304 (S30400), 304L (S30403), 321 (S32100), 347 (S34700), 316 (S31600), 316L (S31603), and 317L (S31703), are widely misunderstood.

Cleaning and pickling are intended to remove foreign materials and permit the surface to equilibrate or homogenize, effecting an electrochemically uniform surface for resistance to localized corrosion. Passivation has been defined as a reduction in the rate of the anodic reaction of an electrode involved in corrosion, as by formation of a protective oxide film.

As previously noted, stainless steels are self-passivating on exposure to air and moisture as the result of their high chromium content (i.e., $>11\%$). Passivation treatment consists of augmenting the naturally occurring protective oxide film by chemical or electrochemical methods. The enhanced passive film is somewhat thicker and more persistent than that formed by autopassivation. For example, in some iron-sensitive organic chemicals (e.g., acetoacetanilide, glyoxal), stainless steel vessels are specially passivated and repassivated at intervals thereafter whenever chromophoric corrosion products are detected in the chemical product. This practice suggests a gradual diminution in thickness (or localized weakening of) the passive film with time in some services. Passivating is accomplished by treatment with oxidizing solutions, such as warm dilute nitric acid (not nitric-HF), ammonium persulfate (which is very safe to handle), dilute hydrogen peroxide, and citric acid–sodium nitrate mixtures.

Table A2.1 of ASTM A 380, Part II, recommends passivation with 20–50% nitric acid at 49–71 °C (120–160 °F) for 10–30 min or 21–38 °C (70–100 °F) for 30–60 min for dull or nonreflective surfaces. In the latter case, a 10% caustic–4% potassium permanganate solution is suggested for smut removal after the final rinse. A 20–40% nitric acid with 2–6% sodium dichromate is suggested for bright-machined or polished surfaces, instead of straight nitric acid. Such solutions are also used for precipitation-hardening (PH) grades, 12–14% chromium grades, and high-carbon martensitic grades, such as Type 440 (S44000). However, chromates are not compatible with pharmaceutical products. (Note: In chemical plants, large tanks or vessels are usually passivated with 10–15% nitric acid at 66 °C (150 °F) for about 4 h.) A 1% citric acid–1% sodium nitrate solution at 21 °C (70 °F) is discussed in ASTM A 380, Part III. The dairy industry has used a 0.5–1.0% nitric acid for 10 min at 65–82 °C (150–180 °F).

A special sequence of alkaline–acid–alkaline (A-A-A) treatment is recommended for free-machining grades (8). Because oil or grease interferes with passivation and because attack on sulfur or selenium inclusions in Type 303 (S30300) and its variants can leave minute pits with residual corrodents on the metal surface, the machined parts are (*1*) degreased for 30 min in 5% sodium hydroxide at about 70–80 °C (160–180 °F) and rinsed, (*2*) immersed in 20% nitric plus 3 oz. per gallon sodium dichromate at 50–60 °C (120–140 °F) for 30 min and rinsed, and finally (*3*) neutralized in the original caustic solution.

Because oil or grease can interfere with self-passivation by preventing ready access of water and oxygen to the metal surface, degreasing may be required. Such contamination occurs primarily from machining operations and should be removed by degreasing with nonchlorinated solvents before passivation is attempted. (Note: Chlorinated solvents may introduce inorganic chloride ions, Cl^-, by hydrolysis or by thermal degradation, and are proscribed.)

Note that nonoxidizing solutions, such as ammonium citrate, hydroxyacetic–formic acid mixtures, and EDTA (ethylenediaminetetraacetic acid) instead of a nitric acid passivation are described in ASTM A 380, Part III. The dairy industry uses a 0.5–1.0% phosphoric acid solution at 65–82 °C (150–180 °F) for 10 min. These nonoxidizing solutions used to remove iron or iron oxides only equilibrate the surface, facilitating autopassivation. They do not actually enhance the film as do the true passivation procedures.

Sanitizing

As a final step (or between batch operations), it is frequently desirable or even necessary to sterilize stainless steel equipment. Sodium or calcium hypochlorite is most frequently utilized, but the oxidizing capacity and chloride content pose serious potential dangers for pitting or crevice corrosion. The conditions and time of exposure and the proper steps to flush the hypochlorites from the system, including low spots, is of paramount importance. Serious failures have occurred in numerous stainless steel installations when adequate measures to preclude attack during treatment or by residual hypochlorites were not taken. Treatment with ozone or dilute hydrogen peroxide, as well as by ultraviolet light, seems to provide useful alternatives.

STANDARD OPERATING PROCEDURES

To summarize, the recommended sequence of treatment of fabricated stainless steel equipment for pharmaceutical applications should embrace the following steps:

1. A pretreatment with solvent or detergent to remove oil, grease, and the like
2. A rinse with water of suitable quality (deionized water being preferred)
3. Necessary pickling or other cleaning operations
4. Passivation
5. Water rinsing
6. Sanitizing
7. Final water rinse; drain and dry

A proper understanding of service conditions and requirements is essential if the several possible cleaning, pickling, and passivation treatments are to be successfully employed for their respective end purposes.

The most valuable resources for practical information on corrosion and corrosion prevention of stainless steels are as follows:

- NACE International (standards and publications), Houston, Tex.
- The Nickel Development Institute (AISI, INCO, and NiDI publications), Toronto, Ontario, Canada
- ASTM International (specifications), Barr Harbor, Pa.
- The Materials Technology Institute of the Chemical Process Industries, Inc. (publications), St. Louis, Mo.

BIBLIOGRAPHY

1. *Metals & Alloys in the Unified Numbering System*, SAE HS-1086, ASTM DS-56 E, 6th ed., American Soc. for Testing and Materials, Philadelphia, 1993.
2. C.P. Dillon, *Corrosion Resistance of Stainless Steels*, Dekker, New York, 1995.
3. C.P. Dillon, D.W. Rahoi, and A.H. Tuthill, *BioPharm* **5**, 3–6 (1992).
4. C.P. Dillon, *Corrosion Control in the Chemical Process Industries*, 2nd ed., Publ. 45, MTI, St. Louis, Mo., 1994.
5. *Guidelines for Preventing Stress Corrosion Cracking in the Chemical Process Industries*, Materials Technology Institute of the Chemical Process Industries, St. Louis, Mo., 1985.
6. P.H. Banes, *Bioprocess Eng.* (1993).
7. D.C. Coleman and R. Evans, *Pharm. Eng.* **11** (1991).
8. T. DeBold, *Machine and Tool Blue Book*, November 1986.

STATIC MIXING, IN FERMENTATION PROCESS

RADU Z. TUDOSE
Technical University Gh.Asachi Iaşi
Iaşi, Romania
MARIA GAVRILESCU
Research Centre for Antibiotics
Iaşi, Romania

KEY WORDS

Fermentation
Heat transfer
Hydrodynamics
Mass transfer
Nonmechanical bioreactors
Non-Newtonian systems
Static mixer

OUTLINE

INTRODUCTION

The intensification of transfer processes in bioreactors using techniques that do not require additional energy costs is of great interest in biotechnology. Traditionally, most bioprocesses, especially aerobic fermentations, are conducted in stirred-tank bioreactors, which require considerable power input to attain adequate oxygen transfer rates. Besides the conventional mechanically agitated fermenters, other types of gas–liquid contactors are used, as pneumatic or jet bioreactors with or without a loop. Advantages of these nonmechanically agitated bioreactors are known to include efficient use of energy for aeration, simple design, and construction. To improve transfer processes in these bioreactors, considerable efforts have been made to study all the essentials of the major relationship for flow, mass and heat transfer over plates, cylinders, baffles, and spheres as well as through sieves and trays. These are the simplest and the oldest contacting bodies, and also are the basic shapes for other contacting devices of practical interest. Both theoretical and experimental studies on flow around these internals together with heat and mass transfer approaches are available (1–4).

Recently, new constructive types of internals known as static mixers or motionless mixers have been developed that supply spectacular effects concerning transfer rate increase. The first of these unconventional mixers was developed for viscous fluids about 30 years ago (5). After a slow start, the development of static mixers has been proceeding since 1970. Static mixers are applied in many operations in heterogeneous systems that require larger contact areas, thermal homogenization, and mass transfer. Today, these devices placed in column bioreactors are increasingly considered for use in fermentation industries, sometimes replacing the traditional mechanically agitated bioreactor design. Static mixers are suitable particularly for use in shear-sensitive cell cultures because of their lower shear rates, compared to those in mechanically agitated bioreactors. They proved to consider some specific criteria in the design of a bioreactor system for a definite process: population damage, energy distribution, and oxygen demand of a culture. The relatively numerous investigations and works that refer to static mixers are concerned with their influence on different process transfer parameters not only in fermentation systems, most of them being performed in model systems, in monophasic and polydispersed laminar or turbulent flow (usually gas–liquid dispersions), using low and viscous liquids, with Newtonian or non-Newtonian rheological behavior.

Static mixers frequently require no additional space (in-line assembly) and can be used as compact inserts or with a distance between the elements. They act on hydrodynamic conditions for turbulence development, the increase of contact areas, or the intensification of driving forces, separately or simultaneously. In all cases, the energy for process intensification is extracted from the fluid flowing through the mixers.

STRUCTURAL TYPES AND CONSTRUCTION

Static mixers can be of numerous constructive types, made of tubes, sheets, helixes, or wire matrix. They can be manufactured by steel, ceramic, or plastics. Their effect depends on the construction of the mixing element and the capacity to produce continuous separation, distribution, and reuniting the fluid stream. Besides their multiple advantages, static mixers often have the difficulty of a complicated structure. At the present time, most of the problems in the construction of static mixers have already been solved, and a number of different types of static mixing systems are being manufactured on an industrial scale, so that about 30 types of static mixers are known today. One of the most important moments in static mixing presentation was the ACHEMA '79 Exhibition in Frankfurt/Main (5). This section discusses some of the most interesting, familiar and, at the same time, most structurally dissimilar static mixers.

Static Mixers from Tubes or Short Pipes

Static mixers from tubes or short pipes generally affect the spatial distribution of the fluid by twisting, distorsion, and expansion of fluid elements in specially designed flow arrangements. For instance, the Ross-ISG static mixer (Aachener Misch- und Knetfabrik) is made by elements turned through 90°, placed one behind other. Over the cross-section of an element there are four obliquely sloped channels twisted with an angle of 120° (Fig. 1**a**), forming a tetrahedral space between elements. The PRM mixer (Prematechnik, Frankfurt/Main) consists of a tubular packet containing obliquely arranged partial tube sections (Fig. 1**b**), which change the direction of the fluid stream (5). Another static mixer used in laminar flow conditions, especially to equalize the distribution of the flow residence time, consists of six guiding holes, directed in the direction of flow (Fig. 1**c**). A portion of the fluid stream, which flows through the central section in front of the device, has a

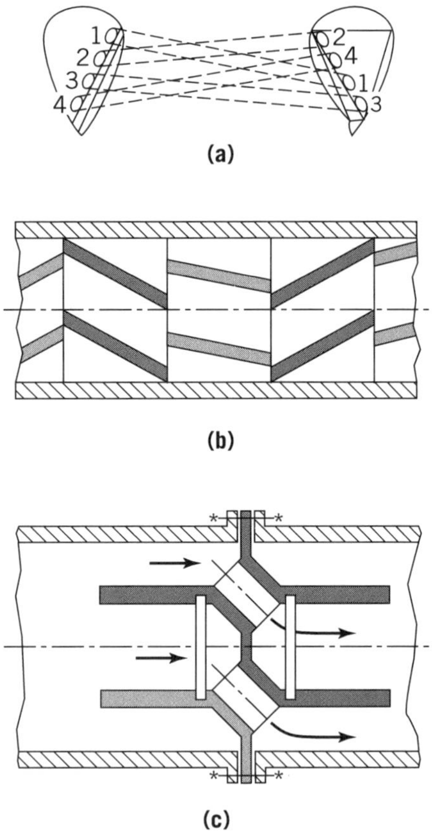

Figure 1. Static mixers made from tubes or short pipes. *Source*: Reprinted from Ref. 5, with permission from VCH Verlagsgesellschaft mbH, and from Ref. 7, with permission from Wiley-VCH.

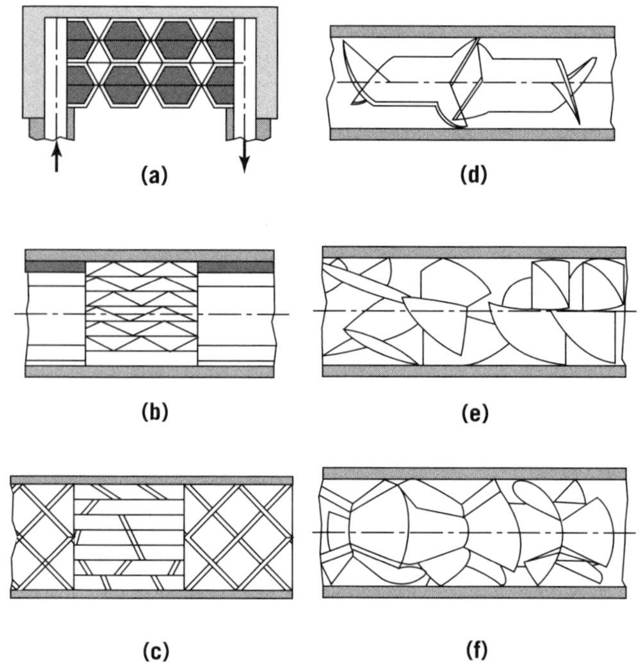

Figure 2. Static mixers made from plates and sheets. *Source*: Reprinted from Ref. 5, with permission from VCH Verlagsgesellschaft mbH and from Ref. 7, with permission from Wiley-VCH.

higher mean velocity and crosses the channels that direct it to a zone with a low mean velocity. Other parts of the fluid stream that flow near the pipe wall cross over the channels that direct them to the central zone of the pipe behind the static mixer. Therefore, the positions of the fluid streams that flow with different velocities are inverted (6–8).

Static Mixers of Plates or Sheets

The effects of these static mixers depend on the continuous cutting, subdividing, distribution, and reuniting the stream of fluid. The PSM mixer (Petzhold, Frankfurt/Main) (Fig. 2**a**) contains rectangular and conical channels combined with conical holes resulted from a special arrangement of individual plates. The fluid flows from the entry to the exit of the static mixing element, being subdivided through a central hole, a conical space created by tangentially arranged slots and furthermore through other channels and openings (5). Sulzer mixers (Sulzer Brothers Ltd., Switzerland) are adequate for both laminar and turbulent regimes. They are constructed in many different ways (SMV, Fig. 2**b**; SMX, Fig. 2**c**) from stacks of corrugate sheets with open crossing channels. The SMX mixer consists of strips inclined to one another that form frameworks with the axis perpendicular to the column axis. The mixing elements are arranged successively in the column turned through 90° one to another. The SMLX mixer type is constructed from corrugate sheets with a distance between the strips increasing in the flow direction. The crossing angle of the sheets is 60°. These static mixers made of corrugate sheets can be mounted in packed form in compact arrangement or with distance between packets in different nonmechanical bioreactors (9). The Komax mixer (Long Beach, Calif.) contains triangular-shaped channels, made by the bent ends of the slotted sheets that form the mixing element (Fig. 2**d**). The Koch mixer (10,11) contains a numer of in-line elements mounted alternately at right angles to each other. The mixing elements are made from layers of corrugate metal sheets joined one to another, forming opened intersecting channels. C_1 and C_2 mixer types (12) (Figs. 2**e** and **f**, respectively) developed in the Central Institute of Physical Chemistry of the German Academy of Sciences and the Pyrapak 480 type developed at the CLG Leipzig-Grimma were also manufactured out of metal sheets.

Helical Static Mixers

The Kenics mixer (Ott Vertriebsgeselschaft, Leonberg) is made of counterclockwise and clockwise rotational elements twisted at 180°, being mounted in tubes alternating left and right with perpendicular leading edges (Fig. 3**a**). A particular form of this helical static mixer consists of an unbroken continuous helical surface that may have the diameter of a screw identical or smaller than the inner di-

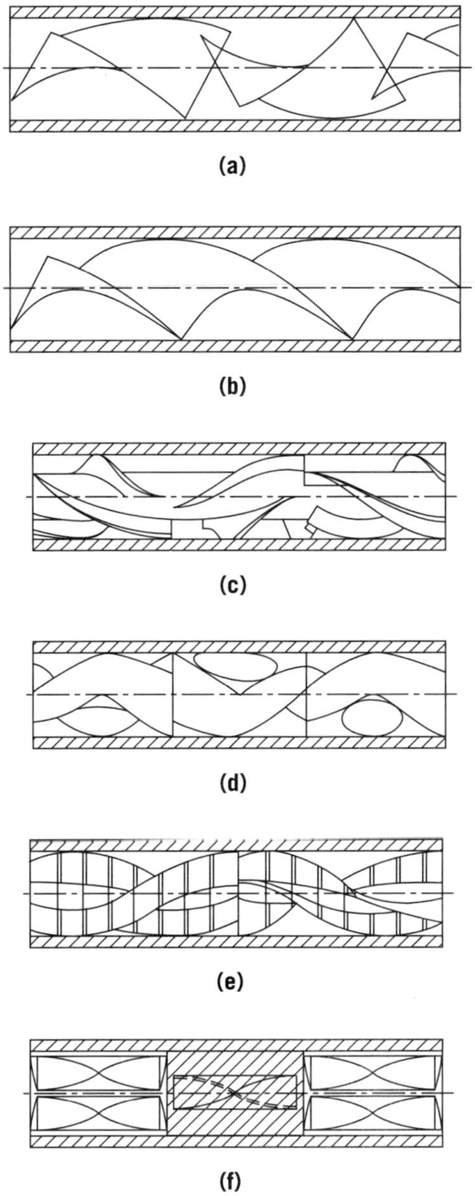

Figure 3. Helical static mixers. *Source*: Reprinted from Ref. 5, with permission from VCH Verlagsgesellschaft mbH, and from Ref. 7, with permission from Wiley-VCH.

ameter of the tube (13) (Fig. 3**b**). Helical static mixers can be disposed in columns as in-line elements or in packed form (14). The Erestat mixer (Yxzet, Dr. Karg GmbH, Affalterbach) consists of four baffles twisted in an alternate manner arranged in a left- or right-handed arrangement on an inner support (Fig. 3**c**). This assembly creates various flowing channels. The N-shaped mixer (Bran and Lübbe, Norderstadt) consists of twisted mixing elements (Fig. 3**d**) that determine variable cross-sectional areas, owing to the formation of variable channels. The Lightin type (Lightin, Rochester, NY) combined in-line mixers obtained by assembling three-bladed elements to form helixes with different senses of rotation (Fig. 3**e**). The HI mixer (Toray, Tokyo) is constructed by helixes twisted at 180°, fitted into holes performed parallel to the axis in a solid cylinder (Fig. 3**f**).

Wire Matrix Turbulators

The Heatex inserts (15) consist of a central wire core onto which are wound a series of wire loops. Each loop is inclined at a certain angle to the core. These inserts are pulled into a tube, the loops being so inclined as to be oversized for the tube and come into a close contact with the tube wall in the form of an arc (Fig. 4). A degree of resistance is offered as the insert is pulled into the tube, indicating that some wall contact is achieved and that the insert is self-supporting. Heatex inserts may have different loop densities, characterized by the number of tube wall contacts achieved per unit insert length.

EFFECTS OF STATIC MIXERS ON MOMENTUM TRANSFER

Flow Structure

Static mixers produce variations in the flowing cross-sectional area that result in modifications of velocity profiles. The fluid is split into individual streams at the open intersecting channels of the mixing element. At each intersection, a partial quantity is sheared off into a crossing channel and stagnant zones are usually eliminated. Also, most of the static mixers divide and reunite continually the main fluid stream and give rise to the formation of elementary layers of fluid. The number of n elementary layers or partial streams produced in laminar flow over a static mixer resulted through assembly of N mixing elements can be calculated from the following expression (16):

$$n = ca^{N} \tag{1}$$

where c and a are parameters characteristic for the static mixer in discussion that are usually given in the literature (16). In the laminar regime, the flow profiles are modified by cutting and twisting, displacement and distortion, or separation and expansion. In the turbulent flow, additional effects occur (5). For instance, the fluid streams are alternating directed from the center to the wall of the column

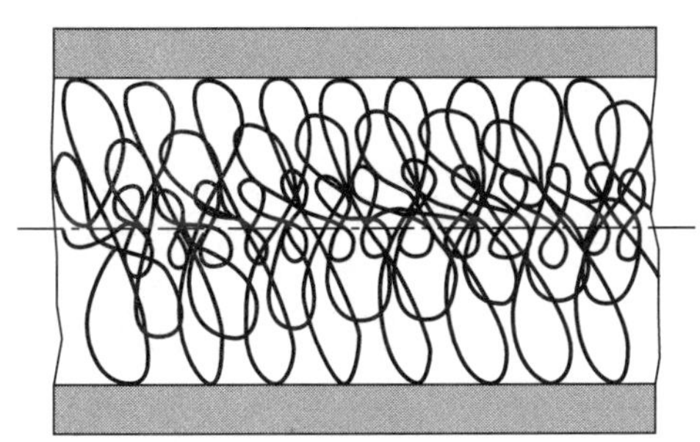

Figure 4. Static mixers of wire matrix. *Source*: Reprinted from Ref. 7, with permission from Wiley-VCH.

and from the wall to the center in the Ross-ISG mixer (Fig. **1a**) or in the static presented in Fig. **1c**. The presence of Heatex inserts (Fig. 4) causes a tube wall–fluid disturbance, and radial mixing is expected to be produced as the fluid flows through the mesh of loops nearer the pipe axis (15). In a Kenics mixer (Fig. 3**a**) a tracer injected below the first element in a certain point can perform a helical rotation strictly geometrically and reach the element edge in an opposite point situated on the same diameter, or it flows along a shortened path after a distortion and arrives at the end of the helix in a intermediate position on the diameter (5). The flow pattern in a Sulzer SMV mixer was presented by Pahl and Muschelknautz (5) using a model that consists in two triangular opened ducts crossing at 90° (Fig. 5) forming two intersected channels. Tracer introduced at the entrance of the channel I flows up to the intersection of the channels, and then about two-thirds of the tracer stream takes the shortest path flowing into duct II owing to the displacement that is produced at the crossing point. The trajectory of the tracer is also distorted; the fluid that exists along the left cathetus of the triangle representing the cross-section of the duct I shifts to the right side of duct I at the exit, performing a rotation. Also, the stream that starts from the opposite position at the entrance of duct I is greatly distorted at the crossing point and flows into both ducts I and II. In gas–liquid dispersions, a reasonably uniform distribution of the gas velocity over the cross-section of a column with static mixers is one of the favorable results. An equalizing effect is achieved by the resistance to flow, which forces the gas to spread out gradually in the direction of flow. This fact is very important in bioreactors with viscous non-Newtonian fermentation fluids, where the bubbles coalesce in the sparged vertical columns, leading to the formation of large bubbles, even at very low gas flow rates. In absence of static mixers, these bubbles rose with large velocities and their size distribution is controlled by bubble coalescence. The velocity profile of a fluid flowing through a tube containing motionless mixers is also considerably altered. There are some works in the literature that give information about radial and axial velocity profiles in the presence of static mixers (13,17,18). The axial and radial velocity profiles in a Kenics mixer change periodically owing to the use of alternating elements of right- and left-hand pitch that results in more intensive mixing of fluids. The screw having the form of a continuous helical surface divides the inside of the tube into two parallel channels, which causes a secondary motion of the fluid in the tube (13). In these motionless mixers, circulating flow occurs in the r-θ plane (r and θ being radial and tangential coordinates, respectively), which is caused by a slow twist in the partitioning chord so that it forms a helix. The actual velocity profiles can be calculated using a complex procedure (17), but an approximation can be made, leading to the following expressions:

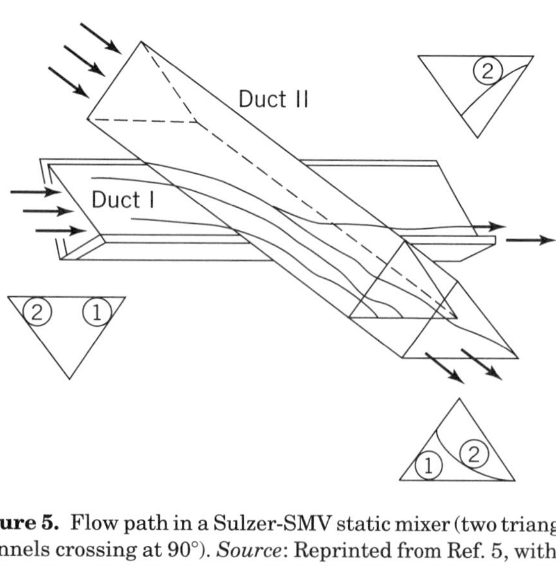

Figure 5. Flow path in a Sulzer-SMV static mixer (two triangular channels crossing at 90°). *Source*: Reprinted from Ref. 5, with permission from VCH Verlagsgesellschaft mbH, and from Ref. 7, with permission from Wiley-VCH.

$$v_r = \frac{dr}{dt} = \beta r(1 - r^{\upsilon}) \sin 2\theta \tag{2}$$

$$v_\theta = \frac{d\theta}{dt} = \beta[2 - (2 + \upsilon)r^{\upsilon}] \sin^2\theta \tag{3}$$

where

$$\upsilon = \left(\frac{11}{3}\right)^{0.5} - 1 \tag{4}$$

and β is a parameter that governs the intensity of circulation (17).

Pressure Drop

The changes in the velocity field caused by the static mixers cause modifications in the pressure loss as well. The pressure drop in static mixers depends very strongly on the geometrical arrangement of the packed elements, the nature of the flowing fluids, and the flow regime (19,20). A simple form representing the pressure loss equation can be referred to the empty tube, together with a geometrical correction factor (5,7,13). For Newtonian media, the equation for calculating the pressure drop is in the following form:

$$\frac{\Delta p_M}{\rho v_{SL}^2}\frac{2D}{L} = f_D(Re_D, \text{geometry}) \tag{5}$$

The mean liquid superficial velocity $\mathbf{v}_{SL}$ can be calculated on the basis of the empty tube cross-section as follows:

$$v_{SL} = \frac{4\dot{V}_L}{\pi D^2} \tag{6}$$

In laminar flow, $f_D\, Re_D$ = constant. Rearranging equation 5, a relationship is obtained whose denominator becomes the pressure drop for laminar flow through an empty pipe:

$$\frac{\Delta p_M}{\Delta p_L} = \frac{\Delta p_M}{32\eta v/D^2} \tag{7}$$

Defining a pressure drop factor as

$$z = f_D \frac{Re_D}{64} \tag{8}$$

it follows that

$$\Delta p_M = z \Delta p_L \tag{9}$$

In Figures 6 and 7, the pressure drop and friction factors for different static mixers are presented (5,7). The plot

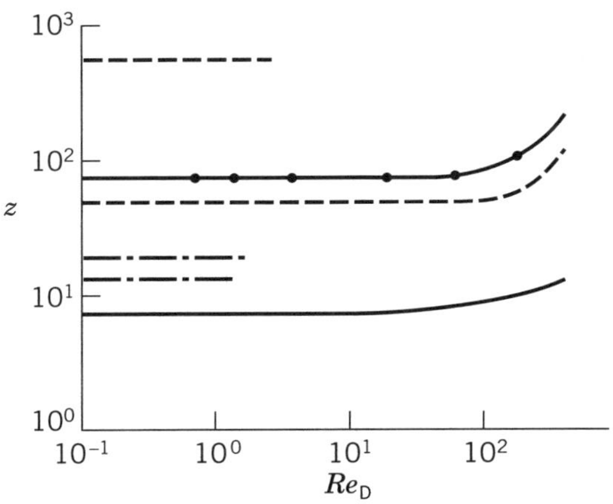

Figure 6. Relative pressure drop for different static mixers versus Reynolds number. — = Kenics; —·— = Erestat; —··— = N-shaped; — — — = HI; —□—□— = Sulzer-SMX; - - - - - - = Ross-ISG. *Source*: Reprinted from Ref. 5, with permission from VCH Verlagsgesellschaft mbH, and from Ref. 7, with permission from Wiley-VCH.

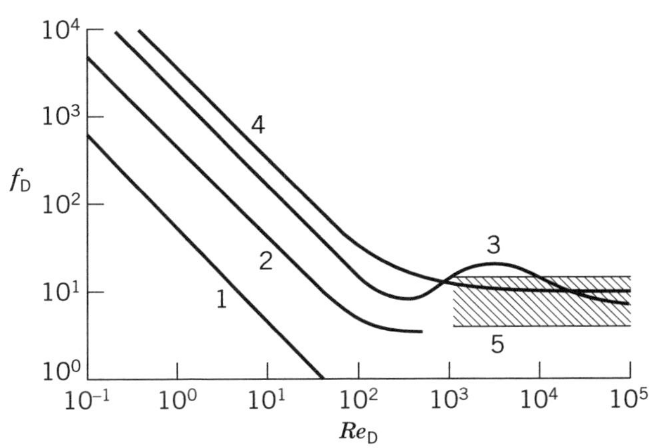

Figure 7. Friction factor as a function of Reynolds number for some static mixers. 1 = empty pipe; 2 = Kenics; 3 = HI; 4 = Sulzer-SMX; 5 = Sulzer-SMV. *Source*: Reprinted from Ref. 5, with permission from VCH Verlagsgesellschaft mbH, and from Ref. 7, with permission from Wiley-VCH.

$f_D - Re_D$ could be also used for comparing static mixers in the turbulent flow regime, since f_D is then approximately independent of the Reynolds number (Fig. 7). Compared with an empty tube, the pressure drop in static mixers is 7 to 200 times greater for laminar flow and 100 to 600 times greater for turbulent flow. Kenics mixers show the smallest pressure drop, and the Ross-ISG affords the largest pressure drop (5,16). Also, the pressure losses in a tube containing continuous screws calculated using equation 5 are lower than those corresponding to a tube containing Kenics mixers for the same value of the parameter h/d. The following analytical expressions were developed for the different types of helical static mixers (13). For Kenics type (for $15 \le Re_D \le 1{,}200$):

$$f_D = \frac{213.5 + 224.0(h/d)^{-1}}{Re_D} + [-0.549 + 4.775(h/d)^{-1}] \tag{10}$$

For static mixer with a continuous screw surface for $15 \le Re_D \le 90$,

$$f_D = \frac{191.3 + 229.0(h/d)^{-1}}{Re_D} \tag{11}$$

for $90 \le Re_D \le 5{,}000$,

$$f_D = \frac{79.8}{Re_D^{0.839 - 0.454(h/d)^{-1}}} \tag{12}$$

With increasing of h/d, the pressure loss decreases for any given type of fittings (Fig. 8). The specification of notation is given in Table 1.

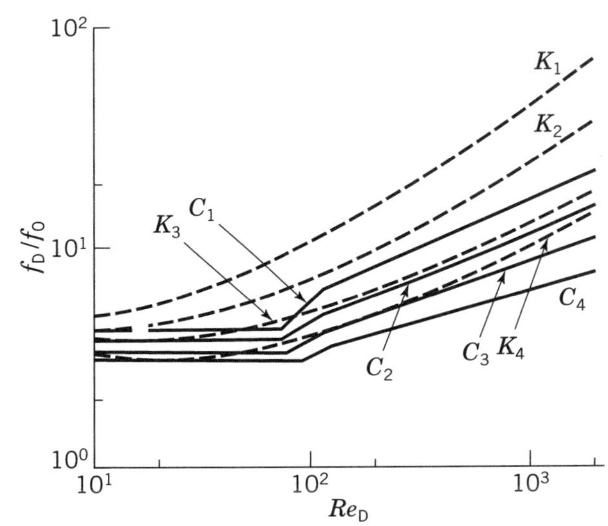

Figure 8. Dependence of the ratio between the friction coefficient in a tube with a screw f_D and that in the same tube without a screw f_0, on the Reynolds number, Re_D, and h/d (specification of C and K in Table 1). *Source*: Reprinted from Ref. 13, with permission from Elsevier Science, and from Ref. 7, with permission from Wiley-VCH.

Table 1. Geometrical Characteristics and Notations for Helical Static Mixers

Type of screw	Notation	h/d	s/d	l/d
Continuous screw	C_1	2.28	0.095	46.85
	C_2	2.95	0.095	47.57
	C_3	3.85	0.095	47.57
	C_4	4.71	0.095	47.69
Kenics	K_1	2.02	0.095	48.26
	K_2	3.03	0.095	49.14
	K_3	4.05	0.095	47.86
	K_4	5.02	0.095	47.86

Source: Reprinted from Ref. 7, with permission from Wiley-VCH.

In Newtonian liquids, the presence of wire matrix turbulators caused an increase in pressure drop by factors up to 20 compared with the empty tube case (15). In non-Newtonian fluids, the pressure drop in a static mixer assembly can be correlated as follows:

$$\Delta p_M = \frac{A}{Re_D} + \frac{B}{Re_D^m} \tag{13}$$

where the values of A, B, and m are dependent on rheological behavior of the liquid phase (7,21). Olivier and Oosterhuis (22) studied the pressure drop in the laminar flow regime for a Sulzer-SMV-type in the xanthan fermentation and found that the pressure drop is approximately 15 times greater than that in an empty tube. Fan et al. (23) studied the pressure drop in a gas–liquid system in a bubble column packed with Sulzer-Koch mixing elements and observed that the ratio of the pressure drop through the packed column to that of the empty one was slightly greater than 1. Data referring to the pressure drop in a column equipped with a static mixer of the Kenics type were obtained for a single (gaseous) phase and two-phase countercurrent flow, respectively (14,24). For one mixing element, the pressure drop, which is a function of the Reynolds number for the gaseous phase, Re_G, is assessed as follows:

$$\Delta p' = \frac{\Delta p_M}{N} \tag{14}$$

The pressure drop in the system was calculated using the gas dynamic pressure for the single phase as

$$\Delta p' = f_{Do}\rho\frac{v_{SG}^2}{2} \tag{15}$$

and for the diphasic flow:

$$\Delta p'_{TP} = f_{DTP}\rho\frac{v_{SG}^2}{2} \tag{16}$$

The plots show three domains with different slopes (Fig. 9). The friction factor also differs in the domains. For a single phase flow, it was correlated as follows:

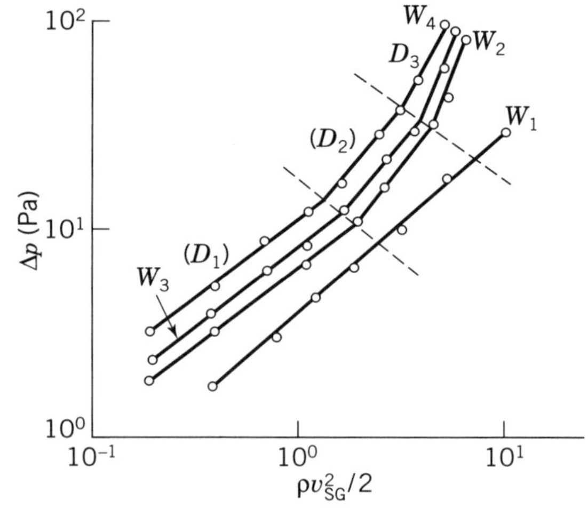

Figure 9. Pressure drop through a single static mixing element as a function of the dynamic pressure of the gaseous phase for different spraying intensities. $W_1 = 0$; $W_2 = 60.93$ m^3/m^{2}h; $W_3 = 81.24$ m^3/m^{2}h; $W_4 = 101.56$ m^3/m^{2}h. *Source*: Reprinted from Ref. 7, with permission from Wiley-VCH.

$$\text{Domain } D_1{:}\quad f_{Do} = 12.86Re_G^{-0.1708} \tag{17}$$

$$\text{Domain } D_2{:}\quad f_{Do} = 1.039Re_G^{0.1062} \tag{18}$$

$$\text{Domain } D_3{:}\quad f_{Do} = 0.9742Re_G^{0.1137} \tag{19}$$

For a diphasic flow, the following relation was established (24):

$$\frac{\Delta p'_{TP}}{\Delta p'} = B'^{W(1+b'Re_G)} \tag{20}$$

The values of the spraying intensity, B', and of the parameter b', are different in the three domains of the curves (Table 2) (24).

For a two-phase flow, Shah and Khale (21) applied a correction factor for the manometer reading, assuming the gas holdup to be proportional with $\dot{V}_G/(\dot{V}_G + \dot{V}_L)$. For an empty tube, the correction factor was found to be 0.83. They presented their data in the form of a Lockhart-Martinelli correlation for three types of static mixers, Kenics, Sulzer, and Komax, for two-phase non-Newtonian liquid–air systems in the following form:

$$\Delta p_{TP} = \varphi_L^2\Delta p_L = \varphi_G^2\Delta p_G \tag{21}$$

$$\varphi_L^2 = 1 + (c'/X) + (1/X^2) \tag{22}$$

$$X = (\Delta p_L/\Delta p_G)^{0.5} \tag{23}$$

with $c' = 2.5 \pm 0.3$ for glycerol solution–air and $c' = 3.27 \pm 0.1$ for CMC solution–air diphasic systems, respectively. Generally speaking, the pressure drop in the gas–liquid systems decreases as the gas flow rate increases (10).

Gas-Holdup

In tower bioreactors with or without a loop (bubble columns, airlifts, jet contactors), liquid flow and gas holdup

Table 2. The Values of Parameters *B′* and *b′* from Equation 20

Domain	B'	b'
1	1.0150	$-5.596 \cdot 10^{-5}$
2	1.00925	$2.2348 \cdot 10^{-5}$
3	$0.947 + 2.6217 \cdot 10^{-4}$ W	$-8.5243 \cdot 10^{-9}$ W^2 + $8.262 \cdot 10^{-7}$ W − $1.4499 \cdot 10^{-4}$

Source: Reprinted from Ref. 7, with permission from Wiley-VCH.

are of basic importance in the determination of their performance (25). Because microbial aerobic cultures often behave as non-Newtonian fluids, much attention was directed toward the performance of these gas–liquid contactors equipped with static mixers, especially when they handle non-Newtonian fermentation liquids. The rheological behavior of biofluids plays an important role in bubble coalescence and breakup, because the large velocities of coalesced bubbles lead to gas holdup decreasing (26). The presence of static mixers in diphasic flow diminishes the bubble coalescence rates owing to their shear effect. Therefore, the average bubble size should decrease and gas holdup is enhanced (7,10) (Fig. 10). The results of MacLean et al. (27) confirm these findings, suggesting that the motionless mixers help to retain more gas in the column. Similar effects can be obtained in triphasic sparged reactors (28). By using Ross-LPD and Sulzer SMV static mixers, larger gas holdup values resulted, compared with the empty contactor (Fig. 11). Kenics mixers mounted in an airlift bioreactor contribute to increase gas holdup values during the batch fermentation of a yeast culture (29). Gas holdup in downcomer (annular section) decreased because of a lower liquid recirculation rate, which implied a more intense gas disengagement at the upper part of the bioreactor, reducing the quantity of the recirculated gas. In the region of greater airflow rates (the riser), gas holdup values are, however, improved. This behavior had a favorable influence on oxygen transfer rates (29,30). The presence of static mixers does not change the correlation between gas holdup and gas superficial velocity, which was established in the following form:

$$\varepsilon_G = a v_{SG}^{b} \tag{24}$$

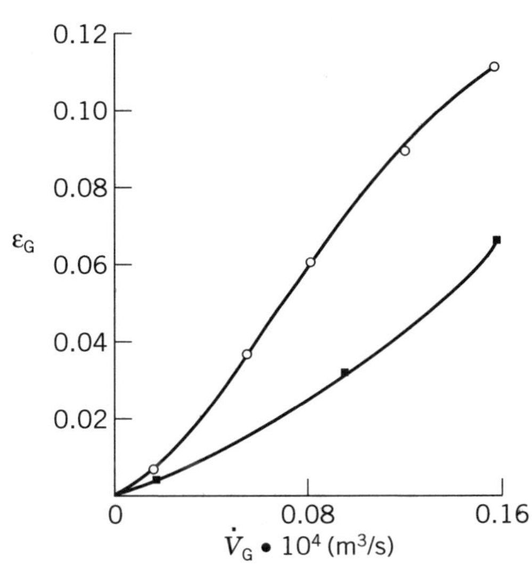

Figure 10. Gas holdup in a Koch mixer column in comparison to the witness one. ● = without static mixers; ○ = with static mixers. *Source*: Reprinted from Ref. 7, with permission from Wiley-VCH.

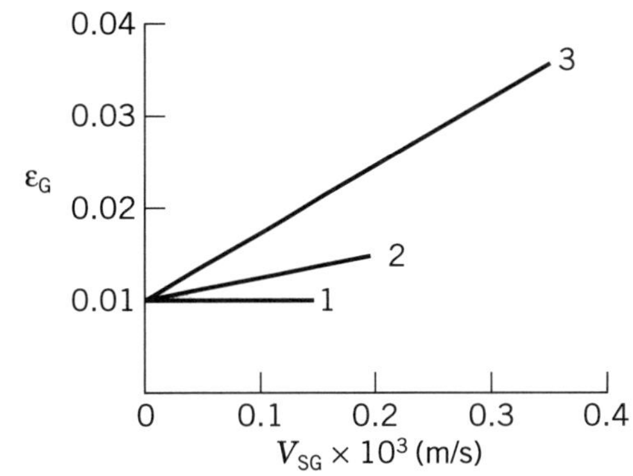

Figure 11. Gas holdup in a triphasic sparged reactor. 1 = witness reactor, 2 = Ross-LPD static mixers; 3 = Sulzer-SMV static mixer. *Source*: Reprinted from Ref. 7, with permission from Wiley-VCH.

For example, static mixers manufactured through assembly of corrugated stainless steel sheets and mounted in the riser section in an external-loop airlift bioreactor can contribute to gas holdup intensification at any given gas flow rate and liquid phase properties, relative to bioreactor configuration without static mixers (Fig. 12), also proved by the modified values of the parameters a and b in equation 24. This effect is found in low-viscous Newtonian and viscous non-Newtonian liquids as well (Table 3) (25). It is evident from Table 3 that static mixers show an increased efficiency in more viscous liquids. Also, they affect the values of the coefficient a in equation 24 in a greater measure than those of b. Therefore, larger values of gas holdup in columns with static mixers resulted because of smaller bubble sizes and hindering action of the packing, which contribute to increasing the gas phase residence time (21,31).

Liquid Velocity

The presence of static mixers changes the liquid circulation rates, because they act as supplementary resistances in the flow path. Besides changes in gas flow pattern, the presence of static mixers in gas–liquid contactors having a net liquid flow (continuous bubble columns, airlifts, con-

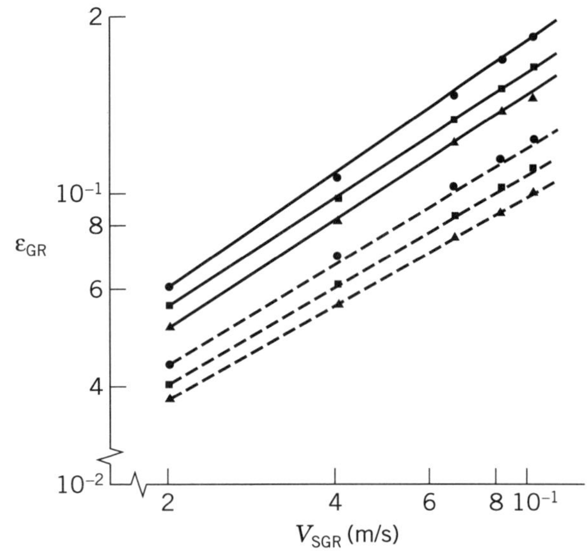

Figure 12. Gas holdup in an external-loop airlift bioreactor without (--------) and with (▬▬) static mixers, in non-Newtonian starch solutions. ● = 0.5% w/v × 2.5 w/v; ▲ = 4.0 w/v. *Source*: Reprinted from Ref. 25, with permission from Springer-Verlag, Germany.

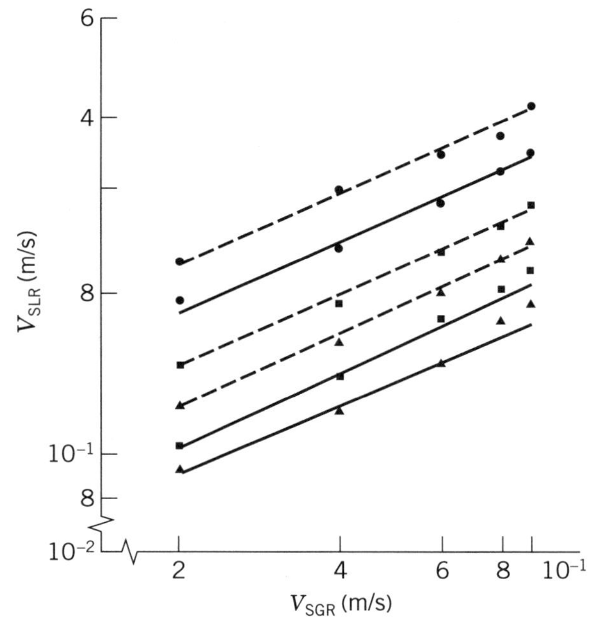

Figure 13. Liquid superficial velocity versus gas superficial velocity in an external-loop airlift bioreactor without (--------) and with (▬▬) static mixers. Symbols as in Figure 12. *Source*: Reprinted from Ref. 25, with permission from Springer-Verlag, Germany.

Table 3. Ratios Between Parameters *a* and *b* in Equation 24 in the Configuration with and without Static Mixers for an External-Loop Airlift Bioreactor

Ratio between parameters in equation 24	K (Pa s^n)		
	0.061	0.438	3.518
a_M/a	1.600	1.845	1.911
b_M/b	1.085	1.159	1.134

Source: Reprinted from Ref. 25, with permission from Springer Verlag.

Table 4. Ratios between the Parameters *f* and *m* in Equation 25 in an External-Loop Airlift Bioreactor with and without Static Mixers, in Non-Newtonian Liquids

Ratio between parameters in equation 25	K (Pa s^n)		
	0.061	0.438	3.518
f_M/f	0.829	0.841	0.856
m/m	0.997	0.994	0.994

Source: Reprinted from Ref. 25, with permission from Springer Verlag.

tactors with a liquid jet) changes the liquid circulation velocities, their effects being complementary to those of several other parameters, including gas superficial velocity, gas holdup, liquid phase viscosity, pressure drop along the flow path, reactor geometry. For instance, the static mixers mounted in an external-loop airlift bioreactor diminish the liquid superficial velocity in the riser section both in Newtonian and non-Newtonian liquids. The parallel lines in the diagram presented in Fig. 13 show that in all cases the liquid velocity increases with gas superficial velocity. This behavior can be expressed by the following mathematical expression as:

$$v_{SL} = f v_{SG}^{m} \tag{25}$$

In non-Newtonian liquids, it was found that the presence of static mixers affects f and m values, dependent on consistency index values (25) (Table 4).

The relationship between the ratio of f values with and without static mixers, respectively, and the consistency index in an airlift contactor can be represented by the following relationship (25):

$$\frac{f_M}{f} = 0.85K^{0.008} \tag{26}$$

The effects of the consistency index on the liquid velocity could be considered as being less significant proving that static mixers are effective in viscous liquids. Popovic and Robinson (32) found that the circulating liquid velocity in the riser section of an airlift reactor is dependent mainly on the downcomer-to-riser cross-sectional areas ratio, A_D/A_R, the apparent viscosity, η_{ap} and gas superficial velocity, as described by the following correlation:

$$v_{SLR} = c' v_{SGR}^{0.32} (A_D/A_R)^{0.97} \eta_{ap}^{-0.39} \tag{27}$$

Considering that equation 27 correlates well a large number of experimental data, it was found that the differences in the values of v_{SLR} should be treated as particular cases of diminuation up to 40% of the geometrical parameter A_D/A_R, as a result of the presence of static mixers (25).

Mixing

As shown, homogenization in static mixers is attained by means of motionless elements using the flow energy of the fluids. The mixing effect depends on the continuous separation, distribution, and reuniting streams of material (5,7). The energy requirements of static mixers when used for mixing are lower than those of conventional dynamic mixers and, moreover, they are of simple construction. In the case of two components, the two quantities are related to each other by the volumetric flow rate ratio, $\dot{V}_2/\dot{V}_1$. The degree of mixing may be described using the variation coefficient $s/\bar{C}_V$ or the relative standard deviation s/τ_0 (5). For component 1, the following expression held:

$$s_1/\bar{C}_V = \sqrt{\dot{V}_2/\dot{V}_1}\, s_1/\tau_0 \tag{28}$$

where

$$s_1^2 = \frac{1}{n-1}\sum_{i=1}^{n}(C_{V_i} - \bar{C}_{V_1})^2 \tag{29}$$

$$\tau_0 = [\bar{C}_{V_1}(1 - \bar{C}_{V_1})]^{0.5} \tag{30}$$

$$\bar{C}_{V_1} = \frac{1}{n}\sum_{i=1}^{n} C_{V_{1,i}} \tag{31}$$

$$C_{V_{1,i}} = \dot{V}_{1,i}/(\dot{V}_{1,i} + \dot{V}_{2,i}) \tag{32}$$

The degree of mixing is strongly dependent on the static mixer type and normalized mixing length, L/D (Fig. 14). It can be seen that the Sulzer-SMX mixer and HI-mixer require shorter homogenization lengths. Also, the degree of mixing is dependent on the entry point of fluids to be mixed as well as by the operation scale (5). Static mixers achieve rapid mixing of fluids with low energy requirements, and the fluids may have either similar or different viscosity. Operation in fermentation processes involves very viscous fluids, and the flow will be predominantly in the laminar region for both liquid and gaseous phases. Suitability of static mixers to mix fluids of different viscosities was investigated in the literature (12,22,33,34).

Mixing behavior is usually described by mixing time, t_M, which is defined as the time required to achieve a definite degree of homogeneity after a trace pulse has been injected into the reactor (35). The mixing time is highly dependent on the presence of static mixers (34). In gas–liquid dispersions, the mixing time is dependent to a great extent on the gas superficial velocity and the presence of static mixers and to a low extent on the liquid phase properties. In Figure 15 the mixing time in water and starch solutions is plotted against superficial velocity. With static mixers, mixing times at a given superficial velocity are essentially

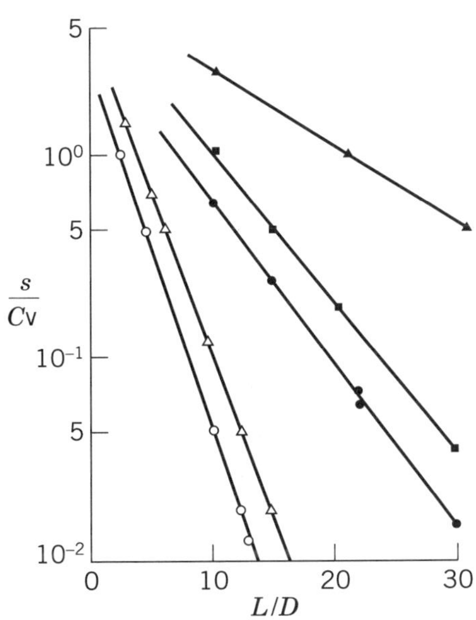

Figure 14. Variation coefficient as a function of the normalized mixing length for various static mixers. ● = Kenics; ○ = Sulzer; ■ = Komax; ▲ = Lightin; △ = HI. $\bar{C}_{V_1}$ = 0.1; $Re_D < 1$. *Source*: Reprinted from Ref. 5, with permission from VCH Verlagsgesellschaft mbH, and from Ref. 7, with permission from Wiley-VCH.

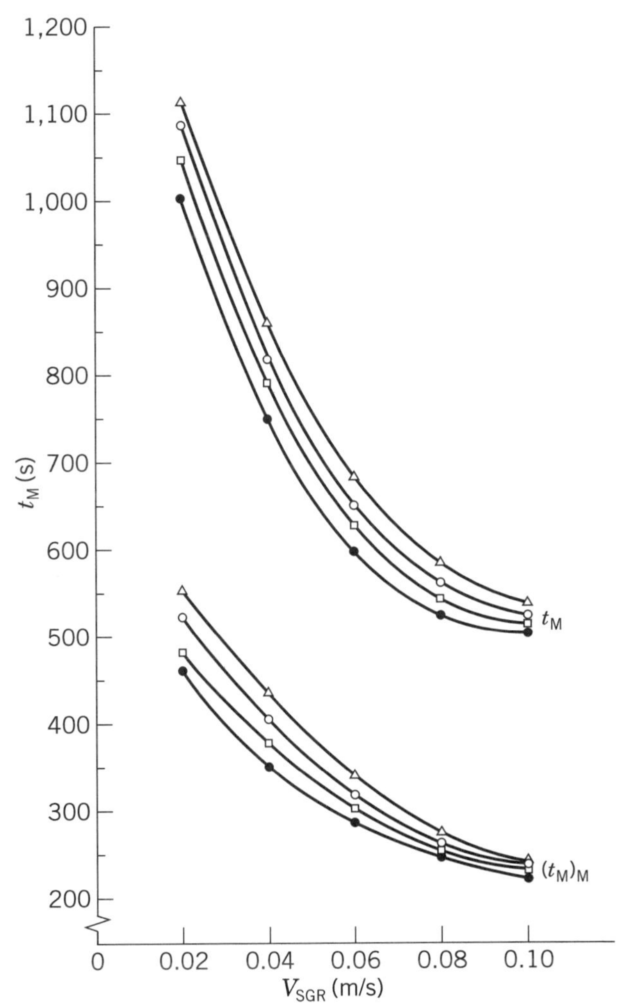

Figure 15. Mixing time versus gas superficial velocity in an external-loop airlift bioreactor without and with static mixers, operating with starch non-Newtonian solutions as liquid phase. ● = water; Starch solutions: □ = 0.5% w/v; ○ = 2.5% w/v; △ = 4% w/v. *Source*: Reprinted from Ref. 34, with permission from Wiley-VCH.

up to 50% shorter than in the reactor without static mixers. The dependence of mixing time on gas superficial velocity can be expressed by the following equation:

$$t_{\mathrm{M}} = kv_{\mathrm{SG}}^{\mathrm{q}} \quad (33)$$

The presence of static mixers does not affect q very much, but significantly influences k. Mixing times increase with increasing pseudoplasticity, but not very greatly (Fig. 15). In non-Newtonian systems, the consistency index had no significant influence on these parameters in the presence of static mixers (Fig. 16).

Other properties of a suitable static mixer are good radial mixing on as short as possible flow path and as low as possible axial mixing. Some static mixers contribute to diminish axial dispersion and influence the residence time distribution in laminar monophasic liquid flow, dependent on the Reynolds values (6). In the presence of static mixer seen in Figure **1c**, the pulse response at the tube exit indicates a smaller axial dispersion (Fig. 17) (6). Also, the dead zones are diminished, and the distribution function of the elements of fluids with ages less than the residence time indicates an evident equalization of the residence times in laminar flow (Fig. 18) (6). In turbulent flow regimes, this effect is not very important. A good radial mixing of a static mixer does not imply in all cases also low axial mixing. The individual static mixers display a different dependence of the axial dispersion on the flow rate of the phases both for single and two-phase flow system. A good static mixer from the standpoint of axial dispersion in single-phase flow does not necessary display a low axial dispersion in two-phase flow (12). Data on axial mixing for the individual types of static mixers are usually missing in the literature. Fialova et al. (12) presented the value of the Péclet number given by Sulzer for its SMX product (*Pé* > 110, d = 0.1 m; L/D = 22). These properties can be ascertained by experimental means.

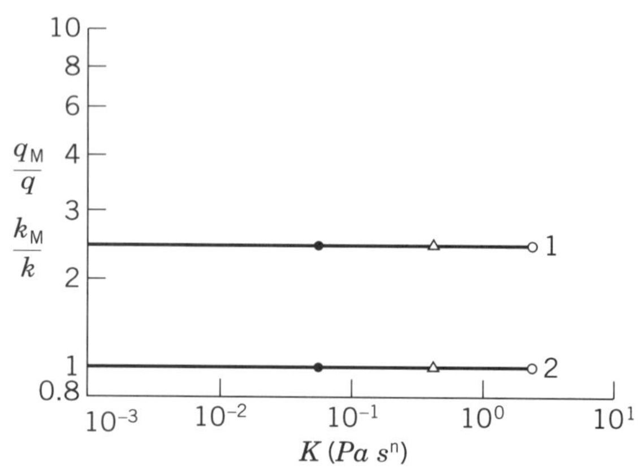

Figure 16. The influence of the consistency index K on the parameters in equation 33. (1 = k_M/k; 2 = q_M/q. Symbols as in Figure 15. *Source*: Reprinted from Ref. 34, with permission from Wiley-VCH).

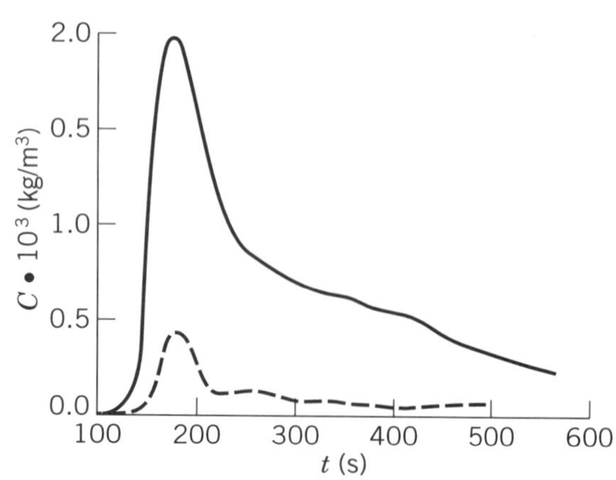

Figure 17. Pulse response at the exit of a tube equipped with the static mixer in Figure **1c**, - - - - - - = without static mixer; — = with static mixer.

MASS TRANSFER IN PRESENCE OF STATIC MIXERS

The application of the static mixers built in nonmechanically bioreactors, especially in aerobic bioprocesses, should lead to attain the most extensive transport rate of oxygen into the liquid phase at the low possible energetic requirement. The influence of static mixers on hydrodynamic behavior has an important impact on mass and heat transfer characteristics and transfer rate intensification. The improvement of the oxygen mass transfer in bioreactors depends on a number of factors, such as bubble formation, bubble size distribution, bubble shear, gas holdup, rate of coalescence, and breakup. Also, bubble or drop coalescence behavior in gas–liquid dispersions, especially in viscous fluids, demonstrates the necessity for redispersing the gaseous or liquid phase to prevent the reduction of interfacial

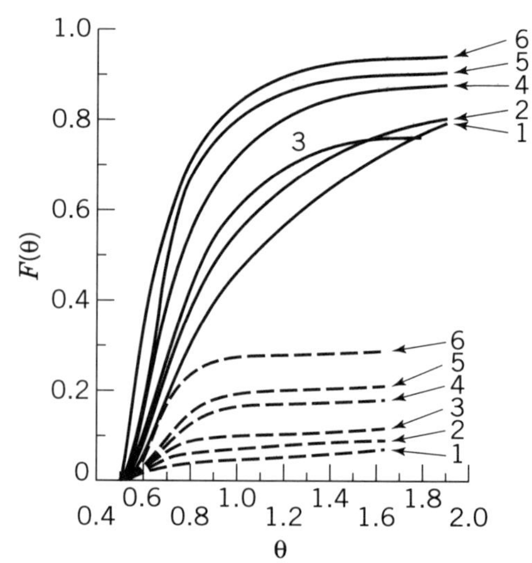

Figure 18. The distribution function of the fluid elements with the age less than the residence time versus Reynolds number in the tube without and with the static mixer from Figure **1c**. 1 = Re_D = 630; 2 = Re_D = 950; 3 = Re_D = 1,260; 4 = Re_D = 1,580; 5 = Re_D = 1,890; 6 = Re_D = 2,2000. - - - - - - = without static mixer; — = with static mixer.

areas and mass transfer rates. A general solution to the mass transfer problems that appear in gas–liquid systems can be the use of static mixers installed at different intervals for increasing the degree of the gas dispersion and mass transfer rates (7,33,36). Usually, the intensity of oxygen transfer is characterized by its volumetric mass transfer coefficient, *Kla*. Static mixers improve *Kla* at any given gas flow rate relative to the bioreactor configuration without static mixer (7). The mass transfer rates are very high during bubble formation, and the values of mass transfer coefficient vary significantly being associated with large surface renewal rates. Coalescence and break-up, bubble formation, bubble shear, and turbulent liquid surface are examples of conditions where mass transfer coefficient is large. The introduction of static mixers has two important roles in *Kla* enhancement:

1. Bubble redispersion produces increased values of interfacial areas, a.
2. The interfacial turbulence during bubble coalescence and redispersion increases *Kl*.

At low gas flow rates, the latter mechanism becomes important (37). Small bubble size can be obtained with a homogeneous bubble distribution over the column cross-section and with a narrow size distribution. However, small rigid bubbles less than 2 mm in diameter have frequently been found to have much smaller mass transfer coefficients (38). Several studies were presented concerning the influence of the Sulzer type (22,33,36,39), Koch (10,11,20,24,40), Kenics (29,30) on *Kla* in bubble columns and external-loop airlift bioreactors in waterlike fluids as well as in non-Newtonian pseudoplastic media. Static mixers significantly enhance the mass transfer rates in waterlike fluids (Fig. 19). In non-Newtonian systems, the improvement in *Kla* due to static mixers depends on the rheological parameters. Stejskal and Potucek (30) and Chisti et al. (39) mounted static mixing elements in the risers of internal- and external-loop airlift reactors that

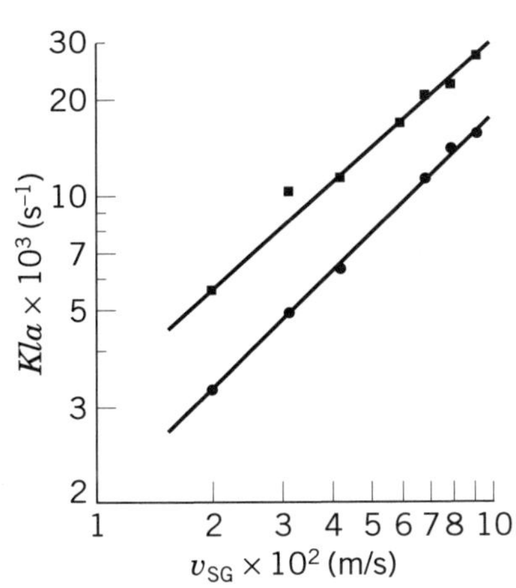

Figure 19. Effects of static mixers on the mass transfer coefficient in aqueous salt solutions. ● = without static mixers, ■ = with static mixers. *Source*: Reprinted from Ref. 7, with permission from Wiley-VCH.

enhanced mass transfer in viscous pseudoplastic CMC aqueous solutions that simulated the rheological behavior of filamentous biofluids. In pseudoplastic liquids, the thicker the fluid was, that is, the higher its value of the consistency index, K, the greater the effect of the static mixer was on *Kla* (36,39). Usually, the accumulation of biomass or product in the culture medium leads to increased viscosity and non-Newtonian flow characteristics that contribute to diminish *Kla* (26). This leads to a drastic decrease in oxygen transfer capability, and the aerobic culture eventually becomes oxygen limited (33,41). The significant decrease in *Kla* is mainly attributed to the formation of large bubbles and a decrease in specific interfacial areas that can be counterbalanced by the static mixers presence. The values of *Kla* can be correlated with the gas superficial velocity, v_{SG}, in tower bioreactors as

$$Kla = uv_{SG}^{t'} \tag{34}$$

The presence of static mixers affects u and t' values. They do not affect t' very much, but significantly enhance u dependent on the consistency index (Fig. 20) (36). The degree of *Kla* enhancement depends on the fluid viscosity and the pattern of fluid circulation through the reactor. The results of Hsu et al. (20) also show that mass transfer coefficient in a bubble column in the presence of Koch type static mixers is improved. A comparison of the oxygen transfer efficiency in terms of kilograms of oxygen transferred per kilowatt hour consumed at low flow rates indicated that the Koch mixed column was almost twofold as efficient as the empty column. However, at high gas flow rates, the efficiency was about the same for the two columns. Also, the ratio Kla/P_G can be used to characterize the efficiency of energy utilization for mass transfer per unit volume of the liquid phase or dispersion. Potucek (29) found that an airlift bioreactor with static mixers showed higher Kla/P_G values for absorption of oxygen in water or polyacrylamid solutions in the region of gas velocities of $v_{SGR} = 0.01 - 0.06$ m/s. Also, in the batch yeast cultivation, Kla/P_G was larger than in the witness bioreactor; gas superficial velocities ranged from $v_{SGR} = 0.03 - 0.06$ m/s, whereas with $v_{SGR} = 0.01 - 0.03$ m/s, lower values of Kla/P_G were obtained. As energy costs increase, more capital should be allocated to motionless mixers that increase the mass transfer efficiency. Olivier and Oosterhuis

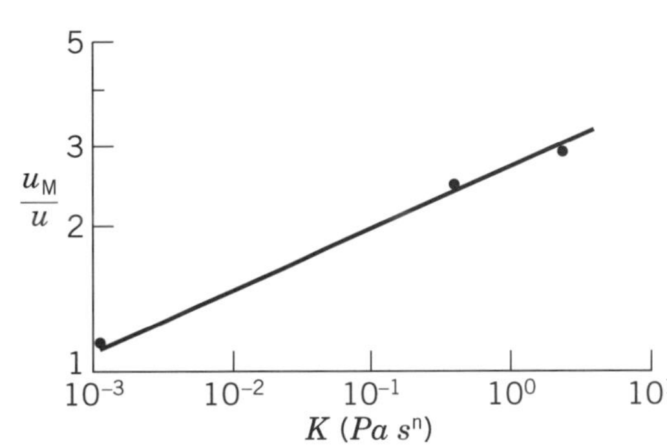

Figure 20. Variation of the ratio u_M/u with the consistency index.

(22) considered that by keeping liquid and gas velocities as well as the volume ratio of empty tube parts and internals constant, the oxygen transfer would be higher and nearly independent on the scale. They also considered that oxygen transfer takes place not only in the internals, but also in the empty tube parts, and correlated the overall oxygen transfer coefficient as (22):

$$Kla = \frac{(Kla\ V_M)_{\mathrm{internals}} + (Kla\ V_E)_{\mathrm{empty\ tube}}}{(V_M)_{\mathrm{internals}} + (V_E)_{\mathrm{empty\ tube}}} \quad (35)$$

The results obtained in different gas–liquid contactors show larger interfacial areas in the systems with static mixers (42) (Fig. 21). Data also indicate that the static mixer contactors are superior to other nonmechanically contacting devices, regarding mass transfer coefficient. Hsu et al. (20) investigated mass transfer in a bubble column with Koch mixers comparative to that with sieve trays. They found a different dependence of *Kla* on gas superficial velocity:

$$\text{for KOCH mixed column:} \quad Kla \approx v_{\mathrm{SG}}^{0.95} \quad (36)$$

$$\text{For sieve-tray bubble column:} \quad Kla \approx v_{\mathrm{SG}}^{0.89} \quad (37)$$

Also, data presented by Atkinson and Mavituna (42) show that in static mixed gas–liquid contactors a larger range of high *Kla/V* values was obtained, compared to those resulted in other gas–liquid contacting devices (Fig. 22).

The optimization of the number of motionless mixing elements and their positioning can enhance the mass transfer rates.

HEAT TRANSFER USING STATIC MIXERS

The role of the static mixers in heat transfer to flowing homogeneous or heterogeneous or heterogeneous fluids consists in fetching fresh fluid to the surface of the heat exchanger by promoting secondary flow, mixing, and turbulence (15). Most heat exchanger performances depend on hydrodynamics, and there are evident advantages that heat transfer may be enhanced by using built-in elements that change the velocity fields inside the tubes. Several studies investigated the heat transfer enhancement by using twisted strip inserts (43), helical static mixers (44,45), and wire matrix (15). The static mixers mostly used for heat transfer enhancement were helical static mixers, which promote a secondary motion in the fluid, increasing the heat transfer between the fluid and the tube wall by convection.

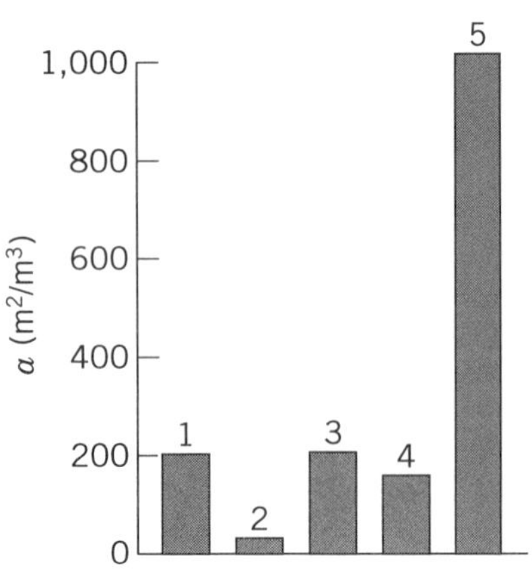

Figure 21. Comparison of specific interfacial areas in different gas–liquid contacting devices. 1 = Baffled agitated tank; 2 = bubble column; 3 = packed tower; 4 = plate tower; 5 = static mixers. *Source*: Data from Ref. 42.

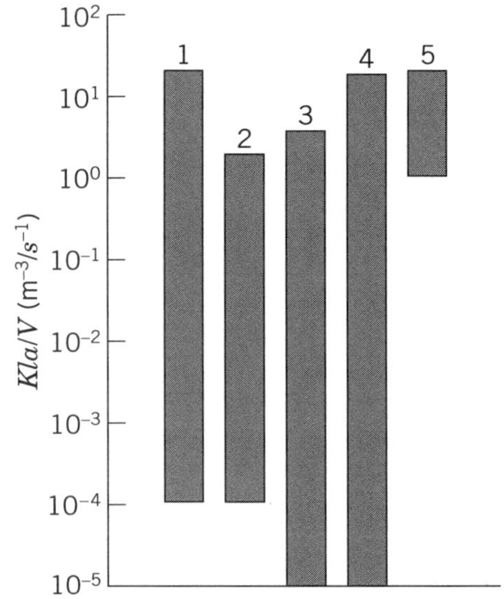

Figure 22. Comparison of different gas–liquid contacting devices regarding *Kla/V* values (significance of numbers as in Figure 21. *Source*: Data from Ref. 42.

The heat transfer performances of a given piece of equipment is usually estimated in terms of a Nusselt number, *Nu*, which is then related to the fluid properties and the Reynolds number, *Re*, characterizing the fluid environment. For instance, the individual heat transfer coefficient α_1 for the inside surface of a tube equipped with a helical static mixer is assessed from the following expression (44):

$$Nu_1 = \frac{\alpha_1 D}{\lambda_1} = \Phi(Re_1, Pr_1, h/d, \eta \mathrm{W}_1/\eta_1) \quad (38)$$

with

$$Re_1 = \frac{D v_1 \rho_1}{\eta_1} \quad (39)$$

$$Pr_1 = \frac{c_{p1}\eta_1}{\lambda_1} \quad (40)$$

The overall heat transfer coefficient, K_T, of a tubular heat exchanger is related to the individual coefficients as follows:

$$K_T = \frac{1}{\frac{1}{\alpha_1}\frac{D_e}{D} + \frac{D_e}{2\lambda_s}\ln\frac{D_e}{D} + \frac{1}{\alpha_2}} \quad (41)$$

The values of K_T in a double-tube heat exchanger with hot water flowing through the annular space and cold water–glycerol solutions circulating through the inner tube containing helical static mixers were determined using the individual heat transfer coefficients calculated with the following equations (44): For the tube ($90 \leq Re_1 \leq 5800$; $6 \leq Pr_1 \leq 5800$);

$$Nu_1 = 1.86\left(Re_1 Pr_1 \frac{d}{L_T}\right)^{0.33}\left(\frac{\eta_1}{\eta_{W_1}}\right)^{0.14} \quad (42)$$

For the annular space ($Re_2 = 1520$):

$$Nu_2 = 1.02 Re_2^{0.45} Pr_2^{0.50}\left(\frac{D_e - D}{L_T}\right)^{0.40} \cdot \left(\frac{D_e}{D}\right)^{0.80} Gr^{0.05}\left(\frac{\eta_2}{\eta_{W_2}}\right)^{0.14} \quad (43)$$

Lecjacks et al. (44) reported the following relationship for helical elements of continuous and Kenics type:

$$Nu_1 Pr_1^{-1/3}\left(\frac{\eta_{W_1}}{\eta_1}\right)^{0.14} = a_1 Re_1^{n_1} \quad (44)$$

with

$$a_1 = \Phi_a\left(Re_1, \frac{h}{D}\right)$$

and

$$n_1 = \Phi_n\left(Re_1, \frac{h}{D}\right)$$

Also, they extended their investigations in the turbulent flow regime, which allowed them to assess the favorable effect of the built-in elements on heat transfer, especially in the laminar regime, for low values of h/d in Kenics static mixer. In the continuous helical static mixer, an increase in friction factor and heat transfer coefficient of 4.4% and 5.3% was obtained, whereas in the Kenics insert this increase was of 19.3% and 10.2%, respectively, comparative with the empty tube (45). Therefore, the heat transfer coefficient was found to be dependent on the flow conditions and geometrical properties of the internals. Olivier and Aldington (15) found a noticeable heat transfer enhancement by factors up to five in round tubes containing wire matrix turbulators, comparative to the empty tube, in laminar regime. Also, the wires caused the falling of the effective viscosity in non-Newtonian solutions at high liquid flow rates. Static mixers are used for thermal homogenization when a breakdown can occur on the heat transfer surface, especially in viscous media. Shear-thinning fluids should give good heat transfer/pressure drop behavior in the presence of tube inserts because the viscosity is reduced in the higher shear rates that are obtained. Static mixers are insensitive and nonresponsive to temperature. They can be well sealed against the surroundings. Maintenance and wear are small, and they are very economical, because they do not require additional space (in-line disposed).

SCALE-UP CONSIDERATIONS

Static mixers are used for many monophasic or gas–liquid, two-phase operations, which involve momentum, mass, and heat transfer. They are largely used in biotechnology and other applications such as processing of natural gases, wastewater treatment, dissolution of gases, continuous mixing of mutually soluble fluids of different viscosities, and so on. It is widely agreed that static mixers are suitable to mix fluids of very widely different viscosities, being especially effective for use with highly viscous fluids. They also could remove any mistakes made by the equipment designed for heat and mass transfer.

As shown in the fermentation field, factors taken into consideration on the choice and the design of an equipment for a definite process, besides the general factors, should include the population safety and damage, the uniform distribution of the phases and energy in phase transfers, the oxygen demand of the aerobic cultures, and so on (4,7). Static mixers can fulfill these objectives. Their action gives rise to elementary fluid layer formation. The maximum thickness of an elementary layer is given by the following expression (16):

$$\delta = \frac{D}{n} = \frac{D}{ca^{N}} \quad (47)$$

As in the case of any apparatus, the static mixer behavior is tested first in small-scale equipment and then designed for large-scale operation, on the basis of the experimental data acquired for the model. For geometric similarity between the model and the actual mixer, the same degree of mixing means that (16):

$$\delta_m = \delta_a \quad (48)$$

or

$$\frac{D_m}{ca^{N_m}} = \frac{D_a}{ca^{N_a}} \quad (49)$$

Boss and Czastkievicz (16) introduced a linear scale-up coefficient, expressed as follows:

$$S = \frac{D_m}{D_a} \quad (50)$$

when equation 49 becomes

$$N_a = N_m + \frac{lgS}{lga} \quad (51)$$

This reasoning is valid in the laminar flow regime. For

constant physicochemical properties of the fluids that flow through the model and the actual mixer, the changes in Reynolds number for the actual mixer relative to the model can be attributed to modifications occurring in the flow rate of the fluids or in the diameter of the static mixer. The relationships between the volumetric flow rates and Reynolds numbers in the model and the actual mixers, respectively, are as follows (16):

$$\dot{V}_a = \frac{z_m}{z_a} \frac{S^3}{1 + \lg S/N_m lga} \dot{V}_m \tag{52}$$

$$Re_a = \frac{z_m}{z_a} \frac{N_m}{N_a} S^2 Re_m \tag{53}$$

These general equations, applicable to scale-up of mixing processes in static mixers, are inferred starting from the pressure drops associated with laminar flow of a Newtonian fluid through circular pipes, which are taken into account by the pressure drop coefficient z, defined in equation 9. Boss and Czastkievicz (16) gave several examples of scale-up calculations for some types of static mixers, namely Kenics, Ross-ISG, and Sulzer.

CONCLUDING REMARKS

An overview of the published results on transfer processes in fluid flow through static mixers has been presented in the preceding pages. The velocity fields inside a tube containing motionless mixers can be modified using static mixers. As a result, a significant enhancement in mixing, mass, and heat transfer are achieved. Static mixers can be used in chemical and biochemical processes to perform a rapid mixing of monophasic, diphasic, and poliphasic systems with low energy requirements, and the fluids may have similar or different viscosities. The energy for mixing is extracted from the fluid flowing through the mixers. Static mixer effects result from specially designed flow arrangements, cutting and twisting, displacement and distortion, or separation and expansion. Significant efforts are devoted by specialists for developing new static mixers and optimal strategies for using minimum number of static mixers located at different positions in tubes, to ensure greater efficiency on transfer processes intensification, especially mixing and oxygen transfer in aerobic fermentation processes.

NOMENCLATURE

A	Cross-sectional area [m^2] (1 in^2 = 6.4516.10^{-4} m^2; 1 ft^2 = 9.290304.10^{-2} m^2)
a	Specific interfacial area [m^2/m^3]
c_p	Specific heat at constant pressure [J/kg °K] (1 ft pdl °R^{-1} lb^{-1} = 0.1672255 J/kg °K)
D	ID of the tube [m] (1 in = 2.54.10^{-2} m; 1 ft = 3.084.10^{-1} m)
D_e	Outside tube diameter [m]
d	Diameter of static mixing element [m]
f_D	Friction factor
Gr	Grashof number
h	Height (length) of a static mixing element [m]
K	Consistency index in power-law rheological model [Pa s^n] (1 lb s ft^{-2} = 1.488163944.10 Pa s)
K_T	Overall heat transfer coefficient [W/m^2 °K] (1 lb s^{-3} °R^{-1} = 0.816466266 W/m^2 °K)
L	Length of the tube [m]
N	Number of static mixing elements
Nu	Nusselt number
n	Number of elementary layers produced in laminar flow over a static mixer
Pr	Prandtl number
Δp	Pressure drop [Pa] (1 lb in^{-2} = 6.89476.10^{-3} Pa)
Re	Reynolds number
Re_D	Reynolds number in a tube with inner diameter D
t_M	Mixing time [s]
V	Volume [m^3] (1 ft^3 = 2.83168 10^{-2} m^3)
$V\cdot$	Volumetric flow rate [m^3/s] (1 ft^3 s^{-1} = 2.83168.10^{-2} m^3/s)
$\mathbf{v}_S$	Superficial velocity ($\mathbf{v} = 4V\cdot/\varphi D$) [m/s] (1 ft s^{-1} = 3.048 m/s)
α	Individual coefficient of heat transfer [W/m^2 °K]
ϵ_G	Gas holdup
δ	Maximum thickness of an elementary layer [m]
λ	Thermal conductivity [W/m °K] (1 ft lb s^{-3} °R^{-1} = 0.248858918 W/m °K)
η	Dynamic viscosity [Pa s]
ρ	Density [kg/m^3] (1 lb ft^{-3} = 1.60185 10 kg/m^3)

Indices

a	Referring to an actual static mixer
D	Downcomer section of an airlift reactor
E	Referring to the empty tube
G	Gaseous phase
L	Liquid phase
M	Referring to static mixer presence
m	Referring to model static mixer
o	Referring to a single phase
R	Riser section of an airlift reactor
s	Referring to solid surface
T	Referring to heat transfer
TP	Referring to diphasic flow
W	Referring to wall temperature
1	Fluid 1
2	Fluid 2

BIBLIOGRAPHY

1. B.H. Chen, *Ind. Eng. Chem. Process Des. Dev.* **9**, 20–24; 121–126 (1970).
2. C.J. Slattery, *AlChE J.* **16**, 345–352 (1970).
3. H. Brauer and D. Sucker, *Int. Chem. Eng.* **18**, 367–380 (1978).

4. V.D. Prokopenko and U.E. Viesturs, *Acta Biotechnol.* **6**, 325–338 (1986).
5. M.H. Pahl and E. Muschelknautz, *Chem.-Ing.-Tech.* **52**, 285–291 (1980).
6. R.Z. Tudose and A. Moise, in *Proc. of the First Natl. Symposium on Inventing* Sept. 7–9, Iaşi, Romania, 1984, pp. 209–212.
7. M. Gavrilescu and R.Z. Tudose, *Acta Biotechnol.* **15**, 3–26 (1995).
8. Pat. RO 73630 (1979), R.Z. Tudose, F. Vitan, and L. Bujor (to Technical University of Iaşi).
9. Pat. RO 98240 (1989), M. Ionescu, R.V. Roman, M. Gavrilescu, A. Sauciuc, A. Pintilie, A. Pascal, S. Matache and T. Rez (to Chemical-Pharmaceutical Research Institute of Bucharest).
10. K.H. Hsu, L.E. Erickson, and L.T. Fan, *Biotechnol. Bioeng.* **19**, 247–265 (1977).
11. J.R. Gutierrez and L.E. Erickson, *Biotechnol. Bioeng.* **20**, 487–501 (1978).
12. M. Fialova, K.H. Redlich, and K. Winkler, *Collect. Czech. Chem. Commun.* **51**, 1925–1932 (1986).
13. Z. Lecjaks, I. Machak, and J. Sir, *Chem. Eng. Process.* **18**, 67–72 (1984).
14. Pat. RO 83269 (1983), R.Z. Tudose, A. Bacaoanu, and L. Bujor (to Technical University of Iaşi).
15. D.R. Olivier and R.W.J. Aldington, *Chem. Eng. Res. Des.* **66**, 555–565 (1988).
16. J. Boss and W. Czaskievicz, *Int. Chem. Eng.* **22**, 362–367 (1982).
17. E.B. Naumann, *Chem. Eng. J. (London)* **47**, 141–148 (1991).
18. R. Daraktschiev, *Int. Chem. Eng.* **30**, 222–227 (1990).
19. F. Gross-Roll, *Int. Chem. Eng.* **20**, 542–553 (1980).
20. K.H. Hsu, L.E. Robinson, and L.-T. Fan, *Biotechnol. Bioeng.* **27**, 499–514 (1975).
21. N.F. Shah and D.D. Khale, *AIChE J.* **38**, 308–310 (1992).
22. A.A.P.C. Olivier and N.M.G. Oosterhuis, in D.G. Bobichon and J. Florent eds., *8th International Biotechnology Symposium*, Société Française de Microbiologie, Paris, 1998, pp. 388–409.
23. L.T. Fan, H.H. Hsu, and K.B. Wang, *J. Chem. Eng. Data* **20**, 26–32 (1975).
24. R.Z. Tudose and F. Vitan, *Rev. Chim. (Romania)* **32**, 1003–1006 (1981).
25. M. Gavrilescu and R.Z. Tudose, *Bioprocess Eng.* **16**, 93–99 (1997).
26. M. Gavrilescu and R.V. Roman, *Acta Biotechnol.* **14**, 27–36 (1994).
27. G.T. MacLean, L.E. Robinson, K.H. Hsu, and L.T. Fan, *Biotechnol. Bioeng.* **19**, 493–505 (1977).
28. L. Rusnak and R. Vlađea, *Rev. Chim. (Romania)* **42**, 597–601 (1990).
29. F. Potucek, *Collect. Czech. Chem. Commun.* **54**, 3213–3219 (1989).
30. J. Stejskal and F. Potucek, *Biotechnol. Bioeng.* **18**, 1552–1572 (1976).
31. K.B. Wang and L.T. Fan, *Chem. Eng. Sci.* **33**, 945–952 (1978).
32. M. Popovic and C.W. Robinson, *Biotechnol. Bioeng.* **32**, 301–312 (1988).
33. M. Gavrilescu and R.V. Roman, *Acta Biotechnol.* **15**, 323–335 (1995).
34. M. Gavrilescu and R.V. Roman, *Acta Biotechnol.* **16**, 145–153 (1996).
35. O. Levenspiel, *Chemical Reaction Engineering*, John Wiley, New York, 1972.
36. M. Gavrilescu, R.V. Roman, and A. Sauciuc, *Biotechnol. Bioequip. (Bulgaria)* **6**, 60–64 (1992).
37. Y. Kawase and M. Tsujimura, *Biotechnol. Bioeng.* **44**, 1115–1121 (1994).
38. P.H. Calderbank, in N. Blakebrough ed., *Biochemical and Biological Engineering Science*, Academic Press, London, 1967, pp. 101–180.
39. Y. Chisti, M. Kasper, and M. Moo-Young, *Can. J. Chem. Eng.* **68**, 45–50 (1990).
40. C.H. Lin, B.S. Fang, C.S. Wu, H.Y. Fang, T.F. Kuo, and C.Y. Hsu, *Biotechnol. Bioeng.* **18**, 1552–1572 (1976).
41. M. Gavrilescu, R.V. Roman, and V. Efimov, *Acta Biotechnol.* **13**, 59–70 (1993).
42. B. Atkinson and F. Mavituna, *Biotechnical Engineering and Biotechnology Handbook*, Stockton Press, New York, 1991, p. 711.
43. A. Date, *Int. J. Heat Mass Transfer* **17**, 845–859 (1974).
44. Z. Lecjaks, I. Machac, and J. Sir, *Int. Chem. Eng.* **27**, 210–217 (1987).
45. D. Burfoot and P. Rice, *Chem. Eng. Res. Des.* **62**, 128–132 (1984).

See also Mass transfer.

STERILIZATION, AIR. See Filtration, air; Sterilization-in-place.

STERILIZATION-IN-PLACE

P.T. Noble
Fluor Daniel GmbH
Wiesbaden, Germany

KEY WORDS

Clean steam
Dead legs
Microbiological control
Overkill
Ozone
Steam-in-place
Steam traps
Validation
Vaporized hydrogen peroxide
Vent filters

OUTLINE

Introduction
Applications
Steam-in-Place Technology
  Validation Issues

INTRODUCTION

Sterilization-in-place (SIP) refers to the practice of sterilizing process equipment or materials in process at their place of use or installed location. Sterilization can be simply defined as a procedure for rendering a system free of living organisms. SIP is a very important technology, needed for microbiological control during production. Although the process of microbial sterilization is a well-studied area of science, SIP varies widely in its interpretation and industrial application. Steam is usually employed as the sterilizing agent, although many sterilizing methods are conceivable. The term SIP in some places is used loosely for a sanitization process whereby the goal is not necessarily achievement of a high assurance level of sterility (SAL).

APPLICATIONS

SIP supplements or replaces the use of process equipment whose purpose is to sterilize materials such as autoclaves, sterilization tunnels, and ovens. Introduction of the equipment into a sterilization chamber often is not practical. SIP is also preferred when several components can be sterilized together via SIP rather than separately treated and subsequently connected under aseptic conditions.

Clean piping systems, such as a water for injection (WFI) supply ring, will usually be sterilized in place before start-up and after service interruptions. Here, the goal is to begin or resume microbiological control but not achieve a specified sterility assurance level. Fermentors, especially those used to culture genetically altered cells, need to be sterilized before innoculation, which can be achieved to a high level of assurance by SIP. Fermentation media are sometimes first introduced into the fermentor and then sterilized with the fermentor. Aseptic manufacture of parenteral drugs and devices requires sterile process equipment, which is usually achieved by SIP. For example, modern filling equipment often undergoes an SIP of the filling lines and nozzles. It is also a current Good Manufacturing Practice (cGMP) standard that lyophilizers for parenteral products be outfitted with an SIP cycle to sterilize the drying chamber and process connections that are open to the product. Finally, as part of biocontainment practice (especially when using recombinant or hazardous organisms), processing equipment may receive an SIP after use but before opening to inactivate any residual microorganisms and prevent their release into the environment.

STEAM-IN-PLACE TECHNOLOGY

The use of saturated steam for sterilization has been relied on for many years. It is natural that SIP would evolve from this experience. One of the first publications on the use of steam for SIP was provided by Myers and Chrai (1) and later reviewed by Agalloco (2). The preference for steam as the sterilizing agent can also be attributed to its advantages in heat transfer, which are elaborated in the following.

Sterilization in general is a kinetic process and is a subject of sufficient breadth for textbooks (3). It follows first-order kinetics for individual microbial populations, whereby the rate constant is species specific as well as being dependent on the sterilizing conditions. When only water is present at the saturated steam condition, the temperature alone is sufficient for completely defining these conditions and hence the rate constant for a particular organism. The rate constants are exponentially dependent on temperature, and the dependencies are significant for temperature variations as small as 1 °C.

For scientific studies, sterilization processes are usually quantified in terms of the D and F values. The D value provides the kinetic information and is simply related to the rate constant. It is species and temperature dependent, and its value is the time in minutes for a 10-fold reduction in the population. The F value gives the time in minutes for a desired reduction in the population, as calculated by $F = (\log N_0 - \log N)\,D$, where N_0 is the starting count and N is the desired count.

The F_0 value has become a standard for defining and comparing sterilization processes. It is based on the sterilization kinetics that have been found for heat-resistant organisms, such as *Bacillus stearothermophilus*. The F_0 value can be calculated from thermal data using the following integral equation, $F_0 = \int 10^{(T-121.1)/10}\,dt$, where T is the sterilization temperature in degrees Celsius and t is the sterilization time in minutes. SIP operations typically should demonstrate an F_0 value in the range of 8 to 20 min, depending on the application.

Steaming conditions other than with pure saturated steam have been found by experience to be less effective (3). Apparently, the presence of both phases, liquid and gas, is important because superheated steam at the same temperature is not as lethal. Also, the presence of air reduces sterilization effectiveness. If air is present, and the sterilizing pressure is kept constant, then the contribution of air to the total pressure lowers the partial pressure of the steam and hence its saturation temperature. The lower temperature has a direct effect on the kinetics. If an outside heating source raises the temperature, the system departs from saturated steam conditions; that is, water is present then only as superheated steam. If the total pressure is raised, so that the steam saturation temperature is maintained, such as with an overpressure autoclave cycle, then the kinetics predicted with saturated steam are followed.

Validation Issues

Important validation issues arise with SIP. The general problem is the estimation of sterility, but the additional process complexities associated with an SIP operation often add problems of control for the process parameters. Validation is needed to provide assurance that the probability of surviving organisms is small enough to be of insignificant risk to the operation. Because the initial bioburden (number and type of microorganisms) can only be partially known from past monitoring experience, kinetic models cannot predict with accuracy the outcome of SIP. Measurements of sterility are in general destructive tests; that is, they violate the integrity of the enclosed sterile system.

Accurate temperature and pressure measurement and control is of special concern for steam-in-place processes. It is especially important to know the location of the coldest points in the system to estimate conservatively the sterilization effect. An accurate measurement of pressure is needed to verify saturated steam conditions. Validation testing include extensive thermal measurements, the use of biological indicators (e.g., spore strips) and often sampling methods via swabbing or media contact. Such testing should yield consistent, favorable, and reproducible results to give such assurance.

Reflecting the uncertainties and risks associated with a sterilization process, regulatory bodies, such as the U.S. Food and Drug Administration (FDA) (4), have specified relatively severe process parameters to meet for general use. These conditions are often termed overkill. In the cited reference, 121.5 °C is to be maintained with saturated steam for 20 min, or equivalency to these conditions must be demonstrated as by achieving the same F_0 value. Using the formula provided, this value would be $F_0 = 21.9$ min.

Process Description

A technical description and a theoretical modeling of the steam-in-place process can be found in Ref. 5 and is relied on here. A steam-in-place process follows the general operations of an autoclave, which have the following cycle steps:

1. Air purge
2. Heat-up
3. Sterilization holding period
4. Displacement and collapse of the steam with sterile air

SIP of a vessel or piping system usually accomplishes the air purge step by displacing the air with steam. The air exits the system through drain and vent openings, which are usually equipped with steam traps.

A residual presence of air is to be avoided because it introduces a poorly controllable heat transfer resistance, which is described next. Its contribution to the total gas pressure also lowers the effective steam pressure and hence saturation temperature, which affects directly the sterilization kinetics. Therefore, this first step is critical to the reproducibility and success of the operation. Unfortunately, displacement of air with steam is often not very effective. Some autoclaves utilize a series of vacuum pump-outs followed by steam repressurization to remove the air from the chamber. Vacuum pump-outs are seldom part of an SIP operation unless an efficient vacuum system is already present for process needs, as with freeze dryers.

Convective displacement of air with steam always results in some mixing of air with the steam. In branched piping systems with poor flow distribution or in large vessels, this mixing can be extensive. In theory (5), turbulent steam flow is desirable because the additional kinetic energy helps to break up any air pockets and transport the air (with steam) out. The air purge step can be successful when the piping design meets appropriate criteria (to be described later) and restrictions to flow are minimized such that maximal steam flow can be reached.

Using steam to heat an enclosure combines both heat and mass transport phenomena. In the gas phase, a comparable situation (condensation of vapor in the presence of a noncondensable gas) has been analyzed and modeled for psychrometric calculations (6). Geometric complications prevent an exact analysis of a typical SIP application. Additionally, steam-in-place must consider transport in the condensed phase as well as heat transport in the walls of the system.

The combined heat and mass transport created by steaming yields important advantages for heat transfer. When saturated steam contacts a colder surface, it condenses, and the resulting void created by the collapse of water vapor molecules into a condensed phase drives more steam to the region of the colder surface. The pumping effect of the condensation helps bring steam heat to the colder regions of the enclosure and thereby improves the heat transfer. Reliance on latent heat also reduces temperature drops, which are otherwise needed for sensible heat transfer.

During the heat-up step, the system pressure is allowed to build up to the steam supply pressure by restricting any vents or drains or via the functioning of the installed steam traps. Any trapped air left will be largely swept to the cold boundaries by the dragging effect of the moving steam. Thus, the residual air will not be uniformly distributed but will be concentrated near the colder boundaries. As a stagnant layer, through which the steam must pass to condense on the walls, residual air can bring a significant resistance to heat transfer. Enclosures with a large volume-to-surface ratio, such as tanks, are particularly prone to interference from trapped air. In Ref. 5, temperature drops of the order of 1 °C are predicted from residual amounts of air equivalent in some cases to less than 0.1% of the original volume. Such exquisite sensitivity is possible in piping dead legs (to be described later) and is greatly amplified when the equipment is not thermally insulated. Insulation reduces heat transport requirements and thereby mitigates the sweeping of air into stagnant boundary layers.

Condensate removal during steam-in-place is a serious concern. Condensate forms on all cold surfaces and will subcool if allowed to accumulate. Because of its low thermal diffusivity, liquid water can maintain a substantial temperature difference between the steam phase and the equipment walls. Condensate layers will collect at all low points. Flow in these layers to the next steam trap is usually laminar. A one-dimensional analysis (5) is useful to predict order-of-magnitude effects. In an insulated pipe, a condensate layer 1 mm in depth will have a temperature drop of the order of 0.5 °C. Without insulation, only a 0.2-mm depth would show such a temperature drop.

The boundary layer phenomena discussed here are not usually observable without extensive temperature monitoring, which may only be performed during validation as a thermal mapping activity. From this study the coldest locations are identified. Thermal data from these locations are then used to define a sterilization holding period to achieve equivalence to the overkill process, or a predefined F_0. In normal operations only one or a few sensors, which for practical reasons sometimes cannot be positioned at the coldest locations, are used to monitor temperature. Temperature nonuniformities bring an uncertainty into the estimation of the sterilization effect; their existence demonstrates the presence of heat transfer resistances, usually resulting from boundary layer phenomena, which are not under direct control. Statistical analysis of temperature data during process validation will be used to add additional time to the holding period to provide the desired degree of sterility assurance.

On completion of the sterilization process, the steam supply is shut, and the equipment can be cooled either actively or passively before use. The collapse of residual steam will ultimately create a vacuum in the enclosure, which is to be avoided because a vacuum increases the risk of recontamination of the system. Therefore, the steam is usually displaced with sterile-filtered process air while there is still overpressure present.

Equipment Design

Process requirements for steam-in-place operations have determined to a large degree the special equipment design specifications associated with bioprocesses. The sterilization conditions usually chosen are 121 °C or greater, resulting in a saturated steam pressures above 1 bar gauge. Materials of construction and construction methods must be suitable for these conditions.

Materials of Construction. 316L stainless steel is usually chosen because of its strength and chemical resistance to the high-purity steam used. Vessels need to be pressure rated for these conditions and protected with safety relief valves or rupture disks.

Specifications for surface finishes and construction methods have the purpose of reducing the potential for the buildup of soil and deposits on the equipment surfaces. Surface deposits provide an additional barrier to the transport of heat and steam. They may also introduce an uncontrolled bioburden level. A high degree of polish will be required; electropolished is preferred. Corners and joints should meet with a generous radius of curvature. Ideally, sealing methods would be seamless. For this reason, welding is preferred, but the welds must be polished and passivated. Seiberling (7) provides general specifications of equipment to receive SIP.

All low points inside the enclosures must have drains; these are needed to drain the condensate and are also required to effectively clean the equipment so as to prevent soil buildup. Thermal insulation of the equipment is needed to reduce the temperature variations in the system. If the equipment is located in a clean room, the insulation must be clad with a housekeeping skin, so that particles from the insulation do not enter the clean rooms and the surfaces can be maintained clean. Stainless steel sheathing is usually used on vessels, and plastic is easier to install on piping. The enclosed insulation must not chemically release agents that would attack the equipment, such as chlorides, which are known to attack stainless steel.

Piping Design. Piping particularly should be thermally insulated because of its large surface-to-volume ratios. All piping must be sloped to drain condensate. Because the mass transport of steam and condensate are driven by heat transport and gravity, respectively, and not by large pressure gradients, it is important in the piping design to remove any uncertain flow paths. A clear example of this concern can be made with parallel flow arrangements. If two or more identical items such as heat exchangers are steamed in parallel, it is likely that one item will not receive the proper treatment because of poor steam distribution. In an extreme case, one item may stay filled with air or become blocked with accumulated condensate, effectively stopping the flow of steam.

Dead legs are to be avoided, but this is not usually possible. Dead legs are commonly understood to be sections of piping with no outlet. Connections that are closed with a valve during SIP are typical dead legs. Instrument ports can be another. Dead legs also need to be sloped to allow for draining. Depending on their length, flow of steam in and air out can be largely a diffusive process. Dependence on diffusion means some air is always trapped in dead legs during SIP, contributing to the difficulty of their sterilization (8). General design guidelines (7,9) limit dead legs in length by a scale factor expressed in terms of pipe diameters, but this rule cannot be defended in theory (5). The best guideline is to minimize their length because they will usually be the most difficult locations to sterilize. Practically, dead legs can easily be limited to four pipe diameters with commonly available pipe fittings.

Dead leg–free valves are commercially available (Fig. 1); they function by using the pipe wall on one side as the valve seating surface. The seating design employs a flexible diaphragm or membrane, which is pushed into the flow path to press against the seating surface. Such valves belong to a type, commonly referred to as diaphragm or membrane valves, which are preferred for equipment that receives SIP or clean-in-place (CIP). Such valves have no dynamic seals for moving parts such as valve stems that may enter or leave the sterile side of the valve. Dead leg–free valves are only such on one side of the valve. On the other side, the valve body itself is a short dead leg.

Instrumentation and Monitoring. The critical parameters to monitor are temperature and pressure. By measuring both, one can confirm that saturated steam conditions are present. Instrumentation must be of a sanitary design, which follows the general considerations given earlier. Temperature is usually measured with resistance temperature detectors (RTD) because their signal can be recorded, they are robust, and they can be accurately cali-

Figure 1. Dead leg–free valve. *Source*: Reproduced with permission of SÜDMO Schleicher AG, Riesbürg, Germany.

brated. A calibration accuracy of no more than ±0.5 °C is recommended because such variations significantly affect the sterilization effectiveness.

Clean Steam Supplies. If the SIP application is for parenteral pharmaceutical products, high-purity steam, commonly known as clean steam, must be specified. Product contact surfaces in such applications must meet only strictly controlled media. Clean steam is usually defined by the purity of its condensate, and this must usually have WFI quality. Normal plant steam will contain more dissolved solids and potentially also some suspended debris. From these discussions, the steam supply should in any case be free of air, which is normally the case even with plant steam.

The steam supply system (boiler and distribution headers) normally is at higher pressure than the desired steam sterilization pressure. The steam connection to the equipment will then include a pressure-regulating valve to reduce the pressure. This design is preferred because the supply system then has reserve to supply large amounts of steam, without a serious loss of steam pressure. Too large a pressure reduction should not be expected from the pressure regulation valve, however. When too large a reduction is made just before entering the equipment, the steam acquires significant superheat. Superheated steam is known to be less effective as a sterilizing agent and as a heat transfer medium. Its presence affects the sterilization kinetics but is not easily observable. An industrially based guideline is to design the pressure reduction for 2:1 (10).

Sizing the steam supply for a steam-in-place is not straightforward, and there is no consensus in the industry. The greatest demand for steam occurs during the air purge and subsequent heat-up phases. The demand during the air purge is really determined by the restrictions to flow, often created largely by the vents and drains, as is discussed in the next section. During the heat-up step, the capacity of the steam supply could limit the heat-up rate, but the actual rate is not normally considered a critical variable.

Steam Traps and Vents. From the previous process description, it is clear that these components are critical to a successful steam-in-place operation. Steam traps are devices that are designed to not pass steam but rather air and condensate. There are many types available (11), and some models have been created especially for the biotech industry (Fig. 2). These devices operate as valves that are self-actuating, can sense by physical property differences whether they are passing steam, air, or water, and thus open or close appropriately. For steam-in-place applications, it is critical that they remove air and condensate. However, the steam trap designs that have evolved from other industries have the major goal of conserving steam. The types commonly selected for biotechnological applications allow some accumulation of condensate to occur upstream and may bring a significant restriction to the flow of air out of the system.

As a consequence, pharmaceutical companies have in the past dispensed with the use of steam traps and employed their own simpler solutions. The simplest solution

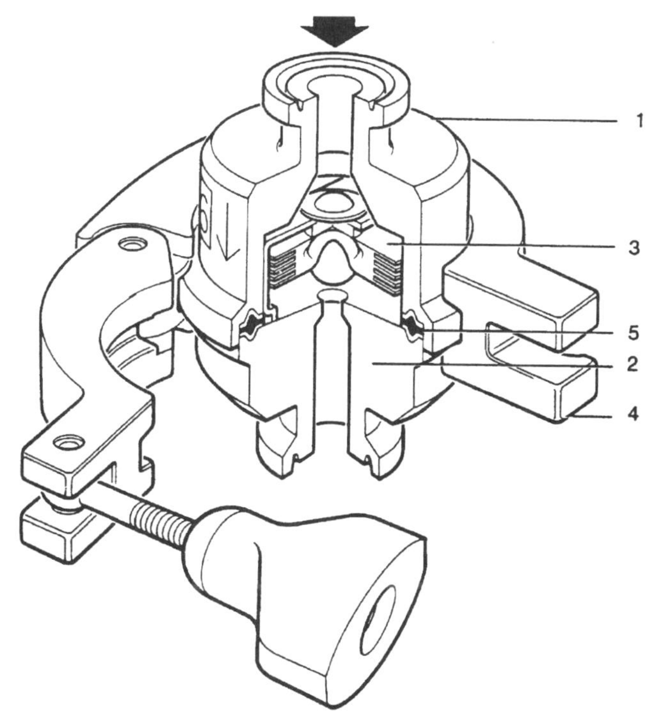

Figure 2. Sanitary blanced pressure steam trap. *Source*: Reproduced with permission of Spirax Sarco GmbH, Konstanz, Germany.

is to install manual valves instead of the steam traps, which must then be manually adjusted during the course of the procedure. During the air purge step, the valves may be fully opened. They are subsequently throttled, but only so far that steam continues to emerge, that is, condensate does not back up. A more automated approach is to employ restriction orifices instead (Fig. 3). During the air purge step, the orifice is bypassed with an open isolation valve to allow a relatively unrestricted flow of air and condensate. During the subsequent steps in the process, the isolation valve is closed, forcing remaining flow through the orifice. The orifice must be custom sized to ensure is large enough to allow all condensate to freely drain. Correctly sized, the orifice will be continuously passing (and wasting) steam and will prevent any buildup of condensate.

It is usual now to encounter sanitary steam traps of the type illustrated in Figure 2. Provided one realizes their limitations, successful steam-in-place operations can be performed with them. Typically, the steam traps are placed a significant piping distance from the limits of the system to be sterilized. The limit is usually an isolation valve, which will close immediately on completion of the operation. The piping downstream of this valve should then be noninsulated to allow for a sufficient condensation of steam and subcooling of condensate to keep the steam trap open. Steam traps must also be selected for the ability to pass air, and a sufficient number must be planned to allow for a successful air purging operation and condensate removal.

Filters. Sterilizable enclosures are almost always equipped with sterile filters, acting as barriers for sterile exchange of air or other media with the outside. Sterile vent filters are needed for pressure equalization, and process liquids are introduced though sterilizing liquid filters. Sterile filter installations are often sterilized together with the enclosure, although there is also no consensus on this practice. The filter element and enclosure may also be sterilized separately, but this raises concerns about ensuring that the separately sterilized components can subsequently be connected together aseptically.

The steaming-in-place of filters is not straightforward (12). Of special concern is the passage of steam, air, and condensate through the filter element. Also, the pressure drop across the filter element is limited by design. These concerns result in special piping designs and controls to ensure that both sides of the filter receive sterilizing conditions. It is not practical to attempt only a sterilization of the future sterile side of the filter, because without flow through of steam the filter will not be air purged successfully.

When steaming-in-place an enclosure, the vent filter will not be employed as the major vent for the air purge operation nor will it be used as the connection to the steam supply because it is typically too limited in flow capacity. It usually receives a secondary source of steam from the vessel during the sterilization process. Vent filter elements are typically nonwetting, which allows them to be quickly functional after steaming. A sterilizing liquid filter is sometimes steamed through to the enclosure (i.e., becomes the source of steam for the enclosure). Because of its limited ability to pass steam, the enclosure downstream is limited in size, often only piping. Liquid filters do not easily pass air after steaming because their hydrophilic pores become blocked with condensate.

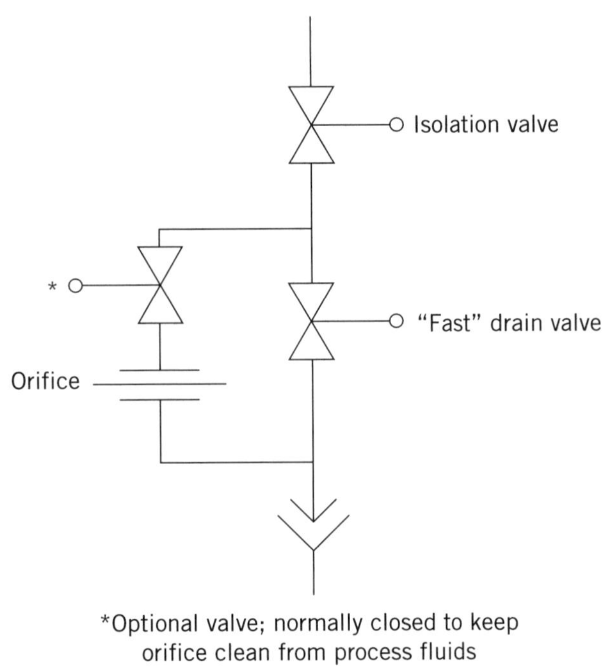

Figure 3. Use of orifices in place of steam traps.

ALTERNATIVE SIP TECHNOLOGIES

Equipment design requirements for steaming-in-place constrain opportunities for applying SIP, which raises interest in alternative technologies that are less demanding on mechanical designs. By employing chemical agents, sterilizing conditions can be reached without elevated temperatures or pressures. Sterilizing methods using radiation are not usually considered for SIP applications because the distribution of radiation throughout an extended enclosure is not practical.

When a new alternative SIP technology is chosen, a large amount of development work follows to make it a reliable industrial process. Much of this work involves scientific studies of the sterilization effect on the diverse microflora potentially present. A large amount of work is needed to develop monitoring and testing procedures for the validation and subsequent monitoring studies. Chalumeau has outlined the concerns raised with alternative sterilization methods in general (13). Some representative SIP methods receiving industrial attention are now used as examples for the technical issues encountered with alternative methods.

SIP with Superheated Water

The application of superheated water is an obvious extension to the use of saturated steam. This method can rely largely on the scientific knowledge base of the thermal sterilization effect with saturated steam. Häggström has reported its effectiveness (14), but it has not been widely practiced. The advantage of using superheated water over steam is that it can be pumped and reheated to distribute the thermal conditions needed. Häggström claims that no steam traps or vents are needed and that a more uniform temperature distribution is attained.

A theoretical analysis (15) points to a fundamental weakness of the technology when compared with steaming-in-place. Heat transport by conduction through water is less effective than the combined heat and mass transport occurring in steam. The analysis considered heat transport in dead legs of hot WFI systems, where significant temperature drops of the order of 10 °C were predicted for typical dead legs. Although acceptable for WFI operation, such temperature variations are very unattractive for SIP. The advantages of superheated water have no benefit for the dead legs in the system, which are typically the most difficult to SIP. Also, to properly heat an enclosure with superheated water, it must be completely filled, or the headspace must be purged of air as in a classical steam-in-place process; this means that steam traps and vents are still required, although fewer would be needed. When the technical features are considered that would be needed to prepare and then discard large amounts of superheated water, it is not surprising that this method has not gained popularity.

SIP with Chemical Agents in a Liquid Phase

Chemical agents used for desinfection or sanitization are sometimes considered for SIP. Sanitizing solutions are not generally useful as sterilizing agents for SIP. To be useful they would have to have biocidal activity against the full spectrum of microorganisms yet be relatively noncorrosive and safe to handle. Commercially available agents that approach this ideal are often termed cold sterilants but still require many hours of exposure to achieve sterilizations.

In general, one must design a system to uniformly distribute the agent, monitor the operation for the proper exposure, and finally remove the agent, so that the equipment can be put back in operation. Monitoring of the operation must consider all parameters that can significantly affect the chemical reactions that determine the sterilization kinetics. Temperature may be important, but the concentration of the agent certainly is. Other material components or chemicals present can potentiate or interfere with the desired reactions. Unless a sensor is available, routine sampling and subsequent chemical analysis are required to verify sterilizing conditions.

High-purity water systems, such as those claiming to be sterile, have been reported using ozone as a circulating agent (16). Ozone has many advantages as a sterilant. It is extremely effective even at low concentrations because of its high reactivity, and it is easily generated electrically from ambient oxygen. In water systems, its decomposition product is also oxygen, which is usually not considered a contaminant. Before use of the media, residual ozone is catalytically decomposed with the aid of ultraviolet radiation. Widespread acceptance of this agent, even for simple microbiological control, has not been reached because of some underlying problems. First, it is an extremely aggressive agent that will degrade the piping system, especially gaskets. Second, it is difficult to remove it to levels sometimes desirable for end use. (See analyses in Ref. 15. Finally, penetration of chemical agents such as ozone into dead legs can be a difficult problem. Where diffusion dominates the transport, the relatively low liquid-phase diffusivities coupled with the low bulk concentrations result in very slow penetration of agent into dead legs (15).

At another extreme, caustic (sodium hydroxide) at relatively high concentrations can be considered as a sterilizing agent. Caustic is often used as a cleaning agent, such as in a CIP operation. If the concentration is high enough, it can be used as a sterilizing agent. Validation of the sterilization effect can be difficult because of the lack of scientific knowledge regarding such use of this agent with microorganisms. Also, the agent must be washed from the installation with large amounts of sterile water. However, it can be appropriate if the goal is to inactivate a biological agent left in equipment or in a waste stream before opening to the environment. Here, the removal of caustic after the operation is not critical; it will be subsequently removed in a washing step.

SIP with Chemical Agents in a Gaseous Phase

The use of gaseous chemical agents is probably the most active area of interest for alternative SIP technologies. Unlike liquid-phase agents, gaseous agents are easier to distribute and remove. There is also a long history of use, and consequently a significant knowledge base, with some of the biocidal gases. Formaldehyde and ethylene oxide have long been used for room and equipment disinfections in hospitals. As with all chemical agents, their use requires measurements of all important reaction conditions for the sterilization. One must also determine their reactivity with any system components and the ultimate fate of their decomposition products.

Both formaldehyde and ethylene oxide have become restricted in their use because of their toxicity and their activity as carcinogens. Ethylene oxide also has physical hazards; not only is it flammable, its chemical instability can lead to explosive decompositions. Recently, vaporized hydrogen peroxide as a sterilizing agent has been developed (17) to the point of a commercially available generator for SIP applications (Fig. 4). Its commercial success has been spurred by the active interest in the use of isolator technology for aseptic processing. Haas (18) has reviewed isolator design issues and listed gaseous agents that have been attempted for their SIP. At present, there does not appear to be a more promising chemical sterilant than hydrogen peroxide, which is likely to become the method of choice for pharmaceutical applications.

The AMSCO VHP™ systems have a basic three-step cycle: dehumidification, sterilization, and sterilant removal. Their commercial systems perform these functions on an attached isolator or enclosure. Often the isolator has operating components that assist the process. The reader is directed to the vendor for specific details on the operation of the commercial AMSCO VHP™ systems; the basic physical and chemical phenomena are considered here.

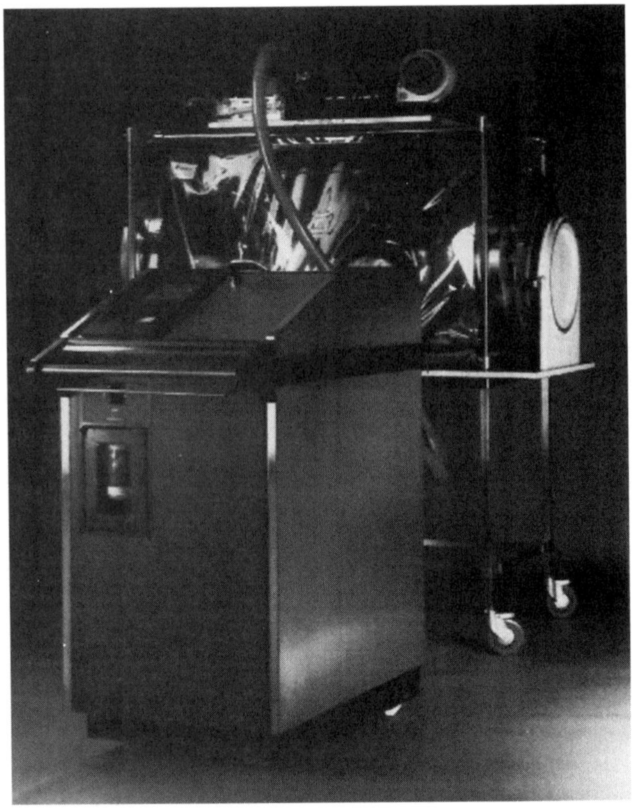

Figure 4. Vaporized hydrogen peroxide generator. *Source*: Reproduced with permission of AMSCO Scientific, Apex, North Carolina.

Hydrogen peroxide solutions have long been recognized for their disinfecting value. The greater lethality of vaporized hydrogen peroxide is a more recent discovery (17). H_2O_2 is an unstable material, available commercially in aqueous solutions up to 70%. High concentrations of H_2O_2 require proper handling because it can explosively decompose or cause fires (19). Its decomposition products are water and oxygen, and it will react with any oxidizable substance, particularly organic materials. Aqueous solutions of H_2O_2 are corrosive, but they have been found to be relatively benign to materials when only vapors are present (17).

A relatively safe solution concentration of 30% is usually the raw material consumed by a H_2O_2 generator. H_2O_2 has a lower vapor pressure than water, and considerable water vapor will also be generated when a 30% solution is vaporized. The water vapor released with the H_2O_2 limits the gas concentration of H_2O_2 achievable because water can easily reach saturation conditions in air. When water condenses, H_2O_2 is also lost from the gas phase because it will dissolve readily in the condensate. SIP with vaporized hydrogen peroxide must therefore control not only the concentration of H_2O_2 but also water in the air when the solution is vaporized. The first step in such an SIP is usually to dehumidify the enclosure so that reproducible initial conditions are established. The temperature of the isolator and the air within is also important because this has a strong effect on the relative humidity. The kinetics of the sterilization reaction do not appear to be strongly a function of temperature.

During the sterilization step, the H_2O_2 reagent is vaporized, and the vapors are introduced into the chamber. The chamber must be leak-tight to prevent emissions and exposures to personnel. Forced convection is needed to distribute the vapors throughout the enclosure and ensure a uniform exposure to, and concentration of, H_2O_2. To maintain the gas concentration, additional H_2O_2 must usually be generated for the duration of the sterilization to replace H_2O_2 that has decomposed. The sterilization step is timed to achieve the desired level of assurance of sterility. This activity on a closed system results in a steady rise in both water vapor content and air temperature. The commercial generators are successful in part because they continuously remove water vapor from the returning gas stream, as they are generating more H_2O_2 in the supply stream. They are thus able to achieve and maintain the necessary H_2O_2 concentrations needed for a rapid sterilization (0.5–2 mg/L) while avoiding reaching water saturation conditions.

As the air temperature rises during the sterilization, a temperature difference develops between the air and the enclosure and its fixtures. Saturation conditions are likely to occur at these surfaces, and process development efforts must focus on avoiding this condition by parameter adjustment. SIP with gas requires good gas circulation to achieve a uniform exposure. Modern isolator designs (18) usually have the necessary vigorous air circulation needed because they employ it to also remove airborne particles via HEPA filtration. Fixtures inside an isolator may have restricted spaces where air circulation is ineffective. Transport of biocidal gas into these spaces then is dependent on diffusion, and dead leg effects for transport can be expected.

Graham et al. (17) reported sterilization conditions could be reached up to 60 pipe diameters into a dead leg with a pulsed cycle, which included a vacuum pulldown between cycles. The vacuum pulldown ensures that transport into the dead leg is not entirely dependent on diffusion because breaking the vacuum with sterilant creates a pressure difference for convection up the tube. Most isolators are not designed for vacuum conditions, and reliance on diffusion reduces this impressive penetration result.

The analysis of diffusional transport of H_2O_2 into dead legs can be modeled analogously to the transport of ozone into dead legs (15). In the gas phase, diffusional transport is much more rapid, but the decomposition kinetics of H_2O_2 are also larger than for ozone. The diffusivity of H_2O_2 in air is reported to be 0.188 cm^2/s at 60 °C. For such gas systems, the temperature dependence of the diffusivity should be of the order of $T^{1.5}$. This corresponds to a value of 0.167 cm^2/s at 35 °C. In comparison to ozone in water, such gas-phase diffusivities are four orders of magnitude larger. The defining parameter for penetration is the diffusivity divided by the rate constant for decomposition (15). Because H_2O_2 decomposition is observable in terms of minutes and ozone decomposition in hours, one would expect that the penetration of H_2O_2 into dead legs would be dramatically better than predicted for ozone in water (15). An industrial guideline to use for limiting dead leg lengths with H_2O_2 has not yet evolved but would be strongly affected by the specific conditions created by the enclosure, the sterilization cycle, and the material of the dead leg, and hence not generally useful.

Finally, it must be realized that the SIP of an enclosure or isolator has the main goal of sterilizing the exposed surfaces. The air in the enclosure is usually recirculated through a sterilizing HEPA filter to keep microbes out of the air, and it therefore does not have the major bioburden. Surfaces must be clean. Microorganisms encased in dirt or a biofilm will not be exposed to the gas-phase H_2O_2 concentration because of transport resistance through the film and decomposition reactions occurring with the film. Reactive surface materials will in general reduce the local concentration of agent at the surfaces.

After the sterilization step, a H_2O_2 removal step is needed before normal operations can resume. The AMSCO VHP™ systems employ a catalytic reactor in the recirculating air stream to the enclosure to accelerate the decomposition and removal of H_2O_2. Plastic components are generally porous to H_2O_2, and thus a significant length of the cycle time often is devoted to removing the outgassing amounts of H_2O_2. This phase is again a diffusional process; but now in a solid phase and hence relatively slow.

The permeability of plastics to biocidal gases, notably ethylene oxide, has been relied on to sterilize materials sealed in plastic bags via exposure from the outside. This method has also been tested successfully with H_2O_2, but a vacuum pulsed cycle is needed (17); this step is not normally considered an SIP operation. A sterilization chamber, similar in concept to an autoclave, would be needed to alternatively evacuate the air and introduce H_2O_2.

BIBLIOGRAPHY

1. T. Myers and S. Chrai, *J. Parenter. Sci. Technol.* **35**, 8–12 (1981).
2. J. Agalloco, *J. Parenter. Sci. Technol.* **44**, 253–256 (1990).
3. K.H. Wallhäuser, *Praxis der Sterilisation Desinfektion-Konservierung*, 4th ed., G. Thieme Verlag, Stuttgart, Germany, 1988.
4. FDA, *Code of Federal Regulations Title 21*, Ch. 1, part 606.60 (Nov. 18, 1975).
5. P.T. Noble, *Biotechnol. Prog.* **8**, 275–284 (1992).
6. R.B. Bird, W.E. Stewart, and E.N. Lightfoot, *Transport Phenomena*, Wiley, New York, 1960.
7. D. Seiberling, in W.P. Olson and M.J. Groves eds., *Aseptic Pharmaceutical Manufacturing*, Interpharm Press, Prairie View, Ill., 1987, pp. 247–314.
8. J.H. Young and B.L. Ferko, *J. Parenter. Sci. Technol.* **46**, 117–123 (1992).
9. C.E. Meyrick, *Pharm. Eng.* **9**, 20–27 (1989).
10. B.D. Reeks, *British Parenteral Society Tutorial Booklet No. 2: The Validation of Steam Sterilisers*, P.S. Swindon, Wiltshire, 1990, p. 11.
11. D. Coleman and P. Smith, *Pharm. Eng.* **12**, 8–14 (1992).
12. T. Myers and S. Chrai, *J. Parenter. Sci. Technol.* **36**, 108–112 (1982).
13. H. Chalumeau, *J. Parenter. Sci. Technol.* **47**, 9–15 (1993).
14. M. Häggström, *Biotechnol. Forum Eur.* **3**, 164–167 (1992).
15. P.T. Noble, *J. Parenter. Sci. Technol.* **48**, 108–112 (1994).
16. C. Nebel and W.W. Nezgod, *Solid State Technol.* **10**, 185–193 (1984).
17. G. Graham, J. Rickloff, and J. Dalmasso, in PDA, Inc., *Proceedings of the International Congress*, Basel, Switzerland 1992, pp. 32–51.
18. P. Haas, *Pharm. Technol.* **2,** 26–38 (1995).
19. J. MacKenzie, *Chem. Eng.* **6**, 85–90 (1990).

See also BIOREACTORS, AIR-LIFT REACTORS; FILTRATION, AIR; PROCESS VALIDATION; SCALE-UP, STIRRED TANK REACTORS.

SUSPENSION CULTURE, ANIMAL CELLS

JOHN R. BIRCH
Lonza Biologics PLC
Berkshire, United Kingdom

KEY WORDS

BHK cells
Bioreactor
Cell damage
Chinese hamster ovary cells
Continuous (perfusion) suspension culture
Mammalian cell
Monoclonal antibodies

OUTLINE

INTRODUCTION

Mammalian cells can be distinguished by their requirement to grow attached to a surface (anchorage dependence) or in free suspension. The ability to grow in suspension is frequently associated with cell lines that demonstrate an "immortal" or infinite lifespan phenotype. Suspension culture systems are preferred for most large-scale manufacturing processes because scale-up is more straightforward. Relatively homogeneous conditions can be achieved in a suspension bioreactor, allowing efficient monitoring and control of key process parameters.

Suspension culture technology for animal cell culture started in the 1950s with the demonstration that several types of cell could be grown in simple agitated systems such as tumbling tubes and shaken flasks (1,2). By the end of the decade methods had been developed for growing cells in magnetically stirred spinner vessels (3,4) and in fermenter vessels similar to those used for microorganisms (5,6). By the 1960s pilot plant reactors at scales of hundreds of liters were in operation (7).

The initial drive to develop an industrial process based on mammalian cell suspension culture came from the need to produce very large volumes of vaccines against foot-and-mouth disease (FMD). Processes were developed using baby hamster kidney (BHK) cells growing in stirred-tank reactors up to 3,000 liters (8–10). Subsequently stirred-tank reactors of up to 8,000-L were used for the production of interferon α from human Namalwa cells (9). The industrial application of animal cell culture has increased dramatically over the last 20 years, driven by the need to produce monoclonal antibodies and recombinant proteins in addition to vaccines. It is the demonstration that these products can be made safely in immortal cell lines that has made possible their large-scale manufacture. Very large scale suspension processes (up to 10,000-L) have been described for the production of proteins such as tissue plasminogen activator (tPA) (11).

CELL TYPES USED FOR LARGE-SCALE PRODUCTION IN SUSPENSION CULTURE

A very wide range of cell types can be grown in suspension culture. The most important cells for industrial processes are Chinese hampster ovary (CHO), BHK, hybridomas, and mouse myelomas; these are discussed in more detail later. Other cell types grown on a large scale include insect cells (for example, see Ref. 12), especially for the production of research materials. Human cell lines such as the human kidney 293 cell line are becoming increasingly important for the production of virus vectors for gene therapy (13).

CHO Cells

The cell type most commonly used for recombinant protein production is the CHO line. CHO cells have been used to produce a wide range of therapeutic proteins (hormones, growth factors, thrombolytics, blood clotting factors, immunoglobulins). The choice of the CHO cell is based on several factors: compatibility with efficient gene expression systems leading to good productivity, ability to carry out important posttranslational modifications of proteins, and freedom from detectable pathogenic agents. In addition the cell type can be grown in large-scale suspension bioreactors. CHO cells can also grow as attached cultures, and in fact, growth in suspension generally requires a period of adaptation after the production cell line has been created. This requirement for adaptation, which can take several weeks, or months, can be circumvented by using host cells for gene transfection that have been preadapted to grow in suspension (14). Kurano et al. (15) isolated several anchorage-independent sublines of CHO, one of which grew in suspension even in static flasks. A different approach was taken by Renner et al. (16) who demonstrated that expression of recombinant cyclin E (a cell cycle regulator) in CHO cells prevented surface attachment and additionally permitted growth in protein-free medium.

BHK Cells

BHK cells have been used for large-scale production of foot-and-mouth disease vaccine because of their ability to propagate the virus and their capacity to grow in large-scale suspension culture (10). Rabies vaccine for veterinary use is also manufactured in BHK cells (17). BHK cells are also used (albeit less frequently than CHO) for production of recombinant proteins such as factor VIII (18).

Hybridomas and Myeloma Cells

Rodent monoclonal antibodies are typically produced in hybridoma cells, which can readily be grown in suspension culture (see, for example, Ref. 19). Increasing therapeutic use is being made of genetically engineered antibodies that have been "humanized" to reduce their immunogenicity in humans. Engineered antibodies can be made in CHO cells (20), but rodent myeloma cells (e.g., the mouse NSO line) are also frequently used (21), and like hybridomas, these can be grown in large-scale reactors (22).

Cells such as CHO and BHK, which can grow as attached cultures, have a tendency to form aggregates when grown in suspension (23). Aggregation can be reduced by rigorous adaptation to suspension culture and by design of culture media. Boraston et al. (24) showed that aggregation of CHO cells was reduced by adjusting the inorganic salt and amino acid composition of the culture medium. When developing cell lines for suspension culture, careful screening in an appropriate system (e.g., shake-flask culture) is essential. Brand et al. (25) compared the specific production rate in attached and in suspension culture of six closely related CHO cell lines, all making the same monoclonal antibody. Production rates were higher, lower, or the same depending on the particular cell line. The same authors found similar results with NSO myeloma cells; measurement of specific production rate in static culture for clones from the same transfection was of limited value in predicting results in suspension culture.

SUSPENSION CULTURE REACTORS

Stirred Reactors

The most commonly used systems for suspension culture are based on stirred reactors used in both batch and perfused configurations (Table 1). Typically reactors are stainless steel with height-to-diameter ratios in the range 1:1 to 3:1. In some cases mammalian cell bioreactors have been developed by retrofitting microbial vessels, usually by changing the agitators and aeration system. Backer et al. (19) described the retrofitting of 150- and 1,300-L reactors. Rushton impellers were replaced with marine propellers, and the agitator drives were altered to allow operation between 25 and 250 rpm. Air was sparged through a sintered stainless steel sparger with 10-μm pore size. Garnier et al. (12) describe the retrofitting of a 150-L microbial bioreactor (height-to-diameter ratio 3). They used two large pitched-blade impellers (45°), three surface baffles, and a polypropylene porous sparger (80-μm pore size). Agitation was controlled at 60–120 rpm. Nienow et al. (29) found that Intermig impellers improved mixing times in a mammalian reactor compared with the Rushton turbines commonly used in microbial reactors. Impellers may be directly driven via a shaft or indirectly driven through a magnetically coupled drive (30,31). Magnetic drives remove the potential for microbial ingress through mechanical shaft seals. The operation of mammalian reactors differs from that of microbial reactors in one very important respect: power inputs are very low to minimize the risk of damage to cells. For an 8,000-L reactor it has been reported (29) that a maximum agitator speed of 1.5 rev/s is used, giving a maximum power input of 30 W/m^3. This is around two orders of magnitude less than the value typically achieved in microbial fermentations. Similarly very low aeration rates are used in animal cell reactors compared with microbial cultures. This combination of low power input and low aeration rate means that particular attention has to be paid to the design of vessels to ensure good bulk mixing and adequate mass transfer of gases (especially oxygen and carbon dioxide). Nienow et al. (29) describe a study of mixing characteristics and oxygen uptake rates in an 8,000-L reactor and discuss ways of improving reactor performance.

Airlift Reactors

The airlift reactor is used less commonly than the stirred reactor for cell culture and was originally developed for microbial fermentation. Sparging of air or other gas mixtures is used to aerate the culture and to provide mixing. Mixing is achieved by separating the vessel into riser and downcomer sections using a draught tube, divider plate, or external recirculation loop. The presence of gas bubbles in the riser section creates a density difference between riser and downcomer that causes the fluid to circulate, providing mixing. Airlift reactors tend to have larger height-to-diameter ratios than stirred reactors (typically as high as 10:1). Their advantage is simplicity (no mechanical agitators) combined with good mass transfer and low shear characteristics (32,33). The basic design features of airlift reactors have been described in reviews (34,35). Airlift reactors have been used for the culture of BHK and human lymphoblastoid cells (36), insect cells (37), and at scales up to 2,000 L for a variety of cell types including CHO, hybridomas, and myelomas (38,39) (Fig. 1). Growth and productivity profiles for a GS-NSO cell line making a recombinant antibody in an airlift reactor are shown in Figure 2 (from Ref. 40).

Operational Considerations

While the type of vessel used for suspension culture is usually a stirred reactor (occasionally airlift), the mode of reactor operations can vary significantly. Production methods have been developed based on batch, fed-batch, and continuous (perfusion) systems. The choice of process will be dictated by several factors. Batch and fed-batch culture have the advantage of simplicity, reliability, flexibility, and reduced timescales for development and validation (compared with continuous systems). On the other hand continuous systems may have significant economic advantages, particularly for high-volume products, and thorough cost modeling is necessary to determine the most appropriate route for a given product. Griffiths (41) has reviewed the benefits of continuous processes.

Batch and Fed-Batch Culture

Batch culture is very commonly used in industry. Cells and medium are added to the reactor, and the culture is harvested after an appropriate period of culture. The time of harvest will depend on the kinetics of product accumulation and, in some cases, on the stability of product within the reactor. A commonly used variation on this type of culture is repeated batch or "draw and fill" culture. This is a batch culture in which a proportion of the culture is retained in the reactor as an inoculum, and fresh medium is added. This allows repeated batch growth cycles within the reactor, obviating the need to repeatedly clean, sterilize,

Table 1. Examples of Industrial-Scale Mammalian Cell Suspension Culture Processes

Product	Cell Line	Reactor	Reference
Foot-and-mouth disease vaccine	BHK 21	3,000 L	Pullen et al. (9)
Interferon	Namalwa	8,000 L	Pullen et al. (9)
Tissue plasminogen activator (tPA)	CHO	10,000 L	Lubiniecki et al. (11)
Factor VIII	BHK 21	Perfused 500-L reactor with cell recycle system	Bödeker et al. (18)
Factor VIII	CHO	2,500-L batch refeed	Adamson (26)
Monoclonal antibodies	Hybridoma/myelomas	Perfused 500-L spin filter reactor	Deo et al. (27)
Monoclonal antibody	NSO myeloma	2,000-L fed-batch reactor	Ray et al. (22)
Monoclonal antibody	Hybridoma	1,300-L batch, draw and fill	Backer et al. (19)
Monoclonal antibody	Myeloma	2,000-L airlift reactor	Brown et al. (28)

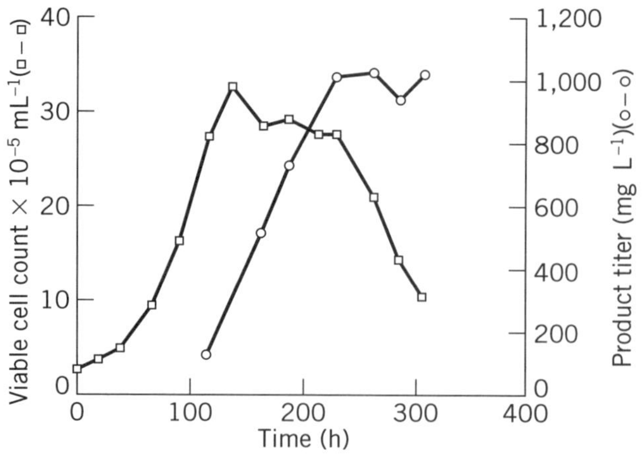

Figure 1. Growth and productivity of a GS-NSO cell line making a recombinant antibody in airlift culture. *Source*: From Ref. 40.

Figure 2. A 2,000-liter airlift bioreactor.

and reinoculate the vessel. Backer et al. (19) described such a system, which was operated for several months at a time, for monoclonal antibody production in a 1,300-L reactor. Every 84 h, the reactor was harvested leaving, 10% of the culture in the reactor as inoculum for the next culture.

A further commonly used variant on batch culture is fed-batch culture in which small volumes of selected nutrients are fed to the culture during the growth cycle to improve productivity. Bibila and Robinson (42) and Bibila et al. (43) describe the optimization of a feeding strategy for production of monoclonal antibodies. The use of a multinutrient feed to increase cell culture longevity resulted in antibody titers in the range of 1 to 2 g/L. A similar feeding strategy led to a 10-fold increase in productivity for large-scale myeloma cultures producing recombinant mouse and rat growth hormones (44). Ray et al. (22) described the fed-batch culture of GS-NSO cells making a humanized monoclonal antibody in a 2,000-L stirred-tank bioreactor, using a supplemental feed containing glucose and amino acids. Xie and Wang (45–48) have developed a feeding strategy based on a stoichiometric model of the demands of nutrients for growth and product formation. The strategy is aimed at preventing nutrient depletion while reducing the formation of toxic products of metabolism such as ammonia and lactate (by avoiding excess concentration of the nutrient precursors of these products). Application of this strategy in 2-L fed-batch cultures of hybridoma cells (45) resulted in very high cell densities (1.7×10^7 cells/mL), extended culture span (550 h), and a final monoclonal antibody concentration of 2.4 g/L. This approach clearly demonstrates the potential for improving the productivity of suspension culture processes. It may ultimately be possible to use on-line measurement to estimate nutrient utilization rate and feed stoichiometric amounts of nutrient to support growth and reduce accumulation of toxic metabolites. Zhou et al. (49) demonstrated such a system using on-line measurement of oxygen uptake rate to automatically control the feed to a 500-mL hybridoma culture. Maximum cell concentrations were increased from 2×10^6/mL in batch culture to 1.36×10^7/mL in the fed culture. Kurokawa et al. (50) used on-line measurement of glucose and glutamine, using HPLC, to control the concentration of these nutrients in fed batch culture at low levels (0.2 and 0.1 g/L, respectively). This strategy reduced the accumulation of lactate and ammonia and gave an approximately twofold increase in cell growth and antibody production.

Continuous Culture Systems

Continuous (perfusion) suspension culture systems are now being used for large-scale manufacturing operations. For a discussion of perfusion culture technology and the engineering issues involved in designing and operating such systems, the reader is referred to the review by Ozturk (51). Perfusion culture uses a continuous supply of culture medium to the reactor and a continuous offtake of spent culture fluid containing product. By retaining cells in the reactor (by preventing them leaving the vessel or by an external recycle device), very high cell densities and consequently very high product throughput can be achieved. The earliest device of this kind was the spin filter culture developed by Himmelfarb et al. (52). A spinning filter within the reactor allowed cell retention, and with continuous perfusion of culture medium, cell densities approaching 10^8/ml were achieved.

Several groups have described production systems based on the spin filter principle. Avgerinos et al. (53) described a 20-L perfused bioreactor for production of a plasminogen activator from CHO cells. Deo et al. (27) described the design and operation of a 500-L perfused spin filter reactor for monoclonal antibody production. They note that continuous perfusion generally gives an approximately 10-fold greater volumetric productivity in comparison with batch and fed-batch processes. Hence less reactor capacity is required. At an antibody titer of 400 to 500 mg/L the 500-L reactor is capable of producing greater than 1 kg of antibody per week, which they estimate would require a 5,000- to 10,000-L bioreactor operating in batch or fed-batch mode. Successful operation of spin filter systems depends on careful optimization of the filter system to prevent fouling (27,54).

Other cell retention devices have also been described in the literature. Roth et al. (55) describe an external vortex flow filtration device used with perfused 3-L cultures of CHO, HEK 293, HeLa, and Sp2 cell lines. Hülscher et al. (56) describe the use of a settling tank for cell retention in a perfused 5.4-L airlift loop reactor. Pui et al. (57) have developed an acoustic resonator to cause cell aggregation and sedimentation back into the culture. At industrial scale Bödeker et al. (18) (Table 1) have used a perfused 500-L bioreactor with an external cell retention/recycle system to manufacture recombinant factor VIII from BHK cells. This is apparently the first approved biopharmaceutical from recombinant mammalian cells made using perfusion culture. The authors discuss the advantages of this process for factor VIII production, which requires the generation of greater than one million liters of harvest. These advantages are a high degree of culture control compared with batch operation (operates in steady-state conditions) and reduction in reactor size (hence easier to operate, clean, and sterilize). The main disadvantage reported was the extended process validation required. This validation included process stability, stability of product characteristics, and genetic stability of the cell line over the duration of the culture, which was up to 185 days. Several full-length campaigns were run to establish process consistency.

An alternative approach to continuous culture is to use a chemostat in which growth rate is controlled by the supply of a single limiting nutrient. The advantage of this system is that true physiological steady-state conditions can be maintained indefinitely, allowing very precise control of culture conditions, and this can be extremely valuable in a variety of research applications for studying cell physiology (see for example Refs. 58–60). To date, however, the technique has not been applied in industrial processes.

Process Monitoring and Control

The key parameters that are usually monitored and controlled in industrial-scale suspension cultures are pH, temperature, and dissolved oxygen. There is also increasing interest in the use of devices, which are now available, that allow real-time monitoring of biomass concentration (61). Such devices will allow improved monitoring and control of fermentations, particularly where fed-batch and perfusion strategies are adopted.

pH Control. Mammalian cell growth metabolism and productivity can be profoundly influenced by even small changes in pH of the culture. Ozturk and Palsson (62) found that the growth rate for a hybridoma was optimum at pH 7.2 but that the specific antibody production rate increased twofold at pH values below 7.2. They also showed that glucose and glutamine uptake rates and production rates for lactate and ammonia increased at higher pH values. Wayte et al. (63) reported that differences in pH of as little as 0.1 unit could profoundly affect cell growth and/or productivity of hybridoma and myeloma cell lines. The effects observed were cell line specific. pH may also influence product characteristics, especially glycosylation. Borys et al. (64) showed that pH affected the glycosylation pattern and specific expression rate of a recombinant lactogen protein made in CHO cell culture. Hence it is essential to establish the optimum pH for a given cell line and to measure and control pH precisely. pH is typically controlled at a value in the range 7.0 to 7.5. Most cell culture media rely on a bicarbonate-CO_2 buffering system. Hence when a bioreactor is sparged with air the pH can be controlled by injecting a controlled flow of CO_2 on demand to maintain a constant pH value. It is frequently found that lactic acid accumulates to levels that cannot be controlled by sparging with air and CO_2; it is then common practice to control the pH by addition of a sterile base solution such as sodium bicarbonate (19) or sodium carbonate (29). Care is needed in designing bioreactors to avoid pH gradients during addition of base solution. Regions of high pH may arise, especially if base is added to a poorly mixed zone (often the upper region) in both stirred reactors (29) and airlift reactors (63). The problem can be minimized by adding base to a well-mixed region of the reactor.

Dissolved Oxygen Concentration. Animal cells have an absolute requirement for oxygen. Their are several ways in which it can be provided to cultures, but the most common method is to bubble or "sparge" air or oxygen into the reactor. A less commonly used method involves gas exchange across gas permeable membranes inside or outside the reactor (51,65). Dissolved oxygen concentrations can be measured in situ with a dissolved oxygen electrode and controlled automatically by adjusting the flow rate of air or oxygen to the reactor. The optimum dissolved concentration varies from one cell line to another, although most cell types will grow over quite a wide range of dissolved oxygen concentrations. Boraston et al. (58) found that growth of a hybridoma cell was relatively unaffected over the range 8–100% air saturation. Ozturk and Palsson (62) showed that growth rates of hybridoma cells were not influenced at dissolved oxygen concentrations in the range 20–80% air saturation. Specific death rates were lowest in the range 20–50% air saturation. Nienow et al. (29) found that a CHO cell line producing recombinant interferon γ grew well and had similar productivity over the range 20–80% saturation, but growth and productivity were reduced at 5 and 100% saturation. The oxygen uptake rate of animal cells is typically in the range 0.05 to 0.5 mmol $O_2/10^9$ cells/h (66). It is of course necessary to ensure that oxygen transfer into the culture is sufficient to meet the uptake rate of the cells, and this becomes an increasingly important issue as higher cell densities are achieved through improvements in process design. Factors influencing oxygen transfer include height-to-diameter ratio for the vessel, sparger and impeller designs, and gas flow rates. Tramper (67) has pointed out that gradients of dissolved oxygen concentration may occur in large bioreactors. It is not clear what impact such gradients have in practice, but they should be borne in mind and minimized when designing reactors. Aeration in cell culture bioreactors has been reviewed in detail by Aunins and Henzler (65).

Dissolved Carbon Dioxide Concentration. In addition to CO_2 added to the culture as part of the pH buffering system, account must be taken of CO_2 generated by cell metabolism. Although systems for in situ monitoring and control of CO_2 are not yet commonly used, it is useful to measure CO_2 levels off-line, for example, using a clinical blood gas analyser. In recent years it has become apparent that in some large-scale processes CO_2 can accumulate to levels that inhibit cell growth and recombinant product formation (12,65–69). It has been suggested (65) that CO_2 accumulation may be the factor that will ultimately limit the reactor scale for some processes. Kimura and Miller (70) point out that inhibition may be caused by an effect of CO_2 on intracellular pH (even if the culture pH is controlled) and may also be caused by an increase in osmolarity caused by the additional base required to maintain pH.

Gray et al. (69), working with CHO cells in a high density perfused process, showed that productivity of a recombinant product was maximized when CO_2 was maintained in the range 30–76 mm Hg. Levels greater than 105 mm Hg resulted in inhibition of growth and productivity. CO_2 concentration may also influence glycosylation of recombinant proteins (70). Aeration systems designed for efficient oxygen transfer will not necessarily be efficient at stripping out CO_2. Gray et al. (69) found that sparging with microbubbles of pure oxygen gave very efficient oxygen transfer but was inefficient for CO_2 removal because the bubbles dissolved before reaching the surface of the culture. Sparging with large bubbles (2–3 mm diameter) improved CO_2 removal while retaining adequate oxygen transfer. Zhou et al. (44), working with GS-NSO cell lines making recombinant mouse and rat growth hormones in a 250-L fed batch reactor, also found that use of a sintered steel frit delivering pure oxygen gave good oxygen transfer, but CO_2 accumulated to 200 mm Hg, depressing cell growth. By using a less efficient sparger (12-mm × 1.5-mm orifice ring sparger) CO_2 accumulation was reduced to a noninhibitory level.

Temperature. Temperature is typically controlled at around 37 °C. Bloemkolk et al. (71) studied the effect of temperature on growth and antibody production of a hybridoma in suspension culture; 37 °C was the optimum temperature for cell growth and antibody production. At 39 °C growth was completely inhibited. In some instances the optimum temperature for product accumulation may be below 37 °C (72).

Culture Media for Suspension Culture

In general media for suspension culture are similar to those used for other processes, with the exception that an agent to protect against mechanical damage is frequently added (see later section on mechanical damage). For reactors that are aerated by sparging it may also be necessary to use an antifoam agent, which is usually a silicone emulsion (65,73). In some instances changes in nutritional requirements have been observed in suspension as opposed to static culture. For example, Murakami (74) showed that a myeloma cell line in suspension culture had a requirement for a phospholipid component that was not required in static culture.

Significant progress is being made in the optimization of media composition as well as of nutrient feeds for fed-batch culture (see earlier section). Brown et al. (28) described how an iterative program of medium development gave a six- to sevenfold improvement in productivity of antibody from a recombinant NSO myeloma cell line. Addition of a feed gave a further twofold increase in productivity. Jo et al. (75) increased maximum cell concentrations significantly for several cell lines by increasing the nutrient concentrations in RPMI 1640 medium.

There is an increasing drive to remove animal-derived raw materials from cell culture processes to reduce the risk of inadvertently introducing adventitious agents. Encouraging progress has been made, and there are now several examples in the literature of industrially important cell lines being grown in chemically defined protein-free medium. Keen (76) described the growth of myeloma and hybridoma cell lines in a protein-free medium in shake flask culture. Zang et al. (77) showed that CHO cell lines making recombinant proteins could be grown in protein-free culture medium.

PREVENTION OF CELL DAMAGE IN BIOREACTORS

Two mechanisms have been identified that can potentially cause cell damage in bioreactors; the first is the interaction of cells with bubbles in sparged reactors, and the second is the stresses created by impellers at high agitation rates in stirred reactors. In practice, given the low impeller speeds used in animal cell reactors, interaction with bubbles is the only likely source of damage in aerated cultures (65,78). The problem is readily overcome by incorporation of a protective polymer such as the nonionic surfactant Pluronic F68 into the culture medium. For a review of protective additives see Ref. 78. The role of these polymers is not totally clear. They may act by preventing cells from attaching to bubbles, by altering the way in which bubbles burst when they disengage at the surface of the culture, or by adsorbing to the surface of cells and strengthening them (65,79,80).

CONCLUSIONS

Large-scale animal suspension cell cultures have become an extremely important source of biopharmaceutical products, and with the steadily increasing number of biotechnology products in clinical trial (81), this trend seems set to continue. New areas of medical development, for example in gene therapy, give rise to new process development challenges that will lead to new uses for large-scale suspension culture. The requirement for large volumes of some animal-cell-derived pharmaceuticals drives the need to develop processes that are highly productive and cost efficient. Currently two approaches are being adopted: the use of very large bioreactors (to benefit from economies of scale), and the use of continuous (perfusion) culture using high-density cell culture at smaller scale. Advances in reactor design and operations are being complemented by the use of efficient gene expression technology and by continuing improvement in the design of culture media, especially chemically defined media.

BIBLIOGRAPHY

1. O. Owens, M.K. Gey, and G.O. Gey, *N.Y. Ann. Acad. Sci.* **58**, 1039–1055 (1954).
2. W.R. Earle, E.H. Schilling, J.C. Bryant, and V.J. Evans, *J. Natl. Cancer Inst.* **14**, 1159–1171 (1954).
3. W.R. Cherry and R.N. Hull, *Anat. Rec.* **124**, 483 (1956).
4. W.F. McLimans, E.V. Davis, F.L. Glover, and G.W. Rake, *J. Immunol.* **79**, 428–433 (1957).
5. W.F. McLimans, and F.R. Giardinello, *J. Bacteriol.* **74**, 768–771 (1957).
6. D.W. Ziegler, E.V. Davis, W.J. Thomas, and W.F. McLimans, *Appl. Microbiol.* **6**, 305–310 (1958).
7. G.E. Moore, P. Hasenpusch, R.E. Gerner, and A.A. Burns, *Biotechnol. Bioeng.* **10**, 625–640 (1968).
8. P.B. Capstick, R.C. Telling, W.G. Chapman, and D.L. Stewart, *Nature* **195**, 1163–1164 (1962).
9. K.F. Pullen, M.D. Johnson, A.W. Phillips, G.D. Ball, and N.B. Finter, *Dev. Biol. Stand.* **60**, 175–177 (1985).
10. P.J. Radlett, T.W.F. Pay, A.J.M. Garland, *Dev. Biol. Stand.* **60**, 163–170 (1985).
11. A. Lubiniecki, R. Arathoon, G. Polastri, J. Thomas, M. Wiebe, R. Garnick, A. Jones, R. van Reis, and S. Builder, in R.E. Spier, J.B. Griffiths, J. Stephenne, P.J. Crooy, eds., *Advances in Animal Cell Biology and Technology for Bioprocesses*, Butterworth, London, 1989, pp. 442–449.
12. A. Garnier, R. Voyer, R. Tom, S. Perret, B. Jardin, and A. Kamen, *Cytotechnology* **22**, 53–63 (1996).
13. I. Nadeau, A. Garnier, J. Côté, B. Massie, C. Chavarie, and A. Kamen, *Biotechnol. Bioeng.* **51**, 613–623 (1996).
14. M.S. Sinacore, T.S. Charlebois, S. Harrison, S. Brennan, T. Richards, M. Hamilton, S. Scott, S. Brodeur, P. Oakes, M. Leonard, M. Switzer, A. Anagnostopoulos, B. Foster, A. Harris, M. Jankowski, M. Bond, S. Martin, and S.R. Adamson, *Biotechnol. Bioeng.* **52**, 518–528 (1996).
15. N. Kurano, C. Leist, F. Messi, C. Gandor, S. Kurano, and A. Fiechter, *J. Biotech.* **16**, 245–258 (1990).
16. W.A. Renner, K.H. Lee, V. Hatzimanikatis, J.E. Bailey, and H.M. Eppenberger, *Biotechnol. Bioeng.* **47**, 476–482 (1995).
17. T.W.F. Pay, A. Boge, F.J.R.R. Menard, and P.J. Radlett, *Dev. Biol. Stand.* **60**, 171–174 (1985).
18. B.G.D. Bödeker, R. Newcomb, P. Yuan, A. Braufman, W. Kelsey, in R.E. Spier, J.B. Griffiths, and W. Berthold eds., *Animal Cell Technology, Products for Today, Prospects for Tomorrow*, Butterworth-Heinemann, Oxford, 1994, pp. 580–583.
19. M.P. Backer, L.S. Metzger, P.L. Slaber, K.L. Nevitt, and G.B. Boder. *Biotechnol. Bioeng.* **32**, 993–1000 (1988).
20. C.R. Wood, A.J. Dorner, G.E. Morris, E.M. Alderman, D. Wilson, R.M. O'Hara, R.J. Kaufman, *J. Immunol.* **145**, 3011–3016 (1990).
21. C.R. Bebbington, G. Renner, S. Thomson, D. King, D. Abrams, and G.T. Yarranton, *BioTechnology* **10**, 169–175 (1992).
22. N.G. Ray, R. Rivera, R. Gupta, and D. Mueller, in M.J.T. Carrondo et al. eds., *Animal Cell Technology*, Kluwer Academic, Dordrecht, 1997, pp. 235–241.
23. J.L. Moreira, P.M. Alves, J.G. Aunins, and M.J.T. Carrondo, *Appl. Microbiol. Biotechnol.* **41**, 203–209 (1994).
24. R. Boraston, C. Marshall, P. Norman, G. Renner, and J. Warner, in R.E. Spier, J.B. Griffiths, and C. Macdonald eds., *Animal Cell Technology: Developments, Processes and Products*, 1992, pp. 424–426.
25. H.N. Brand, S.J. Froud, H.K. Metcalfe, A.O. Onadipe, A. Shaw, and A.J. Westlake, in R.E. Spier, J.B. Griffiths, and W. Berthold eds., *Animal Cell Technology Products of Today, Prospects for Tomorrow*, Butterworth-Heinemann, Oxford, 1994, pp. 55–60.
26. R. Adamson, *Ann. Hematol*, **68**, S9–S14 (1994).
27. Y.M. Deo, M.D. Mahadevan, and R. Fuchs, *Biotechnol. Prog.* **12**, 57–64 (1996).
28. M.E. Brown, G. Renner, R.P. Field, and T. Hassell, *Cytotechnology* **9**, 231–236 (1992).
29. A.W. Nienow, C. Langheinrich, N.C. Stevenson, A.N. Emery, T.M. Clayton, and N.K.H. Slater, *Cytotechnology* **22**, 87–94 (1996).
30. J. Cameron and E.I. Godfrey, *Biotechnol. Bioeng. Symp.* **4**, 821–835 (1974).
31. N.B. Finter, in A.S. Lubiniecki ed., *Large Scale Mammalian Cell Culture Technology*, Marcel Dekker, New York and Base, 1990, pp. 1–14.
32. H. Katinger and W. Scheirer in R.E. Spier, and J.B. Griffiths eds., *Animal Cell Biotechnology*, vol. 1, Academic, London, Orlando, 1985, pp. 167–193.
33. J.R. Birch, K. Lambert, R. Boraston, P.W. Thompson, S. Garland, and A.C. Kenney, in J.M.C. Duarte, L.J. Archer, A.T. Bull, and G. Holt eds., *Perspectives in Biotechnology*, Plenum, New York and London, 1987, pp. 97–109.
34. K.C. Merchuk and M.H. Siegel, *J. Chem. Tech. Biotechnol.* **41**, 105–120 (1988).
35. U. Onken and P. Weiland, in *Adv. Biotechnol. Processes* **1**, 67–95 (1983).
36. H.W.D. Katinger, W. Scheirer, and E. Krömer, *Ger. Chem. Eng.* **2**, 31–38 (1979).
37. B. Maiorella, D. Inlow, A. Shauger, and D. Harano, *BioTechnology* **6**, 1406–1410 (1988).
38. W.R. Arathoon, and J.R. Birch, *Science* **232**, 1390–1395 (1986).
39. M. Rhodes and J. Birch, *BioTechnology* **6**, 518–523 (1988).
40. J.R. Birch, J. Bonnerjea, S. Flatman, and S. Vranch, in J.R. Birch and E.S. Lennox eds., *Monoclonal Antibodies: Principles and Applications*, Wiley-Liss, New York, 1995 pp. 231–265.
41. J.B. Griffiths, in R. Sasaki and K. Ikura eds., *Animal Cell Culture and Production of Biologicals*, Kluwer Academic, Dordrecht, 1991, pp. 401–410.
42. T.A. Bibila and D.K. Robinson, *Biotechnol. Prog.* **11**, 1–13 (1995).
43. T.A. Bibila, C.S. Ranucci, K. Glazomitsky, B.C. Buckland, and J.G. Aunins, *Biotechnol. Prog.* **10**, 87–96 (1994).
44. W. Zhou, T. Bibila, K. Glazomitsky, J. Montalvo, C. Chan, D. Di Stefano, S. Munshi, D. Robinson, B. Buckland, and J. Aunins, *Cytotechnology* **22**, 239–250 (1996).
45. L. Xie and D.I.C. Wang, *Cytotechnology* **15**, 17–29 (1994).

46. L. Xie and D.I.C. Wang, *Biotechnol. Bioeng.* **43**, 1164–1174 (1994).
47. L. Xie and D.I.C. Wang, *Biotechnol. Bioeng.* **43**, 1175–1189 (1994).
48. L. Xie and D.I.C. Wang, *Biotechnol. Bioeng.* **51**, 725–729 (1996).
49. W. Zhou, J. Rehm, and W.-S. Hu, *Biotechnol. Bioeng.* **46**, 579–587 (1995).
50. H. Kurokawa, Y.S. Park, S. Iijima, and T. Kobayashi, *Biotechnol. Bioeng.* **44**, 95–103 (1994).
51. S. Ozturk, *Cytotechnology* **22**, 3–16 (1996).
52. P. Himmelfarb, P.S. Thayer, and H.E. Martin, *Science* **164**, 555–557 (1969).
53. G.C. Avgerinos, D. Drapeau, J.S. Socolow, J.-I. Mao, K. Hsiao, and R.J. Broeze, *BioTechnology* **8**, 54–58 (1990).
54. V.M. Yabannavar, V. Singh, and N.V. Connelly, *Biotechnol. Bioeng.* **40**, 925–933 (1992).
55. G. Roth, C.E. Smith, G.M. Schoofs, T.J. Montgomery, J.L. Ayala, and J.I. Horwitz, *Pharmaceutical Technology*, October 1997, and *BioPharm* **10**, 30–35 (1997).
56. M. Hülscher, U. Scheibler, and U. Onken, *Biotechnol. Bioeng*, **39**, 442–446 (1992).
57. P.W.S. Pui, F. Trampler, S.A. Sonderhoff, M. Groeschl, D.G. Kilburn, and J.M. Piret, *Biotechnol. Prog.* **11**, 146–152 (1995).
58. D.K. Robinson, and K.W. Memmert, *Biotechnol. Bioeng.* **38**, 972–976 (1991).
59. P.M. Hayter, E.M.A. Curling, M.L. Gould, A.J. Baines, N. Jenkins, I. Salmon, P.G. Strange, and A.T. Bull, *Biotechnol. Bioeng.* **42**, 1077–1085 (1993).
60. R. Boraston, P.W. Thompson, S. Garland, and J.R. Birch, *Dev. Biol. Stand.* **55**, 103–111 (1984).
61. K. Konstantinov, S. Chuppa, E. Sajan, Y. Tsai, S. Yoon, and F. Golini, *Tibtech* **12**, 324–333 (1994).
62. S.S. Ozturk, and B.O. Palsson, *Biotechnol. Prog.* **7**, 481–494 (1991).
63. J. Wayte, R. Boraston H. Bland, J. Varley, and M. Brown, *Genet. Eng. Biotechnol.* **17**, 125–132 (1997).
64. M.C. Borys, D.I.H. Linzer, and E.T. Papoutsakis, *BioTechnology* **11**, 720–724 (1993).
65. J.G. Aunins, and H.-J. Henzler, in G. Stephanopoulos ed., *Biotechnology*, vol. 3, *Bioprocessing*, VCH, Berlin, 1993, pp. 219–281.
66. R.J. Fleischaker, and A.J. Sinskey, *Eur. J. Appl. Microbiol. Biotechnol.* **12**, 193–197 (1981).
67. J. Tramper, *Cytotechnology* **18**, 27–34 (1995).
68. R. Kimura and W.M. Miller, *Biotechnol. Bioeng.* **52**, 152–160 (1996).
69. D.R. Gray, S. Chen, W. Howarth, D. Inlow, and B.L. Maiorella, *Cytotechnology* **22**, 65–78 (1996).
70. R. Kimura, and W.M. Miller, *Biotechnol. Prog.* **13**, 311–317 (1997).
71. J.W. Bloemkolk, M.R. Gray, F. Merchant, and T.R. Mosmann, *Biotechnol. Bioeng.* **40** 427–431 (1992).
72. J.P. Mather, and M. Tsao, in A.S. Lubiniecki ed., *Large-Scale Mammalian Cell Culture Technology*, Marcel Dekker, New York and Basel, 1990, pp. 161–177.
73. L.A. van der Pol, D. Bonarius, G. van de Wouw, and J. Tramper, *Biotechnol. Prog.* **9**, 504–509 (1994).
74. H. Murakami, T. Edamoto, K. Shinohara, and H. Omura, *Agric. Biol. Chem.* **47**, 1835–1840 (1983).
75. E.-C. Jo, D.-I. Kim, and H.M. Moon, *Biotechnol. Bioeng.* **42**, 1218–1228 (1993).
76. M.J. Keen, *Cytotechnology* **17**, 193–202 (1995).
77. M. Zang, H. Trautmann, C. Gandor, F. Messi, F. Asselbergs, C. Leist, A. Fiechter, and J. Reiser, *BioTechnology* **13**, 389–392 (1995).
78. T. Papoutsakis, *Tibtech* **9**, 316–324 (1991).
79. A. Handa-Corrigan, A.N. Emery, and R.E. Spier, *Enzyme Microb. Technol.* **11**, 230–235 (1989).
80. Z. Zhang, M. Al-Rubeai, and C.R. Thomas, *Enzyme Microb. Technol.* **14**, 980–983 (1992).
81. J.R. Birch, *Eur. J. Parent Sci.*, Biotechnology Special Issue, pp. 3–10 (1997).

See also MAMMALIAN CELL BIOREACTORS.

SWEETENERS. See ASPARTAME.

T

THERMAL UNFOLDING, PROTEINS

LI SHI
DAVID B. VOLKIN
Merck Research Laboratories
West Point, Pennsylvania
GAUTAM SANYAL
Astra Research Center Boston
Cambridge, Massachusetts

KEY WORDS

Protein folding
Protein pharmaceuticals
Protein stability
Secondary structure
Tertiary structure
Thermal denaturation
Unfolding kinetics

OUTLINE

INTRODUCTION

Recombinant therapeutic proteins have emerged as important pharmaceuticals for a wide variety of indications in a number of clinical disciplines. These include immunoregulatory proteins, peptide and protein hormones, growth factors, and immunoglobulins. Developing stable formulations of proteins offers challenges related to the inherent instability of their native three-dimensional structures that must usually be maintained for biological function (1,2). The pharmaceutical aspects of developing stable protein formulations include designing ways to preserve their chemical and conformational integrity as measured by sensitive, stability-indicating assays.

Protein inactivation can occur by either or both of two distinct pathways: chemical degradation, including hydrolytic and proteolytic clipping, and physical inactivation, including global unfolding, a more subtle local conformational change involving a functionally critical site. Protein stability studies must include assays that can quantitatively determine the extents of physical and chemical changes as well as changes in biological activity that may result from perturbations introduced by storage, handling, and formulation conditions. A well-designed study using various biophysical and biochemical techniques is essential to monitoring protein structure and activity as a function of formulation processing. Protein conformational stability studies under various stress conditions have been widely applied in protein formulation research. The techniques include both thermal, pH, and denaturant induced unfolding. In this article, we discuss specifically the application of thermal unfolding studies to formulation development of protein-based pharmaceuticals.

MECHANISMS OF PROTEIN UNFOLDING

The native or functional structure of a protein is a minimum energy state that is held together by an intricate balance of a multitude of covalent and noncovalent interactions. Perturbations of these interactions can lead to unfolding of the native conformation and will usually lead to the loss of biological activity. In general, there are two major types of noncovalent interaction contributing to the stability of a folded protein structure: hydrophobic interactions between apolar side chains that prefer to exclude water and electrostatic or polar interactions mediated by water. The latter includes hydrogen bonding and salt bridges. Although from a macroscopic point of view a protein molecule may appear to be in a completely folded state, the so-called native protein conformation is in fact in a fast exchange between folded or native (N) and unfolded (U) states in a dynamic equilibrium, with an overwhelming majority of the population being in the folded state. The equilibrium constant, K, describing the population distribution in these two states, is strongly dependent on the environment of the protein and is correlated to the Gibbs energy, ΔG, of the protein molecule through relation $\Delta G = -RT\ln K$.

Protein unfolding occurs when the balance of forces between the intramolecular interactions in the protein and the protein's interaction with its environment is disrupted. This may be a result of perturbation of normal water structure around the protein. The conformational stability of most proteins is sensitive to temperature. Increased temperatures lead to increased thermal motions in proteins, which, in turn, perturb intermolecular interactions as well as solvent–protein interactions. When the environment of a protein is altered because of changes of temperature, pressure, pH, or solvent polarity, the equilibrium balance between unfolded and native states is altered, and a new equilibrium is reached in which the unfolded state may be favored. However, this population redistribution is not significantly noticeable until an apparent conformational transition is approached in the unfolding process.

Disruption of Secondary and Tertiary Structure

It is important to understand the mechanism of the disruption of protein secondary and tertiary structures when the protein is exposed to stress conditions. The secondary structure of a protein refers to the spatial arrangement of amino acid residues that are near one another in the linear sequence (3–6). Some of these steric relationships manifest a regular repeat pattern, giving rise to periodic structures. The α-helices, β-sheets, turns, and random coils are elements of secondary structure. Tertiary structure refers to the overall three-dimensional architecture of the spatial arrangement of amino acid residues, including those that may be far apart in the linear sequence (3–6). This spatial arrangement is also referred to as the packing assembly of secondary structure units.

The net stabilization free energy that maintains the native folded structure of a globular protein is in the order of only 2 to 20 kcal/mol. Therefore, a few kcal of additional energy gained from the interaction changes of the protein with the environment can result in the destabilization of the protein molecule resulting in either local or global conformational changes (7). Information about obtaining the ΔG value of a protein is available (8).

It is necessary to point out that thermodynamic analysis of the free energy of protein conformational change is limited to reversible conformational change or unfolding, because the relation $\Delta G = -RT\ln K$ is only valid when the mass balance is retained for all species involved in the conformational change. For a thermally induced reversible unfolding transition, the dependence of ΔG on T (and hence the dependence of K on T) is given by $\Delta G = \Delta H - T\,\Delta S$. ΔH and ΔS are, respectively, the enthalpy and entropy differences between U and N states. The values of ΔH and ΔS are dependent on temperature through a positive heat-capacity change, ΔCp. The positive ΔCp is thought to be related to the hydrophobic exposure of apolar side chains as a result of unfolding of globular proteins.

Two-Step versus Multistep Unfolding

The simplest model for the unfolding of a protein is the two-state model in which transition from N state to U state occurs in a cooperative process, that is, in a single unfolding transition step N ↔ U (4,9,10). A typical two-state unfolding of a protein follows a sigmoidal shaped unfolding curve, as shown in Figure 1 (for a review see Ref. 11). The N state is generally considered to have a narrow distribution of conformations, which are relatively ordered (low in entropy), whereas the U state involves a broader distribution of microscopic conformational states, which are relatively disordered (high in entropy). In recent years, protein biochemists have focused on the multistep unfolding model, in which protein unfolding proceeds from N to U by passing through one or more intermediate (I) states:

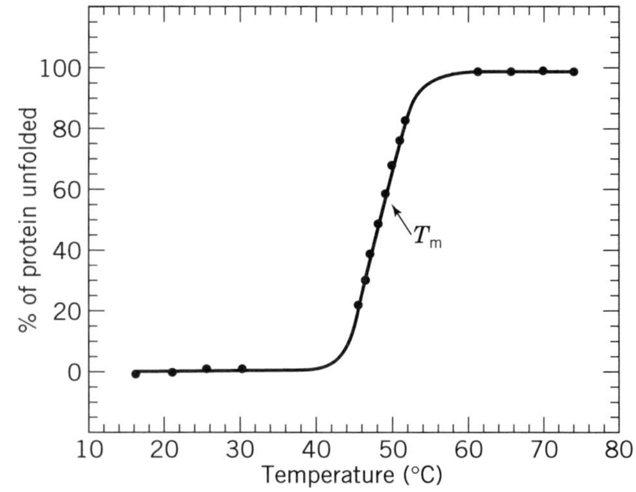

Figure 1. Two-state protein unfolding curve: Ribonuclease T1 thermal unfolding as measured by specific optical rotation at 295 nm. Data were adapted from reference 10 with permission.

$$N \leftrightarrow I \leftrightarrow \;\ldots\; \leftrightarrow U \qquad (1)$$

If an I state is stable in the unfolding process, both N ↔ I and I ↔ U transitions can be characterized biophysically from equilibrium and kinetic unfolding studies (12–14). Several features of a partially unfolded state as a folding intermediate have been identified:

1. Relatively well-defined secondary structure (15,16) folded in a compact globular shape (17)
2. A non-cooperative folding/unfolding pathway (18)
3. Weak tertiary contacts and substantial solvent exposure of hydrophobic groups (18,19)
4. Relatively high affinity for hydrophobic ligands and tendency to aggregate (17,19)

For a multidomain protein, each individual domain may unfold sequentially or simultaneously (13).

Protein unfolding can lead to either reversible or irreversible loss of the folded structure. By definition, proteins denatured by a reversible pathway can regain their native structures and biological functions. An example of this process is the refolding of a protein after solubilization from inclusion bodies using denaturing conditions. Irreversible denaturation often leads to protein aggregation and results in loss of biological function (20,21).

APPLICATION OF THERMAL UNFOLDING STUDIES TO DEVELOPMENT OF STABLE PROTEIN FORMULATIONS

The unfolding of a protein during storage could be induced by perturbations such as elevated temperatures, surface adsorption, and physical agitation. A successful formulation of a protein pharmaceutical would require the development of a dosage form in which the physicochemical properties and biological activity of the protein are maintained during storage. The design of a pharmaceutical formulation should be biocompatible, possess pharmaceutical elegance, and be suitable for commercial scale-up. The development of a stable formulation requires a thorough understanding of the causes and mechanism of inactivation, including both physical and chemical pathways. In approaching this goal, thermal unfolding studies are frequently used in the following ways. Intrinsic conformational stability data such as unfolding transition temperature can be obtained. If the transition is reversible, thermodynamic information such as ΔCp can be derived, and the results obtained from these accelerated stability measurements can often be extrapolated to make reasonable predictions about stability at different temperatures. Thermal unfolding studies can be used to investigate the mechanism of inactivation. The inactivation process is generally accelerated at an elevated temperature. Potential stabilizers can be screened by observing their effects on the thermal unfolding transition. An excipient that provides protection against temperature-induced denaturation or inactivation of a protein is often effective in prolonging its shelf life at refrigeration temperature (22). For example, an unstabilized respiratory syncytial virus (RSV) preparation lost most of its infectivity at −86 °C within 2 to 3 weeks, at 37 °C within 24 h, at 56 °C within 3 min. This study aided the development of a stabilized RSV formulation that was stable at −86 °C for 3 years, at 37 °C for 3 days, and at 56 °C for 6 min (23).

Compared to pH or denaturant-induced unfolding studies that also address the conformational stability of a protein, thermal unfolding studies are more direct and avoid the use of nonnative solution conditions (denaturants). In most cases, proteins and vaccines are likely to be exposed during handling to temperatures higher than the temperature usually recommended for their stable storage (e.g., 2 to 8 °C). Monitoring biological activity and structure of proteins as a function of increasing temperatures is the most convenient method to perform an accelerated stability study, although cold denaturation and freezing-induced aggregation of proteins have been studied as well (24–28). A major benefit of the accelerated stability studies is to expedite the process of screening suitable excipients for development of a stable formulation.

There have been many recent developments in using the approaches of stabilizing proteins by stabilizing the folded state, destabilizing the unfolded state, and altering the kinetics of unfolding. The factors that affect stability of a folded state versus the unfolded state as well as factors affecting rates of folding and unfolding play a role in maintaining stability (29). More detailed information on approaches to minimize inactivation and develop stable formulations of proteins are available (1,4,20,22,28–33).

PROTEIN STABILITY EVALUATED BY MEASUREMENT OF THERMAL UNFOLDING TRANSITION

The denaturation transition temperature (T_m) has been widely used as a reference to evaluate the thermal stability of proteins and polynucleotides (34–43). T_m is defined as the temperature at the midpoint of protein unfolding transition (Figs. 1 and 2). At T_m, 50% of the total protein is unfolded (1,10). Therefore, for a reversible two-state unfolding process, at T_m, the unfolding equilibrium constant K is 1 and the Gibbs free energy ΔG is 0. Based on the complete unfolding curve, a two-state or multistate unfolding process can be suggested. For well-resolved, reversible, multiple transitions, the transition midpoint between states and the apparent unfolding ΔG can be determined for each transition under various conditions (37). The fractions of different conformational states at equilibrium can also, in such ideal cases, be determined from the unfolding transition curves (10,37,44).

There are several methods to determine T_m of protein unfolding with reasonable accuracies. (1) The first derivative of the unfolding curve yields T_m as the sharp peak in the middle of the unfolding transition (Fig. 2**a**). (2) A plot of K values against temperature yields T_m at the temperature at which $K = 1$ (Fig. 2**b**). (3) A plot of ΔG value against temperature yields T_m at the temperature at which $\Delta G = 0$ (Fig. 2**c**).

An increase in T_m is indicative of structural stabilization of a protein in solution. The stabilization or destabilization effect of an excipient can be studied by measuring the change in T_m upon the addition of the excipient. For example, addition of heparin, a sulfated polysaccharide, to a solution of acidic fibroblast growth factor (aFGF) causes a dramatic increase in the T_m for aFGF unfolding, as measured by circular dichroism (CD) and fluorescence spectroscopy (31,45). A related study shows that the binding of heparin results in about 2.5 kcal/mol of stabilization for the native form of aFGF (46). Another example is the differential scanning calorimetric studies of the addition of osmolytes such as glycine, sarcosine, *N,N*-dimethylglycine, and betaine to ribonuclease A and lysozyme. The data indicate a 22 to 23 °C increase in T_m of the proteins upon the addition of the osmolytes (43). For ribonuclease, the 22 °C increase in T_m corresponds to an increase in stabilization free energy of 7.2 kcal/mol. Changes in T_m as a function of pH, salt, and osmolyte concentrations have been used to monitor thermal stability of proteins under different conditions (43,47,48). The T_m of tissue-type plasminogen activator (TPA) in phosphate buffer is approximately 66 °C. The addition of arginine (the stabilizer selected for the formulation) shifts the T_m up to 71 °C (28). Binding of nucleotides and substrates plays an important role in the thermal stabilization of a 70-kDa heat shock protein. Addition of nucleotide (ATP and/or ADP) or substrate protein (unfolded protein or peptide) to the solution increased the T_m of DnaK by 10 to 20 °C (38,49).

Protein unfolding can facilitate aggregation or chemical degradation (Scheme 1). Where U is fully or partially unfolded protein, I_d is the protein inactivated by degradation and I_a is the protein inactivated by aggregation. Unfolding could be a reversible process, but inactivation is normally an irreversible process. Study of the correlation of protein unfolding T_m with the rate of protein inactivation could provide useful information in understanding the mechanism of the inactivation process (7,33). In the case of reversibly denaturing of protein, the shift of T_m upon the addition of excipient is often related to the difference be-

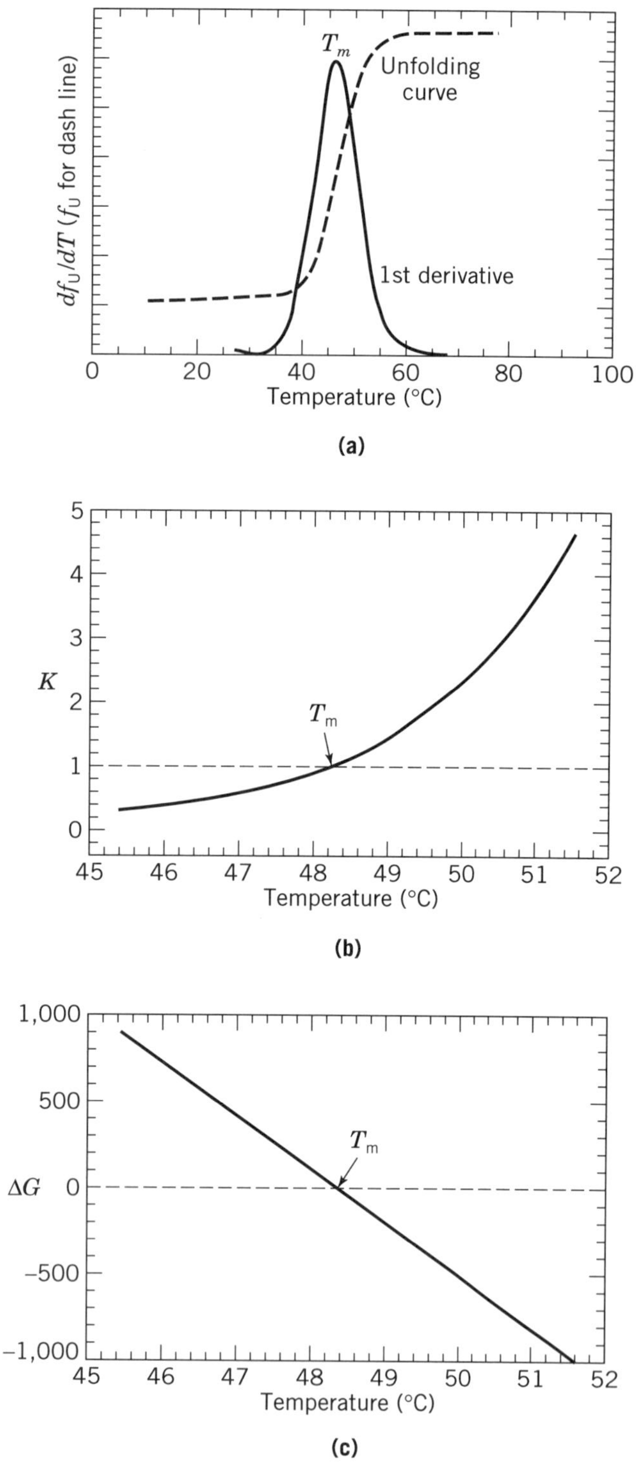

Figure 2. Methods for determining T_m value from protein unfolding curves (the plots were based on data shown in Figure 1) A) f_U (% protein unfolded) and df_U/dT versus T; B) K versus T; C) ΔG versus T. **(a)** First derivative of the fraction of unfolded protein as a function of temperature. **(b)** Equilibrium constant versus temperature. **(c)** Gibbs free energy versus temperature.

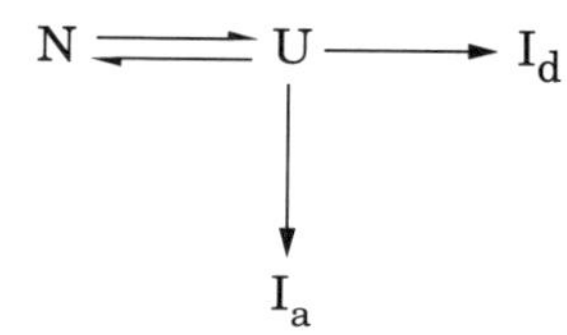

Scheme 1. Protein unfolding related inactivation pathways.

tween the binding affinities of the excipient for native and denatured states (50). For example, heparin increases the T_m of antithrombin III by 7 °C through a preferential binding to the native state of the protein (51).

Protein preparations are well known to be more stable in the dried state. Conformational flexibility is believed to be restricted in the dry state because interactions with water molecules are minimized. However, unfolding of proteins during the process of freeze-drying has been observed with FTIR spectroscopy. Many sugars and even some polymers, such as poly(vinylpyrrolidone) have been found to inhibit lyophilization-induced unfolding (24,52).

Many proteins irreversibly aggregate as they start to unfold with increasing temperature, making it difficult to obtain a complete thermal unfolding curve, and therefore the unfolding data can not be used to calculate thermodynamic parameters such as ΔG (31). As an alternative means for obtaining thermodynamic parameters, unfolding by urea or guanidine hydrochloride is frequently used, provided that these processes are reversible. These denaturants help to maintain the solubility of unfolded forms of these proteins. For the same reason, when performing the thermal unfolding studies of such proteins, a relatively low concentration (0.2 to 1 M) of chemical denaturants such as urea or guanidine hydrochloride could be added to the protein solution to prevent fully or partially unfolded protein from falling out of the solution because of aggregation (53). When this is the case, the contribution of the added denaturant to the apparent unfolding T_m should be taken into consideration. Normally, unfolding should be carried out at a series of denaturant concentrations, and T_m and other thermodynamic parameters can be obtained by extrapolation to zero denaturant concentration. A two-dimensional protein unfolding (thermal unfolding profile under various guanidine hydrochloride concentration conditions) study aimed at elucidation of biomolecular energetics has been carried out by Straume (54).

PROTEIN STABILITY EVALUATED BY KINETICS OF TEMPERATURE-INDUCED PROTEIN INACTIVATION OR UNFOLDING

Another approach to quantitative characterization of the stability of a protein is to study the protein unfolding kinetics and inactivation rate (29,55,56). Rapid unfolding, aggregation, or inactivation at high temperature indicates a less-stable protein product. A formulation that can slow down these denaturation and inactivation processes at elevated temperatures may be able to increase the stability of the protein product at storage temperature.

Arrhenius Relationship

For most proteins, unfolding can be simplified as a first-order kinetic process displaying single-exponential unfolding kinetics. The equilibrium unfolding curve can be predicted from the rate constant assuming a two-state kinetic model. The unfolding rate can be expressed by the decrease in population of the native structure with time as

$$-d[\mathrm{N}]]/dt = k[\mathrm{N}] \qquad (2)$$

where [N] is the population of the structure remaining native at time t during unfolding, and the proportionality constant, k, is the apparent rate constant of the unfolding. The population of the native structure remaining can be expressed as

$$[\mathrm{N}] = [\mathrm{N}]_0 \exp(-kt) \qquad (3)$$

where $[\mathrm{N}]_0$ is the population of native structure at time zero. Arrhenius in 1889 suggested that a reasonable equation for relating the rate constant, k, with an activation energy term (Ea) for the forward part of a reversible reaction is

$$d\ln k/dT = Ea/RT^2 \qquad (4)$$

where R is the gas constant, and T is temperature. At temperature T, the above equation can be expressed as

$$k = A\exp(-Ea/RT) \qquad (5)$$

where A is a frequency factor or preexperiential factor that can be determined by performing a multimeasurement of k at various temperatures (assuming A and Ea are independent of temperature). Equation 5 shows that for a reaction with a positive activation energy, the reaction rate increases with temperature.

The goal of a stable formulation is to minimize unfolding and other degradative pathways. According to equation 5, this can be achieved by either decreasing storage temperature or increasing the activation energy (Ea) for unfolding. The apparent activation energy, Ea, of unfolding determines the rate of the transition between the folded and unfolded states. This can be affected by altering solution conditions. Stabilizing excipients will increase the activation energy of unfolding and will, therefore, reduce the rate of unfolding under the formulation condition. The calculation of Ea requires measurement of a structural or functional property of the protein that directly reflects unfolding at several different temperatures. The Arrhenius plot of ln k versus $1/T$ should be linear (the slope of which yields the activation energy), and this may be true only over a limited temperature range. For example, the amount of aggregated protein might increase beyond a certain temperature to the point of causing interference with rate measurements. A biphasic Arrhenius plot in the absence of aggregation would indicate either a change in the pathway by which temperature affects the protein structure or the breakpoint that occurs at the temperature at which pathway changes. Measurement of unfolding activation energy has been applied to protein stability studies (22,28).

Experimental Approaches

Careful considerations must be given to the selection of methods for measuring the unfolding rate. Far-UV CD detects changes in secondary structure, but subtle and functionally important changes might remain undetected. Near-UV CD and tryptophan fluorescence measurements are well suited to monitor tertiary structural changes. The active sites of proteins, including receptor-binding sites, catalytic sites, and antigen-binding sites, would often be perturbed (locally) before a global unfolding occurs. In such cases, the unfolding rate will be slower than inactivation rate. A good strategy is to use a number of different techniques, for example, CD and fluorescence spectroscopy to monitor the unfolding rate and hydrodynamic methods to monitor rate of aggregation.

APPLICATION OF ACCELERATED PROTEIN STABILITY STUDIES TO FORMULATION DEVELOPMENT

Although the expiration dating of a biological product is based on real-time and real-temperature storage stability data, accelerated stability studies of biological products under stress conditions usually constitute an important part of the stability data package (57). It is common to determine the degradation rate of a chemical compound under a stress condition, for example, an elevated temperature (37 °C or higher), and then extrapolate the results to a lower (storage) temperature (22). When exposed to sufficiently elevated temperatures, all proteins will lose biological activity within limited a time.

Thermal stability studies include two main approaches: temperature-induced conformational change and kinetics of unfolding and degradation at elevated temperatures. The first approach is to follow the unfolding transition temperature to test, for example, effects of various excipients on T_m. The second approach is to accelerate the process that may result in the destabilization of the protein samples. At elevated temperatures, protein unfolding rate increases and most chemical interactions become faster compared to lower temperatures (37). Under these conditions, various formulations could be screened for their ability to inhibit the inactivation process. Under accelerated stability study conditions, the intrinsic stability of protein pharmaceuticals is often decreased dramatically. The addition of various excipients or stabilizers may enhance the stability of the protein pharmaceuticals under the same condition. For example, a properly buffered solution of acidic FGF (aFGF) complexed with heparin is stable at 5 °C for 1 year, but the protein is inactivated when stored at 30 °C for 2 to 4 months. Addition of 0.15 mM EDTA to the formulation enhanced storage stability at 30 °C to be roughly equivalent to that at 5 °C, and more than 70% of the protein mass remained after 1 year of storage at 30 °C (58).

Questions have been raised about the validity of accelerated protein stability studies in defining stable formulations. There are two simple but critical questions that deserve attention. One is the question of reliability of extrapolating results obtained at elevated temperatures to a lower storage temperature, such as 5 °C (22). Another question is how high the temperature for an accelerated condition could be pushed to. So far, there is no generally correct answer to the first question. The answer to the second question is that for quantitative extrapolation of kinetic data to low temperatures, the accelerated studies should be kept well below the lowest protein unfolding transition temperature (T_m). Pearlman and Nyguyen (59) were able to successfully extrapolate data obtained in an accelerated stability study at 40 °C down to 5 °C.

If protein unfolding is a first-order kinetic process, the unfolding activation energy could be obtained by carrying out the unfolding measurement at two or more different temperatures, and then using the temperature coefficient of kinetics to calculate the unfolding rate of the protein at storage temperature. Recently, Eberlein et al. (60) used a turbidity test for 15 h at 30 to 40 °C by UV spectrophotometry to predict storage stability of recombinant human bovine fibroblast growth factor (rhbFGF) with 95% confidence. Figure 3 shows a theoretical calculation for different activation energies of a first-order inactivation process using the Arrhenius equation. The results indicate that the storage stability increases as a function of activation energy of the unfolding and inactivation process. The higher the activation energy, the longer the shelf life is expected to be.

TECHNIQUES USED IN THERMAL UNFOLDING STUDIES

The spectral property of a protein molecule is sensitive to the molecular environment and the mobility of its chromophores. Far-UV CD (180–250 nm) and amide I Fourier transform infrared (FTIR) (1600–1700 cm^{-1}) are convenient spectroscopic tools for monitoring changes in the secondary structures of proteins (61,62). Fluorescence and near-UV CD (250–320 nm) are used to monitor environmental changes around aromatic chromophores of proteins and can often detect subtle changes in tertiary structure even in the absence of a global structural change in a protein (1,44,63,64). Differential scanning calorimetry (DSC), which measures changes in heat capacity, provides a direct method for monitoring thermal stability of proteins (43,65–67). Changes in hydrodynamic properties of proteins, such as aggregation, often induced at elevated temperatures, are monitored by techniques such as dynamic light scattering (DLS), small-angle X-ray scattering, analytical ultracentrifugation, and size exclusion chromatography (SEC). Specific applications of some of the more commonly used techniques that readily provide useful information are described in this section.

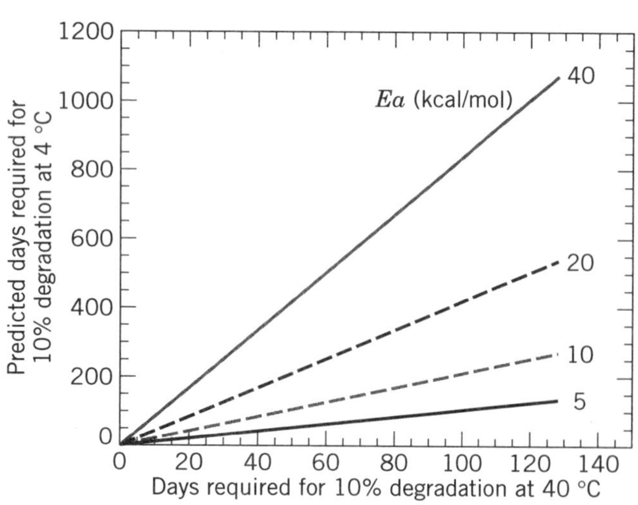

Figure 3. Theoretical prediction of 4 °C shelf-life time based on stability data at 40 °C.

Spectroscopic Methods

Tryptophan and tyrosine residues are the most commonly used intrinsic aromatic chromophores and fluorescence probes of proteins. Phenylalanine fluorescence exhibits a low quantum yield and has not found much use in protein unfolding studies. Tryptophan fluorescence spectral maxima of proteins typically range from 315 to 345 nm, and a fully water-exposed tryptophan manifests maximal fluorescence around 355 nm. The intrinsic fluorescence of tryptophan residues of most proteins is very sensitive to changes in the polarity of their immediate environments. Protein unfolding is usually accompanied by a change in the fluorescence (emission) spectrum of tryptophan, caused by increased exposure of the tryptophan residue(s) to solvent. The spectral maxima of tyrosine fluorescence are limited to a relatively narrow range (~300 to 310 nm), and they often do not manifest significant shifts upon unfolding. However, changes fluorescence intensity can provide valuable information. In proteins containing tryptophan and tyrosine residues, energy transfer from tyrosines to tryptophans can occur, and this process may be affected because of unfolding-induced changes in the distances between these residues. For most proteins containing tyrosine and tryptophan residues, excitation at or near 280 nm (close to the absorption maxima for both residues) results in tryptophan fluorescence. Fluorescence from the phenolic side chain of tyrosines in proteins is often quenched because of different mechanisms, including energy transfer to tryptophan. However, in a number of proteins, tyrosines contribute significantly to the fluorescence spectrum in the 300 to 350 nm wavelength range. To selectively observe tryptophan fluorescence in such proteins, excitation at the red edge of the tryptophan absorption spectrum (typically 295 to 300 nm) is used (68). Although this reduces the sensitivity of excitation (and consequently of emission), the differential absorption between tryptophan and tyrosine is high at these wavelengths where tyrosine absorption is minimal. For aFGF in its properly folded form, the fluorescence intensity of its single tryptophan residue is dramatically quenched, presumably because of electrostatic interactions with charged residues, and can only be observed, with a very low quantum yield, upon red edge excitation (69). When excited at 280 nm, folded aFGF manifests tyrosine fluorescence and no detectable tryptophan fluorescence. As the temperature is raised and the protein starts to unfold, tryptophan fluorescence begins to emerge with a maximum wavelength characteristic of solvent-exposed tryptophan. Changes in tyrosine and tryptophan fluorescence intensities were simultaneously monitored as a function of temperature to study thermal unfolding of aFGF (55,69). These measurements have helped identify a number of polyanionic stabilizers of aFGF.

The relatively simple measurements of steady-state intrinsic fluorescence spectra of proteins, as described above, can provide a wealth of information about protein unfolding and aid in developing stabilizing formulations of proteins. Extrinsic probes such as anilinonaphthalene sulfonic acid (ANS) and (p-toluidiny)naphthalene sulfonic acid (TNS) that noncovalently bind to hydrophobic surfaces have also been widely used in protein unfolding studies, especially in the context of studying partially unfolded intermediates (70). These and other applications of fluorescence spectroscopy to unfolding and thermal stability studies, such as nonradiative energy transfer measurements, have been recently reviewed (32). It should be noted that light-scattering artifacts make it difficult for fluorescence spectroscopy to be used for protein formulations that are not clear solutions or that contain large aggregates or particles. Front face fluorometry or alternative detection geometries, coupled with careful light scattering corrections, can still yield valuable information in such cases (31).

Near-UV CD signals in the 250 to 300 nm range result from asymmetric environment experienced by aromatic amino acid side chains and in the tertiary structures of folded proteins. Unfolding creates increased mobility and decrease or loss of induced asymmetry. This technique is, therefore, complementary to fluorescence in monitoring subtle changes in tertiary structure of proteins. Phenylalanine, tyrosine, and tryptophan are usually well resolved. For guanidine-hydrochloride-induced unfolding of a chimeric protein, TP40, with potentially chemotherapeutic activity, near-UV CD, and fluorescence of tryptophan residues produced comparable unfolding transitions (63). The sensitivity of this technique, in terms of sample concentration requirement, is lower than that of fluorescence. Furthermore, the extraction of quantitative information from near-UV CD spectra is often cumbersome. Theoretically, a quantitative description of a protein's tertiary structure based on its near-UV CD spectrum is attainable, but it is rarely done because detailed information of transition dipole moment orientations, interactions, and environments are not available.

Far-UV CD spectroscopy is frequently used as a powerful technique to monitor changes in the protein's secondary structure that might arise from thermal and other kinds of perturbation. Each element of the secondary structure of a protein's peptide backbone, that is, α-helices, β-sheets, turns, and random coils, manifests different but overlapping spectral features. Protein unfolding results in a decrease in the amount of ordered secondary structure, for example, α-helices and β-sheets, which, for thermal unfolding, can be conveniently monitored by measuring the CD signal at one or more wavelengths as a function of temperature. For an accurate description of the thermal unfolding transition, careful baseline correction may be necessary. This has been achieved by simultaneously monitoring, as a function of temperature, the background signal at a wavelength where proteins do not manifest CD (71,72).

Amide I FTIR spectroscopy is complementary to CD in monitoring the secondary structure of the peptide backbone. The biggest advantage of FTIR, especially in the context of protein formulation, is that the protein sample does not have to be in a clear solution state. Samples in different physical states can be studied, including solids, gels, suspensions, and solutions. Different sample geometries may be used, including transmission, attenuated total reflection, and diffuse reflectance. The choice of the experimental design often depends on the physical state of the sample being used (62). For example, lyophilized samples can be directly studied in the dry form using the diffuse reflectance technique. Although KBr pellets of dry protein samples have often been used to make transmission measurements, it should be recognized that KBr is a chaotropic agent and may cause perturbation of the protein structure, although chances for protein unfolding are reduced in the dry state. The effect of freeze-drying on structures of proteins has been studied by FTIR spectroscopy and potential stabilizers against freezing-induced structural damage have been identified (52). Infrared spectroscopy has also been used to evaluate the effectiveness of formulations in protecting the secondary structural integrity of proteins both in solution and in dried states. This necessitates making quantitative comparisons of the overall similarity of infrared spectra in the conformational sensitive amide I region. The study of temperature-induced protein unfolding using FTIR is generally performed by following spectral change with temperature at a fixed absorption band (e.g., amide I or I′) or wavelength (e.g., 1,640 cm^{-1}). The plot of FTIR intensity at a fixed wavelength versus temperature should yield the protein unfolding transition temperature (73). Detailed discussions on the application of FTIR to study protein conformational stability and thermal unfolding are available elsewhere (62,74–81). Protein aggregation as a result of thermal unfolding can also be detected by appearance of bands around 1,610 and 1,685 cm^{-1}, as demonstrated for chymotrypsinogen (82). The major disadvantages of FTIR are (*1*) In solution, water signal interferes with the amide I signal and must be carefully subtracted or moved away from the spectral range of interest by dialyzing the sample into D_2O. (*2*) Deconvolution of the rather broad amide I band can, to some extent, be subjective. (*3*) There is significant spectral overlap between peptide bonds in random coil and α-helical conformation. (*4*) The sensitivity for aqueous protein samples is approximately two orders of magnitude lower compared to far-UV CD. As for all the biophysical techniques described here, FTIR should be used for monitoring changes in conformation as a result of a perturbation, such as temperature, and should not be viewed as a technique to determine structure contents in absolute terms. To this end, second-derivative FTIR spectroscopy has been shown to be a valuable comparative tool, and second derivatives of amide I spectra have been particularly useful in monitoring structural changes.

It should be noted that temperature-induced protein unfolding can often be monitored by simply following UV absorbance changes in the far-UV (peptide bond absorption) and near-UV (aromatic side chains) regions. This was demonstrated for ribonuclease A and equinatoxin II by Fink and Painter (83) and Poklar et al. (84). However, this is not as commonly used a method as fluorescence and CD because (*1*) the magnitude of UV absorbance change between folded and unfolded protein states is usually much smaller than the change observed by fluorescence and CD and (*2*) the potential light scattering contribution resulting from protein aggregation that often accompanies unfolding can introduce significant inaccuracy. The resolution of near-UV absorbance spectroscopy, which is inherently a low-resolution technique, can be significantly improved by using second-derivative analysis (85). In addition, UV spectrophotometry offers a simple method to monitor temperature-induced protein aggregation. This method, which is usually called the turbidity assay and frequently used in accelerated thermal stability studies, takes advantage of the increase in total optical density resulting from light scattering as large protein aggregates are formed. The intensity of light scattering is proportional to the inverse fourth power of the wavelength of light so that greater sensitivity in measuring light scattering is achieved at lower wavelengths. Typically, a satisfactory balance between measurable light-scattering intensity and minimal protein absorbance is accomplished in the 300 to 350 nm wavelength range (4). The turbidity assay was one of the methods used to study thermal stabilization of basic and acidic FGF (45,56).

The need to use multiple techniques to study thermal unfolding of proteins cannot be overemphasized. When different techniques yield different T_m values, the pathway of unfolding most likely involves one or more partially folded intermediates. For aFGF in the presence of the stabilizer, heparin, fluorescence, CD, and DSC measurements yielded apparent T_m values of 62 °C, 72 °C and 71 °C, respectively (31). This suggests that local unfolding in tertiary structures precede a more global unfolding in which the ordered secondary structure of the peptide backbone is disrupted. For another potentially therapeutic protein, TP4O, produced as chimera of a truncated form of *Pseudomonas exotoxin* and transforming growth factor-α, tryptophan fluorescence, far-UV CD, and DSC gave apparent T_m values of 42, 55, and 48 °C. In this case also, a local unfolding in one or more parts of the tertiary structure precedes global unfolding. The in vitro cell killing activity of the protein was found to be irreversibly lost at 50 °C, and the hydrodynamic radius was increased to 12 nm, compared to 4 nm at 23 °C as measured by dynamic light scattering, suggesting aggregation (71,86).

Calorimetry

Differential scanning calorimetry (DSC) is another frequently used method in protein structural stability studies. Changes in heat capacity as a function of temperature, obtained from DSC measurements, provide a direct description of the thermal unfolding transitions of proteins. A good example (Fig. 4) is the study of the effects of various

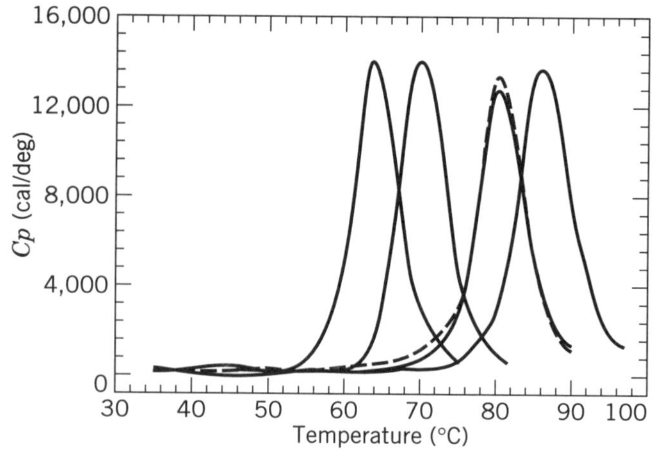

Figure 4. DSC determination of sarcosine effect on the thermal unfolding transition (T_m) of ribonuclease A. Endotherms from lower to higher T_m represent sarcosine concentrations of 0.0, 2.0, 5.5, and 8.2 M. Both of first scan and the rescan are shown for sample with 5.5 M sarcosine (solide and dashline). Data were adapted from reference 47 with permission.

excipients on thermal unfolding transitions of ribonuclease A (43,47,48). The rescanned heat flow profile indicates that thermal unfolding of ribonuclease A is a reversible process (43). In theory, calculation of thermodynamic parameters such as Gibbs free energy can only be made for reversible transitions (66,87). Thermal unfolding of proteins is often irreversible because of aggregation of the unfolded or partially unfolded states. Nevertheless, in protein formulation studies, potential stabilizers can be screened by monitoring their effects on the heat flow profile of a protein scanned from low to high temperature. For example, DSC studies provided critical data in support of stabilizing effects of heparin and osmolytes against thermal unfolding and (irreversible) aggregation of aFGF and lysozyme (43,45,48).

Hydrodynamic Measurements

Unfolding of a protein is usually associated with an apparent increase in its hydrodynamic size and changes in sedimentation and diffusion coefficients. In addition, if the partial or fully unfolded protein states are unstable and form aggregates, the overall hydrodynamic properties of the protein in the solution will change. The techniques for monitoring changes in hydrodynamic properties of proteins include capillary electrophoresis (CE), SEC-HPLC, analytical ultracentrifugation, dynamic light scattering (DLS), and small-angle X-ray scattering (SAXS).

With the recent introduction of a compact and easy-to-use analytical ultracentrifuge that offers both optical absorption and interferometric optics for real-time monitoring of sedimentation, interest has grown exponentially in using equilibrium and velocity sedimentation techniques to measure self-association of proteins (88–90). Equilibrium sedimentation offers a thermodynamically rigorous way of determining molecular weights. The sedimentation properties (sedimentation coefficient) of a protein sample are controlled by several physical factors of the protein and protein complex with other solution components. These factors include protein mass, geometry, and density and are affected by protein oligomerization, aggregation, unfolding, and degradation. Both velocity and equilibrium sedimentation experiments can provide information about protein hydrodynamic properties (33,71). Velocity sedimentation experiments can be performed in only a few hours, although diffusion coefficient is required for obtaining a semiquantitative measure of mass change caused by aggregation.

Based on the difference between the electrophoretic mobility of the native protein and that of the unfolded one, the temperature-induced unfolding transitions of proteins can be analyzed by CE. This technique offers the advantage of very high sensitivity, although it has not yet found wide usage in protein unfolding studies. Ishihama et al. (91) used an in-column incubation method to monitor thermal unfolding of nanogram quantities protein. The thermodynamic parameters of the unfolding transition process were obtained by analyzing the temperature dependence of the electrophoretic mobilities. More information on the application of CE in protein analysis can been found in literature (91–93).

SEC allows separation of proteins based on their molecular shape and size. An unfolded or aggregated protein generally possesses a larger hydrodynamic size than the native protein and is eluted out of the column earlier (i.e., with a smaller elution volume) than the native protein. This method has been used to estimate molecular sizes of proteins under different denaturing conditions (13,63,94). The application of various signal detection techniques in protein gel–filtration analysis has been discussed (32).

Static light scattering (SLS), DLS, and SAXS from protein solutions can all be applied to the measurement of changes in the molecular sizes of proteins induced, for example, by thermal unfolding. SLS measurements normally permit reliable data of molecular weights and radii of gyration for relatively small particles since light scattering of large particles is a very complex function of particle shape and sensitive to detection angle. DLS measures the time-dependent intensity fluctuations of scattered light. The correlation of these fluctuations is related to the diffusion coefficient of the scattering particle, a quantity linked to the hydrodynamic radius through the Stokes-Einstein equation (95). SAXS is a very useful, but underused, technique for characterizing size, compactness, and shape of protein molecules in solution. In addition to determining the overall radius of gyration and the maximum dimension, SAXS can quantitatively describe the globular conformation of a protein molecule, as well as denaturant- and ligand-induced conformational changes in solution (96–98). Detailed discussions on the applications of SLS, DLS, and SAXS on protein conformational change and unfolding are available in the literature (27,32,99–102).

Kinetic Techniques

For a complete description of the thermal unfolding kinetics, a fast time resolution may be necessary. Stopped-flow instrumentation with absorption and fluorescence optics allows studies of unfolding kinetics with a dead time of about 2 ms. Stopped-flow CD instrumentation is also available. The instrumentation of time-resolved CD (TRCD) has been reviewed by Lewis (103). Time-resolved FTIR spectroscopy (104) can complement kinetic CD studies in identifying the faster phases of folding or unfolding of a protein's secondary structure. Techniques for studying fast protein structural change at an atomic level include hydrogen–deuterium exchange in combination with multidimensional NMR, electron spray mass spectrometer, and FTIR spectroscopy (105,106). These techniques monitor the spectral and mass changes induced by exchange of amide protons with deuterium during the folding or unfolding process.

CONCLUSION

Thermal unfolding studies provide formulation scientists a way to evaluate the intrinsic stability of pharmaceutical proteins and to screen excipients that might lead to stabilizing formulations during storage, handling, and administration. The data obtained from thermal unfolding studies can support the design and execution of the formulation processing steps leading to pharmaceutical products. Because temperature is one of the most practically important parameters in testing stability of a protein from small- and large-scale production, detailed information of a protein's conformational stability at various temperatures is critically needed. A rational approach to stable formulation development requires a thorough understanding of the mechanism of protein inactivation. Normally, thermal unfolding experiments provide an understanding of both the physical and chemical events leading to protein inactivation, thereby opening up a faster path to development of stabilizing formulations for protein-based pharmaceuticals and vaccines.

BIBLIOGRAPHY

1. M.C. Manning, K. Patel, R.T. Borchardt, *Pharmacol. Res.* **6**, 903–918 (1989).
2. R.D. Schmid, *Adv. Biochem. Eng.* **12**, 41–118 (1979).
3. C.R. Cantor and P.R. Schimmel, *Biophysical Chemistry, Part I: The Conformation of Biological Macromolecules*, Freeman, New York, 1980, pp. 7–12.
4. R.A. Copeland, *Methods for Protein Analysis: A Practical Guide to Laboratory Protocols*, Chapman & Hall, New York, 1994.
5. T.E. Creighton, *Protein: Structure and Molecular Properties*, Freeman, New York, 1984, pp. 159 and 220.
6. L. Stryer, *Biochemistry*, 3rd ed., Freeman, New York, p. 31, 1988.
7. D.B. Volkin and C.R. Middaugh, in T.J. Ahern and M.C. Manning eds., *Stability of Protein Pharmaceuticals, part A, Chemical and Physical Pathways of Protein Degradation*, Plenum, New York, 1992, pp. 109–134 and 215–247.
8. D. Shortle, *J. Biol. Chem.* **264**, 5315–5318 (1989).
9. C.N. Pace, *CRC Crit. Rev. Phys. Chem.* **38**, 463–488 (1975).
10. B.A. Shirley, in T.J. Ahern, and M.C. Manning eds., *Stability of Protein Pharmaceuticals, part A, Chemical and Physical Pathways of Protein Degradation*, Plenum, New York, 1992, pp. 167–194.

11. C.G. Anfinsen and H.A. Scheraga, *Adv. Protein Chem.* **29**, 205–301 (1975).
12. P.A. Jennings and P.E. Wright, *Science*, **262**, 892–896 (1993).
13. R.D. Palleros, L. Shi, K.L. Reid, and A.L. Fink, *Biochemistry* **32**, 4314–4321 (1993).
14. F.M. Hughson, P.E. Wright, R.L. Baldwin, *Science* **249**, 1544–1548 (1991).
15. D. Atkinson and D.M. Small, *Annu. Rev. Biophys. Chem.* **15**, 403–456 (1986).
16. R.T. Nolte and D. Atkinson, *Biophys. J.* **63**, 1221–1239 (1993).
17. D.L. Barbeau, A. Jonas, T. Teng and A.M. Scanu, *Biochemistry* **18**, 362–369 (1979).
18. J.A. Reynolds, *J. Biol. Chem.* **251**, 6013–6015 (1976).
19. J.M. Donovan, G.B. Benedek, and M.C. Carey, *Biochemistry* **26**, 8116–8125 (1987).
20. J.L. Cleland, M.F. Powell, and S.J. Shire, *Crit. Rev. Ther. Drug Carrier Systems* **10**, 307–377 (1993).
21. L.R. De Young, A.L. Fink, and K.A. Dill, *Acc. Chem. Res.* **26**, 614–620 (1993).
22. Y.J. Wang and M.A. Hanson, *J. Parenter. Sci. Technol.* **42**, S3–S26 (1988).
23. C.K. Gupta, J. Leszcynski, R.K. Gupta, and G.R. Siber, *Vaccine* **14**, 1417–1420 (1996).
24. J.F. Capenter, S.J. Prestrelski, T.J. Anchordoguy, and T. Arakawa, in J.L. Cleland and R. Langer eds., *Formulation and Delivery of Proteins and Peptides*, American Chemical Society, 1994, pp. 137–147.
25. A.T. Da Poian, A.C. Oliveira, and J.L. Silva, *Biochemistry* **34**, 2672–2677 (1995).
26. B.M. Eckhardt, J.Q. Qeswein, and T.A. Bewley, *Pharm. Res.* **8**, 1360–1364 (1991).
27. I. Nishii, M. Kataoka, F. Tokunaga, and Y. Goto, *Biochemistry* **33**, 4903–4909 (1994).
28. R. Pearlman and T.J. Nguyen, *Pharm. Pharmacol.* **44**, 178–185 (1992).
29. A. Shaw and R. Bott, *Curr. Opin. Struct. Biol.* **6**, 546–550 (1996).
30. S.N. Timasheff, in B.A. Shirley ed., *Protein Stability and Folding: Theory and Practice*, Humana, Totowa, N.J., 1995, pp. 219–270.
31. D.B. Volkin and C.R. Middaugh, in R. Pearlman and Y.J. Wang ed., *Formulation, Characterization, and Stability of Protein Drugs*, Plenum, New York, 1996, pp. 181–217.
32. D.B. Volkin, G. Sanyal, C.J. Burke, and C.R. Middaugh, in M. Akers and S. Nail ed., *Development and Manufacture of Protein Pharmaceuticals*, Plenum, in press.
33. D.B. Volkin and A.M. Klibanov, in T.E. Creighton ed., *Protein Function: A Practical Approach*, Oxford Press, Oxford, U.K., 1989, pp. 1–24.
34. D.R. Bachvarov, G.G. Markov, and I.G. Ivanov, *Int. J. Biochem.* **19**, 963–971 (1935).
35. P. Gunning, T. Mohun, S.Y. Ng, P. Ponte, and L. Kedes, *J. Mol. Evol.* **20**, 202–214 (1984).
36. E. Odermatt, J. Risteli, V. van-Delden, and R. Timpl, *Biochem. J.* **211**, 295–302 (1983).
37. C.N. Pace, B.A. Shirley, J.A. Thomson, in T.E. Creighton ed., *Protein Structure: A Practical Approach* IRL, Oxford, U.K., 1989, p. 311.
38. R.D. Palleros, L. Shi, K.L. Reid, and A.L. Fink, *J. Biol. Chem.* **269**, 13107–13114 (1994).
39. T. Porumb, P. Yau, T.S. Harvey, and M. Ikura, *Protein. Eng.* **7**, 109–115 (1994).
40. L.R. Scolnick, Z.F. Kanyo, R.C. Cavalli, D.E. Ash, and D.W. Christianson, *Biochemistry* **36**, 10558–10565 (1997).
41. O. Slodowski, J. Bohm, B. Schone, and B. Otto, *Eur. J. Biochem.* **202**, 1133–1140 (1991).
42. H. Takagi, T. Takahashi, H. Momose, M. Inouye, Y. Maeda, H. Matsuzawa, and T. Ohta, *J. Biol. Chem.* **265**, 6874–6888 (1990).
43. M.M. Santoro, Y. Liu, S.M.A. Khan, L. Hou, and D.W. Bolen, *Biochemistry* **31**, 5278–5283 (1992).
44. M. Eftink, *Biophys. J.* **66**, 482–501 (1994).
45. P.K. Tsai, D.B. Volkin, J.M. Dabora, K.C. Thompson, M.W. Bruner, J.O. Gress, B. Matuszewska, M. Keogan, J.V. Bondi, and C.R. Middaugh, *Pharm. Res.* **10**, 649–659 (1993).
46. C.J. Burke, D.B. Volkin, H. Mach, and C.R. Middaugh, *Biochemistry* **32**, 6419–6426 (1993).
47. C.Q. Hu, J.M. Sturtevant, J.A. Thamson, R.E. Erickson, and C.N. Pace, *Biochemistry* **31**, 4876–4882 (1992).
48. Y. Liu and J.M. Sturtevant, *Biochemistry* **35**, 3059–3062 (1996).
49. L. Shi, D.R. Palleros, and A.L. Fink, *Biochemistry* **33**, 7536–7546 (1994).
50. J.A. Schellman, *Biopolymers* **15**, 999–1000 (1976).
51. T.F. Busby, D.H. Atha, and K.C. Ingham, *J. Biol. Chem.* **256**, 12140–12147 (1981).
52. S.J. Prestrelski, N.T. Tedeschi, T. Arakawa, and J.F. Carpenter, *Biophys. J.* **65**, 661–671 (1993).
53. C.N. Pace, *CRC Crit. Rev. Biochem.* **3**, 1–43 (1975).
54. M. Straume, *Methods Enzymol.* **240**, 530–568 (1994).
55. M.K. Eidsness, K.A. Richie, A.E. Burden, D.M. Kurtz Jr., R.A. Scott, *Biochemistry*, **36**, 10406–10413 (1997).
56. Y.J. Wang, Z. Shahrokh, S. Vemuri, G. Eberlein, I. Beylin, M. Busch, *Pharm-Biotechnol.* **9**, 141–180 (1996).
57. Food and Drug Administration (FDA), *Fed. Regist.* **61**, 36466–36469 (1996).
58. D.B. Volkin, A.M. Verticelli, K.E. Marfia, C.J. Burke, H. Mach, and C.R. Middaugh, *Biochim. Biophys. Acta* **1203**, 18–26 (1993).
59. R. Pearlman and T.J. Nguyen, in D. Marshak and D. Liu eds., *Therapeutic Peptides and Proteins: Formulation, Delivery and Targeting*, Cold Spring Harbor Laboratory, Cold Spring Harbor, New York, 1989, pp. 23–31.
60. G.A. Eberlein, P.R. Stratton, and Y.J. Wang, *PDA J. Pharm. Sci. Technol.* **48**, 224–230 (1994).
61. M. Bloemendal and W.C. Johnson, Jr., in J.N. Herron, W. Jiskoot, and D.J.A. Crommelin ed., *Physical Methods to Characterize Pharmaceutical Proteins*, Plenum, New York, 1995.
62. C.R. Middaugh, H. Mach, J.A. Ryan, G. Sanyal, and D.B. Volkin, in B.A. Shirley ed., *Protein Stability and Folding: Theory and Practice*, Humana, Totowa, N.J., 1995, pp. 137–156.
63. J.O. Gress, D. Marquis-Omer, C.R. Middaugh, and G. Sanayal, *Biochemistry* **33**, 2620–2627 (1994).
64. W. Jiskoot, V. Hlady, J.J. Naleway, and J.N. Herron in J.N. Herron, W. Jiskoot, and D.J.A. Crommelin, eds., *Physical Methods to Characterize Pharmaceutical Proteins*, Plenum, New York, 1995, pp. 1–53.
65. J.I. Boye, I. Alli, and A.A. Ismail, *J. Agric. Food. Chem.* **45**, 1116–1125 (1997).

66. K. Gekko, *J. Biochem.* **91**, 1197–1204 (1982).
67. P.L. Privalov and N.N. Khechinashvili, *J. Mol. Biol.* **86**, 665–684 (1974).
68. J.R. Lakowicz, *Principles of Fluorescence Spectroscopy*, Plenum, New York, 1983.
69. R.A. Copeland, H. Ji, A.J. Halfpenny, R.W. Williams, K.C. Thompson, W.K. Herber, K.A. Thomas, M.W. Bruner, J.A. Ryan, D. Marquis-Omer, G. Sanyal, R.D. Sitrin, S. Yamazaki, and C.R. Middaugh, *Arch. Biochem. Biophys.* **289**, 53–61 (1991).
70. K. Kuwajima, *Proteins: Struct., Funct., Genet.* **6**, 87–103 (1989).
71. G. Sanyal, D. Marquis-Omer, J.O. Gress, and C.R. Middaugh, *Biochemistry* **32**, 3488–3497 (1993).
72. G. Sanyal, D. Marquis-Omer, J.O.L. Waxman, H. Mach, J.A. Ryan, J.O. Gress, and C.R. Middaugh, *Biochim. Biophys. Acta.* **1249**, 100–108 (1995).
73. M. Menedez, M. Gasset, J. Laynez, C. Lopez-Zumel, P. Usobiaga, and E. Toper-Petersen, *Eur. J. Biochem.* **234**, 887–896 (1995).
74. J. Bandeker, *Biochim. Biophys. Acta.* **1120**, 123–143 (1992).
75. A. Dong, S.J. Prestrelski, S.D. Allison, and J.F. Carpenter, *J. Pharm. Sci.* **84**, 415–424 (1995).
76. K. Kato, T. Matsui, and S. Tanaka, *Appl. Spec.* **41**, 861–864 (1987).
77. H. Fabian, C. Schultz, D. Naumann, O. Landt, O. Hahn, and W. Saenger, *J. Mol. Biol.* **232**, 967–981 (1993).
78. I.E. Holzbaur, A.M. English, and A.A. Ismail, *Biochemistry* **35**, 5488–5494 (1996).
79. A. Nabet, J.M. Boggs, and M. Pezolet, *Biochemistry* **33**, 14792–14799 (1994).
80. S. Prestrelski, G.M. Fox, and T. Arakawa, *Arch. Biochem. Biophys.* **293**, 314–319 (1992).
81. I.H.M. Van Stokkum, H. Linsdell, J.M. Hadden, P.I. Haris, D. Chapman, and M. Bloemendal, *Biochemistry* **34**, 10508–10518 (1995).
82. A.A. Ismail, H. Mantsch, and P.T.T. Wong, *Biochim. Biophys. Acta* **1121**, 183–188 (1992).
83. A.L. Fink and B. Painter, *Biochemistry* **26**, 1665–1671 (1987).
84. N. Poklar, J. Lah, M. Salobir, P. Macek, and G. Vesnaver, *Biochemistry* **36**, 14345–14352 (1997).
85. H. Mach, G. Sanyal, D.B. Volkin, and C.R. Middaugh, in Z. Sharokh, V. Shuzky, J.L. Cleland, S.J. Shire, and T.W. Randolph eds., *Therapeutic Protein and Peptide Delivery*, ACS Symposium Series 675, 1997. pp. 186–205.
86. G. Sanyal, D. Marquis-Omer, and C.R. Middaugh, in R. Pearlman and Y.J. Wang eds., *Formulation, Characterization and Stability of Protein Drugs: Case Histories*, Plenum, New York, 1996, pp. 365–392.
87. A. Tanaka, J. Flanagan, and J. Sturtevant, *Protein Sci.* **2**, 567–576 (1992).
88. J. Behlke, *Eur. Biophys. J. Biophys. Lett.* **25**, 319–323 (1997).
89. H. Durchschlag, P. Zipper, G. Purr, and R. Jaenicke, *Colloid Polymer Sci.* **274**, 117–137 (1996).
90. P.G. Varley, A.J. Brown, H.C. Dawkes, and N.R. Burns, *Eur. Biophys. J.* **25**, 437–443 (1997).
91. Y. Ishihama, Y. Oda, N. Aksakawa, and M. Iwakura, *Anal. Sci.* **13**, 913–938 (1997).
92. Z. Deyl and R. Struzinsky, *J. Chromatogr.* **569**, 63–122 (1991).
93. T.A. Van De Goor, *Pharm-Biotechnol.* **7**, 301–327 (1995).
94. Y. Endo, H. Nagai, Y. Watanabe, K. Ochi, and T. Takagi, *J. Biochem. (Tokyo)* **112**, 700–706 (1992).
95. D.E. Koppel, *J. Chem. Phys.* **57**, 4814–4820 (1972).
96. M. Kataoka, J.F. Head, B.A. Seaton, and D.M. Engelman, *Proc. Natl. Acad. Sci. U.S.A.* **86**, 6944–6948 (1989).
97. M. Kataoka, Y. Hagihara, K. Mihara, and Y. Goto, *J. Mol. Biol.* **229**, 591–596 (1993).
98. L. Shi, M. Kataoka, and A.L. Fink, *Biochemistry* **35**, 3297–3308 (1996).
99. S. Beretta, L. Lunelli, G. Chirico, and G. Baldini, *Appl. Optics* **35**, 3763–3770 (1996).
100. J.M. Flanagan, M. Kataoka, D. Shortle, and D.M. Engelman, *Proc. Natl. Acad. Sci. U.S.A.* **89**, 748–752 (1992).
101. M. Kataoka, I. Nishii, T. Fujisawa, T. Ueki, F. Tokunaga, and Y. Goto, *J. Mol. Biol.* **249**, 215–228 (1995).
102. E.E. Lattman, K.M. Fiebig, and K.A. Dill, *Biochemistry* **33**, 6158–6166 (1994).
103. J.W. Lewis, R.A. Goldback, D.S. Kliger, X. Xie, R.C. Dunn, and J.D. Simon, *J. Phys. Chem.* **96**, 5243–5245 (1992).
104. D. Reinstadler, H. Fabian, J. Backmann, and D. Naumann, *Biochemistry* **35**, 15822–15830 (1996).
105. J. Backmann, H. Fabian, and D. Naumann, *FEBS Lett.* **364**, 175–178 (1995).
106. J. Backmann, C. Schultz, H. Fabian, U. Hahn, W. Saenger, and D. Naumann, *Proteins* **24**, 379–387 (1996).

See also DENATURATION, PROTEINS, SOLVENT MEDIATED; PROTEIN AGGREGATION, DENATURATION.

THERMOLYSIN

SATOSHI HANZAWA
Tokyo Research Laboratory
Hayakawa, Japan
SHUN-ICHI KIDOKORO
Sagami Chemical Research Center
Kanagawa, Japan

KEY WORDS

Aspartame
Site-directed mutagenesis
Thermolysin

OUTLINE

INTRODUCTION

Thermolysin (EC 3.4.24.4) is a thermostable metalloprotease produced by *Bacillus thermoproteolyticus* Rokko (1). It is the most thermostable protease currently produced for industrial use. Thermolysin and similar microbial metalloproteases require a metal ion, usually Zn^{2+}, for their catalytic activity. Formerly, they were called neutral proteases because their optimum pH is around pH 7. They also have a common property of substrate specificity in hydrolyzing peptide bonds involving an amino group of hydrophobic and bulky amino acid residues such as Leu or Phe (2). Among these proteases, thermolysin has been extensively studied regarding its structure and catalytic mechanism (3–6). Currently, we can easily obtain structural data of thermolysin from the Protein Data Bank. Although metalloproteases have been produced not only for laboratory use but also for industrial use, the market size has been small, because their substrate specificity is narrower compared to that of other proteases such as serine proteases or cysteine proteases. Their use has been restricted to hydrolysis of proteins that cannot be hydrolyzed by other proteases, for example, that of barley protein, which contains a specific inhibitor toward serine protease (7). Recently, however, thermolysin has been used extensively in aspartame (APM) production at Holland Sweetener Company (HSC), a joint venture of TOSOH (Japan) and DSM (the Netherlands). APM, L-α-aspartyl-L-phenylalanine methyl ester, is an artificial sweetener that is 200 times sweeter than sucrose. In APM production, thermolysin plays an important role because it catalyzes the synthesis of an APM precursor, *N*-benzyloxycarbonyl-APM (Z-APM), from Z-L-Asp and L-Phe methyl ester (L-PheOMe).

$$\text{Z-L-Asp} + \text{L-PheOMe} \underset{\text{thermolysin}}{\rightarrow} \text{Z-L-Asp-L-PheOMe}$$

Because the coupling reaction is the key step in APM production, improvement of this reaction is highly desired. Immobilization of thermolysin has been attempted as one of the improvements to establish efficient recycling of thermolysin (8,9). We sought to improve thermolysin activity by site-directed mutagenesis, that is, by protein engineering, because such activity enhancement not only results in the reduction of enzyme consumption but also circumvents the enzyme recovery step, if enzyme cost is reduced to a negligible amount. It thus also can affect the fixed cost of APM production. In this article, we explain the results of our study to improve thermolysin.

ENHANCEMENT OF THERMOLYSIN ACTIVITY

Before introduction of our improvement study, several other reports regarding activity enhancement of thermolysin are described herein (Table 1). It is well known that enzyme activity is affected by additives in the reaction mixture, such as salts and organic solvents. Thermolysin is also affected by additives, which results in either enhancement or reduction of the activity. Thermolysin activity is enhanced by almost all salts consisting of monovalent anions and cations (10,11). In the case of alcohol, *n*-pentanol at concentrations below saturation enhances activity while *n*-butanol is inhibiting (12). The effect of alcohol is highly dependent on physicochemical properties such as log P or ϵ (13). Pressure is known to be another factor that enhances activity (14). The mechanism of the activation by salts, organic solvents, or high pressure is not clear, although a slight structural change of the enzyme by salts has been observed spectrophotometrically (10). We thought that activity enhancement by these methods on an industrial scale is not feasible from the economical point of view, because expensive equipment would be required to avoid corrosion by salts or explosion of solvents or to resist high pressure. Additionally, such additives result in the increase of industrial wastes.

Kubó et al. (15) demonstrated activity enhancement of thermolysin for casein hydrolysis (Table 1). Mutant enzymes can be expected to be more effective for industrial use compared to salts, organic solvents, or high pressure. However, activity for casein hydrolysis does not correlate with that for hydrolysis or synthesis of dipeptide derivatives as described later. Two thermolysin mutants, Y110W and Y211W, have not been evaluated for their activity for Z-APM hydrolysis or synthesis.

IMPROVEMENT BY PROTEIN ENGINEERING

As described, thermolysin activity can be enhanced by additives. However, we preferred improvement by protein engineering because the use of thermolysin mutants is more cost effective. The strategy, method, and results of our study are described next.

Strategy

In recent protein engineering studies, thermal stability of proteins can be enhanced by introducing or reinforcing one or more of several physicochemical intramolecular interactions, such as hydrogen bonding, S–S bridges, and so forth. Most attempts to modify the activity, however, have resulted in suppression of biological activity. To find a novel strategy for producing more active enzymes, we conducted a "protein engineering cycle" consisting of three elements—design, synthesis, and evaluation—as shown in the following scheme:

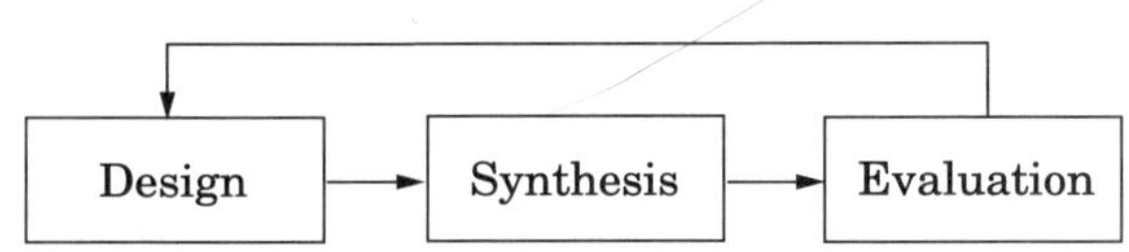

In this scheme, feedback from evaluation to design is highly important. Feedback can revise our working hypothesis for the structure–function relationships. Working with this cycle will not only give us novel useful proteins

Table 1. Activity Enhancement of Thermolysin by Additives, Pressure, and Site-Directed Mutagenesis

Additive	Condition	Substrates	Reaction	Relative activity[a]	Reference
NaCl	4 M	Z-APM	Hydrolysis	9.0	10
NaCl	2.7M	Z-APM	Synthesis	4.2	10
NaCl	4 M	FaGLa	Hydrolysis	13	11
N-Pentanol	60% saturation	Z-FFM	Synthesis	4.7	12
Pressure	1,000 psi	FaGLA	Hydrolysis	4.5	14
Mutation	Y110W	Casein	Hydrolysis	1.6	15
Mutation	Y211W	Casein	Hydrolysis	1.5	15

Z-APM, *N*-benzyloxycarbonyl-aspartame; Z-FFM, Z-Phe-PheOMe; FaGLa, Fa-Gly-Leu-Ala.
[a]Relative activity compared with activity in absence of the additives.

as products but also gives us worthwhile knowledge about the mechanisms underlying the relationship between protein structure, properties, and functions.

Although enzymes do not change the thermodynamic equilibrium between starting materials and products, they accelerate reaction rates by stabilizing the transition state of the reaction. This stabilization is realized enthalpically or entropically, and the Gibbs energy of the transition state is decreased. It is well known that one of the stabilization forces for the transition state is electrostatic interaction (including hydrogen bonding) between an enzyme and its ligand. Several ionizable amino acids such as Asp, Glu, and His play an important role in this stabilization. The equilibrium between the protonation and deprotonation states of these residues, which is mainly determined by electrostatic interaction, controls enzyme activity. Usually the change of protonation state from its active form completely eliminates activity. The pH profile of enzyme activity is generally thought to reflect the protonation state. Therefore, changes in electrostatic potential at the active site may affect both the stability of the transition state and the protonation equilibrium. The optimum pH for the activity of thermolysin is known to be around 7 (1). The synthesis of Z-APM is, however, carried out industrially under a more acidic pH because the methyl ester of the substrate, phenylalanine methyl ester, and the product are hydrolyzed rapidly above pH 7 (16). If the optimum pH can be shifted to the acidic pH region, enzymatic activity in this region may be enhanced. To accomplish this we decided to change the electrostatic potential around the active site.

Design for Higher Activity

There are two residues, Glu 143 and His 231, and zinc ion, Zn^{2+}, in the active site of this enzyme (5,6). We decided to change first the electric charge of Gln 119 to affect the electrostatic potential at Glu 143 (17,18). Gln 119 is known to form a hydrogen bond to the side chain of Ser 103. When Gln 119 is replaced by other amino acids, that hydrogen bond is broken, thereby affecting dynamic properties and stability of the enzyme.

Synthesis of Thermolysin Mutants

The site-directed mutagenesis at the 119th site was performed by the overlap extension method (18,19) using the polymerase chain reaction. The mismatched primer and its complementary sequence were synthesized and used as inside primers, which anneal to the same segment of *npr*M gene coding for thermolysin. The mismatches lead to random sequence alteration in the 119th amino acid residue. The amplified DNA fragment containing random mutation was digested with *Bam*HI and *Sph*I, and substituted for the same region of the native *npr*M gene in pUBTZ2. The *npr*M gene was cloned from *Bacillus stearothermophilus* MK232, which was isolated by TOSOH from soil to provide a protease for APM production (20).

In the early stages of investigation of this metalloprotease, it was believed that its primary structure was different from that of thermolysin at the 37th and the 119th sites (21), because the amino acid sequence of thermolysin was reported to be Asp at the 37th site and Glu at the 119th site by Titani et al. (4). More recent studies, however, revised the sequence of thermolysin as Asn at the 37th and Gln at the 119th (22,23), exactly the same as the metalloprotease from *npr*M gene. The *npr*M gene was first cloned by Kubo et al. (20) in *Bacillus subtilis* MT-2 whose genotype is *trp*C2 *leu*C7 *hsd*R *hsd*M *Npr*$^-$ (24). Expression plasmid pUBTZ2, which is a shuttle vector between *Escherichia coli* and *B. subtilis*, was constructed from pUC9, pUB110 (25), and a fragment of pMK4 (26) involving *npr*M. The thermolysin mutants were expressed in *B. subtilis* MT-2 by cultivation in 2 L broth (10 g of yeast extract, 20 g of Bacto peptone, and 5 g of NaCl in 1 L of H_2O) containing 20 μg/mL kanamycin at 37 °C for 20 h in a culture flask. The mutants were recovered by ammonium sulfate precipitation of culture supernatant and further purified by two-step column chromatography with Butyl Toyopearl 650 M (Tosoh, Japan) and TSK gel G2000SW (Tosoh, Japan).

Evaluation of Thermolysin Mutants

The amino acids at the 119th site were confirmed by sequencing the plasmid DNA. The purity and the isoelectric point were determined by gel isoelectric focusing using Phast System (Pharmacia, Sweden). The *pI* value of three mutants, namely Lys mutant (Q119K), Arg mutant (Q119R), and Met mutant (Q119M), is 5.3 while that of the other mutants and the wild type (WT) is 5.0. The difference in pI was thought to result from a change in one electric charge of the mutant, judging from the *pI* calculation of the amino acid composition.

The hydrolytic activities were evaluated by measuring hydrolysis of *N*-(3-{2-furyl}acryloyl)-glycyl-L-leucine amide (FaGLa) and *N*-(3-{2-furyl}acryloyl)-L-aspartyl-L-phenylalanine methyl ester (Fa-APM). The theoretical curve of an integral Michaelis–Menten equation (27) fitted well with all experimental data within experimental error (10^{-3} optical density). The decrease in absorbance at 345 nm was monitored by a spectrophotometer (UB-35; Jasco, Japan) connected to a thermostat circulator (EL-15; Taitech, Japan) and a personal computer (PC9801; NEC, Japan) for data acquisition. The stored data were fitted with an integral Michaelis–Menten equation by a nonlinear least-squares method to determine k_{cat} and K_m. All measurements were done under a solution condition of 200 mM Tris-maleate buffer with 10 mM $CaCl_2$ at 37 °C. The concentrations of substrate and enzyme were about 1–2 mM and 1 nM, respectively.

Protein concentration was determined by photometry with an extinction coefficient of 1.67×10^4 OD g/mL. The computer program was described in BASIC on PC9801 and FORTRAN with a program package SALS on a workstation (Titan 750; Kubota Computer, Japan). Fa-APM was synthesized using *N*-(3-{2-furyl}acryloyl)-L-Asp and L-PheOMe following a modified method of Oyama et al. (28). To check the electrostatic effect of the 119th site on activity, we selected the potentially positively charged residues, His, Lys, and Arg, and the potentially negatively charged residues, Glu and Asp. In contrast to the expectations based on the electrostatic effect, hydrolytic activity of all mutants was higher than that of WT, indicating that the main reason for the activity enhancement is not the electrostatic effect. We therefore decided to generate all mutants at this site.

The hydrolytic activity toward FaGLa of the mutants depending on pH is presented in Figure 1. The mutant Q119E exhibited the highest activity, five times higher than WT. It should be noted that the hydrolytic activities of almost all these mutants exceeded that of the WT, indicating that the amino acid at this position is selected by nature to show the lowest activity.

We then substituted Ala and Gly for Ser 103, expecting to cause the disruption of the hydrogen bond between this site and the 119th site. These two mutants, S103A and S103G, showed activity 3.0 and 2.6 times higher, respectively, than that of the WT. The similar increase in activity observed as a result of amino acid substitution at the 103rd site indicates that disruption of hydrogen bond is the main cause of activity enhancement.

We found low K_m values for another substrate, Fa-APM, and this enabled us to determine k_{cat} and K_m values separately. The $1/K_m$ values of the mutants showed a strong correlation to k_{cat}/K_m value, while k_{cat} values showed little change from the WT (18). This result indicates that the mutation increases the affinity of the substrate so that it enhances k_{cat}/K_m for FaGLa and Fa-APM (18). The hydrogen bond of Gln 119 seems to fix the part of the β sheet from the 112th to 128th residues to the other strand. Judging from the three-dimensional structure, the strand from the 112th to 118th residue is considered to contact the pep-

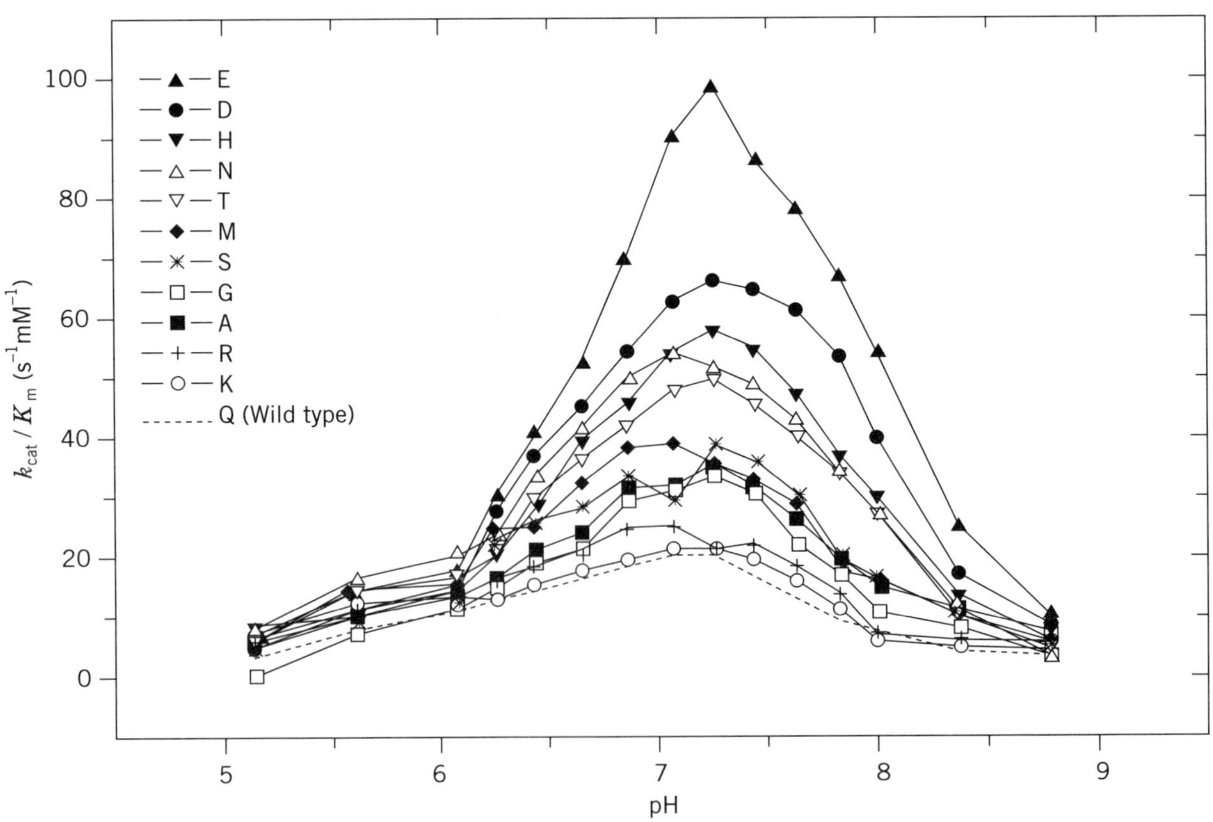

Figure 1. pH dependence of the hydrolytic activity of thermolysin and its mutants at the 119th position in 200 mM Tris-maleate buffer with 10 mM $CaCl_2$ at 37.0 °C.

tide chain of the substrate directly. The flexibility of the β strand may be increased by breakage of the hydrogen bond, thereby causing an increase in affinity of the binding site to the substrate. When flexibility is increased, it is usually observed that the thermal stability of a protein is decreased (18). In this case, a negative correlation between activity and stability is seen. If we assume that the activity is enhanced by flexibility, then there is negative correlation between flexibility and stability. Two mutants, Q119K and Q119R, however, markedly deviated from the negative correlation. From the shift of the *pI* values of these mutants, the Lys and Arg at the 119th site were supposed to be positively charged. The positive charge at this site may suppress the activity. It is reasonable that this change in electric charge affects the activity because this site is selected as a site that affects the electrostatic potential of Glu 143. Measurement of *pI* values of Q119D and Q119E showed the Asp and Glu at this site to be neutral (protonated) around pH 5.0.

From these results, we concluded that flexibility around the substrate-binding site plays an important role in enhancing the activity. The electrostatic potential also influences the change of activity. Measurement of the synthetic activity of these mutants, however, disclosed that synthetic activity of the 119th series mutants did not exceed that of WT (see Table 2). Under the conditions for the coupling reaction, the concentration of the substrates is very high so that the K_m value, which represents the affinity to substrate, does not affect the reaction rate. The important factor for enzymatic activity under the synthetic conditions, therefore, is not the K_m value but the k_{cat} value.

Reiterating the Protein Engineering Cycle

The 213th and 227th sites were then selected as two of the sites possibly most effective in changing the electrostatic potential of His 231 at the active site. Mutations of *npr*M at the 213th and 227th sites were performed by the M13 phage mutagenesis method (29).

Table 2. Activity of Thermolysin Mutants with Z-APM and Casein as Substrates

	Z-APM[a]		
Thermolysin mutant	Synthesis (kat/mol): pH 6.0	Hydrolysis (kat/mol): pH 7.0	Casein[b] (U/mg): pH 7.2
Wild type	0.06 (1.0)	3.5 (1.0)	9,200 (1.0)
Q119E	0.07 (1.1)	5.6 (1.6)	4,000 (0.4)
L144S	0.09 (1.5)	6.3 (1.8)	4,600 (0.5)
D150H	0.12 (2.0)	6.3 (1.8)	9,200 (1.0)
D150W	0.26 (4.3)	11.9 (3.4)	7,400 (0.8)
N227H	0.09 (1.5)	7.4 (2.1)	7,400 (0.8)
L144S-D150H	0.14 (2.4)	6.4 (1.8)	2,700 (0.3)
L144S-N227H	0.11 (1.8)	10.5 (3.0)	3,800 (0.4)
D150H-N227H	0.21 (3.5)	11.3 (3.2)	9,800 (1.1)
D150W-N227H	0.45 (7.5)	22.6 (6.5)	6,700 (0.7)
L144S-D150H-N227H	0.30 (5.0)	21.5 (6.1)	3,200 (0.4)
L144S-D150W-N227H	0.63 (10.5)	37.4 (10.7)	1,900 (0.2)

Note: Numbers in parentheses are activity relative to wild type.
[a]The activity for Z-APM, that is, the synthesis or hydrolysis of 1 mol Z-APM in 1 s, was defined as 1 katal (kat).
[b]The activity for casein, that is, the release of acid-soluble peptides corresponding to the absorbance of 1 μg of tyrosine in 1 min, was defined as 1 unit (U).

In the case of Asp 213 mutants, four mutations, D213N, D213H, D213K, and D213R, were introduced in thermolysin. The hydrolytic activity of these mutants against FaGLa became lower than that of the WT, which led to a negative correlation between the activity and *pI* value of the mutants (30). Considering the three-dimensional structure and electrostatic calculations, it was concluded from these experimental data that the contribution of the pK shift of residues other than the His 231 and the 213th residue must be taken into account in explaining the change in activity.

The mutant N227H, in which a positively charged residue His is introduced instead of Asn 227, showed higher hydrolytic and synthetic activities (Table 2). In contrast, in the mutant N227D, where a negatively charged residue, Asp, is introduced at this site, lower hydrolytic and synthetic activities were seen. The positive charge enhanced enzymatic activity at this site, while the positive charge at the 213th site suppressed the activity. The mechanism of enhancement and suppression of activity by electrostatic interaction seems to be complex and cannot be explained by a simple model. Further theoretical and experimental research is necessary to explore this process.

The Asp 150 was replaced by neutral Asn or by positively charged His so as to influence the electrostatic potential of the Zn ion at the active site. The mutations were performed by PCR (31). In this case the k_{cat}/K_m for hydrolysis of Fa-APM decreased (both k_{cat} and K_m are increased) below that of WT. However, these mutants, D150N and D150H, respectively, exhibited increased activity for Z-APM synthesis (Table 2).

As shown in the Q119 series mutants, flexibility around the active site may have a great effect on the activity. To modify the flexibility, we searched for cavities in the protein molecule using its three-dimensional structure. Figure 2 shows the density map of a cross section of thermolysin; the darker and lighter portions represent higher and lower densities, respectively. As can be seen, the central part of the N-terminal domain has lower density than the surrounding part, clearly indicating the presence of a cavity. Computer graphics suggested that the size of the cavity can be easily modified by replacing several amino acids such as Leu 144. This cavity was also remarked by Vriend and Eijsink in stabilization of thermolysin-like metalloprotease from *B. stearothermophilus* CU21 (32), although the Phe mutant at Leu 144 (L144F), designed to fill the cavity, did not show an increase in thermal stability. We introduced mutation at the 144th site by PCR. It was found that several mutants, such as the Ser mutant (L144S), showed higher hydrolytic and synthetic activities than WT (Table 2).

Based on the three-dimensional structure, several single-point mutants having higher synthetic activity for Z-APM were prepared. We combined these single mutations to obtain more active enzymes, and their synthetic activities are shown in Table 2. We obtained fivefold higher synthetic activity for a triple mutant, L114S-D150H-N227H (33).

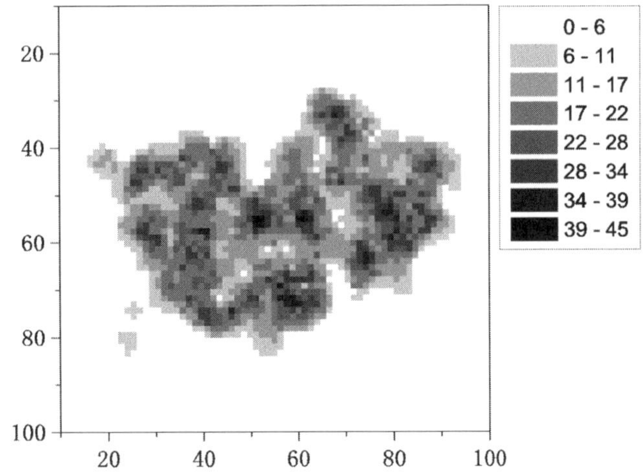

Figure 2. Cross section of average number–density of thermolysin. The density is represented by shading, the highest density being represented by black. *N*-terminal domain is plotted on the left.

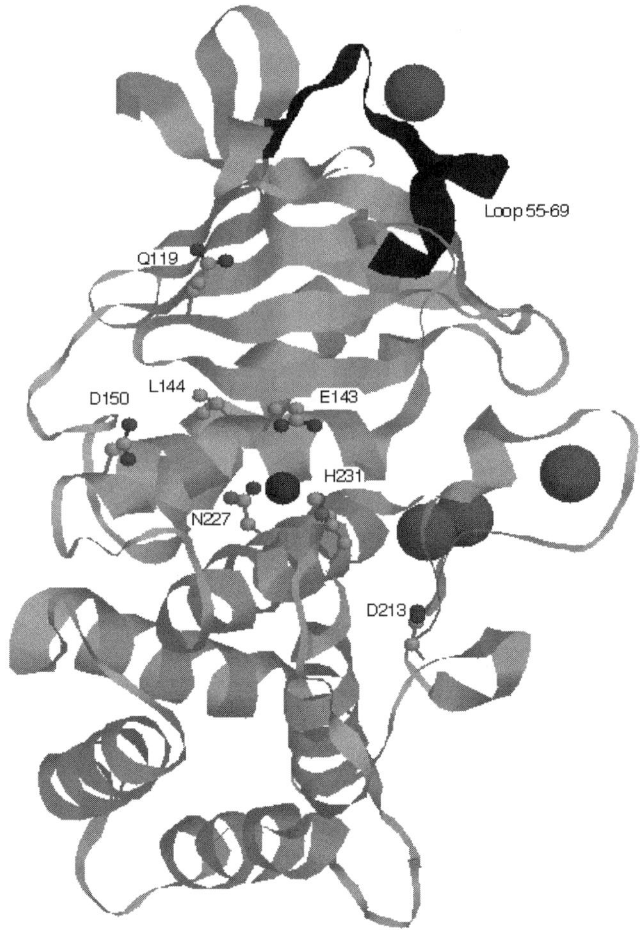

Figure 3. Three-dimensional structure of thermolysin. The peptide chain is represented by a ribbon model. The small ball in the center of this model represents the zinc ion, which is essential for catalytic activity. The four larger balls represent the calcium ion required for stabilization of the three-dimensional structure. E143 and H231 are essential for the catalytic activity. Mutated residues are located as shown here. Loop 55–69 is the dominant region in the stability of thermolysin (38).

Activity for the Aspartame Precursor Z-APM

The activity of the thermolysin mutants for Z-APM synthesis is described in detail in this section. Because mutants at Gln 119, Leu 144, Asp 150, and His 227 sites, specifically the Q119E, L144S, D150H, and N227H mutants, exhibited higher activity for FaGLa or Fa-APM hydrolysis, we examined their activity for Z-APM synthesis. Positions of these sites are indicated in Figure 3. Z-APM synthetic activity was evaluated with 0.1 M Z-L-Asp and 0.1 M L-PheOme as substrate in 0.1 M Tris-maleate buffer at pH 0.6 at 35 °C. After 30 min incubation, the reaction was terminated by addition of EDTA. Z-APM in the mixture was measured by a HPLC system (Tosoh, Japan) equipped with an ODS column (Nacali Tesque, Japan). Z-APM hydrolytic activity, the reversal of Z-APM synthesis, was also analyzed according to the modified method of Inouye (10) by monitoring absorbance at 224 nm at 35 °C with 1 mM of Z-APM as substrate in 0.1 M Tris-HCl buffer (pH 7.0) containing 10 mM $CaCl_2$.

L144S, D150H, and N227H exhibited enhancement of activity both for synthesis and hydrolysis of Z-APM while Q119E only exhibited enhancement in the hydrolysis (see Table 2). The greatest effect on the activation was obtained by D150H wherein synthetic activity was 2-fold higher than for WT; activities of both L144S and N227H were only 1.5-fold higher. The combination of L144S and another mutation enhanced the activity by simple addition of the effects of each mutation. The activity of the double mutants L144S-D150H and L144S-N227H were 2.5-fold and 2-fold higher than WT, respectively. However, when D150H and N227H were combined, the resulting activity was higher than the simple addition of the effect of each mutation. For example, if the enhancing effect were caused by simple addition, the activity of the double mutant D150H-N227H would be 2.5-fold higher than WT, instead of the observed 3.5-fold higher activity (Table 2). The triple mutant, L144S-D150H-N227H, exhibited the highest activity, which was 5-fold higher than WT. We called this thermolysin mutant TZ-1, and its efficiency was further evaluated by small-scale production under the production conditions of the HSC plant; see also ASPARTAME.

These mutants did not always exhibit enhanced activity for other substrates. For example, the activity for casein hydrolysis, analyzed following the method of Endo (1), was decreased in two mutants, Q119E and L114S, while the activity of D150H and N227H was the same as that of the WT (Table 2).

The kinetic parameters of Z-APM synthesis were examined at various concentrations of Z-Asp and 800 mM of L-PheOMe. The reaction rate increased with increasing concentration of Z-Asp up to 100 mM and then declined. This inhibition of the synthetic reaction at high concentration of Z-Asp was also reported by Nagayasu et al. (34) when the reaction was performed with immobilized thermolysin in MES buffer saturated with ethyl acetate. Nagayasu reported that the K_m value of the reaction was 2.8 mM and 250 mM for Z-Asp and PheOMe, respectively, and the K_i value for Z-Asp was 100 mM. Although we have not calculated the K_i value of the condensation reaction, k_{cat} and K_m values of the reactions were calculated from Lineweaver–Burk plots at the range of Z-Asp concentration in which no inhibition is observed (Fig. 4a). The kinetic parameters for Z-Asp are summarized in Table 3.

In our results, the K_m of WT for Z-Asp was 13 mM, which is four to five times higher than the results obtained by Nagayasu et al. (34), apparently because of differences in reaction conditions such as solvent, concentration of substrates, or immobilization of enzyme. L144S decreased the K_m value significantly with slight increase of k_{cat} value, D150H increased both k_{cat} and K_m, and N227H increased k_{cat} without affecting K_m. These results, except for the k_{cat}/K_m value of D150H, were similar to the kinetic parameters obtained when Fa-APM is the substrate. In Z-APM synthesis, the k_{cat}/K_m value of D150H increased while there was a decrease for Fa-APM hydrolysis. This change was caused by the greater effect of this mutation on K_m for Fa-APM than that for Z-APM and the difference of the Fa group and Z group at the N-terminal of APM. The Fa group is less hydrophobic than the Z group.

When k_{cat}/K_m values for Z-Asp of TZ-1 and WT were compared, the efficiency of the former was seven times higher. We were not able to estimate the K_m value for PheOMe in Z-APM synthesis because the Lineweaver–Burk plot at 100 mM Z-Asp and variable PheOMe passed through the graph origin (Fig. 4b), suggesting that the K_m values for PheOMe were greater than 900 mM under this reaction condition.

Optimization of Amino Acids at the 150th, 144th, and 227th Sites

Because D150H exhibited the greatest activity in the complete set of mutants tested, we further examined the substitution of other amino acids for Asp 150. The mutations were introduced by PCR using a random primer. In the complete set of 19 mutants at this site, the Trp mutant (D150W) exhibited the highest activity both for synthesis and for hydrolysis of Z-APM (Table 2) (35), about four times higher than the activity of the WT. The activity of the series of the mutants apparently exhibits correlation with hydrophobic parameters (33), such as the π value (36) or ΔH value (37) of substituted amino acids. From the analysis of three-dimensional structure, we assumed that the enhancement of activity might result from reinforcement of the hydrophobic interaction between the Z group of the substrate and the amino acids at the 150th site.

We also evaluated the substitution of other amino acids for Leu 144 and Asn 227 by random mutation. In the 12 variants of the mutants obtained at the 227th site, that is, Asp, Glu, His, Lys, Arg, Thr, Ala, Met, Trp, and Leu mutants, only N227H exhibited apparent activity enhancement. We could not find any relationship between activity and the physicochemical properties of amino acids at the 227th site. The activation mechanism of N227H is not yet clear, but it is believed that the electrostatic effect caused by the positive charge of His might be the main cause of the activity enhancement. In case of the 144th site, 10 variants of mutants, that is, the Glu, Asn, Gln, Ser, Thr, His, Cys, Ala, Tyr, Phe, and Trp mutants, were obtained. Although this series of mutants exhibited almost the same activity for Z-APM synthesis, activity for the hydrolysis of Z-APM exhibited a negative correlation with the hydrophobic properties of the substituting amino acids. This result seems to support our hypothesis described in the previous section; that is, the activity is affected by the flexibility of the enzyme molecule, which thereby influences the cavity in the center of thermolysin. Hydrophilic amino acids at the 144th site appear to contribute not only to cavity expansion but also to weakening hydrophobic interactions within the thermolysin molecule.

Finally, the triple mutant obtained, L144S-D150W-N227H (named TZ-5), exhibited about 10-fold higher activity for Z-APM synthesis and hydrolysis (Table 3) (35). D150W apparently enhanced the k_{cat} value of the reaction without an effect on K_m value. The k_{cat}/K_m value of TZ-5 was 11 times higher than that of WT.

STABILIZATION OF THERMOLYSIN

Recently, Van den Burg et al. (38) reported remarkable stabilization of thermolysin-like metalloprotease in boiling conditions (100 °C) by site-directed mutagenesis at eight sites (Table 4). This protease (TLP-ste) from *B. stearothermophilus* CU21 differs from thermolysin at 44 sites in its sequence and is less stable than thermolysin. Five mutations, A4T, T56A, G58A, T63F, and A69P, are substitutions of thermolysin-type residues for TLP-ste. In these mutations, G58A was originally suggested by Imanaka et al. (39). In a previous study, Van den Burg et al. (40) replaced almost all residues of TLP-ste by thermolysin-type amino acids and found that substitutions in the N-terminal domain are more effective than those in the C-terminal domain. They presented a hypothesis that local denaturation at the 55–69 loop in the N-terminal domain determines thermal denaturation of TLP-ste (41,42). The region corresponding to the 55–69 loop in thermolysin is shown in Figure 3. The three mutations, S65P, G8C and N60C, were rationally designed to fix the flexible region by substitution of Pro (42) and by introduction of an S–S bridge between the 8th and the 60th sites (38). Thus, the resultant 8-point mutant exhibited a half-life 170 times longer than thermolysin at 100 °C. These mutations did not affect the specific activity of TLP-ste (38). This result suggests that a highly active and extremely stable thermolysin mutant is possibly produced when their mutations are combined with our mutations. Three of these mutations, S65P, G8C, and N60C, can be introduced into thermolysin. Other mutations that stabilize thermolysin (15) are also summarized in Table 4. Thermolysin will be further stabilized when this mutation is combined with others.

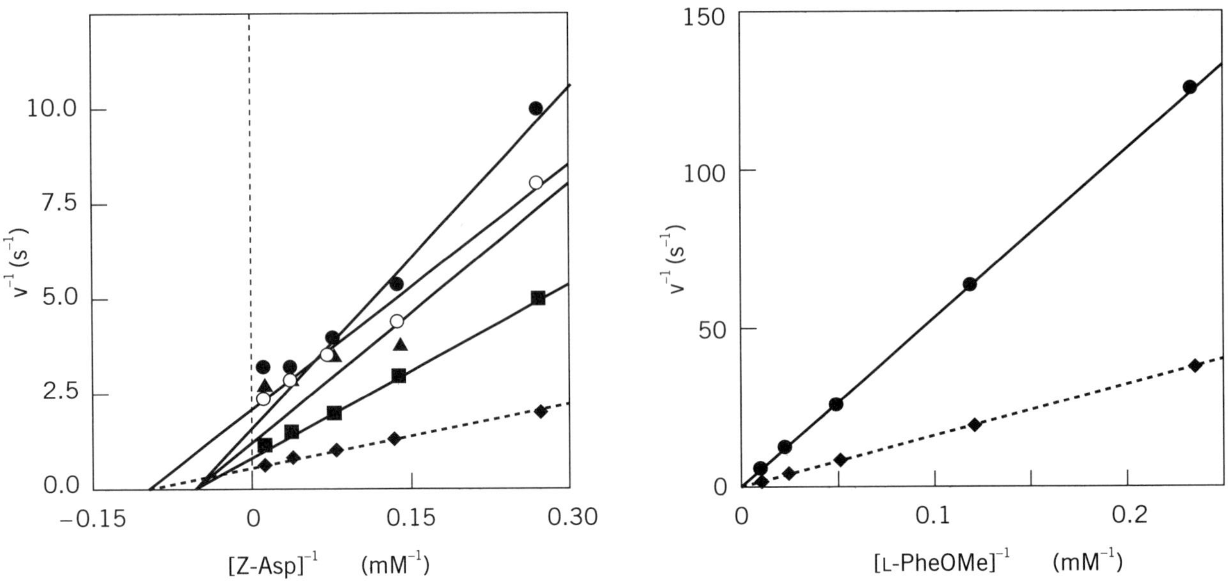

Figure 4. Lineweaver–Burk plots of the condensation reaction of Z-L-Asp (**a**) and L-PheOMe (**b**) catalyzed by thermolysin mutants. The reactions were performed at a constant concentration of L-PheOMe and various concentrations of Z-L-Asp (**a**) or at constant Z-L-Asp and various L-PheOMe (**b**) at pH 6.0 and 35 °C. Closed circles, open circles, squares, triangles and rhombuses represent wild type, L144S, D150H, N227H, and L144S-D150H-N227H, respectively.

CONCLUSION

In designing thermolysin mutants, we started by influencing the electrostatic environment around the active site. It was found during the protein engineering cycle that two additional factors influence thermolysin activity, namely, molecular flexibility and the hydrophobic interaction between the N-protective group of the substrate and the side chain of the 150th site of the enzyme. We were able to locate sites that effectively influence the activity of thermolysin, although some of the mutants at the sites do not enhance the activity, the reasons being only partly inter-

Table 3. Kinetic Parameters for Z-APM synthesis

Thermolysin variant	k_{cat} (s^{-1})	K_m (mM)	k_{cat}/K_m (s^{-1} mM^{-1})
Wild type	0.4	13	0.03
L144S	0.4	9	0.04
D150H	0.9	16	0.06
D150W	1.5	12	0.12
N227H	0.6	13	0.05
L144S-D150H-N227H	1.6	8	0.20
L144S-D150W-N227H	3.0	9	0.33

The kinetic parameters were determined by Lineweaver–Burk plots at 3–100 mM Z-Asp and constant L-PheOMe concentration (800 mM).

Table 4. Stabilization of Thermolysin and Thermolysin-Like Metalloprotease from *Bacillus stearothermophilus* CU21 by Site-directed Mutagenesis

Enzyme	Mutation	Half-life[a] (min)	$T_{1/2}$[b] (°C)	Incubation time/temp.	Reference
TLN	Y93S/G	+36/24	—	80 °C	15
TLP-ste[c]	A4T	—	+1.8	30 min	32
TLP-ste	T56A	—	+1.5	30 min	32
TLP-ste	G58A	—	n.d.	30 min	32,39
TLP-ste	T63F	—	+6.2	30 min	32
TLP-ste	A69P	—	+5.6	30 min	32
TLP-ste	S65P	—	+4.2	30 min	40
TLP-ste	6 points[d]	—	+23.5	30 min	41
TLP-ste	8 points[e]	+169	—	100 °C	38

TLN, Thermolysin.
[a]Difference in half-life between wild-type TLN and mutants of TLN or P-TLP-ste during incubation at indicated temperature.
[b]Difference in temperature required for 50% denaturation between mutants and wild type.
[c]TLP-ste indicates the metalloprotease from *B. stearothermophilus* CU21.
[d]6 points is the combination of A4T, T56A, G58A, T63F, A69P, and S65P.
[e]8 points is the combination of G8C and N60C in the 6-points mutant. The mutated sites are indicated according to the sequence alignment by Vriend and Eijsink (32).

preted (30). Further studies, integrating up mutations and down mutations, must be carried out to establish a more reliable design principle. Apart from the theoretical aspect, we found it quite successful to enhance the synthetic activity of thermolysin by 10-fold. The production cost of APM will be lowered when the thermolysin mutant is applied in the actual industrial process.

ACKNOWLEDGMENTS

The improvement of thermolysin activity was supported by Holland Sweetener Company. We are grateful to Drs. Hiromasa Nagao, Toshio Miyake, Atsuo Aoyama, Takashi Yoneya, Ken-ichi Kai, Seigo Ooe, Yoshikazu Tanaka, You-ichiro Miki, and Kimiko Endo for preparation and evaluation of thermolysin mutants. We thank Dr. Kakidani and Dr. Dionisio-Sese for their critical reading of this manuscript. We would like to express special thanks to Dr. Kiyotaka Oyama, Mr. Ichiro Hiraki, Mr. Akira Tokuda, and Prof. Akiyoshi Wada for the organization of this work.

BIBLIOGRAPHY

1. S. Endo, *J. Ferment. Technol.* **40**, 346–353 (1962).
2. K. Morihara, *Biochem. Biophys. Res. Commun.* **26**, 656–661 (1967).
3. K. Titani, M.A. Hermodson, L.H. Ericsson, K.A. Walsh, and H. Neurath, *Biochemistry* **11**, 2427–2435 (1972).
4. K. Titani, M.A. Hermodson, L.H. Ericsson, K.A. Walsh, and H. Neurath, *Nat. New Biol.* **238**, 35–37 (1972).
5. P.M. Colman, J.N. Jansonius, and B.W. Matthews, *J. Mol. Biol.* **70**, 701–724 (1972).
6. M.A. Holmes, and B.W. Matthews, *J. Mol. Biol.* **160**, 623–639 (1982).
7. K. Aunstrup, in A.H. Rose ed., *Microbial Enzymes and Bioconversions*, Academic Press, New York, 1980, pp. 50–114.
8. K. Oyama, S. Irino, and N. Hagi, in K. Mosbach ed., *Methods in Enzymology*, vol. 136, *Immobilized Enzyme and Cells, Part C*, Academic Press, New York, 1987, pp. 503–516.
9. K. Nakanishi, A. Takeuchi, and R. Matsuno, *Appl. Microb. Biotechnol.* **32**, 633–636 (1990).
10. K. Inouye, *J. Biochem. (Tokyo)* **112**, 335–340 (1992).
11. K. Inouye, S. Lee, and B. Tonomura, *J. Biochem. (Tokyo)* **315**, 133–138 (1996).
12. M.N. Alam, K. Tadasa, and H. Kayahara, *Biotechnol. Lett.* **18**, 45–50 (1996).
13. M.N. Alan, K. Tadasa, T. Maeda, and H. Kayahara, *Biotechnol. Lett.* **19**, 1129–1133 (1997).
14. H. Kitano and N. Ise, *Biotechnol. Bioeng.* **31**, 507–510 (1987).
15. M. Kubo, Y. Mitsuda, M. Takagi, and T. Imanaka, *Appl. Environ. Microbiol.* **58**, 3779–3783 (1992).
16. B.E. Homler, in L.D. Stegink and L.J. Filer, Jr. eds., *Aspartame Physiology and Biochemistry*, Dekker, New York, 1984, pp. 247–262.
17. S. Kidokoro, Y. Miki, K. Endo, and A. Wada, in A. Pullman and B. Pullman eds., *Modeling of Biomolecular Structures and Mechanisms*, Kluwer, Dordrecht, 1995, pp. 399–407.
18. S. Kidokoro, Y. Miki, K. Endo, A. Wada, H. Nagao, T. Miyake, A. Aoyama, T. Yoneya, K. Kai, and S. Ooe, *FEBS Lett.* **367**, 73–76 (1995).
19. S.N. Ho, H.D. Hunt, R.M. Horton, J.K. Pullen, and L.R. Pease, *Gene* **77**, 51–59 (1989).
20. M. Kubo, K. Murayama, K. Seto, and T. Imanaka, *J. Ferment. Technol.* **66**, 13–17 (1988).
21. M. Kubo and T. Imanaka, *J. Gen. Microbiol.* **134**, 1883–1892 (1988).
22. EP Pat. 0418625A1 (September 6, 1989), R. Marquardt, R. Hilgenfeld, and R. Keller (to Hoechst).
23. Y. Miki, *J. Ferment. Bioeng.* **77**, 457–458 (1994).
24. M. Fujii, M. Takagi, T. Imanaka, and S. Aiba, *J. Bacteriol.* **154**, 831–837 (1983).
25. T.J. Gryczan, S. Contente, and D. Dubnau, *J. Bacteriol.* **134**, 318–329 (1978).
26. M. Yamada, M. Kubo, T. Miyake, R. Sakaguchi, Y. Higo, and T. Imanaka, *Gene* **99**, 109–114 (1991).
27. I.A. Nimmo and G.L. Atkins, *Biochem. J.* **141**, 913–914 (1974).
28. K. Oyama, K. Kihara, and Y. Nonaka, *J. Chem. Soc. Perkin Trans. II*, 356–360 (1981).
29. M.A. Vandeyer, M.P. Weiner, C.J. Hutton, and C.A. Batt, *Gene* **65**, 129–133 (1988).
30. Y. Miki, S. Kidokoro, K. Endo, A. Wada, T. Yoneya, A. Aoyama, K. Kai, T. Miyake, and H. Nagao, *J. Mol. Catal. B Enzym.* **1**, 191–199 (1996).
31. R. Higuchi, B. Krummel, and R. K. Saiki, *Nucleic Acids Res.* **16**, 7351–7367 (1988).
32. G. Vriend and V. Eijsink, *J. Comput.-Aided Mol. Des.* **7**, 367–396 (1993).
33. EP Pat. 0616033A1 (March 17, 1993), H. Nagao, T. Yoneya, T. Miyake, K. Kai, S. Kidokoro, Y. Miki, K. Endo, and A. Wada (to Holland Sweetener Company and Sagami Chemical Research Center).
34. T. Nagayasu, M. Miyanaga, T. Tanaka, T. Sakiyama, and K. Nakanishi, *Biotechnol. Bioeng.* **43**, 1108–1117 (1994).
35. WO Pat. 9516029 (June 15, 1995), Y. Tanaka, Y. Miyake, S. Hanzawa, S. Oe, S. Kidokoro, Y. Miki, K. Endo, and A. Wada (to Sagami Chemical Research Center and Holland Sweetener Company).
36. J. Fauchere and V. Pliska, *Eur. J. Med. Chem. Chim. Ther.* **18**, 369–375 (1983).
37. R. Wolfenden, L. Andersson, P.M. Cullis, and C.C.B. Soughgate, *Biochemistry* **20**, 849–844 (1981).
38. B. Van den Burg, G. Vriend, O.R. Veltman, G. Venema, and V.G.H. Eijsink, *Proc. Natl. Acad. Sci. U.S.A.* **95**, 2056–2060 (1998).
39. T. Imanaka, M. Shibazaki, and M. Takagi, *Nature* **324**, 695–697 (1988).
40. O.R. Veltman, G. Vriend, P.J. Middlehoven, B. Van den Burg, G. Venema, and V.G.H. Eijsink, *Protein Eng.* **9**, 1181–1189 (1996).
41. V.G.H. Eijsink, O.R. Veltman, W. Aukema, G. Vriend, and G. Venema, *Struct. Biol.* **2**, 374–379 (1995).
42. H. Hardy, G. Vriend, O.R. Veltman, B. Van der Vinne, G. Venema, and V.G.H. Eijsink, *FEBS Lett.* **317**, 89–92 (1993).

THERMOPHILIC MICROORGANISMS

PAULA M. HICKS
ROBERT M. KELLY
North Carolina State University
Raleigh, North Carolina

KEY WORDS

Archea
Continuous culture
Extreme thermophiles
Hyperthermophiles
Phylogenetic diversity
Pyrococcus
Thermoacidophiles
Thermophiles
Thermotoga

OUTLINE

INTRODUCTION

It has been 25 years since it was first reported by Brock et al. (1) that microbial life was indeed possible above 80 °C and in that time the number and diversity of known thermophilic, or heat loving, microorganisms has continued to expand. The so-called extreme thermophiles, defined here as organisms having an optimal growth temperature of at least 75 °C, can be found in a variety of geothermal sites ranging from terrestrial hot springs to seeps associated with volcanic activity to deep sea hydrothermal vents (2). The most thermophilic of these will be referred to as *hyperthermophiles* (growth at temperatures of 90 °C or above), some of which prefer to grow at temperatures of 100 °C or higher (3). Much has been accomplished in developing protocols for culturing these novel organisms in laboratory settings (4,5). As such, their physiological and metabolic characteristics have been examined (6,7), and numerous highly thermostable proteins have been purified and studied (8,9). Efforts have also begun to address the genetic manipulation of certain extreme thermophiles, not only to probe the intracacies of their replication apparatus, but also to develop cloning and expression systems that are functional at elevated temperatures (10,11).

The driving force to study extremely thermophilic microorganisms is based on both scientific and technological possibilities. The origin of life on earth has been proposed to be traced to this group of microorganisms (12), and prospects for extraterrestrial life have been fortified by the increasing variety of inhabitable environments on this planet, which were long thought to be barren of life. If one accepts the premise that among this group are the most primitive of all known life forms (13), the templates for all existing metabolic strategies and bioenergetic processes can be identified and examined for clues to not-yet-understood cellular phenomena. Fundamental insights into the basis for biomolecular stability, particularly proteins, have been sought through the study of extreme thermophiles (8,14). Encouraged by the incredible utility of DNA polymerases from extreme thermophiles in the polymerase chain reaction (15), commercial interest in enzymes from these organisms is high. Biocatalyst instability often has been a major drawback in many existing applications, and it also diminishes enthusiasm for untried enzymatic routes to new products. Extreme thermophiles represent a fertile source of robust biocatalysts that may change this perspective.

Bioprocessing has proved to be an enabling expertise in the study of extreme thermophiles and will be central to the eventual use of these organisms or their products for commercial purposes. The earliest efforts in this field of inquiry were facilitated by the development of unusual bioreactors and cultivation protocols to provide biomass for characterization (16,17). Imaginative approaches for using extreme thermophiles, ranging from mixed communities of viable cells to thermostable enzymes catalyzing multiple step reactions, will be required in future years. As such, bioprocess engineering and technology will have to accommodate the anticipated application of extreme thermophiles and their constituent biomolecules.

In this review, the physiological and metabolic characteristics of extremely thermophilic microorganisms are discussed with regard to bioprocessing issues that arise in their cultivation. The physiological and phylogenetic diversity of currently known members of this group is covered in addition to efforts to define the untapped microbial content of geothermal locales. In a separate review in this volume, the characteristics of enzymes from extreme thermophiles are examined.

OBTAINING EXTREMELY THERMOPHILIC MICROORGANISMS FROM GEOTHERMAL SITES

A variety of natural biotopes associated with geothermal activity have yielded extremely thermophilic microorganisms (18–20). Samples from these sites (in the form of hot water, sediments, or solid rock or even from intestines of animals inhabitating these sites [21]), inoculated into various types of liquid growth media at temperatures below approximately 120 °C and pH between 0 and approaching 11, can contain one or more culturable species of extreme thermophiles. In fact, it has been demonstrated that samples from geothermal sites can contain any number of microorganisms whose cultivation in a laboratory setting may be difficult at best (22,23). Obtaining samples containing extreme thermophiles from geothermal biotopes can be as simple as drawing liquid into a syringe at the edge of a hot spring, as has been done at Yellowstone National Park (19), or as difficult as having to dive in a submersible several kilometers into the ocean and then expertly manipulate external sampling equipment to procure material that cannot be contaminated by the cold surrounding seawater (3).

Although no connection has been found between the degree of difficulty in sampling or remoteness of location to the isolation of the most unusual organisms, sites that have unique characteristics are continually sought. Table 1 lists some examples of geothermal locations from which extreme thermophiles have been isolated and some characteristics of these sites. As more geothermal locales are investigated, it has become clear that similar organisms may be isolated from locations that have very different physicochemical characteristics and are geographically distant. For example, members of the hyperthermophilic genus *Thermococcus* have been found in shallow marine settings (36), deep sea vents (37), and terrestrial surface waters (32) and petroleum reservoirs of various depths (38). The ubiquity of certain extreme thermophiles may reflect evolution under similar environmental conditions or a common source manifested through global volcanic activity.

Enrichment cultivation of samples taken from geothermal sites to isolate extreme thermophiles follows methodology used for less thermophilic microorganisms, with several key exceptions (18,19,39). For aerobic thermoacidophiles, typically growing optimally at temperatures of approximately 75 °C and pH 2.0, iron pyrite or elemental sulfur may be added to the medium to support chemolithotrophy (1). Many of these organisms grow better in the presence of added complex media such as yeast extract, although the nutritional requirements met by particular complex media are often unclear. Most hyperthermophiles, which grow above 90 °C, are anaerobic and thus require careful attention to ensure that oxygen is eliminated from the media. Fortunately, the reduced solubility of oxygen at such high temperatures somewhat simplifies anaerobic technique to the extent that glove boxes are usually unnecessary. Enrichment media for hyperthermophiles typically contain an array of trace elements, basic salts, oligosaccharides, complex substrates (e.g., yeast extract, tryptone, etc.), and elemental sulfur (18). Adams et al. demonstrated the importance of tungsten as a trace element for several hyperthermophiles that incorporate this metal into specific oxidoreductases (40). For example, a threefold stimulation of cell density levels of one hyperthermophile, *Pyrococcus furiosus*, was noted in the presence of trace amounts of tungsten (41). Although there have been reports of chemically defined growth media for certain hyperthermophiles, such media often result in sig-

Table 1. Examples of Geothermal Sites Yielding Extreme Thermophiles

Site	Location	Geothermal characteristics	Examples of extreme thermophiles found
Yellowstone National Park	Wyoming, Montana, Idaho	Varied; includes alkaline and acid hot springs	*Sulfolobus acidocaldarius* (1)
Vulcano Island	Mediterranean Sea, off western coast of Italy	Geothermally heated sediments and shallow marine hot springs	*Pyrococcus furiosus* (24), *Staphylothermus* sp. (11), *Thermococcus* sp. (25)
Juan de Fuca Ridge	Off coast of Washington, Oregon	Deep sea hydrothermal vents	*Pyrococcus endeavori* (ES4) (26)
Macdonald Sea Mount	Polynesia	Erupting submarine volcano	*Pyrodictium occultum* (27), *Archaeglobus fulgidus* (28), *Pyrococcus furiosus* (24), *Thermococcus celer* (25)
Icelandic hot springs	Various locations in interior Iceland	Varied; includes alkaline and acidic hot springs	*Desulfurococcus mucosus* (29), *D. mobilis* (29)
East Pacific Rise	21 °N, depth 2,610 m	Deep-sea hydrothermal vent	*Methanococcus jannaschii* (30)
Oil reservoirs	3,000 m below the bed of the North Sea	Deep oil-bearing strata	*Archaeglobus* and *Thermococcus* species (31)
Shallow hot springs	New Zealand (central North Island)	Thermal pools, samples taken from 1-m depth	*Thermococcus* ANI (32)
Solfatara fields	Various terrestrial locations; many in Iceland and Italy	Acidic, sulfur-rich	*Acidianus brierleyi, A. infernus* (33), *Thermofilum pendens* (34)
Guaymas Basin	Off the coast of Guaymas, Mexico in Gulf of California	Deep-sea hydrothermal vent	*Methanopyrus kandleri* (35), *Methanococcus* strains, *Pyrococcus* sp., *Archaeoglobus profundus* (18)

nificantly lower growth yields (42). Thermal lability of many medium components is always a concern (43), especially for amino acids, such as glutamine and asparagine (42), certain vitamins, and some monosaccharides. Maltose and cellobiose, however, serve as excellent carbon and energy sources for *P. furiosus* growing at 98 °C, although glucose does not.

Extreme thermophiles can be very difficult to establish in pure, laboratory culture, whether the inoculum is directly from isolation efforts or even from storage. This is especially true for some of the hyperthermophilic, sulfur-reducing chemolithotrophs, such as *Pyrodictium occultum* (44), which grow to very low cell densities and are highly sensitive both to shearing forces and the presence of various metals (Schicho and Kelly, unpublished observation). Problems with culture stability can be encountered; cultures that have transferred successfully for long periods of time will abruptly stop growing and may require reinoculation from stock cultures.

Cultures can be stored in various ways. In some cases, extreme thermophiles will remain viable for years in liquid media. Attempts to freeze-dry cultures for long-term storage have met with mixed success (Rinker and Kelly, unpublished data); this may have something to do with the unusual cell membrane characteristics that members of the archaea possess. Lyophilization has been used successfully for some hyperthermophilic archaea, although there are also reports of limited storage times for organisms stored in this way (45). *P. furiosus* (24) can be stored in dimethyl sulphoxide (DMSO) for extended periods in glass capillary tubes over liquid nitrogen, although plastic cryotubes were found to be too permeable to oxygen and thus ineffective for extended storage (45,46).

The inability to use standard microbiological techniques on solid media has been a limitation in establishing genetic systems for extreme thermophiles. However, plating techniques for *Thermotoga* species (47,48), *Aquifex pyrophilus* (49), *P. furiosus* (50), and *Sulfolobus* strains (51) have been reported. Methodology to carry out mutagenesis experiments for the hyperthermophilic pyrococci has been developed and used to produce a uracil auxotrophic mutant of *Pyrococcus abyssi* (52).

BIODIVERSITY OF EXTREMELY THERMOPHILIC MICROORGANISMS

There are now more than 50 different species of extreme thermophiles that are classified into 25 genera and 11 orders (2); these are summarized in Table 2. The pioneering work of Carl Woese and co-workers, using a 16S rRNA-based phylogenetic tree, helped to distinguish the domain Archaea (which contains most extreme thermophiles) from the domains Bacteria and Eucarya (13). On this basis, new extremely thermophilic isolates can be compared to those already discovered on the basis of 16S phylogeny, in addition to distinguishing them by physiological differences. By using both phylogenetic and physiological yardsticks, a framework for differentiating among the many extreme thermophiles that are known and are newly isolated can be established. This is important not only for scientific reasons, but because it also helps directing attention to a certain species that may contain particular enzymes of interest or has desirable physiological characteristics for application as a whole cell.

Phylogenetic Diversity

Using the sequence of 16S rRNA as a chronometer, evolutionary trees showing the relationships among extreme thermophiles can be developed, and such sequenced-based maps of biodiversity have proved to be valuable tools in studying these microorganisms (23). Figure 1 shows one such representation. The extreme thermophiles include two genera that group with the Bacteria: *Aquifex* and *Thermotoga*. The rest are of the extreme thermophiles are found within one of two archaeal kingdoms: the Crenarchaeota and the Euryarchaeota. It should be pointed out that the Archaea include members that are not thermophilic, such as certain methanogens and the halophiles.

In some cases, it can be difficult to sort out members within a given extremely thermophilic genus based strictly on 16S RNA sequence. Within the genus *Thermococcus*, for example, differences of only a few bases may exist between species that have otherwise distinguishing physiological characteristics (83). Whether these differences are in phenotype or genotype are not completely clear, and this issue will remain unresolved until the genomes of the species being compared are examined. In view of the fact that 16S sequence alone is not sufficient in some cases for phylogenetic differentiation of microorganisms, more detailed bases for such differentiation have been proposed, including using DNA polymerase restriction fragment length polymorphisms (RFLP) (102) and using 16S/23S rRNA spacer regions (103). Of course, the best approach is to combine elements of phylogeny with classical approaches to taxonomy and growth physiology.

Physiological Diversity

The extreme thermophiles are a physiologically diverse group of microorganisms that share a common dependence on high temperatures for growth. Within this group are fermentative anaerobes, methanogens, extreme acidophiles, and to a lesser extent thus far, nitrate reducers and sulfate reducers. Some details on these classifications are provided in this section.

Fermentative Anaerobes. Geothermal sites, both terrestrial and marine, have yielded an array of fermentative microorganisms that use a variety of organic carbon and energy sources and reduce sulfur as a consequence of growth. Most of these fermentative heterotrophs require sulfur for growth, with the exception of some species in the genera *Pyrococcus, Thermococcus* and *Thermotoga*. Of these fermentative anaerobes, species of the archaeal genus *Pyrococcus* and of the bacterial genus *Thermotoga* have been the most carefully studied. As such, species from each of these will be discussed in more detail to illustrate the current level of understanding the physiology of extremely thermophilic fermenters.

Pyrococcus furiosus, A Hyperthermophilic Archaeon. Members of the genus *Pyrococcus* typically ferment peptides and saccharides to CO_2, H_2, and fatty acids (24). Alanine is also produced in significant amounts in the absence of sulfur (104). Hydrogen sulfide will be produced in copious amounts, if a polysulfide source (e.g., elemental sulfur) is present in the growth medium (105). *Pyrococcus furiosus*, the type strain in this genus, was first isolated from geothermally heated sediments near Vulcano Island, Italy (24), and has been the subject of a number of physiological, enzymatic, and bioenergetic studies (6,7). *P. furiosus* seems to grow best on α-linked polysaccharides, such as starch and glycogen (106), and the disaccharides maltose (41) and cellobiose (104). Growth on β-linked polysaccharides larger than cellobiose has not been reported. Although *P. furiosus* will grow on pyruvate (107), it will not grow on glucose even though it has been shown to take up and metabolize this substrate (108). Defined media for this organism have been difficult to develop (105), but one such successful attempt has been reported (43). The significance of sulfur in the bioenergetics of this organism is not completely clear. Schicho et al. (109) showed that sulfur addition led to a twofold improvement in maximum growth yield of *P. furiosus* grown in continuous culture on maltose. Ma et al. (110) showed that the hydrogenase from *P. furiosus* also catalyzes sulfide formation from polysulfide, implicating this enzyme in both sulfur and nonsulfur growth. It was later shown that *P. furiosus* produces significant amounts of alanine in the absence of sulfur (104), which is seemingly an energetically inefficient alternative to acetate formation. This organism may use sulfur as a more productive electron acceptor in that alanine formation decreases in its presence. It remains to be seen why *P. furiosus* is capable of growth in the absence of sulfur, which is an unusual feature among hyperthermophilic heterotrophs.

There has been progress in elucidating the central metabolic pathways of *P. furiosus*, and this has served as a basis for examining other heterotrophic hyperthermophiles (6,7). Studies on glycolysis have shown that this microorganism possesses a version of the Embden-Meyerhof-Parnas (EMP) pathway, which has some interesting alterations compared to the one found in bacteria (Fig. 2). For example, this pathway involves ADP-dependent kinases, glucokinase, and phosphofructokinase (7). Although ADP and ATP have similar thermodynamic properties in regard to the free energy released upon hydrolysis, it is not clear why AMP-forming enzymes are involved (7). Another interesting feature in this modified pathway is the presence of a tungsten-based glyceraldehyde-3-phosphate–ferredoxin oxidoreductase (111), which appears to oxidize glyceraldehyde-3-phosphate to 3-phosphoglycerate directly. Yet another apparent difference that appears to be present in all archaea is the direct conversion of acetyl-CoA to acetate by means of an acetyl-CoA synthetase (ADP forming) (112). The transfer of reducing equivalents generated by glycolysis have been traced to several potential sinks. From ferredoxin (113), electrons may go to the formation of H_2, when growth is in the absence of sulfur, or to polysulfides that can be formed upon nucleophilic attack of elemental or colloidal sulfur (105). Ferredoxin seems to supplant the need for large pools of $NAD^+/NADH$ in *P. furiosus*, which may be a strategy to offset the thermal lability of this coenzyme.

P. furiosus is an attractive candidate for enzymological studies because of its rapid growth rate (doubling time of $\sim$40 minutes, in some cases), its ability to grow in the absence of sulfur, and the cell densities that can be achieved (more than 10^9 cells/mL) (43). Thus, biomass yields of over a kilogram have been reported from 600-L fermentations (8). A number of enzymes spanning most known classifications have been purified from *P. furiosus* and characterized; these are shown in Table 3. An effort is underway to sequence the entire genome of this organism, which will a valuable tool in further examining its physiological and metabolic characteristics (141).

Thermotoga maritima and Thermotoga neapolitana, Hyperthermophilic Bacteria. The genus *Thermotoga* contains obligately anaerobic bacteria with optimum growth temperatures of up to about 80 °C, with some species capable of growth up to 90 °C (98). Species of *Thermotoga* have been found in high-salt, marine environments and low-salt, terrestrial settings. A sheathlike outer structure envelopes these rodlike organisms and balloons over the ends, hence the designation *toga* (66,98). During the stationary phase, cells may become spherical but remain within the *toga*. Nutritionally, *Thermotoga* species are very diverse and can use peptides, simple sugars, and polysaccharides, such as starch (98), xylan (142), and galactomannan (143), as carbon and energy sources.

Thermotoga maritima (98), the type species, has been studied to the greatest extent in terms of its physiological and metabolic characteristics, and *Thermotoga neapolitana* (99) has also been studied in some detail. *T. maritima* ferments sugars to CO_2, acetate, lactate, and H_2 by a standard EMP pathway (144,145); hydrogen sulfide appears to replace H_2 as a product in the presence of polysulfide. Alanine is also produced from sugar fermentation (146). For *T. neapolitana*, sugar transport was found to involve substrate level phosphorylation (147), which is presumably also the case for *T. maritima* species. *T. maritima* and *T. neapolitana* appear to be inhibited by H_2 produced during fermentation; this inhibition is minimized or avoided in the presence of elemental sulfur or if paired with a complementary extremely thermophilic methanogen (148). Both *T. maritima* and *T. neapolitana* are able to reduce thiosulfate to sulfide, which appears to improve growth rates and yields (146). *T. maritima* has been shown to be motile and migrates at a speed that is proportional to temperature (149).

Because *T. maritima* and *T. neapolitana* can be cultured in the absence of sulfur and are nutritionally diverse, these species have been examined for the presence of an array of enzymes (Table 4). In particular, several glycosyl hydrolases have been characterized from these and other Thermotoga species. These include amylases (176), xylanases (142), mannanases (143), galactosidases (143,152), and glucanases (150). The localization of these activities is interesting because some have been found to be associated with the toga (176).

Methanogens. Biotic methane generation is characteristic of many anaerobic geothermal environments, even at

Table 2. Extremely Thermophilic Microorganisms ($T_{opt} \geq 75$ °C)

Order	Genus	Species	Habitat	DNA G + C	T_{opt} [T_{max}]	Growth mode	Carbon source	Metabolic notes	Ref.
				Archaea					
Pyrodictiales	*Pyrodictium*	*abyssi*	Marine	60	97 [110]	Hetero	Cbh, pep	Fermentative	53
		occultum	Marine	62	105 [110]	Auto/hetero	Cbh, pep, CO_2	Forms pyrite in fermenter	27,54
		brockii	Marine	51	105 [110]	Auto	CO_2	Stimulated by YE	54
	Hyperthermus	*butylicus*	Marine	55.6	98 [108]	Hetero	Pep	Fermentative	55
	Thermodiscus	*maritimus*	Marine	49	88 [98]	Hetero/auto	Complex, CO_2	Fermentative?	56
Sulfolobales	*Acidianus*	*infernus*	Terrestrial	31	88 [95]	Auto	CO_2	Facultative aerobe	33
		ambivalens	Terrestrial	32.7	80 [87]	Auto	CO_2	Facultative aerobe	57
	Sulfolobus	*acidocaldarius*	Terrestrial	39	75 [80]	Auto/hetero	Cbh, Pep, CO_2	S^0, Fe^{2+D}; O_2, Fe^{3+A}	1,58
		solfataricus[a]	Terrestrial	38	87 [87]	Hetero	Cbh	Aerobe	58,59
		islandicus	Terrestrial	—	—	Hetero	Pep, Cbh	Inhibited by virus SIRV	59
		shibatae	Terrestrial	34.6	81 [86]	Auto/hetero	Cbh, pep, CO_2	10-mM phosphate inhibits growth	60
	Metallosphaera	*sedula*	Terrestrial	45	75 [80]	Auto/hetero	CO_2, rich organics	Oxidizes metal sulfides	61
		prunae	Terrestrial	46	75 [80]	Auto/hetero	CO2, Pep, rich organics	H_2S, pyrite, sphalerite, chalcopyrite, H_2^D	62
	Stygiolobus	*azoricus*	Terrestrial	38	80 [89]	Auto	CO_2	$S^0 \rightarrow H_2S$	63
	Desulfurolobus	*ambivalens*	Terrestrial	31.7	81 [87]	Auto	CO_2	Facultative aerobe	64
Desulfurococcales	*Desulfurococcus*	*mucosus*	Terrestrial	51.3	85 [97]	Hetero	Pep	Fermentative	29
		mobilis	Terrestrial	50.8	85 [95]	Hetero	Pep	Fermentative	29
		saccharovorans		50	[97]	Hetero		S^0 respiring	56
		amylolyticus			[97]				65
		S	Marine	52.0	85 [90]	Hetero	Pep	Growth strongly S^0 dependent	66
		SY	Marine	52.4	90 [95]	Hetero	Pep	Growth strongly S^0 dependent	66
	Staphylothermus	*marinus*	Marine	34.5	92 [98]	Hetero	Pep	Requires S^0	67
Thermoproteales	*Pyrobaculum*	*aerophilum*[a]	Marine	52	100 [104]	Auto/hetero	CO_2, Pep, acetate	Dissimilatory nitrate reduction	68
		islandicum	Terrestrial	46	100 [102]	Auto/hetero	Pep, CO_2	S^0, thiosulfate, sulfite, cystine[A]	69
		organotrophum	Terrestrial	46	102 [102]	Hetero	Pep	Obligate organotroph	69
	Thermofilum	*librum*	Terrestrial	57	80 [95]	Hetero		Sulfur respiring	56
		pendens	Terrestrial	57	88 [95]	Hetero	Pep	Sulfur respiring	34
	Thermoproteus	*tenax*	Terrestrial	56	88 [97]	Auto/hetero	Cbh, Pep, CO_2	Requires H_2S and S^0	70
		uzoniensis	Terrestrial	56.5	90 [97]	Hetero	Pep	Fermentative	71
		neutrophilus	Terrestrial		85 [97]	Auto/hetero	CO_2, acetate	Reductive citric acid cycle	56,72
Unclassified		FC89	Marine	31.2	85 [88]	Auto	CO_2	Hydrogenotrophic methanogen	73
		TY	Marine	42.0	88 [90]	Hetero	Pep, Cbh	Fermentative	74
		TYS	Marine	46.0	88 [90]	Hetero	Pep, Cbh	Fermentative	75
		GB-4	Marine	48.1	90 [92]	Hetero	Pep, Cbh	Not S^0 dependent	74
		GB-6	Marine	35.2	90 [92]	Hetero	Pep, Cbh	Not S^0 dependent	74
		NA10	Terrestrial	37.7	75 [85]	Hetero	Cbh	Ferments cellulose	75

	Caldococcus	*litoralis*	Marine	41	88 [100]	Hetero	Pep	Fermentative	76
	Aeropyrum	*Pernix*	Marine	67	90-5 [100]	Hetero	Pep	Strict aerobe	77
Thermococcales	*Pyrococcus*	*furiosus*[a]	Marine	38	100 [103]	Hetero	Cbh, pep	S^0 not required	24
		GB-D	Marine	39.5	95 [103]	Hetero	Pep, aa	S^0, cystine[A]	74
		AL-2	Marine	41	100 [108]	Hetero	Pep	Barophilic	78
		woesei	Marine	38	100 [105]	Hetero	Cbh, Pep	S^0 respiration	79
		abyssi GE5	Marine	44.7	96 [105]	Hetero	Pep, aa	Fermentative	80
		ES4	Marine	55.0	90-9 [110]	Hetero	Cbh, Pep, aa	S^{0A}	26
	Thermococcus	*celer*	Marine	56.6	87 [93]	Hetero	Pep	Fermentative	25
		litoralis A3	Marine	40	85 [98]	Hetero	Cbh, pep	Fermentative	36
		stetteri	Marine	50.2	76 [98]	Hetero	Cbh, pep	Grows on pectin	81
		litoralis NSC	Marine	39.1	88 [98]	Hetero	Cbh, Pep	Produces exopolysaccharide, forms biofilm	82,42
		barossii	Marine	60	82.5	Hetero	Cbh, Pep	—	83
		AN1	Terrestrial	46.2	75–80 [85]	Hetero	Cbh, Pep	S^0, L-cystine^	32
		chitonophagus	Marine	46.5	85 [93]	Hetero	Cbh, Pep	Grows on chitin	84
		peptonophilus OG-1	Marine	52	85 [100]	Hetero	Pep	Rifampicin inhibits growth	85
		profundus	Marine	52.5	80 [90]	Hetero	Cbh, Pep	Requires sulfur	86
		fumicolans ST557	Marine	54.5	85 [103]	Hetero	Pep, aa	Cystine, polysulfide^	87
		ES1	Marine	58.6	82 [91]	Hetero	Rich organics	Fermentative	37
		alcaliphilus AEDII2	Marine	42.4	85 [90]	Hetero	Pep	Fermentative	88
		AL-1	Marine	51	90 [94]	Hetero	Pep	Barophilic	89
		DT1331	Marine	52.3	80 [93]	Hetero	Pep	S^0 required	90
Archaeoglobales	*Archaeoglobus*	*fulgidus* VC16[a]	Marine	46.0	83 [95]	Auto/hetero	Cbh, Pep, CO_2	Oxidizes lactate completely to CO_2	31
		fulgidus Z	Marine	45.0	75–80 [90]	Auto/hetero	Cbh, Pep, CO_2	Sulfate[A], thiosulfate[A], sulfite[A]	28
		profundus	Marine	41.0	82 [90]	Hetero	Cbh, Pep	Mixotrophic	91
Methanopyrales	*Methanopyrus*	*kandleri*	Marine	60	98 [110]	Auto	CO_2	S^0 inhibitory	35
Methanococcales	*Methanococcus*	*jannaschii*[a]	Marine	31.4	85 [91]	Auto	CO_2	Requires selenium	30
		CS-1	Marine	31.4	85 [94]	Auto	CO_2	Formate[D]	92
		AG86	Marine	33	85 [92]	Auto	CO_2	Requires selenite or tungstate	93
		igneus	Marine	31	88 [91]	Auto	CO_2	Inhibited by H_2S	94
Methanobacteriales	*Methanothermus*	*sociabilis*	Terrestrial	33	88 [97]	Auto	CO_2	Ammonia as nitrogen source	95
		fervidus	Terrestrial	33	83 [97]	—	—	—	96
	Methanobacterium	*thermoautotrophicum*[a]	Terrestrial	—	75	Auto	CO_2	Does not use formate	97
					Bacteria				
Aquificiales	*Aquifex*[a]	*pyrophilus*	Marine	40	85 [95]	Auto	CO_2	Tolerates 6% O_2	49
Thermotogales	*Thermotoga*	*maritima*[a]	Marine	46	80 [90]	Hetero	Cbh	Fermentative	98
		neapolitana NSE	Marine	41.3	77 [95]	Hetero	Cbh, Pep	Inhibited by sulfide	99
		sp ZB	Marine	46	[85]	Hetero	Cbh	Fermentative	100
	Thermosipho	*africanus*	Marine		75 [77–80]	Hetero/auto	Pep, CO_2	Requires cysteine	101

Notes: [A] = electron acceptor; [D] = electron donor; Cbh = carbohydrates; Pep = peptides; hetero = heterotroph; auto = autotroph; aa = amino acid.

[a]Genomes corresponding to sequencing projects finished or in progress.

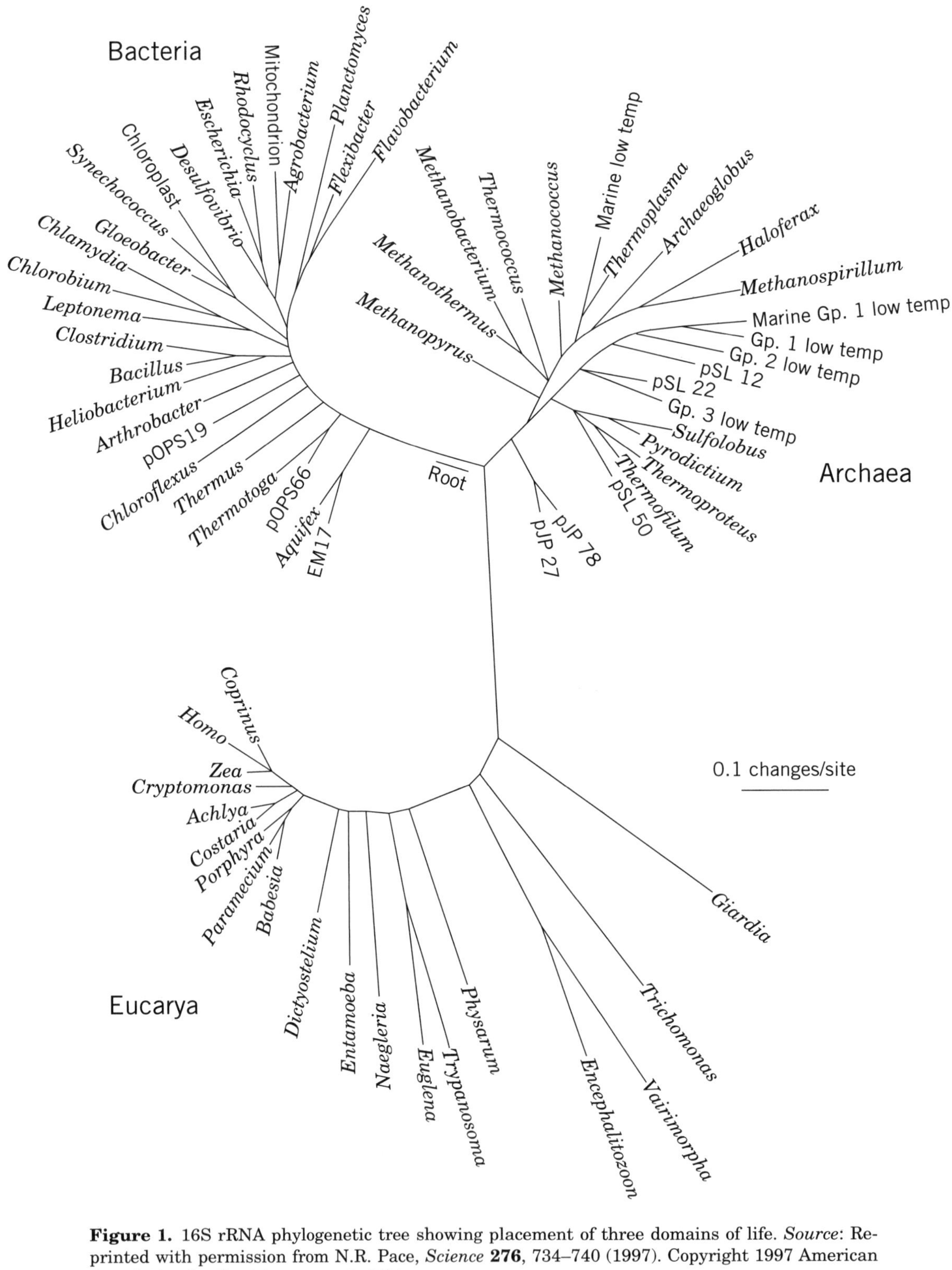

Figure 1. 16S rRNA phylogenetic tree showing placement of three domains of life. *Source*: Reprinted with permission from N.R. Pace, *Science* **276**, 734–740 (1997). Copyright 1997 American Association for the Advancement of Science.

the most extreme temperatures. Methanogenic microorganisms have been isolated from such environments. For example, *Methanococcus jannaschii* was isolated from the base of a deep-sea white-smoker chimney on the East Pacific Rise and found to grow optimally at 85 °C with a doubling time of 26 min (30). Subsequent work with this organism found it to be barophilic with elevated pressures, even higher than those associated with its isolation habitat, promoting both increased growth rate and methane production (177). *M. jannaschii* grows well when

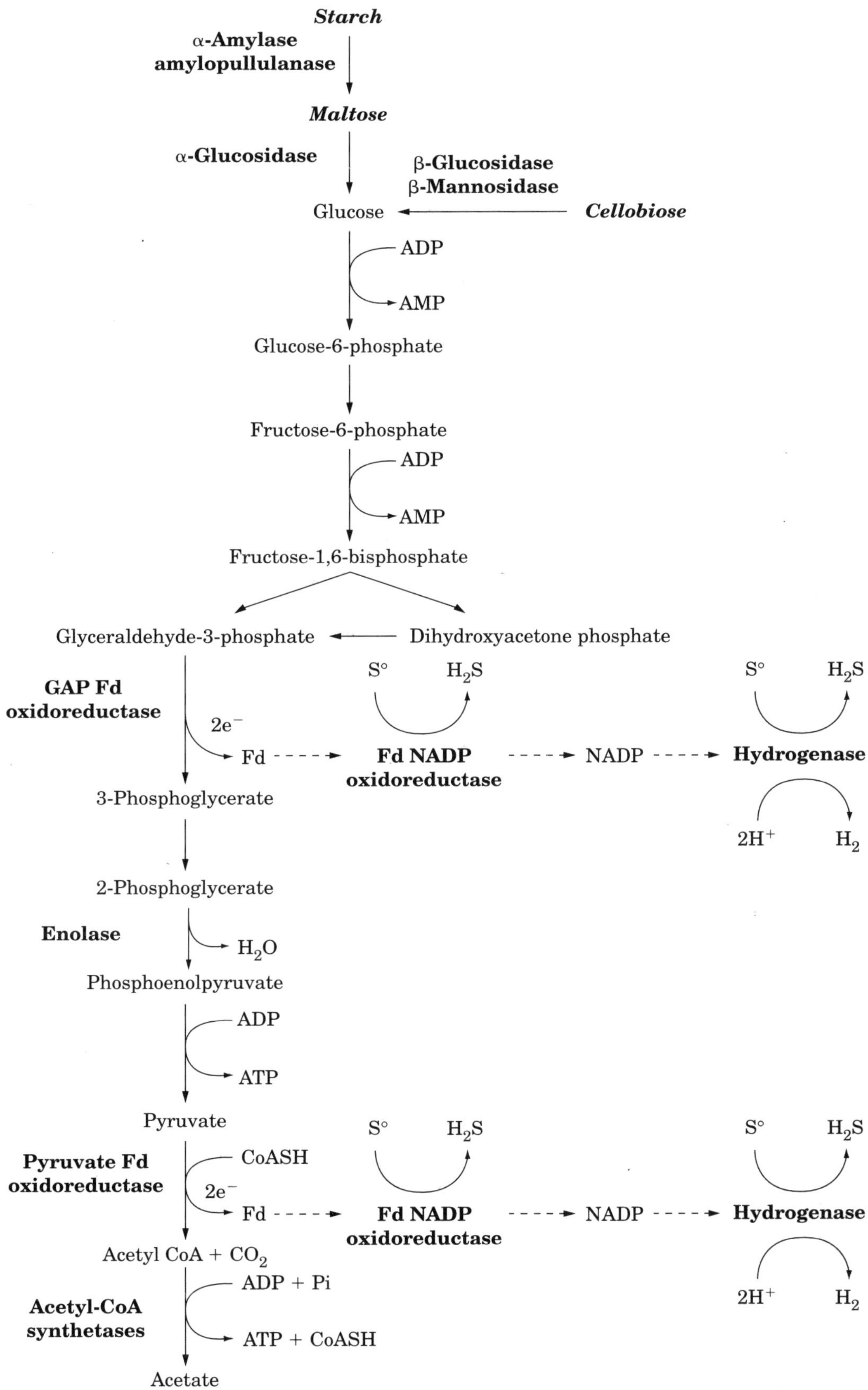

Figure 2. Proposed pathway for glucose oxidation and production of acetate, H_2, and H_2S in *Pyrococcus furiosus.* Growth substrates are shown in bold and italics Fd = ferredoxin. *Source:* Modified from Ref. 111.

Table 3. Enzymes Purified from *Pyrococcus furiosus*

Enzyme	References
Hydrolases	
α-Glucosidase	114
Amylopullulanase	106
Protease PfpI-H (S66)	115
Protease PfpI-C1, PfpI-C2	116,117
α-Amylase	118
β-Glucosidase	119,120
Prolyl oligopeptidase (putative)	121
β-Mannosidase	120
Proteasome	122
Protease pyrolysin	123
Hydrogenases / Redox Proteins	
Hydrogenase/sulfhydrogenase	124
Ferredoxin	113
Rubredoxin	125
Glutaredoxin (putative)	126
Thioredoxin	127
Oxidoreductases	
Aldehyde Fd oxidoreductase	128
Glyceraldehyde-3-Pi Fd oxidoreductase	111
Indolepyruvate Fd oxidoreductase	129
Pyruvate Fd oxidoreductase	130
2-Ketoisovalerate Fd oxidoreductase	131
Ferredoxin NADP oxidoreductase	132
Dehydrogenase	
Glutamate dehydrogenase	133
Isomerases / invertases	
Sucrose α-glucohydrolase	134
Nucleic acid modifying enzymes	
DNA polymerase	135
Synthetases	
Acetyl CoA synthetase (I + II)	136
Citrate synthase (putative)	137
Kinases	
Glucokinase	7
Other	
P1 protein	138
Enolase	139
Aromatic transaminase	140

Fd = ferredoxin.

Table 4. Enzymes purified from Thermotoga species

Enzyme	Species	References
Hydrolases		
endo-β-1,4-Glucanase	*maritina*	150
endo-Glucanase	*neapolitana*	151
exo-β-1,4-Glucanase	*maritima*	150
β-Mannanase	*neapolitana*	143
β-Mannosidase	*neapolitana*	143
α-Galactosidase	*neapolitana*	143
β-Galactosidase	*maritima*	152
β-Glycosidase	FjSS3-B.1	153
β-Xylosidase	FjSS3-B.1	153
α-Arabinofuranosidase	FjSS3-B.1	154
exo-glucanase	FjSS3-B.1	155
endo-1,4-β-D-Xylanases (2)	*thermarum*	156
β-Glucosidase (cloned)	*maritima*	152
endo-Xylanase A (cloned)	*neapolitana*	157
endo-Xylanase	*neapolitana*	151
endo-Xylanase A and B	*maritima*	142
endo-Xylanase A	FjSS3-B.1	158
Enolase	*maritima*	159
Hydrogenases / dehydrogenases		
Hydrogenase	*maritima*	160
D-Glyceraldehyde-3-phosphate	*maritima*	161
Dehydrogenase (cloned)		
L-Lactate dehydrogenase (cloned)	*maritima*	162
Oxidoreductases / redox proteins		
Pyruvate ferredoxin oxidoreductase	*maritima*	163
Ferredoxin	*maritima*	164
Nucleic acid modifying proteins		
RNA polymerase	*maritima*	165
DNA gyrase	*maritima*	165
DNA polymerase	sp.	166
RNAse P (putative from sequence)	*maritima, neapolitana*	167
RuvB (cloned)	*maritima*	168
RecA protein (cloned)	*maritima*	169
Isomerases		
Triosephosphate isomerase	*maritima*	170
Xylose (glucose) isomerase	*maritima, neapolitana*	171,172
Other		
Phosphoglycerate kinase	*maritima*	170
Omp alpha	*maritima*	173
Chemotaxis proteins	*maritima*	174
4-α-Glucanotransferase (cloned)	*maritima*	175

cocultured with hydrogen-producing extreme thermophiles (148,178), which may reflect its ecological role. A novel group of rodlike methanogens have also been isolated from marine geothermal systems that do not appear to belong to the three main methanogenic linkages defined by 16S rRNA phylogeny. *Methanopyrus kandleri*, the type species of the novel methanogenic genus, grows optimally at 98 °C and up to 110 °C and has some unusual cell wall features, including a new type of pseudomurein containing ornithin (35). Both *M. jannaschii* and *M. kandleri* exhibit rapid growth rates under optimal conditions (less than 1 h doubling times) relative to many mesophilic methanogens, which suggests dynamic interspecies hydrogen transfer in geothermal systems (148). To date, only hydrogenotrophic methanogens have been found among the extreme thermophiles. The isolation of a rapidly growing, acetate-using species would be interesting in view of attempts to accelerate mesophilic and moderately thermophilic anaerobic digestion processes.

Extreme Thermoacidophiles. Brock's pioneering work in investigating microbial life in hot springs in Yellowstone National Park revealed the existence of thermoacidophiles that grew in temperatures in excess of 70 °C and at pH under 2.0 (19). Subsequently, a number of thermoacidophilic genera have been described (Table 2), and organisms that grow at a pH below 1.0 have been reported (179). The metabolism of these organisms is complex and may involve the mixotrophic utilization of both organic carbon and CO_2. Extreme thermoacidophiles typically require some complex organic carbon source, but it is not always clear as to what components of this source are being used to fulfill carbon and energy requirements. Many thermoacidophiles oxidize reduced metal and inorganic sulfur species and have, thus, been evaluated for use in the desulfurization of coal and the recovery of metals from ores (180).

Sulfolobus acidocaldarius was first isolated from Locomotive Spring (pH 2.4, 83 °C) in Yellowstone National Park and subsequently found to be distributed widely in hot, acidic geothermal environments (1). Although the pH optimum is very acidic, *S. acidocaldarius* was also found to grow at a pH up to 5.8. Another member of the order Sulfolobales, *Metallosphaera sedula* grows optimally at pH 2.0 and 75 °C (61). Unlike *S. acidocaldarius*, this organism does not use saccharides as a carbon or energy source, but will grow on peptide-based media supplemented with ferrous iron or elemental sulfur. *M. sedula* will solubilize metals from ores very rapidly, even though its optimal doubling time is more than 5 hours (61).

The identification of an HSP60-like heat shock protein (designated TF 55) from *Sulfolobus shibatae* (181) has led to studies focusing on heat shock in the extreme thermoacidophiles. Although survival rates for *S. shibatae* at supraoptimal temperatures could be increased by thermal acclimation (181), similar experiments with *M. sedula* revealed that thermotolerance was fleeting, and no residual thermotolerance was observed once cultures were returned to normal growth temperature ranges (182). Thermal stress in *M. sedula* appears to result in damage to the proton pumping system or leakage of protons caused by membrane damage. This caused the cytoplasmic pH to drop from around 6 to lethal levels of 4 or below upon extended exposure to supraoptimal temperatures (183).

Other Extremely Thermophilic Physiologies. ***Sulfate Reduction.*** Although the reduction of sulfate to sulfide appears to be a physiological feature of less thermophilic microorganisms, the hyperthermophilic archaeon *Archaeglobus fulgidus*, isolated from Italian geothermal systems, will use sulfate, sulfite, and thiosulfate, but not sulfur, as electron acceptors (31). This organism is a facultative autotroph, growing heterotrophically on a variety of complex media or only on H_2, CO_2, and thiosulfate. *A. fulgidus* grows optimally at 83 °C.

Nitrate Reduction. Nitrate reduction, although rare to this point among the extreme thermophiles, has been observed to occur in both a hyperthermophilic archaeon and a hyperthermophilic bacterium. *Pyrobaculum aerophilum*, isolated from a marine water hole in Ischia, Italy, grows optimally at 100 °C and pH 7.0 by dissimilatory nitrate reduction forming dinitrogen as a product (68). This same organism also grows microaerophically on organic and inorganic compounds by aerobic respiration. In either growth mode, elemental sulfur was inhibitory. Similarly, a hyperthermophilic bacterium, *Aquifex pyrophilus*, isolated from Icelandic hot springs, a strict autotroph, also was found to grow by nitrate reduction or under low concentrations of oxygen (49). Molecular hydrogen, thiosulfate, and elemental sulfur were used as electron donors, forming either sulfuric acid or H_2S, depending on whether growth was aerobic or anaerobic. For *A. pyrophilus*, growth was optimal at 85 °C, with a doubling time of 75 min (49). Of further interest, *A. pyrophilus* represents the most deeply rooted bacterial genus on the 16S rRNA phylogenetic tree (49).

CULTIVATION METHODOLOGY FOR EXTREMELY THERMOPHILIC MICROORGANISMS

The cultivation of extremely thermophilic microorganisms for physiological studies or for generation of biomass has relied on methodology previously developed for mesophiles (138). Modifications have been made to equipment and protocols to accommodate the requirement of high temperature and other extreme conditions, such as pressure and pH. In some cases, specialized bioreactors and approaches have been developed to deal with the unusual bioprocessing conditions. For example, because most of these organisms either reduce (anaerobes) or oxidize (aerobes) sulfur, the highly corrosive conditions created by high levels of hydrogen sulfide or sulfuric acid make materials of construction an important issue. Gaseous products and substrates associated with these organisms also include H_2 and CH_4, so explosive conditions can exist. In any case, many extremely thermophilic microorganisms have been successfully cultivated, in some cases producing in excess of a kilogram of biomass (wet) from a single fermentation (184). Table 5 illustrates some of the reactor types used in the bioprocessing of extreme thermophiles.

Batch Cultivation

Batch cultivation of extreme thermophiles has been done in a variety of ways and on different scales. Hungate tubes or serum bottles are often used for small scale culture, which are made anaerobic by addition of reducing agents and degassing procedures when required. Aerobic thermoacidophiles are often grown at pH of approximately 2.0, making contamination problems minimal, although the cultures must be oxygenated in some way. Details on the types of media used and protocols for establishing cultures have been described elsewhere (4,5,18,89).

Table 5. Bioreactors for Growth of Extreme Thermophiles

Reactor characteristics	Operating mode(s)	Examples of organisms grown	Volume (L)	Operation
Serum bottle	Batch	Most extreme thermophiles	~0.1	Sparge with inert gas before inoculation for anaerobes
Glass resin flask; with 4-5 ports; condenser	Batch, fed-batch, continuous	*Pyrocccus furiosus* *Thermococcus litoralis* *Thermococcus barossii* *Thermotoga neapolitana* *Methanococcus jannaschii* *Metallosphaera sedula*	1–10	Sparge with inert gas H_2S scrubbed with alkali for S^0 growth Can be agitated with magnetic mixing bar or impeller
Glass fermenter; with stainless steel heating mantle; condenser; baffles	Batch, fed-batch	*Pyrocccus furiosus* *Thermococcus litoralis* *Thermococcus barossii* *Thermotoga neapolitana* *Methanococcus jannaschii*	4–18	Sparge with inert gas H_2S scrubbed with alkali for S^0 growth
Ceramic-lined fermenter	Batch	Most extreme thermophiles	50–500	Sparge with inert gas
Stainless steel fermenter	Batch	*Sulfolobus* species *Pyrococcus furiosus* *Thermococcus litoralis* *Thermotoga maritima* *Pyrococcus endeavori* (ES4)	100–600	Sparge with inert gas Mostly non-S^0 growth Carefully clean after S growth to prevent corrosion Can be used for themoacidophile cultivation
High pressure reactor (hyperbaric)	Batch	*Methanococcus jannaschii* *Pyrococcus endeavori* *Sulfolobus acidocaldarius*	~1	Use He for gas phase + substrate gases
High pressure reactor (hydrostatic)	Batch	*Pyrococcus endeavori* (ES4)	~0.5	Flexible gold bag used to contain sample Can be subsampled
Gas-lift reactor	Fed-batch, continuous	*Pyrococcus furiosus*	2	Low hydrodynamic stress
Dialysis membrane reactor	Batch, fed-batch	*Pyrococcus furiosus*	1.2	Both outer and inner chambers are flushed Very high cell yields

Fermenters with modifications for high-temperature operation have be used to culture anaerobic extreme thermophiles. Glass vessels with working volumes up to approximately 20 L (0.7 ft^3) must also have provision for condensing large amounts of water that vaporize because of the high operating temperatures. Although mixing is important to suspend many medium components that are sparingly soluble (e.g., polysaccharides, elemental sulfur), many cultures are sensitive to even modest shearing forces. Also, glass fermenters that have heating mantels or baffle cages that are metallic can be susceptible to corrosion. In some cases, enough metal is leached from stainless steels to inhibit the growth of certain hyperthermophiles.

Growth at High Pressure

Many extremely thermophilic archaea have been isolated from ocean depths of up to 3,000 m (pressures of about 200 atm) and from within temperature gradients that go from 350 °C to 4 °C over distances much less than 1 m. Nonetheless, most isolates are enriched at atmospheric pressure or low hyperbaric pressure (less than 3 atm) and temperatures below 115 °C (78) for experimental convenience. It has been shown that there is a trend toward barotolerance and barophily at pressures greater than those encountered in the native environment (185), and this has spurred efforts to understand the interaction of pressure and temperature as parameters influential to microbial growth (186). It should be noted that special precautions must be used at high pressures to ensure that changes in growth or gas production are actually coupled to increased pressure and not a side effect related to other factors, such as protonation and deprotonation or gas solubility differences (185).

For hydrostatic pressure, a hydraulic fluid must be used to provide overpressures; often distilled water or growth media is sufficient for these purposes (78). Pressure is controlled manually during heating by adjusting hydraulic fluid volume. Hydrostatic pressure bioreactors may be as simple as a syringe or a metal cannister (89). To ensure proper sampling and mixing within the hydrostatic chamber, more sophisticated systems have been developed. A compressible gold bag, fitted with sampling ports and placed in an hydrostatic chamber within a heating mantel, has been used to culture extreme thermophiles (16). This system can be operated semicontinuously by adding and removing material through syringe pumps and rocked to promote mixing. Using such a system, Reysenbach and Deming (78) found that two hyperthermophilic strains showed tolerance to pressures up to 440 atm, and in one case, an improvement in growth rate was observed. In addition, cells grown under pressure appeared to have more uniform size and were less irregular in shape compared to the same cells grown at low pressure (78).

In a hyperbaric pressure reactor, pressure is transmitted to the culture by the gas phase. Miller et al. (187) developed a hyperbaric pressure reactor that was constructed from a transparent sapphire cell, allowing visual observation of the culture medium during growth experiments. The reactor and a recirculating pump were both contained in an oven to facilitate experiments with extreme thermophiles. One version of this system could be operated at pressures of over 100 bar, with a 10-mL working volume. A magnetically operated pump was used to sparge inert gases through the medium, providing some degree of agitation and promoting thermodynamic equilibrium between vapor and liquid phases. The culture medium was supplied directly from a glove box into the cell, and a gas chromatograph, used to measure volatile products of growth, was directly interfaced to the reactor. The sapphire cell could be replaced with a 167-mL stainless steel vessel, which was then suitable for pressures up to 1,000 atm (177). With the larger volume system, a more powerful pneumatic pump was necessary to achieve desired flow rates to the reactor. Growth of the extremely thermophilic methanogen *M. jannaschii* in the 10-mL hyperbaric bioreactor demonstrated that increases of pressure from 7.8 to 100 bar resulted in accelerated production of biomass and methane and extended the maximum growth temperature from 90 to 92 °C (187). In the stainless steel reactor, the high-temperature limit for methanogenesis was increased to 98 °C by hyperbaric helium pressures up to 750 atm (177).

Continuous Culture

In fed-batch or continuous culture systems used for the growth of extreme thermophiles, the same considerations must be taken as those described for batch systems. Internal parts are typically glass or Teflon-coated to avoid corrosion. Media feedstreams are heat sterilized before use, or in some cases, in situ, given the normally high growth temperatures of these microorganisms. Inorganic feedstreams are filtered through a 0.2-μm porous filter to avoid contamination. Continuous culture systems have been described in some detail for *M. sedula* (182) and *P. furiosus* (41,108). Figure 3 illustrates both a schematic of and an actual continuous culture system.

The gas-lift bioreactor is an alternative type of continuous culture and has been demonstrated for the extreme thermophile *P. furiosus* (50,189). A protocol for setup and operation of an anaerobic gas-lift reactor is described in the work by Sharp and Raven (89). As mentioned previously, some hyperthermophiles are sensitive to high shear rates during growth. Gas-lift systems characteristically have low shearing forces in conjunction with high rates of mass transfer and are well suited for cultivation of extreme thermophiles (189). Such reactors for extreme thermophile growth have been constructed of glass components, with a gas distribution tube for inert gas (filtered with 0.2 μm) supply. The glass vessel can be heated through a heat transfer fluid-containing jacket; an oversized condensor is used to recover water vapor from the exhaust. Both pH and temperature can be controlled continuously. Experiments by Raven et al. (189) have shown that nitrogen is the preferred inert gas to use for sparging, allowing higher cell densities under otherwise equivalent conditions with other gases. The optimal dilution rate for cell mass production of *P. furiosus* was found to be 0.4 h^{-1}, at which point the bioreactor generated more than 1.5 g (wet weight) cells l^{-1} (189). A defined medium was optimized for *P. furiosus*, resulting in reported cell densities of 10^{10} mL^{-1} (50).

Dialysis Membrane Reactor

The impetus for using a membrane bioreactor is twofold: carbon and energy sources can be supplied directly to the cells through the membrane, and inhibitory or toxic end products of fermentation can be removed through the membrane, improving cellular growth rate and cell yield. The dialysis reactor described by Krahe et al. (43) is composed of two reactor chambers of unequal volumes, separated by a membrane. Both of these compartments can be sparged and agitated. The chamber that houses the cells is typically several times smaller than the chamber that contains the dialysing medium. Using complex media for both chambers, a 100-fold increase (as compared to serum bottles) in cellular concentration of *P. furiosus* has been reported, along with a 10-fold higher cell density at stationary phase where cell lysis can be significant and lead to loss of whole cells for subsequent recovery steps.

NEW DEVELOPMENTS IN THE ISOLATION AND STUDY OF EXTREMELY THERMOPHILIC MICROORGANISMS

The recognition that the number of extreme thermophiles isolated to date is but a small fraction of those inhabiting geothermal environments has become apparent from analysis of 16S rRNA signatures from geothermal samples (23). Although phylogenetic diversity may not correlate well with physiological diversity, it nonetheless illustrates the untapped potential that exists for new types of microorganisms that have proved difficult to isolate into pure culture.

Molecular phylogeny has been the basis for recent methods to isolate extreme thermophiles. Single cultures of archaea containing 16S rRNA signatures, previously identified from in situ analysis of samples from hot springs, were obtained by the use of optical tweezers (190,191). This method makes use of an inverse microscope and an infrared laser to pull a single cell with a particular 16S rRNA signature into a separate section of a capillary and then put it into cultivation media, where it can be cloned and characterized (192). If effective, this short-circuits the often laborious methods of serial dilution, from which the result might be an organism that has already been isolated and characterized.

Whole genome sequencing efforts have thus far focused to a significant extent on extreme thermophiles. Table 2 indicates which of the genomes of extreme thermophiles are currently being sequenced or are finished. The Institute for Genomic Research (TIGR) maintains a list available on-line of current sequencing projects. In general, extreme thermophiles are an attractive target for sequencing

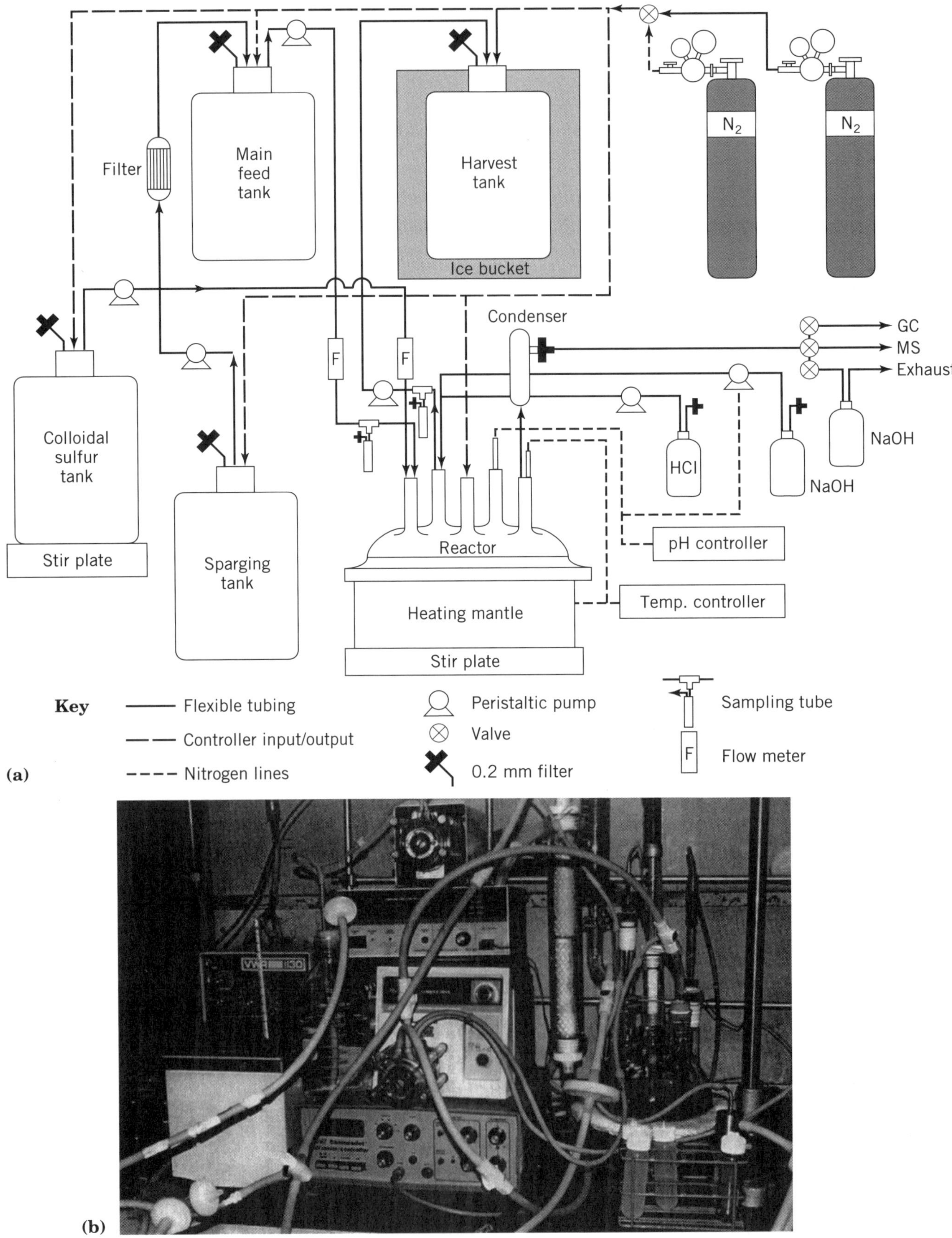

Figure 3. (**a**) Schematic and (**b**) Actual continuous culture setup for growth of hyperthermophiles. *Source*: Schematic modified from Rinker et al. (188). Photo and schematic courtesy of K.D. Rinker, North Carolina State University.

efforts given their evolutionary significance, their commercial potential, and the fact that their genomes are relatively small. The first genome sequence for an extremophile to be completed was that of *M. jannaschii* (193), undoubtedly many more will soon follow. Not only does this present an important new approach to assessing metabolic diversity, but it provides a tool to study the regulation of key pathways within a given organism and a means to study genetic, biochemical, and physiological differences and similarities among this group. The explosive increase in sequence information provides yet another challenge in that mechanisms to use these data must be developed.

SUMMARY

This review illustrates the progress that has been made during the past two decades in exploring the basis for life at extremely high temperatures. Bioprocessing has made important contributions to this effort in developing ways in which extreme thermophiles can be grown for purposes of biomass generation and physiological study. The remaining challenges include developing methods for culturing extreme thermophiles that have, thus far, not been cultured under laboratory conditions; this includes establishing mixed cultures that might be used for particular biotransformations, such as bioremediation. In the future, bioprocessing may even find a role in determining the prospects for life on the subsurface of other planets and moons that may harbor extreme thermophiles not unlike those inhabiting volcanic subsurface environments on earth. The development of bioreactors mimicking such environments will be an important contribution in this regard.

BIBLIOGRAPHY

1. T.D. Brock, K.M. Brock, R.T. Belly, and R.L. Weiss, *Arch. Microbiol.* **84**, 54–68 (1972).
2. K.O. Stetter, *FEMS Microbiol. Rev.* **18**, 149–158 (1996).
3. K.O. Stetter, *ASM News* **61**, 285–290 (1995).
4. F.T. Robb, A.R. Place, K.R. Sowers, H.J. Schreier, S. DasSarma, and E.M. Fleischmann eds., *Archaea: A Laboratory Manual (Thermophiles section)*, Cold Spring Harbor Laboratory Press, New York, 1995, pp. 1–51.
5. K.D. Rinker, C.J. Han, M.W.W. Adams, and R.M. Kelly, in *ASM Manual of Industrial Microbiology and Biotechnology*, 1997, in press.
6. R.M. Kelly and M.W.W. Adams, *Antonie van Leeuwenhoek* **66**, 247–270 (1994).
7. S.W.M. Kengen, A.J.M. Stams, and W.M. de Vos, *FEMS Microbiol. Rev.* **18**, 119–137, (1994).
8. M.W.W. Adams, F.B. Perler, and R.M. Kelly, *Bio / Technol.* **13**, 662–668 (1995).
9. C. Leuschner and G. Antranikian, *World J. Microbiol. Biotechnol.* **11**, 95–114 (1995).
10. C. Aagaard, I. Leviev, R.N. Aravalli, P. Forterre, D. Prieur, and R.A. Garrett, *FEMS Microbiol. Rev.* **18**, 93–104 (1996).
11. W. Zillig, D. Prangishvilli, C. Schleper, M. Elferink, I. Holz, S. Albers, D. Janekovic, and D. Gotz, *FEMS Microbiol. Rev.* **18**, 225–236, (1996).
12. C.R. Woese, *Microbiol. Rev.* **51**, 221–271 (1987).
13. C.R. Woese, O. Kandler, and M.L. Wheelis, *Proc. Natl. Acad. Sci. U.S.A.* **87**, 4576–4579 (1990).
14. M.W.W. Adams and R.M. Kelly, *Chem. Eng. News* **73**, 32–42 (1995).
15. F.B. Perler, S. Kumar, and H. Kong, *Adv. Protein Chem.* **48**, 377–435 (1996).
16. R.M. Kelly and J.W. Deming, *Biotechnol. Prog.* **4**, 47–62 (1988).
17. D.S. Clark and R.M. Kelly, *CHEMTECH* **20**, 641–648 (1990).
18. J.A. Baross and J.W. Deming, in D.M. Karl ed., *Deep-Sea Hydrothermal Vents*, CRC Press, Boca Raton, 1995, pp. 169–217.
19. T.D. Brock, *Thermophilic Microorganisms and Life at High Temperatures*, Springer-Verlag, New York, 1978.
20. E. Blöchl, S. Burggraf, G. Fiala, G. Lauerer, G. Huber, R. Huber, R. Rachel, A. Segerer, K.O. Stetter, and P. Völkl, *World J. Microbiol. Biotechnol.* **11**, 9–16.
21. S.C. Cary, M.T. Cottrell, J.L. Stein, F. Comacho, and F. Desbruyeres, *Appl. Environ. Microbiol.* **63**, 1124–1130 (1997).
22. S.M. Barnes, R.E. Fundyga, M.W. Jeffries, and N.R. Pace, *Proc. Natl. Acad. Sci. U.S.A.* **91**, 1609–1613.
23. N.R. Pace, *Science* **276**, 734–740 (1997).
24. G. Fiala and K.O. Stetter, *Arch. Microbiol.* **145**, 56–61 (1986).
25. W. Zillig, I. Holz, D. Janekovic, W. Schäfer, and W.D. Reiter, *Syst. Appl. Microbiol.* **4**, 88–94 (1983).
26. R.J. Pledger and J.A. Baross, *J. Gen. Microbiol.* **137**, 203–211 (1991).
27. R. Huber, P. Stoffers, J.L. Cheminee, H.H. Richnow, and K.O. Stetter, *Nature* **345**, 179–181 (1990).
28. G. Zellner, E. Stackebrandt, H. Kneifel, P. Messner, U.B. Sleytr, E.C. De Macario, H.-P. Zabel, and K.O. Stetter, *Syst. Appl. Microbiol.* **11**, 151–160 (1989).
29. W. Zillig, K.O. Stetter, D. Prangishvilli, W. Schäfer, S. Wunderl, D. Janekovic, I. Holz, and P. Palm, *Zbl. Bakt. Hyg., I. Abt. Orig.* **C3**, 304–317 (1982).
30. W.J. Jones, J.A. Leigh, F. Mayer, C.R. Woese, and R.S. Wolfe, *Arch. Microbiol.* **136**, 254–261 (1983).
31. K.O. Stetter, *Syst. Appl. Microbiol.* **10**, 172–173 (1988).
32. K.U. Klages and H.W. Morgan, *Arch. Microbiol.* **162**, 261–266 (1994).
33. A. Segerer, A. Neuner, J.K. Kristjansson, and K.O. Stetter, *Int. J. Syst. Bacteriol.* **36**, 559–564, (1986).
34. W. Zillig, A. Gierl, G. Schreiber, S. Wunderl, D. Janekovic, K.O. Stetter, and H.P. Klenk, *Syst. Appl. Microbiol.* **4**, 79–87 (1983).
35. M. Kurr, R. Huber, H. Konig, H.W. Jannasch, H. Fricke, A. Trincone, J.K. Kristjansson, and K.O. Stetter, *Arch. Microbiol.* **156**, 239–247 (1991).
36. A. Neuner, H.W. Jannasch, S. Belkin, and K.O. Stetter, *Arch. Microbiol.* **153**, 205–207 (1990).
37. R.J. Pledger and J.A. Baross, *Syst. Appl. Microbiol.* **12**, 249–256 (1989).
38. G.S. Grassia, K.M. McLean, P. Glenat, J. Bauld, and A.J. Sheehy, *FEMS Microbiol. Ecol.* **21**, 47–58 (1996).
39. H.W. Jannasch, C.O. Wirsen, and T. Hoaki, in F.T. Robb, A.R. Place, K.R. Sowers, H.J. Schreier, S. DasSarma, and E.M. Fleischmann eds., *Archaea: A Laboratory Manual*,

Cold Spring Harbor Laboratory Press, New York, 1995, pp. 9–13.

40. A. Kletzin, S. Mukund, T.L. Kelley-Crouse, M.K. Chan, D.C. Rees, and M.W.W. Adams, *J. Bacteriol.* **177**, 4817–4819 (1995).
41. R.N. Schicho, L.J. Snowden, S. Mukund, J.B. Park, M.W.W. Adams, and R.M. Kelly, *Arch. Microbiol.* **159**, 380–385 (1993).
42. K.D. Rinker and R.M. Kelly, *Appl. Environ. Microbiol.* **62**, 4478–4485 (1996).
43. M. Krahe, G. Antranikian, and H. Markl, *FEMS Microbiol. Rev.* **18**, 271–285 (1996).
44. A.K. Parameswaran, C.N. Provan, F.J. Sturm, and R.M. Kelly, *Appl. Environ. Microbiol.* **53**, 1690–1693 (1987).
45. H. Connaris, D. Cowan, M. Ruffett, and R.J. Sharp, *Lett. Appl. Microbiol.* **13**, 25–27 (1991).
46. J. Carlsson, G.P.D. Granberg, G.K. Nyberg, and M.B.K. Edlund, *Appl. Environ. Microbiol.* **37**, 383–390 (1979).
47. S.E. Childers, M. Vargas, and K.M. Noll, *Appl. Environ. Microbiol.* **58**, 3949–3953 (1992).
48. O.T. Harriott, R. Huber, K.O. Stetter, P.W. Betts, and K.M. Noll, *J. Bacteriol.* **176**, 2759–2762 (1994).
49. R. Huber, T. Wilharm, D. Huber, A. Trincone, S. Burggraf, H. Konig, R. Rachel, I. Rockinger, H. Fricke, and K.O. Stetter, *Syst. Appl. Microbiol.* **15**, 340–351 (1992).
50. N.D.H. Raven and R.J. Sharp, *FEMS Microbiol. Lett.* **146**, 135–141 (1997).
51. D.W. Grogan, *J. Bacteriol.* **171**, 6710–6719 (1989).
52. L. Watrin and D. Prieur, *Curr. Microbiol.* **33**, 377–382 (1996).
53. R. Pley, J. Schipka, A. Gambacorta, H.W. Jannasch, H.F.R. Rachel, and K.O. Stetter, *Syst. Appl. Microbiol.* **14**, 245–253 (1991).
54. K.O. Stetter, H. Konig, and E. Stackebrandt, *Syst. Appl. Microbiol.* **4**, 535–551 (1983).
55. W. Zillig, I. Holz, D. Janekovic, H.-P. Klenk, E. Imsel, J. Trent, S. Wunderl, V.H. Forjaz, R. Coutinho, and T. Ferreira, *J. Bacteriol.* **172**, 3959–3965 (1990).
56. K.O. Stetter, in T.D. Brock ed., *Thermophiles: General, Molecular and Applied Microbiology*, Wiley, New York, 1986, pp. 40–74.
57. T. Fuchs, H. Huber, S. Burggraf, and K.O. Stetter, *Syst. Appl. Microbiol.* **19**, 56–60 (1996).
58. W. Zillig, K.O. Stetter, S. Wunderl, W. Schulz, H. Priess, and I. Scholz, *Arch. Microbiol.* **125**, 259–269 (1980).
59. W. Zillig, A. Kletzin, C. Schleper, I. Holz, D. Janekovic, J. Hain, M. Lanzendorfer, and J.K. Kristjansson, *Syst. Appl. Microbiol.* **16**, 609–628 (1994).
60. D. Grogan, P. Palm, and W. Zillig, *Arch. Microbiol.* **154**, 594–599 (1990).
61. G. Huber, C. Spinnler, A. Gambacorta, and K.O. Stetter, *Syst. Appl. Microbiol.* **12**, 38–47 (1989).
62. T. Fuchs, H. Huber, K. Teiner, S. Burggraf, and K.O. Stetter, *Syst. Appl. Microbiol.* **18**, 560–566 (1995).
63. A.H. Segerer, A. Trincone, M. Gahrtz, and K.O. Stetter, *J. Bacteriol.* **41**, 495–501 (1991).
64. W. Zillig, S. Yeats, I. Holz, A. Bock, M. Rettenberger, F. Gropp, and G. Simon, *Syst. Appl. Microbiol.* **8**, 197–203 (1986).
65. E.A. Bonch-Osmolovskaya, A.I. Slesarev, M.L. Miroshnichenko, T.P. Svetlichnaya, and V.A. Alexeyev, *Microbiologyia* **57**, 78–85 (1985).
66. H.W. Jannasch, C.O. Wirsen, S.J. Molyneaux, and T.A. Langworthy, *Appl. Environ. Microbiol.* **54**, 1203–1209 (1988).
67. G. Fiala, K.O. Stetter, H.W. Jannasch, T.A. Langworthy, and J. Madon, *Syst. Appl. Microbiol.* **8**, 106–113 (1986).
68. P. Volkl, R. Huber, E. Drobner, R. Rachel, S. Burggraf, A. Trincone, and K.O. Stetter, *Appl. Environ. Microbiol.* **59**, 2918–2926 (1993).
69. R. Huber, J.K. Kristjansson, and K.O. Stetter, *Arch. Microbiol.* **149**, 95–101 (1987).
70. W. Zillig, K.O. Stetter, W. Schäfer, D. Janekovic, S. Wunderl, I. Holz, and P. Palm, *Zbl. Bakt. Hyg., I. Abt. Orig.* **C2**, 205–227 (1981).
71. E.A. Bonch-Osmolovskaya, M.L. Miroshnichenko, N.A. Kostrikina, N.A. Chernych, and G.A. Zavarzin, *Arch. Microbiol.* **154**, 556–559 (1990).
72. S. Schäfer, C. Barkowski, and G. Fuchs, *Arch. Microbiol.* **146**, 301–308 (1986).
73. F. Canganella and W.J. Jones, *Curr. Microbiol.* **28**, 299–306 (1994).
74. H.W. Jannasch, C.O. Wirsen, S.J. Molyneaux, and T.A. Langworthy, *Appl. Environ. Microbiol.* **58**, 3472–3481 (1992).
75. M. Taya, H. Hinoki, T. Yagi, and T. Kobayashi, *Appl. Microbiol. Biotechnol.* **29**, 474–479 (1988).
76. V.A. Svetlichnyi, A.I. Slesarev, T.P. Svetlichnaya, and G.A. Zavarzin, *Mikrobiologiya* **56**, 831–838.
77. Y. Sako, N. Nomura, A. Uchida, Y. Ishida, H. Morii, Y. Koga, T. Hoaki, and T. Maruyama, *Int. J. Syst. Bacteriol.* **46**, 1070–1077 (1996).
78. A.-L. Reysenbach and J.W. Deming, *Appl. Environ. Microbiol.* **57**, 1271–1274 (1991).
79. W. Zillig, I. Holz, H.-P. Klenk, J. Trent, S. Wunderl, D. Janekovic, E. Imsel, and B. Haas, *Syst. Appl. Microbiol.* **9**, 62–70 (1987).
80. G. Erauso, A.-L. Reysenbach, A. Godfroy, J.-R. Meunier, B. Crump, F. Partensky, J.A. Baross, V. Marteinsson, G. Barbier, N.R. Pace, and D. Prieur, *Arch. Microbiol.* **160**, 338–349 (1993).
81. M.L. Miroshnichenko, E.A. Bonch-Osmolovskaya, A. Neuner, N.A. Kostrikina, N.A. Chernych, and V.A. Alekseev, *Syst. Appl. Microbiol.* **12**, 257–262 (1989).
82. S. Belkin and H.W. Jannasch, *Arch. Microbiol.* **141**, 181–186 (1985).
83. G.D. Duffaud, O.B. d'Hennezel, A.S. Peek, A.-L. Reysenbach, and R.M. Kelly, *Syst. Appl. Microbiol.* **21**, 40–49 (1998).
84. R. Huber, J. Stohr, S. Hohenhaus, R. Rachel, S. Burggraf, H.W. Jannasch, and K.O. Stetter, *Arch. Microbiol.* **164**, 255–264 (1995).
85. M. Gonzalez, C. Kato, and K. Horikoshi, *Arch. Microbiol.* **164**, 159–164 (1995).
86. T. Kobayashi, Y.S. Kwak, T. Akiba, T. Kudo, and K. Horikoshi (1994).
87. A. Godfroy, J.-R. Meunier, J. Guezennec, F. Lesongeur, G. Raguenes, A. Rimbault, and G. Barbier, *Int. J. Syst. Bact.* **46**, 1113–1119 (1996).
88. M. Keller, F.-J. Braun, R. Dirmeier, D. Hafenbradl, S. Burggraf, R. Rachel, and K.O. Stetter, *Arch. Microbiol.* **164**, 390–395 (1995).
89. R.J. Sharp and N.D.H. Raven, in P.M. Rhodes and P.F. Stanbury eds., *Applied Microbial Physiology: A Practical Approach*, IRL Press, Oxford, 1997, pp. 23–51.
90. Y.S. Kwak, T. Kobayashi, T. Akiba, K. Horikoshi, and Y.B. Kim, *Biosci. Biotech. Biochem.* **59**, 1666–1669 (1995).

91. S. Burggraf, H.W. Jannasch, B. Nicolaus, and K.O. Stetter, *Syst. Appl. Microbiol.* **13**, 24–28 (1990).

92. W.J. Jones, C.E. Stugard, and H.W. Jannasch, *Arch. Microbiol.* **151**, 314–318 (1989).

93. H. Zhao, A.G. Wood, F. Widdel, and M.P. Bryant, *Arch. Microbiol.* **150**, 178–183 (1988).

94. S. Burggraf, H. Fricke, A. Neuner, J. Kristjansson, P. Rouvier, L. Mandelco, C.R. Woese, and K.O. Stetter, *Syst. Appl. Microbiol.* **13**, 263–269 (1990).

95. G. Lauerer, J.K. Kristjansson, T.A. Langworthy, H. Konig, and K.O. Stetter, *Syst. Appl. Microbiol.* **8** (1986).

96. K.O. Stetter, M. Thomm, J. Winter, G. Wildgruber, H. Huber, W. Zillig, D. Janecovic, H. Konig, P. Palm, and S. Wunderl, *Zbl. Bakt. Hyg., I. Abt. Orig.* **C2**, 166–178 (1981).

97. P.J.L. Derikx, G.A.H. de Jong, H.J.M. Op den Camp, C. van der Drift, L.J.L.D. van Griensven, and G.D. Vogels, *FEMS Microbiol. Ecol.* **62**, 251–258 (1989).

98. R. Huber, T.A. Langworthy, H. Konig, M. Thomm, C.R. Woese, U.B. Sleytr, and K.O. Stetter, *Arch. Microbiol.* **144**, 324–333 (1986).

99. S. Belkin, C.O. Wirsen, and H.W. Jannasch, *Appl. Environ. Microbiol.* **51**, 1180–1185 (1986).

100. G. Zellner and H. Kneifel, *Arch. Microbiol.* **159**, 472–476 (1993).

101. R. Huber, C.R. Woese, T.A. Langworthy, H. Fricke, and K.O. Stetter, *Syst. Appl. Microbiol.* **12**, 32–37 (1989).

102. F.B. Perler, M.W. Southworth, D.G. Wilbur, and D. Wallace, in F.T. Robb, A.R. Place, K.R. Sowers, H.J. Schreier, S. DasSarma, and E.M. Fleischmann eds., *Archaea: A Laboratory Manual*, Cold Spring Harbor Laboratory Press, New York, 1995, pp. 139–147.

103. J. DiRuggiero, J.H. Tuttle, and F.T. Robb, *Mol. Mar. Biol. Biotechnol.* **4**, 123–127 (1995).

104. S.W.M. Kengen and J.M. Stams, *Arch. Microbiol.* **161**, 168–175 (1994).

105. I.I. Blumentals, M. Itoh, G.J. Olson, and R.M. Kelly, *Appl. Environ. Microbiol.* **56**, 1255–1262 (1990).

106. S.H. Brown, H.R. Costantino, and R.M. Kelly, *Appl. Environ. Microbiol.* **56**, 1985–1991 (1990).

107. T. Schäfer and P. Schonheit, *Arch. Microbiol.* **155**, 366–377 (1991).

108. I.A. Usenko, L.O. Severina, and V.K. Plakunov, *Microbiology* **62**, 272–277 (1993).

109. R.N. Schicho, K. Ma, M.W.W. Adams, and R.M. Kelly, *J. Bacteriol.* **175**, 1823–1830 (1993).

110. K. Ma, R.N. Schicho, R.M. Kelly, and M.W.W. Adams, *Proc. Natl. Acad. Sci. U.S.A.* **90**, 5341–5344 (1993).

111. S. Mukund and M.W.W. Adams, *J. Biol. Chem.* **270**, 8389–8405 (1995).

112. T. Schaefer, M. Selig, and P. Schoenheit, *Arch. Microbiol.* **159**, 72–83 (1993).

113. S. Aono, F.O. Bryant, and M.W.W. Adams, *J. Bacteriol.* **171**, 3433–3439 (1989).

114. H.R. Costantino, S.H. Brown, and R.M. Kelly, *J. Bacteriol.* **172**, 3654–3660 (1990).

115. I.I. Blumentals, A.S. Robinson, and R.M. Kelly, *Appl. Environ. Microbiol.* **56**, 1992–1998 (1990).

116. S.B. Halio, I.I. Blumentals, S.A. Short, B.M. Merrill, and R.M. Kelly, *J. Bacteriol.* **178**, 2605–2612 (1996).

117. S.B. Halio, M.W. Bauer, S. Mukund, M.W.W. Adams, and R.M. Kelly, *Appl. Environ. Microbiol.* **63**, 289–295 (1997).

118. K.A. Laderman, B.R. Davis, H.C. Krutzsch, M.S. Lewis, Y.V. Griko, P.L. Privalov, and C.B. Anfinsen, *J. Biol. Chem.* **268**, 24394–24401 (1993).

119. S.W.M. Kengen, E.J. Luesink, A.J.M. Stams, and A.J.B. Zehnder, *Eur. J. Biochem.* **213**, 305–312 (1993).

120. M.W. Bauer, E.J. Bylina, R.V. Swanson, and R.M. Kelly, *J. Biol. Chem.* **271**, 23749–23755, (1996).

121. K.A. Robinson, D.A. Bartley, F.T. Robb, and H.J. Schreier, *Gene* **152**, 103–106 (1995).

122. M.W. Bauer, S.H. Bauer, and R.M. Kelly, *Appl. Environ. Microbiol.* **63**, 1160–1164 (1997).

123. R. Eggen, A. Geerling, J. Watts, and W.M. de Vos, *FEMS Microbiol. Lett.* **71**, 17–20 (1990).

124. F.O. Bryant and M.W.W. Adams, *J. Biol. Chem.* **264**, 5070–5079 (1989).

125. P.R. Blake, J.B. Park, F.O. Bryant, S. Aono, J.K. Magnuson, E. Eccleston, J.B. Howard, M.F. Summers, and M.W.W. Adams, *Biochemistry* **30**, 10885–10895 (1991).

126. A. Guagliardi, D. de Pascale, R. Cannio, V. Nobile, S. Bartolucci, and M. Rossi, *J. Biol. Chem.* **270**, 5748–5755 (1995).

127. C.A. Raia, S. D'Auria, and M. Rossi, *Biosens. Bioelectron.* **10**, 135, (1995).

128. S. Mukund and M.W.W. Adams, *J. Biol. Chem.* **266**, 14208–14216 (1991).

129. X. Mai and M.W.W. Adams, *J. Biol. Chem.* **269**, 16726–16732 (1994).

130. E.T. Smith, J.M. Blamey, and M.W.W. Adams, *Biochemistry* **33**, 1008–1016 (1994).

131. J. Heider, X. Mai, and M.W. Adams, *J. Bacteriol.* **178**, 780–787 (1996).

132. K. Ma and M.W.W. Adams, *J. Bacteriol.* **176**, 6509–6517 (1994).

133. F.T. Robb, J.B. Park, and M.W.W. Adams, *Biochim. Biophys. Acta* **1120**, 267–272 (1992).

134. H.R. Badr, K.A. Sims, and M.W.W. Adams, *Syst. Appl. Microbiol.* **17**, 1–6 (1994).

135. K.S. Lundberg, D.D. Shoemaker, M.W.W. Adams, J.M. Short, J.A. Sorge, and E.J. Mathur, *Gene* **108**, 1–6 (1991).

136. X. Mai and M.W.W. Adams, *J. Bacteriol.* **178**, 5897–5903 (1996).

137. J.M. Muir, R.J. Russell, D.W. Hough, and M.J. Danson, *Protein Eng.* **8**, 583–592 (1995).

138. R.M. Kelly, S.H. Brown, I.I. Blumentals, and M.W.W. Adams, in M.W.W. Adams and R.M. Kelly eds., *Biocatalysis at Extreme Temperatures*, American Chemical Society, Washington, D.C., 1992, pp. 23–41.

139. M.J. Peak, J.G. Peak, F.J. Stevens, J. Blamey, X. Mai, Z.H. Zhou, and M.W. Adams, *Arch. Biochem. Biophys.* **313**, 280–286 (1994).

140. G. Andreotti, M.V. Cubellis, G. Nitti, G. Sannia, X. Mai, M.W. Adams, and G. Marino, *Biochim. Biophys. Acta* **1247**, 90–96 (1995).

141. J.P. Huelsenbeck and B. Rannala, *Science* **276**, 227–232 (1997).

142. C. Winterhalter and W. Liebl, *Appl. Environ. Microbiol.* **61**, 1810–1815 (1995).

143. G.D. Duffaud, C.M. McCutchen, P. Leduc, K.N. Parker, and R.M. Kelly, *Appl. Environ. Microbiol.* **63**, 169–177 (1997).

144. C. Schröder, M. Selig, and P. Schönheit, *Arch. Microbiol.* **161**, 460–470 (1994).

145. M.W.W. Adams, *FEMS Microbiol. Rev.* **15**, 261–277 (1994).
146. G. Ravot, B. Ollivier, M.-L. Fardeau, B.K.C. Patel, K.T. Andrews, M. Magot, and J.-L. Garcia, *Appl. Environ. Microbiol.* **62**, 2657–2659 (1996).
147. M.Y. Galperin, K.M. Noll, A.H. Romano, *Appl. Environ. Microbiol.* **62**, 2915–2918 (1996).
148. V. Muralidharan, K.D. Rinker, I.S. Hirsh, E.J. Bouwer, and R.M. Kelly, *Biotechnol. Bioeng.* **56**, 268–278 (1997).
149. M.F. Gulch, D. Typke, and W. Baumesiter, *J. Bacteriol.* **177**, 5473–5479 (1995).
150. K. Bronnenmeier, A. Kern, W. Liebl, and W.L. Staudenbauer, *Appl. Environ. Microbiol.* **61**, 1399–1407 (1995).
151. J.D. Bok, S.K. Goers, and D.E. Eveleigh, *ACS Symp. Ser.* **566**, 54–65 (1994).
152. J. Gabelsberger, W. Liebl, and K.H. Schleifer, *FEMS Microbiol. Lett.* **109**, 131–138 (1993).
153. L.D. Ruttersmith and R.M. Daniel, *Biochim. Biophys. Acta* **1156**, 167–172 (1993).
154. L.D. Ruttersmith, R.M. Daniel, and H.D. Simpson, *Ann. N.Y. Acad. Sci.* **672**, 137–141 (1992).
155. L.D. Ruttersmith and R.M. Daniel, *J. Biochem.* **277**, 887–890 (1991).
156. A. Sunna, J. Puls, and G. Antranikian, *Biotechnol. Appl. Biochem.* **24**, 177, (1996).
157. V. Zverlov, K. Piotukh, O. Dakhova, G. Velikodvorskaya, and R. Borris, *Microbiol. Biotechnol.* **45**, 245–247 (1996).
158. H.D. Simpson, U.R. Haufler, and R.M. Daniel, *J. Biochem.* **277**, 413–417 (1991).
159. H. Schurig, K. Rutkat, R. Rachel, and R. Jaenicke, *Protein Sci.* **4735**, 228–236 (1995).
160. A. Juszczak, S. Aono, and M.W. Adams, *J. Biol. Chem.* **266**, 13834–13841 (1991).
161. A. Tomschy, R. Glockshuber, and R. Jaenicke, *Eur. J. Biochem.* **214**, 43–50 (1993).
162. R. Ostendorp, W. Liebl, H. Schurig, and R. Jaenicke, *Eur. J. Biochem.* **216**, 709–715 (1993).
163. J.M. Blamey and M.W. Adams, *Biochemistry* **33**, 1000–1007 (1994).
164. J.M. Blamey, S. Mukund, and M.W. Adams, *FEMS Microbiol. Lett.* **121**, 165–169 (1994).
165. P. Palm, C. Schleper, I. Arnold-Ammer, I. Holz, T. Meier, F. Lottspeich, and W. Zillig, *Nucleic Acids Res.* **21**, 4904–4908 (1993).
166. H.D. Simpson, T. Coolbear, and R.M. Daniel, *Ann. N.Y. Acad. Sci.* **613**, 426–428 (1990).
167. J.W. Brown, E.S. Haas, and N.R. Pace, *Nucleic Acids Res.* **21**, 671–679 (1993).
168. J. Tong and J.G. Wetmur, *J. Bacteriol.* **178**, 2695–2700 (1996).
169. J.G. Wetmur, D.M. Wong, B. Ortiz, J. Tong, F. Reichert, and D.H. Gelfand, *J. Biol. Chem.* **269**, 25928–25935 (1994).
170. H. Schurig, N. Beaucamp, R. Ostendorp, R. Jaenicke, E. Adler, and J.R. Knowles, *EMBO J.* **14**, 442–451 (1995).
171. S.H. Brown, C. Sjøholm, and R.M. Kelly, *Biotechnol. Bioeng.* **41**, 878–886 (1993).
172. C. Vieille, J.M. Hess, R.M. Kelly, and J.G. Zeikus, *Appl. Environ. Microbiol.* **61**, 1867–1875 (1995).
173. A.M. Engel, Z. Cejka, A. Lupas, F. Lottspeich, and W. Baumeister, *EMBO J.* **11**, 4369–4378 (1992).
174. R.V. Swanson, M.G. Sanna, and M.I. Simon, *J. Bacteriol.* **178**, 484–489 (1996).
175. W. Liebl, R. Feil, J. Gabelsberger, J. Kellermann, and K.H. Schleifer, *Eur. J. Biochem.* **207**, 81–88 (1992).
176. J. Schumann, A. Wrba, R. Jaenicke, and K.O. Stetter, *FEBS Lett* **282**, 122–126 (1991).
177. J.F. Miller, N.N. Shah, C.M. Nelson, J.M. Ludlow, and D.S. Clark, *Appl. Environ. Microbiol.* **54**, 3039–3042 (1988).
178. E.A. Bonch-Osmolovskaya and K.O. Stetter, *Syst. Appl. Microbiol.* **14**, 205–208 (1991).
179. C. Schleper, G. Puehler, I. Holz, A. Gambacorta, D. Janekovic, U. Santarius, H.-P. Klenk, and W. Zillig, *J. Bacteriol.* **177**, 7050–7059 (1995).
180. G.J. Olson and R.M. Kelly, *Biotechnol. Prog.* **2**, 1–15 (1986).
181. J.D. Trent, J. Osipiuk, and T. Pinkau, *J. Bacteriol.* **172**, 1478–1484 (1990).
182. C.J. Han, S.H. Park, and R.M. Kelly, *Appl. Environ. Microbiol.* **63**, 2391–2396 (1997).
183. T.L. Peeples and R.M. Kelly, *Appl. Environ. Microbiol.* **61**, 2314–2321 (1995).
184. M.W.W. Adams, in F.T. Robb, A.R. Place, K.R. Sowers, H.J. Schreier, S. DasSarma, and E.M. Fleischmann eds., *Archaea: A Laboratory Manual*, Cold Spring Harbor Laboratory Press, New York, 1995, pp. 47–49.
185. G. Bernhardt, A. Disteche, J. Rainer, B. Koch, H.D. Ludemann, and K.O. Stetter, *Appl. Microbiol. Biotechnol.* **28**, 176–181 (1988).
186. J. Konisky, P.C. Michels, and D.S. Clark, *Appl. Environ. Microbiol.* **61**, 2762–2764 (1995).
187. J.F. Miller, E.L. Almond, N.N. Shah, J.M. Ludlow, J.A. Zollweg, W.B. Streett, S.H. Zinder, and D.S. Clark, *Biotechnol. Bioeng.* **31**, 407–413 (1988).
188. K.D. Rinker, Ph.D. Thesis, North Carolina State University, Raleigh, N.C., 1998.
189. N. Raven, N. Ladwa, D. Cossar, and R. Sharp, *Appl. Microbiol. Biotechnol.* **38**, 263–267 (1992).
190. A. Ashkin, J.M. Dziedzic, and T. Yamane, *Nature* **330**, 769–771 (1987).
191. A. Ashkin and J.M. Dziedzic, *Science* **235**, 1517–1520 (1987).
192. R. Huber, S. Burggraf, T. Mayer, S.M. Barns, P. Rossnagel, and K.O. Stetter, *Nature* **376**, 57–58 (1995).
193. C.J. Bult, O. White, G.J. Olsen, L. Zhou, R.D. Fleischmann, G.G. Sutton, J.A. Blake, L.M. FitzGerald, R.A. Clayton, J.D. Gocayne, A.R. Kerlavage, B.A. Dougherty, J.F. Tomb, M.D. Adams, C.I. Reich, R. Overbeek, E.F. Kirkness, K.G. Weinstock, J.M. Merrick, A. Glodek, J.L. Scott, N.S.M. Geoghagen, and J.C. Venter, *Science* **273**, 1058–1073 (1996).

See also ENZYMES, EXTREMELY THERMOSTABLE.

THERMOPHILIC/THERMOSTABLE ENZYMES.

See ENZYMES, EXTREMELY THERMOSTABLE; THERMOPHILIC MICROORGANISMS.

THREONINE. See AMINO ACIDS, PRODUCTION PROCESSES.

TRANSFER PHENOMENA IN MULTIPHASE SYSTEMS IN MIXING VESSELS

RODICA-VIORICA ROMAN
Chemical Pharmaceutical Research Institute
Iaşi, Romania

KEY WORDS

Biosynthesis of antibiotics
Hydrodynamics
Mass Transfer
Modified agitators

OUTLINE

INTRODUCTION

The dispersion of gases and solids in liquids is one of the most commonly applied processes of chemical engineering, and the fully baffled stirred vessel is one of many types of equipment used for this purpose. The hydrodynamic regime of the impeller controls the mixing performance of the whole reactor, as is reflected in phases of homogenization, the intensity of local energy dissipation distribution in the region of severe shear conditions, the retained gas fraction, bubble size, solid concentration distribution, mass transfer, and heat transfer (1–3). Good understanding of the dispersion mechanism of the impeller is indispensable for a reliable prediction of the behavior of the multiphase system flow in the reactor as a whole.

In multiphase processes carried out in stirred vessels, interphase mass transfer is often an important step and may be rate determining for the overall process. In aerobic fermentation systems, the oxygen transfer from the gas phase to the surface of the microbial cells is of primary importance, especially at high cell densities when microbial growth is likely to be limited by the availability of oxygen in the liquid phase (4–6). For many years, a set of radial flow Rushton disk turbine impellers of roughly one-third of the standard reactor diameter has been considered the optimum design for the multiphase system mixing, including fermentation processes. Mass transfer cannot be improved by increasing the specific power consumption or the aeration rate above certain limits, both of which lead to increased overall costs, but only by a modification of the blades of the impellers or by the development of impellers that allow a better bulk mixing with a minimum power consumption.

It has been reported (7) that a simple modification of the blades of a Rushton turbine through increase in the blade height simultaneously with perforation of the blade surface could significantly diminish power consumption and improve oxygen transfer efficiency in either a non-biological system or in an antibiotic biosynthesis process.

DESCRIPTION OF THE MODIFIED RUSHTON TURBINE AGITATORS

The modified blades were obtained by increase in the conventional (standard) blade height simultaneously with perforation of the blade surface. Five modified blade turbine types (denoted TP1–TP5) were investigated; these are characterized by the fact that the number of perforations and their size varied and that the filled surface of the modified blades is equal to the blade surface of the standard Rushton turbine (TR) (Fig. 1) (7,8).

The application of perforations to the modified blades decreases the pathway between successive shearing actions on the liquid streams, increasing the number of times that the fluid is sheared. This effect has direct implications on local bubble size distribution in a gas–liquid mixture, which develops as a result of the balance between the complementary processes of dispersion and coalescence. The surface fraction of the perforations, defined as the ratio between the area S_G of the perforations on the surface and the full surface area S_C of the blade is in the range of 0 to 0.588. The modified blade turbines do not require any modification in the electrical motor and drive assemblies and are simple to manufacture. The experiments were carried out in a transparent vessel of inside diameter $D = 0.25$ m, with a flat bottom and four symmetrical wall baffles of width $w = 0.1\,D$. Fluid mixing was carried out by conventional and modified blade turbines, positioned singly (1TR, 1TP1–1TP5) or doubly (2TR, 2TP1–2TP5) on the same shaft.

MULTIPHASE SYSTEM HYDRODYNAMICS

Gas–Liquid Dispersing Characteristics

The hydrodynamic regimes were defined by reference to the distribution of the gas phase in the vessel. For an air–water system, the flow regimes were detected through visual observation of the zone close to the outer edge of the TR and TP1–TP5 turbines (8). In the single-turbine configurations, the three classical regimes (9,10) were observed: flooding (F), loading (L), and complete dispersion (CD). For

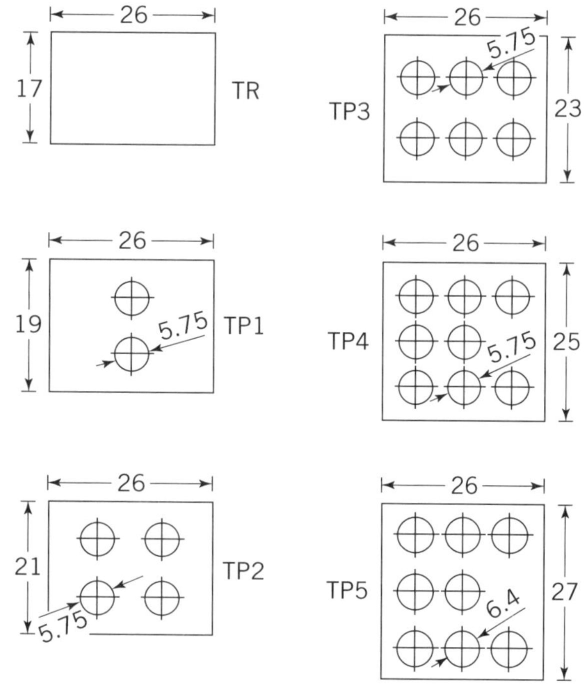

Agitator	TR	TP1	TP2	TP3	TP4	TP5
S_G/S_C	0	0.117	0.235	0.353	0.470	0.588

Figure 1. Schematic representation of the standard (TR) and modified (TP1–TP5) blades of the Rushton turbine. (*Source*: From Ref. 8, with permission.)

a double-turbine system, two hydrodynamic configurations were defined in the upper zone: ineffective dispersion (DI) and effective dispersion (DE), respectively; the transition between the hydrodynamic regimes corresponds to the rotational speed at which the bubbles have a radial velocity sufficient to reach the vessel wall. The gassed power consumption, P_g, is usually determined by varying the operational parameters: the volumetric flow rate, Q_g, and the impeller speed, N.

The curve representing the ratio P_g/P as related to the gas flow number presents maximum and minimum points when the volumetric gas flow rate is kept constant. Each of these extreme points corresponds to a gas–liquid hydrodynamic regime in the impeller region. The power curves change notably when the power consumption is expressed as a power number. Interesting results were obtained by plotting the gassed power number $(\mathrm{Ne})_g$ against the impeller speed N, when the aeration rate Q'_g, expressed as volumetric flow rate of gas per minute/volume of liquid (vvm) was kept constant. These results are presented in Figures 2 and 3 for mixing systems having one or two turbines, respectively. Table 1 explains the symbols used in Figures 2 and 3.

In the range of low N values, the power number for a single turbine decreased with increasing impeller speed until a certain point: subsequently, it increased suddenly in the case of the standard blade turbine and slowly in the case of the modified blade turbines. The impeller speed at which the transition between these two zones occurred was very close to the transition impeller speed between flooding and loading of the turbine. For two turbines on the same shaft, clear delimitation between the two zones of the impeller speed was not apparent (Fig. 3). The $(\mathrm{Ne})_g$ values decreased slowly with increasing impeller speed until the agitation system exceed the flooding state of the bottom turbine. In the range of large N values, when the gas dispersed uniformly through the entire vessel section, the gassed power number became independent of the impeller speed, as it was a function of aeration rate and geometrical configuration of the mixing system only. The variation in the power number depending on the surface fraction of the perforations, which indicates the geometry of the mixing system, is presented in Figure 4.

The range of power number for an individual Rushton turbine reported in the literature (11) is $\mathrm{Ne} = 4.8 - 6.3$ for $Q'_g = 0$. In our case, the value $\mathrm{Ne} = 5.5$ for a single Rushton turbine is in good agreement with these data. The Ne values for Rushton turbines are dependent on a scale (12), and for a mixing system in which the geometrical symplexes deviate from the standard values, the introduction of a correction factor of the power number is recommended (13,14). The Ne and $(\mathrm{Ne})_g$ values for all the modified blade turbines are smaller than those for the Rushton turbine, as a result of the change in fluid flow around and over the blades and of the minimization of the drag forces associated with the motion of blades such that the energy losses due to form drag are very low. These results imply that it is possible to rotate the modified blade turbines much faster than the standard Rushton turbine at an equal specific power consumption. The TP3 modified blade turbine, with the surface fraction S_G/S_C of the perforations equal to 0.353, has the smallest gassed and ungassed power number values, these being reduced by approximately 50% in comparison with the standard Rushton turbine. The increase in the surface fraction of the perforations over $S_G/S_C = 0.353$ is not justified on the basis of the power consumption.

For double turbines, the power number values are approximately twice those obtained for the single turbines, irrespective of the turbine type, at constant aeration rate Q'_g. This shows that, in the case of two identical turbines, the power consumption P_g is twice that obtained using a single turbine: this is possible because, if Q'_g is kept constant, the volumetric gas flow rate, Q_g, $(\mathrm{m}^3\ \mathrm{s}^{-1})$ is 60% higher when the liquid volume is higher (Table 1). Also, the power consumption of two TP3 modified turbines is reduced by 50% in comparison with the double standard Rushton turbines. Because the surface of each modified blade increases simultaneously with the S_G/S_C ratio, it is to be expected that a larger gas volume is retained in the ventilated cavities behind the modified blades and dispersed afterward. Therefore, the modified blades have a greatly increased gas-handling capacity: the TP3 turbines handle more than 35% more gas than do the standard Rushton turbines at the same specific ungassed power consumption, before flooding or ineffective dispersion.

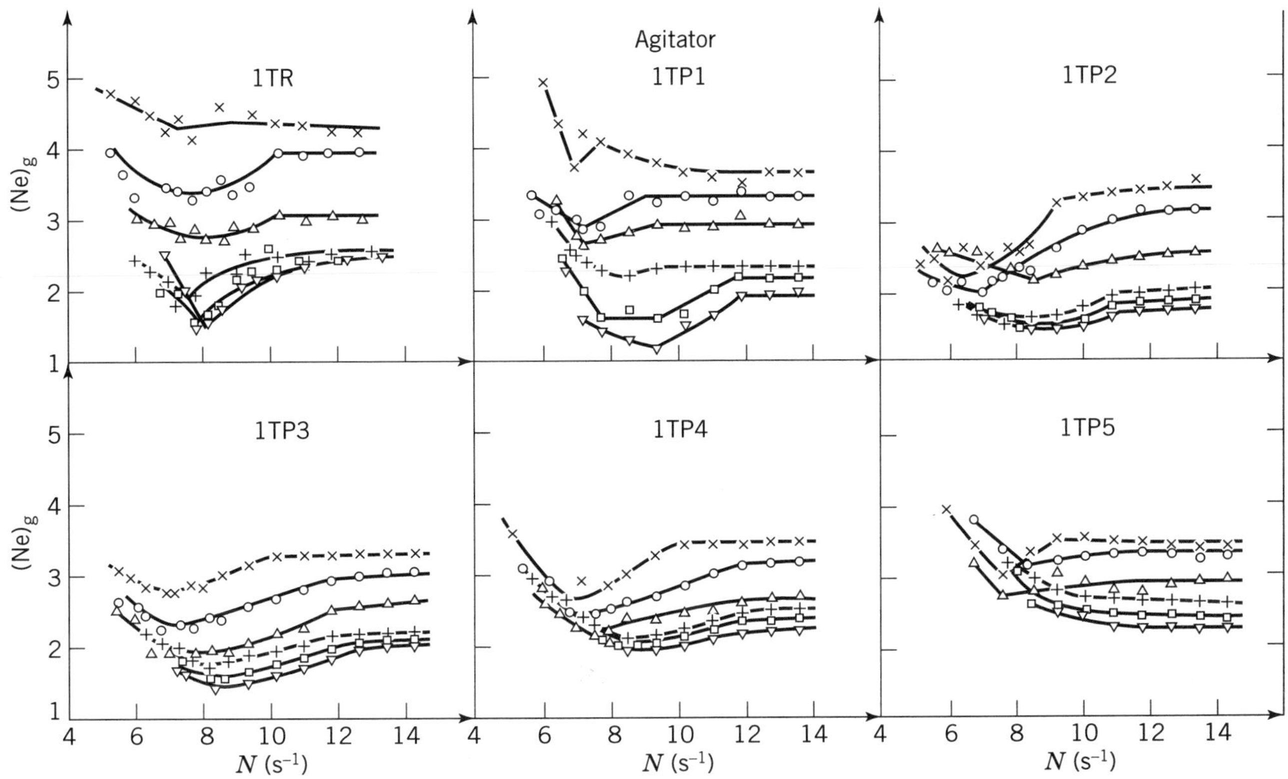

Figure 2. Variation in the gassed power number $(Ne)_g$, depending on the impeller speed (N) at various gas flow rates for single standard and modified Rushton turbines (symbols are defined in Table 1). (*Source:* From Ref. 8, with permission.)

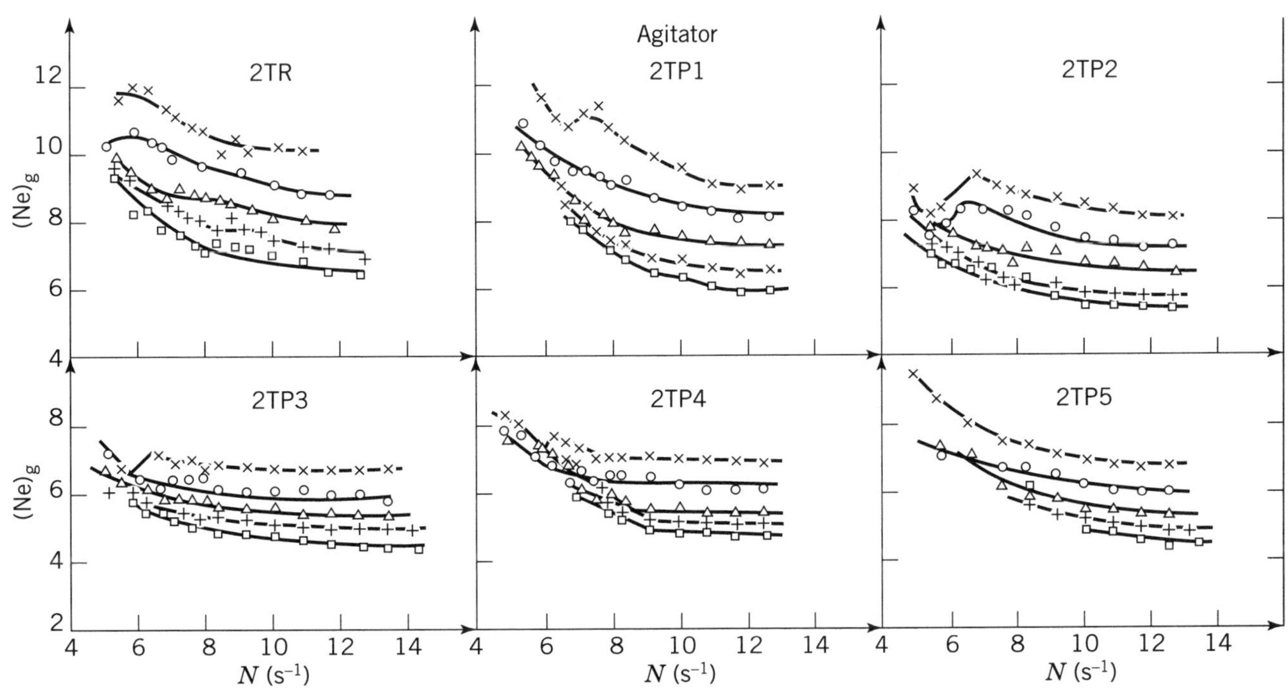

Figure 3. Variation in the gassed power number $(Ne)_g$ depending on the impeller speed (N) at various gas flow rates for double standard and modified Rushton turbines (symbols are defined in Table 1). (*Source*: From Ref. 8, with permission.)

Table 1. Symbols used in Figures 2 and 3

Symbol	Aeration rate, Q'_g (vvm)	Volumetric gas flow rate, $Q_g\ 10^4$ (m^3/s) One turbine (V = 0.0125 m^3)	Two turbines (V = 0.020 m^3)
×	0.30	0.625	1.0
○	0.50	1.042	1.666
△	0.75	1.555	2.500
*	1.00	2.083	3.33
□	1.25	2.611	4.166
▽	1.50	3.125	—

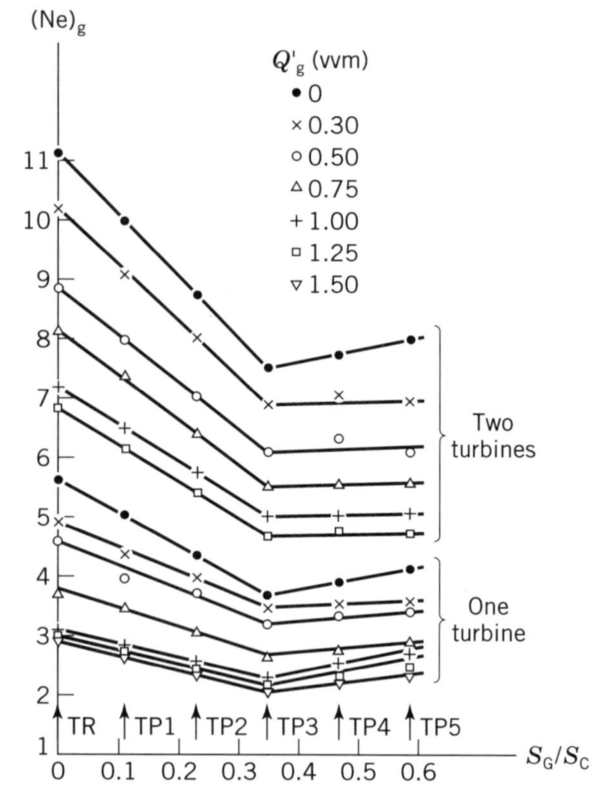

Figure 4. Variation in the power number depending on the surface fraction of the perforations (S_G/S_C) at various aeration rates (Q'_g). (*Source*: From Ref. 8, with permission.)

Suspension of Solid Particles in Liquid

Three suspension states can be defined, as follows (15):

1. Incomplete suspension, in which a part of the solid phase is deposited on the bottom of the vessels
2. Complete suspension, in which all particles are in suspension
3. Homogeneous suspension, in which the particle concentration is uniform throughout the vessel

Many earlier investigations concentrated on the impeller speed for complete suspension, when the maximum surface area available for processing is achieved (15–18). The most widely used criterion to describe solids suspension in mechanically agitated vessels is the just-off-the-bottom suspension condition, first described by Zwietering (19).

In this section, an attempt is made to understand the effects of the physical and rheological properties of the suspensions and number and turbine types on the complete suspension speed and on the power dissipation at complete and homogeneous suspension states of the solid particles (20). The solid particles used are $CaCO_3$, a cation exchange resin (Amberlite IR-120 type), and an anion exchange resin (Amberlite IRA-410 type). A range of density differences $\Delta\rho = (\rho_s - \rho_L)$ between 150 and 1,200 kg/m^3 was examined. The ion exchange resin particle shape is spherical, the particle size distribution being measured by a microscopic method. The change in particle size during the experimental run was negligible. The surface-to-volume (Sauter) mean diameter of the particles, d_p, was calculated for the ion exchange resin particles. In the $CaCO_3$ suspension, the mean diameter of the particles was 15 μm (21). The particle concentration, X, was between 20 and 150 kg/m^3 for each particle type used. The physical properties are given in Table 2. The suspensions used are pseudoplastic in behavior. The rheological properties of suspensions depend on the particle size and the concentration of solids. Rheological data of the suspensions were obtained with a rotational cylinder viscosimeter (Rheotest-2.1) and were fitted according to the power law model (22). The rheological parameters, consistency index, K, and flow behavior index, n, are also presented in Table 2. The definition of complete suspension speed, N_{js}, was taken as the speed at which no particles were visually observed to remain at rest on the vessel bottom for more than 1 or 2 s (19). The last particles to be suspended were located near the center of the vessel base and also behind the baffles. The experimental relationships between complete suspension speed and solid concentrations are shown in Figure 5 for each of the five particle types and for modified and standard turbines positioned singly or doubly on the same shaft.

For the same operation conditions, the N_{js} values for the TP3 modified turbine are slightly higher than those for the standard turbine. Also, Figure 5 shows that, independent of the agitator type, N_{js} actually is increased when two turbines are used, as was also observed by Armenante (23). Although the phenomenon of solid suspension off the vessel bottom is largely dominated by the lower impeller, the presence of two turbines modifies the flow pattern in the vessel.

In all geometrical systems, the N_{js} values are found to increase with solid concentration. This increase of the N_{js} value may result from an increase of probability to find particles at rest on the bottom as the number of particles in suspension increases. It was observed that at N_{js} the concentration of the solid particles in the zone near the bottom is considerably higher than the average value in the vessel. The physical background of the concentration effect is the increased power dissipation by presence of solid particles in the turbulent field (24). The particles can interfere with the turbulent field via two routes. (*1*) Because of their inertia, particles do not follow the motion of the liquid. Consequently, the kinetic energy of a particle

Table 2. Physical and Rheological Properties of Solid Particles

		Physical parameters				Rheological parameters	
Particle type	Symbol	d_p (μm)	ρ_s (kg/m^3)	X (kg/m^3)	ρ_{sp} (kg/m^3)	K (Pa s^n)	n (−)
$CaCO_3$	S1	15	2,200	20	1,026	0.105	0.834
				50	1,065	0.118	0.815
				100	1,130	0.133	0.812
				150	1,195	0.186	0.751
Amberlite IR-120	S2	680	1,350	20	1,008	0.135	0.760
				50	1,020	0.149	0.741
				100	1,040	0.164	0.729
				150	1,060	0.204	0.721
Amberlite IR-120	S3	860	1,350	20	1,008	0.140	0.811
				50	1,020	0.150	0.789
				100	1,040	0.220	0.710
				150	1,060	0.382	0.609
Amberlite IRA-410	S4	600	1,150	20	1,002	0.194	0.730
				50	1,005	0.244	0.686
				100	1,010	0.355	0.611
				150	1,015	0.530	0.570
Amberlite IRA-410	S5	1,000	1,150	20	1,002	0.235	0.662
				50	1,005	0.292	0.651
				100	1,010	0.428	0.568
				150	1,015	0.753	0.553

Source: Ref. 20.

changes continuously. Acceleration of a particle occurs at the expense of energy from the turbulent eddies, whereas deceleration leads to energy dissipation in the form of heat. (*2*) At high solids concentration, the particles will also interfere with each other (collison and/or hindrance). In a similar way, this will lead to increased dissipation of the energy of the turbulent eddies. Thus, for the point of complete suspension to be maintained, a higher solid concentration requires a higher power input to the systems.

Figure 6, where the power consumption expressed as power number for complete suspension condition $(\mathrm{Ne})_{js}$, is plotted against the Reynolds number Re, demonstrates that in the vessel equipped with the modified blade turbines, single or double, the $(\mathrm{Ne})_{js}$ values are approximately 30% lower than those in the standard vessel. In these plots, $(\mathrm{Ne})_{js}$ and Re numbers are calculated using the properties of the suspensions. The $(\mathrm{Ne})_{js}$–Re representation based solely on liquid properties is inadequate (25), because at the highest solids concentration laminar or transition range appear to persist up to very high Reynolds numbers. By using a suitable apparent viscosity for the solid–liquid suspension, it is possible to condense all the data in the adequate range. In our case, the Reynolds number values for complete suspension condition are in the transition range ($\mathrm{Re} < 1000$); the $(\mathrm{Ne})_{js}$–Re dependency has a negative slope.

The main reason for the smaller power number of modified blade turbines is the change of suspension flow around and over the blades. The fluid flow around the blades of the standard Rushton turbine was studied by Vant' Riet (26). Dead zones behind each turbine blade are formed. From these dead zones with the stagnation lines, the trailing vortex pair originates as a combined effect of the vortex motion at the rear of the vertical inner edge of the blade and the wrapping of the vortex sheets behind both horizontal edges. In the case of the modified blade turbines, the fluid jets penetrating the perforations, eliminate the stagnation lines and change the conditions of the trailing vortex formation.

When all the observation are combined, the following correlation is evident between complete suspension speed and the physical and rheological properties of the solid particles:

$$N_{js} = c d_p^{0.2} \Delta\rho^{0.45} X^{0.13} K^{0.1} \qquad (1)$$

The exponents of d_p, $\Delta\,\rho$, and X are essentially the same as those reported by Zwietering. The correlation, equation 1, includes the consistency index K as a pseudo-viscosity of the suspensions. The value of the dimensionless constant, c, is specific for each of the geometrical configurations studied. The influence of the turbine type and number of the turbines on N_{js} can be expressed by incorporation of two dimensionless groups, namely $(1 + S_G/S_C)$ and $n_p/6$ (S_G/S_C is the surface fraction of the perforation and n_p is the number of blades of the mixing system). Using nonlinear least squares regression, the following correlation was established:

$$N_{js} = 1.1 d_p^{0.2} \Delta\rho^{0.45} X^{0.13} K^{0.1} (1 + S_G/S_C)^{0.2} (n_p/6)^{0.25} \qquad (2)$$

In the turbulent range ($\mathrm{Re} > 1000$), when $N > N_{js}$, the particle concentration is uniform throughout the vessel, and power number in solid–liquid system, $(\mathrm{Ne})_s$, is constant.

In the vessel equipped with a standard turbine, the maximum value of the Reynolds number is 1,800; for the modified blade turbine, the maximum Reynolds number is

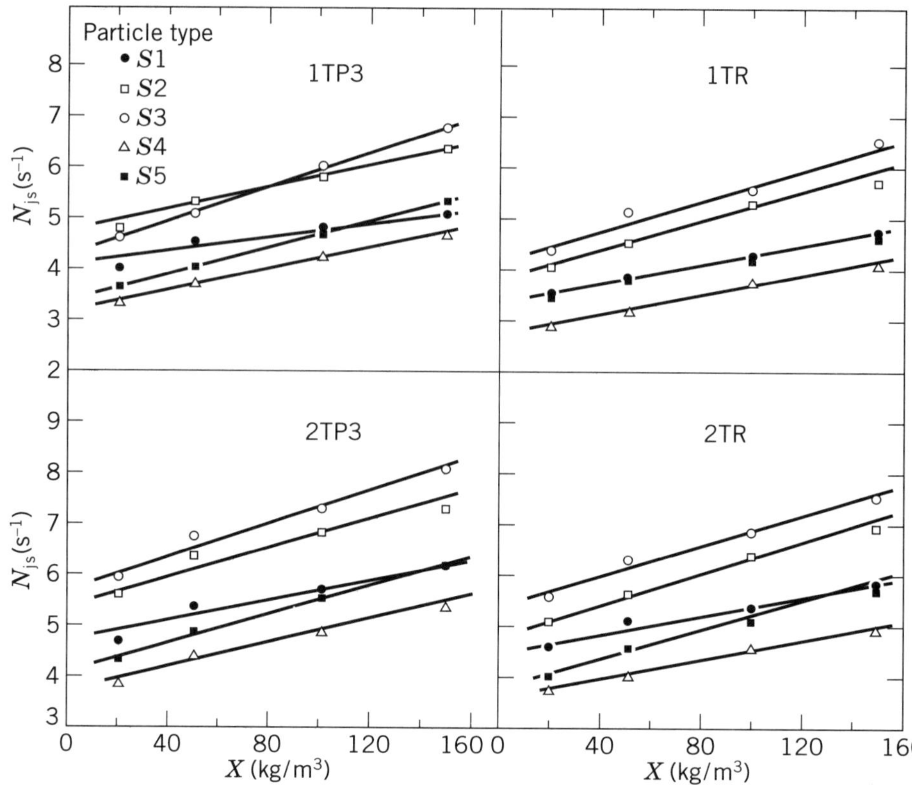

Figure 5. Effect of solid concentration on complete suspension speed for standard and modified Rushton turbines, positioned singly or doubly on the same shaft. S1–S5, particle types (see Table 2). (*Source*: From Ref. 20, with permission.)

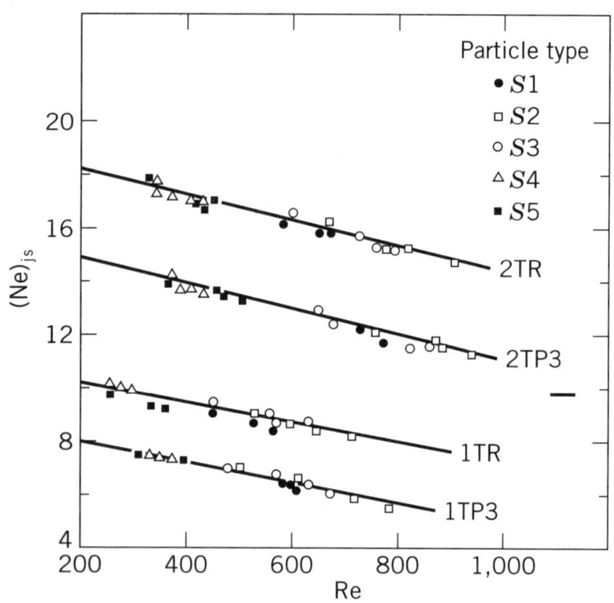

Figure 6. Relation between the power number and Reynolds number (Re) at the complete suspension state, for standard and modified Rushton turbines, single or double. (*Source*: From Ref. 20, with permission.)

2,400. This is possible because the modified blade turbine rotates much faster than the standard turbine at an equal specific power input (11). The relation between the mean values of the power number and the physical properties of the solid particle, in the turbulent range, is

$$(\mathrm{Ne})_3 = 5.6175(n_p/6)(1 + S_G/S_C)^{-1.3867} \cdot (1 + 10 d_p^{0.2} \Delta\rho^{-0.45}) \tag{3}$$

The estimated error using equations 2 and 3 lies within ±10%.

Effect of Particles on Gas–Liquid Hydrodynamics

Establishment of the major interactions between gas dispersion and particle suspension mechanisms is a basis for examining in detail the effect of the major variables on the minimum agitator speed for particle suspension under gassed conditions and on the minimum agitator speed for gas dispersion to all parts of the vessel in the three-phase systems.

The manner in which the presence of the particles affects the gas dispersion is inferred from the variation of the relative power consumption P_{gs}/P_s (P_{gs} and P_s are power consumption in gas–liquid–solid and solid–liquid systems, respectively), with gas flow rate, Fl, which is associated with different stages of cavity growth behind the impeller blades. In the curves presented in Figure 7, for various volumetric gas flow rates and concentrations of the S5 type solid particles for 1TR and 1TP3 turbines, the transition to complete gas dispersion is obtained when a minimum occurs in P_{gs}/P_s and the cavity sizes are maximum (27). For the 1TR turbine, the maximum values of P_{gs}/P_s are between 0.5 and 0.3 for all solid concentrations and depend on the volumetric gas flow rate. The 1TP3 turbine shows a behavior similar to that of 1TR turbine but only at $X = 20$ kg/m³ and $X = 50$ kg/m³. At $X = 100$ kg/m³ and $X = 150$ kg/m³, the minimum values of the P_{gs}/P_s for

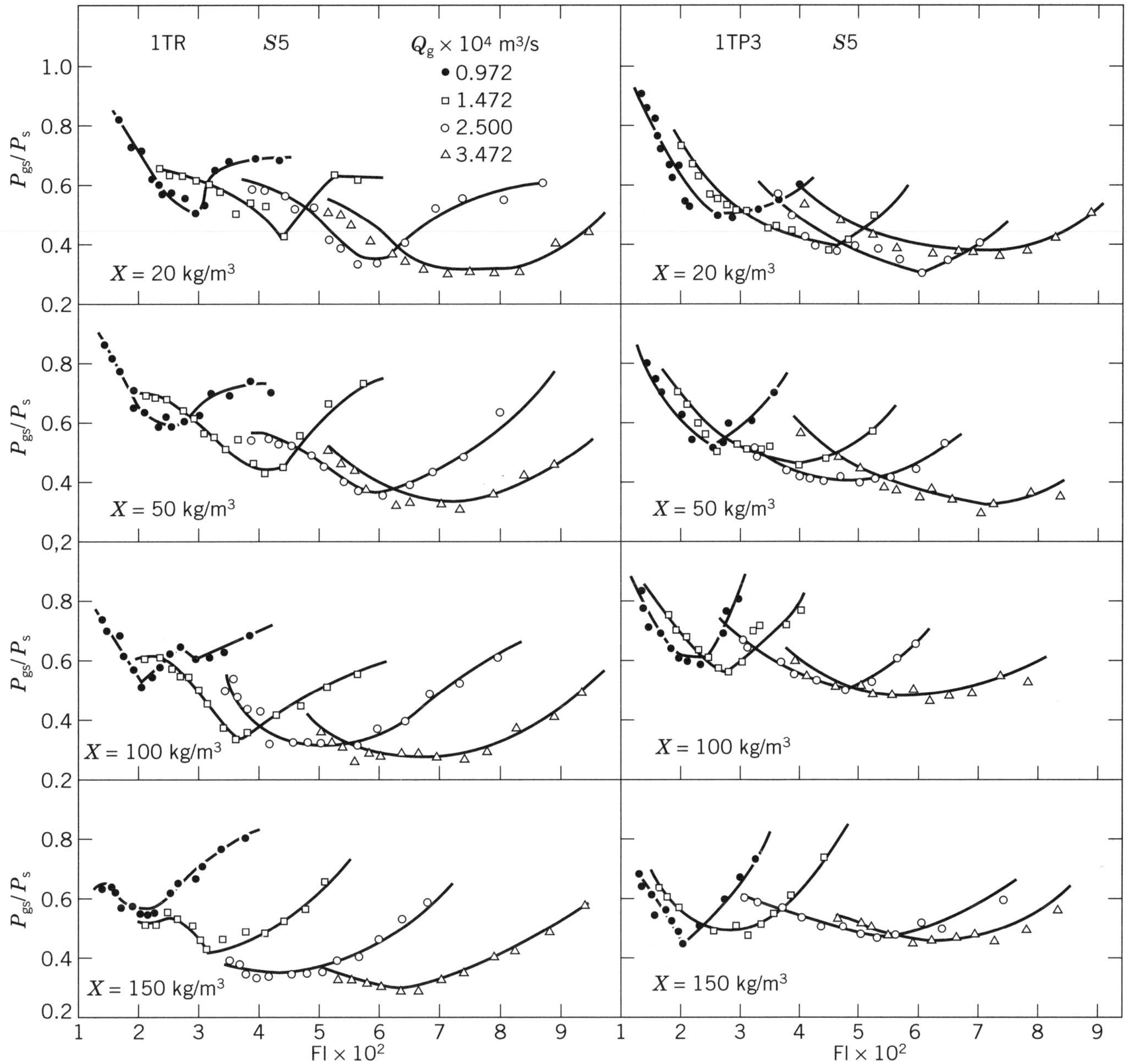

Figure 7. Relation between relative power consumption (P_{gs}/P_s) and gas flow number (Fl) for the S5 particles using single standard and modified turbines. (*Source*: From Ref. 27, with permission.)

1TP3 are approximately 0.5, irrespective of the volumetric gas flow rate. This confirms previous observations (8) as regards the fact that the modified blades have a greatly increased gas-handling capacity.

Effect of particle concentration on power number in a three-phase system, $(\mathrm{Ne})_{gs}$, at constant superficial gas velocity is presented in Figure 8 for the S2 type ion exchange resin, in modified turbines, single or double. A decrease in power number at $N < N_{jsg}$ (jsg is the moment of complete solid suspension) is due to the "false bottom" formed by the particles effectively reducing impeller clearance at low speeds. As the agitator speed N tends to N_{jsg} (as indicated in Fig. 8), the particles were suspended, the false bottom removed, and the power number tended to the common value. The power number generally increased as particle concentration increased before complete particle suspension had been obtained.

In the case of the 1TP3 turbine, the moment of the complete gas dispersion N_{CD} shifts into the range of smaller gas flow rates when the particle concentration increases. For two turbines (2TP3), the variation $(\mathrm{Ne})_{gs} = f$ (Fl) showed the following transition points: the transition to effective dispersion of the gas at the upper impeller (minimum point of the each curve) and also the flooding–loading transition of the bottom impeller (maximum point of each curve). The transition at complete gas dispersion

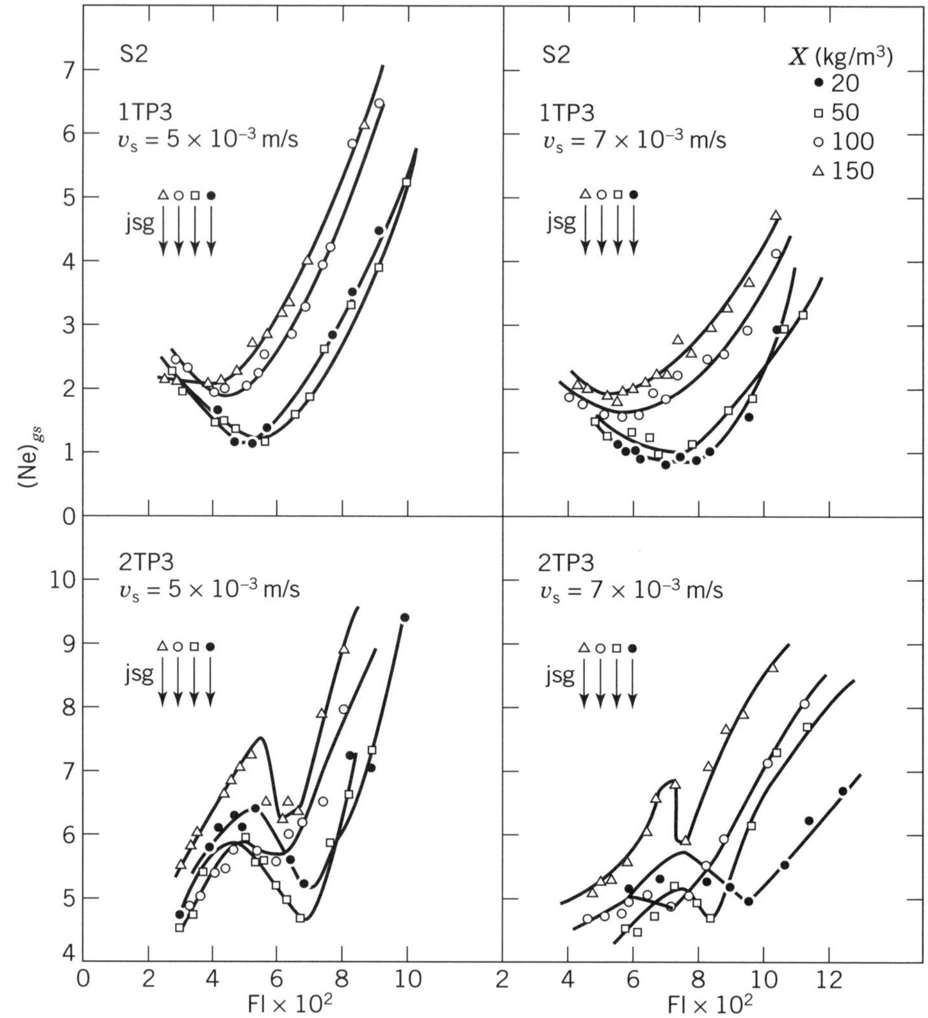

Figure 8. Variation of the power number with gas flow number for the S2 particles using modified turbines, single and double. (*Source*: From Ref. 27, with permission.)

does not show in Figure 8, probably because, once the particles are suspended, the increasing suspension density in the impeller region (and of the power consumption) compensates the decrease in power consumption because of large cavity formation.

The examination of the experimental data in correlation with the Reynolds number shows the following (27):

- For Re < 1,000, mixing is in the transition range, in which the $(Ne)_{gs}$–Re dependence has a negative slope.
- In the range 1,000–1,400 of the Reynolds number, the $(Ne)_{gs}$ values for 1TR attain a minimum point more pronounced for the lower solids concentration. In the case of the 1TP3 turbine, the variation of $(Ne)_{gs}$ with Re is negligible, as a result of the change in fluid flow around and over the perforated blade. The formation of the ventilated cavities behind the modified blades was influenced by the impact of the jets that come through the blade perforations.
- For Re > 1,400, $(Ne)_{gs}$ becomes independent of the Reynolds number, but depends on particle and agitator types. The mixing is in the turbulent range and the phases are uniformly dispersed within the entire vessel section.

The relation between the power number and the physical and rheological properties of solid particles is presented in Figure 9 for the 1TR turbine at various aeration conditions and for 1TP3, 2TP3, and 2TR turbines at constant aeration rate. For the same operation conditions, the modified blade turbines are more efficient, the power number $(Ne)_{gs}$ being reduced by 50% in comparison with the standard Rushton turbine.

Effect of Gas on Solid–Liquid Hydrodynamics

The effect of gas on solid–liquid hydrodynamics in mixing vessels was studied to determine the agitation speed required to just completely suspend all the particles under gassed conditions, N_{jsg} (28). The introduction of a small amount of gas into a solid–liquid system, for the just-suspended condition, causes slight sedimentation such that, for the same impeller speed, the particles were no longer just suspended (i.e., $N < N_{jsg}$). Further increases in the volumetric gas flow rate led to more sedimentation because the pumping capacity and ability of the impeller to circulate fluid and the power input decrease. Any decrease in pumping capacity and power input will decrease all the parameters that cause particle suspension: drag forces, energy dissipation, and associated turbulent eddies. An in-

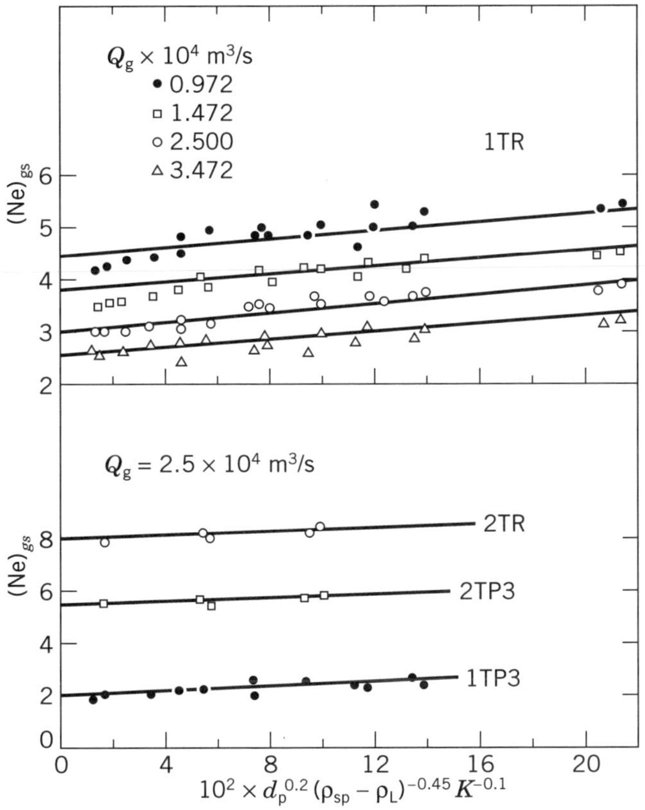

Figure 9. Relation between the power number and the physical and rheological properties of the suspension agitated by various agitator geometries. (*Source*: From Ref. 27, with permission.)

crease of the agitator speed to N_{jsg} ensured complete resuspension. Similarly, under gassed conditions, if $N < N_{jsg}$, a reduction in volumetric gas flow rate gave complete resuspension. In fact, there is a unique combination of gas rate and impeller speed at which the just-suspended condition is achieved and which depends on the agitator type used.

If N_{jsg} is plotted against superficial gas velocity, v_s, the form of the curves is slowly parabolical, for any particle concentration (Fig. 10). At superficial gas velocity of approximately 4×10^{-3} m/s, the slope of the dependence $N_{jsg} = f(v_s)$ decreases, demonstrating that the mechanism of the solid suspension changes. Indeed, this value of the superficial gas velocity is found around a gas flow number Fl = 0.03, when in gas–liquid systems, transition occurs between the clinging cavity regime and the large cavity regime (29,30). The presence of particles has no measurable effect on the formation of the gas-filled cavities.

As showed previously, for complete suspension to be maintained, a higher solid concentration requires a higher power input to the systems. The presence of the gas amplifies the effects mentioned, such that with increasing aeration rate, associated with increasing particle concentration, the agitator speed N_{jsg} must be larger (Fig. 11). The relationship for evaluation of the agitation speed required to just completely suspend all the particles under gassed conditions, N_{jsg}, takes into account the physical and rheological properties of suspensions, the aeration rate, and the equipment configuration (28).

MASS TRANSFER

The mass transfer process from the gas phase to the liquid phase is usually expressed in terms of the volumetric mass transfer coefficient, Kla. The determination methods of the volumetric oxygen transfer coefficient are divided by Rainer (31) into indirect methods, applied without biological system (gassing-out, electrode momentum, sulfite oxidation, carbon dioxide, and glucose oxidase methods), and direct methods (method of gas balance, dynamic method, continuous culture method). The Kla values in mechanically agitated reactors depend on various parameters such as system properties, vessel and stirrer dimensions and arrangement, and operating conditions (32). To examine the behavior of the modified turbines regarding mass transfer, the Kla coefficients were determined in the presence and absence of the solid particles by the sulfite oxidation method.

Gas–Liquid Mass Transfer Characteristics

Figure 12 shows the measured Kla values as a function of gas flow number, Fl, for single and double turbine agitator

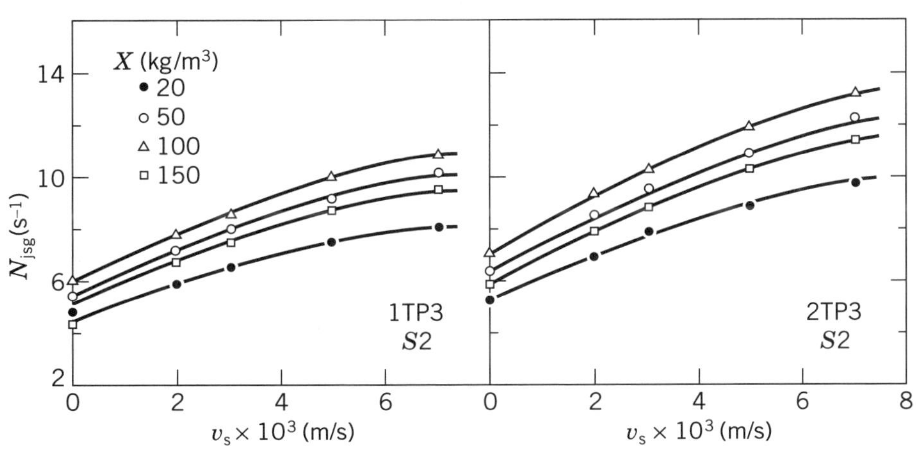

Figure 10. Relation between N_{jsg} and the superficial gas velocity (V_s) for the S2 suspension mixing with modified turbines. (*Source*: From Ref. 28, with permission.)

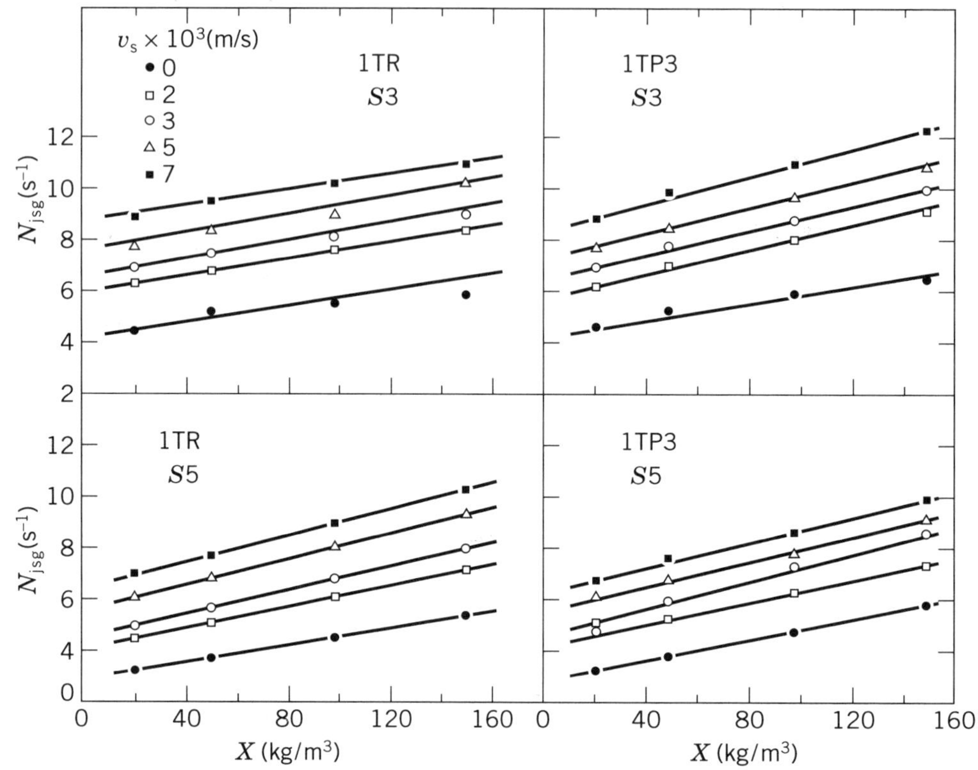

Figure 11. Relation between N_{jsg} (jsg, moment of complete solid suspension) and the solid concentration for various values of the superficial gas velocity. (*Source*: From Ref. 28, with permission.)

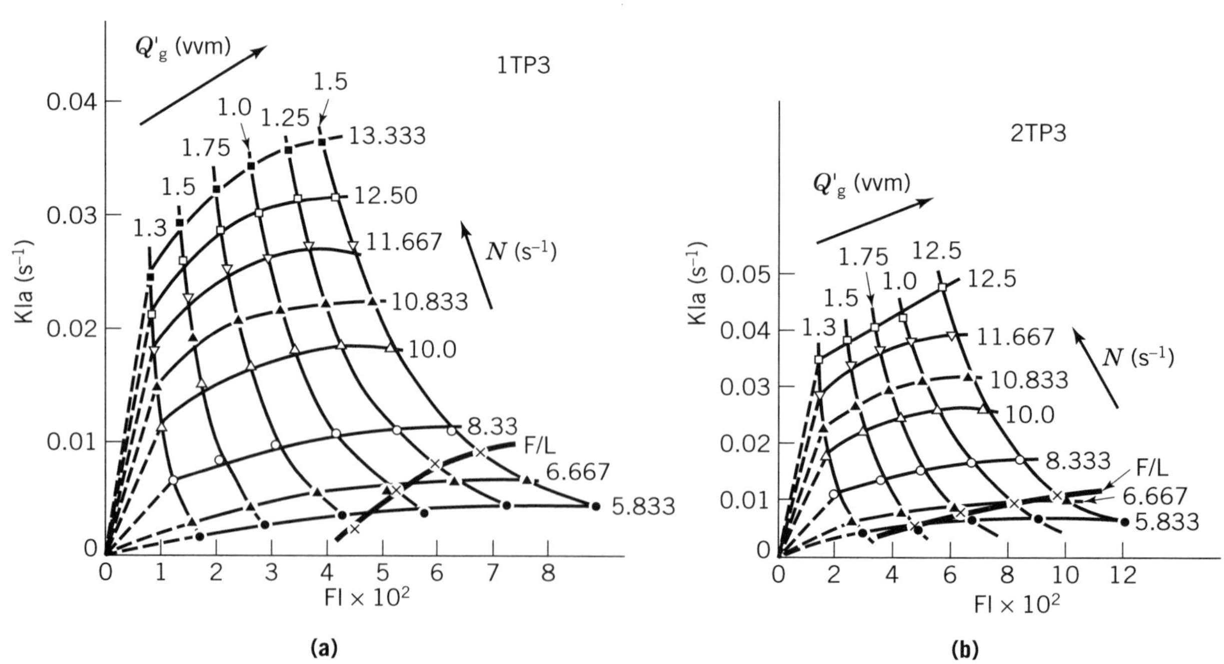

Figure 12. Variation of the volumetric mass transfer coefficient (Kla) depending on the gas flow number: (**a**) single TP3 modified Rushton turbine; (**b**) double TP3 modified Rushton turbines. (*Source*: From Ref. 33, with permission.)

of TP3 type, having as parameter the agitator speed (curves indicated in Fig. 12 by the continual line) or the aeration rate (curves indicated by the dotted line). The same variation, Kla = f(Fl), was also observed for the other modified turbines and the Rushton turbine (33). The dependence of Kla on Fl, when N is constant, is greater at lower gas flow rates (lower gas flow numbers). At higher gas flow numbers, the curves tend to constant Kla values for all agitator speed values and single turbine types (Fig. 12a). In the case of double turbines (Fig. 12b), this tendency of the curves Kla = f(Fl) appears only at values $N < 10\ s^{-1}$. This dependence of Kla on Fl can be explained by the prism of various hydrodynamic regimes, considering the value Fl = 0.03 as the transition point between the vortex-clinging cavity regime and large cavity regime (33). When the Kla values are represented depending on the gas flow number at a constant aeration rate (curves indicated by the dotted line in Fig. 12), these increase slowly once with increasing the rotational speed when the agitator works in flooding regime (below line noted F/L in Fig. 12). When the agitator works in the loading regime (over line F/L), the volumetric oxygen mass transfer coefficient increases considerably with the decreasing gas flow number.

One procedure for evaluating the volumetric oxygen mass transfer coefficient is based on knowledge of the specific power consumption (P_g/V) and the superficial gas velocity (v_s) (34):

$$\text{Kla} = c(P_g/V)^{0.95}(v_s)^{0.67} \tag{4}$$

The relationship between Kla and the product of the variables $(P_g/V)^{0.95}$ $(v_s)^{0.67}$ for all turbine types and their geometrical disposition is linear, the average value of the dimensional constant c being specific for each of the geometrical configurations studied.

The influence of the mixing system geometry on gas–liquid mass transfer is quantified by including, in equation 4, the geometrical elements (n_p/6) and (1 + S_G/S_C), which indicate the number of turbines and their type. The following correlations were obtained for the evaluations of the volumetric mass transfer coefficient:

$$\text{Kla} = 7.2202 \times 10^{-4}(1 + S_G/S_C)^{1.753}(n_p/6)^{-0.126} \cdot (P_g/V)^{0.95})(v_s)^{0.67} \text{ for } S_G/S_C \leq 0.353;\ r = 0.9843 \tag{5}$$

$$\text{Kla} = 2.721 \times 10^{-3}(1 + S_G/S_C)^{-2.73}(n_p/6)^{-0.141} \cdot (P_g/V)^{0.95}(v_s)^{0.67} \text{ for } S_G/S_C > 0.353;\ r = 0.9629 \tag{6}$$

To obtain the same value of the volumetric mass transfer coefficient, the power consumption in the vessel equipped with the TR agitator is higher than that in the vessel equipped with the TP3 agitator. Therefore, the oxygen transfer efficiency E (m^3O_2/W h) (the oxygen volume transferred into the liquid phase for 1 W h of energy consumption for mixing) for one and two TP3 modified Rushton turbines, for any value of the aeration rate (Fig. 13), is more than 30% higher than those obtained with the TR turbines. The TR and TP5 turbines are similar as regards mass transfer efficiency.

Effect of Particles on Gas–Liquid Mass Transfer

It has been observed that the addition of solids has an influence on gas–liquid mass transfer characteristics, even if these particles are not reactive and do not show increased adsorption of dissolved gases. The inert solid particles used in our work are $CaCO_3$ (S1) and cation exchange resin Amberlite IR-120 type (S3) (35). The experiments were carried out in the range of three-phase homogeneous mixing. The presence of the solid phase, with the concentration between 20 and 50 kg/m^3, influences the gas–liquid mass transfer as shown in Figure 14 for the standard (TR) and modified (TP3) turbines, single or double, at various values of superficial gas velocity. All curves show the same general trend.

For $CaCO_3$ particles (S1 solid), a very slight increase of the Kla coefficient is observed until $X = 50$ kg/m^3, while the addition of the cation exchange resin in a proportion of 20 kg/m^3 leads to a pronounced increase of the Kla value, particularly at $N = 11.667\ s^{-1}$ and at higher values of the superficial gas velocity. This trend of the Kla coefficient to increase in the range of small solid concentration was explained by Brehm (36) that at the moderate agitator speed ($N < 15\ s^{-1}$) the solid particles can move independently of fluid elements and break up the gas–liquid interface, which leads to decreasing thickness of the boundary layer. When increasing the solid concentration beyond the values mentioned, the Kla coefficient gradually decreases toward values smaller than those obtained in gas–liquid systems.

Two main factors are responsible for this reduction of the Kla values (37): one factor is the change of the rheological properties of the suspensions and the other is enhanced bubble coalescence, which is caused by turbulence intensity damping. The higher Kla values for mixing with two turbines (Fig. 14) are due to the their higher power consumption for the same operation conditions.

In the three-phase system, the oxygen transfer efficiency, E, decreases with increase of solid concentration, irrespective of the agitator type (Fig. 15), but for mixing with the TP3 modified turbines, the E values are more than 45% higher than those obtained using TR turbines. The experimental data regarding the mass transfer in gas–liquid suspensions demonstrate that the Kla values depend on the physical and rheological properties. The effect of the solid particles on Kla value is given by the group $K^{-0.15}X^{-0.1}$, which includes the consistency index K as a pseudo-viscosity of the suspensions. Therefore:

$$\text{Kla} = 7.2202 \times 10^{-4}(1 + S_G/S_C)^{1.753}(n_p/6)^{-0.126} \cdot (P_g/V)^{0.95}(v_s)^{0.67}K^{-0.15}X^{-0.1} \tag{7}$$

Equation 7 describes the experimental data with an average error of 20%.

PERFORMANCE OF THE MODIFIED AGITATORS IN BIOSYNTHESIS OF ANTIBIOTICS

The transport phenomena in submerged aerobic fermentation involve mixing and mass and heat transfer in three-phase systems consisting of gas bubbles and suspended microorganisms in a liquid phase and are strongly influenced by the rheological properties of the medium used. The rhe-

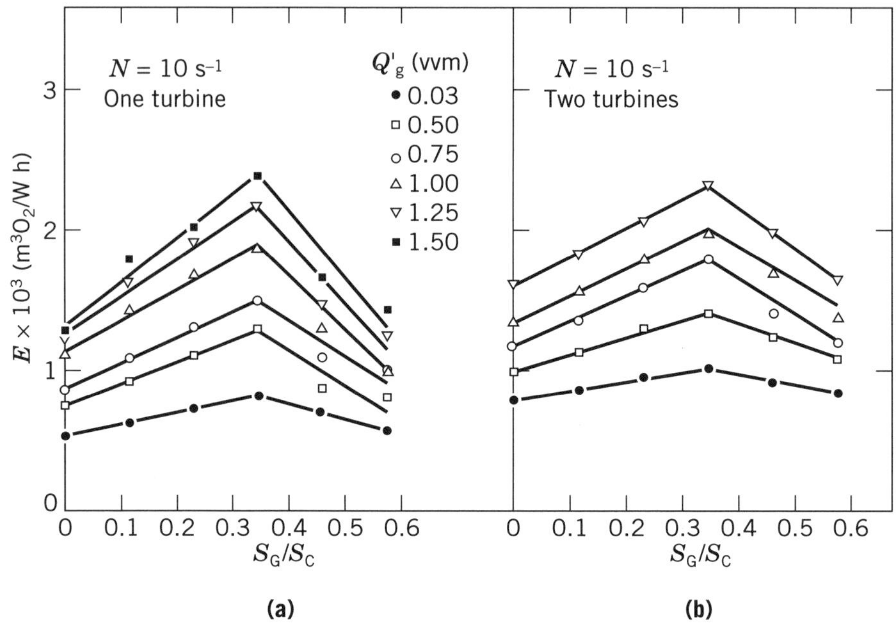

Figure 13. Variation of the oxygen transfer efficiency (E) depending on the surface fraction of the perforation at various values of the aeration rate: (**a**) single turbine; (**b**) double turbines. (*Source*: From Ref. 33, with permission.)

ology of the broth is affected by the composition of the original medium and its modification by the growing culture, the concentration and morphology of the biomass, and the concentration of microbial products. Complex interaction between transport phenomena and reaction kinetics characterizes bioreactors and determines their performance. The modified turbine TP3 with the surface fraction S_G/S_C of perforations equal to 0.353, considered an optimum in dispersion of the gas and solids in liquid, was investigated in antibiotic biosynthesis liquids in semiindustrial and industrial bioreactors.

Biosynthesis of Antibiotics in Semiindustrial-Scale Bioreactors

The experiments were carried out in two 20-m^3 stainless steel stirred tank bioreactors, equipped with three modified Rushton turbines (TP3) (blade width × blade length = 0.306 × 0.225 m), and three witness Rushton turbines (TR) (blade width × blade length = 0.225 × 0.225 m), respectively, having the same diameter and rotating with the same agitator speed (38). The experimental data in the biological systems were obtained using four different microorganisms: *Streptomyces erithreus*, *S. griseus*, *S. noursei*, and *Nocardia mediterranei*, producing erythromycin, streptomycin, nystatin, and rifamycin, respectively. During the four parallel fermentations, the specific power consumption, the volumetric oxygen mass transfer coefficient, the oxygen transfer efficiency, and the antibiotic concentrations were investigated. For each fermentation type, the power consumption for the TP3 agitator is diminished by a factor of 1.18 (streptomycin) to 1.6 (erythromycin) compared to that for the TR agitator (Fig. 16). Therefore, in the viscous mycelial fermentation, the TP3 agitator minimizes the drag forces associated with the motion of the blades, confirming the observations presented for a nonbiological system.

The Kla values, as measured by the dynamic method, decrease during the fermentation processes, depending on the microorganism growing in the liquid (Fig. 17). The fermentation media for *S. erithreus* and *S. noursei*, producing erythromycin and nystatin, respectively, contain starch as a degradable component (22,39), which confers a high viscosity on the initial medium; thus, Kla is smaller at the beginning of biosynthesis, compared to the initial medium for *S. noursei* and *N. mediterranei* without this component. The Kla values obtained under mixing conditions with the TP3 agitator are 25% (streptomycin) to 50% (nystatin) higher than those obtained by using the TR agitator. This effect can be explained by the improvement of the dispersion of the gas bubbles as a result of the impact of the jets that come through the perforations in the TP3 blades onto the relatively surface in the back of large cavities.

Because the TP3 blades are higher by 36% in comparison with the witness blades, the size of the well-mixed, low-viscosity, and high-Kla region is effectively increased. In the micromixing region, excellent bubble breakup and shear thinning of the viscous medium occurs, which results in oxygen transfer improving in accordance with the concepts of Bajpai and Reuss (40). Also, the macromixing region, away from the impeller where the oxygen transfer is relatively poor, is much smaller.

In the investigated biological systems, for 1 W h of energy consumption (for agitation and air compression), the oxygen volume transferred into the liquid phase is approximately 30% higher in the bioreactor equipped with a TP3 agitator (Fig. 18). The oxygen transferred into the medium is consumed mainly in the combustion reaction that takes place within the biomass, on which the antibiotic production depends. The relative antibiotic concentrations measured over the entire course of the fermentations are plotted in Figure 19.

The efficiency of the TP3 agitator was confirmed in that antibiotic production increased up to 35% in nystatin biosynthesis, as much as 9% in rifamycin biosynthesis, and up to 20% in streptomycin biosynthesis as well. No difference in erythromycin production was observed between

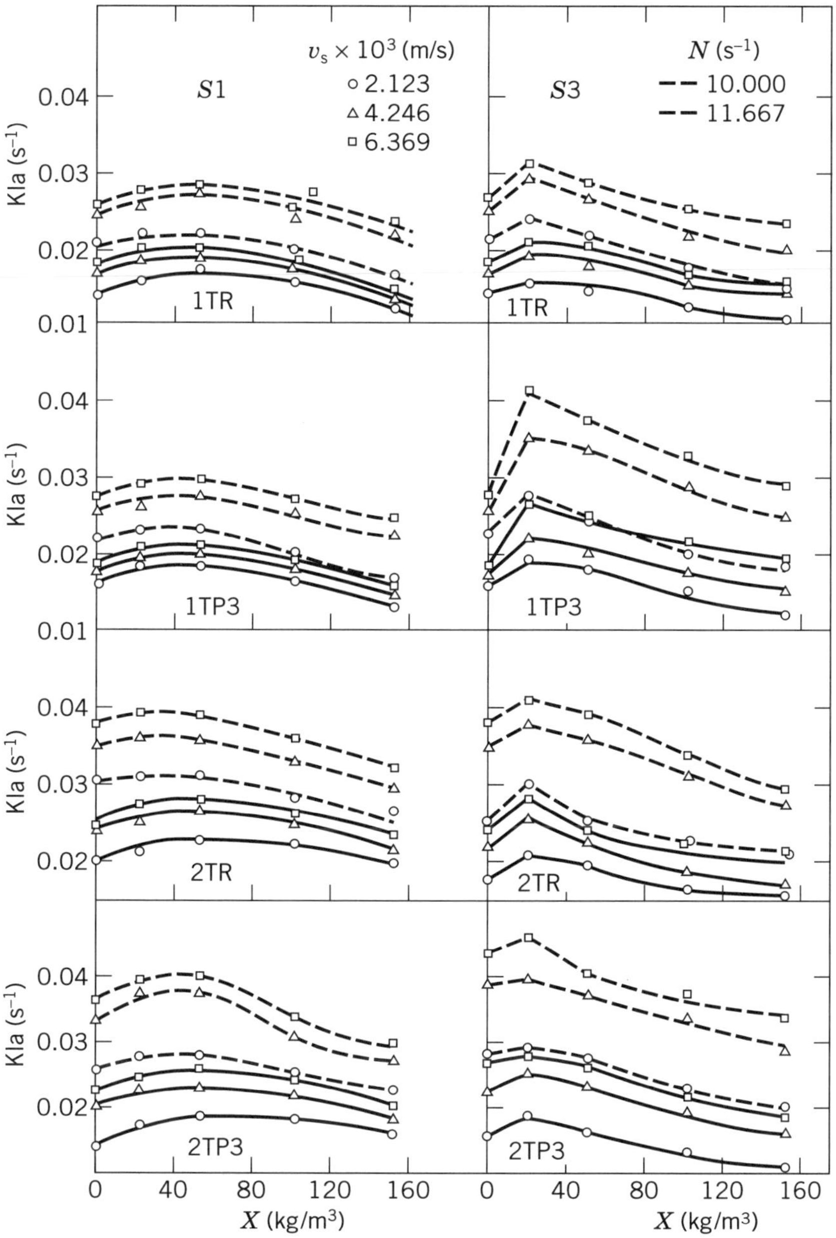

Figure 14. Effect of solid concentration on volumetric mass transfer coefficient for standard and modified Rushton turbine, positioned singly or doubly on the same shaft. (*Source*: From Ref. 35, with permission.)

the two parallel fermentations, but efficiency occurs from the energy savings. Therefore, using the TP3 agitator in the biosynthesis of antibiotics, it is possible to minimize the power consumption and to increase the antibiotic production for the same operating conditions as well as the TR agitator.

Biosynthesis of Antibiotics in Industrial-Scale Bioreactors

In the biosynthesis of antibiotics in semiindustrial-scale bioreactors, it was demonstrated that when the modified agitator has the same diameter and rotates with the same agitator speed as the standard agitator, the power consumption is reduced by a factor of 1.18 to 1.6 (38). Therefore, the use of a modified agitator in biosynthesis of the antibiotics may improve the productivity for a given power input by increasing the TP3 impeller diameter and keeping the agitator speed as high as for the TR agitator, and by increasing the TP3 agitator speed and keeping the impeller diameter as large as for the TR agitator.

The improved agitation performance by modified Rushton turbine agitators corresponding to the two aforementioned cases (if the specific power consumption is constant) was verified experimentally in industrial-scale bioreactors, as follows (41):

- In parallel fermentations of *S. aureofaciens* and *S. rimosus*, producing tetracycline and oxytetracycline, respectively, in two 100-m^3 stainless steel stirred tank bioreactors equipped with four modified impellers (TP3) and four standard impellers (TR), respectively (first case)
- In parallel fermentations of *Penicillium chrysogenum* producing penicillin, in two 63-m^3 stainless steel

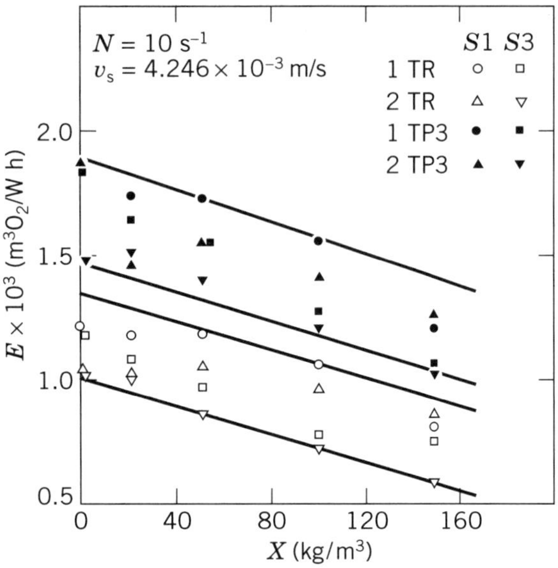

Figure 15. Relation between oxygen transfer efficiency and solid concentration for various agitators. (*Source*: From Ref. 35, with permission.)

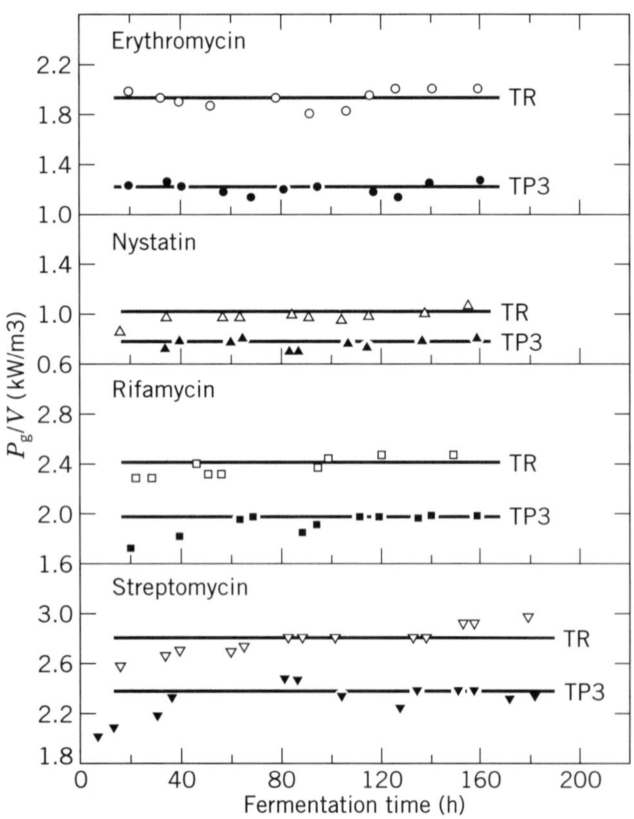

Figure 16. Comparison of TP3 and TR agitators with respect to specific power consumption in the course of processes in the biosynthesis of antibiotics. (*Source*: From Ref. 38, with permission.)

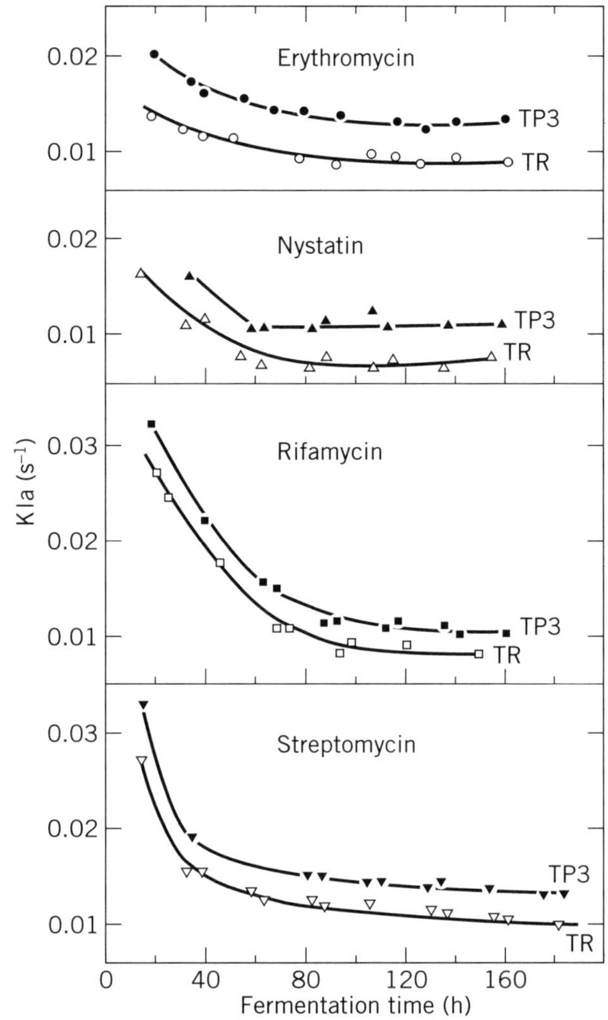

Figure 17. Comparison of TP3 and TR agitators with respect to the volumetric oxygen mass transfer coefficient in the course of processes in the biosynthesis of antibiotics. (*Source*: From Ref. 38, with permission.)

stirred tank bioreactors equipped with three modified impellers (TP3) and three standard impellers (TR), respectively (second case)

The morphology of molds is a function of shear stresses in the bioreactor. The growth rate and partial pressure of oxygen, and the following elements, play a role: the growth of hyphae, the branching of hyphae, the dispersion (breakup) of hyphae caused by the shear stress in the vessel, and the tensile strength of hyphae. The breakup frequency (λ/k_o) is proportional to the frequency that a particle will have in the so-called dispersion zone (V_{disp}), that is, the volume at which the turbulent shear forces are greater than the forces necessary to break the hyphal cell wall (42). The λ/k_o and V_{disp} values for the agitators investigated are presented in Table 3.

Therefore, by using the TP3 agitators, the size of the well-mixed region is effectively increased up to 27% and the breakup frequency of the mycelial hyphae is increased up to 75%. The relative antibiotic production and variation

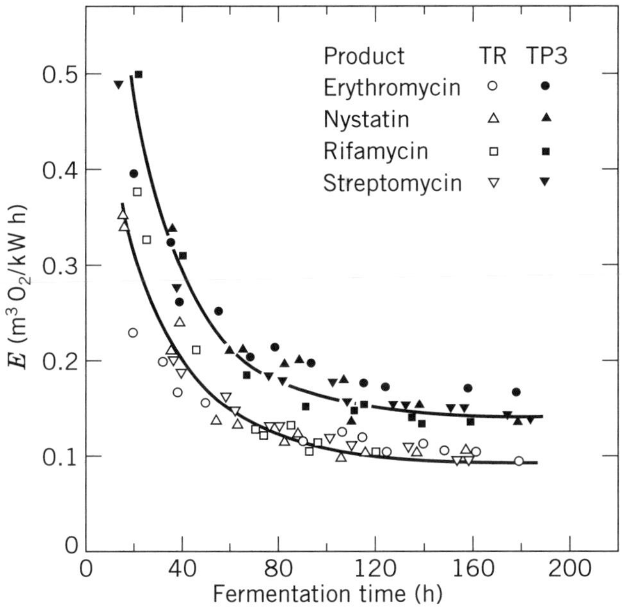

Figure 18. Comparison of TP3 and TR agitators with respect to oxygen transfer efficiency in the course of processes in the biosynthesis of antibiotics. (*Source*: From Ref. 38, with permission.)

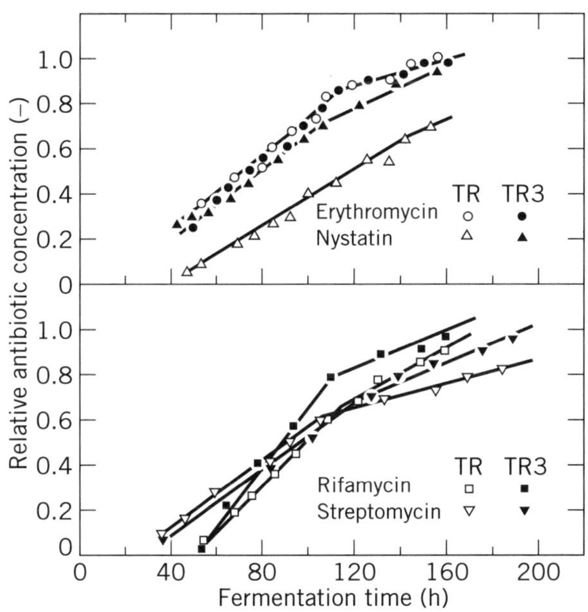

Figure 19. Relative antibiotic concentration profiles during the biosynthesis process using the TP3 and TR agitators in semiindustrial-scale bioreactors. (*Source*: From Ref. 38, with permission.)

of the specific antibiotic production rate, Q_p, during the cultivation time, are plotted in Figures 20 and 21, respectively. The improvement in antibiotic production by more than 30% by using modified agitators can be explained by creating a better hydrodynamic microclimate, and effectively increasing the well-mixed region, oxygen mass transfer, discharge efficiency, and the breakup frequency of mycelial hyphae.

Table 3. Breakup Frequency of the Mycelial Hyphae and the Dispersion Zone of TP3 and TR Agitators

Product	Agitator	V_{disp} (m^3)	$\lambda/k_0 10^2$ (s^{-1})
Tetracycline and oxytetracycline	TR	1.846	1.688
	TP3	3.522	4.808
Penicillin	TR	1.948	2.255
	TF3	2.474	3.937

Source: Ref. 41.

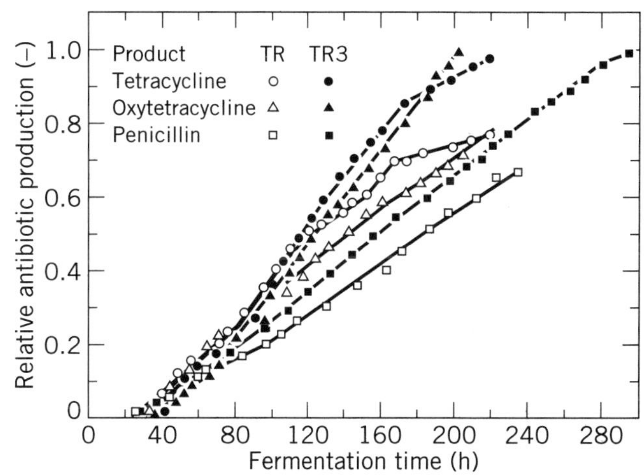

Figure 20. Relative antibiotic concentration profiles during the biosynthesis process using the TP3 and TR agitators in industrial-scale bioreactors. (*Source*: From Ref. 41, with permission.)

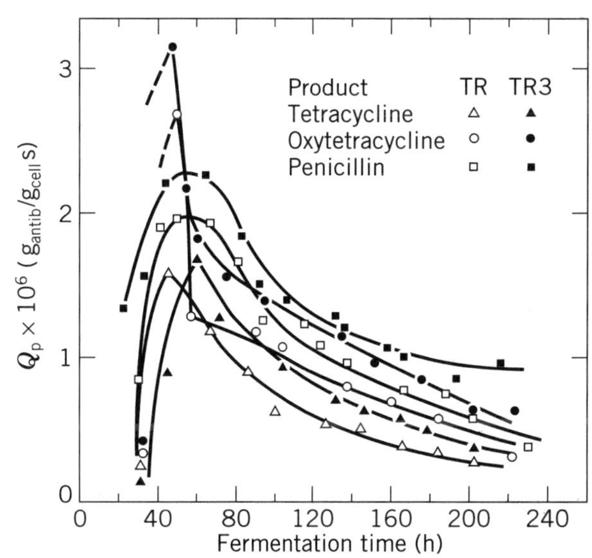

Figure 21. Specific antibiotic production rate profiles during the biosynthesis process using TP3 and TR agitators. (*Source*: From Ref. 41, with permission.)

The Q_p values reach a maximum in the early exponential phase of batch growth, although the cell mass concentration is small. The maximum values of the specific antibiotic production rate and the fermentation time at which these maximum points are reached depend on the type of producing microorganism and the type of agitator. Therefore, the standard and modified agitators create various types of environmental conditions under which the cells appear to adapt by structural and functional changes.

BIBLIOGRAPHY

1. M.M.C.G. Warmoeskerken, Ph.D. Thesis, University of Delft, The Netherlands, 1985.
2. J.M. Smith, in J.J. Ulbrecht and G.K. Patterson eds., *Mixing of Liquid by Mechanical Agitation,* Gordon and Breach, New York, 1985, pp. 139–202.
3. A.W. Nienow and J.J. Ulbrecht, in J.J. Ulbrecht and G.K. Patterson eds., *Mixing of Liquid by Mechanical Agitation,* Gordon and Breach, New York, 1985, pp. 203–237.
4. F. Kargi and M. Moo-Young, in M. Moo-Young ed., *Comprehensive Biotechnology*, vol. 2, Pergamon Press, Oxford, 1985, pp. 5–56.
5. R.V. Roman and R.Z. Tudose, *Rev. Chim.* **46**, 547–564 (1995).
6. R.V. Roman and R.Z. Tudose, *Roum. Chem. Q. Rev.* **4**, 269–293 (1996).
7. R.V. Roman, Ph.D. Thesis, Technical University, Iasi, Roumania, 1997.
8. R.V. Roman and R.Z. Tudose, *Chem. Eng. J.* **61**, 83–93 (1996).
9. R.V. Roman and M. Gavrilescu, *Hung. J. Ind. Chem.* **22**, 87–93 (1994).
10. V. Abrardi, G. Rovero, G. Baldi, S. Sicardi, and R. Conti, *Trans. Inst. Chem. Eng. A* **68**, 516–521 (1990).
11. R.V. Roman and R.Z. Tudose, *Rev. Roum. Chim.* **41**, 309–319 (1996).
12. W. Bujalski, A.W. Nienow, S. Chatwin, and M. Cooke, *Chem. Eng. Sci.* **42**, 317–326 (1987).
13. E.A. Bratu, *Operatii Unitare în Industria Chimică*, vol. 2, Editura Technică, Bucharest, 1985, p. 20.
14. R.V. Roman, M. Gavrilescu, and V. Efimov, *Hung. J. Ind. Chem.* **20**, 155–160 (1992).
15. R.K. Geisler, R.K. Buurman, and A.B. Mersmann, *Chem. Eng. J.* **51**, 29–39 (1993).
16. K. Takahashi, H. Fujita, and T. Yokota, *Chem. Eng. Jpn.* **26**, 98–100 (1993).
17. A. Barresi and G. Baldi, *Chem. Eng. Sci.* **42**, 2949–2956 (1987).
18. K.J. Myers, *Can. J. Chem. Eng.* **72**, 745–748 (1994).
19. T.N. Zwietering, *Chem. Eng. Sci.* **8**, 244–253 (1958).
20. R.V. Roman and R.Z. Tudose, *Bioprocess Eng.* **15**, 221–229 (1996).
21. T. Koloini, I. Plazi, and M. Zumer, *Chem. Eng. Res. Des.* **67**, 526–536 (1989).
22. M. Gavrilescu, R.V. Roman, and V. Efimov, *Acta Biotechnol.* **12**, 383–396 (1992).
23. P.M. Armenante and T. Li, *AIChE Symp. Ser.* **293**, 105–111 (1993).
24. C. Buurman, *I. Chem. E. Symp. Ser.* **121**, 343–350 (1990).
25. M. Greaves and V.Y. Loh, *I. Chem. E. Symp. Ser.* **89**, 69–96 (1984).
26. K. Vant'Riet, Ph.D. Thesis, University of Delft, The Netherlands, 1975.
27. R.V. Roman and R.Z. Tudose, *Bioprocess Eng.* **16**, 135–144 (1997).
28. R.V. Roman and R.Z. Tudose, *Bioprocess Eng.* **17**, 55–60 (1997).
29. R.V. Roman and R.Z. Tudose, *Hung. J. Ind. Chem.* **24**, 161–169 (1996).
30. R.V. Roman and R.Z. Tudose, *Bioprocess Eng.* **17**, 307–316 (1997).
31. B.W. Rainer, *Chem. Biochem. Eng.* **Q4**, 185–196 (1990).
32. H. Oguz, A. Brehm, and W.D. Deckwer, *Chem. Eng. Sci.* **42**, 1815–1822 (1987).
33. R.V. Roman and R.Z. Tudose, *Hung. J. Ind. Chem.* **24**, 185–192 (1996).
34. N.M.G. Oosterhuis, Ph.D. Thesis, University of Delft, The Netherlands, 1984.
35. R.V. Roman and R.Z. Tudose, *Bioprocess Eng.* **17**, 361–365 (1997).
36. A. Brehm, S. Ledakowicz, and R. Kokuun, *Proc. 8th Int. Cong. CHISA*, Praha, 1984.
37. W.M. Lu, R.C. Hsu, and H.S. Chou, *J. Chim. I. Ch. E.* **24**, 31–39 (1993).
38. R.V. Roman and M. Gavrilescu, *Acta Biotechnol.* **14**, 181–192 (1994).
39. M. Gavrilescu, R.V. Roman, and V. Efimov, *Acta Biotechnol.* **13**, 59–70 (1993).
40. R.K. Bajpai and M. Reuss, *Can. J. Chem. Eng.* **60**, 384–392 (1982).
41. R.V. Roman, R.Z. Tudose, M. Gavrilescu, M. Cojocaru, and S. Luca, *Acta Biotechnol.* **16**, 43–56 (1996).
42. J.C. Van Suijdam and B. Metz, *Biotechnol. Bioeng.* **23**, 111–148 (1981).

See also MASS TRANSFER.

TRANSGLUTAMINASE

KATSUYA SEGURO
Food Research and Development Lab
Kawaski, Japan

KEY WORDS

Acyl acceptor substrate
Acyl donor
Acyl transfer reaction
Calcium
Cross-linked proteins
Enzymatic activity
Food proteins
MTG (microbial enzyme)

OUTLINE

INTRODUCTION

Transglutaminase (glutaminyl-peptide γ-glutamyltransferase/R-glutaminyl-peptide: amine γ-glutamyl transferase, EC 2.3.2.13) is an enzyme that catalyzes the acyl-transfer reaction between the γ-carboxyamide group of peptide- and/or protein-bound glutamine as the acyl donor and a variety of primary amines as the acyl acceptors (Fig. 1a) (1). If the side chain of protein-bound lysine acts as the acceptor, transglutaminase cross-links protein molecule(s) inter- and/or intramolecularly through the formation of ϵ-(γ-glutamyl)lysine isopeptide bonds between the γ-carboxyamide groups of glutamine and ϵ-amino groups of lysine in the protein molecule(s) (Fig. 1b). If no amines are present in the reaction system, the water molecule acts the acceptor, and the reaction results in deamidation of the γ-carboxyamide groups of glutamine in the proteins (Fig. 1c). Transglutaminase therefore is responsible for the post-translational modification of proteins as a biocatalyst.

In 1950, Borsook et al. observed the incorporation of radioactive lysine into protein components in the crude extract of guinea pig liver (2). Since then, research on the enzyme has continuously progressed. At present, the activity of most transglutaminases known so far is expressed in a calcium-concentration-dependent manner (3–8). Under physiological conditions in which the calcium catch-release mechanism is well regulated, most transglutaminases are controlled by such a mechanism. Transglutaminase is thus said to be calcium dependent. A typical example of transglutaminase-catalyzed protein cross-linking is readily observed in the final stabilization step of the blood coagulation or wound-healing process in warm-blooded species. In this case, plasma, platelet, and/or placental factor XIIIa (the activated form of blood transglutaminase) plays an important role in fibrin cross-linking, when the thrombin-mediated proteolytic cascade reaction along with elevation of endogenous calcium ion levels is induced by injury. The resulting cross-links are covalent, stable, and resistant to proteolysis, thereby increasing the resistance of fibrin clots to chemical, enzymatic, and physical degradation. The transglutaminase reaction with polyamines results in protein modifications, possibly affecting the biological activity or turnover of the target protein, but not polymer formation. Transglutaminase has a broad specificity for the acyl acceptor substrates (a variety of primary amines, peptide-bound lysine, or polyamines), but a relatively narrow specificity for the acyl donors (peptide-bound glutamine residues) is confirmed (amino acid sequence around the susceptible glutamine residue influences the specificity). Denaturation or proteolysis of a nonreactive protein can convert it to a transglutaminase substrate.

Transglutaminase thus can be used for intentional alteration of food protein functionalities (9). The starting point for the present industrial production of transglutaminase is the discovery of Ando et al. (10); calcium-independent transglutaminase is obtained by fermentation of a nonpathogenic and nontoxicogenic variant of *Streptoverticillium mobaraense*.

DISTRIBUTION OF TRANSGLUTAMINASE IN NATURE

Distribution of transglutaminases is ubiquitous in nature. Most mammalian organs, epidermis, plasma, and extracellular fluids contain transglutaminases in multiformity (6–8). Mammalian transglutaminase exists in five types: keratinocyte transglutaminase (TG_K, or type I), tissue transglutaminase (TG_C, or type II), epidermal transglutaminase (TG_E, or type III), prostate transglutaminase (TG_P, or type IV), and plasma and platelet transglutaminases (factor XIIIa) (Table 1). Each enzyme may have distinct physiological functions in the body; that is, the well-known blood transglutaminase, referred to as factor XIIIa, stabilizes fibrin entanglement by cross-linking; the guinea pig liver transglutaminase is classified as TG_C. Although this guinea pig liver transglutaminase was the first enzyme to be found and has been well characterized, its physiological role was the last to be solved.

Transglutaminases are found in not only in mammals, but also in fish, shellfish, crustaceans, amphibians, plants, and microorganisms. For instance, white croaker, carp, sardine, Alaska pollock, chum salmon, rainbow trout (11), horse mackerel (12), hoki (13), *Crassostrea gigas* (oyster) (14), *Tachypleus tridentatus* (horseshoe crab) (15), tadpoles of *Rana catesbeiana* (bullfrog) (16), *Beta vulgaris* L. (Silver beet), broccoli, spinach, (17), *Medicago sativa* L. (alfalfa) (18), *Helianthus tuberosus* (19), *Pisum sativum* seed (20), *Physarium* (21), *Bacillus* (22), and *Streptoverticillium* (10,23) have been reported to contain transglutaminase ac-

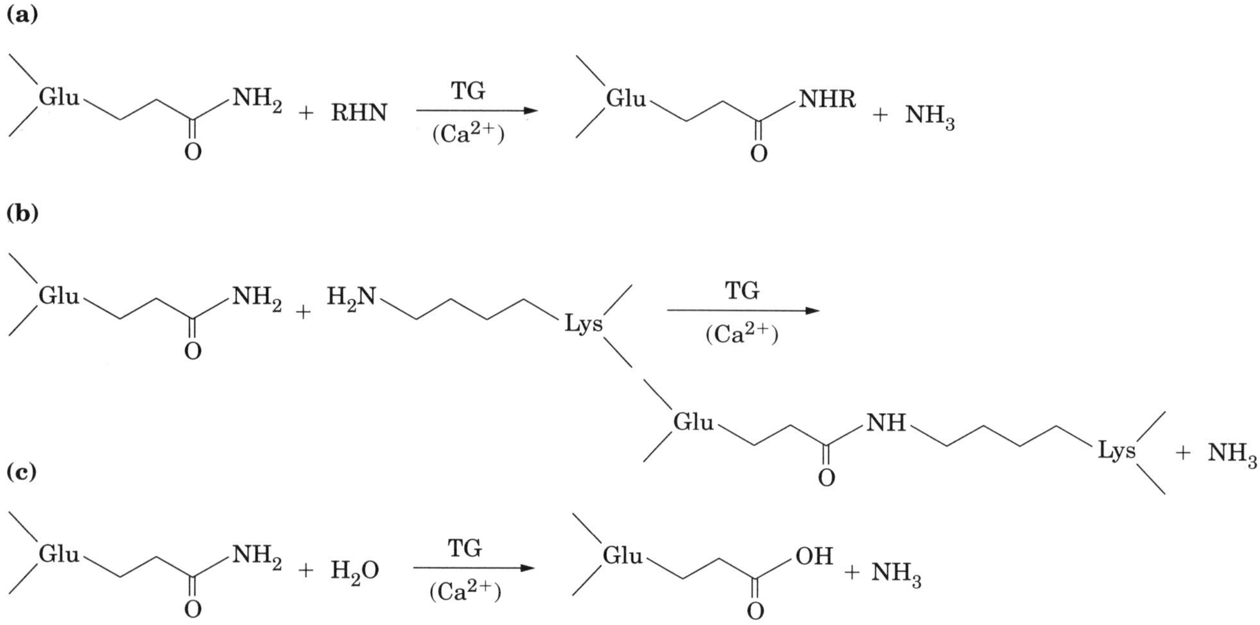

Figure 1. Reaction scheme of transglutaminase activity. (**a**) incorporation of a primary amine into γ-carboxyamide group of peptide-bound glutaminyl residue; (**b**) cross-linking through formation of ϵ-(γ-glutamyl)lysine bond between the γ-carboxyamide group and ϵ-amino group of peptide-bound lysyl residue; and (**c**) deamidation of the γ-carboxyamide group.

Table 1. Physicochemical Properties of Mammalian Transglutaminases

Type	Molecular weight	Subunit structure		Protease activation[a]	GTP effect[b]
Type I, TGK	≈90,000	Monomer	+ fatty acylation	No	None
Type II, TGC	≈8,000	Monomer	+ none	No	Inhibitory
Type III, TGE	≈80,000	Monomer	+ none	Yes[c]	?
Type IV, TGP	≈150,000 (≈71,000 × 2)	Homodimer	+ ?	?	None
Factor XIII					
Plasma	≈350,000	a2b2	+ none	Yes (thrombin)	None
Platelet	≈160,000	a2	+ none	Yes (thrombin)	None

[a]Protease responsible for in vivo activation of transglutaminase is indicated in parentheses.
[b]Effect of guanosine triphophate on transglutaminase activity.
[c]Proteases physiologically involved have not been specified.

tivity, for example, in organs, muscles, or tissue matrices. Although the classification of many enzymes is left undetermined, it is obvious that the enzymatic activity of all the transglutaminases, except microbial enzymes, is expressed only in the presence of calcium ions. The activity of transglutaminases from some microorganisms, such as *Bacillus* and *Streptoverticillium*, is expressed in both the absence and presence of calcium ions. This calcium independence is advantageous for food protein modification because many food proteins are affected by calcium ions. Further, recent investigations revealed that the activity of transglutaminases from oyster and scallop was quite interestingly regulated by both calcium and sodium ions.

ASSAYS FOR TRANSGLUTAMINASE ACTIVITY

There are both quantitative and qualitative assay methods for transglutaminase activity. Quantitative determination can be accomplished by (*1*) measuring the incorporation of any of several radioisotopically labeled primary amines (e.g., [^{14}C]- or [^{15}N]-labeled putrescence, cadaverine, methylamine, and glycine ethyl ester) into caseins, dimethylated caseins, the acetylated B-chain of oxidized insulin, or the glutaminyl peptide derivative, carbobenzoxy(*Z*)-glutaminylglycine; (*2*) measuring the incorporation of fluorescently-labeled primary amines (in most cases, dansylcadaverine) into caseins, dimethylated caseins, the acetylated B-chain of oxidized insulin, or *Z*-glutaminylglycine; (*3*) colorimetric measurement of an enzymatically produced colored γ-glutamylhydroxamate, between hydroxylamine and *Z*-glutaminylglycine, with acidic conditions (Fig. 2); or (*4*) measuring the ammonium released in the acyl transfer reaction (24).

For qualitative determination of the transglutaminases, the detection of either cross-linked proteins by

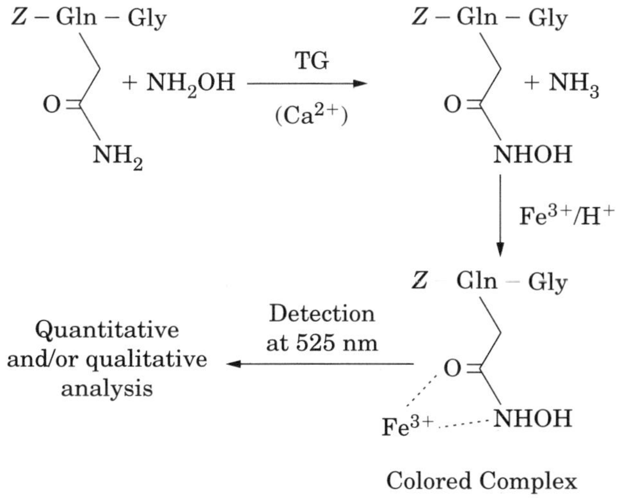

Figure 2. Reaction scheme of hydroxamate assay. Carbobenzoxy(*Z*)-glutaminylglycine and hydroxylamine act as the acyl donor and acceptor substrates. On the addition of ferric chloride/trichloroacetic solution in the acidic pH, the γ-glutamylhydroxamate derivative of *Z*-glutaminylglycine gives a red-colored complex.

(*1*) sodium dodecysulfate polyacrylamide gel electrophoresis (SDS-PAGE) with a reducing agent (e.g., 2-mercaptoethanol or dithiothreitol), or (*2*) test-tube inversion is effective. However, concentrated protein solution (a protein concentration of more than 50 to 100 mg/mL) must be used in the test-tube inversion assay because transglutaminase cross-links the protein substrate in both dilute and concentrate solutions, but a gel that remains at the bottom of the tube can be visually detected only in the concentrated protein solution. The detection of the ϵ-(γ-glutamyl)lysine bond, the enzymatic product, also indicates the presence of transglutaminase (25,26). Further, enzyme-linked immunosorbent assay (ELISA) using either a monoclonal or polyclonal antibody specifical to certain transglutaminases (27) indicates the presence of the enzyme both qualitatively and quantitatively, depending on the conditions of the assay. This method is effective for the detection of denatured transglutaminases in transglutaminase-utilized products.

To detect transglutaminase activity, appropriate assay methods should be carefully chosen. For instance, the sensitivity of the method, the quantity, and the specific activity of transglutaminase may alter the result. Also, substrate specificity, as well as the availability of the detective apparatus and appropriate facilities should be considered. It is known that transglutaminase from the hair follicles of guinea pigs did not recognize hydroxylamine and/or *Z*-glutaminylglycine as a substrate (28). Therefore, this method is not applied for the detection of hair follicle transglutaminases, although the hydroxamate method (hydroxylamine and *Z*-glutaminylglycine) is the most convenient for quantitative and qualitative detection of most transglutaminases.

In the all assays except that for ϵ-(γ-glutamyl)lysine detection, one has to bear in the mind that false results are always a possibility. For instance, some amidases and peptidoglutaminase (EC 3.5.1.43; an enzyme that catalyzes the incorporation of primary amines into the γ-carboxyamide group of glutamines in only short peptides, not in proteins [29]) give a positive result in the hydroxamate assay. Further, various amine oxidases, including lysyl oxidase and peroxidase, may introduce various primary amines in the protein substrates, or cross-link the protein substrates through Schiff base formation by oxidation of amines or the ϵ-amino group of lysine residues (6). Therefore, for final confirmation that the enzyme is truly a transglutaminase, one should check the presence or increment of the ϵ-(γ-glutamyl)lysine bond.

MAMMALIAN TRANSGLUTAMINASE

Transglutaminases in mammalian organs and tissues exit in multiformity depending on their physiological functions and significance. Ordinarily, for all the transglutaminases, including the well-characterized guinea pig liver enzyme and plasma factor XIIIa, the calcium ion requirement for expression of enzymatic activity is absolute (3–8). The primary structure of the five mammalian transglutaminases, namely, keratinocyte transglutaminase (TG_K, or type I), epidermal transglutaminase (TG_E, or type III), tissue transglutaminase (TG_C or type II), and prostate transglutaminase (TG_P or type IV), including plasma and platelet transglutaminase (factor XIIIa), have been established by either cDNA cloning or protein sequencing.

A major physiological role of all the distinct transglutaminases, except tissue transglutaminase, is to cross-link certain protein substrates in accordance with their existence in the organs, plasma, and extracellular fluids. Distinct transglutaminases may recognize the same protein as substrate, but different glutamine residues within the protein act as the acyl donors with different affinity and/or specificity. Their possible physiological functions are reviewed in detail elsewhere (6–8).

Physicochemical Properties

Transglutaminases can be purified by a set of various modes of chromatography (i.e., anion and cation exchange, heparin, blue Sepharose, or antibody-conjugated affinity, hydrophobic, and size-exclusion chromatography). The properties of the purified transglutaminases are summarized in Table 1. The monomeric forms of the four typical mammalian transglutaminases quite resemble one another, and their primary structures differ only slightly, thereby resulting in slightly different molecular weights (Table 1). The primary structures as revealed by cDNA cloning and/or protein sequencing have approximately 670–820 amino acid residues, and approximate molecular weights of the monomeric form of the four transglutaminases stay within the range of 71–92 kDa. TG_K has extra N- and C-terminal regions (about 60 amino acid residues) and fatty acylation through thioester-linked palmitic and myristic acid; hence, TG_K has the largest molecular weight. Other transglutaminases, such as TG_C and TG_E, have a few possible N-linked glycosylation sites and more than 10 cysteine residues, but no glycosylation and disulfide bonding have been observed.

The structural similarity, as expressed by degrees of sequence identity among the four transglutaminases, is 92% between human and rat TG_K's, 80% between human and guinea pig TG_C's, and 75% between human and mouse TG_E's. On the other hand, the structural similarity between each of the four typical transglutaminases ranges from 31% to 37%. This indicates that the degree of sequence identity between species is rather more conserved than that between the types, thereby suggesting that the ancestral genes for each transglutaminase are limited to a few types (8).

Plasma and platelet factor XIII molecules have molecular weights of approximately 360,000 and 150,000, respectively (3,4,30). They are the zymogen of factor XIIIa, an activated form. Plasma factor XIII exists as a tetramer comprising two noncovalently associated a-subunits and two filamentous b-subunits, whereas platelet factor XIII exits as a dimer comprising two noncovalently associated a-subunits. Both plasma and platelet factor XIII are activated by α-thrombin cleavage of the a-subunit into factor XIIIa. In plasma factor XIIIa, b-subunits dissociate from a-thrombin-treated plasma factor XIII (or factor XIIIa) in the presence of calcium ions. In the continued presence of calcium ions, both plasma and platelet factor XIIIa become catalytically active. The a-subunit resembles four typical transglutaminases, with a sequence identity of 43% with TG_K, 36% with TG_C, 34% with TG_E, and 27% with TG_P. The three-dimensional structure of an activated form of factor XIII has been determined (29). Thorough observation of the structure revealed that the catalytic triad residues (cysteine-314, histidine-373, and aspartate-396) are essential for the expression of transglutaminase activity. Such structural information is critical not only for protein chemistry, but also for protein engineering and genetic engineering. The differences therefore are seen mainly in the physicochemical properties of the six transglutaminases from various sources and origins; amino acid sequences around the active site (an important region for expression of the catalytic activity) are highly conserved among all six transglutaminases (Fig. 3).

Enzymatic Properties

Enzymatic properties, such as optimum pH and temperature for the enzymatic activity of the guinea pig liver transglutaminase (TG_C) and factor XIII, have been extensively characterized and reviewed (3–5), whereas those of the remaining transglutaminases have been summarized briefly (7,8). For instance, TG_C can bind guanosine triphosphate (GTP) (32). At the same time, TG_C hydrolyzes GTP, and the TG_C activity is inhibited by GTP. In this context, the possible physiological function of TG_C has been postulated as a G-protein (Gh) in receptor-mediated signal transduction (33), and such a property is unique among the six transglutaminases.

MICROBIAL TRANSGLUTAMINASE

To achieve commercialization of transglutaminase, a constant mass supply is an absolute prerequisite. Among microbial transglutaminases from three microorganisms, *Bacillus, Physarium*, and *Streptoverticillium*, transglutaminase from *Physarium* has a molecular weight of 70,000 (21) (the largest of the three), whereas very few characteristics have been determined for *Bacillus* transglutaminase (22). On the other hand, transglutaminases from *Streptoverticillium mobaraense* var. (MTG) have been well characterized (10,23,34), and their industrial production has been established (see later). MTG is secreted into the culture broth so that cell disruption is unnecessary, and subsequently, easy purification has accelerated commercialization. *Streptoverticillium* transglutaminase was first named "bacterial transglutaminase" (BTG, or BTGase), however, the final taxonomical classification into *S. mobaraense* var. (23) urged the renaming of the enzyme as "microbial" transglutaminase (MTG, or MTGase).

Manufacturing Process

MTG is produced through the fermentation process using *S. mobaraense* var. (Fig. 4). The inoculum for industrial fermenters is built up in three stages: started in a shake flask, then moved via aseptic transfer to a large fermenter, and finally to a larger fermenter. Food-grade raw materials are sterilized and are either batched or continuously

	Residues	Sequence
TGK	368–402	SVPYGQCWVFAGVTTTVLRCLGLATRTVTNFNSAH
TGC	270–304	RVKYGQCWVFAAVACTVLRCLGIPTRVVTNYNSAH
TGE	266–300	PVRYGQCWVFAGTLNTALRSLGIPSRVITNFNSAH
TGP	248–282	PVRFGQCWVFSGVLTTALRAVGIPARSITNFESAH
XIII	281–315	PVRYGQCWVFAGVFNTFLRCLGIPARIVTNYNSAH
MTG	58–92	EWLSYGCVGVTWVNSGQYPTNRLAFASFDEDRFKN

Figure 3. Sequence alignment of the active site region of six types of human transglutaminase. TG_K, keratinocyte enzyme; TG_C, tissue enzyme; TG_E, epidermal enzyme, TG_P, prostate enzyme; XIII, factor XIIIa subunit; MTG, microbial enzyme from *Streptoverticillium mobaraense* var. Underlines indicate residues conserved through all the transglutaminases, except MTG.

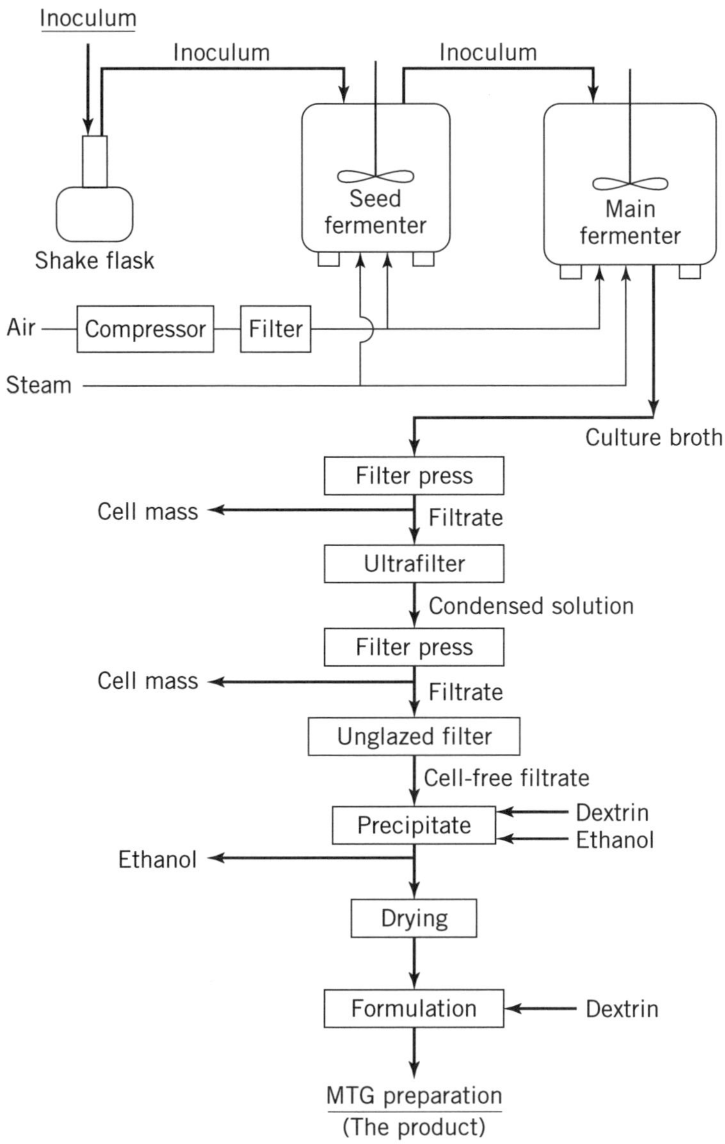

Figure 4. Manufacturing flow of transglutaminase from *Streptoverticillium mobaraense* var.

pumped to the fermenters. The build-up of the inoculum is started when the media are properly prepared and adjusted to suitable growth conditions, and a small inoculum of the pure culture is added. When proper growth is confirmed, the broth is pumped to the next phase. After these serial transfers, the inoculum is then transferred to the main fermenter. Under aseptic conditions, with pH, temperature, aeration and nutrients controlled, the organism begins to produce and excrete the enzyme. When MTG is accumulated to the proper level, the aerobic process is stopped. Thereafter, the whole broth is considered suitably treated, and the fermentation phase is considered completed. Next come the recovery and purification processes. The recovery of MTG from the cultural broth is initiated by the process that removes the organism and spent media. The liquid (broth) that contains the enzyme is subjected to ethanol precipitation, and the fraction containing the enzyme is concentrated by drying in vacuo before it is formulated for the product. All substances added during the recovery and formulating processes are of food grade. The product is then set aside until quality control can assure that it passes all the specifications.

There are various industrial technologies for enzyme recovery, most of which are currently employed. These technologies include the salting-out method, column chromatography, liquid–liquid extraction, ultrafiltration, diafiltration, and so forth. The processes employed in the recovery of MTG are ultrafiltration and solvent precipitation. Following fermentation of the producing organism, the organism is eliminated from the broth by a filter press and then a fine filter apparatus (made of unglazed filter tube). Food-grade dextrin is added to the concentrated enzyme solution for protection of the enzyme, and the enzyme is precipitated by ethanol. Then, the ethanol is removed by centrifugation, and the enzyme is dried in vacuo. The dried enzyme is formulated into the product, whose enzymatic activity is adjusted to 1,000 to 1,150 U/g. The product is kept until the quality control studies are completed.

Fermentation Flow. The pure culture of *S. mobaraense* var is lyophilized; the medium for storage consists of 10% skim milk and 1% monosodium L-glutamate. A lyophilized culture in an ampoule is thawed and suspended in sterile water. After suitable dilution with sterile water, a 0.15-mL portion of the suspension is aseptically inoculated in Venette agar medium containing glucose, peptone, meat extract, yeast extract, and agar. The inoculated media are incubated at 30 ± 1 °C for 10 days. After confirming sufficient mycelium growth and spore formation, the test tubes are examined macroscopically and microscopically. The strain is checked for the standard requirement of MTG activity (hydroxamate assay), protease activity, and pH value. The best strain selected by the above purity test is stored as stock spores at 5 ± 1 °C.

The preculture medium has the following ingredients, in decreasing order of predominance: starch, glycerol, meat extract, yeast extract, potassium dihydrogen phosphate, magnesium sulfate, and antifoam agent. These ingredients are dissolved in water, and the solution is neutralized with sodium hydroxide solution and sterilized at 121 °C for 20 min. Five 2-L shake flasks containing 500 mL of sterile preculture medium are inoculated with a loopful of the spores. The flasks are incubated on a reciprocal shaker at 30 ± 1 °C for 48 h. All of them are inoculated to seed fermenters including sterile preculture medium. The seed fermenter is aerobically incubated at 30 ± 1 °C for about 26 h under a pressure of 0.5 kg/cm^2 with agitation. The whole seed culture broth is transferred to the main fermenter containing a medium described later. The fermenter is conducted aerobically at 30 ± 1 °C for 85–90 h under a pressure of 0.5 kg/cm^2 with agitation. During the fermentation, the pH of broth and the amount of CO_2 in the waste gas are continuously measured; the mycelial growth is also checked. A microscope check for foreign microorganisms is done at 24-h intervals; if any are found, the entire batch is discarded.

The main culture medium has the following ingredients, in decreasing order of predominance: starch, peptone, meat extract, yeast extract, corn steep liquor, potassium dihydrogen phosphate, dipotassium phosphate, magnesium sulfate, and antifoam agent. These ingredients are dissolved in water, and the solution is neutralized with sodium hydroxide solution and sterilized at 121 °C for 30 min.

Purification Flow. The culture broth is passed through a filter press using diatomaceous earth to remove viable cells of *S. mobaraense* var. The obtained filtrate is concentrated to approximately 1/12 volume of the original culture broth and desalted by ultrafiltration. The concentration is filtered using a filter press and a fine-filter apparatus, which is made of an unglazed filter tube. The cell-free filtrate is mixed with ethanol after addition of food-grade dextrin, then stood at 10 ± 2 °C. Ethanol is removed with decantation. The precipitate is washed with ethanol twice. The precipitates are recovered by centrifugation, and the cake is dried in vacuo at 39 ± 1 °C for 22–24 h. The dry powder is ground out and passed through a 60-mesh screen. If necessary, the TG activity can be adjusted to an appropriate uniformity by mixing with a diluent such as dextrin.

Physicochemical Properties

MTG has a molecular weight of approximately 38,000 as determined by SDS-PAGE and gel-permeation chromatography (10). Mass spectrometric analysis indicates that MTG has a molecular weight of 37,869.2 ± 8.8. Amino acid (34) and cDNA (23) sequences reveal that MTG comprises 331 amino acid residues and has a single cysteine residue as a potential active site (Fig. 5). Although two potential glycosylation sites (-Thr-Xxx-Asn-) exist in the primary structure, MTG is a monomeric and simple protein (not a glycoprotein or lipoprotein, etc.). The cDNA structure reveals that MTG has a signal peptide of 18 amino acid residues in its N-terminal in the precursor. In fact, MTG is secreted into the cultural broth during the fermentation. A high isoelectric point (pI 8.9) suggests that MTG is a basic protein.

Enzymatic Properties

MTG has a wide optimum pH (from 5 to 8) for activity. Even at pH 4 or 9, MTG still shows some enzymatic activity. MTG is also stable at wide pH ranges. An optimum temperature for MTG activity was 45–55 °C, and MTG fully sustained its enzymatic activity even after heating at 50 °C for 10 min, both in the presence and absence of reducing agents (Fig. 6). On the other hand, MTG losses its activity within a few minutes on heating to 70°C. MTG expresses its activity even at 10 °C, and it exerts some activity even at near freezing (0 °C).

Expression of enzymatic activity for MTG is totally independent of calcium ions, and in this aspect MTG is unique as compared with other mammalian enzymes. Such calcium independence is useful for the modification of the functional properties of food proteins. Many food proteins, (e.g., caseins, soybean globulins, and myofibrillar proteins) are susceptible to calcium ions and, subsequently, are precipitated easily by calcium ions. The sensitivity toward other cations has also been investigated, and copper, zinc, and lead ions notably inhibited MTG. The results are reasonable because these heavy metals bind to and block the thiol group of the single cysteine residue; meanwhile, MTG was inhibited by thiol-modifying agents (10). For substrate specificity of MTG, to date, most food proteins, such as legume globulins, wheat glutens, egg yolk and white proteins, ovalbumin, actins, myosins, fibrins, milk caseins, α-lactalbumin, and β-lactoglobulin, as well as many other albumins, are cross-linked (35–42).

ALTERATION OF FUNCTIONALITIES OF FOOD PROTEINS BY TRANSGLUTAMINASE

In the formation and maintenance of tertiary and quaternary structures of protein molecules, hydrogen bonds, disulfide bonds, and electrostatic and van der Waal's hydrophobic interactions are essential. Especially in food, not only these bonds, but also other bonds and interactions, particularly those found in connective tissues, are very

1 … 50
DSDDRVTPPAEPLDRMPDPYRPSYGRAETVVNNYIRKWQQVYSHRDGRKQ

51 … 100
QMTEEQREWLSYGC*VGVTWVNSGQUPTNRLAFASFDEDRFKNELKNGRPR

(* printed above C)

101 … 150
SGETRAEFEGRVAKESFDEEKGGQRAREVASVMNRALENAHDESAYLDNL

151 … 200
KKELANGNDALRNEDARSPFYSALRNTPSFKERNGGNHDPSRMKAVIYSK

201 … 250
HFWSGQDRSSSADKRKYGDPDAFRPAPGTGLVDMSRDNIPRSPTSPGEG

251 … 300
FVNFDYGWFGAQTEADADKTVWTHGNHTHAPNGSLGAMHVYESKFRNWSE

301 … 331
GYSDFDRGAYVITFIPKSWNTAPDKVKQGWP

Figure 5. Amino acid sequence of microbial transglutaminase from *Streptoverticillium mobaraense* var. All amino acid residues are represented by one-letter. An asterisk indicates the active center.

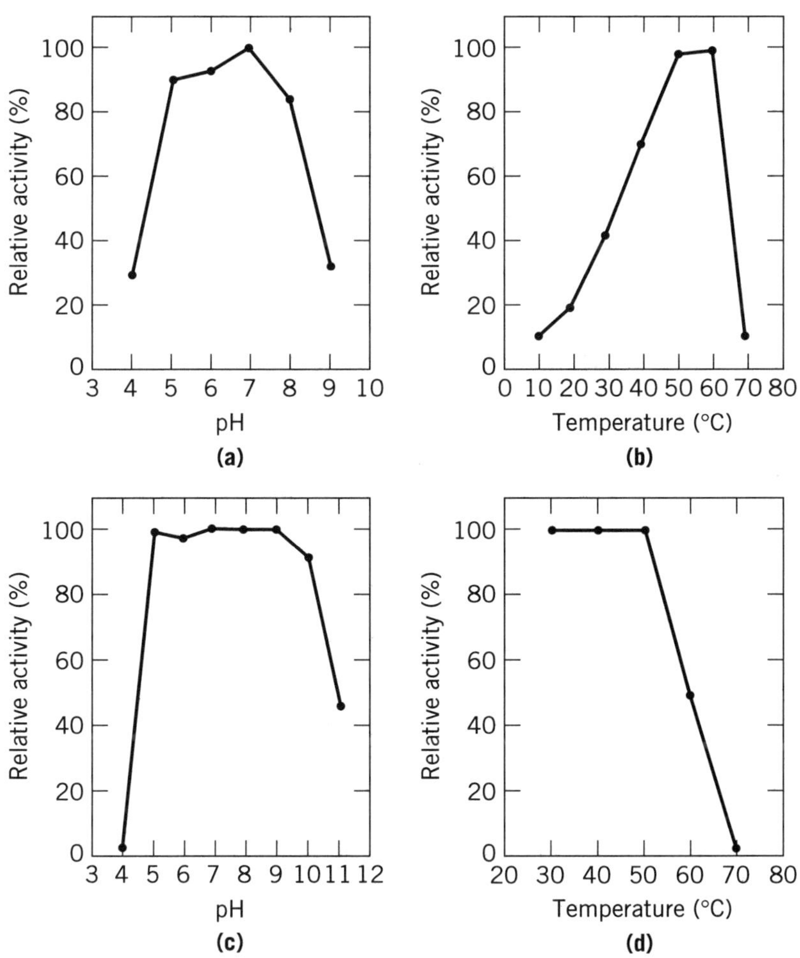

Figure 6. Enzymatic properties of microbial transglutaminase. (**a**) pH-activity profile; (**b**) temperature-activity profile; (**c**) pH stability profile; and (**d**) thermal stability profile.

important in the expression of textures or mouth-feel. Among them, covalent ϵ-(γ-glutamyl)lysine cross-linking is considered noteworthy. An example of such a cross-linking formation is found in the setting phenomenon or "suwari" of salted, ground fish protein pastes in Japanese kamaboko manufacturing. Contribution of such endogenous transglutaminase activity of fish in the suwari process has been well documented (43–45). As demonstrated in the these cases, transglutaminase has been believed to modify functional properties of protein substrates (9). In an attempt to develop new foods with unique protein utilization and processing methodology, we have been interested in the modification of the functional properties of food proteins through formation of ϵ-(γ-glutamyl)lysine bonds by transglutaminase.

Feasibility Study on Protein Modification by Transglutaminases

In the early 1980s, the possibility of modification of functional properties in milk caseins, soybean globulins, whey proteins, and beef, pork, chicken, and fish actomyosins was demonstrated using transglutaminase derived from guinea pig liver (46–59) or bovine plasma factor XIIIa (60,61). When protein solutions are cast on a flat surface, a transparent, water-resistant, and slowly digestible protein film can be prepared. A protein emulsion of the oil-in-water type can also serve as the substrate, and it is gelled by transglutaminase. In these studies, cross-linking between various proteins of different origins, gelation of various proteins by the transglutaminase reaction, as well as incorporation of amino acids or peptides to improve nutritive deficiency was shown. These results indicate that transglutaminase was potentially useful in creating proteins with new, unique functional properties, as previously postulated (9). However, limited supply and the unfamiliarity of guinea pig liver transglutaminase for food use hindered its commercialization.

Protein Cross-Linking

Discovery of MTG drastically accelerated its applicative studies on the alteration of food protein functionalities. MTG gels concentrate solutions of such proteins as soybean protein, milk proteins, gelatin, and beef, pork, chicken, and fish myosins, as does the guinea pig liver enzyme. Subsequently, improvement in solubility, water-holding capacity, and thermal stability was demonstrated. The gelled soybean globulin further hardened by heating at 100 °C, for 15 min, resulting in a protein with new gel properties. Caseins, non-heat-setting proteins, are also gelled by MTG without heating. Gelatin, a cold-setting protein, is also gelled, and in this case, the gelled gelatin no longer melts on heating at 100 °C. In practical application, soybean curd can be prepared by the ordinary process with the aid of MTG. This MTG-treated soybean curd becomes retort resistant in a way that the soybean curd retains its shape and is nonporous even after the retort treatment(s). More than two different proteins can be covalently conjugated by MTG to produce new proteins with novel functionalities, as with guinea pig transglutaminase. For instance, conjugation of caseins or soybean globulins with ovomucin, one of the egg white glycoproteins, improved emulsifying activity (44). Also conjugation of casein and gelatin by transglutaminase yielded novel proteins that were highly soluble in acidic pH regions (41). Practical applications of transglutaminases, including MTG, in food processing and manufacturing are summarized in Fig. 7, and detailed explanations of transglutaminase applications have been given in elsewhere (62–64).

Amino Acids and/or Peptides Incorporation into Proteins

Amino acids and/or peptides can be covalently introduced into substrate proteins by MTG. This improves nutritive values of caseins, soybean proteins, wheat gluten, and so on, whose methionine and lysine, are limiting amino acids. For practical application of amino acid incorporation, all the amino acids, except lysine, must have their α-carboxyl group amidated, esterified, or decarboxylated in order to be substrates of the transglutaminases. On the other hand, lysine with an ϵ-amino group on the side chain is a good substrate for transglutaminase. In the case of peptides, most should contain at least one lysine residue to be a suitable substrate. For instance, lysylmethionine or methionyllysine can be incorporated into caseins to improve the methionine deficiency. Likewise, lysylarginine or arginyllysine can be incorporated into caseins to improve the arginine deficiency in infant prawns. On the other hand, carnosine and β-alanylhistidine can serve as the substrate (65). Carnosine and anserine, a 1-methylated histidine derivative of carnosine, are thus used as inhibitors of protein cross-linking, because both peptides competitively act as the substrates in the transglutaminase cross-linking reaction.

BIOAVAILABILITY OF CROSS-LINKED PROTEINS

Although the scope of application of MTG-catalyzed modification of food proteins is vast, here attention is focused on the nutritional efficiency of the cross-linked proteins. MTG-modified proteins differ from native proteins only in the quantity of the ϵ-(γ-glutamyl)lysine cross-links, all else is the same. However, the digestibility or bioavailability of the ϵ-(γ-glutamyl)lysine moiety in the cross-linked proteins has been intensively investigated.

Distribution of Naturally Occurring ϵ-(γ-Glutamyl)lysine

Distribution of naturally occurring ϵ-(γ-glutamyl)lysine cross-links in raw materials such as animal organs and tissues and in commonly consumed processed foods have been determined by measuring the ϵ-(γ-glutamyl)lysine dipeptide after exhaustive digestion by proteases and peptidases (25,26). Raw food materials, e.g., meats, fish, shellfish, and most processed foods (e.g., kamaboko, ham, stewed beef, fried chicken, grilled pork, and hamburger) contained more or less certain amounts of ϵ-(γ-glutamyl)lysine cross-links (Table 2). However, dairy products (e.g., cheese, milk, and yogurt) scarcely contained these cross-links. The level of ϵ-(γ-glutamyl)lysine cross-links was higher in processed, especially cooked, foods than in raw materials. The ϵ-(γ-glutamyl)lysine cross-link

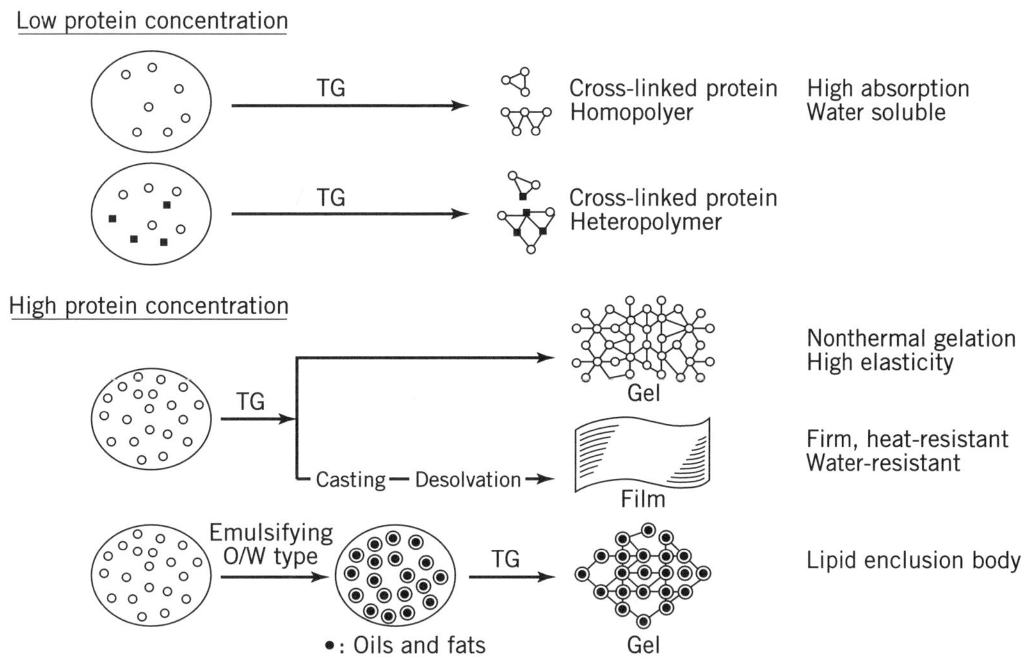

Figure 7. Application of transglutaminase on alteration of food functionalities. Open small circles represent substrate protein molecules. Closed small squares represent other substrate protein molecules. Closed small circles represent oils and/or fats.

Table 2. Distribution of Naturally Occurring ϵ-(γ-Glutamy)lysine Cross-Links

Items		Amount of ϵ-(γ-Glutamyl)lysine[a]
Raw materials	Beef tongue	78.4
	Crab	34.6
	Beef liver	14.5
	Prawn	3.77
	Milk	nd
	Chicken egg	nd
	Tomato	nd
	Apple	nd
	Potato	nd
Processed food	Stewed beef	68.7
	Hamberger	40.9
	Ham	39.4
	Peking duck	31.5
	Bread	15.7
	Cooked rice	11.1
	Popcorn	6.80
	Fried chicken	6.76
	Cheese	nd
	Yogurt	nd

[a]Micromoles of ϵ-(γ-Glutamyl)lysine in 100 g protein of each item.

occurs naturally in eggs of red salmon, lumpfish, herring, sardine, and Alaska pollock (66), all of which have been popularly consumed as caviar in both Western and Eastern countries.

Mechanism of ϵ-(γ-Glutamyl)lysine Formation

It is highly likely that endogenous transglutaminases are responsible in the formation of ϵ-(γ-glutamyl)lysine cross-links in protein during the cooking process, because a variety of living organisms contain intrinsic, endogenous transglutaminase in their tissues and organs. During cooking, the temperature elevation in food materials is in fact very slow, such that the endogenous transglutaminases have some time to exert their enzymatic activity before heat inactivation takes place. Thus cooked or processed foods may contain larger amounts of ϵ-(γ-glutamyl)lysine cross-links than raw materials due to transglutaminase activity. That dairy products contained no ϵ-(γ-glutamyl)lysine cross-links may support an idea that milk itself does not contain any endogenous transglutaminase activities.

Besides the catalysis of the transglutaminases, heating has been reported to induce the formation of ϵ-(γ-glutamyl)lysine cross-links in chemical dehydration between the γ-carboxyl group of glutamate and ϵ-amino group of lysine (67–69). In this respect, humans have been ingesting the ϵ-(γ-glutamyl)lysine moiety since the discovery of cooking. Consequently, the safety of the ϵ-(γ-glutamyl)lysine moiety has been unintentionally and automatically demonstrated.

Degradation In Vitro of ϵ-(γ-Glutamyl)lysine

Ordinarily mammalian gastrointestinal digestive enzymes cleave proteins into amino acids, but leave the ϵ-(γ-glutamyl)lysine cross-link uncleaved after ingestion of cross-linked proteins. The uncleaved ϵ-(γ-glutamyl)lysine dipeptide may be absorbed through intestinal brush-border membranes to the kidney. As a factor responsible for the degradation of the ϵ-(γ-glutamyl)lysine cross-link, a kidney enzyme, γ-glutamylamine cyclotransferase, has been reported to cleave the ϵ-(γ-glutamyl)lysine dipeptide form. This reaction yields a free lysine and 5-*oxo*-proline (synonym: pyroglutamate) from the dipeptide (70). Because lysine is an essential amino acid for humans and other mammals, the presence of such a pathway would help satisfy nutritional requirements. On the other hand, 5-*oxo*-proline should be converted to glutamate by an ATP-dependent enzyme, 5-*oxo*-prolinase. If a large amount of the cross-linked protein was ingested, a large amount of ATP would be consumed in the kidney. Such a large ATP consumption might overload kidney function, however, recently, γ-glutamyltransferase (γ-glutamyltranspeptidase; EC 2.3.2.2, (71)), present mainly in the intestinal brush-border membranes, kidney, and blood, has been found to cleave the ϵ-(γ-glutamyl)lysine dipeptide into a lysine and glutamate (72). This reaction uses no ATP to generate glutamate. Thus, the ϵ-(γ-glutamyl)lysine moiety, could be effectively metabolized into lysine by the two enzymes.

Degradation In Vivo of ϵ-(γ-Glutamyl)lysine

Many researchers have demonstrated that the ϵ-(γ-glutamyl)lysine dipeptide could be metabolized in rats, and that lysine was integrated in rat tissues (73–76). On the other hand, the bioavailability of the ϵ-(γ-glutamyl)lysine cross-linked moiety in the cross-linked proteins has not yet been experimentally demonstrated. This is because it would be unrealistic and impractical to prepare large-scale cross-linked proteins for animal feeding by using the guinea pig liver transglutaminase. With abundantly available MTG, cross-linked caseins in kilogram scale were fed to rats to evaluate the nutritive value of lysine in the ϵ-(γ-glutamyl)lysine moiety. Lysine generated in the rats has been proved to be nutritionally beneficial in the rats using MTG-catalyzed cross-linked wheat glutens (77). It is thus suggested that the ϵ-(γ-glutamyl)lysine moiety in the cross-linked caseins is cleaved, and the lysine in the moiety metabolically utilized in the body.

CONCLUSION

As stated earlier (9), transglutaminases have been found to drastically alter protein functionalities through the formation of ϵ-(γ-glutamyl)lysine cross-links in many food proteins. Vast application of the transglutaminase modification on food proteins can bring forth such promising results as novel foods and innovative processing techniques. Further, the discovery of transglutaminase from *S. mobaraense* var. (MTG) has accelerated applicative studies, and

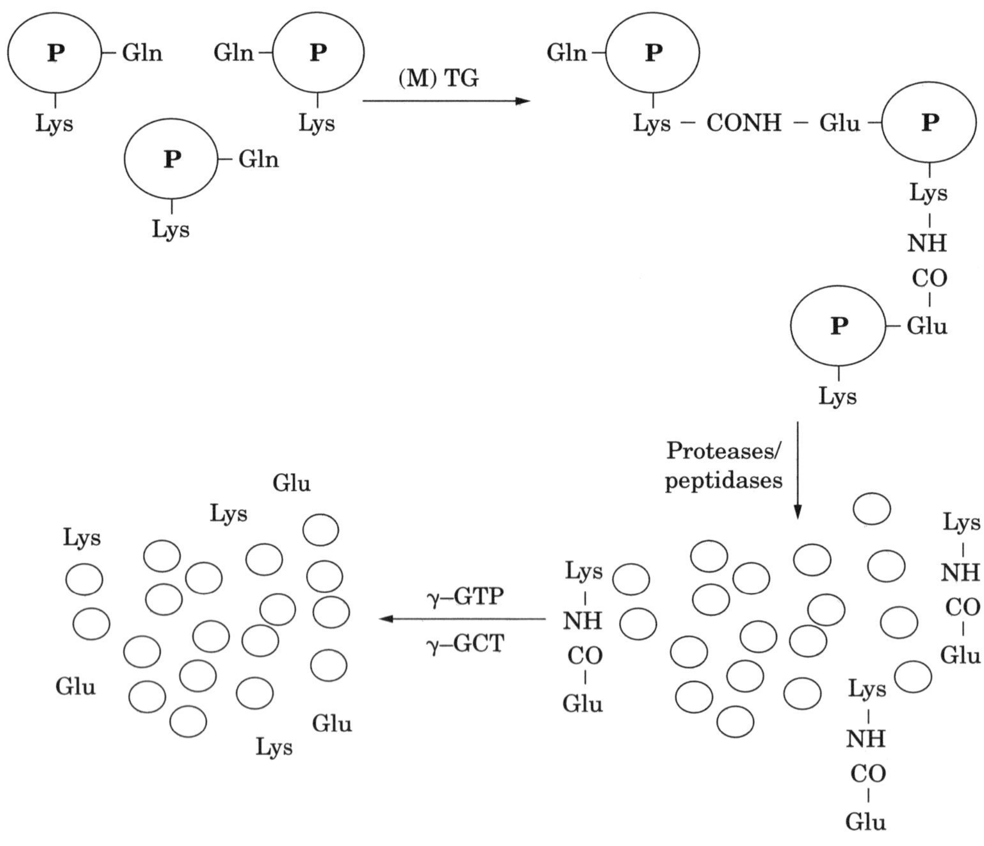

Figure 8. Reaction scheme of the digestion of transglutaminase-catalyzed cross-linked proteins. A large circle with *P* represents a protein or peptide; a small circle represents a single amino acid produced by digestion of proteases and/or peptidases; Glu-CONH-Lys, ϵ-(γ-glutamyl)lysine dipeptide; (M)TG, (microbial) transglutaminase; γ-GTP, γ-glutamyltransferase; γ-GCT, γ-glutamylamine cyclotransferase.

consequently, MTG can be used in surimi and sausage manufacturing in Japan.

With respect to the bioavailability of the ϵ-(γ-glutamyl)lysine moiety in the cross-linked proteins, many studies have been done, and it has been demonstrated that the ϵ-(γ-glutamyl)lysine dipeptide that usually can be produced from the cross-linked protein after exhaustive digestion by gastrointestinal digestive proteases and peptidases can be cleaved by both γ-glutamylamine cyclotransferase and γ-glutamyltransferase, yielding a free lysine. The lysine in the cross-linked proteins has been shown to be metabolizable in the body. The nutritive value of the lysine of the cross-linked protein has been proven using wheat gluten as a lysine-deficient model protein in rats (Fig. 8). Finally, with the evidence that the safety of ϵ-(γ-glutamyl)lysine moiety has been proved by long-established, habitual intakes of cooked foods by humans, the overall effectiveness and safety of protein modification by transglutaminase have been shown.

BIBLIOGRAPHY

1. Nomenclature Committee of the International Union of Biochemistry and Molecular Biology, *Enzyme Nomenclature*, Academic Press, San Diego, Calif., 1992, p. 201.
2. H. Borsook, C.L. Deasy, A.J. Haagen-Smit, G. Keighley, and P.H. Lowy, *J. Biol. Chem.* **184**, 529–543 (1950).
3. J.E. Folk and S.I. Chung, *Adv. Enzymol.* **38**, 109–191 (1973).
4. J.E. Folk, J.S. Finlayson, *Adv. Protein Chem.* **31**, 1–133 (1977).
5. J.E. Folk, *Adv. Enzymol. Relat. Areas Mol. Biol.* **54**, 1–56 (1983).
6. L. Lorand and S.M. Conrad, *Mol. Cell Biochem.* **58**, 9–35 (1984).
7. C.S. Greenberg, P.J. Birckbichler, and R.H. Rice, *FASEB J.* **5**, 3071–3077 (1991).
8. D. Aeschlimann and M. Paulsson, *Thromb. Haemost.* **71**, 402–415 (1994).
9. J.R. Whitaker, *Food Proteins—Improvement through Chemical and Enzymatic Modification*, American Chemical Society; Washington, D.C., 1977, pp. 95–105.
10. H. Ando, M. Adachi, K. Umeda, A. Matsuura, M. Nonaka, R. Uchio, H. Tanaka, and M. Motoki, *Agric. Biol. Chem.* **53**, 2613–2617 (1989).
11. H. Araki and N. Seki, *Nippon Suisan Gakkaishi*, **59**, 711–719 (1993).
12. Y. Kumazawa, K. Seguro, M. Takamura, and M. Motoki, *J. Food Sci.* **58**, 1062–1064 (1993).
13. I. Kimura, M. Sugimoto, K. Toyoda, N. Seki, K. Arai, and T. Fujita, *Nippon Suisan Gakkaishi* **57**, 1389–1396 (1991).
14. K. Sano, Y. Kumazawa, H. Yasueda, K. Seguro, and M. Motoki, *Fifth International Conference on Transglutaminase and Protein Crosslinking Reactions: Book of Abstracts*, 1996, p. 65.
15. F. Tokunaga, M. Yamada, T. Miyata, Y.L. Ding, M. Hiranaga-Kawabata, T. Muta, and S. Iwanaga, *J. Biol. Chem.* **268**, 252–261 (1993).
16. M. Ando, Y. Nagata, and R. Hammerschlag, *Zool. Sci.* **8**, 721–727 (1991).
17. M. Signorini, S. Beninati, and C.M. Bergamini, *J. Plant Physiol.* **137**, 547–552 (1991).
18. S.A. Margosiak, M.E. Alavarets, D. Louie, and G.D. Kuehn, *Plant Physiol.* **92**, 88–96 (1990).
19. D. Serafini-Fracassini, S.D. Duca, and P. Torrigiani, *Plant Physiol. Biochem.* **27**, 659–669 (1989).
20. I. Icekson and A. Apelbaum, *Plant Physiol.* **84**, 972–974 (1987).
21. J.D. Klein, E. Guzman, and G.D. Kuehn, *J. Biocteriol.* **174**, 2599–2605 (1992).
22. M.V. Ramanujam and J.H. Hageman, *FASEB J.* **168**, A3630 (1990).
23. K. Washizu, K. Ando, S. Koikeda, S. Hirose, A. Matsuura, H. Takagi, M. Motoki, and K. Takeuchi, *Biosci. Biotechnol. Biochem.* **58**, 82–87 (1994).
24. H. Waelsch and M.J. Mycek, *Methods Enzymol.* **5**, 833–838 (1962).
25. M. Griffin, J. Wilson, and A. Lorand, *Anal. Biochem.* **124**, 406–413 (1982).
26. H. Sakamoto, Y. Kumazawa, H. Kawajiri, and M. Motoki, *J. Food Sci.* **60**, 416–419 (1995).
27. T. Ohtsuka, K. Seguro, and M. Motoki, *J. Food Sci.* **61**, 81–84 (1996).
28. S.I. Chung and J.E. Folk, *Proc. Natl. Acad. Sci. U.S.A.* **69**, 303–307 (1972).
29. Nomenclature Committee of the International Union of Biochemistry and Molecular Biology, *Enzyme Nomenclature*, Academic Press, San Diego, Calif., 1992, p. 427.
30. T.J. Hornyak, P.D. Bishop, and J.A. Shafer, *Biochemistry* **28**, 7326–7332 (1989).
31. V.C. Yee, L.C. Pedersen, I.L. Trong, P.D. Bishop, R.E. Stenkamp, and D.C. Teller, *Proc. Natl. Acad. Sci. U.S.A.* **91**, 7296–7300 (1994).
32. Y. Takeuchi, P.J. Birckbichler, M.K. Patterson, Jr., and K.N. Lee, *FEBS*, **307**, 177–180 (1992).
33. H. Nakaoka, D.M. Perez, K.J. Beak, T. Das, A. Husain, K. Misono, M.J. Im, and R.M. Graham, *Science* **264**, 1593–1596 (1994).
34. T. Kanaji, H. Ozaki, T. Takao, H. Kawajiri, H. Ide, M. Motoki, and Y. Shimonishi, *J. Biol. Chem.* **268**, 11565–11572 (1993).
35. M. Nonaka, H. Tanaka, A. Okiyama, M. Motoki, H. Ando, K. Umeda, and A. Matsuura, *Agric. Biol. Chem.* **53**, 2619–2623 (1989).
36. M. Nonaka, H. Sakamoto, S. Toiguchi, H. Kawajiri, T. Soeda, and M. Motoki, *J. Food Sci.* **57**, 1214–1218 (1992).
37. Y.P. Huang, K. Seguro, M. Motoki, and K. Tawada, *J. Biochem.* **112**, 229–234 (1992).
38. H. Sakamoto, Y. Kumazawa, S. Toiguchi, K. Seguro, T. Soeda, and M. Motoki, *J. Food Sci.* **60**, 300–304 (1995).
39. I.J. Kang, Y. Matsumura, K. Ikura, M. Motoki, H. Sakamoto, and T. Mori, *J. Agric. Food Chem.* **42**, 159–165 (1994).
40. E. Kim, M. Motoki, K. Seguro, A. Muhlrad, and E. Reisler, *Biophys. J.* **69**, 2024–2032 (1995).
41. M. Nonaka, Y. Matsuura, and M. Motoki, *Biosci. Biotechnol. Biochem.* **60**, 131–133 (1996).
42. A. Kato, T. Wada, K. Kobayashi, K. Seguro, and M. Motoki, *Agric. Biol. Chem.* **55**, 1027–1031 (1991).
43. G.G. Kamath, T.C. Lanier, E.A. Foegeding, and D.D. Hamann, *J. Food Biochem.* **16**, 151–172 (1992).
44. K. Sato, Y. Tsukamasa, C. Imai, K. Ohtsuki, Y. Shimizu, and M. Kawabata, *J. Agric. Food Chem.* **40**, 806–810 (1992).
45. J. Wan, I. Kimura, M. Satake, and N. Seki, *Fisheries Sci.* **60**, 107–113 (1994).

46. K. Ikura, T. Kometani, M. Yoshikawa, R. Sasaki, and H. Chiba, *Agric. Biol. Chem.* **44**, 1567–1573 (1980).
47. K. Ikura, T. Kometani, R. Sasaki, and H. Chiba, *Agric. Biol. Chem.* **44**, 2979–2984 (1980).
48. K. Ikura, M. Yoshikawa, R. Sasaki, and H. Chiba, *Agric. Biol. Chem.* **45**, 2587–2592 (1981).
49. K. Ikura, M. Goto, M. Yoshikawa, R. Sasaki, and H. Chiba, *Agric. Biol. Chem.* **48**, 2347–2354 (1984).
50. K. Ikura, K. Okumura, M. Yoshikawa, R. Sasaki, and H. Chiba, *Agric. Biol. Chem.* **49**, 1877–1878 (1985).
51. N. Nio and M. Motoki, *J. Food Sci.* **48**, 561–566 (1983).
52. N. Nio, M. Motoki, and K. Takinami, *Agric. Biol. Chem.* **48**, 1257–1261 (1984).
53. N. Nio, M. Motoki, and K. Takinami, *Agric. Biol. Chem.* **49**, 2483–2286 (1985).
54. N. Nio, M. Motoki, and K. Takinami, *Agric. Biol. Chem.* **50**, 851–855 (1986).
55. N. Nio, M. Motoki, and K. Takinami, *Agric. Biol. Chem.* **50**, 1409–1412 (1986).
56. M. Motoki, K. Seguro, N. Nio, and K. Takinami, *Agric. Biol. Chem.* **50**, 3025–3030 (1986).
57. N. Nio, M. Motoki, and K. Takinami, *Agric. Biol. Chem.* **51**, 237–239 (1987).
58. M. Motoki, H. Aso, K. Seguro, and N. Nio, *Agric. Biol. Chem.* **51**, 993–996 (1987).
59. M. Motoki, H. Aso, K. Seguro, and N. Nio, *Agric. Biol. Chem.* **51**, 997–1002 (1987).
60. L. Kurth and P.J. Rogers, *J. Food Sci.* **49**, 573–576, 589 (1984).
61. C.D. Backer-Royer, F. Traore, and J.C. Meunier, *J. Agric. Food Chem.* **40**, 2052–2058 (1992).
62. P.M. Neilsen, *Food Biotechnol.* **9**, 119–156 (1995).
63. Y. Zhu, A. Rinzema, J. Tramper, and J. Bol, *Appl. Microbiol. Biotechnol.* **44**, 277–282 (1995).
64. H. Singh, *Trends Food Sci. Technol.* **8**, 196–200 (1991).
65. J. Wan, I. Kimura, and N. Seki, *Fisheries Sci.* **61**, 968–972 (1995).
66. Y. Kumazawa, H. Sakamoto, H. Kawajiri, K. Seguro, and M. Motoki, *Fisheries Sci.* **62**, 331–332 (1996).
67. R.S. Asquith, M.S. Otterburn, and W.J. Sinclair, *Angew. Chem. Int. Ed.* **13**, 514–520 (1974).
68. R.F. Hurrell, K.J. Carpenter, W.J. Sinclair, M.S. Otterburn, and R.S. Asquith, *Br. J. Nutr.* **35**, 383–395 (1976).
69. M. Otterburn, M. Healy, and M. Sinclair, *Adv. Exp. Med. Biol.* **86B**, 239–262 (1977).
70. M.L. Fink, S.I. Chung, and J.E. Folk, *Proc. Natl. Acad. Sci. U.S.A.* **77**, 4564–4568 (1980).
71. A. Meister, S.S. Tate, and O.W. Griffith, *Methods Enzymol.* **77**, 237–253 (1981).
72. K. Seguro, Y. Kumazawa, T. Ohtsuka, H. Ide, N. Nio, M. Motoki, and K. Kubota, *J. Agric. Food Chem.* **43**, 1977–1981 (1995).
73. G. Raczynski, M. Snochowski, and S. Buraczewski, *Br. J. Nutr.* **34**, 291–296 (1975).
74. P.A. Finot, F. Mottu, E. Bujard, and J. Mauron, *Nutritional Improvement of Foods and Food Proteins*, Plenum, London, 1978, pp. 549–570.
75. K. Iwami and K. Yasumoto, *J. Sci. Food Agric.* **37**, 495–503 (1986).
76. M. Friedman and P.A. Finot, *J. Agric. Food Chem.* **38**, 2011–2020 (1990).
77. K. Seguro, Y. Kumazawa, C. Kuraishi, H. Sakamoto, and M. Motoki, *J. Nutr.*, in press.

See also BIOCATALYSIS DATABASES.

TRANSIENT EXPRESSION SYSTEMS

A.R. BERNARD
H.D. BLASEY
Serono Pharmaceutical Research Institute
Geneva, Switzerland

KEY WORDS

Baculovirus
BHK
COS
DNA transfection
HEK293EBNA
Insect cell culture
Large-scale transient expression
Mammalian cell culture
Recombinant protein
Semliki Forest virus
Sf9
Virus expression vectors

OUTLINE

OVERVIEW

Historically, the search for new genes and new gene functions has been the major driving force for the development of several powerful transient expression systems such as direct DNA transfection into mammalian cells, recombinant virus infection, injection of *Xenopus* oocytes with mRNA, or complete in vitro–coupled transcription–translation. In particular, functional expression of large cDNA libraries has allowed the cloning of many new genes. The analysis of the gene products, however, has lagged behind because of the lack of efficient, quick, and easy ways of producing sufficient amounts of the gene product to be studied. On another front, methods for stable expression of genes in mammalian cells have been available for many years and presently form the basis for the development of biopharmaceutical production in cell culture. One of the major drawbacks of stable expression lies in the time required for the construction of an optimal, stable recombinant cell line. Transient expression of human or mammalian genes has received considerable interest in recent years for its use in producing recombinant proteins, mainly for research but potentially also for clinical purposes. The ability to rapidly and reproducibly express one or several recombinant genes in a eucaryotic cell environment and to produce large amounts of the gene product now complements the conventional set of biotechnological tools. This article reviews the transient expression methods that can be used on a large scale for the production of recombinant proteins of pharmaceutical interest for in vitro biological, biochemical, animal, and human studies. The various methods presently available can be classified according to the vector used: plasmid DNA, insect viruses, or mammalian viruses. Several of the vectors used in transient expression are also used in gene therapy applications, but we will restrict the scope of this review to the production of recombinant proteins in cell cultures.

There are several key features of transient expression systems:

- Simplicity, in particular in the construction of the expression vector
- Intrinsic genetic stability, because the gene is constantly maintained in the expression host
- Speed, that is, the time required to obtain the first quantities of the gene product
- Application to a wide range of host cells (except in the case of the baculovirus system)
- Suitability to parallel processing, that is, the ability to study several genes or mutants at the same time

All of the transient expression methods have historically been designed for small-scale operations (Petri dish or tissue culture flasks). However, the emphasis of recent developments has been on the scale-up and application of these laboratory methods to large-scale production systems, and all methods can now be operated at pilot bioreactor scale.

DIRECT DNA TRANSFECTION

Introduction

Direct DNA transfection is a widely used system to efficiently produce a desired gene product at levels of 1–3 mg/L within a few weeks. Both genomic DNA and cDNA can be expressed using the classical COS host cell line. This allows the generation of large amounts of protein material with mammalian posttranslational modifications (glycosylation, disulfide bridges, carboxylation, amidation). The material produced can then be utilized for studying the structure and function of the gene product. In parallel, COS cells are routinely used for studying gene expression regulation and for expression cloning. The cells are readily available from public collections (ATCC, ECACC) and are efficiently transfected. Cultivation of these adherent cells does not pose any major problem, even for large-scale cultures. These properties have contributed to the wide popularity of this expression system.

COS cells are derived from green monkey kidney cells (CV-1) transformed with an origin-defective SV40 virus. Historically, the cell line was developed by Yakov-Gluzman (1) to allow the generation of SV40 virus mutants. He developed three cell lines: COS-1, COS-3, COS-7. The cells express high levels of the wild-type SV40 large tumor (T) antigen, an early gene product of 100 bp, which binds to the vector DNA at the origin of replication and allows the host-cell polymerase to run repetitive cycles of DNA replication. As a consequence, plasmids carrying the SV40 origin are maintained in the cell at high copy number (10^4–10^5 copies per cell), giving rise to abundant mRNA yielding high-level recombinant gene expression. COS cells are known to express for up to 1 week posttransfection, with peak protein production between 48 and 72 h posttransfection.

Lusky and Botchan in 1981 (2) were the first to use the large T antigen expressing COS in combination with an SV40 origin–containing plasmid for the identification of DNA transcription sequences. In 1982, the COS cell expression system was used for the first time by Rose and Bergmann to express cell surface and secreted proteins such as vesicular stomatitis virus glycoprotein (3). Subsequent applications by numerous other investigators demonstrated that the COS cell expression system generally yields biologically active proteins. The main factors influencing the expression level in this system are the expression plasmid construct and the transfection efficiency.

Today, many SV40 T antigen–dependent vectors (e.g., ATCC 37192, 37193, 53100) are available for gene expression with COS cells. Typical plasmids for COS transient expression carry the following major components (Fig. 1):

1. The SV40 origin for plasmid replication
2. Appropriate promoter, enhancer, and polyadenylation sequences
3. A prokaryotic origin of replication
4. A prokaryotic selectable marker
5. Unique restriction sites for target gene subcloning

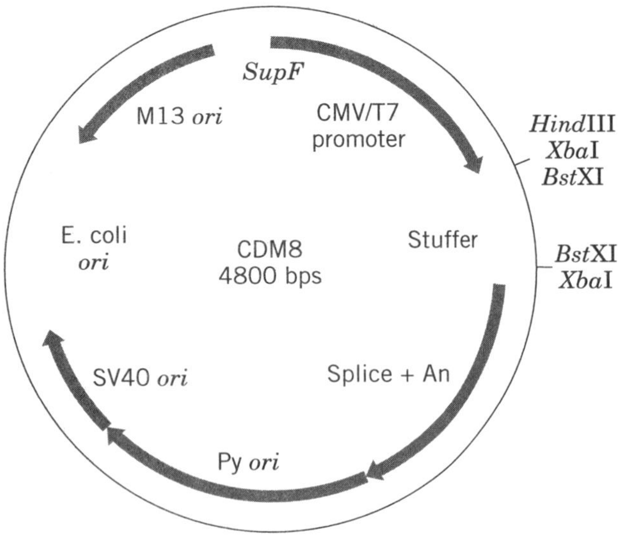

Figure 1. Typical COS expression vector. CDM8, a typical plasmid for transient expression in COS cells. The plasmid consists of CMV/T7 RNA polymerase promoter for high-level gene expression; (SV40 *ori*) SV40 and (Py) polyoma origin of replication; M13 origin of replication; a stuffer with unique restriction sites for subcloning; *E. coli* origin of replication; prokaryotic genetic marker (*SupF*); and splice and An, eukaryotic transcription regulatory elements. *Source*: Reprinted from Ref. 4, with permission of *Nature*.

The Epstein–Barr virus (EBV) also gave rise to a popular expression system based on vectors carrying the replication origin region (OriP) from EBV. EBV–transformed cell lines expressing the EBV nuclear antigen 1 (EBNA-1) replicate transfected plasmids and express the subcloned gene in a similar way to the COS cell–based system. Both genomic and cDNA sequences can be expressed in this system. EBV vectors replicate episomally in the nucleus and, under certain conditions, can be maintained by the cells upon prolonged passage, thus providing a way of obtaining either transient or stable expression of the recombinant protein. In addition, incorporation of the coding sequence for EBNA-1 on the expression vector yields a self-replicating expression plasmid that is autonomously maintained under selection pressure.

By maintaining transfected cells under selective pressure with hygromycin, 293 cells expressing EBNA-1 can be passaged for up to six months while still expressing recombinant proteins at a high level. For example, human placental alkaline phosphatase, tissue plasminogen activator, vascular endothelial cell growth factor, hepatocyte growth factor, prolactin, neurotrophins, and soluble tumor necrosis factor receptors were obtained at levels of 0.2–10 μg/mL (5).

Basic Principles and Protocols

COS Cells and Transfection. COS cell lines are available from the ATCC (COS-1:ATCC CRL 1650, and COS-7: ATCC CRL 1651). They are normally grown adherent onto tissue culture treated matrices in DMEM with 10% FCS and are split by trypsinization before reaching confluence to avoid aggregate formation. ATCC recommends handling the cells in accordance with Biosafety Level 2. Various transfection methods are available to introduce plasmids into mammalian cells. Calcium phosphate precipitation, DEAE dextran, lipofection, and electroporation are commonly used transfection procedures that are well established in cell culture laboratories. In addition, other methods have been described, such as receptor-mediated endocytosis, protoplast fusion, polylysine, and polybrene-mediated transfection. Direct DNA injection into cells with the gene gun (bombardment with DNA-coated gold particles) is used in special applications where cells are difficult to transfect or for certain electrophysiology studies.

Calcium Phosphate. DNA uptake by mammalian cells can be achieved by exposing cells to a calcium phosphate–DNA coprecipitate. The precipitate is formed within 20–30 min by diluting a $CaCl_2$/DNA mix into 2× HBS. It is then added to the cells in culture, and some protocols describe replacing the medium after about 10 min with fresh medium. In all cases, the transfection mixture is removed from the culture within 1 day after precipitate addition and replaced by fresh culture medium, allowing 2–3 days for gene expression.

Calcium phosphate transfection is also applicable to cells grown in suspension. After a wash with PBS, the cells are resuspended and incubated in a solution containing the calcium phosphate–DNA coprecipitate. After fresh medium top-up and a 1-day incubation, the cells are then pelleted and transferred into fresh medium. Several parameters for transfection with calcium phosphate are critical, including the time allowed for formation of the calcium phosphate–DNA coprecipitate and the pH at which this operation is performed. Scaling-up this method requires that the standard protocols be adapted to reduce or avoid the incubation, washing, and separation steps, because these steps are not compatible with large volumes.

Lipofection. DNA can be delivered to cells by liposome-mediated transfection. Positively charged liposomes are formed with various cationic lipids, such as for *N*-[1(2,3-diolexyloxy)propyl]-*N*,*N*,*N*-trimethylammonium chloride (DOTMA), and interact with DNA to produce DNA–liposome complexes. These complexes are adsorbed to and subsequently fuse with the lipid bilayer of cell membranes, delivering DNA into the cytoplasm.

The experimental protocols are relatively simple: plasmid DNA is diluted into serum free medium and mixed with preformed cationic liposomes. For the actual transfection, cells are overlaid with this mixture for several hours. Finally, fresh medium is added, and the culture is continued for a few days during which expression of the target gene is normally observed. No developments in large-scale transfections have been reported to date, probably because of the high costs of the liposome starting material required for lipofection. However, lipofection has been successfully carried out in spinner systems (Blasey et al., unpublished observation), which indicates the potential for scale-up.

DEAE-Dextran. One of the most widely utilized methods for the transfection of COS cells is the DEAE-dextran transfection method. The method is based on the formation of a complex between DNA and high molecular weight DEAE-dextran, followed by the natural uptake of this complex by the cells.

COS cells are first grown up to 50–70% confluence, and medium is exchanged for the transfection medium, which is based on DMEM medium supplemented with 10% NuSerum (Collaborative Research, USA). Normal serum is substituted by NuSerum because of the formation of toxic precipitates. The transfection medium also contains DNA at about 2 μg/mL, DEAE-dextran and chloroquine. Chloroquine is believed to prevent DNA degradation induced by the acidic pH of the endosomes during transport. A DMSO shock is also normally applied to the cells to increase the transfection efficiency. Cells are finally plated out and 24 h after transfection are subjected to either a medium exchange, or a split. The cells or the secreted product are then harvested 24–96 h posttransfection. The major drawback of this method with respect to scale-up consists of the repeated medium changes and washing steps. In addition, the exact timing for the DMSO shock needs to be empirically determined for each transfection.

Electroporation. Transfecting cells by electroporation is achieved by imposing a short, powerful electrical field, (i.e., a pulse) onto the cells. The pulse is generated by discharging a capacitor, and as a consequence, small pores are created into the cell membrane, allowing DNA to enter the cell by diffusion.

Practically, cells are grown to submaximal density, pelleted by centrifugation, and suspended into a conductive medium (e.g., normal culture medium or PBS) together with the plasmid DNA. The mixture is transferred to a sterile electroporation cuvette built with two electrodes embedded in two opposing vertical walls. The electrical pulse is applied, and the cells are then diluted back into fresh medium. Following are the parameters that determine the transfection efficiency:

- The field strength (V/cm), defined by the voltage applied and the distance between the cuvette electrodes
- The pulse shape, which can be square or with a logarithmic decay
- The time constant, which is directly related to the type of the electroporation cuvette
- The type of conductive medium and the cell density
- The concentration of DNA
- The number of electrical pulses applied

The optimal settings of these parameters (Table 1) vary between different cell types and other factors such as the pre- and posttransfection incubation times with the DNA, the temperature at which the electroporation is performed, and the physiological state of the cells. Electroporation has been used to transfect COS cells at the liter scale, however, the process was carried out in batch mode (i.e., by electroporating a large number of cuvettes), and this operation mode clearly limits further scale-up. Commercially available continuous electroporation systems are available, but results using these systems have not been reported.

Results

Transient Expression Scale-Up with COS Cells. Large-scale transient expression with plasmids carrying the SV40 origin includes several steps, shown in Figure 2: production of large DNA quantities, scale-up of cell growth, transfection, and protein production.

Production of Large Quantities of DNA. After transformation of competent *Escherichia coli* cells, colony expansion, and propagation in fermentors, the required gram amounts of plasmid DNA can be obtained. Plasmid DNA is first extracted by conventional alkaline lysis and then purified by selective DNA adsorption onto commercial resins (Qiagen, Germany). One of the main goals of plasmid purification is to achieve a low final endotoxin level, which is essential for a high transfection efficiency.

Culture Scale-Up. COS cells, can be grown up to large quantities by any of several appropriate methods, such as tissue culture flasks, roller bottles, microcarrier cultures, or Cellcube, but clearly, only the last two provide an efficient means of large-scale cell production.

Large-Scale Transfection. COS cells are collected, suspended at high density (10^8 cells/mL) in 0.5 mL of electroporation buffer (see earlier) containing 80 μg plasmid DNA/mL, transferred into electroporation cuvettes (interplate distance of 4 mm), and electroporated (Biorad Gene Pulser) with 2 pulses at 260 V, 960 μF, and infinite resistance.

Protein Production. The transfected culture is allowed to recover for 1 day in FCS-containing medium, which is then exchanged for low-protein, serum-free medium for the protein production phase. Daily medium exchanges allow one to continue the culture for up to 10 days posttransfection. A typical run yields several milligrams of pure protein (6).

Alternatively, COS cells can be grown up in rollerbottles, and the culture is overlaid with a DEAE transfection mix. After applying a DMSO shock, medium is added, and the production is carried out for several days. This method was reported for producing, at 800-mL scale, 1–2 mg/L of the scFv::IgC-kappa fusion protein (7).

Scale-Up of the Calcium Phosphate Transfection with HEK Cells. HEK cells grown in stirred systems can be transfected by the calcium phosphate method. HEK cells previously adapted to suspension growth are propagated in a stirred bioreactor. Once the cell density reaches 3×10^5 cells/mL, the calcium concentration in the medium is elevated to 10 mM. After 30 min transfection is initiated by addition of 20 mL/L of the calcium phosphate/DNA precipitate (at 500 μg DNA/L); 4 h later, calcium phosphate is diluted by simple addition of fresh culture medium, thus avoiding a separation step. In a typical example, 0.5 mg/L recombinant tPA was produced during a 5-day production period (8).

BACULOVIRUS–INSECT CELL EXPRESSION SYSTEM

Introduction

Baculovirus: Overview. The baculovirus–insect cell expression system is based on the infection of insect cell lines

Table 1. Typical Electroporation Conditions (used in our laboratory)

Cell line	CHO Ta	HEK293EBNA	COS M6
Voltage (V)	230	230	260
Capacitance (μF)	960	960	960
Volume (mL)	0.25	0.25	0.5
Resistance (ω)	∞	∞	∞
Cuvette: plate distance (mm)	4	4	4
Plasmid (μg/cuvette)	40	40	30
Electroporation buffer	Basal medium with 10% FCS	Basal medium with 10% FCS	Hanks buffered salt solution
Number of pulses	2	1	1–2
Typical results (% positive)	35–40%	35–40%	40–70%

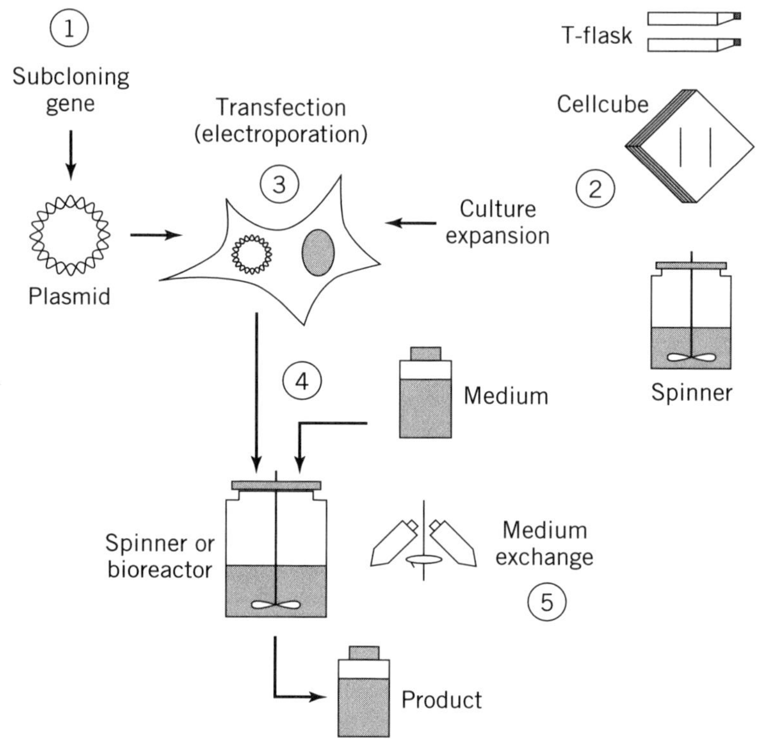

Figure 2. Large-scale COS expression. (1) The gene to be expressed is inserted into the multiple cloning site of the vector. COS cells are grown up in large amounts (2) and then transfected with the plasmid (3) by electroporation. After transfer into a spinner or bioreactor (4) the transfected culture resumes growth and produces the recombinant protein (4). Medium changes (5) extend culture life and allow long-term production (up to 10 days).

with a eucaryotic DNA virus: *Autographa californica* nuclear polyhedrosis virus (AcNPV). A powerful expression vector was originally derived from AcNPV by Summers (9) and Miller (10). It is helper independent and leads to very high expression of heterologous proteins. The expression system exploits strong viral promoters of nonessential genes such as the polyhedrin gene. It is also particularly well suited for coexpression of multisubunit proteins or complex, multicomponent assemblies such as viral capsids. The system is extremely safe because the virus does not replicate in mammalian cells and, in its nonoccluded form, cannot survive in the environment.

Virus Structure and Natural Life Cycle. AcNPV, a member of the Baculoviridae family is a large virus that infects mostly insects. The capsids are 40–50 nm in diameter and 200–300 nm in length. This large size explains in part why large insertions can be made into the viral DNA, a double-stranded circle of approximately 130 kb. The natural life cycle starts from polyhedra or occlusion bodies where virion particles are embedded into a crystalline protein matrix of polyhedrin (29 kDa). These polyhedra are ingested as part of their normal diet by insect larvae. The alkaline pH of the larva's mid-gut dissolves the polyhedrin matrix, releasing up to 100 virion particles per occlusion body. These particles fuse with the mid-gut epithelial cell plasma membranes, migrate to the nucleus where they replicate, and generate nonoccluded (NOV) or budded virus (BV). These budded virus particles are transported by the larva hemolymph to other tissues where they initiate secondary infection sites. In the secondary infection, virus penetrates the cells by receptor-mediated adsorptive endocytosis, a process that is also used during in vitro infection of cells in culture. The virus then infects other susceptible cells, is transported to the nucleus, and enters into a second round of replication that produces NOV again but also, later, occlusion bodies (OB), which close the cycle (Fig. 3). The polyhedrin present in the occlusion bodies can represent up to 50% of the total cellular protein at the peak of production.

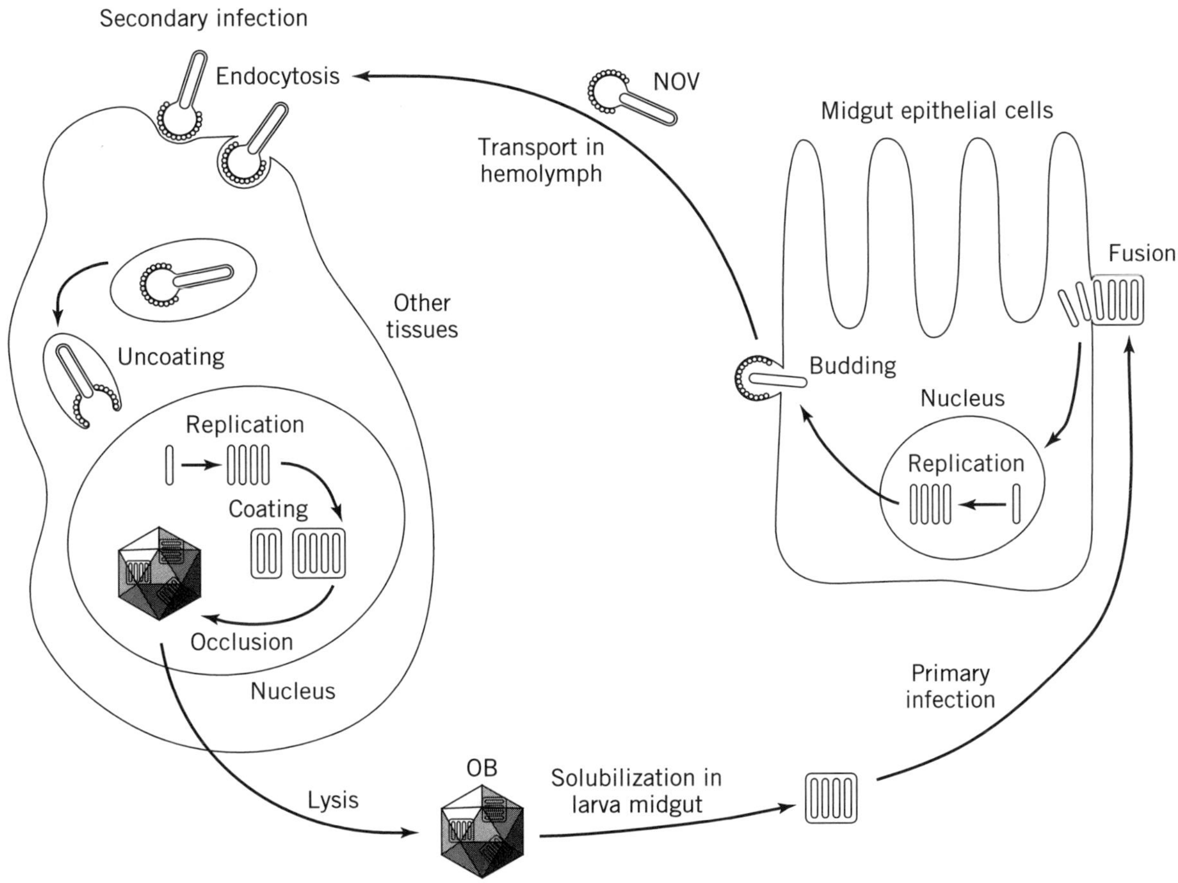

Figure 3. Baculovirus life cycle. The natural life cycle begins with primary infection: the insect larva ingests polyhedra or occlusion body (OB) particles as part of its normal diet. The alkaline pH of the larva's midgut dissolves the polyhedrin matrix. The released virion particles fuse with the midgut epithelial cell plasma membranes, migrate to the nucleus where they start replicating, and generate budded or nonoccluded (NOV) virus particles. NOVs are transported by the larva hemolymph to other tissues where they initiate secondary infection sites. In the secondary infection, virus penetrates the cells by endocytosis, uncoats, is transported to the nucleus, and enters into a second round of replication, which again produces OBs and closes the cycle.

Basic Principles and Protocols

Cell Lines and Media. Several lepidopteran cell lines are highly susceptible to AcNPV, but, two are most commonly used. One is derived from the fall armyworm, *Spodoptera frugiperda* (Sf9 clone), and is available from ATCC (CRL1711), the other is isolated from the cabbage looper, *Trichoplusia ni*, and commonly designated Tn5 (available as High Five cells from Invitrogen Corp.) (11). There are in fact, almost as many variants of these cell lines as laboratories working with them, and extreme caution should be taken when reporting or comparing results without a detailed knowledge of the particular cell line employed. All cell lines have very limited requirements for growth in tissue culture flasks (27 °C, no CO_2, adaptable to serum-free media), and most of them can be readily adapted to suspension culture.

Both cell lines can be used for the expression and production of recombinant proteins, but Sf9 is better suited for generating virus stocks and performing plaque assays because of its natural ability to stick to plastic, its higher susceptibility to the virus, and its ability to produce more virus particles per cell (Fig. 4).

Several media have been developed extensively for efficient growth and expression with the two cell lines described. Serum-containing media such as TNM-FH, IPL-41, and TC-100 are presently used less and less at the production stage, whereas serum-free and even protein-free media are available from commercial sources (JRH Biosciences, Gibco-BRL, Bio-Whittaker). Serum-free media offer clear advantages for large-scale manufacturing, particularly in the case of secreted proteins. When using serum-free media, cells should not be passaged for long periods of time (not more than 40–50 passages) to avoid gradual loss of susceptibility to the virus. Frozen stocks should thus be established to allow rapid and frequent expansion.

Transfer Vectors and Recombinant Virus Construction. In order to produce a recombinant baculovirus, and because the baculovirus genome is too large to be manipulated di-

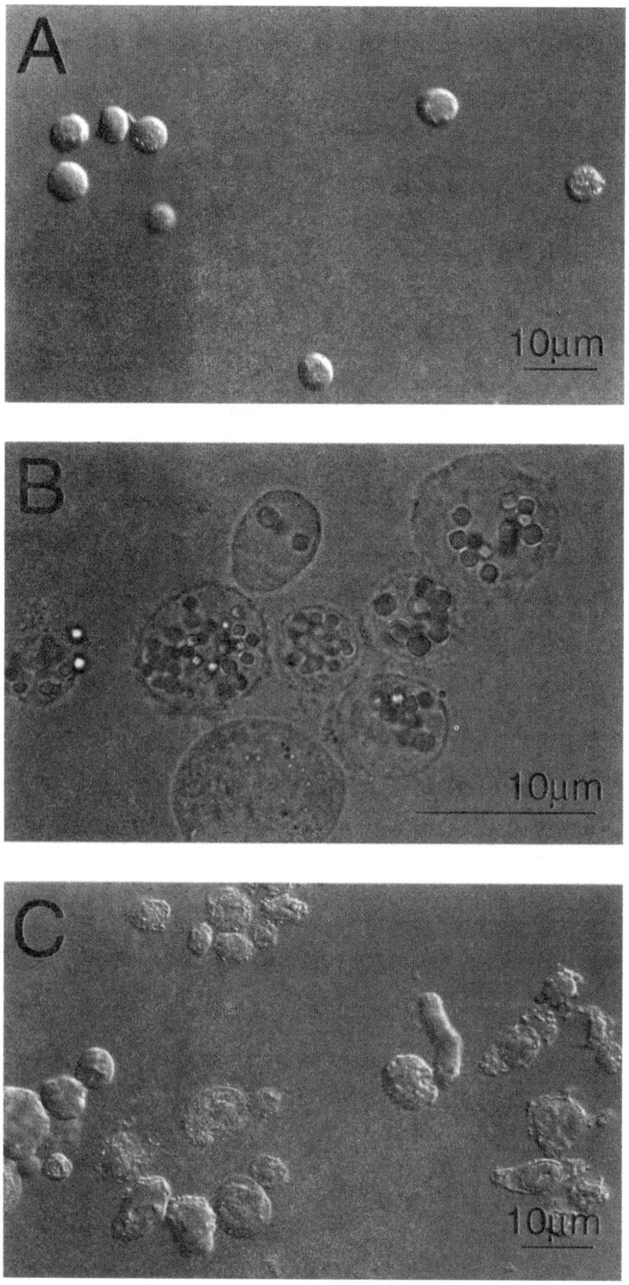

Figure 4. Noninfected Sf9 cells (**a**) and Sf9 cells infected with either WT (**b**) or with a recombinant virus (**c**). Noninfected cells appear round with smooth membranes; WT virus–infected cells contain many occlusion bodies in the nucleus, and recombinant (polyhedrin negative) virus–infected cells display an irregular, sometimes "sausage" shape with rough membranes.

rectly by restriction enzymes, the gene of interest needs to be inserted into a transfer vector (Figs. 5 and 6). Two different types of transfer vector are routinely used, depending on the recombination mechanism chosen:

Transfer Vectors for Recombination in Cells. Once assembled, the transfer vector generally contains one of the strong viral promoters, such as polyhedrin or p10, followed by the recombinant gene coding sequence and transcription termination signals and is flanked on both sides by AcNPV DNA sequences. These sequences allow homologous recombination with wild-type, linearized DNA. A wide variety of transfer vectors of this type have been constructed over recent years (12,13) and offer the possibilities of varying promoters, of multiple gene inserts, and of insertion of tags, fusion, partners, secretion signals, and other helpful features aimed at facilitating detection, expression, and purification.

The transfer vector and wild-type baculovirus DNA are then cotransfected into cells, and the foreign gene coding cassette is inserted into the viral DNA by homologous recombination. Because homologous recombination occurs at a low frequency (1–5%), the virus preparation needs to be purified by plaque assay in order to isolate a pure recombinant virus. It is possible to dramatically increase the proportion of recombinant over wild-type viruses (up to 90%) by using linearized and partially deleted viral DNA, as in the BaculoGold™ system (Pharmingen) (13), or in the MaxBac™ kit (Invitrogen Corp.), but this approach still does not eliminate the requirement for virus plaque purification.

Transfer Vectors for Recombination in E. coli. A newly developed system, BAC-TO-BAC, permits the rapid engineering of the recombinant virus in *E. coli* (14). It is based on site-specific transposition of the DNA cassette into the full viral DNA maintained in *E. coli* as a baculovirus shuttle vector (bacmid). This bacmid contains replication and kanamycin resistance functions and behaves like any common small plasmid in *E. coli*. In addition, a *lacZα* region with a target site for transposition (mini *att*Tn7) was inserted into the bacmid, which allows transposition of the recombinant cassette from the transfer vector. The transposition machinery is provided in trans by a helper plasmid (tet resistant). This strategy allows fast, easy, and reliable selection of recombinant bacmid colonies on chromogenic substrate–containing plates. Most importantly, it eliminates the need for plaque purification of the recombinant virus. The recombinant bacmid DNA is prepared from the positive *E. coli* colonies and used to infect insect cells in culture in order to regenerate a pure recombinant virus stock.

Amplification and Titration. Titration of the virus is commonly performed by a standard plaque assay technique with Sf9 cell monolayers in presence of serum-containing media. This technique is laborious but gives the most accurate estimate of the number of infectious particles per unit volume (or plaque forming units) in the virus stock.

Table 2. Insect Cell Growth Parameters

Cell line	Medium	Td (h^{-1})	Max density (cells/mL)	Ref
Sf9	IPL41 + FCS	0.02	$5\ 10^6$	8
Sf9	SF900II	0.035	$1.5\ 10^7$	19
Sf9	TNMFH + FCS	0.032	$4.7\ 10^6$	1
Sf9	Excell 400	0.034	$2.9\ 10^6$	2
Tn5	Excel 401	0.030	$4\ 10^6$	3
Tn5	Excel 405	0.038	$6\ 10^6$	20
Tn5	SF-1	0.035	$5\ 10^6$	5

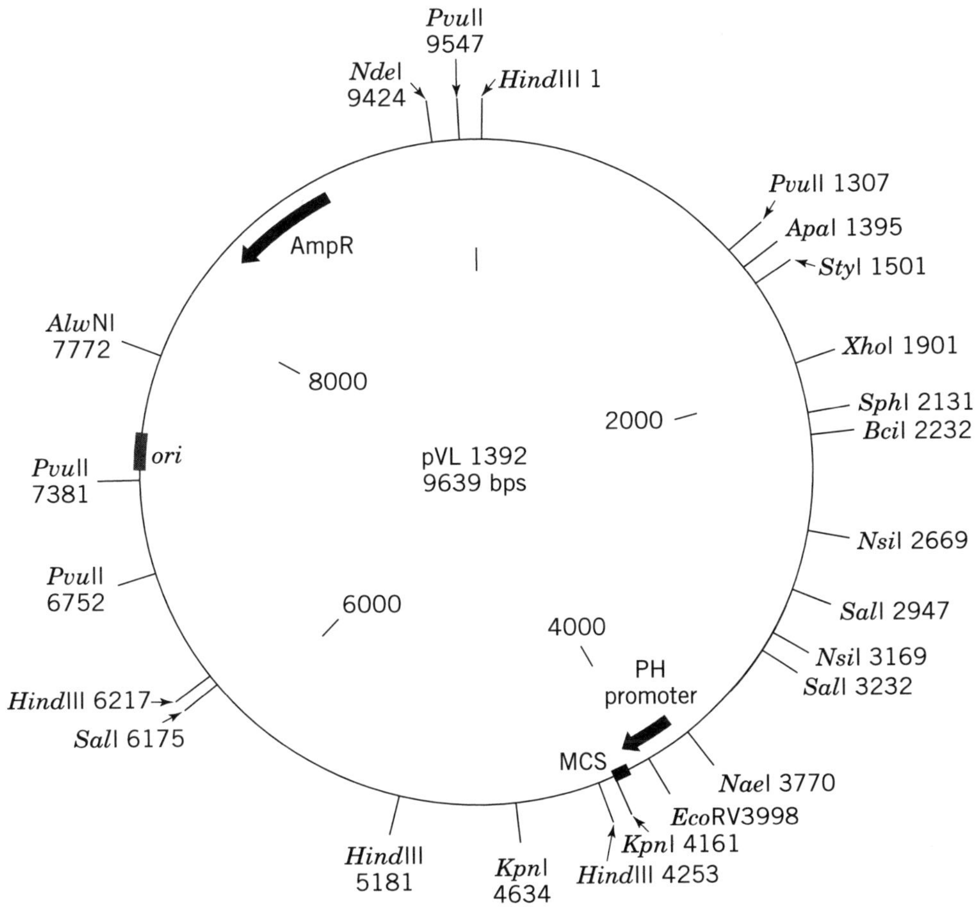

Figure 5. pVL1392 transfer vector. The pVL1392 baculovirus transfer vector was built from pUC8 with its origin of replication (*ori*) and ampicillin resistance gene (*Amp*ʳ). It contains the complete polyhedrin gene promoter sequence (PH promoter) but lacks part of the polyhedrin coding sequence. A multiple cloning site (MCS) region has been inserted 37 nucleotides downstream of the ATH polyhedrin start codon, which has been changed into an ATT. In a variant of this vector, pVL1393, the MCS contains the same restriction sites in the opposite orientation. *Source*: Reprinted from Ref. 12, with permission of Pharmingen, San Diego.

Once obtained in a pure form, the virus isolate must be expanded to larger volumes if required for manufacturing on a large scale. This can be accomplished by infection of a culture at a low multiplicity of infection (MOI = 0.01–0.1) so as to reduce the risk of producing defective interfering particles (DIPs) (15,16) and by harvesting the supernatant of the culture at the end of the infection phase (4 days). Titers between 10^7 and 10^8 PFU/mL are commonly obtained at this stage. Virus stocks can be stored at 4 °C, in the dark, for several months, although titers may be significantly reduced upon prolonged storage. A small aliquot can also be kept frozen at −70 °C for long-term storage.

Growth and Infection. Small-scale growth of the host insect cells and infection are easily obtained in stationary tissue culture flask cultures at 27 °C (15,17). Large-scale cultures can be performed with cells in suspension in spinner flasks or in bioreactors equipped with appropriate controls for temperature, dissolved oxygen, agitation, and a pH around 6.2 (18). Growth and product yields are fairly insensitive to bioreactor design but highly sensitive to reactor operation mode and other process parameters such as dissolved oxygen, cell density, and viability at onset of infection, MOI, and nutrient concentrations.

Insect cells have been known for a long time to be particularly sensitive to mechanical stress, in particular at the gas–liquid interface formed during the bubble-bursting process. This explains many of the initial failures in trying to grow insect cells in airlift bioreactors. This problem can be alleviated by supplementing the growth medium with appropriate surfactants such as Pluronic F-68 at a concentration of 0.02%, which prevents cells from being adsorbed on air bubbles.

Normally growth occurs with a doubling time of less than 24 h, and cell density reaches a plateau at 10^7 cells/mL in conventional media (Table 2). Infection is started by simple addition of a virus stock to the culture in full exponential growth (1–2 × 10^6 cells/mL) and at high viability as determined by the Trypan blue exclusion method. A

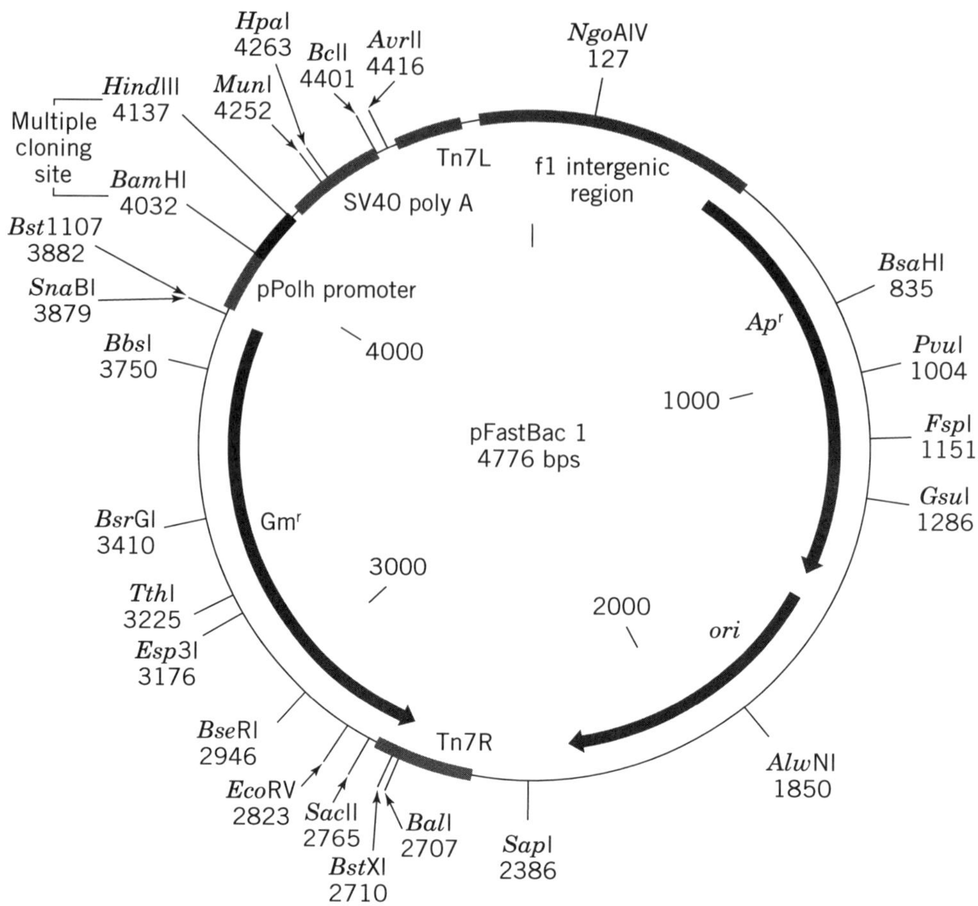

Figure 6. pFast-Bac1 transfer vector. The pFastBac1 vector contains the pUC origin of replication (*ori*), the ampicillin resistance gene (Ap^r), and transposon sequences (Tn7L, Tn7R) to produce recombinant bacmid DNA in *E. coli*. For selection and expression in insect cells, a gentamicin resistance gene (Gm^r), the polyhedrin promoter, SV40 polyadenylation signal, and a multiple cloning site are present. *Source*: reproduced from the GIBCO-BRL 1997–1998 product catalog, with permission from Life Technologies, Inc.

high MOI (5–10) guarantees a fast and uniform infection of all cells and maximizes the final product yield, but the amount of virus should not exceed 10% of the culture volume to reduce the addition of cell debris and spent medium. (Fig. 5).

The level of nutrients present in the medium at the time of infection is probably one of the most important parameters of the process. The nutritional state of the cells conditions the success of the infection and the final product yield, and the lack of a proper control of this parameter explains most observations relating the final expression level to the cell density. It has been clearly demonstrated by several groups that infection could actually be carried out at high cell densities (above 10^7 cells/mL), provided that the nutrient level was kept stable by either one or a combination of strategies, such as medium replacement, concentrated nutrient addition, and perfusion. (Fig. 6).

Another key element in large-scale insect cell processes is oxygen transfer, due to the high natural specific consumption rate (around 5×10^{-17} M/cell/s) by the insect cells. In the early infection process oxygen transfer becomes even more problematic because the cell's requirement can double in the first 24 h of infection. Thus, sparging pure oxygen, increasing agitation rates, and gas flow rates are the most efficient means of maintaining a stable dissolved oxygen concentration (set point between 20 and 50% of air saturation) throughout the whole process.

Results

Timing of Expression. The timing of expression is directly related to the promoter used in the construction of the recombinant virus. The kinetics of accumulation of the recombinant protein in the culture is a result of two inverse phenomena: translation of the recombinant gene from the viral DNA inside the infected cells, and progressive cell death due to viral induced cell lysis. Peak expression is usually observed between 48 and 72 h postinfection when the target gene is under control of very late promoters such as polyhedrin or p10. It may be advantageous, however, to drive expression from earlier promoters, such as the basic protein or *pcor* promoters, which leads to max-

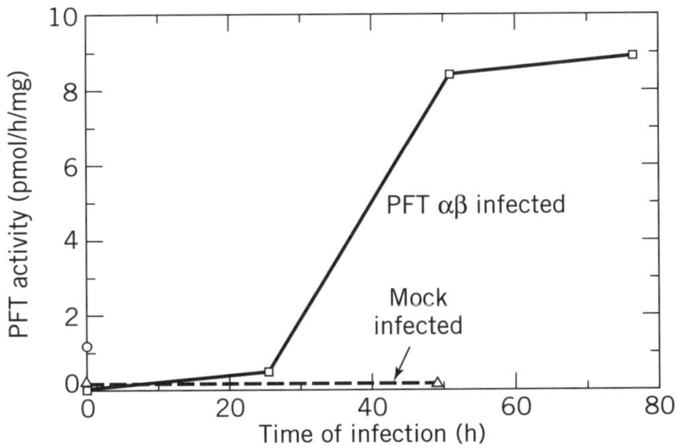

Figure 7. Time course of PFT production by baculovirus–infected Sf9 cells. Multiple cultures of Sf9 cells (4 × 10^6) were either infected with no virus (i.e., mock infected) (○), infected with wild-type virus (□), or infected with recombinant baculovirus containing the coding regions for both the a- and b-subunits of the rat brain PFT (△). Following infection, cells were incubated for the indicated times, whereupon they were harvested by centrifugation and soluble extracts prepared. The extracts were assayed for PFT activity as previously described (21). *Source*: Reprinted from Ref. 21, with permission of *Journal of Biological Chemistry*.

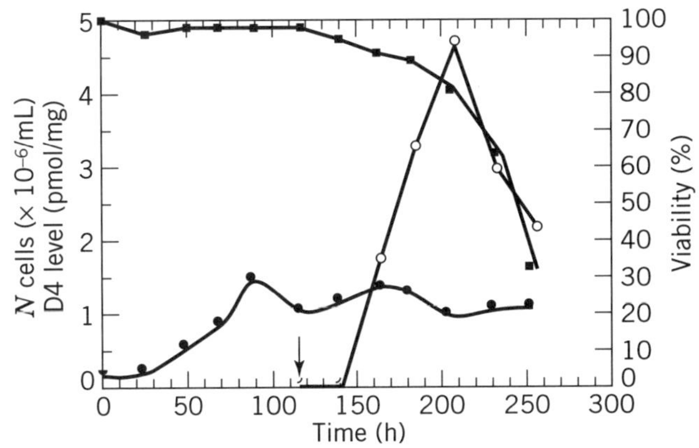

Figure 9. Expression of human dopamine D4 receptors in Sf9 cells. Sf9 cells were propagated in 15-L bioreactors at 27 °C, in TC100 medium (Gibco-BRL) supplemented with 10% FCS and 0.1% Pluronic F68 (BASF, Germany). Dissolved oxygen was controlled at 50% air saturation by pure oxygen sparging on demand. Once the culture reached a cell density of 1.5 × 10^6/mL (•) with a viability above 95% (■), fresh medium was added together with an aliquot of the virus stock so as to obtain a cell density of 1.0 × 10^6/mL and an MOI of 10 (arrow). Every 24 h postinfection, a sample was withdrawn for receptor binding activity measurements (○) with [^{3}H] spiperone, as previously described (24).

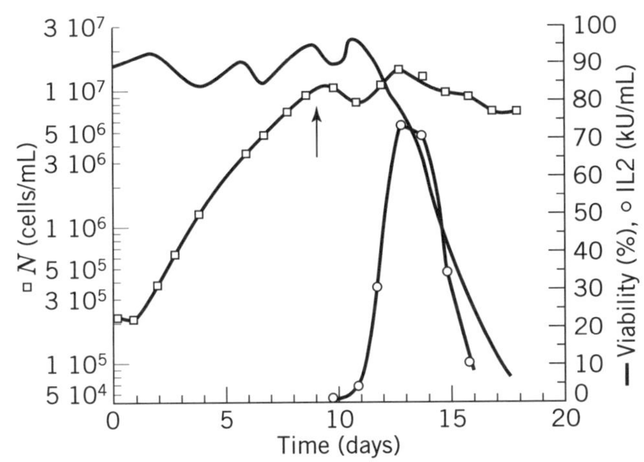

Figure 8. High-density perfusion culture of Sf-21 cells in serum-free Ex-Cell 401 medium. After 9 days, cells were infected with 1.8 × 10^8 AcNPV/pVL30 corresponding to a MOI of 0.015 (arrow). Maximum productivity was reached 4 days postinfection at a viability of 80%. *Source*: Adapted from Ref. 50, with permission.

imal expression earlier when cell viability is still above 80%.

Expression Level. A high expression level of recombinant proteins is commonly achieved with the baculovirus system; however, it is heavily dependent upon the nature and specific characteristics of each protein and each construct. Clearly, the expression level is also affected by the type of promoter used to drive expression of the foreign gene. The highest levels reported are in the range of several hundred micrograms of active protein per milliliter of culture, as in the case of expression of *β*-galactosidase in monolayer cultures (10). Some cytoplasmic or secreted proteins can also be produced in the 100-mg/L range in large-scale cultures (21,22) (Figs. 7 and 8), but more complex proteins such as multisubunit complexes and single or multiple transmembrane proteins are usually produced at lower levels, typically 0.1–5 mg/L (Fig. 9) (24).

Posttranslational Events. One of the strong arguments for the baculovirus is the ability of insect cells to process nascent polypeptides in a way that is similar to that mammalian cells, and most mammalian proteins produced by this system are virtually identical to their native equivalent in terms of their biochemical or biological activity. It is well established, however, that N-glycosylation performed by the most commonly used cell lines, Sf9 or Tn5, is significantly different from mammalian N-glycosylation, with most proteins being processed to the high-mannose structure steps. However, glycosylation by other specialized insect cell lines such as *Estigmene acrea* (25) appears to be more closely related to that of mammalian cell lines. This area is of course of considerable interest and is still the subject of intensive research.

Due to the availability of several promoters, and because large sequences of DNA can be inserted in the viral genome, the baculovirus is particularly well suited to the production of multisubunit complexes. Mammalian viral capsids or viruslike particles can also be reconstituted inside the cell, and this property is being exploited for the development of novel cattle and human vaccines.

Conclusion and Perspectives

The baculovirus system has shown to be a very useful tool for the production of numerous recombinant mammalian

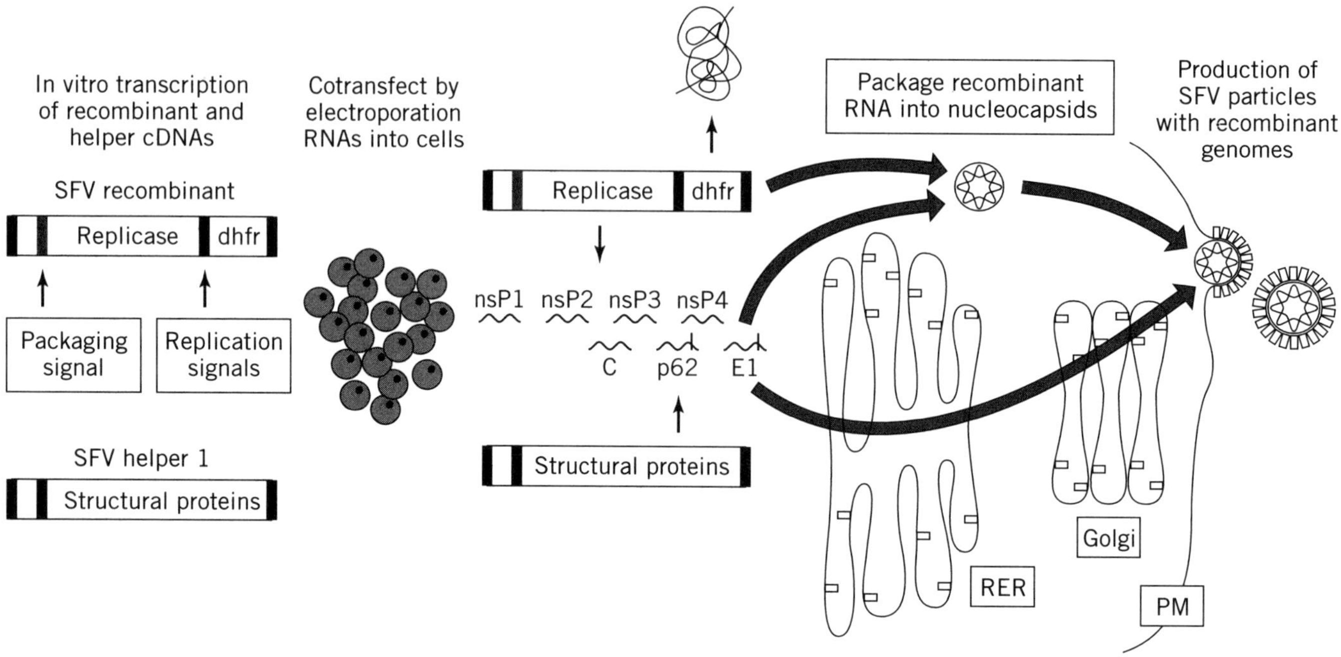

Figure 10. In vivo packaging of recombinant SFV RNA transcripts into infectious particles. After cotransfection of RNA transcripts from the expression and the helper vector, the replicase, coded on the recombinant vector, drives RNA replication and transcription for both plasmids. Structural genes from the helper are translated into proteins. While the capsid protein stays cytosolic and binds RNA to form the nucleocapsid, the membrane proteins undergo posttranslational modifications. E1 and p62 are dimerized in the endoplasmic reticulum and are transported to the plasma membrane where virus budding occurs. Because the viral packaging signal is found only on the recombinant RNA, selective packaging of the pSFVdhfr occurs and as a result yields nonreplicative infectious SFV particles. *Source*: Reprinted from Ref. 26, with permission of Nature-Biotechnology, N.Y.

proteins for research, animal vaccination, and clinical applications. Several biological products produced with this system are being tested in therapies or preventive vaccination, and it seems likely that at least some of them will soon obtain approval for marketing. The major advantages of the system are high-level expression, safety, speed, coexpression of several proteins, easy scale-up, and low cost, which explain its widespread use in particular for the production of research material. Future work is needed to improve control over posttranslational events, which is the only area of difference with more conventional mammalian cell–based expression systems.

SEMLIKI FOREST VIRUS

Introduction: Alpha Virus–Based Expression Systems

Alphaviruses such as the Sindbis virus (SIN) or the Semliki Forest virus (SFV) are very small enveloped viruses containing single-stranded RNA of positive polarity. The 12-kb RNA is capped at the 5′ terminus and polyadenylated at the 3′ terminus. SIN and SFV are the most-studied alphaviruses, and both have been used for the expression of various heterologous genes. Upon infection, the 5′ two-thirds of the RNA is translated into polypeptides, which are processed into four nonstructural proteins (nsP1–nsP4). The nonstructural proteins include transcriptase, which produces a negative-strand copy of the RNA genome, serving as a template for the generation of positive-strand RNA copies. The 3′ third of the genome codes for the structural proteins (i.e., for the capsid C and two membrane proteins, p62 and E1). After translation, the capsid protein C binds to genomic RNA within the cytoplasm. The membrane proteins are translocated into the endoplasmic reticulum, where they form heterodimers and allow virus budding from the cell surface. The released virus can then be taken up by receptor-mediated endocytosis and infect a wide range of mammalian cells.

Semliki Forest Virus Expression System

The broad host range of the SFV expression system has made it a particularly attractive system for functional studies and investigation of the biological role of proteins. It was originally developed by Liljeström and Garoff (26), who split the SFV genome into two plasmids: the cloning vector, pSFV1, which carries the replicase genes; and the pSFV-helper, which encodes the structural proteins of the virus (Figs. 10 and 11). In order to increase the level of biological safety, they later also introduced a mutation into the spike protein–encoding gene, generating pSFV-helper II (27). The mutation prevents the natural proteolytic processing of the spike protein into an active protein, which

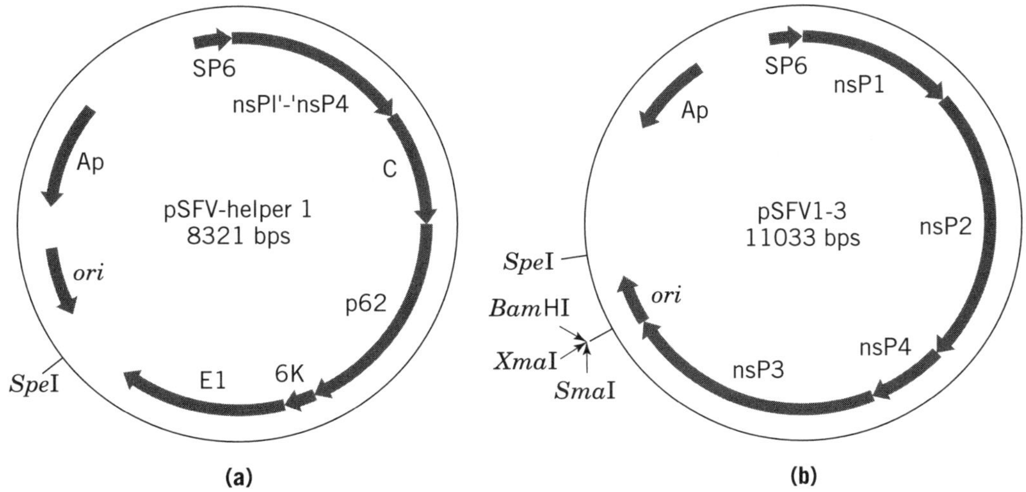

Figure 11. pSFV Helper1 (**a**) and pSFV1-3 plasmids (**b**). Construction of the SFV expression plasmids was done by deleting most of the coding region for the structural proteins (C, p62, 6K, E1) from the cDNA clone, which spans the complete SFV genome. The vectors retain the promoter region of the 26S subgenomic RNA (at the end of the nsP4 region) and the last 49 amino acids of the E1 protein as well as the complete noncoding 3′ end of the SFV genome. Downstream of the 26S promoter, a polylinker (*Bam*III, *Xma*I, *Sma*I) was introduced. The SP6 promoter, which is followed by the replicase (nsP1-nsP4), allows in vitro generation of full-length SFV RNA transcripts. The eukaryotic origin of replication (*ori*) and antibiotic resistance (Ap) allow for plasmid production in *E. coli*. A unique *Spe*I restriction site was inserted for plasmid linearization. The pSFV-helper1 represents the SFV genome cDNA being deleted of most of the nonstructural region but carrying all the structural genes (C, p62, 6K, E1). The pSFV-helper2 codes for a cleavage-deficient variant of the spike protein. *Source*: Reprinted from Ref. 26, with permission of Nature-Biotechnology, N.Y.

mediates entry into the cell. However, the processing can be obtained by chymotrypsin cleavage, thus restoring full infectivity of the virus stock (27). High expression levels are obtained due to the extremely efficient SFV 26S promoter used and also to the high number of copies of RNA (200,000) present in each infected cell. For example, levels up to 25% of total cell protein have been reported for β-galactosidase (26), and several million neurokinin 1 receptors per cell (28). Because the virus can be used to infect almost any mammalian cell, recombinant proteins can be produced with posttranslational modifications based on the specific processing capacities of the host cell selected (Table 3).

Infection and protein expression properties. SFV can infect a broad range of host cells; however, the level of expression differs between cell lines: BHK, CHO, and COS cells yield higher levels of expression than other cells such as NSO, RPMI 8226, and HeLa. This was clearly demonstrated for different types of proteins: a multitransmembrane receptor (neurokinin 2 receptor), a secreted protein (soluble placental alkaline phosphatase), and a cytosolic protein (β-gal).

At least in BHK cells it is known that the optimal pH is different for SFV infection (pH 6.9) and protein production (pH 7.3). The MOI also clearly modulates the expression level (30): a higher MOI results in elevated expression levels.

Transfecting cells with RNA generated in vitro from the cloning vector also leads to RNA replication and subsequent expression of the subcloned gene, a method that should be applicable to the expression of oncogenes and other hazardous genes. However, infection with recombinant packaged SFV particles is still a more efficient way to deliver recombinant RNA inside the cells.

Basic Principles and Protocols

Virus Generation. After subcloning the gene of interest into the multiple cloning site of pSFV, the plasmid is linearized to prepare for in vitro transcription and RNA production. BHK cells are then cotransfected with the recombinant and helper RNAs (Fig. 12), which initiates the in vivo packaging process. It is important to note that the packaging signal is present only on pSFV-1 and is absent from the pSFV-helper vector. As a consequence, infectious but nonreplicative recombinant virus particles are produced, and a high titer (approximately 10^9 virus particles/mL) is harvested 24 h after transfection.

Protein Expression. As a first step, BHK cells are grown to confluence in T-flasks. The virus is activated by incubation with chymotrypsin, which is inactivated after 20 min by addition of aprotinin. Virus is transferred into the culture to give a MOI of 30 for infection. Product can be detected by pulse labeling after 4–6 h postinfection while the host cell protein synthesis is dramatically reduced or shut down and cell growth is arrested. Maximum product concentration is usually reached 1 day postinfection but

Table 3. SFV-Expressed Recombinant Proteins

Protein group	Protein	Ref
Enzymes	β-Gal (up to 25% of cell protein)	26
	Mouse dehydrofolate reductase	26
	Chicken lysozyme	26
	Terminal deoxynucleotide transferase	29
	Human cyclooxygenase 2	30
	Rat and human catechol *O*-methyltransferase	31
	Wild-type and mutant thyroid peroxidase (TPO)	32
Receptors	Dopamine receptor 3 ($>1 \times 10^6$ receptors/cell)	33
	Human neurokinin receptor 1 receptor (2×10^6 receptors/cell)	28
	Human neurokinin receptor 2 receptor	29
	Human transferin receptor	33
	5HT-3 receptor	26
	5HT-1D alpha receptor	34
	Adenosin A3 receptor	35
	Human prostaglandin receptor E4	35
	Dopamin 3 receptor mutant	36
	Human P2×1, rat P2×2	37
	Human P2×1, rat P2×2, rat 2×3, rat 2×4	38
	GABA subunits	39
		40
Virus	Human papilloma type 16 capsid protein	41
	Hamster measles virus protein HNT-PI F and Ph26 H	42
	Influenza virus nucleoprotein	43
	HIV gp120, gp160	44
	Moloney murine leukemia virus spike protein	45
	Vesicular stomatitus virus glycoprotein G	46
	GB virus-C E2 envelope protein	47
	GP4 structural glycoprotein of Lelystad virus	48
	Respiratory-cynytial virus glycoprotein-G	49
	Murine leukemia virus envelopes	50
Others	Rab 8	51
	Rab 12, 22, 24	52
	VIP-21caveolin	53
	Yeast syntaxin Sso2p	54
	Rabbit pIgR	55
	Human APP	56
	Plasminogen activator inhibitor type 2	57
	Mitochondrial precurser protein	58
	APOE isoforms	59
	P35/p40 IL-12 heterodimer	60

may vary depending on the proteins expressed. The culture is harvested by centrifuging the cells after detachment.

Semliki Forest Virus Scale-Up Results

BHK cells are grown in single cell suspension to 10^6 cells/mL using a stirred-tank reactor equipped with a Vibromix stirring system and oxygenation by silicone tube gassing (Chemap/MBR, Switzerland). This agitation method proves to be optimal for minimizing cell aggregate formation, and the oxygenation method is preferred over direct gas sparging into the bulk of the liquid so as to eliminate any aerosol formation.

Before infection, 90% of the medium is exchanged by external cross-flow filtration. The virus is activated separately and the culture infected by addition of the activated virus stock. At 16 h postinfection the culture is harvested by centrifugation and the recombinant protein purified.

Recent results of large-scale protein production with SFV have recently been described:

- Cyclooxygenase-2: 16 mg microsomal protein per 10^9 cells, activity: 3,942 pg $PGE_2/\mu g$ microsomal protein (30)
- 5HT3 receptor: 1–2 mg 5HT3 receptor protein per liter culture, or 25–50 pmol binding sites per cell (19)

Commentary

The SFV expression system has quickly found wide acceptance due to the high level of expression achievable combined with the short time scale for production. It is preferentially used to express receptors and intracellular products.

Virus generation for large-scale infection is still laborious, in particular the RNA generation and transfection of cells with the two RNAs. This constitutes the main difference with other viral-mediated expression systems.

Biological safety of the SFV expression system has been given significant attention. The virus is nonreplicative and only conditionally infective. However, because the culture

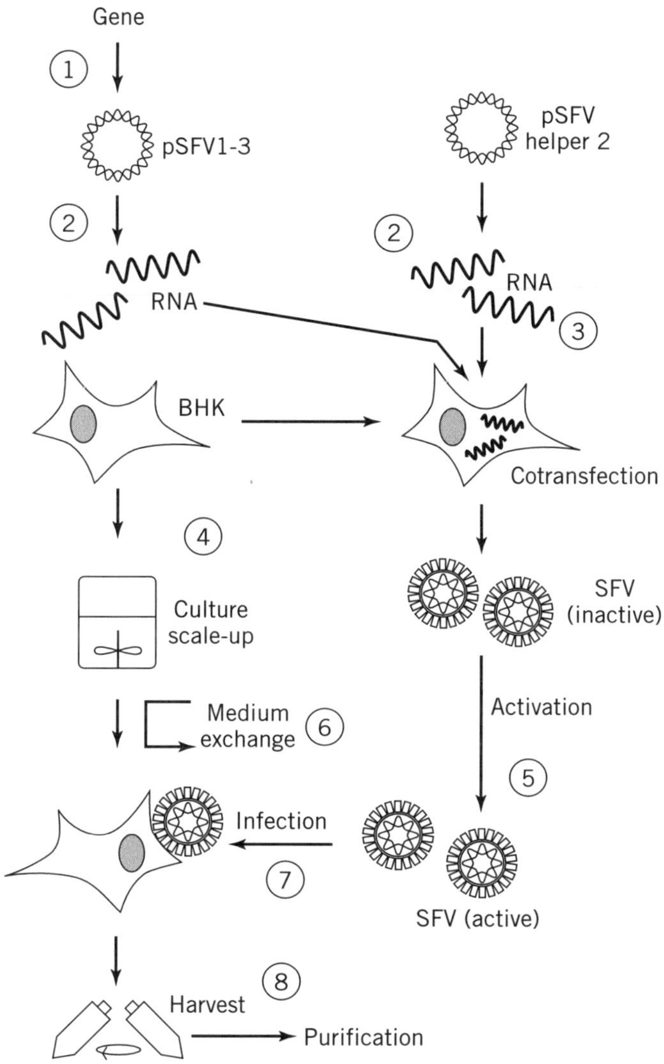

Figure 12. SFV infection process. Large-scale recombinant protein production with the Semliki Forest virus system. The gene to be expressed is subcloned (1) into the recombinant vector pSFV1-3. In vitro transcribed RNA (2) from the recombinant plasmid pSFV1-3 and the pSFV-helper-2 are cotransfected (3) into BHK cells; virus is harvested 24 h later. BHK cells are grown up to 10^6 cell/mL as suspension culture in a mechanically stirred reactor (4). Virus is activated (5), the culture medium is exchanged (6), and the culture infected (7). Cells are harvested (8) 1 day after infection by centrifugation, and the protein is purified.

at the time of harvest potentially contains active virus, it is recommended that this expression system and the handling of products obtained by this technology be operated under safety category two or higher.

OTHER VIRAL-MEDIATED EXPRESSION SYSTEMS

Adenovirus

Introduction. Isolated in the early 1950s, adenoviruses have recently received a renewed interest for their use as vectors for DNA delivery to cells in vivo in gene therapy applications, as potential vaccine vectors, or as expression vectors for heterologous genes in vitro. In this section, the latter application of adenoviruses is reviewed.

Adenovirus Life Cycle. The adenovirus life cycle starts with an early phase in which regions E1–E4 and L1 of the linear double-stranded DNA (36 kb total length) are actively transcribed from the early (E) and major late promoters (MLP), respectively (61). Viral replication begins 8 h after infection, and host transcription as well as protein synthesis are gradually shut off as the cycle enters the late phase of infection. In this phase, late transcripts from the MLP, L2–L5 appear together with structural proteins. The MLP promoter drives high levels of expression of these late proteins, which can account for up to 50% cellular protein. The cycle also leads to cell death and to the release of high virus titers (10^8–10^9 PFU/mL).

Expression Vectors. The extensive knowledge accumulated over the years on adenovirus transcription regulation allowed the engineering of adenovirus vectors (Ad) modified for the expression of heterologous genes (61). The first modification consists in deleting the early regions E1 and E3. Whereas E3 is not essential to viral replication in cell cultures, E1 deletion renders the virus replication incompetent, and this function must then be provided in trans by the host cell, as is the case for 293 cells, which constitutively express the E1 protein. The second modification consists in introducing an expression cassette in place of the deleted E1 region. In the cassette, the recombinant gene is placed under control of an additional (or ectopic) MLP and follows the tripartite leader and splicing signal sequences, thus taking advantage of the strong natural positive regulation of the promoter. Alternatively, the recombinant can be placed under control of an exogenous promoter such as CMV (62) or the natural MLP, but in this latter case, the recombinant virus must then be treated as a helper-dependent virus. Very recently, a new series of vectors, pAdBM5, was constructed (63) with significant improvements in the ectopic MLP region by insertion of enhancer sequences that stimulate the promoter's transcriptional activity. The fact that the virus can be propagated in suspension cell cultures is also of considerable advantage over other similar processes.

Results. Several heterologous proteins have been expressed from Ad vectors at high levels in 293 cells. Most constructions have been done in the Ad5 serotype background, and with a few exceptions, such as DHFR, CAT, factor IX or the vitamin D3 receptor, all were made for expressing viral proteins. SV40 large T antigen was expressed at high levels and could be purified to homogeneity, Herpes simplex virus type 2 ribonucleotide reductase subunit R2 was expressed in 293 cells at 4–5% of cellular proteins, and the polyomavirus middle T antigen and the hepatitis B surface antigen were also expressed at high levels in the same cells. The pAdBM5 vector has been reported to yield even higher levels of expression (10–20% of cellular

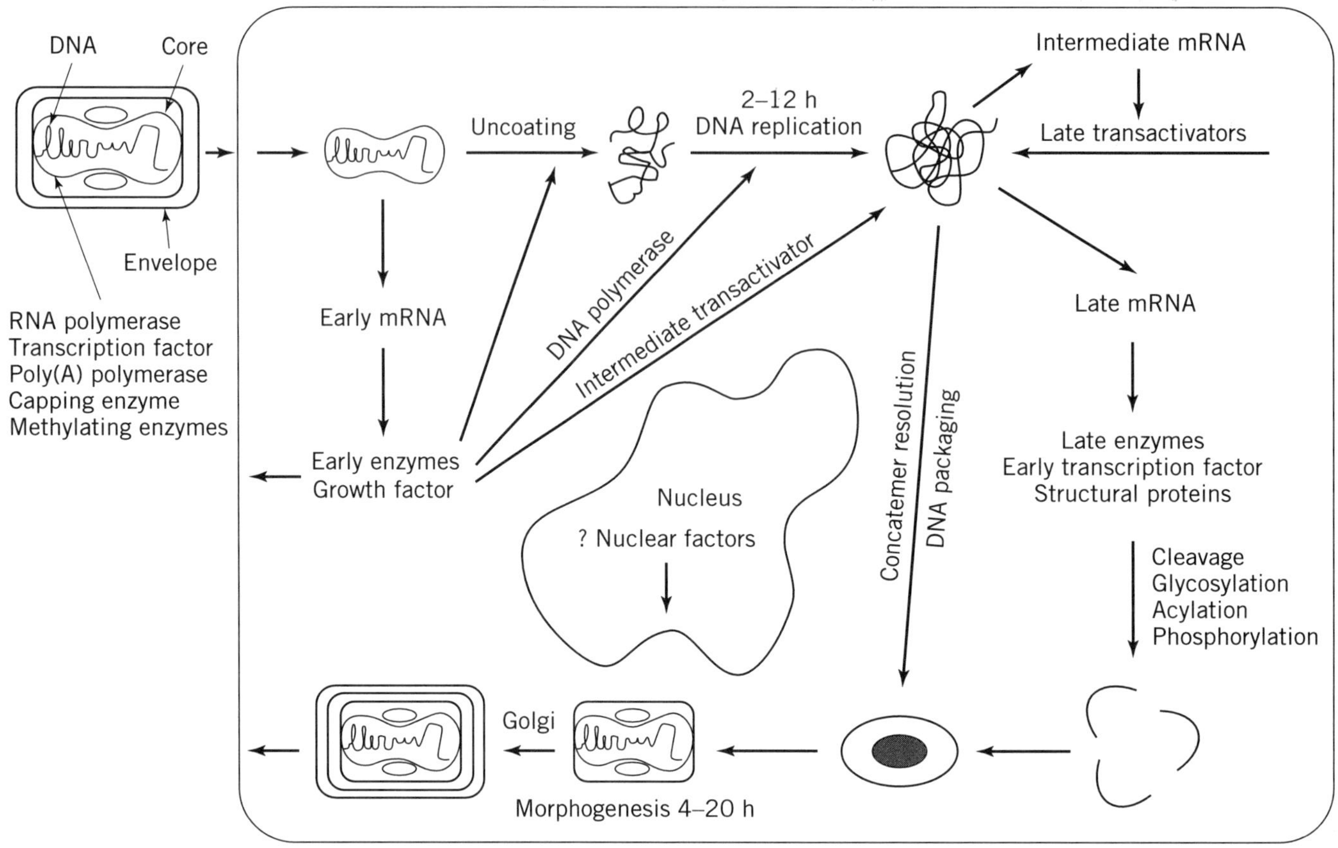

Figure 13. Replication cycle of vaccinia virus. After entry of vaccinia virus into cells, early genes are expressed, leading to secretion of several proteins (including a growth factor), uncoating of the virus core, and synthesis of DNA polymerase (and other replication proteins), RNA polymerase subunits, and transcriptional transactivators of the intermediate class of genes. After DNA replication, intermediate mRNAs are made, some of which encode late transcriptional transactivators, leading to expression of late genes. The latter encode structural proteins, enzymes, and early transcription factors, which are packaged in the assembling virus particles. Some of the mature virions are wrapped in Golgi-derived membranes and are released from the cell. The bold arrows indicate the products that exit the cell. *Source*: Reprinted from Ref. 66, with permission of Raven Press, N.Y.

Table 4. Vaccinia Large-Scale Production

Product	System	Cells	Company	Ref.
HIV-1 rgp160	150-L microcarrier	BHK	Pasteur Merieux	67
HIV-1 gp160	1200-L microcarrier	VERO	Immuno AG	68
Human protrombin	1200-L microcarrier	VERO	Immuno AG	69

proteins), with a tyrosine phosphatase (PTP1C) used as a model in bioreactor-scale cultures of 293 cells (64,65).

As is normally expected, posttranslational processing of Ad-derived recombinant proteins has been shown to be identical to the native protein.

Vaccinia Virus

Vaccinia Virus Characteristics. The vaccinia virus, an *Orthopoxivirus*, belongs to the family of Poxviridae. It has a double-stranded DNA of nearly 200,000 bp and replicates in the cytoplasm of the host cell (Fig. 13). The lipoprotein envelope contains the core structure, including an RNA polymerase, a transcription factor, capping and methylating enzymes, and a poly(A) polymerase.

Infection of cells with the vaccinia virus leads to the suppression of host cell protein synthesis, and cytopathic effects occur. Up to 5,000 virus particles are produced per cell, which, depending on the virus strain, will stay cell associated or can be found in the culture medium.

Table 5. Transient Expression Systems

	Time scale	Advantages	Disadvantages	Applications	Yields
			DNA vectors		
Large T–expressing cell lines (e.g., COS)	3–4 weeks	Established technology Simple to execute Scale up to liter scale demonstrated	Large DNA and cell quantities required for transfection Large-scale transfection Short expression time after transfection (1–2 weeks), overgrowth by non producers Safety category 2–3	Functional and structural studies	0.1–3 mg/L Up to 10 mg protein
HEK293 EBNA	4–8 weeks	Simple to execute Culture scale-up under selective pressure Transfection requires only small quantities of DNA and cells Stable under selection Potential selection of clones Establishing seed: bank possible	Safety category 2	Functional and structural studies, target validation, preclinical studies	0.1–10 mg/L 100 mg protein
			Viral vectors		
Baculovirus	6–8 weeks	Established technology Easy scale-up Safe viral system	Lytic system Nonmammalian glycosylation	Functional and structural studies, target validation, preclinical studies	0.5–30 mg/L gram quantities
Semliki Forest virus	2–6 weeks	Exceptionally short time from gene to product High yields Very broad host range	Safety category 2 for activated virus Limited virus scale-up Lytic system	Functional and structural studies, target validation	Up to 8×10^6 cell Up to 100 mg
Vaccinia	2–8 weeks	Easy scale-up to very large scale Virus simple to scale-up Broad viral host range	Pathogenic: safety category 2 Lytic system	Functional and structural studies, target validation, preclinical studies; vaccines	Around 0.1–5 mg Gram quantities

Vaccinia Virus for Heterologous Expression. For recombinant protein expression, the gene of interest is placed under the control of a vaccinia promoter. Late promoters are preferred for yielding high-level protein expression. Integration of the foreign gene can be accomplished by one of three methods: homologous recombination, direct cloning, or construction of a hybrid of vaccinia with the bacteriophage T7 RNA polymerase.

Homologous recombination is performed by (*1*) subcloning cDNA into a plasmid next to a vaccinia promoter, flanked by nonessential vaccinia gene sequences; (*2*) recombining into the vaccinia genome (cell infection with wild-type vaccinia virus and transfection of recombinant plasmid); and (*3*) screening for recombinant virus.

Direct cloning is performed by inserting the cDNA into a unique restriction site of the vaccinia virus genome, which is packaged after transfecting helper cells (fowlpox virus–infected cells). High-level expression can also be achieved with the vaccinia virus–T7 RNA polymerase hybrid system. This system relies on the cytoplasmic expression of the bacteriophage T7 RNA polymerase (e.g., by a stable transfected cell line). The recombinant gene is inserted into a vaccinia virus under the control of the T7 RNA polymerase promoter. To provide the T7 polymerase, either constitutively expressing cells are infected or cells are cotranfected with a polymerase expressing vaccinia virus.

Advantages of the vaccinia virus system are (*1*) the wide host range of mammalian cells that can be infected, (*2*) the mammalian posttranslational modification machinery, (*3*) the high expression levels, and (*4*) easy production of the virus. A disadvantage of the system is the safety level required when working with vaccinia.

The vaccinia system has been successfully scaled up to very large scale (Table 4).

CONCLUSION

Transient expression technologies have been available for many years and have been applied to many different objectives. Recent novel applications cover the use of these technologies for producing recombinant proteins in amounts sufficient for biochemical and biological studies (100–1,000 mg). Table 5 summarizes the essential features of each technology. A first step common to all technologies presented here is the large-scale preparation of the expression vector (plasmid DNA, mammalian or insect recombinant virus). Speed is the principal motivation (starting from cDNA, the protein is produced within 10–15 days) in contrast to more classical technologies such as stable cell lines, which require selection and amplification of the recombinant gene. In addition, novel routes to pharmaceutical discovery such as human genome sequencing create a serious need for high throughput, medium-scale (10–100 mg of protein) expression systems in order to quickly characterize and evaluate the function of novel gene products. The eukaryotic environment in which most of these transient expression systems operate ensures that elaborate posttranslational modifications are properly carried out and that the protein product obtained provides an authentic representation of its native human counterpart.

ACKNOWLEDGMENTS

We are thankful to Dr. Tom Kost for suggestions and corrections of the manuscript and to Chris Hebert for graphic work.

BIBLIOGRAPHY

1. Y. Glutzman, *Cell* **23**, 175–182 (1981).
2. M. Lusky and M. Botchan, *Nature* **27**, 279–288 (1981).
3. J.K. Rose and J.E. Bergmann, *Cell* **30**, 753–762 (1982).
4. B. Seed, *Nature* **329**, 840–842 (1997).
5. G. Cachianes, C. Ho, R.F. Weber, S.R. Williams, D.V. Goeddel, and D.W. Leung, *Biotechniques* **15**, 255–259 (1993).
6. H.D. Blasey, J.P. Aubry, G.J. Mazzei, and A.R. Bernard, *Cytotechnology* **18**, 183–192 (1995).
7. R. Ridder, S. Geisse, B. Kleuser, P. Kawalleck, and H. Gram, *Gene* **166**, 273–276 (1995).
8. M. Jordan, C. Koehne, and F. Wurm, *Cytotechnology* **26**, 39–47 (1998).
9. G.E. Smith, M.D. Summers, and M.D. Fraser, *Mol. Cell. Biol.* **3**, 2156–2165 (1983).
10. G.D. Pennock, C. Shoemaker, and L.K. Miller, *Mol. Cell. Biol.* **4**, 399–406 (1984).
11. T.J. Wickham, T. Davis, R.R. Granados, M.L. Shuler, and H.A. Wood, *Biotechnol. Prog.* **8**, 391–396 (1992).
12. V.A. Luckow and M.D. Summers, *Biotechnology (N.Y.)* **6**, 47–55 (1988).
13. M.D. Gruenwald and M.S. Heitz, in *Baculovirus Expression Vector System: Procedures and Methods Manual*, Pharmingen, San Diego, 1996, pp. 1–73.
14. V.A. Luckow, S.C. Lee, G.F. Barry, and P.O. Olins, *J. Virol.* **67**, 4566–4579 (1993).
15. D.R. O'Reilly, L.K. Miller, and V.A. Luckow, in *Baculovirus Expression Vectors: A Laboratory Manual*, Oxford University Press, Oxford, 1994, pp. 1–347.
16. M. Kool, J.W. Voncken, F.L. van Lier, J. Tramper, and J.M. Vlak, *Virology* **183**, 739–746 (1991).
17. M.D. Summers and G.E. Smith, in *A Manual of Methods for Baculovirus Vectors and Insect Cell Culture Procedures*, Texas A&M University, College Station, Texas, 1988, pp. 1–56.
18. A.E. Proudfoot, L. Goffin, M.A. Payton, T.N.C. Wells, and A.R. Bernard, *Biochem. J.* **318**, 437–442 (1996).
19. R. Hovius, A.-P. Tairi, H. Blasey, A.R. Bernard, K. Lundström, and H. Vogel, *Protein Sci.* (1997).
20. F.L. Graham and A.J. Van der Eb, *Virology* 456–467 (1973).
21. W.J. Chen, J.F. Moomaw, L. Overton, T.A. Kost, and P.J. Casey, *J. Biol. Chem.* **268**, 9675–9680 (1993).
22. V. Jäger and A. Kobold, *Biotechnol. Tech.* **9**, 435–440 (1995).
23. V. Jäger, A. Kobold, C. Köhne, S.M. Deutschmann, E. Grabenhorst, E. Karger, and H.S. Conradt, in R.E. Spier, J.B. Griffiths, and W. Berthold, eds., *Animal Cell Technology: Products for Today, Prospects for Tomorrow*, Butterworth-Heinemann, Oxford, 1994, pp. 207–211.
24. A. Mills, B. Allet, A.R. Bernard, C. Chabert, E. Brandt, C. Cavegn, A. Chollet, and E. Kawashima, *FEBS Lett.* **320**, 130–134 (1993).

25. R. Wagner, H. Geyer, R. Geyer, and H.D. Klenk, *J. Virol.* **70**, 4103–4109 (1996).

26. P. Liljeström and H. Garoff, *Biotechnology (N.Y.)* **9**, 1356–1361 (1991).

27. P. Berglund, M. Sjoberg, H. Garoff, G.J. Atkins, B.J. Sheahan, and P. Liljeström, *Biotechnology (N.Y.)* **11**, 916–920 (1993).

28. K. Lundström, A. Mills, G. Buell, E. Allet, N. Adami, and P. Liljeström, *Eur. J Biochem.* **224**, 917–921 (1994).

29. V. Ciccarone, J. Jessee, P. Berglund, and P. Liljeström, *Focus* 103–105 (1993).

30. H.D. Blasey, K. Lundström, S. Tate, and A.R. Bernard, *Cytotechnology* **24**, 65–72 (1997).

31. K. Lundström, J. Tenhunen, C. Tilgmann, T. Karhunen, P. Panula, and I. Ulmanen, *Biochim. Biophys. Acta* **1251**, 1–10 (1995).

32. H. Bikker, F. Baas, and J.-J.M. De-Vijlder, *J. Clin. Endocrinol. Metab.* **82**, 649–653 (1997).

33. K. Lundström, A. Mills, E. Allet, K. Ceszkowski, G. Agudo, A. Chollet, and P. Liljeström, *J. Receptor Signal Transduction Res.* **15**, 23–32 (1995).

34. P. Werner, E. Kawashima, J. Reid, N. Hussy, K. Lundström, G. Buell, Y. Humbert, and K.A. Jones, *Mol. Brain Res.* **26**, 233–241 (1994).

35. M. Patel, C. Harris, and K. Lundström, *Drug Dev. Res.* **40**, 35–40 (1997).

36. F.H. Marshall, K. Patel, K. Lundström, J. Camacho, S.M. Foord, and M.G. Lee, *Br. J. Pharmacol.* **121**, 1673–1678 (1997).

37. K. Lundström and M.P. Turpin, *Biochem. Biophys. Res. Commun.* **225**, 1068–1072 (1996).

38. A.D. Michel, K. Lundström, G.N. Buell, A. Surprenant, S. Velara, and P.-P.A. Humphrey, *Br. J. Pharmacol.* **118**, 1806–1812 (1996).

39. K. Lundström, A. Michel, H. Blasey, A.R. Bernard, R. Hovius, H. Vogel, and A. Surprenant, *J. Receptor Signal Transduction Res.* 115–126 (1997).

40. G.H. Gorrie, Y. Vallis, A. Stephenson, J. Whitfield, B. Browning, T.G. Smart, and S.J. Moss, *J. Neurosci.* **17**, 6587–6596 (1997).

41. P. Heino, J. Dillner, and S. Schwartz, *Virology* **214**, 349–359 (1995).

42. K.B. Hummel, J.A. Vanchiere, and W.J. Bellini, *Virology* **202**, 665–672 (1994).

43. X. Zhou, P. Berglund, H. Zhao, P. Liljeström, and M. Jondal, *Proc. Natl. Acad. Sci. U.S.A.* **92**, 3009–3013 (1995).

44. N.L. Paul, M. Marsh, J.A. McKeating, T.F. Schulz, P. Liljeström, H. Garoff, and R.A. Weiss, *AIDS Res. Hum. Retroviruses* **9**, 963–970 (1993).

45. M. Suomalainen and H. Garoff, *J. Virol.* **68**, 4879–4889 (1994).

46. M.M. Rolls, K. Haglund, and J.K. Rose, *Virology* **218**, 406–411 (1996).

47. T.J. Pilot-Matias, R.J. Carrick, P.F. Coleman, T.P. Leary, T.K. Surowy, J.N. Simons, A.S. Muerhoff, S.L. Buijk, M.L. Chalmers, G.J. Dawson, S.M. Desai, and I.K. Mushahwar, *Virology* **225**, 282–292 (1996).

48. J.-J.M. Meulenberg, A.P. Van-Nieustadt, Z.A. Van-Essen, and J.-P.M. Langeveld, *J. Virol.* **71**, 6061–6067 (1997).

49. J. Meanger, I. Peroulis, and J. Mills, *Biotechniques* **23**, 432–434, 436 (1997).

50. F.-K.N.-A.S.-J. Lebedeva-I, *J. Virol.* **9**, (1998).

51. M.J. de Hoop, V.M. Olkkonen, E. Ikonen, E. Williamson, C. von Poser, L. Meyn, and C.G. Dotti, *Gene Therapy* **1** (Suppl 1), S28–31 (1994).

52. V.M. Olkkonen, P. Dupree, I. Killisch, A. Lutcke, M. Zerial, and K. Simons, *J. Cell. Sci.* **106**, 1249–1261 (1993).

53. A.M. Fra, E. Williamson, K. Simons, and R.G. Parton, *Proc. Natl. Acad. Sci. U.S.A.* **92**, 8655–8659 (1995).

54. J. Jantti, S. Keranen, J. Toikkanen, E. Kuismanen, C. Ehnholm, H. Soderlund, and V.M. Olkkonen, *J. Cell. Sci.* **107**, 3623–3633 (1994).

55. M. de Hoop, C. von Poser, C. Lange, E. Ikonen, W. Hunziker, and C.G. Dotti, *J. Cell. Biol.* **130**, 1447–1459 (1995).

56. B. De Strooper, M. Simons, G. Multhaup, F. Van Leuven, K. Beyreuther, and C.G. Dotti, *EMBO J.* **14**, 4932–4938 (1995).

57. P. Mikus, T. Urano, P. Liljeström, and T. Ny, *Eur. J. Biochem.* **218**, 1071–1082 (1993).

58. A. Huckriede, A. Heikema, J. Wilschut, and E. Agsteribbe, *Eur. J. Biochem.* **237**, 288–294 (1996).

59. S. Aleshkov, C.R. Abraham, and V.I. Zannis, *Biochemistry* **36**, 10571–10580 (1997).

60. J. Zhang, P.C. Asselin, F. Bex, J. Bernrd, J. Chehimi, F. Willems, A. Caignard, P. Berglund, P. Liljeström, A. Burny, and S. Chouaib, *Gene Therapy* **4**, 367–374 (1997).

61. F.L. Graham and L. Prevec, *Mol. Biotechnol.* **3**, 207–220 (1995).

62. A.R. Fooks, E. Schadeck, U.G. Liebert, A.B. Dowsett, B.K. Rima, M. Steward, J.R. Stephenson, and G.W.G. Wilkinson, *Virology* **210**, 456–465 (1995).

63. B. Massie, J. Dionne, N. Lamarche, J. Fleurent, and Y. Langelier, *Biotechnology (N.Y.)* **13**, 602–608 (1995).

64. I. Nadeau, A. Garnier, J. Cote, B. Massie, C. Chavarie, and A. Kamen, *Biotechnol. Bioeng.* **51**, 613–623 (1996).

65. A. Garnier, J. Cote, I. Nadeau, A. Kamen, and B. Massie, *Cytotechnology* **15**, 145–155 (1994).

66. B. Moss, in B.M. Fields, D.M. Knipe, R.M. Chanock, M.S. Hirsch, J.L. Melnick, T.P. Monath, and B. Roizman eds., *Virology* Raven Press, New York, 1990, pp. 2079–2111.

67. G. Pialoux, J.L. Excler, Y. Riviere, G. Gonzalez-Canali, V. Feuillie, P. Coulaud, J.C. Gluckman, T.J. Matthews, B. Meignier, M.P. Kieny, P. Gonnet, I. Diaz, C. Meric, E. Paoletti, J. Tartaglia, H. Salomon, S. Plotkin, The Agis Group, and L'Agence Nationale de Recherche sur le Sida, *AIDS Res. Hum. Retroviruses* **11**, 373–381 (1995).

68. N. Barrett, A. Mitterer, W. Mundt, J. Eibl, M. Eibl, R.C. Gallo, B. Moss, and F. Dorner, *AIDS Res. Hum. Retroviruses* **5**, 159–171 (1989).

69. F.G. Falkner, P.L. Turecek, R.T. MacGillivray, W. Bodemer, F. Scheiflinger, S. Kandels, A. Mitterer, O. Kistner, N. Barrett, J. Eibl, and F. Dorner, *Thromb. Haemost.* **68**, 119–124 (1992).

See also Chinese hamster ovary cells, recombinant protein production; Expression systems, *E. coli*; Expression systems, mammalian cells; Insect cell culture, protein expression; Insect cells and larvae, gene expression systems; *Pichia*, Optimization of protein expression.

TRANSPORT, MICROBIAL SOLUTE UPTAKE

R.E. Marquis
University of Rochester
Rochester, New York

KEY WORDS

Antiporters
ATPases
Biofilms
Energy recycling transport
Permeability
Phosphoenolpyruvate
Proton-motive force
Sugar phosphotransferase
Symporters
Transmembrane solute transport

OUTLINE

INTRODUCTION

This article covers transport processes in microorganism of industrial importance. General aspects of transport and uptake of solutes are considered, but as far as possible, examples involving industrial organisms are presented. Figure 1 presents a summary diagram reviewing the major types of transport systems in bacterial cells. Recent reviews from a less applied perspective are cited in the bibliography.

Cell Structure and Solute Movements

Solute transport across membranes plays major roles in the physiology of industrial organisms and may determine which solutes are transformed into products or if the organisms can function under the rigors of industrial process environments. Gram-positive bacteria generally have a single cytoplasmic or protoplast membrane surrounded by a thick cell wall with an elastic peptidoglycan network containing various other polymers and forming a heteroporous molecular sieve that restricts entry of large molecules into the cell. Fungal and plant cells also have thick, heteroporous walls but do not contain the peptidoglycans that are peculiar to eubacteria. The wall is a coarse sieve, whereas the membrane is a finer sieve. In fact, generally only very small solutes pass through the cytoplasmic membrane unless the membrane has specific transport proteins that catalyze movements into or out of the cell. Walled cells generally have what is called osmotrophic nutrition and only small nutrients, for which there are transport systems in the cytoplasmic membrane, can be transformed through the actions of cytoplasmic enzymes. However, bacteria and fungi are noted for secreting hydrolytic enzymes that catalyze breakdown of polymers in the environment. The released fragments then pass through the cell wall to be transported across the cytoplasmic membrane into the cell. Many of these exoenzymes are of major industrial importance, although their primary function for the producing cell is to provide nutrients.

Gram-negative eubacteria generally have two membranes, an inner or cytoplasmic membrane bounding the protoplast, and an outer, cell wall membrane. Both membranes are barriers to solute uptake. The outer membrane is a less-restrictive barrier but does largely prevent movement of larger molecules into the cell unless there are specific receptors for the molecules. Movement even of smaller molecules commonly involves protein porins, which form channels across the membranes. The inner membrane is the equivalent of the cytoplasmic membrane of Gram-positive bacteria and contains transport catalysts for movement of solutes into and out of the cytoplasm. The space between the two membranes contains a thin layer of peptidoglycan, lipoprotein molecules that bridge the peptidoglycan layer and the outer membrane and a variety of enzymes and binding proteins, some of which are involved in solute transport. Small hydrophilic solutes may pass through porins in the outer membrane and be bound by specific binding proteins in the periplasm between the membranes. Binding protein–solute complexes may then interact with specific transport proteins in the cytoplasmic membrane to catalyze transport of the solute into the cell and release of the binding proteins back into the periplasm.

Archeae, which are becoming more and more prominent in biotechnology, have cytoplasmic membranes with ether-linked lipids rather than the ester-linked lipids of other organisms and contain phytanols rather than the more usual types of fatty acids of eubacterial and eukaryotic cells. Various protective envelope structures external to the membrane have been described, but none contains the peptidoglycans characteristic of most eubacterial cells. The membrane transport systems of the *Archeae* are generally similar to systems in eubacterials cells but with differences in the molecular details of the components. Since many *Archeae* are adapted to function in what are considered to be extreme environments, their transport systems are also adapted to function at, for example, high temperatures,

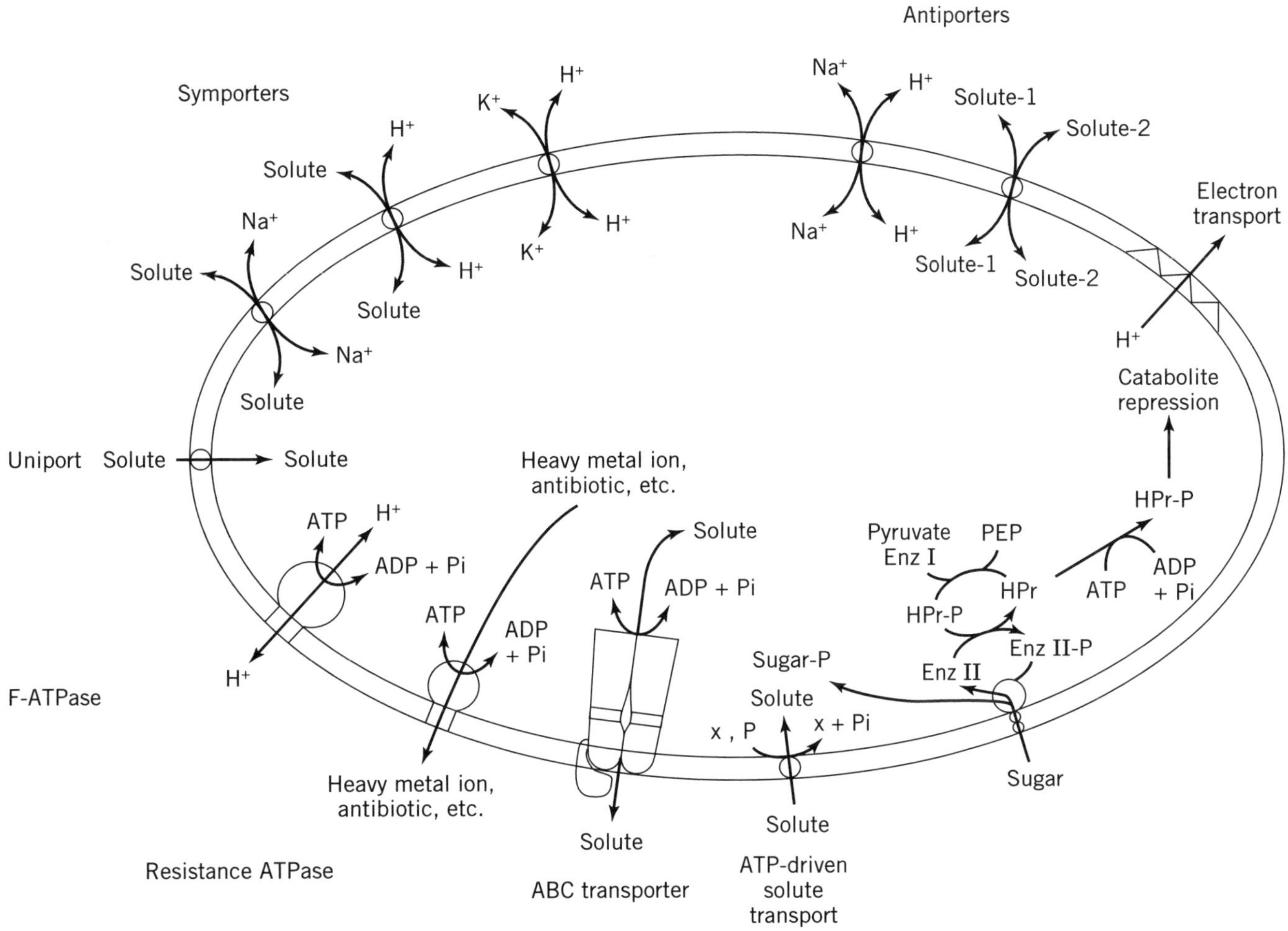

Figure 1. Solute transport systems.

low pH values, very high salt concentrations or low oxidation-reduction potentials.

Eukaryotic organisms used in industrial microbiology include yeasts and other fungi, but also animal cells used, say, for production of monoclonal antibodies, and increasingly, plant cells. Most eukaryotes have multiple membranes that regulate solute transport into and out of the cells and also into and out of the various organelles within in the cells. Indeed, these organisms nearly all have a mix of prokaryotic and eukaryotic characteristics. For example, their major energy-transferring organelles, mitochondria and chloroplasts, are derived from prokaryotic endosymbionts and so have modified prokaryotic membranes and transport systems.

Transport in Biofilms

Special problems of solute transport may be encountered in biofilms, which are important in many biotechnological applications ranging from food fermentations to sewage treatment. Entry of solutes into and out of these consortia of organisms associated with a surface is generally controlled by diffusion across the surface of the biofilm. The situation is often one in which there is flow of a fluid over the biofilm and passage of solues from the fluid across the surface of the film into the polysaccharide matrix and thence into the cells. The passage of solute generally follows the Fick diffusion equation, $dS/dt = PA(\Delta C)$, where dS/dt is the change in the amount of solute in the biofilm per unit of time, P is the permeability coefficient (which has units of distance divided by time), A is the surface area for diffusion, and ΔC is the difference in solute concentration between the interior and exterior of the biofilm surface. Theoretical aspects of antibiotic diffusion into biofilms have been reviewed by Stewart (1), and his analyses can be applied to other solutes.

The biofilm may be mainly exposed to air or other gases. However, the film is still covered by a liquid, and passage, say of O_2, is from the gas phase into the liquid phase and thence into the biofilm. O_2 movement into the film is limited, and biofilms generally contain a diversity of aerobic, facultative, and anaerobic organisms with stratification of the anaerobes away from the surface. Many or most biofilms are penetrated by aqueous channels that may help to deliver solutes to lower parts of the consortium. Nutrients may be leached from solid substrata by biofilms, but again, the limitation on leaching is commonly surface-limited diffusion.

Biofilms have long played key roles in industrial processes ranging from vinegar production to sewage treatment. Our appreciation of them is growing, and they are likely to play even more prominent roles in the future. As mentioned, they allow for wide biodiversity in populations, and this diversity is key in many types of biocatalysis involving cometabolism. They also allow for coexistence of genetically engineered organisms that otherwise would not be competitive. They serve also to mute adverse environmental influences so that members of the community have time to undergo phenotypic or even genotypic adaptations in response to environmental stresses. Certainly, it is well know that cells in biofilms are far more resistant to biocides than are cells in open cultures. However, biofilms do present special challenges in terms of supply of O_2 and other solutes and diffusion of products away from the consortium.

PRIMARY TRANSPORT SYSTEMS

In cells, the primary mechanisms for energy transfer from fuel to function include substrate-level phosphorylations, such as those of glycolysis and other fermentative pathways, and so-called chemiosmotic processes first elucidated by Peter Mitchell. The hypothesis of chemiosmotic energy coupling was based on two fundamental insights. The first involved realization that biological membranes allow for charge separation across a hydrophobic barrier and a conversion of chemical bond energy to electrical energy. During respiration or photosynthesis, protons are moved out of the cell or organelle, such as a mitocondrion or chloroplast, to create a gradient of pH and charge across the membrane. The potential energy of the charge separation was defined by Mitchell in terms of the proton-motive force (Δp) expressed in volts or millivolts

$$\Delta p = \Delta \psi + z\Delta \mathrm{pH}$$

where $\Delta \psi$ is the electrical potential across the membrane (measured experimentally or calculated by use of the Nernst equation), ΔpH is the pH difference across the membrane, and z is equal to 2.3 RT/F with R being the gas constant, T the Kelvin temperature, F the Faraday constant, and the multiplier 2.3 is for conversion of natural log to base 10 log.

The Δp potential energy can then be transferred to synthesis of ATP, to solute movements across the membrane, or to other endergonic processes such as motility. Mitchell's other major insight was that there was associated with the membrane an ATPase that could catabolize synthesis of ATP from ADP and inorganic phosphate in association with proton movements back into the cell. Thus, the electrical energy associated with charge separation across the membrane could be converted back to chemical bond energy. These enzymes are generally called F-ATPases and have the peculiarities that they can operate reversibly and do not become phosphorylated during their catalytic cycles. They are the major agents for chemiosmotic ATP synthesis in respiring and photosynthetic organisms. For example, in the average human, they catalyze synthesis of some 70 kg of ATP per day. The equivalent enzymes of *Archeae* are generally called A-ATPases, and they are able to couple charge separation across the cell membrane to ATP synthesis.

F-ATPases serve major functions even in nonrespiring, fermentative organisms. For example, fermentation commonly results in acid production, and the F-ATPases act to move protons from dissociated acids out of the cytoplasm to develop ΔpH across the cell membrane with the interior alkaline relative to the exterior.

It is now apparent that chemiosmotic energy transfer can involve sodium ions (ΔpNa) and sodium-transporting ATP synthases (2). The ATPases responsible for this type of transfer are distinct from the enzymes responsible for proton-based energy transfer. Proton-based systems are more common than Na^+-based systems, but the latter are widely distributed among prokaryotes.

SECONDARY TRANSPORT SYSTEMS

ATP-Dependent Transport

F-ATPases play a key role in acid-base physiology by transporting protons out of the cytoplasm to reduce internal acidification. This role is clearly important in many industrial fermentations in which acid is produced (e.g., in vinegar production) or in which there is an acidification phase before production of another metabolite (such as in the butanol fermentation by *Clostridium acetobutylicum*). For the bacteria involved in a wide range of processes from yogurt production to acid leaching of copper ores, F-ATPases are needed for the organisms to operate under acidic conditions. Brewing or wine making also are examples in which acidification precedes ethanol production. Yeasts are eukaryotes, and their plasma membrane ATPases, which are not of the reversible F-type, are important for moving protons out of the cell. Also, the vacuole-ATPase (V-ATPase) of the organism appears to be important for acid tolerance. The V-ATPases are multisubunit enzymes with hydrophilic and hydrophobic membrane-embedded components involved in proton pumping or other ion movements. The actions of V-ATPases are not reversible, and so they do not serve for chemiosmotic ATP synthesis. The yeast must then depend on glycolysis for ATP under anaerobic conditions of ethanol production or on the F-ATPases of their mitochondria under aerobic growth conditions.

In bacteria, ATPases other than F-ATPases also are involved in ion movements across the cell membrane. These ATPases are often what are called P-ATPases because they become phosphorylated during their catalytic cycles. Among the best known are heavy-metal-transporting ATPases (3) that play major roles in the resistances of many bacteria to heavy metals such as cadmium. Bacteria also have multisubunit ATPases for the transport of mineral ions such as K^+, Na^+, and CA^{2+}.

ATP can also serve as the energy source for transport of macromolecules across cell membranes, either into the cell or out of the cell, as reviewed by Palmen et al. (4). An example in the dairy industry concerns metabolism of proteins such as casein in manufacture of cheese and related products. As described by Poolman et al. (5) for *Lactococcus lactis*, the proteolytic pathway involves initial cleavage of proteins by proteinases outside of the cell membrane, in this case mainly the so-called Prt-P proteinase. The cleavage may release free amino acids, which are then transported into the cell by standard amino acid transport systems, mainly energized by Δp or by antiport systems. The products of Prt-P hydrolysis are a wide variety of peptides. Di- and tripeptides can be transported into the cell by means of specific transport systems with relatively broad specificities energized by Δp (Dtp-P transport system) or by ATP (Dtp-T system). The peptides are then broken down in the cytoplasm to amino acids. Oligopeptides can be transported into the cell by the oligopeptide transport system, which is an example of ATP binding cassette (ABC) system consisting of two membrane-spanning proteins, two ATP-hydrolyzing subunits, and a binding subunit associated with the outer face of the cytoplasmic membrane. Oligopeptides are then hydrolyzed in the cytoplasm to amino acids by a variety of peptidases. Oligopeptides are the major source of nitrogen for the organism when it is growing in milk. ABC transporters commonly work in the opposite direction to catalyze movement of polymers out of the cell and so are important for many industrial processes such as production of polysaccharides or enzymes and various other proteins.

Phosphoenolpyruvate-Dependent Transport

The phosphotransferase system (PTS), or as it is now more commonly called, the phosphoenolpyruvate:sugar phosphotransferase system, plays major roles in sugar transport in facultative bacteria. However, it appears that the system may be more widespread among organisms and may play major regulatory roles, especially in catabolite repression (6). The biochemistry of catabolite repression has been elucidated for Gram-positive bacteria such as *Bacillus subtilis*. Repression involves an HPr(Ser)-kinase, which can be activated by fructose-1,6-bisphosphate or gluconate 6-phosphate. The kinase can phosphorylate the HPr protein of the PTS at serine-46. The HPr-P can then combine with CcpA (the repressor-activator protein mediating catabolite repression and glucose activation) and fructose-1,6-bisphosphate to form a complex that binds to the cis-acting DNA sequence called the catabolite repression element (CRE) to inhibit transcription. Thus, many Gram-positive bacteria are able to regulate metabolism in response to rises and falls in cytoplasmic levels of glycolytic intermediates. In Gram-negative organisms, catabolite repression involves cyclic AMP and cyclic-AMP-receptor protein, as described in standard textbooks. Catabolite repression is important in many industrial microbiology processes and allows for sequential utilization of mixed substrates, often to allow mainly growth initially and product formation subsequently, for example, in many antibiotic fermentations.

A major function of the PTS is uptake of sugars. The system has common elements, specifically HPr protein and enzyme I, which catalyze transfer of phosphate from phosphoenolpyruvate to a histidine moiety on HPr. The PTS has elements designed for specific sugars, notably IIC proteins or peptides, which are the actual permeases embedded in the cytoplasmic membrane for catalysis of sugar transport across the cell membrane and phosphoryl transfer from enzyme IIB to the incoming sugar. For many of the industrial lactic acid bacteria, the PTS is the major agent for sugar uptake, although nearly all PTS-positive bacteria also have alternative systems for sugar uptake, especially if sugar fermentation plays a major role in their physiology. Thus, for example, dairy lactic acid bacteria may take up milk sugar lactose by a PTS or by transport systems energized either by Δp or ATP.

Proton-Motive Force–Dependent Transport

Coupling of Δp to transport is a very direct means for energy transfer. This coupling is dependent on what is often referred to as *the energized membrane*. The advantage to the organism in direct coupling is obvious, and direct coupling occurs not only for solute transport but also for functions such as motility. However, there can also be disadvantages because the energized state of the membrane is very sensitive to environmental changes. Energizers such as ATP or PEP generally are not released from cells unless there is major damage to membranes. However, Δp can be discharged by a variety of mild insults including exposure to weak acids, which can enter the cell in the protonated state and dissociate within the cytoplasm to diminish the ΔpH component of Δp. This sort of discharge can occur to a degree when weak acids such as formate, acetate, or lactate are produced metabolically. The discharge then puts additional energy demands on the organism, requiring additional catabolism for growth and maintenance of cell functions. The uncoupling may actually be an advantage in many fermentation processes, permitting high production of metabolic products with less production of cells. The discharge of ΔpH can also be of advantage in applications such as the use of fluoride to inhibit acid production by the plaque bacteria that cause dental caries. Fluoride acts in the protonated form, HF, to bring extruded protons back into the cell and thereby to reduce acid tolerance and the capacities of cariogenic bacteria to produce acid at low pH values (7).

Δp-dependent systems occur widely for uptake of small solutes. For example, uptake of many sugars is coupled to proton reentry. Thus entry of galactose, lactose, rhamnose, fucose, and so forth can be coupled through symport proteins to proton movements back into the cell. Many organisms also have PTS systems for the same sugars. We have found for oral streptococci that use of a system for galactose uptake coupled to proton movements, rather than the PTS, actually reduces acid tolerance of the organisms (8).

Δp-coupled uptake and excretion of a wide variety of organic solutes is known to involve proton symport or antiport. For example, lysine is taken up by *L. lactis* through a proton symport mechanism. In the same organism, the transport systems for Leu/Ile/Val, Ala/Gly and Ser/Thr are also dependent on Δp (9). Citrate uptake is commonly coupled to movements of protons back into the cell. Antibiotic resistance in many organisms may have its base in transport systems that move antibiotics out of the cell. Perhaps the best example is that of tetracycline efflux systems in a number of organisms involving antiport of tetracycline out and protons or K^+ in. Multidrug efflux pumps occur also in a variety of organisms and have recently been reviewed by Nikaido (10).

Now that the specifics of Na^+-based energetics are well established (2), it is clear that organisms such as *Propionigenium modestum* depend on ΔpNa, rather than ΔpH, for coupling transport of various solutes to movements of Na^+. In any organism, there can be a mix of Δp and ΔpNa coupling, but in general in the biological world, Δp coupling is predominant.

In general, the biopolymers within cells have a net negative charge and a need for cations to neutralize the excess charge. The potential across the cell membrane ($\Delta\psi$) reflects a nonequilibrium state in metabolizing cells. However, it is a component of Δp that can serve to energize transport. Generally, it is considered that $\Delta\psi$ can be converted readily to ΔpH, and so it is often difficult to separate $\Delta\psi$-driven processes from ΔpH-driven processes. It should be appreciated that a large fraction of the cations within cells, predominantly K^+, are retained simply for charge neutralization. If the cell membrane is damaged in a walled organism, such as a eubacterial cell, there is an efflux of solutes, but there is also a retention of cations as counterion for polymers that cannot exit readily through the porous wall. Of course, when the membrane is damaged, the major agent for selectivity of solute entry is damaged, and cation exchange can occur readily. Movements of mineral ions across membranes by symport and antiport play major roles in cell physiology. A good example is in the capacities of many organisms, predominantly bacteria, to function at alkaline pH values. The most common basis for this ability is H^+/Na^+ antiporters that move protons into the cell to reduce the cytoplasmic pH value in association with movement of Na^+ out.

Energy Recycling Systems

Microorganisms often are able to use transport systems for energy conservation. One of the early proposals for this sort of conservation involved glycolysis by dairy streptococci (11). Glucose is catabolized by these organisms to yield lactate. At higher pH values, lactate excretion is electrogenic because more than one proton is excreted per lactate anion. This excretion then results in Δp across the cell membrane, which can be converted chemiosmotically to chemical energy in the form of ATP through the action of F-ATPases.

Another example pertinent to industrial fermentations is that of the malolactic fermentation carried out by *Oenococcus oenos* and related organisms and important in production of wine, cider, and a variety of fermented beverages. Malate is transported into the cell where it is decarboxylated by the malolactate enzyme to yield CO_2 and lactate, which are then moved out across the membrane in association with protons to create Δp (12). The use of the conserved energy in the Δp can then allow for some 25 to 50% increase in growth yields of the organisms.

Symporters and Antiporters Not Coupled to Δp or ΔpNa

As already reviewed, the energetic coupling of Δp or ΔpNa to solute movements into and out of cells commonly involves mechanisms of symport and antiport. Thus, the potential energy of a gradient of protons or of sodium ions is translated into kinetic energy of movement of other solutes, or the electrical potential across the cell membrane may be the basis for movement of counterions into the cell to reduce the potential. Other types of symporters and antiporters may not be so directly coupled to the energized state of the membrane. A good example is the ornithine/arginine antiport system of many organisms that use the arginine deiminase system (ADS) for ATP synthesis or for neutralization of cytoplasmic protons through production of ammonia (13). The ADS involves three enzymes, arginine deiminase, ornithine carbamyl transferase, and carbamate kinase, plus an antiporter. ATP is synthesized in association with the carbamate kinase–catalyzed reaction. Ornithine is produced in the reaction catalyzed by ornithine carbamyl transferase. It exchanges with arginine so that substrate is brought into the cell, and product is excreted without the need to make use of the ATP produced by the system. This exchange allows for operation with a net gain in ATP, and many organisms can grow with arginine as fuel without need for respiration. In fact, oxygen often is repressive for the ADS. CO_2 is produced by the system, and it can diffuse out of the cell. The other main product, NH_3, reacts with protons within the cytoplasm and serves to maintain acid-base balance in many organisms.

TRANSPORT ASSAYS

Various methods used to assay transport of molecules into cells have been reviewed recently (14), with distinction made between assays primarily for determining permeability of a cell or cell component to a particular compound, and those for assessing the activities of transport systems for active movement of solutes across the cell membrane. Also, methods are reviewed for assessing Δp from its two components, ΔpH and $\Delta\psi$, and for assessing osmotic responses of cells.

In practical situations, cell permeability often is key for entry of solutes into cells. For example, gases are hydrophobic and generally pass readily through hydrophobic membranes. In industrial fermentation in which O_2 must be supplied to aerobic or facultative organisms, the major barrier to be overcome in supplying the gas is at the boundary between the gas phase and the liquid phase. O_2 is relatively insoluble in aqueous solution, and so a high concentration cannot be built up in the aqueous phase of a culture. For a rapidly respiring microbial population, supplying sufficient O_2 can be nearly impossible. Efforts are made then to increase the surface area for diffusion of O_2 into the culture by sparging. Also, pure oxygen may be used to increase supplies. Still, it is common for the cells to be starved for O_2, which passes readily from the medium into the cell as it becomes available. Metabolically produced gases, such as CO_2 and H_2, also pass readily across the cell membrane so that cells do not build up stores of the gases. In fact, gas bubbles generally do not form within cells unless special structures are produced by the cell, such as the gas vesicles used for flotation by certain aquatic microorganisms (15).

Perhaps the most important solute that enters cells without specific transport mechanisms is water. All cells are highly permeable to water, even bacterial spores; and when cells are mixed with water labeled with tritium or deuterium, equilibration of the labeled water generally occurs in seconds. Indeed, osmotic equilibration of water can occur in milliseconds. Other small solutes may also penetrate rapidly. In fact, labeled urea or glycerol is commonly used to assess the total water space of cells because of their ready movement into the total water volume of cells. Recently, the roles of ubiquitous major intrinsic proteins (MIP) in movements of water and neutral solutes across membranes have been studied (16). Hydrophobic solutes may also readily enter biological membranes, especially if they are of low molecular weight. Higher molecular weight hydrophobic solutes also may enter the membrane readily but may remain there and cause membrane dysfunction (17). Mention was made earlier of the effects of organic weak acids on membranes related to movements of the protonated forms into and through membranes (18).

For measurements of passive permeability of cells to solutes, thick suspensions of cells can be used, and the disappearance of solute from the suspending medium assayed. The uptake process is generally not saturable at high solute concentrations, although cells may passively concentrate solutes, especially at low concentration, because of passive binding, for example, binding of cations to negatively charged polymers in the cell wall, membrane, or cytoplasm. For assays of active transport, much less dense suspensions of cells are generally used. Cells and suspending medium may then be separated by centrifugation or by membrane filtration. In some cases, there is no need to make the separation. For example, in assays of proton movements into cells, the pH electrode detects protons outside the cell but not those within the insulating membrane. Thus, as protons move into the cell, the pH value measured by the electrode rises. From the level of pH rise and knowledge of the volumes of cells and suspending medium, the flow of protons across cell membranes can be calculated. When transport depends on metabolism, provision must be made for such metabolism, which may supply ATP, serve to energize the membrane, or produce solutes that may exchange with other solutes in the medium, for example, in arginine/ornithine antiport. In many cases, it is difficult to allow for adequate metabolism, and for example, measurements of movements of protons or other solutes into cells may be affected very much by the experimental conditions. Maintenance of Δp is particularly sensitive to changes in experimental conditions. The cell membrane is at least moderately permeable to protons, and cells must work to maintain ΔpH. If the proton-extruding functions of the F-ATPase are upset, or if proton permeability of the membrane is increased, say, by weak acids, then ΔpH declines and the energized state of the membrane declines. Similarly, $\Delta\psi$ may also be affected adversely, say, by agents such as gramicidin or lantibiotics that enhance membrane permeability to ions.

REGULATION OF TRANSPORT

Transport systems may be regulated in association with the catabolic systems they serve. The best-known example is that of the *lac* operon. The operon contains the gene *lacY* for lactose permease. The operon in *Escherichia coli* is derepressible by β-galactosides and also controlled by catabolite repression involving cyclic AMP and cyclic-AMP-receptor protein. Thus, the permease system is synthesized only when needed for transport of β-galactosides into the cell and maintained at a very low level when β-galactosides are not present. However, genes for transport proteins need not be in the same operon, or even the same regulon, as the genes for metabolism of the transported substrates. An example pertinent to industrial microbiology involves the inducible transport systems for citrate in organisms such as *Lactobacillus rhamnosus* (19). This organism and related organisms are important for flavor development in fermented products through production of diacetyl and acetoin. These products are produced in association with increased consumption of citrate or pyruvate from the environment. The transport system for citrate is repressed in the presence of glucose but becomes derepressed when glucose is exhausted. Thus, flavor enhancement follows derepression of citrate transport after the main phases of growth and acid production.

As might be expected, key enzymes such as the F-ATPases, are mainly constitutive. There is adaptive regulation of F-ATPase synthesis, but the excursion from minimum to maximum levels is small. Thus, for example, for *Streptococcus mutans* and *Enterococcus hirae*, the range is only three to four times the minimal level (20). These organisms depend on their F-ATPases for maintaining acid-base balance, and so appear not to be able to tolerate repression of synthesis of the enzymes to low values. However, upregulation of synthesis of the enzymes does result in increased acid tolerance, and this sort of adaptive tolerance occurs widely among bacteria. It is of importance in many industrial processes in allowing a variety of organisms to operate under adverse conditions.

One of the best examples pertinent to a wide variety of industrial organisms involves the adapation to adverse osmotic conditions of reduced water activity. Adaptation involves pooling of compatible solutes such as K^+, glycine betaine, proline, or polyols and generally involves upregulation of transport systems and enhanced synthetic capacities (21).

INDUSTRIAL EXAMPLES AND APPLICATIONS

Current knowledge of the molecular details of many of the transport systems of industrially important microorganisms is sufficiently advanced to allow for genetic engineering of improved strains, although to date the major work has been on engineering organisms so that they have systems for export of proteins coded by transferred genes. The range of possibilities for advantageous manipulation of transport systems is wide. It is generally considered that organisms that are adapted to oligotrophic environments have transport systems geared for obtaining nutrients from very dilute solutions. Part of their strategy for survival involves formation of biofilms to take advantage of the associations of molecules with interfaces. However, they also need to have transport systems with low K_m (Michaelis–Menten coefficient) values. Thus, they are a source for genes for high-affinity transport systems. The possibilities for developing improved strains include extending the range of substrates that can be used by any one organism. For example, most strains of *Saccharomyces* yeasts used for ethanol production are not very versatile in the range of sugars they can utilize. *Kluyveromyces fragilis* is able to metabolize the sugar lactose, and so there is a possibility of developing hybrids that will allow *Saccharomyces* organisms to use milk sugar for alcohol production. The need to extend the metabolic range of an organism is evident also in development of organisms to transform various xenobiotics. Generally, the xenobiotic must be transported before it can be metabolized. Fortunately, many organisms have two separate transport systems for, say, an amino acid: one highly specific and one with lower specificity. Often xenobiotics can be taken up by the system with lower specificity. However, there are still possibilities for engineering high-affinity systems to the level that they favor the xenobiotic over a related natural compound.

Other examples include work on a variety of organisms for antibiotic production. Generally, producing organisms need to be resistant to the antibiotics they produce, and for example, resistance to tetracyclines can depend on efflux systems. Resistance to heavy metals also commonly depends on efflux systems, often involving efflux ATPases. Mention has already been made of bacterial adaptation to acid conditions and the role of proton-transport systems in the adaptation. However, adaptation to alkali environments in certain foods or in industrial processing also depends very much on transport systems, often K^+/proton antiport or Na^+/proton antiport (22).

Another area in which there has been a great deal of activity concerns excretion of proteins by various genetically engineered organisms. Often simply transferring the genes for synthesis of a particular protein from one organism to another does not result in a useful producing strain. In fact, the transfer can be lethal if the protein cannot be excreted.

ACKNOWLEDGMENT

The work of the author has been supported by grants (P01-DE11549 and R01-DE06127) from the National Institute of Dental Research and by the Center for Aseptic Processing and Packaging Studies, which is a U.S. National Science Foundation Industry/University Cooperative Research Center.

BIBLIOGRAPHY

1. P.S. Stewart, *Antimicrob. Agents Chemother.* **40**, 2517–2522 (1996).
2. P. Dimroth, *Phil. Trans. R. Soc. Lond. B* **326**, 465–477 (1990).
3. G. Ji and S. Silver, *J. Indust. Microbiol.* **14**, 61–75 (1995).
4. R. Palmen, A.J.M. Driessen, and K.J. Hellingwerf, *Biochim. Biophys. Acta* **1183**, 417–451 (1994).
5. B. Poolman, E.R.S. Kunji, A. Hagting, V. Juillard, and W.N. Konings, *J. Appl. Bacteriol.* **79**, 65S–75S (1995).
6. M.H. Saier, Jr., S. Chauvaux, J. Deutscher, J. Reizer, and J.-J. Ye, *Trends Biochem. Sci.* **20**, 267–271 (1995).
7. R.E. Marquis, *Can. J. Microbiol.* **41**, 955–964 (1995).
8. W.A. Belli and R.E. Marquis, *Oral Microbiol. Immunol.* **9**, 29–34 (1994).
9. B. Poolman, A.J.M. Driessen, and W.N. Konings, *Microbiol. Rev.* **51**, 498–508 (1987).
10. H. Nikaido, *J. Bacteriol.* **178**, 5853–5859 (1996).
11. R. Otto, A.S.M. Sonnenberg, H. Veldkamp, and W.N. Konings, *Proc. Natl. Acad. Sci. U.S.A.* **77**, 5502–5506 (1980).
12. T. Henick-Kling, *J. Appl. Bacteriol.* **79**, 29S–37S (1995).
13. A.J.M. Driessen, D. Molenaar, and W.N. Konings, *J. Biol. Chem.* **264**, 10361–10370 (1989).
14. R.E. Marquis, in P. Gerhardt ed., *Methods for General and Molecular Bacteriology*, American Society for Microbiology, Washington, D.C., 1994, pp. 587–599.
15. A.E. Walsby, *Microbiol. Rev.* **58**, 94–144 (1994).
16. J.H. Park and M.H. Saier, Jr., *J. Membrane Biol.* **153**, 171–180 (1996).
17. J.J. Sikkema, A.M. de Bont, and B. Poolman, *J. Biol. Chem.* **269**, 8022–8028 (1994).
18. W.A. Belli, D.H. Buckley, and R.E. Marquis, *Can. J. Microbiol.* **41**, 785–791 (1995).
19. R.M. de Figueroa, I.L. Benito de Cárdenas, F. Sesma, F. Alvarez, A.P. de Ruiz Holgado, and G. Oliver, *J. Appl. Bacteriol.* **81**, 348–354 (1966).
20. W.A. Belli and R.E. Marquis, *Appl. Environ. Microbiol.* **57**, 1134–1138 (1991).
21. L.N. Csonka and A.D. Hanson, *Ann. Rev. Microbiol.* **45**, 569–606 (1991).
22. R.J. Rowbury, Z. Lazim, and M. Goodson, *J. Appl. Bacteriol.* **22**, 429–432 (1996).

ADDITIONAL READING

W.N. Konings, B. Poolman, and H.W. van Deen, *Antonie van Leeuwenhoek* **65**, 369–380 (1994).

P. Mitchell, *Chemiosmotic Coupling and Energy Transduction*, Glynn Research, Ltd., Bodmin, England, 1968.

B. Poolman and W.N. Konings, *Biochim. Biophys. Acta* **1183**, 5–39 (1993).

See also MASS TRANSFER; TRANSFER PHENOMENA IN MULTIPHASE SYSTEMS IN MIXING VESSELS.

TRYPTOPHAN. See AMINO ACIDS, PRODUCTION PROCESSES.

TYROSINE PHENOL-LYASE

HIDEHIKO KUMAGAI
Kyoto University
Kyoto, Japan

KEY WORDS

Enzymes
Gene cloning
L-DOPA
β-Tyrosinase
L-Tyrosine

OUTLINE

INTRODUCTION

Tyrosine phenol lyase (TPL) is a pyridoxal 5′-phosphate (PLP)–dependent multifunctional enzyme catalyzing a degradation reaction of L-tyrosine into phenol, pyruvate, and ammonia. TPL is produced by enterobacteria grown in the medium containing L-tyrosine as an inducer. A crystalline preparation of the enzyme was obtained from two bacterial strains, *Escherichia intermedia* and *Erwinia herbicola*, and their properties were elucidated. The genes of TPL (*tpl*) were cloned, and their nucleotide sequences were determined. The molecule of TPL consists of four identical subunits, each with a molecular mass about 50 kDa, and binds four molecules of PLP per tetramer. The three-dimensional structure of apoenzyme form of TPL (apo TPL) has been refined at 2.3-Å resolution. TPL biosynthesis is regulated transcriptionally, and the DNA binding regulator proteins, TyrR and CRP, are working in this regulation system.

The enzyme catalyzes the synthesis reaction of 3,4-dihydroxyphenyl-L-alanine (L-DOPA) from pyruvate, ammonia, and pyrocatechol in a one-step reaction. L-DOPA is an effective agent in the treatment of Parkinson's disease.

Kakihara and Ichihara (1) reported that phenol is produced from L-tyrosine by an intestinal bacteria, not through stepwise degradation, but through primary fission of the side chain. The tyrosine-inducible enzyme responsible for this conversion was named β-tyrosinase and elucidated to be PLP dependent. The recommended name of this enzyme is tyrosine phenol lyase (deaminating) (EC 4.1.99.2).

Apparently homogeneous preparations of TPL were prepared from cells of *Escherichia intermedia* and *Erwinia herbicola* grown on media supplemented with L-tyrosine. The crystalline preparations of the enzyme catalyze a series of α,β-elimination (equations 1 and 2), β-replacement (equations 3–5), and racemization (equation 6) reactions. The reverse of the α,β-elimination reaction to synthesize L-tyrosine or L-DOPA (equations 7 and 8) was also found to be catalyzed by the crystalline preparations of the enzyme (2).

$$\text{L-Tyrosine} + H_2O \rightarrow \text{pyruvic acid} + NH_3 + \text{phenol} \quad (1)$$

$$\text{L-Serine} \rightarrow \text{pyruvic acid} + NH_3 \quad (2)$$

$$\text{L-Tyrosine} + \text{pyrocatechol} \rightarrow \text{L-DOPA} + \text{phenol} \quad (3)$$

$$\text{L-Serine} + \text{phenol} \rightarrow \text{L-tyrosine} + H_2O \quad (4)$$

$$\text{L-Serine} + \text{pyrocatechol} \rightarrow \text{L-DOPA} + H_2O \quad (5)$$

$$\text{L-Alanine} \rightarrow \text{DL-alanine} \quad (6)$$

$$\text{Pyruvic acid} + NH_3 + \text{phenol} \rightarrow \text{L-tyrosine} + H_2O \quad (7)$$

$$\text{Pyruvic acid} + NH_3 + \text{pyrocatechol} \rightarrow \text{L-DOPA} + H_2O \quad (8)$$

The bacterial cells containing this enzyme catalyze the synthesis of L-DOPA from pyruvic acid, ammonia, and pyrocatechol in significantly high yields, and this reaction is used in industrial production of L-DOPA (3).

PROPERTIES OF TPL

TPLs have been purified and crystallized in the laboratory (4,5,6) from cells of *Escherichia intermedia* and *Erwinia herbicola* grown in a medium supplemented with L-tyrosine by a procedure including cell disruption by ultrasonic oscillation, ammonium sulfate fractionation, protamine sulfate treatment, DEAE-Sephadex column chromatography, hydroxyapatite column chromatography, Sephadex G-150 gel filtration, and crystallization. Photomicrographs of the crystalline holoenzyme prepared from *Erwinia herbicola* are shown in Figure 1.

Crude and partially purified preparations of the TPL were preserved at 5 °C in 0.01 M potassium phosphate buffer, pH 6.0, containing 5 mM mercaptoethanol. Variations in pH in the range of pH 6.0 to 7.0 had little effect on enzyme stability. Outside this range, however, the rate of inactivation increased rapidly. Solutions of highly purified enzyme were less stable than those of partially purified preparations. For example, a sample of the enzyme from *Escherichia intermedia*, with a specific activity of 1.94, lost to 6 to 10% of its initial activity after standing for 2 days at 5 °C in 0.01 M potassium phosphate buffer, pH 6.0, containing 5 mM mercaptoethanol.

The crystalline enzyme preparations could be stored at 0 to 5 °C as suspensions in 10 mM potassium phosphate buffer, pH 6.0, containing 5 mM mercaptoethanol and 60% saturated ammonium sulfate. The specific and total activity of the enzyme remained constant for periods of more than 1 month. Preservation in this fashion constituted the best storage procedure.

Figure 1. A photomicrograph of the crystalline holo TPL prepared from *Erwinia herbicola*.

On heating, the crystalline enzyme was rather stable up to 40 °C, but above 50 °C, it was rapidly inactivated.

Recrystallized enzyme preparations gave a single band on acrylamide gel electrophoresis. The molecular weight of the enzymes was estimated about 200,000, and that of the subunit is about 50,000. The gene cloning and subsequent nucleotide sequencing gave the molecular mass of the subunit 51,439 Da for *Escherichia intermedia* TPL and 51,364 Da for *Erwinia herbicola* TPL (7,8). These data indicate that TPL has homotetramer structure.

The enzyme has been shown to have cofactor requirements (5). Crystalline TPLs prepared by the above procedure showed negligible activity when assayed in the absence of PLP; they are purified as the apoenzyme. Half the maximum enzymic activity was obtained at the respective PLP concentrations of 1.3×10^{-6} M and 1.5×10^{-6} M for the enzymes from *Escherichia intermedia* and *Erwinia herbicola*. On association with PLP, pronounced absorption maxima appeared at 340 and 430 nm. Absorption in the 410 to 440 nm region is characteristic of many PLP enzymes and has been attributed to a hydrogen bound azomethine of PLP. The absorption maxima of holo TPLs were not appreciably shifted by variations in pH between 6.0 and 9.0.

The crystalline enzymes from *Escherichia intermedia* and *Erwinia herbicola* showed a similar requirement for K^+ and NH_4^+ for their maximum activities. However, Na^+ was inactive.

The amount of PLP bound by the apoenzyme was determined by chemical and spectrophotometric titration methods. The results obtained from these two procedures indicated that there are two PLP binding sites per molecule of enzyme. The crystallographic studies showed the existence of sulfate residue at the binding site of phosphate residue of PLP, suggesting the prevention of PLP binding to the binding site. The proper number of PLP molecules bound is estimated to be four per tetramer.

The catalytic properties of the enzyme have been determined with the crystalline TPL from *Escherichia intermedia*. The enzyme catalyzed a series of α,β-elimination, β-replacement, recemization, and reverse reaction of α,β-elimination (equations 1 to 8).

α,β-Elimination Reaction

TPLs catalyze the conversion of L-tyrosine into phenol, pyruvate, and ammonia in the presence of PLP (3,4). Pyruvate formation was also observed with β-chloro-L-alanine, *S*-methyl-L-cysteine, L- and D-serine, and L- and D-cysteine.

L-Alanine competitively inhibits pyruvate formation from L-tyrosine, *S*-methyl-L-cysteine, and L-serine. Phenol, pyrocatechol, and resorcinol also inhibit pyruvate formation, but the inhibition was a mixed type. The α,β-elimination reactions proceed optimally around pH 8.2.

β-Replacement Reaction

TPL catalyzed the synthesis of L-tyrosine from L- or D-serine and phenol in the presence of PLP. β-Chloro-L-alanine, *S*-methyl-L-cysteine, and L-cysteine replaced L- or

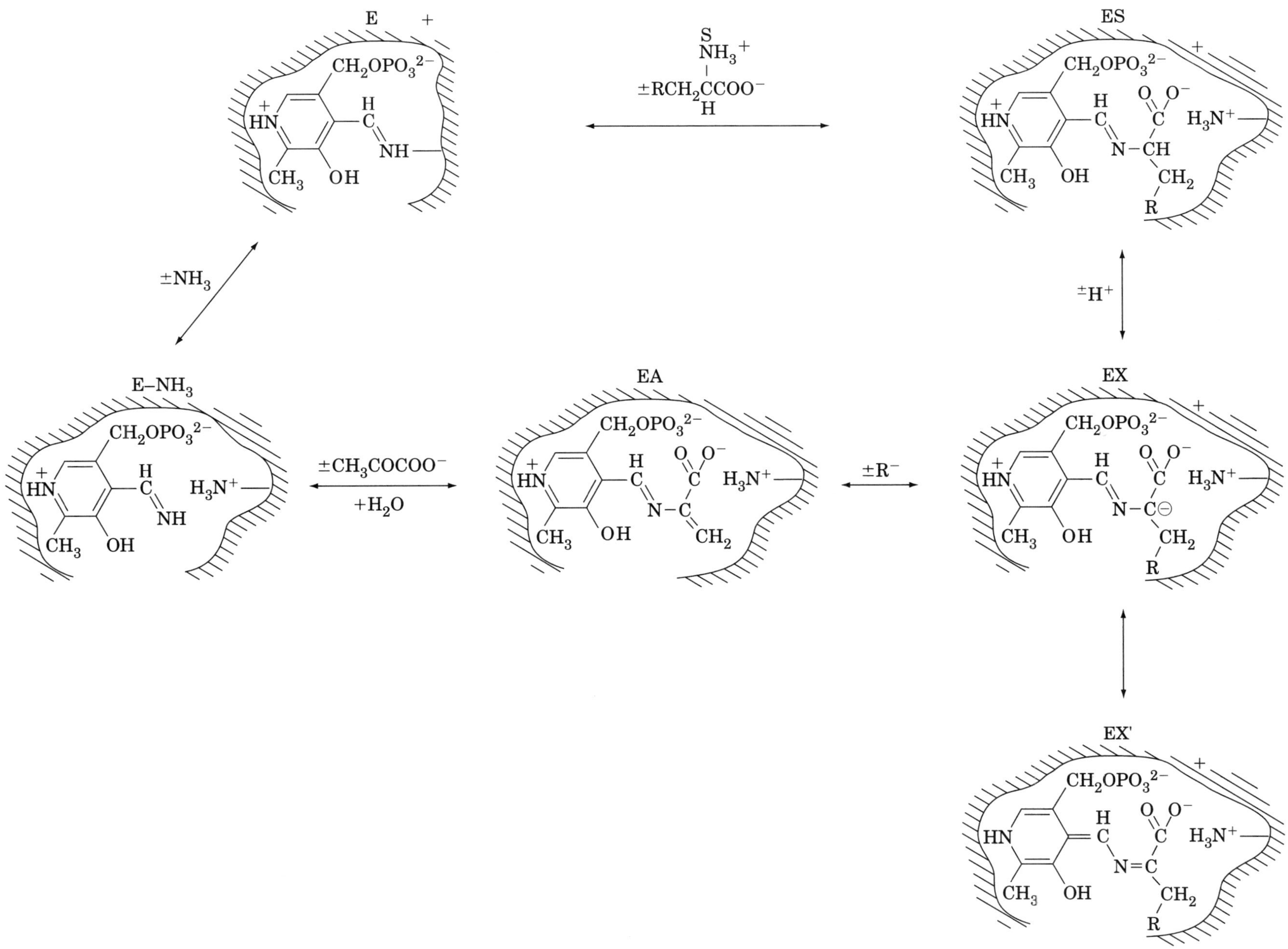

Figure 2. Schematic representation of the mechanism for the reactions catalyzed by TPL. Different species of the enzyme and enzyme-substrate complexes are represented.

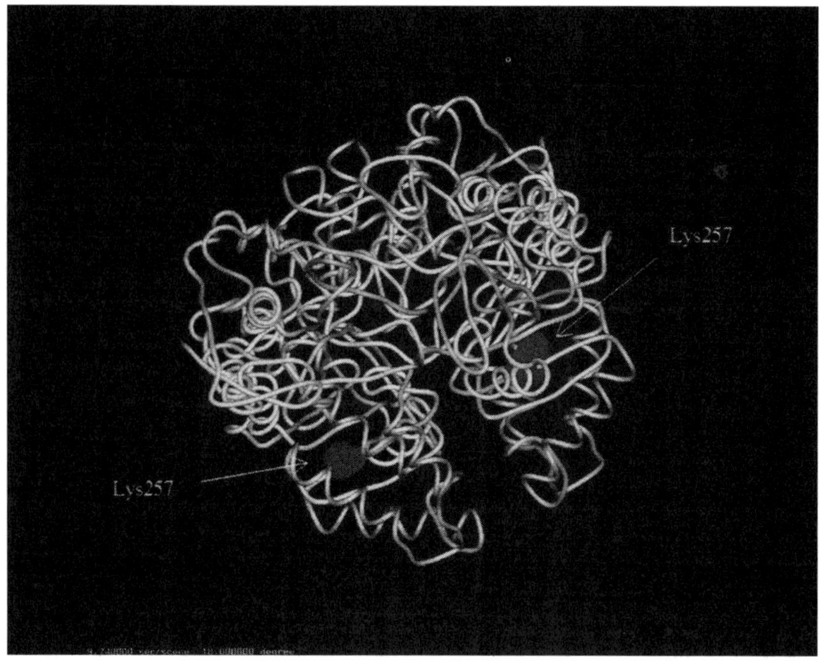

Figure 3. Three-dimensional structure of TPL from *Erwinia herbicola* estimated by computer modeling. TPL has a homotetramer structure, and only dimer structure is shown in this figure. Lysine 257 residues where coenzyme PLP bind are shown by arrows.

D-serine in the synthesis of L-tyrosine (3). When pyrocatechol, resorcinol, pyrogallol, and hydroxyhydroquinone were added to the reaction mixture in place of phenol, L-DOPA, 2,4-dihydroxyphenyl-L-alanine (2,4-L-DOPA), 2,3,4-trihydroxyphenyl-L-alanine (2,3,4-L-TOPA), and 2,4,5-trihydroxyphenyl-L-alanine (2,3,5-L-TOPA) were, respectively, synthesized. These β-replacement reactions proceed optimally at pH 8.0 to 8.5.

Racemization Reaction

The conversion of D- or L-alanine into the racemate is catalyzed by TPL in the presence of PLP (3). The K_m value for L-alanine was 2.6 mM and the maximum velocity of the racemization reaction was 0.05 μmol/min per milligram of protein. The optimum pH for this racemization reaction is between 7.2 and 7.5.

Synthetic Reaction

The synthesis of L-tyrosine from phenol, pyruvate, and ammonia is catalyzed by TPL in the presence of PLP (2,3). This synthesis proceeds as a function of the concentration of phenol, pyruvate, and ammonia. K values were obtained by using three substrate kinetics (3). The K_m values for phenol, pyruvate, and ammonia calculated were 1.1 mM, 12 mM, and 20 mM, respectively, and the maximum velocity was 3.3 μmol/min per mg of protein, which is about 1.5 times higher than that of L-tyrosine degradation by the reaction of equation 1.

When pyrocatechol and resorcinol were, respectively, added to the reaction mixture in place of phenol, L-DOPA and 2,4-L-DOPA were synthesized. Cresols (m- and O-) and chlorophenols (m- and O-) also replaced phenol to synthesize, respectively, methyl-L-tyrosine (2- and 3-) and chloro-L-tyrosine (2- and 3-) (3).

REACTION MECHANISM

The crystalline preparations of TPL from *Escherichia intermedia* and *Erwinia herbicola* catalyze a variety of α,β-elimination, β-replacement, racemization, and synthetic reactions (3). These reactions are partially explained by adopting the general mechanism for pyridoxal-dependent reactions proposed by Braunstein and Shemyakin (9) and by Metzler et al. (10). This mechanism is shown in Figure 2, with a modification in which PLP in the enzyme is shown as an internal azomethine. The enzyme-bound α-aminoacrylate (EA) is the key intermediate in this proposed mechanism. The α,β-elimination reaction is considered to proceed via steps

$$E + S \rightarrow ES \xrightarrow{-H^+} EX \xrightarrow{-R^-} EA \xrightarrow{+H_2O} E + \text{pyruvate} + NH_3$$

The β-replacement reaction proceeds via steps

$$E + S \rightarrow ES \xrightarrow{-H^+} EX \xrightarrow{-R^-} EA \xrightarrow{+R'^-} EX' \xrightarrow{+H^+} ES' \rightarrow E + S'$$

and the racemization reaction proceeds via steps

$$E + L - S \rightarrow E \cdot L - S \xrightarrow{-H^+} EX \xrightarrow{+H^+} E \cdot DL - S \rightarrow E + DL - S$$

The synthetic reaction of L-tyrosine seems to proceed via the reversal steps of α,β-elimination. The stereochemistry and reaction mechanism of TPL were investigated and reported (11).

GENE CLONING AND STRUCTURE ANALYSIS

TPL genes from *Escherichia intermedia* (12), *Cirobacter (Escherichia) freundii* (7), and *Erwinia herbicola* (8,13) were cloned and their nucleotide sequences were determined. The amino acid sequences deduced from the nucleotide sequences showed the following homology: 100% identical between *Citrobacter freundii* and *Escherichia intermedia*; 90.6% between *Escherichia intermedia* and *Erwinia herbicola*. The PLP binding site of TPL is determined as Lys 257 (8). The three-dimensional structure of apo TPL of *Escherichia intermedia* was analyzed by X-ray crystallography at 2.3 Å resolution and reported after refinement with primary amino acid sequence of the *tpl* from *C. freundii*. The three-dimensional structure of apo TPL was estimated from that of *Escherichia intermedia* by computer analysis (Fig. 3).

REGULATION OF TPL BIOSYNTHESIS

Induction and repression of TPL of *Erwinia herbicola* were regulated on the transcriptional level; mRNA of TPL was increased by the addition of tyrosine and decreased by the addition of glucose in the medium. The 5′ flanking region of its gene, *tpl*, was analyzed and its transcriptional start point was determined. TyrR box and operator-like region were also found in this region (14). TyrR box is a typical binding site of DNA where a regulator protein TyrR binds and regulates transcription of structure genes of enzymes or transporters responsible for aromatic amino acid biosynthesis or transportation.

BIBLIOGRAPHY

1. Y. Kakihara and K. Ichihara, *Med. J. Osaka Univ.* **3**, 497–507 (1953).
2. H. Yamada, H. Kumagai, N. Kashima and H. Torii, *Biochem. Biophys. Res. Commun.* **46**, 370–374 (1972).
3. H. Yamada and H. Kumagai, in D. Perlman ed., *Advances in Applied Microbiology*, vol. 19, Academic Press, New York, 1975, pp. 249–288.
4. H. Kumagai, H. Yamada, H. Matsui, H. Ohkishi and K. Ogata, *J. Biol. Chem.* **245**, 1767–1772 (1970).
5. H. Kumagai, H. Yamada, H. Matsui, H. Ohkishi and K. Ogata, *J. Biol. Chem.* **245**, 1773–1777 (1970).
6. H. Kumagai, K. Kashima, H. Torii, H. Yamada, H. Enei, and S. Okumura, *Agric. and Biol. Chem.* **36**, 472–482 (1972).
7. Y. Kurusu, M. Fukusima, K. Kohama, M. Kobayashi, M. Terasawa, H. Kumagai and H. Yukawa, *Biotechnol. Lett.* **13**, 762–772 (1991).
8. H. Suzuki, K. Nishihara, N. Usui, H. Matsui and H. Kumagai, *J. Ferment. Bioeng.* **75**, 145–148 (1993).
9. A.E. Braunstein and M.M. Shemyakin, *Biokhimiya* **18**, 393–411 (1953).
10. D.E. Metzler, M. Ikawa, and E.E. Snell, *J. Am. Chem. Soc.* **76**, 648–652 (1954).
11. M.M. Palcic, S.-J. Shen, E. Schleicher, H. Kumagai, S. Sawada, H. Yamada and H.G. Floss, *Z. Naturforsch.* **42C**, 307–318 (1987).
12. S. Iwamori, S. Yoshino, K. Ishiwata, and N. Makiguchi, *J. Ferment. Bioengineer.* **72**, 147–151 (1991).
13. S. Iwamori, T. Oikawa, K. Ishiwata, and N. Makiguchi, *Biotechnol. Appl. Biochem.* **16**, 77–85 (1992).
14. H. Suzuki, T. Katayama, K. Yamamoto, and H. Kumagai, *Biosci. Biotech. Biochem.* **59**, 2028–2032 (1995).

V

VACCINE TECHNOLOGY

Michael L. Dekleva
Merck & Co., Inc.
West Point, Pennsylvania

KEY WORDS

Adjuvant
Immunization
Vaccine
Validation

OUTLINE

INTRODUCTION

Disease Applications for Vaccines

Vaccines are suspensions of live, killed, or attenuated microorganisms such as bacteria or viruses or components of them. They are designed for human or veterinary use to prime the immune system to produce antibodies or cellular responses that ward off future challenge by pathogenic organisms or their toxins (1–4). *Antibodies* are a group of related proteins that bind specifically to the molecules that induce their formation, *antigens*, and in doing so help facilitate their destruction and clearance from the body. Some molecules, such as carbohydrates, lipids, and small molecules, are themselves poorly immunogenic and must be coupled to proteins to more effectively stimulate an immune response. Although numerous examples exist of vaccines against diseases of bacterial origin (*Haemophilus influenzae, Corynebacteria diphtheria, Streptococcus pneumoniae*, etc.), they are particularly useful prophylactic treatments for viral pathogens, against which antibiotics are ineffective (measles, mumps, rubella, hepatitis B, hepatitis A, etc.). In addition, vaccines have been produced to effectively counteract bacterial toxins such as diphtheria (*Corynebacterium diptheria*) (3) and tetanus (*Clostridium tetanii*) (3) by producing a benign form of the toxin, referred to as a toxoid, by heat or chemical treatment (phenol, formaldehyde, etc.). Vaccines may also offer protection against certain cancers. Hepatocellular carcinoma resulting from chronic hepatitis B infection can be effectively eliminated by preventing the initial infection through vaccination (5–8). Likewise, evidence suggests that certain stomach cancers may be preventable through vaccination with *Helicobacter pylori* preparations (9), or some cervical cancers by vaccination with human papillomavirus vaccines (10,11).

Overview of Preparation

The success of vaccines is most significantly influenced by five factors: (*1*) the nature and dose of the immunogen, (*2*) the adjuvant or excipients used in the formulation, (*3*) the immunization schedule, (*4*) the route of administration, and (*5*) the immune status of the person being vaccinated, influenced by age and disease (2). The vaccine preparations themselves may consist of killed whole bacterial cells or live attenuated or inactivated viral preparations. Alternatively, subunit vaccines are those that rely on recognition of specific components of pathogens by the immune system, instead of the whole microorganism, to elicit a response. These components may include capsular polysaccharides (e.g., vaccines against *Haemophilus influenzae* type b or *Streptococcus pneumoniae*) (12–14), viral coat proteins (15), or other structural components of the organism that can be recognized and responded to by the immune system. Also, recombinant DNA technology provides a means to express subunits of pathogenic bacteria or viruses. For example, a highly effective vaccine against hepatitis B infection has been made by expressing the hepatitis B viral coat protein surface antigen in yeast (15), thereby eliminating the need to the handle infectious blood from which the vaccine was originally isolated. Modern techniques for cell culture, fermentation, isolation, and formulation allow the manufacturer of vaccines of unparalleled purity, specificity, and safety. The high levels of purity offer the advantage of low reactigenicity, with reduced pain, swelling, redness at the site of injection, or systemic effects such as fever. Higher purity presents new challenges, however, in that the highly pure components are typically poorer immunogens than their whole cell counterparts (16–19), driving active research into other agents that can be coadministered (adjuvants) to boost their effectiveness.

Routes of Administration

Vaccines are typically administered intramuscularly or subcutaneously but may be designed, as appropriate, for administration by other routes (e.g., oral polio vaccine). Active research programs exist to study alternate methods for vaccine administration, because most infections are acquired naturally by mucosal routes, either orally, nasally, or genitally, and parenteral vaccines are often poor inducers of mucosal immunity (2). Mucosal vaccines, however, must be designed to overcome the normal rapid clearance, and oral vaccine candidates must be designed to survive the enzymes and pH extremes associated with the gastrointestinal tract (2).

Immunization Strategies

Vaccination strategies include inoculation of individuals who are, because of lifestyle or circumstance, susceptible to a particular pathogen. Alternatively, widespread immunization of a demographic group may be desirable to protect entire populations. This is accomplished in the United States and many other countries through compulsory vaccination of schoolchildren, health care workers, and those serving in the armed forces. This approach has been particularly effective in eliminating the threat of endemic disease in children, such as polio and measles (20,21).

Schedules for vaccine administration vary from product to product and often from one country to the next (4,22). For full and lasting immunity to be achieved, repeated exposure to an antigen may be required (23–25). Therefore, most vaccines are administered through regimens of two to four doses. Even after serum antibody titers drop below detectable levels, immune memory results in a rapid and substantial *anamnestic* response, whereby antibody titers rise rapidly upon revaccination or subsequent challenge with the targeted disease entity (23,25). Table 1 lists a number of currently available vaccines and their administration schedules (2).

It is now widely accepted that the prophylactic prevention of disease through vaccination is an effective and cost-effective means to disease management (4). Smallpox has been eradicated through rigorous worldwide public vaccine programs, and many other diseases have been brought under control, fueling tremendous public optimism in our ability to alleviate morbidity and mortality associated with the many bacterial, viral, and parasitic infections that continue to plague us (20,21). Further applications toward the prevention of reproductive fertility and cancer treatment are also being explored. It is critical at this juncture that continued research into vaccine development and production provide the means to further reduce vaccine costs and make compliance to vaccine schedules easier through the advent of combination vaccines (to reduce required injections) or alternate delivery systems.

EARLY VACCINE HISTORY

Vaccine technology enjoys a rich history with brilliant contributors (26). The first successful vaccine was demonstrated by Jenner in the late 1700s. Jenner had observed that milk maids, who often harbored lesions from benign cowpox infections, were immune to smallpox. He found that a person inoculated with the fluid from a cowpox lesion would be protected from disease if later challenged with fluid from a person with smallpox. The word vaccine is derived from the Latin word *vaccinus*, which means "of or pertaining to cows."

Koch was responsible for advancing the concept that each infectious disease is caused by a single, unique pathogenic species, and whose work with anthrax, cholera, and tuberculosis shook the medical community of his time. Likewise, Pasteur's flamboyant style and flare for publicity popularized the notion, through work with cholera, anthrax, and rabies, that many diseases of animals and people could be prevented or cured. And so early successes with smallpox, cholera, anthrax, and rabies fueled an enthusiastic search for cures and preventative measures for such scourges as diphtheria, tetanus, and polio. Early vaccines for the first time offered hope that the infectious diseases responsible for widespread morbidity and mortality could be prevented.

METHODS FOR VACCINE MANUFACTURE

The earliest vaccine preparations were isolated from tissue fluids obtained from infected people or animals, beginning with Jenner's cowpox vaccine against human smallpox. Through the middle of the 20th century all viral vaccines originated from live or killed whole virus, derived from infected organs or tissues of animals or embryonated hens' eggs (15). Modern methods required for the preparation of vaccines are as varied as the vaccines themselves, tailored to the unique chemical and physical properties of the antigen being cultured or isolated (27–29). Because the course of vaccine treatment is confined to just a few highly potent doses, administered in low volumes (typically a milliliter or less), manufacturing scales for vaccines are small relative to many therapeutic products requiring higher doses or longer courses of treatment.

Propagation of Live, Attenuated Viral Vaccines

The basis for live, attenuated viral vaccines is that by repeatedly subculturing the virus through appropriate host cell lines (e.g., chicken embryo or monkey kidney cell), its virulence can be sufficiently attenuated to avoid a fulminant infection upon inoculation, while still preserving the antigens required to elicit protective immunity. Because viruses are obligate intracellular parasites, they require viable eukaryotic cells for propagation (30–32). The following paragraphs will provide further details on cell culture lines and cell culture techniques that are used in the preparation of vaccines.

Primary Cell Lines. Vaccines have been manufactured by culturing viruses in a variety of animal organs or tissues in vivo or primary cells in vitro. These include embryonated chicken or duck eggs and primary cells obtained from a variety of different animals. Primary cells refer to cells obtained directly from tissue sources and transferred

Table 1. Pharmacokinetic Parameters of Vaccines

Vaccine	Recommended age for first vaccination[a]	Interval for vaccination schedule (mo)	Route	Onset (wk)	Duration (y)	FDA-defined risk factors in pregnancy[b]
		Short-acting vaccines (<2 y)				
Cholera	5 y	0–0.5; every 6 mo	SC or IM (PO) available	2	0.5	C
Influenza	6 mo	Annually	SC	2	1	C
Japanese encephalitis	3 y	0–0.25–0.5; 4 y interval	SC	8	0.5–1	C
Pertussis, accellular[c]	6 wk	0–1–2	IM	4	>1.5	ND
		Moderately short-acting vaccines (2–5 y)				
Haemophillus influenzae b	6 wk	0–2–4–10	IM	16[d]	>4	C
Meningococcus	2–6 y	Single dose	SC	2	3	C
Polio, inactivated	6 wk	0–1–6	SC	6–24	5	C
Typhoid	3 y	4 doses over 1 wk	PO, fasting	1	5	C
		Moderately long-acting vaccines (5–10)				
Diphtheria	6 wk	0–1–2–8; 6 y	Deep SC or IM	8	>10	C
Hepatitis A	2 y	0–1–7	IM	3	5–10	ND
Hepatitis B	2 mo	0–1–6	IM	4–8	5–7	C
Pneumococcus	2 y	Single dose	SC or IM	2–3	5–8	C/X
Varicella	12 mo	0–1	SC	6	5–10	C (D in first trimester)
		Long-acting vaccines (>10 y)				
Measles	9 mo	Single dose; 6 y interval	SC	ND	>21	X
Mumps	15 mo	Single dose; 6 y interval	SC	4	15	X
Polio, live	2 mo	0–2–16	PO	1	>12	C/X
Rubella	12 mo	Single dose	SC	2–4	Lifelong	X
Smallpox	12 mo	Single dose	ID with bifurcated needle	ND	30	X
Tetanus	6 wk	0–1–2–8; 6 y interval	SC or IM	4	12	C
Yellow fever	9 mo	10 y interval	SC	2	30–35	D
		Duration not clear				
Rabies	12 mo	0–0.25–1	ID or IM; postexposure IM	1–2	ND	C
Tuberculosis (BCG)	3 mo	Single dose	ID	2–3	ND	C

Source: Ref. 2.
Note: BCG = bacille Calmette-Guerin; FDA = U.S. Food and Drug Administration; ID = intradermal; IM = intramuscular; mo = month(s); ND = no data available; PO = oral; SC = subcutaneous; wk = week(s); y = year(s).
[a]Younger children may receive some of the vaccines.
[b]C = animal studies have revealed adverse effects on the fetus (use only if potential benefit justifies the potential risk to the fetus); D = positive evidence of human fetal risk; X = human beings have demonstrated fetal abnormalities or evidence of fetal risk, based on human experience.
[c]Duration is still under investigation.
[d]Also dependent on age.

for the first time into an in vitro growth environment (32). Examples include chick fibroblasts cultured from chicken embryos for the production of measles and mumps vaccines. Cell lines for viral propagation grow either as attachment-dependent or, occasionally, suspension cultures (32).

To produce measles and mumps vaccines, those viruses are grown and attenuated by passage through cultures of chicken embryo cells. Chicken embryos are harvested from eggs, obtained from special isolated flocks, and then minced and treated with disaggregating enzymes such as trypsin, collagenase, hyaluronidase, and pronase. The chick fibroblast cells released by this process are attachment dependent, requiring solid surfaces for growth. The successful commercial manufacture of viral vaccines in attachment-dependent cell systems rely on the establishment and maintenance of healthy cell monolayers. Selection of the appropriate growth and maintenance media are critical, as are careful attention to nutrient depletion, waste accumulation, and changes in pH over time (33).

Diploid and Continuous Cell Lines. Serially passaged diploid strains of cultured human cells were first described in the 1960s by researchers at the Wistar Institute (34). The concept of using human diploid cell lines for vaccine preparation was hotly debated in the 1960s, for fear that cancer-causing DNA or human viral agents might be unknowingly coadministered with the vaccine. Extensive karyological (i.e., chromosomal) characterization and thorough searches for viral contaminants were necessary to ensure that cultures were free of exogenous infectious agents before the first diploid cell product, poliomyelitis vaccine, was licensed in the United States (11). Human diploid cell lines (WI-38 or MRC-5) have since been used to produce a number of licensed vaccine products against poliovirus, adenovirus types 4 and 7, rubella (German measles) virus, rubeola (measles) virus, rabies, hepatitis A, and varicella virus (chickenpox) (30–32). A distinguishing feature of the human diploid cells such as WI-38 or MRC-5 is that they have a finite life span, reaching senescence after 40 to 60 population doublings (32). In contrast, continuous cell lines exhibit no such constraints and divide indefinitely. An example of a nonhuman continuous cell line that has been used successfully for the manufacturing of vaccines such as polio, rabies, and influenza is Vero (32,35). Vero cells are a continuous monkey kidney cell line (35).

Serially cultured or continuous cell lines are advantageous in that each new production batch is derived from a uniform master cell bank, characterized to be free of contaminating infectious agents, contaminating proteins, or nucleic acids. Proper maintenance of a master cell bank is critically important for the consistency of cell culture products. They are usually stored as small aliquots in liquid nitrogen and tested at intervals for chromosomal (karyology) and cell integrity to establish long-term stability (32).

Cell Culture Reactors. Cylindrically shaped roller bottles have been used successfully to establish monolayers of chick fibroblasts on their inner surfaces, which can be monitored microscopically through the clear plastic. The cell sheets are continuously bathed by a growth medium contained within the bottle during slow axial rotation on special roller racks. After the cell sheets are established, they are infected by introduction of the specific virus. After incubation, the cells and virus may be harvested (e.g., varicella). In the case of rubella, viral fluids may be harvested at approximately 2- to 3-day intervals, through 10 to 12 harvest cycles. Although millions of doses have been successfully manufactured using roller bottles, capacity is limited by the space that a large number of roller bottles requires and by the time-consuming and often labor-intensive manipulations for harvesting and pooling the viral fluids from them. Robotics may reduce the labor burden associated with the latter and have been successfully used by Merck for the manufacture of their chickenpox vaccine, Varivax, in roller bottles (33).

Multidisk reactors offer an alternative to roller bottles for attachment-dependent cell lines (36). They consist of 10-L stainless steel reaction vessels containing approximately 100 parallel titanium disks that slowly rotate through growth media and provide solid surfaces for cell attachment and monolayer formation. In addition to their improved capacity, the stainless steel reactors are jacketed for temperature control. Monolayers cannot be microscopically monitored, however. Another alternative for cultivating attachment-dependent cells was described in 1979 (37) and consists of stacked polystyrene unit trays (NUNC single-tray unit). Again, cell monolayers can be established, infected, and harvested more efficiently than with roller bottles (33). Other solid supports for attachment-dependent cell lines include ceramic or membrane (hollow fiber) solid supports, and microcarriers. Microcarriers may be composed of a variety of materials such as gelatin, polystyrene, cellulose, polyacrylamide, dextran, or glass, but usually are approximately 200 μm in diameter. They offer advantages of improved process control and scaleability because, like suspension cultures, they can be cultured in large-scale bioreactors. They are, however, susceptible to shear stress and physical deterioration and so must be suspended with low-shear agitators (38). For further reading on cell culture techniques for attachment-dependent cell lines and suspension cultures, please refer to *Large-Scale Mammalian Cell Culture Technology* (39,40).

Purified and Killed Whole Viral Vaccines

Examples of purified whole viral vaccines includes those to protect against polio, influenza, and hepatitis A viruses.

For the large-scale preparation of a killed polio vaccine, viral infected Vero cells have been grown on microcarriers in fermenters (35). Purification of the whole virus vaccine is as follows (36). The crude virus suspension is clarified by filtration, then concentrated 250-fold by hollow-fiber ultrafiltration (100,000 Da MW cutoff), which retains >99% of the virus. The virus is further purified by DEAE-Sephadex chromatography and then sterile filtered through a 0.22-μm filter before formaldehyde inactivation.

Traditionally, methods for influenza vaccine manufacture have always involved growth of the virus in embryonated chicken eggs. Recently, a manufacturing process for a formaldehyde-inactivated whole virus vaccine against influenza virus was developed using Vero cells in a microcarrier-based fermentation system (35). Supernatants from microcarrier cultures are clarified before benzonase treatment to digest Vero cell DNA. After formaldehyde inactivation, the product is concentrated by ultrafiltration (200,000 Da MW cutoff) and further clarified by protamine sulphate precipitation. Further purification is accomplished by continuous-flow zonal centrifugation of a 0 to 50% sucrose gradient, followed by ultrafiltration and sterile filtration.

Development of the hepatitis A vaccine began with demonstration that formaldehyde-inactivated virus extracted from the livers of infected marmosets would induce protective antibodies in susceptible marmosets (41). During the years that followed, various cell culture and purification schemes were developed for hepatitis A vaccine preparations (42,43). VAQTA® is one example (44,45), and is a highly purified, formaldehyde-inactivated hepatitis A viral vaccine. The virus is harvested from MRC-5 cells grown in Costar Cubes (46,47) after lysis with Triton X-100, followed by nuclease treatment of the lysate to digest nucleic acids. Concentration of the product is accomplished by adsorption on an anion exchange column and then PEG precipitation. Solvent extraction of the resuspended PEG pellet contributed significantly to the purification, as did the anion exchange and SEC steps that followed. The product is subsequently formaldehyde inactivated and adsorbed onto aluminum hydroxide.

Purified Subunit Vaccines

The high levels of purity achieved by subunit vaccines have the potential to increase the specificity and significantly reduce the number or degree of adverse reactions associated with less-pure preparations. Their goal is to present the patients' immune systems with the specific feature, be it capsular polysaccharide in the case of *Haemophilus influenzae* or *Streptococcus pneumonia*, or viral coat proteins in the case of hepatitis B vaccines, conserved in all variants of a given species such that specific antibodies produced against it will neutralize the whole organism upon subsequent invasion. These components may be isolated from the organism itself, or expressed by a suitable host through recombinant DNA technology.

Manufacturing processes for subunit vaccines typically have six steps in common (1,48,49), which include (*1*) cell multiplication, either through fermentation or cell culture; (*2*) phase separation to remove cells, if the product is extracellular, or to harvest and wash cells if the product is intracellular; (*3*) product recovery; (*4*) purification; (*5*) polishing; and (*6*) adjuvanting and filling.

Cell Multiplication. The source culture for a subunit vaccine varies, depending on the nature of the product. Examples where components of the pathogens themselves are purified to produce vaccines include *Streptococcus pneumoniae, Haemophilus influenzae* type b, *Neisseria meningitidis, Clostridium tetanii, Corynebacteria diptheria*, or *Bordatella pertussis*. These bacteria are usually grown in batch culture on complex nutrients, such as yeast extract and soy peptone, with addition of specific growth factor supplements, such as hemin, that may be required for growth (50). As a general rule, defined or minimally complex media are sought to minimize purification demands. The antigens of interest, from the above examples, may be capsular polysaccharides (*Streptococcus pneumonia, Haemophilus influenzae*), or protein toxins that are chemically or thermally inactivated (tetanus, diphtheria, or pertussis). Capsular polysaccharide vaccines are sometimes adjuvanted by chemically conjugating them with protein components of other bacteria. Examples are Hib conjugate vaccines in which the capsular polysaccharide of *Haemophilus influenzae* type b is chemically coupled to the outer membrane protein complex of *Neisseria meningitidis*, diptheria toxoid, or tetanus toxoid (12,14). And although the current pneumococcal vaccine is formulated using 23 types of *Streptococcus pneumoniae* (1, 2, 3, 4, 5, 6B, 7F, 8, 9N, 9V, 10A, 11A, 12F, 14, 15B, 17F, 18C, 19F, 19A, 20, 22F, 23F, and 33F), new pneumococcal polysaccharide–protein conjugate vaccines are being developed to improve T-cell-dependent antibody responses, improve long-term immunologic memory, and otherwise enhance immune responses in young children, the elderly, and other groups at risk for pneumococcal infection (13).

Alternatively, proteins representing key antigenic determinants from a target pathogenic microorganism can be recombinantly expressed in bacterial, yeast, or mammalian cell cultures. Important factors influencing the choice of a host expression system are freedom from adventitious agents (51,52), the scientific and organizational experience with a given system, efficiency of product expression, the need for specific glycosylation patterns for full efficacy, or the desirable purification characteristics. Proteins expressed in *Escherichia coli* are not glycosylated and accumulate intracellularly in inclusion bodies that require denaturation, then refolding to return the protein to its native conformation. Proteins expressed in yeast may be either excreted or accumulated intracellularly as viruslike particles, depending on the construct. Mammalian host systems express proteins extracellularly (17). Yeast and mammalian cell host systems, because they are eukaryotic, are often considered when consistent patterns of glycosylation are important (53). Expression systems are also chosen for their scaleability, so that conditions studied in the laboratory can be more easily applied to full-scale manufacturing. Growth media vary depending on the specific host system and on the recombinant expression system. For example, stable maintenance of a plasmid-encoded protein in yeast or bacteria may depend on a culture medium deficient in a particular amino acid, where the plasmid compensates or a specific host auxotrophy (54,55). Alternatively, the coding region for a particular protein product may be regulated by an inducible promoter, such that product expression is triggered by addition of the appropriate chemical or nutritional component (54,55).

Frozen master cell banks (−70 °C), consisting of clonal isolates tested to ensure purity and stability of expression, are subpassaged to generate a working cell bank. In this way, batch-to-batch consistency and purity can be ensured.

Stirred tank reactors, in volumes of hundreds to thousands of liters, are commonly used to cultivate bacterial or recombinant cultures. The bioreactors are jacketed for control of temperature and sparged to provide appropriate levels of oxygenation. Monitoring may be on-line or off-line, by periodic sampling and analysis, for such parameters as biomass, carbon-source (e.g., glucose) consumption, pH, temperature, dissolved oxygen, etc. Valuable information can also be attained on culture physiology by monitoring head space gases for carbon dioxide evolution or oxygen consumption or volatile organic by-products (e.g., ethanol, in the case of yeast cultures). In addition to proper maintenance of master and working cell banks, equipment maintenance and careful attention to fermentation raw materials are essential to ensure reproducible fermentation results (50,56).

Phase Separation. After the final production phase of cell culture, it is usually necessary to separate cells from the remainder of the fermentation broth. This may be accomplished either by continuous centrifugation or microfiltration. If the product is intracellular, the removal of residual culture media components or antifoaming agents from the whole cells is a useful method to reduce the number of contaminants that must be removed by subsequent downstream steps. Regardless of whether products are intracellular or extracellular, care must be taken to avoid cell disruption by excessive shear or osmotic stress to prevent cells from spilling their contents. In the case of intracellular product, this would result in yield losses, and for extracellular products, it would result in the cells' intracellular contents adding to the total number of impurities that must be removed by subsequent downstream steps.

Product Recovery. Figure 1 depicts the typical flow of downstream processing from recovery through purification. Intracellular products are liberated by cell disruption, which is typically accomplished using either bead mills or continuous homogenizers. Bead mills rely on mechanical shear, whereas continuous homogenizers create fluid shear by forcing the cell slurry through a small opening at high pressure (50). Bacterial inclusion bodies require solubilization with chaotropes such as urea or guanidine HCl, followed by renaturation in appropriate buffers (1,27). Virus-like particles in yeast may be tightly associated with intracellular membranes, such as endoplasmic reticulum, requiring detergents for liberation (54). Removal of cell debris is often accomplished by filtration or centrifugation. Where concentration of the desired product is necessary, membrane ultrafiltration is common.

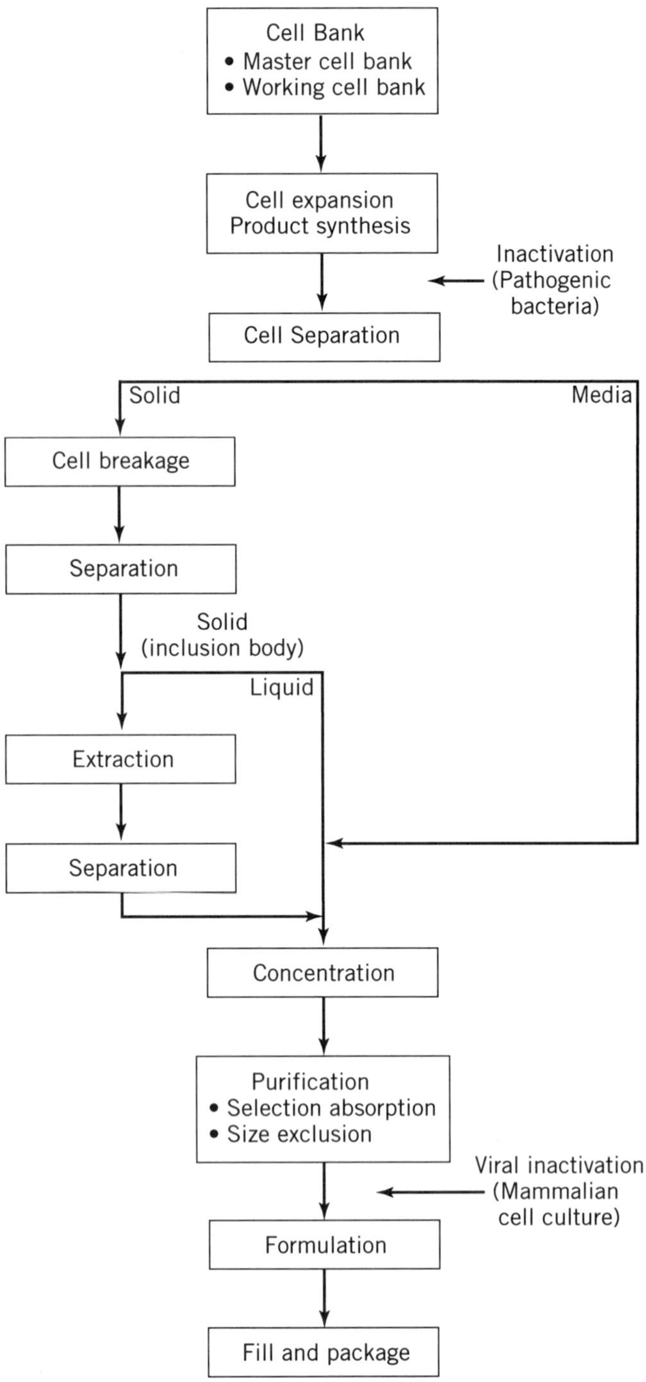

Figure 1. Generic process flow: microbial fermentation or cell culture products.

Purification. Numerous choices exist to accomplish purification, and selection of the most appropriate technologies must be grounded in an understanding of the physical, chemical, and biochemical properties of the starting material and key impurities (1,48,49). Some of the options are outlined in Table 2 (27–29,57). Because yield losses occur at each purification step, the number of steps should be kept to a minimum. This will improve overall recoveries and reduce costs by keeping resource requirements, including raw materials, testing costs, and labor, to a minimum (27–29,48,49,58).

Manufacture of Hepatitis B (Subunit) Vaccines: The original hepatitis B vaccine (HEPTAVAX B®), licensed in 1981, was a subunit vaccine isolated from the blood of hepatitis B infected donors (15). Although derived from infected blood, the final product consisted of a highly purified preparation of 22-nm particles containing HBV surface antigen, subunits of the HBV virus. These particles are expressed in copious amounts by patients suffering from acute infection or in hepatitis B carriers. Although these

Table 2. Downstream Operations for Large-Scale Recovery and Purification of Vaccines

Operation	Physicochemical Property
Cell breakage	
Bead Milling	Liquid/solid shear
Homogenization	Pressure gradient
Cell separation and product recovery	
Filtration	Particle size
Microfiltration	Particle size
Centrifugation	Sedimentation velocity
Precipitation	Solubility
Two-phase extraction	Partition coefficient
Purification	
Adsorption	Hydrogen bonds, polarity, dipole moments, etc.
Ion exchange	Charge
Affinity chromatography	Biological affinity
Hydrophobic interaction	Hydrophobicity
Ultrafiltration	Molecular size
Gel filtration	Molecular size
Reverse-phase liquid chromatography	Hydrophobic and hydrophilic interactions

subunits themselves are noninfectious, they are effective in stimulating a strong immune response and provide lasting protection against hepatitis B infection.

The isolation process consisted of (*1*) defibrination of the plasma with calcium; (*2*) ammonium sulfate precipitation, rate zonal centrifugation (sucrose gradient), and isopycnic banding (sodium bromide); (*3*) pepsin digestion; (*4*) urea denaturation/renaturation; (*5*) SEC; and (*6*) formaldehyde treatment. The process not only resulted in greater than 99% purity, but also ensured safety from extraneous infectious viruses through the sequence of viral inactivation steps—pepsin digestion, urea denaturation, and formaldehyde treatment (15).

Problems associated with the manufacture of plasma-derived hepatitis B vaccines included limitations in the supply of source plasma and a manufacturing process that was state-of-the-art at that time, but still laborious and requiring a year or more to complete each batch. In addition, although HEPTAVAX B® had an excellent safety and efficacy profile in millions of people around the world (59), it was not quickly accepted because of growing concerns over AIDS that might be present in the same infectious plasma that served as the source for the vaccine (59). These concerns led to strong public and government pressure to develop a recombinant hepatitis B vaccine for human use, of which RECOMBIVAX HB® was the first. This recombinant hepatitis B vaccine is prepared for human use by expression of the hepatitis B surface antigen (HBsAg) in a recombinant yeast strain that is batch cultured in stirred-tank fermenters. Cells are separated from the fermentation broth, concentrated, and washed with buffer (to remove media components and antifoam) by microfiltration. Cells are lysed to release the HBsAg, which is accumulated intracellularly by the yeast cells, by passage through a high-pressure homogenizer. Phenylmethylsulfonyl fluoride, a protease inhibitor, is added immediately before lysis to prevent proteolytic degradation. Triton X-100 detergent is added to liberate the HBsAg from tightly associated membrane components, and the product is recovered using a microfiltration system. Cell debris and unbroken cells are retained by the filter and eventually discarded, and the HBsAg is recovered in the filtrate. The product stream is ultrafiltered against a 100-kDa molecular weight hollow-fiber filter cartridge to clear small molecular weight contaminants and concentrate the product. Residual detergent is removed by recirculation through polystyrene beads (XAD-4) followed by adsorption of the antigen on colloidal silica (Aerosil). The HBsAg is eluted with warm borate buffer, which interacts with the silica surface to facilitate its release. Final polishing is accomplished by hydrophobic interaction chromatography, using butyl agarose, followed by treatment with thiocyanate to promote disulfide cross-linking of the protein components of the HBsAg particle. Afterward, the product is sterile filtered through a 0.22-mm membrane. The product is adjuvanted by coprecipitation with aluminum hydroxide (60).

Process Ruggedness. Once promising vaccine candidates are identified through research, and their safety and efficacy is proved through clinical studies, strong medical and financial drivers motivate a rapid introduction of the product into manufacturing. Once the manufacturing facility is constructed and qualified, and the final process is established, a consistency series of three to five lots serves to confirm that product profiles are unchanged through the final process and facility scale-up. Data from these full-scale runs support product licensure. Therefore, at the time of licensure, a considerable body of information will have been amassed at laboratory and pilot scale, but full-scale manufacturing data are limited. Because of the speed through which final process scale-up and manufacturing introduction are accomplished, abundant opportunities generally exist to improve manufacturing efficiencies and overall productivity after product introduction. Such changes must be approved by regulatory agencies before incorporation, supported by data confirming that the product's safety, purity, or potency are not diminished.

It is important to be aware of common pitfalls in new process introductions that might later interfere with a consistent product output. Process ruggedness is a key consideration. For example, a single lot of a key reagent, such as a fermentation raw material or chromatographic resin, may suffice to prepare numerous test lots at laboratory scale, and there may be a strong temptation to show process consistency by using a single lot of a key reagent during the manufacturing demonstration. It is absolutely crucial, however, to identify those critical raw materials early in process development and challenge the process with different vendor lots and from multiple vendors, if possible, to better understand the impact that those changes will have later on in manufacturing when much larger quantities will be consumed and when lot-to-lot variations will stress process consistency. Critical process parameters must also be identified early and the process challenged around them to better understand failure limits. In that way, unplanned process deviations associated with equipment failures or operator error can be more easily addressed, and rapid decisions can be made on the quality of a particular lot. Also, the more thorough the understanding of which steps are truly critical to achieving product quality goals, the less likely that a deviation at a noncritical step will result in product discard (52,61,62).

ADJUVANTS

The term *adjuvant* is derived from the Latin word *adjuvare*, which means to help. Adjuvants are used in conjunction with vaccines to aid in eliciting a rapid, high-level, and long lasting humoral or cellular immune response to the target antigen. Their significance is increasing because of the development of more highly purified subunit and synthetic vaccines that are more specific but inherently are less immunogenic. In addition, such vaccines tend to be very expensive to produce, so an effective adjuvant should also allow for the use of less antigen to stimulate the desired level of immunity. Ideally, an adjuvant should (*1*) enhance the ability of an antigen to stimulate a strong and persistent immune response in all age groups targeted for vaccination, whether newborns or older adults; (*2*) allow lower or fewer doses of antigen to be used; (*3*) be biodegradable; (*4*) not itself be immunogenic; (*5*) not interact with the antigen to cause its deterioration; and (*6*) be nontoxic and without side effects (63). An important issue for the development of vaccines today for human use is striking a balance between stimulation of a strong immune response and toxicity or other undesirable side effects. Those side effects may be localized or systemic. Local reactions, such as pain, swelling, and redness, can be a result of depot formation (localization of an antigen), which is also a primary mechanism of action of some adjuvants, such as mineral compounds, oil–water emulsions, liposomes, or biodegradable polymeric microspheres. These geltype or emulsion-based-adjuvants increase the half-life of the antigens associated with them, which might otherwise be cleared too rapidly from the body for an effective immune response to be mounted. Alternatively, systemic reactions may develop, depending on the type and amount of cytokines that might be elicited after stimulation of the immune system.

Adjuvant History

In 1916, LeMoignic and Pinoy first described the use of mineral oil and lanolin as an adjuvant emulsion in a killed *Salmonella* vaccine (64). In the 1920s, Ramon showed that immune responses to tetanus and diphtheria toxoids were enhanced when used in conjunction with other compounds such as agar, tapioca, lecithin, starch oil, saponin, or even breadcrumbs (65,66). Later, in 1926, Glenny described the use of aluminum compounds as adjuvants (19,67). In the early 1930s and 1940s, Freund developed his water-in-oil emulsion containing killed mycobacteria that became known as Freund's complete adjuvant. The reactigenicity was reduced significantly by elimination of the mycobacterial component. This modified formulation became known as Freund's incomplete adjuvant and was successfully used in several vaccine formulations (19). In 1956, Johnson et al. reported that lipolysaccharide endotoxins (LPS) from Gram-negative bacteria had adjuvant activity (19,68). This work provided the foundation for further studies into the use of detoxified LPS, or subcomponents such as lipid A (19,69). In 1974, Lederer et al. identified muramyl dipeptide (MDP) as the smallest adjuvant-active component of mycobacteria (19,70). Since then, more than 300 compounds have been investigated in search of those retaining the potent adjuvant activity but without unwanted side effects such as pain, redness, swelling, or fever.

Types of Adjuvants and Mechanisms of Action

Adjuvants are a chemically diverse group of substances that can be categorized based on chemical properties and physical or chemical properties (19). Adjuvants have also been characterized by mechanism of action (19,63,71), including (*1*) causing depot formation at the site of injection (e.g., mineral compounds, oil-based adjuvants, liposomes [72], oil adjuvants, biodegradable polymeric microspheres of >10 mm, etc.); (*2*) delivery vehicles for antigens to target specific immune-competent cells such as macrophages or dendritic cells (e.g., liposomes, oil adjuvants, biodegradable polymer microspheres <10 mm, nonionic block polymer surfactants); or (*3*) acting as immunostimulators (e.g., Freund's complete adjuvant, muramyl depeptide, lipopolysaccharide, lipid A, monophosphoryl lipid A, pertussis toxin, etc.) to facilitate production of relevant lymphokines. Antigens can also be modified to enhance immunogenicity (adjuvant effect) through (*1*) polymerization of the protein constituents using aldehydes (tetanus and diphtheria toxoids); (*2*) cationization of proteins (diphtheria toxoid); or (*3*) chemical conjugation of polysaccharide antigens to carrier proteins (e.g., *Haemophilus influenzae* vaccines), which may provide T-cell epitopes to the coupled antigen or simply increase its effective molecular weight.

The only adjuvants used routinely in human vaccines are aluminum salts (16,19), either aluminum hydroxide or aluminum phosphate (19). These have a long history of safety and efficacy, since it was discovered in the 1930s that suspensions of alum-precipitated tetanus and diphtheria toxoids had a much higher immunogenicity than the soluble toxoids, and their use became common with diphtheria-tetanus-pertussis (DTP) vaccines. Billions of doses have been safely administered. Despite their reputation for safety, alum adjuvants do have some drawbacks. They have occasionally been associated with subcutaneous nodules or allergic reactions. They cannot be frozen or lyophilized without collapse of the gel structure. Aluminum-based adjuvants are primarily effective in eliciting a humoral as opposed to cell-mediated response (16). This is a significant limitation when considering their use with vaccines targeting a T-cell response, such as against intracellular viral or parasitic agents such as AIDS or malaria, respectively (19,63,71).

Alum has several forms, including aluminum hydroxide and oxyhydroxides, aluminum phosphate gels (73–76) and precipitated alum [$KAl(SO_4)_2 \cdot 12H_2O$]. Preparation of vaccines or toxoids with aluminum adjuvants is either by coprecipitation of the antigen with the aluminum-based compound $(KAI(SO_4)_2 \cdot 12H_2O$ added to the product (76) or adsorption onto preformed aluminum compounds. Currently, most vaccines are manufactured by the adsorption methods. Their mechanism of adjuvant activity includes (*1*) depot formation at the site of injection, from which antigen is slowly released; (*2*) stimulation of immune-competent cells through complement activation and induction of eosinophils and macrophages; and (*3*) efficient uptake of aluminum-adsorbed antigens due to their particulate nature and optimal size (<10 mm) (21).

Research continues to contribute to a better understanding of adjuvant molecular mechanisms, the biochemistry of the immune response to various diseases, and how to selectively invoke classes of cytokines or immune-competent cells to best suit an antigen's specific purpose (i.e., humoral or cellular response) (19,71). In addition, controlled release of vaccines through the use of biodegradable microspheres may provide a mechanism for single-dose vaccines by either slow or intermittent antigen release. This is particularly important in some developing countries where contact between health care workers and the vaccine recipients for booster doses may be difficult to achieve (77,78).

REGULATORY CONSIDERATIONS FOR VACCINES

Published Federal Guidelines

Vaccine manufacture in the United States is regulated by the arm of the Food and Drug Administration (FDA) known as the Center for Biologics Evaluation and Research (CBER). In addition, the FDA's *Code of Federal Regulations* is a codification of the general and permanant rules published in the *Federal Register* by the executive departments and agencies of the federal government. Title 21, parts 600 to 699, was written with certain vaccines in mind (pertussis, typhoid, anthrax, cholera, polio, measles, mumps, and rubella), but it has evolved to include guidance for the manufacture and testing of all vaccines and many other biological products. Included are sections on personnel responsibilities and training, facilities, equipment, and animal care, record maintenance and retention, labeling, packaging, and shipping.

Parts 210 and 211 of Title 21 provide manufacturers of human drugs, veterinary drugs, and medical devices with guidance regarding current good manufacturing practices (cGMPs). The cGMPs are comprehensive and encompass manufacturing methods, facilities, controls, packaging, storage, and installations for drug operations. Currently, licensure of a vaccine product entails licensure of both the product and procedures by which it was made and tested, packaged, labeled, and distributed (product license) as well as the facility and equipment in which it was made (establishment license). The latter includes air handling systems; water and air supply and testing; personnel, product, and equipment flows; materials of construction for equipment; interior and exterior materials of construction; and cleaning and decontamination procedures. Government inspections are held before product approval and at a minimum of every 2 years thereafter to ensure compliance with existing regulations. Minor adverse inspection observations are presented by the inspectors, at the conclusion of an inspection, to the responsible head of the facility, on form 483. Other mechanisms exist for the FDA to prevent continued distribution or manufacture of products if more severe compliance violations are observed. In any event, the observations from FDA inspections are published and provide manufacturers with information on emerging trends within the FDA (79,80).

Process Validation

Regulations governing vaccine manufacture have evolved and continue to do so. In recent years, that evolution has led to an ever-increasing emphasis on validation, which is a formal method to demonstrate that a given process can accomplish its intended goals consistently and reproducibility. The need for validation of sterilization and cleaning procedures has gained widespread acceptance to ensure that equipment in direct contact with manufactured products is free from extraneous chemical or microbial contamination and from product-to-product or batch-to-batch carryover. In addition, testing procedures used to assess potency, and purity must also be validated.

Increasing emphasis is being placed by the FDA on process validation (81). The goals of process validation are to ensure the consistency of consecutive batches of vaccine using a particular process (82). Each process consists of a series of steps, referred to as unit operations, executed in sequence by a prescribed "recipe." Each of those operations has a specific goal. In the case of fermentation or cell culture, the goal is to increase cell density and expression of the particular antigen of interest. Downstream steps may include release of intracellular antigens through cell disruption, filtration or centrifugation steps, salt or alcohol precipitations, filtration steps to remove small molecular weight contaminants or exchange buffers, or chromatographies to remove contaminants such as unwanted proteins or nucleic acids. Additionally, chemical steps may be required achieve an optimal conformation for enhanced immunogenicity or adjuvant binding. Regardless of the steps, each serves a particular purpose to improve the purity or maintain or enhance the potency of the particular antigen of interest.

To accomplish process validation, a formal document or protocol is prepared delineating those steps critical to the overall success of the process, the intended goals of each of those steps, and the analytical methods required to monitor the consistency of the process to achieve its intended goal. Expectations are established early through laboratory- and pilot-scale runs, and then confirmed upon scale-up to the final manufacturing scale. Those equipment settings necessary to achieve a consistent outcome are referred to as critical process parameters. Analytical measurements of process performance are referred to as critical quality attributes. Prerequisites to process validation include appropriate controls over raw materials; qualification and maintenance of utilities and equipment; calibration and preventative maintenance of scales, gauges, and other mechanical and electronic monitoring devices; comprehensive quality control programs for analytical testing and established procedures for batch review and release; unambiguous manufacturing instructions approved by manufacturing and quality control personnel; and programs for comprehensive training. In addition, validation of equipment cleaning and sterilization are necessary.

Filter validation is another emerging arm of process validation. Goals include insurance that filter membranes, housings, and support components do not leach objectionable chemicals into the product stream during manufacturing and that the filters are sufficiently rugged to stand up to process conditions without integrity failure. Critical process filters are tested before and after use for integrity to provide greater assurance that they performed properly toward their intended goal.

Included among the overall benefits of process validation are the greater understanding of process variables, allowing the manufacturer to more reliably supply customers with the highest quality product with fewer discarded batches.

FUTURE DIRECTIONS FOR VACCINE TECHNOLOGY

Research into vaccine technology continues on many fronts to provide safe, efficacious, and affordable vaccine products. Many at-risk individuals living in Third World and endemic areas do not have ready access to health care, so compliance to the multidose regimens required to achieve full protection may be difficult. Also, cold chains may not exist for those vaccine preparations that must be kept refrigerated to maintain proper levels of potency. Improvements in the thermal stability of vaccine preparations is advantageous (83).

One approach to improve coverage that is being adopted by many vaccine manufacturers is to combine antigens into single preparation so that immunity to two or more diseases can be achieved with a single injection (84). Time-release polymeric preparations are also being explored even for monovalent preparations to achieve repeated antigenic stimulation of the immune system without the need for multiple injections. Other research initiatives seek to eliminate the need for needles for delivery of vaccines through the use of oral vaccines (85,86), aerosolizers for inhalation (87), or needleless devices that use compressed gases at high pressure to deliver transdermal doses of vaccine preparations (88).

A fairly recent development in vaccine technology may offer several advantages over more conventional preparations in terms of manufacturing ease, potency, and flexibility in terms of delivery systems and disease targets. Observations in the early 1990s that plasmid DNA-encoding antigenic proteins can elicit an immune response when injected directly into animals has sparked intensive research and tremendous excitement as well as concern over long-term safety (89). With DNA vaccines, the coding region for a specific antigenic protein is incorporated into plasmids that are amplified in a host bacterial strain, such as *Escherichia coli*, then purified. The bacterial hosts for the coding plasmids are easily scaleable, and generic purification schemes may be possible because the product would always be plasmid DNA. After injection, the naked plasmid DNA yields protein expression in its native conformation, to which both humoral and cell-mediated antibody responses are mounted (90). The DNA constructs can be customized to coexpress both the coding region for the antigen as well as an immune modulator to enhance the immune response or increase its specificity (91,92). The tremendous amount of research in the past few years has led to progress in the development of vaccines against disease targets associated with bacterial, viral, and parasitic infections and provides hope for combating autoimmune and inflammatory diseases (93) as well as cancer.

BIBLIOGRAPHY

1. M. Canales, A. Enriquez, E. Ramos, D. Cabrera, H. Dandie, A. Soto, V. Falcon, M. Rodriguez, and J. De la Fuente, *Vaccine* **15**, 414–422 (1997).
2. S. Gizurarson, *Clin. Pharmacokinet.* **30**, 1–15 (1996).
3. E.H. Relyveld, B. Bizzini, and R.K. Gupta, *Vaccine* **16**, 1016–1023 (1998).
4. R.E. Spier, *Vaccine* **15**, 1163–1164 (1997).
5. J.E. Maynard, M.A. Kane, and S.C. Hadler, *Rev. Infect. Dis.* **2**, S574–S578 (1989).
6. C.N. Shapiro and H.S. Margolis, *Epidemiol. Rev.* **12**, 221–227 (1990).
7. P. Tiollais and M.-A. Buendia, *Sci. Am.*, 116–123 (April 1991).
8. A.J. Zuckerman, *Baillieres Clin. Gastroenterol.* **4**, 775–788 1990.
9. M. Marchetti, M. Rossi, V. Giannelli, M.M. Giuliani, M. Pizza, S. Censini, A. Covacci, P. Massari, C. Pagliaccia, R. Manetti, J.L. Telford, G. Douce, G. Dougan, R. Rappuoli, and P. Ghiara, *Vaccine* **16**, 33–37 (1998).
10. D. Bonn and J. Bradbury, *Lancet* **351**, 810 (1998).
11. R.W. Tindle, *Immunol. Res.* **16**, 387–400 (1997).
12. J. Eskola and H. Kayhty, *Vaccine* **16**, 1433–1438 (1998).
13. C-J. Lee, S.D. Banks, and J.P. Li, *Crit. Rev. Microbiol.* **18**, 89–114 (1991).
14. J.B. Robbins and R. Schneerson, *J. Infect. Dis.* **161**, 821–832 (1990).
15. M. Hilleman, in W.W. Ellis ed., *Hepatitis B Vaccines in Clinical Practice*, Dekker, New York, 1993, pp. 17–39.
16. C.R. Alving, B. Detrick, R.L. Richards, M.G. Lewis, A. Shafferman, and G.A. Eddy, *Ann. N.Y. Acad. Sci.* **690**, 265–275 (1993).
17. R. Elelman, in M.M. Leving, G.C. Woodrow, J.B. Kaper, and G.S. Cobon eds., *New Generation Vaccines*, Dekker, New York, 1997, pp. 173–192.
18. E.H. Relyveld, *Dev. Biol. Stand.* **65**, 131–136 (1986).
19. F.R. Vogel, *Ann. N.Y. Acad. Sci.* **754**, 153–160 (1995).
20. J.-L. Excler, *Vaccine* **16**(14/15), 1439–1443 (1998).
21. A.R. Hinman, *Vaccine* **16**(11/12), 1116–1121 (1998).
22. *Weekly Epidemiological Record: WHO* **70**, 221–228 (1995).
23. L.S. Marsano, D.J. West, I. Chan, T.M. Hesley, J. Cox, V. Hackworth, and R.N Greenberg, *Vaccine* **16**(6), 624–629 (1998).
24. *Morbidity and Mortality Weekly Report* **38**(13), 205–227 (1989).
25. D.J. West and G.B. Calandra, *Vaccine* **14**(11), 1019–1027 (1996).
26. P. De Kruif, *Microbe Hunters*, Harcourt Brace Jovanovich, New York, 1954.
27. S.V. Ho, in M.R. Ladisch, R.C. Willson, C.-d.C. Painton, S.E. Builder eds., *Protein Purification from Molecular Mechanisms to Large-Scale Processes*, American Chemical Society Press, 1989, pp. 14–34.
28. E.N. Lightfoot, in M.R. Ladisch, R.C. Willson, C.-d.C. Painton, S.E. Builder eds., *Protein Purification from Molecular Mechanisms to Large-Scale Processes*, American Chemical Society Press, 1989, pp. 35–51.
29. R.C. Willson and M.R. Ladisch, in M.R. Ladisch, R.C. Willson, C-d.C. Painton, S.E. Builder eds. *Protein Purification From Molecular Mechanisms to Large-Scale Processes*, AChS Press, 1989, pp. 1–13.
30. L. Hayflick, *Dev. Biol. Stand.* **70**, 11–26 (1989).
31. L. Hayflick, S.A. Plotkin, and R.E. Stevenson, *Dev. Biol. Stand.* **68**, 9–17 (1987).
32. M.E. Wiebe and L.H. May, in A.S. Lubiniecki ed., *Large-Scale Mammalian Cell Culture Technology*, Dekker, New York, 1990, pp. 147–160.
33. A.Y. Elliott, in A.S. Lubiniecki ed., *Large-Scale Mammalian Cell Culture Technology*, Dekker, New York, 1990, pp. 207–216.
34. L. Hayflick and P.S. Moorhead, *Exp. Cell Res* **25**, 585–621 (1961).
35. O. Kistner, P.N. Barrett, W. Mundt, M. Reiter, S. Schober-Bendixen, and F. Dorner, *Vaccine* **16**, 960–968 (1998).

36. A.L. Van Wenzel, G. Van Stienis, C. Hannik, and A. Cohen, *Dev. Biol. Stand.* **41**, 159–168 (1978).
37. O. Molin and C.G. Heden, *Progress in Immunological Standardization*, vol. 4, Karger, New York, 1969, pp. 106–110.
38. Skoda, A. Hormann, O. Spath, and A. Johansson, *Dev. Biol. Stand.* **42**, 121–126 (1979).
39. S. Reuveny, in A.S. Lubiniecki ed., *Large-Scale Mammalian Cell Culture Technology*, Dekker, New York, 1990, pp. 271–323.
40. A.S. Lubiniecki ed., *Large-Scale Mammalian Cell Culture Technology*, Dekker, New York, 1990.
41. P.J. Provost and M.R. Hilleman, *Proc. Soc. Exp. Biol. Med.* **159**, 201–203 (1978).
42. D.B. Strader and L.B. Seeff, *Drugs* **51**, 359–366 (1996).
43. A.J. Zuckerman, *Baillieres Clin. Gastroenterol.* **4**, 775–788 (1990).
44. A.J. Hagen, C.N. Oliver, and R.D. Sitrin, *Biotechnol. Prog.* **12**, 406–412 (1996).
45. Eur. Pat. 0302692B (1992), J.A. Lewis, E.A. Emini, and M.A. Armstrong.
46. Pat. Appl. WO 94/03589 (February 17, 1994), R.A. Aboud, J.G. Aunins, B.B. Buckland, P.A. DePhillips, A.J. Hagen, J.P. Hennessey Jr., B. Junker, J.A. Lewis, C.N. Oliver, C.J. Orella, and R.D. Sitrin.
47. M.E. Armstrong, P.A. Giesa, J.P. Davide, F. Redner, J.A. Waterbury, A.E. Rhoad, R.D. Keys, P.J. Provost, and J.A. Lewis, *J. Hepatol.* **18**, S20–S26 (1993).
48. J.A. Asenjo, in J.A. Asenjo ed., *Separation Processes in Biotechnology, Bioprocess Technology*, vol. 9, Marcel Dekker, New York, 1990, pp. 3–16.
49. G.J. Prokopakis and J.A. Asenjo, in J.A. Asenjo ed., *Separation Processes in Biotechnology, Bioprocess Technology*, vol. 9, Marcel Dekker New York, 1990, pp. 571–600.
50. P.F. Stanbury and A. Whitaker, *Principles of Fermentation Technology*, Pergamon Press, New York, 1983.
51. A.J. Darling and J.J. Spaltro, *BioPharm* October, **9**, 42–50 (1996).
52. A.S. Lubiniecki, M.E. Wiebe, and S.E. Builder, in A.S. Lubiniecki ed., *Large-Scale Mammalian Cell Culture Technology*, Marcel Dekker, New York, 1990, pp. 515–541.
53. C.F. Goochee, M.J. Gramer, D.C. Andersen, J.B. Bahr, and J.R. Rasmusen, *Biotechnology* **9**, 1347–1355 (1991).
54. J. Fu, W.J. VanDusen, D.G. Kolodin, D.O. O'Keefe, W.K. Herber, and H.A. George, *Biotechnol. Bioeng.* **49**, 578–586 (1996).
55. S.M. Kingsman, A.J. Kingsman, M.J. Dobson, J. Mellor, and N.A. Roberts, *Biotechnol. Genet. Eng. Rev.* **3**, 377–416 (1985).
56. J.D. Kittle and B.J. Pimentel, *BioPharm*, 48–51 (October 1997).
57. R. Scopes, *Protein Purification: Principles and Practice*, Springer-Verlag, New York, 1982.
58. S.M. Wheelwright, *Protein Purification: Design and Scale up of Downstream Processing*, Carl Hanser Verlag, Munich, 1991, pp. 10–25.
59. D.P. Francis, P.M. Feorino, S. McDougal, D. Wrfield, J. Getchell, C. Cabradilla, M. Tong, W.J. Miller, L.D. Schultz, F.J. Bailey, W.J. McAleer, E.M. Scolnick, and R.W. Ellis, *JAMA* **256**, 869–872 (1986).
60. R.D. Sitrin, E.D. Wampler, and R.W. Ellis, in R.W. Ellis ed., *Hepatitis B Vaccines in Clinical Practice*, Dekker, New York, 1993, pp. 83–101.
61. R. Kieffer and L. Torbeck, *Pharmaceutical Technol.*, 66–76 (June 1998).
62. N.M. Lugo, *BioPharm* **11**, 18–26 (1998).
63. R.K. Gupta and G.R. Siber, *Vaccine* **13**, 1263–1276 (1995).
64. LeMoignic and Pinoy, *Comp. Rend. Soc. Biol.* **79**, 352–354 (1916).
65. G. Ramon, *Bull. Soc. Centr. Med. Vet.* **101**, 227–234 (1925).
66. G. Ramon, *Ann. Institut Pasteur* **40**, 1–10 (1926).
67. A.T. Glenny, C.G. Pope, H. Waddington, and U. Wallace, *J. Pathol. Bacteriol.* **29**, 38–45 (1926).
68. A.G. Johnson, S. Gaines, and M. Landy, *J. Exp. Med.* **103**, 225–246 (1956).
69. E. Ribi, *J. Biol. Res. Mod.* **3**, 1–9 (1984).
70. F. Ellouz, A. Adam, R. Ciorbaur, and E. Lederer, *Biochem. Biophys. Res. Commun.* **59**, 1317–1325 (1974).
71. L. Rong, E. Tarr, and T.C. Jones, *Clin. Infect. Dis.* **21**, 1439–1449 (1995).
72. G. Gregoriadis and A.T. Florence, *Drugs* **45**, 15–28 (1993).
73. R.P. Weissburg, P.W. Berman, J.L. Cleland, D. Eastman, F. Farina, S. Frie, A. Lim, J. Mordenti, M.R. Peterson, K. Yim, and M.F. Powell, *Pharm. Res.* **12**, 1439–1446 (1995).
74. H.S. Warren, F.R. Vogel, and L.A. Chedid, *Annu. Rev. Immunol.* **4**, 369–388 (1986).
75. H.S. Warren and L.A. Chedid, *CRC Crit. Rev. Immunol.* **8**, 83–101 (1988).
76. S. Shirodkar, R.L. Hutchinson, D.L. Perry, J.L. White, and S.L. Hem, *Pharm. Res.* **7**, 1282–1288 (1990).
77. R.K. Gupta, B.E. Rost, E. Relyveld, and G.R. Siber, in M.F. Powell and M.J. Newman eds., *Vacine Design: The Subunit and Adjuvant Approach*, Plenum Publishing, New York, 1995, pp. 229–248.
78. R.K. Gupta, P. Griffin, Jr., An-Cheng Chang, R. Rivera, R. Anderson, B. Rost, D. Cecchini, M. Nicholson, and George Siber, in S. Cohen and A. Shafferman eds., *Novel Strategies in the Design and Production of Vaccines*, Adv. Exp. Med. Biol., vol. 397, Plenum Press, New York, 1996, pp. 105–113.
79. J.G. Dickenson ed., *Dickinson's FDA Review*, Ferdic, Camp Hill Pa., 1998, pp. 1–24.
80. *BioPharm* **11**(8), 8–51 (1998).
81. W.K. Hubbard, *Current Good Manufacturing Practice: Proposed Amendment of Certain Requirements for Finished Pharmaceuticals, Food and Drug Administration* 21 CRF Parts 210 and 211, Docket No. 95N-0362, RIN 0910-AA45, (March 29, 1996).
82. N.M. Lugo, *BioPharm* **11**, 18–26 (1998).
83. A. Robbins and P. Freeman, *Sci. Am.* November, 126–133 (1988).
84. F.E. Andre, *Biologicals* **22**, 317–321 (1994).
85. H.S. Mason, T.A. Haq, J.D. Clements, and C.J. Arntzen, *Vaccine* **16**(13), 1336–1343 (1998).
86. C.O. Tacket, H.S. Mason, G. Losonsky, J.D. Clements, M.M. Levine, and C.J. Arntzen. *Nat. Med.* **4**, 607–609 (1998).
87. A.R. Brown, D.W. George, and D.K. Matteson, *Vaccine* **15**, 1165–1173 (1997).
88. L. Jodar, T. Aguado, J. Lloyd, and P.-H. Lamber, *Gen. Eng. News* **18**, 6, 40, 45 (1998).
89. D.M. Klinman, M. Takeno, M. Ichino, M. Gu, G. Yamshchikov, G. Mor, and J. Conover, *Springer Semin. Immunopathol.* **19**, 245–256 (1997).
90. J.J. Donnelly, J.B. Ulmer, and M.A. Liu, *Life Sci.* **60**, 163–172 (1997).

91. S. Langermann, *Nat. Med.* **4**, 547–548 (1998).
92. H.L. Robinson and C.A.T. Torres, *Semin. Immunol.* **9**, 271–283 (1997).
93. I.A. Ramshaw, S.A. Fordham, C.C.A. Bernard, D. Maguire, W.B. Cowden, and D.O. Willenborg, *Immunol. Cell Biol.* **75**, 409–413 (1997).

VALIDATION. See CLEANING, CLEANING VALIDATION; GOOD MANUFACTURING PRACTICE (GMP) AND GOOD INDUSTRIAL LARGE SCALE PRACTICE (GLSP); PROCESS VALIDATION; VACCINE TECHNOLOGY.

VENT GAS ANALYSIS

DAVID POLLARD
PETER SALMON
Merck and Co., Inc.
Rahway, New Jersey

KEY WORDS

Fermentation
Growth kinetics
Magnetic sector
Mass spectrometer
Online monitoring
Quadrupole
Respiratory quotient
Vent gas

OUTLINE

Introduction
Methods of Vent Gas Analysis
- Thermal Conductivity
- Flame Ionization
- Gas Chromatography
- Electrochemical Sensors
- Infrared Absorption
- Paramagnetism
- Mass Spectrometry
- Vacuum System
- Sample Inlet
- Ionization
- Ion Separation
- Detectors
- Spectral Analysis
- Suitability of Mass Spectrometry

Installation and Operation
- Fermentor Sampling Location
- Sample Delivery and Pretreatment
- Sample-Handling Manifolds
- Sample Flow Control
- System Monitoring

Parameter Calculation
- Well-Mixed Liquid- and Gas-Phase State
- Large-Scale Mixing Models
- Aqueous Concentration Estimation for Volatile Organic Compounds
- Growth Kinetics
- Mass Balancing and RQ

Bibliography

INTRODUCTION

The monitoring of exhaust gas is an ideal universal method for fermentation characterization because most biological processes involve exchange or consumption of gases and volatile compounds. Its noninvasive application does not compromise the sterility of the fermentor, providing a clear advantage over in situ probes, which contact the broth. The monitoring of gas concentrations, such as carbon dioxide and oxygen, in the entry inlet and outlet gas streams and incorporated into liquid and gas balances makes it possible to quantify rates of gas utilization on production and respiratory quotient. The use of these data in conjunction with the fast response of the analyzers allows on-line real-time monitoring of the physiological state of the fermentation, including growth kinetics, and substrate consumption and product formation information. This utilization has made the technology invaluable in research and manufacturing for applications such as fermentation optimization, batch-to-batch reproducibility, on-line troubleshooting, process decision making, and inoculation criteria (1,2). The aqueous concentration of organic volatile compounds, such as ethanol, butanol, and methanol, can also be determined from vent gas analysis and have been used to optimize productivity in industrial processes (3,4).

METHODS OF VENT GAS ANALYSIS

A variety of methods are available for monitoring gases. However, only the methods most applicable for monitoring gases of interest to bioprocesses are described here.

Thermal Conductivity

The sensor for thermal conductivity is typically a resistance temperature detector whereby gas is passed over a heated filament and the magnitude of cooling is measured by the change of resistance of the filament as measured by a Wheatstone bridge (5). For a single flowthrough unit, the gas flow rate must be kept constant. Units with a bypass have two flows; one path houses the sensor with two heated coils. The heated gas is transported from upstream to downstream coils, changing their resistance. The bridge circuit with constant current input gives an output voltage in direct proportion to the difference in resistance, with heat input and specific heat assumed constant. For detectors used in gas chromatography, the heat transfer of the sample gas is compared with the transfer of the carrier gas. The rate of transfer depends on the thermal conductivity of the gas, which is proportional to the square root of molecular weight.

The Wheatstone bridge balances the current through the sample element to restore its temperature to that of the reference. Tungsten resistance wires are used for ambient and low temperature; metallic oxides are used for high temperature (5). The reference and sample detectors are housed in a single metal block with a high heat capacity to minimize errors from transient heating and cooling (6). Instruments are calibrated for the gas to be used and can have a fast response time of 1 s, detection threshold of 30 ppm, and accuracy of 2%. Carbon dioxide is often measured with this technique as the thermal conductivity of CO_2 is about two-thirds that of oxygen and nitrogen. The bypass units where the bulk of the flow bypasses the sensor are less sensitive to gas flow rate fluctuations but have larger response times, 20 s compared with 1 s. Although this technology has been used for many years it does not provide accuracy or long-term reproducibility. It is sensitive to gas flow rate, temperature, and mechanical vibration. Contamination of the sample cell can lead to drift, and frequent calibration is required of these inexpensive devices. Sample pretreatment is often required to remove contaminants by condensation and drying.

Flame Ionization

In this method the sample gas is burned in a hydrogen/air-induced flame. An electrical voltage is applied across two electrodes to remove the resulting ions from the flame. The current created across the two electrodes is a measure of the ion concentration (5). The detector has high linearity and sensitivity (ppb) with a short response time. It is efficient for a variety of organic components, but no response occurs for O_2, N_2, H_2, CO_2, Ar, H_2O, NH_3, or simple organics such as formic acid and formaldehyde. This characteristic makes the method universal for organic analysis by gas chromatography, especially because contaminants, water, air, and ammonia are not detected. However, its application for vent gas analysis is thus limited to ethanol and organic volatiles.

Gas Chromatography

Gas chromatography is the phase separation of gas component by adsorption equilibrium differences between stationary and mobile phases (6). The stationary phase is coated on small solid particles (typically silica or polystyrene) that are packed or coated onto the walls of highly efficient fused silica columns (several meters in length). The particles have an affinity for the chosen gas analysis to achieve separation of gas components based on equilibrium differences. The system requires a carrier gas to transport the gas sample through the stationary phase (typically oxygen-free dry helium), a sample injection system, column temperature controller, and a detector. Thermal conductivity and flame ionization detectors are most commonly used, as previously described. An analysis time of about 5–30 min is typical for each sample, which is much longer than other simpler, faster, and less expensive methods. However, gas chromatography has been used to determine H_2, CO_2, O_2, N_2, and organic volatiles.

Electrochemical Sensors

Amperometric methods can measure oxygen concentration in the vent gas using technology similar to the probes used for dissolved oxygen tension. The Clark principle incorporates the separation of electrodes from the broth using a semipermeable membrane (silicone or polytetrafluroethylene, PTFE) (7,8). A polarization voltage (600–800 mV) reduces oxygen at the cathode (Pt or Au) to peroxide and hydroxyl ions. The resulting current between cathode and anode (Ag/AgCl) is proportional to the oxygen partial pressure. Cl ions are consumed and supplied by the electrolyte (KCl). The probe relies on diffusion between gas, membrane, and electrolyte, which results in a slow response time delay, usually about 10–100 s to reach 63% of total change in dissolved oxygen level. The oxygen diffusion across the membrane results in a temperature-sensitive process. Equally, electrolyte depletion and water loss into the gas stream and membrane fouling make the system prone to drift. Thus, the technology is rarely used for vent gas monitoring.

Zirconia ceramic sensors have been developed for oxygen detection. Thimble-shaped, nonporous, oxygen ion-conducting electrolyte (yttrium-stabilized zirconia) separates two platinum electrodes and the reference from the measurement gas (9,10). The sample gas passes through the center of the sensor tube, contacting the anode platinum electrode, while the reference gas passes between the cathode electrode and the cylinder outer wall. The electrolyte, which separates the two gas phases, is a good conductor of oxygen ions because the ceramic lattice has vacancies for oxygen from the substitution of tetravalent zirconia ions by trivalent yttrium ions. When the oxygen sensor is exposed to a gas mixture, oxygen molecules absorb onto the electrode surface. This changes the molecule concentration on sensor electrode, which in turn changes the charged oxygen vacancy concentrations inside the solid zirconia electrolyte. An electrostatic field is generated from the differences in charge density between the charged oxygen vacancies inside the zirconia electrolyte and electrons in the electrode. The electromotive force (emf) generated across the electrochemical zirconia cell is proportional to the concentration of the adsorbed oxygen. The electrochemical reaction is temperature sensitive and takes place between 600 and 800 °C, requiring good temperature control. Temperature elevation can catalyze a reaction between platinum anode and combustible gases in the sample, resulting in measurement of net oxygen concentration rather than total oxygen concentration (6). For cost-effectiveness as opposed to accuracy, this sensor may be applicable as it is inexpensive, robust, and simple to operate, with accuracy about 1%, suitable for fermentation monitoring (11).

Alternative sensors have been developed for volatile organic species based on solid-state technology. The technology is mostly proprietary, but the sensors are often tin (IV) oxide (SnO_2) and have been demonstrated for ethanol monitoring of fermentation vent gas (12). The gas of interest is detected using two chemically inert electrical conductors of the same composition interfacing with a solid electrolyte. The gas diffuses into the electrolyte and alters its impedance by a change in the charge density. These inexpensive sensors offer good sensitivity but have slow response times.

Infrared Absorption

The bonds of nonsymmetrical molecules vibrate and stretch from bombardment with infrared radiation, resulting in energy absorption. The specific vibration or rotation of a particular bond only takes place at a given frequency. For infrared adsorption, the molecule must be exposed to radiation of the precise frequency at which the molecular motion occurs. Thus, the fundamental frequencies of vibration and rotation for a particular compound can be used to identify it from a mixture. For fermentation analysis, carbon dioxide is the only gas of interest that can be detected by infrared (IR) because oxygen and nitrogen contain only one element. Most commonly used analyzers are nondispersive, which excludes most wavelengths apart from a few narrow bands of radiation. They are less selective than dispersive analyzers but more sensitive, simple, and reliable (13). Sources of radiation include quartz iodide lamps, tungsten filaments, and nichrome wires (6). A specific interference filter assembly, a monochromator, is chosen to filter out all wavelengths except for a desired wavelength band. The resulting radiation is a narrow band based around the mean of the desired wavelength (4,200–4,400 nm for CO_2). Alternatively, a cell containing the gas of interest is used to isolate the desired wavelength source.

In a typical optical arrangement, such as the Luft analyzer, an optical beam splitter is used to split the single radiation source to two beams, that pass into two separate chambers (Fig. 1). One chamber houses the sample cell and the other the reference cell. Both chambers contain a gas that absorbs at the desired wavelength, and the detector gas is usually the same as the desired gas for analysis. The sample cell is a flowthrough device with inlet–outlet ports designed to have zero dead legs. Its material depends on the corrosivity of the sample, usually stainless steel or Teflon, and the sample window is usually quartz or sapphire (6). Differences of IR absorption occur between the two chambers, reference and sample, generating pressure imbalances. This imbalance distorts a diaphragm that separates two chambers in the detector and forms part of a condenser microphone. A chopper wheel, a perforated or segmented disk, is used to block the radiation from one chamber and to allow radiation to reach the detector through the other chamber. The alternate blocking of the reference and sample chambers allows each radiation to be sampled at the appropriate side of the detector. The alternating flow of radiation produces an oscillating effect on the diaphragm. The amplitude of the oscillations is measured by the capacitance of the condenser microphone, which is correlated to the concentration of absorbed gas. A variation of this analyzer is to use a microflow sensor to replace the capacitance detector. Fluctuating radiation flow, generated by the chopper, is detected by a flow cell. The cell measures the temperature difference between two heated elements in the flow path. The temperature difference produces a corresponding difference of electrical resistance using a Wheatstone bridge that is correlated to the sample gas concentration.

Other alternative detectors include a thermocouple detector made of blackened gold foil to which two wire leads made of dissimilar metals are attached (6). The metals are selected to provide large thermoelectric emf. As the gold foil absorbs radiation, an emf forms between the gold foil and the metal. The strength of the emf is a function of the radiant power absorbed. The entire device is enclosed to minimize heat loss. Pyroelectric detectors absorb IR radiation, which raises the temperature of the ferroelectric crystal sensing element, typically Pd-$LiTaO_3$ (14), or $Pb(Zr_{52}Ti_{48})O_3$ (15,16). Spacing between the crystalline lattice changes as heat is absorbed, which affects the electric polarization of the element (14). Two electrodes are used, one of which is transparent to radiation, and the crystal is mounted between the electrodes. The sample gas concentration is correlated from the voltage required to balance the polarization charge of the crystal.

IR analyzers have been used to monitor CO_2 and organic volatiles such as ethanol for vent gas analysis (17), but they are sensitive to interference from other components of the gas such as water. The accuracy of IR analyzers is reasonable for fermentation processes, typically about 1% of the range, but are prone to drift (as much as 1% full scale per day for both the zero and span). The cal-

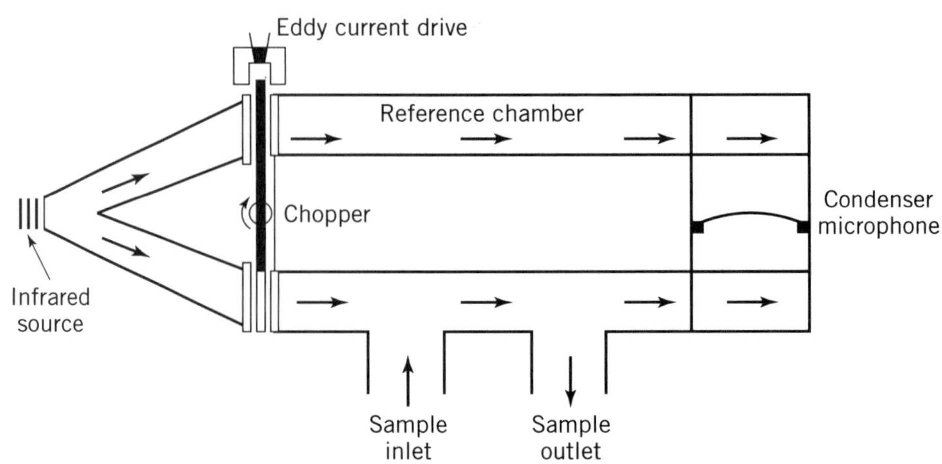

Figure 1. Luft nondispersive infrared analyzer with microphone sensor.

ibration curve is typically nonlinear, which gives added difficulty to frequent recalibration required to counteract drift. However, the 0–90% response time of 0.5–5 s is reasonably fast compared with other technologies already discussed.

Paramagnetism

Gases with unpaired electrons, such as oxygen, generally have high magnetic susceptibility. The unpaired electrons spin around the nuclei in a unbalanced manner. When a magnetic field is applied, the atoms orientate themselves in the orientation of the direction of the field, which has enabled specific instruments to be made for oxygen analysis. The principal instrument, a thermomagnetic analyzer, uses the principle that the magnetic susceptibility is proportional to the difference between sample temperature and the Curie point of 80 °C for oxygen (6). In a typical analyzer, gas flows through a heated magnetic field so that the oxygen becomes less magnetic. Cooler oxygen, which is more magnetic, pushes the hot oxygen, creating a magnetic wind. The magnitude of the flow determines the temperature of the coils, which changes the coil resistance as measured by a Wheatstone bridge. These instruments are sensitive to temperature variations and the thermal properties of the sample gas. Severe calibration shift and slow response times have also been reported (18).

An alternative analyzer is the magnetodynamic analyzer, which measures the mechanical deflection of a rigid test body (dumbbell) caused by the paramagnetic properties of oxygen. The dumbbell consists of two diamagnetic fused quartz spheres (2 mm in diameter) filled with nitrogen and mounted on a rod that is suspended by stiff quartz or platinum–iridium fiber in the cell chamber. A nonuniform magnetic field is applied to the gas sample in the chamber. The oxygen of the sample is attracted toward the most intense magnetic field, and the spheres are displaced until the force is balanced by the torque on the suspended fiber. The resulting torque is proportional to the difference of magnetic susceptibility between test body and surrounding gas. A light ray reflected by a mirror, mounted on the test body, transmits the angular deflection, which is directly proportional to the oxygen concentration in the chamber. Some instruments employ a feedback winding on the dumbbell with an electromagnet to maintain zero deflection, and the oxygen concentration is calculated from the current required. Unlike the thermomagnetic unit, the magnetic analyzer is not affected by temperature or thermal conductivity of the sample. However, care must be taken in orientation of the unit as it is sensitive to vibration. The unit is also sensitive to pressure as the amount of O_2 present in the sensing chamber, rather than the oxygen concentration, determines the balance of magnetic forces (6).

Another type of instrument is the Quincke-type analyzer, which measures the pressure difference when two gases with different oxygen concentrations come together in the magnetic field. The reference gas (N_2, O_2, air) flows through two restrictions so the gas entering either side of the sample chamber has equal flows (Fig. 2). One section of the reference gas is subjected to a pulsating magnetic field, and if oxygen is present then a pressure difference is created between the two reference flows. The Luft capacitance sensor has been used as a detector. Alternatively, Siemens AG developed an analyzer with a microflow detector in a connecting channel that detects the pulsing flow, producing a current proportional to the oxygen concentration (Fig. 2). The separation of the sample from the detector reduces the possibilities of drift and instrument failure from contamination or corrosion of parts.

These paramagnetic analyzers have found to be susceptible to vibration, pressure, and sample conditions with respect to moisture content and flow rate, resulting in frequent calibration to compensate for drift (2). Nevertheless, Quincke-type analyzers have been found to be more robust and have been widely used, including at the Merck & Co., Inc., production facilities. The units are priced similarly to IR analyzers but are more expensive than zirconia sensors. However, they provide only single-component analysis, with susceptibility to moisture and high sample flow requirement, and have poor accuracy (1–2% of range), with slow response times (5–100 s) (19). The units require calibration from drift, the frequency of which depends on care on calibration, maintenance of the machine, and sample treatment (20).

Mass Spectrometry

Mass spectrometry allows simultaneous measurement of both air gases (Ar, O_2, N_2, and CO_2) and volatile substances (including ethanol, acetone, and acetic acid) (21,22) present in the fermentor headspace. A mass spectrometer with multipoint analysis and fast response times (seconds) allows many fermentors and different processes to be monitored on the same instrument. The analysis involves admitting a gas stream to an ion source and then ionizing the molecules by electron bombardment under a high vacuum. The ions produced are subsequently separated, according to their mass-to-charge ratio, by either a magnetic sector or a quadrupole mass filter and detected sequentially. The systems are expensive but offer long-term accuracy and precision, with flexible analysis type and with sub-ppm detection limits and minimum operator intervention.

Vacuum System

The analyzer chamber of a vacuum system is kept under high vacuum (between 10^{-6} and 10^{-8} mmHg) to minimize collisions between particles. At this pressure, the mean free path of particles will be long enough, 50 m as opposed to <0.01 m at atmospheric pressure, and the number of collisions small enough for good analysis (6). Particle collision reduces instrument resolution and sensitivity by increasing kinetic energy distribution of the ions, inducing fragmentation or preventing ions from reaching the detector (23). The vacuum is created in two stages: a roughing pump, which reduces the vacuum to 10^{-3} mmHg, then a diffusion or turbomolecular pump, to obtain pressure as low as 10^{-9} mmHg. The roughing pump is usually a reciprocating mechanical pump lubricated with low vapor pressure oil to prevent oil contamination of the mass spectrometer (6). The diffusion pump uses an oil, often polyphenol

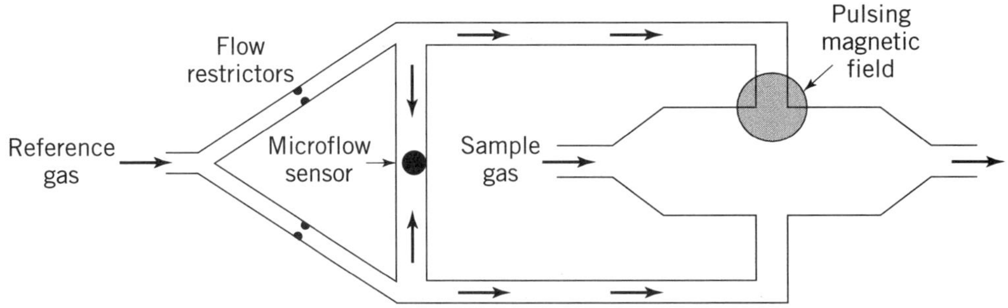

Figure 2. Quincke-type oxygen analyzer with capacitance sensor developed by Siemens AG.

ether, to force molecules out of the vacuum chamber. Oil is heated at the bottom of the pump, causing it to vaporize, and water-cooled coils at the top of pump then allow the oil to cool and condense. Large oil molecules force the smaller molecules down into the pump via momentum exchange, and smaller gaseous molecules exit through the mechanical pump at about 200 L/s. The disadvantage is that diffusion pumps are selective because noble gases are pumped at reduced rates (24). Turbomolecular pumping is mechanical; molecules are forced into the fins of the turbomolecular pump by momentum exchange. Fins revolve at about 10,000 rpm and produce an effective pumping capacity of about 400 L/s. Turbomolecular pumps are often preferred to diffusion types because they cause very low contamination by hydrocarbons and high vacuum is obtained very quickly on start-up. However, turbomolecular pumps are noisy, although they can be muffled to some degree, and highly mechanical, requiring regular servicing.

Sample Inlet

The capillary inlet design is most often used for mass spectrometer sampling, although some systems use a membrane inlet. The membrane material is typically silicone rubber or Teflon, which protects the vacuum integrity of the instrument and separates the analyzer from the atmosphere and contaminates such as foam or water. The membrane inlet has mostly been used for on-line in situ liquid analysis of organic volatiles in the fermentor (21,24–26) and has limited use for vent analysis because the membrane increases the response time to about 30 s. The most popular design is the capillary inlet on which a small representative sample flows from the main sample gas stream, which is continually venting, and "leaked" into the instrument, typically at 20 mL/min (18,27). The constant flow of gas stream from sample line to vent and to the leak of the analyzer means that line flushing can be ignored. The capillary is heated at 80–120 °C to preserve the vapor state of the sample. The rotary pump draws the sample by creating a pressure drop through the capillary. At the outlet, before the ionization chamber, the molecules cross a fritted glass or metal filter to further reduce pressure (13). The response times of such an inlet system can be about 0.5 s, achieved by low dead volumes and maintaining the whole system at 80 °C (28). Volatile components in the gas stream can contaminate the stainless steel capillary by adsorption and desorption, causing larger response times and memory effects from long transition times through the capillary (19,29). These factors can be reduced by using a capillary (diameter, 3 mm; length, 1–3 cm) lined with borosilicate glass that is heated resistively through a stainless steel outer sheath to 150 °C.

Ionization

The most widely used technique is electron impact ionization, where gas passes into an electron beam, resulting in electron ejection and a degree of fragmentation (2). Electrons ejected from a heated tungsten filament are accelerated through an electric field at 70 V to form an electron beam. The vacuum system draws molecules through the sample leak into the ionization chamber. Neutral molecules pass freely through the negatively charged ion repeller in front of the leak and are thermally desorbed into the electron beam, transferring kinetic energy to the sample molecule. The ionization (electron ejection) results in parent ions with the same mass to charge to MW with 6 eV of excess energy (Fig. 3). The excess energy in the molecule leads to some degree of fragmentation to molecular ions, fragment ions, and neutral fragments. A negative ionization step (electron capture) also occurs but is 100 times less efficient than electron ejection. Electron bombardment is most stable and reliable but is complicated by excessive fragmentation. Other soft ionization techniques have been developed: fast atom or ion bombardment, matrix-assisted laser desorption–ionization, and electrospray ionization, which produce only a few fragments. These methods give greater sensitivity and mass range but have primarily been developed for large molecules and solids analysis (23).

Ion Separation

Several methods are available to separate the ions after bombardment: quadrupole filter, magnetic analyzer, time-of-flight detectors, and Fourier transform detectors. The aim is to separate the ions of mass m from the ions of m', which is most commonly done by quadrupole filter or magnetic analyzer because these are easy to operate and compact (30).

The quadrupole filter is a set of four cylindrical rods, typically 125–300 mm long and 6–16 mm in diameter, held in a square array to which an RF and DC potential are applied (24,31). Two rods are connected and two are kept

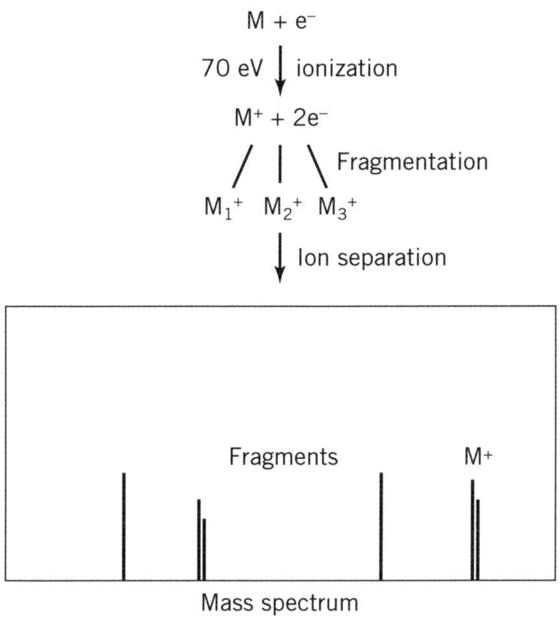

Figure 3. Ionization of molecules (M) producing fragment ions (M_1^+, M_2^+, and M_3^+).

180 degrees out of phase (Fig. 4). This forms crossed, oscillating electric fields perpendicular to the line of travel of the ions between the source and the detector. If the oscillating motion of the ion is out of resonance, the ions strike the walls and are not detected. Applying a certain voltage and RF can control the trajectory of a known mass-to-charge ratio to selectively strike the detector (30). Scanning RF and DC voltages from zero allows ions to be transmitted through the quad filter sequentially in order of their m/z values with constant resolution allowing multiple ion analysis. Resolution attained is determined by the accuracy of control of the parameters U, V, f, and ω radius within the quad rods (30).

The magnetic analyzer uses a magnetic field to separate the different ions. A moving charge passing through a magnetic field will experience a force and travel in a circular motion with a radius of curvature depending on the m/z of the ion. A magnetic analyzer separates ions according to their radii of curvature, and therefore only ions of given m/z will be able to reach a point detector at any given magnetic field (Fig. 5) (30). In a single detector system, an electromagnet is used to adjust the magnetic field to detect any m/z peak of interest (2). Alternatively, a multicollector system is set up whereby each ion fragment is measured by a small number of Faraday bucket collector assemblies, located along the focal plane according to which particular signals are of interest. Single detector systems are more common because they allow the combined use of the secondary electron multiplier that is too large to be used in conjunction with multiple detector systems.

Quadrupole systems are known to be faster, less expensive, and smaller than magnetic sector instruments. Yet the resolution is lower, easily contaminated, and less stable over time, making the magnetic sector a superior choice for industrial on-line monitoring (19,30). The response times of both systems are comparable because the magnetic hysteresis associated with magnetic sectors has been eliminated using laminated magnets. The precision from magnetic sectors comes from the flat-topped peak profiles of ion current versus magnetic field strength that they exhibit (Fig. 5). Thus, peak center and peak height can be precisely determined. Any minor drifting that may occur from temperature fluctuations has no effect, which results in high stability, accurate peak jumping, and long periods between recalibration (up to several weeks for fermentation application). Absolute accuracy is less than 0.05% for O_2 and less than 0.1 for N_2. Quadrupole filters exhibit a Gaussian-type peak profile and are more susceptible to drift (see Fig. 4) and temperature variation. Reduced accuracy and precision result in more frequent calibration. Insufficient resolution leads to an abundance sensitivity problem whereby peak discrimination is difficult at low concentrations with adjacent peaks. Experience with mass spectrometers at Merck during the past two decades favors the magnetic sector as opposed to quadrupoles. VG Gas Analysis Systems manufacture both quadrupoles and magnetic sectors and have shown improved accuracy and drift by a factor of 2 for magnetic sectors over quadrupole systems (27,29). Nevertheless, a considerable number of useful applications for fermentation vent gas analysis have been demonstrated using quadrupole systems (32,33).

Ion separation is still the limiting factor for mass spectrometry, however, because soft ionization produces larger molecular weight ions. Quads with electron ionization have been used since the 1950s but have found new utility when coupled with electron spray systems for solution analysis; however, this method has not been applied to vent gas analysis (23).

Detectors

Faraday cup and secondary electron multiplier are the two common detectors used for mass spectrometry. In the Faraday cup, ions pass the beam-defining slit of the cup and strike a metal dynode plate of a secondary emitting material, such as BeO, GaP, or CsSb, that induces several secondary electrons to be ejected and temporarily displaced. This temporary emission of electrons induces a current in the cup and provides a small amplification of the signal when an ion strikes the cup (23). The signal is measured by a electrometer and is directly proportional to the ion concentration. For the single-collector system, with a single amplifier and auto gain switching, high precision can be maintained over the widest range. In this arrangement any such drift is equal for all components and can therefore be compensated for by using any internal standard (2). The secondary electron multiplier (SEM) is a series of dynodes (about 10) maintained at ever-increasing potentials. Ions strike the dynode surface, resulting in emission of secondary electrons that are attracted to the next dynode where again secondary electrons are generated. A cascade of electrons is produced, transferring from plate to plate in a cycloidal flow path between two glass plates.

The signal amplification or current gain of an electron multiplier is 10^6 with a lifetime of 1–2 years, limited by

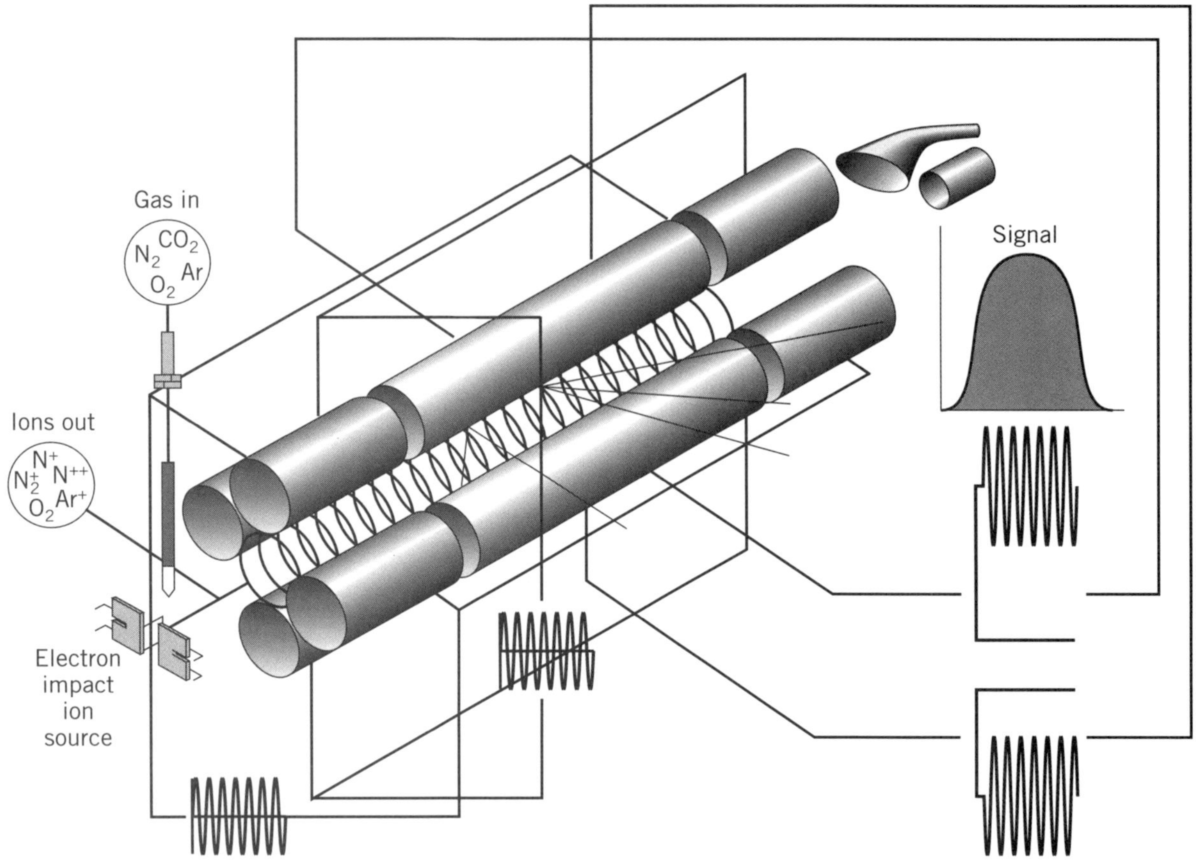

Figure 4. Quadrupole mass filter configuration and Gaussian-type profile curve for the signal, ion current versus electric field strength (VG Gas Analysis Systems Ltd.).

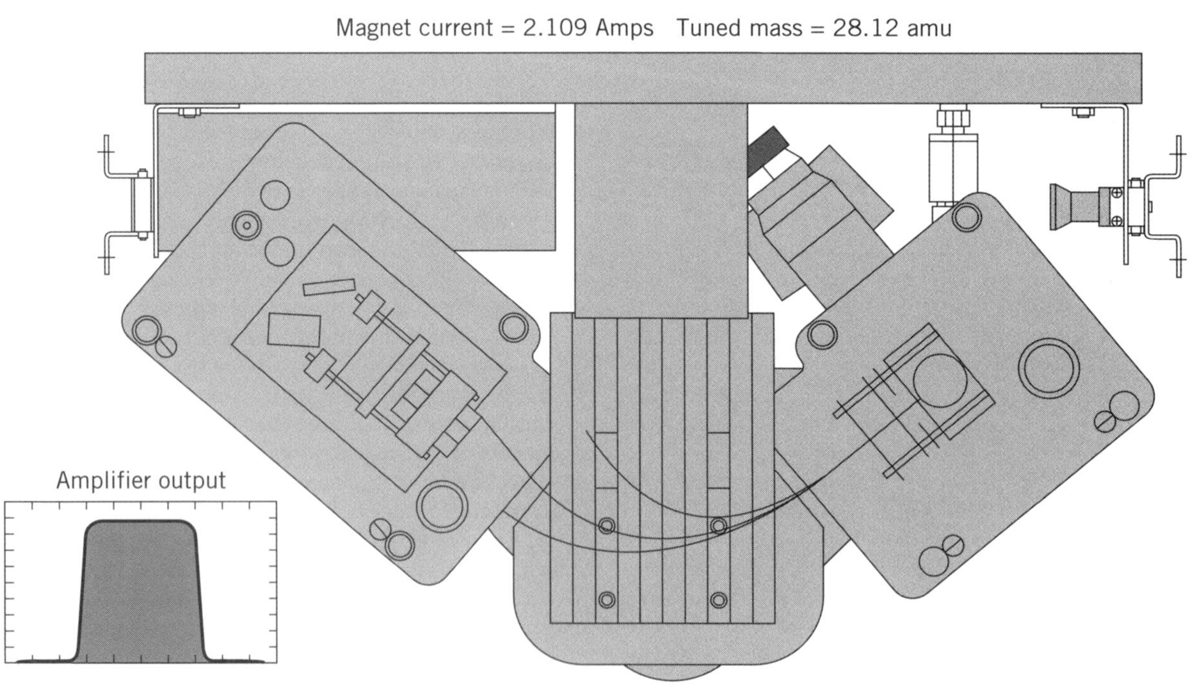

Figure 5. Magnetic sector mass spectrometer showing ion trajectory through magnetic field and the flat-top type profile of the ion current versus magnetic field strength (VG Gas Analysis Systems Ltd.).

surface contamination from incident ions or poor vacuum (19). The Faraday cup is considered rugged, reliable, and typically capable of measurements to 1–10 ppm concentration levels (28). The SEM detector offers greater sensitivity (sub-ppm) with a faster response but has poorer long term stability and is generally less accurate (2). SEM can be sensitive to repeated air exposure; background noise increases when the gain decreases, and it is far less rugged than the Faraday cup (13). In a single-collector arrangement with either a magnetic sector or quadrupole system, it is common to have remotely switchable Faraday cup–SEM assemblies. The appropriate detector is selected by the software, allowing all species to be monitored from 100% to parts per million (24). For example, a Faraday cup is used for air components and a secondary system for volatiles such as ethanol, acetone, or butandiol (22).

Spectral Analysis

To analyze a mixture of ion fragment intensities, with n components, a system of linear equations must be solved to calculate individual component concentrations $c_1, c_2 \ldots c_n$. The ion current signals, S_1, for each mass/charge is considered to be the sum of the contributions of all components, which are proportional to the concentration of each component:

$$\begin{aligned} s_1 &= c_1 m_{11} + c_2 m_{12} + \ldots + c_q m_{1q} \\ &\vdots \\ s_p &= c_1 m_{p1} + c_2 m_{p2} + \ldots + c_q m_{pq} \end{aligned}$$

or

$$\vec{s} = \mathbf{M}\vec{c}$$

where s is the signal, M is the fragmentation pattern or sensitivity factor matrix determined by calibration with individual pure components, and c is the desired concentration (28). Most species produce several peaks due to fragmentation or double ionization. For instance, nitrogen has peaks at nominal m/z values 28, 14, 29, and 15. Thus, a set of linear equations is overdetermined and solved using an iterative route of multidimensional least squares technique (34). The best fit is then made between the observed spectrum and that which can be synthesized following small adjustments of the species concentration. These adjustments use the known fragmentation patterns of various species. Therefore, all the mass spectral information is made available by the single detector instrument, which is most effectively used in achieving the concentration analysis.

Suitability of Mass Spectrometry

A typical magnetic sector mass spectrometer instrument has excellent repeatability of $\pm 0.1\%$ of range at calibration, zero and gain drift less than 1% of range per month, and a response time of 1 s (1,19,28). The limiting step of the response time becomes the passage through the sampling system rather than the mass spectrometer itself. The fast analysis times (seconds) mean that many fermentors and indeed several different processes can be monitored simultaneously. This method provides a cost-effective device for multifermentor sampling despite the high purchase cost of an individual unit. Maintenance is usually associated with contamination of the inlet device and filament and ion pump replacement. However, this is offset by the rapid analysis for polar organic species, excellent long-term accuracy and precision, plus flexibility of analysis. These features make mass spectrometry the most flexible and versatile instrument for vent gas analysis.

INSTALLATION AND OPERATION

Careful consideration must be made in designing a sampling system for a vent gas system, which usually involves the multiplexing of one analyzer to a variety of fermentors. A sample handling system incorporates gas transport from the fermentor, followed by pretreatment before passing through the multipoint sampling manifold into the analyzer. The design of such a system must minimize the response time of sample analysis. The gas must take the shortest route from the fermentor to the pretreatment steps through to the analyzer. The amount of sample pretreatment required can be process specific. Most applications require removal of moisture and contaminating species from the vent gas, which involves steps such as filtering and heating before the analyzer. The materials of construction for transport lines, line purging, sample flow rate, and minimizing tubing lengths must all be considered.

Fermentor Sampling Location

The position of a gas sampling point from the fermentor requires care and to some degree depends on the species for analysis. If fermentor overpressure is to be used for gas transport to the mass spectrometer, then it is pertinent to have the sample port between the headplate and the control valve. The vent lines of fermentors often contain a condenser, double-piped steam heater, or vent filters that are steam dried from a jacket. These techniques can remove volatile compounds from the exhaust gas, so the sampling point should be upstream of these devices if organic volatile analysis is desired. The removal of water via the condenser may also affect carbon dioxide because the solubility of this species is high. Nevertheless, moisture must be removed before the analyzer, so sampling after the condenser may be advantageous.

If fermentor operation requires the use of a vent filter for containment, then sampling must be completed after the filter unless a separate vent gas filter is installed. Foaming and liquid entrainment can occur in the vent line. Some prevention of entrainment into the sampling system can be made by facing the sample port downstream of the gas flow (34). Gas has to make a 180-degree turn to the sample port, while tending to exclude liquid as its inertia prevents it from turning. Foam detection systems can also prevent sample line contamination. The foam sensor, using capacitance or conductivity between two elements, usually probe and tank wall or two elements in the probe, can be mounted on the fermentor headplate; this can be part of an automatic defoamer addition system on foam detection. A second probe can be mounted in the vent gas line to close the off-gas sample valve on foam detection.

Sample Delivery and Pretreatment

Commonly, the fermentor headplate overpressure is used to transport gas to the analyzer. However, for laboratory-scale vessels a gas membrane or piston pump may be required. A 0.1-bar pressure is sufficient for a 100-m length of standard 6-mm nylon tubing (or ¼-in tubing) (28). The tube can be stainless steel, nylon, or preferably PTFE for volatile samples. The tubing is heated to between 60 and 160 °C to ensure compounds in the vapor phase, which avoids removal via condensation; this can be done electrically, by steam, or by jacketed hot water (29). For volatile analysis, heating the tube to about 150 °C and the use of Teflon are essential for fast responses times (35). Adsorption–desorption effects of volatile components with stainless steel tubing walls can hinder response times (>5 min) and cause memory effects.

The vent gas is often passed through a (0.2-μm) hydrophobic filter (PTFE) as a final particulate protector before the analyzer. For unheated lines, moisture is removed by condensation. Knockout pots are used in combination with a coalescing filter whereby large drops form and settle by gravity (34). A graded porosity hydrophilic matrix is used with a drain layer to direct coalescing liquids into a sump at the base of the housing. Automatic draining of the pot is ideal via solenoid actuation periodically or on liquid detection. If further moisture removal is required, various technologies are available, including small concentric tubes or coils in shell heat exchangers with refrigerated coolant. Other systems include the Peltier effect cylindrical gas cooler with a sealed tube containing a pure fluid. A wick lines the inner walls, which ensures fluid return by capillary action. One end of the tube is heated, liquid evaporates, and the gas flow displaces the steam toward the end of the tube. The steam expands and condenses back to the heated end (13). Desiccants such as anhydrous calcium chloride, alumina, or silica gel can be used for further sample drying, but this is rarely necessary.

Sample-Handling Manifolds

Most industrial applications utilize one mass spectrometer for vent gas analysis of an array of fermentors (>10). This requires a fast manifold system free of cross-contamination and leaks to automatically switch between fermentor samples and calibration gases. A typical system has the sampling streams from all tanks under permanent flow to guarantee instant sampling availability and avoid line purging (Fig. 6). Three-way valves can eliminate dead legs, and use of two valves per sample stream allows gas to flow to drain rather than contaminate other streams if both valves fail. Solenoid valves are often used as they are inexpensive and compact compared to air-actuated ball valves. Rotameters can be used to verify flow at the vent side of the three-way valve. The manifold is mounted in a cabinet with heat control to a temperature above the process temperature or to 110 °C required for organic volatiles. Multistream rotary valves have been designed that are actuated by an electric stepper motor for cross-contamination sampling (27,28). The unit (designed by VG Gas Analysis Systems Ltd.) can be configured for 16, 32, 48, or 64 sample streams, connected to the periphery of a circular flat distribution plate, that vent into a cylindrical housing with a common exhaust (Fig. 7). The electric stepper motor drives the rotary arm to position at a sample stream, and sample flows directly into the sample transfer tube. Cross-contamination of other gases is inhibited by a spring-loaded face sealing sleeve (2).

The response time of the manifold system can be reduced by minimizing tubing lengths. The time required to flush the lines should be four times the gas residence time (2). This is estimated using plug flow, but experimental measurements are preferred because of backmixing through valves and regulators.

Sample Flow Control

The signal from a gas analyzer is proportional to the volumetric concentration of the component of interest so that sample pressure will directly influence the results. This can be remedied by measuring internal standards (typically N_2, or Ar), or by measuring the concentrations of all other components, and converting to mole fractions. Alternatively the pressure can be measured and the data corrected using the ideal gas law or by controlling the pressure at the fixed value (34). Pressure control can be overcome by simply venting the instrument outlet to atmosphere. An upstream needle valve can be used to reduce the pressure and to control sample flow rate through instrument. Locating the valve as far upstream as possible reduces the lag time, as the gas velocity is increased after the pressure reduction. However, this leaves a 2–3% variation from atmospheric pressure. If this is an unacceptable error, then an absolute backpressure regulator must be employed on the outlet (34). Temperature is controlled within the analyzer but it is useful to control temperature of the manifold cabinet by an electric heater. A temperature greater than 60 °C will prevent condensation, unless moisture has been removed, and minimize adsorption of compounds.

System Monitoring

Water can be detected in a manifold system by a U-shaped liquid trap with a sensor to detect water in the sample lines. The low electrical conductivity of water in the sample means that the sensor must monitor either the dielectric constant or refractive index of the stream or even thermal dissipation from a self-heated thermistor or electric coil. The sample flow rate (>1 mL/s) can be controlled by a simple needle valve and rotameter. A thermal sensor and auto control valve will ensure sample flow rate is adequate and stable. Leak testing by pressuring the system from instrument air and closing the valves. Alternatively, the pressure can be monitored while the sample system is evacuated if this step is part of the sampling sequence. Diagnostic packages supplied with most analyzers will alarm for problems associated with the capillary, filament, vacuum, and power supply.

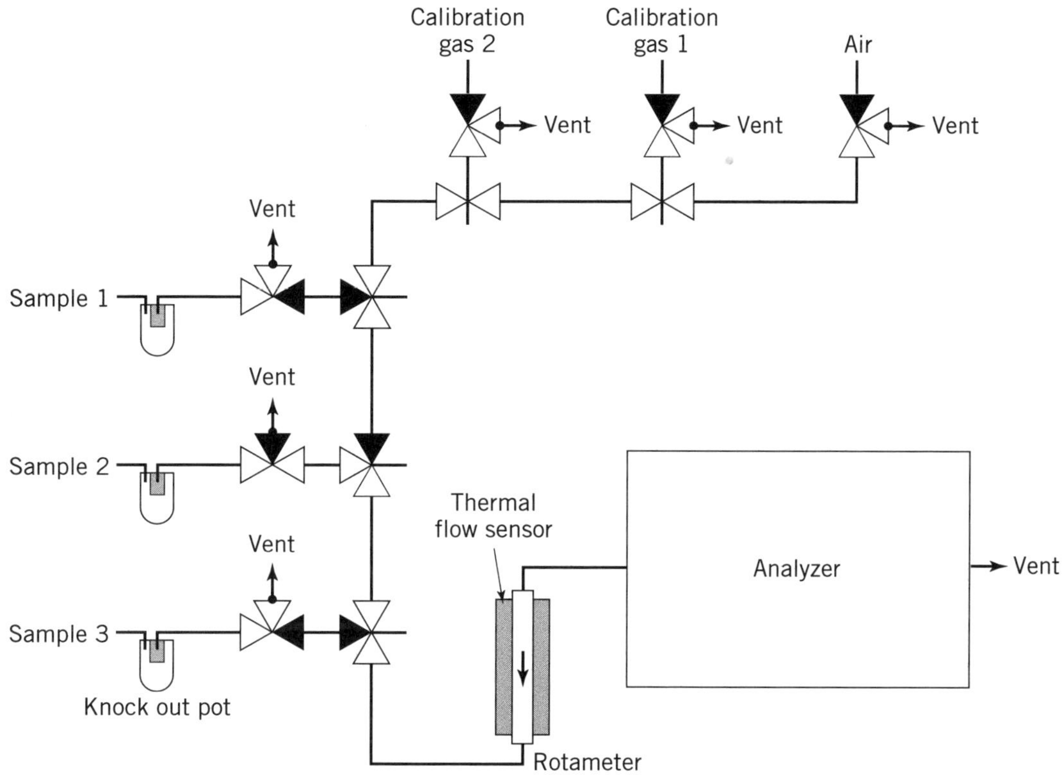

Figure 6. Multistream sample switching system. Solid segments indicate valve-closed position.

Process Measurements Required for Variable Calculation. Measurements of pressure, volume, air flow, and dissolved oxygen are required. To calculate the vent gas variables that are useful to fermentation (oxygen uptake rate [OUR], carbon evolution rate [CER], and respiratory quotient [RQ]). Broth volume can be determined by load cell or differential pressure transducers, air flow rate by a mass flow meter, and dissolved oxygen concentration by a polarographic probe whose response should be checked systematically by varying the headspace pressure or degassing with N_2. Strain gauge, piezoresistive, capacitance, or resonant wire transducer can be used for pressure measurement.

Data Acquisition. For a number of fermentors connected to a multistream manifold, such as a RMS, fast turnover of samples is required although measurements must be accurate. Therefore data acquisition must occur after a suitable line purge time. Multiple measurements can subsequently be made and filtered or smoothed to reduce noise (2). A typical approach is to use an algorithm to produce a stable reading (34). A linear exponential filter is used where each successive measured value c_t is used to update a smoothed estimate as $\hat{c}_t = \kappa c_t + (1 - \kappa)\hat{c}_{t-1}$ where k is a filtering parameter. The trend in the smoothed data is estimated as $\hat{S}_t = \lambda(\hat{c}_t - \hat{c}_{t-1}) + (1 - \lambda)\hat{S}_{t-1}$, where λ is another parameter. A typical algorithm acquires at least $1/\kappa$ values and continues until the absolute value of the trend $\hat{S}_t$ is less than a target tolerance, δ, indicating that a stable result has been obtained. The values of κ, λ, and δ depend on the rate of data acquisition and on instrument performance (34).

Vent gas analysis processes gases on a dry gas basis whereas the actual gases from the fermenter are saturated with water vapor. Standard handbooks can be used to correct these data (36). Most vent gas analysis is expressed as a mole fraction basis. If this is not the case, then data should be adjusted using the ideal gas law to standard conditions of temperature and pressure to allow mass balances to be performed.

Good precision is achieved if signals are noise free with no drift. However, the combination of vent gas data with online data such as air flow rate and dissolved oxygen for mass balancing introduces noise. These errors of measurement can dramatically enlarge errors in the data. Low-pass filters can be used to smooth the data, and many acquisition systems have built-in analog filters for this purpose.

PARAMETER CALCULATION

Vent gas analysis data are only useful when configured to quantify consumption rates and metabolic factors, which requires equation balances for the gas and liquid phases. When completed, the vent gas variables enable characterization of the microbial physiology of the fermentation broth.

Well-Mixed Liquid- and Gas-Phase State

The following balance equations for species i in the gas and liquid phases are written assuming well-mixed phases in

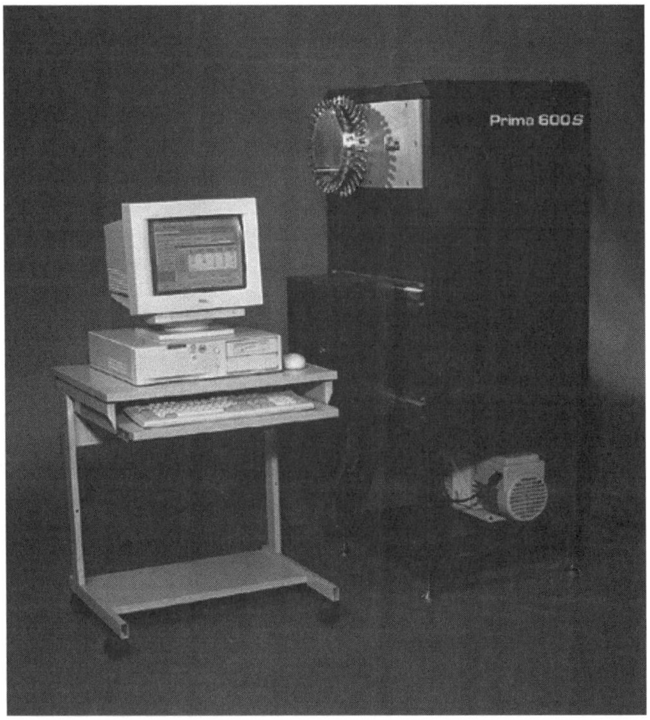

Figure 7. Magnetic sector mass spectrometer unit from VG Gas Analysis Systems Ltd. fitted with the rotary motor sample device.

the fermentor; this assumes that concentrations are uniform throughout the liquid phase, and that mole fractions are uniform throughout the gas phase. The assumption is accepted for small-scale stirred reactors but is less valid at large scale, >100 L. Axial volumetric concentration gradients of species in the gas phase as a consequence of the hydrostatic pressure gradient have been observed at large scale, which leads to alternatives for large-scale vessels, discussed in the next section. For low pressure and dilute concentrations of interest, the ideal gas law and the simplest form of Henry's law (relating equilibrium gas and liquid-phase concentrations through a simple proportionality constant) can be used. The equations have been given by Heinzle and Dunn (2) and modified by Salmon and Buckland (34). It is assumed that the fermentor is isothermal and has a uniform cross-sectional area.

$$\frac{dV_G C_{G_{e,i}}}{dt} = G_o C_{G_{o,i}} - G_e C_{G_{e,i}} - k_{L_i} a V_L (C^*_{L_{e,i}} - C_{L_{e,i}}) \quad (1)$$

$$\frac{dV_L C_{L_{e,i}}}{dt} = L_o C_{L_{o,i}} - L_e C_{L_{e,i}} + k_{L_i} a V_L (C^*_{L_{e,i}} - C_{L_{e,i}}) - V_L \Sigma r_i \quad (2)$$

where $C_{G_{o,i}}$, $C_{G_{e,i}}$, $C_{L_{o,i}}$ and $C_{L_{e,i}}$ are the gas- and liquid-phase concentrations, subscripts *o* and *e* refer to the influent and effluent, and subscripts *G* and *L* refer to the gas and liquid phases, respectively, V_G and V_L are the gas- and liquid-phase volumes, G_o, G_e, L_o, and L_e are the influent and effluent gas and liquid flow rates, $k_{L_i}a$ is a mass transfer coefficient, $C^*_{L_{e,i}}$ is the liquid-phase concentration that would be in equilibrium with $C_{G_{e,i}}$, and Σr_i is the total consumption rate of species *i* in the liquid phase. Standard conditions of pressure and temperature (typically $P_{ref} = 1$ atmosphere, $T_{ref} = 25$ °C) have been used for gas-phase concentrations, volumes, and flow rates.

An expression can be written for $C^*_{L_{e,i}}$, liquid-phase concentration, using Henry's law and integrating over the fermentation broth volume (34):

$$C^*_{L_{e,i}} = (RT_{ref}/H_i P_{ref})(P_h + V_L \rho_L g/2A) C_{G_{e,i}} \quad (3)$$

where H_i is the Henry coefficient, P_h is the headspace pressure, ρ_L is the liquid-phase density, and A is the cross-sectional area of the fermentor. It is usual to assume that the liquid-phase supply, removal, and accumulation terms in equation 1 are negligible for gases such as oxygen that have low aqueous solubility. The gas-phase accumulation term in equation 1 is also ignored because gas holdup is small and changes in oxygen utilization occur slowly. However, this simplification may not be valid for rapidly growing cultures (37) or when sudden metabolic changes occur (38), in which case the calculated results will lag actual events in the fermentor. Using the discussed assumptions, equations 1 and 2 can be added and rearranged to obtain an expression for $OUR = r_{O_2}$, the net oxygen utilization rate:

$$OUR = G_o(C_{G_{o,O_2}} - G_e/G_o C_{G_{e,O_2}})/V_L \quad (4)$$

An expression for the mass transfer coefficient, in a dimensionless form, can also be made from equation 1:

$$\alpha_{O_2} = \frac{k_{L_{O_2}} a V_L}{G_o} = \frac{C_{G_{o,O_2}} - G_e/G_o C_{G_{e,O_2}}}{\beta_{O_2}(1 + \gamma/2) C_{G_{e,O_2}} - C_{L_{e,O_2}}} \quad (5)$$

where $\beta_{O_2} = R\ T_{ref} P_h / H_{O_2} P_{ref}$, and $\gamma = V_L \rho_L g/AP_h$. Values of the Henry coefficient, H_{O_2}, can be obtained from the literature (39). The aqueous oxygen concentration, $C_{L_{e,O_2}}$, found in equation 5, can be measured on line using a polarographic probe calibrated to read $\beta_{O_2}(1 + \gamma/2) C_{G_{o,O_2}}$ before inoculation. The constant, 1, assumes that the difference of oxygen solubility in the fermentation medium compared to water is negligible. This assumption is reasonable as the constant can be estimated for most fermentation broths (39,40) and a value of 0.945 is achieved for a typical broth containing, salts, yeast extract, and glucose.

Inert gas species (such as Ar or N_2) are monitored to compensate for gas flow changes caused by the presence of moisture and volatiles in the exhaust gas. The net utilization for the inert gas is zero and consequently $G_e/G_o = C_{G_{o,inert}}/C_{G_{e,inert}}$ by using the simplifications for oxygen.

For carbon dioxide the situation is more complex because its high solubility (H_{CO_2} of 34 L bar/mol compared with H_{O_2} of 856.9 L bar/mol) means the liquid-phase accumulation cannot be ignored. Also, carbon dioxide reacts with water to give biocarbonate at pH > 6.5.

$$CO_2 + H_2O \underset{k_b}{\overset{k_f}{\rightleftarrows}} H_2CO_3 \overset{K_a}{\leftrightarrow} H^+ + HCO_3^- \leftrightarrow 2H^+ + CO_3^{2-} \quad (6)$$

The hydration of CO_2 to H_2CO_3 is a slow rate constant; however, its concentration is low (<0.3% of CO_2 present) (41) because it is almost instantaneously dissociated to HCO_3^-. At the pH range 4–8 for fermentation processes, the concentration of carbonate ions is negligible, complexing of CO_2 with amine groups can be ignored, and the reaction of dissolved CO_2 with OH^- is negligible (42). Bicarbonate accumulation increases with pH, and at pH 6.3 the concentration can be equal to the dissolved CO_2. Thus, the basic liquid-phase equation 2 must be modified to include accumulation of bicarbonate:

$$\frac{dV_L C_{L_{e,CO_2}}}{dt} = L_o C_{L_{e,CO_2}} - L_e C_{L_{e,CO_2}} + k_{L_{CO_2}} a V_L (C^*_{L_{e,CO_2}} - C_{L_{e,CO_2}}) + V_L(CER - r_{HCO_3^-}) \quad (7)$$

$$\frac{dV_L C_{L_{e,HCO_3^-}}}{dt} = L_o C_{L_{o,HCO_3^-}} - L_e C_{L_{e,HCO_3^-}} + V_L r_{HCO_3^-} \quad (8)$$

where $CER = -r_{CO_2}$ is the net carbon dioxide evolution rate from the biomass and $r_{HCO^-}3$ is the net rate of bicarbonate formation given by

$$r_{HCO_3^-} = k_f C_{L_{e,CO_2}} - k_b/K_a\, C_{L_{e,H^+}}\, C_{L_{e,HCO_3^-}} \quad (9)$$

Values for the constants k_f, k_b, and k_a can be obtained from the literature (43). Rearrangement of equation 1 gas-phase balance, and neglecting accumulation terms as before can produce the following balance:

$$k_{L_{CO_2}} a V_L (C^*_{L_{e,CO_2}} - C_{L_{e,CO_2}}) = G_o C_{G_{o,CO_2}} - G_e C_{G_{e,CO_2}} \quad (10)$$

and solved for the liquid-phase CO_2 concentration,

$$C_{L_{e,CO_2}} = \left\{\frac{1}{\alpha_{CO_2}}\frac{G_e}{G_o} + \beta_{CO_2}(1 + \gamma/2)\right\} C_{G_{e,CO_2}} - \frac{1}{\alpha_{CO_2}} C_{G_{o,CO_2}} \quad (11)$$

where $\alpha_{CO_2} = k_{L_{CO_2}} a V_L / G_0$, and $\beta_{CO_2} = RT_{ref} P_h / H_{CO_2} P_{ref}$. Henry coefficient H_{CO_2} estimates are found in the literature (39). Combining and rearranging equations 7, 8, and 10 gives

$$CER = \frac{1}{V_L}\left\{\frac{dV_L(C_{L_{e,CO_3}} + C_{L_{e,HCO_3^-}})}{dt} - L_o(C_{L_{e,CO_2}} + C_{L_{e,HCO_3^-}}) + L_e(C_{L_{e,CO_2}} + C_{L_{e,HCO_3^-}}) - G_o(C_{G_{o,CO_2}} - G_e/G_o C_{G_{e,CO_2}})\right\} \quad (12)$$

The evaluation of $C_{L_{e,HCO_3^-}}$ can be considered in three algorithms of complexity (34). Equations 8 and 12 can be solved progressively as the fermentation proceeds using current and past data. This practice enables monitoring of CO_2 evolution under dynamic conditions, so long as the rate of data acquisition by the analyzer is sufficiently rapid. This may be a problem when acid is introduced intermittently for pH control because the CO_2 concentration in the vent gas will increase for a short period after each addition (44). A second approach can be applied if the time constant for process changes is greater than 5 min. In this case, the hydration reaction can be assumed to be at equilibrium, yielding $C_{L_{e,HCO_3^-}} \approx (k_f/k_b) K_a 10_{pH} C_{L_{e,CO_2}}$. Substituting this expression into equation 12 allows the value of CER to be calculated directly if the rate of change of the dissolved CO_2 concentration is estimated using current and past data (34). Finally, it is common practice to ignore the transient terms so long as changes of pH and CER are slow. Heinzle et al. (45) demonstrated that including dynamic changes of CO_2 and HCO_3^- in a simulation only produced relative errors <2% in the CER.

Large-Scale Mixing Models

The assumption of ideal mixed gas and liquid phases is reasonable for tanks less than 100 L. However, upon scale up to production scale (100– 100,00 L), axial gradients and large mixing times are observed and so previous equations for oxygen transfer rate, OTR, and OUR based on the well-mixed system may not be valid. Hydrostatic pressure gradients in tall fermentors will cause large differences in oxygen solubility C_L from liquid surface to tank bottom. The dissolved oxygen concentration in a stirred tank can vary linearly from 5.4 mg/L at the surface to 12.1 mg/L at a depth of 10 m. At the bottom of the tank the oxygen solubility for a given gas composition is twice as a consequence of the doubled hydrostatic pressure, as shown by Henry's law. This change will double the maximum oxygen transfer rate, and, hence, dissolved oxygen concentration, oxygen solubility, gas holdup, and transfer parameters may vary with depth in a tall reactor.

Limited experimental and theoretical data are available on concentration in homogeneities or gradients within large fermentors, but radial and axial gradients of pH, conductivity, and dissolved oxygen have been observed (46,47). Residence time distribution data using liquid and gas transfers have characterized the mixing to produce a variety of models (48–51). Various models have been determined for large-scale mixing, where axial backmixing has been described by dispersion theory or a tank in series with recirculation based on the measurement of residence time distribution. Other models include compartmentalization of the reactor into zones, such as two well-mixed impeller compartments and the remainder for the rest of the tank, with transport between compartments based on the pumping power of impellers (46).

The model behavior is described through simulations of the coupled material balance equations for oxygen transfer rate, mixing, and kinetics. The importance of backmixing and hydrostatic pressure on scale-up was demonstrated by using a three-phase multimixed compartment model (50). The flow conditions of these models fall between extremes of well-mixed and plug flow. Simplification for large-scale reactors is to assume that the liquid phase is well mixed but that the gas phase moves in plug flow (no backmixing) from inlet to outlet (51). The gas-phase concentration of species i was described by Salmon and Buckland (34) using this equation:

$$\frac{dC_{G_i}}{dx} - \alpha_i\{\beta_i[1 + \gamma(1 - x)]C_{G_i} - C_{L_{e,i}}\} = 0 \qquad (13)$$

where α_i, β_i, and γ are defined in the previous section. For the gas-phase concentration of species i, a closed-form solution to this equation was used to calculate $C_{G_{e,i}} = C_{G_i}(1)$ from the inlet condition of $C_{G_i}(0) = C_{G_{o,i}}$. Salmon and Buckland (34) used this solution to produce a comparison between the estimates of equilibrium between liquid and gas phases at two scales, 0.15 and 60 m^3 (where differences are solely due to hydrostatic pressure), and between the two models of well-mixed phases and a gas phase with plug flow. For a given mass transfer coefficient, the predicted liquid-phase concentrations were closer to equilibrium with the effluent gas for the plug flow model than with the well-mixed system. This difference was accentuated with the large-scale tank (where hydrostatic pressure is significant). This conclusion seems reasonable because the mass transfer coefficient required to achieve a desired liquid-phase concentration is lower for a production vessel than for a small laboratory fermentor. Thus, for a given mass transfer coefficient the gas-phase and liquid-phase concentrations in a large vessel will be closer to equilibrium than those in a small vessel. This agrees with the observation that the dissolved carbon dioxide in a large tank can be calculated by presuming equilibrium with the vent gas, whereas this assumption is not appropriate for small fermentors.

Aqueous Concentration Estimation for Volatile Organic Compounds

The liquid-phase concentration for an organic species can be calculated as for $C_{L_{e,CO_2}}$ (equation 11) assuming that the system is well mixed with respect to the gas and liquid phases and that the inlet gas-phase concentration is zero (34):

$$C_{L_{e,i}} = \beta_i\left[\frac{1}{\alpha_i\beta_i}\frac{G_e}{G_o} + (1 + \gamma/2)\right]C_{G_{e,i}} \qquad (14)$$

As $\alpha_i\beta_i \rightarrow \infty$, the liquid-phase concentration approaches $\beta_i(1 + \gamma/2)\,C_{G_{e,i}}$, which is the concentration in equilibrium with the gas phase at the average fermentor pressure. Equilibrium constants can be attained for any volatile compound under most conditions (22) because most compounds have a high affinity for the liquid phase and hence a low Henry's coefficient (ethanol 10^{-2} bar/L/mol) compared to gas. Camelbeeck et al. (29) have experimentally confirmed this conclusion by demonstrating that changes in the rates of agitation and aeration have no influence on the vent gas measurements.

The ability to monitor a specific organic compound depends on concentration and volatility and analyzer sensitivity. Oeggerli and Heinzle (22) compared partial pressures above an aqueous solution at a concentration of 0.1 g/L at 30 °C. Compounds were considered analyzable if the partial pressure was greater than 1 μbar; hence, ethanol at 22.24 μbar and acetone at 181 μbar could be monitored (oxygen, $>10^6$ μbar). However 2,3-mesobutanediol was considered undetectable as the partial pressure was 0.0174 μbar, and its limit of detection was found to be 5 g/L with a quadrupole mass spectrometer.

Calibration is usually achieved by adding a known amount of compound to the fermentor before inoculation and measuring concentrations from liquid samples off line by HPLC or GC. Most volatiles require a SEM for accurate detection as volatiles are in the ppm level in the headspace (22,29).

Growth Kinetics

Growth rate kinetics including yield of biomass from oxygen consumed, Y_{x/O_2}, from carbon dioxide evolution, Y_{x/CO_2}, and the specific growth rate (52,53) can be simply determined using vent gas analysis:

$$Y_{x/O_2} = \frac{d(\mathrm{TOC})/dt}{dx/dt} \qquad (15)$$

$$X = Y_{x/O_2}\,(\mathrm{TOC}) \qquad (16)$$

where TOC is the total oxygen consumed estimated from the summation of the OUR versus time and X is the biomass concentration from oxygen. Linear regression from a curve of biomass concentration (determined off line) versus TOC and total carbon dioxide evolved (TCE) can then lead to the respective yields factors for biomass from oxygen and carbon dioxide. The specific growth rate, μ, can be determined by

$$\mu = \frac{d(\mathrm{TOC})/dt}{\mathrm{TOC}} = \frac{\mathrm{OUR}}{\mathrm{TOC}} \qquad (17)$$

where the Y_{x/O_2} is assumed to be constant. Good agreement can be made between specific growth rates as calculated on line using TOC data and off-line data from experimental dry cell weight (53). This calculation provides a useful on-line tool for analyzing growth rate during process and development. This analysis is especially important for secondary metabolites in which the understanding of growth and production rates are critical to process optimization, often involving substrate feeding. For example, the control of specific growth is very important for the synthesis of avermectin by *Streptomyces avermitilis* (53). The impact of decrease of temperature by a few degrees at 12 h into the fermentation clearly showed a dramatic decrease in specific growth rate compared to the control, as measured by the on-line method (Fig. 8).

Mass Balancing and RQ

Vent gas analysis data can be used to describe the substrate utilization and production kinetics of a fermentation via mass balancing (2,34). A typical fermentation could involve $CH_{h_S}N_{n_S}O_{o_S} + aNH_3 + bO_2 \rightarrow Y_{X/S}CH_{h_X}N_{n_X}O_{o_X} + Y_{P/S}CH_{h_P}N_{n_P}O_{o_P} + cCO_2 + dH_2O$, where biomass $CH_{h_X}N_{n_X}O_{o_X}$ and product $CH_{h_P}N_{n_P}O_{o_P}$ are formed from substrate $CH_{h_S}N_{n_S}O_{o_S}$. The coefficients a, b, c, and d can be solved from elemental balances for carbon, hydrogen, nitrogen, and oxygen. Coefficients for oxygen and carbon dioxide are $b = (\gamma'_S - Y_{X/S}\gamma'_X - Y_{P/S}\,\gamma'_P)/4$ and $c = 1 - Y_{X/S} - Y_{P/S}$, where $\gamma'_i = 4 + h_i - 3n_i - 2o_i$ is the so-called generalized degree of reduction for compound

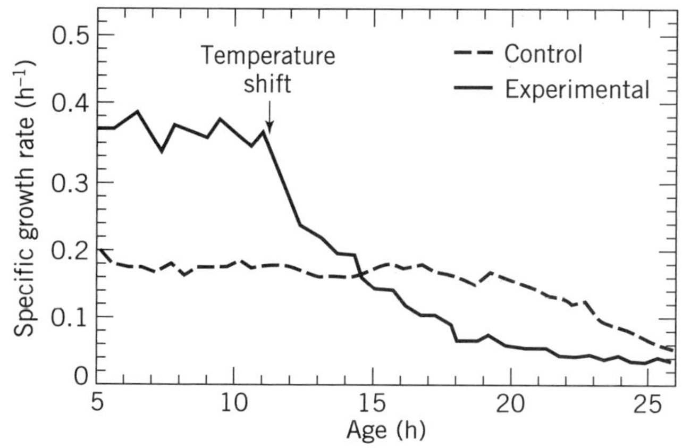

Figure 8. Impact of temperature shift on estimated specific growth rate (on line) of *Streptomyces avermitilis* at 800 L (53).

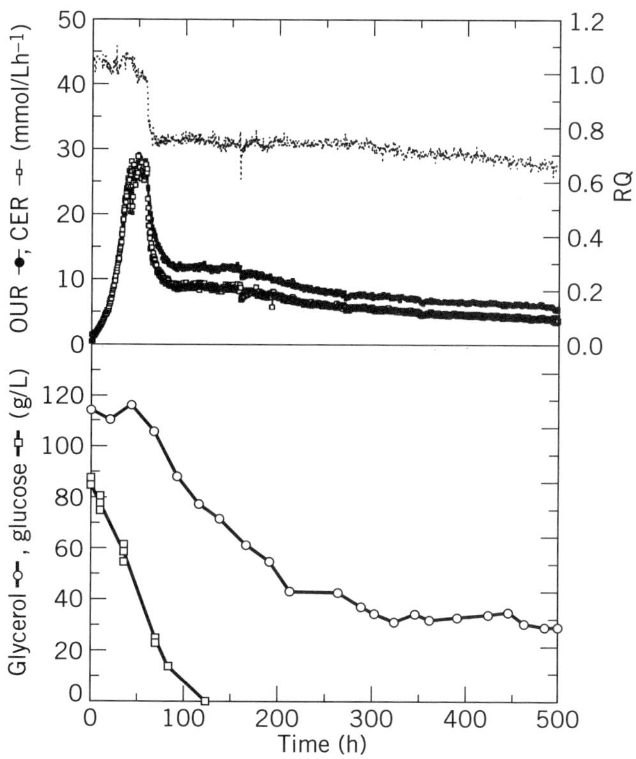

Figure 9. Comparison of vent gas analysis to off-line substrate data for a fungal fermentation at 1,600 L.

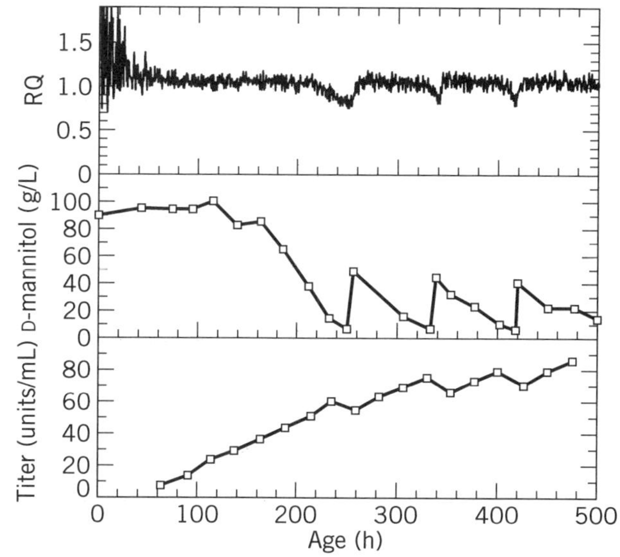

Figure 10. Use of on-line RQ measurements to trigger nutrient additions for secondary metabolite production.

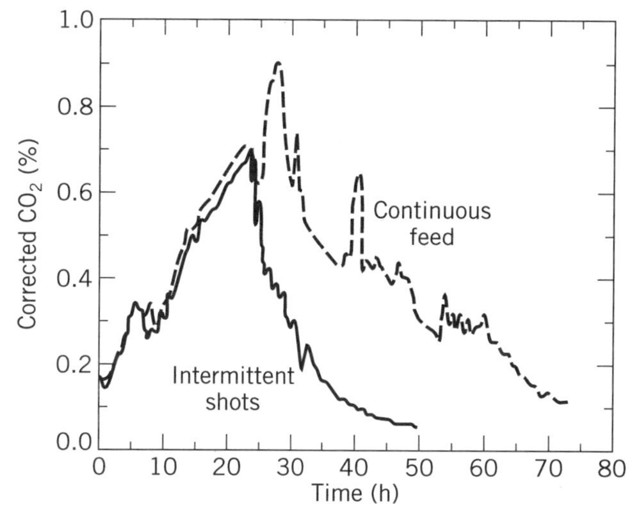

Figure 11. Comparison of the vent gas CO_2 concentration for *Nocardia* sp. between continuous and intermittent shot process for addition (54).

$CH_{h_i}N_{n_i}O_{o_i}$. The value of μ' for biomass is consistently about 4.2, and the values for some common substrates and products can be found in the literature (8). An expression can then be made for the respiratory quotient, RQ, the ratio between the rates of CO_2 evolution and oxygen utilization, as $RQ = 4(1 - Y_{X/S} - Y_{P/S})/(\gamma' s - Y_{X/S}\,\gamma'_p)$. Thus, for a given fermentation the RQ, defined from vent gas analysis, CER/OUR, can be used with off-line analysis of biomass, carbon, nitrogen, and production concentrations to define the yield coefficients for substrate and production (32). Elemental balancing requires high-precision gas analysis with 0.01% accuracy because a 1% error relative offset calibration error for oxygen can result in an error for product estimation >50% (2).

For industrial applications, the unusual mutant strains have complex metabolism, of which little is understood. It is thus difficult to define stoichiometric equations for a process, limiting the use of elemental analysis. Nevertheless, vent gas analysis is a valuable tool for studying process kinetics and metabolism. For example, during a fungal fermentation the RQ was found to switch from unity to 0.7, indicating the change in metabolism from glucose to glycerol (Fig. 9). The reduction of growth rate, as shown by the OUR profile, and commencement of glycerol utilization indicated the start of secondary metabolite production. In another fungal fermentation, the depletion of D-mannitol and subsequent reduction in volumetric productivity corresponded with the RQ reduction from 1 to 0.8 (Fig. 10). Shot additions of D-mannitol were made in response to the change in RQ, which prolonged the fermentation and increased production.

During the microbial hydroxylation of simvastatin (a cholesterol-lowering drug) by a *Nocardia* species, high concentrations of the substrate inhibit production (54). Maintaining a low concentration of the substrate using a controlled continuous feed strategy as opposed to a shot process enabled yield improvements of the biotransformation. The carbon dioxide profile was used to monitor process performance (Fig. 11) where the continuous substrate feed prolonged the decrease of carbon dioxide after 30 h compared to the rapid decrease observed with the shot processes. Therefore, combining vent gas data, including RQ, with off-line analysis of substrate utilization and productivity rates, along with an understanding of biochemical pathways, can enable process monitoring and feedback control.

BIBLIOGRAPHY

1. B. Buckland, T. Brix, H. Fastert, K. Gbewonyo, G. Hunt, and D. Jain, *Biotechnology* **3**, 982–988 (1985).
2. E. Heinzle and I.J. Dunn, in H.-J. Rehm, G. Reed, and K. Schugerl eds., *Biotechnology: Measuring, Modelling, and Control*, VCH Verlagsgesellschaft, Weinheim, 1991, pp. 27–74.
3. S. Aiba, S. Nagai, and Y. Nishizawa, *Biotechnol. Bioeng.* **18**, 1001–1006 (1976).
4. C. Fieschki, K.M. Egan, T. Ritch, R.A. Koski, and M. Jones, *Biotechnol. Bioeng.* **29**, 1113–1118 (1987).
5. J. Sevcik, *Detectors in Gas Chromatography*, Elsevier, New York, 1976, pp. 39–101.
6. G.D. Nichols, *On-Line Process Analyzers*, J. Wiley, New York, 1988, pp. 3–136.
7. Y.H. Lee and G.T. Tsao, *Adv. Biochem. Eng.* **9**, 30–42 (1979).
8. B. Atkinson and F. Mavituna, *Biochemical Engineering and Biotechnology Handbook*, Stockton Press, New York, 1991, pp. 1032–1038.
9. A.D. Brailsford, M. Yussouff, and E.M. Logothetis, *Sens. Actuators B* **44**, 321–326 (1997).
10. N. Hara and D.D. MacDonald, *J. Electrochem. Soc.* **144**, 4158–4160 (1997).
11. K.-D. Anders, W. Muller, R. Kammeyer, and T. Scheper, *Biotechnol. Tech.* **6**, 97–104 (1992).
12. H.M. Muhlemann and H.R. Bungay, *Biotechnol. Tech.* **7**, 575–578 (1993).
13. M.-N. Pons, *Bioprocess Monitoring and Control*, Carl Hanser Verlag, Munich, 1992, pp. 1–63.
14. A. Mandelis and C. Christofides, *Physics, Chemistry, and Technology of Solid State Gas Sensor Devices*, Wiley, New York, 1993, pp. 254–308.
15. Y. Taniguchi and K. Murakami, *Electron. Commun. Jpn.* **79**, 86–96 (1996).
16. C.C. Chang and C.S. Tang, *Sens. Actuators A* **65**, 171–174 (1998).
17. L. Rottenbacher, L. Behlau, and W. Bauer, *J. Biotechnol.* **2**, 137–147 (1985).
18. K. Schaefer and M. Schultis, in E. Henizle and M. Reuss eds., *Mass Spectrometry in Biotechnological Process Analysis and Control*, Plenum, New York, 1987, pp. 39–48.
19. A.T. Dadd, *Am. Biotechnol. Lab.* **8**, 35–38 (1990).
20. J.R. Swartz and C.L. Cooney, *Process Biochem.* **13**, 3–7 (1978).
21. S. Chauvatcharin, T. Seki, K. Fujiuyama, and T. Yoshida, *J. Ferment. Bioeng.* **79**, 264–269 (1995).
22. A. Oeggerli and E. Heinzle, *Biotechnol. Prog.* **10**, 284–290 (1994).
23. G. Suizdak, *Mass Spectrometry for Biotechnology*, Academic Press, New York, 1996, pp. 4–55.
24. H. Degn, R.P. Cox, and D. Lloyd, *Methods Biochem. Anal.* **31**, 165–194 (1985).
25. S. Bohatka, G. Langer, J. Szilagyi, and I. Berecz, *Int. J. Mass Spectrom. Ion Phys.* **48**, 277–280 (1983).
26. R.P. Cox, in E. Henizle and M. Reuss eds., *Mass Spectrometry in Biotechnological Process Analysis and Control*, Plenum, New York, 1987, pp. 63–74.
27. J.P. Camelbeeck, D.M. Comberbach, J. Goossens, and P. Roelants, *Biotechnol. Tech.* **2**, 183–188 (1988).
28. M.J. Winter, in E. Henizle and M. Reuss eds., *Mass Spectrometry in Biotechnological Process Analysis and Control*, Plenum, New York, 1987, pp. 17–38.
29. J.P. Camelbeeck, D.M. Comberbach, M. Orval, J.O. Petre, and P. Roelants, *Biotechnol. Tech.* **5**, 443–452 (1991).
30. E.R. Schmid, in E. Henizle and M. Reuss eds., *Mass Spectrometry in Biotechnological Process Analysis and Control*, Plenum, New York, 1987, pp. 7–16.
31. C.D. Bartman, in E. Henizle and M. Reuss eds., *Mass Spectrometry in Biotechnological Process Analysis and Control*, Plenum, New York, 1987, pp. 49–62.
32. E. Heinzle, *J. Biotechnol.* **25**, 81–114 (1992).
33. S.J. Coppella and P. Dhurjati, *Biotechnol. Bioeng.* **29**, 679–689 (1987).
34. P. Salmon and B.C. Buckland, in P.M. Rhodes and P.F. Stanbury eds., *Applied Microbial Physiology*, Oxford University Press, New York, 1997, pp. 131–163.
35. H. Heinzle, *Adv. Biochem. Eng. Biotechnol.* **35**, 1–46 (1987).
36. R.H. Perry, D.W. Green, and J.O. Maloney, *Chemical Engineers Handbook*, McGraw-Hill, New York, 1984.
37. S. Aiba and H. Furuse, *Biotechnol. Bioeng.* **36**, 534–538 (1990).
38. R.M. Dekkers, *Proceedings of the First IFAC Workshop on Modelling and Control of Biotechnical Processes*, Pergamon, Oxford, 1982.
39. A. Schumpe, G. Quicker, and W.D. Deckwer, *Adv. Biochem. Eng.* **24**, 1–38 (1982).
40. M. Popovic, H. Niebelschutz, and M. Reuss, *Eur. J. Appl. Microbiol.* **8**, 1–15 (1979).
41. J. Smith, S.W. Davison, and G.F. Payne, *Biotechnol. Bioeng.* **35**, 1088–1101 (1990).
42. P.N.C. Royce and N.F. Thornhill, *AIChE J.* **37**, 1680–1686 (1991).
43. H.S. Harned and B.B. Owen, *The Physical Chemistry of Electrolyte Solutions*, Reinhold, New York, 1958.
44. P.K. Yegneswaran, M.R. Gray, and B.G. Thompson, *Biotechnol. Bioeng.* **36**, 92–96 (1990).
45. E. Henizle, A. Oeggerli, and B. Dettwiler, *Anal. Chim. Acta* **238**, 101–115 (1990).
46. N.M.G. Oosterhuis and N.W.F. Kossen, *Biotechnol. Bioeng.* **26**, 546–550 (1984).

47. G. Larsson, M. Tornkvist, E. Stahl Wernersson, C. Tragardh, H. Noorman, and S.-O. Enfors, *Bioprocess Eng.* **14**, 281–289 (1996).
48. H.M. Ruffer, A. Petho, K. Schugerl, A. Lubbert, A. Ross, and W.D. Dekwer, *Bioprocess Eng.* **11**, 145–152 (1994).
49. A. Lubbert, in H.-J. Rehm, G. Reed, and K. Schugerl eds., *Biotechnology: Measuring, Modelling, and Control*, VCH Verlagsgesellschaft, Weinheim, 1991, pp. 107–148.
50. M. Reuss and R. Bajpai, in H.-J. Rehm, G. Reed, and K. Schugerl eds., *Biotechnology: Measuring, Modelling, and Control*, VCH Verlagsgesellschaft, Weinheim, 1991, pp. 299–348.
51. S. George, G. Larsson, K. Olsson, and S.-O. Enfors, *Bioprocess Eng.* **18**, 135–143 (1998).
52. D.W. Zabriskie and A.E. Humphrey, *AIChE J.* **24**, 138 (1978).
53. K. Gbewonyo, D. Jain, G. Hunt, S.W. Drew, and B.C. Buckland, *Biotechnol. Bioeng.* **34**, 234–241 (1989).
54. K. Gbewonyo, B.C. Buckland, and M.D. Lilly, *Biotechnol. Bioeng.* **37**, 1101–1107 (1991).

VINEGAR, ACETIC ACID PRODUCTION

HEINRICH EBNER
Linz, Austria
SYLVIA SELLMER-WILSBERG
Bad Honnef, Germany

KEY WORDS

Acetic acid bacteria
Bacteriophage
Cell recycling
Classification
Continuous submerged fermentation
Frings acetator
Semicontinuous submerged fermentation
Surface fermentation
Vinegar production

OUTLINE

INTRODUCTION

Worldwide, acetic acid is called vinegar, if obtained by oxidative fermentation of ethanol-containing solutions by acetic acid bacteria. Although vinegar does not entirely exclude diluted chemically produced acetic acid in every country of the world, this term is used here to define acetic acid that is produced from ethanol by a primary microbial metabolism, the so-called acetic acid fermentation or vinegar fermentation.

BASES OF ACETIC ACID FERMENTATION

Acetic acid fermentation (acetous fermentation) is an oxidative fermentation by which solutions of ethanol are oxidized to acetic acid and water by acetic acid bacteria using atmospheric oxygen. The oxidation proceeds according to the basic equation

$$C_2H_5OH + O_2 \rightarrow CH_3COOH + H_2O$$
$$G^\circ = -455 \text{ kJ mol}^{-1}$$

The alcohol-containing solution is called mash. Its alcohol concentration is given in percent by volume. Usually it also contains some acetic acid expressed in grams of acetic acid per 100 mL (% w/v). The sum of ethanol (vol%) and acetic acid (g 100 mL^{-1}) is called total concentration because the sum of these rather incommensurable values gives the maximal concentration of acetic acid that can be obtained by complete fermentation (1). The quotient of the total vinegar concentration produced over the total mash concentration indicates the concentration yield.

RAW MATERIALS FOR ACETIC ACID FERMENTATION

All mashes must contain ethanol and water and provide the acetic acid bacteria with nutrients.

Alcohol

By far the largest percentage of vinegar is alcohol vinegar, which is produced from diluted purified ethanol. Other common names for the same product are white vinegar or spirit vinegar. It is customary in almost all countries to denature the ethanol serving as raw material for the vinegar industry. In the United States, this is mostly done with ethyl acetate, and in most European Countries, with alcohol vinegar.

Mashes obtained by alcoholic fermentation of natural-sugar-containing liquids also serve as raw materials. The designation of vinegar is according to the particular raw material used: wine vinegar is produced by acetic acid fermentation of grape wine; cider vinegar is produced from fermented apple juice; malt vinegar is the undistilled product of alcoholic and subsequent acetous fermentation of an infusion of barley or other cereals, the starch of which was converted by malting; rice vinegar is made from saccharified rice starch, followed by alcoholic and acetous fermentation.

Water for Processing

The water used for the preparation of mashes must be clear, colorless, odorless, bacteriologically safe, and without sediments or suspended particles. It must also be free of chlorine, ozone, other chemical compounds, and damaging bacteria.

Nutrients

Most natural raw materials do not require the addition of extra nutrients. However, apple cider is usually low in nitrogenous compounds, but this can be corrected by the addition of ammonium phosphate. Some grape wines also require the addition of ammonium phosphate for an optimal fermentation. In rare cases, all nutrients discussed next must be added for the production of wine vinegar.

For the production of alcohol vinegar, a mixture of the required nutrients was developed. When producing vinegar with up to 15% acetic acid, the acetic acid bacteria definitely require glucose and potassium, sodium, magnesium, calcium, ammonium as ammonium phosphate, sulfate, and chloride. The following trace minerals are needed: iron, manganese, cobalt, copper, molybdenum, vanadium, and zinc. These additions are sufficient for an optimal acetic acid fermentation.

Commercially available mixtures contain supplements such as dried yeast extract in order to restart a fermentation more quickly if stopped by a disturbance such as a power failure. Principally, nutrients should be added sparingly in order to exert a selection pressure directing to a low requirement for nutrients.

PRODUCTION BY ACETIC ACID FERMENTATION

Microorganisms

Overview and Basic Problems of Classification. The microorganisms oxidizing ethanol to acetic acid are commonly called acetic acid bacteria. This special primary microbial metabolism at low pH of the surrounding medium differentiates them from all other bacteria. Acetic acid bacteria are polymorphous. Cells are Gram-negative, ellipsoidal to rod shaped, straight or slightly curved, 0.6 to 0.8 μm long, occurring singly, in pairs or in chains. There are nonmotile and motile forms with polar or peritrichous flagella. They are obligately aerobic. Some produce pigments, some cellulose.

Numerous attempts to classify acetic acid bacteria were carried out as summarized in the work by Ebner and Follman (1). A major contribution to the exact taxonomy of acetic acid bacteria were the DNA-rRNA hybridization studies of Gillis and De Ley (2). A first conclusion from this work was that the two groups *Gluconobacter* and *Acetobacter*, as classified earlier, were closely related groups, justifying their union in the family *Acetobacteraceae*. Although in 1984 a set of 32 features was given to differentiate the genera *Gluconobacter*, *Acetobacter*, and *Frateuria* (3), not all of them are necessary for a satisfactory identification.

The feature of acetate overoxidation identifies the strains used in high-acid fermentations as members of the genus *Acetobacter*, and thus separates them from the genera *Gluconobacter* and *Frateuria*.

Required Properties of Industrially Used Strains. Considering the interest of many microbiologists in acetic acid bacteria, it is surprising that the available information about these bacteria has had so little effect on vinegar production. Working strains were isolated, but lost again, demonstrating the difficulties to isolate and grow *Acetobacter* strains from industrial vinegar fermentations on solid media. Considerable progress was achieved by propagating bacteria from fermenters in Japan on a special double-layer agar. The bacteria isolated on this medium have been described as *Acetobacter polyoxogenes* (4). However, *A. polyoxogenes* was not available from the Japan Collection of Microorganisms because of problems of propagation and preservation.

A new species in the genus *Acetobacter*, for which the name *Acetobacter europaeus* was proposed (5), has been isolated and characterized in pure culture from vinegar fermentations at high acidity in Germany and Switzerland. All investigated strains isolated from submerged fermenters and trickle generators had very low (0 to 22%) DNA–DNA similarities with the traditional strains of *Acetobacter* and *Gluconobacter*. The phenotypical differentiation on the species level of the genus *Acetobacter* is rather difficult, because different strains of a single species do not necessarily utilize the same carbon source. A useful and significant criterion for the distinction of *Acetobacter europaeus* from the other *Acetobacter* species is the strong tolerance to acetic acid of 4 to 8% in AE-Agar and the absolute requirement of acetic acid for growth.

Independent of all these problems and difficulties, the first and foremost interest of the vinegar industry is to use a strain of acetic acid bacteria that tolerates high concentrations of acetic acid and total concentration, requires small amounts of nutrients, does not overoxidize the acetic acid formed, and yields high production rates. At present, it also seems to be necessary to make this strain phage resistant (6).

Industrially Used Strains. The vinegar industry has always worked with acetic acid bacteria that, in most cases, were not derived from pure cultures. The striking fact that common microorganisms used on a large scale for industrial vinegar production were not properly described and characterized in taxonomic terms is caused by the difficulty to cultivate them on semisolid media.

Industrial submerged vinegar fermentations are started by inoculation with inoculation vinegar, microbiologically undefined remains from previous fermentations. Acetic acid fermentation can be carried out for years without interruption or decrease in efficiency or yield, if suitable conditions are chosen to allow a permanent selection of strains tolerating high acidity at a minimum of nutrient concentration. Such cultures are delivered to vinegar factories in all parts of the world by Heinrich Frings GmbH & Co KG, Bonn, Germany.

Plasmid Profile Analysis for Characterization of Acetobacter Strains. Plasmids are circular, extrachromosomal DNA molecules that can be extracted from bacteria by a sophisticated technique. The plasmid profile of a strain is very specific. Plasmids can determine certain properties of bacteria, such as phage resistance or citrate metabolism. The potential of this technique for vinegar fermentations was investigated (7–9). The majority of the acetic acid bacteria contain 1 to 8 plasmids from 1 to more than 17 MDa. Plasmid profile analysis has been applied to characterize the *Acetobacter europaeus* strains and the microflora of industrial vinegar production plants (10). It was shown that the microflora in submerged fermentations consists of one predominant strain, whereas in trickling generators the microflora is quite heterogenous, containing a mixture of several *Acetobacter europaeus* strains with quite different plasmid profiles. This is probably the reason for the susceptability of submerged fermentations to bacteriophage attack. In conclusion, the determination of plasmid profiles from submerged vinegar fermenters represents a new and powerful technique to characterize the microflora of such fermentations and to answer questions that could not have been approached until now, such as the stability, origin, and identity of the prevailing microflora.

Recent investigations have shown that the plasmid profiles of these strains are able to rest stable over a period of at least 5 years and are not changing their phenotypic features during this time. Complete sequencing of the 16S rDNA genes have shown the phylogenetic position of *Acetobacter europaeus* to be close to *Acetobacter xylinum* and well separated from seven other *Acetobacter* species. Acetic acid bacteria together with *Acidiphilium* species and *Rodopila globiformis* represent a distinct acidophilic branch in the α-subclass of *Proteobacteria*.

Electrophoretic protein patterns correlated with the percentage of DNA similarity of DNA–DNA hybridization proved to be very useful for the delineation of the strains from industrial vinegar fermentations at the species/subspecies level (11,12).

Genetic Improvement of *Acetobacter* strains. To improve strains of *Acetobacter* and *Gluconobacter* genetically, recombinant DNA techniques are considered to be useful. Host–vector systems and an efficient transformation method have been developed. The preparation of spheroplasts and the development of a spheroplast fusion method have been described (13). A strain growing at 37 °C and 4% acetic acid and a strain unable to grow at 35 °C and 5% acetic acid were spheroplasted and fused. Fusion products were received growing at 37 °C on agar plates containing 5% acetic acid, on which both parent strains were unable to grow. Some more genetic transformation systems have been described (14–16).

When summarizing the present technique, Fukaya (17) states that recombinant DNA technology can contribute to advances in the basic knowledge of acetic acid bacteria and additionally provide new tools for improving them. But to apply recombinant DNA technology to improving acetic acid bacteria even more, basic research concerning gene expression, protein engineering, and other fundamental studies are required.

Bacteriophages. *Acetobacter*-specific bacteriophages demonstrated by electron microscopy in submerged and trickling fermenters are said to be responsible for fermentation problems in industrial vinegar fermenters (18–21). In trickling generators, fermentation is sometimes slowed down, but hardly stops completely. In submerged fermentations, complete loss of productivity has been observed. Only three morphologically different *Acetobacter* phage types were described until 1992, when a great variety in phage types, regarding the dimensions of the phage heads and tails was reported (20). All phages isolated from vinegar fermentations in Germany, Austria, and Denmark showed long contractile tails and isometric heads belonging to Bradley's group A (22) and the family of *Myoviridae* (23), respectively. Although not all phages described have actually been shown to be infective for defined *Acetobacter* strains because of a lack of a suitable technique, their occurrence in high numbers (10^6 to 10^9 mL^{-1}) in disturbed fermentations suggests that bacteriophages may actually be the source of fermentation problems.

Biochemistry

During acetic acid fermentation, ethanol is almost quantitatively oxidized to acetic acid. Concentration yields between 95 and 98% are common, and the remainder is mainly lost in the effluent gas. At the same time, a suitable C_6-carbon source (preferably glucose) is oxidized. End products of this oxidation are CO_2 and H_2O.

Ethanol. Ethanol oxidation is performed by two sequential reactions catalyzed by alcohol dehydrogenase (ADH) and aldehyde dehydrogenase (ALDH), located at the outer surface of the cytoplasmic membrane. Their function is linked to the respiratory chain of the organism (24,25).

Alcohol dehydrogenase has been purified from several acetic acid bacteria and shown to contain pyrroloquinoline quinone (PQQ) and heme C as prosthetic groups (26–28). Extensive purification and characterization of the enzymes involved in ethanol oxidation of *Acetobacter aceti*, *A. rancens*, and *G. suboxydans* have been reported (26–32). It has been shown that ADH of acetic acid bacteria donates electrons directly to ubichinone in the cytoplasmic membranes. Thus, the ethanol oxidase respiratory chain of acetic acid bacteria is made of only three membranous respiratory components, alcohol dehydrogenase, ubichinone, and terminal oxidase (33).

It has been confirmed that ADH and ALDH of acetic acid bacteria had the same prosthetic group, PQQ, as their glucose dehydrogenase (30). These quinoproteins, glucose, ADH, and ALDH are involved in sugar oxidation that links them to an electron transport system in the membrane of oxidative bacteria. ADH and ALDH from *A. polyoxogenes* sp. nov. have been purified (34,35). This strain originated from a vinegar factory showing higher acid productivity and higher tolerance to acetic acid than strains isolated from natural sources. In contrast to the ADH and ALDH obtained from *A. aceti* and *G. suboxydans*, the enzymes of *A. polyoxogenes* were quite stable, and this high stability seems at least to be partially involved in the ability of *A. polyoxogenes* to produce high concentrations of acetic acid.

Sugar. For the breakdown of sugars, *Acetobacter* is equipped with the hexose monophosphate (HMP) pathway and the TCA cycle (36), whereas glycolysis is either absent or very weak. The enzymes of the Entner-Doudoroff pathway are present in *Gluconobacter* and in *Acetobacter xylinum*. *Gluconobacter* and *Acetobacter* are known for their direct oxidative capacity on sugars, alcohols, and steroids (37). For further details see the work by Ebner and Follman (1). The biochemistry of the ketogenic activities of acetic acid bacteria have been reviewed (38).

Acetate. Strains of *Acetobacter* are able to oxidize acetate as well as lactate either in the presence or in the absence of ethanol by using enzymes of the tricarboxylic cycle.

Carbon Dioxide. *Acetobacter* needs CO_2 for growth (39). CO_2 is incorporated into cell substances, where approximately 0.1% of the cell carbon is derived from CO_2. A very small but measurable portion of acetic acid too is derived from CO_2 (40).

Nitrogen. If an organic carbon source is present, a number of acetic acid bacteria can use ammonium as sole nitrogen source. Some strains need the presence of amino acids; others need cofactors such as vitamins or purines (41). No essential amino acids are known.

Growth Factors. Growth factor requirement depends on the carbon source supplied (42). Some strains require *p*-aminobenzoic acid, niacin, thiamin, or pantothenic acid as growth factors (43). Some *Acetobacter* strains do not need vitamins in the presence of glucose (44). But the combined addition of glutathione with Na glutamate has a cumulative effect on the growth of *A. aceti* (45).

PQQ, the novel prosthetic group in acetic acid bacteria, also shows an interesting growth-stimulating effect for various microorganisms (45–47). One of two PQQ effects was the stimulation of both the growth rate and the total cell yield with only a trace amount of PQQ as an essential growth factor. The second effect of PQQ observed universally was a marked reduction of the lag phase in microbial growth, but no increase of the growth rate in the subsequent exponential phase or in the total cell yield at the stationary phase.

Physiology

Oxygen Demand and Total Concentration. During commercial use of many aerobic microorganisms it is common practice to permit the inoculum to remain in a fermenter for some time up to several days without aeration before inoculation into the main fermenter. However, application of this procedure is impossible in the case of acetic acid fermentation.

Extensive tests have been conducted (48,49) concerning the damage to acetic acid bacteria caused by an interruption of aeration as a function of total concentration, acetic acid concentration, rate of fermentation, and length of interruption of aeration. The longer the interruption of aeration and the higher the total concentration of the mash, the more marked is the effect. At a total concentration of 5%, an interruption of aeration for 2 to 8 min leads to the same damage as an interruption for 15 to 60 s at total concentrations of 11 to 12%. At constant total concentrations, damage increases with increasing concentration of acetic acid and with increasing fermentation rate. The specific production of cell material is extremely low (1).

Experiments conducted later (50) confirmed these results. It was found that there was a sharp drop in the oxidation rate of ethanol as well as in the activity of enzymes involved (ADH and ALDH) after an interruption of oxygen supply, with an acetic acid concentration of 6%. ADH activity in the membrane fraction was highly stable in the culture broth containing 4% or less acetic acid. But the purified enzyme was very unstable even in the culture broth of 2% total concentration. It was suggested from this fact that the ADH has a membrane component contributing to its own stability at acetic acid concentrations of less than 4%, which becomes gradually liable to damage with further increasing acidity. But, the relationship between the ADH system and oxygen has not been elucidated at all.

Under conditions of industrial acetic acid production, the microorganisms undergo considerable stress caused by a high concentration of acetic acid and limited oxygen supply. A high utilization of oxygen up to 70% can be achieved in industrial production. A sparing use of air is very important because of the volatility of ethanol and acetic acid. However, the use of pure oxygen or highly oxygen-enriched air may easily damage the bacteria (48).

In 1985, it was found that at a total concentration of 13%, the ATP pool and the growth rate show a reverse behavior (51). The energy charge (average value of intracellular ATP, ADP, and AMP) showed a rather high value of 0.84 at the beginning of the fermentation. During fermentation, although the acetic acid concentration increased from 9 to 12%, the energy charge decreased to 0.7. After interruption of aeration for 45 s, the energy charge dropped to 0.58, and after several weeks of storage the charge of the inoculum was only 0.50.

Lack of Ethanol. Acetic acid bacteria are also damaged if a vinegar fermentation is carried out to the point where all the ethanol has been oxidized and the addition of fresh ethanol-containing mash is delayed. This is analogous to the damage resulting from an interruption of oxygen supply and also depends primarily on the total concentration and the duration of ethanol lack.

Specific Growth Rate. In 1953, the specific growth rate of acetic acid bacteria was calculated for the first time (52). In semicontinuous fermentations there was no decrease of the specific growth rate at increasing acetic acid concentrations, but at increasing total concentrations the specific growth rate decreased from 0.49 h^{-1} at 5% to 0.16 h^{-1} at 11%.

In continuous culture at a total concentration of 12%, the specific growth rate decreased from 0.027 h^{-1} at 4.5% ethanol to 0.006 h^{-1} at 1% ethanol. It should be mentioned that the specific growth rate obtained in continuous fermentations with constant concentrations of acetic acid and ethanol are less favorable than those obtained under semicontinuous conditions. Batch fermentations carried out with *Acetobacter suboxydans* (53) showed at a total concentration of 7.5% a specific growth rate of 0.30 h^{-1}, a result that is in good agreement with other studies (51).

Specific Product Formation. In 1949, a specific product formation was found of 21 g of acetic acid per gram of bacterial dry substance per hour in semicontinuous tests at a total concentration of 10% (54). This very high value was independent of the acetic acid concentration of 4.5 to 7.2%. In 1972, a strong dependence between specific product formation and acetic acid concentration in continuous fermentations at a total concentration of 7% was found (55). The maximum specific product formation was 15 g g^{-1} h^{-1}. The specific product formation follows a growth-associated mechanism as long as the acetic acid concentration is below 3%. At higher acetic acid concentrations up to 7%, an additional nongrowth-associated term must be considered. The highest value obtained with cell recycle in continuous culture of *A. suboxydans* was 11.5 g g^{-1} h^{-1} (53).

Changes in Concentration and Temperature. Charging of a fermentation must be done with mash at the same total concentration and under rapid mixing because a locally formed strong concentration gradient damages the bacteria. The next cycle starts without any lag phase. At the same time, temperature has to be kept constant because a repeated change in temperature causes a similar result (1).

Overoxidation. Overoxidation is the undesirable oxidation of acetic acid to CO_2 and H_2O that has to be prevented by avoiding lack of ethanol and keeping the total concentration at a high level. In trickle fermenters, this is much more difficult than in submerged fermenters (1).

Industrial Processes

Submerged Vinegar Fermentation ***Semicontinuous and Continuous Fermentations.*** For the production of vinegar with more than 12% and up to 15% acetic acid (56), the process has to be carried out in a semicontinuous manner; each fermentation cycle taking about the same time as the preceding and following cycles. The starting concentration of each cycle is 7 to 10% acetic acid and about 5% ethanol. When an alcohol concentration between 0.05 and 0.3% has been achieved in the fermentation liquid, a quantity of vinegar is discharged from the fermenter. Refilling with new mash of 0 to 2% acetic acid and 12 to 15% alcohol leads to the starting concentrations for the new cycle mentioned earlier. Discharging must be carried out quickly in order to avoid complete alcohol depletion. Charging must be done slowly under constant fermentation temperature and rapid mixing.

Since the middle of 1994, it has been possible to produce vinegar of up to 19% acetic acid in a modified single-stage process. Starting with similar concentrations as mentioned earlier, alcohol is slowly added at a constant alcohol level of 2 to 3% in the fermentation liquid, creating a corresponding increase of the total concentration. Addition of alcohol is stopped when the desired total concentration has been reached. When alcohol approaches zero, a part of the fermentation liquid is discharged and replaced by mash with lower total concentration to bring the acetic acid concentration and total concentration back to starting conditions, thereby enabling the bacteria to multiply faster. Later, total concentration is increased again by adding alcohol.

Continuous fermentation is only possible up to a maximum of 9 to 10% acetic acid because the specific growth rate of the bacteria decreases with decreasing ethanol concentration. To obtain high yields, the fermentation must be carried out at a low alcohol concentration.

Two-stage Processes. In the canning industry, vinegar with a very high percentage of acetic acid is in demand. The vinegar industry is interested in producing vinegar of high acidity in order to save storage and transport costs. Therefore, a two-stage process for the production of vinegar with more than 15% acetic acid was developed (57). During a submerged fermentation with a total concentration below 15% in a first fermenter, alcohol is added slowly to increase the total concentration up to about 18.5%. After the acetic acid concentration has reached 15%, about 30% of the fermenting liquid is transferred into a second fermenter. The first fermenter is resupplied with new mash of lower total concentration. In the second fermenter, the fermentation continues until the alcohol is almost depleted. The whole quantity of the finished vinegar is discharged. The fermentation liquid in the first fermenter is supplied with alcohol at the appropriate time and later divided again.

Since the end of 1993, vinegar of more than 20% acetic acid has been produced using this process. An automatic control for this process has been described (58).

In 1981, a similar two-stage process for the production of vinegar with more than 20% acetic acid was described (59,60). The second fermentation stage is carried out at a reduced temperature of only 18 to 24 °C, whereas in the first fermentation stage, the normal temperature range of 27 to 32 °C is used.

In 1984, another two-stage process was claimed (61) with semicontinuous fermentation in at least two fermenters, membrane filtration of the mash from the semicontinuous stage, using lower temperatures at higher acetic acid concentrations, and with the membrane-filtered vinegar to be used as a second end product.

The Frings Acetator®. Frings Acetator® is the most common equipment for the production of all kinds of vinegar. At the end of 1996, 642 Acetators, continuously improved, enlarged and automated, were in operation all over the world with a total production of 1,524 × 10^6 L/year of vinegar with 10% acetic acid. Energy consumption is about 400 W L^{-1} of fermented ethanol with decreasing tendency; yield is up to 98%.

Commercial sizes of Acetators are for the fermentation of up to 3,600 L of pure ethanol in 24 h.

The different processes demand excellent performance of the fermenter, especially of its aeration system. The Frings aerator (62), as shown in Figure 1, is self-aspirating; no compressed air is needed. The hollow rotor is installed on the shaft of a motor mounted under the fermenter, connected to an air suction pipe and surrounded by a stator. It pumps liquid that enters the rotor from above outward through the channels of the stator that are formed by the wedges, thereby sucking air through the

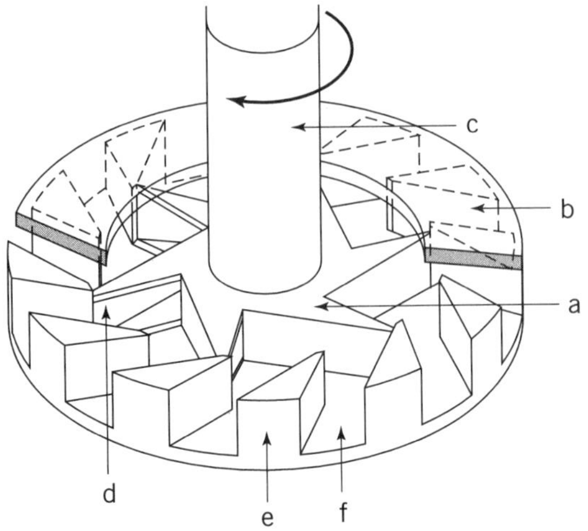

Figure 1. The Frings Aerator. a = hollow rotor; b = stator; c = air suction pipe; d = openings for air exit; e = wedges to form the channels; f = channels to form the beams of air–liquid emulsion.

openings of the rotor and creating an air–liquid emulsion that is ejected outward at a given speed. This speed must be chosen adequately so that the turbulence of the stream causes a uniform distribution of the air over the whole cross-*section of the fermenter.

Figure 2 shows this situation in a fermenter containing for demonstration a small amount of water immediately after the start of the aeration. Air is aspirated through the air suction pipe from the top, and the formed emulsion is spread out in beams at the bottom of the fermenter where they reach the wall of the fermenter somewhat outside of the cooling coils. Seconds later, the uniform aeration of the liquid takes place.

Because foam formation cannot always be avoided, the Acetator is equipped with a mechanical defoamer. From 1966 to 1991, the following equipment (63) was used. The rotor, turning in a spiral housing, separates liquid and gas and returns the liquid into the lower part of the fermenter. Only during discharging of the finished vinegar a part of this returned liquid is pumped out of the fermenter. Because this liquid causes foam formation, the defoamer must operate most of the time of a fermentation cycle. In an improved version (64), the separated liquid portion is not pumped back into the fermenter, but into a small collection vessel to be mixed with the vinegar end product later. Lysis of harmed vinegar bacteria releases foam-causing surface-active substances. By means of the total elimination of the separated liquid, further foam formation is avoided. Minimized foam formation with a shorter operating time of the defoamer and consequently a lower power uptake and a negligibly small volume of the separated liquid are the result.

The Frings Alkograph® (65), an automatic instrument for measuring the amount of ethanol in the fermenting liquid, has been in use since 1965. Small amounts of liquid flow through the analyzer continuously, at first through a heating vessel and then through three boiling vessels. The boiling temperature of the incoming liquid is measured in the first boiling vessel. While alcohol is distilled off continuously from the second and the third boiling vessel, the higher boiling point of the liquid from which ethanol has been removed is measured in the third boiling vessel. The difference in temperature corresponds to the ethanol content and is recorded automatically. Because the flow through the vessels takes some time, there is a delay of about 15 min between the beginning of the inflow of a sample and the appearance of the correct value on the recorder. Because alcohol concentration is decreasing slowly during fermentation, this delay has no disadvantage on fermentation control.

Figure 2. How the Frings Aeration System works. Acetator filled for demonstration with a small amount of water immediately after the start of the aeration. Air is aspirated through the suction pipe from the top and the formed emulsion is spread out in beams at the bottom of the fermenter, reaching the wall of the fermenter somewhat outside of the cooling coil.

To carry out the single-stage semicontinuous process at a defined alcohol content, a contact in the Alkograph activates the vinegar discharge pump. As soon as a preset level has been reached, the mash pump starts adding fresh mash. This pump is controlled by the fermentation temperature in order to refill under constant temperature. The pump is stopped when the desired level is reached and an automatic cooling system is activated. A fermentation cycle takes 24 to 48 h.

The Frings Alkograph® has worked with high reliability for about 30 years in 231 Acetators around the world. New biosensors together with modern membranes permitted the development of a device to measure alcohol without any time delay; the Frings Alkosens®. A device for measuring volatile constituents of the fermentation medium was described (66). A membrane serves to filter these volatile constituents from the fermentation liquid into a carrier medium, and they are recorded by an electronic sensor. Because the measurement relies on diffusion, ambient temperature has to be kept always at the same level. Thus, the alcohol probe is situated in the fermentation broth. The alcohol value displayed at the analyzer is influenced by the concentration of acetic acid, and, therefore, new calibrations are needed at different total concentration levels. This disadvantage has been overcome recently by a new system, which incorporates a multilayer membrane. New sensor technologies and more sophisticated computer equipment permit a fully automated process control of continuous, semicontinuous as well as single-stage or two-stage high-strength fermentations—preparation of mash included.

Further Trends in Submerged Processes. A computerized experimental laboratory system was applied 1992 (67) to submerged acetic acid fermentation with industrial *Acetobacter* strains at about 10% of final acidity using automated semicontinuous cycles during a period of one year. Based on these data, a mathematical model was developed to predict the time course of a cycle, even in the case when at the beginning of the cycle a lag phase occurred because of a lack of substrate at the end of the foregoing cycle. The parameters substrate, product, and biomass, however, were not sufficient to characterize the physiological status of the culture, since the behavior of the microorganisms also depends on their past. RNA concentration per unit of cell mass was found to be an internal key parameter that showed a reasonable agreement of the experimental data and the mathematical model.

In 1989, the possibilities of on-line measurement of alcohol concentration in vinegar production were summarized (68). In addition to the Frings Alkograph® the Alcolyzer®, a Japanese instrument based on the same principle, immobilized microbial sensors, and semiconductor alcohol sensors were described.

Experiments for the continuous production of acetic acid by submerged fermentation with continuous removal of acetic acid by electrodialysis, continuous supply of ethanol and concentrated nutrient solution, and control of ethanol and pH were reported (69–71). The reason for the continuous removal of acetic acid from the reaction mixture was to alleviate the inhibitory effect of the produced acetic acid. Ethanol concentration was kept constant at 1%, acetic acid concentration at approximately 2%. Without dialysis and pH control, acetic acid production already stopped after 2 days at approximately 4%, whereas with dialysis it lasted 30 days.

Other Submerged Fermentation Processes. Other processes of little practical significance and abandoned processes have been described (1).

Continuous Culture and Cell Recycle. Much work was invested (53,72) to increase the productivity of the acetic acid fermentation by continuous culture and cell recycle.

It must indeed be admitted that at first sight the productivity in an Acetator does not seem to be very high. The Acetator 3600® produces during 24 h 4,000 kg acetic acid with a concentration of up to 15% in a fermenting volume of 100 m^3, that is, 1.6 g L^{-1} h^{-1} acetic acid in average or about 2.4 g L^{-1} h^{-1} at the maximum fermentation rate. The aeration rate is 0.1 vvm only; the nutrient consumption is the same as described in Physiology.

Kinetic studies (72) clarified the dependence of the bacterial growth rate on ethanol concentration, on acetic acid concentration, and on concentration of dissolved oxygen. The optimum ethanol concentration is 23 g L^{-1}, the optimum acetic acid concentration is 36 g L^{-1}, and the optimum oxygen concentration is 2.5 − 3 mg L^{-1}. The productivity of acetic acid formation mainly depends on the growth rate but also on the cell concentration. Therefore, the maximum productivity of 6 to 6.6 g L^{-1} h^{-1} acetic acid can be obtained at 32 g L^{-1} ethanol, and 60 to 66 g L^{-1} acetic acid.

The synergistic effects of acetic acid and ethanol on the growth of *Acetobacter* sp. at low concentrations showed similar results (73).

At first sight, the potential productivity is about 4 times higher than presently achievable and could even be further increased by recycling of the bacteria. However, this high degree of productivity can only be achieved at a narrow concentration range. A slight change of ethanol, acetic acid, or oxygen concentration results in an immediate return to lower values.

For technical applications, a second fermenter in series to the fermenter, operating at optimal fermentation conditions, would be necessary to complete the fermentation. To keep a high oxygen concentration in the fermenting liquid at the higher fermentation rate, a high aeration rate would be necessary with recovery of ethanol and acetic acid from the exhaust gas to obtain high yields, and probably addition of oxygen would be necessary. Higher nutrient consumption as well as increasing foam formation would result. The cell recycle system would need a continuous microfiltration avoiding a lack of oxygen and require an automatic control of cell concentration. But all the expenditure in process control only allows the production of the same amount of vinegar, containing not more than 10% acetic acid, in a smaller fermenter. Apparently, the demand for high productivity does not necessarily indicate the optimal developments for future production.

An attempt was made to enhance continuous acetic acid production in a laboratory fermenter equipped with a hollow fiber filter module by increasing the concentration of dissolved oxygen (74,75). The maximum acetic acid concentration achieved was 5%. The highest productivity obtained with yeast extract as nutrient was 107 g L^{-1} h^{-1}. The concentration of dissolved oxygen was already lower than the critical concentration of 1 ppm, although pure oxygen was supplied at an aeration rate of 1 vvm. Furthermore, it was difficult to maintain a high concentration of viable cells in the cell recycle culture, because the cells were inactivated by the lethal conditions of acetic acid concentration, oxygen deficiency, and nutritional limitation.

The effect of dissolved oxygen and acetic acid concentration on acetic acid production in continuous culture has extensively been investigated (75). The optimum concentration of dissolved oxygen was found to be about 2 ppm for acetic acid production at low acidity. The specific acetic acid production rate, however, was diminished to complete inhibition when the dissolved oxygen concentration was higher than 2 ppm at an acetic acid concentration of more than 4.5%. This is in good agreement with older findings (48), stating that the use of pure oxygen during fermentation can completely inhibit the acetic acid formation.

Because viability of the bacteria is important for high productivity, the factors affecting viability have been investigated (76). To study the effect of cell bleeding (withdrawing at constant flow rate), serial experiments were performed in a bioreactor with cell recycle and were supported by mathematical analysis. Acetic acid concentrations were up to 5.6%. The maximum value of productivity was 123 g L^{-1} h^{-1}. The bleed ratio and dilution rate at which concentration of acetic acid and viable cells are relatively high can be predicted now.

The authors expect these results to be useful for designing a two-stage culture for acetic acid production using the active cells in the bleed from the first stage in a second-stage vessel beyond a lethal acetic acid concentration.

How far these certainly interesting results will influence the construction of industrial equipment in Japan remains to be seen. The costs of such complicated fermenters may slow down or even prevent a fast conversion to industrial scale, although in Japan the demand for vinegar with 5% acetic acid is relatively high. However, these results will certainly have no influence on the production of vinegar with more than 10% acetic acid. A review of the Japanese development work (77) states that the problem of scale-up remains for further investigation.

Semicontinuous Culture with Cell Recycle. During investigation of semicontinuous cultures with cell recycle, an increase was found in the productivity compared to semicontinuous culture without cell recycle at an acidity of up to 6% (78). Although the concentration of total cells increased in every cycle, the viable cells reached a much lower quasi steady-state value, which was explained by product inhibition. The productivity was 2.9 g L^{-1} h^{-1} without and 5.0 g L^{-1} h^{-1} with cell recycle at 6% final acetic acid concentration, but 14 g L^{-1} h^{-1} at 3% final acetic acid concentration.

In another study (79), semicontinuous fermentations were reported in which the ethanol concentration was kept for a certain time between 20 and 30 g L^{-1} by adding ethanol-enriched medium. At the end of the cycle, ethanol-poor medium was added. Cells were recycled by membrane filtration. The goal was to obtain high concentrations of acetic acid; the highest amount reached was 9%. It was found that the *Acetobacter* cells lost their viability at about 6% acetic acid, but these nonviable cells were still effective for ethanol oxidation. This interesting feature of *Acetobacter* cells had already been reported much earlier (80).

Surface Fermentation. ***Old Surface Fermentation Processes.*** Detailed descriptions of the history of vinegar fermentation were published in 1976 (81). Today, the equipment used in the famous Orleans process for the production of wine vinegar can only be seen in museums.

New Experiments with Surface Fermentation. A new approach to surface fermentation has been tried (82). The authors used a bioreactor with a shallow, horizontal flow of the medium under a bacterial film with a surface of a few hundred cm^2. Liquid depth was up to 10 mm, and the acetic acid concentration of the effluent was 5.76%. The oxygen absorption rate through the microbial film was found to be very high; a kinetic study of acetic acid production was published (83).

A surface fermentation process in a system of serial intercommunicating fermentation vessels, with flow-regulating devices that enforce the flow to approach plug flow, was described (84). The liquid depth under the layer of acetic acid bacteria was about 50 mm. The inflowing liquid had 2% acidity and contained 3.5% alcohol. The acidity of the effluent stabilized at 5% acetic acid. The residence time was about 21 h.

Investigations on continuous surface fermentations in a single vessel with a working volume of 16 L were reported (84). Under optimum conditions with a liquid depth of 100 mm, 2 to 3% acidity and 20 to 30 g L^{-1} h^{-1} alcohol concentration of the inflowing liquid, vinegar with 4.5% acetic acid was produced during 80 days. The flavor of this vinegar was claimed to be better than that of vinegar produced by other processes.

As a consequence of the efforts to produce rice vinegar at low acidity with better quality, surface fermentation is developing in Japan into an automatic fermentation process at lower costs.

Trickling Processes. The old semicontinuous generator process used beech wood shavings, birch twigs, or corn cobs as carrier material for the vinegar bacteria. Large wooden tanks with up to 100 m^3 shavings are still in use. For more details refer to the work by Ebner and Follman (1). Despite the disadvantages of the trickling process, the use of a silicate carrier material with a large specific inner surface and an open pore structure was proposed recently (86).

Experiments with Immobilized Acetic Acid Bacteria. Many Japanese scientists investigated the fermentation of rice or fruit vinegar with the usual concentration of about 5% acetic acid in continuous fermentation of high productivity with immobilized acetic acid bacteria.

A 460-day run in an airlift bioreactor with acetic acid bacteria immobilized on κ-carrageenan gel beads, where living cells were newly released into the reactor, was reported (68). These newly released bacteria showed extremely high growth rates and respiratory activity that they retained for a few generations after leakage.

Polypropylene fibers as carrier material for *Acetobacter* producing rice vinegar of 3% acetic acid were proposed (87,88). With ceramic honeycomb-monolith as carrier in a fixed-bed reactor a maximum of 4% acetic acid was achieved (89). Calcium alginate beads in a fluidized bed reactor were investigated and 3.5% acetic acid obtained (90). The same author (91,92) used this reactor in combination with an ethanol fermenter to produce rice vinegar of up to 6.5% acetic acid from a saccharified rice medium.

A two-column bioreactor of porous ceramics was tested to produce vinegar of an orange wine with up to 8.4% acetic acid (93). If the cells were fixed on woven cotton fabric, vinegar of kiwi fruit and persimmon of 4.5% acetic acid could be produced (94). A new ceramic carrier called Aphrocell with a continuous pore structure was tried (95), but the oxygen supply to this structure was difficult. None of the tested carriers proved to be ideal.

Experimental and theoretical studies on the continuous acetic acid production by immobilized cells in a three-phase fluidized-bed bioreactor at very low acetic acid concentrations were carried out (96). The influence of gel-entrapped and suspended cells, gel size, k_La, and solid hold-up on productivity was studied by theoretical calculations and experiments.

Mori et al (97–99) reported that immobilized acetic acid bacteria are highly oxygen dependent. Therefore, it is advantageous to supply oxygen to a closed and pressurized culture. Because acetic acid inhibits the bacteria, it is considered that multiple reactors in series would be the best solution to obtain an increasing acidity, thereby keeping the dilution rate higher and creating the prolonged residence time necessary. An output acidity of 5.5% was obtained in two fluidized-bed type tabletop reactors in series. In scaling up to a pilot reactor, however, the microbial layer peeled off from the carrier too easily independent of its material. More operational know-how is needed for the practical application of this technology.

Environmental Protection. Today it is possible to produce vinegar without any environmental pollution. The exhaust gas contains alcohol, acetic acid, and ethyl acetate according to their vapor tension at a fermentation temperature of about 30 °C. To obtain high yields, the aeration rate is chosen as low as possible. It is furthermore common practice today to cool the exhaust gas down with cooling water as far as possible, and in order to absorb the rest of the volatiles it can be scrubbed with water that is recycled and used for mash preparation later.

The raw vinegar contains bacteria that have to be filtered off. Today, most modern factories use cross-flow filtration without adding any filter aid. The very small quantity of remaining solids of about 100 g $\times$ 1,000 L^{-1} of vinegar formed by acetic acid bacteria are degraded during wastewater treatment.

Production Volumes

United States. The U.S. vinegar industry ceased years ago to control and to publish vinegar production figures. Therefore, older figures must be quoted here. Production volumes in 1987 were 581 $\times$ 10^6 L of distilled vinegar, 54 $\times$ 10^6 L of cider vinegar and 126 $\times$ 10^6 L of other types of vinegar, all of different strength.

European Union. In 1995, the European Union produced 349 $\times$ 10^6 L alcohol vinegar, 160 $\times$ 10^6 L wine vinegar and 57 $\times$ 10^6 L other kinds of vinegar (all of 10% acetic acid), together 566 $\times$ 10^6 L.

Japan. In 1994, Japan produced 56 $\times$ 10^6 L of rice vinegar, 146 $\times$ 10^6 L of other grain vinegar, 22 $\times$ 10^6 L of fruit vinegar, and 166 $\times$ 10^6 L of other brewed vinegar, all together 390 $\times$ 10^6 L at a concentration of 4 to 5% acetic acid.

Worldwide. The world production (excluding the USSR and China) of vinegar of 10% acetic acid is assumed to be around 2,000 $\times$ 10^6 L per year, or 200, 000 tons of pure acetic acid.

BIBLIOGRAPHY

This article is a shortened and revised republication of H. Ebner, S. Sellmer, and H. Follmann, *Acetic Acid in Biotechnology*, 2nd ed., vol. 6, H.J. Rehm, G. Reed, A. Pühler, P. Stadler eds., Weinheim, VCH, 281–401, 1996, by permission of VCH Publishing Group of May 30, 1996.

1. H. Ebner and H. Follmann, in H.J. Rehm and G. Reed eds. *Biotechnology*. vol. 3, Verlag Chemie, Weinheim, Germany, 1983, pp. 387–407.
2. M. Gillis and J. De Ley, *Int. J. Syst. Bacteriol.* **30**, 7–27 (1980).
3. J. De Ley, J. Swings, and F. Gosele, in N.R. Kreig, G.J. Holt eds., *Bergey's Manual of Systematic Bacteriology*, Williams & Wilkins, Baltimore, 1984, pp. 268–274.
4. E. Entani, S. Ohmori, H. Masai, and K.J. Suzuki, *J. Gen. Appl. Microbiol.* **31**, 475–490 (1985).
5. M. Sievers, S. Sellmer, and M. Teuber, *Syst. Appl. Microbiol.* **15**, 386–392 (1992).
6. A.J. Schocher, H. Kuhn, B. Schindler, N.J. Palleroni, C.W. Despreaux, M. Boublik, and P.A. Miller, *Arch. Microbiol.* **121**, 193–197 (1979).
7. T. Inoue, M. Fukuda, and K. Yano, *J. Ferment. Technol.* **63**, 1–4 (1985).
8. M. Fukaya, T. Iwata, E. Entani, H. Masai, T. Uozumi, and T. Beppu, *Agric. Biol. Chem.* **49**, 1349–1355 (1985).
9. M. Teuber, M. Sievers, and A. Andresen, *Biotechnol. Lett.* **9**, 265–268 (1987).
10. M. Sievers, A. Andresen, and M. Teuber, *Food Biotechnol.* **4**, 555 (1990).
11. M. Sievers, M. Teuber, *J. Appl. Bacteriol. Symp. Suppl.* **79**, 84S–95S (1995).
12. M. Sievers, W. Ludwig, and M. Teuber, *System. Appl. Microbiol.* **17**, 189–196 (1994).
13. M. Fukaya, H. Tagami, K. Tayama, H. Okumura, Y. Kawamura, and T. Beppu, *Agric. Biol. Chem.* **53**, 2435–2440 (1989).
14. H. Okumura T. Uozumi, and T. Beppu, *Agric. Biol. Chem.* **49**, 1011–1017 (1985).
15. M. Fukaya, H. Okumura, H. Masai, T. Uozumi, and T. Beppu, *Agric. Biol. Chem.* **49**, 2083–2090 (1985).
16. M. Fukaya, H. Okumura, H. Masai, T. Uozumi, and T. Beppu, *Agric. Biol. Chem.* **49**, 2407–2411 (1985).
17. M. Fukaya, in Y. Murooka and T. Imanaka eds., *Recombinant Microbes for Industrial and Agricultural Applications*, Decker, New York, 1994, pp. 529–542.
18. M. Teuber, A. Andresen, and M. Sievers, *Biotechnol. Lett.* **9**, 37–38 (1987).
19. W.W. Stamm, M. Kittelmann, H. Follmann, and H.G. Trüper, *Appl. Microbiol. Biotechnol.* **30**, 41–46 (1989).
20. S. Sellmer, M. Sievers, M. Teuber, *System. Appl. Microbiol.* **15**, 610–616 (1992).
21. C. Defives, D. Ochin, J.P. Hornez, and M. Werquin, *Microbiol. Aliments Nutr.* **8**, 77–79 (1990).
22. O.E. Bradley, *Bacteriol. Rev.* **31**, 230–314 (1967).
23. H.W. Ackermann, *Microbiol. Sci.* **4**, 214–218 (1987).
24. M. Ameyama and O. Adachi, *Methods Enzymol.*, **89**, 450–457 (1982).

25. M. Ameyama and O. Adachi, *Methods Enzymol.*, **89**, 491–497 (1982).
26. O. Adachi, K. Tayama, E. Shinagawa, K. Matsushita, and M. Ameyama, *Agric. Biol. Chem.* **42**, 2045–2056 (1978).
27. O. Adachi, E. Miyagawa, E. Shinagawa, K. Matsushita, and M. Ameyama, *Agric. Biol. Chem.* **42**, 2331–2340, (1978).
28. H. Muraoka, Y. Watabe, N. Ogasawara, and H. Takahashi, *J. Ferment. Technol.* **59**, 247–235 (1981).
29. O. Adachi, K. Tayama, E. Shinagawa, K. Matsushita, and M. Ameyama, *Agric. Biol. Chem.* **44**, 503–515 (1980).
30. M. Ameyama, K. Matsushita, Y. Ohno, E. Shinagawa, and O. Adachi, *FEBS Lett.* **130**, 179–183 (1981).
31. R. Hommel and H.P. Kleber, *Proc. 3rd Eur. Congr. Biotechnol.* **1**, 133–137 (1984).
32. H. Muraoka, Y. Watabe, and N. Osagawara, *J. Ferment. Technol.* **60**, 171–180 (1982).
33. K. Matsushita, Y. Takaki, E. Shinagawa, M. Ameyama, and O. Adachi, *Biosci. Biotechnol. Biochem.* **56**, 303–310 (1992).
34. M. Fukaya, K. Tayama, T. Tamaki, H. Tagami, H. Okumura, Y. Kawamura, and T. Beppu, *Appl. Environ. Microbiol.* **55**, 171–176 (1989).
35. K. Tayama, M. Fukaya, H. Okumura, Y. Kawamura, and T. Beppu, *Appl. Microbiol. Biotechnol.* **32**, 181–185 (1989).
36. T. Asai, *Acetic Acid Bacteria*, Univ. of Tokyo Press, Tokyo, Univ. Park Press, Baltimore, 1968.
37. J. De Ley, and K. Kersters, *Bacteriol. Rev.* **28**, 83–95 (1964).
38. M. Kulhanek, in V. Krumpanzl and Z. Rehacek eds., *Modern Biotechnology*, vol. 2, Institute of Microbiology, Czechoslovak Academy of Sciences, Prague, 1984, pp. 614–676.
39. Z.G. Razumovskaya and T.Z. Belousova, *Mikrobiologiya* **21**, 403–407 (1952).
40. O. Hromatka and H. Gsur, *Enzymologia* **25**, 81–86 (1962).
41. G.D. Brown and C. Rainbow, *J. Gen. Microbiol.* **15**, 61 (1956).
42. M.R. Raghavendra Rao and J.L. Stokes, *J. Bacteriol.* **66**, 634–638 (1953).
43. F. Gossele, J. Swings, and J. De Ley, *Zentralbl. Bakteriol. Parasitenkd. Infektionskrankh. Hyg. I Abt. Orig. Cl.* 348–350 (1980).
44. M. Ameyama and K. Kondo, *Agric. Biol. Chem.* **30**, 203–211 (1966).
45. O. Adachi, K. Okamoto, K. Matsushita, E. Shinagawa, and M. Ameyama, *Agric. Biol. Chem.* **54**, 2751–2752 (1990).
46. M. Ameyama, E. Shinagawa, K. Matsushita, and O. Adachi, *Agric. Biol. Chem.* **48**, 2909–2911 (1984).
47. M. Ameyama, E. Shinagawa, K. Matsushita, and O. Adachi, *Agric. Biol. Chem.* **49**, 699–709 (1985).
48. O. Hromatka and H. Ebner, *Enzymologia* **15**, 57–69 (1951).
49. O. Hromatka and W. Exner, *Enzymologia* **25**, 37–51 (1962).
50. H. Muraoka, Y. Watabe, N. Ogasawara, and H. Takahashi, *J. Ferment. Technol.* **61**, 89–93 (1983).
51. A. Hitschmann and H. Stockinger, *Appl. Microbiol. Biotechnol.* **22**, 46–49 (1985).
52. O. Hromatka, G. Kastner, and H. Ebner, *Enzymologia* **15**, 337–350 (1953).
53. F.M. Vera and D.I.C. Wang, *174th Am. Chem. Soc. Mfg.*, Chicago, 1977.
54. O. Hromatka and H. Ebner, *Enzymologia* **13**, 369–387 (1949).
55. A. Mori and G. Terui, *J. Ferment. Technol.* **50**, 70–78 (1972).
56. U.S. Pat. 4,503,078 (Mar. 5, 1985), H. Ebner (to Heinrich Frings GmbH & Co. KG, Bonn, Fed. Rep. of Germany).
57. U.S. Pat. 4,076,844 (Feb. 28, 1978), H. Ebner and A. Enenkel (to Firma Heinrich Frings, Bonn, Germany).
58. U.S. Pat. 4,773,315 (Sept. 27, 1988), A. Enenkel (to Heinrich Frings GmbH & Co. KG, Bonn, Fed. Rep. of Germany).
59. U.S. Pat. 4,282,257 (Aug. 4, 1981), Y. Kunimatsu, H. Okumura, H. Masai, K. Yamada, and M. Yamada (to Nakano Vinegar Co., Ltd., Japan).
60. U.S. Pat. 4,364,960 (Dec. 21, 1982), Y. Kunimatsu, H. Okumura, H. Masai, K. Yamada, and M. Yamada (to Nakano Vinegar Co., Ltd., Tokyo, Japan).
61. U.S. Pat. 4,456,622 (Jan. 26, 1984), J.A. Maselli, and R.O. Horwath (to Nabisco Brands, Inc., Parsippany, N.J.).
62. U.S. Pat. 3,813,086 (May 28, 1974), H. Ebner and A. Enenkel (to Firma Heinrich Frings, Bonn, Germany).
63. U.S. Pat. 3,262,252 (July 26, 1966), H. Ebner.
64. U.S. Pat. 4,997,660 (Mar. 5, 1991), R. Wittler (to Heinrich Frings GmbH & Co. KG, Bonn, Fed. Rep. of Germany).
65. U.S. Pat. 3,290,924 (Dec. 13, 1966) H. Ebner and A. Enenkel (to Heinrich Frings Maschinen-und Apparatebau, Bonn, Germany).
66. U.S. Pat. 4,404,284 (Sept. 13, 1983), M. Heider, J. Hofschneider, W. Schallenberger, F. Nodes, and E. Kempe (to Vogelbusch Gesellschaft m.b.H., Vienna, Austria).
67. D. Hekmat and D. Vortmeyer, *J. Ferment Bioeng.* **73**, 26–30 (1992).
68. A. Mori, *Instrum. Control. Engin. Jpn., Kogyo Gijutsu-sha* 18–24 (1989).
69. Y. Nomura, M. Iwahara, and M. Hongo, *Appl. Environ. Microbiol.* **54**, 137–142 (1988).
70. Y. Nomura, M. Iwahara, and M. Hongo, *J. Biotechnol.* **12**, 317–326 (1989).
71. Y. Nomura, *Hakkokogaku* **70**, 205–216 (1992).
72. J.J. Bänsch and W. Bauer, *BioEngineering* **7**, 26–36 (1991).
73. A. Nanba, A. Tamura, and S. Nagai, *J. Ferment. Technol.* **62**, 501–505 (1984).
74. Y.S. Park, H. Ohtake, K. Toda, M. Fukaya, H. Okumura, and Y. Kawamura, *Biotechnol. Bioeng.* **33**, 918–923 (1989).
75. Y.S. Park, H. Ohtake, M. Fukaya, H. Okumura, Y. Kawamura, and K. Toda, *J. Ferment. Bioeng.* **68**, 315–319 (1989).
76. Y.S. Park and K. Toda, *J. Gen. Appl. Microbiol.* **36**, 221–233 (1990).
77. M. Fukaya, Y.S. Park, and K. Toda, *J. Appl. Bacteriol.* **73**, 447–454 (1992).
78. Y.S. Park, M. Fukaya, H. Okumura, Y. Kawamura, and K. Toda, *Biotechnol. Lett.* **13**, 271–276 (1991).
79. Y.S. Park, K. Toda, M. Fukaya, H. Okumura, and Y. Kawamura, *Appl. Microbiol. Biotechnol.* **35**, 149–153 (1991).
80. H. Ebner, in *Ullmanns Enzyklopädie der Technischen Chemie, 4.* Aufl. Bd. 11, Verlag Chemie Weinheim, 1976, pp. 41–55.
81. H.A. Conner and R.J. Allgeier, *Adv. Appl. Microbiol.* **20**, 81–133 (1976).
82. K. Toda, Y.S. Park, T. Asakura, C.Y. Cheng, and H. Ohtake, *Appl. Microbiol. Biotechnol.* **30**, 559–563 (1989).
83. Y.S. Park, H. Ohtake, and K. Toda, *Appl. Microbiol. Biotechnol.* **33**, 259–263 (1990).
84. Brit. Pat. 1 305 868 (Feb. 7, 1973), Kewpie Jozo Kabushiki Kaisa, Tokyo, Japan.
85. T. Higashide, C. Shimada, Y. Kawamura, M. Hisamatsu, and T. Yamada, *Bull. Fac. Biores. Mie Univ. Tsu* **8**, 43–49 (1992).
86. Eur. Pat. 0523 524 A1 (Jan. 20, 1993), Carl Kühne KG, Hamburg, Deutschland.

87. A. Okuhara, *J. Ferment. Technol.* **63**, 57–60 (1985).
88. A. Nanba, K. Kimura, and S. Nagai, *J. Ferment. Technol.* **63**, 175–179 (1985).
89. M. Kondo, Y. Suzuki, and H. Kato, *Hakkokogaku* **66**, 393–399 (1988).
90. A. Saeki, *Nippon Shokuhin Kogyo Gakkaishi* **37**, 191–198 (1990).
91. A. Saeki, *Nippon Shokuhin Kogyo Gakkaishi* **37**, 722–725 (1990).
92. A. Saeki, *Nippon Shokuhin Kogyo Gakkaishi* **38**, 891–896 (1991).
93. M. Takada and T. Hiramitsu, *Nippon Shokuhin Kogyo Gakkaishi* **38**, 967–971 (1991).
94. S. Yamashita, S. Ohta, and H. Suenaga, *Nippon Shokuhin Kogyo Gakkaishi* **38**, 608–613 (1991).
95. M. Sueki, N. Kobayashi, and A. Suzuki, *Biotechnol. Lett.* **13**, 185–190 (1991).
96. Y. Sun, S. Furusaki, *J. Ferment. Bioeng.* **69**, 102–110 (1990).
97. A. Mori, *Process Biochem.* **20**, 67–74 (1985).
98. A. Mori, in A. Tanaka, T. Tosa, and T. Kobayashi eds., *Industrial Application of Immobilized Biocatalysts*, Dekker, New York, 1992, pp. 291–313.
99. A. Mori, S. Tanaka, N. Matsumoto, and C. Imai, in S. Furusaki, I. Endo, and R. Matsuno eds., *Biochemical Engineering for 2001*, Springer-Verlag, New York, 1992, pp. 441–443.

W

WASTE GAS CLEANING, BIOLOGICAL

CORNELIUS G. FRIEDRICH
UDO WERNER
University of Dortmund
Dortmund, Germany

KEY WORDS

Biofilms
Biofilters
Bioscrubber
Flame ionization detection
Mass transfer
Microbial degradation
Plant Design
Trickle-bed bioreactor
Volatile Organic Carbon
Volatile organic emissions

OUTLINE

INTRODUCTION

Waste air or waste gases of industrial processes may contain volatile organic or inorganic odorous or toxic compounds. These gases are not to be released to the environment, according to national or European legislation, which defines the allowances of the individual compound to be released. Environmental biotechnology including waste gas treatment was first developed in countries using the German language and in The Netherlands. In these countries, waste gas treatment was applied by the industry for elimination of odor emitted from wastewater treatment plants, slaughterhouses, and such facilities. For more than two decades, plants have been operated with increasing efficiency, and their application has been extended to the chemical, pharmaceutical, and food industries. Initially, most information has evolved from these European countries; experience has been compiled and merged into guidelines for design and application of industrial biofilters, trickle-bed reactors, and bioscrubbers. The principles of these plants are shown in Figure 1. These guidelines have been published by the Verein Deutscher Ingenieure (VDI, Association of German Engineers) in German and in English (1,2). The developing technology was subsequently recognized on other continents. Previous articles on biological waste gas cleaning have covered different processes, mass transfer, biological, chemical, and olfactory aspects, and costs, and the reader is referred to comprehensive reviews (3–5) with detailed references therein.

High concentrations of organic or inorganic pollutants are currently eliminated from waste gases by sorptive processes or thermal of catalytic oxidation. Low pollutant concentrations and odorous compounds are also being traditionally eliminated by biological processes. Plant investment and operating costs of biological waste gas cleaning processes are significantly less than alternative physicochemical processes (6). This fact is the major driving force for academic and industrial research to develop biological processes and to extend these for elimination of high concentrations of pollutants from waste gases.

Presently, the three types of plants mentioned (Fig. 1) are currently used for biological treatment of different qualities of waste gases, which later are treated in detail. Biofilters require a substantial area and are filled to a low height with organic matter such as compost or chopped wood, which serves in part as a source for minerals for the attached microorganisms. Biofilters are suited for cleaning low concentrations of pollutants. Trickle-bed reactors require less area, are operated as towers, and are filled with incompressible porous minerals as air–liquid contactor and as support for microbial growth. Bioscrubbers are generally composed of two units—one for scrubbing the pollutants from the gas phase and the other as fermentor for microbial degradation of the pollutants comparable to activated sludge.

Microorganisms, bacteria or fungi, use the organic or inorganic compounds for energy transformation and for biomass formation. Biological waste gas cleaning is an aerobic process and is brought about by aerobic microorganisms either suspended in water or attached to some support material as a biofilm. Process development is required to establish and to maintain optimum conditions for these microorganisms to achieve an optimal elimination of pollutants from air. Microorganisms require different elements to build up biomass, which is composed of proteins, sugar polymers, fatty acids, nucleic acids, and so forth. The

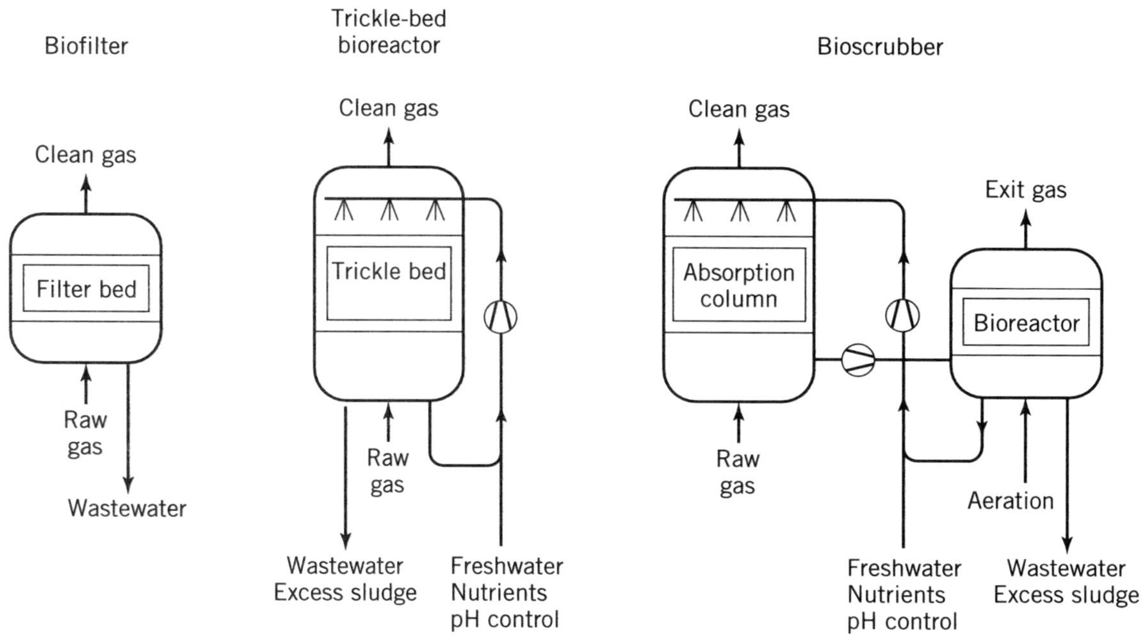

Figure 1. Schematic representation of different types of plants currently used for waste gas treatment.

energy to build up the respective monomers and to polymerize these is derived from the biochemical oxidation of volatile organic matter to carbon dioxide or from gaseous inorganic reduced compounds. Because the microorganisms are present in the aqueous phase, the organic or inorganic substrates must be transported from the gas to the liquid phase. Because biological waste gas cleaning is also an aerobic process, oxygen has to be transported to the liquid film and products have to be transported either to the gas or to the liquid phase.

The degree of elimination of a pollutant and the selection of the process is determined by different factors (Fig. 2): (*1*) the quality of the waste gas determined by the chemical nature of the pollutant, composition of different pollutants, pollutant concentrations, pollutant metabolic accessibility, and the volume of the gas emitted per unit of time; (*2*) the biomass and its physiological state determined by the rate of conversion of the substrates; (*3*) the availability of the substrates determined by the mass transfer of the pollutants and of oxygen from the gas phase to the biofilm and the export of degradation products therefrom (mass transfer depends on the process selected [e.g., biofilter, trickle-bed reactor, or bioscrubber] and on the physicochemical nature of the pollutants to be eliminated); and (*4*) costs of investment and of operation for waste gas cleaning determine the selection of each of these factors by the impact on overall productivity, reliability, and thus profitability.

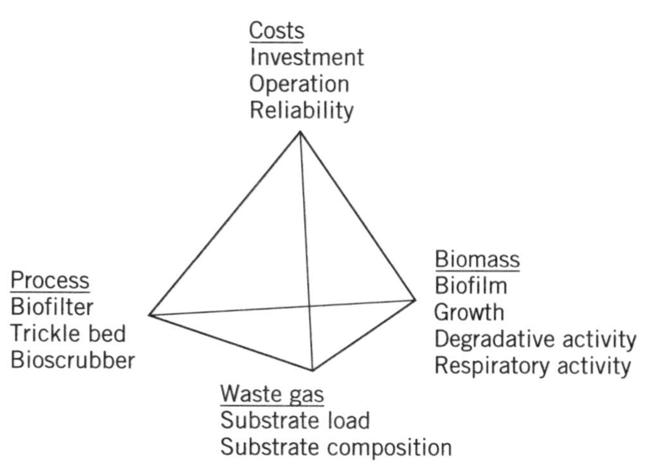

Figure 2. Interrelationships of different parameters crucial for costs and efficiency of waste gas treatment plants.

MICROORGANISMS AND THE BIOFILM

Industrial waste gas cleaning systems are currently operated under nonaxenic conditions and, as a result, contain different populations of bacteria, fungi, and protozoa. Such populations may establish randomly by means of the content of bacteria and bacterial and fungal spores in the surrounding air. To rapidly establish a microbial population in the reactors, these are often inoculated with suspensions of wastewater treatment plants. Because the composition of the waste gas determines the development of microorganisms best fitted to respective conditions, initial efficiency of a plant may be suboptimal. Also, because microorganisms are the crucial phase in biological treatment, they are emitted by the stream of the clean gas that is the subject of hygienic concern.

Establishment of the Microbial Population

Initial colonization of the microbes is the result of adsorption onto the surface of the support material, which occurs almost instantly. Secondary colonization is characterized by the formation of filaments and of extracellular polysaccharides, which together form the biofilm. The kinetics of biofilm formation, its composition, changes, properties, and impact on biocorrosion have been reviewed previously (7). The composition of the species changes with duration of operation. Moreover, under steady-state conditions of the plant, the biofilm represents an ecosystem with grazing protozoa digesting free-living bacteria. Protozoan predation has resulted in a low and constant level of of bacteria, which, however, may not be detrimental to the efficiency of a plant but increase stability of trickle-bed reactors, for example. Enhancement of carbon mineralization up to 22 to 80% and decrease in pressure drop by preventing the extent of clogging have been reported from laboratory waste gas treatment systems. Moreover, most protozoa are not attached to surfaces and could be purged out by flushing the different systems (summarized in Ref. 8).

The use of pure bacterial cultures with defined catabolic properties with the ability to degrade xenobiotic or toxic compounds at known rates has been suggested to be advantageous with respect to waste gas cleaning. Several bacterial species specialized to degrade different solvents have been adsorbed on activated carbon. With suitable bacterial strains, 90% of the solvents could be eliminated from the gas phase, and mass transfer to the biofilm appeared to be limiting the efficiency of the laboratory trickle-bed reactor (9). No reports are presently available for industrial application of monocultures.

Health Considerations

Health hazards for personnel or the surrounding population emerging from waste gas treatment plants have not been reported. However, different bacterial species so far identified from various waste gas treatment plants have been mainly related to the degradative capability of the respective facility. Among these, species of the genus *Pseudomonas* have been identified that are highly versatile in utilizing aromatics and other volatiles. One concern is the release of potentially pathogenic microbes to the environment with the exit gas stream. Some studies have been devoted to this fact. The transport of microbial germs, mainly bacteria and to a smaller extent spores of molds, in the outlet gas of the different full-scale biofilters varied between 10^3 and 10^4 cells/m^3, which is of the same order of magnitude as indoor air (10). Search for fungal spores, the potential causative agent of mycogenic allergies, revealed a minor reduction of spores emitted from clean gas as compared to raw gas from moist biofilters, whereas emission from a dry biofilter temporarily exceeded that determined from raw gas by more than three orders of magnitude (11).

Identification of microorganisms is laborious, time consuming, and costly, and most reports consider the absorption of pollutants rather than the emission of microbes. The technology to detect and to identify specific bacteria within multispecies biofilms, however, is presently available and is used to describe microbial ecosystems. Nucleic acid hybridization techniques and comparative sequence analyses are increasingly being applied to environmental microbiology. More recently, fluorescent dye-labeled oligonucleotide probes have been used for identification of single cells (12).

MICROBIAL GROWTH AND DEGRADATIVE ACTIVITY

Organic or inorganic pollutants of waste gases are substrates for microorganisms. Organic carbon sources have a dual function as substrate for the living cell. Dissimilation of the carbon source characterizes its quality as an energy source to donate electrons for aerobic respiration, which yields the energy required for growth and maintenance of the cell. Carbon is also assimilated by the cell for growth to build up cell constituents. Carbon dioxide and water are the products of the dissimilatory aerobic respiration, and biomass is the product of the resulting assimilation. The quality of a given carbon source as an energy source depends on its degree of reduction and on the type of metabolism, as is shown for dissimilation of one-carbon compounds such as methane, methanol, and formaldehyde (Table 1).

These compounds are oxidized to carbon dioxide, water, and the reductant. Moreover, substitutions of hydrogen by halogens such as methylenechloride cause the release of hydrochloric acid. Degradation of methylmercaptane results in the release of hydrogen sulfide and that of methylamine results in the release of ammonia. The by-product hydrogen sulfide is subsequently oxidized by thiobacteria to sulfuric acid, and ammonia is oxidized by nitrifying bacteria to nitric acid, both resulting in acidification of the environment (Table 1). Therefore, the nature of the pollutant to be mineralized by the microorganisms has a significant impact on the pH of the microenvironment within the plant and on pH-dependent microbial degradative activity.

Table 1. Respiration and Mineralization of Differently Reduced and Substituted One-Carbon Compounds, Nitrification, and Sulfur Oxidation

Respiration

$$4[H]^a + O_2 \rightarrow H_2O$$

Mineralization

$$CH_4 + O_2 \rightarrow CO_2 + 4[H]$$
$$CH_3\text{-}OH + H_2O \rightarrow CO_2 + 6[H]$$
$$CH_3\text{-}NH_2 + 2H_2O \rightarrow CO_2 + NH_3 + 6[H]$$
$$CH_3\text{-}SH + 2H_2O \rightarrow CO_2 + H_2S + 6[H]$$
$$CH_2\text{-}Cl_2 + 2H_2O \rightarrow CO_2 + 2HCl + 4[H]$$
$$CH_2O + H_2O \rightarrow CO_2 + 4[H]$$

Nitrification

$$NH_3 + 2H_2O \rightarrow HNO_2 + 6[H]$$
$$HNO_2 + H_2O \rightarrow HNO_3 + 2[H]$$

Sulfur oxidation

$$H_2S + 4H_2O \rightarrow H_2SO_4 + 8[H]$$

[a] [H] is equivalent to $(H^+ + e^-)$ and represents the reductant hydrogen carrying one electron.

Acidification as a result of the elemental composition of the pollutants as well as the composition and distribution of the different microbial species may require mixing, washing, and pH control of the biofilm attached to the support material or being suspended in the aqueous phase. The different principles of waste gas treatment processes such as biofilter, trickle-bed reactor, or bioscrubber differ in efficiency in mass transfer, capacity to harbor biomass, and costs. Thus, knowledge of the quality of the waste gas with respect to its composition or load is essential for selection of the process.

One crucial criterion in microbial transformations is the overall rate of degradation. The mean residence time of a gas passing certain industrial waste gas treatment plants is occasionally less than a second (2). In this period of time, mass transfer and degradation of the pollutant must occur—and it does.

CELLULAR YIELDS AND MAINTENANCE REQUIREMENTS

The composition of the waste gas usually contains few elements—carbon, hydrogen, or oxygen as in the case of acetone—and other elements (N, P, S, Mg, K, Fe, etc.) are also required to build up biomass. Formation of biomass with constant elemental composition is a function of the organic substrate when each element required for growth is supplied by the environment and expressed as the cellular yield ($Y_{X/S}$) (gram dry cell weight per gram of substrate). In biofilters, some minerals may originate from the support material such as wood, heather, or compost, whereas in other waste gas cleaning plants minerals may be supplied from the added water. Mostly, there will be a severe imbalance in supply with elements required for microbial growth, whether the major pollutant was a hydrocarbon, solvent, hydrogen sulfide, ammonia, or methylenchloride. Imbalance in nutrient supply may originate from the surplus of the carbon source or from the limited supply of minerals, both resulting in two different consequences. First, imbalance of the elemental composition of the biomass may result from storage of different compounds. Excess phosphate supply to certain bacteria results in accumulation of intracellular polyphosphate; excess carbon and energy source may result in either intracellular accumulation of polyhydroxy fatty acids or the formation of extracellular polysaccharides and biofilms. Second, the growth rate may be affected and the requirements for maintenance energy may increase significantly. Maintenance energy is defined as the amount of substrate required by the cell to maintain vital cell functions without growth (equation 1). The concept of maintenance requirements was first described by Pirt (13) and was later extended by Neijssel and Tempest with the definition of overflow metabolism (14).

$$1/Y = m/\mu + 1/Y_{max} \qquad (1)$$

where Y is the cellular yield, Y_{max} the maximum cellular yield, μ the specific growth rate (1/h), and m the maintenance coefficient (g substrate/g dry cell weight h^{-1}).

Overflow metabolism may result in two different phenomena: *(1)* the uncoupling of energy transformation of respiration leading to a low cellular yield and higher rate of oxidation of the substrate to carbon dioxide and *(2)* the formation of polymeric or monomeric by-products. The applicability of these academic findings to waste gas cleaning systems has been subsequently proven successful in laboratory plants (15). However, defined conditions of specific nutrient limitations are susceptible to changing concentrations of volatile organic carbon in the raw gas, which changes the balance in nutrient supply, and thus conditions for optimal mineralization to carbon dioxide and water are difficult to maintain.

ORIGIN, COMPOSITION, AND SENSING OF WASTE GAS COMPONENTS

Waste gas originates from very different industries, from air exchange of buildings, and from various processes. More than 500 biofilters and more than 200 bioscrubbers are currently operated in Germany and in The Netherlands, although outside Europe this technology is less often used. Different industries that apply biological waste gas cleaning as an air pollution control technology are listed in Table 2. The efficiencies of some of these to remove volatile organic carbon or odor are compiled in Table 3. A more comprehensive compilation is given elsewhere (1,2). Most of these facilities emit gases at high flow rates of 1,000 to 120,000 m^3/h but with low concentrations of volatile organic and inorganic compounds. Odor removal is quantified by olfactometry, which is costly and requires reliable calibration as evident from an interlaboratory comparison of odor measurements of n-butanol (25). Recent developments aim at on-line sensors for the quantification of odors. Odor threshold concentrations to be recognized by humans depend on the respective compound and composition of the volatiles. Some threshold concentrations for sensing the odor of chemicals such as dimethylethylamine, phenol, or volatiles of defined composition have been correlated to total organic carbon determined by flame

Table 2. Compilation of Some Industries Applying Biological Waste Gas Cleaning

Biofilter	Trickle-bed reactor	Bioscrubber
Animal rendering	Chemical industry	Aluminum foundry
Bone processing	Concrete slap production	Animal rendering
Chemical industry	Fibrous skin production	Blood processing
Coal mine exit gas	Grease melting plant	Chipboard production
Dump gas	Plastic production	Egg farms
Feces drying	Rayon production	Grease melting plant
Fish smoking	Sewage treatment plants	Grinding wheel production
Fish processing		Printing plant
Foundries		Tobacco industry
Fragrance production		Polyester production
Food production		
Pig farms		
Slaughterhouse		
Tobacco processing		
Varnish production		

Table 3. Compilation of Selected Efficiencies of Different Industrial Waste Gas Cleaning Facilities

Waste gas cleaning plant	Type of reactor	Eliminated pollutant	Total gas flow rate	Raw gas concentration	Clean gas concentration	Refs.
Rendering plant	Tree bark biofilter	Odor/VOC	160 $m^3\ m^{-3}\ h^{-1}$	256 mg organic C/m^3	38 mg organic C/m^3	
				1,700 OU	160 OU	1
Domestic refuse compost	Compost biofilter	Odor/VOC	60 $m^3\ m^{-3}\ h^{-1}$	230 mg organic C/m^3	8–9 mg organic C/m^3	1
Piggery	Peat/twigs biofilter	Odor	141 $m^3\ m^{-3}\ h^{-1}$	50–200 OU	Not detectable	1
Leather processing facility	Compost/chaff biofilter	Solvents	128 $m^3\ m^{-3}\ h^{-1}$	61 g $m^{-3}\ h^{-1}$	41 g $m^{-3}\ h^{-1}$	16
Canning industry	Compost biofilter	Solvents	40 $m^3\ m^{-3}\ h^{-1}$	470 mg organic C/m^3	55 mg organic C/m^3	23
Lacquerworks	Pilot biofilter	Solvents	30–71 $m^3\ m^{-3}\ h^{-1}$	642–1,415 mg/m^3	60–79 mg/m^3	17
Cacao roast house	Biofilter	VOC	75,000 m^3/h	7–615 mg organic C/m^3	1–6 mg organic C/m^3	18
		Odor		49,927–446,000 OU	35–773 OU	18
Printer's color production	Trickle-bed reactor	Toluene/ ethanol	214 $m^3\ m^{-3}\ h^{-1}$	450 mg organic C/m^3	<70 mg organic C/m^3	19
Resin-laden waste gas	Pilot trickle-bed reactor	Formaldehyde	760 $m^3\ m^{-2}\ h^{-1}$	10–30 mg/m^3	0.9–2.7 mg/m^3	20
		Phenol		10–30 mg/m^3	0.1–0.3 mg/m^3	20
		Ammonia		20–90 mg m^3	1.6–7.2 mg/m^3	20
Plastic production	Trickle-bed reactor	Styrene	51,000 m^3/h	0.6 g/m^3	Data not given	21
Lacquer production	Bioscrubber/ biofilter	Solvents	16,000 m^3/h	≤800 mg/m^3	50 mg/m^3	21
Car tunnel exit gas	Pilot trickle-bed reactor	Carbon monoxide	8–23 m^3/h	10–30 ppm	0.8–2.4 ppm	22
		NO_x		0.5 ppm	0.1 ppm	22
		C_xH_y		2 ppm	0.8 ppm	22
Wastewater treatment	Bioscrubber	Hydrogen sulfide	6.000 m^3/h	20 ppm	0.2 ppm	23
Rendering plant	Bioscrubber	Odor	40.000 m^3/h	10,000–70,000 OU	<100 OU	2
Plastic production facility	Bioscrubber	Hydrophobic VOC	10.000 m^3/h	≤200 mg organic C/m^3	10–40 mg organic C/m^3	24

Note: VOC, volatile organic compounds; OU, olfactometric units (cannot be compared from different references).

ionization detection, which is a reliable and available technology for continuous exit gas analysis. Threshold concentrations for odor sensing vary from 0.1 to 4.9 mg organic carbon/m^3 for defined chemicals and from 0.1 to 0.5 mg organic carbon/m^3 for waste gas from food processing and other industries. Usually, the degree of reduction of odor intensity is more effective than the elimination of volatile organic carbon.

Flame ionization detection of volatiles allows quantification of organic carbon but does not differentiate between different components of the waste gas. Thus, the degree of elimination of individual compounds cannot be differentiated. To follow the elimination of different components, identification by gas chromatography is required. This analysis has been used to follow degradation of different solvents in waste gases.

Corrosion of the detecting devices may be observed on on-line quantification of haloorganic compounds such as chloromethanes, chlorobenzenes, chloroethanes, or fluorinated or bromated compounds by flame ionization detection because of formation of the respective hydrohalic acids. Therefore, sampling, gas chromatography, and different detection systems may be advisable.

MASS TRANSFER IN WASTE GAS TREATMENT PLANTS

The principles of mass transfer were first elaborated to understand and to design chemical reaction processes and have been described elsewhere (26,27). However, these principles can be transduced also to biochemical and biological reactions where substrates must be delivered to the microbial cell and metabolic products have to be removed therefrom (3,28,29). This is especially important when a substrate is supplied from the gas phase to the aqueous phase, which is the environment for microbial life and activity.

The principles of mass transfer are applicable to the different processes treated here. However, the differences in design of these plants are based on different principles of gas–liquid contactors for transfer of volatile chemicals of different properties such as hydrophilic or hydrophobic and soluble or insoluble compounds. Therefore, the principles of the transfer of substrates of different physicochemical properties from the gas phase to the biofilm are discussed first, and subsequently mass transfer in biofilters, trickle-bed reactors, and bioscrubbers.

Mass Transfer from the Gas Phase to the Biofilm

Total mass transfer rate (J) from the gas to the liquid phase is driven by the difference in concentration of a given component in equilibrium with the gas phase (c_G) and the actual concentration in liquid phase or the biofilm (c_B), the specific interfacial area (a), and a correlation factor designated mass transfer coefficient (β) as described by equation 2:

$$J = \beta a(c_G - c_B) \qquad (2)$$

The specific interfacial area is generally defined as the quotient of the total area for mass transfer and the total reactor volume.

The mass transfer coefficient is influenced by the different steps for mass transfer from the gas phase to the liquid phase and to the biofilm. These steps are the superpositions of convection and diffusion in both fluid phases and the absorptive transition of the component to the liquid phase and to the biomass, as shown schematically in Figure 3a. The equilibrium in absorption at the transition phase is expressed by Henry's law, equation 3:

$$c_G = Hc_F \qquad (3)$$

where c_G is the partial pressure of the component in the gas and c_F of that in the fluid. A similar correlation holds for the phase-interface liquid or biofilm.

During steady-state operation of a plant, mass transfer rate (J) is in equilibrium with the substrate consumption rate (J_R) of the biomass as expressed in equation 4:

$$J = J_R = K_R a c_B^r \qquad (4)$$

where K_R is the reaction rate of the biomass, a the specific area, c_B the concentration of the component in the biofilm, and r the order of the reaction. Mostly, biological reactions are of zero order when these are independent of the substrate concentration and of first order when the substrate concentration determines the reaction rate. The overall reaction rate is determined by the different steps just mentioned. However, in special cases one step may dominate the process, for example, when the rate of absorption of a pollutant is low as given for chemicals with Henry coefficients >1 MPa or low biological degradation rates. Parameters affecting the capacity of waste gas cleaning facilities are discussed next.

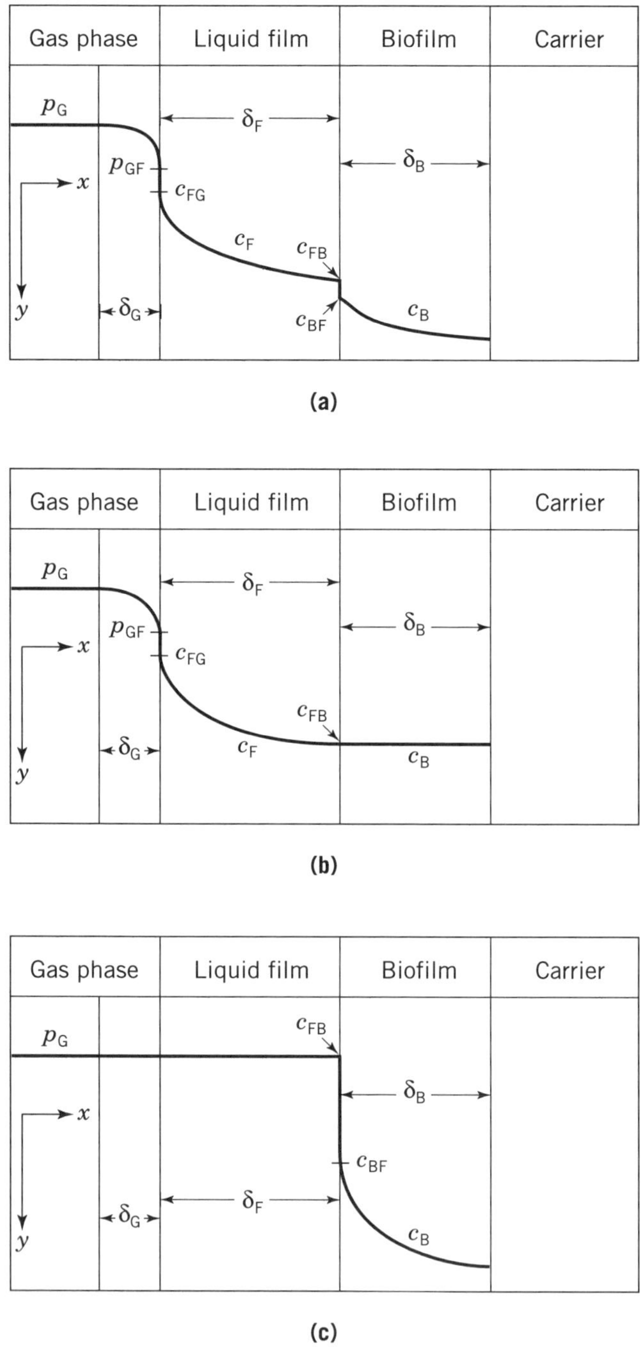

Figure 3. Substrate concentration profiles within the gas phase, the liquid phase, and the biofilm. (**a**) Decrease in concentration caused by biofilm degradative activity and substrate transport by diffusion and convection; (**b**) limitation by mass transfer from the gas phase to the liquid film; (**c**) limitation by degradative activity of the biofilm. p_G, p_{GF}, pollutant partial pressure in the gas phase and at the phase-interface gas–fluid; c_{FG}, c_F, c_{FB}, pollutant concentration at the phase-interface fluid–gas, fluid, and phase-interface fluid–biofilm, respectively; c_{BF}, c_B, pollutant concentration at the phase-interface biofilm–fluid and in the biofilm; δ_G, δ_F, δ_B, boundary layer thickness of the gas phase, fluid, and biofilm, respectively.

Limitations of Reaction Rates

Several examples have been described that demonstrate the significance of single steps on the overall process (27,28,30). The degradative activity of a given biofilm may be limited by suboptimal physiological conditions, which are discussed here, and the mass transfer rates. These conditions apply to the organic or inorganic electron donor for energy transformation, in our case the volatile pollutants, mass transfer rate of the electron acceptor oxygen, the low rate of pollutant degradation by the biofilm, or a combination of these caused by local differences within the reactor.

- Limitation by mass transfer of oxygen from the gas to the liquid phase may be given for a pollutant highly soluble in water, easily transferred from the gas into the biofilm, and rapidly degraded therein (Fig. 3a). Although the oxygen partial pressure of air is 21 kPa, it is barely soluble in water (7.5 mg of oxygen/L at 30 °C), and thus its mass transfer rate competes with that of low concentrations of the pollutants.
- Limitation by mass transfer of the pollutant from the gas to the liquid phase may result from rapid degradation of a pollutant by the biomass. In this case, mass transfer can be enhanced by increase in superficial gas velocity or by reduction of the irrigation rate. Both operations decrease the thickness of the boundary layers δ_G and δ_F (Fig. 3b).
- Limitation by microbial degradation may result either from low degradative activity or from low biomass. Both affect the efficiency of waste gas cleaning similarly. Low degradative activity is given for toxic or xenobiotic pollutants and other persistent chemicals not accessible to the common microbial metabolism (Fig. 3c). Low biomass may result from insufficient supply of mineral nutrients, nonphysiological pH, and support material inadequate to allow sufficient biomass attachment.
- Different limitations, as have been described, may be present in one system at different locations. In a system with a high load of a pollutant, insufficient degradation may take place at the entrance of the reactor where the mass transfer rate is higher than the degradation rate. As the pollutant is removed from the gas stream while passing through the reactor, the concentration in the gas phase decreases. Finally, a limiting mass transfer rate may result as compared to the potential microbial degradation rate.

The efficiency of elimination of a pollutant is clearly dependent on its solubility and mass transfer characteristics. Consequently elimination of a pollutant from the gas phase is linked to the Henry coefficient of the respective compound whether it is eliminated by physicochemical processes or microbial degradation. Such a link is evident for the theoretical and actual efficiency of elimination of different pollutants shown in Figure 4.

The differences in the processes of waste gas cleaning beneficial or deterious to the aspects as stated are discussed in the respective sections.

PARAMETERS FOR PLANT DESIGN

For a given waste gas to be treated, the appropriate type of plant has to be selected, and its size has to be estimated. Therefore, basic characteristics are crucial for plant design. Available area and costs are criteria for choosing between biofilter and trickle-bed reactor, while the hydropathic nature of a pollutant is crucial to decide for or against a bioscrubber. For elimination of pollutants with Henry coefficients >1 MPa, bioscrubbers are usually not chosen.

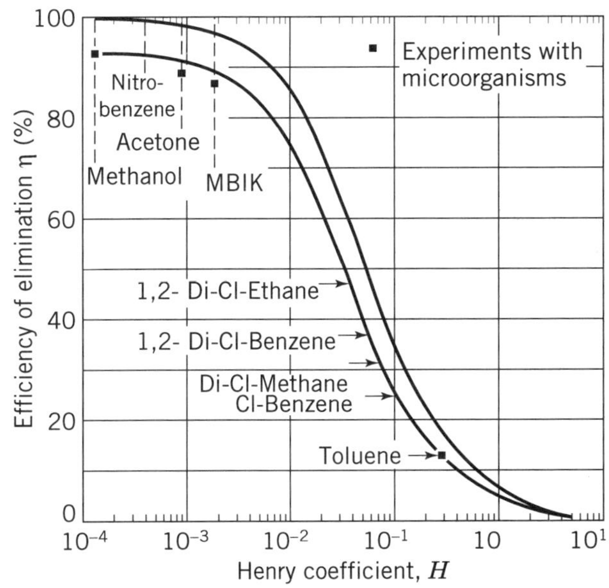

Figure 4. Dependency of the efficiency of elimination of various components on their Henry coefficients. The data are based on a single plate bioscrubber. The upper line represents theoretical efficiency; the lower line gives data of a single-plate bioscrubber. Liquid height on plate = 70 cm, T = 20 °C, u_0 = 0.55 m/s. The dimension of H (MPa) is equivalent to 135.4 ($kg/m_G^3/kg/m_L^3$). *Source*: Redrawn from Ref. 33, with permission.

In general, the mean residence time of the gas and the fluid is crucial for the three types of plants. The residence time is a measure to define the time given for contact of the gas, the liquid, and the biomass to allow mass transfer. The mean residence time (t_m) of a gas or fluid in a reactor with a given volume (V_R) at a given flow rate (V^*) is expressed in equation 5:

$$t_m = V_R/V^* \quad (5)$$

The inverse relation of this mean residence time of the gas defines the volumetric flow rate (τ_V), as expressed in equation 6:

$$\tau_V = V^*/V_R \quad (6)$$

The relation of the flow rate of a gas or fluid to the area of the cross section of the plant (A_R) is defined as area load (τ_A) as expressed in equation 7:

$$\tau_A = V^*/A_R = u_0 \quad (7)$$

where u_0 is the superficial velocity of a gas or fluid, which is the velocity with which they pass the free cross section of the reactor.

In waste gas cleaning, two criteria are important with respect to a pollutant: its concentration and its amount. The amount of a pollutant fed to a plant of a defined volume per unit of time is given by its concentration per volume of gas (c_{in}) and by the gas flow rate, and is expressed as volumetric load (V_L) as shown in equation 8:

$$V_L = (V^*/V_R)c_{in} \quad (8)$$

The pollutant is metabolized in the plant by the biomass and eliminated from the gas phase. Generally, the elimination is equivalent to the degradation of the pollutant, and this term is subsequently used. The overall degradation is defined as volumetric degradation rate (V_D), as expressed in equation 9:

$$V_D = (V^*/V_R)(c_{in} - c_{ex}) \quad (9)$$

The difference in concentration of a pollutant in the feed and the exit gas (c_{ex}) related to the feed concentration characterizes the efficiency of a plant, as expressed in equation 10:

$$\eta = (c_{in} - c_{ex})/c_{in} \quad (10)$$

The efficiency of a plant is determined by various parameters such as the chemical nature of the pollutant, biomass degradative activity, and gas or fluid flow rate. The latter is determined by the resistances within biofilters or trickle-bed reactors given by the type of filling, its density, porosity, and height of the package resulting in pressure loss (26), and may be a significant cost factor for operation. The power (P) required to achieve a certain gas flow rate (V^*) will compensate for the overall pressure loss (Δp) across the reaction zone of a plant, as expressed in equation 11:

$$P = \Delta p V^* \quad (11)$$

Other pressure losses resulting from such components as pipings, nozzles, and orifices are not specific to waste gas cleaning facilities and are not treated here.

The porosity (ε) is a crucial parameter of pressure loss for biofilters or trickle-bed reactors, and is defined by the interstitial volume of a rector. This factor is given by the difference in volume of the empty reactor (V_R) and the volume of the particles (V_P) related to the total reactor volume, as expressed in equation 12:

$$\varepsilon = (V_R - V_P)/V_R \quad (12)$$

TYPES OF WASTE GAS TREATMENT PLANTS

As mentioned, three types of plants are currently used for treatment of industrial waste gases of different origin and composition: biofilter, trickle-bed reactor, and bioscrubber (Fig. 1). Their principles, industrial sizes, capacities, and advantages or disadvantages are outlined next. Certain characteristics increase with the plant used. The applied volumetric flow rate (see equation 6) increases roughly 10-fold as the order of biofilters, trickle-bed reactors, and bioscrubbers. This order reflects not only an inverse relationship of efficiency and size of the plants but also their increasing complexities.

Biofilters

Different designs are currently applied for the construction of biofilters: single bed, multiple bed, and closed bed, either requiring land or installed on top of a roof. Biofilters require an area of 10 to more than 1,000 m^2 filled to a height of 0.5–1.6 m with different materials as support for microbes and for fine distribution of the gas. The filters are operated at rates of 100–120,000 m^3/h. The gas flow rate related to the filter volume (see equation 6) varies from 10 to 580 $m^3\,m^{-3}\,h^{-1}$. The superficial gas velocity of biofilters varies from 0.1 to 8 cm/s. The mean residence time within a filter bed of 1-m height is rather long and amounts to about 10 to 700 s (1). The filling material may be compost, peat, heather, chopped wood, bark, or disrupted root wood to which microorganisms are attached. Biofilters are suited to eliminate odors and volatile organic or inorganic compounds at rather low concentration, as detailed next. The area of the biofilter contains a wall of wood or concrete and a system for distribution of the main stream of the gas. It is practical for the biofilter to be vented from the bottom, that the orifices or slits allow constant gas flow rates over the filter area, and—for larger plants—to allow machines to enter the filter for exchange of the filling.

A waste gas may contain dust particles that are removed by a deduster before feeding into the biofilter. The gas passed through the biofilter should either be humidified to about 95% or the filter should be irrigated to ensure a moist environment as required for microbial activity. The concentration or the composition of the pollutants may vary with the operation of the emitting source, resulting in changes of load. Moderately higher concentrations can be temporarily reduced and a portion of the compounds stored in the filling material according to its adsorptive properties. The compounds are subsequently released therefrom for mineralization by the biomass. Thus, the efficiency of the biofilter can be maintained.

The filling material supplies the attached microorganisms with the minerals necessary for growth. The degradative capacity of the biofilters is limited by the low supply of minerals, which in turn causes a low amount of active biomass present on the support material, its limited buffering capacity for protons or volatiles, and the lack of control of the microbial environment with respect to pH. The filling material is of natural organic matter and subject to microbial degradation at different rates according to the structure, surface, and density. Moreover, environmental conditions may add to the rate of decomposition and loss of mechanical properties of the filling material. Such conditions may be acidification by organic or mineral acids as detailed in Table 1. Decomposition of the filling material results in settling, reduction in pore size, and increase in pressure loss and consequently in increased energy requirements for air compression. Moreover, settling of the filling may result in channeling wherein the gas stream circumvents but does not penetrate portions of lower porosity within the filter bed. This in turn causes reduction of the residence time of the gas and reduction in efficiency of pollutant elimination. Reduction in efficiency of the plant requires exchange of the decomposed filling material, which usually occurs after 1–3 years of continuous operation depending on filling texture, mechanical properties, and degradability and the load of the waste gas to be treated.

Trickle-Bed Reactor

Pollutants with low solubility in water may be eliminated from waste gases by trickle-bed reactors. The advantage of operation of a trickle-bed reactor over biofilters is not only in the increased mass transfer but also in the ease of controlling physiological conditions such as pH and the addition of minerals in the trickling fluid in the bottom of the column. Moreover, trickle-bed reactors require less land than biofilters and are operated in single or multiple stages. They are built as columns filled with incompressible and mostly inert support material for gas–liquid contact and attachment of biomass. The bed heights vary from 1 to 3 m, the reactor volumes from 2 to 160 m^3, and the flow rates from about 3,000 to 85,000 m^3/h. The mean residence time of the gas within the columns is generally less than 4 s, which is considerably shorter than reported for biofilters (see Table 3), and the superficial gas velocities vary from 0.1 to 1.6 m/s. The volumetric gas flow rates are rather divergent and vary from 250 to 4,700 $m^3\ m^{-3}\ h^{-1}$ (2).

The inert support material may be slag, natural porous stones, coke, charcoal, granular clay, ceramics, sintered glass, or plastics and other material proven to be suitable in absorption technology. The filling is trickled from top, and the surface area of the filling provides a major basis for mass transfer from the gas to the fluid phase. Sufficient empirical data are available for choosing geometry, size, and filling for various industrial applications. Moreover, based on the geometry of the fillings, the total surface area of the trickle bed can be calculated. The thickness of the fluid film is proportional to the irrigation rate and inversely related to the total surface area. This in turn allows calculating the mass transfer rate for a given pollutant (27). Also, total biomass can be deduced from the mean thickness of the biofilm. Trickle-bed reactors may be seeded with microorganisms by spraying activated sludge from wastewater treatment plants. The microbes will be selected and the population of the biofilm will be established according to the given pollutants.

A major resistance for mass transfer from the gas phase to the biofilm is the thickness of the fluid film, determined by the rate of irrigation and rate and direction of the gas flow. Therefore, the irrigation rate should be reduced to generate thin films. Depending on the waste gas to be treated, the rates vary from 0.04 to 50 $m^3\ m^{-2}\ h^{-1}$ as reported from laboratory and industrial reactors. However, industrial plants are preferentially operated at $<0.5\ m^3\ m^{-2}\ h^{-1}$ (2). To enhance mass transfer and to reduce diffusional resistance by the liquid film, intermittent irrigation has been applied for elimination of pollutants by a laboratory trickle-bed reactor, which still ensured sufficient moisture for the biofilm (31–33).

Mass transfer may also be increased by increasing the surface area as given with support material of high specific surface. Application of such material results in lower interstitial volume, lower bed volume, and a smaller plant size. However, the thickness of the biofilm will increase with time of operation of the trickle-bed reactor, decreasing or abolishing the small interstitial volumes, which results in increase in pressure loss, clogging, and channel formation. The mean residence time is reduced and the efficiency of the plant declines in eliminating pollutants from the gas stream. The otherwise favorable characteristics of trickle-bed reactors with respect to mass transfer and control of circulation have led to the concept of removing excess biomass by various means: increased irrigation rate to wash off the biofilm (2), periodic pulses of air pressure after flooding, and movement of the trickle bed itself to shear off the biofilm (32). In laboratory systems this procedure of biomass removal has led to continuous operation for more than 2 years with constant degrees of elimination of pollutants. Research data now suggest that the degradative capacities of trickle-bed reactors can be better described mathematically than biofilters and allow conclusions for growth of biomass and of operation of trickle-bed reactors in industrial scale (31,32). Reduction of biomass as a result of a biological equilibrium of biomass formation and biomass degradation has been reported from a pilot trickle-bed reactor after long operation. In this system, elimination of the acidogenic pollutant dichloromethane was reported, involving conversion into biomass, whereas clogging as a result of biomass degradation was not observed (34). Dichloromethane is a substrate with low yield in energy transformation (Table 1) and maintenance requirements (see earlier discussion).

Bioscrubbers

The decision to select bioscrubbers for waste gas cleaning is based on the physicochemical properties of the pollutants and the availability of space for placing the plant. Bioscrubbers are usually smaller than biofilters and consist mostly, but not exclusively, of two units, one for absorption of the pollutant from the gas phase and the other as a fermentor for their mineralization and growth of microorganisms. Bioscrubbers are usually inoculated with activated sludge, and the microbial population will develop according to the volatiles supplied with the gas stream. Waste gas is passed through a spray tower, bubble column, absorber column, or other gas–liquid contactors for absorption of the pollutants. Industrial bioscrubbers presently treat 9,000 to 150,000 m^3 waste gas per hour. Scrubber units are built in heights of 2–4 m and volumes of 10–100 m^3 and exhibit volumetric gas flow rates of 1,000–3,600 $m^3\ m^{-3}\ h^{-1}$. Bioscrubbers are usually operated with superficial gas velocities of 0.8–2.0 m/s and with mean residence times of <3 s. The specific flow rate of the liquid phase to be circulated per area of the absorber unit ranges from 12 to 40 $m^3\ m^{-2}\ h^{-1}$. Some data of industrial plants are compiled in Table 3, and more have been given elsewhere (2). Elimination of hydrophobic pollutants from the gas phase in single-unit bioscrubbers is usually not advisable because of the high volumes of water to be moved. However, single-unit bioscrubbers have been successfully applied to clean industrial volatile pollutants with Henry coefficients of more than 1 MPa (24). Moreover, new developments in part operating in an industrial scale have been reported for application of bioscrubbers that combine both units in one apparatus. These systems consist of special devices for gas or liquid dispersion, special orifices at the bottom of bubble columns that contain the dispersed microorganisms (33), or gassed closed-loop air-lift reactors (35), which, however, await industrial application.

The physiological conditions for the microorganisms such as pH, salinity, and oxygen partial pressure can be well controlled in bioscrubbers, as similarly described for trickle-bed reactors. Also, mineral salts nutrients can be added to the medium to allow microbial biomass to grow. The advantage of bioscrubbers over trickle-bed reactors is in the ease of controlling the environment for microorganisms and mechanically improving mass transfer rates by movement of the liquid according to the need to absorb pollutants from the gas phase. On the other hand, elimination of hydrophobic compounds with low solubilities in water requires movement and dispersion of large volumes of liquids. Bioscrubbers are an interesting alternative to other technologies that have been outlined with respect to the ease of operation, environmental control, and adjustment to changing conditions in the quality of the feed waste gas. Thus, bioscrubbers can be applied for elimination of hydrophobic pollutants (24). A variation of the process uses silicon oil for absorption of hydrophobic volatiles. Problems, however, may arise from disposal of silicon oil–containing sludge (2).

COSTS

Application of biological treatment has been proven useful to economically clean waste gases of different sources. Presently, waste gases with concentrations of less than 1 g of organic carbon/m^3 or of inorganic compounds are treated by biological processes. At these concentrations, biological processes are economically advantageous over alternative processes such as condensation, thermal or catalytic oxidation, or adsorptive and desorptive technologies. These processes are economically feasible at higher concentrations, as detailed in Table 4. Separation of organics from the gas phase by membrane technologies listed in Table 4 is just in its infancy (6,37). Performance of membrane bioreactors for waste gas treatment is presently being examined at a laboratory scale. Reactor performance, however, is hampered by unstable biofilms or by clogging of the liquid channels from excess biomass formation. On solution of these problems, membrane bioreactors are expected to be useful tools for treatment of poorly water-soluble pollutants such as highly chlorinated hydrocarbons (reviewed in Ref. 21).

In a comparative study, costs for investment and for operation were related to treatment by biofilters. Costs of investment include those for the plant and for measuring and controlling devices. Costs of operation include consumables, energy requirements, and costs for personnel (6) (Table 5). From this study it is evident that bioscrubbing is more costly with respect to investment and operation. At present, data on the economy of trickle-bed reactors are not available. However, biofiltration and bioscrubbing are significantly less costly than catalytic oxidation or adsorption (Table 5).

Adsorptive processes aim at the recovery of solvents, for example, whereas thermal or catalytic oxidation aims at the elimination of the pollutants. To make use of the low costs of biological processes for elimination of a pollutant, it is anticipated to extend this technology to waste gases with organic loads significantly larger than 1 g organic carbon/m^3. Two alternatives are presently considered. First, classical engineering processes as specified are applied to eliminate high concentrations or peak concentrations toxic to microorganisms and to combine these with conventional biological processes. This concept was successfully applied recently for elimination of solvents from waste gas of the lacquer industry (38). Second, the inexpensive biological processes are to be extended to treat waste gases with high organic load. The conversion of this concept requires addition of mineral nutrients, which results in formation of high amounts of biomass. Removal of the excess biomass is required to prevent undesired consequences such as clogging in trickle-bed reactors or increase in salt concentration in bioscrubbers or both. However, at laboratory scale high concentrations of biomass allowed the complete (100%) elimination of as much as 4.2 g toluene/m^3 or 8.7 g *n*-butanol/m^3 from the gas stream (39).

Table 4. Processes and Ranges of Concentrations Commonly Used to Eliminate Hydrocarbons from Waste Gases

Process	Concentration (g organic carbon/m^3 waste gas)
Condensation	>250
Membrane separation	>20
Thermal oxidation	6–20
Catalytic oxidation	1–10
Absorption	1–50
Adsorption	1–25
Biological waste gas treatment	<1
Biological treatment with biomass removal[a]	1–10

Source: Data compiled according to Ref. 36, extended. Pilot and laboratory scale.

GENERAL CONSIDERATIONS

Biological waste gas treatment is less costly and requires less maintenance than conventional physicochemical, thermal, or catalytic processes. However, problems may arise from the chemical nature of the compounds to be

Table 5. Relative Costs of Investment and of Operation for Different Waste Gas Treatment Plants

Type of process	Investment costs (%)	Annual costs of operation (%)
Biofilters	100	100
Bioscrubbers	158	196
Catalytic oxidation	182	288
Adsorption	236	365

Source: Altered from Ref. 6, with permission.
The data compare calculated costs of plants for treatment of waste gas with a flow rate of 30,000 m^3 and 6,000 h of operating time.

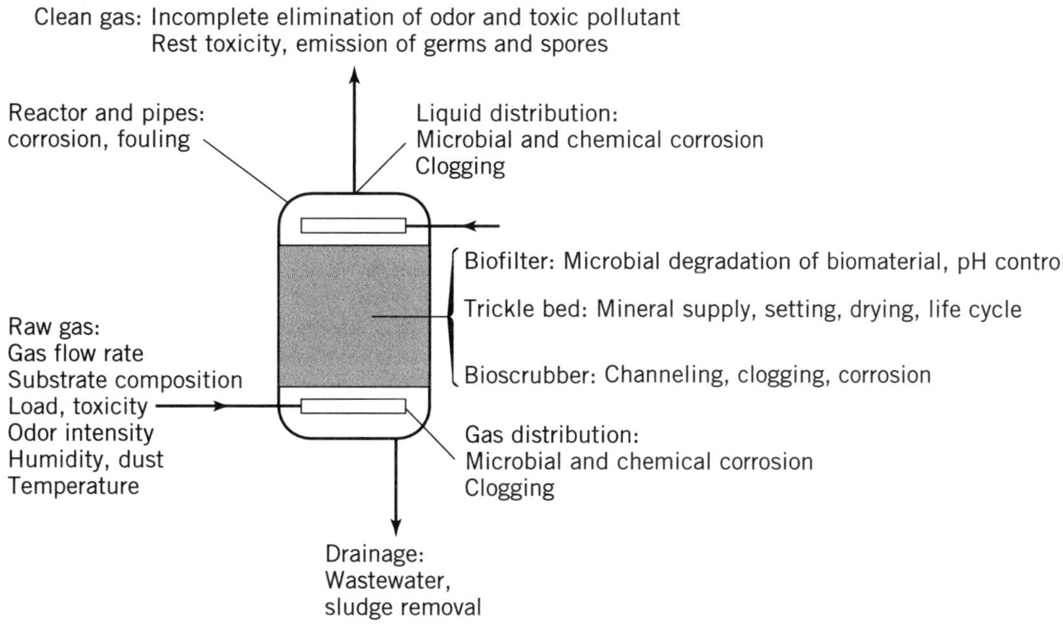

Figure 5. Summary of general problems of biological waste gas treatment plants.

eliminated from waste gases and from the material of which parts of the plant are constructed, such as the walls of the reactor and such parts as piping, orifices, and nozzles (Fig. 5). The waste gases to be treated may contain compounds that are hardly or very slowly biodegradable, such as benzene and xenobiotics, and other compounds toxic to different cell functions, such as solvents, cyanic acid, and hydrogen sulfide. However, successful operation of laboratory-scale waste gas treatment plants to eliminate such compounds point to the applicability of biological treatment to this category of pollutant also.

Different compounds may be converted to organic acids by incomplete oxidation when the mass transfer of oxygen is insufficient. Inorganic acids may be released on microbial degradation of acidogenic compounds, as detailed in Table 1. Acidification, however, not only results in decrease of pH and an unfavorable change of the physiological conditions for the microorganisms but also in corrosion of the material of the plant. The waste gas of rendering plants contains volatile organic sulfur compounds converted to sulfuric acid; this causes severe corrosion of concrete supports of the gas distribution devices of the filter bed. Therefore, such devices have to be exchanged for such components as wooden beams. The same applies for other acidogenic compounds of waste gases. Therefore, the appropriate material for piping and such should be selected according to the expected composition of the waste gas for the three types of plants.

Trickle-bed reactors may especially suffer from clogging and channeling by excess formation of biomass or slime with high loads of the waste gas or extended operation, as was outlined. Therefore, altered processes are presently being examined to extend the application of biological treatment of waste gases loaded with compounds that are slowly degraded or that contain high loads of organic and inorganic pollutants. This application in turn will require devices for the control of the environment for the microorganisms and the removal of excess biomass, and recent developments have been devoted to this need.

BIBLIOGRAPHY

1. VDI-Richtlinie 3477, *VDI/DIN-Handbuch Reinhaltung der Luft*, vol. 6, Beuth Verlag, Berlin, 1991.
2. VDI-Richtlinie 3478, *VDI/DIN-Handbuch Reinhaltung der Luft*, vol. 6, Beuth Verlag, Berlin, 1996.
3. S.P.P. Ottengraf, in H.J. Rehm and G. Reed eds., *Biotechnology*, vol. 8, VCH Verlagsgesellschaft, Weinheim, 1986, pp. 425–452.
4. G. Leson and A.M. Winer, *J. Air Waste Manage. Assoc.* **41**, 1045–1054 (1991).
5. P.L. Bishop and N. Kinner, in H.J. Rehm and G. Reed eds., *Biotechnology*, vol. 8, VCH Verlagsgesellschaft, Weinheim, 1986, pp. 114–176.
6. H. Menig, H. Krill, and T. Jäschke, in W.L. Prins and J. van Ham eds., *Biological Waste Gas Cleaning*, VDI Verlag, Düsseldorf, 1997, pp. 27–58.
7. H.-C. Flemming, *Wasser Abwasser* **132**, 197–207 (1991).
8. H.H.J. Cox and M.A. Deshusses, in W.L. Prins and J. van Ham eds., *Biological Waste Gas Cleaning*, VDI Verlag, Düsseldorf, 1997, pp. 233–240.
9. K. Kirchner, S. Wagner, and H.-J. Rehm, *Appl. Microbiol. Biotechnol.* **37**, 579–587 (1992).
10. S.P.P. Ottengraf and J.H.G. Konings, *Bioprocess Eng.* **7**, 89–96 (1991).
11. C. Huwe and H. Pöhle, *Biofilter* **7**, 47–49 (1997).
12. R.I. Amann, J. Stromley, R. Devereux, R. Key, and D.A. Stahl, *Appl. Environ. Microbiol.* **58**, 614–623 (1992).
13. S.J. Pirt, *Proc. R. Soc. London* **163**, 224–231 (1965).
14. O.M. Neijssel and D.W. Tempest, *Arch. Microbiol.* **107**, 215–221 (1976).

15. S.-M. Wübker and C.G. Friedrich, *Appl. Microbiol. Biotechnol.* **46**, 475–480 (1996).
16. K. Stefan, A. Windsperger, and R. Buchner, *Verfahrenstechnik* **24**, 12–15 (1990).
17. D. Eitner, *Wasser, Luft und Boden* **7–8**, 44–47 (1996).
18. D. Eitner, in K. Fischer ed., *Biologische Abluftreinigung*, Expert, Böblingen, 1990, pp. 55–73.
19. C. Kellner and E. Vitzthum, in W.L. Prins and J. van Ham eds., *Biological Waste Gas Cleaning*, VDI, Düsseldorf, 1997, pp. 61–66.
20. M. Arnold, J. Lehtomäki, and J. Roine, in W.L. Prins and J. van Ham eds., *Biological Waste Gas Cleaning*, VDI, Düsseldorf, 1997, pp. 75–82.
21. H. Herzog and N. Thißen, in W.L. Prins and J. van Ham eds., *Biological Waste Gas Cleaning*, VDI, Düsseldorf, 1997, pp. 83–90.
22. K.-H. Robra, M. Wellacher, F. Kirchmeir, R. Leistentritt, and K. Pucher, in W.L. Prins and J. van Ham eds., *Biological Waste Gas Cleaning*, VDI, Düsseldorf, 1997, pp. 131–140.
23. D. Eitner, in W.L. Prins and J. van Ham eds., *Biological Waste Gas Cleaning*, VDI, Düsseldorf, 1997, pp. 385–395.
24. S. Stockhammer, W. Schäfer-Treffenfeldt, and K. Zetzmann, *Chem. Ing. Tech.* **64**, 148–155 (1992).
25. P. Heeres and H. Harssema, *Gefahrst.-Reinhalt. Luft* **56**, 55–60 (1996).
26. R.H. Perry and C.H. Chilton, *Chemical Engineer's Handbook*, McGraw-Hill, New York, 1987.
27. M. Baerns, H. Hoffmann, and A. Renken, *Chemische Reaktionstechnik*, Thieme, Stuttgart, 1992, pp. 82–97.
28. K. Kirchner, in R. Margesin, M. Schneider, and F. Schinner eds., *Praxis der Biotechnologischen Abluftreinigung*, Springer, Berlin, 1996, pp. 1–16.
29. A. Windsperger and M. Sotoudeh, in R. Margesin, M. Schneider, and F. Schinner eds., *Praxis der Biotechnologischen Abluftreinigung*, Springer, Berlin, 1996, pp. 17–53.
30. R.M.M. Dicks and S.P.P. Ottengraf, in *Biologische Abgasreinigung*, VDI-Berichte, 1104, Düsseldorf, 1994.
31. H. Heits, A. Laurenzis, and U. Werner, in W.L. Prins and J. van Ham eds., *Biological Waste Gas Cleaning*, VDI, Düsseldorf, 1997, pp. 313–320.
32. A. Laurenzis, H. Heits, S.-M. Wübker, U. Heinze, C. Friedrich, and U. Werner, *Biotechnol. Bioeng.* **57**, 497–503 (1998).
33. F. Wolff, *Biologische Abluftreinigung mit Suspendierten und Immobilisierten Mikroorganismen*, Fortschrittsberichte VDI 94, Düsseldorf, 1992, pp. 71–106.
34. R.M.M. Diks, S.P.P. Ottengraf, and S. Vrijland, *Biotechnol. Bioeng.* **44**, 1279–1287 (1994).
35. A. Laurenzis and U. Werner, in W.L. Prins and J. van Ham eds., *Biological Waste Gas Cleaning*, VDI, Düsseldorf, 1997, pp. 115–122.
36. G. Börger, *Chem. Ing. Tech.* **64**, 905–914 (1992).
37. M. Schultes, *Abgasreinigung*, Springer, Berlin, 1996, pp. 209–238.
38. M.W. Reij, J.T.F. Keurentjes, and S. Hartmans, *J. Biotechnol.* **59**, 155–167 (1998).
39. S.-M. Wübker, U. Heinze, and C.G. Friedrich, in W.L. Prins and J. van Ham eds., *Biological Waste Gas Cleaning*, VDI, Düsseldorf, 1997, pp. 241–248.

WASTE TREATMENT, ACTIVATED SLUDGE, CONTROL STRATEGIES

Gustaf Olsson
Lund Institute of Technology
Lund, Sweden

KEY WORDS

Activated sludge
Control
Instrumentation
Operation

OUTLINE

TREATMENT PLANT PROCESSES

The treatment of municipal and some industrial wastewater generally uses a combination of primary, secondary, and tertiary treatment. Primary treatment simply screens and settles large particulates and skims off floating greases and oils. Secondary treatment biologically removes organic carbon, and in newer plants, soluble nitrogen or phosphates. Tertiary treatment attempts to limit the microorganisms and other pathogens in the treated water by membrane filtration or deep-bed filters and some form of disinfection using chlorine, ozone, or ultraviolet light.

The secondary or biological treatment of wastewater uses a number of key processes, unit operations in the process engineering terminology. These include primarily processes for biological reactions and processes for settling and clarification. Chemical precipitation to remove phosphorous is sometimes used just before the settling and clarification. There are many flowsheet configurations, often of a proprietary nature, that attempt to obtain the optimal combination of anaerobic, anoxic, and aerobic conditions for the microorganisms. Commonly used flowsheets are shown in Figure 1. A detailed account of instrumentation, control and automation in wastewater treatment plants has been made in the textbook by Olsson and Newell (1).

GOALS AND INCENTIVES FOR CONTROL

The goal of the operation of a wastewater treatment plant is to satisfy the effluent requirements at all times at minimum cost. The effluent quality and a good performance has to be guaranteed *consistently*, which will have a great impact on instrumentation, control, and automation. Effluent requirements are usually formulated in terms of daily, weekly, or annual averages of suspended solids, biological oxygen demand (BOD), total nitrogen, and total phosphorus. To comply with the limits of the effluent concentrations, certain safety margins are required. The goal of automatic control is to keep the effluent quality consistently good, despite disturbances. This allows the margins to be smaller, thus saving costs.

Criteria

The capital costs are usually the dominant part of the total cost. Traditionally, many plants have been designed with large tank volumes to meet the varying loads to the system. The total cost can be decreased, if the balance between design and operational costs is systematically considered. There are many components of the operational costs, such as consumption of electric energy (for pumping, aeration, etc.), chemicals for phosphorus removal, or polymers for sludge. Obviously, labor costs are important. The balancing of costs is, of course, an important reason for control. Considering the increasing demand of consistent operation and effluent water quality around the clock, however, will probably be the main driving force for plant control and automation.

Driving forces for instrumentation, control, and automation (ICA) in a large number of countries were discussed at an IAWQ symposium in 1993 (2). The major incentives were categorized as

- Effluent quality standards
- Economy
- Increasing plant complexity
- Available tools (sensors, computers, models)

In some countries, there is a water shortage, which in itself is a driving force for improved wastewater treatment, such as for recycling.

In many industrial countries there is today an elaborate requirement for nutrient removal. This has led to the need for quality control based on advanced on-line instrumentation. With more complex processes, the needs for energy conservation and consistent operation are apparent. Ammonium and nitrate on-line instrumentation in the activated sludge was developed and documented in the late 1970s. At that period, the on-line measurements were still limiting the applicability. Very large efforts have been spent on the development of N and P analysers, and there has been a lot of progress in the instrumentation technology. In 1990s, the control is extended to include biological P removal in many plants. Today, commercial applications of nutrient removal control based on advanced instrumentation are common (3). A typical representative layout of nutrient removal plants is the University of Cape Town (UCT) activated sludge plant (Fig. 2).

Disturbances

The essence of the control problem in wastewater treatment is related to disturbances. They are critical to the approach to control and have some specific features:

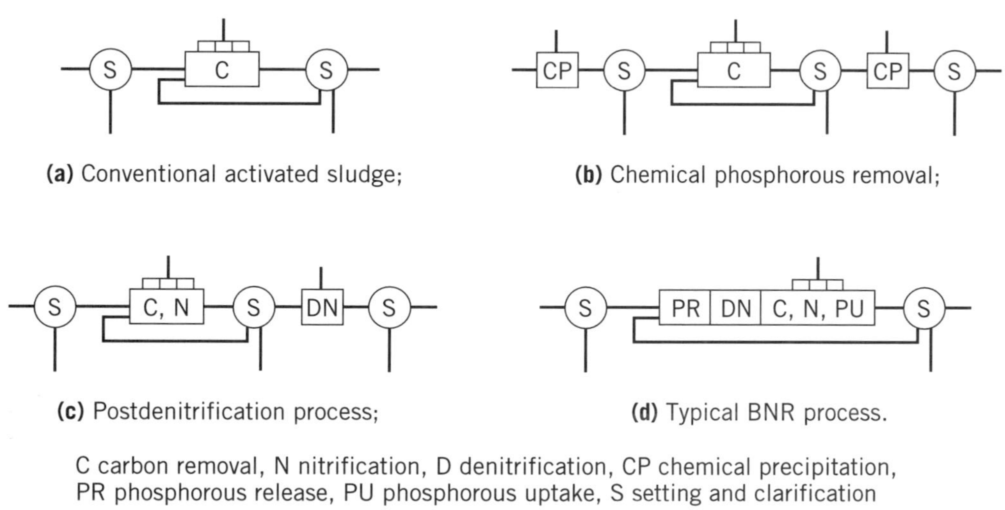

Figure 1. Typical wastewater treatment process flowsheets. (**a**) Conventional activated sludge; (**b**) Chemical phosphorous removal; (**c**) Postdenitrification process; (**d**) Typical BNR process. C = carbon removal; N = nitrification; D = denitrification; CP = chemical precipitation; PR = phosphorous release; PU = phosphorous uptake; S = settling and clarification.

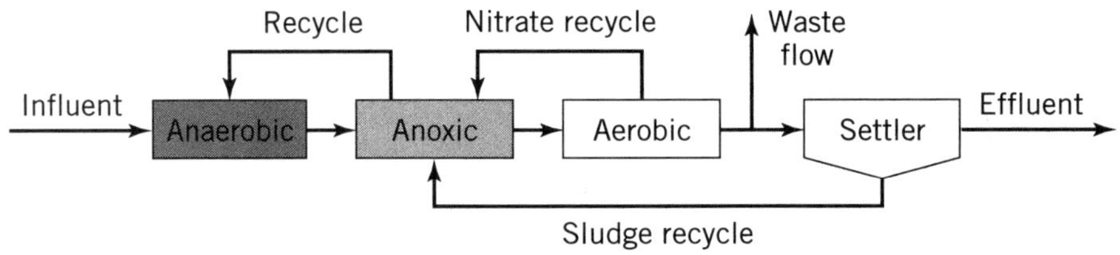

Figure 2. The UCT process, designed to remove organic material, ammonium nitrogen, and phosphorus.

- The daily volume of wastewater treated can be huge.
- The disturbances in the influent are enormous compared to most industries.
- The influent must be accepted and treated; no returning it to the supplier.

Because the disturbances to a wastewater treatment system are significant, the plant is for all practical reasons never in steady state. The disturbances are related to flow rate changes as well as concentration and composition variations. Figure 3 illustrates a typical influent situation in two municipal treatment plants.

The disturbances are not only related to the influent flow characteristics, but may be generated internally in the plant because of operational procedures. For example, poor pumping operations contribute to hydraulic disturbances (4). Clarifier upsets are often caused by inadequate pump control. A sudden flow increase caused by a pump start will influence the flow propagation along the plant within a fraction of an hour. A proper settler operation requires that the influent pumping has to be smooth. Variable speed pumping is a proved technology today and is a crucial component to create a smooth hydraulic operation of the plant. If primary pumps or other recycle pumps are operated on/off in too-large steps, the resulting solids–liquid separation will suffer, as will the effluent water quality.

Concentration disturbances are also generated within the plant itself. For example, sludge thickener overflow is usually returned to the plant inlet. This overflow is usually operated with on/off pumping without considering the influent load. Because the recirculated flow may have a large concentration, it will make a significant contribution to the overall load, both in terms of hydraulics and substrate loading.

It is usually impossible to remove the source of the disturbance, even if this is an important option, especially for industrial effluents, where production control in the industry may be improved. Sometimes the magnitude of the disturbance can be reduced *before* it reaches the plant. An integrated sewer-treatment plant control can attenuate hydraulic disturbances, and some waters can be pretreated in order to avoid harmful effects on the plant, such as neutralization of wastes with high or low pH. The control can also compensate for effects *within* the process. A common example is dissolved oxygen control.

All these disturbances will more or less influence the effluent water quality. An ultimate operational goal is to minimize the influence of disturbances and to maintain consistently a high effluent quality.

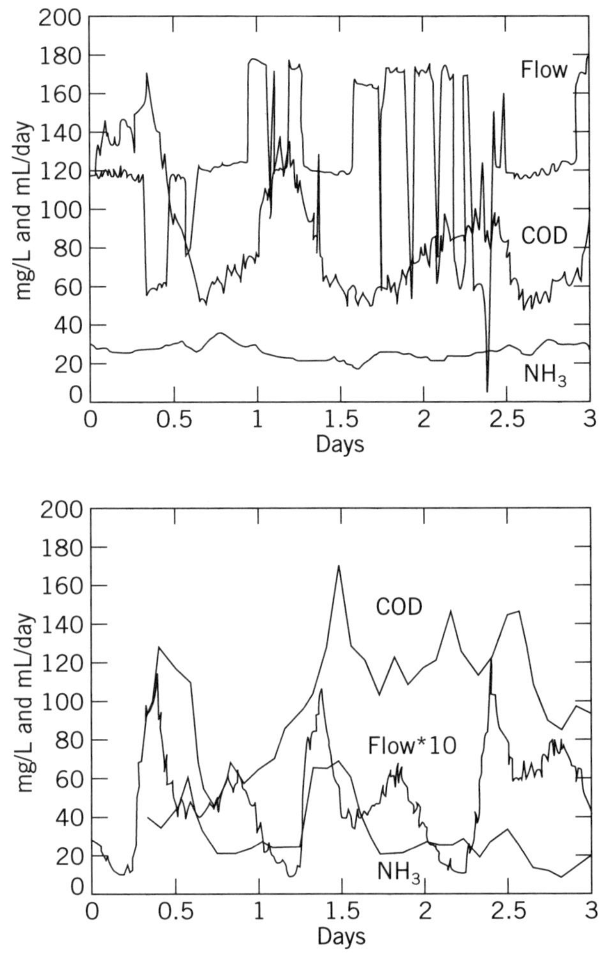

Figure 3. Typical influent disturbances from a large plant in Sweden (**a**) and a small plant in Australia (**b**). The flow rate as well as influent concentrations of chemical oxygen demand (COD) (related to organic content) and total nitrogen have significant variations.

CONTROL AND OPERATIONAL GOALS

It is far from straightforward to implement control in an activated sludge system; some of the reasons are listed here:

- The process has significant nonlinearities limiting the usefulness of simple controllers.
- There is a very wide range of response times, both a problem and a blessing.
- There are many interactions within the process with major recycles.

- There are many external couplings from the sewer to the receiving waters.
- Designers often leave too few manipulated variables and too little control authority.
- The concentrations of nutrients (pollutants) are very small, even challenging sensors.
- Sensors do not exist for many states and are expensive for others.
- The value of the product in the marketplace is remarkably low.

Other attributes why control is far from straightforward include:

- The microorganisms change their behavior and their population distributions.
- The separation of the effluent from the biomass is challenging and easily disturbed.
- Future effluent standards will be tight and based on spot checks.

Disturbance rejection is a major goal of activated sludge control. Large disturbances must be attenuated, and we must do it as early in the process as we can. This will ease the task of controllers later in the process.

The formulation of the goal of a wastewater treatment operation is much more demanding than to define the effluent standards. When a goal or a performance criterion is defined, then the necessary control actions can be analyzed. There are several goals of wastewater treatment, and one has to recognize the different levels and interest groups. The specific goals are set so that they contribute to meeting the more general corporate and community goals. They include at least

- To meet effluent discharge requirements.
- To achieve good disturbance rejection.
- To optimize operation to minimize the operating cost.

The first goal of course includes keeping the plant running. This is the major task for many operations, at least in terms of manpower. It is not a trivial task to maintain advanced instrumentation and equipment, to make sure it runs smoothly, and to detect any faults in its operation. Much of this is traditional automation and process control and is not specific to running wastewater treatment systems (5,6). The environment may be very specific and harsh for much equipment, but that is also the case in other industries as well, so from a strict control point of view the problems are traditional.

However, keeping the plant running and simply recording effluent quality is too often taken as the only goal. This has been and is still largely because of the lack of process knowledge on the part of plant operators. Too many operators of plants rely on the consultant for expertise in both design and operation. To satisfy the effluent quality requirements, it is crucial to ask: is this task solved primarily by design or do instrumentation, control, and automation play significant roles in this part? The traditional answer among designers used to be *no*. When only carbonaceous removal was considered, the plant was considered self-regulating and could take care of load disturbances and still produce a satisfactory effluent. The traditional method was large volumes, excess aeration, and chemical dosage. Consequently, a lot of plants have been oversized. No incentive was given for improved operation, because operational costs were not compared with investment costs. Often they were met from different financial sources.

Because of the significant load variations and disturbances typical for wastewater treatment systems, the plant is hardly ever in steady state. Despite the wide swings of load, the plant still has to *consistently* produce a satisfactory effluent. If the volumes of the tanks are not oversized, this means that some control system has to minimize the influence of disturbances and ensure that the operation is satisfactory. Pressures to fully utilize volumes and the complexities of nutrient removal are making process control essential. This of course includes disturbance attenuation.

Minimizing the operating cost and maximizing the use of available capacity is becoming necessary. Here, control and optimization are the key tools. Costs to be minimized include electrical energy to produce air, chemicals, pumping, and carbon addition.

MANIPULATED VARIABLES

There are quite a few variables to manipulate in an activated sludge process. Still, the possibilities to control the plant in a flexible way are quite limited. In many plants, however, there may be a potential to make a better use of the manipulated variables that are available. We categorize them in the following groups:

- Hydraulic, including sludge inventory variables and recirculations.
- Additions of chemicals or carbon sources.
- Air or oxygen supply.
- Pretreatment of influent wastewater.

There are several other manipulated variables in a plant that are related to the equipment and to basic control loops in the process, such as local flow controllers, level controllers, and so on. They are not included in this discussion. Further details are found in the work by Olsson and Jeppsson (7).

A majority of the manipulated variables are meant to change the hydraulic pattern in the plant. The different flow rates will influence the retention times in the different units. Moreover, the clarification and thickening processes are particularly sensitive to the *rate* of change. The hydraulic flows also determine the interaction between different unit processes. We classify the hydraulic manipulated variables into four major groups:

- Variables controlling the influent flow rate.
- Variables controlling the sludge inventory and its distribution, including return sludge flow rate, waste sludge flow rate and step feed control.
- Internal recirculations within the activated sludge system.

- External recycle streams, influencing the interactions between sludge and liquid processes.

Sequential batch reactors (SBR) offer other control possibilities because of the flexibility associated with working in time rather than in space. Some of the features of SBRs are described in some detail in the work by Irvine-Ketchum (8). The Danish Bio-Denitro and Bio-Denipho processes are not the traditional fill-and-draw process but are alternating processes with continuous discharge (9,10).

Some of the control variables are sometimes called *selection mechanisms* for the viable bacteria in the system (11). They are classified in the groups:

- Electron acceptors (e.g. oxygen or nitrate).
- Substrate.
- Thickening or clarification properties.
- Temperature.
- Growth rate.
- Free swimming organisms.

Not all of these can be manipulated. Only the electron acceptors (mostly oxygen) and (to some extent) the thickening and clarification properties can be controlled (by chemical precipitation). Moreover, substrate can be added (e.g. carbon sources).

IMPORTANT CONTROL LOOPS

Out of the many manipulated variables, we will consider a few. Probably the most important is the control of dissolved oxygen (DO) by using the air supply. Other important measures are the sludge inventory, controlled by the return sludge flow rate, and the waste sludge flow rate.

DO Control

The DO supply is a key control variable for the aerator system, both for carbonaceous and nitrogenous organisms. It influences the operating costs significantly, so there is an incentive to minimize the air consumption. From a microbiology point of view the choice of a proper DO set point is crucial, and too little DO will limit the growth of organisms.

From a process control point of view, DO control is known and proved technology. As long as the air supply is variable, the control of DO is quite straightforward. The tuning of the controller has to be different at different loads because of the inherent nonlinearities, and this can be compensated for by self-tuning regulators, prove in full scale for a long time (12). Another way is demonstrated in pilot scale, where a nonlinear controller can be used to operated over a wide range of loads (13).

The oxygen transfer rate determines how fast gaseous oxygen is transferred into DO. This in turn decides that the DO concentration can be manipulated by air flow changes within fractions of an hour. Changes in the oxygen demand, however, may influence the DO concentration within minutes. Not only the total amount of oxygen but its distribution in space and time are essential for the operation. Some of the considerations are:

- *Total air supply*. The total demand of air is basically determined by the total mass of viable organisms in the system multiplied with the biodegradable substrate concentration. The air supply control has an essential impact on the energy cost for the plant.
- *DO set points*. It is important to be able to select the DO set point. However, little is known about the proper choice of it. Generally, under closed loop DO control, it is made constant. There is, however, some evidence that a time variable set point may be advantageous (12).
- *DO spatial distribution (profile)*. It is a great advantage to be able to individually change the air flow rates along the aerator. This makes it possible to obtain an accurate air flow spatial distribution.
- *Pure oxygen addition*. Some aeration systems are supplied with pure oxygen instead of air. Pure oxygen is now and then used in conventional air systems to add extra DO to the system under extreme load situations.
- *Anoxic/aerobic zones*. Because loading conditions may change significantly on a weekly or longer basis, it is of interest to vary the size of the anoxic reactor volume and change the relative retention times between nitrification and denitrification. Such a control is, of course, easy to realize in SBR systems.

Given an adequate controller, it is straightforward to change the DO set point. However, to know which set point is the best from a microbiological point of view is a much more intricate question, and much remains to be studied before the ideal DO control system has been found. Consider an example of a compromise between competing goals within the same plant. The anoxic unit in the activated sludge system, Figure 2, is used to denitrify the nitrate, that is, to reduce the nitrate to free nitrogen by the oxidation of organic carbon. The dominating organisms in the process are the heterotrophs, a proportion of which reduce nitrate in an (ideally) oxygen-free environment. Therefore, the DO level has to be kept at a minimum. If not, the organisms will use oxygen instead of nitrate as the electron acceptor, and the zone will work at least partly like an aerobic reactor.

There are several disturbances that may violate the anoxic condition. The nitrate recycle stream may have a high DO concentration that will rapidly influence the DO concentration of the anoxic zone, thus the denitrification (DN) rate. It is apparent that from a DN point of view that the DO concentration at the outlet of the aerator has to be kept as low as possible, so that no more oxygen than necessary is recirculated. However, for the nitrification taking place in the aerator, a high DO concentration is required. Consequently, the DO concentration at the aerator outlet has to be a compromise between the needs of nitrification and denitrification.

Sludge Inventory Control

The sludge inventory can be controlled primarily by three manipulated variables: the waste-sludge flow rate, the return-sludge flow rate, and the step-feed flow rates. Manipulation of the waste-sludge flow rate is used to control the total inventory of sludge in the process. Because the total inventory is a function of the total growth rate of organisms, it is used to control the sludge retention time, or the sludge age. It will influence the system in a time scale of several days or weeks.

Manipulation of the return-sludge flow rate is used to distribute the sludge between the aeration basins and the settler units or between the acidogenic and methanogenic reactors in two-stage anaerobic systems. Recycle from the settling stage is an important variable for obtaining the right operating point in the reactors, but seldom useful for the control on an hour-to-hour basis (14). Some systems are supplied with several feeding points for the return sludge. This has a potential for sludge redistribution for certain loads, such as toxic loading.

A *combination* of different recycle may be important. In systems with chemical precipitation, sludge from the secondary settler may be combined with chemical sludge from a postprecipitation settler unit. In that way, the flocculant properties may be influenced, and the chemicals better utilized for phosphorus removal.

By controlling the step feed in an activated sludge plant, the sludge within the aeration basin can be redistributed, given the proper amount of time (15). As a special case of step-feed control, one will obtain a *contact stabilization* structure.

MEASUREMENT AND DIAGNOSIS

Measurement technology is crucial for adequate operation. In particular, on-line analysis is of key interest to control the effluent water quality. It is of great importance to extract maximum information from available instrumentation. Therefore, detection and diagnosis, based on combination of measurements, are models that make up a basis for successful plant operation.

Sensor Technology

Sensor technology is crucial for the success of control and operation, and great progress has been made over the last few years (3). Measurement for control is not necessarily the same as measurements for quality check or for regulatory agencies. The sampling rates have to be consistent with the plant dynamics, and the location of some of the sensors has to be chosen with care, for example, DO probes. Important types of measurements include:

- Flow rates in different plant units.
- Air flow rates and air pressure.
- Sludge flow rates.
- Temperature in water basins and in digesters.
- Gas (e.g., methane) flow rates and temperatures.
- Alkalinity.
- pH.
- DO in different locations.
- Sludge levels.
- Suspended solids in different places.
- BOD, COD, TOC.
- Phosphorus fractions.
- Nitrogen fractions.
- Respiration rate.

There are many instruments that satisfactorily measure flows, levels, pressures, dissolved oxygen in liquids and suspended solids, and dry solids concentrations in sludge. Even if an instrument is deemed to be reliable, proper maintenance is still required. Maintenance requirements will vary greatly between different applications. The development of respirometry has been significant. Within IAWO, task group prepared a scientific and technical report during 1997 (16). Also, advanced N and P analyzers are becoming common in the advanced operation and control of nutrient removal plants (3).

Diagnosis and Estimation

It is clear that more information than the primary sensor signals is required in order to get good operation. The idea of estimation is to mix sensor information with mathematical models. In this way, variables that are not measurable may be reconstructed by combining a mathematical model with sensor information about some other variable. Consider the respiration rate (or the oxygen uptake rate) in an aerator. It can be measured by using respirometers. Alternatively, it can be obtained from measuring the DO concentration and calculation of the DO mass balance. Consider the DO mass balance in a complete mix aerator:

$$\frac{dc}{dt} = d_i \cdot c_i - d \cdot c + k_L a \cdot (c^s - c) - R \qquad (1)$$

where c is the DO concentration; c_i the influent and c^s the saturation DO concentrations, respectively; $k_L a$ the oxygen transfer rate, and R the oxygen uptake rate. The parameters d and d_i are dilution terms. If the oxygen transfer rate $k_L a$ is known, then R can be easily calculated from the DO measurements and the mass balance (equation 1). Usually, however, $k_L a$ must be determined, because it varies slowly with time. Therefore, both the oxygen transfer rate and the respiration rate R to be determined simultaneously from measurements of the DO, preferably under closed loop DO control. This is not trivial, because there is an identifiability problem. A positive change in R can be explained by a similar positive change in $k_L a$ and no unique solution is found (17). However, certain measures can be taken in order to find the parameters uniquely, so that the respiration is found under DO control (13,18,19).

INTEGRATED OPERATION

Because of the combined development of instrumentation, computer technology, operating technology, and the understanding of plant dynamics, the goal is today a plant-wide control. Recognizing the interaction between various unit processes, the overall control of a plant tries to make maximum use of volumes, energy, and other resources.

Because biological nutrient removal (BNR) will become more common, plant complexity will grow. This may be a major driving force for more instrumentation, control, and automation. In a BNR process, there is a delicate balance between carbon, nitrogen, and phosphorus removal. Some important couplings (see Fig. 2) are:

- The phosphorus release has to take place in an anaerobic environment—no oxygen or nitrate can be present. If oxygen-rich water is recirculated, it may impede the P release. If nitrate is allowed to enter the anaerobic zone, the bio-P process is also inhibited.
- Denitrification takes place in an anoxic environment. The various couplings to other unit processes were discussed earlier.
- Both the phosphorus release and the denitrification require a certain carbon concentration. At low loads this may not be sufficient, and an extra carbon source has to be added.

It will become more and more important to understand the impact of load variations. Furthermore, the increased quantities of sludge makes it necessary to develop better control methods for sludge treatment and disposal processes. The drive for a sustainable environment is an incentive to make better use of automation in order to reach such goals.

An efficient operation of a plant has to consider not only the couplings between various unit processes within the plant, but couplings to external processes. Realizing that sewer operation must be coordinated with plant control, there has been a strong development of combined sewer and plant control (20). Disturbances can be detected and controlled as early as possible.

BIBLIOGRAPHY

1. G. Olsson, and B. Newell, *Control of Biological Wastewater Treatment Plants*, 1997, in press.
2. G. Olsson, *Water Sci. Tech.* **28**, 1–7 (1993).
3. IAWO, *Water Sci. Technol.* **33** (1995).
4. G. Olsson, B. Andersson, B.G. Hellström, H. Holmström, L.G. Reinius, and P. Vopatek, *Water Sci. Technol.* **21**, 1333–1345 (1989).
5. G. Olsson, and G. Piani, *Computer Systems for Automation and Control*, Prentice Hall, Englewood Cliffs, N.J., 1992; also in German: *Steuern, Regeln und Automatisteren*, Carl Hanser, München, 1993 (revised and extended revision of the English version).
6. D.E. Seborg, T.F. Edgar, and D.A. Mellichamp, *Process Dynamics and Control*, Wiley, New York, 1989.
7. G. Olsson, and U. Jeppsson, *Forum for Applied Biotechnology (FAB)*, Bruges, Belgium, Sep. 28–30, 1994.
8. R.L. Irvine, and L.H. Ketchum Jr. *CRC Crit. Rev. Environ. Control* **18**, 255–294 (1989).
9. E. Bundgaard, *Proc. of the Int. Workshop on Wastewater Treatment Technology*, Copenhagen, 1988.
10. J. Einfeldt, *Water Sci. Technol.* **25**, 161–168 (1992).
11. W. Gujer, and M. Henze, *Water Sci. Technol.* **23**, 1011–1023 (1990).
12. G. Olsson, L. Rundqwist, L. Eriksson, and L. Hall, in R.A.R. Drake ed., *Advances in Water Pollution Control*, IAWPRC, Pergamon Press, New York, 1985, pp. 473–480.
13. C.F. Lindberg, B. Carlsson, *Water Sci. Technol.* **34**, 173–180 (1996).
14. G. Olsson, *Comprehensive Biotechnology*, Pergamon Press, New York, 1985, pp. 1107–1119.
15. J.F. Andrews, *Water Res.* **8**, 261–289 (1974).
16. H. Spanjers, P. Vanrolleghem, G. Olsson, and P. Dold, *Water Sci. Technol.* **34**, 117–126 (1996).
17. U. Holmberg, in R. Briggs ed., *Advances in Water Pollution Control*, IAWPRC, 1990, pp. 113–120.
18. U. Holmberg, G. Olsson, and B. Andersson *Water Sci. Technol.* **21**, 1185–1195 (1988).
19. B. Carlsson, C.F. Lindberg, S. Hasselblad, and S. Xu, *IAWO 17th Biennial Int. Conf.*, Budapest, Hungary, July 1994.
20. U. Nyberg, B. Andersson, and H. Aspegren, *Water Sci. Technol.* **34**, 127–134 (1996).

WASTEWATER TREATMENT, IMMOBILIZED CELLS

TAKAMITSU IIDA
Niigata University
Niigata, Japan

KEY WORDS

Activated sludge
Adsorption
Biological oxygen demand (BOD)
Entrapment
Immobilization
Immobilized cells
Nitrogen removal
Photo-crosslinked gels
Polyethyleneglycol
Wastewater

OUTLINE

INTRODUCTION

Activated sludge for biological wastewater treatment has been broadly available for a long time. But, water pollution and eutrophication in rivers lakes, and inland seas have resulted from insufficiently treated water. To protect and control these occurrences, it is critical to consider biological oxygen demand (BOD) and nitrogen-phosphorus (N-P) removal in the development of anaerobic–aerobic wastewater treatment systems.

Increasing volumes of wastewater, combined with limited space availability and progressively tightening standards and quality control, require the development of new intensive biotechnologies for wastewater treatment. Recently, there has been a significant increase in the use of immobilization processes for municipal and industrial wastewater treatment. Immobilization processes offer several advantages because of higher volumetric load, increased process stability, compactness, and so on, as shown in Table 1.

The successful application of an immobilized system as biocatalyst relies on the proper choice of the immobilizing materials and reactor type. The most important requirements for wastewater treatment are as follows:

1. High cell mass loading capacity.
2. Simple nontoxic immobilization procedure.
3. High surface-area-to-volume ratio.
4. Mechanical stability.
5. Chemical stability.
6. Reusability.
7. Suitable for conventional reactor systems.
8. Economic feasibility.

An ideal system that possesses all these optimal properties does not exist, and a suitable compromise has to be found.

For wastewater treatment, bioreactors with immobilized microbes have been examined as a combination of activated sludge process for BOD and nitrogen removal and as an anaerobic process for methane production. But, the attempted immobilization to an anaerobic process is scarcely used in a practical scale because the handling of microbes is difficult at the immobilization and the effects of immobilization are not clear (1). The immobilized microbe treatment process clearly is an effective method of aerobic wastewater treatment.

On a laboratory scale, the immobilization of microorganisms has also been applied to the removal of specific compounds such as phenol and its derivatives (2–5), coloring matter (6), and other inorganic compounds such as cadmium (7,8), uranium (9), and hydrogen sulfide (10). Consequently, they can maintain higher biomass concentrations, and a high degree of organic carbon or nitrogen removal, or both, is achieved by these compact systems. Moreover, minimized production of excess sludge is expected by maintaining a low food-to-microorganism ratio.

CLASSIFICATION OF IMMOBILIZATION TECHNIQUES

Whole-cell immobilization is always used in wastewater treatment and may be defined as the physical confinement or localization of intact cells to a certain defined region of space with the preservation of some decomposition activity (11–13). Accordingly, the immobilization technique using biocatalysts is developed to reuse the catalytic activity and for continuous operation in a bioreactor.

The immobilization of microbes for wastewater treatment is broadly classified into four major categories: adsorption of attachment to a surface, entrapment within a porous matrix, containment behind a barrier (semipermeable membrane and microcapsule), and self-aggregation. These techniques can be applied to all wastewater treatments that use a viable cell system.

Cells Attached to a Surface

Attaching cells to a surface is a very popular immobilization technique because it is simple and fast. A suitable adsorbent for attachment should display a high affinity toward microbial cells and yet cause minimal denaturation. A surface-attached process is the immobilization method in which cells are bound to a surface, regardless of the type of binding. Microbial cell walls carry a charge that depends on the pH of wastewater and other environmental conditions. Biofilms resulting from the growth of cells on the surfaces of inorganics such sand or rock are present in conventional industrial processes (14–21), and the self-immobilization of cells (22,23) is a widespread process. But in recent years, intensive treating processes have indicated a need for the development of immobilization for a given process.

Table 1. Advantages and Disadvantages of Immobilized Living Cell Systems

Advantages	Disadvantages
1. Self-proliferation or self-regeneration of immobilized cell systems.	1. The mechanical and chemical stability of some supports materials limits its use (shear, dissolution, and attack by product).
2. Higher productivities because of higher biomass density.	2. Cell growth can destroy the matrix.
3. Tailoring to the most efficient reactor is possible.	3. Diffusion limitations may limit the biodegradation rates.
4. During continuous fermentations, washout can be avoided at high dilution rates.	4. Biocompatibility of immobilization matrices is not sufficient for cells.
5. Better protection from surroundings.	5. Immobilization techniques and materials can be too expensive be used in large-scale processing.

The strength with which the cells are bonded to the support varies, with cell type, support type (24,25), and the variation of operation condition. The industrial use of surface-attached cell systems (biofilm) has been extensively examined in the field of wastewater treatment (26). For strengthening the attachment, porous supports such as porous glass are used (27).

However, in the fluidized-bed type bioreactors for wastewater treatment, inorganic materials are generally not used because of the difficulty in controlling their specific gravity, brittleness, and shortcoming of abrasion-resistance.

Cells Entrapped in a Porous Matrix

The second immobilization technique is entrapment within a porous matrix that is synthesized in situ around the cells. This method is rapidly coming into widespread use (28). The cells are entrapped in a matrix that protects them from the shear field outside the particles. A high degree of cell viability is retained in most entrapment methods. These systems are therefore useful for a variety of growth-related processes. An advantage of the preformed support particles is that they are more resistant to the compression and disintegration in fluidized-bed bioreactors. Cell growth is controlled by the continuous flow of wastewater past the particles (29).

A variety of compounds can be used for cell entrapment, with minimal loss of viability (30). Natural polymers used for immobilization include polysaccharides, such as alginate (31), carrageenan (32), cellulose, and agar, and polypeptides, such as gelatin and collagen. These natural polymers are compatible with microorganisms, but they are not suitable for wastewater treatment because of their assimilation by microorganisms in activated sludge.

Accordingly, only synthetic polymers, such as polyacrylamide (33), poly(vinyl alcohol) (34,35), poly(ethylene glycol), and polypropylene, and hydrophilic photo-cross-linkable resin can be applied to the immobilization of biomass in terms of mechanical strength and durability that are required for industrial usage. For successful immobilization, the hydrogel must be effective in cell viability and function (biocompatible) (36). The method of entrapment can reduce the culture viability. The entrapment in a polyacrylamide gel destroys viability in 50 to 90% of the cells by the toxic acrylamide monomers (37).

Several reviews (28,38,39) offer a thorough treatment of methods and application in immobilized cells. Examples of the various types of polymers are shown in Table 2.

Cells Contained Behind a Barrier

The third major category of cell immobilization is containment behind a barrier. When cell separation from the effluent is required, these systems are highly useful (56).

The barrier that immobilizes cells can be as simple as the liquid–liquid phase interface between two miscible fluids. Mohan and Li (57) immobilized the denitrifying bacteria, *Micrococcus denitrificans*, by emulsifying the cell suspension with surfactant into a hydrocarbon solvent. Cells in the surfactant-enclosed aqueous droplets did not leak from aqueous phase through the organic phase to the cell-free aqueous phase surrounding the droplets. Because the cells are retained behind a semipermeable wall or membrane, this type of system is very similar to microcapsulation in an engineering sense. Nutrients (mainly BOD components) are supplied to, and products are removed from, the cell mass by diffusion. In wastewater treatment, the barrier loses its separation activity during long operations.

Self-Aggregating Cells

Cells naturally aggregate or flocculate and produce extracellular polysaccharides. The large size of the aggregates makes their use possible in reactors designed for immobilized cells, such as packed-bed or fluidized-bed reactors. Artificial flocculating agents or cross-linking agents may be added to enhance the process of aggregation for cells that do not naturally flocculate. Similar mechanisms govern both artificially and naturally induced flocculation (58). The durability of the aggregate will depend on its hydrodynamic interactions with the environment. Many cells produce extracellular polysaccharides, some of which can act to promote the cohesion of a biofilm. These substances might be important determinants of the diffusive resistance in these films.

INDUSTRIAL APPLICATIONS OF IMMOBILIZED CELLS

Much research has focused on industrial applications using immobilized cell technology for wastewater treatment. In an industry setting, entrapment and adsorption are the predominant techniques used for wastewater treatment. The nitrifying process is described in this section, examples include entrapment using PBG gels and adsorption using photo-cross-linked resins.

Gel Entrapment of Microbial Cells

The entrapment of microbes is the most popular immobilization technique for living cells in wastewater treatment. The BOD sensor was developed to abbreviate the long measurement time by entrapping the activated sludge (59,60). In wastewater treatment, immobilization processes should be mild to minimize the damage to the living cells. Entrapping materials must also have proper permeability to allow sufficient diffusion and transport of oxygen and essential nutrients, metabolic waste, and secretory products across the hydrogel network at supplemental culture.

Considering the use of synthetic polymers serving as the synthetic extracellular matrixes for cells, it is important to understand the transport properties through these gels to predict whether nutrients can diffuse into the matrix. Supplemented culture media provide the cells with essential nutrients for growth and development as well as a means of oxygen and metabolic product transport while in vitro (61,62). Traditionally, mass transfer behavior in the gel matrix has often been recognized as a rate-limiting factor in an entrapping cell system. Some of the immobilizing matrixes are highly hydrated, allowing almost unhindered diffusion for products and substrates with molec-

Table 2. Environmental Application of Gel-Entrapped Microorganisms

Substrate	Microorganism	Entrapment method	Reference
Ammonia	*Nitrosomonas europaea*	Ca-alginate	40
		κ-carrageenan	41
	Nitrosomonas europaea	Polyelectrolyte	42
	Paracoccus denitrificans	Complex-stabilized Ca-alginate	
	Thiosphaera pantotropha	Agarose	43
	Scenedesmus quadricauda	κ-Carrageenan	44
Dairy effluents	*Candida pseudotropicalis*	Alginate	45
Nitrate	*Anasistis nidulans*	Collagen	46
	Denitrifying sludge	PVA	34
	Pseudomonas denitrificans	Ca-alginate	47,48
	Thiobacilus denitrificans	Agar/agarose	49
Nitrite	*Nitrobacter agilis*	Ca-alginate	50
		Carrageenan	51
	Nitrobacter sp.	PVA	35
Nitrogen	*Anabaena* CH3	Ca-alginate	52
	Nitrifiers	Polyethyleneglycol resin	53
Phenol	*Candida tropicalis*	Alginate	54
	Pseudomonas sp.	Ca-alginate, polyacrylamide hydrazide PVA-boric acid-Ca-alginate	55,36

ular weights less than 10,000 Da (36). The effective diffusivity of oxygen, glucose, and various ions in gel has been studied by several groups (63–65).

For industrial applications, microbes must be cultivated for entrapment and storage to keep the activity of immobilized microbes after immobilization.

Entrapping Materials. Mass transport through gels is governed by the size and shape of the pores and solute molecules as well as by the nature of the interaction between solute molecule and gel material. The value of the equilibrium coefficient reflects the relative affinity of the solute for the gel relative to its affinity for the surrounding medium. The species concentration in the gel at equilibrium is obtained by mass balance using the known values for the reservoir and gel volumes. The value of K, the partition coefficient, is calculated by the following equation:

$$K = C^{\text{gel}}/C^{\text{bath}} \tag{1}$$

The average pore diameters of gels were estimated from the diffusion of solutes and the hydrodynamic diameter of solute (60).

Biomass decay is an important parameter in the dynamics of the immobilized microbe system. *Nitrosomonas europaea* cells entrapped in a gel can lose their viability during cultivation; they die first in the center of a gel bead but also will die in the outer layers at low substrate loadings. If the substrate concentration is raised, the number of viable microbes near the surface of the gel increases again. The fluorescein diacetate and lissamine-green staining techniques were useful in distinguishing between viable and dead cells in gel beads (66).

Successive nitrification and denitrification are effective biological processes for the removal of nitrogen in the form of ammonia, nitrite, and nitrate from aqueous waste. In this biological nitrogen-removal system, nitrification is generally the rate-limiting step because of the extremely low growth rate of obligate autotrophic nitrifying bacteria. The use of immobilized cells in nitrification overcomes this disadvantage and shortens greatly the hydraulic retention time (HRT) in continuous nitrification.

In a synthetic resin gel such as polyacrylamide, the gel strength decreased about 50 to 60% in the presence of microbes. But the density of microbes in the gel did not influence its strength (67). The immobilization of nitrifying bacteria by using urethane prepolymers of the hydrophilic type exhibited a drastic improvement in the activity yield (68). These immobilized cells were prepared by entrapment using alginate (69–71), polyelectrolite complex (72), polyethyleneimine (73), photo-cross-linkable resin (74), poly(vinyl alcohol) (75), polyethylene (76), and poly(ethylene glycol) (PEG) (77).

Characteristics of Microbes Entrapped in Polyethyleneglycol. The PEG gel was prepared by radical reaction under existing microbes to exhibit about 3 kg/cm^2 of gel strength (78). These gels are nontoxic for microbial cells and show viability. Furthermore, this method is easy industrialized for a large-scale production.

Microorganisms such as nitrifying bacteria grow near the surface of gels (about 60-μm depth). After about a month's cultivation, they have constant activity. The rate of oxygen consumption the parameter of activity, is 60 mg O_2/L gel/h at initially and 800 mg O_2/L gel/h at the end of precultivation. At the steady state, the concentration of microbes in the gel carrier is increased to contain about 10^{10} cells per milliliter of carrier. On the other hand, the concentration of nitrifying microbes in the activated sludge is usually 10^3 to 10^5 cells/mL. Figure 1 depicts nitrifying pellets (about 3 mm^3) and a cross-section of pellets and bacteria in colony.

Nitrification and denitrification of entrapped microbes were scarcely influenced by the operation temperature, and the activity of nitrification was about three times higher than that of the activated sludge, as shown in Figure 2. In Figure 2, the reaction rates are exhibited on the basis of suspended solid for comparing the activity of im-

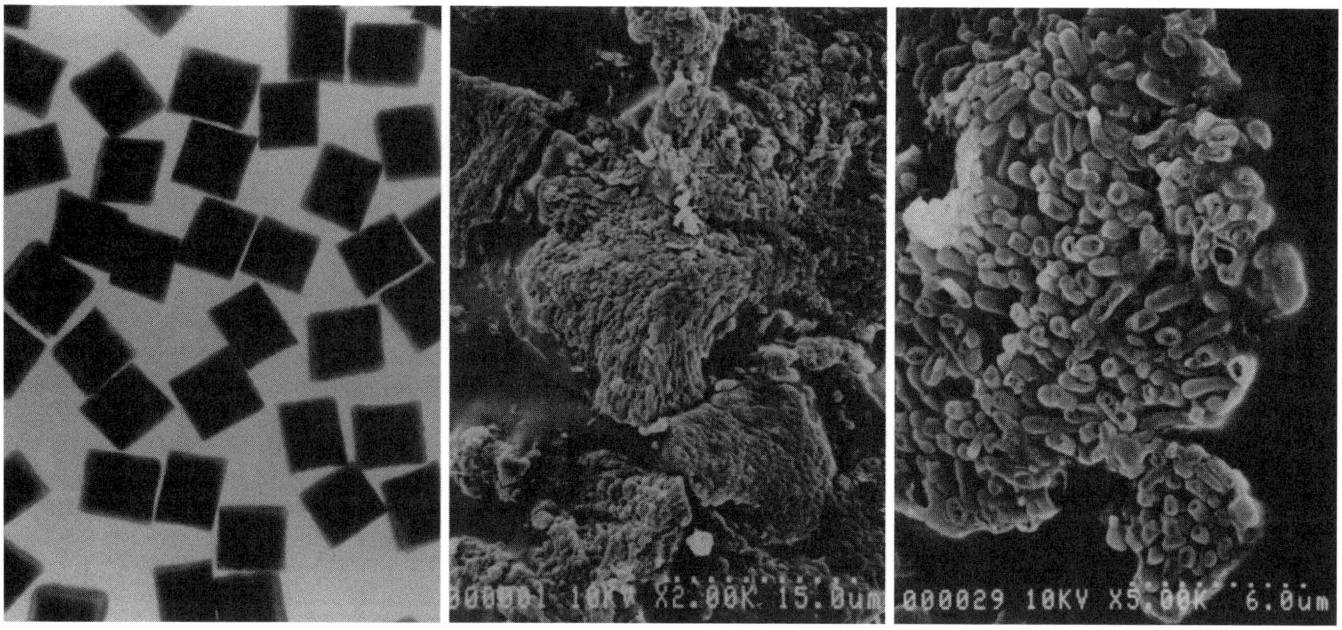

Figure 1. Photographs of nitrifying pellets. Cross-section of pellets and bacteria in colony. (a) Nitrifying pellets of 3-mm cubic gel; (b) Cross-section of pellet by electron scanning micrograph; (c) Bacteria in colony under electron scanning micrograph.

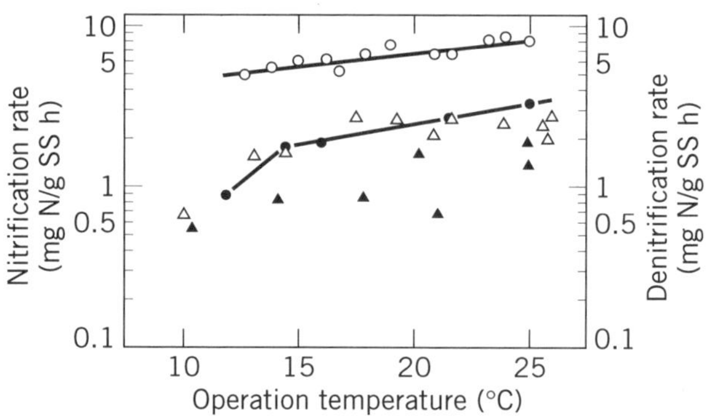

Figure 2. Effect of operation temperature on nitrification rate and denitrification rate of entrapping method against the activated sludge method. Nitrification: ○ = immobilized: △ = activated sludge. Denitrification: ● = immobilized: ▲ = activated sludge. Range of BOD loading: immobilized = 0.1–0.15 (kg/kg SS day). Activated sludge = 0.02–0.08.

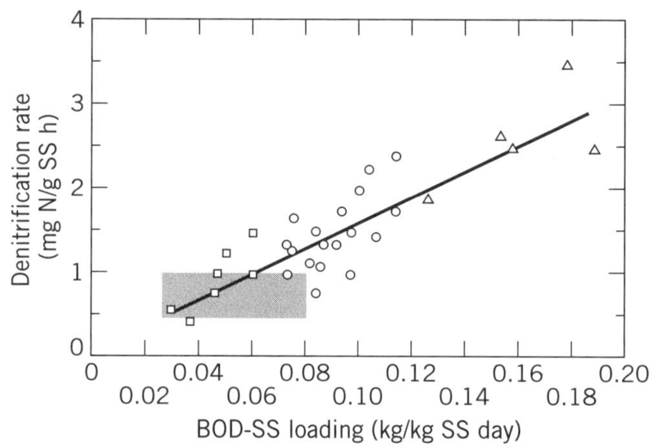

Figure 3. Relationship between denitrification rate and BOD-SS loading in the entrapping method. BOD loading (kg/kg SS day) at 15 °C: □ = 0.02–0.07; ○ = 0.07–0.12; △ = 0.12–0.19. Shaded part indicates the range of the activated sludge method.

mobilization with that of activated sludge. The nitrification rate of precultured immobilized gels was about 100 mg N/L gel/h at 13 °C. Accordingly, the system using entrapped microbes showed a high denitrification rate owing to high BOD loading resulting from short retention time in the nitrification tank (Fig. 3).

Continuous Wastewater Treatment by Entrapped Microbes (79). Figure 4 shows a schematic diagram of a pilot plant using PEG-entrapped gels. It is composed of an initial and final sedimentation tank, a denitrification tank, and a nitrification tank. The volume ratio of denitrification to nitrification is 1.5/1.0, and the gel separator (the net of 1.5-mm mesh) in the nitrification tank is installed for preventing gel bead washout. The mesh size of the gel separator should be half the gel diameter for sufficient separation. Wastewater treatment was recycled between the denitrification tank and the nitrification tank. Retention times are 4.8 and 3.2 h for denitrification and nitrification, respectively. The average total nitrogen removal is represented as

$$\text{Total N removal (\%)} = R/(R + 1) \times 100 \qquad (2)$$

with R as the recycling ratio (−). The total N removal is

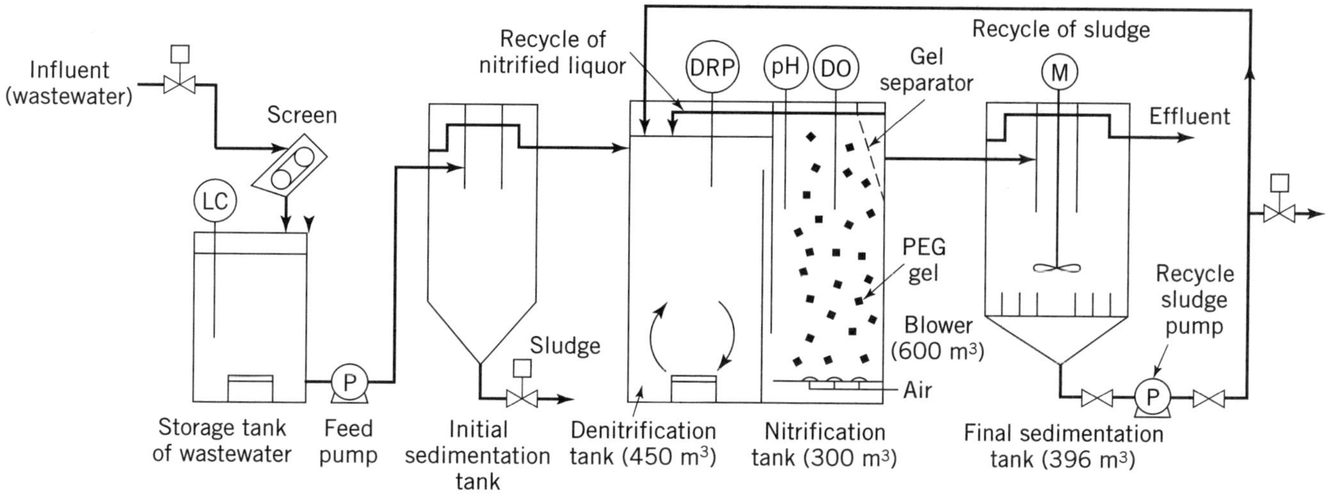

Figure 4. Schematic diagram of wastewater treatment plant with a capacity of 2,250 m^3/day (max, 3,000 m^3/day). DO = dissolved oxygen electrode; ORP = oxidation–reduction potential; pH = pH electrode; LC = level control; P = pump; M = mixer.

increased according to the enlargement of the recycling ratio, but the dissolved oxygen in the denitrification tank is also increased. Consequently, recycling is generally operated at a 1:2 recycling ratio to maintain the anaerobic condition of denitrification and compensate for the increased power cost in nitrification. Gel beads are fluidized by the air blowing at the bottom of nitrification tank.

The packing ratio of entrapped gels in the reactor is usually 5:15 volumetric percent depending on the treatment. In the entrapment method, gels are added at 7.5 volumetric percent to the nitrification tank. The floating carriers are distributed homogeneously. In the entrapment method, specialized equipment and sufficient time are needed to prepare an adequate number of gels for large-scale wastewater treatment.

The result of continuous wastewater treatment in the PEG entrapping method is shown in Figure 5, and the operation conditions of the plant are shown in Table 3. Even in winter nitrification was thoroughly carried out, and BOD as well as suspended solids in the effluent were lower than 10 mg/L. This method needs about 8 h of retention time to eliminate 70% of nitrogen in wastewater. After continuous operation for 2 y, it was concluded that gel beads must have above 1.5 kg/cm^2 compressive strength; therefore, the initial compressive strength of gel absent from microbial cells needs to be over 2.7 kg/cm^2.

For the nitrification, photo-cross-linkable resin was used to obtain an activity of 1.5 kg of N/m^3 gel/day for continuous wastewater treatment (80). On the other hand, the drop in denitrifying activity at the immobilization was restored in the medium containing nitrite or nitrate with sodium acetate as a hydrogen donor.

Industrial Application of Adsorption Method

Systems in which cells are immobilized by adsorption are popular because of the ease of this type of immobilization. The strength with which the cells are bonded to the support varies with cell type and support type (81,82).

The adsorption method differs from the entrapment method, advantages include easy gel formation (no consideration of microbes at cross-linking reaction), mass transfer of gel barrier, good storage stability before usage, and low transportation cost of gels at the industrial scale. A disadvantage is the long induction period for steady wastewater treatment.

The growth rate of nitrifying bacteria is about 10 times less than that of general microbes. The attachment of biomass on the gel surface occurs during the start-up phase. After this phase, the reactor is operated continuously at different hydraulic and surface loading rates. Accordingly, the gels must have high strength and good cell attachment on their surfaces.

A positively charged macroporous cellulose carrier is used for immobilization of nitrifying bacteria by the adsorption method (83). This carrier promotes stable adhesion of bacteria onto the carrier walls and allows transfer of nutrients into the carrier. A porous cellulose carrier was produced from cellulose with a polyethyleneimine coating to form a stable cross-linked structure. Similarly, polyurethane foam (84) and other carriers (85,86) have been used for wastewater treatment.

Advantages of Adsorption Using Photo-Cross-Linked Gels. To use immobilized microbes in wastewater treatment, the preparative method, especially the support material used for the preparation, must be selected from the standpoint of its economy and feasibility on a large-scale operation. For these reasons, photo-cross-linkable resins are promising support materials for the immobilization of microbes.

Although this immobilization process may have various conceivable features, its features are summarized as follows (87):

1. The immobilization procedure (gel formation) is simple because of ultraviolet irradiation.

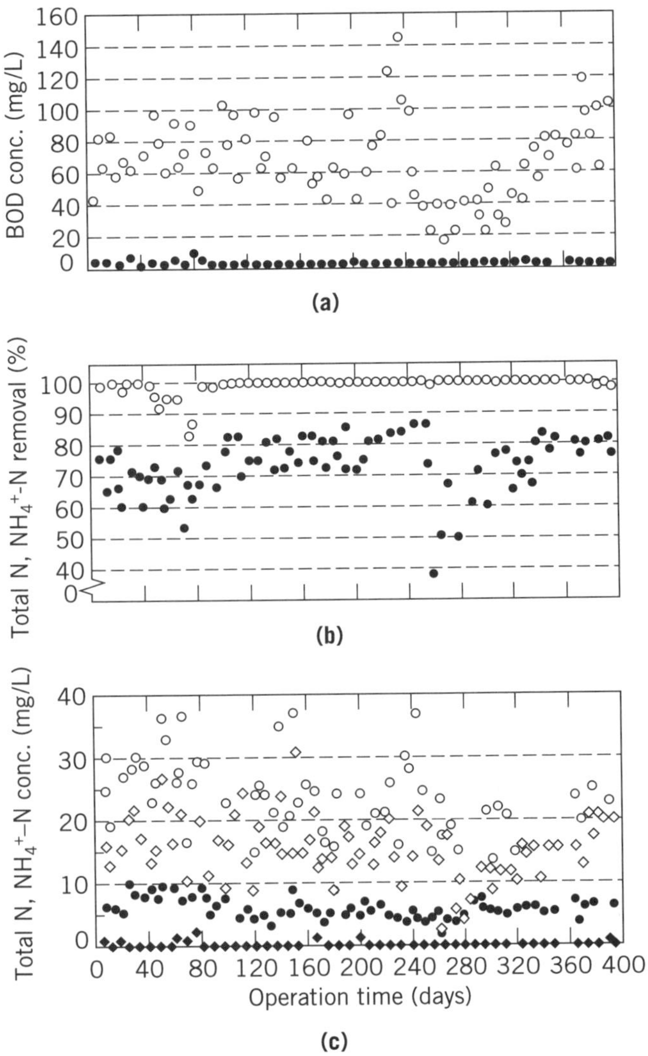

Figure 5. Continuous wastewater treatment using entrapped microbes. (**a**) Influent (mg/L): ○ = BOD; effluent (mg/L) ● = BOD. (**b**) Removal (%): ● = total N removal (%): ○ = NH_4^+-N. (**c**) Influent (mg/L): ^ = total N, ◇ = NH_4^+-N; effluent (mg/L): ● = total N, ◆ = NH_4^+-N.

2. The gel materials have great mechanical strength.
3. The surface of gel beads can be regulated by the properties of photo-cross-linkable resin.
4. The manufacture of photo-cross-linkable resins and the process of gel formation can be easily industrialized.

The photo-cross-linkable resins usually have a main chain of PEG, a mixture of PEG and poly(propylene glycol) (PPG), or poly(vinyl alcohol) (PVA) with photo-cross-linkable ethylenic double bonds. By irradiation of ethylenic double bonds with UV eight, polymerization is induced, forming three-dimensionally-cross-linked gels.

Surface Structure of Gels for Good Cell Adsorption. The sludge used is taken from the recycle line of an activated sludge treatment plant treating wastewater, because organic nitrogen and ammonia nitrogen are present in wastewater. Acclimatization of the sludge is carried out in a batch reactor. The carrier elements are made of hydrophilic photo-cross-linkable resin with a 4-mm bead diameter.

Because the nitrifying bacteria do not flocculate differently from the activated sludge, it is difficult to attach the nitrifying bacteria on the surface of the carrier. The adsorption rate of nitrifying bacteria is maximum at pH 4.8, and nonbiological factors are important at the initial stage of attachment (88). Fluorescence studies of the adsorbing surface clarified that the affinity to water and ζ potentials of the surfaces are important factors for adsorption (89). Among the synthetic polymers, photo-cross-linkable resin has largely resolved the problems with inorganic carriers described in the Introduction. Its surface structure may be suitable for attachment of microbes, controlling the properties of photo-cross-linkable resins.

Several factors affect cell affinity and behavior on hydrogel: the general chemistry of the monomers and the cross-links, hydrophilic and hydrophobic properties, and surface properties involving charged functional chemical groups. Physical properties of the hydrogel also govern the adhesion affinity; therefore, altering pore size and network structure can convert cell adhesion as well as morphology and function.

The attachment of biomass on the charged gel surface occurs during the start-up phase. Designing gel carriers using photo-cross-linkable resin has resulted in good cell growth on their surface with good mechanical strength (90). The surface of PEG-PPG-type hydrophilic resin is most hydrophobic and has an excellent affinity to microbes. This phenomenon is practically confirmed in continuous denitrification, as shown in Table 4.

When observing the surface of PEG-PPG-type photo-cross-linked resin and the results of contact angle, the phase separation of the gel in which hydrophobic domains of about 1-μm diameter separated on the hydrophilic phase of PEG was seen on SEM of the surface (Fig. 6) and with the optical microscope. From these results, the surface of hydrophobic domains formed on hydrophilic phase, the so-called macrophase separation of hydrophilic gel, may be best for the attachment and growth of nitrifying microbes.

The structure of the gel bead is controlled with a combination of PEG and PPG, PVA type, and PEG type. Beads with a diameter of about 4 mm and compressive strength of more than 20 kg/cm^2 can be produced by immobilization (91).

Continuous Wastewater Treatment. Continuous wastewater treatment can be carried out by a process using gel for microbial adsorption, as shown in Table 5 (92–94). Removal performance of the pilot plant with a retention time of 8 to 9 h in the bioreactor is obtained as follows. The average total N of the influent is 31.8 mg/L, and that of the effluent total N is 8.3 mg/L. The average P of the influent is 2.57 mg/L, and that of the effluent total P is 0.60 mg/L.

The nitrification rate of immobilized gels is of zero-order

Table 3. Operation Condition of Plant

Item	Run 1	Run 2	Run 3
Operation period	18	11	18
Infuluent (m^3/day)	2,720	2,280	1,350–3,150
Retention time (h)	6.6	7.9	5.7–13.3
Aeration (Nm^3/h)	610	670	800
Recycling ratio (−)	2.8	2.7	2.0–7.0
MLSS (mg/L)	2,800	1,840	1,880
BOD-SS loading (kg/kg day)	0.070	0.149	0.0965–0.302
BOD loading (kg/m^3 day)	0.196	0.275	0.182–0.570
Total N and SS loading (kg/kg day)	0.0285	0.0535	0.0226–0.0772
Total N loading (kg/m^3 day)	0.0793	0.0980	0.0426–0.146
Recycling ratio of sludge (%)	64	64	47–109
Temp. of water (°C)	23.5	14.0	14.2
Total SRT (day)	14.6	9.0	8.8
Aerobic SRT (day)	5.8	3.6	3.5

Notes: Run 3 = Variation test of loading. MLSS = Mixed liquor suspended solid; BOD = Biological oxygen demand; SS = suspended solid; SRT = sludge retention time.

reaction in which the produced NO_z-N increases directly with the reaction time, and the nitrification rate is about 2.5 kg NO_k/m^2 gel/day, as calculated in Figure 7. In the relationship between the dissolved oxygen (DO) and the nitrification rate coefficient (K_M) of gels, K_M largely depended on the DO. It is considered that the oxygen diffusion through the biofilm on the surface is the rate-determining step for nitrification. The phosphorus uptake increased in proportion to the denitrified nitrogen, and the weight ratio of phosphorus to nitrogen is about 0.75. From these facts, nitrifying microbes are suggested to be phosphorus-accumulating microorganisms (95,96).

Water qualities and flow diagram for metal-finishing wastewater are shown in Table 5 and Figure 8, respectively. Since 1997, a denitrification plant using these gels (nitrifying tank volume, 10,000 m^2) has been operating in Tokyo (Fig. 9).

EVALUATION OF IMMOBILIZATION

The development of processes using immobilized microorganisms has made it possible to eliminate BOD and nitrogen in half the time (6 to 8 h), compared with conventional activated sludge. In addition, these processes maintained constantly high activity during winter. The treatment cost containing the depreciation is 80 to 90% of the activated sludge.

Table 4. Comparison of PEG-PPG/PVA Type Gels and PVA Type Gels

Species of gels	PEG-PPG/PVA type	PVA type
DO (mg/L)	2.8	6.0
NH_3-N value added (kg/m^3 day)	0.29	0.21
Nitrification velocity (kg/m^3 day)	0.21	0.15
Cells attached (mg/L)	6,920	1,620

To gain the advantages of immobilization, reductions are needed in the immobilized gel costs, particularly for large-scale operations, which include raw materials costs, immobilization facility, transportation cost, storage cost, and so on.

Wastewater treatment processes using immobilized

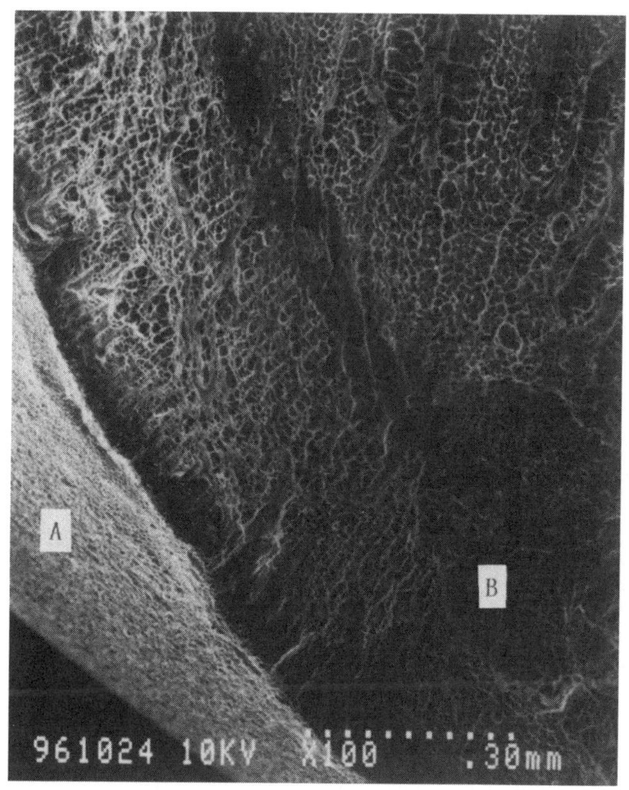

Figure 6. Scanning electron micrograph of photo-cross-linked resin gel. A = Gel surface. Convex is formed by phase separation of hydrophilic and hydrophobic structure. B = Cross-section of gel.

Table 5. Operation Conditions of the Pilot Plant

Item	Step 1	Step 2	Step 3
Influent flow rate (m^3/d)	25.0	25.0	28.6
HRT (h)	8.7	9.0	8.0
Anaerobic tank (m^3)	1.0	1.5	1.5
SRT (day)	16	11–14	8–11
Aerobic SRT (day)	5.7	3.7–4.7	2.7–3.7
Gel diameter (mm)	5.0	5.0	4.2
Charge ratio (%)	20	20	16 -n
DO (mg/L)	1.8–5.0	2.0–5.8	2.8–7.8
Influent BOD (mg/L)	49–151	45–110	48–118
Total N (mg/L)	14–34	13–47	16–45
Total P (mg/L)	1.2–2.9	2.0–5.8	2.8–7.8

HRT = Hydraulic retention time; SRT = sludge retention time; DO = dissolved oxygen.

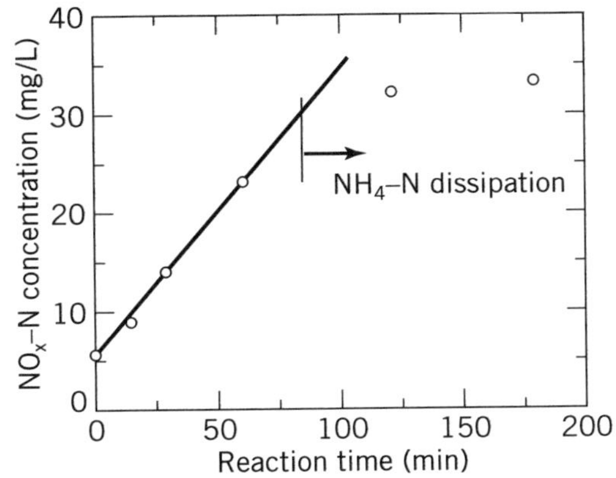

Figure 7. Nitrification rate of adsorpted gels.

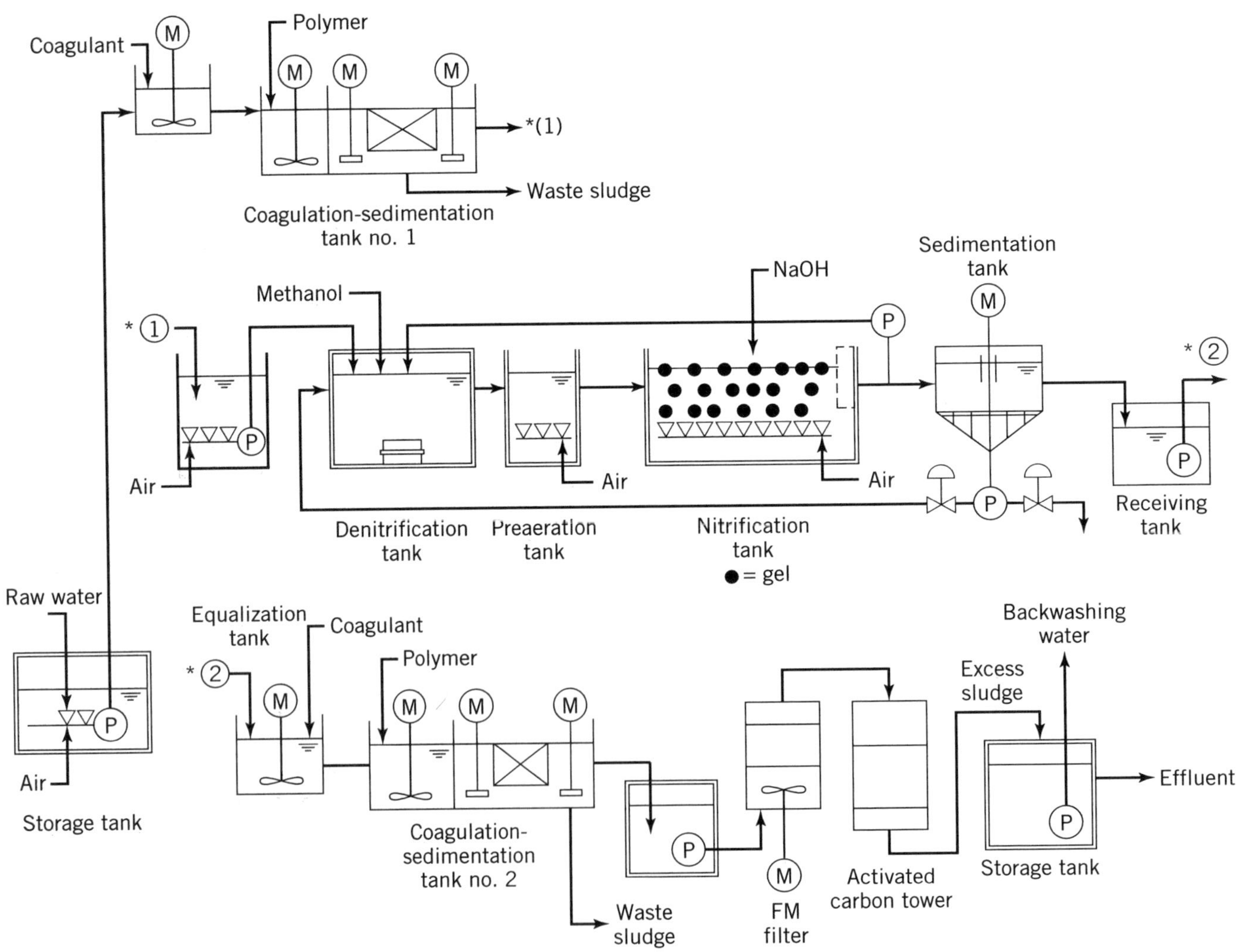

Figure 8. Flow diagram of metal finishing wastewater treatment process.

Figure 9. Photograph of denitrification plant using charged photo-cross-linked resin gels in Tokyo. Nitrifying tank volume, 10,000 m^3, Packing ratio of gel, 10%.

cells are still under development and will be further improved in the future. Particularly, it is expected that carrier gels will regulate their structure and form a particular reaction site in gels or on the surface of gels independent of the surroundings.

BIBLIOGRAPHY

1. J. Poels, P. van Assche, and W. Verstraete, *Biotechnol. Lett.* **6**, 747–752 (1984).
2. H. Bettman and H.J. Rehm, *Appl. Microbiol. Biotechnol.* **22**, 389–393 (1985).
3. A.M. Anselmo, M. Mateus, J.M.S. Cabral, and J.M. Novais, *Biotechnol. Lett.* **7**, 889–894 (1985).
4. H. Keweloh and H. Gruetzmacher, *Appl. Microbiol. Biotechnol.* **31**, 383–389 (1989).
5. J.E. Lin and H.Y. Wang, *J. Ferment. Bioeng.* **72**, 311–314 (1991).
6. T. Ogawa and C. Yatome, *Bull. Environ. Contam. Toxicol.* **44**, 561–566 (1990).
7. L.E. Macaskie and A.C.R. Dean, *Environ. Technol. Lett.* **5**, 177–186 (1984).
8. L.J. Michel, L.E. Macaskie, and A.C. Dean, *Biotechnol. Bioeng.* **28**, 1358–1365 (1986).
9. L.E. Macaskie and A.C.R. Dean, *Enzyme Microbiol. Technol.* **9**, 2–4 (1987).
10. K.L. Sublette, *Appl. Biochem. Biotechnol.* **20/21**, 675–686 (1989).
11. F.K. Steven, B.L. Shari, and R.R. Channing, *Chem. Eng. Sci.* **40**, 1321–1354 (1985).
12. I. Chibata and T. Tosa, *Appl. Biochem. Bioeng.* **4**, 1–9 (1983).
13. J. Klein and F. Wagner, *Appl. Biochem. Bioeng.* **4**, 11–51 (1983).
14. B. Atkinson and H.W. Fowler, *Adv. Biochem. Eng.* **3**, 221–277 (1974).
15. W.G. Characklis and K.E. Cooksey, *Adv. Appl. Microbiol.* **29**, 93–138 (1983).
16. M.S. Switzenbaum, *Enzyme Microb. Technol.* **5**, 242–250 (1983).
17. M. Henze and P. Harremoes, *Water Sci. Technol.* **15**, 1–101 (1983).
18. G. Bitton and K.C. Marshall, *Adsorption of Microorganisms to Surfaces*, Wiley, New York, 1980.
19. K. Elliot, M. O'Connor, and J. Whelan, *Adhesion and Microorganism Pathogenicity*, Pitman Medical, London, 1981.
20. R.C.W. Berkeley, J.M. Lynch, L. Melling, P.R. Rutter, and B. Vincent eds., *Microbial Adhesion to Surfaces*, Ellis Horwood, Chichester, U.K., 1980.
21. D.C. Ellwood, J. Melling, and P. Rutter, *Adhesion of Microorganisms to Surfaces*, Academic Press, London, 1979.
22. B. Atkinson and I.S. Daoud, *Adv. Biochem. Eng.* **4**, 41–124 (1976).
23. B. Metz and N.W.F. Kossen, *Biotechnol. Bioeng.* **19**, 781–799 (1977).
24. M. Fletcher, M.J. Latham, J.M. Lynch, and P.R. Rutter eds., *Microbial Adhesion to Surfaces*, Ellis Horwood, Chichester, U.K., 1980.
25. S.L. Daniels, *Dev. Ind. Microbiol.* **13**, 211–253 (1972).
26. B. Atkinson and I.S. Daoud, *Adv. Biochem. Eng.* **4**, 41–124 (1976).
27. R.A. Messing, C. Ghommidh, and J.M. Navarro, *Biotechnol. Bioeng.* **24**, 1991–1999 (1982).
28. S. Birnbaum, P.O. Larsson, and K. Mosbach, in W.H. Scouten ed., *Solid Phase Biochemistry: Analytical and Synthetic Aspects*, Wiley, New York, 1983.
29. G.M. Black, C. Webb, T.M. Matthews, and B. Atkinson, *Biotechnol. Bioeng.* **26**, 134–141 (1984).
30. I. Chibata and T. Tosa, *Annu. Rev. Biophys. Bioeng.* **10**, 197–216 (1981).
31. J. Tramper and A.W.A. de Man, *Enzyme Microb. Technol.* **8**, 472–476 (1986).
32. J. Tramper and D.R.J. Grootjen, *Enzyme Microb. Technol.* **8**, 477–480 (1986).
33. T. Sumino, H. Nakamura, and N. Mori, in A. Tanaka, T. Tosa, and T. Kobayashi eds., *Industrial Application of Immobilized Biocatalysts*, Dekker, New York, 1992, pp. 377–391.
34. W.R. Vieth and K. Venkatasubramanian, in *Immobilized Microbial Cells*, ACS Symposium Series 106, American Chemical Society, Washington D.C., 1979, p. 1.
35. H. Bettmann and H.J. Rehm, *Appl. Microbiol. Biotechnol.* **22**, 389–393 (1985).
36. H. Tanaka, M. Matsumura, and A.I. Veliky, *Biotechnol. Bioeng.* **26**, 53–58 (1984).
37. N.G. Starostina, K.A. Lusta, and B.A. Fikhte, *Eur. J. Appl. Microbiol. Biotechnol.* **18**, 264–270 (1983).
38. I. Chibata and L.B. Wingard, *Appl. Biochem. Bioeng.*, vol. 4, *Immobilized Microbial Cells*, Academic Press, New York, 1983, p. 355.
39. S. Fukui and A. Tanaka, *Annu. Rev. Microbiol.* **36**, 145–172 (1982).
40. C.G. van Ginkel, J. Tramper, K.Ch.A.M. Luyben, and A. Klapwijk, *Enzyme Microb. Technol.* **5**, 297–303 (1983).
41. R.H. Wijffels and J. Tramper, *Appl. Microbiol. Biotechnol.* **32**, 108–112 (1989).

42. E. Kokufuta, M. Shimohashi, and I. Nakamura, *J. Ferment. Technol.* **65**, 359–361 (1987).
43. C.M. Hooijmans, S.G.M. Geraats, E.W.J. van Neil, L.A. Roberttson, J.J. Heijnen, and K.Ch.A.M. Luyb, *Biotechnol. Bioeng.* **36**, 931–939 (1990).
44. P. Chevalier and J. de la Noue, *Enzyme Microb. Technol.* **7**, 621–624 (1985).
45. P. Chevalier and J. de la Noue, *Biotechnol. Lett.* **7**, 395–400 (1985).
46. S.S. Marwaha, J.F. Kennedy, P.K. Khanna, H.K. Tewari, and A. Redhu, in J.A.M. de Bont, J. Visser, B. Mattiasson, and J. Tramper eds., *Physiology of Immobilized Cells*, Elsevier, Amsterdam, 1990, p. 265.
47. K.C. Chen and Y.F. Lin, *Water Sci. Technol.* **28**, 159–164 (1993).
48. I. Nilsson, S. Ohlson, L. Haggstrom, N. Molin, and K. Mosbach, *Eur. J. Appl. Microbiol. Biotechnol.* **10**, 261–274 (1980).
49. A.B. Salleh and M. Aminuddin, *Pertanika* **4**, 21–24 (1981).
50. J. Tramper and A.W.A. de Man, *Enzyme Microb. Technol.* **8**, 472–476 (1986).
51. T. Willke and K.D. Vorlop, *Symposium Immobilised Cells: Basic and Applications, The Working Party Applied Biocatalysis of the European Federation of Biotechnology*, Noordwijkerhout, The Netherlands, November 26–29, 1995.
52. C.M. Lee, C. Lu, Y.H. Yin, and P.C. Chen, *Symposium Immobilised Cells: Basic and Applications, The Working Party Applied Biocatalysis of the European Federation of Biotechnology*, Noordwijkerhout, The Netherlands, November 26–29, 1995.
53. K. Tanaka, T. Sumino, H. Nakamura, and H. Emori, *Symposium Immobilised Cells: Basic and Applications, The Working Party Applied Biocatalysis of the European Federation of Biotechnology*, Noordwijkerhout, The Netherlands, November 26–29, 1995, pp. 622–632.
54. J. Klein and H. Eng, *Biotechnol. Lett.* **1**, 171–176 (1979).
55. K.Y.A. Wu and K.D. Wisecarver, *Biotechnol. Bioeng.* **39**, 447–449 (1992).
56. A.P. Jarvis and T.A. Grdina, *BioTechniques* **1**, 22/24–27 (1983).
57. R.R. Mohan and N.N. Li, *Biotechnol. Bioeng.* **17**, 1137–1156 (1975).
58. M.W. Tenny and F.H. Verhoff, *Biotechnol. Bioeng.* **15**, 1045–1073 (1973).
59. T. Iida, H. Izumida, Y. Akagi, and M. Sakamoto, *J. Ferment. Bioeng.* **75**, 28–31 (1993).
60. H.L. Rebecca, D.H. Altreuter, and F.T. Gentile, *Biotechnol. Bioeng.* **50**, 365–373 (1996).
61. J. Kein and G. Manecke, *Enzyme Eng.* **6**, 181–189 (1982).
62. D.A. Sirotti and A. Emery, *Biotechnol. Bioeng.* **25**, 1773–1779 (1983).
63. K. Williamson and P.L. McCarty, *J. Water Pollut. Control. Fed.* **48**, 281–296 (1976).
64. J.V. Matson and W.G. Characklis, *Water Res.* **10**, 877–885 (1976).
65. M. Onuma and T. Omura, *Water Sci. Technol.* **14**, 553–568 (1982).
66. E.J.T.M. Leenen, A.A. Boogert, A.A.M. van Lammeren, J. Tramper, and R.H. Wijffels, *Biotechnol. Bioeng.* **55**, 630–641 (1997).
67. T. Sumino, H. Nakamura, and N. Mori, *J. Ferment. Bioeng.* **72**, 141–143 (1991).
68. T. Sumino, H. Nakamura, N. Mori, Y. Kawaguchi, and M. Toda, *Appl. Microbiol. Biotechnol.* **36**, 556–560 (1992).
69. I. Nilsson, S. Ohlson, L. Haggstrom, N. Molin, and K. Mosbach, *Eur. J. Appl. Microbiol. Biotechnol.* **10**, 262–274 (1980).
70. I. Nilsson and S. Ohlson, *Eur. J. Appl. Microbiol. Biotechnol.* **14**, 86–90 (1982).
71. J. Tramper and A.W.A. de Man, *Enzyme Microb. Technol.* **8**, 472–476 (1986).
72. E. Kokufuta, M. Shimohashi, and I. Nakamura, *J. Ferment. Technol.* **64**, 533–538 (1986).
73. S. Cizinska, V. Vojtisek, J. Maixner, J. Barta, and V. Krumphanzl, *Biotechnol. Lett.* **7**, 737–742 (1985).
74. M. Taya, H. Miura, K. Uchiyama, S. Iijima, and T. Kobayashi, *Hakkoukougaku* **66**, 235–240 (1988).
75. Y.F. Lin and K.C. Chen, *Water Res.* **29**, 35–43 (1995).
76. F. Cecen and E. Orak, *J. Chem. Tech. Biotechnol.* **65**, 229–238 (1996).
77. Japan Sewerage Division, Evaluation Report, *Pegasus*, 1994.
78. T. Sumino, K. Noto, T. Ogasawara, N. Hashimoto, and Y. Suwa, *Proceedings of the Water Environment Federation 70th Annual Conference Exposition*, Chicago, 1997, pp. 165–172.
79. T. Sumino, H. Nakamura, N. Mori, and Y. Kawaguchi, *J. Ferment. Technol.* **73**, 37–42 (1992).
80. T. Takatera, H. Izumida, T. Iida, and T. Kobayashi, *Ferment. Bioeng. Conf. Proc.* 1987 p. 161.
81. M. Fletcher, M.J. Latham, J.M. Lynch, and P.R. Rutter, in R.C.W. Berkeley, J.M. Lynch, J. Melling, P.R. Rutter, and B. Vincent eds., *Microbial Adhesion to Surfaces*, Ellis Horwood, Chichester, 1980.
82. S.L. Daniels, *Dev. Ind. Microbiol.* **13**, 211–253 (1972).
83. M. Matsumura, T. Yamamoto, K. Shinabe, and K. Yasuda, *Water Res.* **31**, 1027–1034 (1997).
84. H. Deguchi and M. Kashiwaya, *Water Sci. Technol.* **30**, 143–149 (1994).
85. G. Claus and H.J. Kutzner, *Appl. Microbiol. Biotechnol.* **22**, 289–296 (1985).
86. T.H. Lessel, *Water Sci. Technol.* **23**, 825–834 (1991).
87. T. Iida, *Industrial Application of Immobilized Biocatalysts*, Dekker, New York, 1992, pp. 163–182.
88. Y. Lue, *Colloids Surface* **5**, 213–219 (1995).
89. P.G. Rouxhet and N. Mozes, *Water Sci. Technol.* **22**, 1–16 (1990).
90. M. Kikuta and T. Iida, *RadTech Asia '97 Radiation Curing Conference Proceedings*, Yokohama, 1997, pp. 820–823.
91. U.S. Pat. 4,605,622 (August 12, 1986), E. Hasegawa, T. Iida, M. Sakamoto (to Kansai Paint Co., Ltd.).
92. G.A. Stanley, T.J. Hobley, and N.B. Pamment, *Biotechnol. Bioeng.* **53**, 88–99 (1997).
93. K. Mishima, T. Nishimura, M. Goi, and N. Katsukura, *Water Sci. Technol.* **34**, 137–143 (1996).
94. K. Mishima, T. Yamada, and H. Itoh, *Ebara Time Signal* **170**, 81–85 (1996).
95. T. Kuba, E. Murnleitner, M.C.M. van Loosdrecht, and J.J. Heijnen, *Biotechnol. Bioeng.* **52**, 685–695 (1996).
96. E. Murnleitner, T. Kuba, M.C.M. van Loosdrecht, and J.J. Heijnen, *Biotechnol. Bioeng.* **54**, 434–450 (1997).

WINE PRODUCTION

O. Colagrande
Università Cattolica Sacro Cuore
Piacenza, Italy

KEY WORDS

Alcoholic fermentation
Colloidal dispersion
Enzyme treatments
Glucan
Malolactic fermentation
Organoleptic
Polyphenols
Saccharomyces cerevisiae

OUTLINE

INTRODUCTION

During the past 20 years there have been considerable developments in winemaking techniques, with the introduction of processes and machinery borrowed from other sectors of the food industry. These developments have affected all phases of wine production, wine being the end product of a technological chain that includes the preparation and treatment of the must, fermentation, aging, and bottling. Thus, technological progress has improved the quality of the wine produced. Wine quality is measured by finesse, intensity, and originality in taste and smell and by microbiological and physicochemical stability (1–4).

The main phases of production of white and red wines are illustrated schematically in Figures 1 and 2. Figure 3 shows those technological innovations that may be of practical use. Among these, during the past few years particular attention has been paid to enzyme treatments and to the use of selected fermentation starters (both free and immobilized). Some of these processes have already been applied to production, and others are still in the experimental stage.

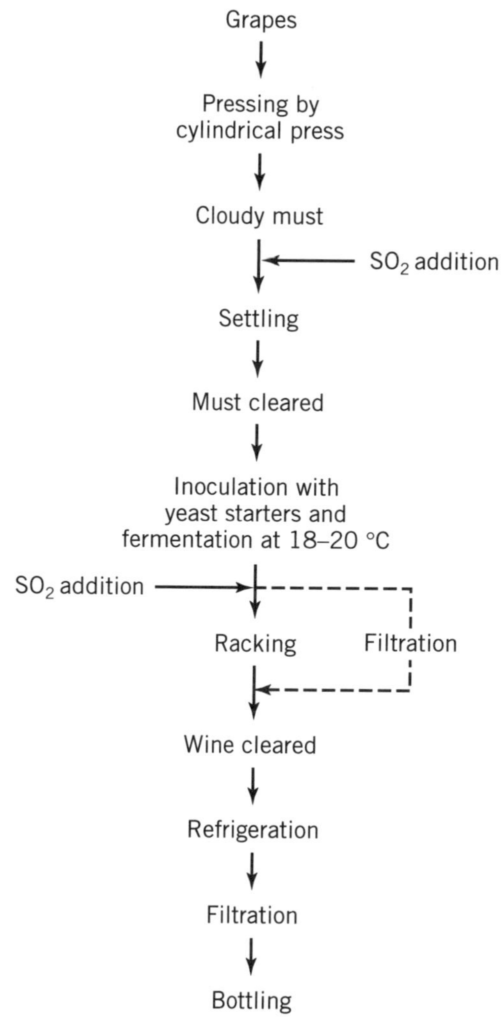

Figure 1. Scheme of a process for white wine preparation (fermentation with skins off). Terms are defined as follows. Pressing by cylindrical press: The operation of applying pressure to grapes to extract the liquid, which allows the juice to separate from the solids (skins, seeds, and stems). Settling: The operation of clarifying must at T = 5 to 10 °C (must can also be clarified by centrifugation or filtration). Refrigeration: The operation of cooling to 0 °C or below, used to hasten the precipitation of potassium bitartrate and to improve the stability of the wine.

ENZYME TREATMENTS

Winemaking is a biotechnological process in which enzymes play a fundamental role and in which the use of exogenous enzyme preparations makes it possible to overcome the problem of insufficient activity of endogenous enzymes in the grapes. The enzymic treatments have a multiple purpose in that the changes made to the grapes, the

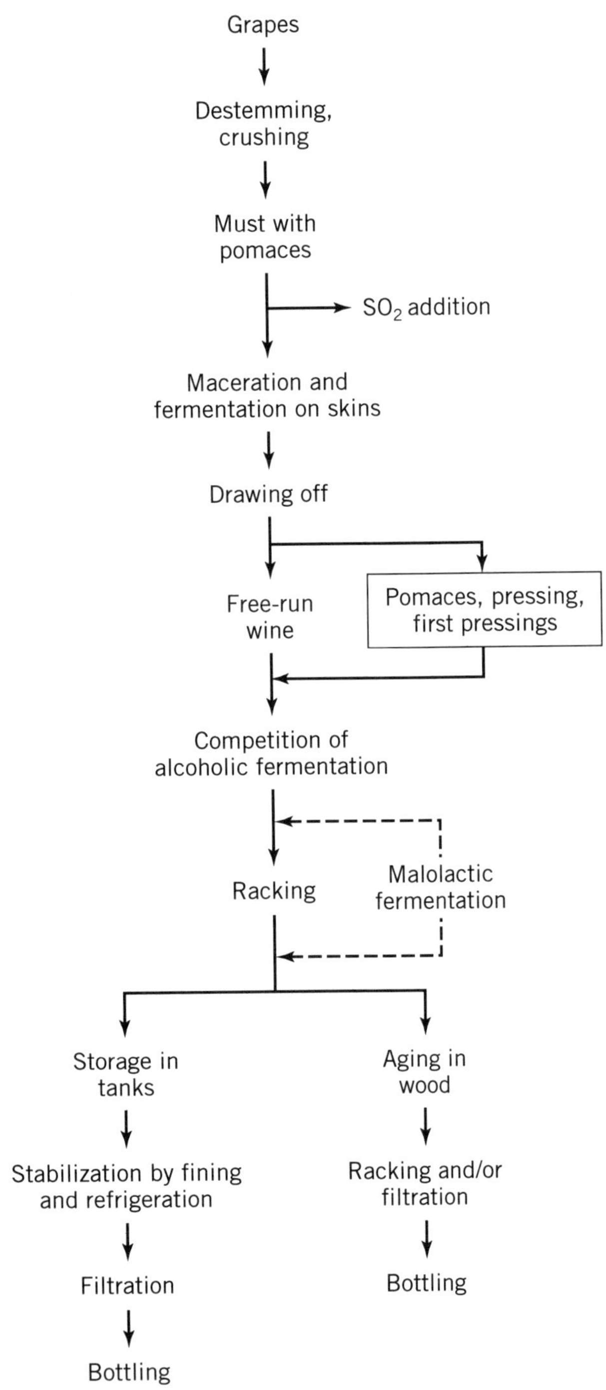

Figure 2. Scheme of a process for red wine preparation (fermentation with skins on). Terms are defined as follows. Destemming-crushing: The operation that mechanically removes the stems from the grapes and liberates the juice. Maceration: Contact of the must with the solid parts (skins and seeds) of the berries to enrich the juice with skin constituents. The maceration conditions (temperature, duration, pumping oversystem) change in relation to the grape variety and to the organoleptic characteristics of the wine produced. Alcoholic fermentation: With natural or selected yeasts. Malolactic fermentation: The biological decomposition of malic acid can be carried out soon after alcoholic fermentation or during storage (aging).

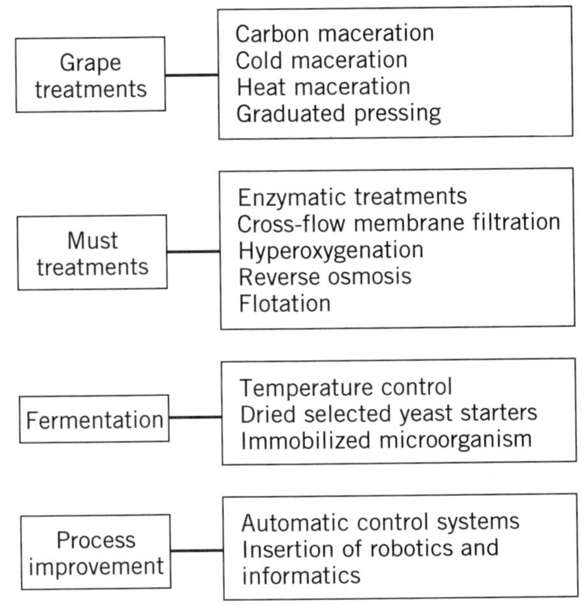

Figure 3. Technological improvement in the enological industry (5).

must, and the wine have an influence on the processes of clarification and filtration, the processes of extraction of color and aroma, and the stability of the wine (6–10). It must, however, be remembered that the use of enzymes in the wine industry, apart from pectolytic enzymes, remains extremely limited for a series of reasons that can be summarized as follows:

- The traditionalism of the wine industry, which is still anchored to classical methods
- Limitations of enzymic activity connected with the composition of the must and the wine (low pH, ethanol concentration, presence of tannins, and a fairly low working temperature compared to the optimum for enzyme preparations)
- Legal restrictions

However, during the past few years, research has suggested the use of enzymes that catalyze the transformation of glucan (glucanase), proteins (peptidase), and polyphenols (laccase, peroxidase, phenolase, and tannase), all of which are compounds that condition the stability and organoleptic characteristics of wines. Particular attention has been paid to glycosidases that are able to free odorous compounds by hydrolyzing terpene glycosides.

Pectolytic Enzymes

Pectolytic enzymes have been widely used for many years, because they act on the pectins. Pectins are complexes of acids (homogalacturonan and rhamnogalacturonan) and neutral polysaccharides (arabinan, galactan, and arabinogalactan) that are found in the must in colloid form. They act as protective colloids, causing problems in the clarification and stabilization (11–13).

Enzymes that can act on the pectic substances are usually classified into two groups, as shown in Figure 4 (14,15): de-esterifiers (pectinmethylesterase) and depolymerases that catalyze both hydrolysis (polygalacturonase) and β-elimination reactions (pectin lyase and pectate lyase). Purified enzymic preparations from cultures of microorganisms normally contain polygalacturonase and lyases associated with hemicellulase and cellulase. However, the presence of pectin methylesterase must not be too great in order to limit the liberation of methanol, which is found esterified with the carboxylic groups of the galacturonic acid in the pectic molecule (16). Pectolytic enzymes are used at vintage time in the preparation of both white and red wines. Italian law permits the use of pectolytic enzymes in the processing of pressed grapes, the must to be used for producing white wine, musts and wines obtained using the thermovinification system, and wines after they have been separated from the pomaces (Ministerial Decree of 8 September 1976, G.U.I.). In the context of numerous experiments that have been carried out during the past few years as well as the development of new commercial products, various authors have evaluated the influence of the use of these enzymes on the extraction of juice, color, and aroma and on the clarification and filtration processes. Ough and Crowell (17) investigated the effect of the use of some exogenous pectolytic enzymes at pressing on the must yields, turbidity, and solids, using different varieties of grapes. The enzyme-treated samples gave better juice yields than the untreated samples. The amount of sediment after overnight settling was less for the controls. After racking and centrifuging, the enzyme-treated juice samples were clearer and contained fewer insoluble solids.

The addition of pectolytic enzymes has a favorable influence on the clarification of musts and, consequently, also has a positive effect on such parameters as viscosity and filterability (18,19). The treatment of crushed grapes with pectolytic enzymes in the production of red wine leads to a more rapid and complete breakdown of the cell wall of the skin. This leads to greater extraction of anthocyanins and, thus, to an increase in color, although the results may vary according the cultivar of grapes used (20).

Rational use of pectolytic enzymes, especially in the case of grapes particularly rich in pectic substances (e.g., muscat) makes it possible to improve the economics of the process (e.g., through reduction in time required, better use of equipment, and reduction in loss of must and wine) without negatively altering or modifying the quality characteristics of the final product (21,22).

β-Glucanase

Glucan is a polysaccharide with a high molecular weight (M_r 10^5 – 10^6), consisting entirely of glucose subunits. It is a (1–3:1–6)-β-D-glucan. It is found in must and wines from grapes infected by *Botrytis cinerea* (23). In these wines, the glucan is of great technological importance because it tends to clog the filters if a quantity is present by forming a glutinous layer on them. This causes considerable problems in clarification and filtration. In order to solve this problem of clogging, an enzyme treatment was set up in the 1980s, based on β-glucanase, which hydrolyzes the glucan excreted by the *Botrytis cinerea* (25).

Experiments carried out by Villettaz (9,26,27) and Wucherpfenning and Dietrich (28) to evaluate the efficacy of β-glucanase have demonstrated the technological and economic importance of this enzyme in improving the filterability of wines produced from grapes affected by *Botrytis*.

β-glucanase has been included in the practices and treatments permitted by the EEC, although the conditions for use have yet to be defined (29,30).

Proteolytic enzymes

In winemaking, treatments are carried out to lower the protein content, because in must and wine, proteins are found in colloidal dispersion, which can become insoluble, forming haze and precipitates (31,32). When this occurs during processing it causes delays, but if it occurs when the wine is already on sale it causes economic loss. This phenomenon occurs predominantly in white wines and only very rarely in red wines. Red wines contain tannins, which are believed to react with the proteins to form insoluble tannoprotein compounds. During fermentation and aging, this leads to a drastic reduction in the protein levels (33). Although the proteins responsible for the instability have not yet been well characterized, it seems that the protein fractions of lower MW (12,600 and 20,000 to 30,000) and lower P_i (4.1 to 5.8) and glycoproteins are the major and most important fractions contributing to protein instability in wines (34). Treatments with proteolytic enzymes have been tried as an alternative to using traditional finings such as bentonite. These enzymes cause hydrolysis of the proteins into amino acids or short-chain peptides that are completely soluble in the wine (35–37); however, intervention with exogenous enzymes is not free from problems. Many of the traditional proteases (papain, pepsin, and trypsin) are not suitable for use in wine; they require high temperatures that are deleterious to wine quality, and they do not have their optimal pH in the range of 3 to 4, which is ideal for wines. For these reasons, studies have mostly been directed toward the use of acid proteases of fungal origin in free and immobilized forms; however, so far the results have not always been positive. Modra and coworkers (38) assessed the action of commercial peptidases on proteins isolated from muscat grapes. This treatment modified the HPLC elution profile of a high relative molecular weight fraction containing protein. The most generally discernible change was a decrease in the M_r 23.000 peak, with a concomitant increase in the 10.000 M_r peak, and a loss of resolution in both. The change in protein profile varied among the peptidases, but none of the enzymes gave complete proteolysis after 7 days. The tests carried out on muscat must confirmed that the peptidases modified the proteins; however, their efficacy was not sufficient to make the wine stable, and treatment with bentonite was required.

Woiwodow et al. (37) tested the action of an acid protease from *Aspergillus niger* immobilized on Sylochrom C-80 on the proteins of white table wines. Experiments were carried out with different quantities of enzyme (2 to

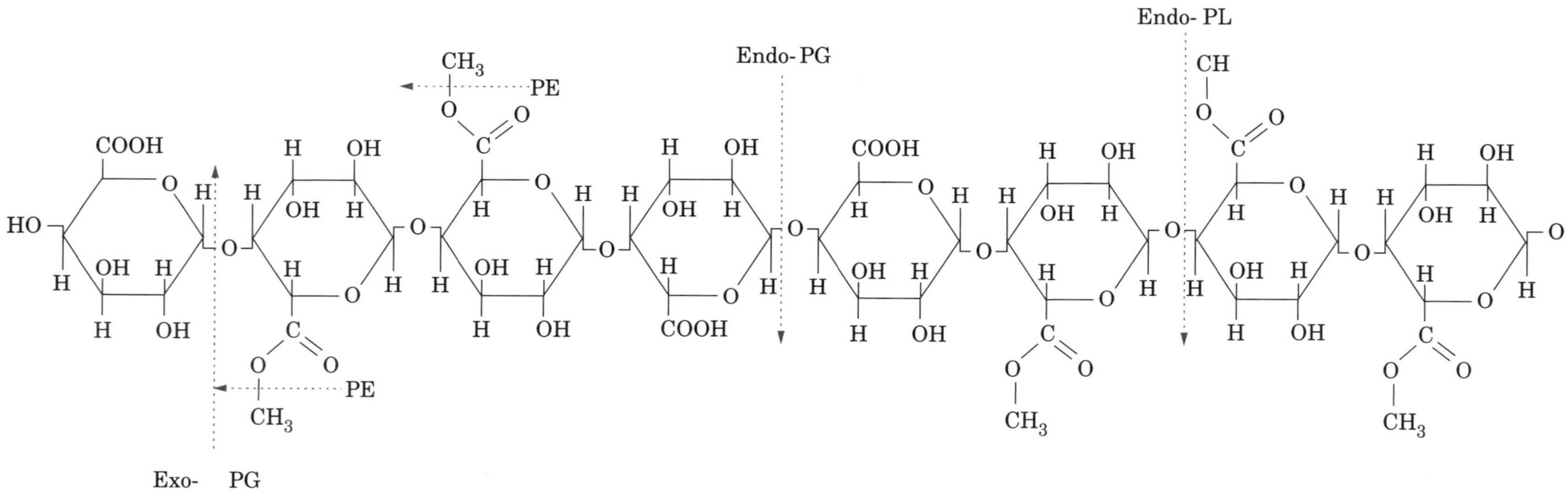

Name	Elective substrate	Polygalacturonases (PG) Mode of action	End products
Pectinesterase (PE)	Exo-PG	Endo-PG	Endopectinlyase (PL)
Pectin	Pectate	Pectate	Pectine
Hydrolysis of ester bond	Hydrolysis of α-1,4 bonds	Random hydrolysis of α-1,4 bonds	
Pectate and methanol	D-galacturonate	Oligogalacturonates	Methyl esters of unsaturated oligogalacturonates

Figure 4. Enzymic degradation of pectins and some enzyme characteristics.

10 g/L) and with different times of contact (2 to 48 h) at a temperature of 23 to 25°C. Complete stability of the wine was reached after prolonged treatment (48 h) with very high quantities of immobilized enzyme (10 g/L). In other tests, electrophoresis of the proteins isolated from the wine before and after treatment showed a diminution in both number and intensity of protein fractions, but the wine was not completely stable. Bakalinsky and Boulton (35) studied the use of food-grade acid protease (from *A. niger*) covalently linked to agarose spheres of 100 μm. Four types of wine were treated at temperatures of 30 and 37 °C in a recirculating flow packed-bed reactor containing 226 mg of dry weight conjugate. At the end of the treatment, the wines were subjected to heat-stability testing. The results showed that the enzyme, although active in all of the wines tested, achieved protein stability only in Riesling wine. The Riesling was the only wine whose pH was adjusted to 3.0 for treatment, whereas the other wines were treated at their natural pH levels (Sauvignon blanc, 3.65; chardonnay, 3.57; Gewürztraminer 3.59). The pH adjustment may have been the cause of the successful stabilization.

That immobilized protease was active in the wine but failed to have any significant effect on the protein stability may indicate that the proteases do not hydrolyze the protein fractions responsible for the instability or that the hydrolysis occurs along with the formation of products that are still unstable. If the potentially unstable wine proteins were complexed with polyphenolic compounds, one would expect them to be more resistant to enzymic hydrolysis. Currently, the use of free or immobilized proteases to stabilize wine requires more information on the nature of the proteins, their interactions with the other components of wine, and the conditions that cause them to become insoluble. Furthermore, sensory evaluation of the treated wine is recommended to assess possible detrimental alterations in flavor caused by the probable increase in peptide content. Small hydrophobic peptides, in particular, are recognized as having a bitter flavor (39).

Enzymes with Specific Action for Polyphenols

Phenols in wine predominantly arise from the grapes and are primarily flavonoid. They are crucial components in wines because of their basic importance to the color and flavor. Flavonoids (catechins, proanthocyanidins) are also responsible for discoloration, turbidity, and flavor changes in wines. Review of the importance of phenolic compounds in grapes and wine has recently been published (40,41). Chemical and enzymic oxidation of phenolic compounds leads to the formation of highly reactive intermediates (quinones) that interact with amino acids to reduce sugars and alcohol and give rise to chromophores and volatile substances (40,42,43). This results in a loss of freshness, anywhere from yellowing to browning of the wine, and turbidity (42,44). A summary of these reactions can be found in Figure 5.

The removal of polyphenols is a common practice in winemaking processes to prevent discoloration and to obtain stabilization against oxidation, especially in white and rose wines (45). Wine technology tends toward the adoption of systems for extraction of the must that limit the solubilization of oxidizable substances (46) and toward the use of finings (gelatin and casein), adsorbants (active carbon), and polyvinylpolypyrrolidone (PVPP) (47,48). The use of these additives in wine production is aimed at reduction of the level of polyphenols and compounds that catalyze their transformation, particularly enzymes and heavy metals. To limit the amount of treatment with chemical-physical agents, it has been suggested that enzymes active against polyphenols (oxydases, hydrolases, and transferases) be used in the prefermentative phase. The aim of this is to set off a process of enzymic oxidation of polyphenols in the must under controlled conditions, so as to destabilize them and accelerate the process of polymerization and flocculation. Enzymes and reactions are shown in Table 1. The possibility of eliminating or achieving reactivity loss of the polyphenols responsible for wine instability has been tested on the bench and on a pilot scale using enzymic preparations containing, respectively, tannase (from *Aspergillus* sp.), phenolase (prepared as an extract or acetoin powder from *Agaricus campestris*), and laccase (from *Polyporus versicolor* in a liquid culture, partially purified) (47,49).

These different enzyme treatments were compared with each other, and some were then compared with conventional systems of stabilization. The polyphenol concentration in wines prepared from the same, but differently processed, pinot grapes showed that laccase was more effective than the other enzymes tested and produced a wine with a more stable color (Table 2). The possibility of using laccase was checked on a pilot scale with the must from Trebbiano grapes. Laccase treatment was compared with clarification carried out with casein, active carbon, and bentonite, with or without sulphur dioxide. The use of the enzyme was shown to be highly effective, preferable or practically identical to the results from traditional processing. The wines treated with laccase had different and better organoleptic maderization-resistant features (47). In the case of muscat and Riesling, the action of laccase (8,300 units/L) was compared with that of fining agents (gelatin/SiO_2) (50). The results obtained with muscat must without SO_2 and Riesling must with SO_2 (150 ppm) (Table 3) shows that the enzymic treatment is highly effective, preferable, or practically identical to traditional processing, especially when using must without sulphur dioxide (Table 3).

These interesting results allow us to conclude that laccase prefermentative treatments make it possible to eliminate instability in white wines caused by oxidizable polyphenols. Obviously, the process of winemaking must include physical (e.g., ultrafiltration) or chemical (e.g., addition of sulphur dioxide, deproteinization) systems that make it possible to eliminate the enzyme. The enzymic treatment coupled with a filtration system (filtration by diatomaceous earth, PVPP, or silica of ultrafiltration) is probably a technique that gives sufficient stability for white wines.

Glycosidases Acting on Precursors of the Aroma

The aroma of grapes includes volatile free odorous substances, especially terpenes (linalool, geraniol, nerol, citro-

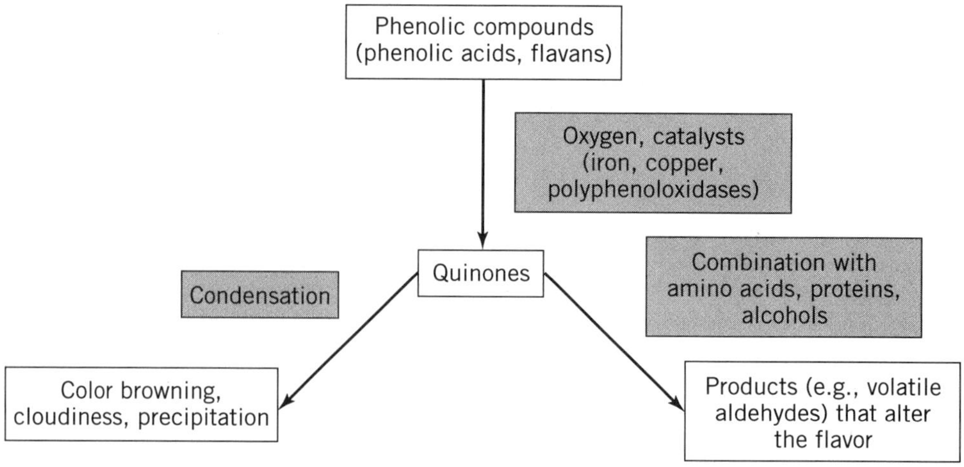

Figure 5. Possible reactions of maderization.

Table 1. Phenolic Enzymes

Oxidases
Peroxidase (EC 1.11.1.7)
Donor (catecol) + H_2O_2, oxidized donor (quinone) + $2H_2O$
o-Diphenoloxidase (catecholoxidase, polyphenoloxidase, phenolase, tyrosinase, cresolase[a]) (EC 1.10.3.1)
2 *o*-Diphenol + O_2; 2 *o*-quinone + $2H_2O$
p-Diphenoloxidase (laccase) (EC 1.10.3.2)
2 *p*-Diphenol + O_2; 2 *p*-quinone + 2 H_2O
Hydrolases
Tannin acylhydrolase (tannase) (EC 3.1.1.20)
Galloylglucose, gallic acid + glucose
Anthocyanase (EC 3.1.1)
Anthocyanidin 3-monoglycoside, anthocianidin + hexose

[a]Monophenol + O_2 + $2e^-$, *o*-diphenol + O^{2-}.

Table 3. Comparison of Untreated and Use of Fining or Laccase Enzyme on Total Phenolics and Catechins of Muscat and Riesling Must

	Untreated	Fining	Laccase
	Muscat (raw)		
Total phenolics	232	184	90
Catechins	43	56	30
Color E_{420mm} (×1,000)	226	99	89
	Riesling (150 ppm SO_2 added)		
Total phenolics	303	282	286
Catechins	29	28	29
Color E_{420mm} (×1,000)	83	53	60

Note: Total phenolics expressed as mg/L gallic acid, and catechins expressed as mg/L *d*-catechin.
Source: From Ref. 48.

Table 2. Polyphenol Content and Color Increases Found in Maderization Test of Pinot Wines Obtained by Various Enzyme Treatments

	No enzyme	Laccase	Phenolase	Tannase
	Polyphenol (mg / L)			
Total	339	310	319	333
Nontannic	267	261	284	284
	Color			
E_{420mm} (×1,000)	92	82	78	64
Increase due to maderization test	121	95	160	156

Source: From Ref. 47.

nellol, α-terpineol, linalool oxide) that play a fundamental role in giving character to certain cultivars, for example, muscat (51–53). In addition, grapes also include compounds called aroma precursors, which are glycosides mainly of linalool, nerol, and geraniol (for example 6-*O*-α-L-arabinofuranosyl-β-D-glucopyranosides, 6-*O*-α-L-rhamnopyranosyl-β-D-glucopyranosides and β-D-apiofuranosyl-β-D-glucopyranosides (54–57). The ratio between bound and free monoterpenols ranges from 1 to 5 in juice grape cultivars of muscat and Riesling and can be as high as 15 in the Gewürtztraminer variety (58,59). From a technological point of view, the existence of such bound forms is interesting because these glycosides, nonvolatile precursors, can give flavoring aglycones when hydrolytic reactions take place. High-temperature acid hydrolysis can be used, but it may lead to extensive rearrangements of the terpenols. A study of the transformations of the terpenic compounds has in fact shown that, as the pH of the aqueous medium diminishes, the hydration and cyclization of linalool increase (60).

Enzymic methods are attractive for increasing the concentration of free flavorants with minimal changes in the natural monoterpene composition. Recent studies have shown that enzymic hydrolysis occurs in two stages, following a sequential mechanism illustrated in Figure 6. In the first stage, the intersugar linkage is cleaved by arabinofuranosidase, rhamnopyranosidase, or apiofuranosidase (depending on the structure of aglycon moiety), and the corresponding monoterpenyl-β-D-glucosides are released. In the second stage, the liberation of monoterpenols takes place after action of a β-glucosidase on the previous monoterpenyl β-D-glucosides (62,63).

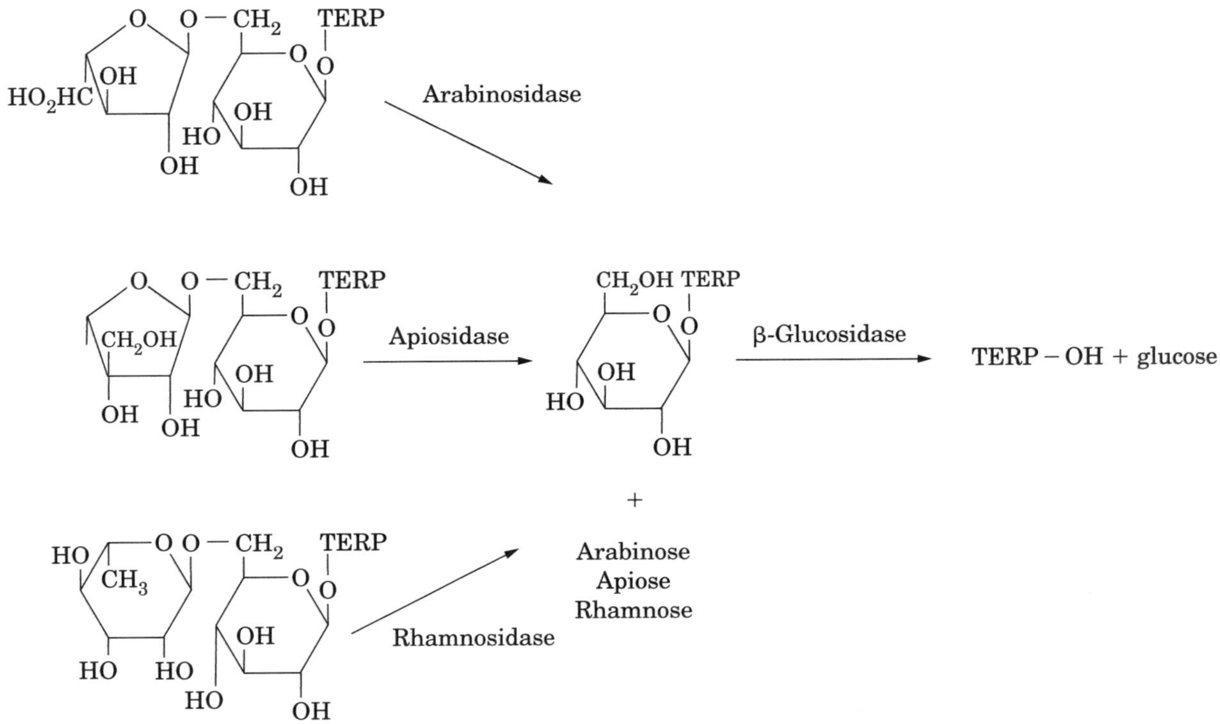

Figure 6. Sequential mechanism of terpene glycosides enzymic hydrolysis (61).

In the search for an enzymatic compound suitable for the hydrolysis of these precursors, different commercial products (pectinase, cellulase, hemicellulase, lypase, and protease) have been tested. Some of the preparations tested have contemporaneously good α-arabinofuranosidase, α-rhamnopyranosidase, and β-glucosidase action, but their efficacy in the must and wine is a function of numerous factors (64). The efficiency of hydrolysis of monoterpenyl β-D-glucosides by β-glucosidases was found to be dependent on the structure of the aglycon and the origin of the enzyme. Moreover, the efficiency of β-glucosidase is influenced by the concentration of glucose and gluconolactone, and in sweet wines, the hydrolysis is interrupted at the stage of the monoglucoside without freeing the odorous monomers. In the case of dry wines, on the other hand, there is a considerable increase of most of the free terpenes (65). Considering that the glycosidases used so far have not given the results hoped for, and that this is principally because of the properties of β-glucosidase, the hydrolytic activities of various plant and fungal β;-glucosidases toward terpene glycosides are being studied. The goal of these studies is to find an enzyme with large aglycon specificity and good stability to low pH as well as tolerance to glucose and ethanol.

USE OF SELECTED CULTURES AND INNOVATION IN THE FERMENTATION PROCESS

Of the two fermentation processes, alcoholic fermentation causes the transformation of sugars into ethyl alcohol and CO_2, whereas malolactic fermentation causes the transformation of malic acid into lactic acid and CO_2. However, both processes produce compounds (organic acid, alcohols, ester, and carbonilic compounds) that influence or determine the organoleptic characteristics of wines and so play a fundamental role in fixing wine quality (3,66–68). For this reason, the microorganisms responsible for these transformations, yeast and bacteria, have for some time been the object of great interest. The work of Pasteur (1886) was followed by numerous studies of the microflora of grapes and of the factors that regulate the equilibria between the various species and their metabolism. In fact, a great variety of microorganisms are present in grapes and musts, and the yeasts responsible for alcoholic fermentation are but a small part of them (69–71). The fermentative yeasts (particularly *Saccharomyces cerevisiae* and related species) generally predominate because the chemical composition of the must is more favorable for their development than for that of other eumycetes and schizomycetes (72,73). However, it is difficult to control alcoholic fermentation that develops naturally, because certain parameters (such as temperature, degree of oxygenation, degree of clarity of the must, and phenomena of microbial antagonism) can modify the natural selection and therefore condition the development and completion of the process. This alters the ratio between the principal (ethanol and CO_2) and secondary (acetic acid, succinic acid, acetoin, 2,3-butanediol, acetaldehyde, glycerol) products.

Because it is impossible to cover in this review even all the main aspects of microbial biochemistry involved in winemaking and the influence of certain environmental conditions on grape must fermentation, textbooks or reviews should be consulted for more detailed information on these topics (66,72,74–80).

Selected *Saccharomyces cerevisiae*

To overcome the problems mentioned earlier, for many years it has been standard practice, especially in the production of white wine, to use dry pure cultures of enologically suitable species (in particular *Saccharomyces cerevisiae* and all its physiological variants) with known properties. This ensures a quicker start to alcoholic fermentation (thanks to the massive addition of yeast), prevents the development of atypical microflora, and therefore eliminates the uncertainty and variation that occur in spontaneous alcoholic fermentation (50,66,76). Selected culture has therefore been the subject of many investigations that have considered the desirable characteristics of wine yeasts and their influence on the composition of wine and its organoleptic characteristics. Numerous properties of yeasts have been studied: some are always desirable, and others are always undesirable (Table 4) (81).

During the past few years, the wine industry has looked with interest at yeast genetics and strain development programs in order to obtain modified yeasts suitable for winemaking technologies that do not impair the flavor and bouquet of the final product. Yeast geneticists have carried out programs of genetic improvement to obtain new yeast strains that possess a high fermentation rate and high ethanol production as well as other relevant winemaking characteristics (82,83). This program has used both classical genetics (e.g., hybridization through conjugation between spores or aploid cells) and more recent biotechnological methods, such as somatic hybridization through fusion of protoplasts induced chemically or physically and transformation through the introduction of fragment of extraneous DNA using molecular vectors. Some reviews on this topic have recently been published by Falcone and Frontali (84) and by Pretorius and van der Westhuizen (85).

Thus far, it has been possible to eliminate foaming capacity of strains thought to be highly suitable for inducing alcoholic fermentation (86,87). Certain wine yeast strains produce proteins that are responsible for the foaming associated with fermentation. Excessive foaming during wine fermentation is an undesirable characteristic of wine yeast strains. The formation of a froth head can result in the loss of grape juice or reduce the capacity of plant equipment, because part of the fermentation vessel may have to be reserved to prevent the froth from spilling over. Foaminess also detracts from the visual appeal of the product (88).

Furthermore, it has been possible to effect genetic marking (89) to differentiate the selected strains from those of the indigenous microflora (always present in musts) and to give the killer phenotype to a strain (90–92) to increase its competitiveness with indigenous flora. The killer phenomenon is related to the ability of some yeasts to secrete toxic proteins that are lethal for so-called sensitive strains (93).

Particular attention is being given to the following possibilities: (*1*) introduction of the flocculent characteristic into yeast strains to facilitate the separation of cells from wine at the end of fermentation (94–97); (*2*) introducing malolactic enzymic activity into yeasts with the object of effecting alcoholic fermentation and malolactic fermentation with a single microorganism (yeast), a modification that would yield obvious technological advantages (98); (*3*) obtaining modified strains for the production of the aroma components (esters, fatty acids, and alcohols) (99); and (*4*) obtaining strains that do not produce H_2S, a component responsible for organoleptic alterations in the wine (100).

Achievements are still modest in the field of genetic improvement of enogical strains because of various factors: (*1*) the complex genetic background of industrial strains (ploidy, homothallism, poor sporulation, etc.) (101–104); (*2*) the impossibility of describing the genetic basis of most enological characteristics and, thus, of elaborating simple screening tests for them (85,105); and (*3*) the presence of indigenous microflora in the must that, unlike other substrata, cannot be subjected to sterilization without altering the wine characteristics.

Table 4. Some Characteristics of *Sacchromyces cerevisiae* Affecting the Winemaking Process

Desirable
Rapid initiation of fermentation
Efficient conversion of grape sugar to alcohol
Ability to conduct even fermentation
Ethanol tolerance
Ability to settle rapidly at the end of fermentation to aid clarification
Fermentation at low temperatures such as 10–14°C
Retention of viability during storage
Sulfur dioxide resistance (sulfur dioxide is normally used in winemaking)
Production of desirable fermentation bouquet and reproducible production of the correct levels of flavor and aroma compounds
Killer factor or resistance to killer toxins
Film-forming capacity
Undesirable
Production of sulfite, hydrogen sulfide, or mercaptan
High production of acetaldehyde, acetic acid (volatile acidity), ethyl acetate, and higher alcohols
foaming capacity
Production of urea, which can result in the formation of ethyl carbamate

Process Innovation in Alcoholic Fermentation

Another line of research that has developed in the field of enology is the search for and optimization of new processes that are able to meet the technological and organizational needs of the industry (greater productivity, reduction in the time needed for the process, reduction in capital and running costs). In this field, we can mention the use of yeasts immobilized in natural nontoxic polymers or contained in a microporous membrane and of the cell-recycle batch fermentation process (CRBF) for the preparation of white wines and sparkling wines. The advantages that are sought are summarized in this section.

Bottle-Fermented Sparkling Wines (Champenois Method). It would be desirable to eliminate the phase of remuage, with a considerable savings in labor and time, thanks to instant sedimentation of the immobilized yeasts by gravity. Sparkling wine produced by the champenois method is characterized by fermentation of wine in the bottle, to which sugars, yeasts and other fermenting agents have been added (Fig. 7). The bottles thus prepared are stacked horizontally in thermoregulated environment (12 to 18 °C) where the secondary fermentation is carried out. Once fermentation has terminated and the wine has been in contact with the yeast for at least a year, the bottles are turned frequently and inclined (an operation referred to as remuage), so that the yeast sediment slides toward the bottle neck and settles on the stopper. The sediment is then eliminated by degorgement. Remuage is both long and arduous in that it requires a large amount of labor over a long period of time (1 to 4 months) and the occupation of large wine cellars. By using immobilized yeasts, it is possible to move directly from the stacks to degorgement with a saving in production costs of about 80% when compared with the traditional method (106).

Production of Sparkling Wine in Tanks (Charmat Process). The possibility of using a high cellular concentration and

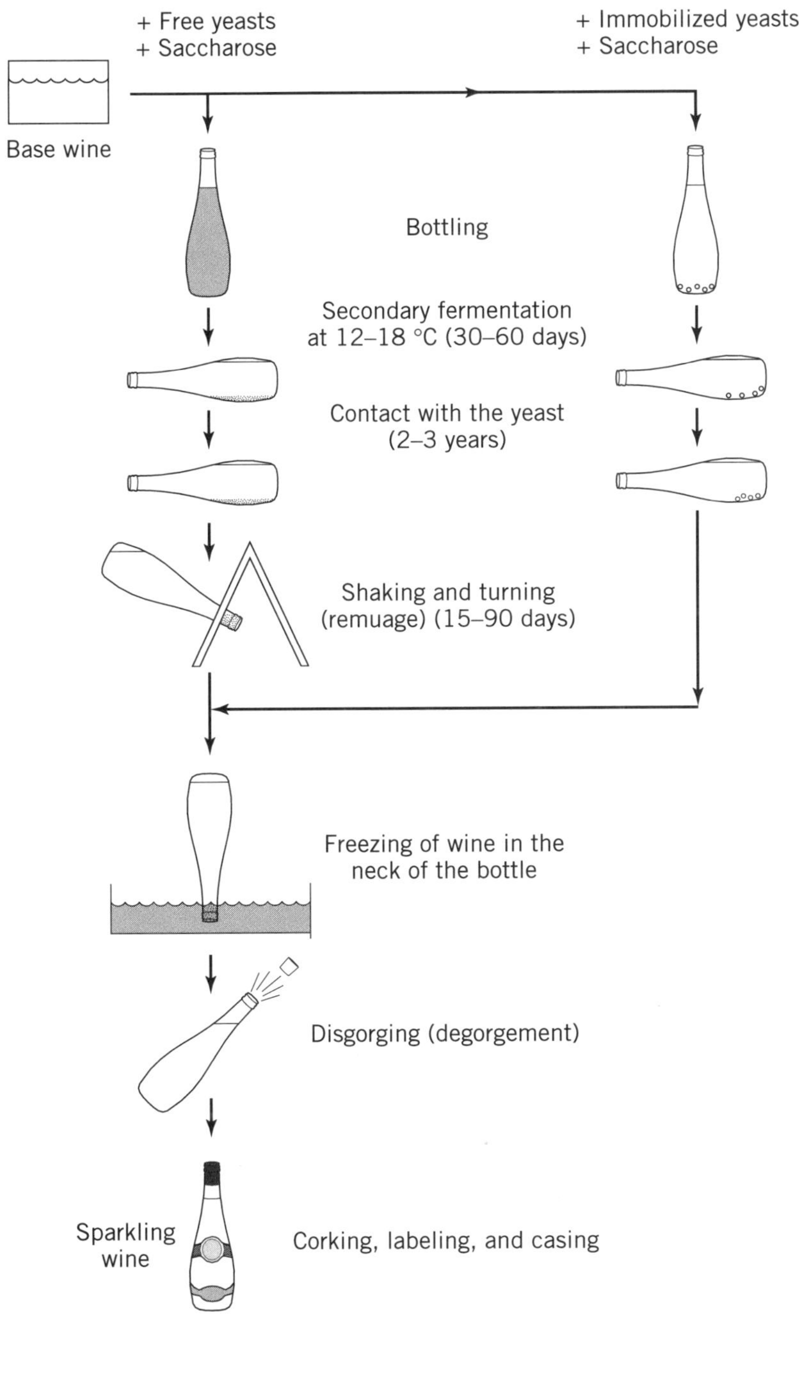

Figure 7. Scheme of the process for bottled sparkling wine preparation with traditional methods and with immobilized yeast.

thus of obtaining a very high rate of reaction so as to achieve a continuous process (107) is another desiderable innovation. This would allow a reduction in the dimension and number of pressure tanks, with a diminution in the costs of running and equipment. In the bulk process for sparkling wine production, the Charmat method (Fig. 8), sugars, other nutrients, and yeasts are added to the base wine, and the second fermentation occurs in large stainless steel sealed tanks. The sparkling wine is then filtered, fined, and bottled. This process is the cheapest for producing sparkling wines, but it still requires a considerable number of expensive tanks.

Use of Yeast Entrapped in Wine Production. The possibility of improving the rate and efficiency of transformation by increasing the concentration of yeast would be an advantage in general wine production. In this way, a linear development of the fermentative process is achieved, excluding the yeast induction phase and exponential growth. This reduces the time required by the fermentative process and the final level of the secondary products of fermentation (acetaldehyde, acetic acid, products of catabolism of the amino acids, etc.) that accumulate in the first phase of fermentation when the biomass is formed (50).

The use of yeasts entrapped in calcium alginate and in other materials has been investigated by Cantarelli (50); Duteurtre et al. (108); Lemonnier and Duteurtre (109); Loureiro (110); Zamorani et al. (111); and researchers at our Institute (106,112–115). This research has made it possible to observe the following:

- Of the various compounds used for entrapment—agar, agarose, calcium alginate, barium alginate, κ-carrageenan, gelatin, pumice stone, porous porcelain, controlled porous glass, and dyalysis membranes—agar and calcium alginate gel are the most suitable for fermentation (110).
- The retention power of the immobilizing matrix is inversely linked to the fermentative capacity of the yeast and to the alginate/yeast ratio (112).
- The presence of structural constituents of alginate (uronic acids) in the wine may be ignored up to bead content equal to 14 g/bottle (106).
- The phosphate and tartaric, malic, and citric acids present in wine do not have a disintegrating effect on the structure of the alginate (106).
- The enrichment of cells that proliferate in nucleic acids, reserve polysaccharides (trehalose and glycogen), and structural polysaccharides (mannan and glucan) is not reflected in a variation in the content of the wine (106).

Base wine
↓ + Saccharose + Yeasts
Secondary fermentation in closed tank at 15–18 °C (15–20 days)
↓
Racking
↓
Refrigeration
↓
Filtration
↓
Bottling, corking, labeling, and casing

Racking, refrigeration, filtration, bottling, and corking are done under isobarometric pressure

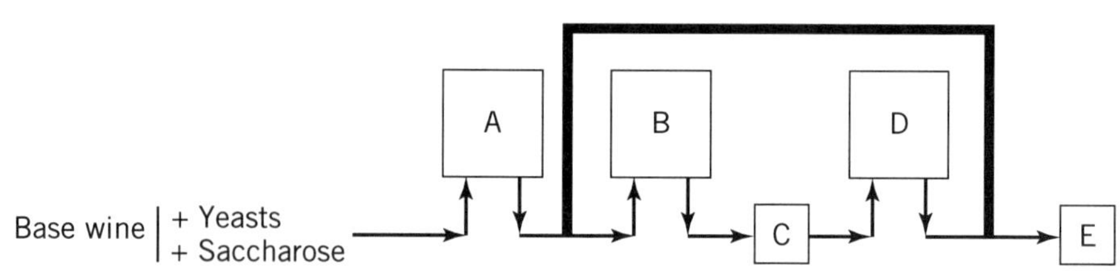

Figure 8. Scheme of a process for sparkling wine preparation by the close tank or Charmat methods. A, closed tank for secondary fermentation; B, tank for refrigeration; C, fieter; D, tank for ready to be bottled; E, isobarometric bottling.

- When the concentrations of alginate and cells in beads are equal, then the smaller the inoculum at draught, the greater the cellular release; however, this release can be avoided by using nonproliferating yeasts (106).

As for the evaluation of the influence of the alginate immobilization technique on the fermentative process, the composition, and the organoleptic characteristics of the bottled sparkling wine, Fumi et al. (114) observed that the use of immobilized cells does not cause any significant change in the second fermentation process. The main components (ethanol, organic acids, total nitrogen, and higher alcohols) are not appreciably different in sparkling wines obtained by immobilized yeast when compared with traditional sparkling wines (Table 5), but some differences are found as regards certain amino acids and some aroma components (Table 6). As shown by tasting, however, these differences do not affect the organoleptic characteristics. Sensory evaluation of sparkling wines obtained with immobilized yeasts versus a control was performed using a duo-trio test and black glasses.

Divies (116) and Duteurtre (108) conducted comparative tests with free and calcium alginate immobilized yeasts for bottled sparkling wine production, and they also noted that there are no variations in the principal chemical or physical characteristics.

Experiments to avoid cell release were performed by coating alginate/yeast beads with a protective cell-free alginate gel layer. These studies were carried out by Duteurtre (108), Fumi et al., (106), and Loureiro (110) for bottled sparkling wine production. A complete bottle fermentation was achieved without cell release, independent of the different conditions adopted—concentration of the alginate, time of gelification, and $CaCl_2$ concentration. However, Ors et al. (117) noted that the double layer must have a minimum critical thickness, equal to 0.2 mm, in order to avoid cellular release.

In an assay where agar cylinders covered with a protective layer of agar without cells were used, equally satisfactory results were not achieved (110).

Although the problem of cell release from the immobilization matrix during fermentation may be solved, there

Table 5. Analytical Characteristics of Sparkling Wine Produced by Free Yeast Cells and Yeast Cells Immobilized in 3% Alginate

	Free yeast	Immobilized yeast
Ethanol (% vol)	12.29	12.29
Sugar (g/L)	2.3	1.7
Ashes (g/L)	1.48	1.48
Ashes alkaline (mequiv/L)	11.0	11.0
pH	3.25	3.21
Titratable acidity (g/L)	6.79	6.93
Volatile acidity (g/L)	0.36	0.34
Tartaric acid (g/L)	1.73	1.82
Malic acid (g/L)	1.47	1.47
Lactic acid (g/L)	2.10	2.06
Citric acid (g/L)	0.20	0.24
Succinic acid (g/L)	0.40	0.44
Pyruvic acid (mg/L)	47.5	49.7
α-Ketoglutaric acid (mg/L)	58	63
Acetaldehyde (mg/L)	71	68
Glycerol (g/L)	5.90	5.60
Color (410 nm)	0.073	0.072
Gauge pressure CO_2 at 20°C (atm)	8.0	8.3
Calcium (mg/L)	66.5	65.5

Source: From Ref. 114.

Table 6. Free Amino Acids and Aroma Components of Sparkling Wines Produced by Free and Immobilized Yeast in 3% Alginate

	Free yeast	Immobilized yeast
Amino Acids		
Histidine (mg/L)	38.60	36.80
Asparagine (mg/L)	167.76	253.62
Arginine (mg/L)	542.11	612.24
Serine (mg/L)	67.83	65.88
Aspartic + glutamic acid (mg/L)	152.39	147.50
Glycine + threonine (mg/L)	88.17	84.16
Alanine (mg/L)	362.0	359.5
γ-Aminobutyric acid (mg/L)	151.2	145.8
Proline (mg/L)	227.03	220.97
Valine (mg/L)	22.75	23.08
Methionine (mg/L)	trace	trace
Tryptophan (mg/L)	trace	trace
Phenylalanine (mg/L)	17.70	18.38
Isoleucine (mg/L)	31.52	31.67
Ornithine (mg/L)	18.58	19.54
Lysine (mg/L)	12.53	14.26
Tyrosine (mg/L)	trace	trace
Total nitrogen (mg/L)	534	528
Esters		
Ethyl butyrate (mg/L)	0.38	0.27
Ethyl caproate (mg/L)	0.53	0.52
Ethyl pyruvate (mg/L)	0.56	0.43
Ethyl lactate (mg/L)	143	113
Diethyl succinate (mg/L)	0.63	0.60
Isobutyl acetate (mg/L)	0.13	0.08
Isoamyl acetate (mg/L)	1.15	0.82
Alcohols		
Methyl alcohol (mL/100 mL of EtOH)	0.03	0.03
1-Propanol (mg/L)	29	31
1-Butanol (mg/L)	1.62	1.23
2-Methyl-1-propanol (mg/L)	55	52
2-Methyl-1-butanol (mg/L)	24	19
3-Methyl-1-butanol (mg/L)	100	94
Hexanol (mg/L)	1.46	1.33
Benzyl alcohol (mg/L)	trace	trace
β-Phenylethanol (mg/L)	17.1	16.7
Acids		
3-Methylbutanois acid (mg/L)	0.55	0.57
Caproic acid (mg/L)	3.70	0.57
Caprylic acid (mg/L)	5.63	4.98

Note: Test with immobilized yeast: yeast dry weight, 20 mg/L gel; 4 g of beads/bottle. Chemicals were analyzed at the end of the fermentation.
Source: From Ref. 114.

are other difficulties to overcome for the industrial application of calcium alginate immobilization. The possibility of supplying wineries with ready-to-use immobilized cells would be a significant step for the implementation of this technique. For this, Loureiro (110) has performed freeze-drying experiments of cells immobilized by gel occlusion. These tests with different polysaccharide gels showed that it was possible to freeze-dry without changing bead shape using either κ-carrageenan or κ-carrageenan plus calcium alginate. However, cell release always occurred, and further studies will be necessary to optimize the thickness and rigidity of the protective bead layer for controlling cell release.

Lemmonier and Duteurtre (109) developed a device consisting of a small cartridge containing two microporous membranes, one hydrophilic and the other hydrophobic, which is filled with yeast and placed in the neck of the bottle. The two membranes hold the yeasts captive, enabling a free and total exchange between the yeasts and the wine. When the cap is opened, the cartridge is automatically ejected. The first results showed that production of sparkling wine with immobilized yeasts occurs in a manner similar to that using traditional methods with free cells. Some differences have been noted regarding protein and polysaccharides (118).

The use of a continuous fermentative system with a calcium alginate immobilized yeast bioreactor was tested by Fumi et al. (115) for the production of sparkling wines in tanks. The experimental system is shown in Figure 9. The

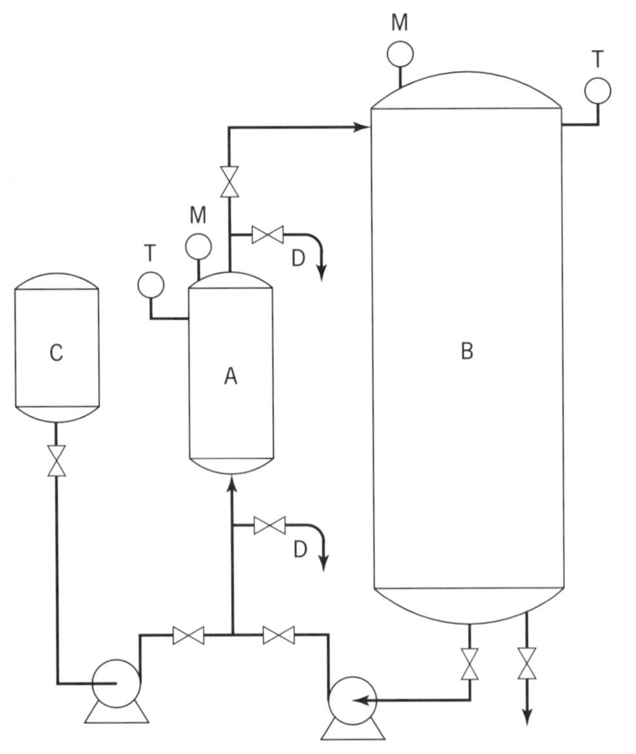

Figure 9. Scheme of the system used in the production of sparkling wine in tanks with immobilized yeast (117); A, bioreactor with immobilized yeast; B, sealed tank; C, syrup tank; D, sampling taps; M, manometers; T, thermometers.

general idea was to run the second fermentation outside the principal stainless steel tank using equipment that could easily be moved from one tank to another. The tests were carried out by placing 2 L of 1.5% alginate and 8% yeast beads (yeast inoculum was 250 g dry weight) or 3 L of 1.5% alginate and 10% yeast (yeast inoculum was 500 g dry weight) distributed or not on five distanced supports of pierced stainless steel, in a 10-L bioreactor. The base wine (9.95% v/v ethanol, 5.61 g/L titratable acidity, pH 3.04) was continuously supplied at a rate of 5 L/h and sucrose syrup at 0.14 L/h. The results showed that a high concentration of immobilized yeasts and a distribution of immobilized yeast mass in the bioreactor greatly affect the fermentation rate, making the process faster than the traditional one (78 h, as compared to 10 to 15 days needed for the traditional process). The sparkling wines have a composition (Table 7) comparable to that of traditional sparkling wines, and sensory evaluation (expert panel of eight tasters) of immobilized test versus control did not show significant differences at a level of $p < 0.05$.

The influence of cell-recycle batch fermentation process on sparkling wines produced in pressure tanks was studied by Ciani et al. (119). They noted a reduction in the time of fermentation as well as an improvement in ethanol productivity and yield. This was linked to the increased inoculum and also to the disappearance of the lag phase, which lasted approximately 150 h in batch. The increase in ethanol yield is probably correlated with a resting cell state highlighted by a lower percentage of new cells produced in steady-state conditions (120). The analytical characteristics of the sparkling wine obtained were not significantly different from those of sparkling wine produced by traditional batch fermentation.

As far as the fermentative process of must is concerned, Cantarelli (50) conducted comparative tests with free and calcium alginate immobilized yeasts. A fluidized or

Table 7. Chemical Composition of Sparkling Wines Obtained with Charmat Methods by Free and Immobilized Yeasts

	Free yeast	Immobilized yeast
Ethanol (v/v %)	10.60	10.62
Sugar (g/L)	4.5	4.7
pH	3.07	3.07
Total acidity (g/L)	5.70	5.78
Volatile acidity (g/L)	0.25	0.28
Tartaric acid (g/L)	2.37	2.28
Malic acid (g/L)	0.58	0.53
Lactic acid (g/L)	1.81	1.85
Succinic acid (g/L)	0.41	0.47
Glycerol (g/L)	4.79	5.08
Acetaldehyde (mg/L)	49	53
Total nitrogen (mg/L)	205	199
Calcium (mg/L)	81	87
n-Propanol (mg/L)	30	30
Isobutyl alcohol (mg/L)	55	62
Amyl alcohols (mg/L)	125	125
Higher alcohols (mg/L)	210	217
OD_{660nm}	>1	0.003

Source: From Ref. 115.

packed-bed reactor consisting of a 200-L tubular fermenter with immobilized yeast beads at various concentrations (maximum 8 g/L) was used for wine production. The results showed that immobilized yeast, compared with free yeast, had almost linear fermentation kinetics without a lag or exponential phase. In addition, organoleptic and chemiometric evaluation of analytical data both showed that there were no significant differences in by-products. Therefore, the final by-product content of wine produced by a fluid-bed fermenter was lower than by a packed one.

Batch fermentation tests on Riesling grape juice were carried out by Fumi et al. (113) using two 50-L working volume fermenters with calcium alginate immobilized biocatalysts (beads with alginate/yeast ratio of 1.5 wt/dry wt, reintroduced into 1% alginate and supported by stainless steel pierced plates vertically disposed) with surface area to volume ratios 1.33 and 0.66, respectively. The immobilized-system kinetic parameters had similar values, and they were not influenced by the different surface area to volume ratios of the biocatalysts. Analytical data showed no detectable differences between the wines obtained by the immobilized system and wines produced traditionally. During the first phases of fermentation with immobilized cells, the must was kept perfectly clean in both bioreactors. Nevertheless, at the end of the fermentation process, a free-cell deposit, caused by yeast leakage from the beads, was found in the bioreactors.

Selected Lactic Acid Bacteria and Innovation in Malolactic Fermentation

Malolactic fermentation is an important factor for many red wines or at least for those high acid wines produced from grapes grown in cool climate (121,122).

This process is effected by a strain of lactic bacteria of the genera *Leuconostoc, Lactobacillus*, and *Pediococcus* (67,122,123) and essentially consists of the degradation of malic acid into lactic acid and carbon dioxide, the consequence of which is a reduction in total acidity (deacidification). In addition, the transformation of malic acid is always accompanied by the fermentation of sugars, which although on a modest scale has the result of increasing aroma components such as acetaldehyde, acetic acid, acetoin, diacetyl, 2,3-butandiol, ethyl lactate, and high alcohols (67,68,122). As a whole, the wine gains in mellowness, roundness, and fullness and becomes more pleasing to the palate (68,123).

The difficulties that hinder growth of lactic bacteria in wine have causes known only in part (pH, ethanol concentration, SO_2 concentration, fermentation temperature, competition between lactic bacteria and other microorganisms, lack of nutrients). Thus, spontaneous biological deacidification of wines is not always possible. To help in the deacidification of wines, it is a common practice to make use of an inoculum of active indigenous bacteria effected through the addition of wines where malolactic fermentation is in process or has just finished. However, this system does not always give good results and is unsuitable for those wines that must maintain their peculiar characteristics unaltered (122).

For quite some time, attempts have been made to induce malolactic fermentation by inoculating the wine with selected lactic acid bacteria (121). For this purpose, lactic bacteria have been isolated from musts and wines and have been selected for their tolerance of ethanol (13 to 14 vol.%) and sulphur dioxide, their capacity to develop in the pH of wine, their capacity to produce small quantities of acetic acid, and their capacity to not alter organoleptic characteristics (121,124,125). Dried or frozen cultures of selected bacteria suitable for malolactic fermentation have been on the market for some years. Research carried out thus far on the efficacy of these preparations has shown that malic acid is broken down completely if the strain is previously reactivated in suitable substrata (67,126–130). Lonvaud-Funel (128) observed that complete breakdown of malic acid is achieved by using an inoculum containing up to 10^7 cells/mL that has been allowed to develop previously for 24 h in either a synthetic substrate (40 g/L glucose, 3 g/L malic acid, 5 g/L of neopeptone, 50 mg/L of Mg^{2+}, 20 mg/L of Mn^{2+}, 50 g/L of yeast extract, pH 4.5) or in a grape juice base substrate.

Some investigators (131–136) have suggested the use of immobilized lactic bacteria or enzymes to convert malic acid into lactic acid.

Spettoli et al. (136) immobilized *Leuconostoc oenos* ML34 in 1.67% calcium alginate gels and tested their ability to convert L-malic acid into L-lactic in a red wine having 10.20% v/v alcohol; 7.90 g/L of titratable acidity; pH 3.15; 21 mg/L of total SO_2, and 1.50 g/L of L-malic acid. The cells, in batch reaction, maintained 56% conversion of L-malic acid after 16 h.

Cuenat and Villettaz (132) immobilized a strain of *Leuconostoc oenos* on alginate (1 vol of 15% dry weight cells in 5 to 12 vol of alginate, 5% (w/v)). A complete degradation of malic acid (from 6.1 to 0.07 g/L) was obtained in a continuous system where 25 L of wines (pH 3.3; 11.3% (v/v) ethanol, 139 mg/L total SO_2) was passed through the columns in 5.5 h.

Naouri et al. (134) compared the malic-acid-degrading ability of a strain of *Leuconostoc oenos* with that of a strain of *Lactobacillus* sp. immobilized in alginate in a high compacting multiphase reactor (HCMR) containing 170 g of alginate beads and 1.41 L of red wines. After bioconversion in the reactor, malic acid was almost completely (82%) degraded by the immobilized *Leuconostoc* strain, whereas incomplete (45%) degradation was obtained with the *Lactobacillus* strain.

Crapisi et al. (131) reported on the efficiency of a continuous-flow bioreactor filled with *Lactobacillus brevis* immobilized on four types of alginate and κ-carrageenan for the control of malolactic fermentation in wine. Bioreactor efficiency seems to be dependent on gel properties ranging from 34 to 45%, depending on the immobilization matrix used.

Attempts to deacidify wine using malolactic enzyme immobilized on gels have not been successful (137) probably because of the inactivity of the enzyme at wine pH and because the required cofactor, NAD, is particularly unstable in wine.

CONCLUSION

Factors limiting the application of standardized biotechnological processes to wine production include the following:

- The seasonality and high variability in grape composition
- The limited annual season of vintage
- The length of the winemaking process for some wines, which delays experimental results
- The variations in the amount of wine production (ranging from a few hundred hectolitres to more than 200.000)

In fact, under the generic term wine there is a diversity and hierarchy of quality that is quite unique among the products of the soil and that is connected to the cultivar, the area of cultivation, and the transformation technologies applied to the grapes. The latter includes the transformation of the must by yeast fermentation, which is certainly the oldest of biotechnologies. The principles of winemaking have been established empirically over the course of several centuries of rigorous and methodical observation. The production of quality wines came before all scientific knowledge.

However, although its origins were empirical, the production of quality wine has, in the last quarter of a century, benefited most from research developments. In this area, biotechnological knowledge may well become more and more important, not only as regards the microorganisms involved in the fermentation process but also because of the use of enzymes in pressing grapes, prefermentation treatments, and refinement of the wines.

Since the development of the use of pectolytic enzymes, the tendency has been to search for enzymic preparations that have precise effects and that allow us to reduce the use of chemical additives. In this field, studies have made it possible to show that there are numerous limitations to the use of enzymes in the field of enology. For example, experiments with enzymes to improve the stabilization of white wines and to increase the aroma of some wines have demonstrated that the composition of the medium and the working conditions are not very favorable to the action of enzymes such as proteases and glycosidases. At present, the use of extracellular yeast enzymes, such as proteases, able to act on the constituents of the must are being studied (138,139). This is obviously of technological interest, because it would make it possible to take advantage of the enzyme activity of the yeast, not only for fermentation but also for stabilization of wines. In addition, the energy liberated during the fermentation process would give temperatures better suited to the activity of the enzymes.

These days a selected yeast is required not only to ensure that undesirable characteristics do not arise but also in order to have a positive effect on the characteristics of the wine. Methods of genetic improvement offer a means of programming and constructing new strains of yeasts. However, certain problems remain when one attempts genetic improvement of the strains, not the last among which is the lack of objective methods of evaluation of the quality of the yeast. The contribution of yeast to the quality of wine is complex because it is the result of interactions between grapes, yeast strains, and technology. In spite of this, the application of recombinant DNA technologies to the improvement of wine strains has undergone rapid development, and although practical applications have not yet been achieved, the basis for future advances has been established.

In the field of bottled sparkling wine production, studies have led to the creation of technologies that permit the use of alginate-immobilized yeasts at a semi-industrial level (108). However, the technology of entrapped yeast for must fermentation requires further optimization of immobilized systems, in the areas of mass transfer, reactor design, and yeast physiology. The most logical strategy for providing the means of increasing productivity without changing the flavor characteristics of fermented products must be found.

Reactors with immobilized lactic acid bacteria have also been shown to be potentially very interesting, but putting them into operation on an industrial scale still requires critical evaluation, not only of the efficiency of the breakdown of malic acid but also of the results on the organoleptic characteristics of the wines.

Every biotechnological advance that is to be accepted in winemaking must allow one to maintain the quality of the product, must take into account the hygienic and legal aspects as well, and must not impair the flavor and bouquet of the final product.

BIBLIOGRAPHY

1. D. Dubourdieu, *Experientia* **42**, 914–921 (1986).
2. A.C. Noble, in H.F. Linskens and J.f. Jackson eds., *Wine Analysis: Modern Methods of Plant Analysis*, vol. 6, Springer-Verlag, Berlin, 1988, pp. 9–28.
3. A. Rapp and H. Mandery, *Experientia* **42**, 873–884, (1986).
4. P. Schreier, *Crit. Rev. Food Sci. Nutr.* **12**, 59–111, (1979).
5. O. Colagrande, *Ann. Fac. Agraria U.C.S.C.* **28**, 99–126 (1988).
6. R.M. Canal-Llauberes, *Rev. Oenol.* **53**, 17–22 (1989).
7. R.M. Canal-Llauberes, *Rev. Franc. Oenol.* **30**, 28–33 (1990).
8. G. Montedoro, *Riv. Sci. Tech. Alim. Nutr. Um.* **6**, 133–144, (1976).
9. J.C. Villettaz, In E. Lemperle and H. Rasenberger, eds., *Proc. of 7th Int. Oenological Symposium, Rome, Italy*, International Association for Modern Winery Technology and Management, Breisach, Germany, 1984, pp. 99–119.
10. A. Zamorani, in C. Cantarelli, G. Lanzarini, eds., *Biotechnology Application in Beverage Production*, Elsevier Applied Science Publishers, New York, 1989, pp. 223–246.
11. M. Feuillat, *Rev. Oenol.* **45**, 7–17 (1987).
12. L. Saulnier, J.M. Brillouet, and M. Montounet, *Conn. Vigne Vin* **22**, 135–158, (1988).
13. L. Usseglio-Tomasset, *Conn. Vigne Vin* **10**, 193–226, (1976).
14. A. Baron, *Rev. Franc. Oenol.* **30**, 21–27 (1990).
15. J.M. Brillouet, L. Saulnier, and M. Montounet, *Rev. Franc. Oenol.* **30**, 43–54 (1990).
16. A. Silva, O. Colagrande, and P. Fontana. In G. Domenichini ed., *Proc. of the 4th Symposium La Difesa Antiparassitaria Nelle Industrie Alimentari e la Protezione Degli Alimenti* Camera di Commercio Industria Artigianato e Agricoltura, Piacenza, Italy, 1989, pp. 389–407.
17. C.S. Ough and E.A. Crowell, *Am. J. Enol. Vitic.* **30**, 22–27, (1979).
18. M.R. Brown and C.S. Ough *Am. J. Enol. Vitic.* **32**, 272–276 (1981).

19. P. Martinière, J.C. Sapis, G. Guimberteau, and J. Ribéreau-Gayon, *Compt. Rendus Acad. Agric. (France)* **59**, 267–273 (1973).
20. M. Castino and P. Bella, *Riv. Vitic. Enol.* **34**, 179–197 (1981).
21. P. Berta, *Vigne Vini* **18**, 37–42 (1991).
22. M. Castino, A. Bosso, and G. Marescalco, *Vini Ital.* **32**, 7–20 (1990).
23. D. Dubourdieu, P. Ribéreau-Gayon, and B. Fournet, *Carbohydr. Res.* **93**, 294–299 (1981).
24. J. Ribéreau-Gayon, P. Ribéreau-Gayon, and G. Seguin, in J.R. Coley-Smith, K. Verhoff, W.R. Jarvis eds., *The Biology of Botrylis*, Academic Press, London, 1980, pp. 251–274.
25. D. Dubourdieu, J.C. Villettaz, C. Desplanques, and P. Ribéreau-Gayon, *Conn. Vigne Vin* **15**, 161–177 (1981).
26. J.C. Villettaz, *Rev. Franc. Oenol.* **27**, 65–72, (1987).
27. J.C. Villettaz, *Rev. Franc. Oenol.* **30**, 59–63, (1990).
28. K. Wucherpfenning and H. Dietrich. *Bull. O.I.V.* **56**, 733–745, (1983).
29. *European Community Official Gazette* 1987, **30**, L86.
30. *European Community Official Gazette* 1988, **31**, L198.
31. M. Feuillat, Ph.D. Thesis, Université de Dijon, France, May 30, 1974.
32. P.R. Ngaba-Mbiahop, Ph.D. Thesis, Oregon State University, 1981.
33. J.C. Hsu and D.A. Heatherbell, *Am. J. Enol. Vitic.* **38**, 6–10 (1987).
34. J.C. Hsu and D.A. Heatherbell, *Am. J. Enol. Vitic.* **38**, 11–16 (1987).
35. A.T. Bakalinsky and R. Boulton, *Am. J. Enol. Vitic.* **36**, 23–29 (1985).
36. B.S. Gaina, N.M. Pavlenko, E.N. Datunashvili, Y.I. Krylova, L.V. Kozlov, V.K. Antonov, *Appl. Biochem. Micro.* **12**, 167–173 (1976).
37. K. Woiwodov, B. Galunsky, S. Djankov, N. Gorinova, and D. Tzakov, *Mitt. Klosterneuburg* **32**, 117–121, (1982).
38. E.J. Modra, P.J. Williams, T.H. Lee, and W. Wallace, in P. Ribéreau-Gayon and A. Lonvaud eds., *Actualités Oenologiques 89, Proc. of the 4th Symposium International d'Oenologie, Bordeaux 1989*, Dunod, Paris, 1990, pp. 217–221.
39. S. Arai, G. Charalambous ed., *The Analysis and Control of Less Desirable Flavors in Foods and Beverages*, Academic Press, New York, 1980, pp. 133–147.
40. J.J. Macheix, J.C. Sapis, and A. Fleuriet, *Crit. Rev. Food Sci. Nutr.* **30**, 441–486 (1991).
41. A.C. Noble, in R.L. Rouseff ed., *Bitterness in Foods and Beverages: Developments in Food Science*, vol. 25, Elsevier, Amsterdam, 1990, pp. 145–158.
42. V.L. Singleton, *Am. J. Enol. Vitic.* **38**, 69–77 (1987).
43. H.L. Wildenradt and V.L. Singleton *Am. J. Enol. Vitic.* **25**, 119–127 (1974).
44. G. Bonaga, U. Pallotta, and K. Syrghi, *Vini Ital.* **32**, 13–30, **32**, 31–38 (1990).
45. U. Pallotta and C. Cantarelli, *Vigne Vini* **6**, 19–46 (1979).
46. O. Colagrande, R. Ottina, and M.D. Fumi, *Vini Ital.* **28**, 43–56 (1986).
47. C. Cantarelli, *Vini Ital.* **28**, 87–98 (1986).
48. C. Cantarelli, G. Giovannelli, and S. Gallizia, in Atti Simposio *Criteri razionali di controllo e di impiego dei Coadiuvanti Enologici*, Università Cattolica del S. Cuore, Istituto di Enologia, Piacenza, Italy, 1989, pp. 55–67.
49. C. Cantarelli, O. Brenna, G. Giovannelli, and M. Rossi, *Food Biotechnol.* **3**, 203–213 (1989).
50. C. Cantarelli, C. Cantarelli and G. Lanzarini eds., in *Biotechnology Application in Beverage Production*, Elsevier Applied Science, New York, 1989, pp. 127–151.
51. P. Ribéreau-Gayon, J.N. Boidron, and A. Terrier, *J. Agric. Food Chem.* **23**, 1042–1047 (1975).
52. P.J. Williams, C.R. Strauss, and B. Wilson *Am. J. Enol. Vitic.* **32**, 230–235 (1981).
53. R. Di Stefano, *Vini Ital.* **130**, 29–43 (1981).
54. R. Di Stefano, *Vigne Vini* **9**, 45–47 (1982).
55. Z. Gunata, C. Bayonove, R. Baumes, and R. Cordonnier, *J. Sci. Food Agric.* **36**, 857–862 (1985).
56. P.J. Williams, C.R. Strauss, and B. Wilson *Phytochemistry* **19**, 1137–1139 (1980).
57. S. Voirin, R. Baumes, S. Bitteur, Z. Gunata, and C. Bayonove, *J. Agric. Food Chem.* **38**, 1373–1378 (1990).
58. P.J. Williams, C.R. Strauss, B. Wilson, R.A. Maay-Westropp, *Phytochemistry* **21**, 2013–2020 (1982).
59. Z. Gunata, C. Bayonove, R. Baumes, and R. Cordonnier, *J. Chromatogr.* **331**, 83–90 (1985).
60. R. Di Stefano, *Riv. Vitic. Enol.* **42**, 11–23 (1989).
61. C. Bayonove, Y.Z. Gunata, J.C. Sapis, and R.L. Baumes, *Rev. Oenol.* **18**, 15–18 (1992).
62. Z. Gunata, S. Bitteur, J.M. Brillouet, C. Bayonove, and R. Cordonnier, *Carbohydr. Res.* **184**, 139–149 (1988).
63. Z. Gunata, S. Bitteur, R. Baumes, J.C. Sapis, and C. Bayonove, *Rev. Franc. Oenol.* **30**, 37–41 (1990).
64. R.E. Cordonnier, Z. Gunata, R.L. Baumes, and C.L. Bayonove, *Conn. Vigne Vin* **23**, 7–23 (1989).
65. Z. Gunata, I. Dugelay, J.C. Sapis, R. Baumes, and C. Bayonove, *J. Int. Sci. Vigne Vin* **24**, 133–144 (1990).
66. R.E. Kunkee and M.A. Amerine, in A.H. Rose and J.S. Harrison eds., *The Yeasts* Academic Press, New York, 1970, pp. 5–71.
67. S. Lafon-Lafourcade, in H.-J. Rehm and G. Reed eds., *Biotechnology*, Verlag Chemie, Weinheim, 1983, pp. 81–116.
68. F. Radler, *Experientia* **42**, 884–893 (1986).
69. A. Martini, F. Federici, and G. Rosini, *Can. J. Microbiol.* **26**, 856–859 (1980).
70. H.J. Phaff, M.W. Miller, and E.M. Mrak, *The Life of Yeasts*, Harvard University Press: Cambridge, Mass., 1978.
71. G. Rosini, F. Federici, and A. Martini, *Microb. Ecol.* **8**, 93–97 (1982).
72. R.E. Kunkee, *Food Microbiol.* **1**, 315–332 (1984).
73. C. Zambonelli, P. Romano, and G. Suzzi in C. Cantarelli, G. Lanzarini, eds., *Biotechnology Application in Beverage Production*, Elsevier Applied Science Publishers, New York, 1989, pp. 17–30.
74. H. Bussey, *Adv. Microb. Physiol.* **22**, 93–122 (1981).
75. C. Geneix, S. Lafon-Lafourcade, and P. Ribereau-Gayon, *C.R. Acad. Sci. Paris* **296**, 943–945 (1983).
76. S. Lafon-Lafourcade, *Experientia* **42**, 904–914 (1986).
77. S. Lafon-Lafourcade, D. Dubourdieu, D. Hadjinicolaou, and P. Ribéreau-Gayon, *Conn. Vigne Vin.* **14**, 97–109 (1980).
78. S. Lafon-Lafourcade, C. Geneix, and P. Ribereau-Gayon, *Appl. Envir. Micribiol.* **47**, 1246–1249 (1984).
79. P. Pfeiffer and F. Radler *J. Gen. Microbiol.* **128**, 2699–2706 (1983).

80. R.B. Wickner, in E.W.J. Strathern and J.R. Broach eds., *The Molecular Biology of the Yeast Saccharomyces: Life Cycle and Inheritance*, Cold Spring Harbor Laboratory, New York, 1981, pp. 415–444.
81. C. Zambonelli, *Accademia Italiana Vite e Vino* **29**, 51–64 (1977).
82. L.F. Bisson, *World Biotech. Rep. Food Process.* **2**, 103–116 (1986).
83. C.V. Rous, R. Snow, and R.E. Kunkee, a leucine *J. Inst. Brew.* **89**, 274–278, (1983).
84. C. Falcone and L. Frontali, C. Cantarelli, and G. Lanzarini eds., in *Biotechnology Application in Beverages Production*, Elsevier Applied Science, London, 1989, pp. 31–39.
85. I.S. Pretorius and T.J. van der Westhuizen, *S. Afr. J. Enol. Vitic.* **12**, 3–31 (1991).
86. R. Eschenbruch, K.J. Cresswell, B.M. Fisher, and R.J. Thornton, *Europ. J. Appl. Microbiol. Biotechnol.* **14**, 155–159 (1982).
87. F. Vezinhet, *Sci. Aliments* **9**, 253–265, (1989).
88. P.C. Molan, M. Edward, and R. Eschenbruch, *Eur. J. Appl. Microbiol. Biotechnol.* **16**, 110–113, (1982).
89. F. Vezinhet, *Rev. Franc. Oenol.* **25**, 45–51, (1985).
90. G.A. Farris, F. Fatichenti, L. Bifulco, L. Berardi, P. Delena, and T. Satta *Biotechnol. Lett.* **14**, 219–222 (1992).
91. S. Hara, Y. Iimura, H. Oyama, T. Kozeki, K. Kitano, K. Otsuka, *Agric. Biol. Chem.* **45**, 1327–1330 (1981).
92. T. Seki, E.H. Choi, and D. Ryu *Appl. Environ. Microbiol.* **49**, 1211–1215, (1985).
93. T.W. Young, in A.H. Rose and J.S. Harrison eds., *The Yeasts*, vol. 2, Academic Press, New York, 1987, pp. 131–164.
94. P. Romano, M.G. Soli, G. Suzzi, and L. Grazia *Appl. Environ. Microbiol.* **50**, 1064–1067 (1985).
95. R.J. Thornton, *Am. J. Enol. Vitic.* **36**, 47–49, (1985).
96. F. Vezinhet and P. Barre, in G. Durand, L. Bobichon, J. Florent, eds., *8th International Biotechnology Symposium Paris 1988*, Abstract A43, p. 101–103.
97. J. Watari, Y. Takata, M. Ogawa, N. Nishikawa, and M. Kamimura *Agr. Biol. Chem.* **53**, 901–903, (1989).
98. S.A. Williams, R.A. Hodges, T.L. Strike, R. Snow, and R.E. Kunkee, *Appl. Environ. Microbiol.* **47**, 288–293, (1984).
99. R. Snow, in J.F.T. Spencer, D.M. Spencer, A.R.W. Smith, eds., *Yeast Genetics: Fundamental and Applied Aspects*, Springer Verlag, New York, 1983, pp. 439–459.
100. C. Zambonelli, M.G. Soli, and D. Guerra *Ann. Microbiol.* **34**, 7–15, (1984).
101. C. Gjermansen and P. Sigsgaard, *Carlsberg Res. Commun.* **46**, 1–11 (1981).
102. I. Herskowitz, *Microbiol. Rev.* **52**, 536–553 (1988).
103. G.C. Stewart, *Can. J. Microbiol.* **27**, 973–990 (1981).
104. R.J. Thornton and R. Eschenbruch, *Antonie van Leeuwenhoek* **42**, 503–509 (1976).
105. M. Polsinelli, in O. Colagrande, ed., *Sviluppi della Biotechnologia Nella Produzione dello Spumante Classico*, Proc. of the 4th Simposio *Internazionale sul Vino, Pavia* Chiriotti Publishers, Pinerolo, Italy, 1990, pp. 19–22.
106. M.D. Fumi, G. Trioli, and O. Colagrande, *Produzione dello Spumante Classico, Proc. of the Simposio Internazionale sul Vino, Pavia*, Chiriotti Publishers, Pinerolo, Italy, 1987, pp. 94–102.
107. J. Tampion and M.D. Tampion, *Immobilized Cells: Principle and Applications*, Cambridge University Press, Cambridge, U.K., 1987.
108. B. Duteurtre, P. Ors, and D. Hennequin, in O. Colagrande, ed., *Sviluppi della Biotechnologia Nella Produzione dello Spumante Classico, Proc. of the 4th Simposio Internazionale sul Vino*, Pavia Chiriotti Publishers, Pinerolo, Italy, 1990, pp. 71–73.
109. J. Lemonnier and B. Duteurtre, *Rev. Franc. Oenol.* **29**, 15–26 (1989).
110. V. Loureiro, in O. Colagrande ed., *Sviluppi della Biotechnologia Nella Produzione dello Spumante Classico Proc. of the 4th Simposio Internazionale sul Vino, Pavia*, Chiriotti Publishers, Pinerolo, Italy, 1990, pp. 74–77.
111. A. Zamorani, A. Crapisi, G. Borin, M. Dell'Eva, G. Versini, and P. Spettoli, *Yeast* **5**, 63–66 (1989).
112. M.D. Fumi, G. Trioli, and O. Colagrande, *Biotechnol. Lett.* **9**, 339–342 (1987).
113. M.D. Fumi, G. Trioli, and O. Colagrande *Proc. 4th Europ. Congr. Biotechnol.*, Amsterdam, June 14–19, Elsevier, 1987, pp. 81–84.
114. M.D. Fumi, G. Trioli, M.G. Colombi, and O. Colagrande *Am. J. Enol. Vitio.* **39**, 267–272 (1988).
115. M.D. Fumi, M. Bufo, G. Trioli, and O. Colagrande *Biotechnol. Lett.* **11** 821–824 (1989).
116. C. Divies, in C. Cantarelli, and G. Lanzarini, eds., *Biotechnology Application in Beverages Production*. Elsevier Applied Science, London, 1989, pp. 153–167.
117. P. Ors, B. Duteurtre, D. Hannequin, in P. Ribéreau-Gayon, A. Lonvaud, eds., *Actualités Oenologiques 89, Proc. of the 4th Symposium International d'Oenologie, Bordeaux 1989*, Dunod, Paris, 1990, pp. 270–274.
118. M. Feuillat, in *Progressi Scientifici e Tecnici Nella Moderna Spumantistica Italiana* Rovato, 1991, pp. 25–40.
119. M. Ciani and G. Rosini, *Biotechnol. Lett.* **13**, 533–536 (1991).
120. G. Rosini, *Appl. Microbiol. Biotechnol.* **24**, 140–143, (1986).
121. C.R. Davis, D. Wibowo, R. Eschenbruch, T.H. Lee, and G.H. Fleet, *Am. J. Enol. Vitic.* **36**, 290–301 (1985).
122. R.E. Kunkee, in A.D. Webb ed., *Chemistry of Wine*, American Chemical Society, Washington D.C., 1974, pp. 137, 151–170.
123. D. Wibowo, R. Eschenbruch, C.R. Davis, G.H. Fleet, and T.H. Lee, *Am. J. Enol. Vitic* **36**, 302–313, (1985).
124. P. Chagnaud, P. Nauri, A. Arnaud, and P. Galzy, *Belgian J. Food Chem. Biotechnol.* **45**, 12–20 (1990).
125. S. Lafon-Lafourcade, in J.G. Carr and C.V. Cutting eds., *Lactic Acid Bacteria in Beverages and Food*, Academic Press, London, 1975, pp. 43–53.
126. A. Joyeux and A. Lonvaud-Funel, *Conn. Vigne Vin* **19**, 149–159 (1985).
127. D.C. Hayman and P.R. Monk, *Food Technol. Australia* **34**, 14–18 (1982).
128. A. Lonvaud-Funel, *Application à l'Oenologie des Progrès Récents en Microbiologie et en Fermentation*, Paris 1988, Programme COMETT, O.I.V., Paris, pp. 91–103.
129. P. Naouri, P. Chagnaud, A. Arnaud, P. Galzy, and J. Mathieu, *J. Biotechnol.* **10**, 135–150, (1989).
130. M. Valade, in *Application à l'oenologie des Progrès Récents en Microbiologie et en Fermentation*, Programme COMETT, Paris, 1988, OIV, Paris, pp. 71–80.
131. A. Crapisi, A. Lante, M.P. Nuti, A. Zamorani, and P. Spettoli, in P. Ribéreau-Gayon, and A. Lonvaud, eds., *Actualités Oenologiques 89, Proc. of the 4th Symposium International d'Oenologie Bordeaux 1989*, Dunod, Paris, 1990, pp. 349–352.
132. Ph. Cuenat and J.CL. Villettaz, *Rev. Suisse Vitic. Arboric, Hortic.* **16**, 145–151 (1984).

133. C. Divies and M.H. Siess *Ann. Inst. Pasteur. Microbiologie Paris* **127**, 525–539 (1976).

134. P. Naouri, P. Chagnaud, A. Arnaud, P. Galzy, and J. Matthieu *J. Wine Res.* **2**, 5–20, (1991).

135. J. Rossi and F. Clementi, *Am. J. Enol. Vitic.* **35**, 100–102, (1984).

136. P. Spettoli, A. Bottacin, M.P. Nuti, and A. Zamorani, *Am. J. Enol. Vitic.* **33**, 1–5, (1982).

137. S. Gestrelius, *Enzyme Eng.* **6**, 245–250 (1982).

138. I. Rosi, L. Costamagna, and M. Bertuccioli, *J. Inst. Brew.* **93**, 322–324, (1987).

139. I. Rosi and M. Bertuccioli, in O. Colagrande ed., *Sviluppi della Biotecnologia Nella Produzione Dello Spumante Classico, Proc. of the 4th Simposio Internazionale sul vino*, Chiriotti Publishers, Pinerolo, Italy, 1990, pp. 43–48.

ADDITIONAL READING

C. Biron, R. Cordonnier, O. Glory, Y.Z. Gunata, and J.C. Sapis *Conn. Vigne Vin* **22**, 125–134, (1988).

C. Bucke, *Methods Enzymol.* **135**, 175–189, (1987).

G.B. Calleja, in A.H. Rose and J.S. Harrison eds., *The Yeasts*, vol. 2, Academic Press, New York, 1987, pp. 165–238.

P. Chatonnet, C. Barbe, R. Canal-LLauberes, D. Dubordieu and J.N. Boidron *J. Int. Sci. Vigne Vin* **26**, 253–269, (1992).

D. Dobourdieu, P. Darriet, P. Chatonnet, and J.N. Boidron in P. Riéreau-Gayon and A. Lonvaud eds., *Actualités oenologiques 89, Proc. of the 4th Symposium International d'Oenologie*, Dunod, Paris, 1990, pp. 151–159.

Y.Z. Gunata, Ph.D. Thesis, Academie de Montpellier-Université des Sciences et Techniques du Languedoc, April 19, 1984.

R. Felix and J.C. Villettaz, in T. Godfrey and J. Reichelt eds., *Industrial Enzymology*, Stockton Press, New York, 1983, pp. 410–421.

Th. Henick-Kling, in H.F. Linskens and J.F. Jackson eds., *Wine Analysis: Modern Methods of Plant Analysis* , vol. 6, Springer-Verlag, Berlin, 1988, pp. 276–316.

S. Lafon-Lafourcade, E. Carre, A. Lonvaud-Funel, and P. Ribereau-Gayon, *Conn. Vigne Vin.* **17**, 55–71 (1983).

G. Lanzarini and P.G. Pifferi, in C. Cantarelli and G. Lanzarini eds., *Biotechnology Application in Beverage Production*, Elsevier Applied Science Publishers, New York, 1989, pp. 189–222.

H.J. Peppler and G. Reed, in H-J. Rehm and G. Reed eds., *Biotechnology*, vol. 7, Verlag Chemie, Weinheim, 1987, pp. 584–586.

V.L. Singleton, in H.F. Linskens and J.F. Jackson eds., *Wine Analysis: Modern Methods of Plant Analysis*, vol. 6, Springer-Verlag, Berlin, 1988, pp. 173–257.

H.J.J. Vuuren and L.M.T. Dicks, *Am. J. Enol. Vitic.* **44**, 99–112, (1993).

See also ANAEROBES; CULTURE COLLECTIONS.

X

XANTHAN GUM

J.-L. Flores Candia
W.-D. Deckwer
GBF-Gesellschaft für Biotechnologische Forschung GmbH
Braunschweig, Germany

KEY WORDS

Drug release control
Emulsifier
Enhanced oil recovery (EOR)
Food additives
Microbial exopolysaccharide
Non-Newtonian fluid
Oil drilling
Thickening agent
Xanthan
Xanthomonas campestris

OUTLINE

INTRODUCTION

A variety of polysaccharides are generated by microorganisms. Examples include alginate, curdlane, dextran, gellan, glucan, pullulan, and xanthan. Also cellulose can be produced by microbes. Among the microbial polysaccharides, xanthan plays a dominant role due to the relative easiness to produce it and as a result of its outstanding properties. Indeed, it has found widespread applications ranging from food and cosmetics additives to enhanced oil recovery. This polymer represents the fastest growing segment of the polysaccharide industry (Table 1). The largest growth is expected to be in food applications (8.3%/year), where demand for natural gums is decreasing. Xanthan gum is an exoheteropolysaccharide produced by the bacterium *Xanthomonas campestris*.

Structure and Properties

The xanthan polymer consists of repeated units of five sugars, namely, two D-glucose, two D-mannose, and one D-glucoronate, and varying amounts of acetate and pyruvate. Its primary structure is shown in Figure 1, based on the results of Jannson et al. (5) and Melton et al. (6), revised according to Hassler and Doherty (7), and confirmed by McIntyre et al. (8).

The polysaccharide backbone, made up of β-[1–4]-linked D-glucose units, is identical to that of cellulose. To alternate D-glucose units at the O-3 position, a trisaccharide side chain containing a D-glucoronosyl unit between two D-mannosyl units is attached. The terminal β-D-mannosyl unit is glycosidically linked to the O-4 position of the β-D-glucoronosyl unit, which in turn is glycosidically linked to the O-2 position of an α-D-mannosyl unit. The terminal D-mannosyl unit of the side chain contains a pyruvic acid moiety as a 4,6-cyclic acetal in variable degree. Finally, the nonterminal D-mannosyl unit is stoichiometrically substituted at O-6 with an acetyl group. Xanthan gum also contains monovalent cations, which can be Na, K, or ½ Ca.

The secondary structure is known to undergo an order–disorder conformational transition at particular salinity

Table 1. Annual Average Growth Rates of the Main Commercial Polysaccharides: Market Analysis and Forecast for the United States

	Annual average growth rate (%)		
Polysaccharide	Food	Petroleum	Cosmetics and toiletries
Starches	2.5	1.5	–
Natural gums			
Guar	2.1	–	–
Algins	2.7	–	3.7
Arabic	2.6	–	3.7
Carrageenan	4.5	–	3.7
Locust bean	1.9	–	3.1
Cellulosics			
CMC	3.5	0.5	3.0
HPG	–	1.5	–
HEC	–	0.5	–
MC	3.5	–	3.7
Xanthan	8.3	2.0	5.9
Pectin	5.2	–	–

Source: Business Communications Co.
Note: CMC, carboxymethylcellulose; HPG, hydroxypropyl guar; HEC, hydroxyethylcellulose; MC, methylcellulose.

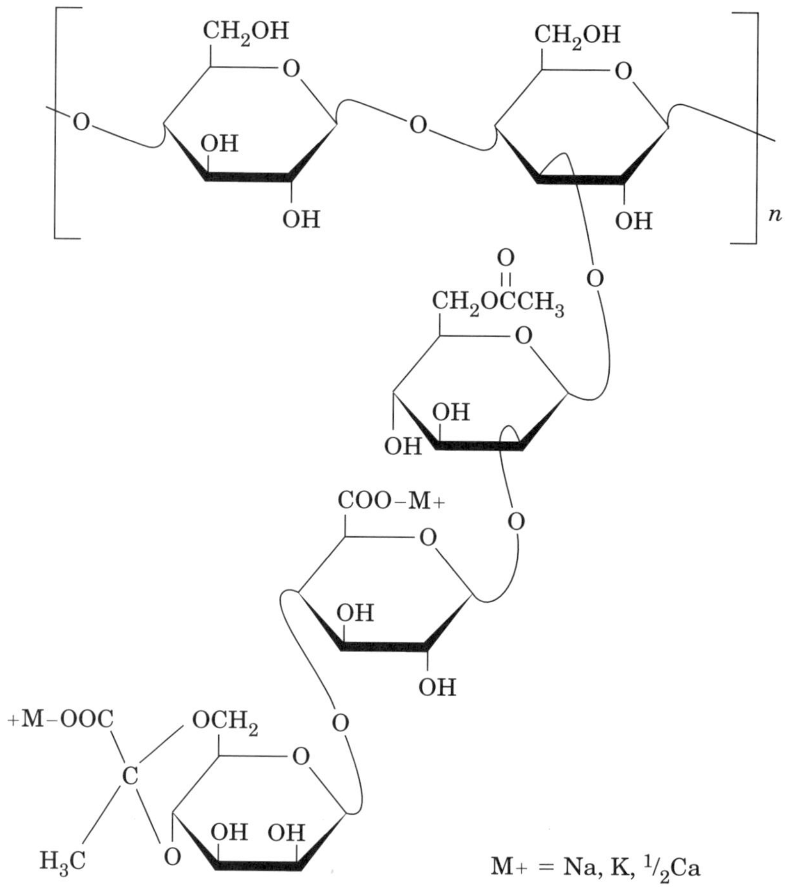

Figure 1. Chemical structure of the xanthan repeating unit.

and temperature levels, depending on the extent of pyruvate substitution (9,10). Most authors explain the stabilization of the ordered form by a noncovalent packing of the side chains along the backbone by H-bonding. X-ray fiber studies have indicated that the polymer adopts a right-handed helical conformation with fivefold symmetry and a helix pitch of 4.7 nm^3 (11). However, the question of whether this polymer dissolves in water as a single (11,12) or a double strand (13,14) has been a matter of controversial discussion. It now seems to be anticipated that xanthan may effectively be a single or a double strand, depending on the fermentation process, the acyl content, the thermal history, and the downstream process (12,15,16).

The most important property of xanthan is its ability to form solutions of high viscosity at rest or at low shear forces, which are highly pseudoplastic, and may display also a viscosity yield value. The viscosity of xanthan solutions is stable over a wide range of salt concentrations (up to 150 g/L NaCl), temperature (up to 90 °C) and pH (2–11). These solution characteristics of xanthan give rise to the functional properties such as thickening and stabilizing ability, which are used in a diverse range of applications. In addition, xanthan shows a synergistic interaction with various galactomannans (17,18), chitosan (19), and β-lactoglobulin (20), which results in enhanced solution viscosity or formation of gels, depending upon the type of mixture.

The rheological behavior of xanthan solution is most conveniently described by the two-parameter model of Ostwald and de Waele (power law)

$$\tau = k\dot{\gamma}^n \quad (1)$$

where τ is the shear stress and $\dot{\gamma}$ the shear rate. From equation 1, the effective viscosity, μ_{eff}, follows as

$$\mu_{\text{eff}} = k\dot{\gamma}^{n-1} \quad (2)$$

The fluid consistency factor, k, and the flow behavior index, n ($\leq$1 in case of the pseudoplastic xanthan solutions), are dependent on xanthan concentration and quality. Xanthan quality is predominantly determined by its molecular mass, but the pyruvate content has also a significant effect, though the influence of the pyruvate content on the viscosity yield is not yet clearly recognized. Flores-Candia (21) observed a sharp increase in viscosity (about a hundred-fold) when the pyruvate content increased above ca. 3% wt. This is in accordance with Cadmus et al. (22) who stated that many industrial applications require a xanthan with a pyruvate content above 3.3% wt. The effect of xanthan molecular mass (pyruvate content $\geq$3.5 wt) on viscosity was studied by Herbst et al. (23) and Peters et al. (24). Figure 2 demonstrates the strong dependency of the vis-

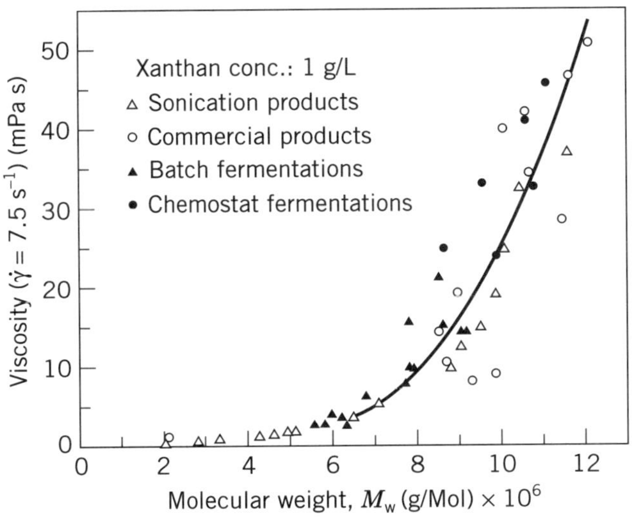

Figure 2. Effect of average molar mass on solution viscosity at a shear rate of 7.5 s^{-1} and a xanthan concentration of 0.1 % wt. *Source*: From Ref. 23.

cosity on the molecular mass of xanthan products of different origin.

Peters et al. (24) developed the following correlations for predicting the rheological parameters of equation 1 from the molar mass, M, (kg $kmol^{-1}$) and polymer concentration, c_p (kg m^{-3})

$$k = 10^{-3} + 2.58 \cdot 10^{-6}\left(\frac{M}{10^6}\right)^{3.69} c_p^{2.72} \qquad (3)$$

$$\frac{1-n}{n-0.071} = 0.0227\left(\frac{M}{10^6}\right)^{2.08} (k - 10^{-3})^{0.47} \qquad (4)$$

The thickening ability is related to the viscosity of a gum solution and can be defined as the resistance to flow of a liquid. Because of its high solution viscosity at low concentration and shear thinning behavior, xanthan is a particularly useful thickener. Xanthan is also used for the stabilization of suspensions, emulsions, and foams. This functional property is enhanced by a high solution viscosity and the existence of a viscosity yield value. Thus, if the suspended particles exert a force less than the yield value, they will be dispersed effectively (25). Again, the shear thinning behavior of xanthan is important because some stabilized products require a low viscosity during their application. Although xanthan does not possess the functional property of gelation, it will form thermoreversible gels with certain galactomannans. Gelation arises from intermolecular associations to form a three-dimensional continuous network that traps and immobilizes the water within it to form a rigid structure that shows resistance to flow under pressure.

BIOSYNTHESIS

Xanthan is produced by the bacterium *Xanthomonas campestris*, isolated originally from cabbage in which it causes black rot disease (26). *Xanthomonas* species are Gram-negative bacteria taxonomically placed in the Pseudomonadaceae family (27). According to the genetic basis of the classification, the average genome size of *Xanthomonas* is 2.5 × 10^9 Da. The percent G + C (guanine + cytosine) content of the DNA ranges from 63 to 71% (28). The cells have a rodlike form, slightly curved with rounded ends, measuring 0.2–0.6 by 0.8–2.9 μm. They occur mostly alone or in pairs, but chains and filaments are also observed. *X. campestris* forms smooth yellow mucoid colonies on solid media, and cells are surrounded by xanthan gum (extracellular polysaccharide). The chemical structure of the yellow pigment was determined to be mono- or dibromoaryl polyene and is probably bound to the cellular membrane. Genetic instabilities that affect quality and yield of xanthan have been observed in *Xanthomonas* cultivation (29). Genetic mutations are associated with the formation of L (large) and S (small) colony types. Whereas L-type colonies produce high xanthan yields with the desired rheological characteristics, the S-type colonies give low product yields with poor quality.

The natural function of xanthan is not fully recognized, but there is strong evidence indicating that the polysaccharide layer, surrounding the microbial cell, protects the microorganisms from various environmental factors. The water-retaining capacity of xanthan, for example, provides the microorganism with a protective layer under dry environmental conditions, thus providing some degree of desiccation resistance (30). Furthermore, recent reports of xanthan yield improvements in the presence of detergents (31) and bacterial-toxic acids (32) reveal that xanthan biosynthesis is induced under stress conditions.

In *X. campestris*, the Entner–Doudoroff pathway in conjunction with the tricarboxylic acid cycle pathway is the predominant mechanism for glucose catabolism (28,33,34). A small portion of glucose is routed via the pentose phosphate pathway (28,35). For glucose uptake, two discrete systems seem to exist (35). The biosynthesis of xanthan, as in most polysaccharide-producing bacteria, utilizes various activated carbohydrate donors to form the polymer on an acceptor molecule. The oligosaccharide repeated units of xanthan are constructed by the sequential addition of monosaccharides from sugar nucleotide diphosphates to an isoprenoid lipid acceptor molecule.

At the same time, acyl substitutes are added from appropriate activated donors. As shown in Figure 3, the xanthan backbone is formed by the successive addition of D-glucose-1-phosphate and D-glucose from two moles of UDP-D-glucose. Thereafter, D-mannose and D-glucoronic acid are added from GDP-mannose and UDP-glucoronic acid, respectively. O-Acetyl groups are transferred from acetyl-CoA to the internal mannose residue, and pyruvate from phosphoenolpyruvate is added to the terminal mannose. This pattern of reactions was elegantly demonstrated by Ielpi and his colleagues (36,37) using permeabilized cells and radioactively labeled precursors. Further, it was also shown that each of these steps, as just described, requires specific substrates and a specific enzyme for completion. If either the substrate or the enzyme is absent, the step is blocked (39–41).

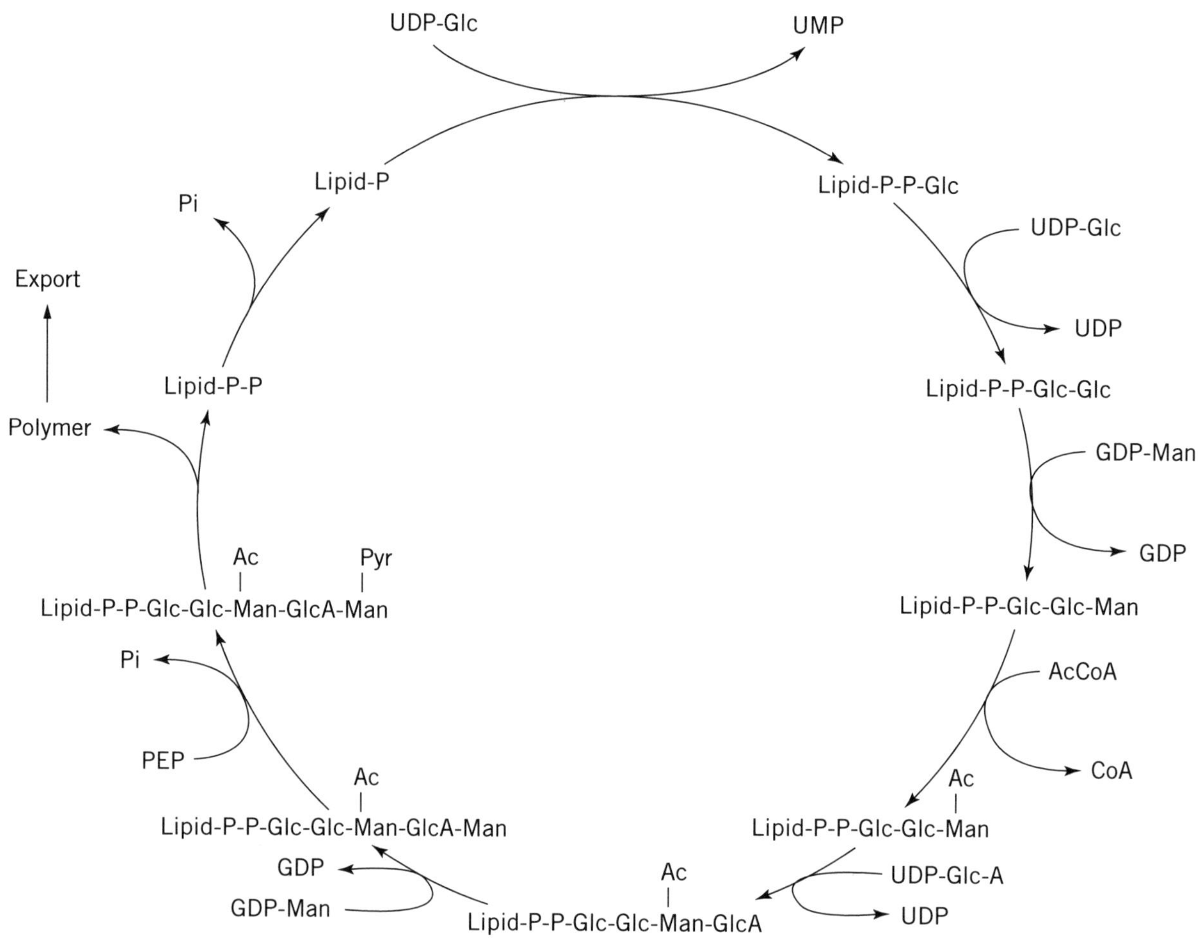

Figure 3. Biosynthesis of xanthan gum from activated precursors of carbohydrates and acyl groups. *Source*: From Refs. 36–38.

It has been suggested that the construction of the exopolysaccharide follows a "tail-to-head" polymerization (36–38,41). After the pentasaccharide repeated unit is formed, oligomers are formed by transfer to other lipid intermediates, gradually increasing the size of the carbohydrate chain. As noticed by Sutherland (38), oligomer construction normally involves the addition of the longer oligosaccharide sequence to the nonreducing terminus of a single repeat unit attached to the isoprenoid lipid diphosphate. The inactive lipid carrier is dephosphorylated to yield isoprenyl phosphate, which can then reenter the biosynthetic sequence.

Although the structure of the repeating units is determined by the sequential transfer of the different monosaccharides and acyl groups from their respective donors by highly specific sugar transferases, the polymerase enzyme responsible for polymerization of the pentasaccharides into a macromolecule of $M > 10^6$ has been shown to be less specific (38). The final stages of exopolysaccharide secretion from the cytoplasmic membrane, passing across the periplasm and outer membrane and finally excreted into the extracellular environment, are much less well specified than the previous biosynthetic events. This mechanism must exist in all polysaccharide-producing bacteria for releasing polymer from the isoprenoid lipid prior to transport to its final destination. The process obviously requires an energy source and may be analogous to the export of lipopolysaccharide to the outer membrane (42), in which ATP is the energy supplier.

Significant advances have been made in genetic research on *X. campestris*. Using transposon mutagenesis, Harding et al. (43) found a cluster of linked genes that encode enzymes for the assembly of the pentasaccharide. This same cluster was also isolated and analyzed by Thorne et al. (44). Vanderslice et al. (45) sequenced the DNA in this region and showed that it contains 12 genes (Fig. 4), which included seven gene products needed for monosaccharide transfer and for acylated repeating units. The genes responsible for xanthan acylation were further corroborated by Marzocca et al. (46). In addition, a recent analysis of the transcriptional organization of the gene cluster indicated that the gum region was mainly expressed as an operon from a promoter located upstream of the first gene, *gumB* (47). Another important contribution to the genetics of *Xanthomonas* is the identification of *xanA* and *xanB* genes (48). These genes were reported to be involved in UDP-glucose and GDP-mannose biosynthesis. Recently, it was also demonstrated that the gene encoding

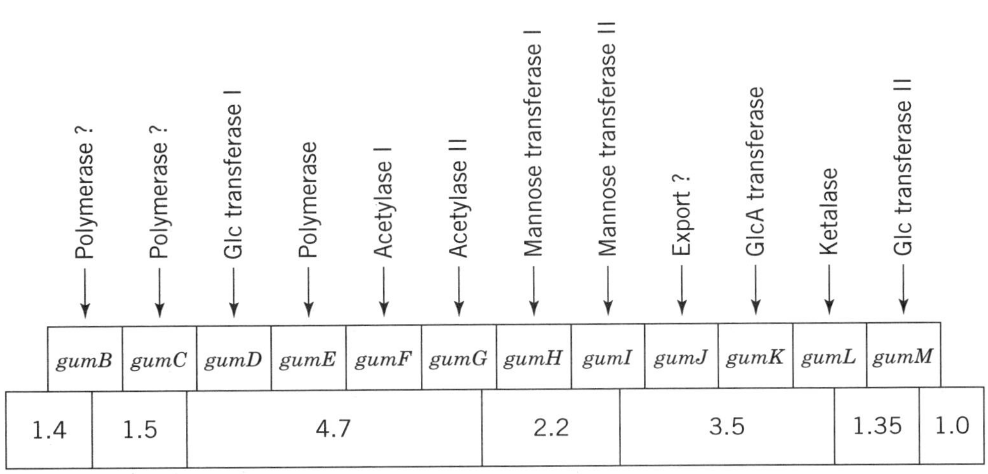

Figure 4. Genetic map of the genome segment of *X. campestris* involved in xanthan production. The *Bam*HI restriction map (bottom) indicates the order and the approximate size in kilobases (kb) of the fragments that comprise the xanthan gene cluster. This cluster has a total DNA length of 16 kb. The genetic map (center) indicates the 12 separate xanthan genes, designated *gumB* to *gumM*. The biochemical functions (top) indicate the enzymatic activity identified for each gene; (?) refers to possible involvement with polymerization or export as indicated; Glc, glucose, GlcA, glucoronic acid. *Source*: Adapted from Vanderslice et al. (45) and Sutherland (38).

UDP-glucose pyrophosphorylase is essential for xanthan synthesis (49). In fact, over 20 relevant genes seem to be identified by genetic analysis and DNA sequencing (50).

BIOPROCESS TECHNOLOGY

Raw Materials

The most efficient xanthan gum producer is *X. campestris*, however other species such as *Xanthomonas phaseolis, Xanthomonas malvacearum, Xanthomonas carotae, Xanthomonas manihotis, and Xanthomonas jugandlis* are known to be able to produce the exopolysaccharide as well. The nutritional versatility of *X. campestris* allows the use of industrial, complex, and synthetic media formulations. In industrial processes, commercial-grade glucose or hydrolized starch is used as the carbon substrate. Sucrose and acid whey are also used. Glucose concentrations up to 5% have been found to give the best product yield; at higher glucose concentrations, xanthan yields decrease. In fact, the maximum level of glucose is limited by the rheology of the fermentation broth. Nitrogen sources, such as corn steep liquor, casein hydrolysates, distiller's dried solubles, yeast, and soybean extracts, can be used, but because of their relatively undefined and variable composition they have been replaced by ammonium salts. Xanthan can indeed be produced in good yield when *X. campestris* is grown in a simple synthetic medium composed of ammonium chloride, glucose, and salts (51). The benefits of using synthetic media are twofold: control of genetic instabilities and improvement of product quality (52). Although the structure of xanthan synthesis is essentially independent of media formulation, there are some differences in acyl substituents and the rheological properties of the polysaccharide, as well as in the product yields (21).

Basic Structure of the Production Process

The design of a xanthan production process depends on the economic requirements to achieve optimum performance with regard to productivity, product concentration, and yield and, at the same time, to fulfill the desired product quality and application specifications. However, this is not a straightforward task because, during the production process, viscosity increases drastically. Therefore, product concentration are usually low and product quality also varies. A compromise between the commercial targets and the limitations inherent to the process must be reached at the industrial level by establishing the relationships between the bacteria's performance, environmental conditions, and equipment design and operation.

Xanthan is produced commercially in a conventional batch process using mechanically agitated vessels (Fig. 5) and a culture of *X. campestris* in a suitable fermentation medium. It is an aerobic process that runs for about 100 h under the following operating conditions: 28–30 °C, pH ~ 7, aeration rate higher than 0.3 vvm, and specific power input for agitation higher than 1 kW/m^3. A product yield approaching 50% is typical in the standard process. The inoculum preparation includes several stages requiring different set of reactors ranging from 10 L for the initial seed culture up to 100 m^3 for production purposes, the need volume being usually enlarged by a factor of 10.

Typical profiles of batch culture of *X. campestris* using synthetic media similar to that used by Davison (53) and the operating conditions mentioned above are shown in Figure 6. The data are plotted as the time courses for cell dried mass (CDM), xanthan, glucose, the rheological parameters k (consistency factor) and n (flow index), the oxygen transfer rate (OTR), and the oxygen partial pressure (pO_2). As the fermentation evolves, cells grow exponentially, resulting in a rapid consumption of the nitrogen

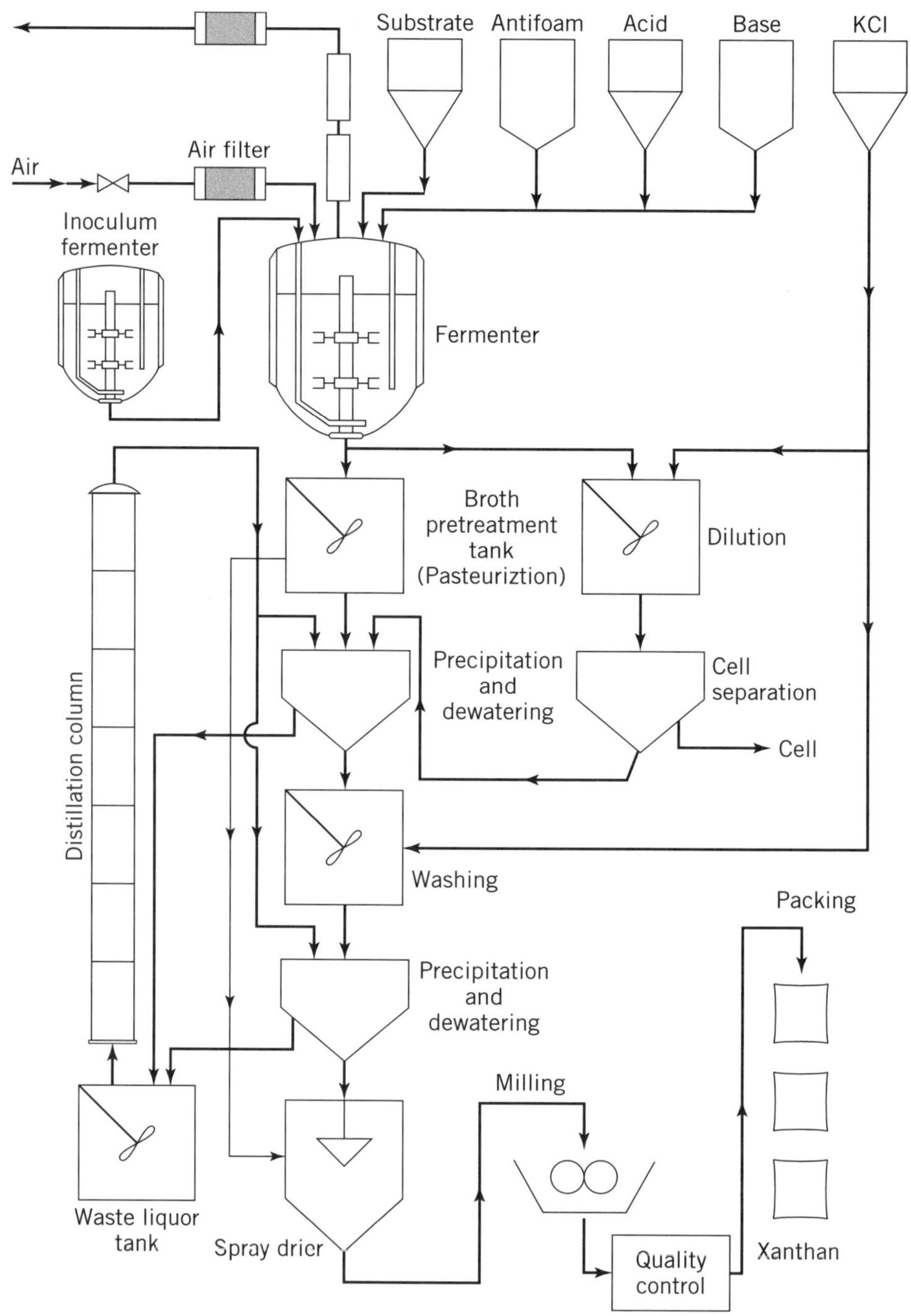

Figure 5. Flowsheet for production of xanthan.

source (not shown). Under N-limitation, cells enter the stationary phase. Xanthan is synthetized in both the growth and the stationary phases, reaching a final concentration of about 20 g L^{-1}. Xanthan biosynthesis ceases as glucose is completely consumed. The consistency factor (k) as a parameter of the rheological properties of xanthan increases from ~15 mPa sn to 30,000 mPa sn. The pO_2 dropped over the initial periods to a minimum value corresponding in time to peak oxygen uptakes rates during the growth phase. Thereafter, both OTR and pO_2 decrease continuously toward the end of the fermentation, reflecting the effect of increasing broth viscosity on the oxygen transfer rate.

After the fermentation stage, xanthan is recovered through a multistep downstream process (Fig. 5). The characteristics of this process are defined by the end product application and the difficulty of removing bacterial cells from viscous culture broths. As pointed out by Smith and Pace (54), the objectives of the recovering process are (*1*) concentration of the fermentation broth to a solid form, which has to be microbiologically stable and easy to handle, transport, and store; (*2*) purification to reduce the degree of nonpolymer solids, such as cells and/or salts, in order to improve the functional performance (odor, color, and taste) of the product; (*3*) deactivation of undesirable contaminating enzymes, such as pectinases and cellulases;

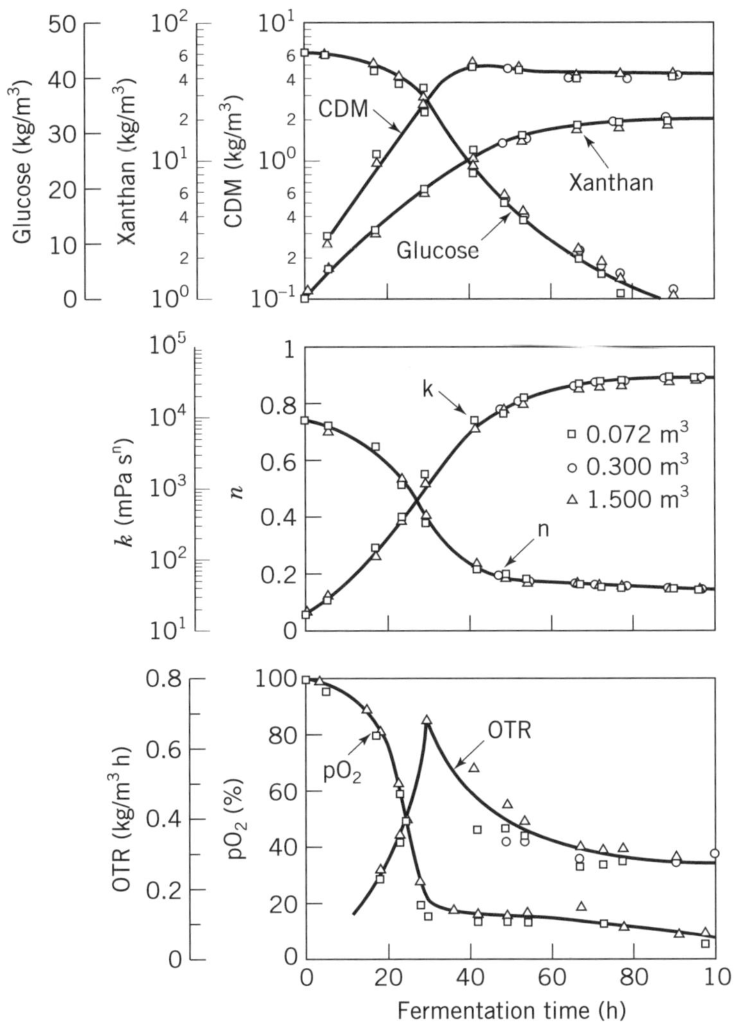

Figure 6. Time profiles of xanthan fermentation in reactors of 72 L, 0.3, and 1.5 m^3 applying the same power input (3.3 kW/m^3) and gas flow rate (7.2 mm s^{-1}). *Source:* Adapted from Herbst et al. (23).

and (*4*) modification of the chemical properties of the product to alter either the application performance or the dispersion and solution rate characteristics of the polymer.

When industrial grade Xanthan is required, the post-fermentation process starts with the pasteurization of the fermentation broth to kill the bacterial cells (Fig. 5). Because *Xanthomonas* are sensitive to temperature and are nonspore formers, complete pasteurization can be readily accomplished. However, caution has to be taken with the temperature of the transition midpoint (T_m) for xanthan in order to avoid thermal degradation of the polymer. The T_m for xanthan is about 100 °C but varies with the total ionic strength and the ratio of the contents of pyruvate and acetate groups (55–57). For xanthan applications where maximum cell removal is required (such as in the food, pharmaceutical, and cosmetics industries), cell separation by centrifugation is facilitated by dilution of the fermentation broth to reduce viscosity. Alternatively, the bacterial cells can also be subjected to chemical or enzymatic digestion (58,59); however, the resultant product may contain cell fragments, and solutions are optically not clear. Such a product may be restricted to applications in the petroleum industry.

After the prerecovery treatment just mentioned, the isolation of xanthan is accomplished by lowering its solubility in order to obtain either a precipitate or a polymer-concentrated phase by either the addition of lower alcohols, or salt or acid, respectively. The more favored method is the addition of ethanol or isopropanol. The added alcohol is subsequently recovered from the spent liquor by distillation (Fig. 5). The addition of salts improves the precipitation by creating a charge reversal effect in which all the available anionic groups are bound with cations of the added salt. The addition of KCl to assist precipitation reduces, for example, the amount of isopropanol required by about 30% (60). Consequently, increasing both the product concentration and the ionic strength prior to precipitation will decrease the amount of alcohol required per mass of product. This has a positive inpact on the economics of the process because the cost of alcohol recovery and the inevitable losses of alcohol during xanthan precipitation represent an important portion of the production cost.

Once the polymer is obtained in a wet solid form, its final purity (quality) and the economics of the next operation units can be determined by dewatering and washing. The xanthan precipitate may be centrifuged or pressed in order to remove free liquid, thereby improving the purity of the product and reducing drying costs. At this stage, the additional washing (Fig. 5) may be carried out to improve the quality of the product by removing as much as possible the particulate matter (cell debris, microgels, organics residues, and pigments). To this end the concentrated form of xanthan is redissolved, washed with water/KCl, precipitated, and dewatered again.

The precipitated cake is then dried in batch or continuous driers under vacuum or with forced air, or with inert gas if the solids contain inflammable solvents. To avoid undesirable effects, such as polymer deterioration, discoloration, and impairment of solubility, the drying temperature must be carefully controlled.

The dried xanthan is then milled to the desired mesh size to control dispersability and dissolution rate as well as to make solids handling easier. The mesh size can range from 40 to 200; the most common mesh sizes are 80 (177 μm) and 200 (74 μm). A range of products with these particle sizes is marketed.

Because an important part of the production costs represents the recovery and purification of xanthan, preparations that have been concentrated directly from the fermentation broth are also commercially available at lower prices. These xanthan concentrates, prepared by tangential flow filtration, contain between 8 and 12% dry polysaccharide to which antibacterial agents are added to prevent deterioration. This form of xanthan, which is specially useful in drilling operations in the oil industry, is available in different industrial grades and can be readily diluted to the required concentrations.

Reactor Designs and Operation Modes

During xanthan fermentation the hydrodynamic behavior of the broth changes substantially as the polymer is produced. The rheology exhibited by the xanthan fermentation culture is complex and includes pseudoplastic flow behavior, yield stress, and viscoelasticity. The effective viscosity increases by about three powers of 10. Under such conditions adequate aeration and mixing have to be provided if high specific xanthan productivity is to be maintained and stagnant zones and hence oxygen limitation avoided. Fermenter and impeller design, therefore, plays a critical role in ensuring good mass transfer and homogeneity and, hence, good product quality. Because the turbulence and shear rate in stirred tank-reactors (STRs) is higher near the impeller than in bulk mixing, optimal design requires the provision of high turbulence to give good gas dispersion and bulk mixing. The conventional Rushton turbine (the most widely used impeller in the fermentation industry) has shown to be disadvantageous because of its high unaerated power number (61) and compartimentalization effects (62). At high xanthan concentration the sorption efficiency (mass of oxygen absorbed at maximum driving force per energy supplied) of an STR with a Rushton turbine drops significantly below that of other reactors (Fig. 7). Consequently, alternative impeller types such as the Intermig, Prochem Maxflo T, and Scaba 6SRGT have been tested (23,64). Among these, the Intermig has shown a four times higher sorption efficiency at high xanthan concentrations, compared with Rushton turbines (23).

Reactor design has also been targeted as a means of improving xanthan production and quality. The plunging jet reactor (65), airlift reactor (66), centrifugal fibrous-bed bioreactor (67), STR with draft tube (68), and bubble column (66), have all been tested for that purpose. In spite of the lower energy requirements and higher oxygen sorption efficiency, the poor mixing in the wall region of the plunging jet reactor is a problem that needs to be solved. The limitation of oxygen supply in the downcomer of airlift reactors has negative effects on the xanthan production rate and product yield (66). Although the design is novel, the recently proposed centrifugal fibrous-bed bioreactor (67) does not show obvious advantages. An STR with a draft tube shows good performance as evaluated by its sorption efficiency (Fig. 7) (68). At low xanthan concentration the bubble column has significantly higher efficiency than the other reactors (Fig. 7). Even at high xanthan concentrations the bubble column exhibited excellent mixing performance; slug flow prevails and inhomogeneities (stagnant zones) are avoided (34,66).

The possibility of running a xanthan bioprocess continuously has been also examined at laboratory scale (69–71). The main features of this mode of operation are (*1*) higher specific productivity, (*2*) no definite limit (in theory) to the length of run because bacterial degeneration could be avoided using synthetic media with reduced levels of sulfur, (*3*) avoidance of time losses in restarting batch runs, and (*4*) near maximum process productivity and a constant product quality, owing to steady-state conditions. One of the most interesting applications of continuous mode operation may be found in the oil industry, where in situ production plants may provide xanthan in different industrial grades that can readily be diluted to the required solution concentration. However, a serious weakness of the process is the low product accumulation, which is around 5–7 g/L at the conditions where maximum productivity and high product quality are favored (dilution rate ~0.075 h^{-1}). This constraint and limitations in avoiding culture contamination are important reasons why industry has not yet adopted this cultivation method.

Unfortunately, fed-batch strategies with glucose feeding (70,72–77) have not brought significant improvement to the process; some resulted in a performance that was even worse than that of batch culture. A common characteristic of these processes is that an important portion of the glucose supplied (up to 60%) remains in the reactor without being utilized. These limitations in fed-batch culture make this process strategy not economically attractive.

Recently, a new fed-batch operation mode was proposed, based on the observation that the N-source inhibits growth of *X. campestris* (21). Therefore, the novel feature of this process is the use of ammonia feeding to control growth. This allows major product accumulation with higher molecular weight during the growth phase (21). The benefits to be expected are as follows: (*1*) larger product yield, (*2*) higher volumetric productivity, (*3*) improvement of product quality and uniformity, (*4*) reduction in capital investment and unit production costs, and (*5*) increase in plant production capacity as compared with the standard batch process (Table 2). The application of this process at the production level may require combined strategies in-

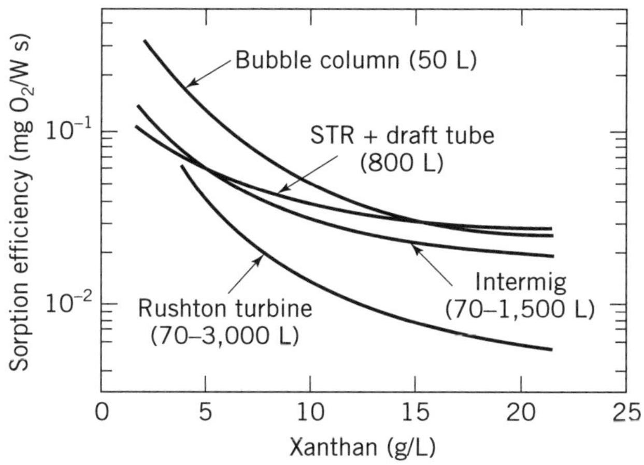

Figure 7. Comparison of the energetic efficiencies of oxygen transfer during xanthan production in various bioreactors and different impeller design. *Source*: From Herbst (63).

Table 2. Comparison of Performance and Costs of Xanthan Cultivation in Batch and Fed-Batch Operation Mode

Parameter	Standard batch	Fed-batch
X_O (g L^{-1})	0.2	0.5
X_f (g L^{-1})	4	5.2
$Product_{(harvest)}$ (g L^{-1})	25.38	30.8
Run time (h)	84	49
Productivity (g L^{-1} h^{-1})	0.3	0.63
$Y_{P/S}$	0.507	0.65
Capital investment cost ($ × 10^3)[a]	5,493	2,930
Production cost per kg ($)[a]	7.28	5.38

Cost analysis has been made taking a production plant capacity of 1,500 tons/year (21).

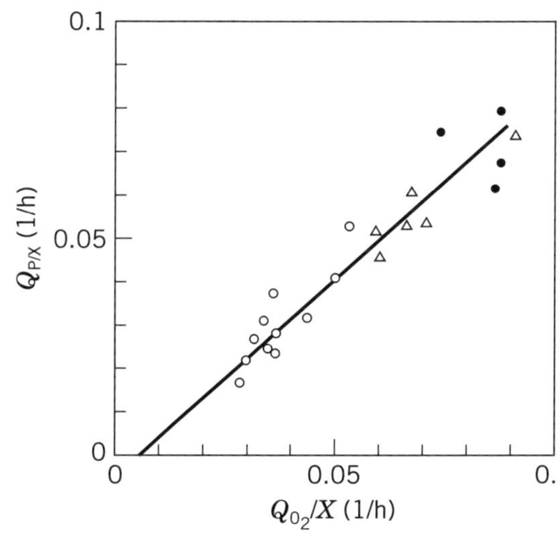

Figure 8. Dependence of the specific xanthan production rate on the specific oxygen uptake rate. Circles refer to runs in an STR at 15 L and 1,500 L, triangle correspond to bubble-column reactor; filled symbols denote no oxygen limitation.

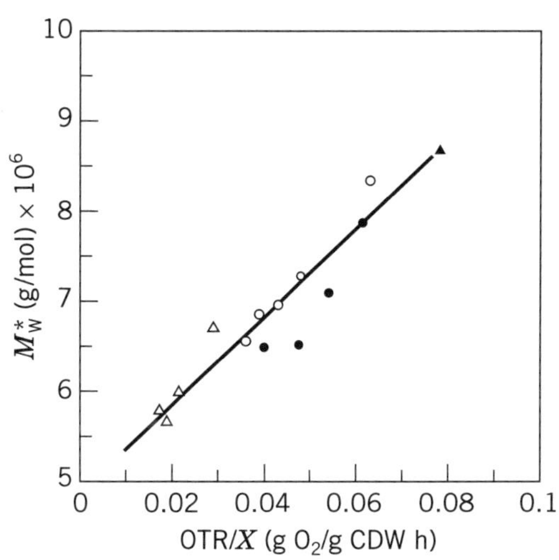

Figure 9. Influence of the specific oxygen transfer rate on the weight-mean molar mass of the xanthan produced.

cluding the use of the Intermig impeller in STRs with a multiple feeding device, or the use of a bubble-column reactor.

Oxygen Transfer and Scale-Up

Transport processes are dependent on scale because the time constants for transport processes increase with scale. Scale-up problems are pronounced in xanthan fermentation because of the highly viscous pseudoplastic flow behavior of the culture broth, which deteriorates the mass transfer, mixing, and heat transfer capabilities of the fermenter. The actual oxygen concentration and the kinetic behavior of the bacteria are dependent on scale owing to the fact that oxygen and other nutrients involved in the bacterial metabolic activity are consumed and have to be supplied by transport processes. Sufficient oxygen supply to the obligately aerobic bacterium is of particular importance. If dissolved oxygen becomes depleted in the culture medium, the specific xanthan productivity of the cells decreases linearly with the specific oxygen uptake rate (78) (Fig. 8). Furthermore, and as shown in Figure 9, a lower molar mass product is formed under oxygen limitation (79). The information in Figures 8 and 9 shows that, among other scale-up criteria, scale-up at constant specific oxygen transfer rate is the most appropriate choice to achieve constant productivity and high product quality. Maintaining a constant volumetric mass transfer coefficient (k_La) in process scale-up is approximately the appropriate condition to guarantee the same oxygen supply rate to satisfy the oxygen demand of the cell population.

Aiming at high product concentration for economic downstream processing, oxygen transfer limitation is likely to occur at the final stage of the fermentation. This applies particularly to industrial fermentations where the specific power input is bound to be low. Any error at the design stage when estimating the oxygen transfer capacity of the fermenter results in a proportional error in the maximum xanthan production rate. Consequently, the volumetric mass transfer coefficient, k_La, must be accurately estimated.

Various approaches toward the estimation of k_La in stirred-tank reactors with viscous pseudoplastic liquid have been proposed in the literature. The flow behavior is commonly described according to the Ostwald-de Waele relationship (equation 1). Because the shear rate in the STR is locally nonuniform, the effective mean shear rate, $\dot{\gamma}_{eff}$, caused by the impeller is described, for example, in terms used by Metzner & Otto (80):

$$\dot{\gamma}_{eff} = AN \quad (5)$$

where A is a constant characteristic of the impeller and N is the sirrer speed. Various expressions for $\dot{\gamma}_{eff}$ in the STR, employed to calculate the effective viscosity, η_{eff}, from equation 2, are listed in Table 3. Equation 6 of Table 3 allows one to predict shear rates at turbine impeller blades, whereas equation 7 applies in the "stirrer zone." Equation 8, is a modification of the Metzner–Otto relationship including a function of the flow behavior index (n) and differs only marginally from the original one (23). Herbst et al. (23) showed that the predicted $\dot{\gamma}_{eff}$, using Calderbank and Moo-young's relationship, is almost two orders of magnitude lower than the shear rate at the stirrer tip proposed by Henzler & Kauling (84). The last predicts a different trend because increasing viscosity is assumed to lower the shear rate. It is obvious that any correlation of the volumetric mass transfer coefficient will be affected by the choice of relationship for the effective shear rate $\dot{\gamma}_{eff}$.

A number of mass transfer correlations exist in the literature, some of them are given in Table 4. The relationships in Table 4 may be divided into those that reflect the effect of the stirring speed directly (equations 10, 11, and 13) and those that take into account the resultant power input (equation 12). The predictive capability of these equations, however, differs significantly (Fig. 10). Based on the results shown in Figure 10, it can be concluded that, the dimensional relationship of Kawase and Moo-Young (87), equation 12, gives the best representation of the experimental data. Herbst et al. (23), working with reactor volumes ranging from 0.015 to 1.5 m^3, demonstrated further that the power input required to achieve a desired mass transfer rate calculated from equation 12 was reduced considerably when Intermig impellers are applied. Figure 11 shows that the experimental data are well described by the relationship of Kawase and Moo-Young (equation 12) when the power input (P/V) was modified by a factor of three (23).

The scale-up criteria of a constant specific oxygen transfer rate was experimentally confirmed to be valid in xanthan bioprocesses by Herbst et al. (23). In order to keep the volume-specific oxygen transfer rate constant, equal power input ($\sim$3.3 kW/m^3) and superficial gas velocity (7.2 mm s^{-1}) in reactors ranging from 0.072 to 1.5 m^3 were employed. Under these operating conditions all the main fermentation performance parameters show an excellent reproducibility independent of the scale. This can be recognized in the results in Figure 6 where all concentrations, the rheological properties (K, n), and the oxygen transfer rate and the dissolved oxygen concentration are reproducible. Moreover, the product quality as evaluated by its molecular weight, the polydispersity index, and the pyruvat content were constant irrespective of the fermenter scale. These results highlight that constant specific oxygen transfer rate is the suitable scale-up criterion for constant xanthan productivity and product quality.

Effects of the Process Design on Product Quality

The high viscosity yield of xanthan at low shear rates and its shear thinning behavior are particularly important during and after processing of many industrial products. Thus, a highly pseudoplastic xanthan is required. This means that, according to the rheology (equation 2) for a determined shear rate, $\dot{\gamma}$, high values of the rheological parameters k and n are desired to achieve a maximum effective viscosity (η_{eff}). However, in real terms, there is no way to vary k and n independently because both appear to be interdependent (see equation 4). The solution of equation 4 is given by equation 14, which exhibits an inverse functional relationship between the flow index, n, and both the xanthan molar mass, M, and the consistency factor, k.

Table 3. Selection of Relationships for the Effective Shear Rate in Stirred Tanks

Reference	Approach	Correlation	Eq.
Wichterle et al. (81)	Semiempirical	$\dot{\gamma} = N(1 + 5.3n)^{1/n}(N^{2-n}d^2\rho_L/k)^{1/(1+n)}$	(6)
Obernosterer and Henzler (82)	Semiempirical	$\dot{\gamma} = \left[0.038 \frac{P}{V_L} \frac{1}{k} \left(\frac{D}{d}\right)^3 \frac{d}{b}\right]^{1/(1+n)}$	(7)
Calderbank and Moo-Young (83)	Semiempirical	$\dot{\gamma}_{eff} = AN\left(\frac{4n}{3n + 1}\right)^{1/(1-n)}$	(8)
Henzler and Kauling (84)	Dimensional analysis	$\dot{\gamma}_{eff} = \left(\frac{P/V}{\mu_{eff}}\right)^{0.5}$	(9)

Table 4. Selection of Correlations for the Volumetric Mass Transfer Coefficient (K_La) in Stirrer Tanks with Viscous (Non-Newtonian) Media

Stirrer	Flow behavior	Shear rate	Correlation	Eq.
Rushton turbine $d/D = 0.4$	$k \leq 13.2$ Pa s^n $n \geq 0.42$	(Eq. 5)	$\frac{k_L a d^2}{D_L} = 0.06\left(\frac{d^2 N \rho_L}{\mu_{eff}}\right)^{1.5}\left(\frac{\mu_{eff}}{\rho_L D_L}\right)^{0.5}\left(\frac{\mu_{eff} u_G}{\sigma}\right)^{0.6}\left(\frac{dN^2}{g}\right)^{0.19}\left(\frac{Nd}{u_G}\right)^{0.32} \cdot [1 + 2(\lambda N)^{0.5}]^{-0.67}$	(10)
Rushton turbine $d/D = 0.27–0.47$	$k \leq 0.005$ Pa s $n = 1$	$10^4\ s^{-1}$	$\frac{k_L a d^2}{D_L} = 2.5 \times 10^{-4}\left(\frac{d^2 N \rho_L}{\mu_{eff}}\right)^{1.8}\left(\frac{\mu_{eff}}{\rho_L D_L}\right)^{0.5}\left(\frac{\mu_{eff}}{\mu_w}\right)^{1.39} \cdot [1 + 100(\lambda N)]^{-0.67}$	(11)
Semitheoretical approach		(Eq. 8)	$k_L a = 0.675 D_L^{0.5} \frac{\rho_L^{3/5}(P/\rho_L V)^{(9+4n)/(10+n)}}{(k/\rho_L)^{(1+n)/2}\sigma^{3/5}}\left(\frac{u_G}{u_b}\right)^{0.5}\left(\frac{\mu_{eff}}{\mu_w}\right)^{-0.25}$ with $u_b = 0.265$ m/s	(12)
6-Blade paddle $d/D = 0.5$	$k \leq 33.2$ Pa s^n $n \geq 0.68$	(Eq. 5)	$\frac{k_L a}{N} = 0.0786\left(\frac{d^2 N \rho_L}{\mu_{eff}}\right)^{0.41}\left(\frac{\rho_L N^2 d^3}{\sigma}\right)^{0.13}\left(\frac{Nd}{u_G}\right)^{0.69}$	(13)

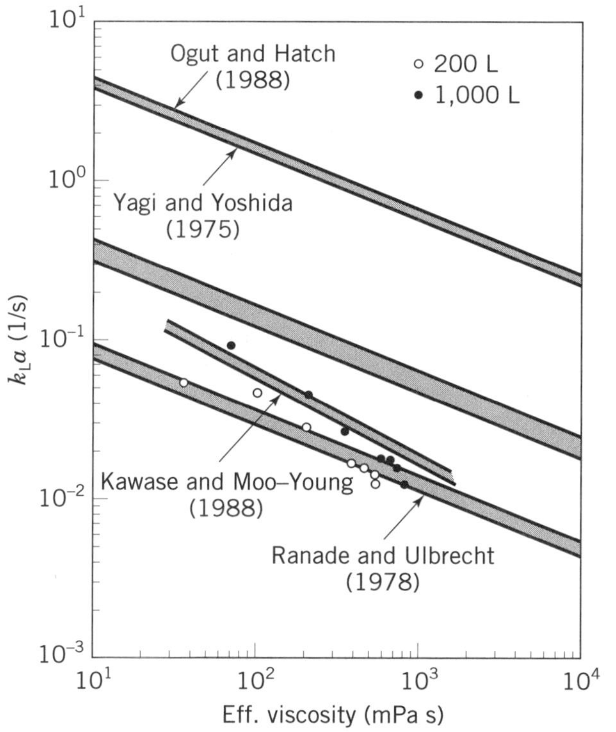

Figure 10. Comparison of experimental data with estimated values of the volumetric mass transfer coefficients using equations 10–13 of Table 4.

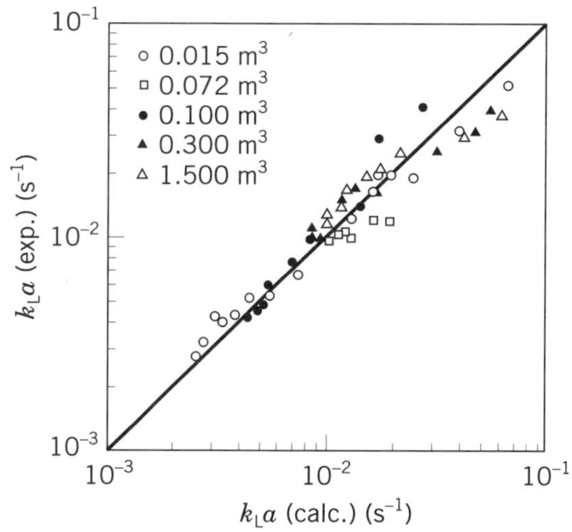

Figure 11. Parity plot for the relationships of Kawase and Moo-Young (82) (equation 12), with experimental data from Herbst et al. (23). Power input in equation 12 was modified by a factor of 3 for Intermig impeller.

$$n = \frac{1 + 0.071 \cdot B}{1 + B}, \text{ with}$$

$$B = 0.0227\left(\frac{M}{10^6}\right)^{2.08}(k - 10^{-3})^{0.47} \qquad (14)$$

As observed in the results in Figure 6, k rises by three orders of magnitude toward the end of the fermentation, while n declines by less than one order of magnitude. This indicates that the consistency factor, k, is a predominant factor in determining the rheological properties of xanthan. The functional dependence of the parameter k on xanthan molar mass and product concentration is well established (see equation 3). Therefore, to achieve high values of k for a given product concentration, conditions of high xanthan molar mass have to be achieved. To this end the empirical correlation of Peters et al. (24) may be considered (equation 15):

$$M = 4.9 \cdot 10^6 + 3.8 \cdot 10^6 \cdot \text{OTR}/Q^*_{O_2} + 2.6 \cdot 10^7 \mu \qquad (15)$$

Equation 15 implies that conditions of no oxygen limitation and high growth rates have to be ensured when xanthan of high viscosity yield to be produced. This rationale was proven to be right by the scale-up results of Herbst et al. (23) mentioned earlier. Furthermore, the media formulation has shown to affect indirectly the mean molar mass of xanthan (21,66). On one hand, when complex media are used, oxygen limitation is encountered at an early stage of the fermentation run, and hence, the mean molar mass of xanthan is impaired (66). On the other hand, high levels of the N-source in synthetic media formulations inhibit the bacterial growth, with the consequent result of low molar mass xanthan synthesis (21). It was further found that the N-source level also affects the pyruvate content of xanthan (21). The same author demonstrated further that the rheological properties of xanthan improve as the degree of pyruvate substitution increases (Fig. 12). This is in accordance with the initial observations of Sandford (86) and Cadmus et al. (52). Thus, the pyruvate content of the polymer should be regarded as an important parameter in order to design, predict, and control the rheological behavior of the fermentation broth, and as regards the product application.

COMMERCIAL APPLICATIONS

5.1 Food uses

Several applications that illustrate xanthan use in the food industry are given in Table 5. First of all xanthan gum is incorporated into foods to control the rheology of the final product. In performing this functional task, the polymer has a major effect on various sensory properties such as texture, flavor release, and appearance, which contribute to the acceptability of the product for consumer use. Because of its solution properties, xanthan can be used to provide a desired functionality at a lower concentration than most gums and, therefore, may be more cost-effective. Additionally, because of its pseudoplastic properties in solution, xanthan has a less "gummy" mouth-feel than gums that exhibit Newtonian behavior. An additional advantage in food formulations is the antioxidative action of xanthan as compared with other polysaccharides and additives (93).

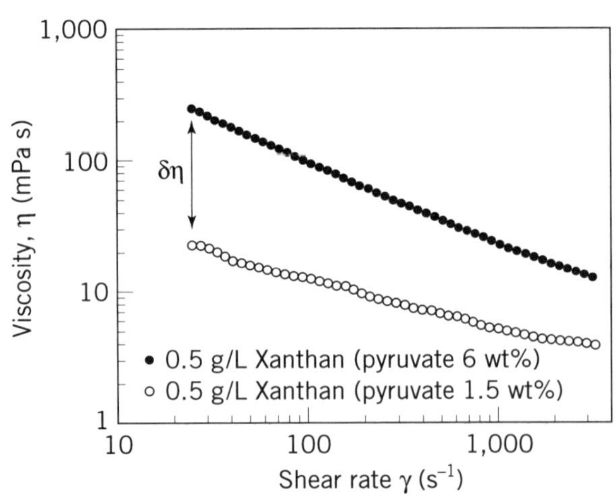

Figure 12. Comparative viscosity curves of 0.5 g/L xanthan in 0.12 M KCl with varying degree of pyruvate substitution. Filled symbols correspond to xanthan with about 6 wt% pyruvate content and M_w of 7×10^6 g/mol; open symbols relate to xanthan with about 1.5 wt% pyruvilation degree and M_w of 6.5×10^6 g/mol (run F7). *Source*: Results from Flores-Candia (21).

Table 5. Main Practical Uses of Xanthan for the Food Industry

Function	Application
Adhesive	Icings and glazes
Binding agent	Pet foods
Coating	Confectionery
Emulsifying agent	Salad dressing
Encapsulation	Powdered flavors
Film formation	Protective coatings, sausage casings
Foam stabilizer	Beer
Gluten substitution and dough processing	Baked goods, pasta
Stabilizer	Ice cream, salad dressings, juice drinks, margarine
Swelling agent	Processed meat products
Syneresis inhibitor, freeze–thaw stability	Cheese, frozen foods
Thickening agent	Jams, sauces, syrups, and pie fillings
Pumping and filling improvement	Processing of canned products

Source: From Refs. 87–92.

Xanthan is also being used in the formulation of reduced-calorie products, representing the second and third generation of fat replacers and stabilizers (87).

Cosmetics and Pharmaceutical Uses

Xanthan gum is used in cosmetics and toiletries as a bodying agent and emulsifier. Personal care products such as shampoos, creams, lotions, make-up, hair care products, and toothpastes can be formulated with xanthan (88,94,95). Additionally, xanthan provides creams and lotions with a good "skin-feel" during and subsequent to application (88). Recently, a new xanthan formulation (Ticaxan) was introduced into the market for that purpose (96).

In the pharmaceutical industry, xanthan is used to suspend pharmaceuticals such as antibiotics and to provide formulations of uniform dosage (88). It can also be used to stabilize emulsified cream formulations containing pharmaceuticals. Presently, xanthan is being evaluated in controlled-release applications, such as in gastrointestinal pharmaceutical applications, as rate limiting membranes for sustained delivery of oral formulations (97,98) and as dissolution and binding agents in tablets and rectal preparations (99,100).

Agricultural and Other Industrial Applications

Xanthan is used in a diverse range of industrial applications, particularly as a suspending or thickening agent. The polymer improves the flowability of fungicides, herbicides, and insecticides by uniformly suspending the solid components of the formulation in an aqueous system or by stabilizing emulsions and multiphase liquid systems (88). The unique rheological properties of xanthan solutions also improve sprayability, reduce drift, and increase pesticide cling and permanence (89,101,102). Recently, various "tolerance exemptions" were issued by the U.S. Environmental Protection Agency for the use of xanthan as a surfactant in pesticide formulations (103). Because of its ability to disperse and hydrate rapidly, xanthan is used in jet injection printing. It is nonpolluting and gives a good color yield. It is also used in several mining processes for removing dissolved metals in a highly efficient process (104). Lately, in the formulation of a new generation of thermoset coatings, xanthan was introduced by Monsanto with the aim of addressing customers' environmental challenges (105).

Petroleum Industry

In the petroleum industry, xanthan gum finds a wide range of applications. Among these are oil drilling, fracturing, pipeline cleaning, workover, and completion. Xanthan's unique solution rheology, its excellent compatibility with salts, and its resistance to thermal degradation have made it useful as an additive to drilling fluids (88,106). The pseudoplasticity of its solutions provides low viscosity at the drill bit, where shear is high, and high viscosity in the annulus, where shear is low. Therefore, xanthan serves a dual purpose in allowing faster penetration at the bit because of the low solution viscosity in that area and, at the same time, suspending cuttings in the annulus because of the high solution viscosity in that area. Hydraulic fracturing is used to effect deep-penetrating reservoir fractures that improve productivity of the well. Xanthan's high solution viscosity at low shear is important for proppant suspension in this operation (107). The pseudoplastic rheology, shear stability, and friction reduction impaired by xanthan provide for maximum delivery of pressure to the formation and minimal friction loss in the tubing and reservoir. A workover and completion fluid is essential for achieving and maintaining maximum well productivity. Xanthan contributes to excellent hole cleaning, suspension, friction reduction, and stability to shear and high temperatures. An important speciality application employs a xanthan gel to remove rust, welding rods, wet slag, and other debris from gas pipelines. This pipeline cleaning method is considered safe, environmentally acceptable for open sea disposal, and cost-effective (88).

Enhanced Oil Recovery

For every barrel of oil actually produced, about another two remain in the ground (108). Therefore, enhanced oil recovery (EOR) may become important in the next decades to meet the increasing energy demand of the future. Xanthan gum can be used in two types of enhanced recovery systems. The principal use is for an improved form of water flooding in secondary oil recovery, a phase in which polymers are used to increase the efficiency of water contact and to displace reservoir oil. The second application is in micellar-polymer flooding as a tertiary oil recovery operation. Polymer-thickened brine is used to drive a slug of surfactant through the porous reservoir rock to mobilize residual oil; the polymer prevents fingering of the drive water through the surfactant ban and ensures good area sweep. In both applications, the function of the polymer is to reduce the mobility of injected water by increasing its viscosity. Evaluation of hundreds of biopolymers has found xanthan to be one of the best candidates for use in EOR operations. The reasons for this are its (*1*) non-Newtonian behavior, (*2*) high viscosity yield even at low concentrations (600–2,000 ppm), (*3*) low sensitivity of viscosity to salinity changes (unlike the polyacrylamides), (*4*) resistance to mechanical degradation, (*5*) stability with respect to higher temperature (up to almost 90 °C).

One of the major limiting factors in the application of xanthan in EOR is that, by the nature of its production method, it is expensive, and the current general market price is in excess of $11/kg (109). In addition, some drawbacks during field operation are associated with poor injectability behavior, adsorption into the rock strata, and plugging problems (110,111). These problems have been attributed to the quality and composition of the polymer (110,112), to water quality (111), and to well configuration (113,114).

MARKET SIZE AND FUTURE DEVELOPMENT

Xanthan gum currently dominates the microbial gum market, which is growing at between 6 and 7% per year (115). It has been suggested that global production exceeds 50,000 tons per year (116). The worldwide xanthan market is valued between $600 and $800 million per year (109,116,117). The European Union market accounted for about $158 million in 1994, and the U.S. market was reported to be about $120 million in 1995. Table 6 shows the U.S. market size and forecast of xanthan consumption by major end uses. Food accounts for about 70% of total consumption; the remainder is spread among petroleum production, cosmetics, and other uses. Overall growth is forecasted at about 7.0% yearly, with food applications providing most of the impetus at 8.3% annually (1). Consumption of xanthan in petroleum is forecasted at a very modest rate (2%), which is due mainly to recovery operations rather than oil well drilling (2). Because of its unique properties, the application of xanthan in cosmetics has been growing successfully at 5.9% per year (82). It should be noted that use in textile production is very nominal because the textile industry in the United States is conservative and only growing with the national economy (3).

An analysis of Table 6 shows that xanthan is positioned well in U.S. markets. Further analysis reveals that xan-

Table 6. Market Analysis and Forecast: Xanthan Consumption in The United States

End use	1993 ($ M)	1995 ($ M)	1998 ($ M)	AAGR (%) 1993–1998
Food	74.4	89.5	111.0	8.3
Petroleum	16.5	17.1	18.2	2.0
Cosmetics and toiletries	9.0	10.2	12.0	5.9
Textiles	1.4	1.4	1.4	0
Other	1.1	1.1	1.7	8.5
Total	102.4	119.1	144.3	7.1

Source: Business Communications Company (1–4,94).
Note: Major xanthan demand according to end use is listed in millions of dollars (M); AAGR, annual average growth rate.

than gum represents the fastest growing segment of the polysaccharide industry (1–4). Most growth is in foods, where demand for natural gums is falling as a result of market share erosion by xanthan (118).

The high growth rate of the xanthan market may explain not only why the major producers have all expanded in recent years (119) but also why Archer Daniels Midland Co. has decided to enter the xanthan market with an estimated 10,000 tons of annual capacity (119,120). Between 1992 and 1994, the global xanthan production capacity doubled in size (119,121). Recently, Nutrasweet Kelco Co., the world's leading producer of xanthan, has announced further expansion (122). In fact, an overcapacity has been created in the xanthan market, which may have two important consequences: (*1*) an accelerated penetration of xanthan into new markets, and (*2*) stiff competition for new polysaccharides. Success may be limited only to those polymers that have a clear advantage in performance or customer perception. Less-expensive polymers that do not perform better may be disadvantaged due to the high costs of incorporating new gum into the end product. An estimate from 1982 suggests that the testing required by the U.S. Food and Drug Administration to gain food clearance approval for a new additive can cost about $2 million and may take up to 8 years (123).

According to a market forecast, the two main factors shaping the xanthan market in the next century are growing xanthan demand in developing countries and worldwide energy demand, which will stimulate EOR operations as reserves decline (124,125). Indeed, the large strategic value of xanthan still remains in EOR operations, whose worldwide production at the beginning of 1996 was estimated to be around 2.2 million barrels/day (126).

The large economic value of xanthan in the EOR market can be assessed, as suggested by Moses (127), by looking at the amount of oil that remains in the earth's crust (Table 7). Thus, it becomes clear that the EOR markets are very large indeed. They are, however, low-price markets that could be developed only through the use of much improved technologies with low production costs. To improve the competitive advantage of xanthan in EOR, more efficient fermentation processes and better control of the polymer quality are required.

Table 7. Approximate Value of Oil Not Economically Recoverable by Present Technology

Source region	Estimated value of oil unrecoverable by present technologies ($ × 10^{12})
Asia-Pacific	0.6
Western Europe	0.8
Middle East	12.4
Africa	1.9
Western Hemisphere	4.1
Former East-Block states	2.9
Total for the whole world	22.7

Source: From Ref. 25.
Note: The estimate was based on the following assumptions: (*1*) a crude oil price of $18.00 per barrel, (*2*) that proven reserves (i.e., reserves regarded as presently recoverable) represent 35% of the original oil-in-place.

Relatively high xanthan costs, some technical drawbacks and the (relatively) low oil price have prevented market entry in recent years. Xanthan usage in EOR is typically 0.4% w/w (about 0.5 kg per barrel of oil), which represents about 25% of the estimated equilibrium oil price ($20 per barrel)—a value that is evidently too high for the oil industry.

A recent survey of EOR shows that polymer flooding is achieving commercial success (126). In this operation, up to 35% of original oil in place has been recovered in the Junggar basin (China). Polymer flooding is being used in several projects in France, Germany, China, India, and Romania. For example, polymer flooding accounts for about 50% of the total EOR operations in China. Even when it is not exactly clear what percentage xanthan is responsible for, the fact that the largest xanthan gum production plant has been brought onstream for oil applications is an indication that xanthan use in EOR is not far away. Analysts in marketing research are convinced that the use of xanthan as well as other microbial gums in EOR will become more prevalent as increasing worldwide energy demand results in rising oil prices.

NOMENCLATURE

a (m^{-1})	Interfacial area per unit volume of liquid
A	Impeller-related factor
b (m)	Height of stirrer blade
C_P (kg m^{-3})	Product concentration
d (m)	Stirrer diameter
D (m)	Reactor diameter
g (m s^{-2})	Gravitational acceleration
k (mPa sn)	Consistency index
k_L (m s^{-1})	Liquid-side mass transfer coefficient
M (kg $kmol^{-1}$)	Weight-mean molecular weight
N (s^{-1})	Stirring speed
n	Flow behavior index
OTR (g L^{-1} h^{-1})	Oxygen transfer rate
P (W)	Total power input
pO_2 (%)	Partial oxygen pressure
Q_{O_2} (g L^{-1} h^{-1})	Microbial oxygen demand
u_b (m s^{-1})	Bubble rise velocity
u_G (m s^{-1})	Volumetric gas flow rate
V (m^3)	Dispersion volume
V_G (k mol^{-1})	Mole volume of ideal gas
x	Molar fraction

Greek Symbols

$\gamma\cdot$ (s^{-1})	Shear rate
η (mPa s)	Viscosity
τ (Pa)	Shear stress
ρ (kg m^3)	Density
λ (s)	Relaxation time
σ (N m^{-1})	Surface tension

Subscripts or Superscripts

L	Liquid phase
eff	Effective
w	Water

BIBLIOGRAPHY

1. Business Communications Company, July, 1994, 00039081.
2. Business Communications Company, July, 1994, 00039094.
3. Business Communications Company, July, 1994, 00039098.
4. Business Communications Company, July, 1994, 00039068.
5. P.E. Jansson, L. Keene and B. Lindberg, *Carbohydr. Res.* **45**, 275–282 (1975).
6. L.D. Melton, L. Mindt, D.A. Rees, and G.R. Sanderson, *Carbohydr. Res.* **46**, 245–257 (1976).
7. R.A. Hassler and D.H. Doherty, *Biotechnol. Prog.* **6**, 182–187 (1990).
8. H. McIntyre, H.J. Ceri, and Vogel, *Starch. Starke* **48** (7–8), 285–291 (1996).
9. A.J. Clarke-Sturman, J.B. Pedley, and P.L. Sturla, *Int. J. Biol, Macromol.* **8**, 355–360 (1986).
10. G. Muller, J. Lecourtier, G. Chauveteau, and C. Allain, *Makromol. Chem. Rapid Commun.* **5**, 203–208 (1984).
11. R. Moorhouse, M.D. Walkinshaw, and S. Arnott, *ACS Symp. Ser.* **45**, 90–102 (1976).
12. G. Muller, M. Anrhourrache, J. Lecourtier, and G. Chauveteau, *Int. J. Biol. Macromol.* **8**, 167–172 (1986).
13. G. Holzward and E.B. Prestridge, *Science* **197**, 757–759 (1977).
14. G. Holzward, *Carbohydr. Res.* **66**, 173 (1978).
15. K.P. Shatwell, I.W. Sutherland, and S.B. Ross-Murphy, *Int. J. Biol. Macromol.* **12**, 71–78 (1990).
16. M. Milas, W.F. Reed, and S. Printz, *Int. J. Biol Macromol.* **18**, 211–221 (1996).
17. K.P. Shatwell, I.W. Sutherland, S.B. Ross-Murphy, and I.C.M. Dea, *Carbohydr. Polymers* **14**, 115–130 (1991).
18. K.P. Shatwell, I.W. Sutherland, S.B. Ross-Murphy, and I.C. M. Dea, *Carbohydr. Polymers* **14**, 131–147 (1991).
19. H. Kumagai, C.H. Chu, T. Sakiyama, S. Ikeda, and K. Nakamura, *Biosci. Biotechnol. Biochem.* **60**, 1623–1626 (1996).
20. D.V. Zasypkin, E. Dumay, and J. C. Cheftel, *Food Hydrocolloids.* **10**(2), 203–211 (1996).
21. J.-L. Flores-Candia, VDI Verlag GmbH, Düsseldorf, (1998).
22. U.S. Pat. 4,394,447 (1983), M. Cadmus and C.A. Knutson.
23. H. Herbst, A. Schumpe, and W.-D. Deckwer, *Chem. Eng. Technol.* **15**, 495–498 (1992).
24. H.U. Peters, I.-S. Suh, A. Schumpe, and W.-D. Deckwer, *Can. J. Chem. Eng.* **70**, 742–750 (1992).
25. A. Parker, P.A. Gunning, K. Ng, and M.M. Robins, *Food Hydrocolloids* **9**(4), 333–342 (1995).
26. J.C. Walker, in *Diseases of Vegetable Crops*, McGraw-Hill, New York, 1952, pp. 128–131.
27. J.F. Bradbury, in *Guide to Plant Pathogenic Bacteria*, CAB International Mycological Institute, Slough, 1986, pp. 198–260.
28. J. Swings, L. Vauterin, and K. Kersters, in Xanthomonas, Chapman & Hall, London, 1993, pp. 121–156.
29. M. Cadmus, C.A. Knutson, A.A. Lagoda, J.E. Pittsley, and K.A. Burton, *Biotechnol. Bioeng.* **20**, 1003 (1978).
30. J.F. Kennedy and I.J. Bradshaw, in *Progress in Industrial Microbiology*, vol. 19, Elsevier, Amsterdam, 1984, pp. 319–365.
31. E. Galindo and G. Salcedo, *Enzyme Microb. Technol.* **19**, 145–149 (1996).
32. J.C. Roseiro, M.E. Esgalhado, A.N. Emery, and M.T. Amaral-Collaco, *J. Chem. Tech. Biotechnol.* **65**, 258–264 (1996).
33. T.R. Jarman and G.W. Pace, *Arch. Microbiol.* **137**, 231–235 (1984).
34. A. Pons, C.G. Dussap, and J.B. Gros, *Biotechnol. Bioeng.* **33**, 394–405 (1988).
35. C. Whitfield, I.W. Sutherland, R.E. Cripps, *J. Gen. Microbiol.* **128**, 981–985 (1982).
36. L. Ielpi, R.O. Couso, and M.A. Dankert, *Biochem. Biophys. Res. Commun.* **102**, 1400–1408 (1981).
37. L. Ielpi, R.O. Couso, M.A. Dankert, *Biochem. Int.* **6**, 323–333 (1983).
38. I.W. Sutherland, in Xanthomonas, Chapman & Hall, 1993, pp. 363–388.
39. J.-L. Tang, C.L. Gough, and M.J. Daniels, *Mol. Gen. Genet.* **222**, 157–160 (1990).
40. U.S. Pat. 4,713,449 (1988), R.W. Vanderslice and P. Shannon.
41. M.R. Betlach, M.A. Capage, D.H. Doherty, R.A. Hassler, N.M. Henderson, R.W. Vanderlice, J.D. Marrelli, and M.B. Wood, in M. Yalpani ed., *Industrial Polysaccharides: Genetic Engineering, Structure / Property Relations and Application*. Elsevier, Amsterdam, 1987, pp. 35–50.
42. G.J. Boulnois and K. Jann, *Mol. Microbiol.* **3**, 1819–1823 (1989).
43. N.E. Harding, J.M. Cleary, D.K. Cabanas, I.G. Rosen, and K.S. Kang, *J. Bacteriol.* **169**, 2854–2861 (1987).
44. L. Thorne, L. Tansey, and T.J. Pollock, *J. Bacteriol.* **169**, 3593–3600 (1987).
45. R.W. Vanderslice, D.H. Doherty, and M. Capage, in *Biomedical and Biotechnological Advances in Industrial Polysaccharides*, Gordon & Breach, New York, 1989, pp. 145–156.
46. M.P. Marzocca, N.E. Harding, E.A. Petroni, J.M. Cleary, and L. Ielpi, *J. Bacteriol.* **173**, 7519–7524 (1991).
47. F. Katzen, A. Becker, A. Zorreguieta, A. Pühler, and L. Ielpi, *J. Bacteriol.* **178**, 4313–4318 (1996).
48. R. Köplin, W. Arnold, B. Hötte, R. Simon, G. Wang, and A. Pühler, *J. Bacteriol.* **174**, 191–199 (1992).
49. C.L. Wei, N.T. Lin, S.F. Weng, and Y.H. Tseng, *Biochem. Biophys. Res. Commun.* **226**, 607–612 (1996).
50. M.O. Martin, *Res. Microbiol.* **145**, 93–97 (1994).
51. M. Cadmus, C.A. Knutson, A.A. Lagoda, J.E. Pittsley, and K.A. Burton, *Biotechnol. Bioeng.* **20**, 1003 (1978).
52. U.S. Pat. 4,394,447 (1983), M. Cadmus and C.A. Knutson.
53. I.W. Davidson, *FEMS Microbiol. Lett.* **3**, 347–349 (1978).
54. I.H. Smith and G.W. Pace, *J. Chem. Tech. Biotechnol.* **32**, 119–129 (1982).
55. K. Hatakenaka, W. Liu, and T. Norisuye, *Int. J. Biol. Macromol.* **9**, 346–348 (1987).
56. P. Foss, B.T. Stokke, and O. Smidsrod, *Carbohydr. Polymers* **7**, 421–433 (1987).
57. F. Callet, M. Milas, and M. Rinaudo, *Int. J. Biol. Macromol.* **9**, 291–293 (1987).
58. U.S. Pat. 3,966,618 (1976), G.T. Colegrove.

59. Br. Pat. 2,065,689 (1981), T.J. Holding and G.W. Pace.
60. A. Margaritis and G.W. Pace, in M. Moo-Young ed., *Comprehensive Biotechnology*, Pergamon Press, Toronto, 1985.
61. A.-W. Nienow and T.P. Elson, *Chem. Eng. Res. Des.* **66**, 5–15 (1988).
62. R. Kuboi and A.W. Nienow, *Chem. Eng. Sci.* **41**, 123–134 (1986).
63. H. Herbst, Ph.D. Thesis, Technical University of Braunschweig, Germany, 1990.
64. A. Amanullah, L. Serrano-Carreon, B. Castro, E. Galindo, and A.W. Nienow, *Biotechnol. Bioeng.* **57**, 94–108 (1998).
65. A. Zaidi, P. Ghosh, A. Schumpe, and W.-D. Deckwer, *Appl. Microbiol. Biotechnol.* **35**, 330–333 (1991).
66. I.-S. Suh, A. Schumpe, and W.-D. Deckwer, *Biotechnol. Bioeng.* **39**, 85–94 (1992).
67. S.-T. Yang, Y.-M. Lo, and D.-B. Min, *Biotechnol. Prog.* **12**, 630–637 (1996).
68. H. Herbst, I.-S. Suh, H.-U. Peters, A. Schumpe, and W.-D. Deckwer, in *DECHEMA Biotechnology Conferences 3*, VCH Verlagsgesellschaft, 1989.
69. H.U. Peters, A. Zaidi, P. Ghosh, A. Schumpe, and W.-D. Deckwer, *DECHEMA Biotechnology Conferences, Annual Meeting of Biotechnologists*, Frankfurt am Main, Germany, May 28–30, 1993.
70. J.C. Roseiro, A.N. Emery, P. Simoes, F. Estevao, and M.T. Amaral-Collao, *Appl. Microbiol. Biotechnol.* **38**, 709–713 (1993).
71. A.-K. Jana and P. Ghosh, *J. Ferment. Bioeng.* **80**, 485–491 (1995).
72. Eur. Pat. 0112661 (1984), G.T. Banks and P.D. Browning.
73. U.S. Pat. 4,828,321 (1981), W.C. Wernau.
74. L. De Vuyst, J. Van Loo, and E.J. Vandamme, *J. Chem. Tech. Biotechnol.* **39**, 263–273 (1987).
75. H. Funahashi, M. Maehara, H. Taguchi, and T. Yoshida, *J. Chem. Eng. Jpn.* **20**, 16–20 (1987).
76. J.C. Roseiro, D.C. Costa, and M.T. Amaral-Collao, *Lebensm. Wiss. Technol.* **25**, 289–293 (1992).
77. K.V. Rajeshwari, G. Prakash, and P. Ghosh, *Lett. Appl. Microbiol.* **21**, 173–175 (1995).
78. H.-U. Peters, H. Herbst, I.-S. Suh, A. Schumpe, and W.-D. Deckwer, in *Proc. 3rd Int. Workshop Recent Dev. Bacterial Polysaccharides: Biomed. Biotechnol. Adv.* pp. 275–281.
79. I.-S. Suh, H. Herbst, A. Schumpe, W.-D. Deckwer, *Biotechnol. Lett.* **12**, 201–206 (1990).
80. A.B. Metzner and R.E. Otto, Agitation of non-Newtonian fluids. *AICHE J.* **3**, 3–10 (1957).
81. K. Wichterle, M. Kadlec, L. Zak, and P. Mitschka, *Chem. Eng. Commun.* **26**, 25–32 (1984).
82. G. Obernosterer and H.-J. Henzler, in *Rheologie und mechanische Beanspruchung biologischer Systeme*, GVC-VDI Tagung, Tübingen, 1987.
83. P.H. Calderbank and M.B. Moo-Young, *Trans. Inst. Chem. Eng.* **37**, 26 (1959).
84. H.-J. Henzler and J. Kauling, *Proc. 5th Eur. Conf. Mixing*, Paper 30, BHRA Cranfield, 1985.
85. Y. Kawase and M. Moo-Young, *Chem. Eng. Res. Dev.* **66**, 284–288 (1988).
86. P.A. Sandford, J.E. Pittsley, C.A. Knudson, P.R. Watson, M.C. Cadmus, and A. Jeanes, *ACS Symp. Ser.* **45**, 192–210 (1977).
87. L.C. Fryer, F.M. Aramouni, and E. Chambers, *J. Food Sci.* **61**(1), 245 (1996).
88. K.S. Kang and D.J. Pettitt, in *Industrial Gums: Polysaccharides and Their Derivatives*, Academic Press, 1993, pp. 341–371.
89. U.S. Pat. 3,717,452 (1973), K.F. Gibsen and A.W. Saddington.
90. M. Satin, *New Sci.* April, 56–59 (1988).
91. I.W. Sutherland, *Food Biotechnol.* **6**, 75–86 (1992).
92. N.M. Edwards, C.G. Biliaderis, and J.E. Dexter, *J. Food Sci.* **60**(6), 1321–1324 (1995).
93. K. Shimada, H. Okada, K. Matsuo, and S. Yoshioka, *Biosci. Biotechnol. Biochem.* **60**, 125–127 (1996).
94. Business Communications Company, July, 1994, 00039104.
95. *Happy Household & Personal Products Industry*, Oct., 33(10), 152 (1996).
96. *Happy Household & Personal Products Industry*, April, 33(4), 136 (1996).
97. J. Sujjaareevath, D.L. Munday, P.J. Cox, and K.A. Khan, *Int. J. Pharm.* **139**(1–2), 53–62 (1997).
98. D.S. Roy and B.D. Rohera, *Pharm. Res.* **13**(9), 320 (1996).
99. J.D. Ntawukulliyayo, S.C. DeSemedt, J. Demeester, and J.P. Remon, *Int. J. Pharm.* **128**(1–2), 73–79 (1996).
100. A. Kondo, T. Irie, and K. Uekama, *Biol. Pharm. Bull.* **19**(2), 280–286 (1996).
101. U.S. Pat. 3,659,026 (1972), H.R. Schuppner.
102. Business Communications Company, July, 1994, 00039109.
103. *Food Chemical News* 12 Feb. **37**(51), 15 (1996).
104. Business Communications Company, November, 1996, 00065453.
105. The Royal Society of Chemistry, Press release, Oct. 23, 1996.
106. Eur. Pat. 611824 (1994), French Institute for Petroleum.
107. U.S. Pat. 5350528 (1994), Weyerhauser.
108. A. Gabriel, in *Microbial Polysaccharides and Polysaccharases*, Academic Press, 1979, pp. 191–204.
109. M. Lerner, *Chemical Market Reporter*, June 24, **26**, (1996), 12–14 (1996).
110. E.I. Sandvik and J.M. Maerker, in *ACS Symp. Series No. 45, Extracellular Microbial Polysaccharides*, 1977, pp. 242–264.
111. U. Grove and D. Kessel, *Chem. Ing. Tech.* **68**, 4196, 413–417 (1996).
112. J. Klein and A. Schumpe, *Chem. Ing. Tech.* **66**(2), 234–238 (1994).
113. D.G. Kessel, *Status Report*, German Institute for Petroleum Research, 1988, p. 41.
114. M. Hassan-Wakkas, Ph.D. Thesis, Technical University of Clausthal, 1995.
115. *Chemical Week*, International Edition, May 31, **156**(21), 47–48 (1995).
116. M. Lerner, *Chemical Market Reporter*, December 11, **248**, 7–14 (1995).
117. *Chemical Marketing Reporter*, June 27, **245**(26 suppl.), SR16–SR18 (1994).
118. *Chemical Marketing Reporter*, Jan. 3, **245**, 17 (1994).
119. B. Banner, *Chemical Marketing Reporter*, **244**(21), 21 (1993).
120. I. Cintron, *Chemical Marketing Reporter*, **243**(15), 7, 16 (1993).
121. *Informations Chimie Hebdo*, Nov. 18, **1192**, 31 (1994).

122. *Chemical Marketing Reporter*, Dec. 11, **284**(24) 7, 14 (1995).
123. M. Gliksman, G.O. Phillips, D.J. Wedlock, and P.A. Williams, in *Progress in Food and Nutrition Science*, vol. 6, *Gums and Stabilisers for the Food Industry*, Pergamon Press, 1982, pp. 299–321.
124. *Bioprocessing Technology*, July, 4–5 (1990).
125. Business Communications Company, November, 1995, 00054955.
126. G. Moritis, *Oil Gas J.* April, 39–61 (1996).
127. V. Moses, in *Biotechnology, the Science and the Business*, Harwood Academic, 1991, pp. 537–565.

See also CULTURE COLLECTIONS; RHEOLOGY OF FILAMENTOUS MICROORGANISMS, SUBMERGED CULTURE.

Y

YEAST, BAKER'S

TADASHI KOMETANI
Toyama National College of Technology
Toyama, Japan
RYUICHI MATSUNO
Kyoto University
Kyoto, Japan

KEY WORDS

Asymmetric reduction
Bubble-column reactor
Enantioselective oxidation
Ethanol
NAD^+
NADPH regeneration
Optically active compound
Optical resolution

OUTLINE

INTRODUCTION

Baker's yeast consists of microorganisms selected for the production of bread. The history of the baker's yeast industry for breadmaking was briefly described in *Encyclopedia of Food Science and Technology* (1). In the modern baker's yeast industry, specially selected strains of *Saccharomyces cerevisiae* are used. Baker's yeast is able to metabolize the commonly available sugars, such as D-glucose, D-mannose, sucrose, and maltose, very readily under both aerobic and anaerobic conditions.

In 1874, the reducing action of fermenting *Saccharomyces cerevisiae* was observed in which hydrogen sulfide was liberated on addition of finely powdered sulfur to a suspension of fresh baker's yeast in sugar solution. The reduction of furfural to furfuryl alcohol, the first transformation of an organic molecule, was then reported. Since the beginning of the 20th century, numerous biotransformations, biodegradations, and fermentations using other microorganisms (or microbial enzymes) in addition to baker's yeast have been studied by biotechnologists and organic chemists (2–6). Organic synthetic chemists have applied the microbial (or enzymatic) transformation to the synthesis of chiral molecules in enantiomerically pure form, and they have used baker's yeast among microorganisms and lipases among enzymes by preference. The reasons for these choices are that baker's yeast is versatile and inexpensive, and its growth does not require the assistance of a microbiologist. In addition, hydrolytic enzymes do not require cofactor regeneration and are easy to use. Thus, a large number of references to scientific studies in this area have been reported to date.

Microbial transformation of unnatural molecules has been used in industry to prepare optically pure compounds. For example, optical resolution by enantioselective hydrolysis with lipase is used in the synthetic process of a medicine such as diltiazem hydrochloride. Conversely, the baker's yeast-mediated transformation is not industrially applied. However, metabolisms including cofactor regeneration in baker's yeast have been thoroughly investigated in the 1990s. Therefore, the baker's yeast-mediated reduction now has the potential for industrial use.

TYPE OF TRANSFORMATIONS

The scientific studies devoted to baker's yeast-mediated transformation have gradually grown since the 1950s. The transformations of various classes of synthetic organic molecules were examined using baker's yeast as a reagent in organic synthesis, and the number of the papers is now more than 500. Those transformations were reviewed from the viewpoint of chemical reaction in three excellent articles published in 1990, 1991, and 1992, respectively (7–9). In these reviews, the uses of baker's yeast in synthetic organic chemistry are fully covered.

The baker's yeast-mediated transformations are mainly classified into four types as follows.

1. Carbon–carbon bond formation
2. Reduction of ketones and carbon–carbon double bonds
3. Oxidation
4. Hydrolysis of esters

The general goal of these transformations mediated by baker's yeast is considered to be the preparation of chiral molecules in enantiomerically pure form, because the catalytic synthesis of optically pure compounds is one of the most exciting chapters in organic chemistry.

Among the reports on baker's yeast-mediated transformations, more than half of the studies are devoted to the asymmetric reduction of prochiral carbonyl compounds, such as cycloalkanones, aliphatic alkanones, sulfur-containing alkanones, cyclic diketones, acyclic diketones, keto esters, and so on. Chiral alcohols are easily obtained using baker's yeast under mild conditions on a laboratory scale. Asymmetric C–C bond-forming reactions have been investigated by many chemists because of their importance as a fundamental reaction and as an attractive method from a synthetic point in organic chemistry.

Organic chemists sometimes use active dry yeast (certain strains of *S. cerevisiae* having higher resistance to drying with good baking activity [10]) instead of baker's yeast (compressed). Biotransformation mediated by active dry yeast also proceeds in the same manner as that by baker's yeast (compressed); therefore, both biocatalysts are not distinguished in this article.

C–C BOND FORMATION

In 1921, the process for the stereospecific preparation of D-ephedrine was developed using baker's yeast (11). This was the first example of the successful combination of chemical and enzymatic reactions. The key intermediate, L-1-hydroxy-1-phenyl-2-propanone, was produced by treating benzaldehyde under fermenting conditions of baker's yeast in the presence of glucose (Fig. 1). In this reaction, benzaldehyde is stereospecifically condensed with a CH_3CO unit to give the α-hydroxy ketone. The unit comes from decarboxylation of pyruvate through the EMP pathway of glucose and reacts with aldehyde. Hitherto, this is only the industrial use of the baker's yeast-mediated transformation in synthetic organic chemistry.

This acyloin-type condensation with baker's yeast has been applied to other aldehydes and α,β-unsaturated aldehydes. Furthermore, unnatural α-keto acids instead of pyruvic acid were used in this condensation reaction with aldehyde. The α-hydroxy ketones obtained in these reactions were further reduced to afford the *erythro*-diols in more than 95% enantiomeric excess (ee).

Another C–C bond formation mediated by baker's yeast is a Michael-type addition. The reaction of α,β-unsaturated ketones with 2,2,2-trifluoroethanol under fermentative conditions afforded the corresponding trifluoromethyl methyl carbinols with more than 90% ee.

ASYMMETRIC REDUCTION OF PROCHIRAL KETONES

The baker's yeast-mediated reduction of prochiral ketones with varying substituents (Me, Et, *n*-Pr, *n*-Bu, Bz) was systematically investigated in the 1960s. The prochiral ketone, with R_L representing a large substituent and R_S a small substituent adjacent to the carbonyl group, was transformed to the secondary alcohol with an *S* configuration. Therefore, it was judged that a hydrogen transfer to the *re*-face would preferably occur in the baker's yeast cells (Fig. 2). However, it was pointed out that this rule, the Prelog rule (12), should be cautiously applied to baker's yeast-mediated reduction, because exceptions to this rule were found.

In asymmetric reduction, its stereoselectivity is practically most important. Unsatisfactory stereoselectivity would be the consequence of two situations (7). The first case is that a single enzyme, catalyzing the reduction of each substrate, cannot recognize its structural features. The second case is that two or more oxidoreductases for each substrate act simultaneously with opposite stereochemical preferences. Although each of the oxidoreductases catalyzes the enantioselective reduction with high enantioselectivity, stereoselectivity of the transformation and the optical purity of the product are the consequence of the activities of those enzymes. This second case was recently found in baker's yeast-mediated reduction. By characterizing oxidoreductases in baker's yeast followed by modifying the structures or procedures, improvement in the enantiomeric excess of the product was accomplished in some reactions. In this article, the successful result for production of (*S*)-ethyl 3-hydroxybutanoate with high ee is described.

Method

Compared with the chemistry of the baker's yeast-mediated reduction, the method, especially for large-scale production, has hardly been investigated. In the 1990s, regeneration systems for nicotinamide coenzymes in the baker's yeast-mediated reduction were extensively investigated, and consequently, an effective method for large-scale production has been developed (13).

The details of the use of baker's yeast in an asymmetric reduction of prochiral ketones were described on a laboratory scale in *Organic Syntheses*. Reductions of acetol (AC) to (*R*)-propylene glycol (PG) (14), of ethyl acetoacetate (EA) to (*S*)-ethyl 3-hydroxybutanoate (E3-HB) (15), and of 2,2-dimethylcyclohexanone-1,3-dione to (*S*)-3-hydroxy-2,2-dimethylcyclohexanone (16) were fully reported (Fig. 3). Thus, asymmetric reduction mediated by baker's yeast was recognized as a useful technique by synthetic chemists.

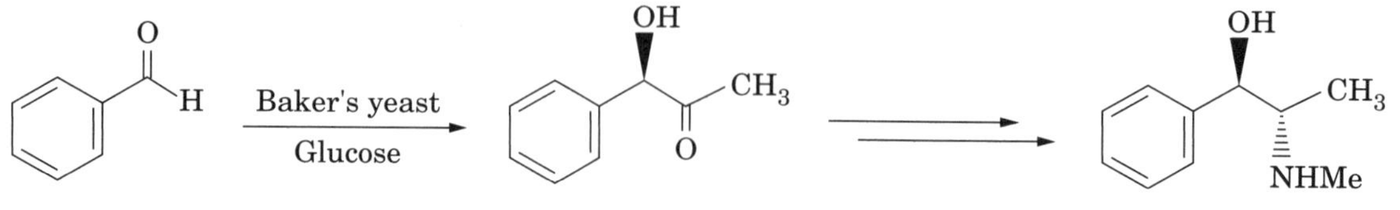

Figure 1. Production of L-ephedrine through baker's yeast-mediated reduction.

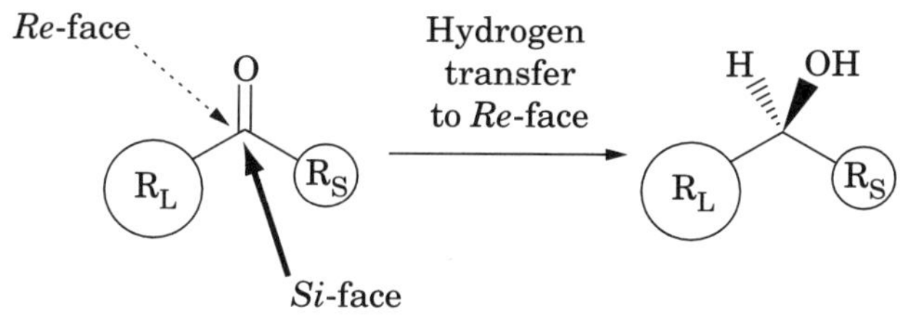

Figure 2. Asymmetric reduction of a prochiral ketone by hydrogen transfer to the *Re*-face.

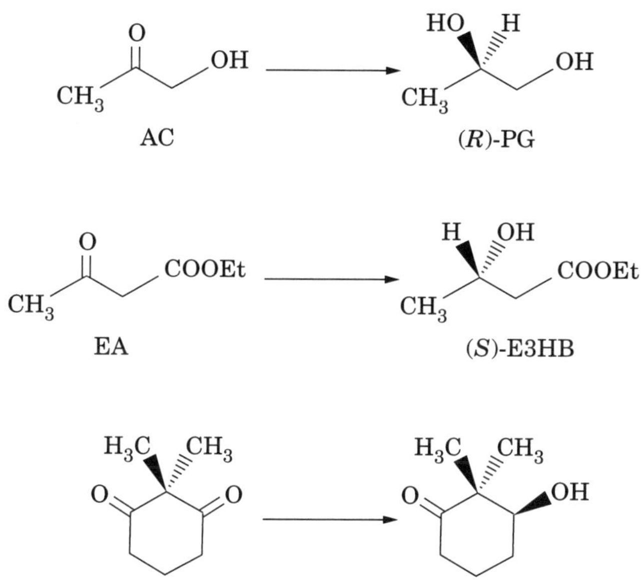

Figure 3. Baker's yeast-mediated reduction fully described in *Organic Syntheses*.

These three bioreductions, and other many similar ones, were done in an aqueous solution containing carbohydrate with stirring, as chemical reactions. Under these conditions, reductions successively proceed consuming carbohydrate, the energy source for coenzyme regeneration. Reduction by carbonyl reductases is coupled with regeneration systems of the reduced form of nicotinamide coenzymes NADPH from the oxidized form $NADP^+$ through glucose oxidation, as shown in Figure 4. However, the use of carbohydrate as the energy source for NADPH regeneration was thought to be inadequate, because the amount of carbohydrate is quite excessive compared with that of the substrate, and the metabolism of the carbohydrate is accompanied by the production of a vigorous foam of carbon dioxide, ethanol, and smelly by-products.

Recently, an efficient method for a baker's yeast-mediated reduction was developed (13). It uses ethanol and oxygen instead of carbohydrate as the energy source for NADPH regeneration. NADPH is regenerated from $NADP^+$ through ethanol oxidation to carbon dioxide, as shown in Figure 5. The asymmetric reduction of several carbonyl compounds by this new method afforded similar results in the chemical yield and optical purity of the product. The method, operating in about 100 mM EtOH under aerobic conditions, is clean and efficient, because no by-product except carbon dioxide was detected, and the energy source can be efficiently used for coenzyme regeneration.

NADPH Regeneration System

In the baker's yeast-mediated reduction, NADPH, which is the hydrogen donor in the reduction, must be successively regenerated from the corresponding oxidized form in the cells. The system responsible for the coenzyme regeneration must be then coupled with reduction of a substrate. Recently, Yamada et al. characterized the enzymes responsible for the NADPH regeneration system using carbohydrate as the energy source with some microorganisms (17). With *S. cerevisiae*, the enzymes were shown to be D-

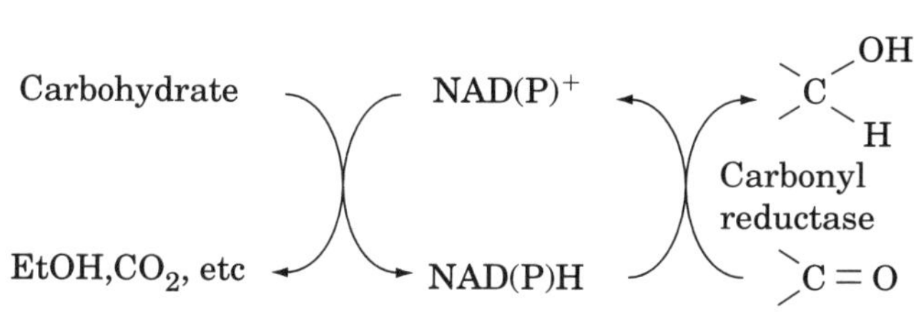

Figure 4. Outline of the reaction mechanism using carbohydrate as the energy source.

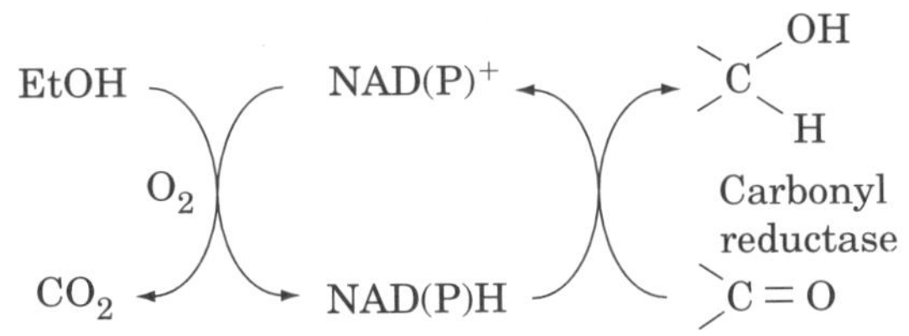

Figure 5. Outline of the reaction mechanism using EtOH as the energy source.

glucose-6-phosphate dehydrogenase and 6-phospho-D-gluconate dehydrogenase in the hexose monophosphate pathway.

In the method using ethanol as the energy source, NADPH is regenerated from $NADP^+$ through the oxidative pathway of ethanol to carbon dioxide, and the system is also efficiently coupled with reduction of a ketone. The enzyme system responsible for NADPH regeneration was investigated by studying the reduction of AC and EA, catalyzed by NADH- and NADPH-dependent carbonyl reductases, respectively. NAD^+ formed during the reduction of AC is reduced to NADH through the oxidative pathway of ethanol to acetate, catalyzed by alcohol dehydrogenase and aldehyde dehydrogenase. On the other hand, $NADP^+$ formed in the reduction of EA is reduced to NADPH through the oxidative pathway of acetate to carbon dioxide. The enzyme responsible for NADPH regeneration must be malate dehydrogenation (decarboxylating) in leaving the TCA cycle into pyruvate.

Asymmetric Reduction of Ethyl Acetoacetate

Reduction of EA by baker's yeast is the most extensively studied transformation, leading to an important chiral building block, (*S*)-ethyl 3-hydroxybutanoate (E3-HB), for the syntheses of the carbapenem thienamycin, (*S*)-(+)-sulcatol, nystatin, and so on. However, the methods described in the literature have been considered to be unacceptable for large-scale production of (*S*)-E3-HB. The reasons are as follows: (*1*) a relatively low concentration, (*2*) tedious isolation and purification processes, and (*3*) a relatively low optical purity. Fortunately, these disadvantages have recently been overcome in developing an efficient system for the baker's yeast-mediated reduction of EA suitable for actual large-scale production.

The first and second disadvantages were simply solved by applying a method using ethanol as the energy source for NADPH regeneration. Because the amount of by-products inactivating baker's yeast cells was considerably diminished as described, they were active longer and, as a result, the product could accumulate in high concentration.

Regarding the third disadvantage, the reported ee values of the product vary from 70 to more than 95%. To realize those results and improve the ee values, purification and characterization of EA reductases in the baker's yeast cells were very helpful. In 1988, two NADPH-dependent oxidoreductases responsible for the reduction of EA in bakers' yeast were purified and characterized (18). One enzyme, with a K_m value for EA of 0.9 mM, catalyzes the reduction to give (*S*)-E3-HB, and the other, with a K_m value for EA of 17.0 mM, to give (*R*)-E3-HB. Therefore, it was supposed that the ee values might easily vary because of the condition of the initial concentration of EA. The former enzyme would mainly participate in the reduction at a low EA concentration, affording high optical purity. Actually, the optical purity was dramatically improved by applying a fed-batch operation of EA to afford more than 99% ee (*S*)-E3-HB (13).

Bubble-Column Reactor

The baker's yeast-mediated reduction of EA and AC were effectively achieved on a large scale using a bubble-column reactor, which had advantages such as decreasing the operating costs due to low energy requirements. Because the aeration rate is the most important factor in this reactor, its effect on the reduction rate was examined. In the case of EA, the specific reduction rate increased according to the increase in the aeration rate, then became constant. Conversely, the specific reduction rate of AC was almost constant regardless of the aeration rate. These results were consistent with coenzyme requirements and their described regeneration mechanisms.

The large-scale production of (*S*)-E-3HB and (*R*)-PG was performed by applying the fed-batch operation for the addition of the substrates and ethanol in a bubble-column reactor containing 10 L of medium (Table 1) (13). Compared with the results by the conventional procedure described in *Organic Syntheses*, the yield and the optical purity were markedly improved, and the procedure could be applied on an industrial scale (14,15). (*R*)-PG is actually used as a chiral building block for preparation of (*S*)-ofloxacin, although it is now industrially produced by catalytic hydrogenation using a chiral *Ru* complex.

Diltiazem Synthesis

Diltiazem hydrochloride, (2*S*,3*S*)-3-acetoxy-5-[2-(dimethylamino)ethyl]-2,3-dihydro-2-(4-methoxyphenyl)-1,5-benzothiazepin-4(5*H*)-one hydrochloride, is a representative calcium channel blocker that has been clinically used as an antianginal and antihypertensive agent in more than 100 countries. In the synthetic process, the key step is the

Table 1. Comparison between EtOH Procedure and the Conventional Procedure Described in *Organic Synthese*

	EtOH procedure (EA reduction)	Conventional (Acetol reductin)
Water	1,000 mL	1,000 mL
Baker's yeast	112 g (77 g)	112 g (100 g)
Energy source	104 g of EtOH (192 g of sucrose)	18 g of EtOH (100 g of sucrose)
Substrate	60 g (15.3 g)	40 g (10 g)
Air	25 L/min	3 L/min
Reaction time	62 h (74 h)	38 h (72 h)
Temperature	30 ° (r.t.)	30 °C (32 °C)
Recovery	40.5 g (10.6 g)	22 g (5.5 g)
Optical rotation	+44.3° (+37.2°)	−17.3° (−15.0°)
ee	99.3% (85%)	98.2% (90%)

introduction of two chiral centers. Optical resolution of a racemic intermediate, lipase-catalyzed enantioselective hydrolysis of methyl 3-phenylglycidate, is now industrially used. Recently, a new efficient synthetic process involving asymmetric reduction by baker's yeast as a key step was reported. The transformation of a racemic ketone, (*RS*)-2-(4-methoxyphenyl)-1,5-benzothiazepin-3,4(2H,5H)-dione afforded only the desired (2*S*,3*S*)-2,3-dihydro-3-hydroxy-2-(4-methoxyphenyl)-1,5-benzothiazepin-4(5*H*)-one (Fig. 6). The difficulty in the industrial application because of its low productivity, caused by low solubility of the substrate in water, was also overcome by changing its physical property. The crystalline substrate was first converted to an amorphous-like solid aggregate by pouring a DMF solution of the substrate into water. After separation, the aggregate was used for the reduction under the usual conditions. Thus, 100 g of the substrate in 1 L of a reaction medium was reduced to produce the chiral intermediate with more than 99% ee in 80% yield (19).

ENANTIOSELECTIVE OXIDATION

Baker's yeast is widely used as a reducing agent in organic synthesis for preparing enantiomerically pure chiral alcohols. One enantiomer can be conveniently prepared by this method, but the other is essentially impossible to prepare with the same system. Several methods for controlling the enantioselectivity of the baker's yeast-mediated reduction have been reported; modifying the structure of the substrate, the reaction conditions, and adding a third reagent. These methods are very interesting but not convenient for large-scale production. Therefore, baker's yeast-mediated oxidation was developed to prepare the antipode of the reduction product on a large scale.

(*R*)-1,2-Alkanediols were found to be easily oxidized to 1-hydroxy-2-alkanones with baker's yeast under aerobic conditions, and (*S*)-1,2-alkanediols, conversely, could not be oxidized under the same conditions. Thus, treatment of racemic 1,2-alkanediols with baker's yeast to oxidize (*R*)-1,2-alkanediols to 1-hydroxy-2-alkanones followed by separation of the 1-hydroxy-2-alkanones afforded the (*S*)-1,2-alkanediols (Fig. 7) (13).

NAD^+ Regeneration

In this biooxidation, NADH-dependent oxidoreductases participating in the asymmetric reduction of 1-hydroxy-2-alkanone catalyze the contrary enantioselective oxidation, and NAD^+ is regenerated from NADH in the yeast cells for the successive oxidation. A preliminary experiment showed that the oxidation did not proceed under anaerobic conditions. Thus, an outline of the oxidation, especially in relation to the condition of oxygen supply, was examined by analyzing the transformation of (*R*)-PG to AC (13). The amounts of oxygen consumed and AC produced increased with increasing k_La, and the ratio was about 1.7 (AC mmol/O_2 mmol). Because 2 mol (*R*)-PG would be theoretically oxidized using 1 mol of oxygen, as shown in Figure 8, this result suggests that oxygen is effectively used during the biooxidation (85% efficiency).

Optical Resolution of 1,2-Alkanediols

Treatment of racemic 1,2-alkanediols with baker's yeast affords (*S*)-1,2-alkanediols and 1-hydroxy-2-alkanones that are easily separated because of a considerable difference in their volatilities. The asymmetric reduction of 1-hydroxy-2-alkanones with baker's yeast affords (*R*)-1,2-alkanediols (13). Thus, the optical resolution of racemic 1,2-alkanediols was conveniently accomplished using only baker's yeast, as summarized in Figure 9.

The oxidative resolution and the asymmetric reduction have been achieved using the bubble-column reactor on a large scale. The specific oxidation rate is sufficiently fast and reaches equilibrium within 24 h, affording the (*S*)-1,2-alkanediols with 79% ee. The aqueous layers after removal of (*S*)-1,2-alkanediols are directly used for baker's yeast-mediated reduction, affording the (*R*)-1,2-alkanediols with over 98% ee.

CHARACTERIZATION OF OXIDOREDUCTASES

There have been several reports on purification and characterization of enzymes from baker's yeast that catalyze the reductions of unnatural substrates. The purposes are to understand the mechanism of the stereochemistry of baker's yeast-mediated transformation and, as a result, to improve the ee of the reduction product by some modifications. Furthermore, some purified reductases would be useful for synthetic purposes.

First, Sih et al. isolated three oxidoreductases that catalyze the reduction of alkyl 4-chloroacetoacetate (20). They reported that stereochemical control by modification of the substrate stems from the difference in relative activities of the D- and L-enzymes depending on the length of the carbon chain in the substrate. Furuichi et al. purified 2-

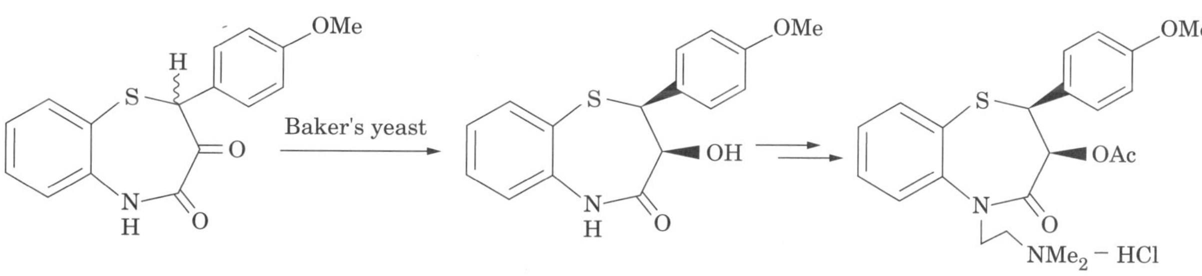

Figure 6. Production of diltiazem hydrochloride through baker's yeast-mediated reduction.

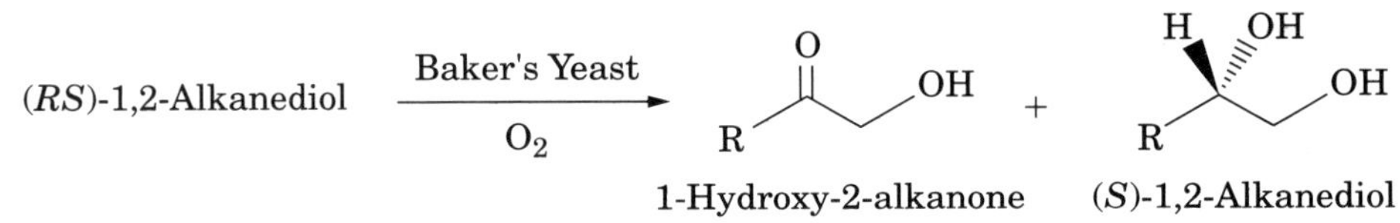

Figure 7. Baker's yeast-mediated oxidation of 1,2-alkanediols.

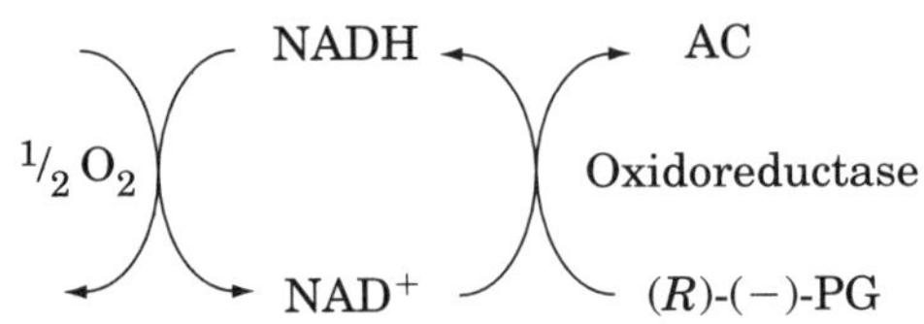

Figure 8. Outline of the reaction mechanism of the yeast-mediated oxidation.

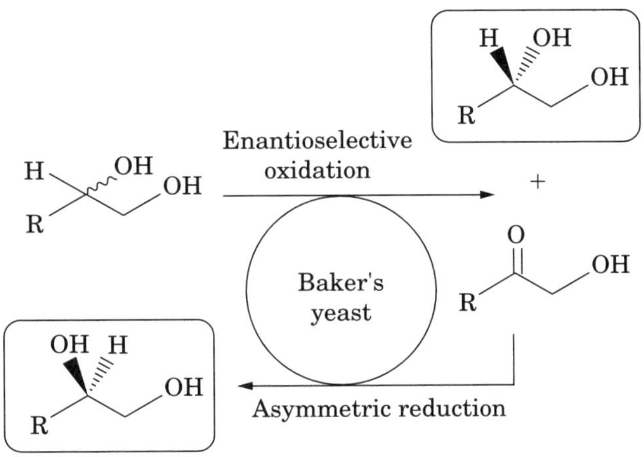

Figure 9. Oxidative resolution of racemic, 1,2-alkanediols with baker's yeast.

methyl-3-oxobutanoate reductase (21) and Heidlas et al. purified two oxidoreductases that catalyze β-, γ-, and δ-keto esters (18). The latter report was very useful for improvement of the ee of the product by baker's yeast-mediated reduction of EA as described. Nakamura et al. isolated four enzymes that catalyze β-keto- and α-alkyl-β-ketoesters and three α-ketoester reductases (22).

IMMOBILIZATION

Immobilized baker's yeast, using polyurethane, calcium alginate, or κ-carrageenan, has been applied to change the reaction conditions (8). Some marked differences have been reported in the stereochemistry, mostly due to a change in the concentration of the substrates. In some cases, immobilization of baker's yeast improved the ee of the products. The immobilized baker's yeast was shown to be active for several weeks, longer than free baker's yeast. However, reductions using immobilized baker's yeast gave lower yields as compared with the analogous reduction with free baker's yeast. Thus, the immobilization did not improve the efficiency of the baker's yeast-mediated transformation for large-scale production.

BIBLIOGRAPHY

1. Y.H. Hui, *Encyclopedia of Food Science and Technology*, Wiley, New York, 1993, pp. 2885–2887.
2. J.B. Jones, C.J. Sih, and D. Perlman, *Application of Biochemical Systems in Organic Chemistry*, Parts 1 and 2, Wiley, New York, 1976.
3. K. Kieslich, *Biotransformations*, Verlag Chemie, Weinheim, Germany, 1984.
4. H. Yamada and S. Shimizu, *Angew. Chem. Int. Ed. Engl.* **27**, 622–642 (1988).
5. C.-H. Wong and G.M. Whitesides, *Enzymes in Synthetic Organic Chemistry*, Pergamon Press, Oxford, 1989.
6. J. Halgas, *Biocatalysts in Organic Synthesis*, Elsevier, Amsterdam, 1992.
7. S. Servi, *Synthesis*, 1990, 1–25, (1990).
8. R. Csuk and B.I. Gläzer, *Chem. Rev.* **91**, 49–97 (1991).
9. E. Santaniello, P. Ferraboschi, P. Grisenti, and A. Manzocchi, *Chem. Rev.* **92**, 1071–1140 (1992).
10. Y.H. Hui, *Encyclopedia of Food Science and Technology*, Wiley, New York, 1993, pp. 2887–2888.
11. C. Neuberg and J. Hirsch, *J. Biochem. Z.* **115**, 282 (1921).
12. V. Prelog, *Pure Appl. Chem.* **9**, 119–130 (1964).
13. T. Kometani, H. Yoshii, and R. Matsumo, *J. Mol. Catal.* **1**, 45–52 (1996).
14. P.A. Levene and A. Walti, *Org. Synth. Coll.* **2**, 545–547 (1943).
15. D. Seebach, M.A. Sutter, R.H. Weber, and M.F. Zuer, *Org. Synth.* **63**, 1–7 (1983).
16. K. Mori and H. Mori, *Org. Synth.* **68**, 56–61 (1989).
17. M. Kataoka, Y. Nomura, S. Shimizu, and H. Yamada, *Biosci. Biotech. Biochem.* **56**, 820–821 (1992).
18. J. Heidlas, K.-E. Engel, and R. Tresal, *Eur. J. Biochem.* **172**, 633–639 (1988).
19. T. Kometani, Y. Sakai, H. Matsumae, T. Shibatani, and R. Matsuno, *J. Ferment. Bioeng.* **84**, 195–199 (1997).
20. W.-R. Shieh, A.S. Gopalan, and C.J. Sih, *J. Am. Chem. Soc.* **107**, 2993–2994.
21. A. Furuichi, H. Akita, H. Matsukira, T. Oishi, and K. Horikoshi, *Agric. Biol. Chem.* **49**, 2563–2570 (1985).
22. K. Nakamura, S. Kondo, Y. Kawai, N. Nakajima, and A. Ohno, *Biosci. Biotech. Biochem.* **58**, 2236–2240 (1994).

See also ANAEROBES; ENERGY METABOLISM, MICROBIAL AND ANIMAL CELLS; FOOD PROCESS ENGINEERING; PROCESS CONTROL, STRATEGY AND OPTIMIZATION; PROCESS MONITORING.

INDEX